Methods in Enzymology

Volume 304

CHROMATIN

METHODS IN ENZYMOLOGY

EDITORS-IN-CHIEF

John N. Abelson   Melvin I. Simon

DIVISION OF BIOLOGY
CALIFORNIA INSTITUTE OF TECHNOLOGY
PASADENA, CALIFORNIA

FOUNDING EDITORS

Sidney P. Colowick and Nathan O. Kaplan

Methods in Enzymology

Volume 304

Chromatin

EDITED BY

Paul M. Wassarman

DEPARTMENT OF CELL BIOLOGY AND ANATOMY
MOUNT SINAI SCHOOL OF MEDICINE
NEW YORK, NEW YORK

Alan P. Wolffe

NATIONAL INSTITUTE OF CHILD HEALTH AND HUMAN DEVELOPMENT
NATIONAL INSTITUTES OF HEALTH
BETHESDA, MARYLAND

ACADEMIC PRESS
San Diego London Boston New York Sydney Tokyo Toronto

This book is printed on acid-free paper. 

Academic Press
a division of Harcourt Brace & Company
525 B Street, Suite 1900, San Diego, California 92101-4495, USA
http://www.academicpress.com

Academic Press Limited
24-28 Oval Road, London NW1 7DX, UK
http://www.hbuk.co.uk/ap/

International Standard Book Number: 0-12-182205-2

PRINTED IN THE UNITED STATES OF AMERICA
99 00 01 02 03 04 MM 9 8 7 6 5 4 3 2 1

Table of Contents

CONTRIBUTORS TO VOLUME 304 ix

PREFACE xv

VOLUMES IN SERIES xvii

Section I. Assembly of Nucleosomes, Chromatin, and Nuclei

1. Preparation of Nucleosome Core Particle from Recombinant Histones — KAROLIN LUGER, THOMAS J. RECHSTEINER, AND TIMOTHY J. RICHMOND — 3

2. Assembly of Defined Nucleosomal and Chromatin Arrays from Pure Components — LENNY M. CARRUTHERS, CHRISTIN TSE, KERFOOT P. WALKER III, AND JEFFREY C. HANSEN — 19

3. Isolation of Minichromosomes from Yeast Cells — JENNIFER A. ALFIERI AND DAVID J. CLARK — 35

4. Chromatin Assembly in *Xenopus* Extracts — DAVID JOHN TREMETHICK — 50

5. *In Vitro* Reconstitution of Nuclei for Replication and Transcription — ALISON J. CROWE AND MICHELLE CRAIG BARTON — 63

6. Preparation/Analysis of Chromatin Replicated *in Vivo* and in Isolated Nuclei — LOUISE CHANG, COLLEEN A. RYAN, CAROLYN A. SCHNEIDER, AND ANTHONY T. ANNUNZIATO — 76

7. Interactions of High Mobility Group Box Proteins with DNA and Chromatin — MAIR E. A. CHURCHILL, ANITA CHANGELA, LINDA K. DOW, AND ADAM J. KRIEG — 99

8. Reconstitution of High Mobility Group 14/17 Proteins into Nucleosomes and Chromatin — YURI V. POSTNIKOV AND MICHAEL BUSTIN — 133

9. Purification and Assays for High Mobility Group HMG-I(Y) Protein Function — RAYMOND REEVES AND MARK S. NISSEN — 155

Section II. Assays for Nucleosome Structure and Function *in Vitro*

10. Cryoelectron Microscopic Analysis of Nucleosomes and Chromatin — JAN BEDNAR AND CHRISTOPHER L. WOODCOCK — 191

11. Electron Microscopy of DNA–Protein Complexes and Chromatin — Jack Griffith, Susan Michalowski, and Alexander M. Makhov — 214

12. Targeted Cross-Linking and DNA Cleavage within Model Chromatin Complexes — Kyu-Min Lee, David R. Chafin, and Jeffrey J. Hayes — 231

13. Base-Pair Resolution Mapping of Nucleosome Positions Using Site-Directed Hydroxy Radicals — Andrew Flaus and Timothy J. Richmond — 251

14. Localization of Specific Histone Regions on Nucleosomal DNA by Cross-Linking — Sergei I. Usachenko and E. Morton Bradbury — 264

15. Restriction Enzymes as Probes of Nucleosome Stability and Dynamics — K. J. Polach and J. Widom — 278

16. Nucleoprotein Gel Assays for Nucleosome Positioning and Mobility — Sari Pennings — 298

17. Assays for Interaction of Transcription Factor with Nucleosome — Qiao Li, Ulla Björk, and Örjan Wrange — 313

18. Assays for Chromatin Remodeling during DNA Repair — Jonathan G. Moggs and Geneviève Almouzni — 333

19. Nuclear Run-On Assays: Assessing Transcription by Measuring Density of Engaged RNA Polymerases — Kazunori Hirayoshi and John T. Lis — 351

Section III. Assays for Chromatin Structure and Function *in Vivo*

20. Mapping Chromatin Structure in Yeast — Philip D. Gregory and Wolfram Hörz — 365

21. Assays for Nucleosome Positioning in Yeast — Michael P. Ryan, Grace A. Stafford, Liuning Yu, Kellie B. Cummings, and Randall H. Morse — 376

22. Mapping DNA Interaction Sites of Chromosomal Proteins Using Immunoprecipitation and Polymerase Chain Reaction — Andreas Hecht and Michael Grunstein — 399

23. Immunological Analysis of Yeast Chromatin — Pamela B. Meluh and James R. Broach — 414

24. DNA Methyltransferases as Probes of Chromatin Structure *in Vivo* — Michael P. Kladde, M. Xu, and Robert T. Simpson — 431

25. Mapping Cyclobutane–Pyrimidine Dimers in DNA and Using DNA–Repair by Photolyase for Chromatin Analysis in Yeast — BERNHARD SUTER, MAGDALENA LIVINGSTONE-ZATCHEJ, AND FRITZ THOMA 447

26. Analysis of *Drosophila* Chromatin Structure *in Vivo* — IAIN L. CARTWRIGHT, DIANE E. CRYDERMAN, DAVID S. GILMOUR, LORI A. PILE, LORI L. WALLRATH, JANET A. WEBER, AND SARAH C. R. ELGIN 462

27. Ultraviolet Cross-Linking Assay to Measure Sequence-Specific DNA Binding *in Vivo* — MARK D. BIGGIN 496

28. DNA–Protein Cross-Linking Applications for Chromatin Studies *in Vitro* and *in Vivo* — DMITRY PRUSS, IGOR M. GAVIN, SVETLANA MELNIK, AND SERGEI G. BAVYKIN 516

29. Chromatin Immunoprecipitation Assays in Acetylation Mapping of Higher Eukaryotes — COLYN CRANE-ROBINSON, FIONA A. MYERS, TIM R. HEBBES, ALISON L. CLAYTON, AND ALAN W. THORNE 533

30. Chromatin Structure Analysis by Ligation-Mediated and Terminal Transferase-Mediated Polymerase Chain Reaction — GERD P. PFEIFER, HSIU-HUA CHEN, JUON-ICHIRO KOMURA, AND ARTHUR D. RIGGS 548

31. Guanine–Adenine Ligation-Mediated Polymerase Chain Reaction *in Vivo* Footprinting — ERICH C. STRAUSS AND STUART H. ORKIN 572

32. Exonuclease III as Probe of Chromatin Structure *in Vivo* — TREVOR K. ARCHER AND ANDREA R. RICCI 584

33. Protein Image Hybridization: Mapping and Positioning DNA–Protein Contacts along DNA — S. BELIKOV, O. PREOBRAZHENSKAYA, AND V. KARPOV 600

34. *In Vivo* Analysis of Chromatin Structure — KENNETH S. ZARET 612

35. Analysis of Nucleosome Positioning in Mammalian Cells — GORDON L. HAGER AND GILBERTO FRAGOSO 626

36. Measurement of Localized DNA Supercoiling and Topological Domain Size in Eukaryotic Cells — PHILLIP R. KRAMER, OLGA BAT, AND RICHARD R. SINDEN 639

37. Fluorescent *in Situ* Hybridization Analysis of Chromosome and Chromatin Structure — WENDY BICKMORE 650

38. Analysis of Nuclear Organization in *Saccharomyces cerevisiae* — MONICA GOTTA, THIERRY LAROCHE, AND SUSAN M. GASSER 663

Section IV. Chromatin Remodeling Complexes

39. Histone Acetyltransferases: Preparation of Substrates and Assay Procedures — CRAIG A. MIZZEN, JAMES E. BROWNELL, RICHARD G. COOK, AND C. DAVID ALLIS 675

40. Analysis of Activity and Regulation of hGcn5, Human Histone Acetyltransferase — NICKOLAI A. BARLEV, JERRY L. WORKMAN, AND SHELLEY L. BERGER 696

41. Purification of a Histone Deacetylase Complex from *Xenopus laevis:* Preparation of Substrates and Assay Procedures — PAUL A. WADE, PETER L. JONES, DANIELLE VERMAAK, AND ALAN P. WOLFFE 715

42. Purification and Biochemical Properties of Yeast SWI/SNF Complex — COLIN LOGIE AND CRAIG L. PETERSON 726

43. Analysis of Modulators of Chromatin Structure in *Drosophila* — PATRICK D. VARGA-WEISZ, EDGAR J. BONTE, AND PETER B. BECKER 742

44. Purification of *Drosophila* Nucleosome Remodeling Factor — RAPHAEL SANDALTZOPOULOS, VINCENT OSSIPOW, DAVID A. GDULA, TOSHIO TSUKIYAMA, AND CARL WU 757

AUTHOR INDEX 767

SUBJECT INDEX 793

Contributors to Volume 304

Article numbers are in parentheses following the names of contributors.
Affiliations listed are current.

JENNIFER A. ALFIERI (3), *Laboratory of Cellular and Developmental Biology (NIDDK), National Institutes of Health, Bethesda, Maryland 20892-2715, and University of California at Davis, Davis, California 95616*

C. DAVID ALLIS (39), *Department of Biochemistry and Molecular Genetics, University of Virginia Health Sciences Center, Charlottesville, Virginia 22908*

GENEVIÈVE ALMOUZNI (18), *Section de Recherche, Unité Mixte du CNRS, Institut Curie, 75231 Paris cedex 05, France*

ANTHONY T. ANNUNZIATO (6), *Department of Biology, Boston College, Chestnut Hill, Massachusetts 02467*

TREVOR K. ARCHER (32), *The University of Western Ontario, London Regional Cancer Centre, London, Ontario N6A 4L6, Canada, and Chromatin and Gene Expression Section, Laboratory of Reproductive and Developmental Toxicology, National Institute of Environmental Health Sciences, National Institutes of Health, Research Triangle Park, North Carolina 27709*

NICKOLAI A. BARLEV (40), *The Wistar Institute, Philadelphia, Pennsylvania 19104-4268*

MICHELLE CRAIG BARTON (5), *Department of Molecular Genetics, University of Cincinnati Medical Center, Cincinnati, Ohio 45267*

OLGA BAT (36), *Institute of Molecular Medicine, The University of Texas, Houston Health Science Center, Houston, Texas 77030*

SERGEI G. BAVYKIN (28), *Englehardt Institute of Molecular Biology, Academy of Sciences of Russia, Moscow 117984, Russia, and Center for Mechanistic Biology and Biotechnology, Argonne National Laboratory, Argonne, Illinois 60439*

PETER B. BECKER (43), *European Molecular Biology Laboratory, 69117 Heidelberg, Germany*

JAN BEDNAR (10), *Biology Department, University of Massachusetts, Amherst, Massachusetts 01003*

S. BELIKOV (33), *Laboratory of Molecular Genetics, Department of Cell and Molecular Biology, Medical Nobel Institute, Karolinska Institute, S-17177 Stockholm, Sweden*

SHELLEY L. BERGER (40), *The Wistar Institute, Philadelphia, Pennsylvania 19104-4268*

WENDY BICKMORE (37), *MRC Human Genetics Unit, Western General Hospital, Edinburgh EH4 2XU, Scotland, United Kingdom*

MARK D. BIGGIN (27), *Department of Molecular Biophysics and Biochemistry, Yale University, New Haven, Connecticut 06520*

ULLA BJÖRK (17), *Department of Cell and Molecular Biology, Karolinska Institute, S-17177 Stockholm, Sweden*

EDGAR J. BONTE (43), *European Molecular Biology Laboratory, 69117 Heidelberg, Germany*

E. MORTON BRADBURY (14), *Department of Biological Chemistry, School of Medicine, University of California at Davis, Davis, California 95616, and Life Science Division, Los Alamos National Laboratory, Los Alamos, New Mexico 87545*

JAMES R. BROACH (23), *Department of Molecular Biology, Princeton University, Princeton, New Jersey 08544*

JAMES E. BROWNELL (39), *ProScript, Inc., Cambridge, Massachusetts 02139*

MICHAEL BUSTIN (8), *Protein Section, Division of Basic Sciences, National Cancer Institute, National Institutes of Health, Bethesda, Maryland 20892*

LENNY M. CARRUTHERS (2), *Department of Biochemistry, The University of Texas Health Science Center at San Antonio, San Antonio, Texas 78284-7760*

IAIN L. CARTWRIGHT (26), *Department of Molecular Genetics, Biochemistry, and Microbiology, University of Cincinnati College of Medicine, Cincinnati, Ohio 45267*

DAVID R. CHAFIN (12), *Department of Biochemistry and Biophysics, University of Rochester Medical Center, Rochester, New York 14642*

LOUISE CHANG (6), *Howard Hughes Medical Institute, Vanderbilt School of Medicine, Nashville, Tennessee 37232*

ANITA CHANGELA (7), *Department of Biochemistry, Molecular Biology, and Cell Biology, Northwestern University, Evanston, Illinois 60208-3500*

HSIU-HUA CHEN (30), *Department of Biology, Beckman Research Institute of the City of Hope, Duarte, California 91010*

MAIR E. A. CHURCHILL (7), *Department of Pharmacology, University of Colorado Health Sciences Center, Denver, Colorado 80262*

DAVID J. CLARK (3), *Laboratory of Cellular and Developmental Biology (NIDDK), National Institutes of Health, Bethesda, Maryland 20892-2715*

ALISON L. CLAYTON (29), *Biophysics Laboratories, University of Portsmouth, Portsmouth PO1 2DT, England, United Kingdom*

RICHARD G. COOK (39), *Department of Microbiology and Immunology, Baylor College of Medicine, Houston, Texas 77030*

COLYN CRANE-ROBINSON (29), *Biophysics Laboratories, University of Portsmouth, Portsmouth PO1 2DT, England, United Kingdom*

ALISON J. CROWE (5), *Department of Molecular Genetics, University of Cincinnati Medical Center, Cincinnati, Ohio 45267*

DIANE E. CRYDERMAN (26), *Department of Biochemistry, University of Iowa College of Medicine, Iowa City, Iowa 52242*

KELLIE B. CUMMINGS (21), *Wadsworth Center, New York State Department of Health, Albany, New York 12208*

LINDA K. DOW (7), *Program in Biophysics and Computational Biology, University of Illinois at Urbana–Champaign, Urbana, Illinois 61801*

SARAH C. R. ELGIN (26), *Department of Biology, Washington University, St. Louis, Missouri 63130*

ANDREW FLAUS (13), *ETH Zürich, Institut für Molekularbiologie und Biophysik, ETH-Hönggerberg, CH-8093 Zürich, Switzerland*

GILBERTO FRAGOSO (35), *Department of Biology, Johns Hopkins University, Baltimore, Maryland 21218*

SUSAN M. GASSER (38), *Swiss Institute for Experimental Cancer Research (ISREC), CH-1066 Epalinges, Switzerland*

IGOR M. GAVIN (28), *Program in Molecular Medicine, University of Massachusetts Medical Center, Worcester, Massachusetts 01606*

DAVID A. GDULA (44), *Laboratory of Molecular Cell Biology, National Cancer Institute, National Institutes of Health, Bethesda, Maryland 20892*

DAVID S. GILMOUR (26), *Department of Molecular and Cell Biology, Pennsylvania State University, University Park, Pennsylvania 16802*

MONICA GOTTA (38), *Department of Genetics, University of Cambridge, Cambridge, England, United Kingdom*

PHILIP D. GREGORY (20), *Institut für Physiologische Chemie der Universität München, D-80336 München, Germany*

JACK GRIFFITH (11), *Lineberger Comprehensive Cancer Center, University of North Carolina at Chapel Hill, Chapel Hill, North Carolina 27599*

MICHAEL GRUNSTEIN (22), *Department of Biological Chemistry, UCLA School of Medicine and the Molecular Biology Institute, University of California at Los Angeles, Los Angeles, California 90095*

GORDON L. HAGER (35), *Laboratory of Receptor and Gene Expression, National Cancer Institute, National Institutes of Health, Bethesda, Maryland 20892-5055*

JEFFREY C. HANSEN (2), *Department of Biochemistry, The University of Texas Health Science Center at San Antonio, San Antonio, Texas 78284-7760*

JEFFREY J. HAYES (12), *Department of Biochemistry and Biophysics, University of Rochester Medical Center, Rochester, New York 14642*

TIM R. HEBBES (29), *Biophysics Laboratories, University of Portsmouth, Portsmouth PO1 2DT, England, United Kingdom*

ANDREAS HECHT (22), *Max-Planck-Institut für Immunbiologie, 79108 Freiburg, Germany*

KAZUNORI HIRAYOSHI (19), *Department of Ultrastructural Research, Institute for Frontier Medical Sciences, Kyoto University, Sakyo-ku, Kyoto 606-8507, Japan*

WOLFRAM HÖRZ (20), *Institut für Physiologische Chemie der Universität München, D-80336 München, Germany*

PETER L. JONES (41), *Laboratory of Molecular Embryology, National Institute of Child Health and Human Development, National Institutes of Health, Bethesda, Maryland 20892-5431*

V. KARPOV (33), *W. A. Engelhardt Institute of Molecular Biology, Russian Academy of Sciences, Moscow 117984, Russia*

MICHAEL P. KLADDE (24), *Department of Biochemistry and Molecular Biology and Center for Gene Regulation, Pennsylvania State University, University Park, Pennsylvania 16802-4500*

JUN-ICHIRO KOMURA (30), *Department of Radiation Research, Tohoku University School of Medicine, Aoba-ku, Sendai 980-77, Japan*

PHILLIP R. KRAMER (36), *Laboratory of Neurochemistry, National Institute of Neurological Disorders and Stroke, National Institutes of Health, Bethesda, Maryland 20892*

ADAM J. KRIEG (7), *Department of Biochemistry, University of Illinois at Urbana–Champaign, Urbana, Illinois 61801*

THIERRY LAROCHE (38), *Swiss Institute for Experimental Cancer Research (ISREC), CH-1066 Epalinges, Switzerland*

KYU-MIN LEE (12), *Department of Biochemistry and Biophysics, University of Rochester Medical Center, Rochester, New York 14642*

QIAO LI (17), *Laboratory of Molecular Embryology, National Institutes of Health, Bethesda, Maryland 20892-5431*

JOHN T. LIS (19), *Section of Biochemistry, Molecular and Cell Biology, Cornell University, Ithaca, New York 14853*

MAGDALENA LIVINGSTONE-ZATCHEJ (25), *Institut für Zellbiologie, ETH-Zürich, Hönggerberg, CH-8093 Zürich, Switzerland*

COLIN LOGIE (42), *Program in Molecular Medicine and Department of Biochemistry and Molecular Biology, University of Massachusetts Medical Center, Worcester, Massachusetts 01605*

KAROLIN LUGER (1), *Institut für Molekularbiologie und Biophysik, ETH Zürich, Hönggerberg, CH-8093 Zürich, Switzerland*

ALEXANDER M. MAKHOV (11), *Lineberger Comprehensive Cancer Center, University of North Carolina at Chapel Hill, Chapel Hill, North Carolina 27599*

SVETLANA MELNIK (28), *A. N. Bach Institute of Biochemistry, Russian Academy of Sciences, Moscow 117071, Russia*

PAMELA B. MELUH (23), *Program in Molecular Biology, Memorial Sloan-Kettering Cancer Center, New York, New York 10021*

SUSAN MICHALOWSKI (11), *Lineberger Comprehensive Cancer Center, University of North Carolina at Chapel Hill, Chapel Hill, North Carolina 27599*

CRAIG A. MIZZEN (39), *Department of Biochemistry and Molecular Genetics, University of Virginia Health Sciences Center, Charlottesville, Virginia 22908*

JONATHAN G. MOGGS (18), *Section de Recherche, Unité Mixte du CNRS, Institut Curie, 75231 Paris cedex 05, France*

RANDALL H. MORSE (21), *Wadsworth Center, New York State Department of Health, and Department of Biomedical Sciences, State University of New York School of Public Health, Albany, New York 12208*

FIONA A. MYERS (29), *Biophysics Laboratories, University of Portsmouth, Portsmouth PO1 2DT, England, United Kingdom*

MARK S. NISSEN (9), *Department of Biochemistry/Biophysics, Washington State University, Pullman, Washington 99164-4660*

STUART H. ORKIN (31), *Division of Hematology/Oncology, Children's Hospital/Howard Hughes Medical Institute, Boston, Massachusetts 02115*

VINCENT OSSIPOW (44), *Laboratory of Molecular Cell Biology, National Cancer Institute, National Institutes of Health, Bethesda, Maryland 20892*

SARI PENNINGS (16), *Department of Biomedical Sciences, University of Edinburgh, Edinburgh EH8 9XD, Scotland, United Kingdom*

CRAIG L. PETERSON (42), *Program in Molecular Medicine and Department of Biochemistry and Molecular Biology, University of Massachusetts Medical Center, Worcester, Massachusetts 01605*

GERD P. PFEIFER (30), *Department of Biology, Beckman Research Institute of the City of Hope, Duarte, California 91010*

LORI A. PILE (26), *Department of Molecular Genetics, Biochemistry, and Microbiology, University of Cincinnati College of Medicine, Cincinnati, Ohio 45267*

K. J. POLACH (15), *Department of Biochemistry, Molecular Biology, and Cell Biology, Northwestern University, Evanston, Illinois 60208-3500*

YURI V. POSTNIKOV (8), *Protein Section, Division of Basic Sciences, National Cancer Institute, National Institutes of Health, Bethesda, Maryland 20892*

O. PREOBRAZHENSKAYA (33), *W. A. Engelhardt Institute of Molecular Biology, Russian Academy of Sciences, Moscow 117984, Russia*

DMITRY PRUSS (28), *Myriad Genetics, Salt Lake City, Utah 84108*

THOMAS J. RECHSTEINER (1), *Institut für Molekularbiologie und Biophysik, ETH Zürich, Hönggerberg, CH-8093 Zürich, Switzerland*

RAYMOND REEVES (9), *Department of Biochemistry/Biophysics, Washington State University, Pullman, Washington 99164-4660*

ANDREA R. RICCI (32), *Department of Biochemistry, University of Western Ontario, London, Ontario N6A 4L6, Canada*

TIMOTHY J. RICHMOND (1, 13), *Institut für Molekularbiologie und Biophysik, ETH Zürich, Hönggerberg, CH-8093 Zürich, Switzerland*

ARTHUR D. RIGGS (30), *Department of Biology, Beckman Research Institute of the City of Hope, Duarte, California 91010*

COLLEEN A. RYAN (6), *Department of Biology, Boston College, Chestnut Hill, Massachusetts 02467*

MICHAEL P. RYAN (21), *Wadsworth Center, New York State Department of Health, Albany, New York 12208*

RAPHAEL SANDALTZOPOULOS (44), *Laboratory of Molecular Cell Biology, National Cancer Institute, National Institutes of Health, Bethesda, Maryland 20892*

CAROLYN A. SCHNEIDER (6), *Salve Regina University, Newport, Rhode Island 02840*

ROBERT T. SIMPSON (24), *Department of Biochemistry and Molecular Biology and Center for Gene Regulation, Pennsylvania State University, University Park, Pennsylvania 16802-4500*

RICHARD R. SINDEN (36), *Center for Genome Research, Institute of Biosciences and Technology, Texas A&M University, Houston, Texas 77030-3303*

GRACE A. STAFFORD (21), *Wadsworth Center, New York State Department of Health, Albany, New York 12208*

ERICH C. STRAUSS (31), *Department of Ophthalmology, Massachusetts Eye and Ear Infirmary, Boston, Massachusetts 02114*

BERNHARD SUTER (25), *Institut für Zellbiologie, ETH-Zürich, Hönggerberg, CH-8093 Zürich, Switzerland*

FRITZ THOMA (25), *Institut für Zellbiologie, ETH-Zürich, Hönggerberg, CH-8093 Zürich, Switzerland*

ALAN W. THORNE (29), *Biophysics Laboratories, University of Portsmouth, Portsmouth PO1 2DT, England, United Kingdom*

DAVID JOHN TREMETHICK (4), *The John Curtin School of Medical Research, The Australian National University, Australian Capital Territory, 2601 Australia*

CHRISTIN TSE (2), *Department of Biochemistry, The University of Texas Health Science Center at San Antonio, San Antonio, Texas 78284-7760*

TOSHIO TSUKIYAMA (44), *Fred Hutchinson Cancer Research Center, Seattle, Washington 98109*

SERGEI I. USACHENKO (14), *Department of Biological Chemistry, School of Medicine, University of California at Davis, Davis, California 95616*

PATRICK D. VARGA-WEISZ (43), *European Molecular Biology Laboratory, 69117 Heidelberg, Germany*

DANIELLE VERMAAK (41), *Laboratory of Molecular Embryology, National Institute of Child Health and Human Development, National Institutes of Health, Bethesda, Maryland 20892-5431*

PAUL A. WADE (41), *Laboratory of Molecular Embryology, National Institute of Child Health and Human Development, National Institutes of Health, Bethesda, Maryland 20892-5431*

KERFOOT P. WALKER III (2), *Department of Biochemistry, The University of Texas Health Science Center at San Antonio, San Antonio, Texas 78284-7760*

LORI L. WALLRATH (26), *Department of Biochemistry, University of Iowa College of Medicine, Iowa City, Iowa 52242*

JANET A. WEBER (26), *Department of Molecular and Cell Biology, Pennsylvania State University, University Park, Pennsylvania 16802*

J. WIDOM (15), *Department of Biochemistry, Molecular Biology, and Cell Biology, Northwestern University, Evanston, Illinois 60208-3500*

ALAN P. WOLFFE (41), *Laboratory of Molecular Embryology, National Institute of Child Health and Human Development, National Institutes of Health, Bethesda, Maryland 20892-5431*

CHRISTOPHER L. WOODCOCK (10), *Biology Department, University of Massachusetts, Amherst, Massachusetts 01003*

JERRY L. WORKMAN (40), *Howard Hughes Medical Institute, Department of Biochemistry and Molecular Biology and the Center for Gene Regulation, Pennsylvania State University, University Park, Pennsylvania 16802*

ÖRJAN WRANGE (17), *Department of Cell and Molecular Biology, Karolinska Institute, S-17177 Stockholm, Sweden*

CARL WU (44), *Laboratory of Molecular Cell Biology, National Cancer Institute, National Institutes of Health, Bethesda, Maryland 20892*

M. XU (24), *Department of Biochemistry and Molecular Biology and Center for Gene Regulation, Pennsylvania State University, University Park, Pennsylvania 16802-4500*

LIUNING YU (21), *Department of Biomedical Sciences, State University of New York School of Public Health, Albany, New York 12208*

KENNETH S. ZARET (34), *Department of Molecular Biology, Cell Biology, and Biochemistry, Brown University, Providence, Rhode Island 02912*

Preface

The last decade has seen remarkable progress in our understanding of the diverse roles of nucleosomes, chromatin, and nuclear organization in the biological activities (e.g., transcription and replication) of DNA. Histones and HMG proteins facilitate the assembly of specific regulatory nucleoprotein structures, whose architecture can be modulated by the recruitment of chromatin remodeling complexes. This dynamic quality of chromatin fulfills essential regulatory functions in transcription and, most probably, also in DNA replication, recombination, and repair.

This volume of *Methods in Enzymology* includes up-to-date procedures employed for the assembly of nucleosomes, chromatin, and nuclei, extending the classical procedures described nearly ten years ago in Volume 170 (*Nucleosomes*) of this series, and should assist in the further investigation of the ways in which the structural dynamics of chromatin contribute to the regulation of transcription, replication, recombination, and repair. Assays for the structure and function of *in vitro* reconstituted chromatin and for defining the organization and characteristics of natural chromosomal material from yeast (*Saccharomyces cerevisiae*), flies (*Drosophila melanogaster*), and frogs (*Xenopus laevis*), as well as from mammalian tissues, are described. Finally, the purification and assay procedures for various chromatin remodeling activities, including histone acetyltransferases, histone deacetylases, and SWI/SNF ATPases, are detailed.

We thank the many authors for their excellent contributions and their patience in dealing with publication schedules. We hope that this volume on *Chromatin* will be useful to many investigators and will find its way to the laboratory bench.

Finally, PMW would like to take this opportunity to remember Dr. Nathan O. Kaplan, his thesis advisor at Brandeis University. Nate was a superb biochemist who imparted the excitement of scientific research to everyone around him. He is missed, but not forgotten by those who knew him.

PAUL M. WASSARMAN
ALAN P. WOLFFE

METHODS IN ENZYMOLOGY

VOLUME I. Preparation and Assay of Enzymes
Edited by SIDNEY P. COLOWICK AND NATHAN O. KAPLAN

VOLUME II. Preparation and Assay of Enzymes
Edited by SIDNEY P. COLOWICK AND NATHAN O. KAPLAN

VOLUME III. Preparation and Assay of Substrates
Edited by SIDNEY P. COLOWICK AND NATHAN O. KAPLAN

VOLUME IV. Special Techniques for the Enzymologist
Edited by SIDNEY P. COLOWICK AND NATHAN O. KAPLAN

VOLUME V. Preparation and Assay of Enzymes
Edited by SIDNEY P. COLOWICK AND NATHAN O. KAPLAN

VOLUME VI. Preparation and Assay of Enzymes (*Continued*)
Preparation and Assay of Substrates
Special Techniques
Edited by SIDNEY P. COLOWICK AND NATHAN O. KAPLAN

VOLUME VII. Cumulative Subject Index
Edited by SIDNEY P. COLOWICK AND NATHAN O. KAPLAN

VOLUME VIII. Complex Carbohydrates
Edited by ELIZABETH F. NEUFELD AND VICTOR GINSBURG

VOLUME IX. Carbohydrate Metabolism
Edited by WILLIS A. WOOD

VOLUME X. Oxidation and Phosphorylation
Edited by RONALD W. ESTABROOK AND MAYNARD E. PULLMAN

VOLUME XI. Enzyme Structure
Edited by C. H. W. HIRS

VOLUME XII. Nucleic Acids (Parts A and B)
Edited by LAWRENCE GROSSMAN AND KIVIE MOLDAVE

VOLUME XIII. Citric Acid Cycle
Edited by J. M. LOWENSTEIN

VOLUME XIV. Lipids
Edited by J. M. LOWENSTEIN

VOLUME XV. Steroids and Terpenoids
Edited by RAYMOND B. CLAYTON

VOLUME XVI. Fast Reactions
Edited by KENNETH KUSTIN

VOLUME XVII. Metabolism of Amino Acids and Amines (Parts A and B)
Edited by HERBERT TABOR AND CELIA WHITE TABOR

VOLUME XVIII. Vitamins and Coenzymes (Parts A, B, and C)
Edited by DONALD B. MCCORMICK AND LEMUEL D. WRIGHT

VOLUME XIX. Proteolytic Enzymes
Edited by GERTRUDE E. PERLMANN AND LASZLO LORAND

VOLUME XX. Nucleic Acids and Protein Synthesis (Part C)
Edited by KIVIE MOLDAVE AND LAWRENCE GROSSMAN

VOLUME XXI. Nucleic Acids (Part D)
Edited by LAWRENCE GROSSMAN AND KIVIE MOLDAVE

VOLUME XXII. Enzyme Purification and Related Techniques
Edited by WILLIAM B. JAKOBY

VOLUME XXIII. Photosynthesis (Part A)
Edited by ANTHONY SAN PIETRO

VOLUME XXIV. Photosynthesis and Nitrogen Fixation (Part B)
Edited by ANTHONY SAN PIETRO

VOLUME XXV. Enzyme Structure (Part B)
Edited by C. H. W. HIRS AND SERGE N. TIMASHEFF

VOLUME XXVI. Enzyme Structure (Part C)
Edited by C. H. W. HIRS AND SERGE N. TIMASHEFF

VOLUME XXVII. Enzyme Structure (Part D)
Edited by C. H. W. HIRS AND SERGE N. TIMASHEFF

VOLUME XXVIII. Complex Carbohydrates (Part B)
Edited by VICTOR GINSBURG

VOLUME XXIX. Nucleic Acids and Protein Synthesis (Part E)
Edited by LAWRENCE GROSSMAN AND KIVIE MOLDAVE

VOLUME XXX. Nucleic Acids and Protein Synthesis (Part F)
Edited by KIVIE MOLDAVE AND LAWRENCE GROSSMAN

VOLUME XXXI. Biomembranes (Part A)
Edited by SIDNEY FLEISCHER AND LESTER PACKER

VOLUME XXXII. Biomembranes (Part B)
Edited by SIDNEY FLEISCHER AND LESTER PACKER

VOLUME XXXIII. Cumulative Subject Index Volumes I–XXX
Edited by MARTHA G. DENNIS AND EDWARD A. DENNIS

VOLUME XXXIV. Affinity Techniques (Enzyme Purification: Part B)
Edited by WILLIAM B. JAKOBY AND MEIR WILCHEK

VOLUME XXXV. Lipids (Part B)
Edited by JOHN M. LOWENSTEIN

VOLUME XXXVI. Hormone Action (Part A: Steroid Hormones)
Edited by BERT W. O'MALLEY AND JOEL G. HARDMAN

VOLUME XXXVII. Hormone Action (Part B: Peptide Hormones)
Edited by BERT W. O'MALLEY AND JOEL G. HARDMAN

VOLUME XXXVIII. Hormone Action (Part C: Cyclic Nucleotides)
Edited by JOEL G. HARDMAN AND BERT W. O'MALLEY

VOLUME XXXIX. Hormone Action (Part D: Isolated Cells, Tissues, and Organ Systems)
Edited by JOEL G. HARDMAN AND BERT W. O'MALLEY

VOLUME XL. Hormone Action (Part E: Nuclear Structure and Function)
Edited by BERT W. O'MALLEY AND JOEL G. HARDMAN

VOLUME XLI. Carbohydrate Metabolism (Part B)
Edited by W. A. WOOD

VOLUME XLII. Carbohydrate Metabolism (Part C)
Edited by W. A. WOOD

VOLUME XLIII. Antibiotics
Edited by JOHN H. HASH

VOLUME XLIV. Immobilized Enzymes
Edited by KLAUS MOSBACH

VOLUME XLV. Proteolytic Enzymes (Part B)
Edited by LASZLO LORAND

VOLUME XLVI. Affinity Labeling
Edited by WILLIAM B. JAKOBY AND MEIR WILCHEK

VOLUME XLVII. Enzyme Structure (Part E)
Edited by C. H. W. HIRS AND SERGE N. TIMASHEFF

VOLUME XLVIII. Enzyme Structure (Part F)
Edited by C. H. W. HIRS AND SERGE N. TIMASHEFF

VOLUME XLIX. Enzyme Structure (Part G)
Edited by C. H. W. HIRS AND SERGE N. TIMASHEFF

VOLUME L. Complex Carbohydrates (Part C)
Edited by VICTOR GINSBURG

VOLUME LI. Purine and Pyrimidine Nucleotide Metabolism
Edited by PATRICIA A. HOFFEE AND MARY ELLEN JONES

VOLUME LII. Biomembranes (Part C: Biological Oxidations)
Edited by SIDNEY FLEISCHER AND LESTER PACKER

VOLUME LIII. Biomembranes (Part D: Biological Oxidations)
Edited by SIDNEY FLEISCHER AND LESTER PACKER

VOLUME LIV. Biomembranes (Part E: Biological Oxidations)
Edited by SIDNEY FLEISCHER AND LESTER PACKER

VOLUME LV. Biomembranes (Part F: Bioenergetics)
Edited by SIDNEY FLEISCHER AND LESTER PACKER

VOLUME LVI. Biomembranes (Part G: Bioenergetics)
Edited by SIDNEY FLEISCHER AND LESTER PACKER

VOLUME LVII. Bioluminescence and Chemiluminescence
Edited by MARLENE A. DELUCA

VOLUME LVIII. Cell Culture
Edited by WILLIAM B. JAKOBY AND IRA PASTAN

VOLUME LIX. Nucleic Acids and Protein Synthesis (Part G)
Edited by KIVIE MOLDAVE AND LAWRENCE GROSSMAN

VOLUME LX. Nucleic Acids and Protein Synthesis (Part H)
Edited by KIVIE MOLDAVE AND LAWRENCE GROSSMAN

VOLUME 61. Enzyme Structure (Part H)
Edited by C. H. W. HIRS AND SERGE N. TIMASHEFF

VOLUME 62. Vitamins and Coenzymes (Part D)
Edited by DONALD B. MCCORMICK AND LEMUEL D. WRIGHT

VOLUME 63. Enzyme Kinetics and Mechanism (Part A: Initial Rate and Inhibitor Methods)
Edited by DANIEL L. PURICH

VOLUME 64. Enzyme Kinetics and Mechanism (Part B: Isotopic Probes and Complex Enzyme Systems)
Edited by DANIEL L. PURICH

VOLUME 65. Nucleic Acids (Part I)
Edited by LAWRENCE GROSSMAN AND KIVIE MOLDAVE

VOLUME 66. Vitamins and Coenzymes (Part E)
Edited by DONALD B. MCCORMICK AND LEMUEL D. WRIGHT

VOLUME 67. Vitamins and Coenzymes (Part F)
Edited by DONALD B. MCCORMICK AND LEMUEL D. WRIGHT

VOLUME 68. Recombinant DNA
Edited by RAY WU

VOLUME 69. Photosynthesis and Nitrogen Fixation (Part C)
Edited by ANTHONY SAN PIETRO

VOLUME 70. Immunochemical Techniques (Part A)
Edited by HELEN VAN VUNAKIS AND JOHN J. LANGONE

VOLUME 71. Lipids (Part C)
Edited by JOHN M. LOWENSTEIN

VOLUME 72. Lipids (Part D)
Edited by JOHN M. LOWENSTEIN

VOLUME 73. Immunochemical Techniques (Part B)
Edited by JOHN J. LANGONE AND HELEN VAN VUNAKIS

VOLUME 74. Immunochemical Techniques (Part C)
Edited by JOHN J. LANGONE AND HELEN VAN VUNAKIS

VOLUME 75. Cumulative Subject Index Volumes XXXI, XXXII, XXXIV–LX
Edited by EDWARD A. DENNIS AND MARTHA G. DENNIS

VOLUME 76. Hemoglobins
Edited by ERALDO ANTONINI, LUIGI ROSSI-BERNARDI, AND EMILIA CHIANCONE

VOLUME 77. Detoxication and Drug Metabolism
Edited by WILLIAM B. JAKOBY

VOLUME 78. Interferons (Part A)
Edited by SIDNEY PESTKA

VOLUME 79. Interferons (Part B)
Edited by SIDNEY PESTKA

VOLUME 80. Proteolytic Enzymes (Part C)
Edited by LASZLO LORAND

VOLUME 81. Biomembranes (Part H: Visual Pigments and Purple Membranes, I)
Edited by LESTER PACKER

VOLUME 82. Structural and Contractile Proteins (Part A: Extracellular Matrix)
Edited by LEON W. CUNNINGHAM AND DIXIE W. FREDERIKSEN

VOLUME 83. Complex Carbohydrates (Part D)
Edited by VICTOR GINSBURG

VOLUME 84. Immunochemical Techniques (Part D: Selected Immunoassays)
Edited by JOHN J. LANGONE AND HELEN VAN VUNAKIS

VOLUME 85. Structural and Contractile Proteins (Part B: The Contractile Apparatus and the Cytoskeleton)
Edited by DIXIE W. FREDERIKSEN AND LEON W. CUNNINGHAM

VOLUME 86. Prostaglandins and Arachidonate Metabolites
Edited by WILLIAM E. M. LANDS AND WILLIAM L. SMITH

VOLUME 87. Enzyme Kinetics and Mechanism (Part C: Intermediates, Stereochemistry, and Rate Studies)
Edited by DANIEL L. PURICH

VOLUME 88. Biomembranes (Part I: Visual Pigments and Purple Membranes, II)
Edited by LESTER PACKER

VOLUME 89. Carbohydrate Metabolism (Part D)
Edited by WILLIS A. WOOD

VOLUME 90. Carbohydrate Metabolism (Part E)
Edited by WILLIS A. WOOD

VOLUME 91. Enzyme Structure (Part I)
Edited by C. H. W. HIRS AND SERGE N. TIMASHEFF

VOLUME 92. Immunochemical Techniques (Part E: Monoclonal Antibodies and General Immunoassay Methods)
Edited by JOHN J. LANGONE AND HELEN VAN VUNAKIS

VOLUME 93. Immunochemical Techniques (Part F: Conventional Antibodies, Fc Receptors, and Cytotoxicity)
Edited by JOHN J. LANGONE AND HELEN VAN VUNAKIS

VOLUME 94. Polyamines
Edited by HERBERT TABOR AND CELIA WHITE TABOR

VOLUME 95. Cumulative Subject Index Volumes 61–74, 76–80
Edited by EDWARD A. DENNIS AND MARTHA G. DENNIS

VOLUME 96. Biomembranes [Part J: Membrane Biogenesis: Assembly and Targeting (General Methods; Eukaryotes)]
Edited by SIDNEY FLEISCHER AND BECCA FLEISCHER

VOLUME 97. Biomembranes [Part K: Membrane Biogenesis: Assembly and Targeting (Prokaryotes, Mitochondria, and Chloroplasts)]
Edited by SIDNEY FLEISCHER AND BECCA FLEISCHER

VOLUME 98. Biomembranes (Part L: Membrane Biogenesis: Processing and Recycling)
Edited by SIDNEY FLEISCHER AND BECCA FLEISCHER

VOLUME 99. Hormone Action (Part F: Protein Kinases)
Edited by JACKIE D. CORBIN AND JOEL G. HARDMAN

VOLUME 100. Recombinant DNA (Part B)
Edited by RAY WU, LAWRENCE GROSSMAN, AND KIVIE MOLDAVE

VOLUME 101. Recombinant DNA (Part C)
Edited by RAY WU, LAWRENCE GROSSMAN, AND KIVIE MOLDAVE

VOLUME 102. Hormone Action (Part G: Calmodulin and Calcium-Binding Proteins)
Edited by ANTHONY R. MEANS AND BERT W. O'MALLEY

VOLUME 103. Hormone Action (Part H: Neuroendocrine Peptides)
Edited by P. MICHAEL CONN

VOLUME 104. Enzyme Purification and Related Techniques (Part C)
Edited by WILLIAM B. JAKOBY

VOLUME 105. Oxygen Radicals in Biological Systems
Edited by LESTER PACKER

VOLUME 106. Posttranslational Modifications (Part A)
Edited by FINN WOLD AND KIVIE MOLDAVE

VOLUME 107. Posttranslational Modifications (Part B)
Edited by FINN WOLD AND KIVIE MOLDAVE

VOLUME 108. Immunochemical Techniques (Part G: Separation and Characterization of Lymphoid Cells)
Edited by GIOVANNI DI SABATO, JOHN J. LANGONE, AND HELEN VAN VUNAKIS

VOLUME 109. Hormone Action (Part I: Peptide Hormones)
Edited by LUTZ BIRNBAUMER AND BERT W. O'MALLEY

VOLUME 110. Steroids and Isoprenoids (Part A)
Edited by JOHN H. LAW AND HANS C. RILLING

VOLUME 111. Steroids and Isoprenoids (Part B)
Edited by JOHN H. LAW AND HANS C. RILLING

VOLUME 112. Drug and Enzyme Targeting (Part A)
Edited by KENNETH J. WIDDER AND RALPH GREEN

VOLUME 113. Glutamate, Glutamine, Glutathione, and Related Compounds
Edited by ALTON MEISTER

VOLUME 114. Diffraction Methods for Biological Macromolecules (Part A)
Edited by HAROLD W. WYCKOFF, C. H. W. HIRS, AND SERGE N. TIMASHEFF

VOLUME 115. Diffraction Methods for Biological Macromolecules (Part B)
Edited by HAROLD W. WYCKOFF, C. H. W. HIRS, AND SERGE N. TIMASHEFF

VOLUME 116. Immunochemical Techniques (Part H: Effectors and Mediators of Lymphoid Cell Functions)
Edited by GIOVANNI DI SABATO, JOHN J. LANGONE, AND HELEN VAN VUNAKIS

VOLUME 117. Enzyme Structure (Part J)
Edited by C. H. W. HIRS AND SERGE N. TIMASHEFF

VOLUME 118. Plant Molecular Biology
Edited by ARTHUR WEISSBACH AND HERBERT WEISSBACH

VOLUME 119. Interferons (Part C)
Edited by SIDNEY PESTKA

VOLUME 120. Cumulative Subject Index Volumes 81–94, 96–101

VOLUME 121. Immunochemical Techniques (Part I: Hybridoma Technology and Monoclonal Antibodies)
Edited by JOHN J. LANGONE AND HELEN VAN VUNAKIS

VOLUME 122. Vitamins and Coenzymes (Part G)
Edited by FRANK CHYTIL AND DONALD B. MCCORMICK

VOLUME 123. Vitamins and Coenzymes (Part H)
Edited by FRANK CHYTIL AND DONALD B. MCCORMICK

VOLUME 124. Hormone Action (Part J: Neuroendocrine Peptides)
Edited by P. MICHAEL CONN

VOLUME 125. Biomembranes (Part M: Transport in Bacteria, Mitochondria, and Chloroplasts: General Approaches and Transport Systems)
Edited by SIDNEY FLEISCHER AND BECCA FLEISCHER

VOLUME 126. Biomembranes (Part N: Transport in Bacteria, Mitochondria, and Chloroplasts: Protonmotive Force)
Edited by SIDNEY FLEISCHER AND BECCA FLEISCHER

VOLUME 127. Biomembranes (Part O: Protons and Water: Structure and Translocation)
Edited by LESTER PACKER

VOLUME 128. Plasma Lipoproteins (Part A: Preparation, Structure, and Molecular Biology)
Edited by JERE P. SEGREST AND JOHN J. ALBERS

VOLUME 129. Plasma Lipoproteins (Part B: Characterization, Cell Biology, and Metabolism)
Edited by JOHN J. ALBERS AND JERE P. SEGREST

VOLUME 130. Enzyme Structure (Part K)
Edited by C. H. W. HIRS AND SERGE N. TIMASHEFF

VOLUME 131. Enzyme Structure (Part L)
Edited by C. H. W. HIRS AND SERGE N. TIMASHEFF

VOLUME 132. Immunochemical Techniques (Part J: Phagocytosis and Cell-Mediated Cytotoxicity)
Edited by GIOVANNI DI SABATO AND JOHANNES EVERSE

VOLUME 133. Bioluminescence and Chemiluminescence (Part B)
Edited by MARLENE DELUCA AND WILLIAM D. MCELROY

VOLUME 134. Structural and Contractile Proteins (Part C: The Contractile Apparatus and the Cytoskeleton)
Edited by RICHARD B. VALLEE

VOLUME 135. Immobilized Enzymes and Cells (Part B)
Edited by KLAUS MOSBACH

VOLUME 136. Immobilized Enzymes and Cells (Part C)
Edited by KLAUS MOSBACH

VOLUME 137. Immobilized Enzymes and Cells (Part D)
Edited by KLAUS MOSBACH

VOLUME 138. Complex Carbohydrates (Part E)
Edited by VICTOR GINSBURG

VOLUME 139. Cellular Regulators (Part A: Calcium- and Calmodulin-Binding Proteins)
Edited by ANTHONY R. MEANS AND P. MICHAEL CONN

VOLUME 140. Cumulative Subject Index Volumes 102–119, 121–134

VOLUME 141. Cellular Regulators (Part B: Calcium and Lipids)
Edited by P. MICHAEL CONN AND ANTHONY R. MEANS

VOLUME 142. Metabolism of Aromatic Amino Acids and Amines
Edited by SEYMOUR KAUFMAN

VOLUME 143. Sulfur and Sulfur Amino Acids
Edited by WILLIAM B. JAKOBY AND OWEN GRIFFITH

VOLUME 144. Structural and Contractile Proteins (Part D: Extracellular Matrix)
Edited by LEON W. CUNNINGHAM

VOLUME 145. Structural and Contractile Proteins (Part E: Extracellular Matrix)
Edited by LEON W. CUNNINGHAM

VOLUME 146. Peptide Growth Factors (Part A)
Edited by DAVID BARNES AND DAVID A. SIRBASKU

VOLUME 147. Peptide Growth Factors (Part B)
Edited by DAVID BARNES AND DAVID A. SIRBASKU

VOLUME 148. Plant Cell Membranes
Edited by LESTER PACKER AND ROLAND DOUCE

VOLUME 149. Drug and Enzyme Targeting (Part B)
Edited by RALPH GREEN AND KENNETH J. WIDDER

VOLUME 150. Immunochemical Techniques (Part K: *In Vitro* Models of B and T Cell Functions and Lymphoid Cell Receptors)
Edited by GIOVANNI DI SABATO

VOLUME 151. Molecular Genetics of Mammalian Cells
Edited by MICHAEL M. GOTTESMAN

VOLUME 152. Guide to Molecular Cloning Techniques
Edited by SHELBY L. BERGER AND ALAN R. KIMMEL

VOLUME 153. Recombinant DNA (Part D)
Edited by RAY WU AND LAWRENCE GROSSMAN

VOLUME 154. Recombinant DNA (Part E)
Edited by RAY WU AND LAWRENCE GROSSMAN

VOLUME 155. Recombinant DNA (Part F)
Edited by RAY WU

VOLUME 156. Biomembranes (Part P: ATP-Driven Pumps and Related Transport: The Na,K-Pump)
Edited by SIDNEY FLEISCHER AND BECCA FLEISCHER

VOLUME 157. Biomembranes (Part Q: ATP-Driven Pumps and Related Transport: Calcium, Proton, and Potassium Pumps)
Edited by SIDNEY FLEISCHER AND BECCA FLEISCHER

VOLUME 158. Metalloproteins (Part A)
Edited by JAMES F. RIORDAN AND BERT L. VALLEE

VOLUME 159. Initiation and Termination of Cyclic Nucleotide Action
Edited by JACKIE D. CORBIN AND ROGER A. JOHNSON

VOLUME 160. Biomass (Part A: Cellulose and Hemicellulose)
Edited by WILLIS A. WOOD AND SCOTT T. KELLOGG

VOLUME 161. Biomass (Part B: Lignin, Pectin, and Chitin)
Edited by WILLIS A. WOOD AND SCOTT T. KELLOGG

VOLUME 162. Immunochemical Techniques (Part L: Chemotaxis and Inflammation)
Edited by GIOVANNI DI SABATO

VOLUME 163. Immunochemical Techniques (Part M: Chemotaxis and Inflammation)
Edited by GIOVANNI DI SABATO

VOLUME 164. Ribosomes
Edited by HARRY F. NOLLER, JR., AND KIVIE MOLDAVE

VOLUME 165. Microbial Toxins: Tools for Enzymology
Edited by SIDNEY HARSHMAN

VOLUME 166. Branched-Chain Amino Acids
Edited by ROBERT HARRIS AND JOHN R. SOKATCH

VOLUME 167. Cyanobacteria
Edited by LESTER PACKER AND ALEXANDER N. GLAZER

VOLUME 168. Hormone Action (Part K: Neuroendocrine Peptides)
Edited by P. MICHAEL CONN

VOLUME 169. Platelets: Receptors, Adhesion, Secretion (Part A)
Edited by JACEK HAWIGER

VOLUME 170. Nucleosomes
Edited by PAUL M. WASSARMAN AND ROGER D. KORNBERG

VOLUME 171. Biomembranes (Part R: Transport Theory: Cells and Model Membranes)
Edited by SIDNEY FLEISCHER AND BECCA FLEISCHER

VOLUME 172. Biomembranes (Part S: Transport: Membrane Isolation and Characterization)
Edited by SIDNEY FLEISCHER AND BECCA FLEISCHER

VOLUME 173. Biomembranes [Part T: Cellular and Subcellular Transport: Eukaryotic (Nonepithelial) Cells]
Edited by SIDNEY FLEISCHER AND BECCA FLEISCHER

VOLUME 174. Biomembranes [Part U: Cellular and Subcellular Transport: Eukaryotic (Nonepithelial) Cells]
Edited by SIDNEY FLEISCHER AND BECCA FLEISCHER

VOLUME 175. Cumulative Subject Index Volumes 135–139, 141–167

VOLUME 176. Nuclear Magnetic Resonance (Part A: Spectral Techniques and Dynamics)
Edited by NORMAN J. OPPENHEIMER AND THOMAS L. JAMES

VOLUME 177. Nuclear Magnetic Resonance (Part B: Structure and Mechanism)
Edited by NORMAN J. OPPENHEIMER AND THOMAS L. JAMES

VOLUME 178. Antibodies, Antigens, and Molecular Mimicry
Edited by JOHN J. LANGONE

VOLUME 179. Complex Carbohydrates (Part F)
Edited by VICTOR GINSBURG

VOLUME 180. RNA Processing (Part A: General Methods)
Edited by JAMES E. DAHLBERG AND JOHN N. ABELSON

VOLUME 181. RNA Processing (Part B: Specific Methods)
Edited by JAMES E. DAHLBERG AND JOHN N. ABELSON

VOLUME 182. Guide to Protein Purification
Edited by MURRAY P. DEUTSCHER

VOLUME 183. Molecular Evolution: Computer Analysis of Protein and Nucleic Acid Sequences
Edited by RUSSELL F. DOOLITTLE

VOLUME 184. Avidin–Biotin Technology
Edited by MEIR WILCHEK AND EDWARD A. BAYER

VOLUME 185. Gene Expression Technology
Edited by DAVID V. GOEDDEL

VOLUME 186. Oxygen Radicals in Biological Systems (Part B: Oxygen Radicals and Antioxidants)
Edited by LESTER PACKER AND ALEXANDER N. GLAZER

VOLUME 187. Arachidonate Related Lipid Mediators
Edited by ROBERT C. MURPHY AND FRANK A. FITZPATRICK

VOLUME 188. Hydrocarbons and Methylotrophy
Edited by MARY E. LIDSTROM

VOLUME 189. Retinoids (Part A: Molecular and Metabolic Aspects)
Edited by LESTER PACKER

VOLUME 190. Retinoids (Part B: Cell Differentiation and Clinical Applications)
Edited by LESTER PACKER

VOLUME 191. Biomembranes (Part V: Cellular and Subcellular Transport: Epithelial Cells)
Edited by SIDNEY FLEISCHER AND BECCA FLEISCHER

VOLUME 192. Biomembranes (Part W: Cellular and Subcellular Transport: Epithelial Cells)
Edited by SIDNEY FLEISCHER AND BECCA FLEISCHER

VOLUME 193. Mass Spectrometry
Edited by JAMES A. MCCLOSKEY

VOLUME 194. Guide to Yeast Genetics and Molecular Biology
Edited by CHRISTINE GUTHRIE AND GERALD R. FINK

VOLUME 195. Adenylyl Cyclase, G Proteins, and Guanylyl Cyclase
Edited by ROGER A. JOHNSON AND JACKIE D. CORBIN

VOLUME 196. Molecular Motors and the Cytoskeleton
Edited by RICHARD B. VALLEE

VOLUME 197. Phospholipases
Edited by EDWARD A. DENNIS

VOLUME 198. Peptide Growth Factors (Part C)
Edited by DAVID BARNES, J. P. MATHER, AND GORDON H. SATO

VOLUME 199. Cumulative Subject Index Volumes 168–174, 176–194

VOLUME 200. Protein Phosphorylation (Part A: Protein Kinases: Assays, Purification, Antibodies, Functional Analysis, Cloning, and Expression)
Edited by TONY HUNTER AND BARTHOLOMEW M. SEFTON

VOLUME 201. Protein Phosphorylation (Part B: Analysis of Protein Phosphorylation, Protein Kinase Inhibitors, and Protein Phosphatases)
Edited by TONY HUNTER AND BARTHOLOMEW M. SEFTON

VOLUME 202. Molecular Design and Modeling: Concepts and Applications (Part A: Proteins, Peptides, and Enzymes)
Edited by JOHN J. LANGONE

VOLUME 203. Molecular Design and Modeling: Concepts and Applications (Part B: Antibodies and Antigens, Nucleic Acids, Polysaccharides, and Drugs)
Edited by JOHN J. LANGONE

VOLUME 204. Bacterial Genetic Systems
Edited by JEFFREY H. MILLER

VOLUME 205. Metallobiochemistry (Part B: Metallothionein and Related Molecules)
Edited by JAMES F. RIORDAN AND BERT L. VALLEE

VOLUME 206. Cytochrome P450
Edited by MICHAEL R. WATERMAN AND ERIC F. JOHNSON

VOLUME 207. Ion Channels
Edited by BERNARDO RUDY AND LINDA E. IVERSON

VOLUME 208. Protein–DNA Interactions
Edited by ROBERT T. SAUER

VOLUME 209. Phospholipid Biosynthesis
Edited by EDWARD A. DENNIS AND DENNIS E. VANCE

VOLUME 210. Numerical Computer Methods
Edited by LUDWIG BRAND AND MICHAEL L. JOHNSON

VOLUME 211. DNA Structures (Part A: Synthesis and Physical Analysis of DNA)
Edited by DAVID M. J. LILLEY AND JAMES E. DAHLBERG

VOLUME 212. DNA Structures (Part B: Chemical and Electrophoretic Analysis of DNA)
Edited by DAVID M. J. LILLEY AND JAMES E. DAHLBERG

VOLUME 213. Carotenoids (Part A: Chemistry, Separation, Quantitation, and Antioxidation)
Edited by LESTER PACKER

VOLUME 214. Carotenoids (Part B: Metabolism, Genetics, and Biosynthesis)
Edited by LESTER PACKER

VOLUME 215. Platelets: Receptors, Adhesion, Secretion (Part B)
Edited by JACEK J. HAWIGER

VOLUME 216. Recombinant DNA (Part G)
Edited by RAY WU

VOLUME 217. Recombinant DNA (Part H)
Edited by RAY WU

VOLUME 218. Recombinant DNA (Part I)
Edited by RAY WU

VOLUME 219. Reconstitution of Intracellular Transport
Edited by JAMES E. ROTHMAN

VOLUME 220. Membrane Fusion Techniques (Part A)
Edited by NEJAT DÜZGÜNEŞ

VOLUME 221. Membrane Fusion Techniques (Part B)
Edited by NEJAT DÜZGÜNEŞ

VOLUME 222. Proteolytic Enzymes in Coagulation, Fibrinolysis, and Complement Activation (Part A: Mammalian Blood Coagulation Factors and Inhibitors)
Edited by LASZLO LORAND AND KENNETH G. MANN

VOLUME 223. Proteolytic Enzymes in Coagulation, Fibrinolysis, and Complement Activation (Part B: Complement Activation, Fibrinolysis, and Nonmammalian Blood Coagulation Factors)
Edited by LASZLO LORAND AND KENNETH G. MANN

VOLUME 224. Molecular Evolution: Producing the Biochemical Data
Edited by ELIZABETH ANNE ZIMMER, THOMAS J. WHITE, REBECCA L. CANN, AND ALLAN C. WILSON

VOLUME 225. Guide to Techniques in Mouse Development
Edited by PAUL M. WASSARMAN AND MELVIN L. DEPAMPHILIS

VOLUME 226. Metallobiochemistry (Part C: Spectroscopic and Physical Methods for Probing Metal Ion Environments in Metalloenzymes and Metalloproteins)
Edited by JAMES F. RIORDAN AND BERT L. VALLEE

VOLUME 227. Metallobiochemistry (Part D: Physical and Spectroscopic Methods for Probing Metal Ion Environments in Metalloproteins)
Edited by JAMES F. RIORDAN AND BERT L. VALLEE

VOLUME 228. Aqueous Two-Phase Systems
Edited by HARRY WALTER AND GÖTE JOHANSSON

VOLUME 229. Cumulative Subject Index Volumes 195–198, 200–227

VOLUME 230. Guide to Techniques in Glycobiology
Edited by WILLIAM J. LENNARZ AND GERALD W. HART

VOLUME 231. Hemoglobins (Part B: Biochemical and Analytical Methods)
Edited by JOHANNES EVERSE, KIM D. VANDEGRIFF, AND ROBERT M. WINSLOW

VOLUME 232. Hemoglobins (Part C: Biophysical Methods)
Edited by JOHANNES EVERSE, KIM D. VANDEGRIFF, AND ROBERT M. WINSLOW

VOLUME 233. Oxygen Radicals in Biological Systems (Part C)
Edited by LESTER PACKER

VOLUME 234. Oxygen Radicals in Biological Systems (Part D)
Edited by LESTER PACKER

VOLUME 235. Bacterial Pathogenesis (Part A: Identification and Regulation of Virulence Factors)
Edited by VIRGINIA L. CLARK AND PATRIK M. BAVOIL

VOLUME 236. Bacterial Pathogenesis (Part B: Integration of Pathogenic Bacteria with Host Cells)
Edited by VIRGINIA L. CLARK AND PATRIK M. BAVOIL

VOLUME 237. Heterotrimeric G Proteins
Edited by RAVI IYENGAR

VOLUME 238. Heterotrimeric G-Protein Effectors
Edited by RAVI IYENGAR

VOLUME 239. Nuclear Magnetic Resonance (Part C)
Edited by THOMAS L. JAMES AND NORMAN J. OPPENHEIMER

VOLUME 240. Numerical Computer Methods (Part B)
Edited by MICHAEL L. JOHNSON AND LUDWIG BRAND

VOLUME 241. Retroviral Proteases
Edited by LAWRENCE C. KUO AND JULES A. SHAFER

VOLUME 242. Neoglycoconjugates (Part A)
Edited by Y. C. LEE AND REIKO T. LEE

VOLUME 243. Inorganic Microbial Sulfur Metabolism
Edited by HARRY D. PECK, JR., AND JEAN LEGALL

VOLUME 244. Proteolytic Enzymes: Serine and Cysteine Peptidases
Edited by ALAN J. BARRETT

VOLUME 245. Extracellular Matrix Components
Edited by E. RUOSLAHTI AND E. ENGVALL

VOLUME 246. Biochemical Spectroscopy
Edited by KENNETH SAUER

VOLUME 247. Neoglycoconjugates (Part B: Biomedical Applications)
Edited by Y. C. LEE AND REIKO T. LEE

VOLUME 248. Proteolytic Enzymes: Aspartic and Metallo Peptidases
Edited by ALAN J. BARRETT

VOLUME 249. Enzyme Kinetics and Mechanism (Part D: Developments in Enzyme Dynamics)
Edited by DANIEL L. PURICH

VOLUME 250. Lipid Modifications of Proteins
Edited by PATRICK J. CASEY AND JANICE E. BUSS

VOLUME 251. Biothiols (Part A: Monothiols and Dithiols, Protein Thiols, and Thiyl Radicals)
Edited by LESTER PACKER

VOLUME 252. Biothiols (Part B: Glutathione and Thioredoxin; Thiols in Signal Transduction and Gene Regulation)
Edited by LESTER PACKER

VOLUME 253. Adhesion of Microbial Pathogens
Edited by RON J. DOYLE AND ITZHAK OFEK

VOLUME 254. Oncogene Techniques
Edited by PETER K. VOGT AND INDER M. VERMA

VOLUME 255. Small GTPases and Their Regulators (Part A: Ras Family)
Edited by W. E. BALCH, CHANNING J. DER, AND ALAN HALL

VOLUME 256. Small GTPases and Their Regulators (Part B: Rho Family)
Edited by W. E. BALCH, CHANNING J. DER, AND ALAN HALL

VOLUME 257. Small GTPases and Their Regulators (Part C: Proteins Involved in Transport)
Edited by W. E. BALCH, CHANNING J. DER, AND ALAN HALL

VOLUME 258. Redox-Active Amino Acids in Biology
Edited by JUDITH P. KLINMAN

VOLUME 259. Energetics of Biological Macromolecules
Edited by MICHAEL L. JOHNSON AND GARY K. ACKERS

VOLUME 260. Mitochondrial Biogenesis and Genetics (Part A)
Edited by GIUSEPPE M. ATTARDI AND ANNE CHOMYN

VOLUME 261. Nuclear Magnetic Resonance and Nucleic Acids
Edited by THOMAS L. JAMES

VOLUME 262. DNA Replication
Edited by JUDITH L. CAMPBELL

VOLUME 302. Green Fluorescent Protein
Edited by P. MICHAEL CONN

VOLUME 303. cDNA Preparation and Display
Edited by SHERMAN M. WEISSMAN

VOLUME 304. Chromatin
Edited by PAUL M. WASSARMAN AND ALAN P. WOLFFE

VOLUME 305. Bioluminescence and Chemiluminescence (Part C) (in preparation)
Edited by THOMAS O. BALDWIN AND MIRIAM M. Ziegler

VOLUME 306. Expression of Recombinant Genes in Eukaryotic Cells (in preparation)
Edited by JOSEPH C. GLORIOSO AND MARTIN C. SCHMIDT

VOLUME 307. Confocal Microscopy (in preparation)
Edited by P. MICHAEL CONN

VOLUME 308. Enzyme Kinetics and Mechanism (Part E) (in preparation)
Edited by VERN L. SCHRAMM AND DANIEL L. PURICH

VOLUME 309. Amyloids, Prions, and Other Protein Aggregates (in preparation)
Edited by RONALD WETZEL

VOLUME 310. Biofilms (in preparation)
Edited by RON J. DOYLE

VOLUME 311. Sphingolipid Metabolism and Cell Signaling (Part A) (in preparation)
Edited by ALFRED H. MERRILL, JR., AND Y. A. HANNUN

VOLUME 312. Sphingolipid Metabolism and Cell Signaling (Part B) (in preparation)
Edited by ALFRED H. MERRILL, JR., AND Y. A. HANNUN

VOLUME 313. Antisense Technology (Part A) (in preparation)
Edited by M. IAN PHILLIPS

VOLUME 314. Antisense Technology (Part B) (in preparation)
Edited by M. IAN PHILLIPS

Section I

Assembly of Nucleosomes, Chromatin, and Nuclei

[1] Preparation of Nucleosome Core Particle from Recombinant Histones

By KAROLIN LUGER, THOMAS J. RECHSTEINER, and TIMOTHY J. RICHMOND

Introduction

The ability to make defined nucleosome core particles (NCPs), or arrays of nucleosomes, from histone proteins expressed in bacteria has several advantages over previously used methods using histones isolated from natural sources.[1–3] Recombinant histones do not contain posttranslational modifications and can be obtained in a highly pure form due to their high expression levels. Site-directed mutants, rare isotypes, or truncations of histones can be expressed, and any combinations of mutants (or histone truncations) can be assembled into nucleosomes. The ability to express all four histone proteins in bacteria has allowed us to develop a method for the mapping of nucleosome position to base pair resolution[4] and has been instrumental in the structure determination of the NCP at high resolution.[5]

Protocols for the overexpression and purification of histones H2A, H2B, H3, and H4, both as full-length proteins and corresponding trypsin-resistant "globular domains" (as defined by Böhm and Crane-Robinson[6]), are given in this chapter. The method for refolding and purification of histone octamer from denatured recombinant histone proteins is described, together with a protocol for the assembly and purification of nucleosome core particle using 146 bp of DNA. The purity and homogeneity of the final core particle preparation is assessed by a high-resolution gel shift assay. A flow chart of the procedures involved in the preparation of "synthetic nucleosomes" is given in Fig. 1.

[1] C. von Holt, W. F. Brandt, H. J. Greyling, G. G. Lindsey, J. D. Retief, J. de A. Rodrigues, S. Schwager, and B. T. Sewell, *Methods Enzymol.* **170,** 431 (1989).
[2] J. O. Thomas and P. J. G. Butler, *J. Mol. Biol.* **116,** 769 (1977).
[3] R. H. Simon and G. Felsenfeld, *Nucleic Acids Res.* **6,** 689 (1979).
[4] A. Flaus and T. J. Richmond, *Methods Enzymol.* **304** [13] (1999) (this volume).
[5] K. Luger, A. W. Maeder, R. K. Richmond, D. F. Sargent, and T. J. Richmond, *Nature* **389,** 251 (1997).
[6] L. Böhm and C. Crane-Robinson, *Biosci. Rep.* **4,** 365 (1984).

0076-6879/99 $30.00

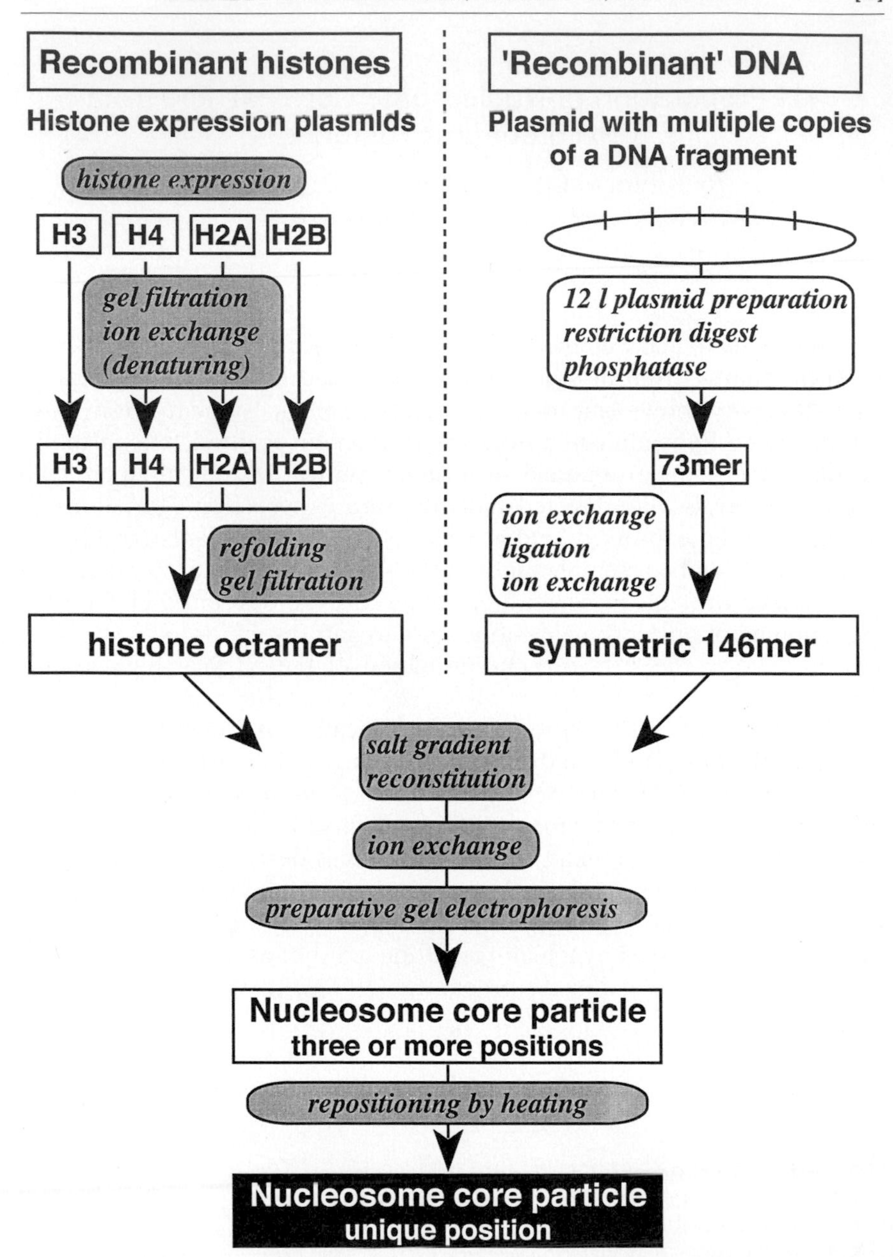

FIG. 1. From histone expression clone to uniquely positioned core particle: the methods discussed in detail in this section are indicated by a shaded box.

Histone Expression and Purification

T7-based expression plasmids for the individual histone proteins and their N-terminally truncated versions have been described previously.[7] Whereas high-level expression of the *Xenopus laevis* histone genes for H2A, H2B, and H3 does not necessitate adaptation of the codon usage to that of *Escherichia coli,* the *X. laevis* gene for histone H4 was only expressed after optimization of the sequence.[7] Expression levels of histones H2A and H2B have been insensitive to the small sequence variation, but histone H3 expression levels differ for sequence variants or mutated genes. H4 expression is sensitive to amino acid substitutions and can drop to an undetectable level for variants with certain single amino acid exchanges. Typical yields for H2A, H2B, and H3 are 50–80 mg of pure protein per liter of cell culture, whereas yields for H4 are four to five times lower, possibly due to degradation by a protease that recognizes the hydrophobic C terminus of proteins and peptides.[8,9] All expressed proteins are found in insoluble form in inclusion bodies.

Histone Expression

The level of histone expression depends strongly on the cell density at which expression is induced. Therefore, the optimal cell density for induction should be tested individually for each histone clone and histone mutant. The time after induction for expression is 90 to 120 min. A typical experiment to optimize expression and procedures for large-scale expression are described.

1. Transfect BL21 (DE3) pLysS cells[10] with 0.1 to 1 μg of the pET-histone expression plasmid[7] and plate on AC agar plates [10% (w/v) Bacto-tryptone, 5% (w/v) yeast extract, 8% (w/v) NaCl, 1.5% (w/v) agar, 100 μg/liter ampicillin, and 25 μg/liter chloramphenicol]. Incubate at 37° overnight.
2. Inoculate 5-ml aliquots of 2× TY-AC medium [16% (w/v) Bacto-tryptone, 10% (w/v) yeast extract, 5% (w/v) NaCl, 100 μg/liter ampicillin, and 25 μg/liter chloramphenicol] with one single colony each from the agar plate. Shake at 37° until slightly turbid (OD_{600} ~0.2). Transfer 0.5 ml of the culture into a sterile Eppendorf tube, add 0.2 ml of sterile glycerol, mix well, and store at −80°. This will serve as the glycerol stock for large-scale expressions. Induce all but one culture by the addition of isopropyl-β-D-

[7] K. Luger, T. Rechsteiner, and T. J. Richmond, *J. Mol. Biol.* **272,** 301 (1997).
[8] D. A. Parsell, K. R. Silber, and R. T. Sauer, *Genes Dev.* **4,** 277 (1990).
[9] K. C. Keiler and R. T. Sauer, *J. Biol. Chem.* **271,** 2589 (1996).
[10] F. W. Studier, A. H. Rosenberg, J. J. Dunn, and J. W. Dubendorff, *Methods Enzymol.* **185,** 60 (1990).

thiogalactopyranoside (IPTG) to a final concentration of 0.2 m*M*. The uninduced sample will serve as a negative control. Incubate for another 2 hr at 37° and analyze expression levels by denaturing gel electrophoresis (18% SDS–PAGE).

3. Restreak the appropriate glycerol culture on an AC agar plate and incubate overnight at 37°. Inoculate 5 ml 2× TY-AC medium and shake at 37° until slightly turbid. Use this starter culture to inoculate 100 ml of 2× TY-AC medium in a 500-ml flask and shake at 37° until slightly turbid. At different values of OD_{600}, induce 5-ml samples from the culture by the addition of 0.2 m*M* IPTG. Note that values obtained by measurements of the optical density at 600 nm depend on the geometry of the spectrophotometer. Start with the experiment as soon as the cell suspension is slightly cloudy.

4. The evening before performing the large-scale expression, restreak the glycerol stock as described earlier and incubate at 37° overnight.

5. The next morning, inoculate five aliquots of 4 ml 2× TY-AC media with one colony each from this plate and shake at 37° for approximately 4 hr. Use the combined starter cultures to inoculate 100 ml of 2× TY-AC media and shake for another 2 hr. The starter culture should not be grown beyond slight turbidity. Inoculate 12 two-liter Erlenmeyer flasks containing 500 ml 2× TY-AC media with 8 ml each of the 100-ml starter culture. Shake at 200 rpm and 37° until the OD_{600} has reached the optimal level for induction, as determined under step 3. Induce by the addition of 0.2 m*M* IPTG (final concentration) and shake for another 2 hr at 37°. Harvest the cells by centrifugation (2000 rpm, 5 min) at room temperature. Resuspend to homogeneity in 100 ml wash buffer (50 m*M* Tris–HCl, pH 7.5, 100 m*M* NaCl, 1 m*M* Na-EDTA, 1 m*M* benzamidine), transfer into a plastic beaker, and flash-freeze in liquid nitrogen.

Note: Cells expressing histone proteins (especially H4) are prone to lysis and should be centrifuged at room temperature. For the same reason, it is not recommended to wash the cell pellet. Resuspend cells well before freezing, as this will improve lysis on thawing. The cells can be stored at −20° for a considerable length of time.

Histone Purification

A three-step purification procedure, which yields up to 1 g of highly pure histone from 6 liters of induced cells, is described. The purification protocol involves the preparation of inclusion bodies, gel filtration under denaturing conditions, and HPLC ion-exchange chromatography under denaturing conditions. If less starting material is used, scale down the protocol for the preparation of inclusion bodies and use a smaller gel filtration column. The purified proteins can be stored as lyophilizates for a considerable length of time, to be used in refolding reactions as described subsequently.

1. Early in the morning, start equilibrating a 5 × 75-cm Sephacryl S-200 high-resolution gel filtration column (Pharmacia, Sweden) with 2 liters of filtered and degassed SAU-1000 buffer (7 *M* deionized urea, 20 m*M* sodium acetate, pH 5.2, 1 *M* NaCl, 5 m*M* 2-mercaptoethanol, 1 m*M* Na-EDTA) at a flow rate of 3 ml/min.

2. Thaw the cell suspension (equivalent to 6 liters of bacterial culture) in a warm water bath. The cell suspension will become extremely viscous as lysis occurs. Transfer into a wide, short measuring cylinder, adjust the volume with wash buffer to 150 ml, and reduce the viscosity by shearing with an Ultraturrax homogenizer. Check whether the mixture is still viscous with a Pasteur pipette and repeat the shearing step if necessary. Centrifuge immediately for 20 min at 4° and 23,000*g*. The pellet contains inclusion bodies of the corresponding histone protein. Wash the pellet by completely resuspending it in 150 ml TW buffer [wash buffer with 1% (v/v) Triton X-100]. Spin for 10 min at 4° at 12,000 rpm. Repeat this step once with TW buffer and twice with wash buffer. After the last wash, the drained pellet can be stored at −20° until further processing.

3. Transfer the pellet to a 50-ml centrifuge tube, add 1 ml of dimethyl sulfoxide (DMSO), and soak for 30 min at room temperature. Mince the pellet with a spatula. Slowly add 40 ml of unfolding buffer [7 *M* guanidinium hydrochloride, 20 m*M* Tris–HCl, pH 7.5, 10 m*M* dithiothreitol (DTT)] and stir gently for 1 hr at room temperature. Resuspend by repeatedly passing through a pipette. Remove undissolved material by centrifugation at 20° and 23,000*g* for 10 min, reextract the pellet, and apply the combined supernatants to the equilibrated gel filtration column using a flow rate of 3 ml/min. Record the elution profile at a wavelength of 280 nm, and collect fractions of suitable size.

Note: Isocyanate can irreversibly modify proteins. Urea-containing solutions that are more than 24 hr old should not be used, and urea stock solutions should always be deionized by passing over Amberlite MB3 (or a similar ion exchange resin) before use. Do not store protein in buffers containing guanidinium hydrochloride or urea for more than 24 hr.

4. Analyze peak fractions by 18% SDS–PAGE. The first peak will contain DNA and larger proteins, whereas the second, much smaller, peak (or shoulder) will contain mainly histone protein. Depending on the separation quality of the column, histones might coelute with the first DNA peak. This can result from the formation of unspecific complexes between cellular

DNA and histones, which sometimes occur despite the presence of 1 *M* salt. For this reason, or if the resolution of the gel filtration column is insufficient, the first fractions of the histone peak might be contaminated with DNA. Before pooling, check the first few histone-containing fractions by UV spectroscopy for DNA contamination, and discard fractions that are contaminated (indicated by a high ratio of OD_{260} to OD_{276}).

5. Pool fractions and dialyze thoroughly against at least three changes of distilled water containing 2 m*M* 2-mercaptoethanol at 4° (dialysis bags with a cutoff of 6–8 kDa are sufficient even for the globular domain of H4, which is the smallest of the constructs). Determine the concentration of the dialyzed sample using the molecular extinction coefficients listed in Table I. Lyophilize and store at − 20°.

6. Dissolve 60–100 mg of lyophilized histone protein in SAU-200 (7 *M* deionized urea, 20 m*M* sodium acetate, pH 5.2, 0.2 *M* NaCl, 5 m*M* 2-mercaptoethanol, 1 m*M* NA–EDTA) and let stand at room temperature for 15 min. Remove insoluble matter by centrifugation. Equilibrate a preparative TSK SP-5PW HPLC column (2.15 × 15.0 cm, Toyo Soda Manufacturing Company, Tokyo, Japan) with SAU-200 buffer. Inject a maximum of 15 mg of protein per run, and elute proteins with the gradients given in Table II [buffer A, SAU-200; buffer B, SAU-600 (7 *M* deionized urea, 20 m*M* sodium acetate, pH 5.2, 0.6 *M* NaCl, 5 m*M* 2-mercaptoethanol, 1 m*M* Na-EDTA)].

7. Analyze the peak fractions by SDS–PAGE. Pool fractions containing pure histone protein, dialyze against water as described in step 6, and lyophilize.

8. Redissolve in a small volume of water, determine the concentration using the values given in Table I, and lyophilize in aliquots suitable for

TABLE I
MOLECULAR WEIGHTS AND MOLAR EXTINCTION COEFFICIENTS (ε) OF HISTONE PROTEINS[a]

	Full-length protein		"Globular domains"		
Histone	Molecular weight	ε (cm^{-1} M^{-1}), 276 nm	Amino acid	Molecular weight	ε (cm^{-1} M^{-1}), 276 nm
H2A	13,960	4050	19 to 118	11,862	4050
H2B	13,774	6070	27 to 122	11,288	6070
H3	15,273	4040	27 to 135	12,653	4040
H4	11,236	5040	20 to 102	9521	5040

[a] Values of ε for full-length and trypsin-resistant globular domains of histone proteins were calculated according to Gill and von Hippel.[15] Molecular weights were determined by a summation of amino acids and were confirmed by mass spectrometry.[7]

TABLE II
SALT GRADIENTS FOR ELUTION OF HISTONE PROTEINS FROM SP COLUMN[a]

H2A, H2B		H3		H4	
t (min)	*B* (%)	*t* (min)	*B* (%)	*t* (min)	*B* (%)
0	0	0	0	0	0
10	0	3	30	5	50
11	40	8	30	10	59
46	100	40	100	40	100
60	100	50	100	50	100
61	0	51	0	51	0

[a] Salt gradients for elution of histone proteins from an SP-5PW HPLC column (2.15 cm × 15.0 cm): the flow rate is 4 ml/min, buffer A is 7 *M* deionized urea, 20 m*M* sodium acetate, pH 5.2, 0.2 *M* NaCl, 5 m*M* 2-mercaptoethanol, 1 m*M* Na-EDTA, and buffer B is 7 *M* deionized urea, 20 m*M* sodium acetate, pH 5.2, 0.6 *M* NaCl, 5 m*M* 2-mercaptoethanol, 1 m*M* Na-EDTA.

subsequent octamer refolding reactions (see later). For example: H3 4.5 mg, H4 3.5 mg, H2A 4.0 mg, H2B 4.0 mg.

Histone Octamer Refolding

The following protocol is used for the refolding of histone octamers from lyophilized recombinant histone proteins. No restrictions have been observed with respect to the source of material or combination of mutants: all possible combinations of recombinant *X. laevis* full-length and globular domain histone proteins, as well as combinations of histone dimer or tetramer from "natural sources" and recombinant histones, can be refolded to functional histone octamer. The method works best for 6–15 mg of total protein. For smaller amounts of protein, a proportionally smaller gel filtration column should be used.

1. Dissolve each histone aliquot to a concentration of approximately 2 mg/ml in unfolding buffer. Pass through a Pasteur pipette repeatedly while rinsing the walls of the tube in the process; do not vortex. Unfolding should be allowed to proceed for at least 30 min and for no more than 3 hr. Remove undissolved matter by centrifugation, if necessary. Determine the concentration of the unfolded histone proteins by measuring OD_{276} (Table I) of the undiluted solution against unfolding buffer. Subtract the reading at 320 nm from the actual OD_{276} value.

2. Mix the four histone proteins to exactly equimolar ratios and adjust to a total final protein concentration of 1 mg/ml using unfolding buffer. Dialyze at 4° against at least three changes of 2 liters refolding buffer (2 *M* NaCl, 10 m*M* Tris–HCl, pH 7.5, 1 m*M* Na-EDTA, 5 m*M* 2-mercapto-ethanol). For a 15-mg setup, use dialysis bags with a flat width of 2.5 cm. The second or third dialysis step should be performed overnight. Histone octamer should always be kept at 0–4° to avoid complex dissociation.

3. Remove precipitated protein by centrifugation. Concentrate to a final volume of 1 ml (use a concentration device suitable for 1- to 25-ml volumes; e.g., Sartorius ultrathimble, Sartorius AG, Göttingen, Germany, or Centricon 10, Amicon AG, Beverley, MA).

4. At the same time, equilibrate the gel filtration column (HiLoad 16/60 Superdex 200 prep grade; Pharmacia, Uppsala, Sweden, with 1.5-column volumes of filtered refolding buffer at a flow rate of 1 ml per minute, at 4°. Load a maximum of 1.5 ml or 15 mg of the concentrated histone octamer and run at a flow rate of 0.6–1.0 ml per minute. High molecular weight aggregates will elute after about 45 ml, histone octamer at 65 to 68 ml, and histone (H2A-H2B) dimer at 84 ml. Separation between histone tetramer and octamer is not complete under these conditions; it is therefore recommended to work with a very slight excess of H2A and H2B if the concentration of histones is ill determined. Yields of pure histone octamer are usually between 50 and 75% of the input material.

Note: Other gel filtration resins with a similar separation range, such as Superose 12 or Sephacryl S-300 (both from Pharmacia) do not give a baseline separation of histone octamer and dimer. Sephadex G-100 (Pharmacia) does not separate histone octamer from high molecular weight aggregates and is therefore not recommended. For all resins tested, significant amounts of histones (or high molecular weight aggregates) may remain attached to the column. Clean the column with NaOH as recommended by the supplier.

5. The histone octamer peak should be symmetric and should not contain shoulders.[7] Check the stoichiometry of the fractions on an 18% SDS–PAGE using purified NCP as a marker (dilute samples of histone octamer fractions by a factor of at least 2.5 before loading onto the gel to reduce distortion of the bands due to the high salt concentration). Pool fractions that contain equimolar amounts of the four histone proteins.

6. Using this protocol, histone octamer will elute at sufficiently high concentrations to be used directly for NCP reconstitution with DNA (see later). Determine the concentration spectrophotometrically (A_{276} = 0.45 for a solution of 1 mg/ml). Alternatively, concentrate to 3–15 mg/ml, adjust to 50% (v/v) glycerol, and store at −20°.

Reconstitution of Nucleosome Core Particles

DNA fragments of the desired length and sequence are obtained by subcloning and amplification of the fragment in pUC-derived plasmids.[11,12] The plasmid is isolated from 12 liters of HB101 cells by a scaled up and modified alkaline lysis procedure, and the fragment is isolated by subsequent cleavage and ligation. Fragments are purified by ion-exchange chromatography using a TSK DEAE-5PW HPLC column (Toyo Soda Manufacturing Company, Tokyo, Japan).

A modification of the salt gradient method described by Thomas and Butler[2,12] is used for the reconstitution of histone octamer with DNA. Histone octamer and DNA are combined at 2 *M* KCl; the gradual reduction of the salt concentration to 0.25 *M* KCl over a period of 36 hr leads to the formation of NCPs. However, as will be described, these preparations still exhibit remarkable heterogeneity with respect to translational and/or rotational position, even if 146 bp of a presumed strong positioning sequence is used.[13]

The procedure works equally well for large and small amounts of NCP (10–0.1 mg) and multiple setups can be dialyzed in one vessel. If amounts smaller than 0.1 mg need to be reconstituted, dialysis buttons (e.g., Hampton Research, Laguna Hills, CA) or the dialysis apparatus described in the accompanying article[4] are useful alternatives.

1. Histone octamer and DNA are combined at 0.9 molar ratio and adjusted to a final DNA concentration of 6 μM. Histone octamer from a glycerol stock should be dialyzed overnight against refolding buffer before it is used in reconstitutions because the required accuracy in the ratio between the histone and the DNA ratio cannot be maintained if pipetted directly from the glycerol stock. Before adding histone octamer to the reconstitution mixture, adjust the salt concentration of the DNA solution to 2 *M* using 4 *M* (or solid) KCl, and add DTT to a final concentration of 10 m*M*. Histone octamer should always be added last to the reconstitution mixture to avoid mixing octamer and DNA at <2 *M* salt concentrations, which might result in the formation of aggregates. Incubate at 4° for 30 min.
2. Prepare 400 ml RB high buffer (2 *M* KCl, 10 m*M* Tris–HCl, pH 7.5, 1 m*M* EDTA, 1 m*M* DTT) and 1600 ml RB low buffer (0.25 *M* KCl, 10 m*M* Tris–HCl, pH 7.5, 1 m*M* EDTA, 1 m*M* DTT) and chill to 4°. Set up the dialysis apparatus as shown in Fig. 2. A peristaltic

[11] R. T. Simpson, F. Thoma, and J. M. Brubaker, *Cell* **42,** 799 (1985).
[12] T. J. Richmond, M. A. Searles, and R. T. Simpson, *J. Mol. Biol.* **199,** 161 (1988).
[13] A. Flaus, K. Luger, S. Tan, and T. J. Richmond, *Proc. Natl. Acad. Sci. U.S.A.* **93,** 1370 (1996).

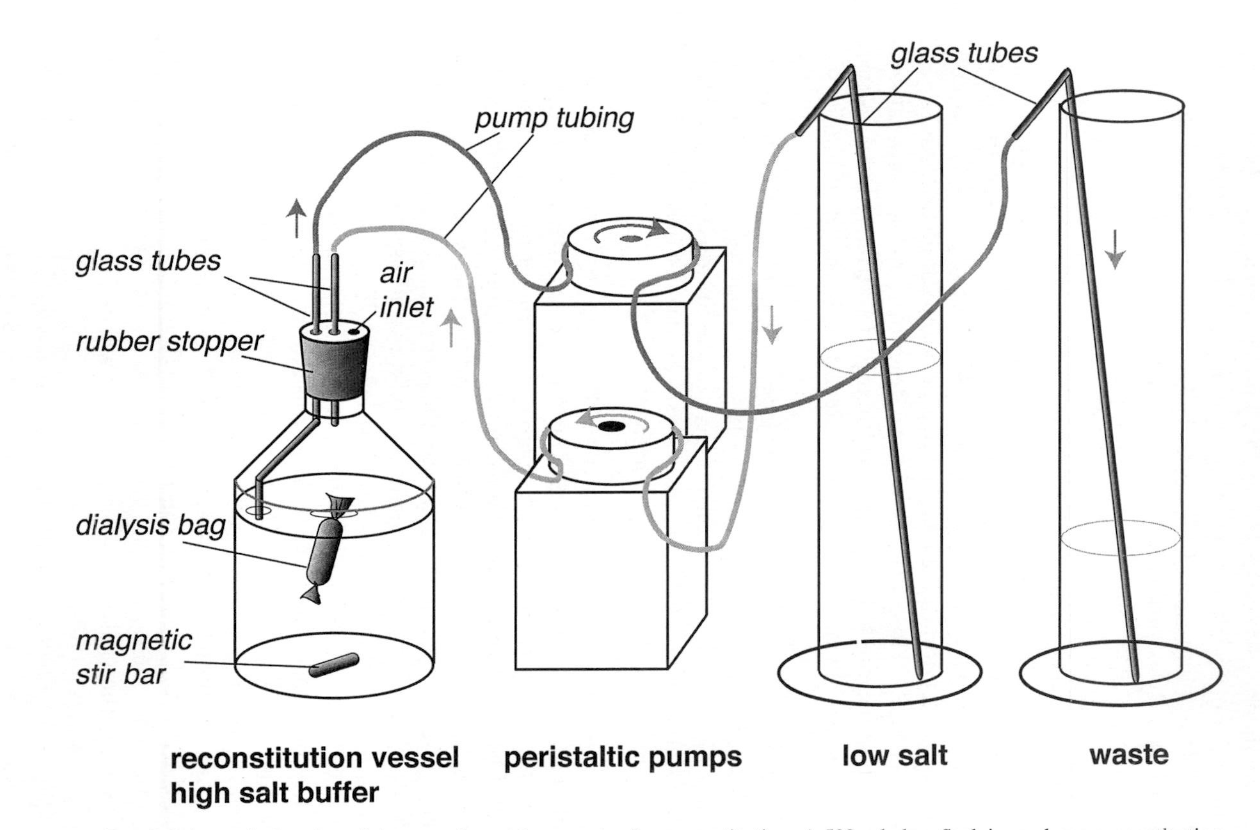

FIG. 2. Schematic drawing of the experimental apparatus for reconstitution. A 500-ml glass flask is used as a reconstitution vessel. Standard glass tubes are bent in the appropriate manner and are connected to the pumps by silicon tubing. The reconstitution vessel contains high salt to start; buffer exchange from high to low salt is achieved by an exponential gradient during a time period of 36 hr.

pump with a double pump head (e.g., Gilson Minipuls 3 peristaltic pump, equipped with tubing with a 2.5-mm inner diameter; Gilson Medical Electronics SA, Villier-leBel, France) or, alternatively, two peristaltic pumps capable of maintaining flow rates of 0.5–1 ml/min can be used. Calibrate the peristaltic pump(s) to a flow rate of 0.7–0.8 ml/min.

3. Dialyze the sample against 400 ml RB high buffer at 4° under constant stirring. Make sure that the dialysis bag can circle freely and rapidly to allow constant mixing of the contents of the dialysis bag. Using the peristaltic pump, continually remove buffer from the dialysis vessel and replace with RB low. Over a period of 36 hr, an exponential gradient toward low salt is generated. After the gradient has finished, dialyze for at least 3 hr against RB low. If the samples are not processed further within the next 24 hr, dialyze against CCS buffer (20 m*M* potassium-cacodylate, pH 6.0, 1 m*M* EDTA). If the core particle is to be purified by preparative gel electrophoresis (see later), dialyze against TCS buffer (20 m*M* Tris–HCl, pH 7.5, 1 m*M* EDTA, 1 m*M* DTT).

Note: Take care to adjust the position of the inlet and outlet tubing as shown in Fig. 2. Use the pump with the apparently faster speed for the removal of buffer. The formation of large amounts of precipitate during dialysis or unexpectedly low yields may result for two reasons: (1) an excess of histone octamer (or histone protein) has been added by mistake. Histone octamer tends to precipitate assembled NCP, and thus any change in the given molar ratio between octamer and DNA will reduce yields significantly. (2) Stalled motion of the dialysis bag creates an uneven salt gradient within the bag itself, resulting in inconsistencies with respect to yield.

Purification and Repositioning of Nucleosome Core Particle

The method just described usually results in less than 80% of the input DNA being assembled into NCPs. Two methods have been developed for the purification of NCPs from aggregates, free DNA, and/or free histones. Both methods have been optimized for NCP with 146 bp DNA, but can be adjusted easily for nucleosomes with different lengths of DNA. The authors use either HPLC DEAE ion–exchange chromatography[12] or gel electrophoresis under nondenaturing conditions[7] for purification principle.

The first method makes use of the fact that free DNA binds tighter to an anion exchange column than NCP, which allows a differential elution from the column by a salt gradient; the second method relies on the different migration of free DNA, NCP, and high molecular weight aggregates on nondenaturing polyacrylamide gels. Both methods alone yield highly pure

NCP preparations; the choice depends on available equipment and on the problem at hand. Ion-exchange chromatography is suitable for large-scale preparations on a routine basis. It has been observed that certain modifications of histone proteins (such as covalently bound heavy atoms or fluorescent labels) might completely alter the binding and elution properties of the NCP. Purification by preparative gel electrophoresis should be preferred if modified samples are to be purified; if a significant amount of material appears as a high molecular weight band in a gel shift assay; or if the histones contain mutations that make the particle prone to salt-dependent dissociation. The two methods can also be used in combination.

Purification by Ion-Exchange Chromatography

1. At 4°, equilibrate a TSK DEAE-5PW HPLC column (2.15 × 15.0 cm, or 7.5 × 75 mm, Toyo Soda Manufacturing Company, Tokyo, Japan) with TES-250 buffer (0.25 *M* KCl, 10 m*M* Tris–HCl, pH 7.5, 0.5 m*M* EDTA). Centrifuge the reconstitution mixture in Eppendorf tubes at 4° and apply the supernatant to the column (a maximum of 10 mg NCP). Samples can be in RB low, TCS buffer, or CCS buffer. Using a flow rate of 4 ml/min, the column is developed with the gradient given in Table III (buffer A, TES-250; buffer B, TES-600: 0.6 *M* KCl, 10 m*M* Tris–HCl, pH 7.5, 0.5 m*M* EDTA) while monitoring the eluate at 260 nm. The gradient needs to be adjusted for optimal separation if core particles containing longer DNA fragments

TABLE III
SALT GRADIENT ELUTION OF NCP FROM DEAE-5PW HPLC COLUMN[a]

NCP146	
t (min)	*B* (%)
0	0
10	0
11	40
46	100
60	100
61	0

[a] The flow rate is 4 ml/minute, buffer A is 0.25 *M* KCl, 10 m*M* Tris–HCl, pH 7.5, 0.5 m*M* EDTA, and buffer B is 0.25 *M* KCl, 10 m*M* Tris–HCl, pH 7.5, 1 m*M* EDTA, 1 m*M* DTT.

are purified or if chromatography is performed at different temperatures or with different salts (NaCl instead of KCl).

2. The heterogeneity of NCP with respect to the relative position of the DNA on the histone octamer (see later) is apparent in the elution profile from the ion-exchange column: NCP elutes from the DEAE column as a major peak with several shoulders (Fig. 3a). The major peak fractions, including the shoulders, are usually pooled without

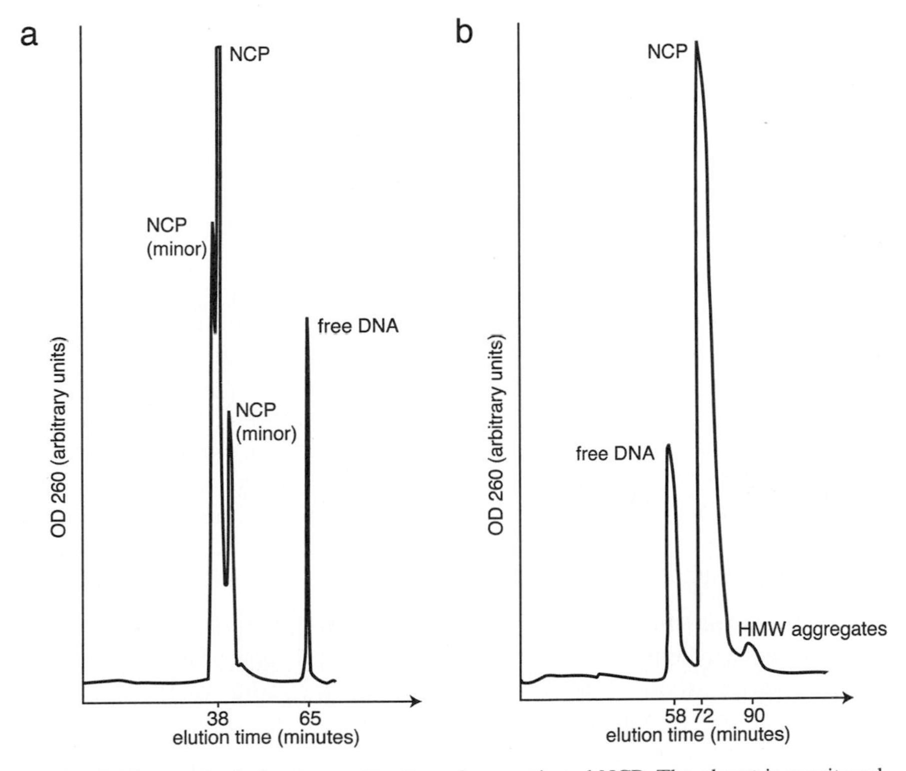

FIG. 3. Two methods for the purification of reconstituted NCP. The eluent is monitored by the absorbance at 260 nm. The approximate elution times are shown for each peak, but are dependent on the flow rate and the geometry of the setup. (a) Purification by HPLC ion-exchange chromatography. NCPs with different rotational positions elute at different times from the column. The relative ratio between main and minor peaks can vary. Major and minor peaks are combined; additional peaks that exhibit baseline separation (not shown) are not included. In some cases, a free octamer can be observed to elute at the beginning of the gradient (not shown). (b) Purification by preparative nondenaturing gel electrophoresis. Note that the three bands observed in Fig. 4 are beyond the resolution range of this method. A good separation from high molecular weight (HMW) aggregates is achieved. An increased elution speed will improve the separation of the peaks, but will also yield a more dilute sample, which might lead to dissociation of assembled NCP.

analysis by nondenaturing PAGE. These peaks are not distinguished by preparative electrophoresis (Fig. 3b).

3. Pool the peak fractions and immediately dialyze against three changes of TCS buffer at 4°. Concentrate to >1 mg/ml and store on ice until further use. For prolonged storage, dialyze against CCS buffer or add 5 m*M* potassium cacodylate at an appropriate pH to prevent microbial growth. Determine the concentration of NCP preparations by measuring the absorbance at 260 nm of a 200- to 500-fold dilution. The yield of this reconstitution and purification method may vary between 20 and 80% of the DNA added to the reconstitution mixture, depending on the sequence of the DNA, the stoichiometry of histones and DNA, and the reconstitution geometry.

Purification by Preparative Gel Electrophoresis

The following conditions have been optimized for the purification of NCP containing 146 bp of DNA. Conditions for preparative gel electrophoresis should be optimized by analytical nondenaturing gel electrophoresis (see later), following the guidelines given in the instruction manual for the Model 491 Prep Cell (Bio-Rad Laboratories, Richmond, CA). The ratio between acrylamide and bisacrylamide, the length of the gel, and the elution speed can alter the relative mobility and the separation of the components greatly. The authors have noticed that the choice of elution buffer and electrophoresis buffer may also influence the relative mobility of the different species. Improved resolution between different peaks is often a tradeoff with high dilution of the sample. Note that the high dilution of NCP during purification might result in a partial dissociation of DNA and octamer.

1. Prepare 20 ml of a 5% polyacrylamide gel mixture (ratio of acrylamide to bisacrylamide 60:1) containing 0.2× TBE and pour a cylindrical gel with an outer radius of 28 mm, an inner radius of 19 mm, and a height of 50 mm. Polymerize overnight and assemble according to instructions given in the manual for the Model 491 Prep Cell. Connect to a standard power supply, UV detector, fraction collector, and a peristaltic pump capable of maintaining flow rates <1 ml/min (e.g., Gilson Minipuls 3 peristaltic pump, Gilson Medical Electronics SA, Villier-leBel, France). Use a dialysis membrane with a 6000–8000 molecular weight cutoff, cut to a circle with a radius of 3 cm. Prerun the gel under constant recircularization of the buffer for 90 min in 0.2× TBE at 4° and 10 W. Record a baseline at 260 nm using TCS buffer as elution buffer.
2. After reconstitution, dialyze NCP against TCS buffer and concentrate. A maximum of 600 μl or 3 mg is mixed with sucrose to a final concentration of 5% (v/v) and loaded on the preparative gel using

a syringe with an attached piece of tubing. Electrophorese at 10 W and elute the complex at a flow rate of 0.7 ml/min with TCS buffer. Recirculate the electrophoresis buffer during the run. Record the elution at an OD of 260 nm, and collect fractions of appropriate size (0.7 to 1 ml). Free DNA will appear first, followed by pure NCP, and finally higher molecular weight aggregates (Fig. 3b).

3. Pool peak fractions, concentrate immediately to at least 1 mg/ml, and store as described earlier.

High-Resolution Gel Shift Assay and Repositioning of Nucleosome Core Particles

Reconstitution on longer DNA fragments usually leads to a heterogeneous population of NCP with respect to the position of the DNA on the histone octamer. Surprisingly, this also holds true for DNA fragments with a limiting length of 146 bp, even if presumed "strong positioning sequences" are used.[7,14] A simple heating step (37–55° for 20–180 min) results in a uniquely positioned NCP preparation for DNA: 145 to 147 bp in length, which is suitable for biochemical studies and crystallization (Fig. 4). Repositioning can be monitored by a high-resolution gel shift assay described later. The incubation time and temperature necessary for repositioning depend on the sequence and the length of the DNA fragment and have to be checked individually for each combination of DNA fragment and histone octamer. For example, a *X. laevis* full-length histone octamer with the 146-bp fragment derived from the 5S RNA gene of *Lytechinus variegatus* is heated for 30 min at 37° for a complete shift,[13] whereas other sequences might require as long as 2 hr at 55°. Shifting to a unique position occurs *without* dissociation of the DNA; an excess of competitor DNA does not exchange onto the histone octamer during the process (K.L., unpublished results, 1995).

The following protocol allows for the separation of NCP with a different translational setting of the DNA by 10 bp (Fig. 4). The ratio between acrylamide and bisacrylamide can completely alter the relative mobilities of different NCP species with respect to the DNA size marker.[7] The choice of gel buffer has only minor effects on the resolution of the different NCP species.

1. Prepare a nondenaturing gel (5% acrylamide with 60:1 acrylamide to bisacrylamide, 0.2× TBE) with the dimensions 20 × 20 × 0.1 cm, using a 10- to 16-well comb.

[14] A. Flaus and T. J. Richmond, *J. Mol. Biol.* **275,** 427 (1998).

[15] S. C. Gill and P. H. von Hippel, *Anal. Biochem.* **182,** 319 (1989).

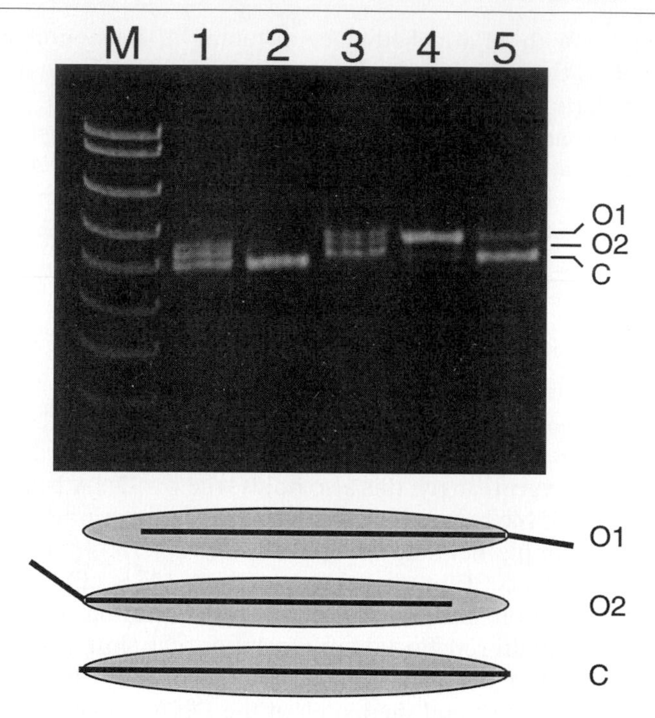

FIG. 4. Nondenaturing gel electrophoresis of NCP reveals multiple positions of the histone octamer on the DNA. Lanes 1 to 3: purified NCP reconstituted from recombinant full length histone proteins and a 146 bp fragment derived from the 5S RNA gene of *Lytechinus variegatus*. Before (lanes 1 and 3) and after (lane 2) heating for one hour at 37°. Lanes 4 and 5: purified NCP prepared from recombinant full-length histone proteins and a 146 bp palindromic DNA fragment derived from human alpha satellite DNA; before (lane 4) and after (lane 5) heating for two hours at 37°. The relative positions of the DNA on the histone octamer in the three bands is shown schematically (shaded ovals: histone octamer, bold line: DNA). Note that NCP containing the asymmetric 5S RNA DNA fragment reconstitutes in two off-centered and one centered rotational position (marked with O1, O2 and C, respectively) whereas NCP containing a palindromic sequence reconstitutes mainly to the off-centered positions. For NCPs containing asymmetric 146-mer DNA, the two off-centered positions can be distinguished; in contrast, they cannot be distinguished in NCPs prepared with a symmetric DNA fragment because the sequence symmetry results in an identical exit angle of the ends of the DNA from the histone octamer.

2. Prerun the gel for at least 3 hr at 4° and 200 V, while constantly recycling the running buffer. Attach a metal plate for better heat distribution during electrophoresis.
3. Shortly before loading the samples, the slots should be rinsed well with 0.20× TBE. Load 1–2 pmol of core particle solution supplemented with 5% (v/v) sucrose, in no more than 10 μl. Traces of bromphenol blue can be added for easier loading.

4. Run the gel for a suitable length of time or until bromphenol blue has reached the bottom of the gel. Recirculate the running buffer at all times.
5. Stain the gel with ethidium bromide. Because free DNA is stained significantly better by ethidium bromide than DNA bound to the histone octamer, a quantification of yields is not possible by this method. Subsequent staining with Coomassie Brilliant Blue sometimes gives better resolution on slightly overloaded gels due to the limited sensitivity of Coomassie Brilliant Blue compared to ethidium bromide.

Acknowledgments

We thank our colleagues, especially Drs. A. Flaus, A. Mäder, and S. Tan, for support, advice, and discussions. The help of R. Richmond, R. Amherd, and J. Hayek in the preparation of histones is gratefully acknowledged.

[2] Assembly of Defined Nucleosomal and Chromatin Arrays from Pure Components

By LENNY M. CARRUTHERS, CHRISTIN TSE, KERFOOT P. WALKER III, and JEFFREY C. HANSEN

Introduction

The structural subunit of chromatin is the nucleosome, which consists of 146 bp of DNA wrapped nearly twice around a core histone octamer. The histone octamer is composed of an H3/H4 tetramer and a pair of H2A/H2B dimers. Linker DNA connects successive nucleosomes to form nucleosomal arrays. Linker histones and nonhistone chromatin-associated proteins (NHCAPs) interact with nucleosomal arrays to form chromatin. Nucleosomal arrays and chromatin undergo complex folding and compaction processes that serve to both regulate the biological activity of the chromatin fiber and condense eukaryotic DNA into chromosomes.[1–4]

[1] K. E. van Holde, "Chromatin." Springer-Verlag, New York, 1988.
[2] J. Widom, *Annu. Rev. Biophys. Biophys. Chem.* **18,** 365 (1989).
[3] A. P. Wolffe, "Chromatin: Structure and Function," 2nd Ed. Academic Press, New York, 1995.
[4] T. M. Fletcher and J. C. Hansen, *Crit. Rev. Eukaryotic Gene Expression* **6,** 149 (1996).

0076-6879/99 $30.00

One way to study the structure/function relationships of nucleosomal arrays and chromatin *in vitro* is to assemble defined model systems from specifically engineered DNA templates and purified protein components. This approach was first made possible by Simpson and co-workers,[5] who created a series of tandemly repeated *Lytechinus* 5S rDNA templates. Because each 5S rDNA repeat binds a single positioned histone octamer after reconstitution and also has a viable class III promoter, this system provided for the first time a way to obtain length and compositionally defined nucleosomal arrays suitable for simultaneous analysis of higher order conformational dynamics and transcriptional activity. Subsequent characterizations of these reconstituted model systems have yielded a wealth of information regarding DNA-directed nucleosome positioning[6,7] and the complex functions of the core histone N termini and linker histones in chromatin condensation,[8–19] as well as the relationships between higher order folding and modulation of transcription by RNA polymerase III.[20–22]

Importantly, it is not trivial to assemble *high-quality* preparations of defined nucleosomal array and chromatin model systems from pure components. Depending on the histone : DNA input ratio, the DNA templates after reconstitution can be subsaturated, saturated, or supersaturated with core histones. Subsaturated templates contain one or more octamer-free "gaps" in the nucleosomal array, saturated arrays have all possible DNA repeats occupied by histone octamers, and supersaturated reconstitutes consist of saturated templates nonspecifically complexed with additional core histones. In addition, further technical problems are encountered when one attempts to assemble stoichiometric amounts of linker histones with nucleosomal arrays to obtain defined chromatin model systems. With this in

[5] R. T. Simpson, F. Thoma, and J. M. Brubaker, *Cell* **42,** 799 (1985).
[6] J. C. Hansen, J. Ausio, V. H. Stanik, and K. E. van Holde, *Biochemistry* **28,** 9129 (1989).
[7] G. Meersseman, S. Pennings, and E. M. Bradbury, *J. Mol. Biol.* **220,** 89 (1991).
[8] F. Dong, J. C. Hansen and K. E. van Holde, *Proc. Natl. Acad. Sci. U.S.A.* **87,** 5724 (1990).
[9] J. C. Hansen, K. E. van Holde, and D. Lohr, *J. Biol. Chem.* **266,** 4276 (1991).
[10] S. Pennings, G. Meersseman, and E. M. Bradbury, *J. Mol. Biol.* **220,** 101 (1991).
[11] M. Garcia-Ramirez, F. Dong, and J. Ausio, *J. Biol. Chem.* **267,** 19587 (1992).
[12] J. C. Hansen and D. Lohr, *J. Biol. Chem.* **268,** 5840 (1993).
[13] P. M. Schwarz and J. C. Hansen, *J. Biol. Chem.* **269,** 16284 (1994).
[14] M. Garcia-Ramirez, C. Rocchini, and J. Ausio, *J. Biol. Chem.* **270,** 17923 (1995).
[15] T. M. Fletcher and J. C. Hansen, *J. Biol. Chem.* **270,** 25359 (1995).
[16] P. M. Schwarz, T. M. Fletcher, A. Felthasuer, and J. C. Hansen, *Biochemistry,* **35,** 4009 (1996).
[17] S. C. Moore and J. Ausio, *Biochem, Biophys. Res. Commun.* **230,** 136 (1997).
[18] C. Tsc and J. C. Hansen, *Biochemistry* **36,** 11381 (1997).
[19] L. M. Carruthers, J. Bednar, C. Woodcock, and J. C. Hansen, *Biochemistry* **37,** 14776 (1998).
[20] J. C. Hansen and A. P. Wolffe, *Biochemistry* **31,** 7977 (1992).
[21] J. C. Hansen and A. P. Wolffe, *Proc. Natl. Acad. Sci. U.S.A.* **91,** 2339 (1994).
[22] C. Tse, T. Sera, A. P. Wolffe and J. C. Hansen, *Mol. Cell Biol.* **18,** 4629 (1998).

mind, protocols are presented for (1) reconstitution of defined nucleosomal arrays, (2) characterization of their extent of histone octamer saturation, (3) binding of linker histones to reconstituted nucleosomal arrays, and (4) enrichment for saturated nucleosomal and chromatin arrays. The protocols are based on our experience with the *Lytechinus* 5S rDNA system consisting of twelve 208-bp repeats, but will be equally applicable to any analogous defined model system assembled from tandemly repeated DNA and pure protein components. The objective of this article is to outline a general strategy for obtaining high-quality preparations of defined nucleosomal array and chromatin model systems assembled *in vitro* from pure components.

Materials, Reagents, and Instrumentation

Materials

Escherichia coli DH5α cells containing plasmid pPOL208-*n*, where the repeat length is 208 bp and *n* is the total number of repeats[23]; whole chicken blood from Pel-Freeze Biologicals (Rogers, AR); HeLa cells (S3 strain) from American Type Culture Collection (Rockville, MD); fetal bovine serum and Joklik's media from GIBCO/BRL (Gaithersburg, MD); low electroosmosis (LE) agarose from Research Organics (Cleveland, OH); Amicon XM50 concentrating filters; and Spectra/Por 2 dialysis tubing (12,000–14,000 molecular weight cutoff) from Spectrum (Houston, TX).

Enzymes

*Eco*RI, *Hha*I, and their respective reaction buffers from Promega (Madison, WI); micrococcal nuclease (MNase) from Worthington Biochemical (Freehold, NJ); trypsin immobilized on DITC-treated glass beads, soybean trypsin inhibitor, and proteinase K from Sigma (St. Louis, MO).

Reagents

Aprotinin, calcium chloride, cesium chloride, dithiothreitol (DTT), ethanol, ethidium bromide, glycerol, leupeptin, sodium acetate, sodium butyrate, sodium chloride, and sodium dodecyl sulfate (SDS).

Buffers

10 m*M* Tris–HCl, 0.25 m*M* Na_2EDTA, pH 7.8 (TE buffer); 10 m*M* Tris–HCl, 0.25 m*M* Na_2EDTA, 2.5 m*M* NaCl, pH 7.8 (TEN buffer); 0.1 *M*

[23] Georgel, B. Demeler, C. Terpening, M. R. Paule, and K. E. van Holde, *J. Biol. Chem.* **268,** 1947 (1993).

potassium phosphate buffer, 5 m*M* sodium butyrate, 2.2 *M* NaCl, 2.5 m*M* DTT, pH 6.8 (hydroxyapatite elution buffer); 40 m*M* Tris–HCl, 1 m*M* Na_2EDTA, pH 7.8 (E buffer); 40 m*M* Tris–acetate, 1 m*M* Na_2EDTA, pH 8.0; 45 m*M* Tris–borate, 1 m*M* Na_2EDTA.

Instrumentation

Quantitative Agarose Gel Electrophoresis Equipment. This technique requires a special electrophoresis apparatus referred to as a multigel (commercially available from Aquebogue Machine Shop, Aquebogue, NY). Other equipment required for quantitative electrophoretic experiments include a high temperature water bath, an oscillating pump with voltage controller, and a digital imaging system (described in detail in Hansen *et al.*[24] and Hansen *et al.*[25]).

Analytical Ultracentrifuge. A Beckman XL-A or XL-I analytical ultracentrifuge is used to obtain sedimentation coefficients (from sedimentation velocity experiments) and molecular mass (from sedimentation equilibrium experiments). Software capable of analyzing sedimentation velocity data by the method of van Holde and Weischet[26] is necessary to obtain diffusion-corrected sedimentation coefficient distributions.[27,28]

Eppendorf Microcentrifuge. This piece of equipment is used to perform differential centrifugation assays of salt-dependent nucleosomal array and chromatin oligomerization.[13,16]

Purification of DNA Templates

The 5S rDNA templates are liberated from the parent pPOL plasmids by digestion with the restriction endonuclease *Hha*I.[5] Purification of up to milligram quantities of the tandemly repeated 5S rDNA insert is best achieved by size-exclusion chromatography,[29] which efficiently separates ≥800-bp 5S rDNA templates ($n \geq 4$) from the ≤400-bp vector fragments obtained after digestion of pPOL plasmids with *Hha*I. Smaller DNA inserts can be purified using either repeated cycles of size-exclusion chromatography or gel electrophoresis/elution techniques.

[24] J. C. Hansen, I. K. Kreider, B. Demeler, and T. M. Fletcher, *Methods* **12,** 62 (1997).
[25] J. C. Hansen, I. K. Kreider, and T. M. Fletcher, *in* "Methods in Molecular Biology: Chromatin Protocols," vol. 119 (P. Becker, ed.), 1999.
[26] K. E. van Holde and W. O. Weischet, *Biopolymers* **25,** 1981 (1978).
[27] J. C. Hansen, J. Lebowitz, and B. Demeler, *Biochemistry* **33,** 13155 (1994).
[28] B. Demeler, H. Saber, and J. C. Hansen, *Biophys. J.* **72,** 397 (1997).
[29] J. C. Hansen and H. Rickett, *Anal. Biochem.* **179,** 167 (1989).

Protocol

1. Purify pPOL208-*n* plasmids by alkaline lysis and banding in CsCl/ethidium bromide gradients as described.[6,30]
2. Digest purified plasmids with 0.03 units of *Hha*I/μg of DNA for 36–48 hr at 37°. Note that these are the optimal conditions for digesting 10–50 mg of plasmid; smaller digestions can be performed more rapidly using greater amounts of *Hha*I.
3. Load 10–15 mg of the *Hha*I digest on a 2.5 × 70-cm column of Bio-Gel A150m (Bio-Rad, Richmond, CA) and elute with TE buffer using the gravimetric flow rate of the column. Other equivalent size exclusion gels can be substituted for Bio-Gel A150m.
4. Collect 4-ml fractions over a 2- to 3-week period. Electrophorese 10-μl aliquots of each fraction on a 1% agarose gel to determine the position of the insert peak.
5. Combine those fractions containing the purified insert, add sodium acetate, pH 5.2, to a concentration of 0.3 *M*, ethanol precipitate the DNA, and store at −20° until needed.

Purification of Native, Trypsinized, and Differentially Acetylated Core Histone Octamers

Core histone octamers were purified from chicken erythrocyte oligonucleosomes using standard protocols.[31] Core histone octamers lacking their N-terminal tail domains were purified from chicken erythrocyte oligonucleosomes that had been treated with immobilized trypsin as described.[18,32] Acetylated histone octamers were purified from sodium butyrate-treated HeLa cells essentially as described[33] with the exception that freshly prepared 2.5 m*M* DTT was added to all the isolation buffers to prevent intermolecular cross-linking of histone H3 during purification. Intact and trypsinized histone octamers, H2A/H2B dimers, and H3/H4 tetramers were purified from the various oligonucleosome fractions by hydroxyapatite chromatography as described previously in detail.[34,35] Purified histone octamers were stored at 4° in hydroxyapatite elution buffer at a

[30] J. Sambrook, E. F. Fitsch, and T. Maniatis, "Molecular Cloning." Cold Spring Harbor Laboratory Press, Cold Spring Harbor, NY, 1989.

[31] T. D. Yager, C. T. McMurray, and K. E. van Holde, *Biochemistry* **28,** 2271 (1989).

[32] J. Ausio, F. Dong, and K. E. van Holde, *J. Mol. Biol.* **206,** 451 (1989).

[33] J. Ausio and K. E. van Holde, *Biochemistry* **25,** 1421 (1986).

[34] R. H. Simon and G. Felsenfeld, *Nucleic Acids Res.* **6,** 689 (1979).

[35] C. von Holt, W. F. Brandt, H. J. Greyling, G. G. Lindsey, J. D. Retief. J. de A. Rodrigues, S. Schwager, and B. T. Sewell, *Methods Enzymol.* **170,** 431 (1989).

final concentration of ~0.40 μg/μl as determined by absorbance at 230 nm. An A_{230} of 4.3 for a l-mg/ml octamer solution was used in all cases.[36] After quantitation, the protease inhibitors aprotinin and leupeptin were added to a final concentration of 10 μg/ml.

Reconstitution of Defined Nucleosomal Arrays Using Simplified Salt Dialysis Method

Numerous variations of the salt dialysis reconstitution method have been developed over the years. The protocol we use has been modified to coincide with key assembly steps that occur during the reconstitution process[9] (see Fig. 1). It is known that as the salt concentration falls below 2 *M* NaCl, the histone octamer dissociates into an H3/H4 tetramer and two H2A/H2B dimers.[37] At 1 *M* NaCl, one H3/H4 tetramer binds to each 5S rDNA repeat.[9] At ~0.8 *M* NaCl, a single H2A/H2B dimer binds to each H3/H4 tetramer–DNA complex, and by 0.6 *M* NaCl nucleosome assembly per se is complete. Below 0.6 *M* NaCl, the assembled nucleosomal arrays are folded until low salt conditions are reached, at which point the nucleosomal arrays adopt an unfolded "beads-on-a-string" state. In view of these results, we first dialyze into 1 *M* NaCl/TE to assemble positioned H3/H4 tetramer arrays, then into 0.75 *M* NaCl/TE to form hexamers. The reconstitutes are then dialyzed immediately against TEN buffer to complete the assembly process as rapidly as possible. Also, the DNA concentration is kept low (≤300 μg/ml) to prevent the need for urea in the reconstitution buffers.

Protocol

1. Prepare salt dialysis buffers as follows: 2.0 liter of TE buffer containing 1.0 *M* NaCl (B1), 0.75 *M* NaCl (B2), and 2.5 m*M* NaCl (B3). Cool to 4° prior to use.
2. Soak dialysis tubing in B1 for ≥30 min.
3. Adjust the template DNA to 2 *M* NaCl. Mix DNA and histone octamers (in hydroxyapatite elution buffer) to final NaCl and DNA concentrations of 2.0–2.3 *M* and 100–300 μg/ml, respectively. Gently mix by inverting the sample after each step. In experienced hands, a ratio of 1.1–1.2 moles of histone octamers per mole of DNA repeat (*r*) usually will yield preparations in which ~50% of the DNA templates are saturated with histone octamers.

[36] A. Stein, *J. Mol. Biol.* **130,** 103 (1979).

[37] T. H. Eickbush and E. N. Moudrianakis, *Biochemistry* **17,** 4955 (1978).

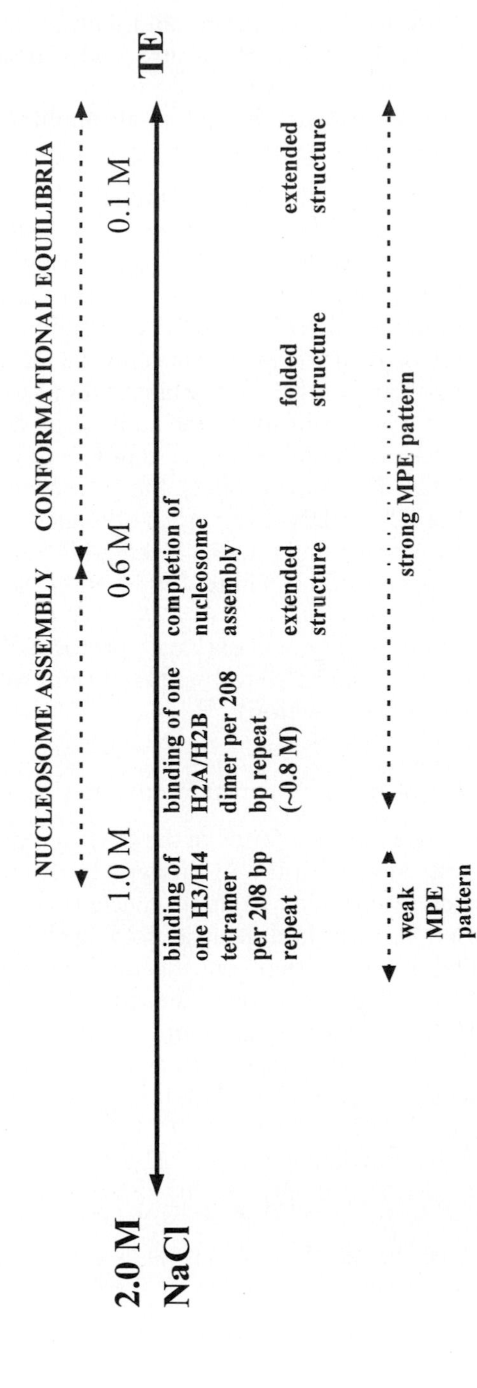

FIG. 1. Schematic illustration of the events that occur during assembly of 208-12 nucleosomal arrays by salt dialysis. MPE, methidium propyl-EDTA · Fe(II).

4. Transfer the mixture to dialysis tubing and dialyze sequentially at 4° against B1 (3 hr), B2 (3 hr), B3 (4 hr), and a fresh change of B3 overnight.
5. After dialysis, add the protease inhibitors aprotinin and leupeptin to a final concentration of 10 μg/ml.

Assembly of trypsinized nucleosomal arrays and H3/H4 tetramer arrays follows the same protocol except that either trypsinized histone octamers or H3/H4 tetramers are mixed with the DNA, respectively. Assembly of acetylated nucleosomal arrays follows this protocol except that the reconstitution buffers also contain 2.5 m*M* DTT and 2.5 m*M* sodium butyrate.

Reconstitution of nucleosomal arrays in which only the H2A/H2B dimer or H3/H4 tetramer components of the nucleosomal subunits were trypsinized (hybrid trypsinized) is achieved using the same protocol except for the step where the histones are added. In the case of hybrid arrays lacking H2A/H2B N termini, intact H3/H4 tetramers and trypsinized H2A/H2B dimers were added to the 208-12 DNA at r = 1.1–1.3 and 2.2–2.6 respectively. Hybrid-trypsinized nucleosomal arrays lacking H3/H4 tail domains were constructed from trypsinized H3/H4 tetramers and intact H2A/H2B dimers using the same strategy.

Importantly, reconstitution by typical exchange protocols[38,39] should be avoided due to complications relating to the inability to control for nonspecific histone deposition (see next section).[7,40]

Determination of Extent of Array Saturation after Reconstitution

The degree of histone octamer saturation of the DNA template achieved after reconstitution is critical to both the quality and the interpretation of subsequent experimental results. For example, nucleosome-free gaps in a subsaturated nucleosomal array inhibit array folding.[13,41] Furthermore, although one never obtains more than one histone octamer per DNA repeat, nonspecific histone deposition onto the underlying nucleosomal arrays occurs readily if the histone : DNA input ratio is too high (see Fig. 2).[12] Thus, it is absolutely essential to accurately assess the extent of saturation of a defined nucleosomal array model system prior to initiating any structural or functional studies. Determination of saturation levels minimally requires analysis by restriction enzyme digestion and quantitative agarose gel electrophoresis, and ideally also by sedimentation velocity in

[38] H. R. Drew and C. R. Calladine, *J. Mol. Biol.* **195,** 143 (1987).
[39] D. Rhodes and R. A. Laskey, *Methods Enzymol.* **170,** 575 (1989).
[40] J. C. Hansen, unpublished observations.
[41] T. M. Fletcher, P. Serwer, and J. C. Hansen, *Biochemistry* **33,** 10859 (1994b).

the analytical ultracentrifuge. One assay that essentially is meaningless for the purpose of establishing reconstitute integrity is micrococcal nuclease (MNase) digestion.

The basis of these conclusions is illustrated by the experiments shown in Fig. 2. The 208-12 DNA template was reconstituted at r = 1.0–1.6 and the resulting reconstitutes were characterized by a number of different assays. Although there are some relatively minor comparative differences in the (MNase) digestion profiles, all of the samples form the characteristic nucleosome "ladder" expected from digestion of positioned nucleosomal arrays (Fig. 2A). It is impossible to establish from these results which samples are closest to being saturated with core histones. Furthermore, even the r = 1.6 appears to be composed of high-quality 12-mer arrays based on MNase digests alone. However, when the same samples are analyzed by sedimentation velocity (Fig. 2B), quantitative agarose gel electrophoresis (Fig. 2C), or *Eco*RI digestion (Fig. 2D), it can be determined in each case (see end of this section) that the r = 1.0 sample is subsaturated, the r = 1.2 sample is nearly perfectly saturated with bound octamers, and both the r = 1.4 and 1.6 nucleosomal arrays are supersaturated with nonspecifically bound core histones. It should be noted that while each of the latter three assays alone provide useful information regarding the degree of saturation, combined they are even more powerful.

Protocol: Determination of Gel-Free Mobility in Low Salt Using Agarose Multigel

Quantitative agarose gel electrophoresis provides information regarding macromolecular surface charge density (μ_0), shape (effective radius, R_e), and flexibility. This information is obtained by measuring electrophoretic mobilities (μ) in 0.2–3.0% aggregate gels (termed "multigels"). This section presents a description for determining the μ_0 of 208-12 nucleosomal arrays. Because the μ_0 provides a direct measure of surface charge *density*,[42] the μ_0 of a nucleosomal array containing one histone octamer per 208 bp of DNA (or any other DNA repeat length) is constant regardless of overall array length. Consequently, this assay provides a sensitive and rigorous assay for the average extent of template saturation.[43] The following procedure supplements detailed protocols presented elsewhere.[24,25,43,44]

1. Pour a 9 or 18 lane multigel with running gels typically in the range of 0.2–1.0% agarose in E buffer. These percentages should define a

[42] O. J. Shaw, "Electrophoresis," pp. 10–12. Academic Press, London, 1969.
[43] T. M. Fletcher, U. Krishnan, P. Serwer, and J. C. Hansen, *Biochemistry* **33,** 2226 (1994a).
[44] P. Serwer, *Methods Enzymol.* **130,** 116 (1986).

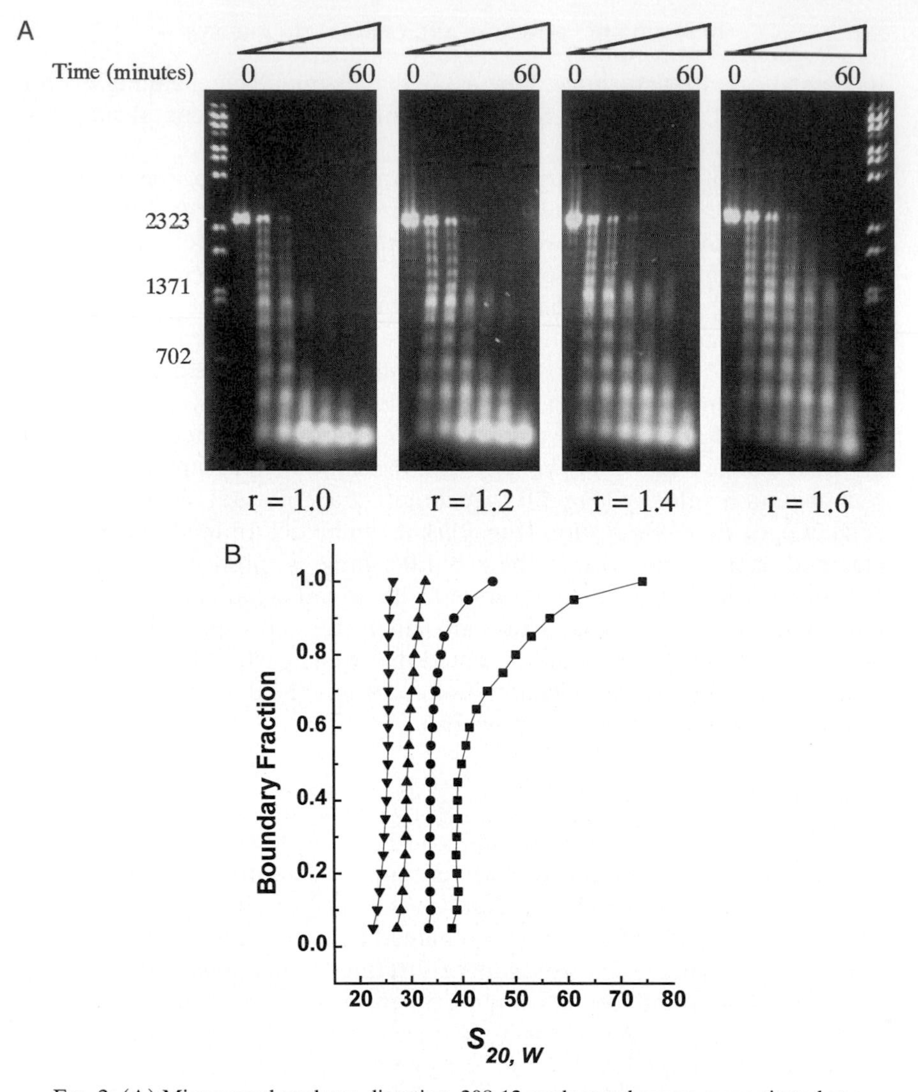

FIG. 2. (A) Micrococcal nuclease digestion. 208-12 nucleosmal arrays reconstituted at r = 1.0, 1.2, 1.4, and 1.6 were digested with 0.05 units of micrococcal nuclease per microgram of DNA in the presence of 1 mM $CaCl_2$. Lanes correspond to 0, 0.5, 1, 2.5, 5, 10, and 60 min of digestion at room temperature for each respective r value. The reactions were quenched by the addition of 1/5 volume of 5% SDS, 25% glycerol, 10 mM Na_2EDTA, 0.3% bromphenol blue, 0.25 μg/μl proteinase K. Samples are incubated at 37° for 30 min to deproteinize the DNA and subsequently electrophoresed for 5 hr at 2 V/cm in a 1% agarose gel buffered with 40 mM Tris–acetate, 1 mM Na_2EDTA, pH 8.0. Bands were visualized under UV illumination after incubation of the gel in ethidium bromide. λ DNA digested with *Bst*EII (M) was used as size markers. (B) Sedimentation velocity analysis. The same reconstitutes were analyzed by sedimentation velocity in the XL-A analytical ultracentrifuge and analyzed by the method of van Holde and Weischet[26] to yield the integral distribution of sedimentation coefficients. Profiles obtained for r = 1.0 (▼), 1.2 (▲), 1.4 (●), and 1.6 (■) are shown. (C) Determination

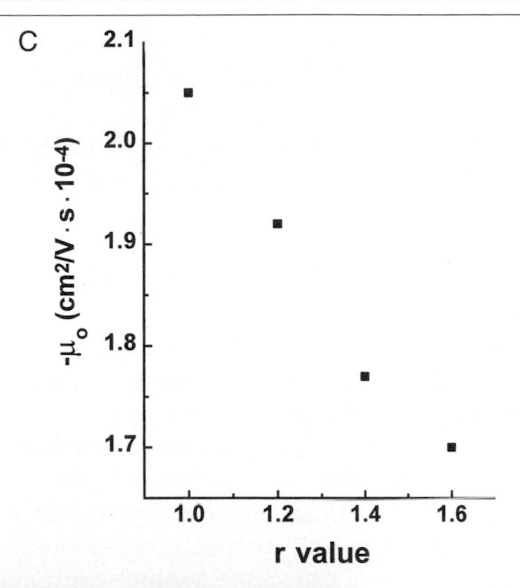

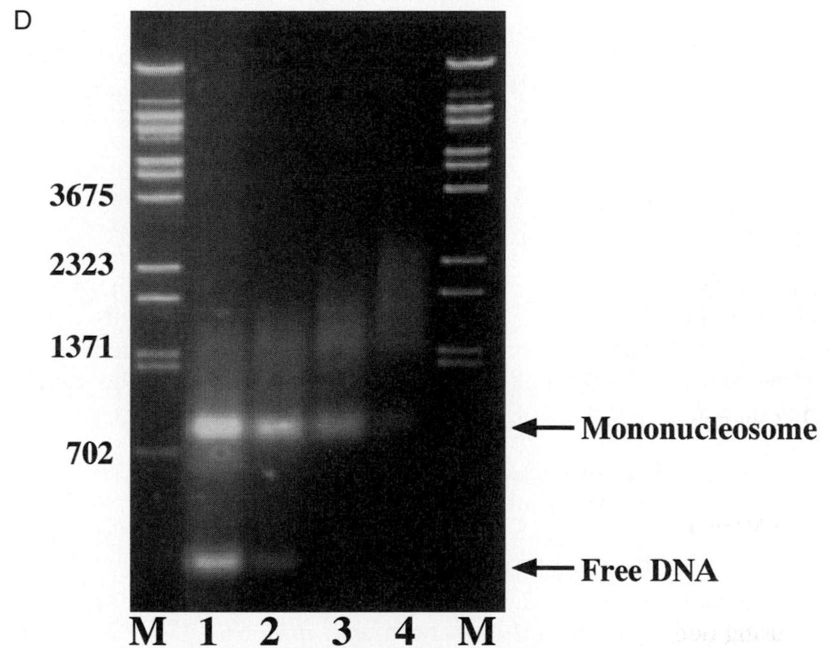

FIG. 2. (*continued*)

of gel-free mobility using quantitative agarose gel electrophoresis. The indicated μ_0 values were obtained as described in the text. (D) *Eco*RI digestion. The same reconstitutes were digested with *Eco*RI and electrophoresed on a 0.8% agarose gel as described in the text. Lanes 1–4 correspond to 208-12 nucleosomal arrays reconstituted at r = 1.0, 1.2, 1.4, and 1.6, respectively. λ DNA digested with *Bst*EII (M) was used as size markers.

region in which a semilogarithmic plot of μ versus agarose concentration is linear, which is essential for accurately determining the gel-free mobility (μ_0').

2. Mix 0.5 μg nucleosomal arrays with 0.5 μg bacteriophage T3 standard (10–15 μl total volume) and load the sample into the wells of the multigel.
3. Electrophorese the samples at 1 V/cm for 8 hr at room temperature using an oscillating pump to circulate the buffer and keep the temperature constant within $\pm 3°$.
4. Stain the multigel with ethidium bromide. Acquire a high-quality photograph and negative of the multigel next to a ruler defining distance (usually centimeters).
5. Digitize the negative (e.g., using a computer interfaced with a digital imaging camera and NIH Image software). Measure the values of μ for both T3 and the nucleosomal arrays (defined as the distance traveled in centimeters from the well to the middle of the bands).
6. Take the tabulated data and create a semilogarithmic plot of μ versus agarose concentration for both T3 phage and nucleosomal arrays. Extrapolate the linear region of the plot to a gel concentration of zero (*y* axis) to obtain the gel-free mobility, μ_0'. The μ_0' is subsequently corrected for electroosmosis and normalized to yield the (μ_0) as described.[25,43]

Protocol: Restriction Enzyme Digestion Assay

The following protocol is described for *Eco*RI digestion of 208-12 nucleosomal arrays.[18] Equivalent assays with analogous tandemly repeated templates will follow the same protocol except that the specific restriction enzyme that cuts in the linker DNA regions of that particular template will be used.

1. Digest ~1 μg of nucleosomal arrays with 10 units of *Eco*RI for 60 min at room temperature in digestion buffer.
2. Quench the reaction by addition of 1/5 volume of 25% glycerol, 10 m*M* Na_2EDTA (pH 8.0).
3. Electrophorese the digested samples in a 0.8% native agarose gel buffered with 45 m*M* Tris–borate. 1 m*M* Na_2EDTA for 3 hr at 1 V/cm to resolve histone-free DNA repeats from nucleosome monomers and larger undigested oligonucleosome fragments.
4. Calculate the percentage of histone-free DNA repeats present in each sample by quantitating the ethidium signal from each gel band using densitometry and multiplying the signal from the nucleosomal

bands by 2.5 to correct for histone quenching of ethidium fluorescence.[45]

Protocol: Determination of Integral Distribution of Sedimentation Coefficients in Low Salt by Sedimentation Velocity in Analytical Ultracentrifuge

Sedimentation velocity experiments performed in an analytical ultracentrifuge are used to directly measure the sedimentation coefficient distributions of nucleosomal array and chromatin samples.[26,27] The sedimentation coefficient distributions in turn provide a sensitive indication of the extent of compaction of each species present in the sample.[27,28] Because the log of the sedimentation coefficient increases linearly with increasing template saturation,[12] sedimentation velocity experiments provide the single most definitive means to determine the extent of saturation of 208-12 nucleosomal or chromatin arrays after reconstitution. The use of the XL-A/XL-I analytical ultracentrifuges has been simplified greatly compared to previous versions of this instrumentation, and as a result this technique has become a very powerful tool in studying defined nucleosomal array and chromatin model systems in the solution state.[27,28]

1. Prepare 400–450 μl of nucleosomal arrays in TEN buffer so the final A_{260} is between 0.6 and 1.0. Equilibrate at room temperature for 1 hr.
2. Operation of the XL-A/XL-I is straightforward and has been described in detail.[24,46] It is important to allow the rotor to equilibrate at the chamber temperature for at least 1 hr before starting the experiment. Choose the fastest speed that allows collection of 8–10 scans (that have completely cleared the meniscus) while maintaining a flat plateau in all scans. During the run, data are acquired digitally by the XL-A/XL-I.
3. Analyze data by the method of van Holde and Weischet[26] to determine the integral distribution of sedimentation coefficients present in the sample (plotted as boundary fraction versus $s_{20,w}$). An extremely powerful, user-friendly public domain computer program that performs the van Holde/Weischet analysis and numerous other analysis methods (Ultrascan II) is available from Dr. Borries Demeler.[47]

In our hands, the best possible initial reconstitution of 208-12 DNA templates will yield samples in which ~50% of the nucleosomal arrays are saturated, ~40% contain one to two nucleosome-free gaps, and <10% are

[45] C. T. McMurray and K. E. van Holde, *Proc. Natl. Acad. Sci. U.S.A.* **83,** 8472 (1986).
[46] Beckman XL-A/XL-I Manual.
[47] http://www.biochem.uthscsa.edu/UltraScan.

slightly supersaturated. Such a preparation is considered to be high quality, i.e., suitable for publication purposes. Identification of the quality of the nucleosomal array preparation achieved after reconstitution in most cases is made easily with the analytical ultracentrifuge; saturated 208-12 arrays sediment at 29-30S in TEN buffer and the fraction of the sample that sediments above and below this value can be determined accurately from the sedimentation coefficient distribution plot (Fig. 2B). If at least half of the array preparation is saturated, the μ_0 will be -1.92×10^{-4} $cm^2/V \cdot sec$ in E buffer[41,43]; the value is more negative if the array is subsaturated[43] and less negative if the array is supersaturated (Fig. 2C). Finally, a sample in which 3–4% of the *Eco*RI digest consists of naked 5S rDNA repeats can also be judged to contain at least 50% saturated arrays.[18,22] This assay also provides a clear indication when the bulk reconstitutes are either subsaturated or supersaturated, as judged by the greater amount of free 5S rDNA and inaccessibility of *Eco*RI-binding sites, respectively (Fig. 2D).

It is important to note that until recently only a few analytical ultracentrifuges were in regular use throughout the world. However, this instrumentation has made a dramatic return and is now much more accessible than in the past. Furthermore, even though it is highly desirable to utilize analytical ultracentrifugation when attempting to reconstitute defined nucleosomal array and chromatin model systems, the extent of template saturation can now be documented rigorously from the combination of μ_0 and *Eco*RI data (Figs. 2C and 2D). In this regard, the multigel apparatus required to measure the μ_0 is relatively inexpensive, making it possible for virtually every chromatin laboratory to exploit the advantages inherent in the analysis of defined nucleosomal array and chromatin model systems if desired.

Selective Enrichment of Saturated Nucleosomal Arrays

Although with regular practice it is possible to achieve the optimal reconstitutes described earlier after the initial attempt, in general this is not an easy task to accomplish. Consequently, a method has been developed for the enrichment of saturated templates, i.e., by manipulating the solubility properties of this system, it is possible to improve on the extent of template saturation achieved after the initial reconstitution. To exploit this protocol to its *fullest*, access to an analytical ultracentrifuge is required (although see comments at the end of this section).

Protocol

1. Follow the reconstitution protocol described earlier, except purposely use $r = 1.3$–1.4 rather than 1.1–1.2. For these purposes, the

initial DNA concentration during reconstitution should be ~300 μg/ml.

2. Perform a sedimentation velocity experiment to determine the fraction of the sample that is supersaturated, and from this estimate the amount of DNA that needs to be added back to the sample to achieve nearly complete saturation. The concentration of both the nucleosomal array sample and the added DNA is critical for accurate adjustment of the saturation level.
3. Dialyze the supersaturated sample from low salt to 2 *M* NaCl in the reverse order described earlier.
4. Add the predetermined amount of DNA template to the initial sample and dialyze into low salt as described previously. Determine the extent of saturation of the nucleosomal arrays using a sedimentation velocity experiment.
5. Repeat steps 3 and 4 until a sample is obtained containing 50–80% saturated nucleosomal arrays as deemed by sedimentation velocity. It may also be necessary to add back core histones, also quantified very carefully, if the nucleosomal arrays become subsaturated during these manipulations.
6. Use an aliquot of the sample (~20 μg) to establish the Mg^{2+}-dependent oligomerization profile of the sample as described.[13,16] Briefly, dilute the sample to an $A_{260} \cong 1.6$ with TEN buffer. Mix the sample and an ~2× concentrated $MgCl_2$ stock solution (40 μl each), incubate at room temperature for 10 min, and centrifuge in a Eppendorf microcentrifuge at 13,500 rpm (16,000*g*) for 10 min. The range of $MgCl_2$ used typically is 2.0–3.0 m*M*. Carefully remove the supernatant and measure the A_{260}. Plot the percentage of the sample that remained in the supernatant ([A_{260} of supernatant/original A_{260} in TEN buffer] × 100) versus $MgCl_2$ concentration.
7. The supersaturated templates oligomerize first. Therefore, determine the $MgCl_2$ concentration required to selectively oligomerize the fraction of the sample that is supersaturated as judged from the sedimentation coefficient distribution plot. Dialyze the sample from step 5 into the $MgCl_2$ concentration determined in step 6 for 2 hr and centrifuge to pellet the supersaturated oligomers. Carefully remove the supernatant and dialyze it for 4 hr against 1 liter of TEN buffer and then overnight against 1 liter of fresh TEN buffer.
8. Analyze the final sample by sedimentation velocity to establish the extent of enrichment of saturated nucleosomal arrays. Confirm the sedimentation results using the μ_0 and restriction enzyme digestion assays.

In the absence of an analytical ultracentrifuge, enrichment can be achieved using a combination of the μ_0 and restriction digestion assays to estimate the initial extent of template supersaturation (see steps 2 and 5). In this case, dialyze the large-scale preparation into 2 m*M* $MgCl_2$ and repeat steps 7 and 8 to establish the final average extent of template saturation.

Incorporation of Linker Histones to Form Chromatin Model Systems

It is fundamentally important to define the complex interactions nucleosomal arrays undergo in "physiological" ionic conditions. However, it is also essential to have the ability to incorporate linker histones and NHCAPs into a chromatin model system in order to study the structure and function of the chromatin arrays. Described is a protocol to incorporate linker histone H5 into saturated nucleosomal arrays that have been selectively enriched. Linker histones were purified following a protocol described previously.[48]

Protocol

1. Adjust the NaCl concentration of a high-quality, preassembled nucleosomal array preparation to 50 m*M*, add purified histone H5 at a molar ratio of 1.4–1.6 per DNA repeat, and incubate on ice for 3 hr.
2. Dialyze the sample against 1 liter of TEN buffer for 4 hr and then overnight against 1 liter of fresh TEN buffer.
3. Analyze by sedimentation velocity in TEN buffer and by quantitative agarose gel electrophoresis in E buffer to determine the sedimentation coefficient distribution and μ_0, respectively. For saturated nucleosomal arrays containing one bound linker histone per nucleosome, the $s_{20,w}$ in TEN buffer is 35–36 and the μ_0 in E buffer is -1.55 ± 0.06 $cm^2/V \cdot sec$.

Although it is possible to achieve a high-quality sample suitable for investigating linker histone effects on chromatin structure and function after the first attempt, the chromatin sample can also be enriched for saturated templates if deemed necessary from the sedimentation coefficient distribution plot and μ_0 value. This is accomplished using essentially the same protocol as that used for nucleosomal arrays except either a lower $MgCl_2$ concentration or 100–150 m*M* NaCl is used to induce oligomerization of the supersaturated chromatin arrays.

[48] M. Garcia-Ramirez, S. H. Leuba, and J. Ausio, *Protein Expression Purif.* **1,** 40 (1990).

Concluding Remarks

The strategies, approaches, and protocols summarized in this article represent the accumulated knowledge derived from 10 years of working with defined nucleosomal array and chromatin model systems. Although at one time it was absolutely necessary to have access to an analytical ultracentrifuge to properly assemble such systems, additional protocols have been presented that allow rigorous assessment of array integrity after reconstitution and that can be performed by virtually any laboratory studying chromatin. As such, it is hoped that this article makes the many advantages of working with defined nucleosomal array and chromatin model systems that have been assembled from pure components more accessible to chromatin researchers.

[3] Isolation of Minichromosomes from Yeast Cells

By JENNIFER A. ALFIERI and DAVID J. CLARK

Introduction

Many of the recent striking advances in the biology of chromatin structure and function concern the insight gained using the tools of genetics in the relatively simple eukaryote, the yeast *Saccharomyces cerevisiae.* Examples include the mechanism of gene silencing as a model for heterochromatin in higher eukaryotes and, more recently, the roles of histone acetyltransferases and deacetylases in gene regulation. However, progress has been limited by the lack of a simple and effective method to purify yeast chromatin containing specific genes or regulatory elements for biochemical analysis *in vitro.*

Several methods for isolation of yeast chromatin in the form of small plasmids have been described in the literature, but they suffer from various problems, reflecting the difficulty of isolating a small amount of relatively pure minichromosome from the huge excess of contaminants (chiefly ribosomes, genomic chromatin, mitochondrial genomes, and yeast double-stranded RNA virus particles, if present). The method of Pederson *et al.*[1] results in highly purified minichromosomes but it is very time-consuming, requiring 4–5 days to complete (because two equilibrium density gradients

[1] D. S. Pederson, M. Venkatesan, F. Thoma, and R. T. Simpson, *Proc. Natl. Acad. Sci. U.S.A.* **83,** 7206 (1986).

0076-6879/99 $30.00

are required), and dissociation of nucleosomal proteins has been observed.[2] More rapid methods result in relatively crude preparations that are contaminated heavily with ribosomes, genomic chromatin fragments, and mitochondrial DNA.[2–6] By far the most elegant method described to date is that of Dean *et al.*[7]: it requires the presence of a *lac* operator in the minichromosome at a nonnucleosomal site such that the minichromosomes can be purified directly from a cell extract using a *lac* repressor–β-galactosidase fusion protein and beads to which antibodies against β-galactosidase are attached. Minichromosomes are eluted using isopropylthiogalactoside (IPTG). The major contaminant is genomic chromatin.[2,7] The method requires a constant supply of antibody and fusion proteins in reasonable quantities. Another drawback is that it is hard to predict where to place the *lac* operator in the minichromosome.

The authors have developed a method for purification of a small yeast minichromosome (2.5 kb) from yeast cells for studies on activated transcription in the context of chromatin. Yeast nuclei are purified and lysed to release minichromosomes that are then sedimented into a sucrose cushion, dialyzed, and concentrated in a filtration device. The procedure takes about a day and a half. Nucleic acid analysis in agarose gels demonstrates that minichromosomal DNA is a major component of the preparation (see later). The method does not require any special reagents and it is easily scaled up (the cost of the lytic enzyme required to destroy the cell wall is probably the major limiting factor).

Design of Plasmid for Use as Minichromosome

A number of factors should be taken into account when designing the plasmid to be used as a minichromosome. The most important is the copy number (i.e., the number of plasmid molecules per cell), which is determined by specific sequences in the yeast plasmid. There are three classes of yeast plasmid: (1) low copy (CEN/ARS) plasmids: they contain an origin of replication (autonomous replicating sequence, ARS) and a centromere (CEN). It is the CEN sequence that confers low copy number (1–3 per cell). (2) ARS plasmids: these are medium to high copy number (20–100 copies/cell) containing just an ARS. Most of these have copy numbers of 20–40, but *TRP1ARS1*, a circularized genomic *Eco*RI fragment, has a copy

[2] S. Y. Roth and R. T. Simpson, *Methods Cell Biol.* **35,** 289 (1991).
[3] C. Martinez-Campa, N. A. Kent, and J. Mellor, *Nucleic Acids Res.* **25,** 1872 (1997).
[4] D. M. Livingston and S. Hahne, *Proc. Natl. Acad. Sci. U.S.A.* **76,** 3727 (1979).
[5] R. G. Nelson and W. L. Fangman, *Proc. Natl. Acad. Sci. U.S.A.* **76,** 6515 (1979).
[6] C. Shalitin and A. Vishlizky, *Curr. Genet.* **9,** 107 (1984).
[7] A. Dean, D. S. Pederson, and R. T. Simpson, *Methods Enzymol.* **170,** 26 (1989).

number of 100–200/cell.[8] A complication is that there are reports that within the same cell population some cells have very high copy numbers whereas others have very low copy numbers. (3) Plasmids containing a 2-μm origin: these contain the special origin of replication found in the endogenous yeast 2-μm plasmid. The plasmid regulates its own copy number at about 60–100 per cell.[9] This type of plasmid would seem to be ideal, but because all of the 2-μm plasmid functions are required to maintain a stable copy number,[9] the cells must also contain endogenous 2-μm circle plasmid (otherwise the plasmid behaves as an ARS plasmid, usually with a lower copy number). The presence of both plasmids in the cell is likely to result in problems with copurification and a risk of recombination between the plasmids resulting in hybrid plasmids. A solution to this problem is to construct a single plasmid that contains the entire 2-μm sequence (6.3 kb),[10] although this would result in a large minichromosome (see later). Another option is to use 2-μm-based vectors containing the *leu2-d* gene as the selection marker (i.e., *LEU2* with a truncated promoter), which results in extremely high copy numbers (200–400) to provide sufficient quantities of the enzyme encoded by *LEU2*, but it is very difficult to transform yeast with these vectors.[10]

The problem with low copy number plasmids is that the yield of minichromosome from reasonable quantities of cells will be very low (we have not attempted to purify CEN/ARS minichromosomes, but it might just be possible). To illustrate this point, it is interesting to calculate the theoretical maximum yields of the different plasmids from 1 liter of culture at an absorbance of 1 at 600 nm (A_{600}), assuming 5×10^7 cells per A_{600} unit: 0.3 μg DNA for a 5-kb CEN/ARS plasmid, 8 μg DNA for a 5-kb ARS plasmid at 30 copies/cell, and 28 μg DNA for a 5-kb 2-μm-based plasmid at 100 copies/cell. Very high copy number minichromosomes are obviously ideal for yield. However, if the objective is to study transcription (as it is in our case), then there is the risk that not all of the copies contain an active gene (a serious problem for studies aimed at correlating the chromatin structure of a gene with its activity).

Another important factor is the size of the plasmid: the larger the plasmid, the more complex its chromatin structure is (i.e., the more nucleosomes there are to map and the more complicated each experiment becomes). However, a larger plasmid with the same copy number will give

[8] V. A. Zakian and J. F. Scott, *Mol. Cell. Biol.* **2,** 221 (1982).

[9] J. R. Broach and F. C. Volkert, *in* "The Molecular and Cellular Biology of the Yeast Saccharomyces," Vol. 1, p. 297. Cold Spring Harbor Laboratory Press, Cold Spring Harbor NY, 1991.

[10] J. R. Broach, *Methods Enzymol.* **101,** 307 (1983).

greater yields in terms of quantity of DNA. Yeast plasmids are often large because they contain bacterial sequences (such as the ampicillin resistance gene), which are unnecessary for propagation in yeast cells. Furthermore, the presence of extraneous sequences might complicate the interpretation of events in the region of interest. Therefore it is better to remove these sequences by transforming the yeast with a gel-purified fragment corresponding to the desired minichromosome DNA that has been circularized by DNA ligase *in vitro*.

We decided to use a *TRP1ARS1*-based plasmid because it combines a high copy number with small size and its chromatin structure has been studied intensively.[2,7,11] *TRP1ARS1* is a circularized genomic *Eco*RI fragment.[8] The DNA of interest should be inserted at the *Eco*RI site because this is the least likely to interfere with *TRP1* and *ARS1* function (it is of course necessary to have a subclone of *TRP1ARS1* in a suitable bacterial vector to perform the required manipulations). However, it should be noted that the copy number of our *TRP1ARS1* minichromosome (in which *CUP1* was inserted at the *Eco*RI site) was much lower than expected (only about 10–25 copies/cell); it appears that the copy number depends on the nature of the insert in *TRP1ARS1*.

Choice of Yeast Strain

In principle, all that is required is a yeast strain with a genotype appropriate for the selection marker present on the minichromosome. Our plasmid is based on *TRP1ARS1* and so a *trp1* strain is necessary. Other commonly used genes such as *URA3*, *LEU2*, and *HIS3* would be reasonable choices. However, there are a few other factors worth taking into account when choosing a yeast strain.

1. Yeast cells contain many proteases which are released on cell lysis (many are localized in the vacuole). These activities can degrade proteins present on the minichromosomes: the "tail" domains of the core histones are particularly susceptible to proteolytic degradation. Therefore we chose strains that have mutations in some of the vacuolar proteases (encoded by the *PEP4*, *PRB1*, and *PRC1* genes). Suitable strains are YPH420[12] (available from ATCC) and BJ5459[13] (available from the Yeast Genetic Stock Center, Berkeley, CA).
2. Haploid versus diploid cells: we have worked only with a haploid strain. If the effects of gene knockouts are to be tested, then use of

[11] F. Thoma, L. W. Bergman, and R. T. Simpson, *J. Mol. Biol.* **177,** 715 (1984).
[12] J. M. Nigro, R. Sikorski, S. I. Reed, and B. Vogelstein, *Mol. Cell. Biol.* **12,** 1357 (1992).
[13] E. W. Jones, *Methods Enzymol.* **194,** 428 (1991).

a haploid strain is a distinct advantage. It is possible that the plasmid copy number is different in haploid and diploid cells.

3. Nature of the mutations in the selection marker to be used: *Saccharomyces cerevisiae* is adept at homologous recombination and one worry is that the plasmid will integrate at a homologous locus in one of the chromosomes, resulting in eventual loss of the plasmid from the culture. The risk can be reduced by minimizing the homology between the plasmid and the chromosomes (i.e., the target size): chromosomal deletion mutations in genes also present in the minichromosome are better than point mutations because the target size is reduced. This argument also applies to the gene of interest. For example, we work with *CUP1*, one of the two genes in the yeast genome that undergoes tandem amplification[14] (and therefore is a large target for recombination); we constructed a strain in which the entire *CUP1* locus was deleted to reduce the risk of plasmid loss by integration.
4. Presence of 2-μm circle plasmid: The method described later for the isolation of minichromosomes will result in the copurification of 2-μm circle plasmid, if present. Most laboratory strains contain this plasmid at high copy number (60–100 per cell; designated cir^+) and it is therefore desirable to cure the strain of the plasmid before transforming it with the minichromosome plasmid. A method to achieve this is described later. However, if the minichromosome plasmid has a 2-μm origin of replication, which is normally employed to ensure high copy number, then its copy number will be reduced to that of ARS-based plasmids because the elevation of the copy number is dependent on functions supplied by the intact 2-μm circle.[9] BJ5459 is cir^+ but YPH420 is cir^0.
5. Presence of LA virus (killer virus): This is a double-stranded RNA virus (for a review, see Wickner[15]) present in some strains, including YPH420 and BJ5459, which contain very large quantities of the virus. Although the method described later for the purification of minichromosomes results in its destruction, it is quite resistant to RNase and it is therefore better if it is not present.

Elimination of Endogenous Yeast Plasmid (2-μm Circle)

The 2-μm circle is difficult to eliminate from yeast cells. Its presence has only a very small negative effect on growth and it confers no known advantage on yeast cells containing it; its continued propagation has been

[14] S. Fogel and J. W. Welch, *Proc. Natl. Acad. Sci. U.S.A.* **79,** 5342 (1982).
[15] R. B. Wickner, *Annu. Rev. Microbiol.* **46,** 347 (1992).

likened to a sexually transmitted infection.[9] Several methods for curing cells of 2-μm circle have been described.[16–19] The method described here is a modification of that of Rose and Broach[16]: overexpression of the 2-μm circle *FLP* gene (which encodes a recombinase) results in an extremely high 2-μm circle copy number and very slow growth. Consequently, the culture is gradually enriched with cells that have spontaneously lost the plasmid. For overexpression, yeast is transformed with a plasmid containing *FLP* linked to a *GAL10* promoter and galactose is used as a carbon source.

YEp51-FLP[16] (the gift of Dr. J. Broach, Princeton University, Princeton, NJ) is an ARS-based plasmid containing the hybrid *GAL-FLP* gene and *LEU2* as a selection marker. We transformed our strain (based on BJ5459) with this plasmid and selected for leu$^+$ colonies with glucose as the carbon source. However, we were unable to obtain a cir^0 strain. The problem was that the difference in growth rates between "fast"-growing colonies and slow ones was small. We hypothesized that the high copy number of *GAL-FLP* also caused slow growth due to the sequestration of the Gal4 protein, which is present in limiting quantitites.[20] Because a single genomic copy of *GAL-FLP* is sufficient to cure the strain,[16] we transferred the *GAL-FLP* gene from YEp51-FLP to pRS416, a low copy (CEN/ARS) vector containing *URA3* as a selection marker [which is easy to counterselect with 5-fluoroorotic acid[21] (5-FOA)] to obtain pRS.GAL-FLP. [The 2022-bp *Xmn*I–*Xba*I *GAL-FLP* fragment from YEp51-FLP was inserted into pRS416 (Stratagene, La Jolla, CA) cut with *Sma*I and *Xba*I.] All of the colonies we tested (17) were cured of 2-μm plasmid. The modified method is described below:

1. Transform yeast to ura$^+$ with pRS.GAL-FLP (or pRS416 as a control) using plates made from synthetic complete (SC) medium lacking uracil (containing glucose as carbon source).
2. Pick a single colony and inoculate 5 ml SC-uracil medium with galactose as a carbon source to induce the *GAL-FLP* gene. The cells grow very slowly relative to cells containing pRS416.
3. When the culture is dense, plate an aliquot of cells on a plate containing SC-uracil with galactose as the carbon source.
4. Pick some fast growing colonies and inoculate 5 ml SC medium. Also, grow some parent strain as a control. Grow cells to high density and store some of each. Prepare genomic DNA from the remainder of

[16] A. B. Rose and J. R. Broach, *Methods Enzymol.* **185,** 234 (1990).
[17] M. J. Dobson, A. B. Fletcher, and B. S. Cox, *Curr. Genet.* **2,** 201 (1980).
[18] A. Toh-e and R. B. Wickner, *J. Bact.* **145,** 1421 (1981).
[19] E. Erhart and C. P. Hollenberg, *J. Bact.* **156,** 625 (1983).
[20] S. M. Baker, P. G. Okhema, and J. A. Jaehning, *Mol. Cell. Biol.* **4,** 2062 (1984).
[21] J. D. Boeke, F. LaCroute and G. R. Fink, *Mol. Gen. Genet.* **197,** 345 (1984).

the cells and analyze the DNA in a 0.7% (w/v) agarose gel containing ethidium bromide. A 2-μm circle DNA is clearly visible as a supercoiled plasmid running at about 3.7 kb relative to linear DNA.
5. Grow cells that have lost 2-μm circle (cir^0) in SC medium and plate on SC plates containing 0.1% (w/v) 5-FOA (PCR Tech., Gainesville, FL) to select for ura^- cells (i.e., those that have lost pRS.GAL-FLP). Pick a single colony.

Transformation of Yeast Cells

Transformation of yeast cells may be achieved by the lithium acetate method[22] or by electroporation.[23] The following protocol is for transformation using the latter method. It is very important to wash the cells very thoroughly with sorbitol to remove ions (which adversely affect the electroporation).

1. Inoculate 5 ml of selective medium or YPD with yeast cells from an agar plate. Incubate overnight at 30°, 300 rpm. By morning, the culture should be quite dense with A_{600} of more than 2.
2. Dilute the cells into 50 ml of the same medium and continue growth until the culture reaches an A_{600} value between 0.5 and 1.0.
3. Prepare fresh, sterile 1 *M* sorbitol (27.33 g in 150 ml; 0.2-μm filter) and cool to 4°.
4. Transfer the cells to a sterile 50-ml Falcon tube and spin cells down (e.g., Sorvall RT6000D at 3400 rpm/4° for 5 min). Discard supernatant carefully.
5. Wash cells twice more with 50 ml of cold 1 *M* sorbitol. Drain pellet with care.
6. Gently resuspend cells in a final volume of about 200 μl 1 *M* sorbitol, using a 1-ml Pipetman, and transfer to a cold, sterile microfuge tube on ice.
7. Distribute cells in 40-μl aliquots in sterile microfuge tubes on ice (one for each transformation plus a control) and add transforming DNA to the cells. Mix well and leave on ice for 5 min. (Use 0.2 μg of a 0.1-mg/ml solution of DNA in TE.)
8. Pipette cells/DNA into a cold electroporation cuvette (Bio-Rad: 0.2-cm electrode gap) and electroporate using the Bio-Rad (Richmond, CA) *Escherichia coli* Pulser set at 1.5 kV (one pulse). Check the time constant afterward (should be 5.6 or less).
9. Add 450 μl 1 *M* sorbitol to the cuvette, mix, and remove cells to a sterile microfuge tube on ice.

[22] J. Hill, K. A. Ian, G. Donald, and D. E. Griffiths, *Nucleic Acids Res.* **19,** 5791 (1991).
[23] D. M. Becker and L. Guarente, *Methods Enzymol.* **194,** 182 (1991).

10. Plate immediately: 10 and 90% of the electroporated cells on appropriate selection plates.

11. Colonies should appear after 2–3 days at 30°. Individual colonies should be picked and restreaked on fresh plates.

Once colonies have been isolated, it is sensible to check that the plasmid is the correct size by Southern blot hybridization. This is particularly true if the transforming DNA was a ligation reaction (to avoid the presence of bacterial DNA sequences in the plasmid; see earlier) because several different ligation products will have been present.

Isolation of Minichromosomes

The challenge is to purify the minichromosomes away from an enormous excess of ribosomes, from mitochondrial DNA, and from large quantities of the double-stranded LA virus (yeast killer virus), if present in the strain chosen. The following method is partly an adaptation of previous methods.[24–26]

Growth of Cells

The following method is based on 1 liter of yeast cells at A_{600} between 1 and 2 (i.e., 1000–2000 units). Grow the cells in appropriate selection medium to late log phase (A_{600} = 1–2) and collect them in GS-A tubes (6000 rpm in the Sorvall GS-A rotor, 5 min, 4°). Drain the pellets thoroughly. The cells can be stored at −80° after freezing on dry ice.

Preparation of Spheroplasts

Spheroplasts are intact yeast cells with their cell walls removed by lytic enzyme. They are much more fragile than cells and will lyse unless an osmotic stabilizer is present, usually sorbitol. Spheroplasting is usually done in the presence of sorbitol only, subjecting the cells to starvation conditions (typically for 10 min to 1 hr at 30°). To avoid this problem, carry out spheroplasting in sterile SC medium (−tryptophan) with 1 *M* sorbitol. The pH is buffered at 8.0 and 2-mercaptoethanol (2-ME) is added to increase the rate of digestion by lytic enzyme.[27] Note that different strains require different amounts of lytic enzyme; the amount required has to be determined empirically for each strain.

[24] R. May, *Z. Allg. Mikrobiol.* **11,** 131 (1971).

[25] C. Szent-Gyorgyi and I. Isenberg, *Nucleic Acids Res.* **11,** 3717 (1983).

[26] S. Y. Roth, M. Shimizu, L. Johnson, M. Grunstein, and R. T. Simpson, *Genes Dev.* **6,** 411 (1992).

[27] J. H. Scott and R. Schekman, *J. Bact.* **142,** 414 (1980).

1. Resuspend the cells in 50 ml spheroplasting medium (with 20 m*M* 2-ME) and warm to 30° with occasional swirling to prevent the cells from sedimenting.
2. Dilute a 40-μl aliquot of cells into 1.2 ml 1% sodium dodecyl sulfate (SDS) and measure the A_{600} (mix well, wait 2 min before reading; should be near 1).
3. Dissolve about 1200 mg yeast lytic enzyme (ICN, Costa Mesa, CA; at about 80,000 U/g) in 6 ml spheroplasting medium (+2-ME). Shake thoroughly to dissolve and spin out insoluble material (Sorvall RT6000D, 3400 rpm, 5 min, 4°).
4. Add lytic enzyme to the yeast cells. Incubate at 30° for up to 1 hr; swirl regularly to prevent settling of the cells. Remove 40-μl aliquots after 5, 10, 20 min digestion etc. and measure the A_{600} in SDS as described earlier. When the A_{600} drops to 5% of the starting value, digestion is complete. (Do not continue with cells that do not reach <15% of the starting value within 1 hr. Try adding more enzyme next time.)
5. Collect the spheroplasts in the Sorvall SS34 rotor: 7500 rpm for 5 min at 4°.
6. Wash the spheroplasts twice with 25 ml cold ST buffer per tube (to remove medium and enzyme) (spin as described earlier).

Solutions Required

1. Spheroplasting medium (-trp): 6.7 g/liter N-base + ammonium sulfate (DIFCO), 2% D-glucose, 0.74 g/liter CSM-trp (Bio101, Detroit, MI), 1 *M* D-sorbitol. Just before making up to volume, add (to avoid the formation of a precipitate): 1 *M* Tris–HCl, pH 8.0, to a final concentration of 50 m*M*. Filter sterilize.
2. ST buffer: 1 *M* D-sorbitol, 50 m*M* Tris–HCl (pH 8.0). Filter sterilize.

Preparation of Nuclei: Lysis of Spheroplasts Using Ficoll and Triton X-100

The next step is to prepare nuclei from the spheroplasts. This represents a major purification step resulting in the elimination of other organelles as well as most of the ribosomes. The spheroplasts are lysed in the presence of 18% (w/v) Ficoll, which protects nuclei from osmotic shock but not the spheroplasts.[24] We also include 0.2% (w/v) Triton X-100 to aid in complete lysis of spheroplasts; specifically, to avoid having to use a Dounce homogenizer (a step sometimes difficult to reproduce). The concentration of Triton should not be increased because of possible lysis of nuclei (the method does work without the addition of Triton). Most protocols call for the addition of divalent cations (magnesium) to help stabilize the nuclei, but we have included spermidine and spermine instead of magnesium, together with EDTA and EGTA, to protect DNA from nucleases, as is usual in protocols for preparing nuclei from cells of higher eukaryotes (adapted from Hewish and Burgoyne[28]). The buffer concentration is fairly high (40 m*M* potassium phosphate) because yeast cell lysates are acidic. Protease inhibitors (AEBSF and leupeptin) are included throughout.

The spheroplast lysate is loaded on a step gradient composed of 65% (w/v) sucrose in the same buffer. The use of 65% (w/v) sucrose is critical because it is sufficiently dense to prevent mitochondria from sedimenting through the step gradient.[29] Other steps commonly used (e.g., 7% Ficoll/20% glycerol) result in a pellet containing both nuclei and mitochondria in about equal amounts. Not only is the mitochondrial genome hard to remove later on (it is very large but it is not packaged into nucleosomes and therefore sediments very slowly), but the mitochondria contain large quantities of ribosomes that are released on lysis. With a 65% sucrose step, very few contaminating mitochondria are present in the nuclear pellet, and the fraction of cosedimenting RNA measured by the A_{260} of extracted nucleic acid (DNA represents less than 2% of total nucleic acids in yeast) is only 10% (about 1.5 mg) of the total (about 14 mg). With a 7% Ficoll/20% glycerol step, 30% of the RNA cosediments with the nuclei (about 4.5 mg).

1. Resuspend spheroplast pellet in 20 ml F buffer. Resuspend on ice thoroughly but carefully.
2. Layer 20 ml lysate over 15 ml of sucrose step buffer in a polycarbonate high-speed tube (with top).
3. Spin at 14,000 rpm for 30 min in the SS34 rotor at 4°. See Fig. 1A. The pellet is crude nuclei. Decant the supernatant carefully. The pelleted material at the interface is mitochondria. Be sure to remove all of it by carefully and thoroughly wiping the inside of the gradient tube using a tissue.

Solutions Required

1. Spheroplast lysis buffer: 18% Ficoll-400, 0.2% Triton X-100, 40 m*M* potassium phosphate, pH 7.0, 2 m*M* Na-EDTA, 0.5 m*M* Na-EGTA, 0.5 m*M* spermidine hydrochloride, 0.15 m*M* spermine hydrochloride. Adjust the pH to 6.50 with phosphoric acid. Just before use, add the following: 10 m*M* 2-mercaptoethanol, 0.1 m*M* AEBSF, 5 μg/ml leupeptin.

[28] D. R. Hewish and L. A. Burgoyne, *Biochem. Biophys. Res. Commun.* **52,** 504 (1973).
[29] D. Rickwood, B. Dujon, and V. M. Darley-Usmar, *in* "Yeast: A Practical Approach" (I. Campbell and J. H. Duffus, eds.), p. 185. IRL Press, Oxford, 1988.

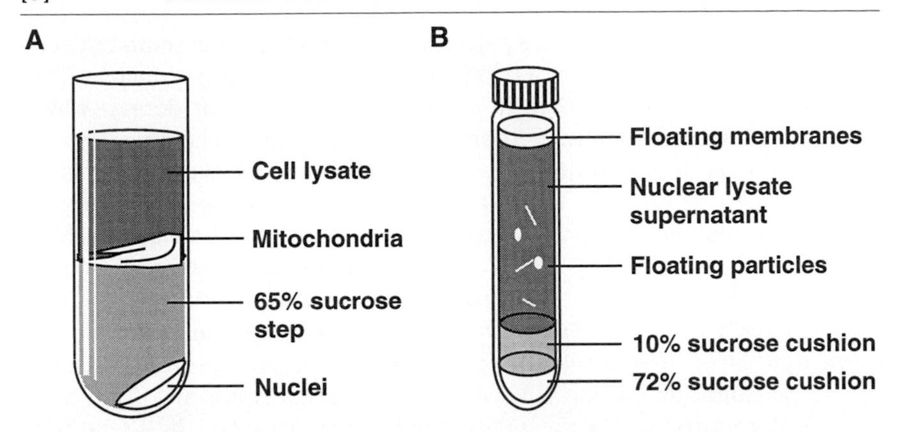

FIG. 1. Preparation of minichromosomes. (A) Preparation of nuclei using a step gradient. The diagram shows the appearance of the 65% sucrose step gradient after spinning. (B) Collection of minichromosomes in a sucrose cushion. The diagram shows the 50Ti tube after spinning; the minichromosomes sediment into the 72% sucrose cushion.

2. Sucrose step gradient buffer: 65% (w/v) sucrose, 40 m*M* potassium phosphate, pH 7.0, 2 m*M* Na-EDTA, 0.5 m*M* Na-EGTA, 0.5 m*M* spermidine hydrochloride, 0.15 m*M* spermine hydrochloride. Just before use, add the following: 10 m*M* 2-mercaptoethanol, 0.1 m*M* AEBSF, 5 μg/ml leupeptin.

Lysis of Nuclei to Release Minichromosomes

Nuclei are susceptible to hypoosmotic shock and can be lysed by resuspension in a low ionic strength buffer. Lysis results in the release of the minichromosomes into the supernatant. The genomic chromatin is undamaged and still attached to nuclear structures and so is easily removed (later) with a short hard spin. The nuclear pellet still contains some ribosomes, which cosedimented with the nuclei (see above), perhaps with the endoplasmic reticulum, which is contiguous with the nuclear membrane. Unfortunately, the ribosomes are also released with the minichromosomes, as are large quantities of LA virus. RNA (about 1.5 mg) is still in very large excess over minichromosome DNA (less than 1 μg). We have tried both gel filtration and sucrose density gradients in an effort to remove the RNA but these methods are only partially successful due to peak overlap with intact or partially disrupted ribosomes, and/or virus particles, depending on the conditions. Therefore, we opted for RNase treatment: ribosomes and virus are digested into much smaller fragments (but only at low ionic strength, where the particles are partially disrupted[30]), which sediment very slowly. The nuclear lysate is left on ice for 30 min to allow nuclear lysis and the RNase digestion to go to completion. The nuclear debris (which includes genomic chromatin) is removed by a hard spin. The minichromosomes are separated from RNase, RNA digestion products, and nuclear membranes by spinning in the ultracentrifuge: the supernatant is layered on top of a small cushion of 10% sucrose placed on top of a small cushion of very dense sucrose (72%) to "trap" the sedimenting minichromosomes. The 10% sucrose layer acts as a "buffer zone" to reduce mixing at the 72% sucrose interface. The spin has to be timed to catch the minichromosomes in the 72% layer. We use 4.5 hr for a 2.5-kb minichromosome; this time should be adjusted for the size of the minichromosome and should be inversely proportional to the DNA length (e.g., use 2.25 hr for a 5-kb minichromosome). Bovine serum albumin (BSA) is added to the cushion solutions to protect against proteases and against loss of minichromosomes to surfaces in the centrifuge tube, during dialysis, and during concentration in the Centricon filter (Amicon, Danvers, MA). Dialysis is necessary to remove the sucrose because the solution is too dense to pass readily through the Centricon filter. Centricon-500 was chosen because its very high molecular weight cutoff (500 K) allows residual RNase and some RNA fragments to pass through and prevents the accumulation of BSA in the sample. We store the minichromosomes frozen in 10% glycerol to limit proteolysis. This does not damage the DNA but we have not rigorously investigated the effects of freezing. Ideally, the minichromosomes should be used in experiments as soon as possible.

1. Resuspend the nuclear pellet thoroughly in 8 ml nuclear lysis buffer and leave 30 min on ice.
2. Spin in the Sorvall SS34 rotor at 10,000 rpm for 5 min at 4°.
3. Prepare a sucrose cushion in a cold 10-ml polycarbonate tube (50Ti): place 1 ml 72% sucrose buffer in the bottom of the tube. Mark the tube at the meniscus with a marker pen. Very carefully overlay 1 ml 10% sucrose buffer. Mark the tube at the meniscus with a marker pen. Then very carefully overlay the nuclear lysate supernatant (which appears cloudy); the solution should reach the neck of the tube. Spin at 50,000 rpm in a Beckman ultracentrifuge in a 50Ti rotor at 4° for 4.5 hr (adjust this time according to the size of the minichromosome DNA; see earlier discussion).
4. After centrifugation there is cloudy material at the top of the tube (presumably nuclear membrane fragments) and there are often some

[30] A. S. Spirin, "Ribosome Structure and Protein Synthesis." Benjamin/Cummings, Menlo Park, CA, 1986.

floating white particles in all layers and perhaps a small pellet (Fig. 1B). Remove and discard the supernatant and most of the 10% sucrose layer. Carefully remove the 72% layer and transfer to a microfuge tube on ice. Wash the bottom of the 50Ti tube with 150 μl dialysis buffer in case some of the minichromosomes pelleted and add this to the 72% layer for dialysis. Spin briefly in the microfuge to remove any insoluble material.

5. Dialyze against 250 ml dialysis buffer overnight to remove sucrose. Use SpectraPor7 tubing (Fisher Scientific, Pittsburgh, PA) (flat width, 12 mm; molecular weight cutoff, 8000 or higher).

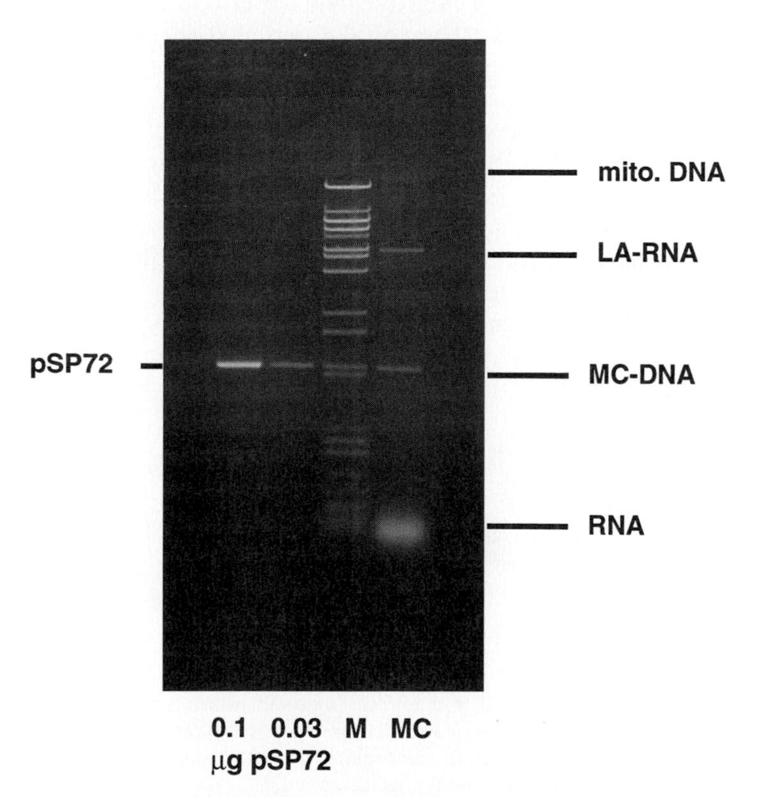

FIG. 2. Nucleic acid content of a typical minichromosome preparation. Nucleic acids from 10% of the preparation were extracted, precipitated, and loaded in a 0.8% (w/v) agarose gel containing ethidium bromide at 1 μg/ml. In this example, the minichromosome is a supercoiled 2.5-kb *TRP1ARS1* plasmid with *CUP1* inserted at the *Eco*RI site. Loaded in separate lanes are 0.03- and 0.1-μg supercoiled pSP72 vector (Promega), which is the same size as the minichromosomal DNA. The size marker (M) is a mixture of a *Bst*EII digest of λ DNA and an *Msp*I digest of pBR322 (New England Biolabs, Beverly, MA).

6. Transfer dialysate to microfuge tubes on ice: spin briefly in a refrigerated microfuge to remove any insoluble material. Transfer supernatant to a washed Centricon-500 unit (Amicon) to concentrate the minichromosomes: spin at 6000 rpm in the SS34 rotor (5000g) at 4° until the volume is reduced to less than 100 μl (about 1.5 hr). The minichromosomes can be washed with fresh buffer, if desired. Collect the minichromosomes (3 min, 3000 rpm).
7. Freeze the minichromosomes on dry ice and store at −20°.

Solutions Required

1. Nuclei lysis buffer: 50 mM Tris–HCl (pH 8.0), 5 mM Na-EDTA. Just before use, add RNase to 0.4 mg/ml (Qiagen, Valencia, CA; DNase-free), 25 μg/ml aprotinin, 0.1 mM AEBSF, 5 μg/ml leupeptin, and 5 mM 2-ME.
2. Sucrose cushion buffers: (1) 72% (w/v) sucrose in 50 mM Tris–HCl (pH 8.0), 5 mM Na-EDTA. (2) 10% (w/v) sucrose in 50 mM Tris–HCl (pH 8.0), 5 mM Na-EDTA. Add AEBSF (0.1 mM), leupeptin (5 μg/ml), 2-ME (5 mM), and 0.1 mg/ml BSA (protease and nuclease free; Calbiochem, La Jolla, CA) to 1 ml of each solution before placing into the centrifuge tube.
3. Dialysis buffer: 10 mM Tris–HCl (pH 8.0), 1 mM Na-EDTA, 10% (v/v) glycerol, 0.1 mM AEBSF, 5 μg/ml leupeptin, 5 mM 2-ME.

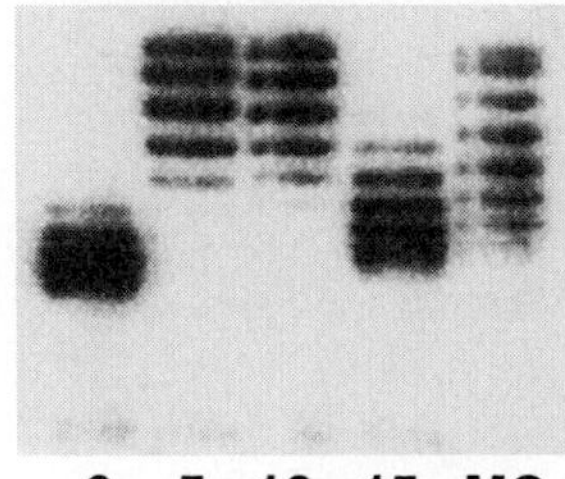

FIG. 3. Counting nucleosomes on a minichromosome by analysis of topoisomers in a chloroquine gel. DNA extracted from a minichromosome preparation was loaded in a 1.2% (w/v) agarose gel containing 10 μg/ml chloroquine. The gel was electrophoresed for 4.5 hr at 40 V with recirculation of the buffer, as described.[31] A Southern blot of the gel was hybridized with radiolabeled probe (a plasmid containing the *TRP1ARS1CUP1* minichromosome sequence) and analyzed using a PhosphorImager (Molecular Dynamics, Sunnyvale, CA). Linking number standards: pSP72 (Promega, Madison, WI) prepared with different numbers of negative supercoils [with excess linking numbers of 0 (relaxed), −5, −10, and −15, referring to the topoisomer at the center of the distribution, as indicated].[31] The minichromosomal DNA (MC) has 12–13 negative supercoils and therefore the minichromosome has 12–13 nucleosomes (as expected).

Analysis of Chromatin

Once isolated, it is important to check that the minichromosomes are intact. The DNA should be supercoiled with very little nicked circle: extract and precipitate 10% of the minichromosome preparation and analyze in an agarose gel containing ethidium bromide (an example is shown in Fig. 2). Use a plasmid DNA of the same size as marker or linearize the minichromosomal DNA by restriction digestion. This also reveals the extent of contamination with other nucleic acids. In the typical preparation shown, there are small amounts of ribosomal RNA, RNA virus, and mitochondrial DNA. Although these contaminants are insignificant relative to the amounts present before fractionation, they nevertheless make it difficult to analyze the protein content of the minichromosomes. The concentration of minichromosomal DNA can be estimated by densitometry using known amounts of plasmid DNA as standards. Preliminary estimates of the copy number of this minichromosome suggest 10–25 per cell and an approximate yield of about 5–15% can be deduced.

A useful test is to measure the number of nucleosomes on the minichromosome by counting the number of negative supercoils present in the DNA after extraction (each nucleosome protects one negative supercoil from relaxation by cellular DNA topoisomerases). This is accomplished using chloroquine gels (which resolve topoisomers according to the chloroquine concentration) and some standards (plasmid DNA of exactly the same size prepared with different degrees of supercoiling). An example is shown in Fig. 3. Detailed protocols can be found in Clark.[31] Further tests of the integrity of the chromatin include mapping the nucleosomes using micrococcal nuclease and mapping DNase I hypersensitive sites. It is certainly an advantage to work with *TRP1ARS1* plasmids because of the wealth of precise mapping data already available.[2,11,32]

Acknowledgments

We thank Ann Dean, Rohinton Kamakaka, Benoit Leblanc, Hugh Patterton, and Chris Szent-Gyorgyi for useful discussions and helpful comments on the manuscript. We thank Dr. J. Broach for providing us with YEp51-FLP.

[31] D. J. Clark, *in* "Chromatin: A Practical Approach" (H. Gould, ed.), p. 139. IRL Press, Oxford, 1998.

[32] F. Thoma and M. Zatchej, *Cell* **55,** 945 (1988).

[4] Chromatin Assembly in *Xenopus* Extracts

By DAVID JOHN TREMETHICK

Introduction

The majority of DNA in eukaryotic cells is packaged by histones, and many nonhistone proteins of unknown function, to form a dynamic structure known as chromatin. Chromatin is a periodic structure made up of repeating, regularly spaced subunits, the nucleosomes. Elegant genetic experiments have clearly demonstrated that histones play a central role in transcriptional control.[1] Histones, via protein–protein interactions or by playing an architectural role, can facilitate or inhibit the transcriptional activation process.[1,2] Moreover, the function of histones themselves may be regulated by protein modifications, which indicates that histones may be targets for cell signaling pathways.[3,4]

To determine the precise molecular details of how chromatin regulates transcription (and how the structure of chromatin itself is regulated), it is essential to mimic the *in vivo* transcriptional activation process *in vitro* within a chromatin context. Indeed, numerous studies employing cell-free protein extracts have reproduced *in vivo* transcription factor–nucleosome interactions *in vitro*.[5–7] Protein extracts prepared from *Xenopus laevis* ovaries have been particularly useful in assembling cloned genes into chromatin *in vitro* for transcription analysis.[6,8]

Xenopus laevis has been used extensively for studying many different aspects of vertebrate development. Many of these studies have used the oocyte as an *in vivo* test tube for the investigation of transcription by microinjecting DNA or RNA into an oocyte.[9] Another advantage of using *Xenopus* toads is that oocytes provide a rich source of nuclear proteins, including chromatin assembly components and transcription factors. The oocyte accumulates such a large pool of proteins because such a store is

[1] M. Grunstein, A. Hecht, G. Fisher-Adams, J. Wan, R. K. Mann, S. Strahl-Bolsinger, T. Laroche, and S. Gasser, *J. Cell Sci.* **19,** 29 (1995).
[2] A. P. Wolffe, *Science* **264,** 1100 (1994).
[3] P. Wade and A. P. Wolffe, *Curr. Biol* **7,** 82 (1997).
[4] S. Y. Roth and C. D. Allis, *Cell* **87,** 5 (1996).
[5] G. Almouzni, M. Mechali, and A. P. Wolffe, *EMBO J.* **9,** 573 (1990).
[6] D. Tremethick, K. Zucker, and A. Worcel,*J. Biol. Chem.* **265,** 5014 (1990).
[7] T. Tsukiyama, P. B. Becker, and C. Wu, *Nature* **367,** 525 (1994).
[8] A. Shimamura, D. Tremethick, and A. Worcel, *Mol. Cell. Biol.* **8,** 4257 (1988).
[9] A. P. Wolffe, *Biochem. J.* **278,** 313 (1991).

0076-6879/99 $30.00

required by rapidly dividing nuclei in the preblastula embryo. The protocol presented here explains the preparation, from *X. laevis* oocytes, of an efficient extract that can assemble the large amounts of chromatin necessary for biochemical analysis. In addition, methods are described for the isolation of key nucleosome assembly components from the oocyte extract.

Preparation of *Xenopus* Oocyte Extracts

Toads, fed with fresh calf liver, are maintained in secured stainless-steel (or appropriate plastic) tanks in tap water (allowed to stand for 24 hr to permit dechlorination) at 20–24°. After surgically removing the ovary, the oocytes are partially dispersed by collagenase treatment. Following this treatment and extensive washing of the oocytes, the oocytes are washed in a low ionic strength extraction buffer. The oocytes are then lysed in the same buffer by a high-speed centrifugation step. Unlike homogenization, this method of lysis does not release proteinases and other degrading enzymes.[10] The resultant supernatant is basically a cytoplasmic extract (3–6 mg/ml). The success of the assembly extract is critically dependent on the presence of large healthy mature (stage 6) oocytes and on the batch of collagenase. It is recommended that a small amount of a specific collagenase batch is first tested, with regard to yielding an active oocyte extract, before purchasing a large batch. Oocyte extracts that have been stored at −70° have remained active for up to 3 years.

Materials and Reagents

10× stock OR2A buffer: 25 m*M* KCl, 825 m*M* NaCl, 50 m*M* HEPES, 10 m*M* Na_2HPO_4, pH 7.6
100× stock OR2B buffer: 100 m*M* $CaCl_2$, 100 m*M* $MgCl_2$
Collagenase (Sigma, St. Louis, MO)
Extraction buffer: 1 m*M* EGTA, 5 m*M* KCl, 1.5 m*M* $MgCl_2$, 10% (v/v) glycerol, 20 m*M* HEPES, 10 m*M* β-glycerol phosphate (disodium salt), 0.5 m*M* dithiothreitol (DTT), 10 μg/ml phenylmethylsulfonyl fluoride (PMSF), 2 μg/ml leupeptin, 2 μg/ml pepstatin, pH 7.5
50-ml conical tubes (Falcon, Becton Dickinson Labware)
SW-41 ultracentrifuge tubes (Beckman Instruments, Palo Alto, CA)

Procedure

1. Anesthetize seven large *Xenopus* toads by hypothermia by placing the animals in an ice bucket (with lid) and covering them with ice for 60–90 min.

[10] G. C. Glikin, I. Ruberti, and A. Worcel, *Cell* **37,** 33 (1984).

2. Remove the ovaries from a toad by making two ventral incisions, in an anterior–posterior direction, on both sides of the abdomen. Tease the ovary out gently using fine-tipped forceps (the toad is then sacrificed by pithing). Place ovaries from the seven toads into a beaker containing 200 ml of OR2 buffer (combine 890 ml of water with 100 ml of 10× OR2A buffer and 10 ml of 100× OR2B buffer). Gently swirl and transfer the ovaries to a second beaker containing another 200 ml of OR2 buffer. Repeat this process until the ovaries are washed with 800 ml of buffer.
3. Following the addition of 0.15 g collagenase (greater than 400 units/mg of protein) to 200 ml of OR2 buffer in a 500-ml flask, add the washed oocytes. Then place the oocytes in a water bath at 25° and shake gently for 2 hr.
4. Wash the collagenase-treated oocytes extensively to remove collagenase, immature oocytes, and connective tissue, with 5 × 200-ml washes of freshly prepared OR2 buffer. For each wash, pour the OR2 buffer down the side of the flask and swirl the flask gently. Decant the buffer when the large oocytes have settled while, at the same time, the small white immature oocytes are still floating.
5. Half fill 50-ml Falcon tubes with the washed oocytes and top with freshly prepared extraction buffer. Invert twice and allow the large oocytes to settle before decanting the buffer. Repeat this washing step three times.
6. Transfer oocytes to a 500-ml beaker containing 100 ml of extraction buffer and proceed to gently load the dispersed oocytes, using the wide end of a Pasteur pipette, and larger clumps of oocytes, using fine-tipped forceps, into 6 SW-41 ultracentrifuge tubes at 4°. Fill the centrifuge tubes to 1 cm from the top. Decant buffer and replace with fresh extraction buffer. This buffer should remain clear as cloudiness is indicative of oocyte lysis.
7. Centrifuge samples at 36,000 rpm for 2 hr at 4°.
8. After centrifugation, draw off the supernatant using a 5-ml syringe connected to a 19-gauge needle. Insert the needle, just above the pellet, gently into the ultracentrifuge tube by using a "twisting motion." Remove the clear supernatant and avoid removing any cell debris and the upper lipid layer. Combine supernatants, add protease inhibitors, and aliquot. Store aliquots at −70°. Alternatively, the oocyte extract can be fractionated further (see later).

Chromatin Assembly Using the Oocyte Extract

The oocyte extract can essentially assemble any piece of DNA (of appropriate size) into chromatin with closed circular DNA being assembled more efficiently than linear DNA. The oocyte extract contains topoisomerase activities, which relaxes supercoiled DNA rapidly prior to the assembly of the plasmid into chromatin. However, it is recommended that supercoiled DNA is first relaxed with purified topoisomerase I prior to being used in assembly reactions. It is also recommended that plasmids less than 4 kb in length be used to ensure efficient assembly.

The chromatin assembly reaction itself is a time-dependent process in which nucleosomes load gradually onto template DNA. While the majority of nucleosomes load onto template DNA within 30 min, the reaction is allowed to proceed for 4–5 hr to ensure the formation of a regular array of nucleosomes. In addition to varying the reaction time, the final nucleosome density can also be varied by altering template DNA concentration or by changing the incubation temperature. For example, the nucleosome density can be decreased by raising the DNA concentration above the optimal concentration for a given extract (it is important to point out that the regular spacing of nucleosomes is lost below a critical nucleosome density). The nucleosomal repeat length of chromatin assembled using the oocyte extract under the conditions described here (at 27°) is approximately 180 bp (Fig. 1B). Increasing the reaction temperature in a gradual manner results in a gradual increase in nucleosome density and a concomitant decrease in the nucleosome repeat of the assembled chromatin.[11] At 37°, the repeat length becomes 160–165 bp.[8]

A critical component of the chromatin assembly reaction is the ATP regeneration system. ATP is required for the ordering of nucleosomes into a regular array, and several ATP-dependent nucleosome spacing activities have been identified.[12–14]

All new oocyte extract preparations are first tested for nucleosome formation by employing a DNA supercoiling assay (Fig. 1A; each nucleosome introduces one negative supercoil into covalently closed circular DNA). Importantly, the authenticity of the assembled chromatin, i.e., whether assembled nucleosomes are organized into a regular array, must then be examined by digesting the chromatin template with micrococcal nuclease. It is also worth noting that although the nucleosome density may be high enough to fully supercoil a circular plasmid template (as assayed in a one-dimensional agarose gel), the density may not be high enough to generate a regular nucleosomal array (Fig. 1; compare Fig. 1C with 1B). Concerning the oocyte extract, occasionally extracts are prepared with low supercoiling activity (or that produce a poor micrococcal nuclease digestion

[11] D. J. Tremethick, unpublished results, 1991.

[12] D. J. Tremethick and M. Frommer, *J. Biol. Chem.* **267,** 15041 (1992).

[13] T. Ito, M. Bulger, M. J. Pazin, R. Kobayashi, and J. T. Kadonaga, *Cell* **90,** 145 (1997).

[14] P. D. Varga-Weisz, M. Wilm, E. Bonte, K. Dumas, and P. B. Becker, *Nature* **388,** 598 (1997).

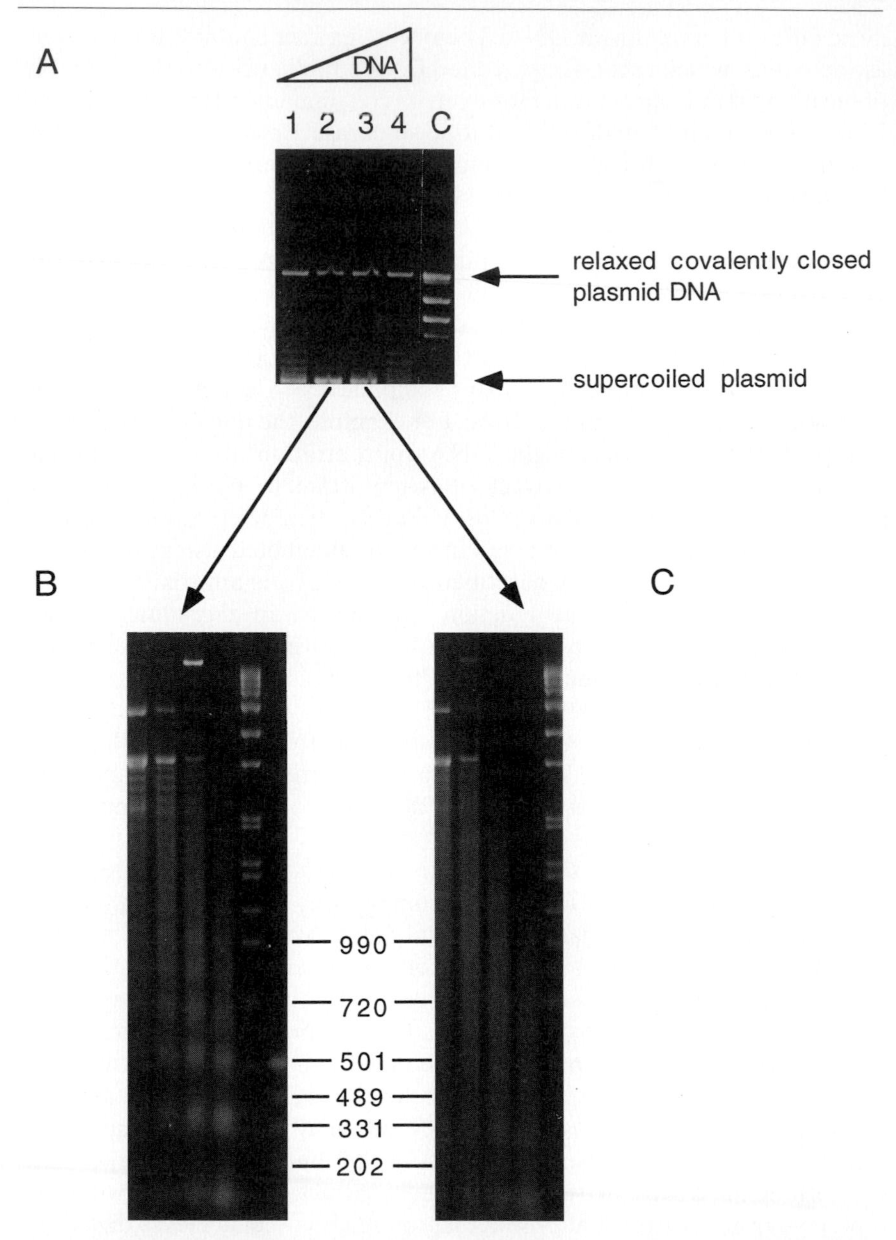

FIG. 1. Supercoiling and micrococcal nuclease analysis of chromatin assembled using unfractionated oocyte extract. Lanes 1–4 (in a total volume of 50 μl) received 50, 150, 300, and 600 ng of relaxed covalently closed plasmid DNA, respectively. Following assembly, template DNA was purified and run on a 1% agarose gel (A). Scaled-up reactions received 150 (B) or 300 (C) ng per 50 μl of reaction mixture. Following micrococcal nuclease digestion, the purified DNA was run on a 1.5% agarose gel.[12]

ladder). In most cases, lowering the pH of the reaction from 7.5 to 7.0 or below can overcome this problem. In addition, optimizing the $MgCl_2$ (by performing a titration between 0.5 to 2.5 m*M*) may improve the quality of the digestion ladder. The following procedure describes the method used to determine the optimal DNA concentration for producing a regular array of nucleosomes.

Materials and Reagents

1. Circular plasmid DNA of interest (relaxed)
2. Extraction buffer: 1 m*M* EGTA, 5 m*M* KCl, 1.0 m*M* $MgCl_2$, 10% (v/v) glycerol, 20 m*M* HEPES, 10 m*M* β-glycerol phosphate, 0.5 m*M* DTT, 10 μg/ml PMSF, 2 μg/ml leupeptin, 2 μg/ml pepstatin, pH 7.5
3. Creatine kinase (Boehringer Mannheim)
4. ATP buffer: 10.0 m*M* $MgCl_2$, 30 m*M* ATP (pH 7.5), 400 m*M* creatine phosphate, 10 m*M* β-glycerol phosphate, 0.5 m*M* DTT, 1 m*M* EGTA, 5 m*M* KCl, 10% (v/v) glycerol, 20 m*M* HEPES, 10 μg/ml PMSF, 2 μg/ml leupeptin, 2 μg/ml pepstatin, pH 7.5

Procedure

1. Add increasing amounts (50 to 500 ng) of circular relaxed plasmid DNA to an appropriate number of Eppendorf tubes in a total volume of 10 μl, on ice, using the extraction buffer as a dilution buffer.
2. Add 5 μl of creatine kinase (1 ng/μl) followed by 5 μl of ATP buffer.
3. Finally, add 30 μl of oocyte extract very slowly to the reaction mix (final volume 50 μl) and incubate for 5 hr at 27°.

Preparation of DEAE Column Assembly Fraction

A potential problem with regard to using the crude oocyte extract for chromatin/transcription studies is that the oocyte extract is not only enriched with chromatin assembly components, but also with transcription factors. The oocyte extract can be fractionated to isolate key nucleosome assembly components and remove many of the proteins involved in the transcription process to yield a more defined *in vitro* system (in addition, this fractionation work has provided important information concerning the mechanism by which DNA is assembled into chromatin *in vitro*[12]). In *Xenopus* oocytes, the two histone pairs exist in two distinct forms. Histones H3/H4 (which are in the diacetylated form) are complexed with carrier proteins known as N1 and N2 to produce a distinct complex.[15] However,

[15] J. A. Kleinschmidt and A. Seiter, *Cell* **29,** 799 (1988).

histones H2A/H2B appear to be mostly in the free form with a small subpopulation being complexed with nucleoplasmin.[12,16] In the first fractionation step, the oocyte extract is passed over a weak anion exchanger (DEAE-Sephacel). Under the conditions used (buffer containing 100 m*M* NaCl), noncomplexed histones H2A/H2B flow through the column whereas the N1,N2–(H3,H4) complex, and H2A/H2B associated with nucleoplasmin remain bound. These complexes are eluted off the column with buffer containing 200 m*M* NaCl.

Materials and Reagents

DEAE-Sephacel (Pharmacia, Piscataway, NJ)
Poly-Prep Chromatography 10-ml columns (Bio-Rad, Richmond, CA)
Column buffer: 1 m*M* EDTA, 1.5 m*M* $MgCl_2$, 20 m*M* HEPES, 0.5 m*M* DTT, 10 μg/ml PMSF, 2 μg/ml leupeptin, 2 μg/ml pepstatin, pH 7.5
Protein assay dye reagent (Bio-Rad)
75 m*M* NaCl extraction buffer: 75 m*M* NaCl, 1 m*M* EGTA, 5 m*M* KCl, 1.5 m*M* $MgCl_2$, 10% (v/v) glycerol, 20 m*M* HEPES, 10 m*M* β-glycerol phosphate, 0.5 m*M* DTT, 10 μg/ml PMSF, 2 μg/ml leupeptin, 2 μg/ml pepstatin, pH 7.5

Procedure

1. Prepare a 1.5-ml DEAE column, ensuring that the column is washed extensively with 100 m*M* NaCl column buffer (containing protease inhibitors).
2. Using a conductivity meter, adjust the oocyte extract to a final conductivity of 0.1 *M* NaCl in column buffer.
3. Very slowly, overnight, load the DEAE column with the oocyte extract (using oocyte extract prepared from 14 toads).
4. After allowing all of the supernatant to run in, wash the column extensively with 10 column volumes using freshly prepared 100 m*M* NaCl column buffer. Once the column is washed, elute chromatin assembly fractions using freshly prepared 200 m*M* NaCl column buffer. Collect 500-μl fractions.
5. Using a standard protein–dye assay, combine the three most concentrated protein fractions. Dialyze against 75 m*M* NaCl extraction buffer, aliquot, and store at $-70°$. Alternatively, use these combined fractions to isolate the N1,N2–(H3,H4) complex (see later).

[16] K. Zucker and A. Worcel, *J. Biol. Chem.* **265,** 14487 (1990).

Purification of Histones H2A and H2B

In contrast to the crude oocyte extract, assembly reactions using the 0.2 *M* DEAE fraction [which contains the N1,N2–(H3,H4) complex] require the addition of histones H2A/H2B to assemble complete nucleosomes. Although histones H2A/H2B purified from the DEAE-Sephacel flowthrough can be used, histones H2A/H2B purified[17] from other sources (e.g., chicken red blood cells or HeLa cells) are used because the oocyte extract contains three H2A variants,[8] which could complicate the interpretation of results from structure/function studies.

Materials and Solutions

Homogenization buffer: 0.2 m*M* EDTA, 20 m*M* Tris, 1 m*M* DTT, 10 μg/ml PMSF, 2 μg/ml leupeptin, 2 μg/ml pepstatin, pH 8.0
Glass Dounce homogenizer.
Hydroxylapatite (Bio-Rad, DNA-grade)
Hydroxylapatite buffer: 0.5 *M* KPO_4, 1 m*M* DTT, 10 μg/ml PMSF, 2 μg/ml leupeptin, 2 μg/ml pepstatin, pH 7.5
Dialysis buffer: 200 m*M* NaCl, 20 m*M* HEPES, 0.1 m*M* EDTA, 10 μg/ml PMSF, 2 μg/ml leupeptin, 2 μg/ml pepstatin, pH 7.5
SS-34 centrifuge tubes
Protein assay dye reagent (Bio-Rad)

Procedure

1. Isolate nuclei from a convenient source of cells[18] using a suitable method[19] and pellet the nuclei by low-speed centrifugation.
2. After weighing nuclei, resuspend nuclei in 5 volumes of homogenization buffer.
3. Homogenize with 10 strokes using a Dounce homogenizer and leave on ice for 30 min.
4. Sonicate twice for 60 sec.
5. Pellet nuclei debris by high-speed centrifugation (14,500 rpm at 4° using SS-34 centrifuge tubes).
6. Remove supernatant and dilute with 9 volumes of fresh 0.6 *M* NaCl-hydroxylapatite column buffer.
7. Load diluted supernatant slowly onto hydroxylapatite column (1 ml of packed resin per 3 mg of chromatin) overnight.

[17] R. H. Simon and G. Felsenfeld, *Nucleic Acids Res.* **6,** 689 (1979)
[18] H. R. Drew and C. R. Calladine, *J. Mol. Biol.* **195,** 143 (1987).
[19] J. D. Dignam, R. M. Lebovitz, and R. G. Roeder, *Nucleic Acids Res.* **11,** 1475 (1983).

8. Wash column with 10 column volumes of 0.6 *M* NaCl hydroxylapatite column buffer.

9. Elute histones H2A/H2B using 0.93 *M* NaCl hydroxylapatite column buffer. Collect 10 fractions, the volume of each fraction being equivalent to one-fourth of the column volume.

10. If required, elute histones H3/H4 using 2.0 *M* NaCl column buffer.

11. Using a standard protein–dye assay, determine which fractions contain protein. Run these fractions on a 18% SDS–polyacrylamide gel to check purity.

12. Combine pure histone fractions and dialyze against dialysis buffer.

Chromatin Assembly Using DEAE Chromatographic Fraction

Like the unfractionated oocyte extract, the 0.2 *M* DEAE fraction assembles regularly spaced chromatin in an ATP-dependent manner. The chromatin assembled has a shortened nucleosomal repeat length of 165 bp (Fig. 2, compare lanes 6–9 with lanes 1–5), indicating that a component(s) responsible for increasing the repeat length from 165 to 180 bp is missing from the 0.2 *M* DEAE fraction. Therefore, this fraction is particularly useful for assembling chromatin with a short repeat length (similar to that observed in lower eukaryotes such as yeast) and for carrying out histone H1 structure/function studies because the addition of histone H1 to the reaction restores the repeat length back to approximately 180 bp.[12]

Initially, new DEAE fractions (in the presence and absence of H2A/H2B, as described in the protocol section) are tested for nucleosome formation by employing the supercoiling assay. A protein titration is performed using the DEAE fraction to determine the optimal protein concentration required for efficient nucleosome formation. Because H2A/H2B is required for the formation of a complete nucleosome, the addition of H2A/H2B enhances the supercoiling reaction. The quality of the assembled chromatin template is then checked by digestion with micrococcal nuclease. Like the unfractionated oocyte extract, it may be necessary to determine the optimal $MgCl_2$ concentration for the assembly reaction.

Materials and Solutions

Circular plasmid DNA of interest (relaxed)

Extraction buffer: 1 m*M* EGTA, 5 m*M* KCl, 1.0 m*M* $MgCl_2$, 10% (v/v) glycerol, 20 m*M* HEPES, 10 m*M* β-glycerol phosphate, 0.5 m*M* DTT, 10 μg/ml PMSF, 2 μg/ml leupeptin, 2 μg/ml pepstatin, pH 7.5

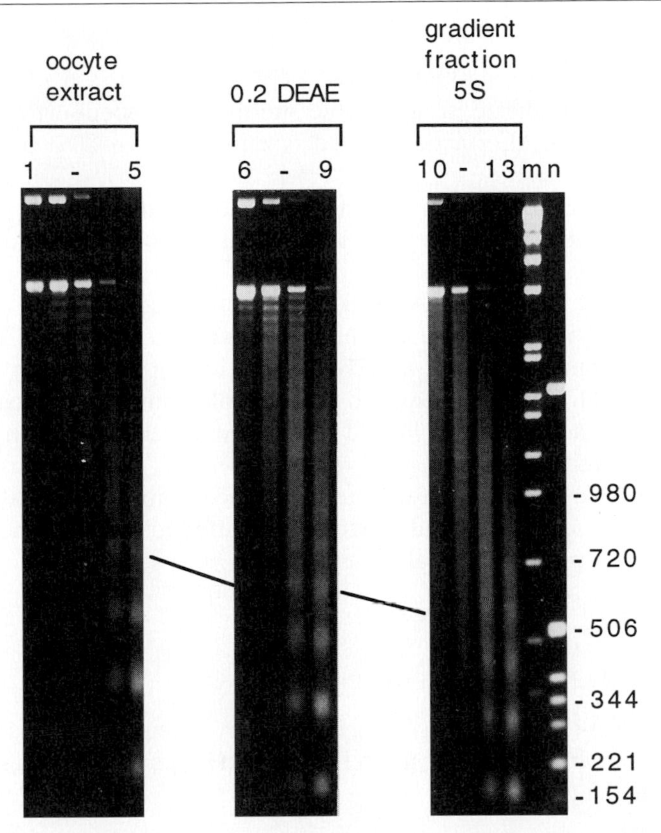

FIG. 2. Micrococcal nuclease digestion analysis of *in vitro*-assembled chromatin. Chromatin was assembled using the unfractionated *Xenopus* oocyte extract (lanes 1–5), a 0.2 *M* NaCl DEAE column fraction (lanes 6–9), and the N1,N2–(H3,H4) complex (lanes 10–13). The assembled chromatin was digested with micrococcal nuclease and the purified DNA was run on a 1.5% agarose gel.[12]

Creatine kinase (Boehringer Mannheim)

ATP buffer: 5.0 m*M* $MgCl_2$, 30 m*M* ATP (pH 7.5), 225 m*M* creatine phosphate, 10 m*M* β-glycerol phosphate, 0.5 m*M* DTT, 1 m*M* EGTA, 5 m*M* KCl, 10% (v/v) glycerol, 20 m*M* HEPES, 10 μg/ml PMSF, 2 μg/ml leupeptin, 2 μg/ml pepstatin, pH 7.5

Pure histones H2A/H2B

Dialysis buffer: 200 m*M* NaCl, 20 m*M* HEPES, 0.1 m*M* EDTA, 10 μg/ml PMSF, 2 μg/ml leupeptin, 2 μg/ml pepstatin, pH 7.5

Topoisomerase I (GIBCO-BRL, Gaithersburg, MD)

Procedure

1. Add increasing amounts of DEAE fraction (4 to 16 μl, with 2-μl increments), in duplicate, to a suitable number of Eppendorf tubes. The volume of the DEAE fraction added is equivalent to 20% of the final volume.
2. To the DEAE fraction, add a volume of creatine kinase (1 ng/μl) followed by the same volume of a ATP buffer equivalent to one-tenth of the final volume.
3. Add pure histones H2A/H2B, at a protein-to-mass ratio of 0.4, to one set of reaction mixtures. To the other set, add 0.2 *M* NaCl dialysis buffer. Depending on the H2A/H2B preparation, in some cases, a higher ratio of H2A/H2B may be required to assemble complete nucleosomes.
4. Add 300 ng of relaxed plasmid DNA slowly (combined with 1.5 units of topoisomerase I).
5. Using dialysis buffer, adust the Na^+ concentration to a final concentration of 85 m*M* and then add extraction buffer to increase the volume of the reaction mix to the final volume.
6. Incubate for 5 hr at 27°.

Isolation of N1,N2–(H3,H4) Fraction

The N1,N2–(H3,H4) complex is isolated by running the 0.2 *M* DEAE fraction on a sucrose gradient; the N1,N2–(H3,H4) complex has a sedimentation coefficient of 5S.

Materials and Solutions

SW-41 ultracentrifuge tubes

5% sucrose buffer: 5% (w/v) sucrose, 23 m*M* KCl, 17 m*M* NaCl, 0.1 m*M* EDTA, 2 m*M* $MgCl_2$, 20 m*M* HEPES, 1 m*M* DTT, 10 μg/ml PMSF, 2 μg/ml leupeptin, 2 μg/ml pepstatin, pH 7.5

20% sucrose buffer: 20% (w/v) sucrose, 23 m*M* KCl, 17 m*M* NaCl, 0.1 m*M* EDTA, 2 m*M* $MgCl_2$, 20 m*M* HEPES, 1 m*M* DTT, 10 μg/ml PMSF, 2 μg/ml leupeptin, 2 μg/ml pepstatin, pH 7.5

Procedure

1. Pour three 11-ml 5–20% sucrose gradients (plus a balance) in gradient buffer using SW-41 ultracentrifuge tubes.
2. Load 500 μl of the combined 200 m*M* NaCl DEAE fraction (total volume 1.5 ml) onto each gradient and centrifuge samples at 40,000 rpm for 24 hr at 4°.

3. After centrifugation, fractionate gradients into approximately 23 fractions of 500 μl each.
4. Run 10 μl of appropriate fractions on a 18% SDS–polyacrylamide gel with standard histones as a control.
5. After silver staining the gel, combine the three gradient fractions with the highest concentration of histones H3 and H4. Routinely, histones H3 and H4 peak in three gradient fractions (usually fractions 16–18 with fraction 1 being the first fraction collected from the bottom of the SW-41 tube). Aliquot the N1,N2–(H3,H4) complex and store at −70°.

Nucleosome Assembly Using N1,N2–(H3,H4) Complex

The isolated N1,N2–(H3,H4) complex combined with purified histones H2A/H2B provides a well-defined nucleosomal assembly system. It is possible to use this system to assemble single positioned nucleosomes on short DNA fragments[20] or assemble plasmid DNA into a nucleosomal template with a wide range of nucleosome densities [by varying the amount of N1,N2–(H3,H4) complex added to the assembly reaction].[6] Assembly reactions receiving these histone components produce a nucleosomal template with a repeat length of 145 bp (Fig. 2, lanes 10–13). Dependent on the experimental aim, longer repeat lengths can be restored by the addition of an ATP-dependent nucleosome spacing activity.[12] Finally, this defined system assembles nucleosomes under physiological conditions and therefore provides a valuable system for transcription studies, e.g., it is possible to add transcription factors to template DNA prior to nucleosome assembly.

Identical to the assays performed with the DEAE fraction, a N1,N2–(H3,H4) complex titration is carried out to determine the optimal protein concentration required for efficient nucleosome formation.[6] The addition of high concentrations of the N1,N2–(H3,H4) complex can inhibit nucleosome formation. After determining the concentration of the N1,N2–(H3,H4) complex required to give the appropriate extent of supercoiling, nucleosome formation is confirmed by micrococcal nuclease digestion analysis.

Materials and Solutions

Circular plasmid DNA of interest (relaxed)

ATP buffer: 30 m*M* ATP (pH 7.5), 135 m*M* creatine phosphate, 10 m*M* β-glycerol phosphate, 0.5 m*M* DTT, 1 m*M* EGTA, 5 m*M* KCl, 10% (v/v) glycerol, 20 m*M* HEPES, 10 μg/ml PMSF, 2 μg/ml leupeptin, 2 μg/ml pepstatin, pH 7.5

[20] K. W. Ng, P. Ridgway, D. R. Cohen, and D. J. Tremethick, *EMBO J.* **16,** 2072 (1997).

Pure histones H2A/H2B
Dialysis buffer: 200 m*M* NaCl, 20 m*M* HEPES, 0.1 m*M* EDTA, 10 μg/ml PMSF, 2 μg/ml leupeptin, 2 μg/ml pepstatin, pH 7.5
Topoisomerase I (GIBCO-BRL)

Procedure

1. Add increasing amounts of the N1,N2–(H3,H4) complex (8 to 24 μl with 2-μl increments), in duplicate, to a suitable number of Eppendorf tubes. The volume of complex added to an assembly reaction is equivalent to 70% of the final volume.
2. To the N1,N2–(H3,H4) complex, add a volume of ATP buffer equivalent to one-tenth of the final volume.
3. Add pure histones H2A/H2B at a protein-to-mass ratio of 0.4 (a protein-to-DNA mass ratio of more than 0.4 may be required if not all of the histone H2A/H2B dimers are functionally active with regard to assembling complete nucleosomes) to one set of reaction mixtures. To the second identical set, add dialysis buffer.
4. Using dialysis buffer, adust the Na^+ concentration to 85 m*M* (if the Na^+ concentration is too high, reduce the creatine phosphate concentration).
5. Add 300 ng of relaxed plasmid DNA plus 1.5 units of topoisomerase I slowly. Incubate for 5 hr at 27°.

Processing of Chromatin Samples for Micrococcal Nuclease and DNA Supercoiling Analysis

Materials and Solutions

Micrococcal nuclease (Boehringer Mannheim)
250 m*M* $CaCl_2$
2.5% sarcosyl (v/v), 100 m*M* EDTA
RNase A (Boehringer Mannheim)
2% SDS (w/v)
Proteinase K (Boehringer Mannheim)
7.5 *M* NH_4SO_4

Procedure

1. On completion of the chromatin assembly reaction, add $CaCl_2$ to a final concentration of 2 m*M*, and then immediately add micrococcal nuclease.[12] Digestions are carried out at 22° (if a supercoiling assay is being carried out, go to step 2).

2. Add 12.5 μl of a 2.5% sarcosyl, 100 mM EDTA solution.

3. Adjust final volumes of samples to 62.5 μl with sterile water.

4. Add 1 μl of RNase A (10 mg/ml) and incubate samples for 30 min at 37°.

5. Next, add 8 μl of a 2% SDS solution and 8 μl of proteinase K (10 mg/ml). Incubate overnight at 37°.

6. Following the overnight incubation, ethanol precipitate the DNA samples by adding 53 μl of 7.5 M NH_4SO_4 followed by 265 μl of ethanol.

7. Following washing DNA samples with 70% ethanol and drying, run the resuspended DNA samples (with appropriate control lanes, see Tremethick *et al.*[6]) on an agarose gel (a 1% gel for supercoiling analysis or a 1.5% gel for micrococcal nuclease analysis).

[5] *In Vitro* Reconstitution of Nuclei for Replication and Transcription

By ALISON J. CROWE and MICHELLE CRAIG BARTON

Introduction

In vitro reconstitution of gene regulation must strike a balance between biochemical accessibility and physiological relevance. Considerable progress toward this goal has been made by utilizing *in vitro* systems that reconstitute higher order protein–DNA structure and thus more accurately reflect the constrictions placed on gene activation *in vivo*. A self-ordered process of *in vitro* chromatin assembly occurs on the incubation of cloned DNA in extracts of embryonic origin, both *Drosophila* and *Xenopus*. Embryonic extracts have been employed to examine the role of chromatin structure in transcription regulation. These extracts are biological warehouses of histones, nucleosome assembly factors, and chromatin remodeling factors that form physiologically spaced nucleosomes on DNA.[1,2]

Organization of more complex nuclear structures *in vitro* relies on the use of *Xenopus* egg extracts that contain not only the chromatin assembly factors just described, but also maternal stores of membrane vesicles.[3,4] Incubation of cloned DNA in these egg extracts activates a stagewise assem-

[1] P. B. Becker, *Sem. Cell Biol.* **6,** 185 (1995).

[2] G. Almouzni and A. P. Wolffe, *Exp. Cell Res.* **205,** 1 (1993).

[3] M. J. Lohka and Y. Masui, *Science* **220,** 719 (1983).

[4] J. J. Blow and R. A. Laskey, *Cell* **47,** 577 (1986).

0076-6879/99 $30.00

bly of functional nuclear organelles.[5–7] These synthetic nuclei are capable of cell cycle function and active protein transport following the encapsulation of chromatin within a bilayer nuclear membrane containing nuclear pores and underlying nuclear lamina.[5,8–10] Initiation of semiconservative DNA replication occurs on formation of an intact membrane envelope. Bidirectional DNA synthesis originates at random DNA sequences and is restricted to a single round.[5,11–17]

Reconstitution of synthetic nuclei provides an accessible, dynamic *in vitro* system that integrates higher-order protein–DNA structure with multiple nuclear functions. Unraveling the respective contributions of complex processes such as DNA replication to gene activation may require the recreation of a nuclear environment *in vitro*. We have coupled synthetic nuclei formation to RNA polymerase II-dependent transcription in order to analyze replication-mediated switches in gene expression.[7,18] As synthetic nuclei formed in the *Xenopus* egg extract are transcriptionally quiescent, it is possible to ascertain the role of DNA replication in mediating changes in chromatin structure in the absence of ongoing transcription. Cellular extracts as a source of RNA polymerase II are added following nuclear formation and DNA replication in order to assess the functional consequences of replication-mediated chromatin remodeling. The power of linking the *Xenopus* egg nuclei assembly system with transcription was initially illustrated by the activation of nucleosome-repressed β-globin templates during DNA replication in the presence of staged erythroid proteins.[7] More recently, we have focused on the α-fetoprotein (AFP) gene as a model for activation of developmentally silenced genes during tumorigenesis by exploiting the coupled nuclei formation–DNA replication–transcription *in vitro* system.

[5] J. Newport, *Cell* **48,** 205 (1987).
[6] P. Hartl, E. Olson, T. Dang, and D. J. Forbes, *J. Cell Biol.* **124,** 235 (1994).
[7] M. C. Barton and B. M. Emerson, *Genes Dev.* **8,** 2453 (1994).
[8] D. J. Forbes. *Annu. Rev. Cell Biol.* **8,** 495 (1992).
[9] D. D. Newmeyer, J. M. Lucocq, T. R. Burglin, and E. M. De Robertis, *EMBO J.* **5,** 501 (1986).
[10] A. W. Murray and M. W. Kirschner, *Nature* **339,** 275 (1989).
[11] M. A. Sheehan, A. D. Mills, A. M. Sleeman, R. A. Laskey, and J. J. Blow, *J. Cell Biol.* **106,** 1 (1988).
[12] J. J. Blow and R. A. Laskey, *Nature* **332,** 546 (1988).
[13] G. H. Leno and R. A. Laskey, *J. Cell Biol.* **112,** 557 (1991).
[14] L. S. Cox and R. A. Laskey, *Cell* **66,** 271 (1991).
[15] A. D. Mills, J. J. Blow, J. G. White, W. B. Amos, D. Wilcock, and R. A. Laskey, *J. Cell Sci.* **94,** 471 (1989).
[16] J. F. X. Diffley, *Genes Dev.* **10,** 2819 (1996).
[17] C. Smythe and J. W. Newport, *Methods Cell Biol.* **35,** 449 (1991).
[18] M. C. Barton and B. M. Emerson, *Methods Enzymol.* **274,** 299 (1996).

Preparation of DNA Templates

Coupling DNA templates to paramagnetic streptavidin beads prior to chromatin or nuclei assembly provides greater flexibility in examining a wide range of gene templates under more biochemically defined conditions.[19,20] A magnetic concentration step gently and easily separates chromatin- and nuclei-assembled DNA from soluble, unbound cellular and egg extract proteins. This step, followed by buffer washes prior to transcription, is essential for the transcription of several gene promoters that are nonspecifically inhibited by the egg extract. A complete change in buffer and/or extract conditions between each step of the experiment is possible. Additionally, a high degree of concentration can be achieved for chromatin and nuclei templates coupled to paramagnetic beads. This bead-coupled DNA has been referred to as "solid phase" by Becker and co-workers.[19,21,22]

Generation of solid-phase DNA templates begins with biotin end labeling of a unique 5′ end. This can be accomplished by digestion with paired restriction enzymes, only one of which generates a 5′ overhanging end containing A's (for biotin dUTP incorporation) or G's (for biotin dCTP incorporation) or by sequential digestion, fill-in, digestion reactions. Labeling of a unique end appears to be critical; coupling of templates labeled on both ends yields templates refractive to transcription and replication. Klenow enzyme and the appropriate combination of deoxynucleotides and biotin-modified deoxynucleotide are used under standard conditions to generate a blunt end containing one or more biotinylated nucleotides.[23] Most frequently in our laboratory, we have digested an 11-kb AFPlacZ plasmid[24] with *Eco*RI and *Cla*I enzymes to generate a 9.0-kb fragment encompassing the AFP enhancer I and promoter sequences fused to the *lacZ* gene. Standard end-labeling reactions containing template DNA (50 μg DNA per 100-μl reaction volume), Klenow enzyme (NEB, Beverley, MA; 5 units per 100-μl reaction volume), 20 μM biotin-21-dUTP (Clontech, Palo Alto, CA), and dATP (20 μM) were performed for 30 min at 37° to generate biotin-labeled *Eco*RI sites. Incorporation of biotinylated nucleotides bearing a linker of 14 carbon residues or greater results in higher

[19] R. Sandaltzopoulos, T. Blank, and P. B. Becker, *EMBO J.* **13,** 373 (1994).

[20] R. Heald, R. Tournebize, T. Blank, R. Sandaltzopoulos, P. Becker, A. Hyman, and E. Karsenti, *Nature* **382,** 420 (1996).

[21] R. Sandaltzopoulos and P. B. Becker, *Nucleic Acids Res.* **22,** 1511 (1994).

[22] E. Ragnhildstveit, A. Fjose, P. B. Becker, and J.-P. Quivy, *Nucleic Acids Res.* **25,** 453 (1997).

[23] F. M. Ausubel, R. Brent, R. E. Kinston, D. D. Moore, J. G. Seidman, J. A. Smith, and K. Struhl, "Current Protocols in Molecular Biology" (K. Janssen, ed.). John Wiley & Sons, Boston, 1987.

[24] B. T. Spear, T. Longley, S. Moulder, S. L. Wang, and M. L. Peterson, *DNA Cell Biol.* **14,** 635 (1995).

efficiency coupling rates, presumably due to a decrease in steric hindrance. Unincorporated deoxynucleotides, biotinylated deoxynucleotides, and small fragments of digested DNA are removed by gel filtration (Chromaspin 1000, Clontech).

Purified biotinylated DNA is coupled to streptavidin-coated paramagnetic beads according to manufacturer's specifications (Dynal Corporation, Oslo, Norway). Coupling efficiency may be highly variable; best results are achieved by employing M-280 Dynal paramagnetic beads and, for fragments greater than 6 kb, kilobase binding buffer (Dynal) in the coupling reaction. Beads are washed twice in an equal volume of binding buffer and then are resuspended in two volumes of binding buffer. An equal volume of biotinylated DNA is added to the resuspended beads to give a 1:1 final dilution. Coupling reactions are incubated at room temperature for 3 hr to overnight on a rotating platform or wheel. Coupled DNA beads are washed three times with 2 *M* NaCl/10 m*M* Tris–HCl, pH 8.0/1 m*M* EDTA and stored in 1× phosphate-buffered saline (PBS) (140 m*M* NaCl/2.7 m*M* KCl/8.1 m*M* Na_2HPO_4/1.5 m*M* KH_2PO_4) at 4°. Rough estimation of the amount of DNA coupled to the beads relies on differences in A_{260} spectrophotometric readings between starting and uncoupled material. A more accurate quantitation is achieved by mixing ethidium bromide (0.5 μg/ml) with a small quantity of coupled beads/DNA and comparing the relative fluorescence of the sample with a series of similarly treated (with ethidium bromide) uncoupled, linearized DNA standards. We routinely couple 20–30 μg of DNA to 1 mg of beads. Transcription and replication reactions are usually performed with 1 μg of DNA (approximately 50 μg of beads) per reaction.

Assembly of Solid-Phase Synthetic Nuclei Templates

Preparation of unactivated, interphase *Xenopus* egg extracts was performed exactly as described previously in detail.[18] The crude cytoplasmic extract of *Xenopus* eggs is clarified by a low-speed spin to yield a supernatant fraction (LSS) competent in formation of synthetic nuclei when combined with cloned DNA templates[5,7,17] and solid-phase DNA.[20] Further fractionation of the LSS by a high-speed spin (200,000 *g*) results in the separation of soluble proteins (HSS) and membrane vesicles. Combination of HSS and membrane vesicles with sperm chromatids or cloned DNA plus glycogen can also assemble functional synthetic nuclei as described previously.[5,7,25,26] However, we have primarily utilized HSS fractions to assem-

[25] P. Hartl, E. Olson, T. Dang, and D. J. Forbes, *J. Cell Biol.* **124,** 235 (1994).
[26] K. S. Ullman and D. J. Forbes, *Mol. Cell. Biol.* **15,** 4873 (1995).

ble chromatin prior to *in vitro* transcription or as a precursor to LSS-mediated assembly of nuclei as described later.

The use of DNA coupled to beads allows one to assemble chromatin and nuclei under optimal conditions for these processes without consideration of parameters that are inhibitory for transcription. With this protocol, a complete change of buffer conditions and protein extracts is possible at each stage of the experiment. We previously found that the LSS fraction and optimal ATP concentrations for the assembly step are highly inhibitory to RNA polymerase II transcription (unpublished observations); however, synthetic nuclei are more robust in nuclear transport and DNA replication when formed in LSS versus the reconstituted HSS plus membrane vesicles.[25] Isolation and purification of solid-phase templates by magnetic concentration and washing permits one to form nuclei under optimal conditions for assembly in LSS and then transcribe the washed templates under optimal transcription conditions.

Assembly of solid-phase AFP DNA into chromatin and nuclei was monitored by fluorescence microscopy (Fig. 1, see color insert). Approximately 1 μg of DNA linked to streptavidin beads (50 μg) was added to approximately 2.5 mg total protein of HSS (Fig. 1, a and b) or 5.0 mg total protein of LSS (Fig. 1, c and d) in a total volume of 100 μl. Parallel reactions were performed with both λ DNA (Fig. 1e) and uncoupled beads (Fig. 1, f and g) in LSS as positive and negative controls, respectively, for nuclei formation. After a 2-hr incubation, 2-μl aliquots were removed and mixed with an equal volume of the DNA-specific fluorescent stain bisbenzimide [Hoechst's stain: 0.8 m*M* KCl/0.15 m*M* NaCl/5 m*M* EDTA/0.15 m*M* PIPES/2.8% paraformaldehyde/220 m*M* sucrose/10 m*M* HEPES/10 μg/ml bisbenzimide (Sigma, St. Louis, MO)]. DNA and nuclear structures were visualized by both phase-contrast (Fig. 1, a, d, and f) and fluorescence (Fig. 1, b, c, e, and g) microscopy. Incubation in HSS for 1 hr at 22° yields chromatin-assembled DNA (Fig. 1, a and b). Paramagnetic streptavidin beads display fluorescence only when coupled with DNA (compare Fig. 1, b, c, and g). The photomicrographs reveal the nonspecific association of beads that tends to occur in the presence of the egg extract (Fig. 1, a, d, and f). This aggregation may reduce the number of templates that are encapsulated within a membrane bilayer during LSS incubation (Fig. 1, c and d). Only the single or smaller groups of beads appear to be enveloped within a membrane (arrow, Fig. 1, c and d) and assume the morphology of synthetic nuclei formed with uncoupled λ DNA (Fig. 1e). Steps taken to reduce aggregation such as gentle rotation during nuclei assembly and replication or inclusion of 0.002% Nonidet P-40 (NP-40) detergent (Sigma) in the reaction may be necessary for optimal nuclei formation. We routinely include NP-40 in the chromatin assembly reactions (0.002%) and during *in vitro* transcription (0.02%).

Elegant studies demonstrating self-organization of microtubules into spindles in the absence of centromeric DNA[20] utilized synthetic nuclei formed from solid-phase DNA incubated in *Xenopus* egg extracts (equivalent to LSS). The morphological and functional characteristics of solid-phase synthetic nuclei were analyzed by electron and immunofluorescence microscopy. These studies revealed a bilayer membrane containing nuclear pores overlying nuclear lamina assembly, which encapsulated the solid-phase DNA. Nuclei were competent in the transport of nuclear localization sequence (NLS) substrate and supported DNA replication.[20]

DNA Replication in Solid-Phase Nuclei

In general, our studies have utilized a two-step process in the assembly of solid-phase nuclei for transcription analysis. The protocol for the initial assembly of solid-phase DNA into chromatin followed by nuclei assembly as diagrammed (Figs. 2A and 3A) was designed with earlier studies by Laskey and co-workers in mind.[15] These experiments revealed that DNA replication by nuclei formed *in vitro* is highly asynchronous in the absence of a preliminary incubation in HSS. Exposure of demembranated sperm chromatid DNA to HSS for 1 hr at 22° mediates decondensation of chromatin without formation of a nuclear membrane.[5,25,27,28] The subsequent addition of LSS leads to membrane formation and DNA replication that is complete in essentially all nuclei in less than 2 hr; this is in contrast to the kinetics observed in LSS with no preincubation where DNA replication is not complete for 4–6 hr.[15]

Prior to synthetic nuclei formation, immobilized AFP templates were preincubated for 1 hr in *Xenopus* egg cytoplasmic fraction (HSS) in an amount determined previously to fully repress transcription (Fig. 2A, chromatin assembly: left-hand side). Reconstitution of chromatin-assembled templates into synthetic nuclei was achieved by the addition of LSS followed by a 2-hr incubation at 22° (Fig. 2A, nuclei assembly and DNA replication: right-hand side). DNA replication in the coupled chromatin assembly (HSS)/nuclei assembly (LSS) system appears to be nearly complete within 1 hr and reaches a plateau by 2 hr (data not shown).

One prediction based on the premise that DNA replication proceeds as a single round of semiconservative synthesis within synthetic nuclei is depicted in Fig. 2A. As solid-phase templates, the biotin end-labeled strand

[27] R. A. Laskey, A. D. Mills, and N. R. Morris, *Cell* **10**, 237 (1977).
[28] A. Philpott, G. H. Leno, and R. A. Laskey, *Cell* **65**, 569 (1991).

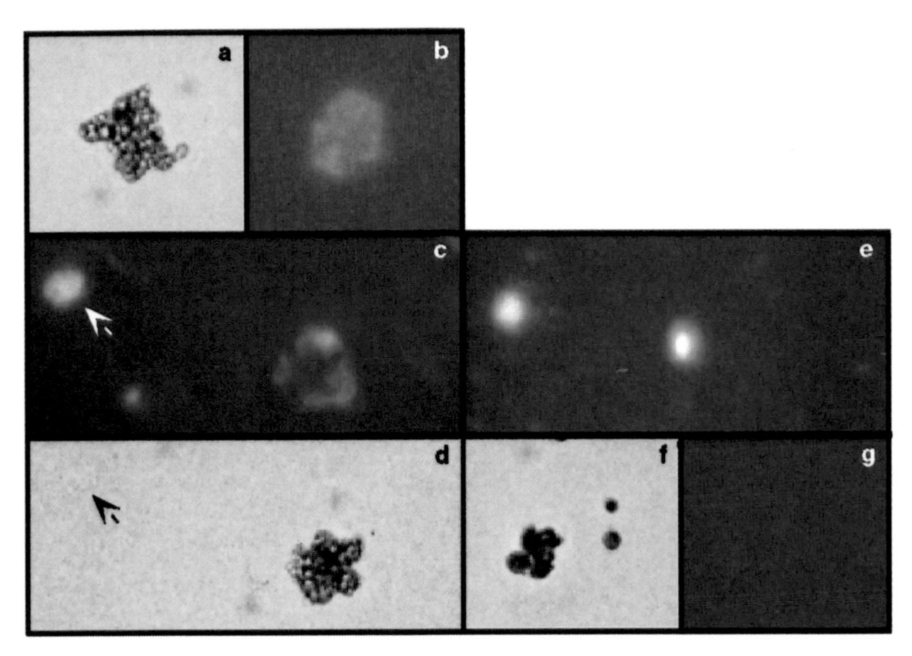

FIG. 1. Solid-phase synthetic nuclei. Bead-coupled DNA was incubated in a *X. laevis* high-speed supernatant (HSS; a, b) for 1 hr or in a low-speed supernatant (LSS; c, d) for 3 hr. Parallel incubations were performed with λ DNA (e) and uncoupled beads (f, g) in LSS. After mixing with bisbenzimide (Hoechst stain), samples were visualized by either phase contrast (a, d, f) or fluorescence (b, c, e, g) microscopy (magnification: $\times 400$). λ DNA is assembled efficiently into synthetic nuclei (e). Uncoupled beads do not fluoresce and do not assemble into nuclear structures (f, g).

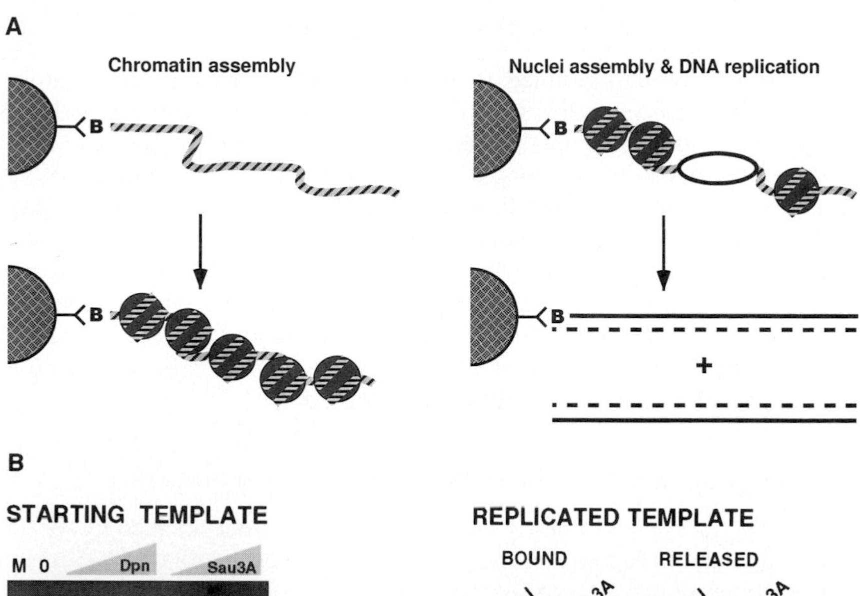

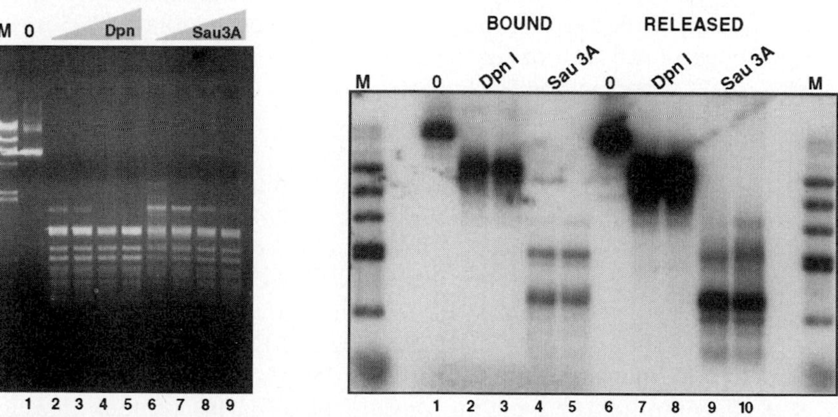

FIG. 2. Replication of solid-phase templates. (A) Diagram of solid-phase template replication. Biotinylated template DNA (wavy line) was coupled to streptavidin-coated magnetic bead (half-circle) through a biotin (B) linkage. Solid-phase templates were assembled into nucleosomes (circles) by incubation with HSS (arrow). Chromatin assembly (left-hand side) was linked to nuclei assembly (right-hand side) by the addition of LSS. The bidirectional replication of solid-phase templates (open circle) results in two populations of DNA: a released double-stranded template and a biotinylated-bound double-stranded template, each containing one template strand (solid line) and one newly replicated daughter strand (dashed line). (B) Replicated DNA is resistant to *Dpn*I cleavage. AFPlacZ template (400 ng) was digested with increasing amounts (0.05, 0.1, 0.5, 1.0 units) of the restriction enzyme *Dpn*I (lanes 2–5) or its isoschizomer *Sau*3AI (lanes 6–9) (left-hand side: starting template). An equivalent amount of bead-coupled template was replicated in Xl LSS in the presence of [μ-^{32}P]dATP (right-hand side: replicated template). After replication, the beads were concentrated and ^{32}P-labeled templates from both the supernatant (containing released templates) and the beads (containing the bound templates) were purified and digested with *Dpn*I or *Sau*3AI as indicated or were left undigested (lanes 1, 6). Digestion reactions contained either 0.1 (lanes 2, 4, 7, 9) or 0.5 (lanes 3, 5, 8, 10) units of enzyme. *Hin*dIII-digested λ DNA molecular weight markers are indicated (M).

of DNA will remain coupled to the beads when copied into a duplex daughter molecule consisting of one strand of newly synthesized DNA (dashed line) and one template strand (solid line). Additionally, the nonbiotinylated template strand, along with its newly synthesized complementary strand, will be released from the bead into the pseudonucleoplasm. Thus, replication of solid-phase, nuclei-assembled templates results in two populations of labeled duplexed DNA: a released fraction and a bound fraction, both of which are assembled rapidly into chromatin (data not shown). Results illustrated in Fig. 2B verify this prediction and further reveal that approximately one-half of the newly synthesized DNA remains bound to the streptavidin bead while one-half is released.

Characterization of Solid-Phase Nuclei Replication Properties

Steady-state levels of newly synthesized DNA can be assayed by continuous labeling during the replication of bead-coupled template (0.75 μg) in the presence of 0.5 μCi of [α-^{32}P]dATP (3000 Ci/mmol, DuPont NEN, Boston, MA) added with unfractionated *Xenopus* egg extract (LSS, 50 μg/μl protein concentration) in a volume of 100 μl (a preincubation with HSS, as described earlier, can also be performed). Replication is terminated after a 3-hr incubation at room temperature by dilution in two volumes of *Xenopus* egg extract buffer (XBII: 100 m*M* KCl/4 m*M* $MgCl_2$/10 m*M* K-HEPES, pH 7.2/100 m*M* sucrose) followed by complete removal of the LSS and buffer to a separate tube (released fraction). After two washes in 100 μl XBII, the bead-bound DNA templates are resuspended in 300 μl XBII (bound fraction). The released and bound DNA fractions are made protein free by incubation with 0.5% SDS/25 m*M* EDTA and proteinase K at a concentration of 1 mg/ml for 1 hr at 37°. This protein digestion step also severs the biotin–streptavidin linkage between the biotinylated template and the paramagnetic bead support. Two phenol–chloroform extractions (1:1, v/v) are performed on the DNA solutions prior to ethanol precipitation and electrophoretic analysis on a 0.8% TBE-agarose gel.

To confirm that the radiolabeled DNA is the result of DNA replication and not due to DNA repair at random sites on the bead-bound DNA, we determined the *Dpn*I resistance of ^{32}P-labeled DNA obtained after the incubation of solid-phase templates in LSS (Fig. 2B). The restriction enzyme *Dpn*I will cleave at its recognition sequence (5′ GATC 3′) only when both adenosines within the double-stranded DNA recognition sequence are fully methylated. Methylation of DNA was achieved *in vivo* by growth in the *Escherichia coli* strain DH5α prior to isolation of covalently closed circular plasmid DNA by cesium chloride centrifugation. The fully methylated AFP DNA was coupled to streptavidin beads and incubated in LSS as described earlier for replication analysis. A single round of DNA replication will yield hemimethylated daughter molecules distributed between bead-bound and released populations.

The restriction enzyme *Sau*3AI recognizes the same 4-bp site and cleaves without DNA methylation specificity or inhibition. Digestion of protein-free starting DNA and replicated DNA by both restriction enzymes in parallel gives a well-controlled comparison of restriction accessibility prior to and following DNA replication. *Dpn*I digestion of the AFP DNA starting template is complete and comparable over the same range of enzyme concentration to the *Sau*3AI pattern, which has the same recognition site but a different cleavage pattern (Fig. 2B, starting template). Following DNA replication, as described earlier, in LSS (total protein concentration of approximately 5 mg) in the presence of [α-^{32}P]dATP, DNA is purified and cleaved with either *Dpn*I or *Sau*3AI enzymes (Fig. 2B, replicated template). Digestions are performed for 2 hr at 37° in 1× REACT 4 buffer (GIBCO-BRL, Grand Island, NY; 20 m*M* Tris–Cl, pH 7.4/5 m*M* $MgCl_2$/ 50 m*M* KCl). Details of the reaction conditions are described in the legend to Fig. 2. The results of this analysis show that the replicated DNA is fully resistant to *Dpn*I cleavage (Fig. 2, lanes 2, 3, 7, and 8). Insensitivity to *Dpn*I digestion is likely the result of single-strand methylation of replicated DNA as the labeled DNA remains fully accessible to *Sau*3AI cleavage (Fig. 2, lanes 4, 5, 9, and 10). The results are similar for both bead-bound and released DNA, demonstrating that radiolabeled DNA synthesis is the result of semiconservative DNA replication rather than fill-in labeling during DNA repair.

Comparison of the gel migration pattern between the starting material and replicated DNA reveals an interesting shift in apparent mobility and size following LSS incubation. We have not determined whether the shift to higher molecular weight migration is the result of concatamer formation during replication or due to the well-known and efficient ability of *Xenopus* egg extracts to ligate DNA.[5]

We have measured the efficiency of DNA replication within solid-phase nuclei under a variety of reaction conditions. This is accomplished easily in the solid-phase system by including a radiolabeled nucleotide during the biotinylation end-labeling reaction, prior to coupling to the paramagnetic bead. For the template, we end labeled an *Eco*RI/*Cla*I-digested fragment with biotin-14-dCTP (Gibco-BRL), [α-^{32}P]dATP, dGTP, and dTTP to generate an asymmetrically labeled template with one biotinylated strand and one radioactive strand. Replication of the radiolabeled DNA coupled to streptavidin beads, in the absence of any further added radioactivity, releases the ^{32}P-labeled strand as one of the newly synthesized daughter molecules. Following the replication incubation period, solid-phase nuclei are isolated by the magnetic concentrator and the supernatant is removed. Solid-phase nuclei are washed twice gently with nuclear dialysis buffer (NDB: 20 mM HEPES, pH 7.9, 100 m*M* KCl, 0.5 m*M* EDTA, 20% glycerol, 2 m*M* DTT). The supernatant is then combined with the two NDB washes and counted in a scintillation counter in order to determine the amount of radiolabeled DNA template that was released. Replication efficiency is expressed as the percentage of radioactivity released relative to the input radioactivity. In general, the replication efficiency determined by this method is 50–75% under conditions that mimic the coupled nuclei assembly–transcription reaction.

Coupling Transcription with Solid-Phase Nuclei Replication

In vitro-assembled synthetic nuclei have been utilized for both RNA polymerase III- and RNA polymerase II-directed transcription.[7,26] Initiation of transcription by RNA polymerase II requires a postassembly addition of nuclear extract as a source of the general transcription factors. In contrast, transcription of cloned tRNA and 5S RNA templates by endogenous RNA polymerase III present in the *Xenopus* egg extract is highly efficient and originated within the membrane-bound organelle.[26] It is highly unlikely that membrane-bound solid-phase nuclei are extant following magnetic concentrator isolation and washing as required for polymerase II reactions.[20] Thus, the coupled nuclei assembly–replication–RNA polymerase II transcription system is best suited for questions addressing the functional consequences of chromatin remodeling during replication or on transport of nuclear proteins rather than as a direct assessment of the structural role that nuclear membrane assembly plays in transcription.

In order to recapitulate the developmental stage-specific repression of AFP transcription and the subsequent loss of repression that occurs in hepatocellular carcinoma, we used the solid-phase chromatin and nuclei transcription system, as diagrammed in Fig. 3A. With this system, we analyzed the ability of various cellular and tissue-specific extracts to establish active or repressed AFP transcription. Cellular extracts (*trans*-acting factors) or dialysis buffer were preincubated for 20 min at room temperature with immobilized bead–DNA templates or, alternatively, added during chromatin assembly or synthetic nuclei formation, as indicated. A *Xenopus* egg cytoplasmic high-speed supernatant fraction (Xl HSS) in an amount previously determined to fully repress transcription (usually 600–800 μg protein) is added to assemble the bead–DNA into chromatin for 1 hr at 22°. Following chromatin assembly, templates are washed by the addition of four volumes of NDB, removal, and a repeat wash in NDB alone. Washed chromatin templates are assembled into synthetic nuclei with *Xenopus* egg

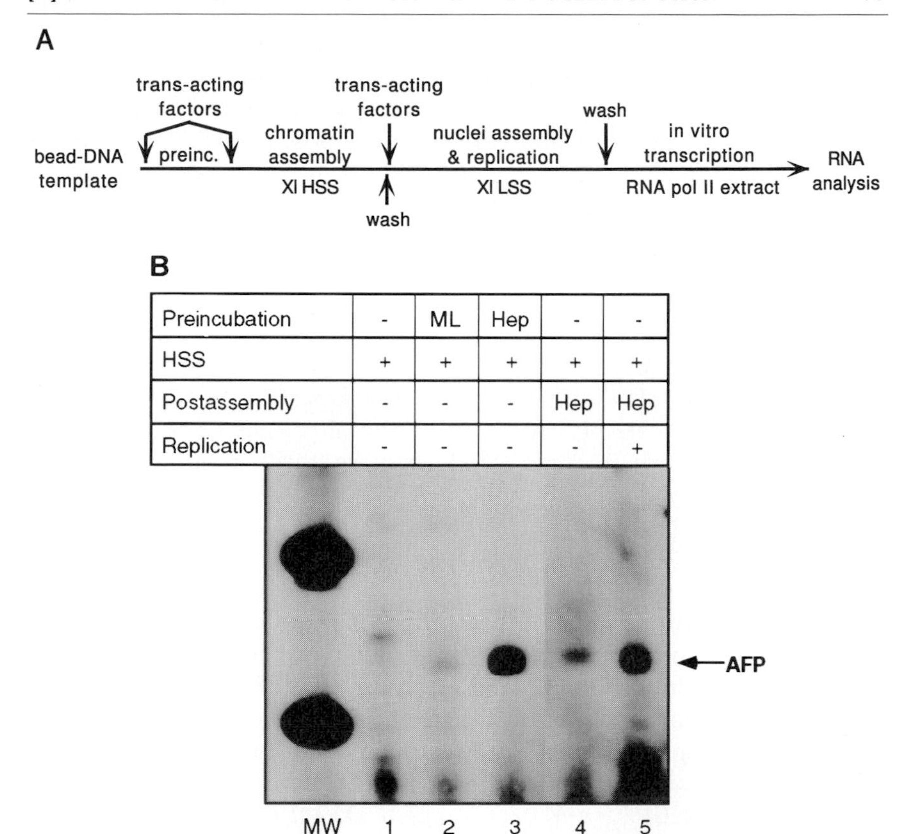

FIG. 3. Coupled DNA replication–transcription system with solid-phase template. (A) Diagram of replication–transcription protocol. Buffer or *trans*-acting factors were added to template DNA coupled to beads during a 20-min preincubation period at room temperature or during chromatin assembly. Chromatin assembly was achieved by incubation of templates with a *X. laevis* high-speed supernatant (Xl HSS) for 1 hr at 22° Chromatin-assembled templates were washed and then assembled further into synthetic nuclei capable of DNA replication by incubation in a *X. laevis* low-speed supernatant (Xl LSS) in the presence of buffer or additional *trans*-acting factors for 2 hr at 22°. Washed templates were transcribed *in vitro* for 1 hr at 30° in the presence of an RNA polymerase II containing extract. The resulting RNA was purified and analyzed by primer extension. (B) Replication-mediated activation of a chromatin-repressed template. AFPlacZ-coupled beads were preincubated in buffer (lanes 1, 4, 5), adult mouse liver extract (ML, lane 2), or HepG2 extract (Hep, lane 3). All templates were chromatin assembled for 1 hr in Xl HSS. Chromatin-assembled templates were washed and incubated in XBII only (lanes 1–3), XBII plus HepG2 extract (lane 4), or assembled further into synthetic nuclei with LSS in the presence of HepG2 extract (lane 5). After a wash step, templates were transcribed in HeLa nuclear extract and RNA detected by primer extension. ϕX174 DNA digested with *Hae*III (GIBCO-BRL) and end labeled with ^{32}P was used as a molecular weight marker (MW).

extract LSS (Xl LSS, total protein concentration of approximately 1.5 mg) along with a second introduction of protein extract or buffer, in a total volume of 50 μl. Once the nuclear envelope is fully formed, replication ensues over the 2-hr incubation period at 22°, as described. The effect of replication on the preprogrammed chromatin structure/transcription state is assessed by *in vitro* transcription analysis. Nuclei-assembled templates are washed three times in NDB prior to transcription exactly as described earlier for replication and structural analyses. Washed templates are then *in vitro* transcribed on addition of an RNA polymerase II-containing HeLa nuclear extract[29] and an NTP/salts/energy-generating mix to give final concentrations of 0.6 m*M* CTP, UTP, GTP, ATP, 5 m*M* $MgCl_2$, 66 m*M* KCl, 5 m*M* creatine phosphate, 10 U/ml of creatine kinase, 9% glycerol, 0.1 m*M* EDTA, and 20 m*M* HEPES (pH 7.9) in a total volume of 50 μl. After a 60-min incubation at 30°, RNA products were purified and analyzed by primer extension and gel electrophoresis as described previously.[30]

We followed this experimental protocol for the transcription analysis presented in Fig. 3B. Cellular extracts are employed as sources of *trans*-acting factors and added either during the preincubation period or after chromatin assembly as indicated. Adult mouse liver extract (ML) is prepared according to Gorski *et al.*[31] Hepatocarcinoma cell extracts (Hep) are prepared from human HepG2 cells as described by Dignam *et al.*[29] with the following minor modifications: Cells are grown to 70% confluence and harvested by scraping into 1× PBS. Washed pellets are resuspended in six times the packed cell volume (PCV) with hypotonic buffer (20 m*M* HEPES, pH 7.9/10 m*M* NaCl/1.5 m*M* $MgCl_2$/2 m*M* DTT). After swelling 10 min on ice, cells are pelleted and resuspended in two PCV with hypotonic buffer containing 0.05% NP-40 prior to Dounce homogenization (Wheaton, Millville, NJ; type B pestle). The remainder of the procedure is performed exactly as described previously.[29] Protein extracts are dialyzed against two changes of 100 volumes dialysis buffer (20 m*M* HEPES, pH 7.9/50 m*M* KCl/0.2 m*M* EDTA/20% glycerol/1 m*M* DTT/0.2 m*M* PMSF) for 2 hr each. The HeLa nuclear extract is prepared exactly as described[23] and is added to each transcription reaction at approximately 150 μg total protein.

Bead–DNA templates are preincubated with NDB (Fig. 3B, lanes 1, 4, 5), adult mouse liver extract (60 μg protein, lane 2), or hepatoma extract (150 μg protein, Fig. 3B, lane 3) in a total volume of 25 μl. Templates are then chromatin assembled in HSS exactly as described earlier. Only proteins

[29] J. D. Dignam, R. M. Lebovitz, and R. G. Roeder, *Nucleic Acids Res.* **11,** 1475 (1983).
[30] M. L. Waterman, W. H. Fischer, and K. A. Jones, *Genes Dev.* **5,** 656 (1991).
[31] K. Gorski, M. Carneiro, and U. Schibler, *Cell* **47,** 767 (1986).

present within the HepG2 hepatoma extract (Hep) are capable of establishing an active chromatin transcription template (Fig. 3B, lane 3).[32] Developmentally repressed staged mouse liver extract (ML) added prior to chromatin assembly yielded DNA templates that are repressed for transcription (Fig. 3B, lane 2) almost as efficiently as nucleosome assembly in buffer alone (Fig. 3B, lane 1), reflecting the inability of the differentiated adult liver to transcribe AFP. Reconstitution of chromatin-assembled templates (washed two to three times in NDB) into synthetic nuclei (Fig. 3B, lane 5) is achieved by incubation with 25 μl unfractionated *Xenopus* egg extract (LSS, 50 μg/μl) in the presence of additional transacting hepatoma factors (150 μg total protein) in a total volume of 50 μl for 2 hr at 22°. The role of replication in these experiments is determined in comparison to parallel postassembly incubations in XBII buffer alone (Fig. 3B, lanes 1–4).

When exposed to replicating chromatin-assembled DNA templates, transactivators present in hepatoma extracts recreate the tumorigenic switch in AFP expression from chromatin repressed to transcriptionally active (compare lane 5 to lane 1 in Fig. 3B). The addition of hepatoma protein extract postchromatin assembly results in a low level of transcriptional activation, implying the existence of chromatin remodelers in the hepatoma extract (Fig. 3B, lane 4). Replication of chromatin-assembled templates in the absence of hepatoma factors yields transcriptionally silent templates (data not shown), indicating that replication/synthetic nuclei assembly alone is insufficient to activate transcription. These studies suggest that DNA replication is required for hepatoma factors to gain full access to their binding sites on nucleosomal DNA in order to establish a highly active transcription template.[32]

Conclusions

By providing a tumorigenic environment in synthetic nuclei, we have succeeded in recreating the aberrant activation of AFP that occurs *in vivo* during liver tumor formation. DNA replication may be necessary to mediate the "derepression" of chromatin-repressed genes by facilitating the competition between *trans*-acting factors and histones during repackaging of the newly replicated daughter strands.[7,33,34] The relative local concentration of activating factors at the time of replication may thus determine whether a given chromatin structure is maintained or converted to an alternate

[32] A. J. Crowe and M. C. Barton, *Methods* **17,** 173 (1999).

[33] A. P. Wolffe and D. D. Brown, *Cell* **47,** 217 (1986).

[34] G. Almouzni, M. Mechali, and A. P. Wolffe, *EMBO J.* **9,** 573 (1990).

conformation. These studies illustrate the potential of using the solid-phase nuclei assembly system to identify the responsible tumorigenic factors as well as the contribution of cellular proliferation to activation of gene expression.

[6] Preparation/Analysis of Chromatin Replicated *in Vivo* and in Isolated Nuclei

By LOUISE CHANG, COLLEEN A. RYAN, CAROLYN A. SCHNEIDER, and ANTHONY T. ANNUNZIATO

Introduction

Once per generation dividing cells must faithfully copy the nucleoprotein complexes that constitute their genetic information. In eukaryotes this process requires histone deposition and nucleosome assembly. Replication-coupled chromatin assembly has perennially been the focus of considerable research effort, and many of the steps required to generate nucleosomes on newly synthesized DNA are now reasonably well understood.[1] Much of this information has been obtained by taking advantage of cell-free systems, in which DNA templates, histone pools, and assembly factors can be manipulated readily.[2,3–6] In addition, the analysis of chromatin replication in whole cell systems has been indispensable in providing insights into histone synthesis and nucleosome assembly and in the ways in which these two vital cell functions are coordinated.[1,7–10] Moreover, the validation of hypotheses derived from *in vitro* experiments will at some point involve verification in living cells.

[1] A. P. Wolffe, "Chromatin: Structure and Function." Academic Press, San Diego, 1995.
[2] C. Gruss and J. M. Sogo, *Bioessays* **14,** 1 (1992).
[3] G. Almouzni and A. P. Wolffe, *Exp. Cell Res.* **205,** 1 (1993).
[4] P. B. Becker, T. Tsukiyama, and C. Wu, *in* "Methods in Cell Biology" (L. S. B. Goldstein and E. A. Fyrberg, eds.), Vol. 44, p. 207. Academic Press, San Diego, 1994.
[5] S. Dimitrov and A. P. Wolffe, *BBA-Gene Struct. Express.* **1260,** 1 (1995).
[6] P. D. Kaufman, *Curr. Opin. Biol.* **8,** 369 (1996).
[7] M. L. DePamphilis and P. M. Wassarman, *Annu. Rev. Biochem.* **49,** 627 (1980).
[8] K. E. van Holde, "Chromatin." Springer-Verlag, New York, 1988.
[9] A. T. Annunziato, *in* "The Eukaryotic Nucleus: Molecular Biochemistry and Macromolecular Assemblies" (P. R. Strauss and S. H. Wilson, eds.) Vol. 2, p. 687. Telford Press, Caldwell, NJ, 1990.
[10] A. T. Annunziato, *in* "The Nucleus" (A. P. Wolffe, ed.), Vol. 1, p. 31. JAI Press, Greenwich, CT, 1995.

0076-6879/99 $30.00

The investigation of nucleosome assembly *in vivo* poses a number of challenges. At any one time most cellular chromatin is not undergoing replication, even in an S-phase cell. Thus the ability to identify those regions of DNA that have been assembled into nucleosomes, and the histones that have been deposited during assembly, requires the use of suitable tracer reagents. Further, while the control of parameters such as replication rate, protein synthesis level, and histone modification state is often desired, and can be achieved within limits, the potential interdependence of these and other features of cell metabolism must be considered (e.g., the protein synthesis inhibitor emetine also preferentially inhibits DNA synthesis on the retrograde arm of the fork[11]). These concerns can be offset by the value of analyzing chromatin biosynthesis in its natural setting. Certain questions, however, may be best addressed using simpler systems.

This article offers biochemical approaches to the investigation of nascent nucleosomes in cultured mammalian cells, with exclusive focus on nuclear chromatin (for a treatment of the application of replicating SV40 minichromosomes to the study of chromatin assembly *in vivo,* the reader is directed to the discussion by Cusick *et al.*[12]). Methods for the radiolabeling, isolation, immunoprecipitation, and detection of chromatin assembled *in vivo* are described. There then follows a discussion of the analysis of chromatin replicated in isolated nuclei, an *in vitro* system that permits the selective detection of parental histones on newly replicated DNA.

Cell Culture and Radiolabeling

In the following section, HeLa S3 cells grown in spinner culture are used as a model system; however, most of the general principles outlined are applicable to a variety of cell types, whether grown in suspension or not. In our laboratory, HeLa S3 cells are maintained at 37° in Eagle's minimal essential medium (Joklik modification; JMEM) containing 5–10% (v/v) donor calf serum (we have found that calf serum from a donor herd is generally more consistent than nondonor serum). The Joklik modification contains no calcium chloride or sulfate and twice the level of dextrose than the unmodified formulation. If desired, the medium can be supplemented with antibiotics (penicillin G at 50 units/ml; streptomycin sulfate at 50 μg/ml). Typically the cells double in ~24 hr, generating sufficient CO_2 to obviate the need for exogenous carbon dioxide. For immunocytochemistry, cells can be transferred to a medium containing calcium chloride (e.g.,

[11] W. C. Burhans, L. T. Vassilev, J. Wu, J. M. Sogo, F. S. Nallaseth, and M. L. DePamphilis, *EMBO J.* **10,** 4351 (1991).

[12] M. E. Cusick, P. M. Wassarman, and M. L. DePamphilis, *Methods Enzymol.* **170,** 290 (1989).

MEM with Earle's salts) in a 5% (v/v) CO_2-equilibrated incubator; they will then attach to a glass or plastic surface.

Labeling DNA

In the context of nucleosome assembly, one can speak of two classes of DNA: newly replicated DNA and "bulk" or total DNA. It is often advantageous to uniformly label bulk chromatin DNA to provide an internal control for the analysis of nascent chromatin. This is done by growing cells in JMEM for one generation with [^{14}C]thymidine (~50 mCi/mmol) at a final concentration of 0.02–0.1 μCi/ml. Note that the culture medium contains no thymidine to compete with the radiolabel.

Newly replicated DNA is pulse labeled using high specific activity [methyl-^{3}H]thymidine (~80 Ci/mmol). The concentration of [^{3}H]thymidine is varied, depending on the duration of the pulse. For very brief pulses (<5 min), radiolabel is added at 50 μCi/ml; longer pulses (5–30 min) can be performed at 5–10 μCi/ml. To reduce costs as well as the total amount of label used, cells are harvested by centrifugation at room temperature (800*g*; 1.5 min) and are concentrated by resuspension in a small amount of prewarmed JMEM. When cells are to be labeled for very brief periods (less than 5 min), they can be concentrated 20-fold; for longer pulses, cells are resuspended in one-fifth the original volume of JMEM. Cells are harvested while in exponential growth (~4 × 10^5 cells/ml) and are allowed to equilibrate for 5 min at 37° following resuspension before adding [^{3}H]thymidine. A circulating water bath is excellent for rapidly bringing cells to the labeling temperature.

For very short pulse times (0.5–1 min) it is often desirable to halt labeling by diluting cells 1:1 in ice-cold nuclear isolation buffer (see later) containing 0.1% sodium azide; this is to prevent continued incorporation during subsequent manipulations.

Labeling Histone Proteins

Because histones are lysine and arginine rich, these are the compounds of choice for radiolabeling. Lysine and arginine are essential amino acids and are present in most tissue culture media. It is therefore necessary to use media free of these amino acids to label nascent histones. Kits are available commercially for the preparation of culture media lacking selected amino acids (e.g., the MEM Select-Amine Kit from GIBCO-BRL Gaithersburg, MD), and many suppliers will also custom produce formulations lacking specified components. It is most efficient to prepare media lacking several amino acids of possible interest (e.g., arginine, lysine, methionine) and add back those that will not be the labeling reagent in a particular experiment.

To effectively radiolabel proteins *in vivo* it is useful to partially deplete endogenous amino acid pools. Cells to be labeled with $[^3H]$lysine are harvested by centrifugation (800*g*; 1.5 min), washed with lysine-free JMEM containing 5% donor calf serum (LF-JMEM), resuspended in LF-JMEM at one-fifth the starting culture volume, and incubated at 37° for 10–15 min. Cells are again harvested and resuspended in the volume of LF-JMEM appropriate for the labeling period (see later). For brief labeling periods (<10 min), dialyzed serum can be used to eliminate competing nonradioactive lysine from the incubation medium.

For pulse labeling, cells are concentrated 20-fold (2- to 10-min pulse) or 5-fold (>10-min pulse) in LF-JMEM, allowed to equilibrate at 37° for 5 min, and incubated with $[^3H]$lysine (~100 Ci/mmol) at 50 μCi/ml. For longer pulse times, $[^3H]$lysine is added at 5–15 μCi/ml; if desired, $[^3H]$arginine (~40 Ci/mmol) can also be added at the same concentration as lysine, using arginine-, lysine-free JMEM.

Long-term labeling of bulk histones with $[^{14}C]$lysine (300 mCi/mmol; ~0.5 μCi/ml) is performed for one to two generations in JMEM depleted 80% in lysine. The simplest way to do this is to add 20% complete JMEM to the lysine-free culture medium. Serum is added at 5–10%.

Metabolic Inhibitors

It is often instructive to eliminate a selected cell process to help determine its role within a pathway of interest. Table I lists various inhibitors, along with the concentrations at which they are effective under the conditions used for radiolabeling DNA and histones *in vivo*. Cells are typically preincubated 5–10 min with inhibitors prior to the addition of radiolabeled compounds.

Preparation of Nuclei and Chromatin

Nuclei are prepared from HeLa cells without the use of detergents, as follows. Cells are harvested by centrifugation in a benchtop centrifuge at room temperature (800*g*; 2 min), washed twice in buffer A (10 m*M* Tris–HCl, 3 m*M* $MgCl_2$, 2 m*M* 2-mercaptoethanol, 5 m*M* sodium butyrate, pH 7.6), and resuspended in buffer A containing 0.5 m*M* phenylmethylsulfonyl fluoride (PMSF), a protease inhibitor, at $\sim 2 \times 10^6$ cells/ml. When phosphoproteins are to be analyzed, buffer A also contains the phosphatase inhibitor microcystin-LR (0.5 μg/ml). After swelling at 4° for 15 min, cells are lysed by homogenization with ~15 strokes of a Dounce homogenizer, using a tight-fitting "B" pestle. Cells are then pelleted at 4° (800*g*; 2 min) and washed twice in buffer A. For the final wash, a known volume is used (typically 10 ml) to permit quantitation of nuclei by means of spectroscopy.

TABLE I
METABOLIC INHIBITORS

Inhibitor	Concentration	Stock solution
Hydroxyurea (DNA synthesis)	10 m*M*	0.1–1 *M* in water
Cycloheximide (protein synthesis)	200 μg/ml	10 mg/ml in water
MG132 (proteasome inhibitor)	40 μM	10 m*M* in dimethyl sulfoxide
Okadaic acid (protein phosphatases 1 and 2A)	500 n*M*	100 μM in 100% ethanol
Sodium butyrate (histone deacetylases)	50 m*M*; 7 m*M*[a]	1 *M* in water
Staurosporine (protein kinases)	300 ng/ml	1 mg/ml in dimethyl sulfoxide
Trichostatin A (histone deacetylases)[b]	10 n*M*	1 mg/ml (3.3 m*M*) in 100% ethanol

[a] As described by L. S. Cousens and B. M. Alberts, *J. Biol. Chem.* **254,** 1716 (1979), 50 m*M* is the concentration of choice to effectively inhibit histone deacetylases *in vivo;* this concentration has been used for the analysis of newly replicated chromatin [C. A. Perry and A. T. Annunziato, *Nucleic Acids Res.* **17,** 4275 (1989); C. A. Perry and A. T. Annunziato, *Exp. Cell Res.* **196,** 337 (1991)]; however, 7 m*M* butyrate is sufficient to generate hyperacetylated histones during longer incubations (>6 hr).

[b] The potential pleiotropic effects of sodium butyrate, particularly during lengthy exposure times [L. C. Boffa, R. Gruss, and V. G. Allfrey, *J. Biol. Chem.* **256,** 9612 (1981)], may be avoided by using trichostatin A (TSA) to inhibit histone deacetylases. TSA is reported to be specific for histone deacetylases and can be used at much lower concentrations than sodium butyrate [M. Yoshida, M. Kijima, M. Akita, and T. Beppo, *J. Biol. Chem.* **265,** 17174 (1990)].

Soluble chromatin is prepared by digesting nuclei with micrococcal nuclease, using either of two methods.

Method 1

Nuclei are resuspended at 40 $A_{260\ nm}$/ml [$A_{260\ nm}$ measured in 1% sodium dodecyl sulfate (SDS)] in buffer A containing 0.5 m*M* $CaCl_2$, and prewarmed at 37° for 5 min. Micrococcal nuclease (Sigma, St. Louis, MO) is then added at a concentration of 1.2 U/ml; digestion generally proceeds for 1–5 min, depending on the relative proportion of mononucleosomes desired. Nuclease digestion is terminated by the addition of EGTA (ethylene glycol bis(β-aminoethyl ether)-*N,N,N′,N′*-tetraacetic acid) to a final concentration of 1 m*M* from a 100 m*M* stock solution (pH 7.6) and by placing the nuclei on ice. After 10 min, nuclei are pelleted in an adjustable speed microcentrifuge at 4° (1000*g*; 5 min) to generate a first supernatant (S1) containing mononucleosomes.[13] The nuclear pellet is then resuspended in 2 m*M* EDTA, pH 7.2 (in the same volume as the S1), cooled on ice for 10 min, and then pelleted in a microcentrifuge (~12,000*g*; 10 min; 4°) to yield a second supernatant (S2).

[13] K. S. Bloom and J. N. Anderson, *Cell* **15,** 141 (1978).

Note: Nucleosomes in the S1 supernatant are depleted of histone H1 and are enriched in transcriptionally active DNA sequences; chromatin in the S2 contains stoichiometric levels of H1 and is depleted in transcribed genes.[13,14] Adjusting the EDTA-supernatant (S2) to physiological ionic strength results in the preferential precipitation of "inactive" chromatin.[15] Thus one consequence of Method 1 is that some sequences may be lost to analysis (or may require separate examination). We therefore frequently use an alternative method that permits the solubilization of 30–50% of nuclear chromatin under more physiological conditions.

Method 2

This method of chromatin solubilization is a modification of that described by Worcel *et al.*[16] Nuclei are isolated in buffer A, as in Method 1. They are then washed once in 10 m*M* PIPES (piperazine-*N,N'*-bis[2-sulfonic acid]), 80 m*M* NaCl, 20 m*M* sodium butyrate, pH 7.0 (PB buffer). Nuclei are resuspended in PB buffer, adjusted to 0.5 m*M* $CaCl_2$ (using a 50 m*M* stock solution), and chilled on ice for at least 5 min. Micrococcal nuclease (Sigma) is then added at 5 U/ml, and the digestion is allowed to proceed on ice for 5–15 min, depending on the percentage of mononucleosomes required. The reaction is halted by adjustment to 2–5 m*M* EGTA, using a 100 m*M* stock solution (pH 7.6). Digested nuclei are kept on ice for 15 min and then pelleted in an adjustable-speed microcentrifuge (800*g*; 4 min; 4°). The first supernatant (S1) is collected and centrifuged again at 12,000*g* at 4° for 5 min to remove any remaining nuclei. The S1 obtained in this manner contains 30–50% of the starting bulk chromatin DNA in a buffer that approaches physiological ionic conditions.

Comments: It is important to keep in mind that soluble chromatin prepared by Method 2 contains potentially active micrococcal nuclease. Therefore either EDTA or EGTA must remain present during all subsequent manipulations of S1 chromatin.

A note on Sigma micrococcal nuclease: Sigma defines a unit of micrococcal nuclease as that amount of enzyme that produces 1.0 μmol of acid-soluble polynucleotides per minute at 37°. For other enzyme suppliers (e.g., Worthington Biochemical Corp., Lakewood, NJ), one unit corresponds to a change in optical density of 1.0 at 260 nm. Because Sigma equates one micromolar unit with ~85 A_{260} units, one Sigma unit equals approximately 85 Worthington units.

[14] R. L. Seale, A. T. Annunziato, and R. D. Smith, *Biochemistry* **22,** 5008 (1983).
[15] J. A. Ridsdale and J. R. Davie, *Nucleic Acids Res.* **15,** 1081 (1987).
[16] A. Worcel, S. Han, and M. L. Wong, *Cell* **15,** 969 (1978).

Liquid Scintillation Counting

It may be helpful to briefly summarize some of the steps involved in measuring radioactivity in samples containing labeled nascent chromatin. To separate soluble chromatin from free (radiolabeled) nucleotides, all samples are precipitated with 10% (w/v) trichloroacetic acid (TCA) prior to scintillation counting. An aliquot of chromatin is placed into a test tube with 0.2 ml of 20 m*M* EDTA containing 50 μg each of bovine serum albumin (Sigma) and DNA (Sigma) as carriers. After mixing well, ~3 ml of cold 10% (w/v) TCA is added, and the samples are chilled on ice for 30 min. The resulting precipitate is collected onto glass fiber disks (24 mm) using a vacuum apparatus, washed twice with cold 10% (w/v) TCA and twice with cold 95% (v/v) ethanol. The filters are placed into glass scintillation vials and dried either at room temperature or in a 60° oven.

Samples containing newly replicated chromatin DNA are often double labeled, e.g., [^{3}H]thymidine for new DNA and [^{14}C]thymidine for bulk DNA. The low energy of the tritium and ^{14}C radionuclides necessitates that chromatin be solubilized off the glass fiber filters prior to counting. If this is not done, differential quenching of the two radioisotopes will give inaccurate results. To do this the dried filters are digested overnight with 0.2 ml of a solubilizing agent (e.g., Solusol from National Diagnostics, Atlanta, GA) containing 5% H_2O (water is essential for the hydrolysis reaction). The digested chromatin can then be suspended in a biodegradable scintillation cocktail that is formulated for aqueous samples and counted using an appropriate double-label program.

Comments: Tissue solubilizers can cause some plastic vials to leak; thus glass vials are preferable for double-labeled samples. Also, the alkaline nature of most solubilizing agents can induce chemiluminescence. This can often be alleviated by adding one drop of concentrated acetic acid to each vial.

Polyacrylamide Gel Electrophoresis

Electrophoresis of Nucleosomal DNA

The electrophoretic analysis of newly replicated chromatin DNA labeled *in vitro* with [^{32}P]TTP is typically performed using agarose gels, followed by autoradiography.[17–21] Because nucleotide triphosphates do not cross intact cell membranes, [^{3}H]thymidine is typically substituted for

[17] G. Almouzni, D. J. Clark, M. Méchali, and A. P. Wolffe, *Nucleic Acids Res.* **18,** 5767 (1990).
[18] G. Almouzni, M. Méchali, and A. P. Wolffe, *EMBO J.* **9,** 573 (1990).
[19] B. Stillman, *Cell* **45,** 555 (1986).
[20] R. Fotedar and J. M. Roberts, *Proc. Natl. Acad. Sci. U.S.A.* **86,** 6459 (1989).
[21] T. Krude and R. Knippers, *J. Biol. Chem.* **268,** 14432 (1993).

[^{32}P]TTP when labeling living cells, as described earlier. However, the use of tritiated compounds necessitates the employment of fluorography to visualize radiolabeled DNA and histones resolved in gels. Many commercially prepared fluorographic agents are now available, and the original PPO (*i.e.,* 2,5-diphenyloxazole)/dimethyl sulfoxide (DMSO) fluorographic method[22,23] also gives excellent results. Because agarose gels dissolve in DMSO,[23] we most often use polyacrylamide gels for studies of nascent DNA labeled *in vivo.* The following method is based on that of Loening,[24] modified to include sodium dodecyl sulfate (SDS) in all buffers. The inclusion of SDS eliminates the requirement to purify nucleosomal DNA prior to electrophoresis. Gels and running buffer contain 40 m*M* Tris–acetate, pH 7.2, 20 m*M* sodium acetate, 1 m*M* Na_2EDTA, and 0.1% SDS.

Stock Solutions

3E buffer: 120 m*M* Tris base, 60 m*M* sodium acetate trihydrate, 3 m*M* Na_2EDTA, pH 7.2 (adjusted with acetic acid)
21% acrylamide: 20% (w/v) acrylamide, 1% (w/v) *N,N'*-methylenebisacrylamide in H_2O
10% (w/v) SDS in H_2O
10% (w/v) ammonium persulfate in H_2O
Sample buffer: 40 m*M* Tris–acetate (pH 7.2), 20 m*M* sodium acetate, 1 m*M* Na_2EDTA, 5% glycerol, 0.01% bromphenol blue, 1% (w/v) SDS.

For One 30-ml, 14 × 15-cm Slab Gel (3.9% Acrylamide). Mix 10 ml 3E buffer, 5.6 ml 21% acrylamide, and 13.95 ml water in a side-arm flask. Degas, add 0.3 ml 10% SDS, 25 μl TEMED, and 150 μl 10% ammonium persulfate. Mix well, pour, and allow to polymerize for at least 1 hr (or overnight). Gels are preelectrophoresed (110 V, 45 min) and then run at 110 V using constant current for at least 1.5 hr (or until the dye runs off).

Comments: This concentration of polyacrylamide can be used to resolve up to octamer-length nucleosomal DNA. To prevent drying during overnight polymerization, gels can be placed in a zip-locked plastic bag containing about 10 ml water; use large stainless-steel clamps (Research Products International Corporation, Mount Prospect, IL) at the base of the plates to raise gels above water in the bag. In preparation for electrophoresis, chromatin samples are adjusted to 10 m*M* magnesium acetate (from a 1 *M* stock solution), precipitated with two volumes of ethanol in a dry ice/ethanol slurry, vacuum dried, and solubilized in SDS sample buffer overnight at room temperature.

[22] W. M. Bonner and R. A. Laskey, *Eur. J. Biochem.* **46,** 83 (1974).
[23] R. A. Laskey and A. D. Mills, *Eur. J. Biochem.* **56,** 335 (1975).
[24] U. E. Loening, *Biochem. J.* **102,** 251 (1967).

Electrophoresis of Histones

Histones are extracted from nuclei or chromatin samples with 0.2 *M* H_2SO_4 overnight at 4°, precipitated on ice with 25% TCA, washed once with ice-cold acetone containing 0.05 *M* HCl and once with acetone alone, and air or vacuum dried. They are then resolved in 18% polyacrylamide gels in the presence of SDS,[25] in acid–urea gels,[26,27] or in Triton–acid–urea gels.[28] The electrophoresis of histones is discussed in detail in a previous volume of this series.[29]

Transfer of Histones to Membranes

Histones that have been resolved in the presence of either acid–urea or SDS are transferred electrophoretically to Immobilon-P membrane (Millipore, Bedford, MA) using the standard procedures of Towbin *et al.*[30] For histones separated in Triton–acid–urea gels, the method of Delcuve and Davie[31] gives excellent results. This latter method involves washing the gel (following electrophoresis) in 50 m*M* acetic acid, 0.5% SDS, and then in 62.5 m*M* Tris–HCl, 5% 2-mercaptoethanol, 2.3% SDS, pH 6.8. Transfer to Immobilon-P is then performed in 25 m*M* (cyclohexylamino)-1-propanesulfonic acid (CAPS), 20% (v/v) methanol, pH 10, with cooling; with this protocol, histones migrate toward the anode.

Chromatin Immunoprecipitation

Antibodies that are specific for defined chromatin proteins have long been used as probes of chromatin architecture and function.[32,33] Detailed procedures for the preparation and application of immunological probes for chromatin analysis have been described in previous volumes of this series.[34–37] The following section addresses the use of antibodies to immuno-

[25] J. O. Thomas and R. D. Kornberg, *Proc. Natl. Acad. Sci. U.S.A.* **72,** 2626 (1975).
[26] S. Panyim and R. Chalkley, *Arch. Biochem. Biophys.* **130,** 337 (1969).
[27] R. Hardison and R. Chalkley, *in* "Methods Cell Biology" (G. Stein, J. Stein, and L. J. Kleinsmith, eds.), Vol. 17, p. 235. Academic Press, New York, 1978.
[28] A. Zweidler, *in* "Methods in Cell Biology" (G. Stein, J. Stein, and L. J. Kleinsmith, eds.), Vol. 17, p. 223. Academic Press, New York, 1978.
[29] R. W. Lennox and L. H. Cohen, *Methods Enzymol.* **170,** 532 (1989).
[30] H. Towbin, T. Staehelin, and J. Gordon, *Proc. Natl. Acad. Sci. U.S.A.* **76,** 4350 (1979).
[31] G. P. Delcuve and J. R. Davie, *Anal. Biochem.* **200,** 339 (1992).
[32] M. Bustin, *Curr. Top. Microbiol. Immunol.* **88,** 105 (1979).
[33] M. Bustin, *Cytometry* **8,** 251 (1987).
[34] M. Bustin, *Methods Enzymol.* **170,** 214 (1989).
[35] C. L. F. Woodcock, *Methods Enzymol.* **170,** 180 (1989).
[36] S. Muller and M. H. Van Regenmortel, *Methods Enzymol.* **170,** 251 (1989).
[37] L. P. O'Neill and B. M. Turner, *Methods Enzymol.* **274,** 189 (1996).

precipitate newly assembled nucleosomes; however, the procedures described are of general usefulness for other chromatin studies.

Immunoprecipitation of Unfixed Chromatin

Core histone–DNA and histone–histone interactions are highly stable, and for many applications it is unnecessary to fix chromatin prior to immunoprecipitation.[38–41] The advantage of using unfixed chromatin is that it eliminates the possibility of epitope masking due to fixation. The following method has been used to examine histone complexes and histone–DNA interactions during nucleosome assembly in HeLa cells.[42–43a]

Solutions

Low salt buffer: 20 m*M* Tris–HCl, pH 8.6, 0.15 *M* NaCl, 1% Triton X-100 (Calbiochem, La Jolla, CA; ultrapure, supplied as a 10% solution)

High salt buffer: 20 m*M* Tris–HCl, pH 8.6, 0.5 *M* NaCl, 2 m*M* EDTA, 1% Triton X-100

Preparation of Protein A–Agarose Beads. Weigh out immobilized protein A beads (e.g., protein A–Sepharose from Pharmacia, Piscataway, NJ), suspend beads at ~0.5 mg/ml in high salt buffer, and invert to hydrate (do not vortex, which can break beads). Centrifuge at ~4000*g* for ~30 sec at room temperature (this can be varied). Repeat twice. Wash beads once in low salt buffer (pelleting by centrifugation) and finally resuspend in low salt buffer to make a 1:1 (v/v) slurry. To coat beads with immunoglobulin G (IgG), collect 50–200 μl of slurry (25–100 μl of beads) by centrifugation and add 50–100 μl of undiluted antiserum. Incubate serum with the beads for 90 min at 37° with constant inversion. Collect the beads by centrifugation and wash three times for 5 min at room temperature with 1 ml low salt buffer containing 1 m*M* freshly added PMSF, using constant inversion. If desired, the beads can then be "blocked" by incubating for 1 hr at room temperature in PB buffer (see Preparation of Chromatin, Method 2) containing 1 mg/ml bovine serum albumin (or 1 mg/ml acetylated BSA, Promega, Madison, WI); for an "acid-soluble" blocking reagent, ubiquitin

[38] S. Druckmann, E. Mendelson, D. Landsman, and M. Bustin, *Exp. Cell Res.* **166,** 486 (1986).
[39] T. Dorbic and B. Wittig, *Nucleic Acids Res.* **14,** 3363 (1986).
[40] T. R. Hebbes, A. W. Thorne, and C. Crane-Robinson, *EMBO J.* **7,** 1395 (1988).
[41] L. P. O'Neill and B. M. Turner, *EMBO J.* **14,** 3946 (1995).
[42] C. A. Perry, C. A. Dadd, C. D. Allis, and A. T. Annunziato, *Biochemistry* **32,** 13605 (1993).
[43] L. Chang, S. S. Loranger, C. Mizzen, S. G. Ernst, C. D. Allis, and A. T. Annunziato, *Biochemistry* **36,** 469 (1997).
[43a] R. Lin, J. W. Leone, R. G. Cook, and C. D. Allis, *J. Cell Biol.* **108,** 1577 (1989).

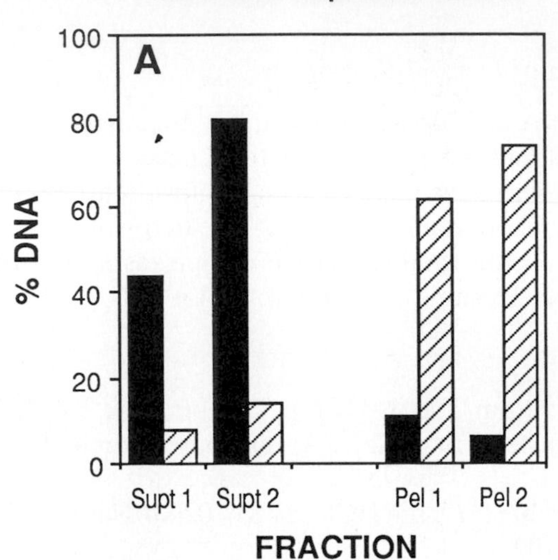
Immunoprecipitation of
Chromatin Replicated in vivo
A
% DNA
100
80
60
40
20
0
Supt 1
Supt 2
Pel 1
Pel 2
FRACTION

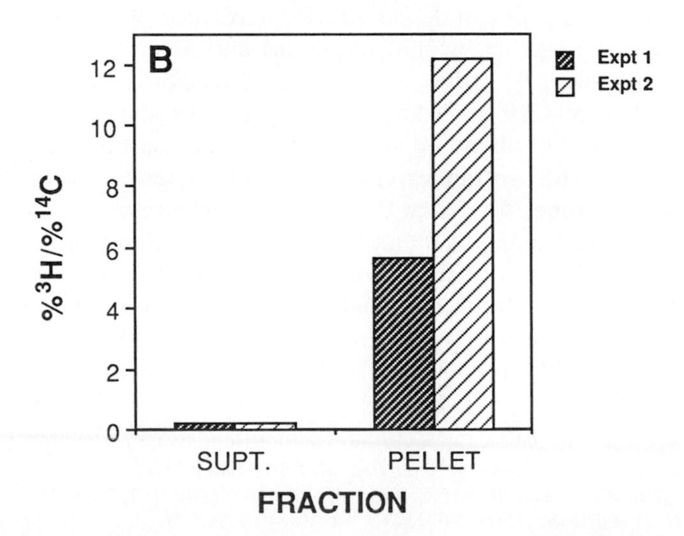
Enrichment for
Chromatin Replicated in vivo
B
Expt 1
Expt 2
$\%^3H/\%^{14}C$
12
10
8
6
4
2
0
SUPT.
PELLET
FRACTION

(Sigma; 100 μg/ml) can also be used. Wash three more times with low salt buffer plus PMSF. Finally, wash the beads once with the buffer in which the sample will be contained.

Immunoprecipitation. Prepare soluble nucleosomes in PB buffer (see Preparation of Chromatin, Method 2). Adjust the sample to contain 1 m*M* PMSF, 0.5 m*M* dithiothreitol (DTT), and 0.5% Triton X-100; bring the pH to ~8.0 using a stock solution of 1 *M* Tris base (usually a few microliters is sufficient), while monitoring with pH paper. Apply 2.5–5 μg of DNA as nucleosomes to 30–100 μl of packed beads. Incubate for 2.5 hr at 37° (or at room temperature, if desired). Recover the beads by centrifugation (4000*g*; 30 sec; room temperature) and save the unbound supernatant. Wash five times for 10 min (constant rotation) at room temperature with high salt buffer, then once with 10 m*M* Tris–HCl, pH 8.6. For liquid scintillation analysis, the unbound supernatant and the beads are precipitated with 10% trichloroacetic acid and treated as described earlier (Fig. 1). To resolve radiolabeled proteins or DNA in gels, the unbound supernatant can be precipitated using the magnesium/ethanol procedure outlined in the electrophoresis section; the beads are dried briefly under vacuum. The unbound supernatant and the immunopellet can then be resuspended in SDS-containing sample buffer and solubilized overnight prior to electrophoresis (Figs. 2 and 3).

It is also possible to use this procedure to immunoprecipitate cytosolic histone complexes. In this case all washes of the immunopellet are performed using low salt buffer only. By using "blocked" beads, this procedure yields minimal nonspecific sticking (Fig. 4). Samples can also be "precleared" using beads treated with control serum prior to immunoprecipitation.

FIG. 1. Immunoprecipitation of acetylated nascent chromatin. (A) HeLa cells were labeled for 30 min with [^{3}H]thymidine in the presence of 50 m*M* sodium butyrate and chased for 15 min without radiolabel in the continuous presence of butyrate (new, ▨). [^{3}H]Nuclei were isolated, mixed with control nuclei from cells prelabeled for one generation with [^{14}C]thymidine under normal conditions (minus butyrate) (old, ■), and digested with micrococcal nuclease in PB buffer; soluble nucleosomes were then incubated with protein A–Sepharose-conjugated "penta" antiserum,[43a] which recognizes acetylated H4 in HeLa cells.[42] The beads were pelleted, yielding a supernatant fraction (Supt), washed, and recovered by centrifugation, yielding a final *pellet.* Data for two independent experiments (numbered 1 and 2) are expressed as the percentage of the total acid-precipitable ^{3}H or ^{14}C radioactivity applied to the beads. (B) ^{3}H/^{14}C ratios in the supernatant and pellet fractions from data in (A) are given. Results: Chromatin replicated in the presence of the deacetylase inhibitor sodium butyrate is preferentially immunoprecipitated relative to control bulk chromatin, using "penta" antibodies. [Reprinted, with permission, from C. A. Perry, C. A. Dadd, C. D. Allis, and A. T. Annunziato, *Biochemistry* **32,** 13605 (1993).]

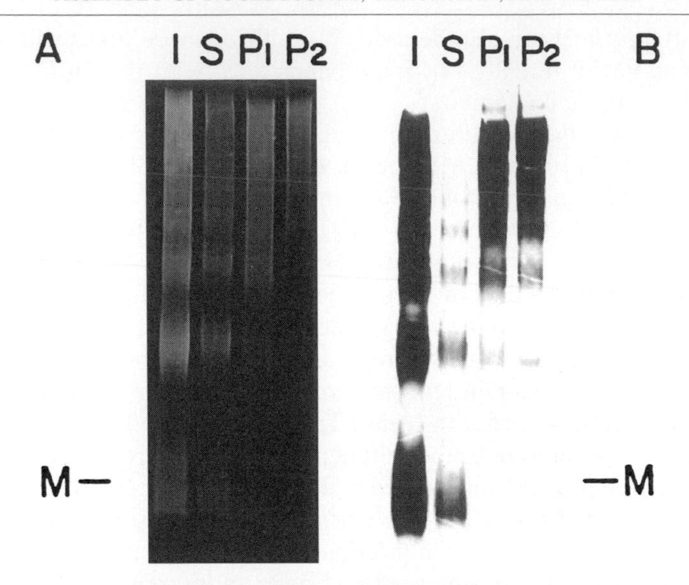

FIG. 2. Electrophoretic analysis of newly replicated DNA in immunofractionated chromatin. Cells were incubated with [^{3}H]thymidine for 30 min in the presence of 50 m*M* sodium butyrate and chased for 15 min in the absence of radiolabel (plus butyrate). Soluble input nucleosomes (lane I) were treated with immobilized "penta" antibodies, yielding supernatant (lane S) and pellet (lane P_1) fractions. The supernatant was also treated a second time with fresh Sepharose-conjugated antibodies, yielding a second immunopellet (lane P_2). Chromatin samples were subjected to polyacrylamide gel electrophoresis and analyzed by ethidium bromide staining (A) and fluorography (B); the position of mononucleosomal DNA is indicated (M). *Note:* Loads have been adjusted to yield optimum resolution in gels; thus, these data do not reflect the absolute percentage of either newly replicated or bulk DNA present in each fraction. [Reprinted, with permission, from C. A. Perry, C. A. Dadd, C. D. Allis, and A. T. Annunziato, *Biochemistry* **32,** 13605 (1993).]

Comments: As has been noted by other investigators,[40–42,44] antihistone antibodies preferentially immunoprecipitate oligonucleosomal fragments and capture mononucleosomes relatively poorly (see Fig. 2); this can be alleviated by using mononucleosomes purified from sucrose gradients as the starting material.

Immunoprecipitation of Fixed Chromatin

For some studies it will be necessary to fix chromatin with a suitable cross-linking agent to prevent the rearrangement of nucleoprotein components. Several methodologies for the immunoprecipitation of fixed chroma-

[44] E. Mendelson, D. Landsman, S. Druckmann, and M. Bustin, *Eur. J. Biochem.* **160,** 253 (1986).

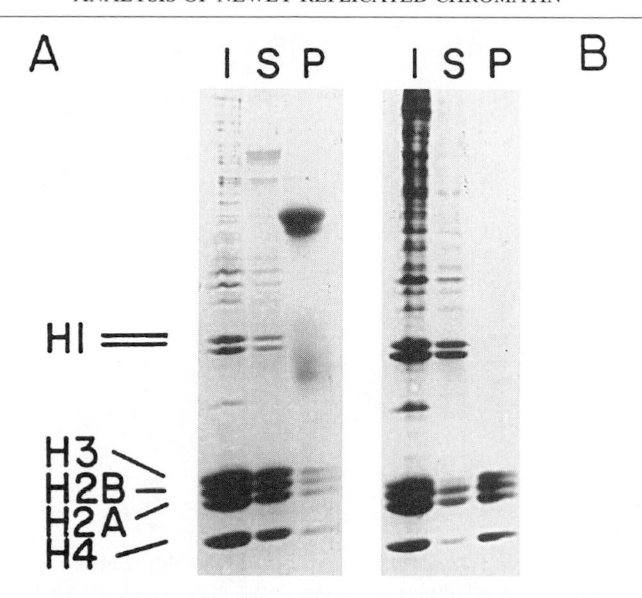

FIG. 3. Immunoprecipitation of newly synthesized histones. Cells were labeled for 30 min with [^{3}H]lysine in the presence of sodium butyrate and then chased for 15 min in the absence of radiolabel (plus butyrate). Soluble nucleosomes (lane I) were treated with immobilized "penta" antibodies, yielding supernatant (lane S) and pellet (lane P) fractions; during these procedures, histone H1 is lost from the immunopellet. Chromatin fractions were subjected to electrophoresis in the presence of SDS, stained with Coomassie Blue (A), and analyzed by fluorography (B). Results: Nucleosomes containing all four nascent core histones are immunoprecipitated, demonstrating that H4 acetylation is integral to the deposition of new H3, H2A, and H2B. [Reprinted, with permission, from C. A. Perry, C. A. Dadd, C. D. Allis, and A. T. Annunziato, *Biochemistry* **32,** 13605 (1993).]

tin have been described.[37,45,46] The following method is essentially that of Dedon *et al.*[47]

A. Formaldehyde Fixation of Cells and Sonication of Chromatin

Solutions

PBS: 10 m*M* NaH_2PO_4, 150 m*M* NaCl, pH 7.4
Triton wash buffer: 0.25% Triton X-100, 10 m*M* EDTA, 0.5 m*M* EGTA, 10 m*M* HEPES [*N*-(2)-hydroxyethylpiperazine-*N'*-(4-butanesulfonic acid)], pH 6.5

[45] D. S. Gilmour and J. T. Lis, *Mol. Cell. Biol.* **5,** 2009 (1985).
[46] M. J. Solomon, P. L. Larsen, and A. Varshavsky, *Cell* **53,** 937 (1985).
[47] P. C. Dedon, J. A. Soults, C. D. Allis, and M. A. Gorovsky, *Anal. Biochem.* **197,** 83 (1991).

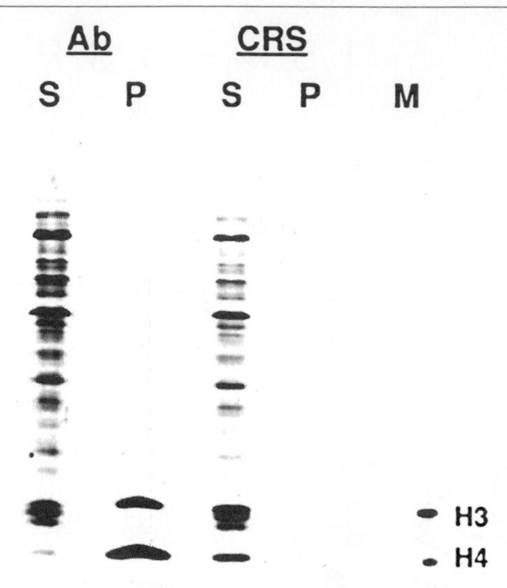

FIG. 4. Nascent cytosolic H3 and H4 form a native preassembly complex. Newly synthesized histones were labeled with [^{3}H]lysine for 5 min in the presence of 50 m*M* sodium butyrate. A cytosolic S100 extract was prepared according to the protocol of Stillman[19] and centrifuged in a 5–20% sucrose gradient as described previously.[43] Fractions containing newly synthesized H3 and H4 were pooled and immunoprecipitated using antibodies that recognize K_5,K_{12}-acetylated histone H4 (Ab) or using control rabbit serum (CRS); the unbound supernatant (S) and the immunopellet (P) were analyzed by SDS–PAGE and fluorography. Lane M contains [^{3}H]acetate-labeled histone markers. Results: The coimmunoprecipitation of cytosolic nascent H3 and H4 indicates that these histones form a soluble complex prior to their stable incorporation into chromatin. [Reprinted, with permission, from L. Chang, S. S. Loranger, C. Mizzen, S. G. Ernst, C. D. Allis, and A. T. Annunziato, *Biochemistry* **36,** 469 (1997).]

Non-Triton wash buffer: 200 m*M* NaCl, 1 m*M* EDTA, 0.5 m*M* EGTA, 10 m*M* HEPES, pH 6.5

Formaldehyde: 37% preserved with 10–15% methanol; discard 2 months after opening bottle

Sonication buffer: 2% SDS, 10 m*M* EDTA, 50 m*M* Tris–HCl, pH 8.1

Dilution buffer: 1% Triton X-100, 150 m*M* NaCl, 2 m*M* EDTA, 20 m*M* Tris–HCl, pH 8.1

Note: Add protease inhibitors to sonication and dilution buffers just before use (PMSF to 1 m*M*; leupeptin to 0.6 μg/ml, pepstatin to 0.8 μg/ml).

Protocol

1. Harvest ~1.5×10^7 HeLa cells and wash twice in tissue culture medium (e.g., JMEM) without added serum.
2. Resuspend cells at 2×10^6 cells/ml in JMEM (without serum) in a 15-ml conical centrifuge tube. Add 37% formaldehyde to a final

concentration of 2% and fix cells at room temperature for 30–40 min. The duration of fixation is adjusted empirically to yield maximum fixation without epitope masking. The cross-linking of histones to DNA can be tested by attempting to acid extract histones from the sonicated cells with 0.2 *M* H_2SO_4, and monitored by PAGE.
3. Pellet cells by centrifugation (800*g*; 2 min; room temperature).
4. Resuspend cells in cold (4°) PBS and transfer to a microcentrifuge tube; pellet again (~1400*g*; 30 sec; 4°).
5. Sequentially wash cells for 10 min in 1.0 ml of each of the following buffers, with rotation at 4°: PBS; Triton wash buffer; and non-Triton wash buffer. Collect cells by centrifugation.
6. Resuspend in sonication buffer at a density of $\sim 1 \times 10^7$ cells/ml.
7. Sonicate cells with 8–11 × 10-sec bursts using a 1/8-in. stepped microtip (Branson Sonifier Model 250). The first burst is at setting 1, the rest at setting 5. Sample is kept in an ice water bath throughout and permitted to cool between bursts.
8. Centrifuge the sonicate at ~12,500*g* for 10 min at 4°; repeat until completely clarified.
9. Collect the supernatant and dilute at least 20-fold with dilution buffer (it is important to dilute the SDS in the sonication buffer to preserve antibody integrity during immunoprecipitation).

Comments: This procedure yields DNA with an average length of ~600–1000 bp. Sonication times can be modified as needed to yield differently sized chromatin fragments.

B. Immunoprecipitation of Fixed Chromatin

Solutions

TSE: 0.1% (w/v) SDS, 1% (v/v) Triton X-100, 2 m*M* EDTA, 150 m*M* NaCl, 20 m*M* Tris–HCl, pH 8.1

High salt TSE: 0.1% SDS, 1% Triton X-100, 2 m*M* EDTA, 500 m*M* NaCl, 20 m*M* Tris–HCl, pH 8.1

TE: 1 m*M* EDTA, 10 m*M* Tris–HCl, pH 8.0

LiCl buffer: 250 m*M* LiCl, 0.1% (v/v) NP-40, 1% (w/v) deoxycholate (sodium salt) in TE, pH 8.0

Protocol

1. Adjust the desired amount of sonicated fixed cells (typically that from $2–8 \times 10^4$ cells) to 1 m*M* PMSF, 0.5 m*M* DTT, and 1 m*M* EGTA. If necessary, adjust pH to 8.0 using 1 *M* Tris base.

2. Prepare protein A–Sepharose beads by hydrating in high salt TSE and washing twice with 1.0 ml high salt TSE and twice with TSE. Resuspend beads as a 1:1 slurry in TSE.
3. Collect the desired amount of beads by centrifugation (typically 25–60 μl of beads) for 30 sec at ~4000*g*. Remove buffer from the beads and add undiluted antiserum (typically 25–50 μl). Rotate at 37° for 2 hr.
4. Wash beads three times with 0.5 ml TSE containing 2 m*M* PMSF at room temperature, rotating for 5 min each time.
5. Add adjusted sample to washed beads. Rotate overnight at 4°.
6. The next day rotate an additional 2 hr at room temperature.
7. Collect the unbound supernatant by centrifugation and save. Wash the immunopellet at 4° in the following buffers, rotating 10 min each time: 2 × 1.0 ml high salt TSE, 1 × 1.0 ml LiCl buffer, and 2 × 1.0 ml TE. At this point samples can be prepared for liquid scintillation counting as described earlier.

Comments: Fixation by this method preserves the accessibility of histone N-terminal tails, as measured by the immunoprecipitation of nascent chromatin with antibodies that recognize acetylated H4 (Table II).

Chromatin Replication and Acetylation in Isolated Nuclei

A. DNA Replication in Isolated Nuclei

For some studies it may be necessary to use a replicating system that is somewhat more defined than that afforded by living cells. An approach that is midway between replication *in vitro* using completely defined components and replication *in vivo* is replication using isolated HeLa cell nuclei. In isolated nuclei there are no new histones available for *de novo* chromatin assembly. Thus one advantage of this system is that it permits the selective examination of parental histones segregated to newly replicated DNA, without the use of protein synthesis inhibitors. The reaction conditions are essentially those of Seale,[48] modified to include all of the ribonucleoside triphosphates, an ATP-regenerating system, and uracil (to inhibit the potential activity of uracil–DNA glycosylase[49]). The reader is also directed to

[48] R. L. Seale, *Biochemistry* **16,** 2847 (1977).
[49] K. Brynoff, R. Eliasson, and P. Reichard, *Cell* **13,** 573 (1978).

TABLE II
IMMUNOPRECIPITATION OF NASCENT CHROMATIN WITH ANTIBODIES THAT RECOGNIZE ACETYLATED HISTONE H4[a]

	Supernatant	Immunopellet
Antiacetylated H4		
Experiment 1		
Unfixed	99.9%	0.1%
Fixed	42.3%	57.7%
Experiment 2		
Fixed	27.3%	72.7%
Control rabbit serum		
Unfixed	99.7%	0.3%
Fixed	99.5%	0.5%

[a] HeLa cells were labeled with 10 μCi/ml [^{3}H]thymidine in the presence of 50 m*M* sodium butyrate for 45 min, collected by centrifugation, and chased for 15 min in JMEM plus butyrate. Cells were then processed and fixed with formaldehyde as described in text (fixed). Newly replicated chromatin was then immunoprecipitated either with antibodies generated against a synthetic peptide representing the K5,K12-acetylated N terminus of histone H4 [L. Chang, S. S. Loranger, C. Mizzen, S. G. Ernst, C. D. Allis, and A. T. Annunziato, *Biochemistry* **36,** 469 (1997)] or with control rabbit serum. Antibodies from different animals were used for Experiments 1 and 2. In some cases, cells were left unfixed but otherwise treated identically (unfixed).

the *in vitro* replication systems described by Krokan *et al.*,[50] Fraser and Huberman,[51] and Shelton *et al.*,[52] based similarly on isolated nuclei.

Reagents and Solutions

5× replication buffer: 250 m*M* dextrose, 60 m*M* $MgCl_2$, 50 m*M* ATP, 5 m*M* EDTA-Na_2, 2.5 m*M* each of CTP, GTP, and UTP, 0.5 m*M* each of dATP, dCTP, and dGTP, 10 m*M* uracil, 250 m*M* phosphocreatine, 300 m*M* HEPES-Na_2, pH 8.0. Prepare 10 ml at time and store as aliquots at −70°. (Note: NTPs and dNTPs are available from several suppliers as 100 m*M* stock solutions)

[50] H. Krokan, E. Bjorklid, and H. Prydz, *Biochemistry* **14,** 4227 (1975).

[51] J. M. K. Fraser and J. A. Huberman, *J. Mol. Biol.* **117,** 249 (1977).

[52] E. R. Shelton, J. Kang, P. M. Wassarman, and M. L. DePamphilis, *Nucleic Acids Res.* **5,** 349 (1978).

10 mg/ml creatine phosphokinase (CPK)
100 m*M* dithiothreitol (store at −20°)
1.0 *M* sodium butyrate in water or 10 μ*M* trichostatin A in ethanol
[^{3}H]TTP (~75 Ci/mmol in Tricine buffer; 2.5 mCi/ml)

Protocol

1. Isolate HeLa cell nuclei in buffer A as described earlier, keep at 4°
2. Begin preparing 0.5 ml of 1× replication buffer by mixing 100 μl 5× buffer, 10 μl of 100 m*M* dithiothreitol (2 m*M* final concentration), 5 μl of 10 mg/ml CPK (100 μg/ml final concentration); if desired, add 1.0 μl sodium butyrate (2 m*M* final concentration) or 0.5 μl TSA (10 n*M* final concentration) to inhibit histone deacetylation.
3. Add 30 μl of [^{3}H]TTP (150 μCi/ml final concentration).
4. Bring to 500 μl with water.
5. Resuspend nuclei at 30 A_{260} units/ml (A_{260} measured in 1% SDS) in replication buffer.
6. Label nuclei for 20 min in a 37° water bath with occasional *gentle* agitation.
7. Collect nuclei by centrifugation (1000*g*, 8 min; 4°), and wash in buffer A prior to micrococcal nuclease digestion, chromatin preparation, etc.

Comments

1. Nuclei should be resuspended in replication buffer as quickly as possible following preparation.
2. It has been demonstrated that histones are segregated to newly replicated DNA in this system, generating segregated nucleosomes of parental origin.[53,54]
3. There is no initiation of DNA synthesis under these conditions, and Okazaki fragment ligation is retarded greatly[48]; the rate of incorporation *in vitro* is ~3–4% the *in vivo* rate.[51,54]
4. If desired, biotinylated dUTP (e.g., biotin-21-dUTP from Clontech, Palo Alto, CA) can be substituted for [^{3}H]TTP to yield nascent chromatin that can be affinity purified with immobilized avidin or streptavidin.[55,56] It is preferable to use a mixture of dTTP and biotin-dUTP (0.075 m*M* each) to yield reasonable incorporation rates. To

[53] R. L. Seale, *Proc. Natl. Acad. Sci. U.S.A.* **75,** 2717 (1978).
[54] C. A. Perry, C. D. Allis, and A. T. Annunziato, *Biochemistry* **32,** 13615 (1993).
[55] M. Shimkus, J. Levy, and T. Herman, *Proc. Natl. Acad. Sci. U.S.A.* **82,** 2593 (1985).
[56] R. Fotedar and J. M. Roberts, *Cold Spring Harb. Symp. Quant. Biol.* **56,** 325 (1991).

label newly replicated DNA, omit dCTP from 5× replication buffer and substitute [^{3}H]dCTP or [^{32}P]dCTP in the reaction mix.

B. Histone Acetylation in Isolated Nuclei

Chromatin-bound histones in isolated nuclei can be acetylated *in vitro* by endogenous acetyltransferases.[57] The coupling of DNA replication and histone acetylation *in vitro* has been exploited to examine the acetylation status of parental histones segregated to newly replicated chromatin, as well as the accessibility of nascent nucleosomes to histone acetylases.[54] The following procedure is derived from that of Vu *et al.*[58] and makes use of an acetyl-CoA-regenerating system.

Solutions

1.0 mM coenzyme A (Boehringer Mannheim, Indianapolis, IN)
10 mg/ml acetyl-CoA synthetase (Boehringer Mannheim; ~3 units/mg protein)
50 mM sodium acetate or
[^{3}H]acetic acid, sodium salt (~4 Ci/mmol in ethanol; typically 10 mCi/ml)

Protocol

1. Prepare 1× replication buffer as described earlier, if desired add 0.1 mM unlabeled TTP (to permit DNA replication) and 2 mM sodium butyrate (or 10 nM TSA) to inhibit deacetylases.
2. Add CoA to a final concentration of 10 μM.
3. Add acetyl-CoA synthetase to a final concentration of 0.04–0.08 units/ml.
4. Add sodium acetate to a final concentration of 500 μM to yield replication/acetylation buffer or
5. Evaporate ethanol off the [^{3}H]acetate (under vacuum) and resuspend in replication/acetylation buffer (minus *non*radioactive acetate) at a final concentration of ~2 mCi/ml to radiolabel acetylated histones.
6. Resuspend isolated nuclei in the 1× replication/acetylation buffer at a final concentration of 30 A_{260} units/ml. Start reactions by bringing to 37°.
7. Stop reactions by cooling on ice; pellet nuclei by centrifugation (1000g; 8 min; 4°)

[57] C. C. Liew and A. G. Gornall, *Fed. Proc.* **34,** 186 (1975).
[58] Q. A. Vu, D. Zhang, Z. C. Chroneos, and D. A. Nelson, *FEBS Lett.* **220,** 79 (1987).

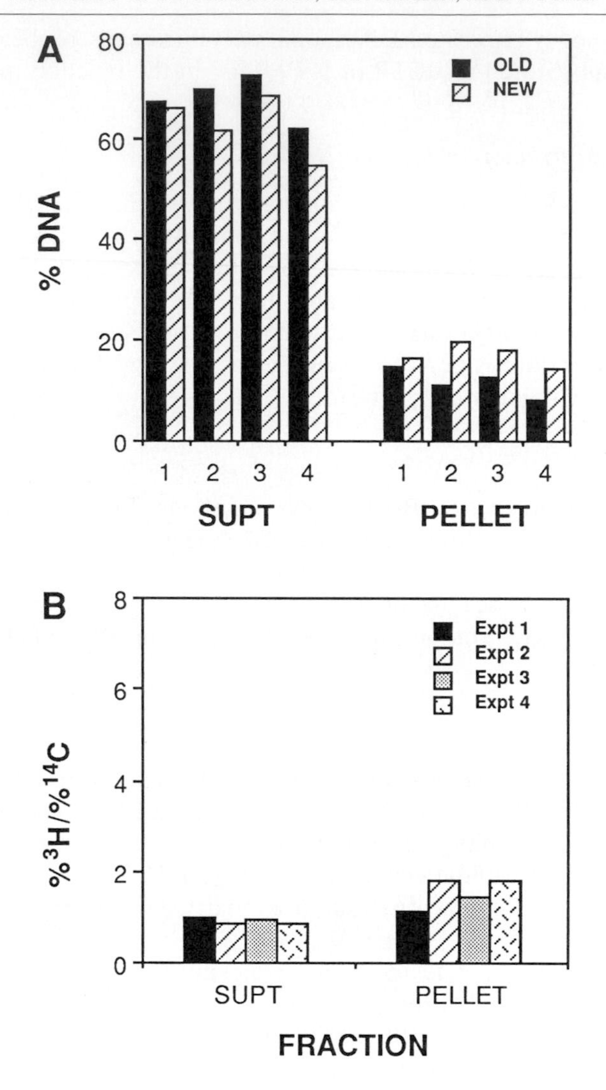

Comments and Applications

1. In the presence of [^{3}H]acetate, this system preferentially acetylates nucleosomal histones H3 and H4 (see marker lane M in Fig. 4); longer exposures also show acetylation of H2A and H2B.
2. [^{3}H]Acetyl-CoA (~5 Ci/mmol) or unlabeled acetyl-CoA may be used as an alternative to the acetyl-CoA-regenerating system; however, this is more expensive.

3. By combining the *in vitro* replication/acetylation system with the immunoprecipitation techniques also described herein, we have shown[54] that parental nucleosomes segregated to nascent DNA are not obligatorily acetylated in advance of the replication fork (Fig. 5); however, segregated histones are susceptible to acetylation during or immediately after DNA synthesis (Fig. 6).

Summary

This article outlined biochemical methodologies for the labeling, detection, and analysis of newly replicated and newly assembled nucleosomes. The isolation of specific vertebrate factors that may be involved in chromatin assembly *in vivo,* such as nucleoplasmin,[59,60] CAF-1,[61] and NAP-1[62] and their counterparts in *Drosophila*[63–65] and yeast[66–69] add a further dimension

[59] R. A. Laskey, A. D. Mills, and N. R. Morris, *Cell* **10,** 237 (1977).
[60] R. A. Laskey, B. M. Honda, A. D. Mills, and J. T. Finch, *Nature* **275,** 416 (1978).
[61] S. Smith and B. Stillman, *Cell* **58,** 15 (1989).
[62] Y. Ishimi, J. Hirosumi, W. Sato, K. Sugasawa, S. Yokota, F. Hanoaoka, and M. A. Yamada, *Eur. J. Biochem.* **142,** 431 (1984).
[63] M. Bulger, T. Ito, R. T. Kamakaka, and J. T. Kadonaga, *Proc. Natl. Acad. Sci. U.S.A.* **92,** 11726 (1995).
[64] T. Ito, J. K. Tyler, M. Bulger, R. Kobayashi, and J. T. Kadonaga, *J. Biol. Chem.* **271,** 25041 (1996).
[65] T. Ito, M. Bulger, R. Kobayashi, and J. T. Kadonaga, *Mol. Cell. Biol.* **16,** 3112 (1996).
[66] Y. Ishimi and A. Kikuchi, *J. Biol. Chem.* **266,** 7025 (1991).

FIG. 5. Segregated nucleosomes are not obligatorily acetylated in advance of the replication fork. (A) Newly replicated chromatin was labeled in isolated HeLa cell nuclei for 20 min in the presence of [^{3}H]TTP and sodium butyrate (new). Soluble nucleosomes (prepared by micrococcal nuclease digestion in PB buffer) were mixed with control chromatin that was prelabeled for one generation with [^{14}C]thymidine (minus butyrate) (old) and immunoprecipitated with immobilized penta antibodies (see Fig. 1), yielding supernatant and pellet fractions. Data for four independent experiments are expressed as the percentage of the total ^{3}H or ^{14}C radioactivity applied to the beads; for Experiment 4, cells were preincubated in butyrate for 1 hr prior to nuclear isolation and replication *in vitro. Note*: [^{3}H]Chromatin was digested with 5 U/ml micrococcal nuclease on ice for 0.75 min (Experiment 1), 1 min (Experiments 2 and 4), or 2 min (Experiment 3); [^{14}C]chromatin was digested for either 10 min (Experiments 1, 3, and 4) or 15 min (Experiment 2). Nascent and bulk chromatin were digested separately to compensate for the increased sensitivity of newly replicated chromatin to nuclease digestion.[54] (B) The ^{3}H/^{14}C ratios in the supernatant and pellet fractions from (A) are given. Results: Unlike chromatin replicated *in vivo* (Fig. 1), nucleosomes segregated to new DNA *in vitro* (in the presence of sodium butyrate) are not preferentially acetylated, as measured by immunoprecipitation with antibodies that recognize acetylated H4. Thus, chromatin is not obligatorily acetylated in advance of the replication fork. [Reprinted, with permission, from C. A. Perry, C. D. Allis, and A. T. Annunziato, *Biochemistry* **32,** 13615 (1993).]

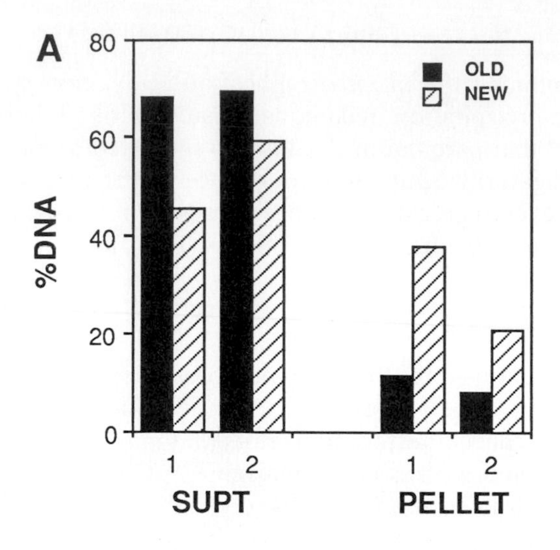

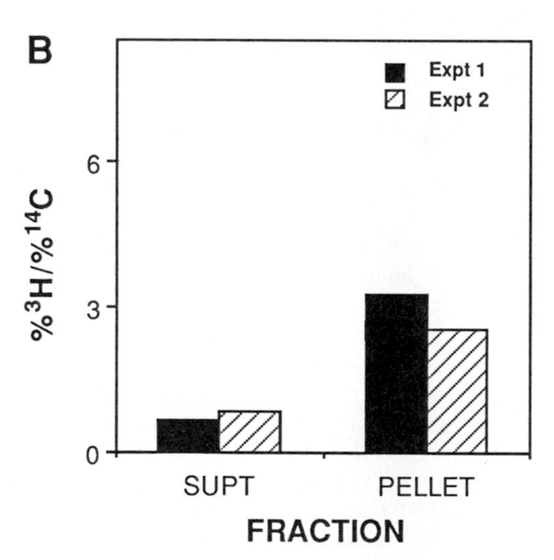

FIG. 6. Segregated nucleosomes can be acetylated during replication. (A) Nascent chromatin was labeled for 20 min with [^{3}H]TTP *in vitro* (new) under conditions in which nucleosomal histones are acetylated (i.e., in replication/acetylation buffer) using nonradioactive sodium acetate. Soluble nucleosomes were mixed with chromatin from control cells prelabeled with [^{14}C]thymidine (old) and immunoprecipitated with penta antibodies, yielding supernatant and pellet fractions. The results of two independent experiments are shown. (B) ^{3}H/^{14}C ratios in the supernatant and pellet fractions of (A) are given. Results: When acetylation accompanies replication, segregated nucleosomes are preferentially immunoprecipitated relative to bulk chromatin (compare with Fig. 5), indicating that newly replicated chromatin is highly accessible to histone acetyltransferases *in vitro.* The preferential immunoprecipitation of nascent DNA can be taken as evidence that acetylation is more pronounced on newly replicated nucleosomes, relative to bulk chromatin. [Reprinted, with permission, from C. A. Perry, C. D. Allis, and A. T. Annunziato, *Biochemistry* **32,** 13615 (1993).]

to the study of nucleosome assembly in living cells. In particular, the ability to genetically manipulate the yeast system, together with the identification of yeast enzymes that acetylate newly synthesized H4,[70,71] will certainly provide exciting new avenues for the investigation of chromatin assembly *in vivo*.

Acknowledgments

We are grateful to Dr. C. David Allis for sharing immunoprecipitation expertise and "penta" antibodies. Work in the authors' laboratory was supported by a grant from the National Institutes of Heath to A.T.A. (GM35837).

[67] P. D. Kaufman, R. Kobayashi, and B. Stillman, *Genes Dev.* **11,** 345 (1997).
[68] S. Enomoto, P. D. McCunezierath, M. Geraminejad, M. A. Sanders, and J. Berman, *Genes Dev.* **11,** 358 (1997).
[69] E. K. Monson, D. Debruin, and V. A. Zakian, *Proc. Natl. Acad. Sci. U.S.A.* **94,** 13081 (1997).
[70] S. Kleff, E. D. Andrulis, C. W. Anderson, and R. Sternglanz, *J. Biol. Chem.* **270,** 24674 (1995).
[71] M. R. Parthun, J. Widom, and D. E. Gottschling, *Cell* **87,** 85 (1996).

[7] Interactions of High Mobility Group Box Proteins with DNA and Chromatin

By MAIR E. A. CHURCHILL, ANITA CHANGELA, LINDA K. DOW, and ADAM J. KRIEG

Introduction

Relationships between transcriptional regulation and chromatin structure have generated much interest in understanding the interactions and regulation of chromosomal proteins such as core histones, histone H1, and nonhistone chromosomal proteins with DNA.[1-3] The proposed functions for many chromosomal proteins are mediated by moderate affinity "nonsequence-specific" DNA binding. Because of this minimal sequence specificity, techniques that were designed originally to study transcription factors and sequence-specific proteins require modifications to facilitate the quanti-

[1] R. D. Kornberg and Y. Lorch, *Annu. Rev. Cell Biol.* **8,** 563 (1992).
[2] H. E. van Holde, "Chromatin." Springer-Verlag, New York, 1989.
[3] A. P. Wolffe, "Chromatin Structure, and Function." Academic Press, London, 1995.

0076-6879/99 $30.00

tative analysis of DNA-binding properties of chromosomal proteins.[4] Electrophoretic methods that can be used effectively to study the nonsequence-specific interactions of the HMG1 class of chromosomal proteins with DNA and nucleoprotein complexes will be described.

High Mobility Group Chromosomal Proteins

The nonhistone high mobility group (HMG) proteins that are abundant in higher eukaryotes have roles in modulating the structure and activity of chromatin.[2,3,5] Of the three distinct families of HMG proteins, the HMG14/17 proteins bind to nucleosomes and are known to associate with transcriptionally active regions in chromatin. Another group, the HMG-I/Y proteins, are found in mammals and act as DNA-binding transcriptional coactivators. The third group of HMG proteins, the HMG1 and 2 (HMG1/2) proteins, function by stabilizing and modulating chromatin structure.[5–7] They may bind to nucleosomes,[8] repress transcription by interacting with the basal level transcriptional machinery,[9] or act as transcriptional coactivators and modifiers of transcription factor specificity.[10] Despite functional differences, the interactions of these three chromosomal protein families in chromatin are all mediated by nonsequence-specific DNA binding.

The archetypes of the HMG1/2 chromosomal protein family are the vertebrate HMG1 and 2 proteins, which contain two copies of a conserved 75 residue DNA-binding motif known as the HMG box[5,11,12] (Fig. 1A). This domain is also found in a number of sequence-specific transcription factors, such as lymphoid enhancer factor (LEF-1) and the mammalian testis determining factor, SRY.[11] The chromosomal proteins bind DNA in a "sequence-tolerant" manner (nonsequence specifically). Interestingly, the degree of specificity, which is defined as the affinity for the specific site/affinity for noncognate DNA (K_{sp}/K_{nsp}), of the site-specific HMG-domain proteins is only 10 to 50, whereas the specificity for other transcription factors is often considerably greater, approximately 100–1000.[13] Specificity

[4] M. Fried and D. M. Crothers, *Nucleic Acids Res.* **9,** 6505 (1981).
[5] M. Bustin and R. Reeves, *Progr. Nucleic Acid Res. Mol. Biol.* **54,** 35 (1996).
[6] C. R. Wagner, K. Hamana and S. Elgin, *Mol. Cell. Biol.* **12,** 1915 (1992).
[7] S. S. Ner and A. A. Travers, *EMBO J.* **13,** 1817 (1994).
[8] K. Nightingale, S. Dimitrov, R. Reeves, and A. Wolffe, *EMBO J.* **15,** 548 (1996).
[9] H. Ge and R. G. Roeder, *J. Biol. Chem.* **268,** 17136 (1994).
[10] V. Zappavigna, L. Falciola, M. Citterich, F. Mavilio, and M. Bianchi, *EMBO J.* **15,** 4981 (1996).
[11] R. Grosschedl, K. Giese, and J. Pagel, *Trends Genet.* **10,** 94 (1994).
[12] A. D. Baxevanis and D. Landsman, *Nucleic Acids Res.* **23,** 1604 (1995).
[13] T. A. Steitz, Structural Studies of Protein–Nucleic Acid Interaction: The Sources of Sequence-Specific Binding. Cambridge Univ. Press, Cambridge, UK, 1993.

differences between these two families reside in regions of the HMG domain for which there are differences in sequence homology and structure (Fig. 1C).[5,12,14]

The experimental methods described in this article make use of a *Drosophila melanogaster* chromosomal protein, HMG-D. HMG-D was isolated as the major HMG-1-like chromosomal protein in *D. melanogaster*[6] and from a screen of a *Drosophila* λgt11 expression library with a DNA region from the *fushi tarazu* upstream element DNA fragment (*ftz*-SAR) under conditions that were favorable for the selection of a sequence-specific protein.[15] HMG-D is a homolog of HMG1 that appears to have a role in embryonic chromatin at a time when the chromatin is transcriptionally silent, but is undergoing rapid rounds of replication.[7] HMG-D contains only a single copy of the DNA-binding domain[6,7] followed by a "tail" region that has a basic motif similar to the C-terminal domain of histone H1 and a C-terminal acidic stretch similar to that seen in HMG1/2 proteins.

Specificity of HMG-Box Proteins

Preliminary analyses of the affinity and specificity of HMG-D were accomplished using electrophoretic mobility shift assays (EMSA) and binding site selection techniques.[4,16] In EMSA of HMG-D with DNA containing the putative "specific site" DNA or any DNA sequence greater than 20 nucleotides in length, a ladder of bands appears (Fig. 2A). Similar results have been obtained for proteins such as the *Escherichia coli* chromosomal protein HU.[17] This banding pattern is the hallmark of a nonsequence-specific binding protein. Further analysis of the DNA specificity of HMG-D was accomplished by application of a binding site selection technique, originally developed by Pollock and Treisman[16] for sequence-specific proteins, to HMG-D (see methods below). This procedure confirmed that HMG-D does not have a specific DNA-binding site and revealed an informative selectivity for AT-rich sequences containing a TG dinucleotide.[18]

DNA Bending by HMG Proteins

The preference for deformable and deformed DNA of HMG-D and other HMG proteins suggests that DNA bending and unwinding are defor-

[14] C. M. Read, P. D. Cary, N. S. Preston, M. Lnenicek-Allen, and C. Crane-Robinson, *EMBO J.* **13,** 5639 (1994).

[15] S. S. Ner, M. E. A. Churchill, S. Searles, and A. A. Travers, *Nucleic Acids Res.* **21,** 4369 (1993).

[16] R. Pollock and R. Treisman, *Nucleic Acids Res.* **18,** 6197 (1990).

[17] E. Bonnefoy and J. Rouviere-Yaniv, *EMBO J.* **11,** 4489 (1992).

[18] M. E. A. Churchill *et al., EMBO J.* **14,** 1264 (1995).

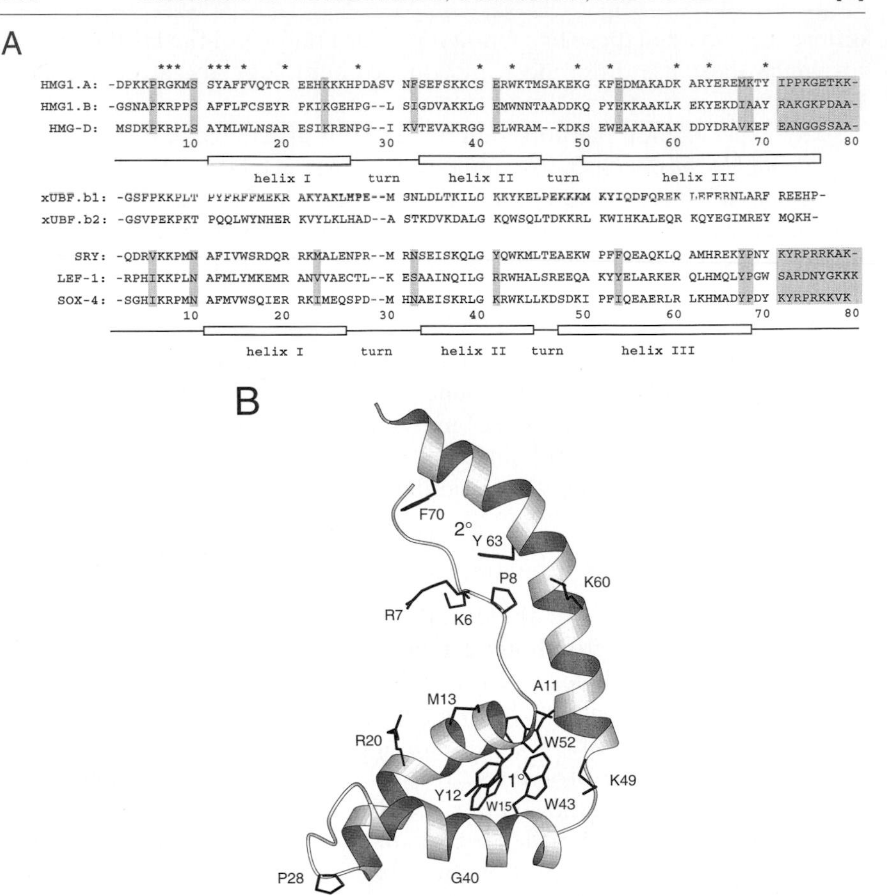

FIG. 1. Sequence structure of the HMG-box domain. (A) Sequence alignment of HMG-1 box A and B, HMG-D, xUBF box 1 and box 2, SRY, LEF-1, and Sox4 based on HMG-D numbering and published sequence alignments.[11,54] Sequences are grouped based on their sequence specificity; the HMG-1 family is the least specific and the transcription factors are the most sequence specific. UBF proteins contain some residues that are SRY-like, 24, 32, 41, and 53, but the N- and C-terminal residues, which form the secondary hydrophobic core, are more similar to the HMG-1 proteins. Asterisks denote residues that are conserved among the HMG protein families and the shaded sequences are those that differ between HMG1 and SRY-type proteins. (B and C) Diagrams of structure of the HMG-D HMG-box domain determined by NMR.[49] (B) Residues that are conserved among members of the HMG domain superfamily are shown.[12,63] (C) Residues that differ between HMG1 and sequence-specific HMG families. The protein regions that make up the primary and secondary hydrophobic cores are shown by 1° and 2°, respectively.

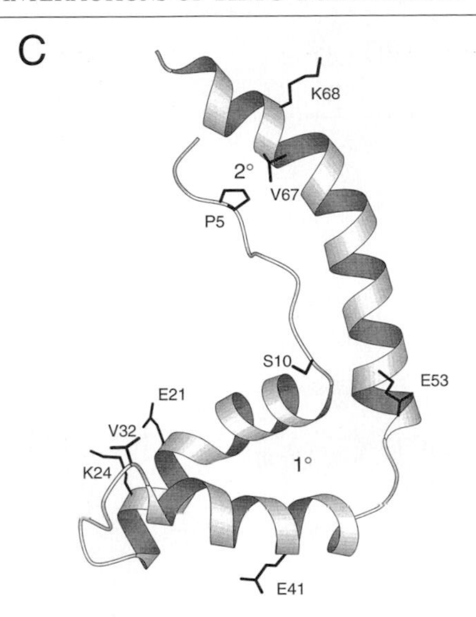

FIG. 1. *(continued)*

mations introduced into DNA by chromosomal HMG protein binding.[18–21] HMG-box proteins also directly induce bends in DNA.[22,23] In order to study the DNA bending of nonsequence-specific proteins, several methods have emerged that are based on the ability of the protein to circularize DNA.[22,23] Ligase-mediated circularization assays are quite sensitive to the degree of DNA bending and unwinding induced by the bound protein.[22,24] In another version of this assay, short DNA fragments, less than 25 bp in length, multimerize and circularize in the presence of the bending protein and ligase.[25] The size of the circles obtained is indicative of the degree and phase of the protein-induced DNA deformation. When this analysis was applied to oligonucleotides of length 10, 15, and 21 bp, HMG-D induced formation of circles as small as 70 bp.[25] However, the helical repeat of DNA bound by HMG-D is predicted to be closer to 11 bp/turn.[26] Consistent with this prediction, for oligonucleotides of length 11 bp, circles as small

[19] D. P. Bazett-Jones, B. Leblanc, M. Herfort, and T. Moss, *Science* **264,** 1134 (1994).

[20] L. G. Sheflin and S. W. Spaulding, *Biochemistry* **28,** 5658 (1989).

[21] S. A. Wolfe, A. E. Ferentz, V. Grantcharova, M. E. A. Churchill, and G. L. Verdine, *Chem. Biol.* **2,** 213 (1995).

[22] T. T. Paull, M. J. Haykinson, and R. C. Johnson, *Genes Dev.* **7,** 1521 (1993).

[23] P. Pil and S. J. Lippard, *Science* **256,** 234 (1992).

[24] J. D. Kahn and D. M. Crothers, *Proc. Natl. Acad. Sci. USA* **89,** 6343 (1992).

[25] D. Payet and A. A. Travers, *J. Mol. Biol.* **266,** 66 (1997).

[26] A. Balaeff, M. E. A. Churchill, and K. Schulten, *Prot. Struc. Func. Gen.* **30,** 113 (1998).

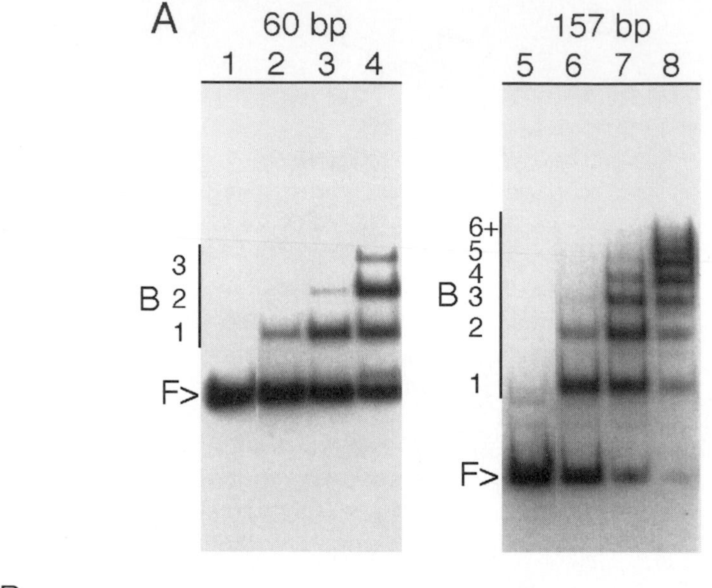

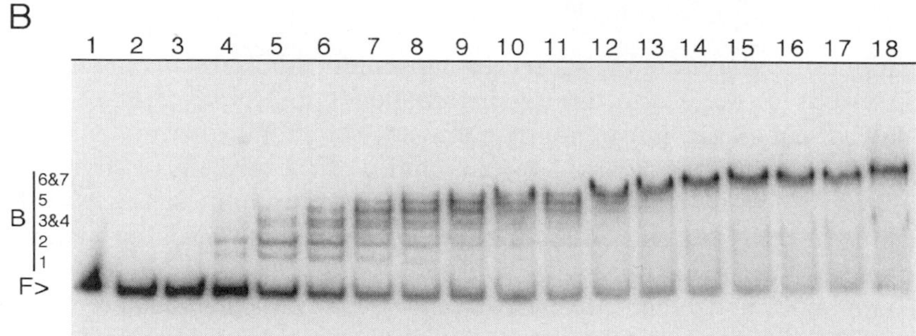

FIG. 2. DNA-binding properties of HMG-D. (A) EMSA of HMG-D bound to DNA of 60 (lanes 1–4) and 157 (lanes 5–8) bp in length illustrating a length dependence of the affinity of HMG-D for DNA. The protein concentrations are 0 (lanes 1 and 5), 25 (lanes 2 and 6), 50 (lanes 3 and 7), and 100 (lanes 4 and 8) n*M*. The free DNA is denoted by F, and B refers to the bound DNA that has been numbered according to the protein–DNA ratio for each complex band. (B) EMSA of HMG-D bound to a 90-bp DNA fragment used for multisite DNA-binding analysis. Lanes 1–18 contain binding reactions with 0, 1, 10, 20, 30, 40, 50, 60, 70, 80, 90, 100, 120, 140, 160, 190, 220, and 250 n*M* HMG-D, respectively. The bound DNA is numbered corresponding to the protein–DNA ratio for each complex band, 1, 2, 3 + 4, 5, and 6+ proteins, respectively. (C) Hill plot using data from B.[37,40] (D) Analysis of the fraction of DNA bound as a function of protein concentration for each of the complexes that could be separated by EMSA, 1, 2, 3 + 4, 5, and 6+. (E) Graph of the fractional saturation of the DNA, ν, versus $\nu/[P]$ of data from B and C. The theoretical binding curve was generated

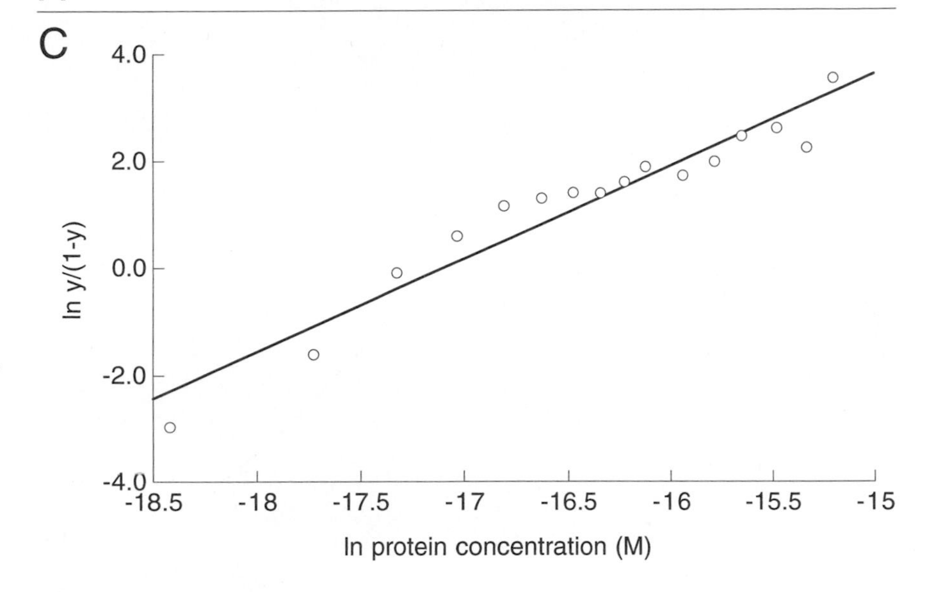

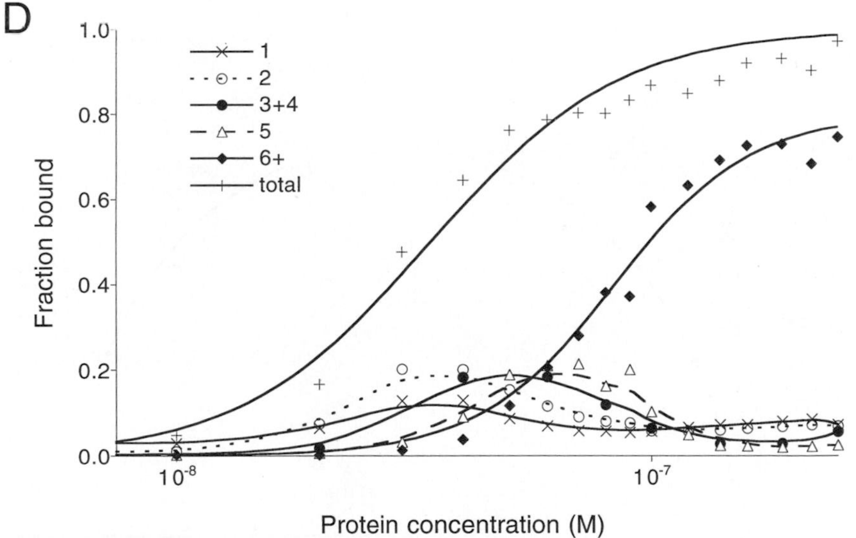

FIG. 2. (*continued*)

using theory describing cooperative ligand binding to an infinite DNA lattice, with a cooperativity factor of 9000, site size of 7 and K_d of 20 μM. (F) Scatchard analysis of the same data illustrates positive cooperativity of binding.[40] The theoretical curve was calculated with K_d of 50 nM and cooperativity of 2.1.

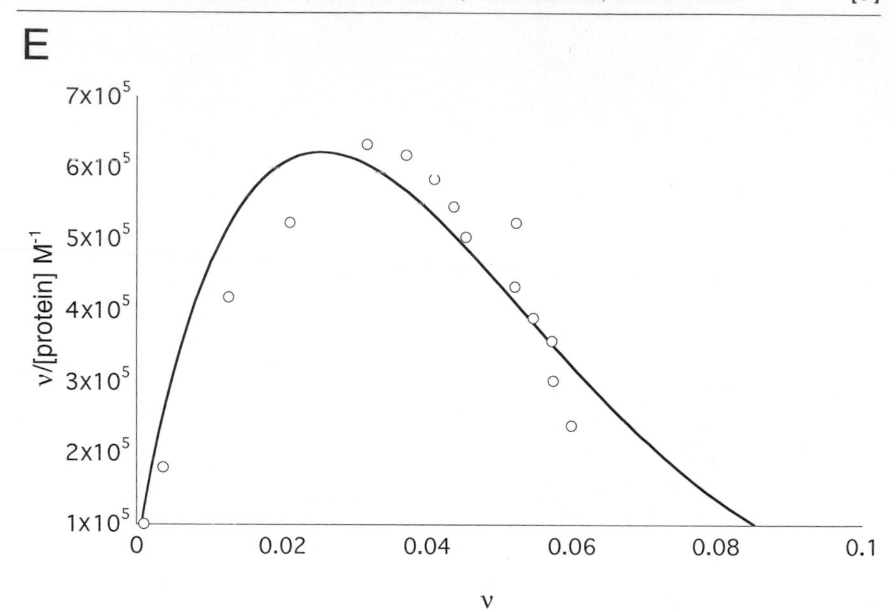

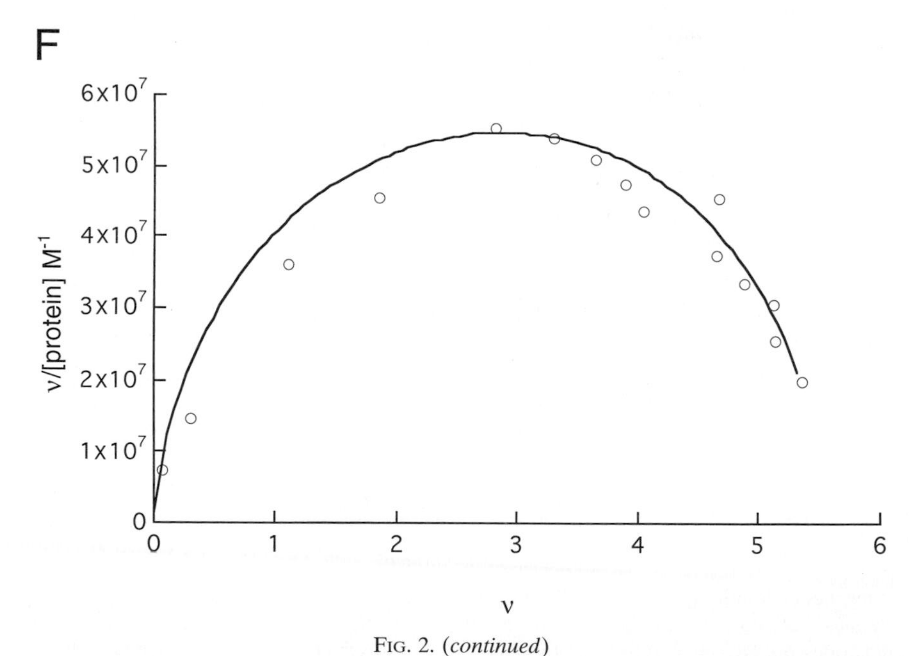

FIG. 2. (*continued*)

as 55 bp were observed for HMG-D and truncated forms of the protein (Fig. 3). The size of the circle also suggests that the bending angle of each HMG-D complex is greater than ~70°. By sampling a range of different size oligonucleotides, it is possible to estimate the net unwinding angle of the complexed DNA, as well as to define the best range of DNA length to use in ligation rate experiments to refine these values.[24]

Use of Electrophoretic Mobility Shift Assay to Study Affinity, Cooperativity, and DNA Bending of Nonsequence-Specific Chromosomal HMG-Box Proteins

The nonsequence specificity of DNA binding by the chromosomal class of HMG-box proteins poses a challenge for those interested in determining thermodynamic parameters describing DNA binding. Approaches traditionally used to study site-specific complexes required modifications in order to obtain quantitative information on equilibrium-binding constants and cooperativity.[27] Thus, elegant methods have been developed for the determination of these parameters for DNA binding of nonsequence-specific proteins,[28,29] especially for fluorescent or fluorescently labeled proteins.[30] In the case of HMG proteins, however, the intrinsic tryptophan fluorescence is not sufficient for the success of these approaches. Therefore, electrophoretic methods have been employed to study binding affinity and cooperativity, structure selectivity, reaction rates, and DNA bending. Restricting the length of the DNA fragment to a single binding site facilitates the determination of the intrinsic DNA-binding constant, and multiple site analysis provides information on the cooperativity of the interaction.

Intrinsic Affinities and DNA-Binding Rates for HMG-D from Single-Site Analysis

The information required to model DNA binding of a multisite DNA-binding protein can be simplified by first determining the values of the intrinsic affinity. For the HMG-D–DNA interaction, the intrinsic affinity may be measured directly for a single DNA site using EMSA, which greatly simplifies the interpretation of the binding curves.[21,31,32] This necessitates the use of short DNA fragments, of 20 bp or less, which bind only one

[27] J. D. McGhee and P. H. von Hippel, *J. Mol. Biol.* **86,** 469 (1974).

[28] C. P. Woodbury and P. H. von Hippel, *Biochemistry* **22,** 4730 (1983).

[29] A. Revzin, Nonspecific DNA–Protein Interactions. CRC Press, Boca Raton, FL, 1990.

[30] T. M. Lohman and D. P. Mascotti, *Methods Enzymol.* **212,** 424 (1992).

[31] D. F. Senear and M. Brenowitz, *J. Biol. Chem.* **266,** 13661 (1991).

[32] L. K. Dow, A. Changela, H. E. Hefner, and M. E. A. Churchill, *FEBS Lett.* **414,** 514 (1997).

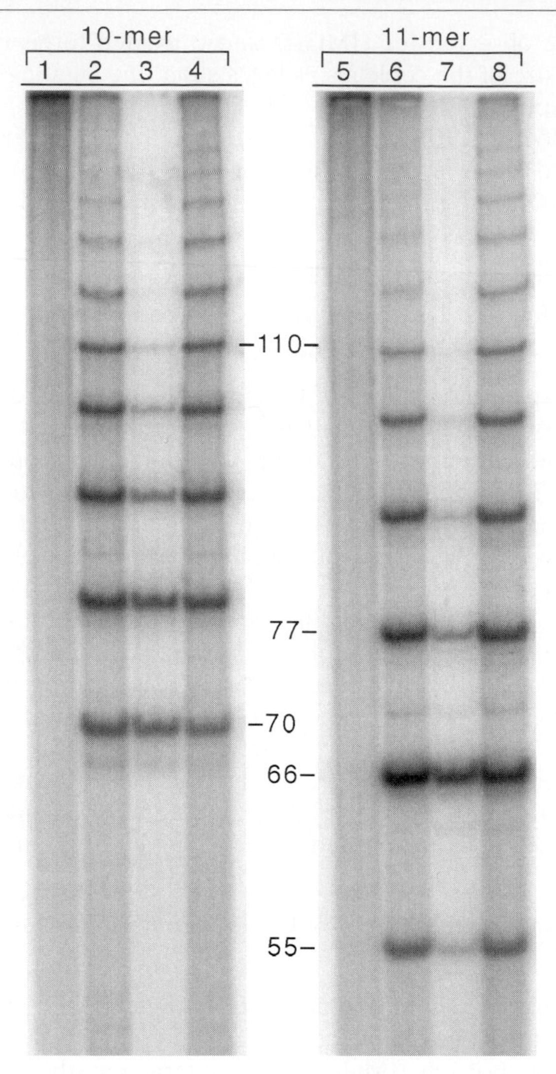

FIG. 3. The effect basic and acidic tail regions of HMG-D on DNA bending using a ligase-mediated circularization assay with HMG-D and truncation mutants. Experiments in lanes 1–4 and lanes 5–8 were performed with DNA of 10 and 11 bp in length, respectively. The DNA fragments have been multimerized and circularized by T4 DNA ligase in the presence of HMG-D (lanes 4 and 8), HMG-D-100 (lanes 3 and 7), or HMG-D-74 (lanes 2 and 6). Lanes 1 and 5 are control experiments, which lack HMG-D. The smallest DNA circles observed, 55 bp, are generated with DNA of length 11 bp, indicating an optimal helical repeat of approximately 11 bp/turn and a bending angle of at least 72° per potential binding site.

TABLE I
STRUCTURE-SPECIFIC DNA FRAGMENTS AND RELATIVE AFFINITIES

Structured DNA	Protein	Reported K_{struct}/K_{lin}	Reference
Holliday junction (17-bp arms)	HMG-D-100	1	18
Cisplatin DNA (24 bp)	HMG-D-100	3	18
Disulfide cross-linked DNA (20 bp)	HMG-D, HMG-D-100	10–15	21
Bulged DNA (21 bp)	HMG-D	~120	25
Circular DNA (75 bp)	HMG-D	~50	25
Circular DNA (75 bp)	HMG-D-74	~1000	25
Holliday junction (~18-bp arms)	HMG1	~100	64
Cisplatin DNA (92 bp)	HMG1	~100	23
Cisplatin DNA (20 bp)	HMGts	230	65
Cisplatin DNA (20 bp)	HMGts boxA	20	65

molecule of HMG-D. The effects of HMG-D and truncated forms of HMG-D binding to two types of DNA fragments, linear and "prebent" DNA fragments, illustrate this approach.

The observed preference of HMG-D for AT-rich DNA sequences that contain a TG dinucleotide was used to design a suitable linear DNA-binding site.[18] This sequence preference indicated that particular structures of DNA that are bent and/or underwound would also be preferred.[33] Several structured DNA probes that have been used for testing the DNA-binding preferences of the chromosomal HMG-proteins are shown in Table I. EMSA of HMG-D-100 binding to a linear DNA fragment and to a "prebent" disulfide cross-linked DNA fragment are illustrated in Figs. 4A and 4B. The binding curves were fit directly with Eq. (1) describing the Langmuir single-site binding isotherm.[31,34]

$$y = \frac{[P]/K_d}{1 + [P]/K_d} \tag{1}$$

The protein binds to each DNA fragment to form a 1:1 complex, and the intrinsic binding affinity, K_d, was calculated to be 11×10^{-9} and $1.4 \pm 0.4 \times 10^{-9}$ M for linear DNA and "prebent" DNA, respectively (Fig. 4C). Therefore, HMG-D-100 has a 10-fold preference for synthetic "disulfide cross-linked" (prebent) DNA molecules over the "unbent" control.[21] HMG-D has preferred binding to other bent DNA fragments such as cisplatin-treated DNA[18] but not synthetic Holliday junction DNA.[18,25,35]

[33] A. A. Travers, *Curr. Opin. Struct. Biol.* **1,** 114 (1991).
[34] T. M. Lohman and D. P. Mascotti, *Methods Enzymol.* **212,** 400 (1992).
[35] J. R. Wisniewski and E. Schulze, *J. Biol. Chem.* **269,** 10713 (1994).

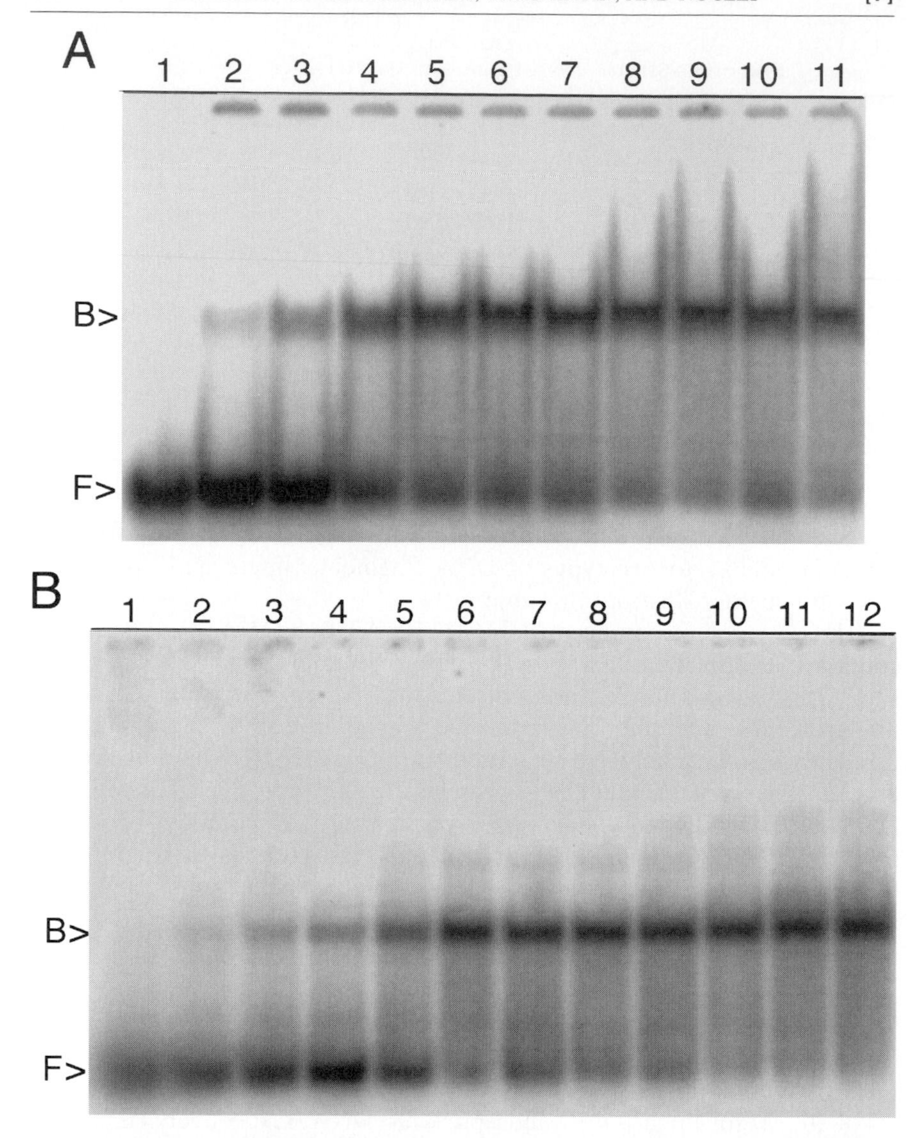

FIG. 4. Binding of HMG-D to short linear and structured DNA fragments. (A) EMSA (7% gel) of HMG-D-100 binding to a 20-bp linear DNA fragment. Lanes 1–11 contain binding reactions of DNA with 0, 0.5, 1.0, 5.0, 7.5, 10, 25, 75, 100, and 500 n*M* HMG-D-100, respectively. Free and bound DNA are denoted by F and B, respectively. (B) EMSA (6% gel) of HMG-D-100 binding to a 20-bp disulfide cross-linked DNA fragment.[21] Lanes 1–12 contain binding reactions with 0, 0.1, 0.3, 0.5, 1.0, 2.5, 3.5, 5.0, 10, 15, 25, and 35 n*M* HMG-D-100, respectively. (C) Binding curves for HMG-D-100 with linear and cross-linked DNA. Equilibrium dissociation constants from single-site-binding isotherms are 10.5 ± 2.0 and 1.4 ± 0.4 n*M* for linear and disulfide cross-linked DNA, respectively. (D) Binding curves illustrating the effect of

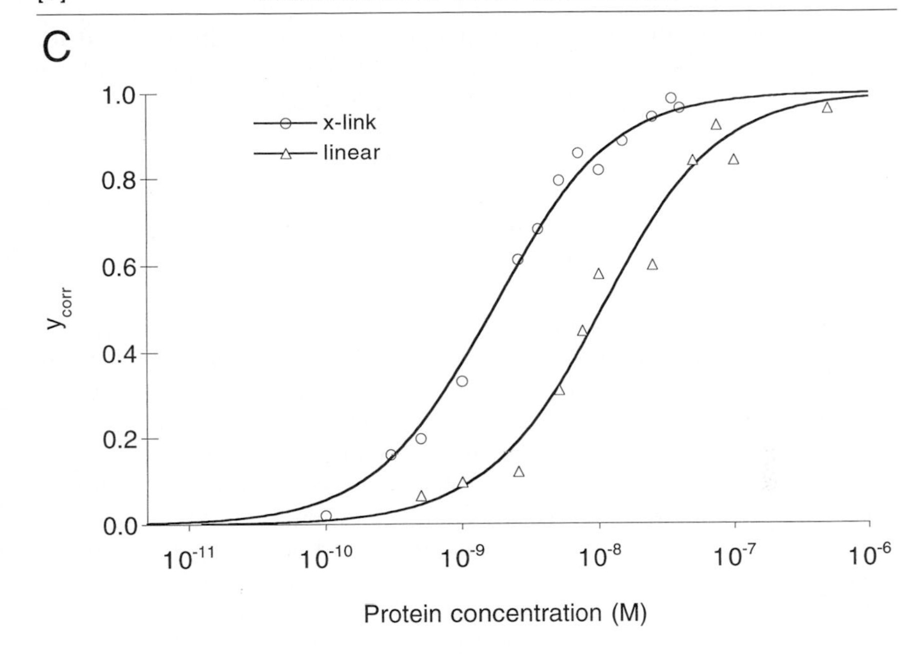

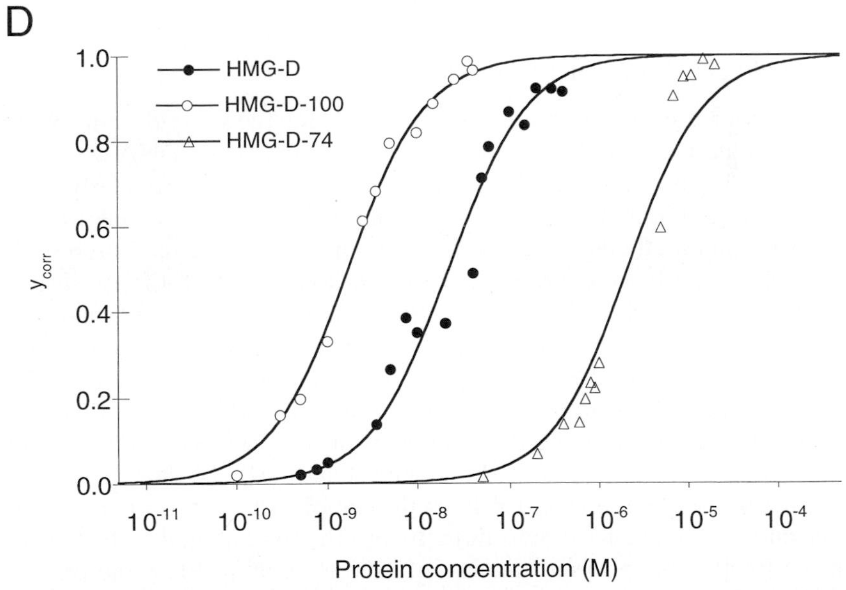

FIG. 4. (*continued*)

basic and acidic regions of HMG-D on binding to the 20-bp disulfide cross-linked DNA. The single-site equilibrium dissociation constants for HMG-D, HMG-D-100, and HMG-D-74 are 24 ± 2.5 nM, 1.4 ± 0.4 nM, and 2.1 ± 2.5 μM, respectively.

Table I illustrates the range of specificities that have been observed for HMG proteins binding to different structured DNA fragments both from single site assays and for structured DNA embedded in longer DNA fragments.

The single-site binding assay is also useful for quantitatively determining the effects of mutations and modifications on DNA binding. HMG-D contains an HMG domain, a basic region, and an acidic tail that all may contribute to the DNA-binding affinity of the protein. Binding curves were obtained for HMG-D, HMG-D-100, and the HMG-domain (HMG-D-74) binding to a disulfide cross-linked DNA fragment using EMSA (Fig. 4D). All of the proteins bind to DNA as a 1:1 complex with intrinsic binding affinities of 2.4×10^{-8}, 1.4×10^{-9}, and 2.1×10^{-6} *M* for HMG-D, HMG-100, and HMG-74, respectively. The basic AK motifs (residues 80–100) are important for high-affinity binding. The reduction in affinity of the protein lacking this region is about 1000-fold, equivalent to a free energy difference of ~4 kcal/mol. Full-length HMG-D binds with an affinity between these values, indicating that the acidic and basic regions modulate HMG-D binding affinity.[35] The precision and reproducibility of this assay are ideal for quantitative analysis of proteins that have even smaller differences in affinity than illustrated by this example.

Rates of HMG Protein DNA Binding

Whether the protein binds to the DNA slowly or rapidly and which rate constant is affected more by differences in the target DNA or protein mutations are properties of the binding kinetics that are particularly useful in the interpretation of binding results obtained for chromosomal proteins. Many elegant methods can be used to obtain quantitative measurement of these rates, such as fluorescence polarization and fast DNase I footprinting.[36] However, a simple electrophoretic method can give values for binding rates that are sufficiently slow (on the order of minutes). This method was developed for the study of transcription factors[4] and is applied here to HMG-D-100 binding to a 16 nucleotide disulfide cross-linked DNA fragment. Figure 5A shows that the association of HMG-D-100 with the DNA fragment is complete within 1 min. Figure 5B shows a measurable dissociation rate of HMG-D-100 from the complex. The first-order dissociation rate constant can be determined from the exponential fit to the plot of the fraction bound as a function of time (Fig. 5C). While the on-rates of HMG-D DNA are too fast to be measured using this method, the highest affinity form of HMG-D, HMG-D-100, has a measurable off-rate and a

[36] V. Petri, M. Hsieh, and M. Brenowitz, *Biochemistry* **34,** 9977 (1995).

complex half-life of approximately 14 min. Prebending the DNA increases the off-rate relative to linear DNA (data not shown). These results suggest that HMG-D can bind and dissociate many times during the course of the binding experiment and that the binding constants calculated were obtained at equilibrium.

Cooperative Interactions of HMG-D Using Multisite Analysis

The DNA-binding ladders observed in Figs. 2A and 2B show that HMG-D binds to multiple sites on DNA. The length dependence of the DNA-binding affinity of HMG-D also suggests a cooperative DNA-binding interaction. Figure 2A illustrates the banding pattern observed for HMG-D bound to two different lengths of DNA derived from the *ftz*-SAR, which is AT rich and contains several TG dinucleotides. The number of complexes correlates with the length of the DNA, such that five complexes are observed for the complex with the 62-bp DNA, at least seven complexes are visible for a 90-bp DNA fragment (Fig. 2B), and more than eight complexes are observed for the 157-bp DNA.

Quantitative analysis of the DNA-binding ladders can be applied in order to determine intrinsic binding affinities and cooperativity. Methods for the study of multiple ligands binding to DNA have been described for the interaction of ligands with specific sites.[37–40] The theory and methodology for determining binding affinity and cooperativity parameters have also been developed for the analysis of large noninteracting and interacting ligands binding to multiple nonspecific sites on an infinite lattice.[27,30,34,41] Both types of analyses have been applied to data obtained from EMSA of HMG-D-100 on the *ftz*-SAR DNA (Fig. 2B).

The Hill plot, the first method of analysis to be discussed, is a standard semiempirical method for analysis of DNA-binding data.[37,40] The Hill plot, as applied to HMG-D data in Fig. 2C, is the quantity $\ln[y/(1-y)]$ graphed as a function of $\ln[P]$, where y is the fraction of DNA bound and $[P]$ is the protein concentration. The slope of a straight line from a linear least-squares analysis of data gave an observed cooperativity of binding of 1.74, and the value of $[P]$ at $y = 0.5$ gave a value of approximately 40 nM for the apparent equilibrium dissociation constant, K_{obs}. Determining the fraction of DNA bound in this way is similar to analyses of data obtained

[37] A. V. Hill, *J. Physiol. Lond.* **40,** 4 (1910).
[38] G. Scatchard, *Ann. N.Y. Acad. Sci.* **51,** 660 (1949).
[39] A. A. Schreier and P. R. Schimmel, *J. Mol. Biol.* **86,** 601 (1974).
[40] C. R. Cantor and P. R. Schimmel, "Biophysical Chemistry." Freeman, San Francisco, 1980.
[41] P. L. deHaseth, T. M. Lohman, and M. T. Record, *Biochemistry* **16,** 4783 (1977).

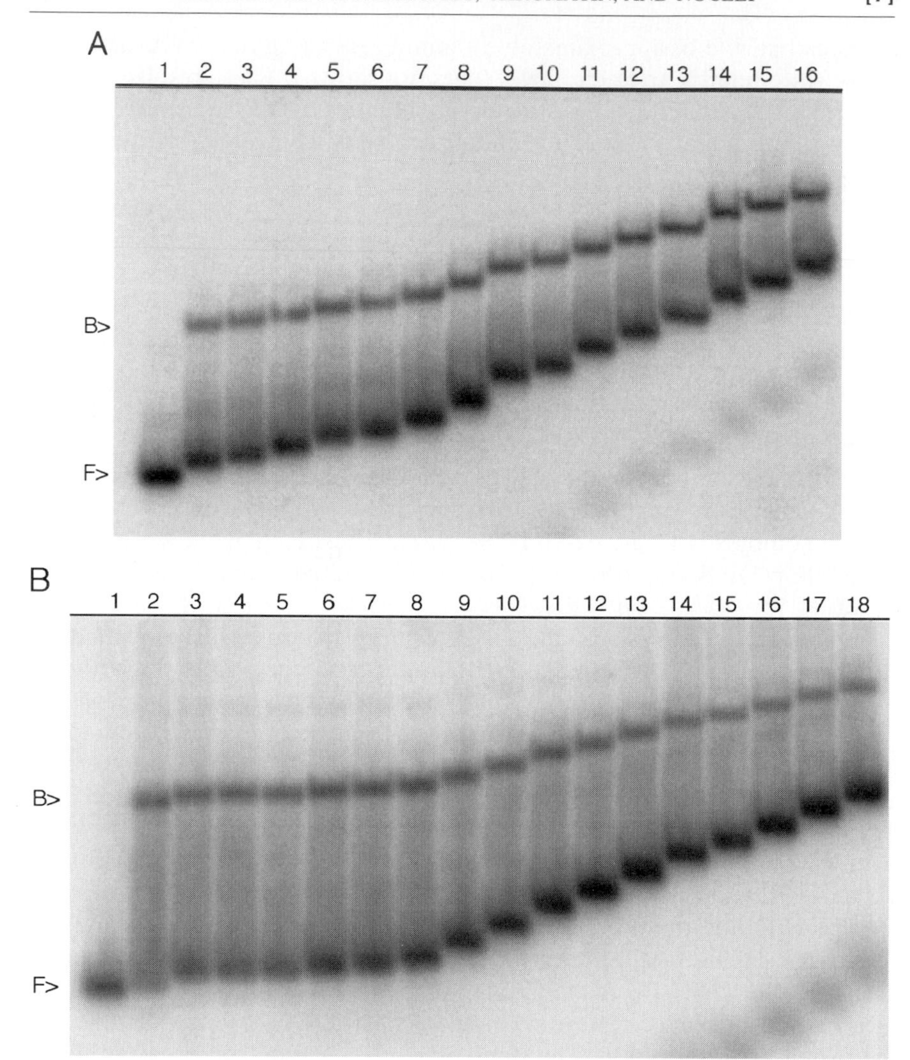

FIG. 5. Rates of chromosomal HMG-D DNA binding. (A) EMSA illustrating relatively fast association rate of HMG-D-100 with DNA. Lane 1 is a control with no protein added. Lanes 2–16 contain binding reactions of DNA with 10 n*M* HMG-D-100 incubated for 0 (<15 sec), 1, 3, 5, 7, 10, 15, 20, 25, 30, 35, 40, 45, 50, and 55 min, respectively. (B) EMSA illustrating a measurable off rate for HMG-D-100 DNA binding. Lane 1 is a control with no protein added and lane 2 is a control with no competitor DNA added. Lanes 3–18 contain binding reactions of DNA with 10 n*M* HMG-D-100 incubated with 100-fold excess of competitor DNA over labeled DNA for 0 (<15 sec), 0.5, 1, 2, 3, 5, 10, 15, 20, 25, 30, 35, 40, 45, 50, and 55 min, respectively. (C) Graph of data obtained from experiments such as B. Curve is fitted with an exponential function describing first-order dissociation kinetics, with a dissociation rate of 5.8×10^{-4} sec^{-1} and a complex half-life of 14 min.

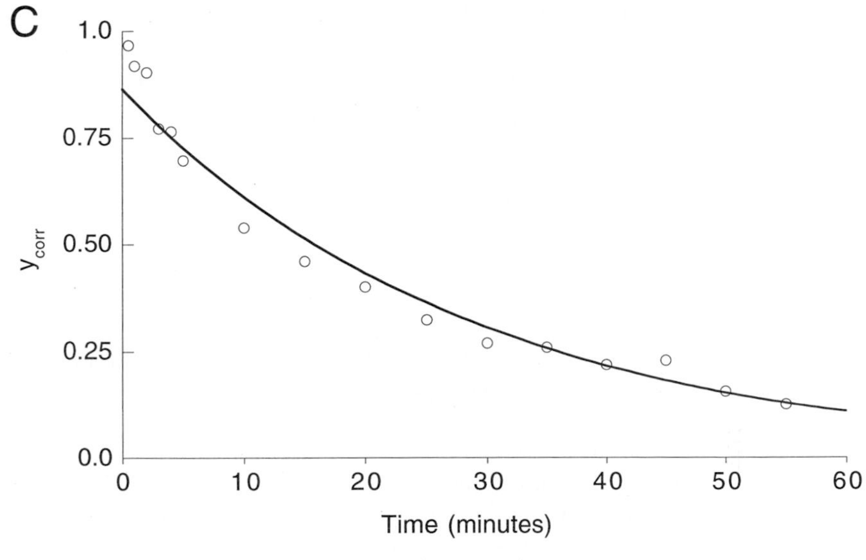

FIG. 5. (*continued*)

from filter-binding experiments,[28] which depend on whether the DNA is bound by protein or not and does not take advantage of knowledge about the degree of saturation of the DNA.[42]

The ability of EMSA to provide information on the saturation of the DNA fragment can be utilized in a second method of analysis. The ladder of bands in Fig. 2B is composed of DNA fragments with different numbers of HMG-D-100 molecules bound; six distinct bands are observed. The fraction of DNA bound with different numbers of proteins ($y_1, y_2, y_3, \ldots, y_m$) can be extracted easily by determining the amount of DNA present in each type of complex separately during the quantitation of digitized gel images. These values, y_1–y_m, can be used to calculate the fractional saturation or binding density of the DNA (ν) using Eq. (2), where i corresponds to the number of the shifted band. For HMG-D, which binds to DNA as a monomer (data not shown), the band number is equal to the number of protein molecules bound to the DNA in band i. The fraction bound (y_i) for each band is plotted as a function of protein concentration in Fig. 2D.

$$\nu = \sum_{i=0}^{m} i(y_i) \tag{2}$$

[42] M. Stros, J. Stokrova, and J. O. Thomas, *Nucleic Acids Res.* **22,** 1044 (1994).

The experimentally determined binding density described earlier allows theoretical models for HMG-D-DNA binding to be evaluated by analysis of the Scatchard plot, $\nu/[P]$ graphed as a function of ν (Fig. 2E). The plot has a downward curvature, indicative of positive cooperativity of binding.[40] The binding affinity and cooperativity parameters can be extracted from such a curve for large ligands binding cooperatively to an infinite lattice by fitting Eq. (3), derived by McGhee and von Hippel,[27] to the HMG-D data. The intrinsic association constant is K_i, ω is the cooperativity, $[P]$ is the total ligand concentration, n is the length of the DNA-binding site and ν is the fractional saturation.

$$\frac{\nu}{[P]} = K_i(1 - n\nu)\left(\frac{(2\omega + 1)(1 - n\nu) + \nu - R}{2(\omega - 1)(1 - n\nu)}\right)^{n-1}\left(\frac{(1 - (n + 1)\nu + R)}{2(1 - n\nu)}\right)^2 \quad (3)$$

where $R = \{[1 - (n + 1)\nu]^2 + 4\,\omega\nu(1 - n\nu)\}^{1/2}$. Even for a range of possible site sizes, intrinsic affinity constants, and cooperativity factors, data from the HMG-D DNA-binding experiment do not fit this model well at low binding densities (Fig. 2E). The intrinsic affinity of DNA binding (K_i) is underestimated ($1 \times 10^{-5}\,M$) and the overall cooperativity is overestimated (9000). Specificially, the calculated curves show that higher $\nu/[P]$ values are expected at lower binding densities than are observed, whereas at higher binding densities the curve fit is reasonable. Interestingly, the EMSA method is intrinsically better at determining the fractional saturation at lower binding densities due to greater resolution of those complexes. At the highest binding densities the fit is better, but the fractional saturation data are not as reliable (for reasons discussed later).

The Scatchard plot can also be analyzed using the theoretical Scatchard model to obtain values of intrinsic affinity and cooperativity. A plot of $\ln[P]$ versus $\ln[(n/\nu\ -)1)]$ gave a straight line for these data with a value for a total number of sites (n) of 6 (not shown). The slope of the line is $-(1/\omega)$ and the y intercept is $\ln K_{obs}$. A cooperativity factor, w, of 2.0 and a K_d of 66 nM were obtained. The theoretical Scatchard curve shown in Fig. 2G, calculated using Eq. (4) with a $K_a = 2 \times 10^7\,M^{-1}$ and cooperativity of 2.1, fits the data quite well:

$$\frac{\nu}{[P]} = \frac{nK_a^{\omega}[P]^{\omega-1}}{1 + K_a^{\omega}[P]^{\omega}} \quad (4)$$

Although the Scatchard analysis overestimates $\nu/[P]$ slightly at lower binding densities, it does so less than the infinite lattice model, and the values used to calculate the theoretical curves are comparable to the intrinsic DNA-binding affinity determined from independent methods, described later. For HMG-D data, the value of intrinsic affinity obtained is consistent with the values that have been determined using single-site binding methods under similar conditions (not shown).

Interpretation of Single-Site and Cooperative DNA-Binding Analyses

Results from single-site quantitative-binding analyses can be useful in identifying complications arising in protein–DNA recognition that may aid in the interpretation of cooperative DNA-binding experiments. When the Hill analysis is applied to binding reactions where a 1 : 1 complex is expected, a cooperativity value greater than 1 may be observed. This phenomenon is not simple to interpret, because by definition a 1 : 1 complex can exhibit no cooperativity. In one case where this cooperativity was observed, the protein sample was not homogeneous, being composed of oxidized and native proteins that bound to the DNA with slightly different affinities. Another possible case involves protein–DNA interactions for which there is internal cooperativity, as defined by Jencks.[43] HMG-box proteins that are composed of two linked HMG-box domains may be modeled by application of this binding theory. Additionally, a complex preequilibrium, such as aggregation or other linked equilibria that could influence the true concentration of free protein available for DNA binding, can also cause unexpected apparent cooperativity. Therefore, single-site analysis is useful even for nonsequence-specific proteins for determining whether a readily physically interpretable binding equilibrium exists before investing effort on cooperative binding analyses.

The Scatchard analysis indicated that the intrinsic affinity of HMG-D was consistent with that calculated in the single-site assay. However, the application of the multiple ligand-binding theory[27] resulted in high cooperativity values and much lower binding constants than were expected from the single-site analyses. The presence of some cooperativity and the transient nature of HMG-D DNA binding illustrated by the binding rate studies suggest that significant reorganization may occur during the long incubation times used in this study, giving the appearance of specific cooperatively interacting sites, which fit the Scatchard model quite well.[38,39] However, consideration must also be given to the limitations of cooperative binding analysis, as observed in this study.

First, there are limitations in the quantitative EMSA method because the resolution of the bands at higher protein : DNA complex ratios decreases. The precise number of proteins capable of binding to a finite length DNA fragment may be difficult to determine accurately at higher binding densities. In this case, seven bands were detectable when the true maximum

[43] W. P. Jencks, *Proc. Natl. Acad. Sci. U.S.A.* **78,** 4046 (1981).

number of proteins bound per DNA fragment could be as high as 9 or 10. This introduces errors in determining the magnitude of the fractional saturation by giving a lower value for the maximum number of proteins bound per DNA fragment. Second, an anticipated difference between the HMG-D experiment and the infinite lattice theory is the finite length of the DNA-binding site used in the EMSA. The theoretical models evaluate the statistical distribution of free binding sites accurately when the length of the lattice relative to the site size is large. However, application of this theory to a finite lattice has been used successfully to calculate the binding affinity and cooperativity values of HU for several shorter DNA fragments from EMSA.[44] Third, the mode of binding of HMG-D to multiple sites on DNA may be more complicated than what has currently been modeled theoretically. For example, super structures may be formed by multimers of these proteins with DNA that may require more than one cooperativity factor to describe the binding interaction.[19,42]

Procedures: Electrophoretic Methods for Studying DNA Recognition by Purified Chromosomal Proteins

Use of EMSA for Chromosomal Protein–DNA Binding

Several aspects of the experimental protocols appear to be critical to obtain reproducible data of high quality for quantitative analysis using EMSA with HMG proteins. Among the most important components for successful affinity measurements are protein homogeneity and DNA purity. It is becoming evident from studies on HMG-box proteins[32,45,46] and the hyperthermophile chromosomal protein Sac7d that minor groove intercalation is a critical component of the molecular recognition mechanism used by these proteins.[47] The intercalating methionine is located on the surface of the concave face of the boomerang shape of HMG-D (Fig. 1B). The oxidation of this residue that may occur during the overexpression of HMG-D in bacteria results in a mixture of oxidized and unoxidized proteins. We have developed a method to obtain homogeneous HMG-D by separating the two forms of the protein.[32] The consequences of oxidation are to decrease DNA binding and add complexity to the single-site binding isotherms so that they are not readily interpretable as 1 : 1 complexes. DNA purity can be as important as protein purity because trace contaminants from DNA preparation and labeling procedures can inhibit DNA binding

[44] E. Bonnefoy, M. Takahashi, and J. Rouviere-Yaniv, *J. Mol. Biol.* **242,** 116 (1994).
[45] C.-Y. King and M. A. Weiss, *Proc. Natl. Acad. Sci. U.S.A.* **90,** 11990 (1993).
[46] J. J. Love *et al., Nature* **376,** 791 (1995).
[47] H. Robinson *et al., Nature* **392,** 202 (1998).

by HMG-D. Commonly used methods to purify DNA are sufficient for this purpose.

Extreme care is required to maintain and measure the fraction bound for the weaker complexes, those with affinities in the range of 1–10 μM. This involves minimizing dissociation of the complex in the gel by decreasing the electrophoresis time. The polyacrylamide concentration and ratios of bisacrylamide to acrylamide are tailored for each set of experimental conditions so that electrophoresis for only 1 hr at 125 V (20 × 20-cm gel) gives sufficient separation of the bound and unbound DNA for quantitative analysis. Gels can be loaded quickly or as the gel is running. Some HMG-D mutants have fast on and off rates, leading to dissociation of the complexes during electrophoresis. Under conditions where dissociation of complexes in the gel is observed, a sufficiently high concentration of protein must be used so that a correction factor based on the degree of dissociation observed at saturating amounts of protein can be applied, as described later.[48] Two other important modifications to the DNA-binding reaction procedure include the use of microcentrifuge tubes (natural polypropylene; Eppendorf Scientific, Westbury, NY) and the method of siliconization of the reaction vials. These details can influence the success and reproducibility of the experiment dramatically.

HMG-D Protein Purification

The gene for HMG-D (full-length 112 amino acids) is subcloned into the *Nde*I and *Bam*HI-digested expression vector, pET13a.[49] Use of the T7 expression system[50] and this kanamycin-resistant expression vector results in stringent selective pressure by the antibiotic necessary with a chromosomal protein, which is toxic to the cells.[49] Variants and mutants of HMG-D are made in pET13a using standard subcloning techniques.[51] HMG-D and its variants are grown in *Escherichia coli* strain BL-21(DE3) (Novagen, Madison, WI) in LB media with 30 μg/ml kanamycin at 37° until an A_{600} of 0.8–1.0 is reached. Expression is then induced by the addition of isopropylthiogalactopyranoside (IPTG) to 250 μg/ml. Cells are harvested after 3–4 hr of further incubation at 37° by centrifugation at 8000 rpm for 10 min and stored at −20°.

The protein purification proceeds either at 4° or on ice with >1.0 mM dithiothreitol (DTT) present until the high-performance liquid chromatog-

[48] B. Hoopes, J. LeBlanc, and D. Hawley, *J. Biol. Chem.* **267,** 11539 (1992).

[49] D. N. M. Jones *et al., Structure* **2,** 609 (1994).

[50] F. W. Studier, A. H. Rosenberg, J. J. Dunn, and J. W. Dubendorff, *Methods Enzymol.* **185,** 60 (1990).

[51] J. Sambrook, E. F. Fritsch, and T. Maniatis, "Molecular Cloning: A Laboratory Manual." Cold Spring Harbor Laboratory Press, Cold Spring Harbor, NY, 1959.

raphy (HPLC) step. Despite the presence of 1 m*M* DTT and/or 10 m*M* methionine included in the buffers, significant oxidation of methionine-13 does occur.[32] The cell pellet from 1 liter of culture is suspended in 15 ml of lysis buffer [50 m*M* Na-HEPES, pH 7.9, 1.0 m*M* EDTA, 1.0 m*M* DTT, 0.6 m*M* phenylmethylsulfonyl fluoride (PMSF), 1.0 m*M* benzamidine, 2 μg/ml leupeptin, 2 μg/ml aprotinin, 1 μg/ml pepstatin A, and 500 m*M* NaCl]. The suspension (on ice) is sonicated for 30 sec, frozen in liquid N_2, and sonicated again. Cell debris is removed by centrifuging the lysate for 15 min at 18,000 rpm at 4°. The supernatant, containing the HMG-D, is dialyzed (two changes) against "HEP" buffer (50 m*M* Na-HEPES, pH 7.9, 1.0 m*M* EDTA, 1.0 m*M* DTT, 0.6 m*M* PMSF, 1.0 m*M* benzamidine, and 50 m*M* NaCl) and then fractionated by precipitation with $(NH_4)_2SO_4$. Full-length and HMG-D-100 (C-terminal truncation at residue 100) proteins precipitate between 65 and 100% $(NH_4)_2SO_4$. The HMG-box domain of the protein (HMG-D-74) precipitates between 60 and 70%. Mutant variants of HMG-D have been observed to precipitate at even lower concentrations of $(NH_4)_2SO_4$. The precipitate is dissolved in and dialyzed against HEP buffer (two changes) and prepared for chromatography.

Isocratic chromatography on a DEAE-Sephacel (Pharmacia) or HQ (Poros) column with HEP buffer (without protease inhibitors) separates the protein from DNA and acidic contaminants. The purest fractions elute in the flow through as determined by sodium dodecyl sulfate–polyacrylamide gel electrophoresis (SDS–PAGE) analysis. When only a single band is observed on the gel, the sample is dialyzed against water and concentrated in a Centricon 3 (Amicon, Danvers, MA) microconcentrator for high performance liquid chemotography (HPLC). If additional bands are present, cation-exchange chromatography using the SP-Sepharose (Pharmacia) or Poros HS column is used to purify the fractions further prior to preparation for HPLC; the protein elutes during a gradient between 0.5 and 1 *M* NaCl. The native and oxidized forms of the protein are separated by reversed-phase (RP)-HPLC (LKB) using a Waters μBondapak C_{18} column. The protein is eluted isocratically at 3.5 ml/min with buffer containing 31% acetonitrile and 69% 0.2 *M* dibasic sodium phosphate, pH 2.3. Fractions corresponding to the two forms of the protein are pooled separately, loaded onto the SP-Sepharose column (Pharmacia, Piscataway, NJ) or Poros HS column (PerSeptive Biosystems, Framingham, MA), and eluted in a gradient of 0.1 to 1 *M* NaCl. Purified protein is pooled, dialyzed, and concentrated. The total yield of unoxidized HMG-D can be as high as 8 mg per liter of culture.

The molecular mass of the protein and the removal of oxidized protein is confirmed by electrospray–ionization mass spectrometry using a Micromass Quattro triple quadrupole mass spectrometer equipped with a Megaflow electrospray ion source. This shows only one species, with the appropriate molecular mass for HMG-D or its variants. The concentration of the protein is determined carefully by ultraviolet (UV) absorption using the extinction coefficient (19.1 A_{280}/mM) based on the amino acid content; the UV absorption of the proteins is the same in TE as 6 M guanidine hydrochloride.[52] The proteins are then diluted in protein dilution buffer, PDB [50 mM HEPES, pH 7.5, 0.1 M KCl, 1 mM EDTA, 1 mM DTT, 100 μg/ml bovine serum albumin (BSA), and 50% glycerol] and stored at −20°.

Binding Site Selection

A polyclonal antiserum against the HMG-D protein (gift from Sarb Ner)[7] is used to immunoprecipitate complexes of 200 nM HMG-D with a DNA fragment containing a 26-bp stretch of randomized DNA sequence flanked by fixed ends for polymerase chain reaction (PCR) priming and subsequent subcloning.[16,18]

CAGGTCAGTTCAGCGGATCCTGT–CG(N)$_{22-26}$GA–GGCGAATT CAGTGCTGCAGC

The immunoprecipitated material is then amplified using PCR and purified by polyacrylamide gel electrophoresis using the crush and soak procedure of Maxam and Gilbert.[53] An excess of poly[d(IC)] (Pharmacia), 1 μg per reaction, is used as a competitor DNA in the binding reactions to ensure the selection of preferred DNA fragments. This process is repeated three times and the amount of DNA precipitated during each cycle increases steadily. After four cycles of precipitation and amplification, the selected fragments are electrophoresed in the presence of the HMG-D on a 4% polyacrylamide gel (40 : 1; acrylamide : bisacrylamide; Bio-Rad, Richmond, CA) for 90 min at 150 V in 0.5× TBE [1×: 100 mM Trizma base (Sigma), 100 mM boric acid, 2 mM ethylenediaminetetraacetic acid (EDTA); electrophoresis grade]. Inserts that retard the mobility of the DNA are isolated, gel purified using the crush and soak procedure,[53] subcloned into the *Bam*HI/*Eco*RI site of a Blue Scribe SK$^-$ vector, and sequenced using standard procedures.[51] Statistical analysis of the probability of occurrence of dinucleotides shows a strand bias and preferences for TG sequences.[18]

Binding Site DNA Preparation

An important aspect of DNA preparation is the final purification of the DNA. DNA fragments, including restriction fragments purified from gels

[52] T. M. Lohman and W. Bujalowski, *Methods Enzymol.* **208,** 258 (1992).
[53] A. M. Maxam and W. Gilbert, *Methods Enzymol.* **65,** 499 (1980).

and short DNA duplexes constructed from purified synthetic oligonucleotides, frequently contain contaminants either from the labeling procedure or purification by gel or chromatographic fractionation that can interfere with the high-affinity binding of some chromosomal proteins. In order to remove trace contaminants, we typically extract the labeled DNA samples with phenol : chloroform : isoamyl alcohol (25 : 24 : 1, v/v) (PCI) and remove the residual phenol using two ether extractions. The DNA is then ethanol precipitated and centrifuged, and the pellet is rinsed with 80% (v/v) ethanol and dried before redissolving the DNA in TE buffer (50 m*M* Tris–HCl, pH 7.9, 1 m*M* EDTA) for storage at −20°.

Restriction Fragments

The probe DNA fragments, 62, 90, and 157 bp, come from the region "400–455" of the *Drosophila fushi tarazu* (*ftz*) upstream element/scaffold associated region (USE/SAR) in plasmids pBSSNP12 and p19Sal12 (gift of Sarb Ner; described previously).[15] The 62- and 90-bp fragments are excised from the p19Sal12 plasmid using *Sal*I and *Bam*HI/*Hin*dIII digests, respectively. The 157-bp fragment is obtained from *Ava*I/*Hin*dIII digests of pBSSNP12. The resulting restriction digests are dephosphorylated using calf intestinal alkaline phosphatase (Promega) and are purified by PCI and ether extraction, ethanol-precipitated, and stored at −20° until use. The quantitated probe DNA (1–3 pmol) is labeled at the 5′ end using [γ-^{32}P]ATP (3000 Ci/mmol; Amersham, Piscataway, NJ) and T4 polynucleotide kinase (GIBCO-BRL, Gaithersburg, MD) and gel purified as described earlier to separate it from the remaining plasmid fragments.[53]

Oligonucleotides

Single-stranded oligonucleotides (AGTTACTGAATTACGCTCAT and TAGAGCGTAATTCAGTAACT) are synthesized commercially. The DNA is purified by Sep-Pak (Waters C_{18}) chromatographic or ion-exchange methods or by polyacrylamide gel purification followed by PCI extraction, ether extraction, and ethanol precipitation. Palindromic sequences are annealed at 90° in 0.1 *M* NaCl and then cooled. The single strands of nonpalindromic sequences are quantitated carefully using UV absorption and electrophoresis techniques to ensure correct duplex formation and to minimize single-stranded DNA.[18] The strands are annealed in a 1:1 molar ratio at ~1 μM concentration in TE buffer by heating at 90° for 2 min followed by slow cooling to room temperature. The DNA is then stored at −20° in TE buffer until use. A 20-bp disulfide cross-linked DNA fragment (kindly provided by Wolfe and Verdine) was labeled as described previously.[54] Other DNA fragments were 5′ end labeled using [γ-^{32}P]ATP and gel purified as described earlier.

Electrophoretic Mobility Shift Assay

Band shift assays are conducted using siliconized Eppendorf vials. To siliconize the reaction vials, the tubes are dipped into a solution of 2% dimethyldichlorosilane (Sigma) in heptane (Sigma) and rinsed immediately with 95% ethanol. The vials are then rinsed three times using deionized water (Milli-Q; Millipore, Bedford, MA) and baked overnight at 100°. Protein dilutions are made from 10 μM or more concentrated stocks of HMG-D in PDB (as required). One microliter of HMG-D (diluted as described earlier) is added to a 9-μl binding reaction composed of between 0.1 and 1 nM radiolabeled DNA, 200 μg/ml BSA, and binding buffer (20 mM HEPES, pH 7.5, 50 mM KCl, 2 mM $MgCl_2$) to give final HMG-D concentrations ranging from 0 to 20 μM. Reactions are incubated at 22° for 45 min. The samples contain 5% glycerol and no added dye. The samples are loaded on prerun (at least 45 min, until constant current is reached) polyacrylamide gels (30 : 1; acrylamide : bisacrylamide) and electrophoresed for 45 min to 1.5 hr at 125 V in 0.33× TBE, dried, and then exposed to film.

Ligase-Mediated Circularization Assay

The DNA oligonucleotides (GCCTATTGAA, GCTTCAATAG, GCCATATTGAA, GCTTCAATATG) used in the circularization assays are synthesized commercially (Operon Technologies, Inc., Alameda, CA) and purified using RP chromatography (Waters Sep-Pak, Waters Corp., Milford, MA). One nanomole of each "top strand" is phosphorylated at the 5′ end using T4 polynucleotide kinase (GIBCO-BRL) with [γ^{32}-P]ATP (3000 Ci/mmol, Amersham) and unlabeled ATP, and 1 nmol of each "bottom strand" is phosphorylated with unlabeled ATP to generate ligatable DNA ends. After purification by PCI extraction and two ether extractions, the DNA is ethanol precipitated, rinsed, and dried before redissolving the DNA in TE buffer. The hot and cold DNA strands are annealed at a 1 : 1 molar ratio at 1 mM concentration in TE buffer by heating at 90° for 2 min followed by slow cooling to room temperature and storage at −20°.

For the circularization assay, 1 μl of 10 μM HMG-D is added to 8 μl of 1 μM double-stranded DNA in ligase buffer (50 mM Tris–Cl, pH 8.0, 5 mM $MgCl_2$, 1 mM DTT) and allowed to incubate for approximately 4 min. One unit of T4 DNA ligase (GIBCO-BRL) is then added and the

[54] H. A. Saroff, *Biopolymers* **33,** 1327 (1993).

reaction is allowed to proceed for 1 hr at 22°. The reactions are quenched by PCI extraction after which the DNA is extracted with ether, ethanol precipitated, dried, and redissolved in 10 μl of water. The samples are digested for 20 min at 37° with exonuclease III (GIBCO-BRL) in 50 m*M* Tris–Cl, pH 8.0, 5 m*M* $MgCl_2$, 1 m*M* DTT to remove linear DNA fragments and subsequently extracted with ether, ethanol precipitated, dried, and redissolved in water and loading buffer (50% glycerol, 1% xylene cyanol, 1% bromphenol blue, w/v). The samples are loaded on a 40-cm 8% (w/v, 30 : 1 acrylamide : bisacrylamide; 1× TBE) native polyacrylamide gel and are electrophoresed for 4.0 hr at 400 V. The gel is dried and exposed to Kodak (Rochester, NY) XAR film and phosphorimaging plates. Control experiments (data not shown) are performed to verify the gel position of the monomer circles.

Time Course of Protein–DNA Binding

Time course experiments are conducted using the same DNA fragments and binding conditions as described earlier for equilibrium binding analysis with a few exceptions. For the on-rate experiments, 20 μl of 100 n*M* HMG-D-100 in PDB is added to a 180-μl solution, which contains 1 n*M* radiolabeled DNA, 200 μg/ml BSA, and binding buffer (20 m*M* HEPES, pH 7.5, 50 m*M* KCl, 2 m*M* $MgCl_2$). At times varying from 0 sec to 60 min, a 10-μl aliquot is removed and loaded directly onto the running gel to capture the protein–DNA complexes formed at that time. For off-rate assays, the reactants (180 μl of 10 n*M* HMG-D-100, 1 n*M* radiolabeled DNA, 200 μg/ml BSA in binding buffer) are incubated for 45 min before the addition of the 20-μl unlabeled competitor DNA, poly[d(IC)], to a concentration of 160 n*M*. At a series of time points after addition of the competitor DNA, aliquots are removed from the vial and loaded directly onto a prerun band-shift gel (8%, w/v, 30 : 1; acrylamide : bisacrylamide in 0.33× TBE) while running. The gels are electrophoresed for 60–75 min at 125 V, dried, and exposed to film.

Data Extraction and Analysis

The dried gels are exposed to a phosphorimaging screen between 10 and 24 hr and subsequently scanned using the Molecular Dynamics Phosphorimager. The integrated areas and background values for individual bands are obtained using the ImageQuant program and manipulated using the Excel (Microsoft) and Kaleidagraph (Abelbeck Software) programs. First, the background value corresponding to the area of each band is subtracted. Then the amounts of DNA bound, DNA_b, and free, DNA_f, are determined, and the fraction bound $y = DNA_b/(DNA_f + DNA_b)$ is calculated and plotted for each protein concentration. The shape of the binding curve is assessed to ensure that a plateau indicative of saturation is observed.

The protein–DNA complexes dissociate somewhat during electrophoresis. In cases of weak binding or where 100% binding is not observed in the gel, but a plateau in the binding curve is reached, data are normalized to give a maximum saturation of approximately 95%. This correction is equivalent to determining and applying the rate of dissociation of the HMG-D–DNA complex within the gel matrix as described previously.[48] Where k_{obs} is the dissociation rate of the complexes in the gel and t is the time of electrophoresis, the correction is given by $y_{corr} = y_{obs}(1 - e^{-tk_{obs}})$.[48] Because the amount of complex dissociation is not constant from gel to gel, the correction is applied to each gel based on the plateau value of y_{obs}, which is typically between 60 and 90%. In cases where no plateau is reached initially, the experiments are repeated using higher protein concentrations until a plateau is observed. For complexes with an affinity as weak as 7 μM, it was possible to obtain saturation with about 50% of the DNA shifted at the binding curve plateau. For the kinetics experiments, a rate of gel dissociation curve can be generated and applied.[48]

Typically, a minimum of three data sets are averaged and fitted using nonlinear least-squares refinement methods implemented in Kaleidagraph with the binding equations described earlier. The curve-fitting errors for single-site binding isotherms are ±15%. The errors for cooperative binding analysis are not estimated as easily, but are in the same range of ±15–20%.

Use of Chromatin Western Blot to Examine Interaction of HMG-D with Chromatin

HMG-D is the most abundant HMG-box protein expressed during *D. melanogaster* embryogenesis.[6] Although numerous functions have been ascribed to HMG-box proteins, little is known about the function or interactions of HMG-D or other HMG-box proteins with DNA and chromatin *in vivo*. HMG-D may associate with nucleosomes in competition with H1 during syncytial nuclear division or may bind to linker DNA in the precellular embryo to facilitate the formation of easily replicated chromatin structures.[7] To address these questions, we used a method that would directly reveal the association of the protein with nucleosomes.

It is possible to separate nucleosome oligomers generated by micrococcal nuclease digestion of chromatin by native gel electrophoresis.[55-57] The length of the DNA is the primary determinant of migratory distance. Mononucleosomes migrate faster than dinucleosomes, etc. However, for a given length of DNA, the molecular weight of the complex is altered by the quantity of protein bound. This is particularly evident for mononucleosomes, which appear as several species, each containing a different complement of proteins and DNA lengths associated with the histone octamer.[55-57] If HMG-D associates with nucleosomes in *Drosophila* chromatin, then its binding may be detected using native gel electrophoresis because a nucleosome bound to HMG-D may migrate differently than nucleosomes bound to histone H1 or other proteins.

At the developmental stage of the embryos used in this example (6 to 12 hr), HMG-D is found at a basal level of about 1 molecule per 50 nucleosomes while H1 has already reached a protein to nucleosome ratio of 1:1. The elegant techniques used to analyze the association of major chromosomal protein–DNA complexes using electrophoretic techniques would fail to show HMG-D–nucleosome complexes because they would be obscured by the more abundant H1-containing chromatosome complexes. If the method is altered, as suggested by Kornberg,[58] to include a Western blot step, then the association of even minor chromosomal proteins with nucleosomes can be detected.

We have developed and applied the chromatin Western blot technique to identify which species of nucleosomes contain HMG-D, using conditions that are as close to the *in vivo* conditions as currently possible. The scheme in Fig. 6 illustrates the procedure for performing the chromatin Western blot on native chromatin. The nuclei are isolated quickly from *Drosophila* embryos of particular ages (6 to 12 hr illustrated here) and are lightly digested with micrococcal nuclease (MNase). The chromatin is extracted directly from the digested nuclei and electrophoresed on a native polyacrylamide gel. The gels are stained, blotted, and probed with various antibodies as described later.

To test the method, antibodies against histone H1 and HMG-D are used to probe Western blots made from chromatin containing both proteins. Figure 7 illustrates results for this procedure as applied to HMG-D and histone H1. The blot for histone H1 clearly shows that histone H1 associates with nucleosome particles. The blot for HMG-D shows that HMG-D also colocalizes with the mononucleosome region of the gel, but the complexes

[55] R. Todd and W. T. Garrard, *J. Biol. Chem.* **252,** 4729 (1977).
[56] A. J. Varshavsky, V. V. Bakayev, and G. P. Georgiev, *Nucleic Acids Res.* **3,** 477 (1976).
[57] S. Huang and W. T. Garrard, *Methods Enzymol.* **170,** 117 (1989).
[58] R. Kornberg, J. W. LaPointe, and Y. Lorch, *Methods Enzymol.* **170,** 3 (1989).

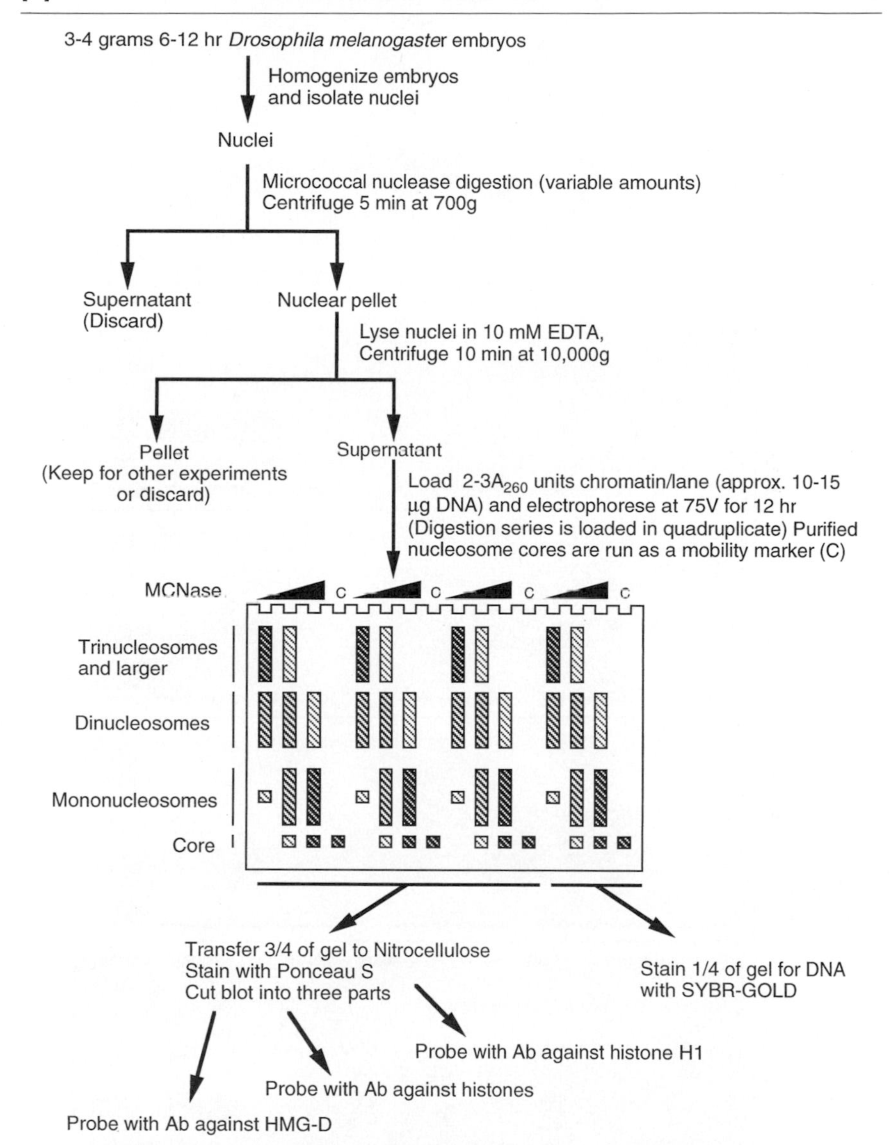

FIG. 6. Scheme for chromatin Western blot analysis of native chromatin samples.

it forms migrate slightly faster and slightly slower than those formed with histone H1. Two-dimensional DNA analysis (not shown) shows that the DNA length in the HMG-D-binding regions was intermediate between nucleosome cores and chromatosomes. This approach is particularly useful

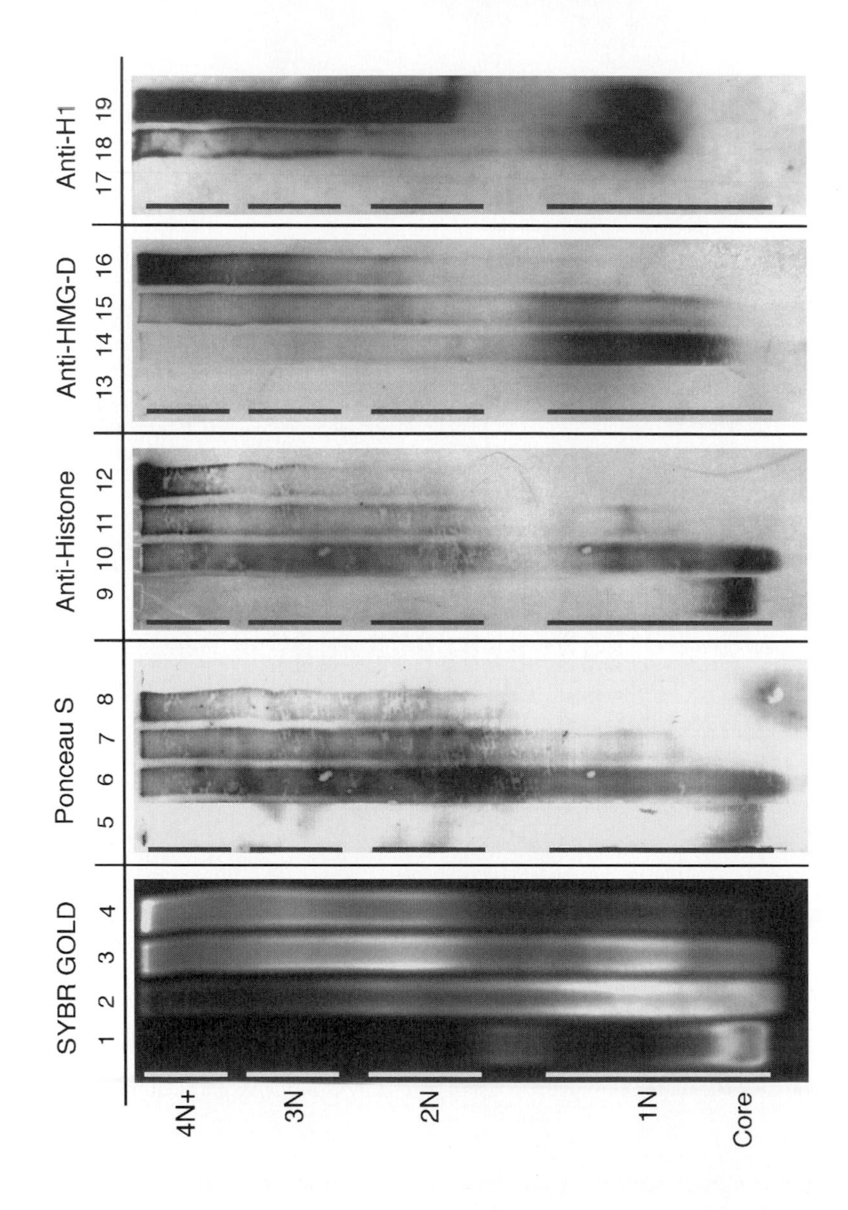
Anti-H1
17 18 19
Anti-HMG-D
13 14 15 16
Anti-Histone
9 10 11 12
Ponceau S
5 6 7 8
SYBR GOLD
1 2 3 4
4N+
3N
2N
1N
Core

for a protein such as HMG-D, which is detected more easily by Western blotting than by gel staining because HMG-D stains very poorly with silver. In conjunction with other *in vitro* control experiments, this approach is powerful because it can reveal associations of even minor chromosomal proteins in a native chromatin environment without prior disruption and reconstitution of the chromatin.[8]

Chromatin Western Blot Procedures

Collection of Drosophila Embryos

Embryos are collected using standard methods[59] from population cages containing Oregon R wild-type flies. Fresh grape plates smeared with yeast paste are placed into population cages for 6 hr, removed, and incubated for 6 hr at 25°. Embryos are washed off the plate onto a Nitex filter, washed with water to remove yeast, weighed, frozen in liquid nitrogen, and stored at −80° until needed for use.

[59] C. D. Shaffer, J. M. Wuller, and S. C. R. Elgin, *Methods Cell Biol.* **44,** 99 (1994).

FIG. 7. Colocalization of histone H1 and HMG-D with nucleosomes using chromatin Western blot analysis of 6- to 12-hr *Drosophila* embryo chromatin. All samples were digested with micrococcal nuclease and electrophoresed, as described in the methods section. The separated gel sections were subjected to different treatments as follow. Lanes 1–4: Nucleoprotein complexes stained with 1× SYBR-GOLD and visualized under UV. Lanes 5–8: Ponceau S-stained nitrocellulose with electrotransferred nucleoprotein complexes. Positions of nucleosome migration were marked, and the stained blot was scanned for visual record. Positions of nucleosome species, mononucleosomes, dinucleosomes, and so on are denoted 1N, 2N, and so on. Lanes 9–12: Western blot using antibody specific for histones. After transfer to nitrocellulose, the blot was probed with 5 μg/ml antihistone (mouse monoclonal) antibody (Boehringer Mannheim). Lanes 13–16: Western blot for HMG-D. After transfer to nitrocellulose, the blot was probed with a 1:2000 dilution of anti-HMG-D polyclonal (rabbit polyclonal; 1:2000 dilution) antibody (gift of Dr. S. Ner). Lanes 17–19: Western blot using histone H1 antibody. Samples were transferred to nitrocellulose and probed with a 1:10,000 dilution of anti-*Drosophila* histone H1 polyclonal (rabbit) antibody (gift of Dr. J. Kadonaga). Lanes 1, 5, 9, 13, and 17 contain approximately 5 μg purified chicken erythrocyte nucleosome cores (CEN) as migration standards. Each lane in the sets of lanes 2–4, 6–8, and 10–12 contains approximately 8 μg DNA as soluble chromatin digested with 0.1, 0.03, and 0.01 units/A_{260} of micrococcal nuclease, respectively. Lanes 18 and 19 (from a different experiment) contain approximately 12 μg of digested chromatin digested with 0.1 and 0.03 units/A_{260} of micrococcal nuclease, respectively.

Nuclear Isolation

Nuclei are isolated on ice using a modification of a protocol published by Shaffer *et al.*[60] Three to four grams of frozen 6- to 12-hr *D. melanogaster* embryos are ground in liquid nitrogen with a ceramic mortar and pestle. The resulting embryo powder is saturated in 3 ml/g of MLB + NP-40 (magnesium lysis buffer + Nonidet P-40: 15 m*M* HEPES, pH 7.3, 60 m*M* KC1, 15 m*M* NaCl, 2 m*M* $MgCl_2$, 0.5% Nonidet P-40, 5 m*M* DTT, 10 m*M* PMSF, 2 μg/ml leupeptin, 2 μg/ml aprotinin, 1 μg/ml pepstatin A). This slurry is homogenized with five strokes of a ground glass homogenizer. The homogenate is transferred to a Dounce homogenizer, disrupted with 10 strokes of pestle B, filtered through two layers of Miracloth (Calbiochem), rinsed with 2 ml MLB + NP-40, overlaid on 1/4 volume of MLBS (MLB with 0.3 *M* sucrose; sucrose cushion), and centrifuged at 1500*g* for 5 min. The nuclear pellet is resuspended in 3 ml MLB + NP-40, homogenized with five strokes of pestle A in the Dounce, and then centrifuged through a sucrose cushion as before. The resulting nuclear pellet is resuspended in 3 ml MLB, transferred to a Dounce homogenizer, and disrupted with five strokes of pestle B. Nuclei are then centrifuged at 700*g* and used in subsequent experiments. If the supernatant remains cloudy, the pellet can be resuspended in MLB and centrifuged through the sucrose cushion again.

Micrococcal Nuclease Digestion of Drosophila Chromatin

Nuclei from *Drosophila* embryos are suspended in 1 ml micrococcal nuclease digestion buffer (15 m*M* HEPES, pH 7.3, 60 m*M* KCl, 15 m*M* NaCl, 1 m*M* $CaCl_2$, 0.25 *M* sucrose).[60] The amount of micrococcal nuclease required for optimal digestion is calibrated by first diluting 0.25–1 μl of nuclear suspension to 100 μl in 0.1 *M* NaOH and then measuring the absorbance at 260 nm. Correction for dilution yielded the absorbance of the solution in units of A_{260}/ml. After the A_{260} determination, the nuclear suspension is divided into four 250-μl aliquots and equilibrated to 15°, and MNase is added according to Table II. For example, if the A_{260} of a 0.25-μl aliquot of nuclei diluted 400× is 0.59, the undiluted nuclei would have an absorbance of 236 A_{260} units/ml; 250 μl of nuclei would thus contain 59 A_{260} units. An intermediate level of digestion can be obtained using 0.03 units MNase/A_{260}. The addition of 1.8 units of MNase would achieve the intended extent of digestion.

Nuclei are then digested for 10 min at 15° with intermittent mixing by gentle agitation of the reaction tubes. Reactions are quenched by the addition of 5 μl of 0.5 *M* EDTA (final concentration of 10 m*M*) followed

[60] C. D. Shaffer, J. M. Wuller, and S. C. R. Elgin, *Methods Cell Biol.* **44,** 185 (1994).

TABLE II
CALIBRATION OF MNase DIGESTION CONDITIONS

Reaction	A_{260}/250 μl	Units MNase/A_{260}	μl MNase/A_{260} (0.1 U/μl)	Total MNase (μl)
1	59	0.01	0.1	5.9
2	59	0.03	0.3	17.7
3	59	0.1	1.0	59.0
4	59	0.3	3.0	177.0

immediately by the chilling of reaction tubes on ice. After the reaction, nuclei are centrifuged at 700*g* for 5 min at 4°. After the supernatant is discarded, nuclei are lysed by rapidly resuspending the pellet in 50 μl 10 m*M* Tris, 10 m*M* EDTA (pH 8). Nuclear debris is removed by centrifuging at 10,000*g* for 10 min at 4°. The supernatant is saved and 1 μl of each reaction is diluted 100× in 0.1 *M* NaOH, and the A_{260} is measured to determine the concentration of chromatin in micrograms per milliliter of DNA. Increasing the lysis volume to 100 μl yields equivalent concentrations of soluble chromatin in the supernatant.

Electrophoresis of Digested Chromatin

Electrophoresis is carried out on a 1 mm × 20 cm × 20 cm 4% (w/v, 19:1) polyacrylamide gel in 0.33× TBE. Only one of the glass plates is siliconized prior to gel casting to ensure adhesion to the second plate. The gel is prerun at 125 V at room temperature until constant current is reached and is then transferred to a 4° cold room for equilibration prior to sample loading. Approximately 5–8 μg DNA as chromatin is loaded per lane. Samples are prepared by adding glycerol to a final concentration of 5% and are loaded as rapidly as possible. Approximately 5 μg of purified chicken erythrocyte nucleosome cores, purified according to Lutter,[61] is loaded as a mobility standard, and marker dyes (Promega) are loaded in the outer lanes, flanking the digestion lanes. Gels are electrophoresed at 75 V until the xylene cyanol marker dye migrates approximately three-fourths of the length of the gel (approximately 12 hr). Provided that the gels are made from the same mixture and electrophoresed at the same time on the same power supply, identical mobility of nucleosomes is observed.

Immunoblotting of Chromatin Gels

The experiment can be performed on multiple gels so that one set of lanes can be stained for DNA. A second set can be blotted and stained

[61] L. C. Lutter, *J. Mol. Biol.* **124,** 391 (1978).

for protein using Ponceau S and then probed with an antibody; the remaining sets of lanes can be blotted to nitrocellulose and probed with various antibodies. Used here are antihistone, anti-HMG-D, and anti-H1 antibodies. After the gels are removed from the electrophoresis apparatus, the boundaries between each set of reactions are traced on the outside of both glass plates. Cut 0.45 μm supported nitrocellulose (MSI) to fit all but one set of reactions and soak in transfer buffer (25 m*M* Tris, 200 m*M* glycine, 0.01% SDS, 20% (v/v) methanol.[62] The gel plates are separated, leaving the gel adhered to the unsiliconized plate, providing a solid surface with which to manipulate the gel without stretching to tearing it. One set of reactions is separated from the others with a razor blade. This gel piece is stained for DNA using 1× SYBR gold stain (Molecular Probes, Eugene, OR) according to the manufacturer's instructions and visualized by UV transillumination.

The remainder of the gel is transferred to nitrocellulose. Whatman (Clifton, NJ) filter paper, soaked in transfer buffer, is placed on top of the part of the gel to be transferred until three layers of buffer-soaked filter paper cover the gel. To remove bubbles, a glass rod or Pasteur pipette is rolled over the gel as each new layer is added. The layers of gel and filter paper are lifted carefully from the glass plate and turned over. The uncovered face of the gel is soaked with additional buffer to ensure equilibration, preventing the gel from adhering to the nitrocellulose after transfer.[62] The precut, preequilibrated nitrocellulose is placed over the gel and the bubbles rolled out. Three layers of buffer-soaked Whatman filter paper are placed over the nitrocellulose as before. The gel sandwich is assembled in a semidry transfer apparatus with the gel on the cathode side and the nitrocellulose on the anode side according to manufacturer's instructions. The transfer is performed at 400 mA for 1 hr at 4°. After transfer, the sandwich is disassembled, taking care to separate the gel from the nitrocellulose. This process is facilitated by streaming transfer buffer between the gel and nitrocellulose with a Pasteur pipette while plying the layers apart.

The blotted nitrocellulose is rinsed in water to remove SDS and then covered in Ponceau S solution (0.1% Ponceau S in 5% acetic acid, w/v) and incubated approximately 10 min with minor agitation. The blot is destained in water. The position at which the chicken erythrocyte nucleosome cores migrated is marked on the blot with a lead pencil. The remaining experi-

[62] G. Jacobsen, *in* "Protein Blotting: A Practical Approach" (D. Rickwood and B. D. Hames, eds.), pp. 53–64. Oxford Univ. Press, New York, 1994.

[63] D. Landsman and M. Bustin, *Bioessays* **15,** 539 (1993).

[64] M. E. Bianchi, M. Beltrame, and G. Paonessa, *Science* **243,** 1056 (1989).

[65] U. Ohndorf, J. P. Whitehead, N. L. Raju, and S. J. Lippard, *Biochemistry* **36,** 14807 (1997).

ments are separated and blocked separately in 100 ml TBS containing 10% nonfat dried milk. The blots are probed according to standard techniques[62] using the appropriate HRP-conjugated secondary antibody under the same incubation conditions. Proteins are detected by chemiluminescence (Super Signal Substrate, Pierce, Rockford, IL).

Acknowledgments

We are grateful to Drs. Scot Wolfe and Greg Verdine for samples of disulfide cross-linked DNA and to Drs. James Kadonaga and Sarb Ner for gifts of antibodies. We appreciate the assistance of Drs. Lori Wallrath, Sarah Elgin, and Chris Doe and his laboratory members in the implementation of the chromatin Western blot experiments. We also thank Frank Murphy and Rebecca Bachmann for contributions and helpful comments on the manuscript and the contributions of past laboratory members, especially Heidi Hefner and Per Jambeck. This work was supported by the Illinois Division of the American Cancer Society, by American Heart Association grants to M.E.A.C., and by Molecular Biophysics NIH Training Grant support of L.K.D. and Cell and Molecular Biology NIH Training Grant support of A.J.K.

[8] Reconstitution of High Mobility Group 14/17 Proteins into Nucleosomes and Chromatin

By YURI V. POSTNIKOV and MICHAEL BUSTIN

High mobility group (HMG) proteins are a ubiquitous and heterogeneous class of nonhistones, which serve as "architectural elements" that modify the structure of DNA and chromatin, thereby facilitating a variety of DNA-related activities in the nucleus of the cell.[1] The HMG14/17 subgroup is the only class of nuclear proteins known to bind specifically to the 146-bp core particle. These proteins bind to the nucleosome core cooperatively and form homodimeric complexes containing either two molecules of HMG14 or two molecules of HMG17.[2] Both HMG14 and HMG17 contact the nucleosomal DNA 25 bp from the end of the core particle and in the two major grooves flanking the nucleosomal dyad axis.[3] Additional specific contacts are made with the amino termini of the core histones.[4] The proteins stimulate various DNA-dependent activities such as transcrip-

[1] M. Bustin and R. Reeves, *in* "Progress in Nucleic Acid Research and Molecular Biology" (W. E. Cohn and K. Moldave, eds.), pp. 35–100. Academic Press, San Diego, 1996.
[2] Y. V. Postnikov, L. Trieschmann, A. Rickers, and M. Bustin, *J. Mol. Biol.* **252,** 423 (1995)
[3] P. J. Alfonso, M. P. Crippa, J. J. Hayes, and M. Bustin, *J. Mol. Biol.* **236,** 189 (1994).
[4] M. P. Crippa, P. J. Alfonso, and M. Bustin, *J. Mol. Biol.* **228,** 442 (1992).

0076-6879/99 $30.00

tion[5–10] and replication,[11] but only in the context of chromatin. Enhancement of the DNA-dependent activities is associated with a decompaction, or unfolding, of the higher order chromatin structure.[7,10] These findings indicate that HMG14/17 proteins are chromatin-specific "architectural elements" that stimulate DNA-dependent activities by unfolding the higher order chromatin structure and facilitating access to the underlying DNA sequence.

This article describes various approaches useful for studies on the structure and function of HMG14/17 proteins in chromatin. These approaches include studies on the binding of HMG14/17 to chromatin subunits, on their reconstitution into *in vitro* assembled chromatin, and on their *in situ* organization in cellular chromatin. The methods described may be suitable for studies on other nonhistone proteins.

Reconstitution of HMG14/17 into Nucleosome Cores

Specific interactions between HMG14/17 and nucleosome core particles can be detected by mobility shift assays (Fig. 1), by hydroxyl radical footprinting (Fig. 2), and by two-dimensional gel analysis (Fig. 3). Two additional methods used for detecting HMG–nucleosome interactions and HMG-induced nucleosome pertubations are thermal denaturation and DNase I footprinting.[4]

The thermal denaturation assay is based on the observation that the binding of HMG14/HMG17 stabilizes the structure of the nucleosome core.[12] In 0.1× TBE buffer (8.9 m*M* Tris, 8.9 m*M* borate, 0.2 m*M* EDTA, pH 8.4), the melting temperature profile of a core exhibits two transitions. The first transition occurs at 62° and corresponds to the melting of the DNA ends at the entry/exit of the core; the second transition around 76° corresponds to the melting of the rest of the DNA. In the HMG–nucleosome complex, both of these structural transitions are shifted to higher temperatures. The shift in the first point is more dramatic than that of the second, reflecting the stabilization of the DNA at the ends of the nucleosome core by these HMG proteins.[4]

[5] M. P. Crippa, L. Trieschmann, P. J. Alfonso, A. P. Wolffe, and M. Bustin, *EMBO J.* **12,** 3855 (1993)

[6] H. F. Ding, S. Rimsky, S. C. Batson, M. Bustin and U. Hansen, *Science* **265,** 797 (1994).

[7] H. F. Ding, M. Bustin and U. Hansen, *Mol. Cell Biol.* **17,** 5843 (1997).

[8] S. M. Paranjape, A. Krumm, and J. T. Kadonaga, *Genes Dev.* **9,** 1978 (1995).

[9] D. J. Tremethick and L. Hyman, *J. Biol. Chem.* **271,** 12009 (1996).

[10] L. Trieschmann, P. J. Alfonso, M. P. Crippa, A. P. Wolffe, and M. Bustin, *EMBO J.* **14,** 1478 (1995).

[11] B. Vestner, M. Bustin, and C. Gruss, *J. Biol. Chem.* **273,** 9409 (1998).

[12] A. E. Paton, S. E. Wilkinson, and D. E. Olins, *J. Biol. Chem.* **258,** 13221 (1983).

High-resolution DNA sequencing gels of a DNase I digest of ^{32}P-end-labeled nucleosome cores have been used to identify a major histone: DNA contacts in the cores. The binding of HMG14/17 proteins to the cores modifies the DNase I-generated histone footprint. In the HMG–core complex, the DNA is more resistant to digestion by DNase I, especially 11 and 22 bp from the ends of the core particle DNA. These results, together with thermal denaturation and protein cross-linking studies, indicate that one of the major sites of interaction between the nucleosomal DNA and these HMGs is located about 25 bp from the end of the nucleosomal DNA.

The following three assays are the most commonly used and provide the most useful information regarding the interactions of HMG14/17 proteins with chromatin subunits.

Gel Mobility Shift Assays

In a solution containing equimolar amounts of core particles and DNA, HMG14/17 proteins form specific complexes with nucleosome cores and nonspecific complexes with DNA. The interaction of the proteins with core particles is ionic strength dependent. At ionic strength close to physiological, the proteins bind to the cores cooperatively, producing complexes containing two molecules of HMG per particle (Fig. 1A). At low ionic strength

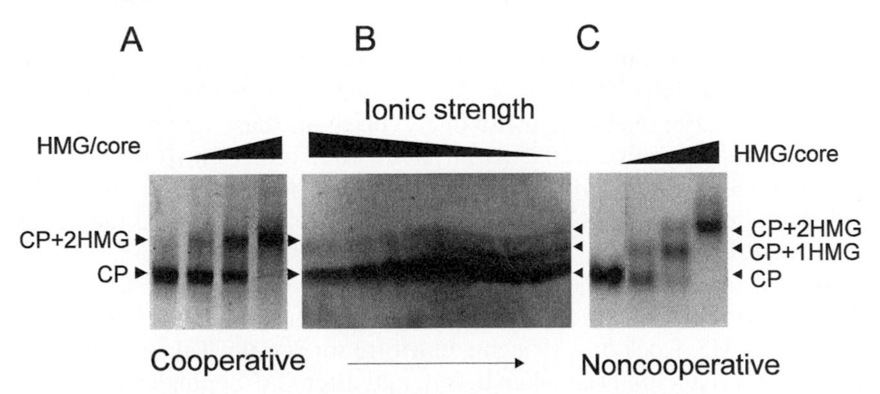

FIG. 1. Mobility shift assay. (A) Cooperative interactions. Core particles (CP) are incubated in 2× TBE with HMG17 protein, and the mixture is separated on 5% native polyacrylamide gels in 2× TBE. (B) A salt-dependent switch between cooperative and noncooperative modes of binding. A 5% native polyacrylamide gel is cast with a transverse linear salt gradient from 0.2× TBE on the right to 2.5× TBE on the left. HMG–nucleosome core complexes in 0.2× TBE are layered on the top. The CP–HMG complexes are formed and equilibrate during the run. Note that the middle band, corresponding to a complex of a nucleosome core with one molecule of HMG17, disappears approximately in the midpoint of the gradient. (C) Noncooperative interactions in 0.2× TBE. The position of the DNA in the gels is visualized with ethidium bromide. Arrows point to the migration of the various HMG–nucleosome core complexes. Under cooperative conditions, CP + 1HMG complexes are not detected.

the proteins bind noncooperatively, producing complexes containing either one or two molecules of HMG per particle (Fig. 1C). The switch from one binding mode to another occurs at an ionic strength of about 1.2× TBE (Fig. 1B). Scanned images of the ethidium bromide-stained gels are suitable for determining the binding constants of HMG14/17 proteins to the core particles. The binding constant for the interaction between two molecules of HMG14/17 and nucleosome cores under cooperative conditions (about $0.75 \times 10^7 M^{-1}$) is 100 times weaker than the binding constant under noncooperative conditions.

Protocol for Mobility Shift Assay for the Binding of HMG14/17 Proteins to Nucleosome Cores

Core Particle Preparation. The nucleosomal core particles were prepared from chicken erythrocyte nuclei according to Ausio *et al.*[13] with minor modifications. Briefly, wash the chicken erythrocytes three times with PBS (20 m*M* sodium phosphate buffer, 140 m*M* NaCl, pH 7.4) at 4°. Lyse the cells with 10 volumes of HBSST [340 m*M* sucrose, 15 m*M* Tris–HCl, 15 m*M* NaCl, 60 m*M* KCl, 10 m*M* dithiothreitol (DTT), 0.5 m*M* spermine, 0.15 m*M* spermidine, 0.1 m*M* phenylmethylsulfonyl fluoride (PMSF), 1% Triton X-100, pH 7.5) at 4° for 20–30 min. Wash the nuclei (spin/resuspend) with HBSS (same buffer without Triton). Store at −20° at about 2 mg/ml.

Wash 4 ml of the nuclei preparation with micrococcal nuclease digestion buffer (10 m*M* Tris–HCl, pH 7.5, 25 m*M* NaCl, 1 m*M* $CaCl_2$), resuspend in the same buffer, and digest with micrococcal nuclease (Sigma, St. Louis, MO; final concentration: 250 units/ml) at 37° for 5 min. Centrifuge at 10,000 rpm in a SS-34 rotor and save the pellet. Resuspend in 0.25 m*M* EDTA and incubate overnight. Centrifuge as described earlier, discard the pellet, and measure OD_{260}.

Add 0.1 volume 5 *M* NaCl to a final concentration of 0.45 *M*. Centrifuge and discard the pellet. Add CM-Sephadex suspension (in 0.5 *M* NaCl) to remove H1/H5. Rock on a swinging platform for 30 min. Take supernatant (nonbound). Run material taken before and after CM-Sephadex treatment on a Laemmli 15% polyacrylamide minigel. If there are no traces of linker histones in the supernatant, proceed. Otherwise, repeat the treatment with CM-Sephadex.

Spin-dialyse the supernatant against micrococcal nuclease digestion buffer using Centriprep concentrators (Amicon, Danvers, MA). Redigest with micrococcal nuclease (100 units/ml) for about 4 min at room temperature. Down-scaled time course analytical digestion is recommended to de-

[13] J. Ausio, F. Dong and K. E. van Holde, *J. Mol. Biol.* **206,** 451 (1989).

termine optimal digestion time for the best yield of core particles. Add 0.5 *M* EDTA to a final concentration 10 m*M*. Keep on ice.

Load the digest (1 ml per 38.5-ml tube) onto 5–20% sucrose linear gradient in nucleosome solution (10 m*M* Tris–HCl, pH 7.5, 10 m*M* NaCl, 1 m*M* EDTA). Centrifuge at 28,000*g* for 20 hr at 4° in a Beckman SW28 rotor. Fractionate by collecting 0.5- to 1-ml fractions. Mix 25 μl from each fraction with 6 μl of freshly prepared loading dye (5% SDS, 100 μg/ml proteinase K, 5× TAE, 10% Ficoll, 1 μg/ml bromphenol blue, 10 μg/ml xylene cyanol). Analyze fractions using a 1.7% agarose gel in TAE. Pool fractions containing mononucleosomes and spin-dialyze in Filtron Macrosep 3 (or Centriprep 3) concentrators. Store at 1 μg/μl at 4°.

HMG–Nucleosome Complexes.[2,14] Routinely, 5 pmol of core particles is incubated with 5–15 pmol of HMG proteins in a volume of 10 μl at 4° for 10 min. For cooperative binding assays, the buffer is 2× TBE; for noncooperative binding assays the buffer is 0.2× TBE. At the end of the incubation, 2 μl of 20% Ficoll-400 (dissolved in H_2O) is dispensed to each tube, and the mixture is loaded onto polyacrylamide gels.

Electrophoretic Separation. 5% native polyacrylamide (acrylamide : bisacrylamide, 19 : 1, v/v) minigel (8 × 6 × 0.075 cm) containing either 2× or 0.2× TBE is prerun for 1 hr at 4°. To monitor the electrophoresis, apply bromphenol blue and xylene cyanol dyes into empty lanes. After loading the samples, the gels are run at a constant current of 10 mA, at 4°, until xylene cyanol reaches two thirds of the gel length. Finally, the bands in the gels are visualized by staining with ethidium bromide.

Hydroxyl Radical Footprinting

Hydroxyl radical cleavage can be used to map contacts between proteins and DNA with the resolution of a single nucleotide.[15] The use of this technique for mapping the interactions between nucleosomes and various proteins, such as HMG14/17, is complicated by the prominent footprint of the histones on the nucleosomal DNA. In the HMG–core complex the histone footprint is modified. The HMG14/17 footprint can be detected on the background of the histone footprint using (1) highly purified nucleosome cores or chromatosomes in which the position of the histone octamer relative to the DNA ends is invariant, (2) high-resolution sequencing gels, and (3) equipment for high-resolution scanning of the sequencing gels. We have purified the chromatin subunits on preparative agarose gels twice. First, they are purified after the sucrose gradient step and then again after the HMG : nucleosome complex is reacted with hydroxyl radicals (see

[14] Y. V. Postnikov, D. A. Lehn, R. C. Robinson, F. K. Friedman, J. Shiloach, and M. Bustin, *Nucleic Acid. Res.* **22,** 4520 (1994).

[15] J. J. Hayes, T. D. Tullius, and A. P. Wolffe, *Proc. Natl. Acad. Sci. U.S.A.* **87,** 7405 (1990).

later). The cleavage pattern is displayed using high-resolution DNA sequencing gels made from several concentrations of acrylamide solution and is analyzed with a Molecular Dynamics computing densitometer with ImageQuant and Microsoft Excel software. Figure 2 depicts densitometric scans of the hydroxyl radical cleavage patterns of nucleosome cores complexed with HMG17 (Fig. 2A) and chromatosomes complexed with HMG14 (Fig. 2B). The scans are compared to those obtained, with the same type of chromatin subunit, in the absence of HMG14/17 proteins.

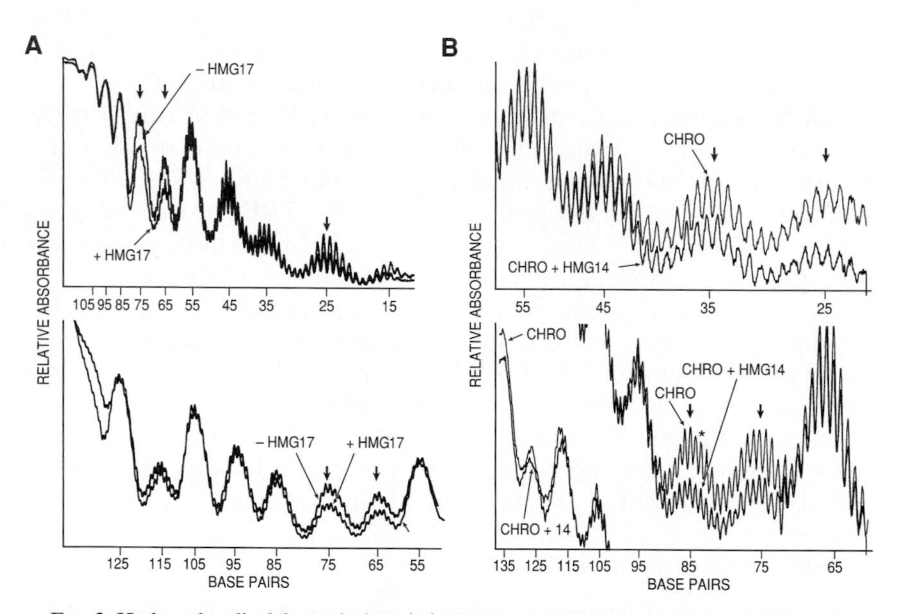

FIG. 2. Hydroxyl radical footprinting. (A). Densitometric analysis of the hydroxyl radical cleavage patterns of nucleosome cores and core–HMG17 complexes. An autoradiograph of either a 10% (top) or a 6% (bottom) polyacrylamide sequencing gel (7 *M* urea) is scanned with a Molecular Dynamics computing densitometer. The upper scan corresponds to the hydroxyl radical cleavage pattern of the core alone and the lower scan corresponds to the cleavage pattern of the core–HMG17 complex. The abscissa denotes the distance (in bp) from the ^{32}P-labeled end of the nucleosome cores. The 10-bp repeat pattern represents the footprint of the core histones in the nucleosome cores. Note that the two scans overlap in most of the positions. Arrows point to the prominent protection conferred by HMG17 centered around positions 25, 65, and 75 bp. (B) Densitometric analysis of the hydroxyl radical cleavage patterns of H1/H5-depleted chromatosomes without (Chro) and with added HMG-14 (Chro + 14). An autoradiograph of an 8% polyacrylamide sequencing gel (7 *M* urea) is scanned with a Molecular Dynamics computing densitometer. The scans are divided into two panels for the best visualization of differences in protection against OH radicals. The abscissa denotes the distance (in bp) from the ^{32}P-labeled end of the chromatosome. The asterisk denotes the dyad axis. Note the prominent protection at the ends (up to about nucleotide 40) and in the major grooves flanking the dyad axis (from position 70 to 90). Adapted from Alfonso *et al.*[3]

Protocol[3]

CHROMATIN PARTICLE PURIFICATION AND LABELING. Core particles and chromatosomes depleted of linker histones are prepared using sucrose gradients as described[13] (see also section on mobility shift assays). The sucrose gradient fraction containing chromatin monomers is dialyzed against 10 m*M* Tris–HCl, 25 m*M* NaCl, 1 m*M* EDTA, pH 7.5, concentrated with Centricon 30 (Amicon), and applied to a 3-mm-thick, 4% (w/v) polyacrylamide gel (acrylamide : bisacrylamide, 20 : 0.69, w/w). The gel is run in TBE buffer at 50 V, 4° for 16 hr. Strips are then taken from the edges of the gel and stained with ethidium bromide to determine the position corresponding to nucleosome cores and chromatosomes. The chromatin subunits are electroeluted from the gel slice with a Strataelutor (Stratagene, La Jolla, CA) at 100 V, for 3 hr, at 4° in 0.5× TBE. The polarity is reversed for 20 sec prior to collecting the samples.

Use siliconized Eppendorf tubes. The particles are end labeled in a 20 μl volume with T4 polynucleotide kinase (Promega, Madison, WI) in 10 m*M* Tris–HCl, pH 8.0, 5 m*M* $MgCl_2$, 2 m*M* dithiothreitol (DTT) for 30 min at 30°. The reaction is stopped by adding 10 μl of 30 m*M* EDTA, pH 8, and diluted 1 : 100 with STE (10 m*M* Tris–HCl, pH 8, 25 m*M* NaCl, 1 m*M* EDTA).

HMG–NUCLEOSOME COMPLEXES. Labeled nucleosome cores or chromatosomes (0.2 μg) are incubated with a two- to fivefold molar excess of HMG14/17 protein (or 0.1% BSA as controls) for 15 min at 4°.

HYDROXYL RADICAL FOOTPRINTING.[3,15,16] It is important that all traces of glycerol, which may be present after the end-labeling reaction, be removed from the incubation mixture as glycerol is a hydroxyl radical scavenger. For a list of buffers compatible with the hydroxyl radical reaction, see Tullius *et al.*[17] For hydroxyl radical footprinting, 20 μl of 1 m*M* $Fe(EDTA)_2$, 20 μl of 10 m*M* sodium ascorbate, and 20 μl of 0.3% (v/v) H_2O_2 are added to 0.2 μg of core particle DNA (or HMG-containing complexes) in 140 μl STE and vortexed vigorously for 15 sec in a 1.5-ml Eppendorf tube. It is important to add the reagents to tube walls and not directly to the solution. The reagents mix and contact the DNA during the vortexing. After 2 min at room temperature, the reaction is stopped with 20 μl of 50% (v/v) glycerol. The reaction mixtures are then loaded onto 3-mm-thick, 4% (w/v) polyacrylamide (acrylamide : bisacrylamide, 20 : 0.69), 0.5× TBE-containing gels and run for 16 hr at 50 V (constant voltage) at 4°. The positions of the protein–DNA complexes are visualized by autoradiography of the wet

[16] J. J. Hayes, D. J. Clark, and A. P. Wolffe, *Proc. Natl. Acad. Sci. U.S.A.* **88,** 6829 (1991).

[17] T. D. Tullius, B. A. Dombrowski, M. E. A. Churchill, and L. Kam, *Methods Enzymol.* **155,** 537 (1987).

gels. Bands are sliced out and the protein–DNA complexes are eluted by crushing the slice and soaking for 16 hr in 0.5–1 ml gel elution buffer [20 m*M* Tris–HCl, pH 7.5, 1 m*M* EDTA, 0.5 *M* ammonium acetate, 0.1% (w/v) SDS]. The DNA is extracted once with phenol/chloroform (1:1, v/v) and precipitated with ethanol. Pellets are washed with 70% ethanol and dried, and the radioactivity is counted. Samples are suspended in TBE/95% formamide (1:1, v/v), heated at 90° for 2 min, and 8 μl containing 20,000 cpm is loaded onto acrylamide (acrylamide:bisacrylamide, 12:1) sequencing gels. The gels are run with a gradient buffer system.[18] Autoradiograms of the fixed gels are analyzed on a Molecular Dynamics computing densitometer with ImageQuant and Microsoft Excel 4.0 software. Minor differences in loading can be corrected by anchoring the scans at positions 55 and 85 bp from the end of the nucleosome core.

Using hydroxyl radicals, we demonstrated that the footprint of HMG14 on the surface of both the nucleosome core and the chromatosome is indistinguishable from the footprint of HMG17. In both types of chromatin subunits, bound HMG proteins protect the DNA from hydroxyl radicals in each of the two major grooves flanking the dyad axis and near the end of chromatin subunit (for nucleosome cores and chromatosomes, 25 and 25–35 bp, respectively).[3]

Immunofractionation of HMG14/17-Containing Nucleosome Cores and Analysis of Protein Content by Two-Dimensional Gel Electrophoresis

Specific antibodies can be used to isolate a subfraction of nucleosome cores associated with a specific protein, such as HMG14/17.[19] Because nucleosome–HMG14 complexes have the same electrophoretic mobility as nucleosome–HMG17 complexes, we use specific antibodies to fractionate the two types of complexes and two-dimensional gel analysis to identify the various proteins in the core–HMG complexes. In these experiments, equimolar amounts of HMG14 and HMG17 proteins are incubated with nucleosome core particles and the resulting complexes are treated with antibodies specific to one of the HMG proteins. These reaction mixtures contain three major components: free nucleosome core particles (CP), core particles complexed with HMG proteins (CP + HMG), and complexes of IgG molecules bound to the CP + HMG complex (CP + HMG + IgG). Electrophoresis on native polyacrylamide gels resolves the three components in the reaction mixture while electrophoresis in the second dimension,

[18] J. Y. Sheen and B. Seed, *Biotechniques* **6,** 942 (1988).

[19] M. Bustin, *Methods Enzymol.* **170,** 214 (1989).

in polyacrylamide gels containing sodium dodecyl sulfate, resolve the various proteins present in each complex[2] (Fig. 3).

Protocol[2]

CORE PARTICLE PREPARATION AND HMG PROTEINS. See previous section on mobility shift assays.

ANTIBODIES TO HMG14/17 PROTEINS. Affinity pure antibodies are obtained with immunoaffinity columns, which are prepared by immobilizing pure recombinant human HMG14 or HMG17 protein on 3M Emphaze Biosupport Medium AB1 (Pierce, Rockford, IL).[2] The antibody preparations are desalted and concentrated with Macrosep (Filtron Tech. Corp.) centrifugal concentrators (30,000 molecular weight cutoff) to approximately 1 mg/ml, aliquoted, and stored at −20°.

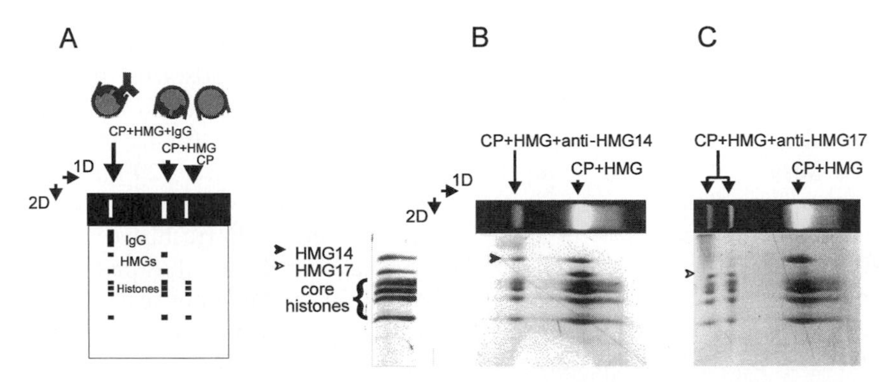

FIG. 3. Two-dimensional polyacrylamide gel analysis of HMG–nucleosome core complexes after antibody-mediated gel supershifts assays. (A) Scheme of the experiment. Core particles are incubated with an equimolar mixture of HMG14 and HMG17 to produce HMG–core particle complexes. Immunoaffinity-pure antibodies to either HMG14 or HMG17 are added to form HMG–core–IgG complexes. The molecular species are separated in the first dimension on native polyacrylamide gels (see Fig. 1A). In this dimension, three types of molecular species can be resolved: free core particles (CP), the CP–HMG complex, and the CP–HMG–IgG complex. Proteins present in each complex are resolved by electrophoresis in SDS-containing gels in the second dimension. (B and C) Experimental results demonstrating that under cooperative binding conditions, HMG14/17 bind to nucleosome cores to form complexes containing either two molecules of HMG14 or two molecules of HMG17. Note that the supershift produced with antibodies to HMG14 (B) contains approximately equimolar ratios of the four core histones and HMG-14 and that HMG17 is not present. The two supershifts produced with anti-HMG17 (C) contain only core histones and HMG17 and are devoid of HMG-14. (C) The two supershifts represent the binding of one or two molecules of IgG to the complex. The gel region containing IgG is not shown. In the first dimension the gel is stained with ethidium bromide. The protein composition in the second dimension is visualized by silver staining of SDS-containing 10–20% gradient polyacrylamide gels. Adapted from Postnikov *et al.*[2]

HMG–CORE COMPLEXES. See previous section on mobility shift assays. For antibody-mediated "supershift," 2–5 pmol of affinity-pure antibodies are added to the binding reaction (i.e., HMG–core complex) and the mixture is incubated for 15 min at 4°.

ELECTROPHORETIC SEPARATION. After electrophoresis in the first dimension (see previous section on mobility shift assays), the lanes of interest are excised and placed on top of the second-dimension gel to determine the protein composition of nucleosome–HMG complexes. The second-dimension running gels are a 10–20% linear gradient of acrylamide (acrylamide : bisacrylamide, 39 : 1, for both light and heavy solutions). A 1-cm-wide stacking gel (5% acrylamide; acrylamide : bisacrylamide, 39 : 1) is layered and polymerized on top of the running gel. The gel strips from the first dimension are embedded and polymerized into a second layer of the stacking gel that is identical to the first stacking gel layer. Throughout the gel the buffer is 0.4 *M* Tris–HCl, pH 8.8. SDS loading dye solution is layered on the top of the gel. The running buffer is 25 m*M* Tris, 190 m*M* glycine, pH 8.3, 0.1% SDS. After electrophoresis the bands are visualized by silver staining. The gel is washed twice for 15 min in deionized water, fixed for 30 min in freshly prepared 10% glutaraldehyde (made in H_2O), and washed at least five times with a large excess of water (20 min each wash). The gel is submerged in the ammoniacal silver solution for 20 min, washed with water for 10 min, and developed with freshly prepared formaldehyde/citrate solution (see Wray *et al.*[20] for these solutions). Commercial kits for silver staining are available, but we have not tested them.

Using this approach we demonstrated that at physiological ionic strength, the interaction between nucleosome cores HMG14/17 leads to the formation of complexes containing either two molecules of HMG14 or two molecules of HMG17. Nucleosome core complexes containing both types of HMG proteins are not detected. Based on additional analysis, we suggested that HMG14/17 proteins induce specific allosteric transitions in the chromatin subunit.[2]

Reconstitution of HMG14/17 Proteins into Chromatin

Studies with purified components in cell-free systems have contributed significantly to the understanding of the effect of chromatin structure on various DNA-dependent activities. The role of HMG14/17 proteins in these processes has been studied with chromatin, which is assembled *in vitro*, in extracts prepared from either *Xenopus* eggs[5,10] or *Drosophila* embryos,[8] and also with *in vivo*-assembled chromatin, with simian virus 40 (SV40)

[20] W. Wray, T. Boulikas, V. P. Wray, and R. Hancock, *Anal. Biochem.* **15,** 197 (1981).

minichromosomes isolated from CV-1 cells.[6,7] In these[10] experimental systems, HMG14/17 stimulated various DNA-dependent activities, such as transcription by both polymerase II[6,8] and III[5,10,21] and replication,[11] but only in the context of chromatin. These activities are not stimulated in "naked" deproteinized DNA. The HMG14/17-mediated simulation of these activities is associated with decompaction of the chromatin template.[7,10] The decompaction occurs without any noticeable change in the histone content or the nucleosomal spacing of the templates. These results suggest that the proteins act as modifiers of chromatin structure rather than polymerase-specific factors.

In considering the methodology for reconstituting HMG14/17 into chromatin, it is important to note that the effect of these proteins varies according to the method by which the chromatin is assembled. With the *in vitro*-assembled chromatin, HMG14/17 stimulate transcription and replication only when added during assembly. With the *in vivo*-assembled, transcriptionally competent SV40 chromatin, HMG14 decompacts the chromatin structure and enhances transcription even when added after assembly, provided that the minichromosomes are compacted by the linker histone H1. Thus, in the *in vivo*-assembled chromatin, HMG14 stimulates transcription by negating the repressive activity of histone H1.

Here we provide a protocol and mention significant methodological considerations for using chromatin assembled in *Xenopus* egg extracts to study the effect of HMG14/17 protein on transcription by polymerase III. It is best to consult the original references for the methodologies used to study the effect of HMG14/17 on replication from chromatin assembled in *Xenopus* egg extracts, on polymerase II transcription from chromatin assembled using *Drosophila* extracts, and on polymerase II transcription from *in vivo*-assembled SV40 minichromosomes.

Xenopus egg extracts contain all the ingredients necessary to assemble chromatin.[22–24] When the DNA added is single stranded, the nucleosome assembly is contingent on the synthesis of the complimentary DNA strand. Starting with single-stranded M13 plasmid templates, the chromatin assembly is completed in approximately 3 hr. Chromatin can also be assembled without DNA synthesis starting with double-stranded plasmids. Using double-stranded M13 plasmids, chromatin assembly is completed in approximately 6 hr. The exogenous addition of MgATP affects the quality of the

[21] N. Weigmann, L. Trieschmann, and M. Bustin, *DNA Cell Biol.* **16,** 1207 (1997).

[22] G. Almouzni, D. J. Clark, M. Mechali, and A. P. Wolffe, *Nucleic Acids Res.* **18,** 5767 (1990).

[23] G. Almouzni and A. P. Wolffe, *Exp. Cell Res.* **205,** 1 (1993).

[24] A. P. Wolffe and C. Schild, *Methods Cell Biol.* **36,** 541 (1991).

resulting template. For a full description of the use of *Xenopus* egg extracts to assemble chromatin, it is necessary to consult additional references.

The various protocols used to study the effect of HMG14/17 on chromatin reconstituted in *Xenopus* egg extracts are presented in Fig. 4. We have found great variability in the *Xenopus* egg extracts with respect to their capacity for supporting chromatin assembly, transcription, and replication. Therefore, each preparation has to be tested for these parameters. Chromatin assembly can be monitored by following the appearance of supercoiled DNA, preferentially by electrophoresis in chloroquine-containing gels, or by direct examination by electron microscopy. The nucleosomal periodicity of the template is assessed by following the kinetics of digestion with micrococcal nuclease. Additional structural information on the assembled chromatin can be obtained by analyzing the kinetics of digestion by various restriction enzymes. The assembled minichromosomes can be purified either by sucrose gradient centrifugation or by Sephacryl 300. The protein content of the minichromosomes is analyzed by polyacrylamide gel electrophoresis or by various immunochemical techniques.

Effect of HMG14/17 on Assembly of Chromatin Template

The effect of HMG14/17 on the rate of replication and chromatin assembly can be monitored by following the appearance of radioactively labeled supercoiled DNA by electrophoresis in polyacrylamide gels (Fig. 5A). With single-stranded plasmids the DNA is labeled by the addition of radioactive deoxynucleotides to the extract. Double-stranded, radioactively labeled DNA can be obtained from the chromatin template assembled in the extract.[21] Supercoiling analysis in gels containing chloroquine (Figs. 5B and 5C) can be used to estimate whether HMG14/17 proteins affect the rate of chromatin assembly and the number of nucleosomes in the minichromosomes.

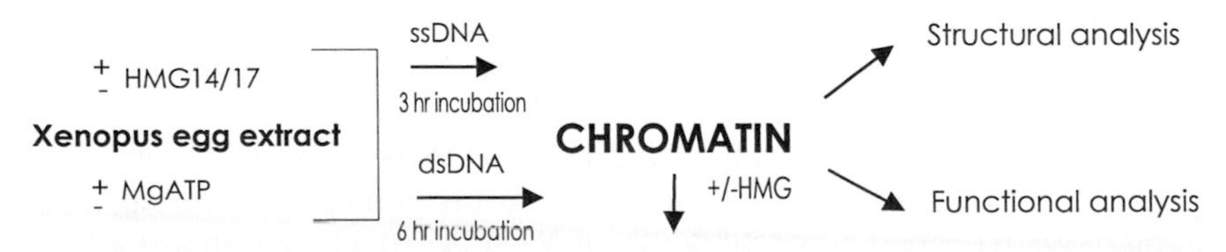

FIG. 4. Outline of the protocols used to reconstitute HMG14/17 into chromatin assembled in *Xenopus* egg extracts. Highly purified single-stranded or double-stranded M13 DNA containing a gene that is transcribed by polymerase III is added to a *Xenopus* egg extract that contains various amounts of exogenously added recombinant HMG14/17 protein. Depending on the type of experiment, the extracts also contain exogenously added MgATP and polymerase III transcription factors. In some experiments the HMG proteins are added to preassembled chromatin.

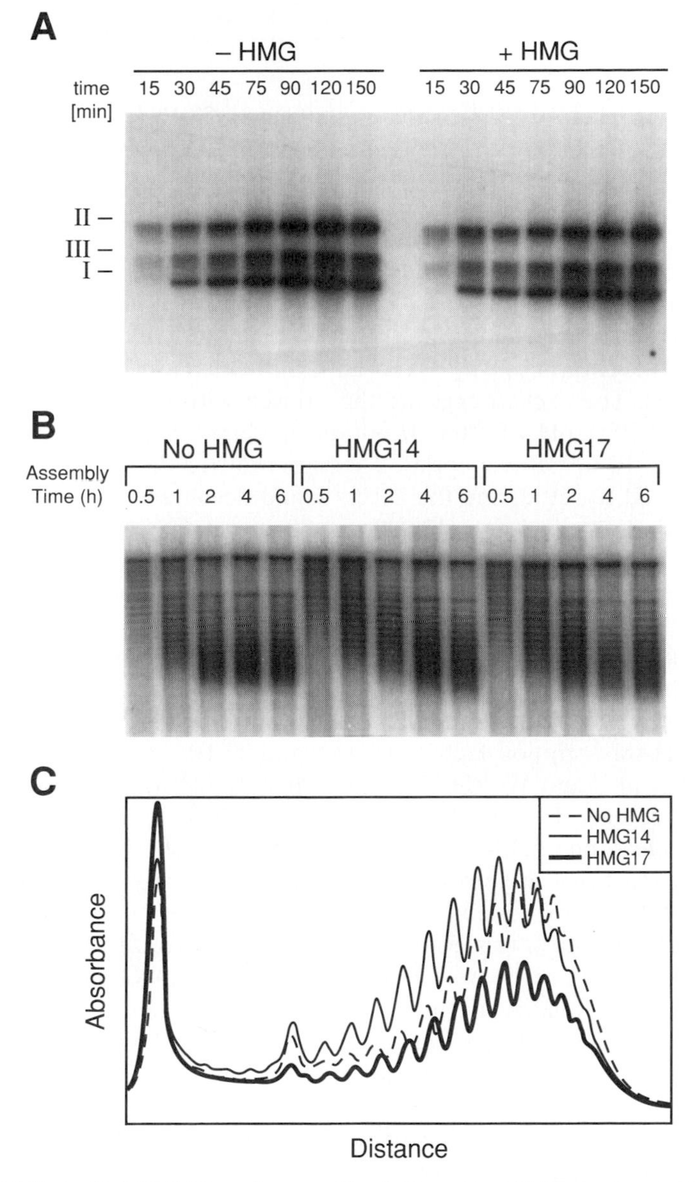

FIG. 5. Effect of HMG14/17 on the assembly of chromatin. (A) Single-stranded DNA and ^{32}P dCTP are added to *Xenopus* egg extracts in either the absence or the presence of exogenously added recombinant HMG17 protein. Samples are removed at the times indicated, and the purified radiolabeled DNA is fractionated on agarose gels. The autoradiograph of the gel depicts the mobility of the supercoiled (form I), linear (form III), and circular (form II) plasmid DNA. (B) DNA supercoiling analysis. Plasmids are assembled into chromatin for various times in either the absence or the presence of either HMG14 or HMG17. The resulting templates are deproteinized, and the DNA topoisomers are resolved on agarose gels containing chloroquine. (C) Densitometric scans of the fully assembled minichromosomes. Adapted from Weigmann *et al.*[21]

Protocol

Xenopus laevis Egg Extract. High-speed extracts from unfertilized *X. laevis* eggs are prepared according to established procedures.[23–25] Briefly, unfertilized eggs are obtained from the *X. laevis* frogs (Xenopus I, Ann Arbor, MI) by injection of human chorionic gonadotropin (Sigma). The eggs are collected in HSB buffer (100 m*M* NaCl, 2 m*M* KCl, 1 m*M* $MgSO_4$, 0.5 m*M* Na_2HPO_4, 2 m*M* $NaHCO_3$, 15 m*M* Tris–HCl, pH 7.6), rinsed several times with ice-cold H_2O, resuspended in HSB buffer, and dejellied with 2% cysteine adjusted to pH 7.9 with 10 *M* NaOH. The dejellied eggs are rinsed several times with HSB (until they are packed at the bottom of the beaker). The packed eggs are then rinsed with cold extraction buffer (20 m*M* HEPES, pH 7.5, 70 m*M* potassium chloride, 1 m*M* dithiothreitol, 5% sucrose, 10 μg/ml leupeptin). The eggs are transferred to a centrifuge tube, allowed to settle, and the excess liquid decanted. The eggs are then disrupted by centrifugation (12,000*g* for 30 min at 4°). The ooplasmic fraction is collected and recentrifuged (150,000*g* for 1 hr at 4°) in a SW55Ti rotor. The supernatant below the lipid layer is collected with a syringe, aliquoted, and stored at −80°.

Chromatin Assembly. All reactions are carried out in either the presence or the absence of pure HMG14/17 proteins.[5,10,21] Prior to the addition of DNA, the reactions are supplemented with 1 μl of an extract enriched in class III transcription factors TFIIIA and TFIIIC (prepared according to Smith *et al.*[26] and Wolffe[27] at approximately 0.3 μg/μl. Routinely, the reactions contain 150 ng of single-stranded DNA or 300 ng of double-stranded DNA and a 15-μl extract in a final volume of 20 μl. Chromatin assembly reactions mixtures are incubated at 22° for either 3 (single-stranded DNA) or 6 (double-stranded DNA) hr.

Topoisomer Analysis. Radiolabeled DNA is extracted from chromatin assembled in the presence of various amounts of HMG and fractionated on an 0.8% (w/v) agarose gel in 1× TPE[28] containing 10 μg/ml chloroquine. Autoradiograms of gels are analyzed on a Molecular Dynamics computing densitometer with ImageQuant and Microsoft Excel 4.0 software.

Effect of HMG14/17 on Sedimentation and Protein Content of Reconstituted Chromatin

The effect of HMG14/17 proteins on the structure of reconstituted chromatin is best studied by isolating the assembled minichromosomes

[25] G. Almouzni and M. Mechali, *EMBO J.* **7,** 665 (1988).

[26] R. C. Smith, E. Dworkin-Rastl, and M. B. Dworkin, *Genes Dev.* **2,** 1284 (1988).

[27] A. P. Wolffe, *EMBO J.* **8,** 1071 (1988).

[28] D. Clark, *in* "Chromatin: A Practical Approach" (H. Gould, ed.), pp. 139–152. IRL, London, 1996.

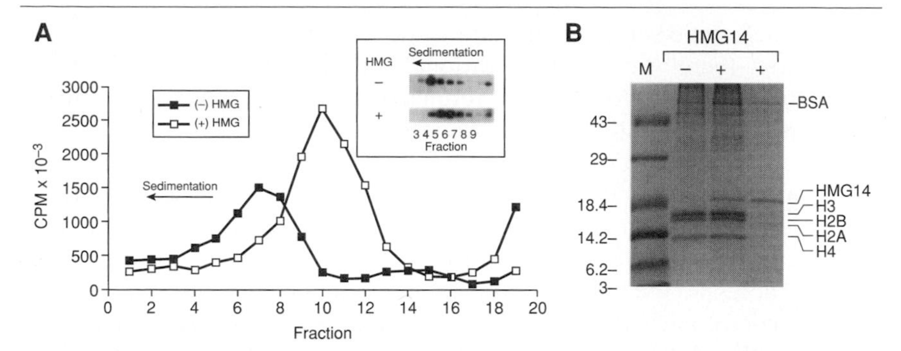

FIG. 6. Effect of HMG14/17 on the sedimentation and protein content of reconstituted chromatin. (A) Sedimentation analysis (fraction volume: 0.5 ml). (Inset) Analysis of the contents of supercoiled plasmid in various fractions from another sucrose gradient (fraction volume: 1.0 ml). (B) Analysis of proteins in purified minichromosomes. Chromatin is assembled from single-stranded DNA in either the absence or the presence of HMG14. Minichromosomes are purified by sucrose gradient centrifugation, and their protein content is analyzed by electrophoresis in polyacrylamide gels containing SDS. M, molecular weight markers; right lane contains purified recombinant HMG14 applied to the gel in a solution containing BSA. Adapted from Trieschmann *et al.*[10] and Wray *et al.*[20]

on sucrose gradients. We have found that the proteins are incorporated stoichiometrically into the chromatin and that they do not displace any of the core histones. Minichromosomes containing HMG14/17 sediment slower that those assembled in the absence of proteins, suggesting that HMG14/17 unfold the minichromosomes[10] (Fig. 6).

Protocol

Radiolabeled minichromosomes assembled in the extract are purified on a sucrose gradient with an SW40 rotor for 2.5 hr at 40,000 rpm. Using a 10–30% (w/v) linear sucrose gradient in 70 m*M* KCl, 20 m*M* K-HEPES, pH 7.5, 1 m*M* EDTA, 0.25% Triton X-100. Minichromosomes recovered from the gradient are pelleted through a 30% (w/v) sucrose cushion in the same buffer. The samples are dissolved in SDS loading buffer and analyzed on 15% polyacrylamide gel containing 0.1% SDS. For topoisomer analysis, radiolabeled DNA is extracted from chromatin and fractionated on an 0.8% (w/v) agarose in 1× TPE buffer.

Micrococcal Nuclease Digestion of HMG14/17-Containing Chromatin

Micrococcal nuclease cleaves chromatin initially in the linker region as an endonuclease and at later stages of digestion in an exonucleolytic attack, gradually reducing the size of the resulting chromatin fragments.[29] During

[29] K. E. van Holde, "Chromatin." Springer-Verlag, New York, 1988.

the exonucleolytic stages of digestion, the histone octamers slide toward each other, thereby exposing additional nucleotides at the ends of the chromatin fragments to further exonucleolytic attack. Therefore, during the course of the digestion with micrococcal nuclease, the size of the oligonucleosomes gradually decreases. HMG14/17 proteins affect the course of digestion in two different ways. They unfold the higher order chromatin structure and at the same time stabilize the structure of the nucleosome core particle. By unfolding the higher order chromatin structure, the proteins facilitate access to the linker region and increase the initial rate of cleavage by micrococcal nuclease. By stabilizing the structure of the nucleosome, they minimize sliding and provide partial protection from exonucleolytic attack. Indeed, as demonstrated in Fig. 7, minichromosomes assembled in the presence of HMG are digested faster than those assembled in the absence of HMG (compare the extent of the large molecular weight smears in lanes 3 to 4 or in lanes 7 to 8). It is also evident that the relative amount of mono- and dinucleosomes generated by the micrococcal nuclease of chromatin assembled in the presence of HMG is larger than in the chromatin assembled in the absence of HMG at early times of digestion, a finding consistent with a faster rate of digestion in the linker region between core particles. Strikingly, when the HMG14 is added to preassembled chromatin, the rate of digestion is slower as is evident from the large molecular weight DNA smears (Fig. 7, lanes 5 and 9). Apparently, when the HMG14/17 is added to preassembled chromatin, some of the protein binds incorrectly to the linker DNA and hinders the accessibility of the micrococcal nuclease to its initial point of attack. We also note that the mobility of the oligonucleosomes is independent of HMG at early digestion times (compare lanes 4 to 5 to 6 or 7 to 8 to 9), suggesting that the proteins do not affect the nucleosomal repeat, i.e., the length of the DNA between adjacent nucleosome cores. However, at later stages of digestion, oligonucleosomes and mononucleosomes derived from HMG-containing chromatin migrate slower than those obtained from HMG-free chromatin (compare lane 16 to lanes 15 and 17). By virtue of their location in chromatin, the proteins stabilize the position of the octamer, minimize sliding, reduce the rate of exonucleolytic attack, and also protect several bases at the end of the particle from digestion. This interpretation[10] is consistent with results of transcription assays and restriction analysis of reconstituted chromatin.

Analysis of the kinetics of chromatin digestion by micrococcal nuclease is a powerful tool to detect the proper assembly of HMG14/17 proteins into folded chromatin. However, if the chromatin is already unfolded, the effect of HMG14/17 is marginal and may not be detectable by this assay.

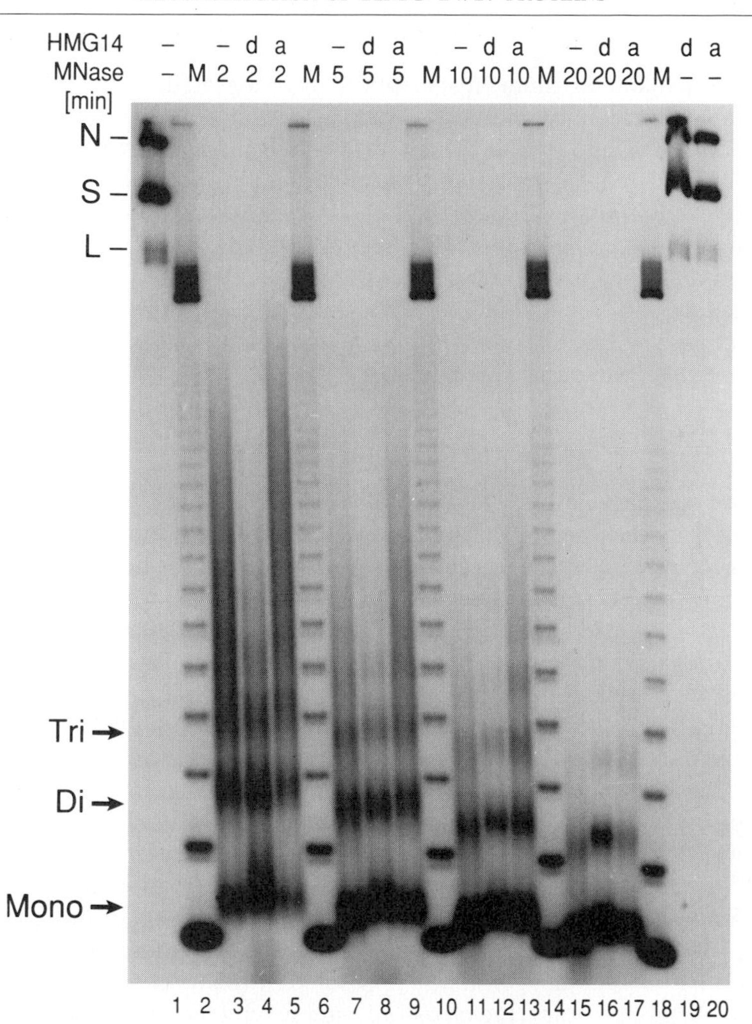

FIG. 7. Effect of HMG14 on the micrococcal nuclease digestion of chromation. Radioactively labeled chromatin assembled in either the absence (lanes 3, 7, 11, and 15) or the presence (lanes 4, 8, 12, and 16, marked as d on top) of HMG14 and chromatin to which HMG14 is added after assembly (lanes 5, 9, 13, and 17, marked as a) are digested with 0.7 U/μl micrococcal nuclease for the time indicated. DNA is purified, fractionated on long 0.8% agarose gels, and exposed for autoradiography. M, 123-bp ladder molecular weight marker. The mobility of supercoiled (S), linear (L), and nicked (N) DNAs is indicated at the left. Adapted from Trieschmann *et al.*[10]

Protocol

The assembled minichromosomes are digested with micrococcal nuclease (Worthington), directly in the egg extract. The digestion is initiated by adding $CaCl_2$ to a final concentration of 1 m*M* and enzyme to a final concentration of 0.5–1.0 U/μl. After various times of digestion at room temperature, aliquots are removed and the digestion is stopped by making the aliquot solutions 20 m*M* Tris–HCl, 20 m*M* EDTA, pH 8, 0.5% (w/v) lithium dodecyl sulfate. The samples are digested with proteinase K (500 μg/ml) at 37° for 1 hr, and the DNA is purified by phenol–chloroform extraction. The DNA fragments are fractionated by electrophoresis in 0.8% agarose gels containing TAE.

Analysis of HMG14/17 Organization in Cellular Chromatin

The organization of HMG proteins in the cell can be visualized using standard immunocytological approaches. However, for biochemical studies, it is necessary to immunofractionate chromatin regions containing these proteins. One of the main questions with this approach, which is inherent to all nucleosome preparations, is whether the purified chromatin preparation indeed represents the entire population present in the nucleus. This question is addressed in detail elsewhere.[19] An additional major challenge in this approach is to isolate the chromatin fragments without significantly altering their "native" state. This consideration is especially significant for HMG14/17 proteins, as it is known that they readily migrate and redistribute among nucleosomes.[30] The following protocol used low ionic strength and, in some cases, cross-linking to overcome these problems. The chromatin is isolated, digested with micrococcal nuclease, and fractionated on sucrose gradients at low ionic strength. In some cases the mono- and oligonucleosomes of defined length are cross-linked prior to immunofractionation. The HMG14/17-containing chromatin is isolated by immunoaffinity chromatography. When necessary, the cross-links are reversed prior to analysis of the protein and DNA in the various chromatin regions.

By immunofractionating oligonucleosomes of defined length and quantitating the ratio of HMG17 to histones in the immunoprecipitates, we were able to demonstrate that nucleosomes containing HMG proteins are clustered into domains, which on average consist of six contiguous HMG-containing nucleosomes.[31] Clustering of architectural proteins may be a general feature of the chromatin fiber. We have defined the size of such a cluster (i.e., the number of contiguous nucleosomes carrying a protein of

[30] D. Landsman, E. Mendelson, S. Druckmann, and M. Bustin, *Exp. Cell Res.* **163,** 95 (1986).
[31] Y. V. Postnikov, J. E. Herera, R. Hock, U. Scheer, and M. Bustin, *J. Mol. Biol.* **274,** 454 (1997).

interest) as "clustering value" and derived an equation for calculating this value.[31] The clustering value "C" (see Fig. 9) is a reliable indicator of the degree of clustering of nucleosomes, which contain a property (i.e., an antigenic site) that can be used to separate them from bulk nucleosomes. Thus, the approach used here, in which we fractionated oligonucleosomes of defined length rather than the total population of micrococcal nuclease-digested nucleosomes, can be used to test the arrangement of nucleosomes containing specific proteins such as histone sequence variants, modified histones or nonhistones, or perhaps unusual nucleotides. Such studies may provide additional insights into the structure–function relation in chromatin.

The experimental protocol outlined in Fig. 8A and the results presented in Figs. 8B and 8C demonstrate how the "C" value is obtained experimentally. The derivation of the equation can be found elsewhere.[31]

Procedure

Antibodies and HMG Proteins. See previous section.

Oligonucleosome Preparation. Chicken erythrocyte nuclei are prepared as recommended.[32] To prevent HMG protein rearrangements[30] and to reduce the effects of higher order chromatin structure on the rate of enzymatic digestion, nuclei are digested mildly with micrococcal nuclease at low ionic strength (15 m*M* Tris–HCl, 20 m*M* KCl, 1 m*M* $CaCl_2$, pH 8.0). Soluble chromatin (about 80% of total chromatin) is fractionated into oligomers using 12–50% linear sucrose gradient ultracentrifugation in NS buffer (10 m*M* NaCl, 10 m*M* Tris–HCl, pH 7.5, 1 m*M* EDTA). DNA from the oligonucleosomes in each fraction is purified by phenol/chloroform, and the lengths of oligonucleosomes in each fraction are determined by agarose gel electrophoresis. Oligonucleosomes are spin-dialyzed (Macrosep 30, Filtron) against NS buffer containing 0.1% Triton X-100 (NST).

Immunofractionation. Oligonucleosomes of defined length (50 μg as DNA) are incubated with 20 μg of either affinity-pure antibodies to HMG-17 or nonimmune rabbit IgG in a volume of 0.5–1 ml at 4° for 30 min in NST buffer. The reaction mixture is then incubated overnight with 10 μl of immobilized protein A on Trisacryl GF-2000 (Pierce) or with 50-μl (bed volume) Dynabeads M-280 coated with sheep anti-rabbit IgG. The Trisacryl beads are washed three times with 100 volumes of NST buffer, and the bound material is released by either 5 volumes 1% SDS or first with 0.35 *M* NaCl followed by 1% SDS. When using Dynabeads, the Triton X-100 in NST is replaced by 0.1% bovine serum albumin.

[32] R. D. Kornberg, J. W. LaPointe, and Y. Lorch, *Methods Enzymol.* **170,** 3 (1989).

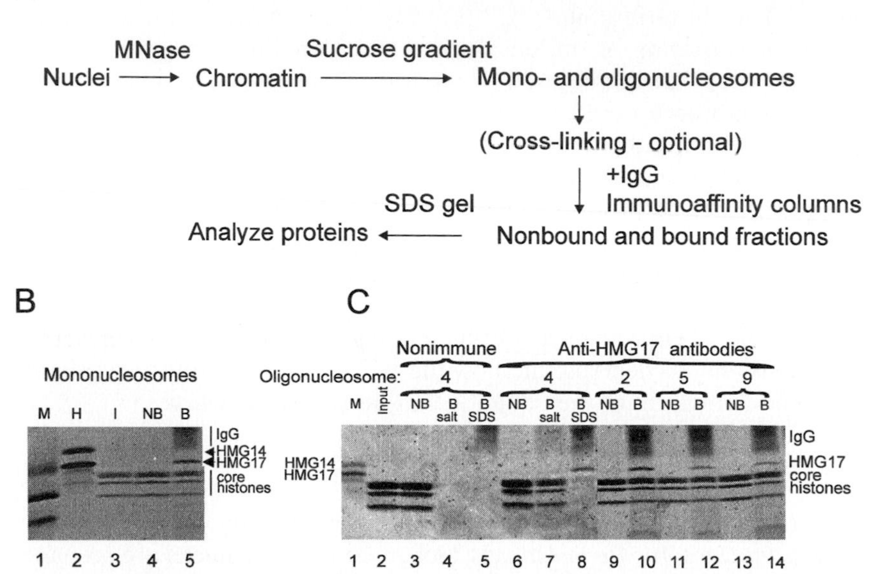

FIG. 8. Isolation of chromatin regions containing HMG14/17 from nuclei. (A) Experimental flow chart. (B) Analysis of proteins in HMG17-containing mononucleosomes. Mononucleosomes are purified by sucrose gradient centrifugation from chicken erythrocytes, and the HMG17-containing nucleosomes are isolated by immunoadsorption with affinity-pure antibodies to HMG17 protein. Lanes 1 to 5 contain, respectively, molecular weight markers (GIBCO-BRL); HMG14/17 markers; input mononucleosome fraction; mononucleosome fraction not bound by anti-HMG17 antibodies; and mononucleosome fraction bound to anti-HMG17 antibodies. Note that in bound mononucleosomes (lane 5), the amount of HMG17 is equivalent to that of the core histones, that all four core histones are present in equivalent amounts, and that HMG14 protein is not detectable. Thus, all the immunoprecipitated mononucleosomes contain two molecules of HMG17. (C) Proteins in oligonucleosomes. Oligonucleosome fractions with average nucleosomal repeats of two (lanes 9 and 10), four (lanes 3–8), five (lanes 11 and 12), and 9 (lanes 13 and 14) are isolated by sucrose gradient ultracentrifugation and subjected to immunofractionation. The specificity of immunoprecipitation is tested using nonimmune rabbit IgG and nucleosomal tetramers (lanes 3–5). Proteins of the tetranucleosomes are found in the nonbound (NB) fraction (lane 3). Treatment of these immunoadsorbents with either 0.35 *M* NaCl (lane 4) or 1% SDS (lane 5) did not release proteins. In contrast, a fraction of the oligonucleosomal tetramers, fractionated similarly using antibodies to HMG17 (lanes 6–8), is retained on the resin. Bound nucleosomes are released by elution with 0.35 *M* NaCl (lane 7). Bound IgG and HMG17 are released subsequently with 1% SDS. Lanes 9 and 10, 11 and 12, and 13 and 14 depict, respectively, proteins in the bound (B) and unbound (NB) fractions of chicken di-, penta-, and nonanucleosome fractions. M, marker recombinant HMG proteins. Lane 2, input, unfractionated chromatin. A quantitative analysis of overloaded gels indicates that the amount of HMG14/17 in chromatin equals 2% of the amount of each of the core histones. HMG14/17 are not visible at the loading levels used in this lane or in the lanes showing proteins in nonbound fractions. Note that the ratio of HMG17 to core histones decreases as the length of the oligonucleosome increases. Adapted from Postnikov *et al.*[31]

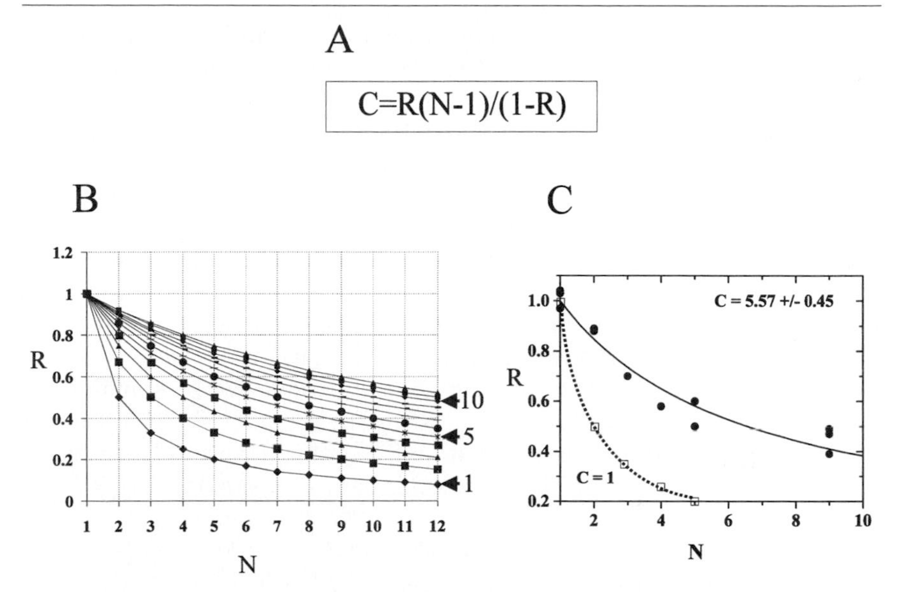

FIG. 9. Calculating the size of the HMG17-containing nucleosome cluster in chromatin. (A) Formula. The size of the HMG-containing nucleosome cluster can be estimated by using the formula $C = R(N - 1)/(1 - R)$, where C is the size of the cluster in nucleosomes, R is the molar ratio of HMG to core histone (obtained from quantitative scanning of the polyacrylamide gels), and N is the length of the oligonucleosome (determined from analysis of the sucrose gradients) subjected to immunofractionation. The deduction of the equation is described in Postnikov *et al.*[31] (B) Theoretical curves for the predicted value of R obtained with all the possible combinations of C and N between 1 and 10. (C) Experimental results obtained by measuring the HMG : histone ratio (R) in oligonucleosomes purified from chicken erythrocytes and mouse thymus. The curve represents the least-square fit to the equation in (A) using RS/1 software (BBN Software Products, Cambridge, MA). The analysis yields a clustering value of 5.57 ± 0.5. The best fit to the experimental points at various N values is indicative of a regular cluster of contiguous HMG17-containing nucleosomes. The broken line depicts the theoretical curve expected from a nonclustered arrangement of HMG-containing nucleosomes in chromatin (i.e., $C = 1$).

Immunofractionation with Cross-Linking. We have found significant batch-dependent variability in the degree of nonspecific binding to the immunoaffinity columns (both protein A–Trisacryl GF-2000 from Pierce and Dynabeads M-280 from Dynal, Inc.). The following version of immunofractionation relies on the total formaldehyde fixation of chromatin prior to immunofractionation and the reversal of cross-links after the nucleosomes containing HMG have been separated from bulk nucleosomes. Although more time-consuming, this procedure works reliably with most commercial immunoaffinity media, including the commonly used protein A–agarose or protein A–Sepharose.

To remove amines from the nucleosome preparation, spin the oligonucleosome fractions (15–20 ml of sucrose gradient fractions) repeatedly in Centriprep centrifugal concentrators with 10,000 cutoff at 5000*g* for 10 min at 4° using HEN buffer (20 m*M* HEPES, pH 7.4, 1 m*M* EDTA, 10 m*M* NaCl). The final concentration of each of the nucleosome fractions should be about 4 mg/ml. If the sucrose gradient is made of HEPES instead of Tris, it is sufficient to just concentrate the fractions.

To cross-link chromosomal proteins and DNA, mix 250 μl of amine-free oligonucleosomes (1 mg), 7 μl 37% formaldehyde (to a final concentration of 1% or 34 m*M*), and incubate for 2 hr at 37°. Put sample on ice and add 25 μl of 1 *M* Tris–HCl, pH 8, to quench the reaction.

The removal of excess formaldehyde is accomplished by spinning the sample in a Microcon 100 K tube at 5000*g* for 5 min at 4° several times, diluting the concentrated material with HEN buffer. Save two 1-μl aliquots for testing the completion of the cross-linking and cross-linking reversal procedures.

To immunoprecipitate the cross-linked material, 100 μl of 3–4 μg/μl oligonucleosome preparation is mixed with 500 μl of AFIP solution (amine-free immunoprecipitation solution, 0.1% SDS, 0.5% Triton X-100, 0.5% DOC-Na, 0.5 *M* NaCl, 50 m*M* Na-HEPES, pH 7.4, 1 m*M* EDTA, 1 m*M* EGTA, 0.1 m*M* PMSF). Add 30 μg of immunoaffinity-pure anti-HMG17 antibodies to the reaction mixture. Use nonimmune IgG instead of anti-HMG antibodies for control immunofractionation. Incubate for 30 min at room temperature.

Take 50 μl of a (1:1, v/v) suspension of protein A–agarose (using a trimmed pipette tip) and add 25 volumes of AFIP solution at room temperature for 5–10 min. The best mixing can be obtained by using a vertically rotating wheel. Spin down at 100*g* for 1 min at room temperature. Discard the supernatant. Add the reaction mix to the prewashed resin. Incubate overnight at room temperature with constant mixing (no magnetic stirrer).

Centrifuge at 100*g* for 1 min. Save supernatant as *nonbound* fraction. Wash twice with 25 volumes of AFIP solution and once with HEN. Add 60 μl 1% SDS and warm the tube at 50° for 2 min with light tapping. Centrifuge at 100*g* for 1 min and collect supernatant as the *bound* fraction.

To reverse the cross-linking, add 10 μl 1 *M* Tris–HCl, pH 8, to both the eluate and an aliquot of the nonbound fraction and heat at 95° for 90 min. Use polymerase chain reaction thin-walled 200-μl tubes and a heated lid. The samples at this point could be analyzed by a SDS–polyacrylamide gel.

Electrophoretic Separation and Analysis of Gels. The proteins are resolved on 10–20% gradient polyacrylamide–SDS gels, 0.7 mm thick, with SDS–Tris–glycine running buffer. The gels are stained with 0.1% Coomassie

Blue R-250 and scanned with a Molecular Dynamics computing densitometer, and the bands are quantitated by ImageQuant software. To evaluate the stoichiometry of Coomassie staining of core histones and HMG proteins, standard curves can be constructed using purified proteins isolated on a RP-HPLC butyl column (Aquapore Brownlee, 250 × 4.6 mm). The HMG/octamer ratio (R) is calculated using the sum of the signals from core histones divided by 4. Typical results of the immunoprecipitation technique are shown in Fig. 8. Calculation of the clustering value (C) is shown in Fig. 9.

Acknowledgments

We thank Drs. M. Bergel, J. Herrera, and J. Wagner for critically reviewing this manuscript.

[9] Purification and Assays for High Mobility Group HMG-I(Y) Protein Function

By RAYMOND REEVES and MARK S. NISSEN

Introduction

The HMG-I(Y) family of "high mobility group" nonhistone mammalian proteins belongs to a group of nuclear proteins collectively known as architectural transcription factors because of their ability to function *in vivo* both as components of chromatin structure and as auxiliary gene transcription factors (reviewed by Bustin and Reeves[1]). The HMG-I(Y) family contains three known members, HMG-I, HMG-Y, and HMGI-C, which, because of their similarities and for convenience, are often referred to simply as the HMG-I(Y) proteins. HMG-I (107 amino acids; ~11.9 kDa) and HMG-Y (96 residues; ~10.6 kDa) proteins are identical in sequence except for an 11 amino acid internal deletion in the latter and are produced by alternative splicing of transcripts from a single gene in both humans and mice. The HMGI-C protein (109 residues; ~12 kDa) has high sequence homology (~50% overall) with the HMG-I and HMG-Y proteins, has the internal deletion of 11 amino acids characteristic of HMG-Y, but is the product of a separate gene in both species. In their capacity as transcription factors, HMG-I(Y) proteins have been implicated in both positive and negative regulation of a number of human genes *in vivo*. In several examples of positive gene regulation, evidence indicates that the HMG-I(Y) proteins

[1] M. Bustin and R. Reeves, *Prog. Nucleic Acids Res. Mol. Biol.* **54,** 35 (1996).

0076-6879/99 $30.00

function by physically interacting with sequence-specific transcription factors to form stereospecific multiprotein complexes, which are necessary for efficient transcription initiation. The ability of HMG-I(Y) proteins to bend, straighten, unwind, and supercoil DNA substrates[2,3] also appears to play a role in gene transcriptional regulation. HMG-I(Y) proteins have also been demonstrated to be a component of the HIV-1 viral preintegration complex in human cells and to be required for the efficient integration of viral DNA *in vitro*.[4] As in the case of their involvement in gene transcriptional regulation, the ability of these proteins to distort DNA structure may also facilitate retroviral DNA integration reactions *in vitro*.

HMG-I(Y) proteins are distinguished from other HMG proteins by their ability to preferentially bind to the minor groove of A·T-rich sequences of duplex DNA both *in vitro* and *in vivo*.[1] Nevertheless, these proteins do not bind to all A·T-rich stretches equally well or with equal affinity, suggesting that they recognize DNA structure rather than sequence.[5,6] Selective substrate structural recognition is also demonstrated by the fact that HMG-I(Y) proteins bind preferentially to non-B form DNA structures found in supercoiled plasmids,[3] to locally distorted regions of DNA located on the surface of nucleosomes,[7,8] to synthetic four-way junction (FWJ) DNAs,[9] and to cisplatin-adducted substrates.[10] Individual HMG-I(Y) proteins have three separate, but very similar, peptide domains (called A·T-hook motifs) that interact specifically with the narrow minor groove of certain A·T DNA sequences, and each of these binding domains is separated by a flexible polypeptide backbone.[5] Free in solution HMG-I(Y) proteins have little, if any, secondary structure but assume specific conformations on binding to DNA substrates.[1,10,11] The recent determination of the solution structure of an HMG-I(Y)–DNA complex by multidimensional nuclear magnetic resonance (NMR) spectroscopy[11] indicates that the highly conserved peptide core of individual A·T-hook DNA-binding motifs has a planar crescent-shape structure resembling the drugs distamycin and ne-

[2] D. Lehn, T. Elton, K. Johnson, and R. Reeves, *Biochem. Intl.* **16,** 963 (1988).
[3] M. S. Nissen and R. Reeves, *J. Biol. Chem.* **270,** 4355 (1995).
[4] C. M. Farnet and F. D. Bushman, *Cell* **88,** 483 (1997).
[5] R. Reeves, T. S. Elton, M. S. Nissen, D. Lehn, and K. R. Johnson, *Proc. Natl. Acad. Sci. U.S.A.* **84,** 6531 (1987).
[6] R. Reeves and M. S. Nissen, *J. Biol. Chem.* **265,** 8573 (1990).
[7] R. Reeves, and M. S. Nissen, *J. Biol. Chem.* **268,** 21137 (1993).
[8] R. Reeves and A. P. Wolffe, *Biochemistry* **35,** 5063 (1966).
[9] D. A. Hill and R. Reeves, *Nucleic Acids Res.* **25,** 3523 (1997).
[10] Unpublished data, 1998.
[11] J. R. Huth, C. A. Bewley, M. S. Nissen, J. N. S. Evans, R. Reeves., A. Gronenborn, and G. M. Clore, *Nature Struct. Biol.* **4,** 657 (1997).

tropsin and the dye Hoechst 33258, ligands that also bind preferentially to the minor groove of A · T sequences due to substrate structural recognition. The ability of HMG-I(Y) proteins to selectively recognize and alter the structure of DNA substrates appears to underlie many of their *in vivo* biological functions.

Isolation and Purification Methods

Salt vs Acid Extraction Procedures

HMG-I(Y) proteins, like other high mobility group proteins,[12] can be isolated from nuclei or chromatin by extraction with 0.3–0.4 *M* NaCl.[13,14] Serendipitously, due to their unusual amino acid composition (having high concentrations of both basic and acidic amino acid residues), these highly charged proteins are also soluble in dilute acids and can be extracted from nuclei, chromatin, or even whole cells and tissues with either 5% perchloric ($HClO_4$) or trichloroacetic (TCA) acids.[12–14] There are advantages and disadvantages to both salt and acid extraction methods. While the salt isolation method is mild and allows for efficient recovery of native, nondenatured HMG proteins, this procedure results in extraction of a very complex mixture of nuclear proteins from which the desired HMG-I(Y) proteins must be subsequently purified. Because the concentration of HMG-I(Y) proteins is usually quite low, or barely detectable, in normal nontransformed cells and tissues,[1,15–17] the salt extraction method is usually reserved for isolating native HMG-I(Y) proteins from cells or tissues that contain high concentrations of these proteins, such as rapidly proliferating cancerous tissues, ascites cells, or transformed cell lines growing in tissue culture[1] (see later). Salt extraction methods also have the disadvantage that they often involve manipulation of large volumes of dilute protein solutions and are frequently accompanied by considerable protein degradation due to the presence of contaminating proteases in the aqueous extracts, thus requiring inclusion of high concentrations of protease inhibitors in the isolation protocol.

Isolation of HMG-I(Y) proteins by dilute acid extraction offers several advantages over salt extraction procedures. First, acid extraction greatly

[12] E. W. Johns and J. Forrester, *Eur. J. Biochem.* **8,** 547 (1969).
[13] T. Lund, J. Holtlund, M. Fredrickesen, and S. G. Laland, *FEBS Lett.* **152,** 163 (1983).
[14] E. W. Johns, ed., "The HMG Chromosomal Proteins." Academic Press, New York, 1982.
[15] T. S. Elton and R. Reeves, *Anal. Biochem.* **157,** 54 (1986).
[16] R. Reeves and T. S. Elton, *J. Chromatogr. Biomed. Appl.* **418,** 73 (1988).
[17] K. D. Johnson, D. Lehn, T. S. Elton, P. A. Barr, and R. Reeves, *J. Biol. Chem.* **263,** 18338 (1988).

simplifies the composition of proteins obtained from isolated nuclei, selectively solubilizing total HMG proteins [HMG1, 2, 14, and 17, in addition to the HMG-I(Y) proteins] plus histone H1-like proteins,[14] thus greatly facilitating the subsequent purification of individual proteins. Second, acid extraction procedures, which usually incorporate a protein precipitation step, can be used to effectively concentrate HMG-I(Y) proteins isolated from large amounts of starting material, thus allowing detection of the low concentrations of these proteins found in normal or nondividing cells. Third, acid extraction minimizes proteolytic degradation of the soluble proteins and also allows direct isolation of HMG proteins from frozen samples during the thawing process. As noted earlier, because the HMG-I(Y) proteins appear to be in random coil configuration while free in solution, exposure of these proteins to dilute acids or even boiling does not adversely affect their structure. Thus, after acid isolation the proteins do not require renaturation to regain their *in vitro* functional activities.[1,6]

Disadvantages of acid isolation methods include a somewhat lower yield of recovered protein than is obtained with salt isolation methods. Also, when whole cells or tissues are used as the starting material, acid extraction results in the solubilization of numerous basic cytoplasmic and mitochondrial proteins (e.g., many ribosomal proteins), as well as protein degradation products, as contaminants in the extract. Extraction with trichloroacetic acid has the added disadvantage that this acid exhibits strong absorbance at 220 nm, thus prohibiting determination of the concentration of the HMG-I and HMG-Y proteins (which lack aromatic amino acids that absorb at 280 nm) by spectrophotometric methods at this wavelength (see later). Extraction of HMG-I(Y) proteins with dilute perchloric acid is, therefore, generally the method of choice. Many HMG-I(Y) isolation protocols combine both salt extraction and acid fractionation procedures to maximize protein recovery and simplify protein purification.[18,19]

Reagents and Extraction Protocols

Salt Extraction Procedures for Native Proteins

Isolated nuclei (or chromatin) from actively growing cells in suspension culture (e.g., human HeLa, mouse Friend erythroleukemia cells) or ascites tumor cells isolated from mice are the best starting materials for salt extraction of native HMG-I(Y) proteins. For cells that grow in suspension, the following nuclear isolation and salt extraction procedures are recom-

[18] G. H. Goodwin, R. H. Nicolas, and E. W. Johns, *Biochim. Biophys. Acta* **405,** 280 (1978).
[19] T. S. Elton and R. Reeves, *Anal. Biochem.* **144,** 403 (1985).

mended. Unless otherwise stated, all solutions are kept at 4° and all extraction steps are conducted at this same temperature.

1. Cells are washed free of serum and other proteins using phosphate-buffered saline (PBS; 137 m*M* NaCl, 2.7 m*M* KCl, 4.3 m*M* $Na_2HPO_4 \cdot 7H_2O$, 1.4 m*M* KH_2PO_4, pH 7.3) by first pelleting the cells by low-speed centrifugation (750*g*, 10 min), resuspending them in 10-pellet volumes of PBS, and then repelleting the cells by centrifugation.
2. The washed cells are resuspended in 2–3 pellet volumes of RSB buffer [10 m*M* NaCl, 3 m*M* $MgCl_2$, 10 m*M* Tris–HCl, 0.5 m*M* phenylmethylsulfonyl fluoride (PMSF), pH 7.4] and incubated on ice for 10 min. Cells are then lysed by the addition of an equal volume of RSB buffer containing 1% NP-40 and homogenization (10–15 strokes) with a loose-fitting Dounce homogenizer. The released nuclei are pelleted by centrifugation and washed twice with RSB buffer alone, followed by pelleting.
3. The nuclei are resuspended in 2–3 volumes of salt extraction buffer (SEB; 0.35 *M* NaCl, 3 m*M* $MgCl_2$, 10 m*M* Tris–HCl, 0.5 m*M* PMSF), and are incubated on ice for 15 min, nuclei are pelleted by centrifugation, and the supernatant is collected. The SEB extraction procedure is repeated and the 0.35 *M* NaCl extracts are pooled. Total salt-soluble proteins can be precipitated from the extract by the addition of 6 volumes of cold-acidified acetone (acetone containing 10 m*M* HCl), incubated on ice for 1 hr, and centrifuged at 20,000*g* for 30 min. The pelleted proteins are then washed carefully (without disturbing the pellet) once with cold acidified acetone, twice with cold acetone alone, and then dried under vacuum and stored at −20°. If desired, HMG-I(Y) protein precipitation during the isolation process can be avoided altogether by the addition of 0.35 g of ammonium sulfate [$(NH_4)SO_4$] per milliliter of salt extract (warmed to room temperature) in order to precipitate proteins other than the HMGs and histone H1. After 1–2 hr at room temperature, the precipitated proteins are pelleted by centrifugation (20,000*g*, 30 min) and the supernatant is collected. The soluble HMG fraction is dialyzed overnight in Spectra-Por 6000 molecular weight cutoff dialysis tubing against several changes of TE solution (10 m*M* Tris–HCl, 1 m*M* EDTA, pH 7.4) to remove salts. The dialyzed samples are centrifuged to remove particulates and the soluble native HMG proteins are concentrated and reduced to a small volume using a membrane concentration apparatus (6000 molecular weight cutoff, preferred) such as the Centricon 3 microconcentrator or Centriprep 3 centrifugation

concentrator (Amicon Corp., Danvers, MA). Concentrated native protein solutions in TE should be stored in small aliquots at −70°. Repeated freeze/thaw cycles of HMG-I(Y) proteins should usually be avoided.

Acid Extraction Procedures

The 0.35 *M* NaCl extract from isolated nuclei or chromatin is the starting point for many acid extraction protocols. The salt extract is made 5% (w/v) in perchloric acid to precipitate the bulk of the salt-soluble nuclear proteins, leaving the HMG proteins and histone H1 fractions in solution. The $HClO_4$-soluble proteins are precipitated by the addition of 6 volumes of cold-acidified acetone, incubated on ice for 1 hr, and centrifuged at 20,000*g* for 30 min. The pelleted proteins are washed carefully, once with acidified acetone and twice with acetone alone, and then dried under vacuum and stored at −20°. Alternatively, trichloroacetic acid can be used to first fractionate the salt extract into a 2% TCA-soluble protein solution[18] from which the HMG proteins are then precipitated by the addition of 100% TCA to the extract until the solution is 25% TCA (final concentration), incubated on ice for 1 hr, and then pelleted by centrifugation and washed as before. In all cases where TCA is used to precipitate proteins, it is important that all traces of residual acid be removed from the pelleted HMG proteins. If several washes of the pellet with acetone fail to remove all traces of TCA, the protein pellet should be solubilized in TE solution and reprecipitated with 6 volumes of acidified acetone, washed with acetone as before, and dried under vacuum before storage at −20°.

Cold 5% perchloric acid can be used to isolate HMG proteins directly from either fresh or frozen cells or tissues. Fresh cells are collected by centrifugation and resuspended in an equal volume of cold 10% $HClO_4$ (to give a final solution concentration of approximately 5% $HClO_4$) and the mixture is homogenized in a Dounce homogenizer (~20 strokes) on ice. The homogenate is incubated on ice for 15 min, centrifuged for 10 min at 20,000*g*, and the supernatant collected. The pelleted residue is then twice reextracted by homogenization with equal volumes of cold 5% $HClO_4$, with all of the supernatants pooled and the proteins precipitated, washed, and dried as described earlier. When frozen samples are used as the starting material, these should be broken into manageable sized fragments, an equal volume of 10% cold perchloric acid added to each sample, and the mixtures homogenized in a precooled electric blender, such as an Omnimixer or Waring blender, until thawed (usually about 3 min for 20 g of material). Blender homogenizations should be conducted in either a 4° cold room or a refrigerated cabinet. After the initial blender homogenization, the samples are reextracted with an equal volume of cold 5% $HClO_4$, with the extracts pooled and processed as described earlier.

Production of Recombinant Proteins

Bacterially produced recombinant wild-type and mutant forms of HMG-I(Y) proteins have proven invaluable for the elucidation of the structure and function of these proteins.[1,11] Although various fusion derivatives of the HMG-I(Y) proteins (e.g., hexahistidine-tagged, hemagglutinin-"tagged," or glutathione *S*-transferase fusion proteins) have been used successfully as aids in purifying recombinant proteins, for many types of experiments, particularly those involving detailed structural analyses,[11] the production of recombinant proteins without such additional amino acid residues is desirable. For this purpose we have found the pET bacterial expression vector system of Studier[20,21] to be particularly useful.[22] Employing standard *in vitro* mutagenesis procedures, site-specific mutations were introduced into the full-length human HMG-I cDNA (clone 7C)[23] so that its protein-coding region could be introduced into the pET-3a[21] expression vector (creating a recombinant vector designated pET7C[22]) for the production of large quantities of bacterially produced recombinant wild-type HMG-I protein. In making such bacterial expression constructs using pET vectors, it is important to include the translation stop codon of the HMG-I cDNA itself as part of the insert so that translation termination occurs at the appropriate endogenous stop site. Otherwise, the termination of recombinant protein synthesis will depend on a 3′ downstream stop codon present in the pET vectors, which will result in the incorporation of several additional amino acids, including a cysteine residue,[21,24] at the C-terminal end of the recombinant protein. Because wild-type human and mouse HMG-I proteins do not contain cysteine residues,[17,23] the incorporation of this pET vector-encoded cysteine residue can, under certain conditions, lead to artifactual dimerization of the recombinant HMG-I proteins[25,26] produced from these vectors.

To produce full-length recombinant human HMG-I protein, the pET7C expression vector[22] is introduced into the double *lon/ompT* protease mutant

[20] A. H. Rosenberg, B. Lade, D.-S. Chui, S. Lin., J. J. Dunn, and F. W. Studier, *Gene* **56,** 125 (1988).

[21] F. W. Studier, A. H. Rosenberg, and J. J. Dunn, *Methods Enzymol.* **185,** 60 (1990).

[22] M. S. Nissen, T. A. Langan, and R. Reeves, *J. Biol. Chem.* **266,** 19945 (1991).

[23] K. D. Johnson, D. Lehn, and R. Reeves, *Mol. Cell. Biol.* **9,** 2114 (1989).

[24] R. Reeves and M. S. Nissen, *J. Biol. Chem.* **268,** 31137 (1993).

[25] D. Thanos and T. Maniatis, *Cell* **71,** 777 (1992).

[26] W. Du, D. Thanos, and T. Maniatis, *Cell* **74,** 887 (1993).

B strain of *Escherichia coli* BL21(DE3)pLysS.[21] This host strain is lysogenic for the bacteriophage DE3, which carries the T7 RNA polymerase gene under the control of the isopropyl-β-thiogalactopyranoside (IPTG)-inducible *lacUV5* promoter. The addition of 0.4 m*M* IPTG to exponentially growing cultures of BL21(DE3)pLysS for 4 hr (37°) induces large amounts of T7 polymerase, which, in turn, efficiently transcribes the target HMG-I cDNA in the pET7C plasmid and produces large amounts of recombinant protein. Following induction, the bacteria are pelleted (10 min at 5000 rpm in a GSA rotor), washed once in PBS, and, at this point, if desired, frozen at −70° for later use. Alternatively, the recombinant proteins can be isolated and purified immediately as described in the following steps.

Steps in Preparation of Crude Recombinant HMG-I(Y)

1. To bacterial pellets obtained from 250 ml of starting culture, add 25 ml of cold 5% $HClO_4$/0.5% Triton X-100 and resuspend the pellets by vigorously vortexing/stirring/pipetting until as fine a colloidal suspension as possible is produced. Incubate the suspension on ice for at least 30 min with occasional shaking. Acid-insoluble material is then removed from the extracts by centrifugation (10 min, 10,000 rpm in a GSA rotor) and the supernatant is collected.
2. Recombinant HMG-I protein is precipitated from the supernatant by adding TCA to a 25% final concentration (i.e., 0.34 volumes of 100% TCA solution is added to the supernatant). Immediately after TCA addition the solution should appear very turbid but it is best to allow it to sit on ice for an hour or more with occasional agitation (the mixture can stand at 4° overnight without adverse effects). Do not use a stir bar for mixing since much of the precipitated recombinant protein will stick to it and be difficult to recover. It is best to perform the precipitation directly in a centrifuge bottle.
3. The precipitated crude recombinant proteins are recovered by centrifugation at 10,000 rpm in GSA rotor for 15 min. Sometimes the protein pellets recovered from bacteria are yellowish and appear to be "oily" and fluid, thus care should be taken not to lose the pellets while decanting the supernatant. Redissolve the protein pellet in a relatively small volume of water and transfer to a 30-ml Corex glass centrifuge tube. Precipitate the proteins as before with TCA (a 30-min incubation on ice is sufficient) and recover them by centrifugation at 10,000 rpm for 10 min at 4°. Discard the supernatant and wash the pellet with acidified acetone and acetone as described earlier. Dry the pellet under vacuum and store at −20°.

Chromatographic Purification Procedures

RP-HPLC Purification of HMG-I(Y) from Cell and Tissue Extracts

Although the HMG-I(Y) proteins can be separated from histone H1 and other HMGs[14,27] by fractionation on CM-cellulose, CM-Sephadex, or HPLC ion-exchange columns, much better resolution and yields can be achieved by use of reversed-phase ion pair HPLC (RP-HPLC) chromatography (Fig. 1).[15] This technique produces denatured protein products but, as noted previously, this is not a problem with HMG-I(Y) proteins because they naturally have little secondary structure when free in solution. If undenatured proteins are desired, native 0.35 *M* salt-extracted proteins can be separated by a modification of the polybuffer exchange chromatography method (see later). Protein concentrations in crude HMG extracts can be determined using either a Bio-Rad protein assay kit (Bio-Rad Laboratories, Hercules, CA) or the Pierce BCA protein assay procedure (Pierce Chemical Co., Rockford, IL) using either histone H1 or purified HMG proteins as reference standards.

RP-HPLC fractions are performed conveniently on a Vydac C_4 silica column (5 μm silica, 300-Å pore size, 4.6 × 250 mm) (Rainin Instruments, Farmingdale, NY).[15] Dried HMG proteins are dissolved in solvent A [0.1% (w/v) trifluoroacetic acid (TFA) in H_2O, pH 2.14] containing 0.1 *M* dithiothreitol (DTT), centrifuged to remove insoluble material, and injected onto the Vydac C_4 column. HMG proteins are eluted at room temperature using a gradient from 0.1% TFA in water (i.e., solvent A) to 0.1% (w/v) TFA in 95% acetonitrile (CH_3CN), 5% H_2O (solvent B) with a flow rate of 1 ml/min. As shown in Fig. 1, RP-HPLC can be used to purify all of the individual HMGs from an "acid-enriched" extract with homogeneous HMG-I protein eluting in a discrete fraction (e.g., lane 4). Proteins are recovered from each chromatography fraction by removing the solvent with a Speed-Vac apparatus (Savant Instruments, Farmingdale, NY).

Because of their unusual amino acid compositions lacking aromatic residues, estimations of the concentration of mammalian HMG-I and HMG-Y proteins in purified chromatography fractions are determined conveniently and accurately by measuring the absorbance of the proteins dissolved in water (or TE) at 220 nm.[6] The extinction coefficient for HMG-I is $\varepsilon_{220} = 73{,}500$ liter/mol · cm or 6.23 ml/mg · cm. For HMG-Y, $\varepsilon_{220} = 67{,}500$ liter/mol · cm or 6.37 ml/mg · cm. In contrast to the HMG-I and HMG-Y

[27] A. Riffe, M. Delphech, F. Levy-Favatier, J.-P. Boissel, and J. Kruh, *Nucleic Acids Res.* **334,** 332 (1985).

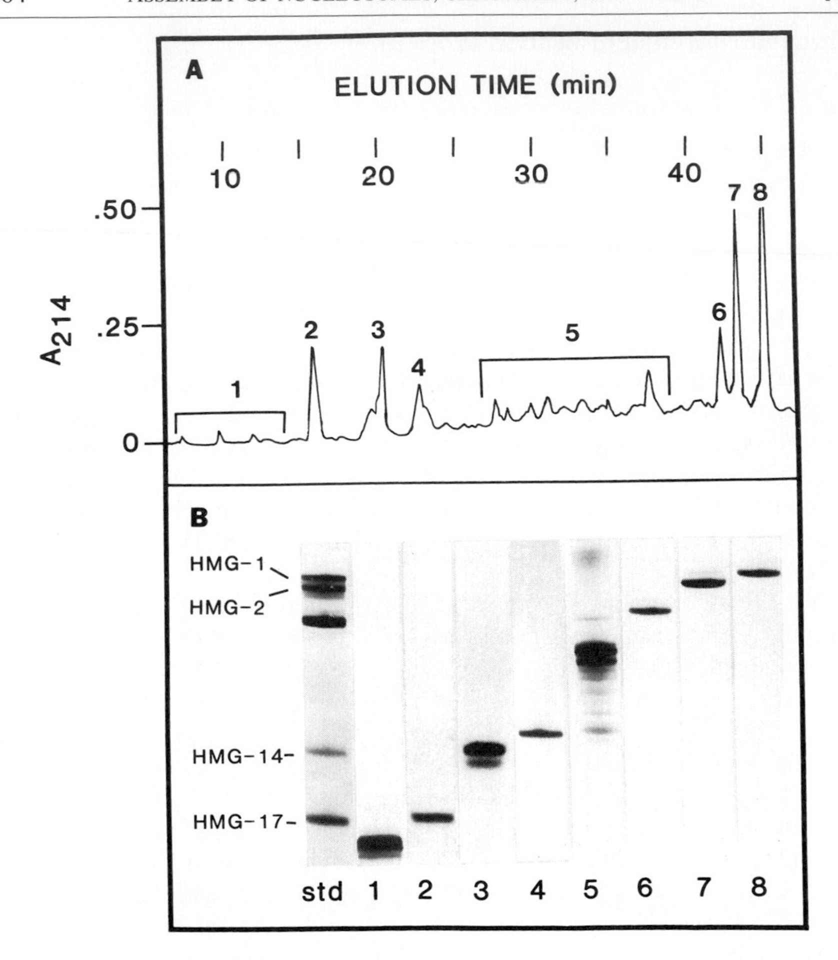

FIG. 1. Ion pair RP-HPLC analysis of Friend erythroleukemic mouse cell HMG proteins. (A) "Enriched" HMG proteins (approximately 100–200 μg) were isolated from Friend cell nuclei,[19] dissolved in solvent A [aqueous 0.1% trifluoroacetic acid (TFA)], injected onto a RP-HPLC column, and eluted with a 15–50% linear gradient of solvent B for 60 min at a flow rate of 1.0 ml/min (gradient slope of 0.58% acetonitrile/min). (B) Acid–urea polyacrylamide slab gel: std, calf thymus HMG1, 2, 14, and 17 and histone H1 proteins; lanes 1–8 correspond to the protein peaks labeled 1–8 in (A). As indicated, homogeneous HMG-I protein elutes off in peak 4 of the gradient. Reproduced from Elton and Reeves,[19] with permission.

proteins, both the human and the mouse HMGI-C protein contains a single tryptophan residue that allows accurate determination of its concentrations in solution by their absorbance at 280 nm. The purity of HMG protein samples can be monitored either by sodium dodecyl sulfate (SDS)–

polyacrylamide gel electrophoresis (SDS–PAGE)[28] or by the acid–urea polyacrylamide gel electrophoretic procedures of Panyim and Chalkley.[29] Unless special precautions are taken, the HMG-I and HMG-Y proteins often comigrate with the unrelated HMG14 and HMG17 proteins on standard 15% SDS–PAGE gels. Thus, the use of acid–urea gels to analyze HMG proteins in crude extracts is recommended[15] (see Fig. 1B). However, SDS–PAGE is suitable for analyzing the purity of either recombinant HMG-I(Y) proteins or chromatographically purified HMG fractions.

PBE94 Polybuffer Exchanger Purification of Nondenatured, "Native" Proteins

The polybuffer exchanger PBE94 resin from Pharmacia (Uppsala, Sweden) has been used by Adachi and colleagues[30] for efficient large-scale chromatographic purification of both nondenatured and acid-extracted HMG-1 and HMG-2 proteins. The PBE94 resin can also be used in anion-exchange column chromatography to efficiently purify nondenatured, native HMG-I(Y) proteins in high yields. The preferred starting material for such native protein purification is the HMG-enriched, $(NH_4)SO_4$-soluble fraction of the 0.35 *M* NaCl extracts from the nuclei of cells producing large amounts of HMG-I(Y) proteins (see earlier). Prior to loading of samples on the PBE94 column (20 × 1.0 cm ID; equilibrated in TE buffer), native HMG samples in TE are centrifuged at 5000*g* for 20 min at 0–4° to remove any debris. The samples (up to about 50 mg of protein in a 3-ml volume[30]) are loaded onto the column and proteins are eluted with a 0–1 *M* NaCl linear gradient in TE buffer. Absorbance is measured at 220 nm, and 4-ml fractions are collected at a flow rate of 10 ml/hr. The HMG-I(Y) proteins elute from the column at around 200 m*M* NaCl. Aliquots of the eluted fractions are analyzed by SDS–PAGE, with appropriate reference standards, to accurately identify those containing the HMG-I(Y) proteins and to determine the purity of the fraction. The identified protein fractions are then dialyzed overnight against TE and concentrated as described earlier. If the identified fractions are contaminated with other proteins, they can be rechromatographed on the PBE94 column with a shallower NaCl gradient to better resolve the contaminants from the HMG-I(Y) proteins.

[28] U. K. Laemmli, *Nature* **227,** 680 (1970)
[29] S. Panyim and R. Chalkley, *Arch. Biochem. Biophys.* **130,** 337 (1969).
[30] Y. Adachi, S. Mizuno, and M. Yoshida, *J. Chromatogr.* **530,** 39 (1990).

Chromatographic Purification of Recombinant HMG-I(Y) Proteins

Recombinant HMG-I(Y) proteins are purified from crude acid-extracted bacterial precipitates by cation ion-exchange chromatography on a Macro-Prep 50-S column (1 × 20 cm; Bio-Rad Laboratories) employing a potassium salt gradient with protein absorbance being monitored at 220 nm.[22] The buffers used with this column are:

Buffer A: 25 m*M* KH_2PO_4, pH 7.0 (use of "HPLC grade" phosphate is recommended, as the "reagent grade" often contains UV opaque materials).

Buffer B: Buffer A plus 1.0 *M* KCl.

The column is initially equilibrated in 5% buffer B (50 m*M* KCl) and is loaded with approximately 2 mg of recombinant protein dissolved in the same buffer. For elution, the KCl concentration is initially raised to 300 m*M* in 5 min (5% buffer B/min) followed by 300–550 m*M* in 70 min (0.36% B/min) at a flow rate of 1.0 ml/min. One-milliliter fractions are collected and their protein content is determined by 15% SDS–PAGE analysis. From these electrophoretic results, determine the fractions to keep, pool them, and dialyze the mixture against several changes of Milli-Q water to remove salt (use 3000 molecular weight cutoff dialysis tubing). Finally, the volume and concentration of the purified recombinant protein in the fractions are determined, the proteins vacuum concentrated to dryness using a Speed-Vac apparatus, and the dried HMG-I(Y) samples stored at −20°.

Figure 2A shows the chromatographic profile of the first passage of recombinant human HMG-I protein over a Macro-Prep 50-S cation-exchange column. It is evident from the profile that the recombinant HMG-I protein elutes as two major peaks, labeled 1 and 2 (Fig. 2). Peptide and amino acid analysis indicates that peak 1 is the full-length recombinant HMG-I protein and that peak 2 is a truncated form of the protein lacking its carboxyl-terminal end (unpublished data). This truncated form of the recombinant human HMG-I protein retains its three A·T-hook DNA-binding motifs and thus binds to DNA in a manner similar to the full-length protein. Thus, in electrophoretic mobility shift assays (EMSAs), when partially purified recombinant HMG-I protein preparations are used that contain both peaks 1 and 2, a doublet band of retarded protein–DNA complexes is observed.[31] As shown in Fig. 2B, when the major fraction of peak 1 is rechromatographed on a second Mono S column, the full-length recombinant HMG-I protein can be purified substantially (>95%) away from the truncated contaminant. If greater purity of recombinant protein is required, the protein in peak 1 can be chromatographed on an RP-HPLC

[31] S. J. Fashena, R. Reeves, and N. H. Ruddle, *Mol. Cell. Biol.* **12,** 894 (1992).

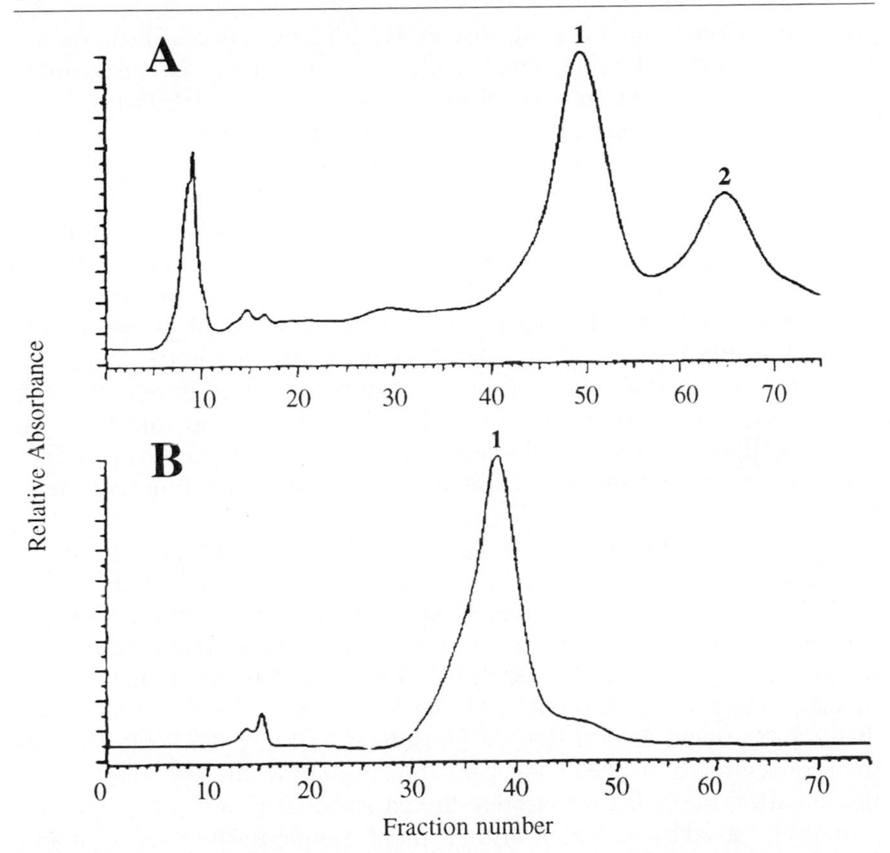

FIG. 2. Purification of recombinant human HMG-I protein by chromatography on a Macro-Prep 50-S cation-exchange column. (A) Chromatographic elution profile of the crude acid-soluble proteins from induced bacteria harboring expression plasmid pET7C containing the full-length human HMG-I cDNA. Peak 1 contains full-length recombinant human HMG-I protein and peak 2 contains recombinant protein that has been truncated at its C-terminal end by residual proteolytic activity in the host bacteria. (B) Chromatographic profile of the proteins in peak 1 of (A) that have been rechromatographed on a Macro-Prep 50-S column under slightly different elution conditions from that shown in (A).

column (see earlier) and the full-length, homogeneous HMG-I protein isolated.

Production and Use of Anti-HMG-I(Y) Antibodies

Antibody Production and Western Blotting Methods

Because of their highly conserved nature and their unusual amino acid composition, it is often difficult to produce specific, high-titer/high-avidity rabbit polyclonal antibodies against HMG-I(Y) proteins.[32] Furthermore, even when such antibodies are raised successfully there is considerable variability between different rabbits in the quality of anti-HMG-I(Y) antibody produced. For example, we have found that while most rabbits can be induced to produce antibodies useful for Western blot assays, few produce a specific, high-titer antibody that can be used effectively in EMSA and immunoprecipitation assays. Nevertheless, such specific polyclonal antibodies have been produced and utilized in both *in vitro* and *in vivo* studies of HMG-I(Y) proteins.[7,31–36] Mononclonal antibodies against the HMG-I(Y) proteins would be useful experimental reagents, but so far, probably due to the difficulties noted earlier, none have been reported in the literature.

Although anti-HMG-I(Y) antibodies can be produced in rabbits when pure protein is cross-linked to *Limulus* hemocyanin and injected with Freund's adjuvant,[32] a more reliable and consistent immunization procedure is to use Ribi adjuvant (MPL + TDM + CWS emulsion; Ribi ImmunoChem, Hamilton, MT). For chemical cross-linking, 1 ml of chromatographically pure HMG-I protein (5 mg/ml in PBS) is added to 1 ml of hemocyanin (5 mg/ml in PBS) and to the mixture (2 ml volume) is *slowly* added, with continuous stirring, 0.5 ml of freshly made 0.5% aqueous glutaraldehyde. The mixture is allowed to cross-link for 1–2 hr at room temperature and the reaction is terminated by adding 0.1 *M* glycine to the reaction, followed by dialysis against 0.1 *M* $(NH_4)_2CO_3$ for 3–4 hr at 4°. The solution is then dialyzed overnight against 0.05 *M* phosphate buffer, pH 7.0 (at 4°), and then concentrated by Speed-Vac centrifugation down to 2 ml volume. For the initial injection, 0.2 ml of cross-linked material (~500 μg of HMG-I protein) is mixed by vortexing with 0.2 ml of complete Ribi's adjuvant and is injected subcutaneously into a rabbit at several different locations. Ten to 14 days later a second injection of 0.2 ml of cross-linked material with 0.2 ml of *incomplete* Ribi's adjuvant is similarly administered, with additional injections of 0.2 ml of pure cross-linked proteins (without any adjuvant) being given at about 2-week intervals until antibody is produced (usually after 3–6 weeks). The anti-HMG-I(Y) serum titer, specificity, and cross-reactivity are monitored during the course of the immunization regime

[32] J. E. Disney, K. R. Johnson, N. S. Magnuson, S. R. Sylvester, and R. Reeves, *J. Cell Biol.* **109,** 1975 (1989).

[33] S. John, R. Reeves, J.-X. Lin, R. Child, J. M. Leiden, C. B. Thomas, and W. J. Leonard, *Mol. Cell. Biol.* **15,** 1786 (1995).

[34] Y. Saitoh and U. K. Laemmli, *Cell* **76,** 609 (1994).

[35] S. R. Himes, L. S. Cole, R. Reeves, and M. F. Shannon, *Immunity* **5,** 479 (1996).

[36] L. T. Holth, A. E. Thorlacius, and R. Reeves, *DNA Cell Biol.* **16,** 1299 (1997).

using standard procedures.[37] For experiments involving *in situ* immunolocalizations, immunopreciptiations, and electrophoretic mobility shift assays, the IgG fraction from the serum is isolated (with a protein A column)[32] and used. Many of the anti-HMG-I(Y) preparations produced in this manner show some cross-reactivity with histone H1.[10] To circumvent this problem, antibody that is specific for only HMG-I(Y) proteins can be obtained by affinity purifying the IgG fraction on an Affi-Gel10 resin (Bio-Rad) linked covalently to purified recombinant HMG-I protein.[37]

For Western blot analysis of HMG-I(Y) proteins using either crude polyclonal sera or partially purified IgG preparations, a stringent washing protocol has been developed (courtesy of Dr. Stefan Dimitrov) that gives a high signal and low background when used in conjunction with standard chemiluminescence detection methods. The antibody reaction and washing procedure, after the usual Western blot transfer of proteins to a membrane and blocking procedures,[37] is as follows.

1. Immediately after the primary antibody incubation [usually about 2 hr at room temperature; anti-HMG-I(Y) rabbit serum or IgG, dilution 1:250] the membrane blot is rinsed in running distilled water for about 5–10 sec to denature and remove unbound serum proteins or IgGs.
2. The blot is then rinsed for 50 min, and then for an additional 5 min, in *each* of the following solutions: PBS containing 0.5 *M* NaCl plus 0.5% Tween 20 and PBS plus 0.5% Tween 20.
3. The blot is incubated in the secondary antibody (horseradish peroxidase-conjugated antirabbit IgG; dilution ~1:5000) for 1 hr at room temperature.
4. Repeat the washes of step 2.
5. Follow standard commercially available chemiluminesence detection protocols using luminol as a substrate (e.g., Supersignal substrate for Western blots from Pierce Chemical Co., or luminol reagent from Santa Cruz Biotechnology, Inc., Santa Cruz, CA).

In Vitro Functional Assays for HMG-I(Y) Proteins

DNase I Footprinting and Directional Substrate Binding

A characteristic feature of HMG-I(Y) proteins that distinguishes them from other HMG proteins is their ability to preferentially bind, with high

[37] F. M. Ausubel, R. Brent, R. Kingston, J. G. Seidman, J. A. Smith, and K. Struhl, "Current Protocols in Molecular Biology," Vols. 1–3. Wiley, New York, 1988.

affinity, to the minor groove of A·T-rich stretches of B-form DNA both *in vitro* and *in vivo*.[1,11] Furthermore, HMG-I(Y) proteins bind to preferred DNA substrates *in vitro* with a distinct directionality, or polarity, that is dependent on the structure, rather than the sequence, of the substrate.[8,10,11] Both of these points are illustrated by the results shown in Fig. 3. For the

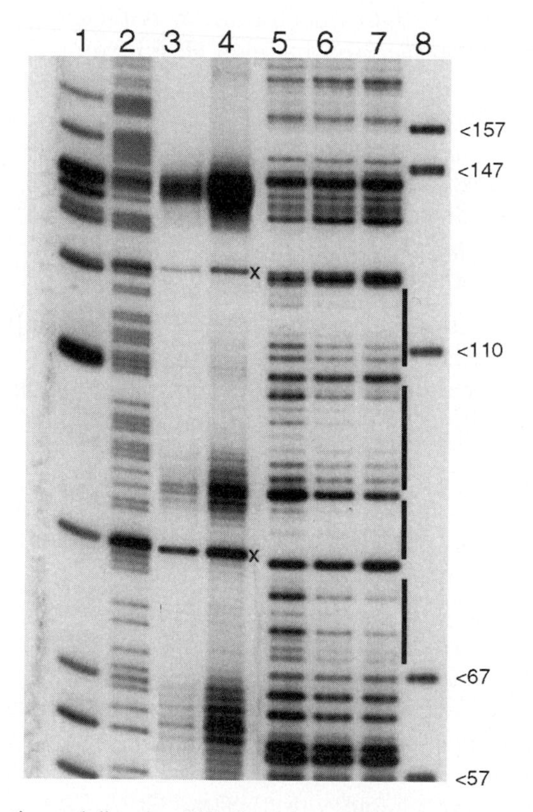

FIG. 3. Footprinting and directional binding of recombinant HMG-IΔE91 protein to A·T sequences on BLT DNA. 1,10-Phenanthroline (OP)-derivatized (lanes 3 and 4), recombinant HMG-IΔE91 protein was bound to ^{32}P-end-labeled DNA and then subjected to different DNA cleavage reactions. In lanes 3 and 4, copper was added to the OP–HMG-IΔE91/DNA complex to induce DNA strand scission by oxidative cleavage,[38] whereas DNase I was added to the complex to enzymatically cleave the DNA in lanes 5, 6, and 7. After the cleavage reactions the products were separated by electrophoresis on a sequencing gel and autoradiography was performed. Lanes: 1, G-specific cleavage reaction of BLT DNA; 2, G + A cleavage reaction; 3 and 6, 10 ng protein; 4 and 7, 50 ng of recombinant protein; 5, no protein added; and 8, molecular weight markers (*Msp*I-digested pBS·SK$^-$ plasmid). The vertical bars next to lane 7 indicate areas of protection against DNase I digestion by the bound protein shown in lanes 6 and 7. An "x" marks the site of artifactual cleavage bands observed in OP–HMG-IΔE91 cleavage reactions in lanes 3 and 4.

footprinting experiments shown in Fig. 3, a carboxyl-terminal deleted form of the human HMG-I protein truncated at glutamic acid residue 91, hence designated HMG-IΔE91, was created by site-specific *in vitro* mutagenesis.[24] This artificially truncated form of the recombinant human HMG-I protein retains its three A · T-hook DNA-binding motifs and thus binds to A · T-rich sequences of DNA in an identical manner to the full-length protein. In addition, because of the use of a translation stop codon in the bacterial expression vector used for its production (see earlier), the recombinant HMG-IΔE91 protein contains a unique cysteine residue at its carboxyl-terminal end that can be used for attachment of a chemical cleavage reagent to determine the polarity/directionality of binding of the protein to a DNA substrate.[38]

Lanes 5–7 of Fig. 3 demonstrate specific binding of the recombinant HMG-IΔE91 protein to A·T-rich stretches of DNA in the 3′-untranslated region of bovine interleukin-2 cDNA (i.e., BLT DNA).[6] For these experiments, increasing concentrations of recombinant protein (0, 20, and 50 ng; lanes 5-7, Fig. 3, respectively) were added to ^{32}P-end-labeled BLT DNA, digested with DNase I at room temperature, and the digestion products fractionated on a 6% polyacrylamide sequence gel. Both the optimal concentration of DNase I and the appropriate digestion time for each reaction were determined empirically in preliminary experiments.[24] As indicated by the vertical bars next to lane 7, the recombinant protein specifically footprinted to the expected A · T stretches in the BLT DNA.[5] In addition, the results shown in lanes 3 and 4 demonstrate that the HMG-IΔE91 protein binds to the BLT DNA with a distinctly polar orientation. For these experiments, the chemical nuclease *N*-(1,10-phenanthrolin-5-yl)iodoacetamide (OP; Molecular Probes, Eugene, OR)[38] was attached covalently to the cysteine residue at the C-terminal end of the recombinant protein, and various concentrations of OP-HMG-IΔE91-conjugated protein (10 and 50 ng; lanes 5 and 6, Fig. 3, respectively) bound to the end-labeled BLT DNA substrate. The DNA/protein complexes were then exposed to conditions that produced oxidation-mediated scission of the DNA by creation of an activated OP–copper complex at the C-terminal end of the protein (i.e., by the addition of 100 μM $CuSO_4$, 0.02% H_2O_2, and 6 mM 3-mercaptopropionic acid to the complexes[38]). After a 5-min reaction at room temperature, the reaction was quenched by the addition of 3 mM 2,9-dimethyl-1,10-phenanthroline, and products of the chemical cleavage were separated on a sequencing gel, as shown in lanes 3 and 4. Directionality of binding of the HMG-1 protein to the DNA substrate is indicated by the

[38] C. Q. Pan, J.-A. Feng, S. E. Finkel, R. Landgraf, D. Sigman, and R. C. Johnson, *Proc. Natl. Acad. Sci. U.S.A.* **91,** 1721 (1994).

strong hydroxyl cleavage sites that are offset in register from the sites of specific footprinting of the protein to A · T sequences on the same substrate (lanes 5–7).[38] The bands marked with an "x" in lanes 3 and 4 are the results of nonspecific, artifactual cleavages intrinsic to the OP cleavage reaction.

Hydroxyl Radical Footprinting

Localization of HMG-I(Y) proteins on DNA or nucleosomal substrates employing hydroxyl radical cleavage reaction has been used as a diagnostic tool to demonstrate that these proteins recognize DNA structure rather than sequence.[8] For hydroxyl cleavage reactions, ^{32}P-end-labeled DNA fragments, either free or bound with HMG-I(Y) protein and/or nucleosomal histones, are processed according to the method of Wolffe and Hayes.[39] A solution of Fe(II) · EDTA is freshly prepared by mixing equal volumes of 1 m*M* $Fe(NH_4)_2(SO_4)_2 \cdot 6H_2O$ and 2 m*M* EDTA. The cutting reaction is initiated by placing the Fe(II) · EDTA solution, 0.12% H_2O_2, and 10 m*M* ascorbic acid (Na salt) on the inner wall of a tilted 1.5-ml Eppendorf centrifuge tube containing the target nucleoprotein complex. The cleavage reagents are mixed gently on the side of the tube and then added to the nucleoprotein complex in the bottom of the tube, and the two solutions are mixed thoroughly by pipetting. The ensuing hydroxyl radical reaction is quenched after an appropriate length of time (empirically determined for each experiment) by the addition of glycerol to 5%. The samples are then digested with proteinase K in the presence of 0.2% SDS, extracted with phenol/chloroform, and precipitated with ethanol. The recovered cleaved DNA is resolved by electrophoresis on a 6% sequencing gel, along with specific DNA markers produced by Maxam–Gilbert cleavage of the target DNA at G residues.[37] The sequencing gels are visualized using a phosphorimager (Molecular Dynamics Corp., Sunnyvale, CA) and cleavage band intensities are quantified and analyzed using ImageQuant (Molecular Dynamics Corp.) and Microsoft Excel 4.0 software. Fourier transformations and other data analyses can be performed using SigmaPlot and PeakFit software programs from Jandel Scientific (Corte Madera, CA).

Fluorescence Competition Binding Assays

The organic dye Hoechst 33258, like HMG-I(Y), binds specifically to the minor groove of A · T-rich DNA and fluoresces when it binds substrate. Thus, competition of wild-type and mutant HMG-I(Y) proteins with Hoechst 33258 for binding to substrate is a convenient and quantitative method for determining the binding constants of these protein ligands

[39] A. P. Wolffe and J. J. Hayes, *Methods Mol. Genet.* **2,** 314 (1993).

for A·T-rich DNA.[6,40] In fluorescence competition experiments, solutions containing 10 mM HEPES buffer, pH 7.5, 50 mM NaCl, 1 mM EDTA, BLT DNA (100 nM expressed as phosphate),[6] and various concentrations of individual HMG-I(Y) proteins are titrated with Hoechst 33258 (Sigma, St. Louis, MO). Final concentrations of Hoechst dye in the titrations range from 0 to 50 nM. Fluorescence of the dye is excited at 354 nm and observed at 450 nm in a fluorescence spectrophotometer (e.g., a Shimadzu RF-540). The change in fluorescence (ΔF) of the sample due to binding of the dye is expressed as the difference of the fluorescence of the test solution and a "blank" containing only buffer and dye.

Binding of Hoechst 33258 to DNA in the presence or absence of competitors can be analyzed using the Michaelis–Menten equation.[6] Curve fitting of experimental data (ΔF versus [dye]) can be performed conveniently using commercially available programs such as the nonlinear regression analysis program Enzfitter[6] or the curve-fitting analysis program SigmaPlot (Jandel Scientific). These programs permit direct determination of the dissociation constant, K_d, of the dye in the absence of competing ligand or the apparent dissociation constant, K_{app}, in the presence of the competing ligand. Using the relationship $K_{app} = K_d(1 + [\text{ligand}]/K_{\text{ligand}})$,[41] the dissociation constant of the competing HMG-I(Y) ligand is determined from a plot of K_{app} versus [ligand]. Representative results from such fluorescence competition experiments are shown in Fig. 4 in which the K_d for the binding of Hoechst dye is determined to be ~9.6 nM and the K_d for HMG-I protein is ~1.02 nM.[6]

Nucleosome-Binding Assays

HMG-I(Y) proteins have been demonstrated to bind specifically to DNA at four locations on the front surface of random sequence nucleosome core particles, at the entrance and exist sites, and in the distorted regions of DNA flanking either side of the dyad axis.[24] Additionally, these proteins also bind to, and can induce rotational changes in, defined sequence DNAs on the surface of nucleosome core particles that have been reconstituted *in vitro*.[8] These studies, together with those demonstrating specific binding of HMG-IY proteins to four-way junction DNA (see later), clearly support the finding of NMR structural studies[11]: that the principal determinant for specific binding of these proteins is substrate structure, not nucleotide sequence. In addition, because none of the other HMG proteins exhibit these same nucleosome core particle binding characteristics, such binding

[40] J. S. Siino, M. S. Nissen, and R. Reeves, *Biochem. Biophys. Res. Commun.* **207,** 497 (1995).
[41] I. H. Segel, "Biochemical Calculations," 2nd Ed. Wiley, New York, 1976.

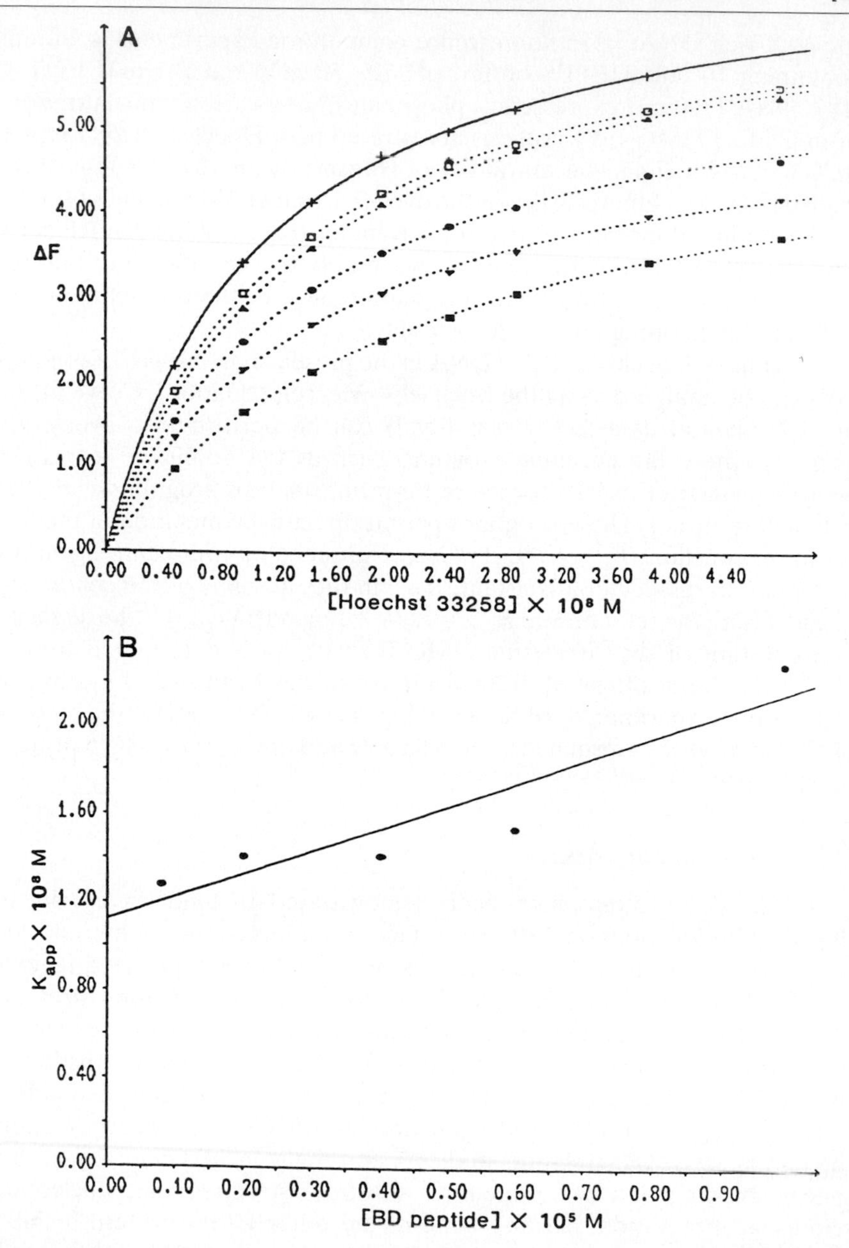
A
5.00
4.00
ΔF
3.00
2.00
1.00
0.00
0.00 0.40 0.80 1.20 1.60 2.00 2.40 2.80 3.20 3.60 4.00 4.40
[Hoechst 33258] × 10⁸ M
B
2.00
1.60
1.20
0.80
0.40
0.00
K_app × 10⁸ M
0.00 0.10 0.20 0.30 0.40 0.50 0.60 0.70 0.80 0.90
[BD peptide] × 10⁵ M

activity constitutes a significant *in vitro* diagnostic assay for the functionality of the HMG-I(Y) family of proteins.

Electrophoresis of HMG-I Nucleosome Core Particle Complexes

The typical starting material for these binding assays is trimmed chicken erythrocyte nucleosome core particles obtained by a modification of the isolation method of Libertini and Small.[42] Complexes of HMG-I(Y) proteins, or control proteins, and nucleosome core particles are analyzed by electrophoretic separation on nondenaturing gels containing 6% acrylamide, 0.2% bisacrylamide, and 0.1× Tris–borate–EDTA (TBE) buffer (90 mM Tris borate, 90 mM boric acid, 2 mM EDTA, pH 8.0). Core particles are mixed with purified HMG-I(Y) proteins in 0.1× TBE, 5% (v/v) glycerol to produce complexes containing from 0.5- to 4(or more)-fold molar excess of protein to core particle. These nucleosome/HMG-I complexes are separated electrophoretically on 1.5-mm polyacrylamide slab gels, at room temperature, with TBE buffer recirculation, as described.[43] The gels are stained with Coomassie Brilliant Blue. With increasing input ratios of protein, the gels should reveal up to four specific HMG-I/nucleosome complexes corresponding to core particles containing one to four bound HMG-I(Y) proteins.[8] In contrast, titration of core particles with HMG14 or HMG17 proteins form only two specific complexes with trimmed core particles,[43] whereas HMG1, HMG2, and other control proteins do not form specific complexes with core particles under these conditions.[8]

[42] L. J. Libertini and E. W. Small, *Nucleic Acids Res.* **8,** 3517 (1980).
[43] A. E. Paton, E. Wilkinson-Singley, and D. E. Olins, *J. Biol. Chem.* **258,** 13221 (1983).

FIG. 4. Fluorescence competition assay for determining the binding constant of HMG-I(Y) protein to A·T-rich DNA. (A) Binding of Hoechst 33258 to A·T-rich BLT DNA in the presence or absence of competing HMG-I protein. The solid curve (+) represents the change in fluorescence (ΔF) observed when BLT DNA is titrated with Hoechst dye in the absence of a competing protein ligand. Dotted curves represent the titration of DNA in the presence of 1.39 nM (Δ), 2.77 nM (■), 4.14 nM (▼), 5.22 nM (●), and 6.89 nM (□) HMG-I protein. In all experiments the BLT DNA concentration was fixed at 100 nM and the dye concentration varied from 0 to 50 nM. K_{app} of dye binding to DNA in the presence of competing protein is determined by nonlinear regression analysis of the ΔF versus [dye] curves. (B) Dissociation constant K_d of the specific DNA–protein interaction is determined from the ratio of intercept to slope of plots of K_{app} versus [HMG-I]. Figure redrawn from Reeves and Nissen,[6] with permission.

Footprinting of HMG-I(Y) Proteins on Nucleosomes Reconsituted in Vitro with Defined Sequence DNAs

HMG-I(Y) proteins bind to and change the rotational setting of defined sequence DNAs that have been reconsituted into monomer and dinucleosomes *in vitro.*[8] In these experiments, radiolabeled, cloned DNA fragments containing A·T-rich segments that are intrinsically bent (phased A·T tracts), flexible (oligo[d(A-T)]), or straight and rigid [oligo(dA)·oligo(dT)] were reconsituted into nucleosomes by either exchange with core particles or dialysis from high salt/urea with purified histone octamers as described in detail by Wolffe and Hayes.[39] Reconsituted monomer and dinucleosomes were incubated with recombinant human HMG-I(Y) protein at various HMG:core particle molar ratios, and the structure of the DNA on the resulting chromatin complexes was analyzed by either DNase I or hydroxyl radical cleavage methods (see earlier). In such analyses, the optimum cleavage conditions for each reagent, and for each reconstituted chromatin preparation, must be determined empirically. After treatment with the cleavage reagents, the reconstituted chromatin particles are immediately isolated from cleavage reagents, free DNA, and any other contaminating materials by electrophoresis on 0.8% agarose (45 m*M* Tris–borate, pH 8.3, 1 m*M* EDTA) native nucleoprotein gels. After electrophoresis, the positions of the reacted monomer or dimer nucleosomes are identified by autoradiography, the appropriate gel fragments are excised, and the chromatin particles are electroeluted. The labeled DNA fragments are recoved from the chromatin particles by protease digestion and phenol/chloroform extraction (1:1, v/v) and precipitated with ethanol. Single-stranded DNA cleavage products are then separated by electrophoresis on either 6 or 8% sequencing gels with Maxam–Gilbert chemical cleavage products of control DNA fragments serving as reference standards.[37] Autoradiograms of the sequencing gels are analyzed on a Molecular Dynamics computing densitometer with ImageQuant, Microsoft Excel, SigmaPlot, and PeakFit software programs as described earlier. The helical periodicities of nucleosomal DNAs are determined quantitatively from patterns of the hydroxyl radical cleavage reactions. Briefly, the area of each DNA cleavage band from the reconstituted chromatin reaction is determined by integration using the ImageQuant program after subtracting the integrals of bands from the control (free DNA) samples corresponding to those in the nucleosomal samples. The resulting values, which represent the amount of cleavage at each nucleotide in the reconstitute sample, are then smoothed and analyzed using the curve-fitting functions of the PeakFit program, and the helical periodicity of the DNA on the nucleosome is determined by calculating the distances, in base pairs, between peaks of cleavage.[8]

DNA Bending and Supercoiling Assays

The ability of HMG-I(Y) proteins to unwind, bend/straighten, or otherwise induce distortions in DNA substrates has been demonstrated by a variety of techniques, including circular dichroism studies,[44] circular permutation assays,[10,45] ligation-mediated ring closure assays,[10] and plasmid supercoiling assays.[4] The latter assays are particularly informative because, as illustrated in Fig. 5, they demonstrate that HMG-I(Y) proteins have the unusual ability (not shared by other HMGs) to introduce positive supercoils in relaxed plasmid molecules at low concentrations of bound protein, whereas at higher input ratios of protein they induce negative supercoils into these same plasmid substrates.

Topoisomerase I-mediated relaxation assays are used to monitor the amount of supercoiling induced by HMG-I(Y) proteins in plasmid DNA substrates.[4] For a typical assay, pure HMG-I, HMG-Y, or various control proteins are added in varying ratios to 2 μg of negatively supercoiled plasmid DNA (pBLT or pUC18) and incubated at 23° for 30 min in 200 μl of 20 m*M* Tris–HCl, pH 8.0, 50 m*M* NaCl, 50 μg/ml serum albumin, 5% glycerol, and 1 m*M* dithiothreitol. Twenty units of avian topoisomerase I (isolated from chicken erythrocytes[4]) is added and the mixture is incubated for 1 hr at 37°. In parallel control reactions, plasmid DNAs are incubated with varying concentrations of either netropsin (which introduces positive supercoils in this assay) or ethidium bromide (EtBr, which introduces negative supercoils in this assay). After the incubation period, the samples are made 0.1% in SDS and digested with 0.1 mg/ml of proteinase K (Sigma) for 30 min at 37° followed by extensive extractions with phenol/$CHCl_3$ and ethanol precipitation of the DNA.

Two-dimensional electrophoresis in native agarose gels, with chloroquine incorporated into the gel in the second dimension, is used to determine both the direction and the extent of ligand-induced supercoiling in plasmid DNAs. DNAs recovered from the topoisomerase I relaxation assays are dissolved in TE buffer and loaded onto 1.5% agarose–TAE (40 m*M* Tris–HCl, pH 8.3, 20 m*M* sodium acetate, 2 m*M* EDTA) gels and are electrophoresed for 3 hr at 6 V/cm at 23°. The gels are then soaked for 1 hr in TAE buffer containing 1.5 μg/ml chloroquine phosphate (Sigma), rotated 90° with respect to the original direction of electrophoretic migration, and the electrophoresis repeated in TAE buffer supplemented with 1.5 μg/ml chloroquine phosphate. The gels are then stained with EtBr and photographed.

[44] D. A. Lehn, T. S. Elton, K. R. Johnson, and R. Reeves, *Biochem. Int.* **16,** 963 (1988).
[45] J. V. Flavo, D. Thanos, and T. Maniatis, *Cell* **83,** 1101 (1995).

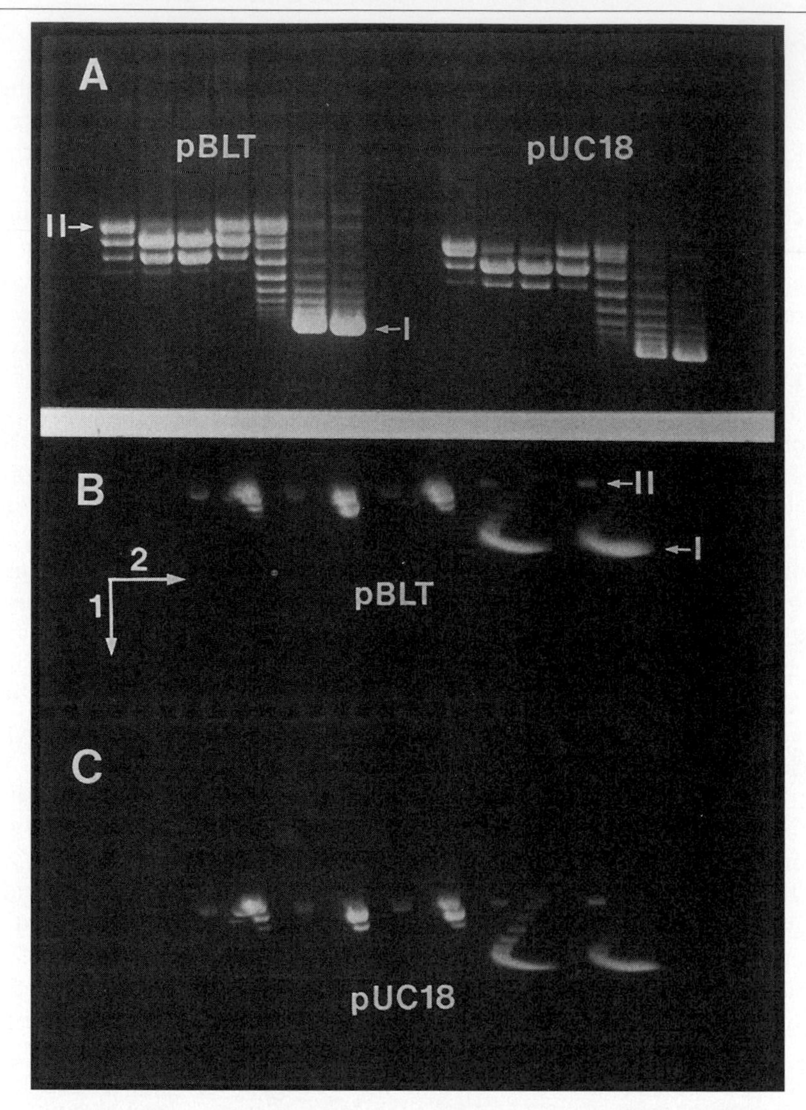

FIG. 5. Effects of recombinant HMG-I protein on the supercoiling of two plasmids, pBLT and pUC18, as detected by a topoisomerase I-mediated relaxation assay. (A) One-dimensional gel showing the results of increasing concentrations of the carboxyl-terminal-deleted protein HMG-IΔE91 on supercoiling of the two plasmids. In each case the protein to nucleotide molar ratios, from left to right, are 0, 0.04, 0.07, 0.11, 0.14, 0.20, and 0.27. (B and C) Two-dimensional gels containing 1.5 μg/ml chloroquine in the second dimension. Protein to nucleotide molar ratios are 0, 0.07, 0.11, 0.22, and 0.27. The migration direction of each dimension is indicated. The upper set of bands (B) shows pBLT DNA and the lower set (C) pUC18 DNA. The relative position of form I (supercoiled) and form II (relaxed) plasmids are indicated. Reproduced, with permission, from Nissen and Reeves.[4]

Chloroquine intercalates into DNA and can be used to distinguish positive and negative supercoils that have been introduced into plasmids during topoisomerase I-mediated relaxation assays. For example, chloroquine introduced into gels during the second-dimension electrophoretic separation will retard the rate of migration of negatively supercoiled plasmids while increasing the rate of migration of relaxed and positively supercoiled plasmids. This effect is clearly seen in Fig. 5. Migration of negatively supercoiled topoisomers tend to the *left* of Fig. 5 (i.e., toward the origin of the second dimension), whereas positively supercoiled topoisomers tend to the *right* (i.e., toward the direction of electrophoretic migration in the second dimension). Figure 5 shows two-dimensional gels of two plasmids, pBLT and pUC18, relaxed in the presence of increasing concentrations (from left to right in the gel) of recombinant human HMG-I protein. At low concentrations of input protein the plasmids initially exhibit positive supercoiling, i.e., the mobility of the topoisomers increases in the second electrophoretic dimension. However, as the input protein concentration increases further, the plasmids become negatively supercoiled as demonstrated by the reduced mobility of the ensemble of topoisomers.

Four-Way Junction (FWJ)-Binding Assays

Consistent with the observation that HMG-I(Y) proteins recognize the structure of the narrow minor groove of A·T-rich sequences in B-form DNA, as well as distorted DNA structures on the surfaces of nucleosomes, these proteins also preferentially recognize other altered DNA structures, such as synthetic FWJ DNAs *in vitro.*[9] Recombinant human HMG-I binds to FWJ DNA with a $K_d \sim 6.5$ nM and can outcompete both histone H1 ($K_d \sim 16$ nM) and HMG-1 ($K_d \sim 80$ nM) in titration competition assays.[9] As shown in Fig. 6, hydroxyl radical footprinting analysis (employing the procedures outlined earlier) demonstrates that the HMG-1(Y) proteins bind preferentially to the central crossover region of FWJs, the same site of binding of both HMG-1 and histone H1 proteins.[10] Such preferential and tight binding to FWJ DNA *in vitro,* which is detected easily by EMSA assays (see later), is a convenient and quantitative method to demonstrate the *in vitro* functional activity of new preparations of HMG-I(Y) proteins and is one we employ routinely for this purpose.

In Vivo Functional Assays for HMG-I(Y)

Acting as architectural transcription factors, HMG-I(Y) proteins have been suggested to be involved in the *in vivo* regulation of more than a dozen mammalian genes. In addition, these proteins have been implicated

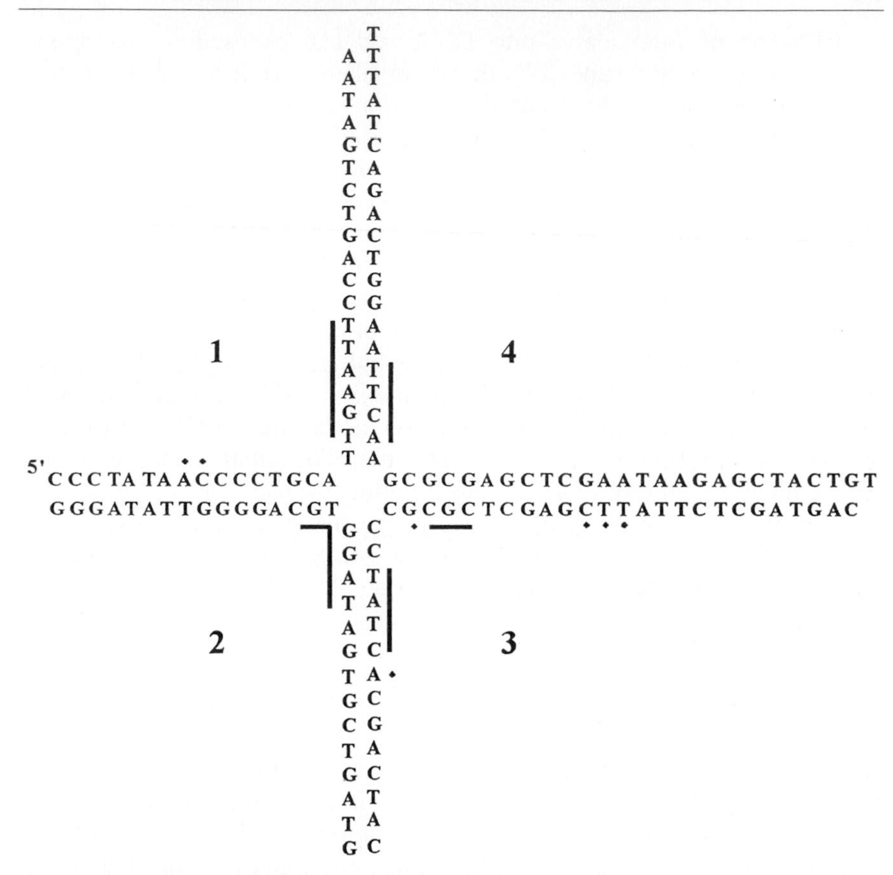

FIG. 6. The sequence and regions of FWJ DNA that recombinant human HMG-I protein protects from hydroxyl radical cleavage while in a 1 : 1 DNA–protein complex. Bars represent the regions of major protection and dots indicate areas of minor protection. Reproduced, with permission, from Hill and Reeves.[9]

in chromosome condensation during the cell cycle, have been demonstrated to interact with the front face of nucleosome core particles *in vivo,* and have been suggested to be involved with alterations in the structure of nucleosomes and chromatin domains *in vivo*[1,7,8,34,46] A number of lines of evidence have been advanced to support these putative *in vivo* functions for HMG-I(Y) proteins. The experimental approach and methods used for some of these *in vivo* assays are outlined, but the reader should go to the original cited literature for more detailed information.

[46] R. Reeves, *Curr. Opin. Cell Biol.* **4,** 413 1992.

In Situ Immunolocalization of HMG-I(Y) Proteins

Polyclonal anti-HMG-I(Y) antibodies have been used in fluorescence *in situ* immunolocalization studies to demonstrate that these proteins are localized on A·T-rich G/Q bands and around the centromere of human and mouse metaphase chromosomes.[32] Treatment of cells with the A·T-binding drugs netropsin or distamycin will displace the HMG-I(Y) proteins from these binding sites on metaphase chromosomes and lead to detectable elongation and decondensation of the chromosomes, suggesting that HMG-I(Y) proteins are involved in changes in chromosome structure during the cell cycle.[10,46,47] Support for such an idea also comes from immunolocalization studies that have colocalized the HMG-1(Y) proteins and topoisomerase II to the A·T-rich scaffold, or "queue," along the backbone of metaphase chromosomes.[34] Furthermore, additional studies with synchronized mammalian cells demonstrate that *in vivo* the HMG-I(Y) proteins are phosphorylated at metaphase by the cell cycle regulating enzyme $p^{34/cdc2}$ kinase.[48] Such phosphorylation markedly decreases the affinity of binding of HMG-I(Y) proteins to A·T-rich DNA *in vitro,*[49] consistent with the notion that these proteins are involved intimately with the dynamics of the chromatin condensation cycle *in vivo.*

Immunofluorescence Reagents and Conditions

In immunolocalization studies, care must be taken during the fixation process not to extract or artificially rearrange the HMG-I(Y) proteins in mammalian cells or chromosomes. Because HMG proteins are soluble in the commonly used dilute acid–methanol mixture used traditionally for the cytological examination of mammalian cells, this fixation method is inadequate. Fixation of mammalian cells with either formaldehyde or strong acidic solutions (such as Bouin's reagent) is therefore preferred.

A convenient protocol for the immunolocalization of HMG-I(Y) proteins on metaphase chromosomes of mammalian cells growing as a monolayer in culture is as follows. Approximately 50% confluent cells in 10 tissue culture dishes (10 cm) are blocked in metaphase with 0.025 μg/ml Colcemid for 90 min. The culture medium is removed from the dishes, and mitotic cells are then selectively dislodged from the dishes by gentle shaking, or pipetting, in PBS (phosphate-buffered saline) solution and collected by centrifugation (200*g* for 5 min, 4°). The cells are fixed immediately by resuspension in cold PBS containing 4% formaldehyde (freshly prepared

[47] M. Z. Radic, T. S. Elton, R. Reeves, and B. A. Hamkalo, *Chromosoma* **101,** 602 (1992).
[48] R. Reeves, T. Langan, and M. Nissen, *Proc. Natl. Acad. Sci. U.S.A.* **88,** 1671 (1991).
[49] M. Nissen, T. A. Langan, and R. Reeves, *J. Biol. Chem.* **266,** 19945 (1991).

from paraformaldehyde) and fixed for 10 min at 4°. The fixed cells are centrifuged for 8 min at 1000 rpm in a cytocentrifuge (Shandon Southern Instruments, Sewickley, PA) onto the surface of a round coverslip, and the coverslip is rinsed three to five times with cold PBS without any fixative. An alternative, and equally effective, method is to centrifuge unfixed mitotic cells in cold PBS onto coverslips followed by immediate fixation with Bouin's fluid (Polysciences, Inc., Warrington, PA), several rinses with distilled water, and, finally, by three rinses with cold PBS (5 min at each step).

An indirect immunofluoresence procedure is usually employed for chromosomal localization of the HMG-I(Y) proteins.[32] Fixed cells or chromosomes on coverslips are soaked for 30 min in 1% bovine serum albumin (BSA) in cold (4°) PBS (PBS/BSA). The cells are then incubated overnight at 4° with the IgG fraction (or affinity purified IgG fraction) of rabbit anti-HMG-I(Y) serum diluted 1:1000 in PBS/BSA. As a control, in parallel experiments, cells should be incubated with antihistone antibodies (Chemicon International, Inc., El Segundo, CA) to immunolocalize the distribution of these proteins on metaphase chromosomes for comparison. After incubation with the primary antibody, the cells are washed with PBS/BSA and incubated at room temperature for 2 hr with biotinylated goat antirabbit IgGs (Sigma) diluted 1:4000 in PBS/BSA. The cells are washed with PBS/BSA followed by PBS alone, and then one of two different indirect immunofluorescence protocols can be used to determine the subcellular localization of the proteins. (A) Cells are incubated at room temperature for 30 min with avidin-FITC (E-Y Laboratories, San Mateo, CA) diluted in PBS as described by the manufacturer, washed for 30 min with PBS, and mounted in glycerol containing *n*-propyl gallate (0.1 μg/ml) as an antiquenching agent and 0.1 μg/ml propidium iodide as a generalized DNA counterstain. (B) Cells are incubated at room temperature for 1 hr with antiavidin DCS (Vector Laboratories, Burlingame, CA) diluted 1:100 with PBS/BSA. After washing for 30 min with PBS/BSA, the cells are incubated with a 1:400 dilution of fluorescein–avidin DCS in PBS/BSA for 1 hr and washed three times with PBS/BSA. Cells are counterstained with 2.5 μg/ml propidium iodide in PBS for 15 sec, rinsed for 2 min in McIlvaine's buffer (pH 7.5), and mounted as described earlier. In either case, the stained cells are observed and photographed with an epifluorescence microscope equipped with filters appropriate for the fluorescent labels used in the experiments.

Cross-linking of HMG-I(Y) Protein to Nucleosomes in Vivo

Investigations using reversible chemical cross-linking reagents, combined with one- and two-dimensional electrophoretic analyses, have shown that the human HMG-I protein is bound to the front face on random sequence nucleosomal core particles in close proximity to histones H3, H2A, and H2B both *in vitro* and *in vivo*.[7] For such *in vivo* cross-linking studies, nuclei of human K562 erythroleukemia tissue culture cells are isolated and resuspended to approximately 1 mg/ml DNA in 300 m*M* sucrose, 50 m*M* triethanolamine, pH 7.4, 3 m*M* $MgCl_2$. Cross-linking is initiated by the addition of freshly prepared 50 mg/ml DSP (dithiobissuccinimidyl propionate; Lamont's Reagent; Pierce Chemical Co.) in acetonitrile to a final concentration of 0.25 mg/ml. The reaction is allowed to proceed for 30 min at 22° and is quenched by the addition of glycine, pH 8.0, to 100 m*M*. Nuclei are pelleted and extracted once with 0.1 *M* NaCl, 0.2 *M* H_2SO_4 and once with 0.2 *M* H_2SO_4 alone. The acid extracts are combined and proteins are precipitated by the addition of 3 volumes of absolute ethanol, collected by centrifugation, washed with acidified acetone, then acetone alone, and dried under vacuum. The protein samples are analyzed by two-dimensional SDS–PAGE employing a 15% acrylamide *nonreducing* first dimension and an 18% *reducing* (to reverse the DSP protein cross-linking) second dimension. The two-dimensional gels are stained for protein mass bands with Coomassie Brilliant Blue. To unambiguously identify protein spots, identical second-dimension gels are processed in parallel for Western blotting by transferring the proteins to a polyvinylidene difluoride membrane (Millipore, Bedford, MA) and probed with anti-HMG-I(Y) antibody as described earlier.

This two-dimensional ("diagonal") SDS–PAGE separation system, in which DSP cross-linked protein products are separated under nonreducing conditions in the first dimension and under reducing conditions in the second dimension to thiolytically cleave the DSP cross-linker, is a sensitive method for determining protein–protein interactions in chromatin.[50] In these two-dimensional separation systems the protein spots lying in vertical lines off the electrophoretic diagonal arise from proteins that were initially cross-linked to each other in the nucleus. Such an analytical system has been used to identify the nearest cross-linkable protein neighbors of the HMG-I(Y) proteins *in vivo*[7] and is a valuable tool for studying the natural structural organization of chromatin within the cell nucleus.

EMSAs, Supershifts, and Ablation Assays with Nuclear Extracts

The first demonstration that HMG-I(Y) proteins participate in gene transcriptional regulation *in vivo* involved, among other experiments, the use of antibodies to supershift and ablate specific HMG-I(Y) containing complexes formed on the promoter region of the mouse lymphotoxin (tumor necrosis factor-β) gene incubated in nuclear extracts from mouse pre-

[50] J. O. Thomas, *Methods Enzymol.* **170,** 549 (1989).

B-cell lines.[31] Similar EMSA antibody supershift/ablation techniques have been used subsequently to demonstrate the involvement of these proteins in the regulation of a number of other genes, including the interleukin(IL)-2, IL-2Rα, IL-4, and granulocyte/macrophage-colony stimulating factor (GM-CSF) genes.[33,35,51] A useful protocol for supershift/ablation assays using nuclear extracts, anti-HMG-I(Y) antibodies, and EMSA electrophoretic separations is as follows.

1. Isolated nuclei from exponentially growing cells are extracted for 20 min on ice with 2 volumes of a buffer containing 20 m*M* HEPES (pH 7.9), 25% (v/v) glycerol, 0.42 M NaCl, 1.5 m*M* $MgCl_2$, 0.2 m*M* EDTA, 0.5 m*M* DTT, and 0.5 m*M* PMSF (freshly added). The resultant extract is clarified by centrifugation in a microcentrifuge (4°) and the extract supernatant is collected. The nuclear extract is dialyzed overnight against buffer "D" [20 m*M* HEPES (pH 7.9), 20% (v/v) glycerol, 0.1 *M* KCl, 0.2 m*M* EDTA, 0.5 m*M* DTT, and 0.5 m*M* PMSF], reclarified by centrifugation, and the total protein concentration determined (see earlier). Aliquots of this nuclear extract can be used immediately or stored at −70° prior to use.
2. Binding assays with nuclear extract are carried out in buffer D that contains 5 ng of ^{32}P-labeled promoter DNA fragment (~10^4 cpm), 2 μg of competitor poly(dG-dC) DNA (Sigma), and 5–10 μg of nuclear extract in a total volume of 25 μl. Following incubation for 15 min at room temperature, the extract and probe are analyzed on 4% polyacrylamide nondenaturing 0.5× TBE gels (acrylamide-to-bisacrylamide ratio, 37.5 : 1). Following electrophoresis at 150 V for 2–2.5 hr at room temperature, they are dried and exposed to a phosphoimager screen and analyzed quantitatively as described earlier.
3. In reactions involving antibody and nuclear extracts, 2.5 μl of purified anti-HMG-I(Y) rabbit IgG is added to the nuclear extract, plus the other components (giving a 1 : 10 dilution of the antibody in a final reaction volume of 25 μl), and the mixture incubated for 15 min at room temperature *prior* to the addition of the labeled probe. After an additional 15-min incubation, samples are analyzed on gels as described earlier. Control reactions consist of the addition of 2.5 μl of preimmune IgGs (or nonimmune rabbit IgGs) to the incubation mixture in place of the immune IgGs. For experiments involving pure protein, 1–15 ng of recombinant HMG-I(Y) protein, 0.5 μg of poly(dG-dC), and 1.25 μg of acetylated BSA (Promega) are added and incubated in buffer D (25 μl final volume) for 15 min at room

[51] J. Kim, R. Reeves, P. Rothman, and M. Boothby, *Eur. J. Immunol.* **25,** 798 (1995).

temperature prior to gel analysis. When antibodies are used in combination with pure recombinant proteins (rather than crude nuclear extracts) in supershift assays, 2.5 μl of immune IgGs is added to the reaction mixture after the formation of a complex of the recombinant HMG-I(Y) protein with the probe, incubated for an additional 15 min at room temperature, and then analyzed as described.

In EMSA supershift/ablation analyses, either poly(dG-dC) or sheared salmon sperm DNA can be used as the nonspecific competitor DNA. Use of poly(dI-dC) as a competitor should be avoided, however, because HMG-I(Y) proteins bind preferentially to this substrate because its structure mimics that of the minor groove of A·T DNA sequences. Conversely, poly(dI-dC) can be used as a competitor in EMSA reactions to eliminate HMG-I(Y) protein contributions to the electrophoretic mobility pattern. Figure 7 shows the results of an experiment in which antibody is observed to both supershift and ablate specific HMG-I(Y) protein-containing complexes formed on the promoter region DNA of the murine lymphotoxin (LT) gene incubated in nuclear extracts.[31]

Binding-Site Mutagenesis and Antisense Transfection Assays to Demonstrate in Vivo HMG-I(Y) Function

HMG-I(Y) proteins often cooperate with other transcription factors to regulate the transcriptional activity of mammalian genes *in vivo.* In many cases, the HMG-I(Y)-binding sites in the promoters of these genes partially overlap the major groove-binding sites for the cooperating sequence-recognizing transcription factors. For example, HMG-I(Y)-binding sites partially overlap those of the NFκB site in the human interferon-β gene promoter,[25] the Elf-1 site in the promoter of the human IL-2 receptor-α gene,[33] several NFAT/octamer purine boxes in the promoter of the murine IL-4 gene,[52] the CD28RE site in the promoter of the human IL-2 gene,[35] and the CK-1/CD28RE site in the promoter of the human GM-CSF promoter.[35] Such binding site overlap often allows for the introduction of *in vitro*-generated mutations in the promoter that interfere with either the minor groove binding of HMG-I(Y) or the major groove binding of the cooperating transcription factor, but not both. Experiments in which these mutated promoters are ligated to reporter genes [e.g., chloramphenicol acetyltransferase (CAT) or luciferase] and the resulting expression vector constructs individually transfected into mammalian cells have demonstrated that both HMG-I(Y) and the major groove-binding transcription factor are required for expression of the reporter gene *in vivo.*[25,33] Complementary experiments

[52] S. Chuvpilo, C. Schomberg, R. Gerwig, A. Heinfling, R. Reeves, F. Grummt, and E. Serfling, *Nucleic Acids Res.* **21,** 5694 (1993).

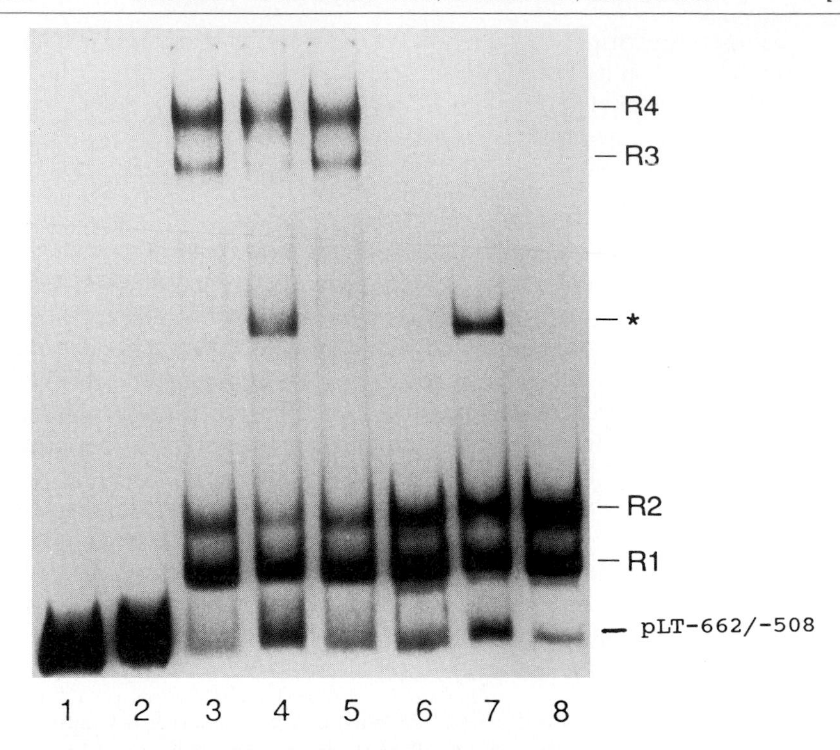

FIG. 7. Anti-HMG-I(Y) antibody specifically alters the EMSA-binding pattern of PD cell nuclear extracts and recombinant human HMG-I to murine lymphotoxin (LT) gene promoter DNA. Autoradiogram of EMSA showing electrophoretic mobility of ^{32}P-labeled LT-662/-508 promoter probe DNA incubated in the presence of PD nuclear extract (10 μg; lanes 3 to 5) or recombinant human HMG-I (15 ng; lanes 6 to 8). Parallel binding assays were performed without antibody (lanes 3 and 6) or in the presence of either anti-HMG-I(Y) IgG (lanes 2, 4, and 7) or control preimmune IgG (lanes 5 and 8). The electrophoretic mobility of free probe (lane 1) was unchanged when incubated with anti-HMG-I(Y) IgG (lane 2). However anti-HMG-I(Y) IgG, but not preimmune IgG, induced a supershift in one HMG-I-containing DNA complex formed both in nuclear extracts (lane 4) and with recombinant HMG-I (lane 7) (indicated by an asterisk) while ablating another larger HMG-I-containing complex (R3) observed only in nuclear extracts. Reproduced with permission from Fashena *et al.*[31]

in which antisense expression vector constructs for either HMG-I(Y) or the major groove-binding transcription factor are transfected separately into cells harboring a wild-type promoter–reporter gene vector have, likewise, demonstrated the requirement of both HMG-I(Y) and the cooperating factor for efficient *in vivo* gene transcription.[25,31,33,35,50] Results of one such antisense expression vector transfection experiment are shown in Fig. 8. In this experiment, increasing concentrations of a plasmid vector expressing

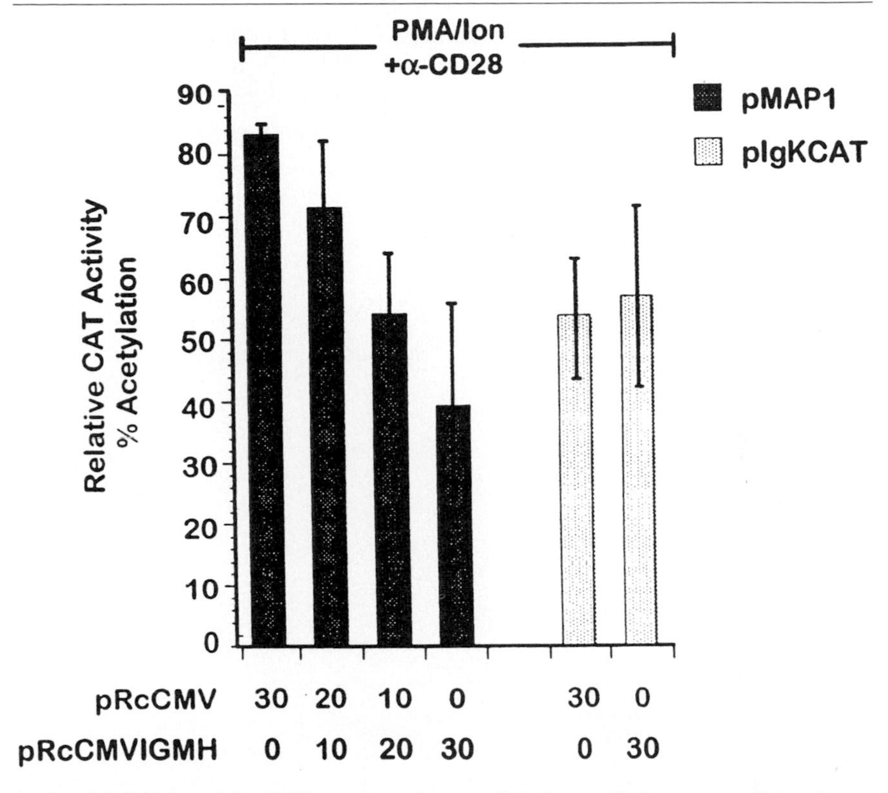

FIG. 8. Inhibition of the CD28 response element of the human IL-2 promoter with antisense HMG-I. Jurkat cells were cotransfected with the expression pMAP1 (5 μg), which contains six copies of the IL-2 CD28 response element cloned upstream of the HSV thymidine kinase promoter in a CAT reporter vector and increasing concentrations (10, 20, and 30 μg) of a plasmid expressing antisense HMG-I RNA (pRcCMVIGMH). Cells were then stimulated with PMA/Ca^{2+} ionophore plus antibody to the CD28 receptor. As a control, a CAT report construct (30 μg) containing a promoter that is not dependent on HMG-I for transcriptional expression (e.g., pIgKCAT) was transfected into Jurkat cells under identical conditions. CAT assays employed 2 μg of cell extract and were quantified using a phosphoimager. Columns represent the mean, and error bars the SEM, of three replicate transfections. Reproduced from Himes *et al.*,[35] with permission.

antisense HMG-I RNA (i.e., pRcCMVIGMH) were transfected into human Jurkat lymphoid cells harboring an expression vector containing multiple copies of the IL-2 CD28RE promoter element ligated to a CAT reporter gene (i.e., pMAP1) and the transfected cells were treated with the IL-2 gene-inducing agents PMA/ionomycin plus anti-CD28 antibody.[35] As shown in Fig. 8, transfection of Jurkat cells with increasing concentrations of the antisense HMG-I(Y) vector resulted in a marked decrease (>50%) of the inducible expression of the IL-2 promoter *in vivo.* In contrast, transfection of the antisense HMG-I(Y) vector into Jurkat cells harboring a control expression vector that is not dependent on HMG-I(Y) for its transcriptional activation (e.g., pIgKCAT) has no effect on the expression of the control vector.

Section II

Assays for Nucleosome Structure and Function *in Vitro*

[10] Cryoelectron Microscopic Analysis of Nucleosomes and Chromatin

By JAN BEDNAR and CHRISTOPHER L. WOODCOCK

Introduction

Of the few techniques available for examining the architecture and three-dimensional (3D) conformation of arrays of nucleosomes, electron microscopy (EM) is the most widely used. It has a resolving power that is quite adequate for imaging nucleosomes and DNA, and a number of methods of contrast enhancement have been developed.[1,2] However, in its conventional mode, EM has a number of important limitations that have severely hampered efforts to obtain structural information about the larger and more compact chromatin assemblies that are presumed to be most relevant to the *in vivo* situation. Conventional EM typically requires the chromatin to be affixed to a flat substrate, dehydrated, and contrasted with heavy metals. These processes not only compromise the 3D conformation, but result in an inability to recognize individual nucleosomes and resolve linker DNA in all but well-dispersed chromatin. The superior contrast afforded by scanning transmission EM has been applied to mononucleosomal samples[3] and has provided resolution similar to that obtainable with cryoelectron microscopy (cryo-EM). However, the technique does not overcome the adsorption-induced loss of 3D information.

Cryoelectron microscopy avoids the aforementioned problems. Although introduced in 1984[4] and used successfully in the field of viruses and protein complexes for more than a decade, its use in the chromatin field has, until recently, been very limited.[5] Additional methodological and instrumental development was required before it could be applied systematically to chromatin research.[1,6] Cryoelectron microscopy in superior to conventional microscopy in that the specimen is neither chemically modified (no fixation, staining, or drying needed) nor adsorbed to a flat surface.

[1] C. L. Woodcock and R. A. Horowitz, *Methods* **12**, 84 (1997).
[2] H. Zentgraf, C.-T. Bock, and M. Schrenk, *in* "Electron Microscopy in Molecular Biology: A Practical Approach" (J. Sommerville and U. Scheer, eds.), p. 81. IRL Press, Oxford/Washington, D.C., 1987.
[3] A. Hamiche, P. Schultz, V. Ramakrishnan, P. Oudet, and A. Prunell, *J. Mol. Biol.* **257,** 30 (1996).
[4] M. Adrian, J. Dubochet, J. Lepault, and A. W. McDowall, *Nature* **308,** 32 (1984).
[5] J. Dubochet, M. Adrian, P. Schultz, and P. Oudet, *EMBO J.* **5,** 519 (1986).
[6] J. Bednar, R. A. Horowitz, J. Dubochet, and C. L. Woodcock, *J. Cell Biol.* **131,** 1365 (1995).

0076-6879/99 $30.00

Instead, it is embedded in a thin layer of vitrified buffer, thus preserving its 3D solution conformation. The resolution obtainable is superior to that of conventional EM, thanks largely to the absence of heavy metal stains or shadows, and the technique can be adapted to examine changes occurring over time spans of the order of milliseconds.[7] From tilted pairs of micrographs, 3D information can be recovered and the configuration of nucleosomes and linker DNA modeled.

However, cryo-EM is not without its disadvantages. The unfixed, unstained specimen is extremely beam sensitive, making low-dose imaging obligatory, and has an inherently very low contrast. Typically, the specimen cannot be observed directly before taking images, and only after developing the negatives can one judge the quality of the specimen and preparation. Despite these considerations, which place a large demand on the skills of the operator, cryo-EM is the technique of choice for obtaining 3D information about macromolecular assemblies and for examining rapid (~msec) conformational changes of such assemblies in solution. In the future, further technical developments in progress, especially the further automation of EM imaging and the wider availability of charge-coupled device (CCD) cameras with large arrays, will make cryo-EM less challenging.

Preparation of Frozen Hydrated Suspension of Chromatin

For a full account of the principles and applications of cryo-EM, the comprehensive review by Dubochet and colleagues[8] is recommended. This article confines its discussion to cryo-EM of chromatin *in vitro,* using frozen hydrated suspensions. Cryoelectron microscopy of chromatin *in situ* is also possible via frozen hydrated cryosections,[9] and although it can provide valuable results,[10] this technique still faces numerous technical difficulties and cannot be considered as a routine addition to the repertoire of the microscopists. Familiarity with the basic principles and methodologies of conventional EM is assumed; for such information the reader is referred to general texts such as Bozzola and Russel[11] and Sommerville and Scheer.[12]

The basic process in the preparation of frozen hydrated suspensions is

[7] J. Berriman and N. Unwin, *Ultramicroscopy* **56,** 241 (1994).

[8] J. Dubochet, M. Adrian, J. J. Chang, J. C. Homo, J. Lepault, A. W. McDowall, and P. Schultz, *Q. Rev. Biophys.* **21,** 129 (1988).

[9] A. W. McDowall, J. J. Chang, R. Freeman, J. Lepault, C. A. Walter, and J. Dubochet, *J. Microsc.* **131,** 1 (1983).

[10] C. L. Woodcock, *J. Cell Biol.* **125,** 11 (1994).

[11] J. J. Bozzola and L. D. Russel, "Electron Microscopy." Jones and Bartlett, Boston, 1991.

[12] J. Sommerville and U. Sheer, "Electron Microscopy in Molecular Biology: A Practical Approach." IRL Press, Oxford/Washington, D.C., 1987.

vitrification: the transformation of a liquid phase into a solid one without the formation of crystals. This is achieved by very rapid cooling and is accomplished by plunging the specimen rapidly into a cryogen at a temperature close to its melting point (usually ethane at −180°). Although the issue of plunging speed is important in the case of bulky specimens, for preparations of frozen hydrated thin film suspensions, speeds between 1 and 2 m/sec are adequate and can be obtained easily by a simple, gravity-driven plunger.

The vitreous ice formed during plunging is unstable and can undergo phase transitions. Depending on the temperature, it will transform into one of two possible crystal forms of ice: cubic or hexagonal.[8] This renders the specimen virtually unusable and will introduce crystallization artifacts. In order to avoid recrystallization, the specimen should always be kept below the −120° devitrification temperature.[13] Under well-controlled conditions, very small crystals of cubic ice can actually aid in high-resolution imaging,[14] but crystallization usually leads to serious degradation of the specimen.

Preparation of Grids and Support Films for Cryoelectron Microscopy

Two basic types of support grid may be used in cryo-EM: bare grids or grids with perforated film. Using the former, the specimen is applied directly to a fine mesh (typically 1000) EM support grid. No film is used to cover the grid surface and a self-supporting thin layer of sample is formed (after blotting) due to the high surface tension of aqueous solutions. The only step in grid preparation is to coat the grid surface with a layer of carbon (ca. 10 nm), which may be accomplished using a simple evaporator (a description of the equipment and protocols for carbon evaporation and other standard EM procedures can be found in, e.g., Bozzola and Russel,[11]). This assures that the grid has homogeneous surface properties. In addition, the carbon may be "activated" by exposure to a glow discharge.[15] Depending on the gases present during the discharge, the surface may be given a net negative or positive charge, and the level of its hydrophilicity altered. As discussed in detail in the section on ice film thickness, the surface properties of the support film influence the spreading of the specimen over the grid surface. The bare grid system is very simple, but it has a major disadvantage in that the specimen is devoid of features that can be used for focusing or astigmatism correction, and there is usually a large variation in specimen film thickness between the center and edges.

[13] M. Blackman and N. D. Lisgarten, *Proc. R. Soc.* **A239,** 93 (1957).

[14] M. Cyrklaff and W. Kuhlbrandt, *Ultramicroscopy* **55,** 141 (1994).

[15] J. Dubochet, M. Groom, and S. Muller-Neuteboom, *in* "Advances in Optical and Electron Microscopy" (V. E. Cosslett and R. Barer, eds.), p. 107. Academic Press, London, 1982.

Although the technique of using grids with perforated film is more laborious than the previous one, it is far superior in that it provides features for focusing and astigmatism and a more even film thickness. EM grids with perforated supporting film are available commercially (e.g., Ted Pella, Inc., Redding, CA), but we do not consider their quality appropriate for a routine work and prepare our own films according the following procedure adapted from Fukami and Adashi.[16]

Reagents needed (all available from Sigma Chemicals, St. Louis, MO) include a 1% (w/w) aqueous solution of benzalkonium chloride (Osvan), a 2% (w/w) aqueous solution of dioctyl sulfosuccinate (Pellex), and a 1% (w/w) solution of cellulose acetate butyrate (Triafol) in ethyl acetate.

Procedure

1. Standard optical microscope glass slides are used as the substrate for plastic holey film formation. They are stored in 90% (v/v) ethanol and cleaned before use by wiping with optical tissue.
2. The slides are immersed in a 1% solution of Osvan for 5 min, rendering their surfaces hydrophobic. They are then dip washed in double-distilled water and slowly withdrawn. If the surface is properly hydrophobic, no water droplets should stick to the slide. If some droplets appear, the slide should be reimmersed into the Osvan solution for additional time and the procedure repeated.
3. The slide is then placed on a clean cold surface for a short time (ca. 15 to 30 sec) so that about 2 cm overhangs the edge. An aluminum block with a flat polished surface precooled to ca. $-20°$ in a freezer or by brief immersion in liquid nitrogen (LN_2) is recommended. Atmospheric moisture will condense on the exposed slide surface and form microdroplets.
4. The slide is removed, held at an angle of approximately 60° from the horizontal, and Triafol solution is applied with a Pasteur pipette uniformly along the upper edge of the slide on the side with condensed water, allowing the solution to run down evenly (see Fig. 1). Excess solution is removed by blotting from the bottom edge.
5. The slide is allowed to air dry after which a "milky" film should appear. The size, density, and distribution of holes are checked under the optical microscope. Optimal hole diameter is between 1 and 3 μm.
6. The slides can be stored for later use for a prolonged period of time (several months or longer) or the film can be floated off and applied to grids immediately.

[16] A. Fukami, and K. Adachi, *J. Electr. Microsc.* **14**(2), 112 (1965).

7. EM grids should be placed on a grid support held under water in a suitable glass container (e.g., glass culture dish 115 × 55 mm). The type of grid is not very important, and virtually any type with mesh greater than 200 can be used. Perforated metal grid support plates about the size of glass slides are available commercially (Ted Pella, Inc.).
8. To detach the film, the slide is dipped into a 1% solution of Pellex for 5 min. This makes its surface hydrophilic again so that water can penetrate between the film and the surface.
9. The slide is dip washed in double-distilled water, the edges scored with tweezers or razor blade, and the film is floated off on the water air interface.
10. Using tweezers, the floating film is maneuvered carefully over the grids and the water level is then lowered either by careful withdrawal by a pipette or, if the water container is equipped with an outlet valve, by carefully regulating the outflow of water (a complete system for film floating is available from Bal-Tec/Technotrade, Manchester, NH). Once the film has settled over the grids, the grid support and grids are removed and air dried.
11. In order to make the plastic film stable under the electron beam, it must be covered with a carbon layer. This can be done in any carbon evaporator with controls sufficient for depositing approximately 10 nm of carbon film.
12. Before use, the surface properties of the holey film can be modified by "activating" the carbon with a glow discharge device[15,17] to leave either a net positive or a net negative charge. Because surface activation usually fades with time, glow-discharged grids should be used within 15 min.
13. A more permanent treatment, which produces a uniform, neutral surface, helps increase the rigidity and stability of the holey film and provides a grainy texture for focusing and astigmatism is to evaporate an additional layer of ca. 5 nm of platinum–carbon (95% Pt–5% C). This is accomplished most easily with an electron beam evaporation apparatus.
14. Optionally, the Triafol plastic film can be removed before use. In the case of platinum-coated films, this step is recommended as it improves the overall stability of the film under the electron beam and accentuates the underfocus granularity for low-dose focusing and astigmatism adjustment and can decrease the mechanical strength of unreinforced carbon significantly. The Triafol film is

[17] C. L. Woodcock, L. L. Frado, G. R. Green, and L. Einck, *J. Microsc.* **121,** 211 (1981).

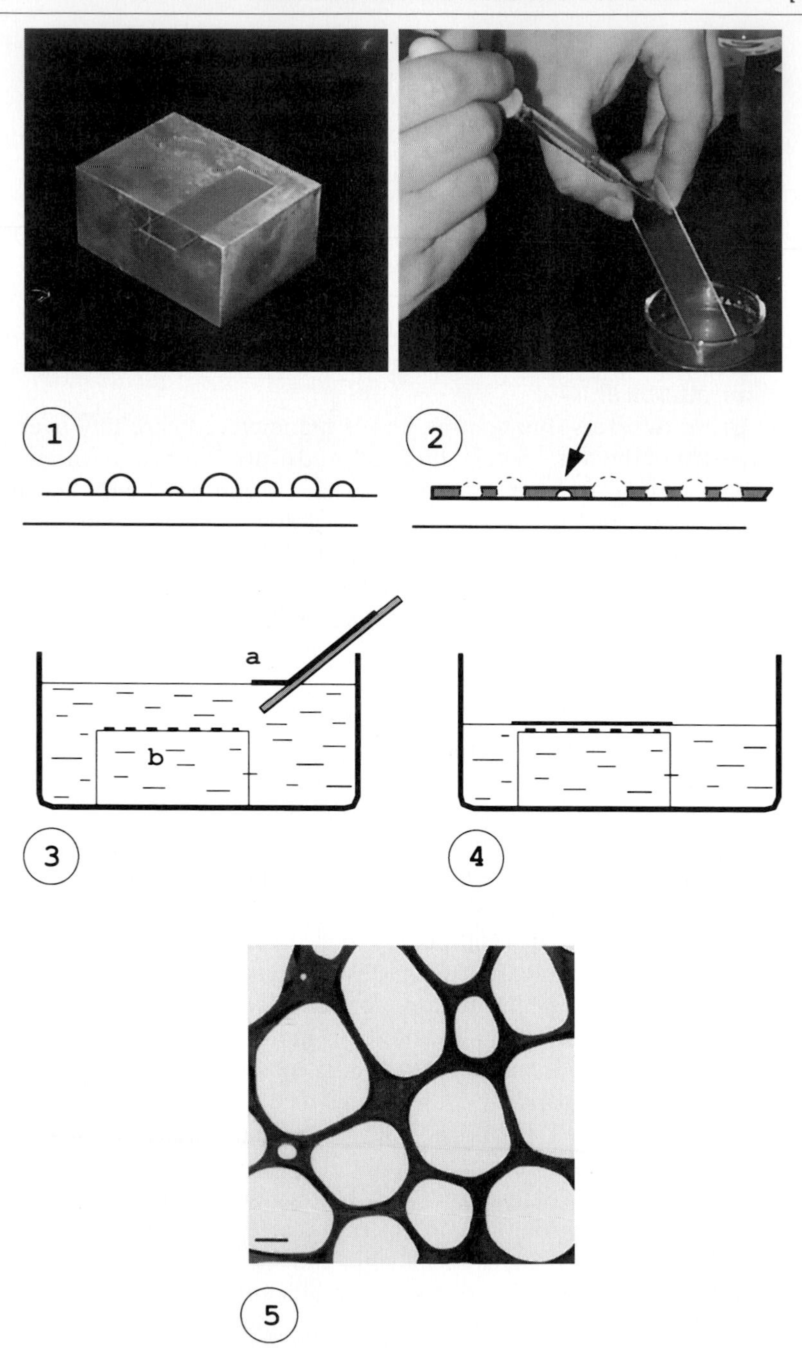
1
2
a
b
3
4
5

removed by dipping grids, held in tweezers, into ethyl acetate for several seconds, then chloroform, and again ethyl acetate.

Remarks on Grid Preparation

Steps 3 and 4 are absolutely crucial, but rather difficult to control. In order to understand how one can influence the result, it is useful to explain the process of holey film formation. Water vapor will condense on the cold glass surface and the size, density, and distribution of the microdroplets will depend on many parameters, with the most important being relative humidity, glass temperature, and time allowed for condensation. Because of the hydrophobic surface, the condensed water will form spherical droplets with a negative wetting angle. The level of humidity and the temperature of the glass will determine how fast and how densely the droplets will form; if they are allowed to grow too large, neighboring droplets will merge, ruining the holey film effect.

Once the plastic Triafol solution is flooded over the surface, it will penetrate in between the droplets, and holes will be left in their place after the water evaporates. For holes to occur, the thickness of the plastic layer has to be smaller than the average diameter of the droplets (see Fig. 1). Another factor is the degree to which the plastic solution penetrates the array of droplets, which depends on the viscosity of the Triafol, a function of its concentration and temperature. Although it is relatively easy to control or change the concentration, the temperature will vary according to the time of contact with the cold substrate. As the Triafol is added from the top, it will become progressively colder (and more viscous) toward the bottom part of the slide; the temperature gradient depends on slide temperature and rate of flow. The latter can be varied by the angle at which the slide is held while the plastic is applied and by the force with which the plastic solution is propelled from the pipette. The number of parameters and the difficulty of controlling them precisely make the reproducibility of the whole process difficult. Optimal conditions on a given day must be

---

FIG. 1. Preparation of EM grids with holey supporting films. (1) A glass slide with a hydrophobic surface is cooled on a precooled aluminum block and water vapor is allowed to condense on its surface, forming droplets with a large wetting angle (close to 90°). (2) A cellulose acetate butyrate solution is flooded over the droplet array, forming a thin plastic film in which holes will be created at the sites of the water droplets. If the film thickness is larger than a droplet diameter, true holes will not be formed (arrow). (3) After making the slide hydrophilic, the film (a) is floated off on a water surface, below which are EM grids arranged on a grid support (b). (4) The water level is lowered and the film is deposited on the grids. (5) An ideal film containing holes ~3 to 5 μm in diameter separated by septae. Bar: 1 μm.

determined empirically, and it can take several hours of trials before a satisfactory film quality is obtained.

From our experience, the ratio of film area to hole area seems to be important. Because the area of support film influences the spreading of the specimen (mostly detrimentally), minimizing the film area by having holes separated only by very thin septae of plastic (i.e., a net) is advantageous. However, net-like films tend to be more unstable in the electron beam, and a compromise, such as that shown in Fig. 1, should be reached.

Preparation of Frozen Hydrated Specimens

The preparation of *in vitro* chromatin specimens for cryoelectron microscopy proceeds as follows (see Fig. 2). A grid with holey supporting film is clamped in tweezers and is mounted on the plunger. The cryogen container is placed in a reservoir of LN_2 and cooled. A weak stream of gaseous ethane (or propane) is then projected through a plastic pipette tip against the bottom of the cryogen container. The gas will liquefy easily, provided

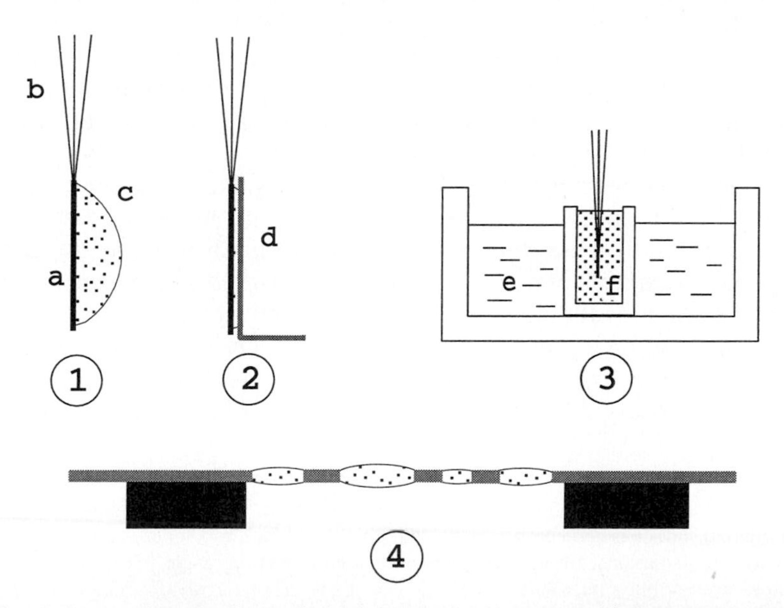

FIG. 2. Preparation of a frozen hydrated thin film suspension. (1) Three microliters of specimen (c) is applied to an EM grid (a) with supporting holey film, held in tweezers (b). (2) Excess specimen solution is blotted away with filter paper (d). (3) The grid is plunged immediately into a container of liquid ethane (f), which is cooled with LN_2 (e). (4) After plunging, the specimen remains embedded in a thin (50–100 nm) self-supporting vitreous layer spanning the holes of the support film (gray areas).

the stream is neither too strong nor too weak. When the container is full, a 3-μl droplet of specimen at a concentration of 50–100 μg/ml is applied to the grid, excess is blotted away by the flat application of filter paper to the grid surface, and the plunger is released immediately, dropping the specimen into the cryogen. The thin film of specimen solution left on the grid after blotting solidifies immediately without the formation of ice crystals.

The tweezers are then unmounted and the grid is transferred into LN_2 before the ethane in the container solidifies. All subsequent operations must be carried out below $-120°$, preferably under LN_2, in order to avoid an increase in temperature and consequent devitrification of the specimen. The grids can be stored under LN_2 practically indefinitely, but immediate use is advised to avoid the buildup of contamination (usually ice microcrystals from the LN_2). Often, a layer of ethane will remain on the grid and will solidify when placed into LN_2. This is of no concern, as after about 10 min it will dissolve away. Propane, however, does not dissolve in LN_2 and has to be removed carefully with precooled filter paper before the grid is placed into the LN_2. This complication makes ethane the preferred cryogen.

Remarks

a. Plunger. The just-described procedure applies to a simple mechanical gravity-driven plunger, with no cryogen temperature control or automated blotting implemented. For more consistency, however, it is recommended to use a plunger with automated blotting, which allows precise control of the blotting time and the pressure of blotting paper on the grid. A basic description of such a system can be found in Cyrklaff *et al.*[18] or Bellare *et al.*[19] Controlling cryogen temperature to just above the melting point prolongs the time available for making specimens (ethane freezes at LN_2 temperature) and allows vitrification of multiple grids with one cryogen filling. We use a plunger apparatus constructed in a private machine shop. To our knowledge, inexpensive cryo vitrification systems are not available commercially, perhaps due to the intrinsic dangers of ethane and propane (see safety note later). A versatile instrument that uses a spring-loaded plunger system is available from Leica Inc. (Deerfield, IL).

b. Blotting and Film Thickness. After appropriate blotting, excess specimen is removed and only a very thin layer (ca. 100 nm) is left spanning the holes of the supporting film. As the typical hole diameter is between 1 and 3 μm, the surface/volume ratio of the layer is very favorable for heat extraction, which explains why the cooling efficiency is very high in such systems and why freezing speeds are sufficient for successful vitrification.

[18] M. Cyrklaff, M. Adrian, and J. Dubochet, *J. Electr. Microsc. Tech.* **16,** 351 (1990).

[19] J. R. Bellare, H. T. Davis, L. E. Scriven, and Y. Talmon, *J. Electr. Microsc. Tech.* **10,** 87 (1988).

The thickness of the layer is affected by several parameters. Most important is the combination of the speed and the extent of blotting. Speed is related to the type of filter paper used, and the use of a medium or slow hardened ashless type of paper (e.g., Whatman, Clifton, NJ; #40, or #52) is recommended. Other parameters influencing thin-layer formation are the surface tension of the specimen solution and the surface properties of the supporting film itself. The possibilities for changing or modifying these are discussed later. An appropriate thickness of the vitreous layer is critical for successful imaging. Thinner films (~50 nm) are ideal for obtaining good-quality micrographs with a relatively high contrast and maximal spatial resolution, but confine the specimen to a smaller volume, consequently increasing the likelihood of distorting the original structure. Thinner films also increase the probability of surface-related artifacts (see later). Thicker films (~80–100 nm), however, provide more space for the specimen, but also increase the proportion of inelastically scattered electrons, which degrades image quality. Thicker films may be used if electron energy filtering or an "intermediate" voltage (200–400 kV) microscope is available; field emission electron guns that provide a highly coherent electron beam are highly advantageous.[20]

c. Evaporation of Water. An important concern is the potential loss of water from the thin film by evaporation. Typical times between blotting and vitrification are between 0.2 sec (for an automatic plunger with mechanical blotting) to about 1 sec for a manual system. During this time, the specimen is exposed to the surrounding atmosphere and, if the specimen temperature is above the dew point for the local atmospheric humidity, water in the specimen solution will evaporate. Even a very slow evaporation rate from thin films can be very significant due to the large surface/volume ratio, increasing the concentration of solutes up to fourfold.[18] In a typical experiment carried out in 50% relative humidity, an approximately twofold increase in solute concentration will occur if no precautions are taken.[21] Possible solutions to this problem are discussed thoroughly in Cyrklaff *et al.*[18] so we will mention only the two simplest. The first option is to carry out the experiment in a cold room, which typically has an atmosphere with relative humidity above 85% and temperature about 4°. This decreases the evaporation rate significantly. The other option is to use a stream of humid air (from a simple humidifier) in the specimen area. If experiments require a precise concentration of components, a more complex apparatus that will assure 100% control over atmospheric humidity around the specimen must be used. A description of such system can be found in Bellare *et al.*[19] In

[20] R. A. Horowitz, A. J. Koster, J. Walz, and C. L. Woodcock, *J. Struct. Biol.* **120,** 353 (1998).
[21] J. Trinick and J. Cooper, *J. Microsc.* **159,** 215 (1990).

climates that vary greatly in humidity during the year, the plunging apparatus should be placed in a Lucite box that can be humidified, and a good quality hygrometer should be used to monitor the conditions before and during specimen preparation. Without this minimal level of humidity control, consistent results will not be attainable.

d. Properties of the Specimen Solution. Sometimes the specimen solution exhibits properties very unfavorable for the formation of layers with appropriate thickness. In such cases, some slight modification using surfactant agents may be used to achieve satisfactory results. In the case of very low surface tension (the specimen has tendencies to form layers that are too thin), the addition of a low concentration (0.01 to 0.05%) of sucrose can be beneficial. The opposite extreme, high surface tensions and consequently poor spreading over the grid surface, may be corrected by the addition of a low concentration of a nonionic detergent. The choice of type and concentration of detergent will depend on the specimen. Fortunately, these cases are rather rare and most of the problems with thin-layer formation can be treated by adjusting the blotting conditions and the delay between blotting and vitrification (in general, the shorter the better). An "activation" of the support film surface, prior to adding the specimen, can be used to increase its hydrophilicity and help promote thinner ice layers. This can be achieved by applying a glow discharge treatment.[15]

e. Safety. Two safety concerns should be mentioned. (1) Liquid ethane is a very efficient cryogen that can easily cause serious skin burns and potentially permanent damage if allowed to contact the eyes. (2) Gaseous ethane can form an explosive mixture with air, and all operations with it should be carried out under a ventilated hood. The use of small cryogen containers (about 10 ml) reduces this potential danger greatly.

Preparation of Chromatin

In most applications, purified fractions of native or reconstituted chromatin will be the starting material; in these cases, care should be taken in designing the purification method. Sucrose gradient fractionation, followed by the removal of sucrose by dialysis, has been a reliable method in our hands,[6] but purification by elution from gels or columns has generally proven unsatisfactory due to the nonchromatin material that almost always contaminates the eluates and makes cryo-EM (which, unlike conventional EM, images all the solution components, not just those that adhere to the support film) very difficult. Additional purification steps, such as sucrose or glycerol gradients, should be included in such cases. The same considerations apply to experiments that are normally carried out in the presence of carriers such as bovine serum albumin or other components that can potentially interfere with cryo-EM images.

The ideal concentration of the material depends somewhat on the size of the chromatin (a lower concentration is required for mononucleosomes than for polynucleosomes), but a starting concentration of 50–100 μg/ml DNA should be used. If fractions need to be concentrated after the final purification step, the use of centrifugation through high molecular weight cutoff membranes is recommended. Amicon 100-kDa units (Centricon 100 and Minicon 100), obtainable through Millipore (Bedford, MA), have proved satisfactory in our hands. A thorough prewash of the membrane with buffer and the use of the lowest gravitational forces possible to achieve a useful flow rate are important in reducing losses that occur due to the sticking of chromatin to the membrane. In all experiments, especially with new types or batches of membrane, the yield should be checked carefully.

The following procedure briefly describes a simple procedure for the preparation of chromatin from chicken erythrocyte nuclei by nuclease digestion. The soluble chromatin obtained has proven to be well suited for cryo-EM and provides a good "test system" for verifying or trouble shooting new batches of support films or other changes.

Buffers

1. Wash buffer: 0.15 *M* NaCl, 5 m*M* PIPES, pH 7.8, 0.5 m*M* PMSF (phenylmethylsulfonyl fluoride)
2. Digestion buffer: 1.0 m*M* PIPES, pH 7.8, 0.2 m*M* $CaCl_2$.

Fresh chicken blood is diluted into a 10-fold excess of ice-cold wash buffer and washed twice at 800*g* for 10 min. Nonidet P-40, (NP-40) detergent is added to the washed erythrocytes from a 10% stock solution to give a final concentration of 0.5%, and the mixture is stirred at room temperature for 5 min. The solution is centrifuged again, and nuclei are purified by washing twice with wash buffer at 1000*g* for 10 min and resuspending in a known volume (similar to the starting volume of blood). The concentration of chromatin–DNA is determined spectrophotometrically by determining the OD_{260} of DNA released from an aliquot of nuclei by diluting a known volume in 0.2 *M* NaOH and agitating thoroughly to disperse and fragment the DNA. One milligram per milliliter of DNA has an absorbance of ~20 OD_{260} and typically 10 ml of blood yields about 50 mg DNA. Convenient nuclease digestion conditions include 5 mg DNA–nuclei (100 OD_{260}) pelleted at 1000*g* and resuspended carefully in 2.0 ml digestion buffer. The mixture is placed in a 37° water bath, and 1 unit micrococcal nuclease (Sigma) is added per 50 μg DNA. After ca. 2 min, digestion is terminated by the addition of EDTA to a final concentration of 2 m*M*. The insoluble chromatin is pelleted, and the supernatant with soluble chromatin is collected and its concentration measured. After dilution with the desired buffer (see later), cryo-EM preparations may be made. Soluble chromatin prepared in this way is stable at 4° for 1 or 2 weeks.

There is principally no restriction on the choice of buffer. Buffers HEPES, PIPES, and Tris have all been used successfully for cryo-EM visualization of chromatin. However, because Tris interacts with glutaraldehyde, our standard procedure of first checking samples using conventional EM cannot be used. Moreover, fixation of the material for cryo-EM (which may occasionally be necessary) is no longer an option. Because the pH of the buffer is temperature dependent, except at pH 7.0, it is advisable to select a buffer that will not show a large pH change during the previtrification cooling period. Although experiments have shown that there is no observable structural change in DNA due to the drop in temperature during vitrification,[22] the situation may vary with the species of ions present. The mobility of small ions in aqueous solutions is several orders of magnitude higher than the typical relaxation times for DNA, and thus it cannot be excluded that some local changes in buffer concentration, and hence pH, could occur. For this reason, buffers with a low temperature constant, such as PIPES or HEPES, are preferred.

In some cases a gentle fixation of the specimen is necessary, especially when experiments in higher ionic concentrations are intended (see later). Fixation with 0.1% glutaraldehyde for 24 hr is used routinely.[23] This treatment does not introduce visible structural changes when compared with unfixed material.[6]

Imaging Conditions

Although the specific operational procedures depend on the type of EM and cryoholder and cryotransfer device in use, the general procedures for cryo-EM apply in all cases. As mentioned earlier, the first task is to insert the specimen into the microscope without raising the temperature to the crystallization point or accumulating contamination in the form of surface ice crystals. Close attention to the proper use of the cryotransfer device and minimal exposure of the grid to the ambient atmosphere are mandatory.

Once inserted successfully in the microscope, it is preferable to wait at least 20 min before attempting to record images. This allows the specimen

[22] J. Dubochet, J. Bednar, P. Furrer, A. Z. Stasiak, A. Stasiak, and A. Bolshoy, *Nature Struct. Biol.* **1,** 361 (1994).

[23] F. Thoma, T. Koller, and A. Klug. *J. Cell. Biol.* **83,** 403 (1979).

and cryoholder temperatures to equilibrate. Because the specimen is held at a low temperature (typically ~−175°) in the microscope, contaminants in the vacuum will tend to condense on it. With cryo-EM, the most significant contamination source is usually water molecules, introduced as ice on the specimen holder and subsequently sublimed into the vacuum system. The effect is a steady increase in ice film thickness and a consequent degradation of image quality. Although modern microscopes are routinely equipped with anticontamination devices, cryo-EM may require an additional system in which blades held at ~−185° are installed very close to the specimen.

After equilibration, the grid is then surveyed at low magnification (~4000 to 6000×) and very low beam intensity to check the distribution of the ice film over holes and to assess their relative thickness. As the chromatin specimen is invisible under these conditions, ice thickness and the absence of contamination are the only criteria in judging the suitability of the site for collecting micrographs. Having selected the most favorable region to examine in detail, the low-dose mode (in which focusing is carried out on a small area of supporting film, after which the beam is shifted automatically to an unirradiated area of ice for image recording) is activated to avoid beam damage.

Specimen contrast depends on the microscope itself, especially the focal length of the objective lens (longer focal lengths generally yield higher contrast), and the parameters selected for imaging. The most important is the accelerating voltage, which is usually selectable between 60 and 100 kV. Lower voltages result in higher contrast, but increased specimen beam damage, whereas higher voltages generally result in weak contrast, but lower beam damage, and better spatial resolution. The final choice will depend on the needs of the experiment. For example, in cases where maximum film thickness is needed to accommodate rather large chromatin assemblies without distortion, higher voltages may be needed to minimize the proportion of inelastically scattered electrons at the expense of contrast.

For all cryo-EM work, microscope settings should be optimized to produce a highly coherent electron beam, as this maximizes the level of phase contrast. Appropriate underfocusing is also essential, with its optimal value depending on the parameters and settings of the microscope, especially the spherical aberration constant (C_s) of the objective lens and the accelerating voltage. Different underfocus levels are needed for "high" and "low" resolution imaging, but will generally be between 1 and 2 μm. Compared with conventional EM of stained specimens, optimal underfocus values for maximal contrast in cryo-EM are rather high and present some problems in image interpretation. This can be partially overcome by a subsequent image-processing operation that calculates the theoretical image change at different spatial resolutions and applies a contrast transfer function (CTF) correction (e.g., Bednar *et al.*[6] and Horowitz *et al.*[20]). CTF correction requires input of the actual defocus value of the micrograph, which can be obtained from the micrograph itself, provided that an area of carbon film is included. This is much more preferable than relying on the defocus value reported by the microscope, as the latter is influenced by any difference between the z plane at the focus site and the image site. If grids with evaporated platinum are used, a precise determination of underfocus is facilitated.

Higher spatial resolution imaging requires smaller underfocus values (~700 nm for a 100-kV accelerating voltage) and is achieved at the expense of contrast. An appropriate starting level of underfocus for obtaining optimal image quality with a specific EM can be selected by examining calculated CTF curves. The underfocus value at which the CTF has its first zero at a spatial resolution of ~2 nm generally provides good conditions for chromatin visualization.

For projects in which 3D information is desired, two or more images of the same area at different tilt angles are required. For tilt pairs, we usually select a 30° separation angle, taking one image at a tilt angle of −15° and the second at 15°. This requires careful attention to the stability of the cryoholder, as image drift is often exacerbated by tilting and careful setting of the z height of the specimen to the eucentric point. It is also important that the image shift between the focus and the image-recording site be in a direction parallel to the tilt axis. As automated specimen tracking and focusing systems[24] become more widely available, this aspect of cryo-EM will become less demanding.

Image Recording

For recording images on film, the high sensitivity of Kodak SO-163 film, developed for maximum density using full-strength D19 developer (Kodak, Rochester, NY) for 12 min at 21° is recommended. Alternatively, images can be recorded using slow-scan CCD cameras coupled to phosphor screens. These have the advantage of providing rapid feedback about specimen and image quality, but generally cover a much smaller area than film. The typical film negative sized $3\frac{1}{4} \times 4$ inches has about 10 times the effective area of a CCD chip with a 1024×1024 array of pixels. When 2048×2048 CCD arrays become more common (and affordable), the advantages of direct digital recording will be much greater. Because most cryo-EM images require computer processing, the digitization of selected areas of negatives

[24] A. J. Koster, H. Chen, J. W. Sedat, and D. A. Agard, *Ultramicroscopy* **46,** 207 (1992).

is necessary and the selection of pixel size should take into account the desired spatial resolution of the data set.

Appearance of Chromatin Embedded in the Thin Layer

A typical cryo-EM image of a chromatin segment prepared by the method described earlier and vitrified in low salt buffer (5 m*M* monovalent ions) is shown in Fig. 3, which was obtained using a Philips CM10 transmission EM high-contrast objective lens, operated at 60 kV. A Model 626-00 cryoholder (Gatan Inc., Pleasanton, CA) was used, and additional decontamination was provided by a Gatan 651N anticontamination device.

The chromatin forms a 3D zigzag pattern of nucleosomes and DNA. As the original 3D orientation of the nucleosomes is preserved, they are seen in a variety of projections, including the familiar *en face* view presented by the majority of nucleosomes in conventional EM images (Fig. 3a). The process of adsorption to a flat substrate required for conventional EM causes nucleosomal arrays to undergo a dramatic change in conformation, resulting in extreme flattening and a preponderance of *en face* views of individual nucleosomes. When observed under "high resolution" cryo-EM conditions, *en face* projections reveal a pear-shaped unit, with the DNA exiting the core particle after completing ~1.78 turns of DNA and subsequently "crossing" at a distance approximately 8 nm from the center of the nucleosome (Fig. 3b; see also Ref. 36). A striking feature of these images is the low electron density of the volume within the two turns of DNA, which is occupied by the core histones, having a molecular mass of ~100 kDa. The very weak electron density produced by the histone core is at least partly attributable to the effect of strong underfocusing on contrast, as, after performing a CTF correction, the relative density of the histone core region increases substantially.[6] A reliable calculation of the expected electron density of the portion of the histone core that lies within the 1.78 turns of nucleosomal DNA will be possible now that the coordinates from crystallographic data on the nucleosome core particle[25] have been released. It seems likely that DNA contributes disproportionately to the contrast of chromatin in cryo-EM, perhaps due to the presence of the relatively heavy phosphorus atoms in the sugar-phosphate backbone.

Another example of the dramatic improvement in resolution afforded by cryo-EM is seen in side projections of nucleosomes, which, under optimal conditions, reveal the two turns of DNA (Fig. 3b).

[25] K. Luger, A. W. Mader, R. K. Richmond, D. F. Sargent, and T. J. Richmond, *Nature* **389,** 251 (1997).

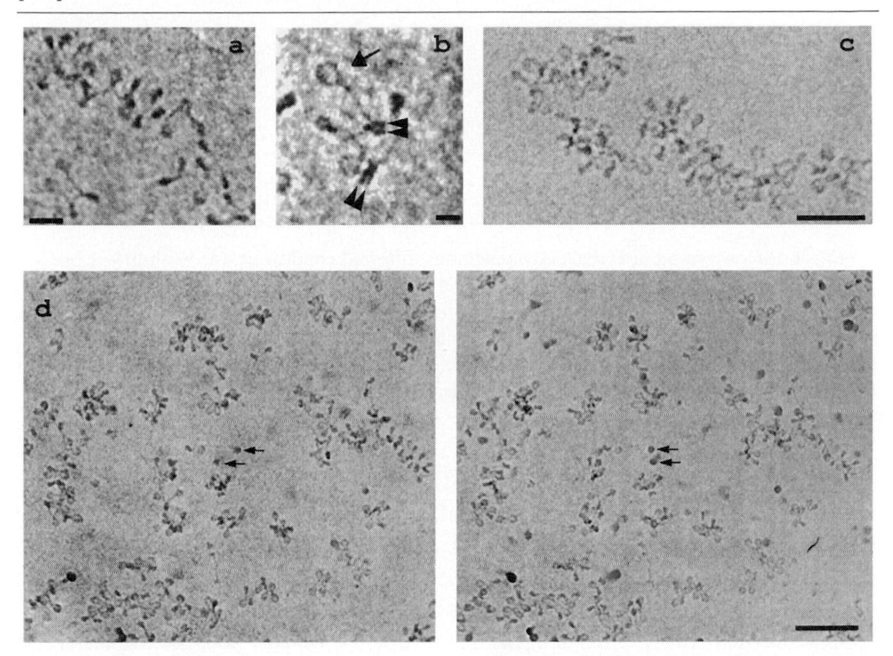

FIG. 3. Appearance of unfixed, unstained chromatin in a thin vitrified layer. (a) A chromatin fiber under low salt conditions (5 m*M* NaCl) documenting the 3D zigzag arrangement of nucleosomes and linker DNA and showing various projections of nucleosomes. (b) Under high resolution conditions, an *en face* projection (arrow) of a nucleosome has a pear-like shape, with entering and exiting DNA crossing at about 8 nm from the nucleosomal center. Side views reveal two distinguishable DNA turns (arrowheads). (c) With increasing salt concentration, the zigzag becomes more compact (example at ~20 m*M* NaCl). (d) Stereo pair of micrographs of chicken neutrophil oligonucleosomes (average ~12 nucleosomes) containing MENT protein.[35] When viewed in stereo, the distribution of the sample throughout the ice layer and the nonplanar zigzag arrangement of nucleosomes can be seen. However, a portion of the particles appears to be "trapped" by the bottom surface of the layer. Ionic conditions correspond to ~20 m*M* monovalent ions. Contaminating ice microcrystals deposited on the opposite layer surfaces (small arrows) can serve as useful reference points for determining layer thickness. The relatively large angular separation of individual images (30°) makes the stereo perception more difficult than a pair with a standard separation (5 to 10°) and also amplifies the depth of the field somewhat. Bars: 30 (a), 10 (b), 50 (c), and 100 (d) nm.

Chromatin compaction is influenced strongly by the ionic strength of the medium.[23,26] When the concentration of monovalent ions reaches ca. 20 to 40 m*M*, the number of nucleosomes per unit length along the fiber axis increases to such an extent that individual linker DNA segments be-

[26] K. E. van Holde, "Chromatin." Springer-Verlag, New York, 1989.

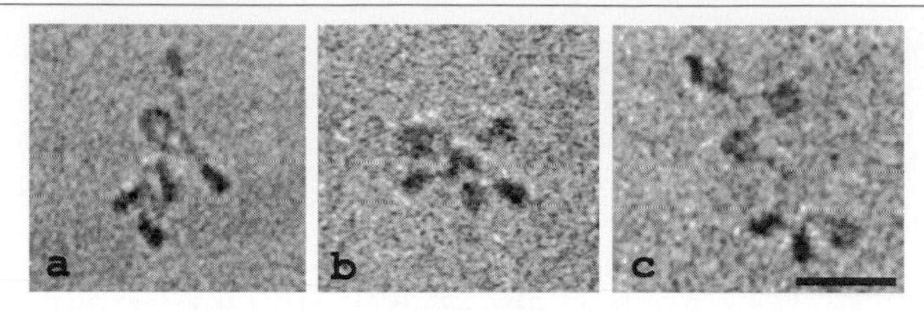

FIG. 4. Reconstituted hexanucleosomes under different conditions. (a) With linker histone present, and at a low concentration (5 m*M* NaCl), an irregular zigzag conformation similar to that seen with native chromatin is observed. (b) At 50 m*M* NaCl, the particles become more compact and it is impossible to distinguish individual linkers and determine their path. This sample was prefixed with 0.1% glutaraldehyde for 24 hr. (c) In the absence of linker histone, the short nucleosomal arrays adopt a more open structure. Bar: 30 nm.

come difficult to resolve and the identification of consecutive nucleosomes is much more difficult from tilt pairs of images. The mode of compaction is observed more easily in a simpler situation, where stretches of chromatin with a smaller number of nucleosomes are visualized. An example of nucleosomes reconstituted onto six tandem repeats of the strong nucleosome positioning sequence found in 5S rDNA[27] imaged in low (5 m*M*) and high (50 m*M*) NaCl is shown in Figs. 4a and 4b. Increased compaction of the hexanucleosome particles is observed, and even in this rather simple case, it is nearly impossible to trace the linker DNA path completely.

The appearance of the nucleosomal chain changes when linker histones are removed. Under these conditions, a high proportion of nucleosomes do not exhibit the "crossing" arrangement of entering and exiting DNA (Fig 4c). In the absence of linker histone, the linker DNA is free to adopt the most energetically favorable conformation and, close to the entry and exit points, its path will be dictated by mutual repulsion of DNA.

Recovery of 3D Information

From two tilted micrographs of an object, it is possible to recover the 3D coordinates of object features and prepare a 3D reconstruction (with certain limitations). If the axis of tilt is selected as one of the coordinate axes (usually y), the coordinates of any point on the object will have the same value of y in both projections. The second (x) coordinate will be different in the two projections and the difference is related to the third (z) coordinate. This very simple reconstruction principle applied to chromatin

[27] R. T. Simpson, F. Thoma, and J. M. Brubaker, *Cell* **42,** 799 (1985).

arrays allows DNA paths and nucleosome centers to be determined. The result is a linear (monodimensional) object in 3D space.

The basic procedure for DNA 3D reconstruction can be found in Dustin *et al.*[28]

1. A tilt pair of micrographs is digitized at the desired resolution, and a system of coordinates is chosen so that the x, y plane corresponds to the plane of the first micrograph and the y axis is parallel to the tilt axis.
2. Image treatment procedures (CTF correction, low-pass or high-pass filtering, etc.) are applied if required or desired to facilitate feature recognition.
3. The digitized micrographs are displayed on a computer screen and mutually aligned along the y direction on the basis of a reference point, which is a well-defined feature, clearly visible on both micrographs. This reference point is taken as the origin of the system of coordinates. In the case of chromatin, the center of a nucleosome can serve as a reference point, and alignment of images can be checked by ensuring that the y coordinates are equal in both projections.
4. With appropriate software, determination of the path of the DNA or the centers of nucleosomes can be made by clicking the mouse along the center of the DNA filaments or other features displayed on the computer screen. The position of points on the first image defines the coordinates (x'_i, y'_i) and also defines the y''_i coordinates for the corresponding points on the second image. The movement of the pointer is then restricted to the x direction at $y'_i = y''_i$.
5. The result is a set of x', y' values for each point in the first tilted image and in x'', y'' in the second one. As the y axis is parallel to the tilt axis, $y'_i = y''_i$ for ith point. A simple calculation then gives the z' coordinate for this point:

$$z'_i = [x'_i \cos(\alpha) - x''_i]/\sin\alpha \tag{1}$$

where α is the angular difference between two tilts.

From the set of 3D coordinates, one can construct a model of the chromatin array; the detailed modeling procedure depends on the available 3D software and computer hardware. The reconstructed DNA trajectories are visualized most usefully as flexible tubes, obtained by attributing a "user defined" thickness representing the appropriate DNA diameter, to

[28] I. Dustin, P. Furrer, A. Stasiak, J. Dubochet, J. Langowski, and E. Egelman, *J. Struct. Biol.* **107,** 15 (1991).

the 1D curve in 3D space. Nucleosomes may be approximated by a sphere, except in cases where the rotational orientation of the nucleosomal disc can be determined, when they can be modeled as cylinders.[6]

In cases where part of a DNA strand is perpendicular to the tilt axis, it is not possible to determine corresponding points on the image pair, and the z coordinates cannot be defined. In these cases, two tilted images are not sufficient to recover 3D information, and a third image tilted along a different tilt axis will be needed. This is technically very difficult at present because the two axis tilt cryoholders available are complicated and difficult to operate. In practice, it is necessary to exclude from analysis objects having a significant amount of DNA perpendicular to the tilt axis.

Potential Artifacts of Cryo-EM

The critical test of any imaging technique is the extent to which it provides a faithful representation of the original specimen, which, in the case of the cryo-EM applications discussed here, is the conformation of chromatin in the chosen solution. We have discussed many of the parameters that need careful attention in order to avoid conditions that may be detrimental to the final image. In addition, there are some conditions whose effects on chromatin are not yet fully analyzed or understood, but which may lead to images that do not reflect the solution conformation.

Cooling Effects

During the brief time between room temperature and vitrification, the sample undergoes rapid cooling, and because the helical repeat of DNA is a function of temperature,[29] this could affect its conformation. For chromatin, a change in the helical repeat of DNA could significantly modify the 3D conformation of both linear and closed arrays of nucleosomes by altering the rotational nucleosome setting[30] and, in the case of covalently closed arrays, by introducing torsional stress. However, it has been shown experimentally[22] that, under normal vitrification conditions, the cooling rate is too rapid to affect the DNA helical repeat. In the worst case, the rapid freezing time limits the effective temperature drop to about 10°.[22]

[29] A. V. Vologodskii, "Topology and Physics of Circular DNA." CRC Press, Boca Raton, FL, 1992.

[30] C. L. Woodcock, S. A. Grigoryev, R. A. Horowitz, and N. Whitaker, *Proc. Natl. Acad. Sci. U.S.A.* **90,** 9021 (1993).

Interface Effects

A more troublesome problem encountered with chromatin is the interaction of the specimen with one or both surfaces of the ice layer. This problem is not confined to chromatin and many other interactions between specimen and air–water interface have been observed and studied.[8,31] These poorly understood interactions are probably caused by the native electrical potential of the air–water interface and its effect on the distribution and type of charges within the thin film of solution. In the case of chromatin, this artifact can be recognized easily, as it causes denaturation, and naked DNA is seen, often in a layer close to the ice surface. Our observations suggest that this effect is related to three factors: salt concentration, specimen concentration, and ice film thickness, which tend to act cooperatively. As mentioned earlier, the optimal specimen concentration range is between 50 and 100 μg/ml. Higher concentrations, especially in combination with ion concentrations above ~20 m*M*, and thinner ice films tend to promote the surface denaturation effect, whereas lower specimen concentrations, thicker ice films, and lower ion concentrations tend to prevent it. Adding a low concentration of sucrose (0.05–0.1%) to the specimen solution can also help considerably. For work with salt concentrations in the range that produce maximal chromatin condensation (~100 m*M* monovalent ions), there are two proven approaches: increase the film thickness to ~150–200 nm and use intermediate voltage microscopy[20] or fix the chromatin at the desired salt concentration and then remove the salt by dialysis. At the level of resolution obtainable with cryo-EM, glutaraldehyde fixation does not appear to induce significant structural changes in chromatin 3D conformation.[6]

An alternative solution, which has not yet been perfected, is to use a protective lipid layer on the thin film surface. This effectively eliminates the air–water interface and the troublesome surface-related artifacts. However, it is difficult to control precisely the amount of lipid that must be added in order to produce a uniform layer and also minimize the appearance of lipid vesicles (which always form in solution) within the ice film. A procedure that has some promise is as follows: just before use add a 2-μl droplet of lipid solution in chloroform to the grid (total amount of lipids ~2 μg) and allow to air dry. When the aqueous specimen is then added, the lipid will form a protective bilayer on both surfaces; although these bilayers will be disturbed during blotting, they will be restored rapidly before vitrification. The most promising lipid is 1-α-phosphatidylcholine,

[31] M. Cyrklaff, N. Roos, H. Gross, and J. Dubochet, *J. Microsc.* **175,** 135 (1994).

which does not interact with DNA or chromatin and readily forms bilayers and micelles.

On some occasions, the attractive forces of the surface are not strong enough to cause chromatin denaturation, but they are sufficient to "trap" the chromatin particles in a region close to one or both interfaces and induce some degree of flattening. This effect is not obvious from the examination of individual micrographs and can only be recognized through a stereoscopic evaluation of tilt pairs (stereo pairs) of micrographs (see Fig. 3). Particles lying in one plane, usually close to the air–ice interface, should not be used for studies of 3D conformation.

Specimen Distribution and Film Thickness

As mentioned earlier, the thickness of the ice film can influence the 3D conformation of specimens that extend more than ~50 nm in any direction, which implies some restrictions for larger arrays of nucleosomes. The small space in the z direction will force them to adopt an orientation with their long axes parallel to the layer and, in the case of long chromatin fibers, their axes will be flattened significantly in the z dimension. However, the thickness of the ice layer is generally large enough to accommodate the 30-nm diameter of chromatin fibers and their internal 3D organization will be preserved.[20]

Thickness also has an impact on the distribution of specimen throughout the grid and within individual holes. Simple diffusion results in more particles being present in thicker parts of the film. Moreover, ice thickness is usually not uniform across the grid and also varies significantly within a single hole. In the latter case, the variation in thickness will be directed by the hydrophilic–hydrophobic properties of the holey film. In general, a hydrophilic film will promote a thicker layer of ice around the perimeter of the hole, causing most of the specimen to be located close to the edge. The opposite effect is seen with hydrophobic films. Platinum–carbon coating of the supporting film is recommended as it produces a rather neutral effect, and large variations in ice film thickness are usually not a problem. The use of lipids as a protective layer complicates the situation, as ice film thickness is usually quite variable.

Conclusions

Cryoelectron microscopy is the technique of choice when chromatin is to be imaged with electrons at the highest possible resolution or when the evaluation of its 3D conformation is needed. However, establishing the technique in a new location requires a substantial investment in instrumentation, and there are similar demands on operator skill and experience. Finding optimal conditions for a given project or sample can require considerably more time and effort than for conventional EM techniques, where fixation is usually mandatory, and the subsequent removal of interfering components from the solution can be carried out without affecting chromatin conformation. Also, the efficiency of the technique (in terms of number of usable images per operator hour) is much lower than with standard EM. In consequence, cryo-EM should not be targeted for experiments where extensive statistics are needed. However, the advantages of being able to image the solution conformation of a 3D macromolecular assembly that tends to collapse when attached to a substrate and dehydrated are very clear. As it becomes possible to reconstitute specific chromatin arrays *in vitro* and study such events as transcription[32] and chromatin remodeling, potentially important applications of cryo-EM are likely to increase dramatically.

Another important potential of cryo-EM that has been exploited for some systems[7,33] but not yet applied to chromatin, is temporal resolution. Reactions can be initialized by the addition of a reagent to the thin film during its descent into the cryogen and the course of the reaction followed with milisecond precision, allowing the course of events to be followed both in 3D and time.

Although some other techniques allow the retrieval of 3D information from chromatin (i.e., atomic force microscopy,[34] NMR, or X-ray crystallography), cryo-EM at present has substantial advantages, the most important of which is the minimal perturbation and hence preservation of the native structure of relatively large arrays of nucleosomes.

Acknowledgment

Supported in part by NIH GM43786 to CLW.

[32] V. M. Studitsky, D. J. Clark, and G. Felsenfeld, *Cell* **83,** 19 (1995).

[33] T. Ruiz, I. Erk, and J. Lepault, *Biol. Cell* **80,** 203 (1994).

[34] S. H. Leuba, G. Yang, C. Robert, B. Samori, K. van Holde, J. Zlatanova, and C. Bustamante, *Proc. Natl. Acad. Sci. U.S.A.* **91,** 11621 (1994).

[35] S. A. Grigoryev and C. L. Woodcock, *J. Biol. Chem.* **273,** 3082 (1998).

[36] J. Bednar, R. A. Horowitz, S. A. Grigoryev, L. M. Carruthers, J. C. Hauser, A. J. Koster, and C. L. Woodcock, *Proc. Natl. Acad. Sci. U.S.A.* **95,** 14173 (1998).

[11] Electron Microscopy of DNA–Protein Complexes and Chromatin

By JACK GRIFFITH, SUSAN MICHALOWSKI, and ALEXANDER M. MAKHOV

Introduction

It has become increasingly clear that chromatin structure plays an important role in the regulation of nuclear function, including transcription and replication. Our understanding of the nature of chromatin and DNA–protein interactions in general can be enhanced by visualizing these complexes using electron microscopy (EM). A growing number of microscopic approaches have become available, including methods employing scanning tips, scanning transmission EM (TEM), and cryo-EM of samples embedded in ice. Conventional electron microscopy, however, remains one of the most reliable, accessible, and easily performed approaches, and in work where imaging needs to be closely coupled with biochemical studies, the attributes of ease and simplicity become critical. Because scanning tip and cryo-EM methods have been described elsewhere, this article focuses on methods and problems related to the use of conventional transmission electron microscopes for the visualization and analysis of chromatin and DNA–protein complexes. A brief summary of the different methods are provided followed by more detailed discussion of problems related to the adsorption of samples to supports, fixation, and the use of negative staining.

General Approaches for Visualizing DNA–Protein Complexes

Surface Spreading Methods

Kleinschmidt Method. Over the past three decades a variety of approaches have been developed for the visualization of DNA and chromatin. The first generally useful method for visualizing DNA was developed by Kleinschmidt and Zahn[1] and was refined by Davis and others[2] and employs a denatured film of cytochrome *c* protein. Here the DNA is mixed with cytochrome *c* and the mixture is either spread onto the surface of a low salt buffer or placed as a drop on a plastic sheet. In either case, some protein denatures and forms a layer of denatured protein at the air–water interface. Denatured protein will bind cooperatively to DNA, increasing

[1] A. K. Kleinschmidt and R. K. Zahn, *Z. Naturforsch.* **14b,** 770 (1959).
[2] R. W. Davis, M. Simon, and N. Davidson, *Methods Enzymol.* **XXI,** 413 (1971).

0076-6879/99 $30.00

its width from 2 nm to as much as 15 nm, and the resulting filament is trapped in the denatured protein film at the interface. Surface tension forces and the inherent stiffness of the filament spread the DNA out, reducing the number of times the DNA crosses itself. The greatly enhanced width means that rough shadow casting methods using platinum–palladium provide very high contrast, which in turn makes it possible to view the samples at magnifications severalfold less than that required for DNA not coated with protein. Variations of the method have used formamide in the spreading buffers to remove the secondary structure of single-stranded DNA or RNA complexes[2] (reviewed by Moore[3]). It is generally considered that the coating of cytochrome *c* will obscure other proteins bound to the DNA and thus these methods are not thought to be useful for chromatin or DNA–protein complexes. However, a review from this laboratory[4] showed that by the careful titration of cytochrome *c*, these methods can be surprisingly useful. For example, gold-labeled antibodies allow the identification of proteins bound to DNA even when obscured by a subsequent layer of cytochrome *c*. For studies using denatured protein films, it has been found that very hydrophobic plastic supports such as parlodion bind the DNA–protein film much better than, for example, carbon. Because plastic films may melt in the electron beam, they are frequently coated with carbon following the metal shadow casting step.

A variation of the Kleinschmidt method used extensively for DNA–protein complexes employs a low molecular weight mild detergent, benzyldimethylalkylammonium chloride (BAC), as a substitute for cytochrome *c*.[5] To enhance contrast, the specimens are stained with uranyl acetate prior to the shadowing with platinum or platinum–palladium. This method has the advantage of a much thinner coating over the DNA and hence protein complexes of moderate size can be distinguished bound to the DNA. The surface forces employed, however, may be strong and could alter or disrupt structures of interest.

Miller Spreading Method. The method developed by Miller and colleagues has been extremely useful in studies of chromatin and was employed by Olins and Olins[6] in their very early visualization of chromatin "nu" bodies. Nuclei are disrupted hypotonically and then stepwise dispersed gently and fixed by low-speed centrifugation through a sucrose gradient containing a mild detergent (e.g., Joy) followed by formaldehyde or glutar-

[3] C. L. Moore, *in* "Electron Microscopy in Biology" (J. D. Griffith, ed.), Vol. 1, p. 67. Wiley, New York, 1981.

[4] R. Thresher and J. Griffith, *Methods Enzymol.* **211,** 481 (1992).

[5] H. J. Vollenweider, J. M. Sogo, and Th. Koller, *Proc. Natl. Acad. Sci. U.S.A.* **72,** 83 (1975).

[6] A. L. Olins and D. E. Olins, *Science* **187,** 1202 (1974).

aldehyde, and finally the material is deposited onto an EM grid placed on a flat shelf at the bottom of the centrifuge tube. Contrast enhancement can be obtained by positive staining or shadow casting. Critical factors for optimal dispersal of chromatin are an alkaline pH (pH 8.5–9) and very low ionic strength. Only a small amount of cellular material and no specialized equipment are required for this method.[7,8] In general the method is excellent for studying disrupted nuclei as contrasted to defined DNA–protein complexes formed *in vitro*.

Direct Mounting Methods

A variety of methods generally termed "direct mounting" have been developed specifically for the visualization and analysis of defined DNA–protein complexes and chromatin. These methods involve the direct adsorption of the sample to a support (usually carbon) that has been modified to bind DNA, followed by dehydration and rotary shadow casting or negative staining. The methods have proven ideal for chromatin and defined complexes of proteins bound to DNA. Figure 1 shows two examples from early studies: a simian virus 40 (SV40) minichromosome and a field of chromatin fibers. The visualization of SV40 minichromosomes[9] provided the first visualization of a true nucleosome and a quantitative link between these particles, which were shown to contain ~160 bp of DNA repeated every ~200 bp and the $(200\text{ bp})_n$ micrococcal nuclease digestion patterns obtained by Hewish and Burgoyne.[10]

When combined with metal shadow casting, these methods provide a reliable way of visualizing chromatin, DNA, and single proteins of ~50 kDa bound to DNA. Samples can be processed at a rate and number that makes such methods compatible with many biochemical studies. For example, the kinetics of growth of DNA loops mediated by the mutS protein has been followed by these methods,[11] and the binding of p53 was shown to be highly specific for Holliday junctions.[12] An older review[13] provides details of the most commonly used method in this laboratory. As a further means of combining EM and biochemical studies, Jett and Bear[14] have developed a method termed snapshot blotting. Here DNA or DNA–protein

[7] O. L. Miller, Jr. and B. R. Beatty, *Science* **164,** 955 (1969).

[8] O. L. Miller, Jr. and A. H. Bakken, *Acta Endocrinol. Suppl.* **168,** 155 (1972).

[9] J. Griffith, *Science* **187,** 1202 (1975).

[10] D. R. Hewish and L. A. Burgoyne, *Biochem. Biophys. Res. Commun.* **52,** 504 (1973).

[11] D. J. Allen, A. Makhov, M. Grilley, J. Taylor, R. Thresher, P. Modrich P, and J. D. Griffith, *EMBO J.* **16,** 4467 (1997).

[12] S. Lee, L. Cavallo, and J. Griffith, *J. Biol. Chem.* **272,** 7532 (1997).

[13] J. D. Griffith and G. Christiansen, *Annu. Rev. Biophys. Bioeng.* **7,** 19 (1978).

[14] S. D. Jett and D. G. Bear, *Proc. Natl. Acad. Sci. U.S.A.* **91,** 6870 (1994).

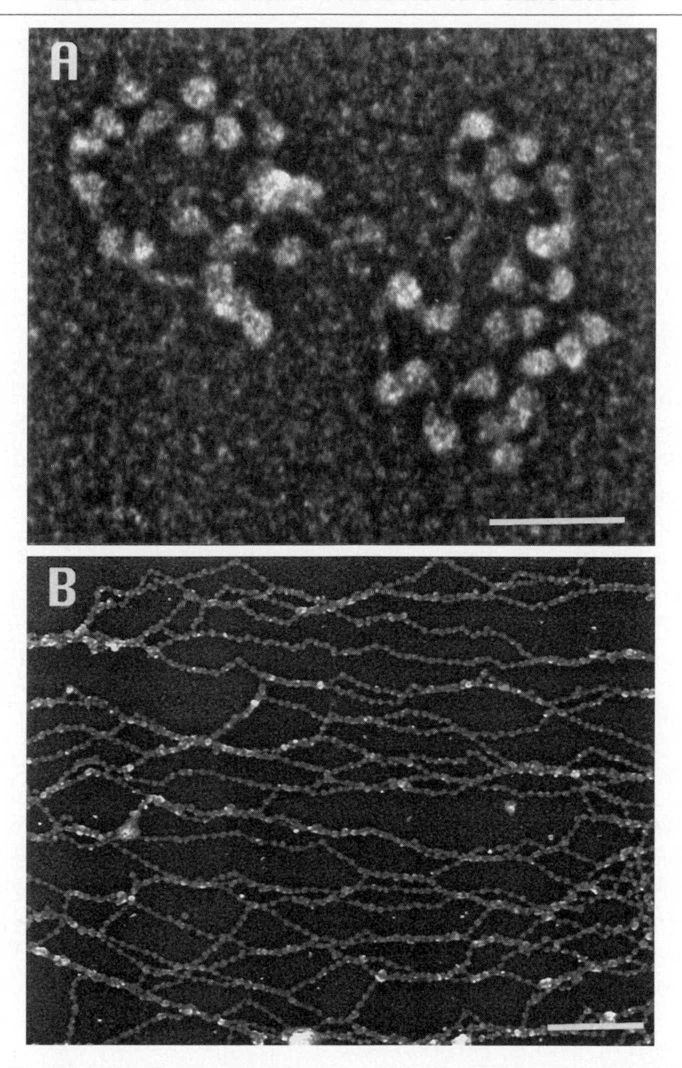

FIG. 1. Visualization of chromatin fibers and minichromosomes. (A) SV40 minichromosomes purified from cells infected by SV40 virus contain a roughly 5-kb circular DNA assembled into 21 nucleosomes formed from cellular histone ocatamer cores. Two SV40 minichromosomes are shown. The sample was prepared by 1% formaldehyde fixation, washed, air-dried, and rotary shadow cast with tungsten. (B) A field of cellular chromatin fibers showing the beaded nucleosomal repeat. Chicken erythrocyte nuclei were lysed, fixed with 0.6% glutaraldehyde, and spread on a glow-charged carbon film in the presence of 2 m*M* spermidine. The sample was washed, air-dried, and rotary shadow cast with tungsten. Shown in reverse contrast. Bar: 50 nm (A) and 500 nm (B).

complexes are electrophoresed in agarose or acrylamide gels as commonly done and the DNA bands are located using ethidium bromide staining. A small well is then cut into the gel just ahead of the band and an EM grid along with a drop of buffer containing spermidine is placed in the well. The sample is then electrophoresed onto the grid, which is then removed, washed, dried, and shadow cast.

Cryo Preparation of DNA–Protein Samples for TEM. Freeze-fracture and freeze-drying methods have been adapted for the analysis of chromatin and DNA–protein complexes.[15,16] In the mica chip method of Heuser,[17] the sample is adsorbed to a slurry of mica chips, which are then frozen rapidly using a liquid helium-chilled block followed by freeze-fracture and deep etching and finally replication with carbon and platinum. This approach has produced very striking images of recA protein complexes with DNA.[18] In a different approach, Bortner and Griffith[19] developed a means of adsorbing DNA–protein complexes to thin carbon supports followed by rapid freezing in liquid ethane and then slow freeze-drying and tungsten shadow casting all in the same ultrahigh vacuum system. The latter method combines advantages of avoiding chemical fixation with the higher resolution afforded by tungsten or tantalum as contrasted to platinum shadowing (see Fig. 2). Both methods, however, are complex and not ideally suited to the high throughput required for combining TEM with biochemical studies of chromatin.

Summary

A large variety of ways are now available for preparing chromatin and DNA–protein complexes for TEM. This article has provided key references to each of these methods and indications of where each is best employed. The following discussions describe in more detail several key areas: preparation of the supports, fixation, and staining that are important to master regardless of which preparative routes are taken.

Preparation of Sample Supports

Preparation of Carbon Supports

A variety of support films have been used for chromatin and DNA–protein complexes. In general, the plastic supports that work well for the

[15] J. T. Finch and A. Klug, *Proc. Natl. Acad. Sci. U.S.A.* **73,** 1897 (1976).
[16] J. Lepault, S. Bram, J. Escaig, and W. Wray, *Nucleic Acids Res.* **8,** 265 (1980).
[17] J. Heuser, *J. Mol. Biol.* **169,** 155 (1983).
[18] J. Heuser and J. Griffith, *J. Mol. Biol.* **210,** 473 (1989).
[19] C. Bortner and J. Griffith, *J. Mol. Biol.* **215,** 623 (1990).

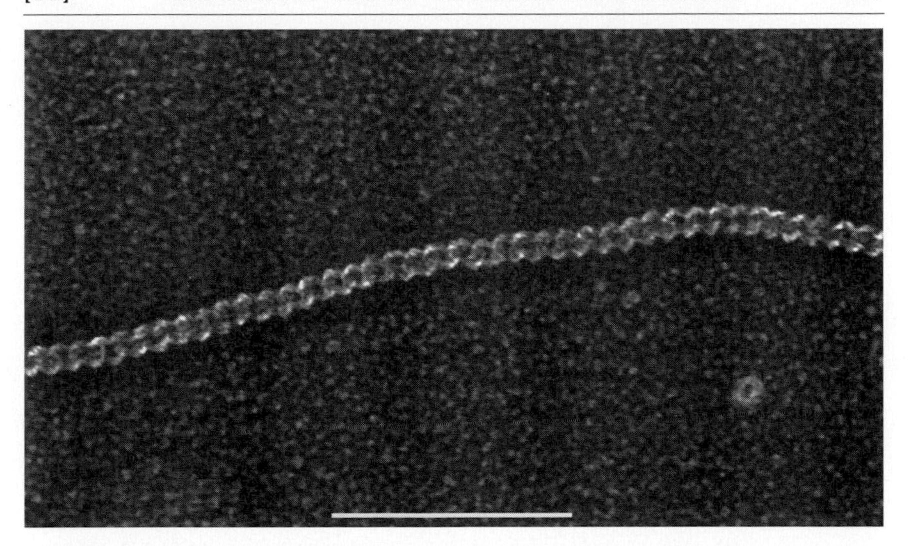

FIG. 2. Rapid freezing/freeze-drying of a recA protein DNA filament. The recA protein was assembled onto double-stranded DNA in the presence of the ATP analog ATPγS. Filaments were adsorbed without fixation to a glow-discharged carbon support in the presence of 2 mM spermidine, washed briefly, frozen in liquid ethane, and then slowly freeze-dried in a fully cryopumped vacuum evaporator. The sample was then rotary shadow cast with tantalum after freeze-drying without opening the vacuum chamber. Shown in reverse contrast. Bar: 100 nm.

Kleinschmidt method are too thick and rough surfaced and are subject to melting in the microscope beam to be useful for the direct mounting of chromatin to be imaged by shadow casting or negative staining. While supports of silicon monoxide, aluminum, or aluminum–beryllium[20,21] and other low molecular weight elements have been tried, carbon prepared by evaporation onto a smooth surface is uniformly used. Carbon provides a number of advantages. It is relatively electron transparent, it is extremely strong, it is thermally stable, and its coefficient of thermal expansion is low so that even very thin films withstand significant beam doses. Carbon films are also very easy to make and store. Their preparation is described in Griffith[22] and Goodhew[23] as well as in classic EM reference texts. In brief, sheets of mica are cleaved to expose a clean surface, placed in a vacuum evaporator, and carbon rods are heated to evaporation with a high current.

[20] J. M. Sogo, R. Portmann, P. Kaufmann, and T. Koller, *J. Microsc.* **104,** 187 (1975).
[21] C. L. F. Woodcock, L.-L. Y. Frado, G. R. Green, and L. Einck, *J. Microsc.* **121,** 211 (1981).
[22] J. D. Griffith, *Methods Cell Biol.* **7,** 129 (1973).
[23] P. J. Goodhew, *in* "Practical Methods in Electron Microscopy" (A. M. Glauert, ed.), Vol. 11, p. 160. Elsevier, New York, 1985.

Once coated, the mica sheets can be stored for years. To prepare the EM support, a small square of the carbon-coated mica is lowered at a 30° angle onto the surface of a water trough to float the carbon free. Copper mesh grids are placed on the carbon for 1 hr and the grids are then removed using a beaker over which a drum of Saran wrap has been formed (see Griffith[22]). This results in the surface of the carbon that was against the mica being the surface onto which the sample will be adsorbed.

Treatment of Supports to Facilitate Sample Adherence

While nearly all of the procedures begin with the preparation of carbon supports, there have been a large number of approaches taken to make the carbon bind DNA or proteins. The fundamental problem is that carbon evaporated onto mica is extremely hydrophobic and unreactive. DNA does not bind readily and, if it does, it forms collapsed toruses on dehydration because it is bound so poorly to the support. The most common initial step is to expose the carbon films to a glow discharge in a vacuum, which is usually combined with some further step.[24,25] Glow charging is carried out by placing the supports in a vacuum of ~200 millitorr and then connecting one of the electrodes of the chamber to a high voltage source such as the ballast of a neon tube. This is done for 30 to 60 sec and the treatment renders the carbon hydrophilic for at least several hours. Analysis of carbon films treated in this way using surface spectroscopy revealed that the major charged species generated were carboxyl groups.[26] In all of the various methods that follow, it is assumed that highly distilled water is used.

One of the most common methods to further facilitate the binding of DNA or chromatin to the glow-treated supports involves the inclusion of small molecule ions. In the early studies from this laboratory it was found that DNA adsorbed much more efficiently to the supports if NaCl was present at a concentration of 0.1 to 0.5 *M*. Others observed that lower concentrations of magnesium (10 to 50 m*M*) had the same effect. In the method developed in our laboratory and which we have found to be most generally useful, the sample is mixed with a buffer containing 0.15*M* salt and 2 m*M* spermidine. The salt keeps the spermidine from condensing the DNA and the spermidine molecule that harbors a positive charge at each end of the chain provides a bridge between the negatively charged DNA phosphates and the negatively charged carboxyl groups on the carbon support.

[24] U. Aeibi and T. D. Pollard, *J. Electr. Microsc. Tech.* **7,** 29 (1987).
[25] E. Namork and B. V. Johansen, *Ultramicroscopy* **7,** 321 (1982).
[26] J. Griffith, unpublished results, 1990.

Polylysine. Williams[27] described a method of postglow treatment that has been used extensively employing polylysine. Formvar–carbon or carbon-covered grids are glow treated and then a drop of polylysine in water at 0.3–1.0 μg/ml is placed on the grid for 30 sec. The drop is removed by touching the edge of the grid with a vacuum-connected aspirator freshly made from a Pasteur pipette flame—drawn to less than a 0.5-mm bore diameter. The residual liquid on the grid can be inspected under a 10 power microscope as it dries; the surface should be so hydrophilic that drying takes several seconds, with the trailing edge of the liquid film showing one to two complete orders of interference colors. No filter papers of any kind are used, either for water purification or for removal of the liquid drop from the grid surface. After the residual polylysine is dried the grids are ready for use and remain so for at least a few days if they are kept in a tightly covered container. A concern with this method is that polylysine is a protein and binds DNA avidly. Thus a rougher background on the support can be expected and some polylysine clustering on the DNA might be confused for the binding of other proteins.

Amylamine and Pentylamine. Dubochet and colleagues[28] have combined glow discharge treatment and exposure to charged ions by carrying out the glow charging in an atmosphere of amylamine or pentylamine. In their method, carbon-covered grids are evacuated in a chamber to about 1 millitorr, amylamine or pentylamine is introduced, and the vacuum is adjusted to 200–500 millitorr. Pumping is then resumed until a vacuum of 20 millitorr is attained, and glow charging is carried out for 10 sec. Glow discharging in the presence of amylamine appears to provide a slightly higher adsorption, but this may not be suitable negative staining (may interfere with the spreading of the stain) and care must be taken to avoid the fumes as they are quite toxic. A similar method has been developed by Bazett-Jones and Ottensmeyer[29] employing ethylenediamine. The carbon-coated films can also be discharged in pure nitrogen at a pressure of 190 millitorr within a 2 hr period before use.[30]

Alcian Blue. A variety of treatments have been described that do not necessarily require an initial glow discharge treatment. These involve exposure to the dyes Alcian blue or ethidium bromide. In the method employing Alcian blue, a 0.2% (w/v) stock solution of Alcian blue 8GX in 3% acetic acid is diluted in redistilled water to make 0.002% Alcian blue and filtered

[27] R. C. Williams, *Proc. Natl. Acad. Sci. U.S.A.* **74,** 2311 (1977).

[28] J. Dubochet, M. Ducommun, M. Zollinger, and E. Kellenberger, *J. Ultrastruct. Res.* **35,** 147 (1971).

[29] D. P. Bazett-Jones and F. P. Ottensmeyer, *J. Ultrastruct. Res.* **67,** 255 (1979).

[30] P. V. C. Hough, I. A. Mastrangelo, J. S. Wall, J. F. Hainfeld, M. N. Simon, and J. L. Manley, *J. Mol. Biol.* **160,** 375 (1982).

through a 0.22-mm filter.[31] Carbon-covered grids are floated carbon side down for 5 min on 0.002% Alcian blue at room temperature. The excess dye is washed off by floating on redistilled water for 10 min. The grids are blotted dry on filter paper and used within the next hour.

Ethidium Bromide. In a replica-based method employing ethidium bromide immediately before use, ethidium bromide is dissolved in water, filtered, and added to a DNA sample at a final concentration of 50–250 μg/ml of the dye.[32] Droplets of 0.1–0.15 ml of this solution are placed on fresh parafilm and are kept for 10–15 min at room temperature. A piece of freshly cleaved mica is then touched to the surface of solution. The mica is then washed in water for several hours. The samples are then dehydrated with ethanol, air dried, and platinum–carbon replicas prepared. Carbon coated and glow-discharged supports can be treated with similar solutions of ethidium bromide, washed, dried, and used for mounting DNA and DNA–protein complexes. Similar results were obtained with propidium diiodide and actinomycin D when they were used in place of ethidium bromide.[32] The carbon films can be treated with a 50-μg/ml solution of bacitracin into 1 m*M* ammonium acetate, pH 6.7–6.9, for 30 sec.[33]

Methods and Problems of Fixation

General

Most all methods for visualizing DNA–protein complexes and chromatin have employed chemical fixation prior to EM preparative steps. Fixation is very useful in stopping the progress of DNA–protein reactions and interactions and in stabilizing complexes so that they can be purified further by, for example, gel chromatography prior to adsorption to EM supports. Fixation clearly helps protect protein complexes from denaturation during exposure to dehydrating agents or air-drying. Although fixation can introduce structural changes, in our experience these changes are secondary to changes that will be seen if fixation is not employed. Even when samples are prepared by rapid freezing and freeze-drying methods, it is often necessary to fix the samples so that excess protein and buffer can be removed prior to rapid freezing and freeze-drying. We now deal with two central concerns: what is the best method of fixing and monitoring the fixation process and what artifacts may be encountered due to fixation?

[31] P. Labhart and T. Koller, *Eur. J. Cell Biol.* **24,** 309 (1981).

[32] T. Koller, J. M. Sogo, and H. Buard, *Biopolymers* **13,** 995 (1974).

[33] C. W. Gray, *in* "Methods in Molecular Biology" (C. Jones, B. Mulloy, and A. H. Thomas, eds.), Vol. 22, p. 13. Humana Press, Totowa, NJ, 1994.

Different reagents have been used for cross-linking chromatin. Formaldehyde, glutaraldehyde, various imido esters and dithiobis(succinimidyl propionate) cross-link between lysine residues with a linker of various lengths. Carbodiimide cross-links mainly between lysine and aspartyl or glutamyl residues, and tetranitromethane and ultraviolet light between tyrosine residues.[34–36] Formaldehyde, glutaraldehyde, and dimethyl suberimidate are commonly used fixatives for EM analysis of chromatin and will be discussed next.

Glutaraldehyde and Formaldehyde

Glutaraldehyde and formaldehyde are often used to fix chromatin and other DNA–protein complexes for EM. Glutaraldehyde effectively cross-links protein and does not react with naked DNA,[37–39] whereas formaldehyde cross-links protein to DNA,[40,41] but is not as efficient at stabilizing proteins. Protein is cross-linked to DNA by formaldehyde at 0.05%; however, the cross-linking of histone octamers does not occur at lower than 0.5% concentration.[42] Formaldehyde induces cross-links through the formation of monomethyloyl derivatives of amines and ultimately methylene bridges between amine groups. A number of reactive groups on DNA bases can be cross-linked to nearby lysines on the histones by this mechanism. It has been shown that formaldehyde forms bonds with exocyclic amino groups of adenosine, guanosine, and cytosine and with CO—NH grouping in purine and pyrimidine heterocycles.[43,44]

In some cases a combination of the two fixatives may be used to prevent denaturation of the protein while stabilizing the interaction between the protein and DNA to prevent slippage from the binding site. It must be emphasized strongly that because all of these fixatives react with primary amines, it is important to avoid buffers that contain reactive groups, such as Tris. If, for example, the sample is buffered in 10 m*M* Tris, the addition

[34] J. Barnues, E. Querol, P. Martinez, A. Barris, E. Espel, and J. Lloberas, *J. Biol. Chem.* **25,** 11020 (1983).

[35] K. S. Lee, M. Mandelkern, and D. M. Crothers, *Biochemistry* **20,** 1438 (1981).

[36] M. Suda and K. Iwai, *J. Biochem.* **86,** 1659 (1979).

[37] A. M. Glauert, *in* "Fixation, Dehydratation and Embedding of Biological Specimens." North Holland, Amsterdam, 1975.

[38] D. Hopwood, *Histochem. J.* **4,** 267 (1972).

[39] D. Hopwood, *Histochem. J.* **7,** 267 (1975).

[40] D. Brutlag, C. Schlehuber, and J. Bonner, *Biochemistry* **8,** 3214 (1969).

[41] R. Chalkley and C. Hunter, *C. Proc. Natl. Acad. Sci. U.S.A.* **72,** 1304 (1975).

[42] B. T. Sewell, C. Bouloukos, and C. von Holt, *J. Microsc.* **136,** 103 (1984).

[43] M. Ya. Feldman, *Prog. Nucleic Acid. Res. Mol. Biol.* **13,** 1 (1973).

[44] J. F. Walker, *in* "Formaldehyde," 3rd Ed. Reinhold, New York, 1964.

of 1% formaldehyde (250 m*M*) will obliterate the buffer, release acid, and likely result in a very uncontrolled reaction. HEPES or phosphate buffers provide good alternatives.

The concentration of the fixative and the length of time and temperature of fixation are parameters that need to be considered (see later). A good starting point is the use of 0.6% glutaraldehyde alone in a HEPES-buffered solution for 5 min at 37°, or 10 min at room temperature or on ice. If formaldehyde is used, we have employed 1% formaldehyde (freshly diluted from a 37% stock solution following heating of the stock to 90° for 5 min) for 15 min on ice. The resulting structures are then stabilized further with 0.6% glutaraldehyde (freshly diluted from an unheated 25% stock solution) for 10 more min on ice.

Dimethyl Suberimidate

This compound is a bifunctional amino group reagent that has been used to cross-link chromatin and other DNA–protein complexes.[36] It appears that nucleosome size and DNA unwinding are not affected by this fixative.[45] Chromatin in cross-linking buffer [10 m*M* HEPES, 0.1 m*M* EDTA, 1 m*M* phenylmethylsulfonyl fluoride (PMSF), and 100 m*M* NaC1, pH 8.0] is placed in dialysis tubing with dimethyl suberimidate and dialyzed against cross-linking buffer for 30 min. The procedure is repeated once followed by dialysis against 0.1 m*M* Tris and 0.1 m*M* EDTA, pH 7.8.[35]

Optimizing Fixation

The fixation regime employed for any particular DNA–protein complex should be optimized prior to a detailed study as this may vary greatly. For example, *Escherichia coli* RNA polymerase is released from DNA on exposure to low temperature and thus fixation on ice is a poor choice, whereas other complexes have been found to be stabilized by chilling on ice. The best approach is to test a matrix of fixation times (1 to 30 min), temperatures (4, 21, 37°), and (1) glutaraldehyde only, (2) formaldehyde with glutaraldehyde, or (3) formaldehyde first followed by glutaraldehyde. EM visualization will then provide clues into which method provides the most uniform appearing samples. Gel mobility shift analysis may also be used to optimize fixation. If the protein forms a specific complex on a DNA fragment that can be detected by gel mobility shift analysis, then different fixations should be carried out (using a nonprimary amine buffer). Following fixation, sodium dodecyl sulfate (SDS) is added to 1% and the sample is placed at 37° for 5 min prior to electrophoresis. Adequate fixation is

[45] H. M. Wu, N. Dattagupta, M. Hogan, and D. M. Crothers, *Biochemistry* **18,** 3960 (1979).

determined to be a condition under which the mobility shift is the same as that observed under native conditions and is resistant to challenge by subsequent SDS treatment.

Artifacts of Fixation

Treatment of DNA–protein complexes or chromatin with chemical fixatives would be expected to alter their structure. This was documented by Heuser and Griffith[18] in a study of recA protein filaments on DNA. When the filaments were prepared by slam freezing onto a liquid helium-chilled silver block without prior glutaraldehyde fixation, very striking regular helical filaments (see Fig. 2) were observed in the platinum replicas that had dimensions close to those seen in samples prepared by negative staining without fixation. When the same filaments were treated with glutaraldehyde under standard conditions and then prepared in parallel by slam freezing and deep etching, filaments were observed but the helical repeat was much less obvious and the fine detail was absent. This argues that fixation may compact or collapse the fine structure of protein complexes. In our hands this loss is usually acceptable as it allows one to follow very long DNA strands and map the location of infrequent or unusual protein complexes on the DNA. Once such structures have been characterized using regimes employing fixation and shadow casting, then fine structure studies using negative staining (see later) without fixation or fast-freezing/freeze-drying may reveal the detail lost due to fixation.

Can fixation generate complexes on DNA that were not present prior to fixation? We have yet to encounter a clear example using relatively mild fixations (0.6% glutaraldehyde treatment for 5 to 10 min). When bovine serum albumin was dissolved at concentrations spanning a 20,000-fold range from 1 μg/ml to 20 mg/ml, fixed with 0.6% glutaraldehyde, and then diluted to 1 μg/ml and mounted on the EM support, the particles present were monomer sized irrespective of the concentration at which they were fixed, arguing against fixation-induced oligomerization under these conditions.[46] It has been observed with the binding of the mutS protein at the site of a G/T mismatch that the protein complexes present were larger when high concentrations of mutS protein were present in the incubations. This may be due to the presence of loosely associated mutS layering onto the specifically bound protein rather than the creation of larger complexes generated by fixation.

Occasionally, fixation may interfere with the binding of a protein to DNA. In the case of fos and jun proteins binding to their specific site on

[46] J. Griffith, unpublished results, 1990.

DNA, fixation was found to release the proteins from the DNA as seen both by EM and by gel-shift analysis. It is likely that the proteins possess one or more lysine groups critical for their binding to DNA and that the glutaraldehyde treatment inactivates these groups before the protein can be fixed to the DNA.[47]

Negative Staining of Chromatin and DNA–Protein Complexes

Most of the methods just described utilize metal shadow casting as a means of contrast enhancement. These methods have been described in great detail elsewhere[13,22] and are generally routine. The resolution from shadow casting is limited by a combination of the amount of metal deposited, the inherent size of the metal grains formed on the support, and the possible degradation of the sample by the very high temperature of the impinging metal atoms. In contrast, negative staining, which is a very long-standing high-resolution method for studies of proteins, protein complexes, and bacteriophage, has found only limited use for chromatin or DNA–protein complexes. It offers nonetheless many advantages. First, it is carried out routinely without exposing the sample to fixation, and the glass formed by the stain may help protect the sample from dehydration. Very small amounts of the sample are needed and the time required for the staining step is minimal. The resolution afforded is potentially much higher, and penetration of the stain into valleys and holes in a protein complex may reveal interior detail. However, air-drying from a negative stain film that only partly covers a protein is likely to give rise to flattening. The achievable resolution of negatively stained samples is usually about 2 nm and at best 1.4 nm. Superior resolution has been reported for negative stain–glucose mixtures from electron crystallography.[48] Finally, great advances have been made in the instruments themselves, which facilitate the acquisition of the highest resolution-stained images. These include automatic film exposure, the use of cooled CCD cameras, and computer-controlled minimal beam exposure systems. Digital image analysis and reconstruction software used to quantify and enhance fine detail in the images are becoming readily available.

Negative staining has not seen as much use for chromatin or DNA–protein complexes because of the general belief that DNA cannot be visualized easily by staining alone. Although DNA is visualized much more easily by shadow casting, particularly if the DNA is long, DNA can be seen by

[47] J. Griffith, unpublished results, 1990.

[48] N. A. Kiselev, M. B. Sherman, and V. C. Tsuprun, *Electr. Microsc. Rev.* **3,** 43 (1990).

negative staining as illustrated in Fig. 3. Negative staining is a relatively simple method resulting in high-resolution images that are particularly useful for the detailed visualization of protein structure.

General Approach

Thin carbon support films prepared for negative staining are exposed routinely to a glow discharge and are used immediately. The samples are adsorbed to the surface of the carbon support for 30 sec to 1 min. Some compounds that may be present in the DNA–protein complexes will interfere with staining. These include carbohydrates, phosphates, detergents, and oxidizers and reducers at significant concentration. High concentrations of salt can also interfere. In this case the grids are washed after adsorption but before staining by floating them on the surface of water or buffer containing low salt concentrations or without salt. HEPES buffer (10–20 m*M*) with or without 0.1–1.0 m*M* EDTA may be used. After washing, the grids are commonly stained for about 1 min. It should be noted that glycerol is commonly included at 20–50% concentration in purified protein preparations. Unfortunately, even 1% (v/v) glycerol will interfere. Thus glycerol should first be reduced by dilution or dialysis. Negative-stained samples are very sensitive to radiation damage and care must be taken to avoid excess beam exposure.

Uranyl Acetate and Uranyl Formate. The most commonly used stains for protein and DNA–protein complexes are uranyl acetate and less often uranyl formate. Generally, solutions of 1–2% uranyl acetate in water (unbuffered) are used for negative staining. The uranyl cation can also bind phosphate and carboxyl groups and may stain DNA positively if the carbon film is hydrophobic. In addition, if the samples are not absorbed first to the carbon support film, the acidity of the uranyl acetate may produce undesirable precipitation or aggregation of protein molecules from solution. Mixtures of uranyl acetate and glucose are relatively unstable in the electron beam, causing pronounced specimen "bubbling" unless the stain–glucose film is thin.[49]

Phosphotungstic Acid. For these stains, aqueous solutions of 1.5–2% (w/v) phosphotungstic acid (PTA) are employed. The pH of PTA is adjusted with NaOH or KOH to pH 7.0. For stabilizing DNA–protein or protein–protein complexes, 0.015% glucose may be added to the PTA solution. PTA usually gives a lower resolution than uranyl acetate, but has been used successfully for the analysis of chromatin.[50] PTA is relatively

[49] J. R. Harris and R. W. Horne, *Micron* **25,** 5 (1994).

[50] C. L. Woodcock, H. Woodcock, and R. A. Horowitz, *J. Cell Sci.* **99,** 99 (1991).

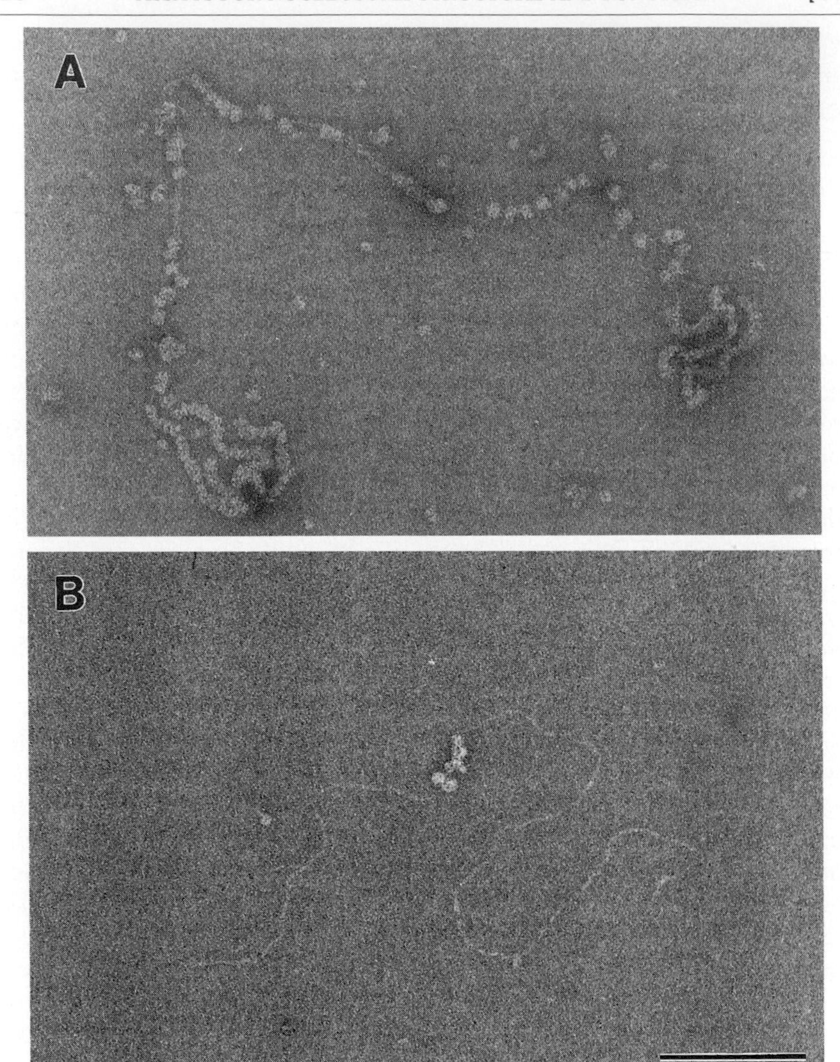

FIG. 3. Examples of DNA–protein complexes stained negatively with 2% uranyl acetate in water (A). The UL9 and ICP8 proteins of herpes simplex virus type I carry out a concerted unwinding of linear double-stranded DNA containing 3′ single-stranded tails. This begins by the binding of UL9 to the single-stranded ends followed by its migration into the central double-stranded region and the replacement of UL9 on the tails by the ICP8 protein. (B) The human TRF1 protein binds to the repeated sequence TTAGGG at telomeres. Here an insert of 27 such repeats is present at the center of a linear 2.5-kb DNA and can be seen bound by hTRF1 protein. Bar: 60 nm (A) and 100 nm (B).

insensitive to beam damage as compared to other stains and was selected for the tomographic reconstruction of chromatin based on negative staining.[51,52]

Uranyl acetate and PTA together have been used as effective stains in the analysis of chromatin. After the adsorption of DNA–protein complexes to carbon supports, the samples were stained with uranyl acetate, washed with water, and stained with PTA.[21]

Methylamine Tungstate. Methylamine tungstate was developed as a stain with both hydrophobic and hydrophilic components.[53] Solutions of 1.5% methylamine tungstate in water, pH 7.0, are usually used.[50] This stain has the interesting property of emphasizing arrays of nucleosomes, perhaps by being excluded from the contact region between nucleosomes.[50] Negative staining was attained by the addition of 0.25% methylamine tungstate to the carbon film for 1 min, washing twice with water, and air drying.[54]

Ammonium Molybdate. Solutions of 2–4% (w/v) ammonium molybdate in water, pH 8.0–8.5, have been used for negative staining. Ammonium molybdate usually gives clear images, but the stain layer is also quite deep. One effect of the deep stain is to minimize the contrast change at the edges of chromatin fibers, giving the appearance that they are substantially thinner than other stains. Ammonium molybdate is also less reactive than other stains and therefore generally more satisfactory as an isotonic negative stain, thus preserving the molecular structure of biological samples.

Aurothioglucose. Solutions of 2% aurothioglucose (ATG), a nonionic sugar derivative (unbuffered), are used. ATG provides very clear images of individual nucleosomes. However, the use of this stain is restricted by its extreme beam sensitivity, requiring the use of a low-dose system. An alternative method of applying aurothioglucose involves mixing the chromatin with ATG (2% final concentration) and placing the mixture directly on the grid.[50] Cadmium thioglycerol is as beam sensitive as aurothioglucose and has been used by Jakubowski *et al.*[55] Dodeca-*m*-bromohexatantalum diacetate has been used to stain DNA and DNA–protein complexes.[56] Methylamine vanadate[57] and sodium silicotungstate[49] were used for the staining of protein and protein complexes.

[51] C. L. Woodcock, B. F. McEwen, and J. Frank, *J. Cell Sci.* **99,** 107 (1991).

[52] C. L. Woodcock and B. F. McEwen, *Eur. J. Cell Biol.* **51,** 45 (1988).

[53] A. C. Faberge and R. M. Oliver, *J. Microsc.* **20,** 241 (1974).

[54] K. J. Neil, R. A. Ridsdale, B. Rutherford, L. Taylor, D. E. Larson, M. Glibetic, L. I. Rothblum, and G. Harauz, *Nucleic Acids Res.* **24,** 1472 (1996).

[55] U. Jakubowski, R. Hegerl, H. Formanek, S. Volker, U. Santarius, and W. Baumeister, *Inst. Phys. Conf. Ser. 93* **3,** 381 (1988).

[56] T. Koller, A. G. Harford, Y. K. Lee, and M. Beer, *Micron* **1,** 110 (1969).

[57] J. F. Hainfeld, D. Safer, J. S. Wall, M. Simon, B. Lin, and R. D. Powell, *in* "Proc. 52nd Ann. Meet. Microsc. Soc. Amer." (G. W. Baily and A. J. Garratt-Reed, eds.), p. 132. San Francisco Press, San Francisco, 1994.

It has been shown that chemical compounds containing only light elements (potassium aluminum sulfate, ammonium borate, sodium tetraborate) can function as negative stains.[58]

Surfactants. It has been suggested that the use of some surfactants during specimen preparation help control the quantity of negative stain surrounding molecular complexes. Among these compounds are glucopyranoside, sucrose, trehalose, glycerol, thioglycerol polyethylene glycol (1500–10,000), bacitracin, octadecanol, and octylglucoside.[49]

Octadecanol was found to facilitate the formation of thin films of negative stain around biological structures adsorbed on a hydrophobic carbon film. About 0.5 μl of a solution of octadecanol (0.015% in hexane) is spread on the surface of the stain droplet just before the excess stain is blotted off with filter paper. Octadecanol treatment allows one to use only a 0.2–0.3% solution of a negative stain, 3–10 times less than what is used in conventional techniques.[59] It has been shown that trehalose provides considerable protection to protein and DNA–protein complexes during the drying of negatively stained specimens. Some reduction in the excessive density imparted by uranyl acetate around large assemblies is also achieved. In the presence of 1% (w/v) trehalose, it is desirable to increase the concentration of the negative stain to 5% (w/v) for ammonium molybdate and to 4% for uranyl acetate to produce satisfactory image contrast. In general, the ammonium molybdate–trehalose negative stain is more satisfactory than the uranyl acetate–trehalose combination because of the greater electron beam sensitivity of the uranyl negative stain.[49,60]

Summary

This article focused on a number of aspects of the preparation of chromatin and other DNA–protein complexes for conventional transmission EM that are critical for success but may not have been addressed in a single chapter before. These include the importance of optimizing fixation, the generation of active supporting supports, and the use of negative staining as a means of obtaining higher resolution detail than can be garnered from shadow casting methods.

[58] W. H. Massover and P. Marsh, *Ultramicroscopy* **69,** 139 (1997).

[59] C. N. Gordon, *J. Ultrastruct. Res.* **39,** 173 (1972).

[60] J. R. Harris, M. Gerber, W. Gebauer, W. Wernicke, and J. Markl, *J. Microsc. Soc. Am.* **2,** 43 (1996).

[12] Targeted Cross-Linking and DNA Cleavage within Model Chromatin Complexes

By KYU-MIN LEE, DAVID R. CHAFIN, and JEFFREY J. HAYES

Introduction

It has become clear that a complete understanding of nuclear processes that involve DNA such as transcription, replication, repair, and recombination will require a comprehensive understanding of how these processes occur in a chromatin environment.[1,2] Such considerations must therefore include questions of how chromatin structures are accommodated by the molecular machines that carry out these processes and, in many cases, how chromatin has been functionally integrated in a process or its control mechanisms.[3,4] The answers to such questions will require a more complete understanding of the essential elements of chromatin structure than is currently available.

The basic building block of the chromatin fiber is a 250-kDa structure known as the nucleosome.[1] This subunit contains about 200 bp of DNA, two copies each of the four core histones, and generally a single molecule of a linker histone. The DNA within this complex is partially wrapped in two 80-bp turns around a spool formed by the core histones. The core histones and the central 146 bp of DNA in tightest association with these proteins form a structure resistant to micrococcal nuclease digestion referred to as the nucleosome core.[1] Details of the organization of DNA and the proteins within the nucleosome core have been well documented.[5,6] The remainder of the DNA within the nucleosome subunit (~40 ± 40 bp) links nucleosome cores together to form a continuous string of these subunits. Strings of nucleosomes are further coiled or compacted into fibers approximately 30–40 nm in diameter.[7] Unfortunately, details of this compaction, such as the conformation of the linker DNA within the 30-nm fiber or the packing of nucleosome cores within the fiber, have not been

[1] K. E. van Holde, "Chromatin." Springer-Verlag, New York, 1989.
[2] A. P. Wolffe, "Chromatin Structure and Function." Academic Press, London, 1995.
[3] G. Felsenfeld, *Cell* **86,** 13 (1996).
[4] M. Grunstein, *Nature* **389** (1997).
[5] G. Arents, R. W. Burlingame, B. W. Wang, W. Love, and E. N. Moudrianakis, *Proc. Natl. Acad. Sci. U.S.A.* **88,** 10148 (1991).
[6] K. Luger, A. W. Mader, R. K. Richmond, D. F. Sargent, and T. J. Richmond, *Nature* **389,** 251 (1997).
[7] F. Thoma, T. Koller, and A. Klug, *J. Cell Biol.* **83,** 403 (1979).

0076-6879/99 $30.00

elucidated.[8,9] These fibers are organized into even higher order structures to form the metaphase chromosome.[2]

Most of the structural information available about chromatin concerns the nucleosome. However, several important questions still remain concerning the details of this basic subunit. A major area of interest concerns the molecular interactions of the histone tail domains. These domains are not resolved in the studies cited earlier but are necessary for the formation of a stable 30-nm chromatin fiber.[10,11] Tails within chromatin are known to make molecular contact with both DNA and protein.[12,13] These complex interactions play crucial roles in moderating the functional state of the chromatin fiber.[2,4] Tail function is mediated primarily via the posttranslational modification of residues within these domains and thus the tail domains are important end points of signal transduction within the nucleus.[14,15] However, the exact nature of the complicated interactions of the core histone tail domains remains unknown.

Another persistent question concerns the location and mode of association of linker histone. Linker histones bind to the surface of the nucleosome and direct the formation of the 30-nm fiber. These proteins have a short N-terminal tail region, a long and highly basic C-terminal tail region (~80–100 residues), and a globular domain (~80–90 residues) that directs structure-specific recognition and binding of the protein to the nucleosome.[1,16] However, exactly where on the surface of the nucleosome and how the globular domain interacts with this complex are in much debate.[17] A classical model suggests that the globular domain interacts with two strands of DNA near the dyad of the nucleosome.[16,18] A second model, based on footprinting and cross-linking results, suggests that this domain binds to protein and DNA at a location inside the superhelical DNA gyre.[19,20]

[8] C. L. Woodcock, S. A. Grigoryev, R. A. Horowitz, and N. Whitaker, *Proc. Natl. Acad. Sci. U.S.A.* **90,** 9021 (1994).

[9] K. van Holde and, J. Zlatanova, *J. Biol. Chem.* **270,** 8373 (1995).

[10] M. Garcia-Ramirez, F. Dong, and J. Ausio, *J. Biol. Chem.* **267,** 19587 (1992).

[11] J. Allan, N. Harborne, D. C. Rau, and H. Gould, *J. Cell Biol.* **93,** 285 (1982).

[12] J. Hansen, *Chemtracts: Biochem. Mol. Biol.* **10,** 56 (1997).

[13] D. G. Edmonson, M. M. Smith, and S. Y. Roth, *Genes Dev.* **10,** 1247 (1996).

[14] J. E. Brownell and C. D. Allis, *Curr. Opin. Genet. Dev.* **6,** 176 (1996).

[15] P. A. Wade, D. Pruss, and A. P. Wolffe, *Trends Biochem. Sci.* **22,** 128 (1997).

[16] J. Allan, P. G. Hartman, C. Crane-Robinson, and F. X. Aviles, *Nature* **288,** 675 (1980)

[17] C. Crane-Robinson, *Trends Biochem. Sci.* **22,** 75 (1997).

[18] Y. B. Zhou, S. E. Gerchman, V. Ramakrishnan, A. Travers, and S. Muyldermans, *Nature* **395,** 402 (1998).

[19] J. J. Hayes, *Biochemistry* **35,** 11931 (1996).

[20] D. Pruss, B. Bartholomew, J. Persinger, J. J. Hayes, G. Arents, E. N. Moudrianakis, and A. P. Wolffe, *Science* **274,** 614 (1996).

We have devised a chemical approach to identify the molecular interactions made by the histone tail domains and linker histone domains within chromatin complexes. This approach involves the construction of site-specific chemically modified proteins that are then used to probe the positional relationships within the complex.[21] These techniques can be used with any chromatin complex that can be reconstituted *in vitro.* We demonstrate uses of these techniques to probe interactions of several core histone tail domains and a linker histone globular domain within an isolated nucleosome complex.

Construction of Cysteine-Substituted Histones

The reactivity of various chemical reagents to cysteines has provided a useful tool for the structural analysis of protein–DNA interactions.[21] Cysteine residues can be placed strategically at specific positions within a protein and the free sulfhydryl can be subsequently modified with structural probes such as cross-linking or DNA-cleaving reagents. This enables the modified protein to act essentially as a chemical probe and allows for versatile, high-resolution mapping of protein–DNA interactions under a number of different conditions or structural contexts. Histones are especially good candidates for this technique as most native proteins are completely devoid of cysteine residues. This methodology has been applied to chromatin complexes to map specific histone–DNA interactions within the nucleosome.

Site-Directed Mutagenesis and Cloning of Cysteine-Substituted Histone Genes

Standard polymerase chain reaction (PCR) mutagenesis is used to substitute a single residue for cysteine within the region of interest. Usually either a single round PCR reaction with two primers or two rounds of PCR reaction with three or four primers will be sufficient to construct a single cysteine-substituted protein. If the codon to be changed is near the end of the coding region, the codon change may be incorporated via one round of PCR reaction with two primers at either end and the cysteine codon incorporated into one of the primers. If the change is more central to the protein sequence, then a three primer technique with two rounds of PCR reaction is used. An initial PCR reaction is done using a primer at one end of the coding region (either 5′ or 3′ end) and a complementary internal primer that contains the codon change. Once this reaction is done, the product from this PCR reaction will serve as a primer for a second PCR

[21] P. S. Pendergrast, Y. Chen, and R. H. Ebright, *Proc. Natl. Acad. Sci. U.S.A.* **89,** 10287 (1992).

reaction. A second PCR reaction containing the product from the first reaction and a primer at the other end of the coding sequence will ultimately generate a fragment with a cysteine codon substituted in the midportion of the coding sequence. An alternate approach used when the former method fails employs a four primer method and two rounds of PCR. This involves two PCR reactions initially: one reaction has the 5′ end primer and an internal primer with the codon change whereas the other contains the 3′ end primer with an internal primer that overlaps the previous internal primer. PCR products from these two reactions are then combined with the original 5′ and 3′ end primer for a second PCR reaction to generate the changed coding sequence.

The coding sequence is then incorporated into an expression vector such as the pET expression vector system (Novagen). We have subcloned coding sequences for each of the core histones into pET3a and pET3d bacterial expression vectors. In order for the insert and vector to be ligated, both DNAs must be digested with the same restriction endonucleases. Equimolar amounts of insert DNA and pET vector DNA are ligated by adding 400 units of T4 DNA ligase (New England Bio-Labs, Beverly, MA) and incubating at 14° overnight. Next, the efficiency of the ligation is checked by transforming a small amount of the ligation sample into *Escherichia coli* (DH5α) and plating them on ampicillin LB plates. Several colonies are analyzed by placing a single colony into 3–5 ml of LB–ampicillin medium and growing the culture at 37°. The plasmid DNA from these cultures is isolated by standard DNA miniprep techniques and is typically digested with the original restriction endonucleases used for ligation. Observation of the DNA fragment corresponding to the original insert indicates that the fragment has been cloned into the expression vector. Plasmids that contain the correct inserts are then used to transform *E. coli* BL21(DE3) for overexpression.

Overexpression and Purification of Cysteine-Substituted Histones

The pET plasmid containing the insert is transformed into BL21(DE3) cells using standard procedures.[22] A 200-ml culture of LB amp medium is inoculated and grown at 37° to an optical density of 0.6 at 595 nm wavelength. Overexpression is induced by adding isopropylthiogalactoside (IPTG) to a final concentration of 0.2 m*M* and growing at 37° for approximately 2–4 hr. Before proceeding, it is recommended that a small amount of the culture be checked for overexpression of the protein of interest. This is done by pelleting down a small aliquot of the culture, resuspending the

[22] J. Sambrook, E. F. Fritsch, and T. Maniatis, "Molecular Cloning: A Laboratory Manual." Cold Spring Harbor Laboratory Press, 1989.

pellet in SDS–PAGE loading dye, incubating the sample at 90° for >10 min, and running the sample on an SDS–PAGE protein gel. The cells in the entire culture are pelleted by centrifugation at 4000*g* for 15 min at 4° and can be stored frozen. If expression is detected, then the large cell pellet is resuspended in 10 ml of TE buffer (10 m*M* Tris–HCl, pH 8.0, 1 m*M* EDTA) and lysozyme and Triton X-100 are added to a final concentration of 0.2 mg/ml and 0.2%, respectively. The suspension is then incubated for 30 min at room temperature. The mixture is diluted twofold with 2 *M* NaCl in TE to a final concentration of 1 *M* NaCl and is transferred to an oakridge centrifuge tube on ice. The bacterial slurry is sonicated for 6 min total with two 3-min sonications (Branson 250 Sonifier, Danbury, CT; 30% duty cycle, power: 4). Sonication tends to increase the temperature of the sample quickly, which could induce denaturation and proteolysis of the proteins. The sample must therefore be cooled continuously on ice during sonication. Allow several minutes between sonication runs to keep the sample as cold as possible. After sonication, the cell debris is pelleted by centrifugation at 10,000*g* for 30 min at 4°. Supernatants are diluted twofold with TE buffer to bring the NaCl concentration to 0.5 *M* and mixed with 12.5 ml of a 50% suspension of Bio-Rex 50-100 (Bio-Rad, Richmond, CA) mesh resin for 1–2 hr at 4°. Linker histones and most other proteins will bind directly to the Bio-Rex resin, but core histone proteins must first be incubated with their dimerization partner proteins (i.e., H2A with H2B) before they will bind to the resin. We have found that the individual core histones do not bind well to the Bio-Rex resin.[23] This behavior could be due to the fact that histone H2A and H2B are completely unfolded when separated from each other.[24] After heterodimerization the proteins will elute off the column at approximately 1 *M* NaCl.[25]

After 1–2 hr the resin is collected in a plastic 10-ml disposable chromatography column and the flowthrough fraction is collected in a 50-ml conical tube and stored at −80°. The column is washed with 2 (20 ml) column volumes of 10 m*M* Tris–HCl, pH 8.0, containing 0.6 *M* NaCl. All wash fractions are collected in a 15-ml conical tube and stored at −80°. The bound proteins are eluted with two 5-ml elutions of 10 m*M* Tris–HCl, pH 8.0, containing 1.0 *M* NaCl, collected in separate 15-ml conical tubes, and stored at −80°. After elution, the column is washed with 5 ml of 10 m*M* Tris–HCl, pH 8.0, containing 2.0 *M* NaCl. The 2.0 *M* elution fraction is collected in a 15-ml conical tube and stored at −80°.

[23] K.-M. Lee and J. J. Hayes, *Proc. Natl. Acad. Sci. U.S.A.* **94,** 8959 (1997).
[24] J. J. Hayes and K.-M. Lee, *Methods* **12,** 2 (1997).
[25] V. Karantza, A. D. Baxevanis, E. Freire, and E. N. Moudrianakis, *Biochemistry* **34,** 5988 (1995).

Modification of Cysteine-Substituted Histones with Chemical Probes

Cysteine-substituted histones readily undergo spontaneous oxidation to form disulfide bonds. Therefore the proteins have to be treated with reducing agents [dithiothreitol (DTT), 2-mercaptoethanol] before modifying with the chemical probes. After reduction, these reagents must be removed quantitatively. The cross-linking reagent APB (4-azidophenacyl bromide) modifies the free sulfhydryl group of cysteine through a maleimide linkage so it is not susceptible to dissociation once it is attached to the protein. In contrast, the cleavage reagent EPD (EDTA-2-aminoethyl 2-pyridyl disulfide) is linked to the protein through a disulfide bond and therefore trace amounts of reducing agents should be avoided completely after modification.

Reduction of Cysteine-Substituted Histones

One milliliter of the protein (~500 μg) of interest is incubated in a 1.5-ml Eppendorf tube with 50 m*M* DTT final concentration for 1 hr on ice. The protein is then transferred to a 15-ml tube and diluted fourfold with TE (10 m*M* Tris–Cl, pH 8.0, 1 m*M* EDTA), which dilutes the NaCl concentration to 250 m*M* NaCl. One milliliter of a 50% slurry of Bio-Rex (Bio-Rad) 100–200 mesh chromatography resin is added and mixed for 2 hr at 4°. The slurry is collected in a 10-ml plastic disposable chromatography column (Bio-Rad) and the flowthrough fraction is collected. The column is washed with about 5 bed volumes of wash buffer (0.5 *M* NaCl, TE) and 20 μl of the wash sample is removed into a separate Eppendorf tube for later analysis on a 15% SDS–PAGE. The reduced histones are eluted with 1 ml of 1 *M* NaCl and TE into a clean Eppendorf tube. Typically, two separate 1.0 *M* NaCl elution steps are performed and collected separately. Again, 5 μl of the fractions is removed for SDS–PAGE analysis and the samples are frozen in dry ice immediately. A final elution with buffer containing 2.0 *M* NaCl buffer will ensure that all of the protein has been eluted from the column. The fractions are stored at −80° or modified directly with the chemical probes. The protein content of each aliquot obtained from the elution fractions is checked on a 15% SDS–PAGE. After separation, the gel is stained in Coomassie Blue stain (45% methanol, 10% acetic acid, v/v, and 2.5 mg/ml Coomassie Brilliant Blue R250) for approximately 1 hr at room temperature and destained with 45% methanol, 10% (v/v) acetic acid until the background of the gel is clear. Note that for proteins to be subsequently modified with metal chelator-based reagents such as EPD, EDTA should be absent from all the buffers described previously.

Modification of Cysteine-Substituted Proteins with Cross-Linking Reagent APB and DNA Cleavage Reagent EPD

Except for a few details, the modification of cysteine-substituted proteins with APB and EPD is essentially the same. About 200 μg of reduced cysteine-substituted protein can be modified with 100 μM APB in the dark at room temperature for 1 hr. For EPD, a 1.1-fold molar excess of EPD is added to 60 μl (~60 μg) of reduced protein and incubated for 1 hr at room temperature in the dark. APB is a photoactivatable reagent so exposure to light should be minimized. The modified protein and the remaining unmodified protein are frozen in dry ice and stored at −80°. If linker histones are being modified with APB or EPD, excess reagents should be removed. This is accomplished most easily by performing one more round of Bio-Rex chromatography identical to that presented earlier except that 1 ml of the 50% slurry (fine mesh) is added to the protein solution. In addition, the slurry is then poured into a minicolumn made from a 1-ml pipette tip fitted with a small amount of glass wool at the opening. Washing and elution are the same as described earlier except all elution volumes are scaled according to the resin bed volume. Note that neither excess APB nor EPD need be removed after core histone modification because they are dialyzed away during nucleosome reconstitution.

Postmodification labeling with [^{14}C]NEM (*N*-[ethyl-1-^{14}C]maleimide) can be used to quantitatively determine the extent of modification with APB and EPD. [^{14}C]NEM is also a cysteine-specific reagent that attaches to the free sulfhydryl group through a maleimide linkage. Approximately 0.25–0.5 μCi of [^{14}C]NEM is added into a 10-μl (~2 μg) aliquot of reagent unmodified and modified H2A/H2B dimer. The histones are incubated at room temperature for 30 min and then quenched with 1 μl of 0.1 *M* DTT. The protein samples are loaded directly onto an 18% SDS–PAGE gel for core histones and a 12% gel for linker histones. After separation, the gel is stained with Coomassie Blue dye for 1 hr and destained overnight. The gel is dried onto a piece of Whatman (Clifton, NJ) filter paper and exposed to Kodak (Rochester, NY) Biomax film or a phosphorimager plate. The efficiency of APB or EPD modification is determined by quantitating the amount of [^{14}C]NEM labeling in modified versus unmodified samples.

In Vitro Reconstitution of Nucleosomes

Radioactive End Labeling of DNA Fragments

Site-specific mapping of protein–DNA interactions requires using either 5′ or 3′ end-labeled DNA. Plasmid DNA is cut with the appropriate restriction enzyme, which produces the site that is being labeled. Once labeling is done, the linear plasmid is cut again to generate the singly end-labeled DNA fragment to be used for reconstitution. Approximately 5 μg of plasmid DNA or ~0.5 μg of gel purified DNA fragment is digested with the appropriate restriction enzyme and precipitated. The DNA is resuspended in phosphate buffer and treated with 1 μl alkaline phosphatase (Boehringer Mannheim) for 1 hr at 37° to dephosphorylate the DNA. SDS is added into the reaction (0.01%), and the solution is phenol-extracted twice and precipitated. The DNA is resuspended in 10 μl TE, and 2.5 μl of 10× T4 polynucleotide kinase buffer is added. [γ-^{32}P]dATP (50 μCi) is then added and the volume is adjusted to 24 μl with water. The reaction is started by adding 10 units of T4 polynucleotide kinase and is incubated for 30 min at 37 °. The kinase reaction is stopped with 200 μl of 2.5 *M* ammonium acetate and 700 μl of cold 95% ethanol. The DNA is pelleted in a microfuge for 30 min at room temperature and the DNA pellet is washed briefly with ice-cold 70% ethanol and dried in a Speed-Vac concentrator. The DNA is dissolved in 30 μl of TE buffer and digested with a second restriction endonuclease that generates the appropriate sized DNA fragment. The labeled DNA fragment can then be isolated from a 6% native polyacrylamide gel. After separation, the gel is wrapped in Saran wrap and exposed to film for ~1 min, which is sufficient to detect the specific band containing the labeled DNA. The use of fluorescent markers on the gel surface allows the alignment of the gel so that the band of interest is excised more easily from the polyacrylamide gel. The gel slice is placed into a clean Eppendorf tube and crushed with an Eppendorf pestle and then 700 μl of TE buffer is added. The labeled fragment can be eluted passively overnight at 4° or in about 1 hr at 37°. The sample is then equally divided and pipetted into two Series 8000 microfuge filtration devices (Marsh, Rochester, NY). This may require snipping the end of the pipette tip to allow transfer of the larger gel chunks. The crushed gel mix is spun for 30 min in a microfuge to elute the DNA. The DNA is precipitated and dissolved in TE buffer. Add enough TE buffer so that the labeled DNA is approximately <5000 cpm/μl. Storing labeled DNA in a more concentrated form is not advised as autodegradation of the DNA may take place. DNA can be stored for several weeks at approximately 5000 cpm/μl.

In Vitro Reconstitution of Nucleosomes

The method described here for the reconstitution of nucleosomes allows for large quantities of homogeneous nucleosomes to be prepared in approximately 12 hr. These nucleosomes have been shown to have the same structural characteristics of nucleosomes prepared from native chromatin.[26] They have also been shown to bind linker histone in a physiologically relevant manner as tested by several assays.[27] Although histones bind DNA without sequence specificity, DNA containing the *Xenopus borealis* 5S rRNA gene has been shown to position the nucleosome uniquely on the DNA.[28] These "5S" nucleosomes are used primarily for mapping experiments.

Approximately 8 μg of unlabeled calf thymus DNA, 200,000–400,000 cpm of singly end-labeled *X. borealis* 5S ribosomal DNA, purified chicken erythrocyte core histone protein fractions (H2A/H2B and H3/H4), 160 μl of 5 *M* NaCl (2.0 *M* final), and TE buffer are mixed in a final volume of 400 μl. Purified histone proteins are often obtained in two fractions: H2A/H2B and H3/H4.[29] Thus, it is possible to replace either the H2A/H2B dimer or the H3/H4 tetramer with APB- or EPD-modified forms for core histone mapping experiments. In addition to the total histone : DNA ratio, the ratio between the two substituents of core histones (H2A/H2B and H3/H4) must be adjusted empirically to yield maximum octamer–DNA complexes. Often, a small amount of a subnucleosomal band containing $(H3/H4)_2$ tetramer–DNA complexes is observed because many competitor DNA bind H2A/H2B dimers independent of octamer formation.[24] The assembly of $(H3/H4)_2$ tetramer–DNA complexes by omitting histones H2A and H2B from the reconstitution should be done as a control for the identification of subnucleosomal bands observed by nucleoprotein gel electrophoresis and to guard against the misidentification of a ditetramer–DNA complex (i.e., two $[H3/H4]/_2$ tetramers on one DNA)[30,31] as a true nucleosome complex. Ideally, about 50% of the labeled DNA will be assembled into the tetramer while the rest remains as naked DNA. After identification of the tetramer complexes, several reconstitutions containing increasing amounts of H2A/H2B can be prepared to determine empirically the precise amounts of these histones needed to form homogeneous octamer–DNA complexes.[24] The empirically determined input of H2A/H2B may be slightly different than that predicted by the theoretical mass ratio between these proteins and the $(H3/H4)_2$ tetramer within the nucleosome core (1 : 1). After combining all the components of the reconstitution, the mixture is placed into a 6000–8000 molecular weight cutoff dialysis bag. All subsequent dialysis steps are for 2 hr at 4° against 1 liter of dialysis buffers unless

[26] J. J. Hayes, T. D. Tullius, and A. P. Wolffe, *Proc. Natl. Acad. Sci. U.S.A.* **87,** 7405 (1993).
[27] J. J. Hayes and A. P. Wolffe, *Proc. Natl. Acad. Sci. U.S.A.* **90,** 6415 (1993).
[28] D. Rhodes, *EMBO J.* **4,** 3473 (1985).
[29] R. H. Simon and G. Felsenfeld, *Nucleic Acids Res.* **6,** 689 (1979).
[30] P. G. Stockley and J. O. Thomas, *FEBS Lett.* **99,** 129 (1979).
[31] C. M. Read, J. P. Baldwin, and C. Crane-Robinson, *Biochemistry* **24,** 4435 (1985).

specified. The first dialysis buffer is 10 m*M* Tris–HCl, pH 8.0, 1.2 *M* NaCl, 1 m*M* EDTA. Subsequent dialysis is carried out with fresh buffer containing 1.0, 0.8, and 0.6 *M* NaCl. The procedure is completed with a final dialysis against TE buffer overnight. Nucleosomes at this stage can be used for H1 gel-shift experiments or for cross-linking experiments with APB as EDTA does not interfere with these methods. However, EDTA does interfere with EPD cleavage thus experiments with proteins such as transcription factor IIIA, which requires Zn^{2+} ions, thus two additional dialysis steps are required. First, the reconstitutes are dialyzed against 10 m*M* Tris–HCl, pH 8.0, for several hours to remove the EDTA. A second dialysis against fresh 10 m*M* Tris–HCl, pH 8.0, removes trace amounts of EDTA and prepares the samples for chemical mapping with EPD.

Binding Single Cysteine-Substituted Linker Histones to Reconstituted Nucleosomes

The exact amount of each mutant linker histone protein to add to the hydroxyl radical reactions needs to be determined empirically. This is accomplished by adding increasing amounts of the mutant protein to a fixed mass of reconstituted nucleosomes (typically ~5000 cpm) and analysis by nucleoprotein gel shift.[27] Several methods can be used for the incorporation of linker histones into reconstituted mononucleosomes. The method described here involves the direct addition of linker histones to mononucleosomes. The binding reaction typically contains 50 m*M* NaCl and 5% (v/v) glycerol. Linker histones are properly folded in low salt solutions in the presence of DNA.[32] Indeed, we find that linker histones can be mixed directly to nucleosomes in either 5 or 50 m*M* NaCl solutions and these proteins bind in a physiologically relevant manner.[27] The binding reactions are incubated for 15 min at room temperature and the complexes are separated on a 0.7% agarose, 0.5× TBE gel. After drying the gel, it is exposed to autoradiography film to determine the amount of protein necessary for complex formation. In addition, several other assays indicating that the correct binding of linker histones to DNA has occurred have been performed.[27]

Site-Directed Photo-Cross-Linking of Histone–DNA Interactions

Site-directed cross-linking is a high-resolution mapping technique for protein–DNA interactions and can be applied to various complexes.[21] For histone–DNA mapping within the nucleosome, two variations of the

[32] D. J. Clark and J. O. Thomas, *J. Mol. Biol.* **187,** 569 (1986).

method are presented. One involves nucleosomes prepared with a sequence-specific DNA (DNA sequence from the 5S rDNA of *X. borealis*) and the other prepared on random sequence DNA or mixed sequence DNA. The advantage of using a sequence-specific DNA for the cross-linking experiments is that the result is much clearer due to the uniformity of the sequence and the well-defined nucleosome position. Determination of cross-linking with random sequence DNA is done to check for probe reactivity artifacts due to sequence-dependent structures within the DNA restriction fragment and bias toward cross-linking to a particular base type. Here we describe the details of attaching the reagent APB to a specific position within a histone protein (see earlier discussion), assembly of the protein–DNA complex, and irradiation of the complex to cause cross-link formation between the protein and the DNA. The cross-linked DNA is then treated with heat and alkali to induce base elimination of the cross-linked base, which results in a strand break at the cross-linked position, which, in turn, is analyzed on sequencing gels.

UV Cross-Linking of Nucleosomes Reconstituted with a Defined Sequence Restriction Fragment

One hundred microliters of reconstituted nucleosomes containing a single APB-modified histone (approximately 100,000 cpm) is combined with 10 μl of 25% glycerol. The sample is loaded on a native 0.7% agarose/0.5× TBE gel and run at 120 V for 3 hr. The core histone nucleosome complex is identified by exposing the wet gel to X-ray film (Kodak XR) for 3 hr at 4°. The agarose slice containing the octamer histone complex is cut out and UV irradiated for 20 sec with a 365-nm light source. Although 20 sec is sufficient for cross-linking histone–DNA interactions, the time of irradiation for other proteins should be determined empirically first. UV irradiation for over 5 min should be avoided because it damages the DNA and causes dissociation of the protein–DNA cross-link (K.-M. Lee and J. J. Hayes, unpublished results). Analysis of the amount of cross-linked species produced over time is done by irradiating samples of reconstituted nucleosome solutions directly in Eppendorf tubes laid directly on the UV light box and separating cross-linked species on SDS–PAGE as described later. However, instead of exposing the wet gel, dry the gel before exposure. We have found that cross-linking within the tubes approximates cross-linking within the gel slice.

The irradiated gel slice is placed in a Spin-X microfuge filter device and put in dry ice for 30 min. The frozen gel slice is spun at maximum speed in a microfuge for 30 min at room temperature to elute the DNA. The volume of the eluate is reduced in a Speed-Vac concentrator and the DNA ethanol is precipitated with 1/10 volume of 3 *M* sodium acetate and 2.5 volumes of cold 95% ethanol. (To minimize the loss of sample it is possible to go directly from this step to the base elimination step.) Continuing with the procedure described later can reduce the background associated with the cross-linking signal. However, proceeding with these steps entails a significant loss of sample and is not necessary when there is a strong enough signal on the sequencing gel.[23] The sample is dissolved in 50 μl of TE, and the DNA complexes are separated on a 6% resolving and a 4% stacking SDS–PAGE gel for 4 hr at constant 200 V. The gel can also be run overnight at low voltage to keep the gel cool. After the run the cross-linked species is identified by exposing the wet gel to X-ray film (Kodak XR) for 2 hr at 4°. The band corresponding to the core histone–DNA cross-linked product is cut out and crushed with a microfuge pestle. The crushed gel is soaked in 500 μl of TE buffer and mixed overnight. The gel/buffer mixture is then transferred to Spin-X filter tubes (Costar) and pelleted for 30 min in a microfuge at 14,000 rpm to elute the DNA from the crushed gel. The volume of the DNA sample is reduced in a Speed-Vac and then ethanol precipitated. Proceed to the base elimination step of the histone–DNA cross-linking after this step.

Figure 1 shows that protein–DNA cross-linking obtained within a nucleosome complex using the procedure just described is dependent on specific incorporation of the photoactivatable probe and UV irradiation. To quantitate the amount of cross-linking obtained, reconstituted nucleosomes irradiated with UV light were treated with SDS and the constituents were separated by SDS–PAGE. Components separated on the gel were then detected by autoradiography. An important control shows that nucleosomes reconstituted with radioactively end-labeled DNA and *wild-type* histones yield only a single band on the autoradiograph of the gel that corresponds to uncross-linked naked DNA (results not shown). Further, a single band is still obtained after UV irradiation of these complexes (Fig. 1, lane 1). However, nucleosomes containing an APB-modified H2A yield a higher molecular weight complex on the gel that is dependent on UV irradiation of the sample (Fig. 1, compare lanes 2 and 3). Moreover, irradiation of a nucleosome containing a wtH2A that had been incubated with APB in the same manner as the cysteine mutant did not yield a higher molecular weight product on the gel (results not shown). Densitometric analysis reveals that the efficiency of cross-linking is approximately ~5%, as expected.[21]

UV Cross-Linking of Nucleosome Core Particles Containing Random DNA Sequences

Nucleosomes reconstituted with random DNA sequences and a modified core histone are prepared as per the following procedure. Approxi-

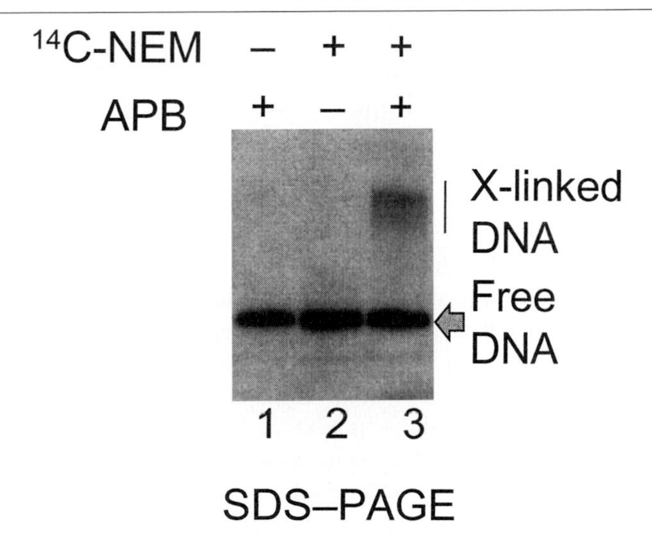

FIG. 1. Protein–DNA cross-linking is dependent on APB modification and UV irradiation. Nucleosomes containing labeled DNA and either APB-modified or unmodified H2A were irradiated with UV light, products were separated by SDS–PAGE, and the gel autoradiographed. Lane 1, UV irradiated wild-type nucleosomes; lane 2, unirradiated APB-modified nucleosomes; and lane 3, APB-modified and UV-irradiated nucleosomes. The positions of uncross-linked naked DNA and protein–DNA cross-linked products are indicated.

mately 40 μg of core histones (20 μg each of H2A/H2B and H3/H4) and 50 μg of calf thymus DNA are mixed together, and a salt dialysis nucleosome reconstitution is performed as described previously except without labeled DNA as the nucleosomal DNA will be labeled directly after the nucleosome cores are prepared. One of the most important steps in this experiment is to determine empirically the optimum amount of micrococcal nuclease digestion required for nucleosome core particle preparation. A preliminary digestion experiment varying the amount of nuclease to digest samples containing ~200 ng (DNA mass) of reconstituted material is therefore performed. These digests are analyzed by stopping the digestion with 5 m*M* EDTA and 0.2% SDS and running the samples on an 8% nondenaturing acrylamide gel and selecting the amount that produces a clean 146-bp core particle band.

After reconstitution, 40 μl of the reconstitute (~5 μg of nucleosomes) is digested in preparative scale with micrococcal nuclease by adding 5 μl of 30 m*M* $CaCl_2$ and 5 μl of approximately 0.2 units of micrococcal nuclease and incubating at 37° for 10 min. The micrococcal digest is stopped with 5 μl of stop mix (25% glycerol and 0.25 *M* EDTA). The entire sample is loaded on a 0.7% agarose/0.5× TBE gel and is separated for 3 hr at constant

120 V. After separation, the gel is directly UV irradiated for 20 sec with a 365-nm light source and stained with ethidium bromide for 1 hr. The nucleosome core particle band is identified and cut out from the gel. This step requires the use of a UV light source to identify the nucleosome core particle. Therefore, it is important to excise the band corresponding to the nucleosomal complex as fast as possible (approximately 5 sec of exposure). The gel slice is placed in a Spin-X filter tube and frozen on dry ice for 30 min. The protein–DNA complex is eluted by centrifugation for 30 min at 14,000 rpm at room temperature, and the volume of the eluate is reduced in a Speed-Vac concentrator and ethanol precipitated. The pellet is dissolved in 43 μl of H_2O and then 5′ end-labeled by adding 5 μl of 10× polynucleotide kinase buffer, 10 μCi of [γ-^{32}P]ATP, and 10 units of polynucleotide kinase. The sample is incubated for 30 min at 37° and ethanol precipitated. The DNA is dissolved in 40 μl of TE buffer. Approximately 50,000 cpm of the sample is loaded on a 6% SDS–PAGE gel and separated for 4 hr at a constant 200 V. The cross-linked core histone–DNA complex is identified by exposing the wet gel to X-ray film (Kodak XR) for 2 hr. The band corresponding to the histone–DNA complex is cut out from the gel and crushed as before. The crushed gel is resuspended in 500 μl of TE buffer and mixed overnight to elute the DNA. The gel/buffer mixture is transferred to a Spin-X tube and pelleted for 30 min in a microfuge at 14,000 rpm to elute the DNA from the crushed gel. The volume of the sample is reduced in a Speed-Vac concentrator and ethanol precipitated. The pellet is resuspended in 48 μl of TE, and 1 μl of 2% SDS and 1 μl of 15 mg/ml proteinase K are added and incubated at 37° for 1 hr to digest away the protein cross-linked to the DNA. The sample is loaded directly onto an 8% (w/w) nondenaturing polyacrylamide gel (1:19, bisacrylamide:acrylamide, w/w) and run for 2 hr at constant 200 V. The labeled DNA band is identified by autoradiography for 2 hr as before and the acrylamide gel slice containing the DNA band is cut out. The gel slice is crushed as described previously, resuspended in 500 μl of TE buffer, and mixed overnight. The gel/buffer mixture is transferred to a Spin-X filter tube and pelleted for 30 min in a microfuge at 14,000 rpm to elute the DNA from the crushed gel. The volume is reduced in a Speed-Vac concentrator and the DNA ethanol is precipitated. The DNA pellet is washed with cold 70% (v/v) ethanol and dried. Proceed to the base elimination step.

Identification of the Site of Cross-Linking

The procedure for elimination and cleavage of the cross-linked or modified bases within the DNA is based on the work of Pendergrast *et al.*[21] and the same for sequence-specific DNA and random sequence DNA. The

DNA pellet is dissolved in 100 μl of buffer (25 mM ammonium acetate; 0.1 mM EDTA; 2% SDS) and layered with 100 μl of mineral oil. The sample is incubated at 90° for 30 min and then 5 μl of 2 M NaOH is added and incubated an additional 60 min at 90°. The bottom layer (aqueous sample) is removed from the tube and 5 μl of 2 M HCl and 100 μl of 20 mM Tris–HCl, pH 8.0, are added into the sample to stop the elimination reaction. The DNA is ethanol precipitated and the pellet is dried in a Speed-Vac concentrator. The DNA is resuspended in 4 μl of neat formamide and analyzed on sequencing gels (see later).

Examples of Site-Directed Cross-Linking

We present results from the cross-linking of core histone H2A with DNA within two differently prepared nucleosomes. Figure 2 shows the results from a cross-linking experiment in which residue 12 of core histone H2A was modified with APB within a nucleosome reconstituted with the 5S DNA fragment (see earlier discussion). Cross-linking was carried out and data analyzed as described. Analysis of the cleavage pattern from these cross-linked complexes when the top strand of 5S DNA is 5′ end labeled reveals a single, strong signal at ~39, about 40 bp away from the dyad axis of symmetry within the 5S nucleosome (located near the start site for

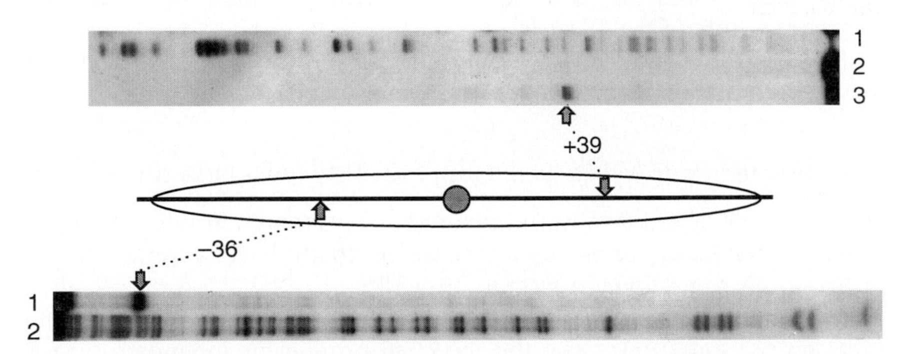

FIG. 2. Location of cross-links between APB-modified H2AA12C and nucleosome DNA. Nucleosomes were reconstituted with the 215-bp *Eco*RI–*Dde*I fragment 5S DNA fragment and core histones including either APB-modified H2AA12C or wtH2A. Cross-linking was carried out and the DNA analyzed on sequencing gels as described in Materials and Methods. (*Top*) Samples from nucleosomes reconstituted with the *Eco*RI–*Dde*I DNA fragment 5′ end-labeled at the *Eco*RI (labels the "top" strand). Lane 1, G-specific sequencing reaction markers; lanes 2 and 3, cross-linking within nucleosomes containing wild-type H2A and APB-modified H2AA12C, respectively. (*Bottom*) As in the top gel except nucleosomes were reconstituted with the *Eco*RI–*Dde*I DNA fragment 5′ end-labeled at the *Dde*I site (labels the "bottom" strand). Lane 1, G-specific markers; lane 2, cross-linking within nucleosomes containing APB-modified H2AA12C.

transcription at +1) (Fig. 2 top, lane 3). A control lane derived from nucleosomes containing only wild-type histones treated in the same way reveals only low-intensity background signals (Fig. 2 top, lane 2). Cross-links made to the bottom strand of 5S DNA in nucleosomes containing this mutant are shown in the bottom gel in Fig. 2. Clearly, a strong signal is detected at approximately position −36 within 5S DNA. Note that the two sets of cross-links detected on complementary strands are approximately symmetry related about the dyad axis of the nucleosome. Controls in which the modified H2AA12C-APB/H2B dimer was added directly to the labeled DNA and irradiated shows that cross-links are scattered throughout the 5S DNA fragment (K.-M. Lee and J. J. Hayes, unpublished results).

The next sample is from cross-linking of the nucleosome on random sequence DNA and involves the same H2AA12C protein (Fig. 3). Nucleosome core particles were isolated by preparative nucleoprotein gel electrophoresis and irradiated, and the positions of cross-links were identified as described previously. A single site of cross-linking is found in cores containing APB-modified H2AA12C at a position approximately 40 bp away from the dyad axis of symmetry (Fig. 3, lane 2), in excellent agreement with results from the 5S nucleosome. Note that because of the size heterogeneity associated with the core particle preparation the cross-linking signal is distributed over a larger range of DNA sizes than is observed in previous experiments employing 5S nucleosomes. Further, because of the random sequence nature of the DNA fragments, bands differing by one nucleotide in length are not resolved on the sequencing gel in samples subjected to the base-elimination chemistry (Fig. 3, lanes 1 and 2).

Site-Directed Hydroxyl Radical Mapping of Histone–DNA Interactions

The single cysteine substitution approach presented in this article provides a versatile way of modifying the free sulfhydryl with several different types of DNA-modifying reagents. In addition to the DNA cross-linking reagent, APB, site-directed DNA cleavage can be carried out with the cleaving reagent EPD. Once the modified protein has bound the DNA, cleavage can be induced by hydroxyl radicals produced from a lone metal center within the EPD moiety by Fenton chemistry. At places where the EPD moiety comes into closest contact to the DNA backbone, the hydroxyl radicals will cause a specific break in the DNA. These strand breaks can then be mapped by analyzing the purified DNA on sequencing gels.

The binding reaction is scaled up to include 40,000–50,000 cpm of labeled reconstituted nucleosomes, and enough modified mutant linker histone is added to form H1–nucleosome complexes. Glycerol is added to 0.5% final concentration, and sodium ascorbate and H_2O_2 are added to a

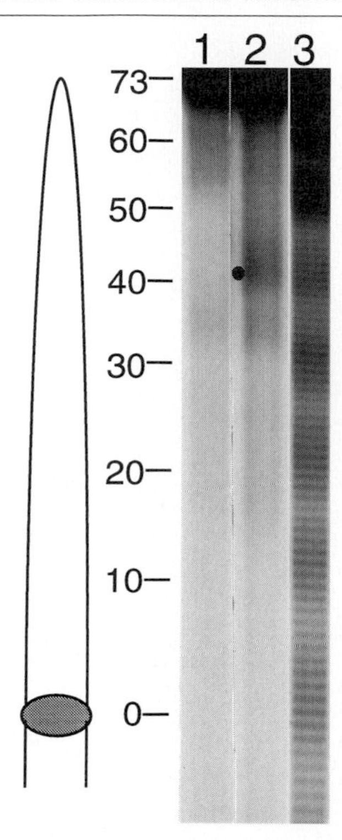

FIG. 3. Location of cross-linked sites with random sequence core particles containing APB-modified H2AA12C. Cross-linked products from core particles were prepared and analyzed by sequencing gel electrophoresis and autoradiography as described in the text. Autoradiograph of products from core particles containing wtH2A (lane 1) and H2AA12C (lane 2). The position of bands is indicated relative to the distance from the dyad axis of symmetry in nucleotides. Only one symmetrical half of the nucleosome core is shown as indicated by the schematic (left). The hydroxyl radical footprint of nucleosome core particles is shown in lane 3 for reference.

final concentration of 1 m*M* and 0.0075%, respectively. The sample is incubated for 30 min at room temperature in the dark. After 30 min, 1/10th volume of 50% glycerol, 10 m*M* EDTA solution is added. The samples are loaded immediately onto a running (90 V) preparative 0.7% agarose/0.5× TBE gel and separated so that the H1–nucleosome complexes are well resolved from tetramer and free DNA bands. Next, the gel is wrapped tightly with Saran wrap so that the gel cannot move within the plastic. Fluorescent markers (Stratagene) are laid onto various portions of the gel for alignment purposes. The wet gel is exposed to film for several hours at 4°. The autoradiograph is then developed and overlaid onto the wet

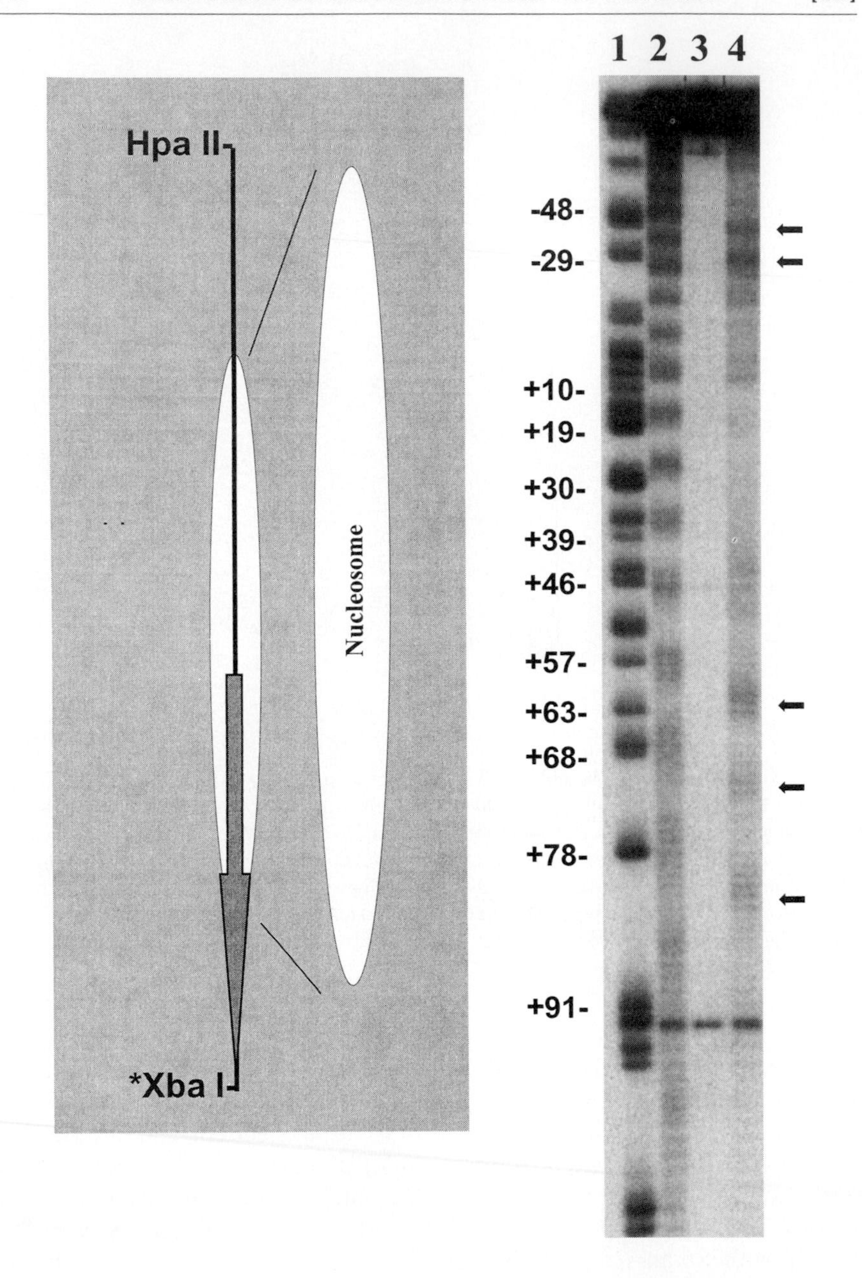
Hpa II
Nucleosome
*Xba I
1 2 3 4
-48-
-29-
+10-
+19-
+30-
+39-
+46-
+57-
+63-
+68-
+78-
+91-

gel, lining up the fluorescent markers. The agarose slice containing the H1–nucleosome complexes or bands of interest are cut out and placed into Series 8000 microfuge filtration devices. The agarose is spun down in a microfuge at maximum speed for 30 min at room temperature. The fluid from the agarose matrix will be collected in the 2-ml centrifuge tube surrounding the filtration device. The fluid is saved to combine later with the second eluate sample. The agarose plug is removed gently from the bottom of the filtration device and placed into a clean Eppendorf tube. The centrifugation device is saved for use later. Using a microfuge tube pestle, the agarose plug is resuspended in 500 μl of 10 m*M* Tris–HCl, pH 8.0, 0.1% SDS. After the agarose is crushed into tiny pieces with an Eppendorf tube pestle, all samples are placed at 4° overnight or for several hours. The crushed agarose is placed into the same centrifugation device and spun down in a microfuge at maximum speed for 30 min at room temperature. The sample is then combined with the previously saved eluate and the DNA is precipitated. The DNA is dissolved in 15 μl of TE buffer.

Example of Site-Directed Cleavage of Nucleosomal DNA by EPD

One example of a site-directed DNA cleavage reaction is presented in Fig. 4. A schematic of the 5S mononucleosome is shown in the left panel. The thick black line represents the 5S ribosomal DNA fragment that contains the transcriptional coding sequence for this gene (gray arrow). The 5S ribosomal gene fragment was used because it contains a nucleosomal positioning sequence that precisely wraps the DNA around the core histones and provides a homogeneous population of nucleosomes. Furthermore, because one major translational position exists within this population of nucleosomes, the precise orientation of the DNA as it wraps around the core histones is known. This enables us to determine base pair resolution,

---

FIG. 4. Location of K59C-EPD cleavages on the coding strand of 5S nucleosomal DNA. Nucleosomes reconstituted with 5S DNA radioactively end-labeled at the *Xba*I site (see schematic) were complexed with unmodified K59C of K59C-EPD. Complexes were incubated with cleavage reagents for 30 min and the DNA cleavages were analyzed by autoradiography of the sequencing gels. DNA from nucleosomes incubated with unmodified K59C protein as a control is shown in lane 3. DNA from nucleosomes incubated with K59C-EPD is shown in lane 4. Positions of K59C-EPD-specific cleavage are highlighted with black arrows. Lane 1 shows a G-specific reaction generated using the same labeled DNA fragment to determine the positions of the K59C-EPD cleavages. The nucleotide positions are indicated to the left of the gel. Lane 2 contains the hydroxyl radical footprint of the 5S mononucleosome. The schematic (left) indicates the orientation of the 5S DNA fragment with respect to the gel and the position of the radioactive labels (star) with regard to the 5S gene-coding sequence (vertical black arrow).

the sites of cleavage by EPD. A singly labeled 5S DNA fragment was incorporated into nucleosomes via the salt dialysis procedure detailed earlier. Labeled mononucleosomes were bound by a mutant, EPD-modified linker histone containing a single cysteine substitution for the lysine residue at position 59, referred to as K59C-EPD. After hydroxyl radical cleavage for 30 min, the protein–DNA complexes were separated on a 0.7% agarose gel, and the labeled DNA fragments corresponding to the H1–nucleosome complexes were purified. These purified DNAs were then analyzed on a sequencing gel (Fig. 4, right). Lane 1 (Fig. 4) represents a specific G reaction generated with the singly labeled 5S DNA fragment to allow precise size determination of the DNA fragments. Mononucleosomes used for this experiment give a characteristic 10-bp protection when hydroxyl radical footprinted in the absence of linker histone, indicating that the 5S DNA has properly formed a histone octamer (Fig. 4, lane 2). No specific cleavages occur when the reaction is run in the presence of mutant but unmodified K59C (lane 3, Fig. 4). In contrast, when hydroxyl radicals were produced from the lone metal center within K59C-EPD-bound nucleosomes, two sets of cleavages are evident (lane 4, Fig. 4). The cleavages at 62, 72, and 82 correspond to the end of the nucleosome where the DNA exits. This result is consistent with previous data suggesting that the linker histone binds the nucleosome at the periphery, tucked inside a superhelical gyre of DNA.[19] A second set of cleavages occurs at −29 and −39. These cleavages occur on the DNA strand directly underneath that of the 62/72/82 cleavages as the DNA comes 360° around the histone octamer. It is possible that amino acid 59 makes close contacts with both strands of the DNA, consistent with the strong cleavages seen at each site. It is also possible that hydroxyl radicals have diffused away from the −29/−39 cleavage site and cleave the DNA in other areas. Inconsistent with this, glycerol, a very good hydroxyl radical scavenger, does not seem to have an effect on the cleavages obtained with K59C-EPD at concentrations known to eliminate hydroxyl radical cleavage (D. R. Chafin and J. J. Hayes, unpublished results).

Sequencing Gel Analysis of APB and EPD Experiments

Maxam–Gilbert G-Specific Reaction

The G-specific reaction used in the Maxam–Gilbert sequencing method provides an easy and quick method to identify the exact location of bases within any known sequence on sequencing gels to base pair resolution. Approximately 20,000 cpm of singly labeled DNA (same DNA used to reconstitute nucleosomes) is mixed with 20 μl of 10× G-specific reaction buffer in a final volume of 200 μl. Add 1 μl of neat dimethyl sulfate (DMS) to the sample, mix immediately, and then spin down briefly in a microfuge (caution: DMS is a potent, volatile carcinogen so use a hood and avoid getting any DMS on the skin or on standard laboratory gloves). Add 50 μl of G reaction stop solution and mix immediately.[22] The DNA is precipitated and resuspended in 90 μl of H_2O. Add 10 μl of piperidine and incubate the sample at 90° for 30 min. The solution is dried to completion in a Speed-Vac. The DNA is dissolved in 20 μl of water and the drying step is repeated. This step is repeated one more time. The DNA is resuspended in 100 μl of TE buffer and stored at 4°.

Sequencing Gel Analysis of DNA Samples

Analysis of DNA samples from both cross-linking and cleavage experiments on sequencing gels is done the same way. Equal amounts of counts for each sample are added as determined with a Geiger counter, including the G-specific reaction, to clean Eppendorf tubes. The sample is then placed into a Speed-Vac concentrator and dried to completeness. The sample is dissolved in 4 μl of formamide loading buffer and incubated at 90° for 2 min to denature the sample. The sample is placed directly onto ice to prevent renaturation. The sample is separated on a 6% polyacrylamide/8 *M* urea sequencing gel running at a constant 2000 V. The gel is transferred after the run to a Whatman 3MM filter paper, dried, and autoradiographed.

[13] Base-Pair Resolution Mapping of Nucleosome Positions Using Site-Directed Hydroxy Radicals

By ANDREW FLAUS and TIMOTHY J. RICHMOND

Introduction

The position of a nucleosome describes the arrangement of the 147 bp of nucleosome core-associated DNA relative to the histone octamer.[1] The particular helical face of this DNA exposed to solvent (rotational phase) can be determined easily using a nucleolytic digestion pattern because nucleosomal DNA has a pitch of approximately 10 bp, allowing maximal accessibility for nucleases to every approximately 10th base pair (e.g., DNase I, free solution hydroxyl radicals[2]). The more rigorous specification of the exact region of DNA in contact with the histone octamer is known as

[1] F. Thoma, *Biochim. Biophys. Acta* **1130,** 1 (1992).

[2] J. J. Hayes, T. D. Tullius, and A. P. Wolffe, *Proc. Natl. Acad. Sci. U.S.A.* **87,** 7405 (1990).

0076-6879/99 $30.00

the translational position.[3] This locates the detailed features of nucleosomal DNA relative to the histone octamer such as points of sharper curvature or interaction with histone tails.[4] The detection of this positioning has been considerably more difficult to achieve experimentally as enzymatic or chemical footprinting can reveal only the outer border of the histone–DNA association, and analysis frequently requires subjective assessment of protection end points for digestion subject to limitations in nuclease accessibility. The most popular method has been using the nucleosomal core protection against micrococcal nuclease,[5] which has the disadvantages that this enzyme (16.8 kDa) is relatively bulky with respect to the nucleosome,[6] has some sequence bias,[7] digests within the core region to create a nonuniform background,[7a] and requires specific reaction conditions such as divalent ions and elevated temperatures. Another approach has been to use the DNA pitch variation at the pseudo dyad axis observed when nucleosomes are footprinted with free solution-generated hydroxy radicals.[8] However, the high-resolution crystal structure of the nucleosome core particle provides no direct explanation for the footprinting observation, and the analysis requires very careful experiments that could be complicated easily by multiple overlapping positions.

Hydroxyl radicals are extremely reactive,[9] with a typical migration of only 10–15 Å in aqueous solution,[10,11] and are applicable under a wide variety of temperatures and ionic strengths. When directed at DNA, they lead to strand scission by attacking the C-1′ and C-4′ carbons of the deoxyribose ring to generate chain breakage resulting in 3′-phosphate or 3′-phosphoglyconate end, respectively, with loss of the nucleotide base.[12,13] These two reaction products are indistinguishable in polyacrylamide gels for DNA lengths greater than 10–20 bp.[12] The site-directed hydroxyl radical method presented here[14] differs from free solution hydroxyl radical methods because it localizes the chelation of the hydroxy radical-generating iron center

[3] F. Thoma, *Methods Enzymol.* **274,** 197 (1996).
[4] K. Luger, A. Maeder, R. K. Richmond, D. F. Sargent, and T. J. Richmond, *Nature* **389,** 251 (1997).
[5] F. Dong, J. C. Hansen, and K. E. van Holde, *Proc. Natl. Acad. Sci. U.S.A.* **87,** 5724 (1990).
[6] K. E. van Holde, "Chromatin." Springer-Verlag, New York, 1989.
[7] J. D. McGhee, and G. Felsenfeld, *Cell* **32,** 1205 (1983).
[7a] M. Cockell, D. Rhodes, and A. Klug, *J. Mol. Biol.* **170,** 423 (1983).
[8] M. S. Roberts, G. Fragoso, and G. L. Hager, *Biochemistry* **34,** 12470 (1995).
[9] B. Halliwell and J. M. C. Gutteridge, "Free Radicals in Biology and Medicine." Clarendon Press, Oxford, 1989.
[10] B. Halliwell and J. M. C. Gutteridge, *Methods Enzymol.* **186,** 1 (1990).
[11] Y. W. Ebright, Y. Chen, S. Prendergrast, and R. H. Ebright, *Biochemistry* **31,** 10664 (1992).
[12] R. P. Hertzberg and P. B. Dervan, *Biochemistry* **23,** 3934 (1984).
[13] A. G. Papavassiliou, *Biochem. J.* **305,** 345 (1995).
[14] A. Flaus, K. Luger, S. Tan, and T. J. Richmond, *Proc. Natl. Acad. Sci. U.S.A.* **93,** 1370 (1996).

to a specific cysteine residue, thus directing the site of attack of the hydroxy radicals. Using the chelating EDTAcyst reagent attached via a disulfide bond to histone H4 residue 47 that lies near the nucleosome pseudo dyad axis, hydroxy radicals create a single major scission per strand separated by 3 bp on the complementary strands. Therefore, the pseudo dyad axis of the nucleosome must pass through the plane of the central of these 3 bp, implying that the nucleosome core includes an *odd* number of base pairs. This has been confirmed subsequently in the high-resolution crystal structure and demonstrates that the method yields translational positions with true base pair accuracy.[14]

Similar site-directed chelating reagents have also been synthesized by other workers[11,15] and there are growing numbers of reports of their use in the study of nucleic acid interactions.

Chemical Synthesis of EDTAcyst(NPS)

The chemical reagent EDTAcyst(NPS) used for site-directed hydroxy radical mapping of nucleosome positions consists of an EDTA-derived chelating group linked by a peptide bond to cysteamine (Fig. 1A).[14] The cysteamine thiol is in an activated disulfide link with the good leaving group 2-nitrophenylsulfenyl and so exchanges readily the EDTA-linked cysteamine onto free thiols such as solvent-exposed protein cysteine residues.[16] The following protocol describes the steps necessary for the chemical synthesis of EDTAcyst(NPS), *S*-(nitrophenylsulfenyl)-cysteaminyl-EDTA (Fig. 1C). These steps use readily available chemicals and simple equipment accessible in any organic chemistry laboratory. The synthesis is achievable by a patient biochemist with basic chemical skills at the practical yields reported below.

1. Protection of cysteine thiol functionality using *tert*-butanol (tBu)[17]: Gently reflux 2.8 g (25 mmol) of cysteamine hydrochloride together with 5 ml *tert*-butanol (53 mmol) and 12.5 ml 2 *M* HCl for 14 hr. Remove excess *tert*-butanol by rotary evaporation and purify the *S*-*tert*-butylcysteamine by recrystallization. Typical yields of *S*-*tert*-butylcysteamine [Cyst(tBu)] are 60% of material judged pure by thin-layer chromatography on silica plates using 4:1:1 (v/v) *n*-butanol:acetic acid:water as the mobile phase (R_f 0.48).

[15] M. R. Ermacora, J. M. Delfino, B. Cuenoud, A. Schepartz, and R. O. Fox, *Proc. Natl. Acad. Sci. U.S.A.* **89,** 6383 (1992).

[16] L. Rydén and J. Carlsson, *in* "Protein Purification" (J.-C. Janson and L. Ryden, eds.), p. 252. VCH Publishers, New York, 1989.

[17] J. J. Pastuzak and A. Chimiak, *J. Org. Chem.* **46,** 1868 (1981).

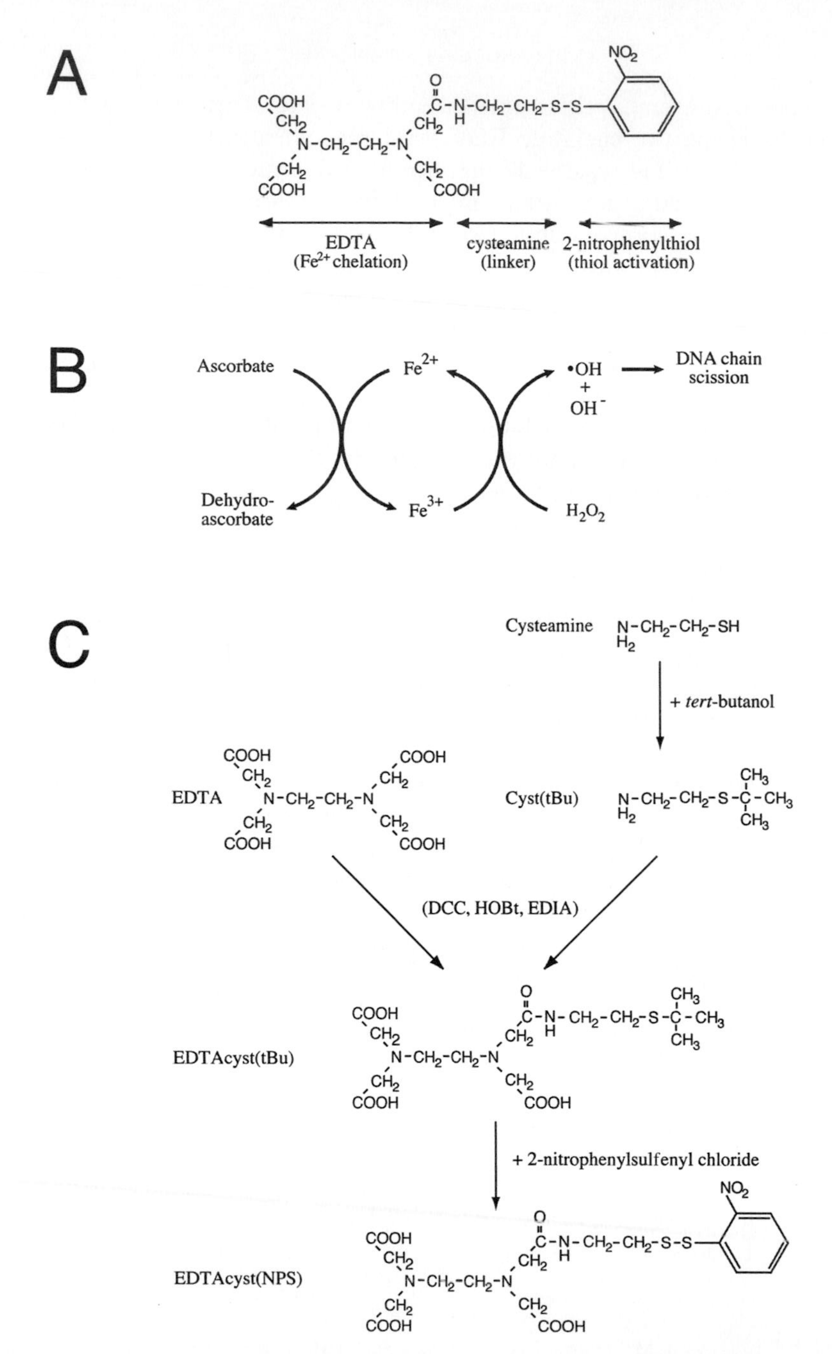

FIG. 1. Chemistry of the site-directed mapping reagent. (A) Structure of EDTAcyst(NPS) reagent. (B) Catalytic cycle for hydroxy radical generation using the Fenton reaction, also known as the Haber–Weiss cycle. (C) Synthetic route for EDTAcyst(NPS).

2. Linkage of *S-tert*-butylcysteamine to EDTA[18]: Chill a mixture of 0.40 g *S-tert*-butylcysteamine (3 mmol), 5.87 g acid-form EDTA (20 mmol), 0.48 g 1-hydroxybenzotriazole (HOBt, 3.3 mmol), 3.5 ml *N*-ethyldiisopropylamine (EDTA, 20 mmol), and 15 ml water-free *N,N*-dimethylformamide in an ice–salt bath ($<-10°$). Add 3.5 ml 1 *M* dicyclohexylcarbodiimide in *N,N*-dimethylformamide (DCC, 3.5 mmol) and allow the reaction to proceed for 14 hr while the ice–salt bath melts. Filter the reaction products through a sintered glass funnel and evaporate the solvent under high vacuum. Resuspend the oily yellow liquid that remains in 10 ml of 0.1% trifluoroacetic acid (TFA) and centrifuge to remove all precipitate. Load the supernatant on a C_{18} reversed-phase column (e.g., 10 × 250 mm Nucleosil 300-7) equilibrated in 0.1% TFA and elute with a gradient from 8 to 24% (v/v) acetonitrile in 0.1% (v/v) TFA over 40 min at 2.5 ml/min. Pool peak fractions and remove solvent using a Speed-Vac or rotary evaporator, resuspend in water, and lyophilize. Typical yields of *S-tert*-butylcysteaminyl-EDTA [EDTAcyst(tBu)] are 20% of material judged pure by thin-layer chromatography on silica plates using 4:1:1 *n*-butanol:acetic acid:water as the mobile phase (R_f 0.68).
3. Exchange of *tert*-butyl protection for 2-nitrophenylsulfenyl (NPS)[17]: Stir 100 mg of *S-tert*-butylcysteaminyl-EDTA (245 μmol) with 70 mg 2-nitrophenylsulfenyl chloride (368 μmol) in 10 ml acetic acid for 3 hr. Remove the solvent by rotary evaporation and resuspend in 400 μl of 80% (v/v) acetonitrile, 0.1% (v/v) TFA. Add 2 ml of 0.1% TFA dropwise and centrifuge to remove precipitate. Load the supernatant on a C_{18} reversed-phase column equilibrated in 20% acetonitrile, 0.1% TFA and elute with a gradient to 56% (v/v) acetonitrile, 0.1% TFA over 20 min at 2.5 ml/min. Pool peak fractions and remove solvent, resuspend in water, and lyophilize. Typical yields of *S*-(nitrophenylsulfenyl) cysteaminyl EDTA [EDTAcyst(NPS)] are 50% of material judged pure by thin-layer chromatography on silica plates using 3:2:4:1 (v/v) *n*-butanol:pyridine:acetic acid:water as the mobile phase (R_f 0.69).
4. Analysis of *S*-(nitrophenylsulfenyl)cysteaminyl EDTA: fast atom bombardment mass spectrometry using 3-NOBA matrix, elemental analysis, and nuclear magnetic resonance has been used to confirm the identity of the reagent. The reagent has a molecular mass of 505 Da and is stable if stored in a dry environment.

[18] J. C. Sheehan and G. P. Hess, *J. Am. Chem. Soc.* **77,** 1067 (1955).

Attachment of EDTAcyst(NPS) to Histone Octamers

A cysteine residue has been introduced at the serine codon of histone H4 by site-directed mutation of the pET-based *Xenopus laevis* sequence H4 expression vector (K. Luger and T. J. Richmond, unpublished results, 1995). This protein is expressed and purified, and the histone octamer is assembled by combining this site-mutated H4 with wild-type sequence H2A, H2B Ala7Pro, and H3 Cys110Ala and then purifying the resulting histone octamer complex. The derivatization of the exposed, unique cysteine-47 on histone H4 can then be carried out by thiol–disulfide exchange of the EDTAcyst onto the thiol functional group. The mutation of the natural cysteine of H3 residue 110 is made to simplify experiments, because although it can be derivatized by the reagent, it does not catalyze any DNA digestion by hydroxy radicals. The H3 Cys110Ala exchange does not cause any detectable change in the structure of the nucleosome (A. Flaus, K. Lunger, and T. J. Richmond, unpublished results, 1995).

1. Prepare histone octamer (xOct) comprising H2A, H2B Ala7Pro, H3 Cys110Ala, and H4 Ser47Cys.[18a] Incubate the desired amount of approximately 15 μM histone octamer in standard buffer (20 mM Tris–Cl, pH 7.5, 2 M NaCl) supplemented with 10 mM dithiothreitol (DTT) for at least 1 hr on ice.
2. Dialyze against at least three changes of 500 ml low pH buffer (5 mM potassium cacodylate buffer, pH 6.0, 2 M NaCl) at 4° for at least 3 hr per step, then measure its concentration spectrophotometrically using E (276 nm, 1 mg/ml) = 0.44, molecular weight 108,000 (K. Luger, T. Rechsteiner, and T. J. Richmond, unpublished results).
3. Prepare fresh EDTACyst(NPS) solution by dissolving 0.5–0.8 mg in 100 mM Tris–Cl, pH 7.5, to a concentration of 10 mM and then add an equal volume of 4 M NaCl for a final concentration of 5 mM EDTAcyst(NPS) in 50 mM Tris–Cl, pH 7.5, 2 M NaCl.
4. Add 1 M Tris-Cl, pH 7.5, to make the dialyzed xOct to 50 mM Tris–Cl, pH 7.5, buffer and then immediately add sufficient 5 mM EDTACyst(NPS) solution for a 75- to 100-fold molar excess. Allow this reaction to proceed overnight at room temperature.
5. Dialyze against three or more changes of 5 mM Potassium cacodylate, pH 6.0, 2 M NaCl at 4° for at least 6 hr per step. Calculate the final concentration of xOct after its dilution by the addition of the reagent solution. Alternatively, measure this spectropho-

[18a] K. Luger, T. J. Rechsteiner, and T. J. Richmond, *Methods Enzymol.* **304,** [1], (1999) (this volume).

tometrically after confirming that all excess NPS chromophore has been removed by monitoring the absorbance at 365 nm using E (365 nm, 1 *M*) = 3000).

6. The derivatized octamer should be stored on ice and is stable for several weeks. Carefully prepare and run 5–20 pmol underivatized and derivatized octamer side by side on a fresh standard 18% (w/w) acrylamide (1:60 bisacrylamide/acrylamide), 1% SDS gel. Because of its increased bulk, derivatized H4 runs slightly more slowly than underivatized protein. Alternative methods to detect derivatization include binding $^{63}Ni^{2+}$ to the reagent before loading on SDS–PAGE and then detecting on X-ray film with the aid of a fluorogenic reagent or separating histone by Triton-acetic acid–urea PAGE.

Assembly of Nucleosomes Containing Derivatized Histone Octamer

The following protocol is a small-scale adaptation of one developed for histone exchange from bulk chromatin onto end-labeled DNA fragments.[19] The stepwise progression in salt reduction is related to the approximate points of equilibration of the H3–H4 tetramer and the H2A–H2B dimer onto DNA.[20] The positions of nucleosomes deposited by this method depend on the salt-dependent equilibration[21] and later thermally driven changes such as nucleosome sliding.[22,23]

1. Temperature equilibrate a suitable dialysis apparatus capable of efficiently dialyzing an appropriate amount of assembly mixture at 4 °. We routinely dialyze 40- to 50-μl samples by placing them in open 4-mm-internal-diameter acrylate tubes projecting from a plate and whose bottom faces are covered with dialysis membrane secured by a rubber O ring. This is placed over an acrylate block containing matching 10-ml volume (20 mm diameter) buffer wells. It is also possible to dialyze <300-μl volumes by carefully sealing small dialysis bags.
2. Prepare samples as a mixture containing 5 μM derivatized xOct with equimolar labeled DNA fragment and with a minimum salt concentration of 1.8 *M* and load samples into the dialysis container, taking care not to pierce the membranes. DNA labeling can be carried out, for example, by 5″ labeling using T4 polynucleotide

[19] D. Rhodes and R. Laskey, *Methods Enzymol.* **170,** 575 (1989).
[20] J.-E. Germond, M. Bellard, P. Oudet, and P. Chambon, *Nucleic Acids Res.* **11,** 3173 (1976).
[21] H. R. Drew, *J. Mol. Biol.* **219,** 391 (1991).
[22] A. Flaus and T. J. Richmond, *J. Mol. Biol.* **275,** 427 (1998).
[23] G. Meerseman, S. Pennings, and E. M. Bradbury, *EMBO J.* **11,** 2951 (1992).

kinase and [γ-^{32}P]ATP or by end filling of a 5′ overhanging restriction end using T4 DNA polymerase and an appropriate [α-^{32}P]dNTP.

3. Dialyze stepwise against 20 m*M* Tris–Cl, pH 7.5, 1 m*M* potassium EDTA with consecutively 2, 0.85, 0.65, and 0.5 *M* potassium chloride per 2-hr step and then for 2 hr against 20 m*M* Tris–Cl, pH 7.5. Finally, dialyze overnight against 20 m*M* Tris, pH 7.5, with 5% buffer-equilibrated Chelex 100 resin slurry.
4. Remove the assembled 5 μM nucleosome samples to prechilled Eppendorf tubes. The nucleosomes can be stored on ice for several weeks.

Analysis of Assembled Nucleosomes

Acrylamide gels can distinguish nucleosome organization at very high resolution, e.g., separating nucleosome core particles differing in position by 10 bp.[22,23] Care is required to thermally equilibrate the gel at 4° before use for optimal resolution.

1. Prepare a 5% acrylamide (1:60 bisacrylamide/acrylamide), 0.2 × TBE gel between 16 cm × 18 cm × 1 mm glass plates with 10-mm-wide sample wells and prerun at 250 V for 3 hr at 4° with buffer recirculation. Allow the gel to temperature equilibrate at 4° for several hours (preferably overnight).
2. Prepare 10–20 μl of sample containing 1–2 pmol assembled nucleosomes at 5% sucrose. Ensure that the loading solution is prechilled before adding nucleosomes.
3. Load and run samples on gel at 250 V for 3 hr at 4° with buffer recirculation. Dry the gel without fixing and expose to X-ray film.

Mapping of Nucleosome Positions

For mapping the position of a nucleosome using site-directed hydroxy radicals, the site-specifically attached, chelating EDTAcyst reagent must be loaded with Fe^{2+} ions. The Fenton hydroxy radical-generating cycle can then be catalyzed at this reactive center by the addition of hydrogen peroxide and using ascorbate as the reducing agent to create a catalytic cycle (Fig. 1B).[24] Because the cycle will continue until the radicals destroy the chelating activity of the reagent either by attack of the chelating EDTA arms or at the sulfur atoms anchoring the reagent, only 5–10% of DNA is apparently digested, even after long digestion times. Furthermore, because

[24] T. D. Tullius, *Nature* **332,** 663 (1988).

the restricted space between the histone octamer and DNA reduces the supply of the hydrogen peroxide and ascorbate substrates for the Fenton cycle, the site-directed mapping reaction is slower than the free solution hydroxyl radical footprinting method, which is also carried out at higher Fe^{3+}/EDTA concentrations. Excess Fe^{2+} ions will bind as counterions to the DNA and cause nonspecific background digestion so it is necessary to avoid any extraneous metal ions that could either generate hydroxy radicals themselves (Cu^{2+}, Ni^{2+}) or release Fe^{2+} by competing with it for binding to the chelating reagent (any divalent or trivalent metal ion). This is the major difficulty experienced in practice and underlines the necessity to acid wash glassware,[25] purify the major buffer components by passing them over Chelex resin,[25] and to use only the highest metal-free grades of reagents with explicit metal contamination specifications. We have used only Fluka (Ronkonkoma, NY) chemicals in all studies to date, with the exception of Sigma (St. Louis, MO) Tris base.

1. Using a standard water aspiration pump, degas two acid-washed Büchner flasks each containing 50 ml of buffer stock and two further acid-washed flasks containing 50- and 100-ml aliquots of water, respectively. Add 39 mg (100 μmol) ammonium ferrous sulfate hexahydrate to the degassed 100 ml of water to make a 1 m*M* solution and continue degassing while gently swirling to dissolve the salt. Remove two aliquots of 987 μl buffer from the 50-ml degassed buffer stock to Eppendorf tubes on ice, labeling one as "buffer" and the other as "buffer + iron." Add 13 μl degassed 1 m*M* Fe salt to the aliquot labeled buffer + iron.
2. Place 10 μl of 5 μM assembled nucleosome into prechilled reaction tubes on ice for each reaction. Add 50 μl from the aliquot labeled buffer to each reaction tube, add 5 μl buffer + iron, and then stand on ice for 15 min.
3. While allowing iron to bind in the nucleosome, add 0.10 g (576 μmol) ascorbic acid to the second 50-ml degassed buffer stock to make a 12 m*M* ascorbate solution and continue degassing while swirling to dissolve. Chill a 1-ml aliquot of the 12 m*M* ascorbate solution on ice and label as "ascorbate." At the same time, add 333 μl of 30% H_2O_2 to the 50-ml degassed water to make a 0.2% solution and swirl gently to mix. Chill a 1-ml aliquot of the 0.2% H_2O_2 on ice and label as "H_2O_2 solution." Prepare for native gel samples by prechilling fresh Eppendorf tubes containing 15 μl 10% sucrose for each reaction. Prepare for zero time point samples by mixing 5 μl "stop solution"

[25] K. M. Schaich, *Methods Enzymol.* **186,** 121 (1990).

(25 m*M* EDTA pH 8.0, 1 *M* thiourea) and 20 μl buffer in fresh Eppendorf tubes for each reaction.

4. After the 15-min equilibration of nucleosome-containing reaction samples (step 2), remove 5 μl of each reaction to tubes for native PAGE analysis and 20 μl of each reaction sample to tubes for the zero time point. Add 20 μl of ascorbate and 20 μl of H_2O_2 solution to each reaction tube in quick succession to start the reactions.
5. At desired time points, remove samples (20–80 μl) and add to fresh Eppendorf tubes containing 0.125 volumes of stop solution. Immediately after adding samples to stop solution, extract once with phenol/CIA [25 parts (v/v) Tris-equilibrated phenol, 24 parts chloroform, 1 part isoamyl alcohol]. If subsequent restriction enzyme digestion is necessary, add 0.15 volumes of precipitation solution (2.18 *M* sodium acetate, pH 5.2, 91 m*M* $MgCl_2$, 0.18 mg/ml carrier tRNA) and 2.5 volumes of absolute ethanol to the extracted samples and centrifuge to precipitate. Aspirate off the supernatant and dry the pellet under vacuum for >15 min and then resuspend DNA pellets in 1-μl formamide dyes (deionized formamide with 0.1% xylene cyanol, 0.1% bromphenol blue, 10 m*M* EDTA) per 1000 cpm as estimated from Cerenkov counting.
6. If a restriction enzyme digestion is necessary subsequent to the reaction, extract the stopped reaction samples once with phenol/CIA and once with CIA [24(v/v) parts chloroform, 1 part isoamyl alcohol] and then ethanol precipitate without carrier. Resuspend the DNA pellet in a suitable buffer and carry out the digestion to completion. Continue processing digested samples from the phenol/CIA extraction as described in step 5.
7. Prepare a Maxam–Gilbert G-specific sequence ladder to allow the identification of DNA sizes. These DMS reaction products have the same free phosphate ends as those generated from hydroxy radical attack. We have found empirically that in the gel-running conditions described later, dideoxy-terminated ladders from standard enzymatic sequencing with incorporated labels (i.e., no 5′-phosphate on primer) can be used as markers because species of equivalent length to hydroxy radical cuts migrate 2 bases slower than the equivalent hydroxyl radical cut fragment in the 60–120 base range, presumably due to the presence in the latter of the 5′-phosphate, the charged 3′ terminus, and the loss of the terminal base moiety.[22]
8. Prepare an 8.3 *M* urea, 6% acrylamide (1:20 bisacrylamide/acrylamide ratio), 0.5× TBE gel between 40 cm × 18 cm × 0.1 mm siliconized glass plates and a 20-well comb and assemble the gel in an electrophoresis box with 1× TBE in the upper and lower reservoirs.

Prerun the gel at 45 W for at least 45 min, continuing until samples have been boiled for 3 min and are ready to be loaded. Immediately load 2–3 μl of each sample per lane and then electrophorese at 45 W until the bromphenol blue marker, which comigrates with 18–20 base

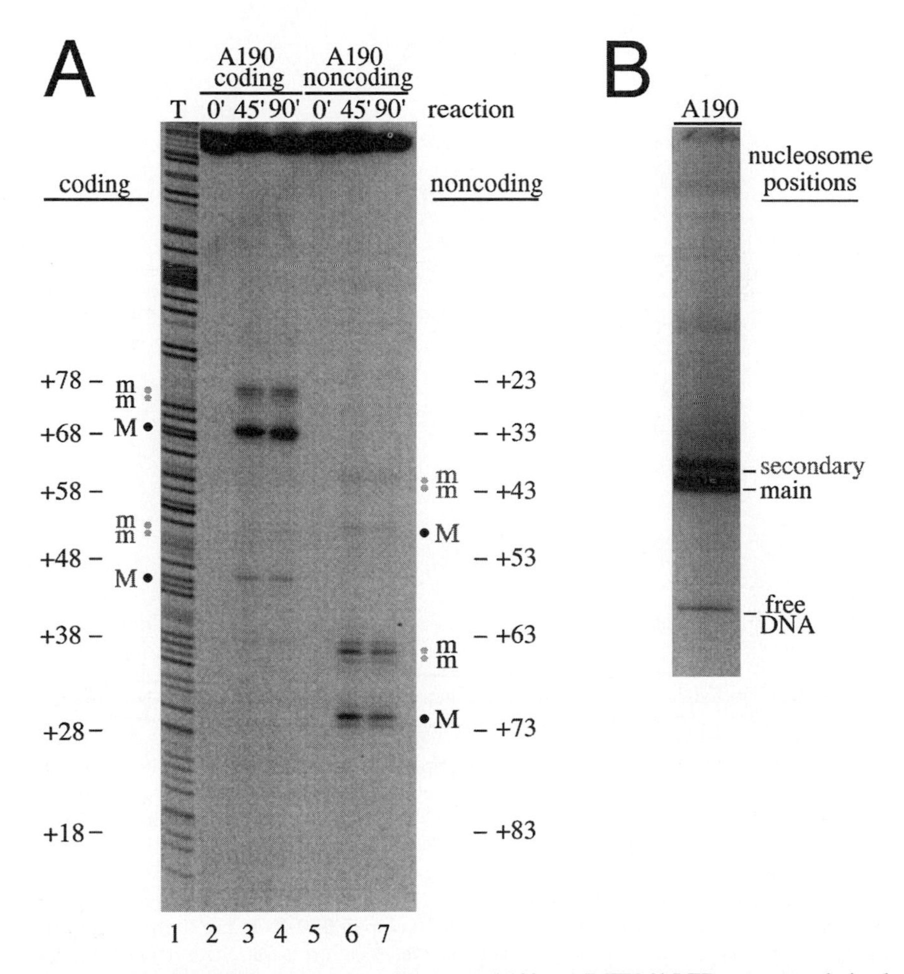

FIG. 2. Mapping of nucleosome positions on A190, a MMTV 3′ LTR promoter-derived DNA fragment spanning base pairs −45 to 145 relative to the transcription start site. (A) Denaturing PAGE of mapping reactions on DNA 5′ labeled independently on coding (lanes 2–4) and noncoding (lanes 5–7) strands. Nucleosomes assembled at 4° and stored at 0° were reacted on ice for the times in minutes shown above each lane. A calibrating dideoxy ladder is shown in lane 1. The single strong cut (M) and weaker (m) cuts are indicated in black and gray for the main and secondary positions, respectively. (B) Native PAGE at 4° of a sample used for mapping in (A) showing the faster migration of the terminally located main species relative to the secondary position.

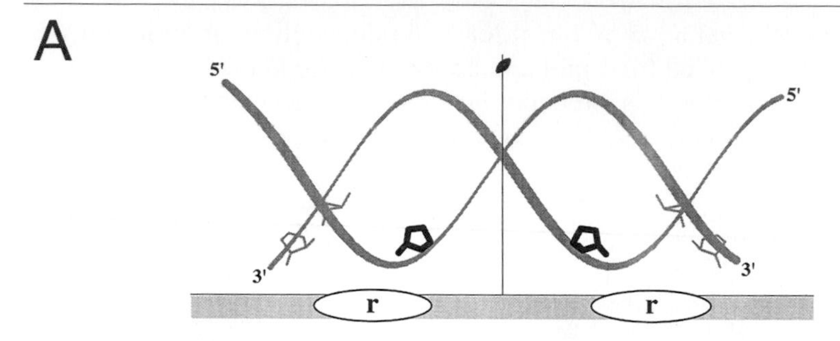

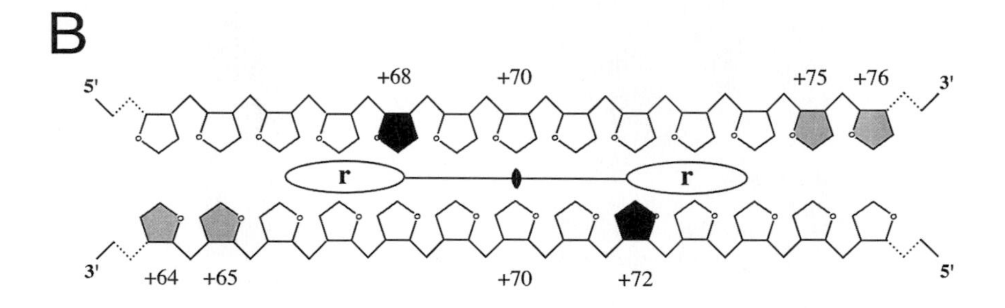

FIG. 3. Diagrammatic representations of mapping analysis. (A) Idealized representation of DNA backbone near the dyad axis of the nucleosome. The single major cut generated on each strand is indicated by the bold black ribose ring. The two weaker cuts 7 and 8 bases in the 3′ direction are indicated as gray ribose rings. The pseudo dyad symmetry axis is indicated. (B) Schematic representation of cuts on complementary strands illustrating the strong (black-filled ribose ring) and weaker (gray-filled ribose rings) cut bases and how the pseudo dyad axis must therefore pass through the plane of the base pair 2 bases in the 3′ direction of the strong cut on each strand. (C) Diagram showing the location of the main (solid oval) and secondary (dashed oval) nucleosome positions for the results shown in Fig. 2.

fragments, reaches the bottom (about 100 min). Fix the gel for 15 min in 10% (v/v) acetic acid, 10% (v/v) methanol, dry, and expose to X-ray film with enhancement screens (e.g., Kodak, Rochester, NY, MR film, or Fuji RX film depending on exposure characteristics) or to a freshly erased phosphorimager plate.

Determining Nucleosome Positions Using Mapping Gel Results

Determining the location of the nucleosomes is sometimes confusing at first because each position yields both a single strong cut and two somewhat weaker ones on each strand (cuts labeled M and m, respectively, in Fig. 2A).[14] Study the idealized DNA backbone in Fig. 3A to satisfy yourself about the structural basis for this pattern and the schematic in Fig. 3B to see how the pseudo dyad location can be determined precisely using nucleosomal dyad symmetry. Figure 3B shows that the pseudo dyad axis of the nucleosome lies 2 bp in the 3′ direction from each strong cut. Using the mapping gel, identify each nucleosome position on each strand by looking for the characteristic single strong cut followed by the weaker ones 7 and 8 bases in the 3′ direction (i.e., migrating more slowly in a gel if DNA was 5′ end labeled). With the aid of the calibration ladder on the gel, determine the location of each strong cutting site for each strand, mark these on a linear printout of the DNA sequence, and determine the pseudo dyad position as halfway between the complementary cuts.

For example, in the mapping shown in Fig. 2A, characteristic patterns with main strong cuts at 68 and 72 occur on coding and noncoding strands, respectively, indicating that the pseudo dyad lies through base pair 70 (Fig. 3B). Weaker characteristic patterns show that there is also a minor position with pseudo dyad at 48 (Fig. 3C). Because the local stereochemistry can differ theoretically for cutting at different sites, the rate of cutting may not be strictly proportional to nucleosome proportional occupancy so estimates of relative occupancy of different positions must be made using native PAGE. Figure 2B shows resolution of the more centrally located, slower migrating secondary 48 position from the more terminal, faster migrating main 70 position, and phosphorimaging of this gel allowed estimation of the relative occupancy as approximately 1:10. Care should be taken in interpreting the meaningfulness of nucleosome occupancies at the extreme end of DNA fragments, such as the 70 nucleosome in Fig. 2A, as they can appear disproportionately when nucleosomes have been allowed to "slide" due to their elevated energetic stability under low salt conditions.[22]

Acknowledgments

We are very grateful for the advice, suggestions, and assistance of colleagues, especially to Drs. T. Rechsteiner, A. Maeder, K. Luger, and S. Tan, to A. Tun-Kyi, K. Berndt, and J. Bushweller for advice on chemistry, and to M. Bumke and R. Richmond for technical assistance. We thank Dr. Carl Deluca for carefully reading this manuscript and Dr. S. Halford for supplying *Eco*RV enzyme.

[14] Localization of Specific Histone Regions on Nucleosomal DNA by Cross-Linking

By SERGEI I. USACHENKO and E. MORTON BRADBURY

Introduction

Chromatin and nucleosomes are labile structures that are maintained by histone–DNA interactions. These interactions are the major factor involved in the modulation of the structure–function relationship of chromatin and nucleosomes, which continue to be studied by a variety of methods. One of the most powerful and informative methods is zero-length covalent protein–DNA cross-linking induced by mild methylation of DNA with dimethyl sulfate.[1] This method allows the direct determination of histone–DNA contacts in nucleosomes and thereby provides significant advantages to the studies of histone–DNA interactions both *in vitro*[2] and *in vivo*.[3,4] Currently, this is the only methodology that provides a direct approach for the localization of specific histone–DNA contacts in native chromosomes within intact nuclei.[5,6] It has also been used successfully to map histone–DNA contacts in isolated chromatin and nucleosomes[7] or nucleosomes reconstituted on specific DNA sequences.[8] The histone–DNA contacts identified by this method for the isolated core particle[5] has been confirmed by X-ray crystallography analysis.[9] This method also facilitates the studies of the overall structure and stability, as well as the dynamic aspects of the nucleosome.[2,5,6] For an overview of this methodology, as well as detailed cross-linking protocols, the interested reader is referred to the general

[1] E. S. Levina, S. G. Bavykin, V. V. Shick, and A. D. Mirzabekov, *Anal. Biochem.* **110,** 93 (1981).

[2] I. M. Gavin, S. I. Usachenko, and S. G. Bavykin, *J. Biol. Chem.* **273,** 2429 (1998).

[3] V. L. Karpov, O. V. Preobrazhenskaya, and A. D. Mirzabekov, *Cell* **36,** 423 (1984).

[4] S. Bavykin, L. Srebreva, T. Banchev, R. Tsanev, J. Zlatanova, and A. Mirzabekov, *Proc. Natl. Acad. Sci. U.S.A* **90,** 3918 (1993).

[5] S. G. Bavykin, S. I. Usachenko, A. I. Lishanskaya, V. V. Shick, A. V. Belyavsky, I. M. Undritsov, A. A. Strokov, I. A. Zalenskaya, and A. D. Mirzabekov, *Nucleic Acids Res.* **13,** 3439 (1985).

[6] S. I. Usachenko, I. M. Gavin, and S. G. Bavykin, *J. Biol. Chem.* **271,** 3831 (1996).

[7] A. V. Belyavsky, S. G. Bavykin, E. G. Goguadze, and A. D. Mirzabekov, *J. Mol. Biol.* **139,** 519 (1980).

[8] D. Pruss and A. P. Wolffe, *Biochemistry* **32,** 6810 (1993).

[9] K. Luger, A. W. Mader, R. K. Richmond, D. F. Sargent, and T. J. Richmond, *Nature* **389,** 251 (1997).

0076-6879/99 $30.00

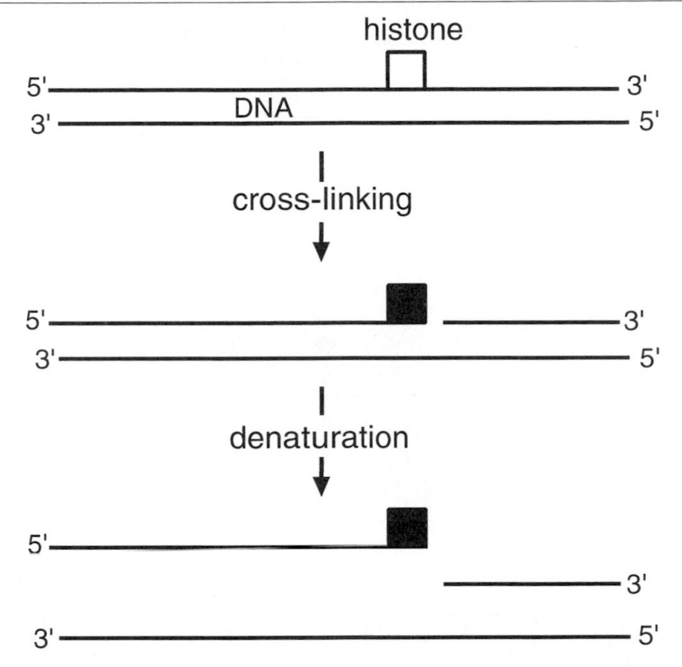

FIG. 1. The scheme for histone–DNA cross-linking induced by mild methylation of DNA with dimethyl sulfate. Cross-linking causes a single-stranded nick in the nucleosomal DNA at the cross-linking site such that only the 5′-terminal DNA fragment is cross-linked to a histone molecule. The length of the cross-linked DNA fragment is the precise distance of a protein cross-linking site from the 5′ end of the nucleosomal DNA.[1]

review on DNA–protein cross-linking[10] where the chemical reactions have also been described. This article describes the application of methodological improvements that have been developed and used successfully for the localization of specific histone regions on the nucleosomal DNA.

Mapping of Histone–DNA Contacts by Cross-Linking

The scheme of histone–DNA cross-linking induced by mild methylation of DNA with dimethyl sulfate is represented in Fig. 1. The length of the cross-linked DNA fragment, which is the precise distance of the protein cross-linking site from the 5′ end of the nucleosomal DNA, can be assessed by two versions of denaturing two-dimensional gel electrophoresis: the "DNA version" and the "protein version."[10] Based on previous experience,

[10] A. D. Mirzabekov, S. G. Bavykin, A. V. Belyavsky, V. L. Karpov, O. V. Preobrazhenskaya, V. V Shick, and K. K. Ebralidse, *Methods Enzymol.* **170,** 386 (1989).

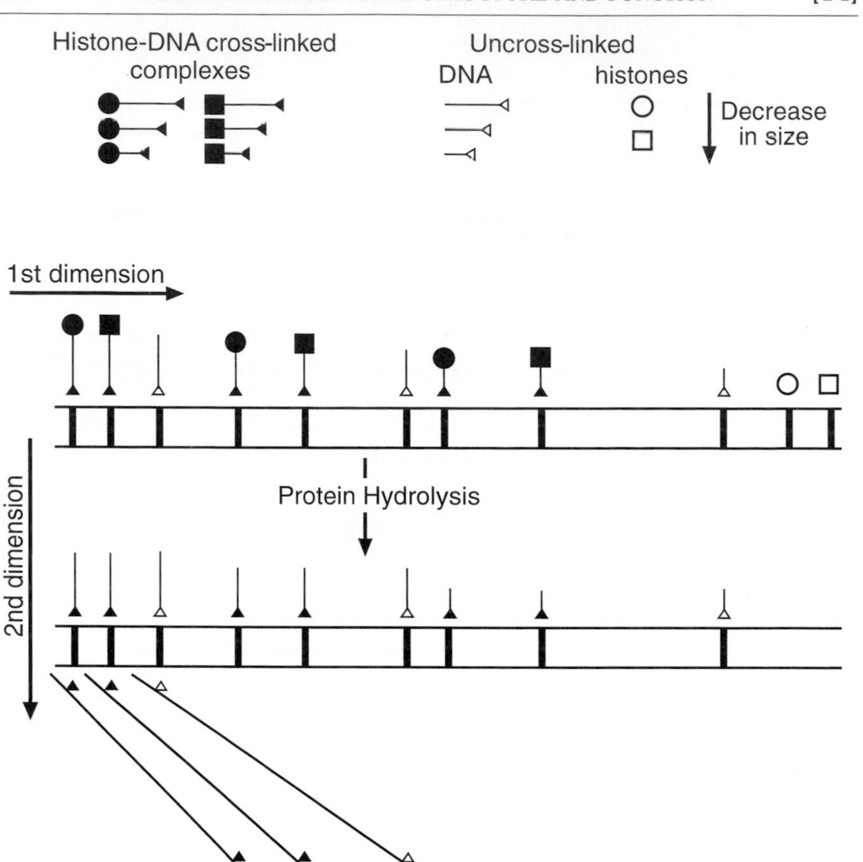

FIG. 2. Schematic presentation of the "DNA version" of two-dimensional gel electrophoresis. In the first dimension of the denaturing SDS gel the mobility of the cross-linked histone–DNA complexes depends on the molecular weight of the histone and the size of the cross-linked DNA fragment. After separation of histone–DNA cross-linked complexes in the first dimension the histones are digested directly in the gel by protease and the released DNA fragments are separated according to their size in a denaturing second-dimension gel. In the first dimension, histones cross-linked to the DNA decrease the mobility of the cross-linked complexes proportionally to the histone size. As a result, after histone digestion, the released DNA fragments that were cross-linked to the different histones fall on different diagonals in

the DNA version is the most informative and reproducible for the direct mapping of histone–DNA contacts.[6,11] However, sometimes the protein version is necessary to complement the DNA version. Another advantage of the DNA version is that the cross-linked complexes are labeled with ^{32}P whereas in the protein version the labeling is with ^{125}I. This article deals with the mapping of histone–DNA contacts using the DNA version of two-dimensional gel electrophoresis, which is represented schematically in Fig. 2.

In order to achieve optimal resolution in the mapping of histone–DNA contacts, a number of factors need to be taken into account. One of the major factors affecting the resolution of histone–DNA contacts in the two-dimensional gel is the separation of the diagonals corresponding to the individual histones. The required separation can be achieved by optimizing the first-dimension gel conditions. Based on our experience, the optimal resolution in two-dimensional gel electrophoresis for most chromatin sources can be obtained with gels containing 15% acrylamide (AA), 0.5% (w/v) *N,N'*-methylenebisacrylamide (MBA) for the first dimension and 15% AA, 0.2% MBA for the second dimension.[11] However, if better separation of a diagonal corresponding to a particular histone is desired, this can be achieved by modifying the concentrations of acrylamide and *N,N'*-methylenebisacrylamide in the first-dimensional gel.

Another factor that can affect the separation of histone–DNA cross-linked complexes in the first dimension and therefore the diagonals corresponding to the individual histones in the second dimension is the distance the cross-linked complexes migrate in the first-dimensional gel. Histone–DNA cross-linked complexes are much bigger than free proteins and their separation is much less efficient in the first-dimensional gel. In order to obtain sufficient separation, we increase the length and running time of the first-dimensional gel. In addition, this prevents undesired background in the second-dimensional gel. For each new source of histone–DNA cross-linked complexes, it is advisable to undertake preliminary experiments to optimize the experimental conditions. For the best results, the separation time of cross-linked complexes in the first-dimensional gel should be estimated by a short preliminary fractionation. This investment of time and

[11] S. I. Usachenko, S. G. Bavykin, I. M. Gavin, and E. M. Bradbury, *Proc. Natl. Acad. Sci. U.S.A.* **91,** 6845 (1994).

the second-dimension gel. Diagonals corresponding to particular histones in the two-dimensional gel are arranged from left to right in the same order as free histones migrate from top to bottom in the first-dimensional gel. The position of the spots within each diagonal indicates the length of the DNA fragments cross-linked to a particular histone. This length can be determined by running DNA fragments of known size in the gel of the second dimension.

effort is necessary to optimize resolution and facilitate the interpretation of results.

Procedure for Short Preliminary Fractionation

1. Purify the cross-linked complexes from uncross-linked proteins by Cetavlon precipitation as described.[10]
2. Label 0.5–1 A_{260} units of cross-linked complexes with ^{32}P by incubation at 37° for 30–45 min in 30 μl buffer [40 mM bicine, pH 9.0, 10 mM $MgCl_2$, 10 mM dithiothreitol (DTT), 0.1 mM spermidine] containing 300 μCi of [γ-^{32}P]ATP and 10 units of polynucleotide kinase. The labeling of cross-linked complexes is more efficient before its purification from uncross-linked DNA.
3. Precipitate the ^{32}P-labeled cross-linked complexes with ethanol, centrifuge for 5 min at 10,000g, and wash the pellet twice with cold 75% ethanol. Purify the ^{32}P-labeled cross-linked complexes from uncross-linked DNA as described[10] and precipitate the ^{32}P-labeled cross-linked complexes with ethanol as described earlier.
4. Dissolve the pellet in 10 μl 0.2% sodium dodecyl sulfate (SDS) and incubate for 5 min.
5. Add 15 μl loading buffer (15% Ficoll-400, 62.5 mM Tris–HCl, pH 6.8, 10 mM DTT, 7 M urea, 1% SDS, 0.1% bromphenol blue), mix thoroughly, and incubate in boiling water for 2–3 min.
6. Load the sample on the first-dimensional gel (300–350 mm long and 0.6–0.75 mm thick) consisting of a separating gel (15% AA, 0.5% MBA, 375 mM Tris–HCl, pH 8.8, 0.1% SDS, 7 M urea) and a concentrating gel (6% AA, 0.2% MBA, 125 mM Tris–HCl, pH 6.8, 0.1% SDS, 7 M urea).
7. Mark the loaded well on the glass plate with a marker pen. Attach the electrophoresis apparatus to an electric power supply (the positive electrode should be connected to the bottom buffer reservoir) and carry out electrophoresis in Tris–glycine electrophoresis buffer for SDS gel[12] at 10–12 mA constant current. Measure the speed of the bromphenol blue migration.
8. Stop the gel when the bromphenol blue reaches the bottom.
9. Place the gel on a flat surface so that the glass with the marked well is on the bottom. To remove the upper glass plate successfully, the following procedure can be helpful. Insert a 25-gauge (or thinner) syringe needle between the gel and the upper glass plate. Insert a plastic spatula between the glass plates close to the inserted needle.

[12] J. Sambrook, E. F. Fritch, and T. Maniatis, "Molecular Cloning." Cold Spring Harbor Laboratory, Cold Spring Harbor, NY, 1989.

Carefully apply upward pressure to allow the entrance of air through the needle between the gel and the upper glass plate. Carefully lift off the upper glass plate so that the gel remains attached to the lower glass plate.

10. Remove spacers from the sides and dry thoroughly with a paper towel moistened with alcohol.
11. Replace the spacers and cover the whole surface of the gel and spacers with enough Saran wrap so that extra Saran wrap can be attached with adhesive tape to the back of the glass plate.
12. Place X-ray film inserted between two pieces of thin paper on the Saran wrap-covered gel and cover with a clean dry glass plate of the same size. Secure the assembled "sandwich" with clamps on the sides.
13. Autoradiograph the gel in a dark cool (4–10°) place. The time for exposure depends on the level of radioactivity incorporated into the cross-linked complex and can be estimated by a series of short exposures. For cross-linked complexes labeled with 300–500 μCi radioactivity, the time of exposure is between 15 and 30 min.

The speed of the fastest migrating band should be calculated from the autoradiograph relative to the speed of bromphenol blue migration (see step 7 in the previous protocol). From this, the time for fractionation of the cross-linked complexes can be determined. The best resolution of cross-linked complexes in the first-dimensional gel is obtained when the fastest migrating band approaches as close as possible (e.g., 3–5 cm) to the bottom of the gel. When the appropriate time for the migration of cross-linked complexes in the first-dimensional gel is determined, two-dimensional gel electrophoresis can be carried out according to the following procedure.

Two-Dimensional Gel Electrophoresis

1. Repeat steps 1–7 in the just-described procedure.
2. Carry out the first-dimensional gel electrophoresis until the fastest migrating band is 3–5 cm from the bottom of the gel (calculated according to the preliminary fractionation).
3. Stop the gel and follow steps 9–13.
4. According to the autoradiograph, cut out the informative part of the gel strip and wash it in at least 50 volumes of buffer containing 62.5 m*M* Tris–HCl, pH 6.8, 0.1% SDS three times for 20 min with gentle shaking to extract urea. Finally, for better polymerization in the second-dimensional gel, soak the strip for another 15 min in the just-mentioned buffer containing 6% acrylamide, 0.08% bis-acrylamide, 0.5 mg/ml ammonium persulfate.

5. Polymerize the second-dimensional gel (400 mm long, 300 mm wide, and 1–1.3 mm thick) consisting of a separating gel (15% AA, 0.2% MBA, 375 m*M* Tris–HCl, pH 8.8, 0.1% SDS, 7 *M* urea) and a concentrating gel (6% AA, 0.08% MBA, 125 m*M* Tris–HCl, pH 6.8, 0.1% SDS). According to our experience, this AA : MBA ratio gives optimal resolution of the DNA fragments in the second dimension.
6. Cast another concentrating gel on top of the already polymerized concentrating gel and immediately insert the first-dimensional gel strip prepared as described in step 4. For the best results it is important that the boundaries between the different sections (concentrating and separating gels) of the second-dimensional gel are very straight and defined. This can be achieved by carefully overlaying 0.1% SDS on the top of each section immediately after casting.
7. Remove any air bubbles between the gel strip and the polymerized concentrating gel by applying light pressure from the top of the strip with a clean thin plastic strip (e.g., spacer from the first-dimensional gel). Carefully overlay the concentrating gel with 0.1% SDS. After polymerization is complete, pour off the overlay and wash the top of the gel twice with 0.1% SDS.
8. Prepare the loading buffer for the second-dimensional gel as follows: Dissolve 3 mg Pronase in 200 μl water, add 700 μl first-dimension loading buffer (see step 6 in the procedure for short preliminary fractionation) without urea and mix thoroughly. Add 200 μl formamide solution of single-stranded DNA fragments (A_{260} 500) of known size (e.g., from DNase I digest of nuclei[13]) as molecular weight markers drop by drop, mixing between drops.
9. Mount the gel in the electrophoresis apparatus and add Tris–glycine electrophoresis buffer for SDS gel[12] to the top and bottom reservoirs covering the top and bottom of the gel.
10. Using a pipette with a narrow tip, carefully apply the prepared loading buffer on the top of the second dimensional gel. Wait for a few minutes until the loading buffer is distributed evenly over the top of the gel.
11. Attach the electrophoresis apparatus to an electric power supply (the positive electrode should be connected to the bottom buffer reservoir) and carry out electrophoresis at 7 mA until the bromphenol blue has migrated through the concentrating gel. Increase the current to 15 mA and carry out electrophoresis until the bromphenol blue reaches the bottom of the gel.

[13] M. Noll, *Nucleic Acids Res.* **1,** 1573 (1974)

12. Remove the concentrating gel and extract SDS and urea by soaking the separating gel in 75% alcohol until it turns white. Then soak for 30 min in water with two to three changes and stain the marker DNA fragments with ethidium bromide (1 μg/ml).
13. Wash the gel in water for 10 min and place on a UV transilluminator. Remove excess liquid from the surface of the gel with a paper towel. Using the UV light, mark the positions on the gel of the ethidium bromide-stained marker fragments by Indian ink dots along each band. Soak the gel in 5% glycerol for 5 min and dry under a vacuum.
14. Place at least two fluorescent tags (e.g., manufactured by Stratagene, La Jolla, CA) on separate corners of the gel. Make a replica of the gel by placing a transparency on the gel and marking the location of the DNA markers and fluorescent tags. Autoradiograph the gel.
15. Superimpose the gel replica with the developed X-ray film, aligning the fluorescent tags. Assess the size of cross-linked DNA fragments on the X-ray film according to the size of DNA markers on the gel replica.

Mapping of Histone–DNA Contacts in Specific Histone Regions

The location of specific histone regions on the nucleosomal DNA can be identified by comparison of histone–DNA contacts before and after digestion with a specific proteolytic enzyme that either cleaves or digests the histone region of interest. The dimethyl sulfate cross-linking method consists of two protocols.[14] The first protocol favors cross-linking through histidines (90%) rather than lysines (10%) and mostly detects the location of histone regions containing histidine. The second protocol provides cross-linking of histones through both lysines and histidines and allows the localization of histone regions containing both amino acid residues. A combination of these two cross-linking protocols and different specific proteases affecting different histone regions allows one to determine the location of these regions on the nucleosomal DNA. This article describes the technique using two proteolytic enzymes, trypsin and clostripain, and cross-linking through histidines. This approach was used successfully to map the contacts of the central (globular) domains of all core histones, the C-terminal domain of histone H2A, and the N-terminal domain of histone H4 on nucleosomal DNA.[11] Trypsin digests all of the histone terminal tails,[15–17] whereas clostri-

[14] G. A. Nacheva, D. Y. Guschin, O. V. Preobrazhenskaya, V. L. Karpov, K. K. Ebralidse, and A. D. Mirzabekov, *Cell* **58,** 27 (1989).

[15] L. Böhm, C. Crane-Robinson, and P. Sautiere, *Eur. J. Biochem.* **106,** 525 (1980).

[16] L. Böhm, G. Briand, P. Sautiere, and C. Crane-Robinson, *Eur. J. Biochem.* **119,** 67 (1981).

[17] L. Böhm, G. Briand, P. Sautiere, and C. Crane-Robinson, *Eur. J. Biochem.* **123,** 299 (1982).

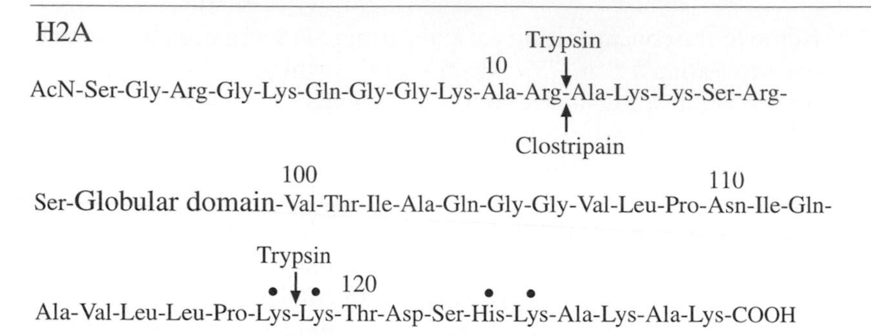

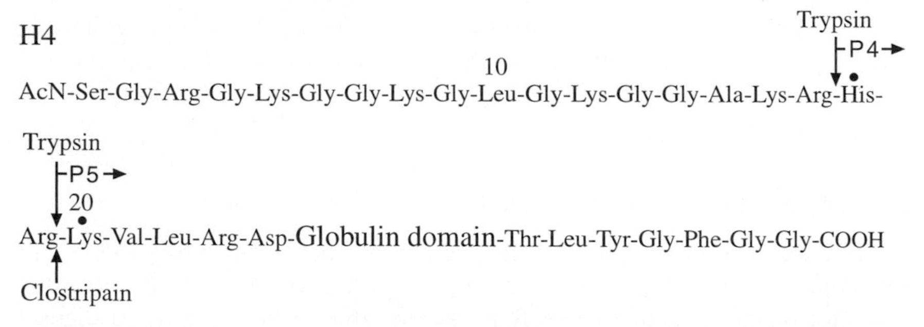

FIG. 3. Partial amino acid sequence of chicken erythrocyte histones H2A and H4. Vertical arrows indicate the boundaries of limited digestion of nucleosomes with trypsin and clostripain. Amino acids in the histone terminal domains that were cross-linked to the nucleosomal DNA are marked by solid circles.

pain digests only the N-terminal tails of the core histones[18] (Fig. 3). To obtain unambiguous results it is very important that the proteolytic reaction is complete so that the major products are not under- or overdigested. Conditions for proper digestion can be established from the results of a preliminary analytical time course digestion. However, the final products of a large-scale digestion must also be tested by gel electrophoresis. The digestion is complete when there are no bands corresponding to the products of under- or overdigestion observed on the gel between the bands corresponding to the major peptides. Under- or overdigested products may also overlap with the major peptides in the gel so that they cannot always be detected in a regular or short SDS gel. However, they can be detected efficiently in a first-dimensional gel, which provides better resolution (Fig.

[18] A. Dumuis-Kervabon, I. Encontre, G. Etienne, J. Jauregui-Adell, J. Mery, D. Mesnier, and J. Parello, *EMBO J.* **5,** 1735 (1986).

4E). This test is important to achieving the appropriate resolution of diagonals corresponding to the major peptides in the two-dimensional gel and can be carried out as follows.

1. Bring the nucleosome preparation to $A_{260} \sim 20$ in 100 μl of appropriate buffer (e.g., 25 m*M* HEPES, pH 7.0).
2. Preincubate for 10–20 min at 25°, add 1 μl trypsin (7 mg/ml), and mix.
3. Incubate at 25° and stop the reaction at 5-, 10-, 20-, 40-, and 60-min time points by taking 12-μl samples and adding each to 1 μl 10 mg/ml N^{α}-*p*-tosyl-L-lysine chloromethyl ketone (TLCK) (Sigma, St. Louis, MO) on ice.
4. Add 2 μl 1% SDS, mix thoroughly, and incubate at room temperature for 5 min.
5. Add 10 μl gel-loading buffer for the first dimension (15% Ficoll-400, 62.5 m*M* Tris–HCl, pH 6.8, 10 m*M* DTT, 7 *M* urea, 1% SDS), mix, and incubate in boiling water for 2–3 min.
6. Load the first-dimension gel (see procedure for short preliminary fractionation) and fractionate the samples at a 10- to 12-mA constant current until the bromphenol blue reaches the bottom of the separating gel.
7. Stain the gel with Coomassie Blue.

A similar test can be carried out under the appropriate conditions with any other specific proteolytic enzyme. For example, clostripain digestion was similar to that described for trypsin by substituting the trypsin digestion buffer with the clostripain digestion buffer and trypsin solution with the clostripain solution. In addition, the stop solution may contain EDTA to ensure the complete inhibition of this Ca^{2+}-dependent enzyme.[18]

Analysis of Histone-DNA Contacts Revealed by Cross-Linking

Analysis of histone–DNA contacts revealed by cross-linking involves the determination of position and an estimation of the intensity of the signals in two-dimensional gels. The position of the signals within each diagonal indicates the length (in nucleotides) of cross-linked DNA fragments, which determines the location of a particular histone on the nucleosomal DNA from the 5′ end. This can be determined by running single-stranded DNA fragments of known size in the gel of the second dimension. The intensity of the signal on the autoradiograph indicates the efficiency of cross-linking at that particular DNA site. This can be estimated by scanning the X-ray films and comparing the relative intensities of the signals within the same diagonal. An example of heavy attenuation of certain signals due to the loss of a major cross-linking site of a histone is demonstrated in the trypsin- and clostripain-digested core particles compared with

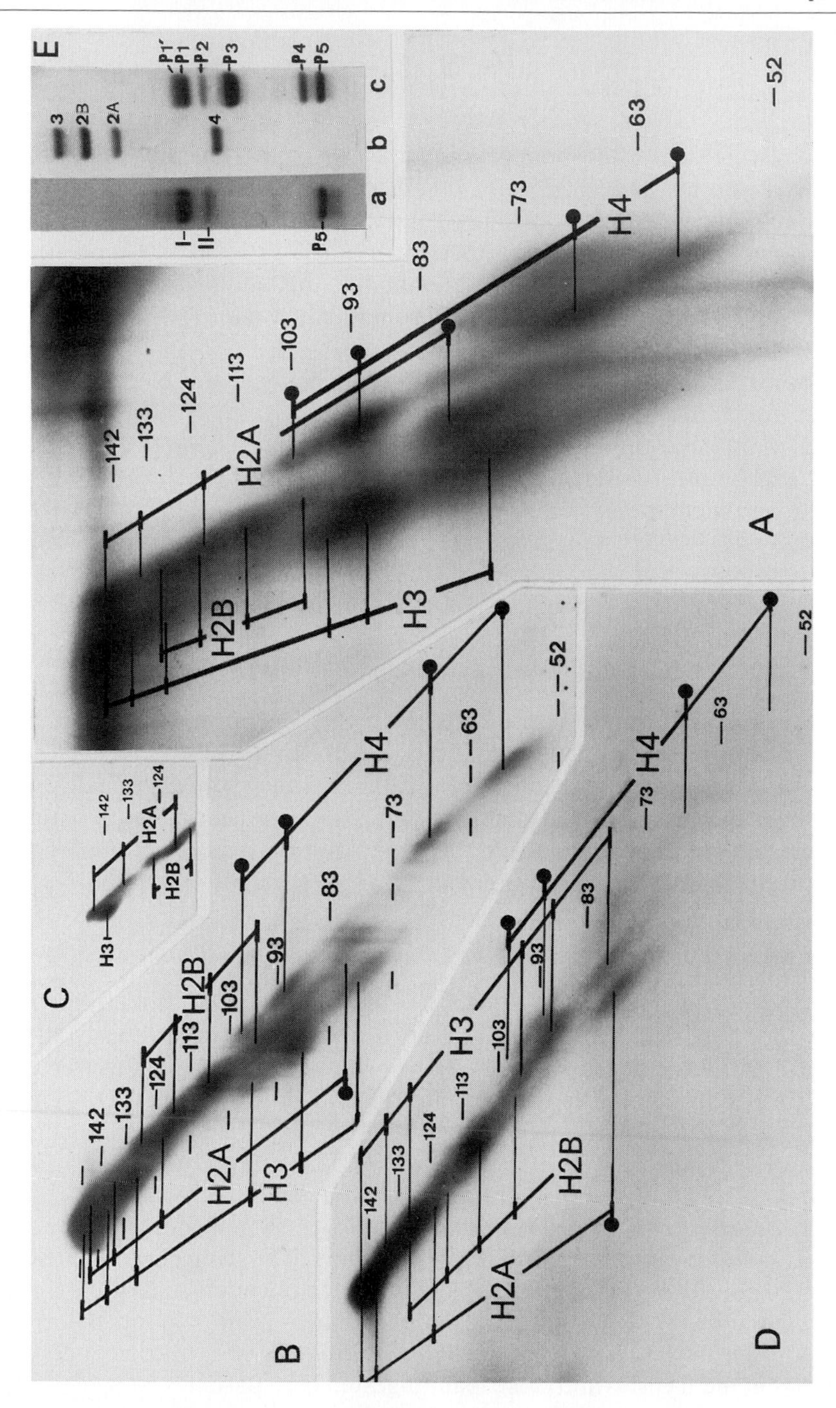
A
B
C
D
E
H2A
H2B
H3
H4
a
b
c
P1
P2
P3
P4
P5
I
II
3
2B
2A
4
142
133
124
113
103
93
83
73
63
52

intact (control) core particles (Fig. 4). Heavy attenuation of the signal corresponding to the peptide of histone H2A around nucleotide 77 was observed for nucleosomes digested with trypsin (Fig. 4B), which removes the histone H2A C-terminal domain containing one of the major cross-linking sites, His123 (Fig. 3). In contrast, this signal did not change in nucleosomes digested with clostripain (Fig. 4D), which leaves the H2A C-terminal domain intact (Fig. 3). Unaltered contacts (e.g., histone H2A around nucleotides 121, 135, and 146) in the trypsin- or clostripain-digested core particles (Fig. 4) indicate the location of the trypsin- or clostripain-resistant central (globular) domain of the histone molecule on the DNA.

Similar heavy attenuation was also observed for signals corresponding to peptides of histone H4. The digestion of nucleosomes with trypsin results in two major peptides of histone H4, peptides P4 and P5,[16] whereas digestion with clostripain results only in peptide P5[18] (Fig. 3). The P4 peptide contains His18, which is the major cross-linking site of histone H4[19] and therefore can be identified by two-dimensional gel-electrophoresis (Fig. 4). However, peptide P5 does not contain this major cross-linking site (Fig. 3) and therefore signals corresponding to this peptide are attenuated heavily in two-dimensional gel electrophoresis (Fig. 4C).

In order to avoid ambiguous interpretation of the results, the cross-linking should be carried out before the enzymatic treatment of nucleosomes because the variation in signal intensity can indicate nucleosome conformational changes[6] as a result of proteolytic digestion. The conditions

[19] K. K. Ebralidse, S. A. Grachev, and A. D. Mirzabekov, *Nature* **331,** 365 (1988).

FIG. 4. Two-dimensional gel electrophoresis of single-stranded ^{32}P-labeled DNA fragments cross-linked to histones in nucleosome core particles. (A) Intact core particle. (B) Trypsin-digested core particle. (C) Shorter exposure of the upper part of B. (D) Clostripain-digested core particle. (E) First-dimensional gel electrophoresis of peptides from clostripain (lane a)- and trypsin (lane c)-digested core particles and histones (lane b) from intact (control) core particles (see text). In two-dimensional gel electrophoresis the first dimension was carried out until the bromphenol blue migrated approximately 1.5 lengths of the separating gel. Positions of the ^{32}P-labeled DNA fragments cross-linked to various histones are arranged on separate diagonals and are indicated by solid lines. The arrangement of the diagonals from left to right in A–D corresponds to the order in which free histones or their peptides migrate from top to bottom in the first-dimensional gel (E). Bands I and II (lane a in E) correspond to peptides of histone H2A/H2B and H3, respectively.[11,18] Peptides P1/P1′, P2, and P3 correspond to histones H3, H2A, and H2B, respectively.[15–17] The dashed lines in A–D show the position in the gel, and the numbers indicate the length of ethidium bromide-stained DNA fragments from DNase I digests of rat liver nuclei. Solid circles indicate the signals of H2A and H4, which are attenuated heavily in trypsin(B)- and clostripain(D)-digested core particles, respectively, compared to intact (control) core particles (A). From S. I. Usachenko, S. G. Bavykin, I. M. Gavin, and E. M. Bradbury, *Proc. Natl. Acad. Sci. U.S.A.* **91,** 6845 (1994), with permission.

for histone–DNA cross-linking in nucleosomes and soluble chromatin have already been described in sufficient detail.[10] However, for cross-linking within intact nuclei, the following precautions must be taken into account.

Histone–DNA Cross-Linking in Nuclei

The main advantage of histone–DNA cross-linking in nuclei is to study histone–DNA contacts in native chromosomes. To prevent changes in nuclear structure and avoid experimental artifacts it is very important to use proper conditions starting with nuclei isolation. The method of dimethyl sulfate cross-linking requires several buffer changes involving the pelleting and resuspension of nuclei. The appropriate procedures, including general hints and troubleshooting for the isolation and handling of each new source of nuclei, have been described previously.[20] In addition, the incubation of nuclei during cross-linking at elevated temperatures (37–45°) for several hours requires efficient protection from endogenous proteases and nucleases. Depending on the source of the nuclei, endogenous proteases can be inhibited by using a cocktail of appropriate protease inhibitors that should first be tested by a simple incubation of the nuclei suspension for 6–8 hr at the temperature used during cross-linking. An alternative way to remove the majority of proteases from any nuclear source is to wash the suspension of nuclei in buffer containing 0.35 *M* NaCl before cross-linking. However, one should keep in mind that some nonhistone proteins can also be extracted during this washing. After this washing the addition of diisopropyl fluorophosphate (DFP) dissolved in dimethyl sulfoxide (DMSO) at 0.5–1 m*M* sufficiently inhibits proteolytic degradation of histone proteins during cross-linking.[5,6]

Another major point of concern during cross-linking within intact nuclei is to preserve nuclei membrane and chromosome integrity. In general, this can be maintained by an appropriate balance of monovalent and divalent cations or polyamines. However, at the elevated temperature during cross-linking, divalent cations activate endogenous nucleases that can introduce nicks in the chromosomal DNA. These nicks release the DNA torsional stress and can affect the chromatin and nucleosome conformation and thereby histone–DNA contacts.[6] In order to prevent the activity of endogenous nucleases during cross-linking, divalent cations should be removed by proper concentrations of EDTA unless some appropriate nuclease inhibitors are included. Because polyamines can be potentially cross-linked to DNA, they should also be excluded. In the absence of divalent cations and polyamines, the nuclear membrane and chromosome integrity can be maintained during cross-linking by using a physiological concentration of

[20] M. Bellard, G. Dretzen, A. Giangrande, and P. Ramain, *Methods Enzymol.* **170,** 317 (1989).

monovalent cations (approximately 150 m*M*). In order to prevent disruption of the nuclei membrane, extreme caution should be followed during centrifugation and resuspension steps. Centrifugation should not exceed 150*g* and resuspension should be started with gentle swirling followed by gentle pipetting with wide-bore pipettes.

Procedure

1. Isolate nuclei using an appropriate procedure for the particular tissue.[20] Bring the nuclei suspension to a concentration no higher than $A_{260} \sim 20$ and maintain this concentration during all washing steps. Wash twice in ice-cold phosphate buffered saline (PBS) containing a final concentration of 0.35 *M* NaCl by gentle resuspension and centrifugation at 150*g* for 5 min at 4°.
2. Wash nuclei as just described twice in regular PBS. If divalent cations were used during the isolation, EDTA (10 m*M*) should be included in the washing buffer.
3. Wash nuclei once and resuspend in 50 m*M* cacodylate–NaOH, pH 7.0, 100 m*M* NaCl, 0.1 m*M* EDTA, 0.5 m*M* DFP.
4. Dissolve $(CH_3O)_2SO_2$ (dimethyl sulfate) in 50 m*M* cacodylate–NaOH, pH 7.2, up to a final concentration of 50 m*M* by extensive vortexing. Add 1/10 volume of this solution to the nuclei suspension, mix gently and thoroughly, and incubate on ice for 18 hr.
5. Wash nuclei twice in PBS containing 1 m*M* EDTA, resuspend in the same solution, add DFP dissolved in DMSO to a final concentration 1 m*M*, and incubate at 45° for 8 hr.
6. Chill on ice, resuspend in 100 m*M* sodium phosphate, pH 7.0, add 1/10 volume of sodium borohydride (10 mg/ml) prepared freshly in ice-cold 100 m*M* sodium phosphate, pH 7.0, and incubate on ice for 30 min in the dark.
7. Dilute the suspension at least two to three times with ice-cold PBS, centrifuge at 150*g* for 5 min at 4° and wash the pellet twice by resuspending in ice-cold PBS.

Cross-linked nuclei prepared with this procedure can be used further for the isolation of chromatin or nucleosomes and subsequent analysis of cross-linked complexes.

Conclusion

The purpose of this article is to provide general guidelines and detail protocols for studying the interactions of specific histone regions with nucleosomal DNA and their involvement in chromatin structure/function relationships. The examples presented here illustrate the applicability of protein–DNA cross-linking and its ability to yield information that cannot be obtained by other techniques, particularly in intact nuclei. This approach can be applied not only to histones but also to the interaction of other proteins with DNA (e.g., different regulating factors) and their possible effects on nucleosome structure. Furthermore, the combination of this approach with other techniques can also be applied to study the structure of nucleosomes containing certain DNA sequences, posttranslational modifications of histones, and different histone subtypes.

Acknowledgments

We are grateful to Jodie L. Usachenko for assistance in the preparation of this manuscript. This work was supported by a grant from the Department of Energy (DE-FG03-88ER60673).

[15] Restriction Enzymes as Probes of Nucleosome Stability and Dynamics

By K. J. Polach and J. Widom

Introduction

Restriction enzymes have several attributes that make them particularly valuable reagents for studies of nucleosome stability and dynamics. (i) Most importantly, restriction enzymes are large in comparison to the diameter of DNA, and in their site-specific complexes they embrace their target sites, wrapping nearly entirely around the DNA circumference. Thus restriction enzymes cannot bind to their DNA target sites at the same instant that those sites are wrapped in nucleosomes. For such enzymes, there is no such thing as a nucleosomal target site that is accessible simply because it faces "out," away from the histone octamer: much of the area of any such site is inevitably occluded by proximity to the histone surface and is further occluded by the adjacent turn or turns of DNA in the same nucleosome. (ii) Restriction enzymes are site specific and therefore allow one to probe accessibility at specific sites while leaving the remainder of the DNA unaffected. By analogy with spectroscopy, they allow one to sensitively detect events on a "black background." An entire population of DNA molecules may be analyzed in the "single hit" regime for accessibility at a particular site. In contrast, when using a relatively nonspecific probe such as DNase I, the great majority of DNA molecules will be destroyed by cleavage

0076-6879/99 $30.00

elsewhere along their length long before cleavage at a particular desired site ever occurs. With such probes, signals obtained from a particular site of interest may come from a tiny fraction of DNA molecules that are not representative of the entire population. (iii) Restriction enzymes can serve as neutral probes, reporting on the existence of spontaneous nucleosome conformational changes that make their DNA target sites transiently accessible rather than driving the formation of such states. This will be true provided that the restriction enzymes are used at (free) concentrations that are small in comparison to their dissociation constants. (iv) Finally, because restriction enzymes are of prokaryotic origin and know nothing of histones or nucleosomes, they eliminate any possibility that unexpected properties specific to eukaryotic transcription factors could possibly account for the accessibility of sites that one thought were buried in nucleosomes.

Routine nucleosome mapping experiments often use restriction enzymes as probes, taking advantage of the inaccessibility of nucleosomal target sites. Sites that are digested readily are considered to be present in linker regions between nucleosomes or in larger nucleosome-free regions, whereas sites that "cannot" be cleaved are considered to possibly be present in nucleosomes (or else in other protein–DNA complexes).

Notwithstanding such routine analyses, we have discovered that sites that are indeed buried in nucleosomes can nevertheless be accessed and cleaved by restriction enzymes. As will be discussed later, this accessibility arises as a consequence of highly frequent but short-lived conformational fluctuations of nucleosomes that expose the wrapped DNA. We term such events "site exposure." In the transiently exposed state, target sites are accessible to restriction enzymes and can be cleaved with a rate that depends on the fraction of time nucleosomes exist in such alternative conformational states: the equilibrium constants for site exposure.

Remarkably, this picture of transient site accessibility, quantified by studies of restriction enzyme digestion kinetics, turns out to be linked theoretically to—and provide a physical explanation for—a set of quite different experiments from many laboratories that investigate the equilibrium binding of various site-specific DNA-binding proteins to nucleosomal target sites. Furthermore, it leads to quantitative prediction of a novel type of cooperativity that arises as a consequence of multiple proteins competing simultaneously against the same histone octamer for binding to nucleosomal target sites.

This article reviews the theoretical analyses of the restriction enzyme digestion experiments and the relationship between the kinetic analyses of restriction enzyme cleavage and equilibrium and cooperative binding studies. It then discusses important issues and controls for experimental studies using restriction enzymes as probes of nucleosome stability and dynamics.

Theory

Site Exposure Model for Restriction Enzyme Digestion of Nucleosomal DNA

Studies using restriction enzymes as probes reveal that nucleosomes *in vitro* exist in a constant dynamic conformational equilibrium, transiently exposing stretches of DNA off their surface so as to make target sites within these DNA stretches accessible to the restriction enzymes[1] (for reviews, see Widom[2,3]).

A formal mechanism consistent with data is illustrated in Fig. 1 and is summarized in Eqs. (1a) and (1b). For simplicity, we make the assumption that sufficient nucleosomal DNA is exposed such that the microscopic rate constants for enzyme action on naked DNA [Eq. (1b)] and on nucleosomal DNA after site exposure [Eq. (1a)] are identical. (At some level, proximity to the histones and other effects cannot be ignored; for example, as will be discussed later, this introduces coupling in the effects of rotational and translational positioning on protein binding. However, as will be seen, this simplifying assumption leads to results that are in broad agreement with results in the literature from very different kinds of experiments, implying that the assumption is valid for real nucleosomes.) Thus

$$N \underset{k_{21}}{\overset{k_{12}}{\rightleftharpoons}} S + E \underset{k_{32}}{\overset{k_{23}}{\rightleftharpoons}} ES \xrightarrow{k_{34}} E + P \tag{1a}$$

and

$$S + E \underset{k_{32}}{\overset{k_{23}}{\rightleftharpoons}} ES \xrightarrow{k_{34}} E + P \tag{1b}$$

for nucleosomal and naked DNA, respectively, where N is the starting nucleosome, S is fully accessible DNA, E is the restriction enzyme, and P are the products of digestion.

Restriction Enzymes as Quantitative Probes for Site Exposure

Restriction enzymes allow for the qualitative detection of site exposure by the observation of cleavage of sites that are known to be present in nucleosomes. Additionally, however, restriction enzymes allow the quanti-

[1] K. J. Polach and J. Widom, *J. Mol. Biol.* **254,** 130 (1995).
[2] J. Widom, *Annu. Rev. Biophys. Biomol. Struct.* **27,** 285 (1998).
[3] J. Widom, *Methods Mol. Biol.,* in press (1999).

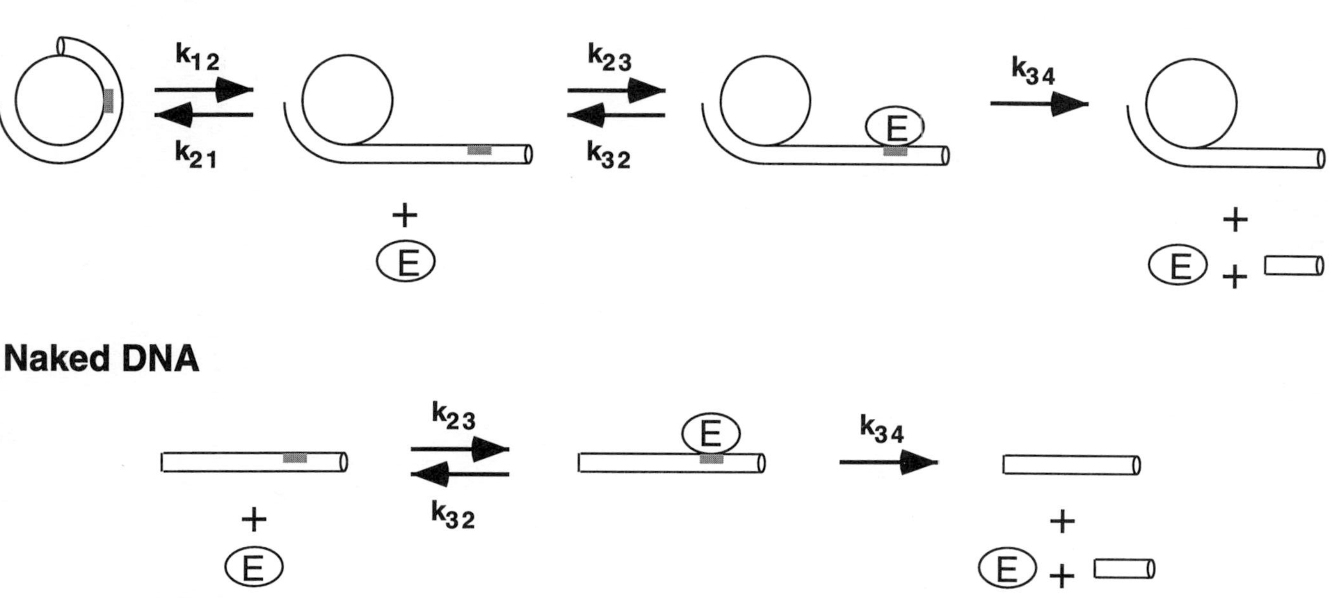

FIG. 1. Site exposure model for the action of a restriction enzyme (E) on a nucleosomal target site (shaded) compared to its action on naked DNA. Restriction enzymes embrace nearly the entirety of the circumference of their DNA target sites, thus target sites present in nucleosomes are accessible only after the DNA is displaced off the histone surface. Real nucleosomes exhibit this behavior, transiently uncoiling stretches of their DNA. The exposed stretches behave similarly to naked DNA, interacting with the restriction enzyme with comparable binding (k_{23}, k_{32}) and catalytic (k_{34}) rate constants. Site exposure is nondissociative and occurs as a rapid preequilibrium; available data suggest that innermost sites are reached by successive uncoiling from the near end.

tative measurement of the equilibrium constant for the site exposure process, K_{eq}^{conf} and its corresponding free energy ΔG_{conf}^{0},

$$K_{eq}^{conf} = k_{12}/k_{21} = e^{-\Delta G_{conf}^{0}/RT} \tag{2}$$

Experimentally, one monitors loss of reactant nucleosomal DNA (D)

$$(D) = (N) + (S) \tag{3}$$

which disappears according to a first-order rate law with an observed rate constant k_{obs},

$$k_{obs} = \frac{-1}{(D)} \frac{d(D)}{dt} \tag{4}$$

Making the steady-state approximation for (S) and for (ES), one obtains

$$k_{obs} = \frac{k_{34}(E)}{K_m} \frac{k_{12}}{k_{12} + k_{21} + \dfrac{k_{34}E}{K_m}} \tag{5a}$$

with

$$K_m = \frac{k_{32} + k_{34}}{k_{23}} \tag{5b}$$

Digestions proceed in the rapid preequilibrium limit[1]

$$k_{21} \gg k_{23}(E) \tag{6}$$

for which Eq. (5) reduces to

$$k_{obs} = \frac{k_{34}(E)}{K_m} \frac{K_{eq}^{conf}}{1 + K_{eq}^{conf}} \tag{7}$$

Studies on nucleosomes reveal that K_{eq}^{conf} varies with location but is always $\ll 1$. If, in addition, experiments are set up such that $(S) \ll K_m$ so that $(E) \approx (E_0)$ (the total concentration of added restriction enzyme), then the observed first-order rate constant for the loss of reactant nucleosomal DNA [Eq. (7)] becomes

$$k_{obs}^{nucleosome} = \frac{k_{34}(E_0)}{K_m} K_{eq}^{conf} \tag{8}$$

If, in separate experiments, naked DNA is digested under identical solution conditions [but possibly with different (E_0)], the reactant naked DNA will disappear with an apparent first-order rate constant given by

$$k_{\text{obs}}^{\text{naked DNA}} = \frac{k_{34}(E_0)}{K_{\text{m}}} \tag{9}$$

Combining Eqs. (8) and (9) yields

$$K_{\text{eq}}^{\text{conf}} = \frac{k_{\text{obs}}^{\text{nucleosome}}/(E_0^{\text{nucleosome}})}{k_{\text{obs}}^{\text{naked DNA}}/(E_0^{\text{naked DNA}})} \tag{10}$$

Thus, one can use restriction enzymes to obtain an experimental measurement of $K_{\text{eq}}^{\text{conf}}$ from the ratio of the two observed rate constants scaled by their respective enzyme concentrations.

Structural Model for Site Exposure

Site exposure appears to occur progressively from an end. Experimentally measured equilibrium constants for site exposure decrease more or less progressively as one moves in from one end, with values of $\sim 10^{-2}$ for sites just inside from one end to $\sim 10^{-5}$ for sites directly over the nucleosomal dyad.[1] Additional evidence consistent with uncoiling from an end is the agreement between the predictions of the site exposure model and experimental observations of cooperative equilibrium binding.[4] The structure of the nucleosome core particle itself[5] is consistent with a picture in which DNA spontaneously and transiently uncoils inward starting from an end.[6] The structure shows the DNA wrapped on the histone surface as making contacts ("bonds") in small patches, approximately every 10 bp, each time that the phosphodiester backbone (minor groove) faces inward toward the octamer. Thus, uncoiling would naturally proceed stepwise, with incremental increases in energetic cost (decreased equilibrium constant for site exposure) associated with each additional ~10-bp segment uncoiled.

Site Exposure: Spontaneous and Rapid Equilibrium

Two different theoretical models, assuming either diffusion-limited or activated processes, lead to the expectation that the "recapture" of transiently exposed DNA will occur with a rate $k_{21} >$(or $\gg$) $10^5\ \text{sec}^{-1}$ (J. Widom, unpublished results, 1996). Given the measured equilibrium constants $K_{\text{eq}}^{\text{conf}}$, this implies that the forward rate of site exposure, k_{12}, will be $>$(or $\gg$) $10^3\ \text{sec}^{-1}$ for exposure of the outermost stretches. Exposure at

[4] K. J. Polach and J. Widom, *J. Mol. Biol.* **258,** 800 (1996).

[5] K. Luger, A. W. Mader, R. K. Richmond, D. F. Sargent, and T. J. Richmond, *Nature* **389,** 251 (1997).

[6] J. Widom, *Curr. Biol.* **7,** R653 (1997).

sites further inside the nucleosome would occur with a reduced overall rate, although if site exposure does indeed occur stepwise as suggested by the structure, each individual successive segment could uncoil with a microscopic rate constant comparable to that for the outermost segment.

Experimentally, as will be discussed later, we find that digestion by restriction enzymes occurs in a rapid prequilibrium limit as implied by the mechanism in Fig. 1 and Eq. (1). Digestion by restriction enzymes is carried out on the minutes to tens of minutes time scale. Studies using other enzymes that also require prior site exposure for their action[7,8] reveal that site exposure sufficient to allow access to the entirety of the nucleosomal DNA occurs on a time scale of seconds or less (possibly much faster).

Site Exposure Model Linking Restriction Enzyme Digestion Kinetics to Simple Equilibrium-Binding Reactions

Measurements of the equilibrium constant for site exposure obtained from the analysis of restriction enzyme digestion kinetics are linked thermodynamically to measurement of the dissociation constant for simple equilibrium binding of proteins to nucleosomal target sites.

Figure 2a illustrates an example in which a site-specific, DNA-binding regulatory protein *R* can bind to a nucleosomal target site if and only if that site is first displaced away from the surface of the histone octamer. We again suppose that sufficient DNA is (transiently) exposed in the nucleosome so as to allow for identical interactions of R with the site-exposed nucleosomal DNA as with naked DNA. Then,

$$N \underset{k_{21}}{\overset{k_{12}}{\rightleftharpoons}} S + R \underset{k_{32}}{\overset{k_{23}}{\rightleftharpoons}} RS \tag{11a}$$

and

$$S + R \underset{k_{32}}{\overset{k_{23}}{\rightleftharpoons}} RS \tag{11b}$$

for binding to a nucleosomal target and naked DNA, respectively. Binding of a regulatory protein to a nucleosomal target sequence will then occur with a net free energy change

$$\Delta G^0_{\text{net}} = \Delta G^0_{\text{conf}} + \Delta G^0_{\text{naked DNA}} \tag{12}$$

[7] R. U. Protacio, K. J. Polach, and J. Widom, *J. Mol. Biol.* **274,** 708 (1997).
[8] J. Widom, *Science* **278,** 1899 (1997).

in which $\Delta G^{\circ}_{\text{naked DNA}}$ is the free energy change for process [Eq. (11b)] and $\Delta G^{\circ}_{\text{conf}}$ is the same free energy cost for site exposure measured from the studies of restriction enzyme digestion kinetics

$$\Delta G^{0}_{\text{conf}} = -RT \ln K^{\text{conf}}_{\text{eq}} = -RT \ln (k_{12}/k_{21}) \tag{13}$$

Expressed in equilibrium constants, a site-specific, DNA-binding protein would bind to a nucleosomal target sequence with an apparent dissociation constant

$$K^{\text{apparent}}_{\text{d}} = K^{\text{naked DNA}}_{\text{d}} / K^{\text{conf}}_{\text{eq}} \tag{14}$$

where $K^{\text{naked DNA}}_{\text{d}}$ is the dissociation constant for binding to naked DNA ($= k_{23}/k_{32}$), and $K^{\text{conf}}_{\text{eq}}$ is the same equilibrium constant for site exposure measured from studies of restriction enzyme digestion kinetics.

Real nucleosomes behave as described by this model and these equations.[1,4] Thus, measurements of equilibrium constants for site exposure based on restriction enzyme digestion kinetics can be used to *predict* the results of simple equilibrium-binding studies; conversely, measurements of apparent dissociation constants for binding to nucleosomal target sites (together with measurements on naked DNA) can be used to predict the results of studies probing site exposure equilibria using restriction enzyme digestion kinetics.

Alternatively, the experimental truth that one finds good agreement between the results from these two very different experiments, which are essentially unrelated to each other except through the site exposure model, provides strong experimental evidence that the nucleosomes are in fact behaving as described by this model.

System-Specific Effects

Despite the generally good agreement between the results from these two very different assays, one may nevertheless anticipate that system-specific, second-order effects exist, which may lead to discrepancies between results from the two experiments. Similarly, one may anticipate the existence of discrepancies between results obtained using either one of these methods with two different probe molecules (restriction enzymes or binding proteins) monitoring accessibility at the same or closely spaced sites.

Figure 2b illustrates a hypothetical case in which the binding of a protein to a nucleosomal target site can occur with no requirement for prior site exposure, simply because the binding protein is small and can bind its (correspondingly small) site with no requirement for site exposure when the site happens to face "out." (Note that because nucleosomes are mobile in physiological conditions, no site can be counted on to face "out" all of

a

Nucleosomes

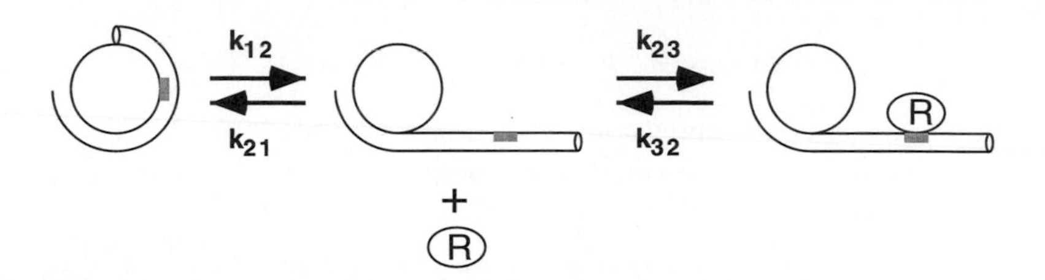

Naked DNA

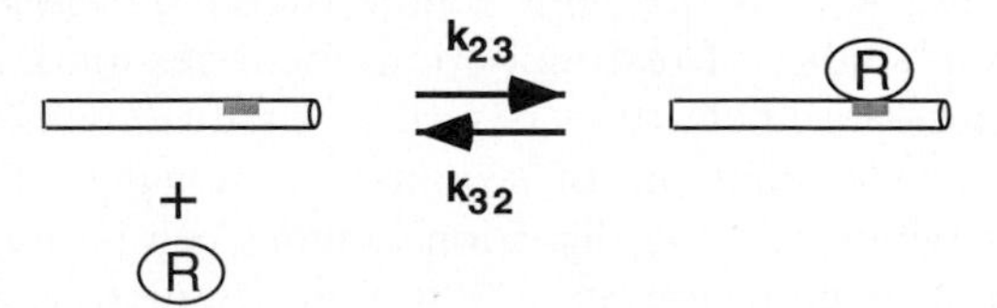

b

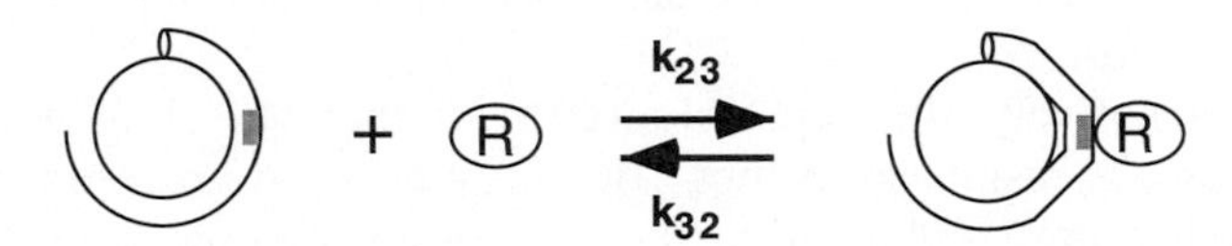

c

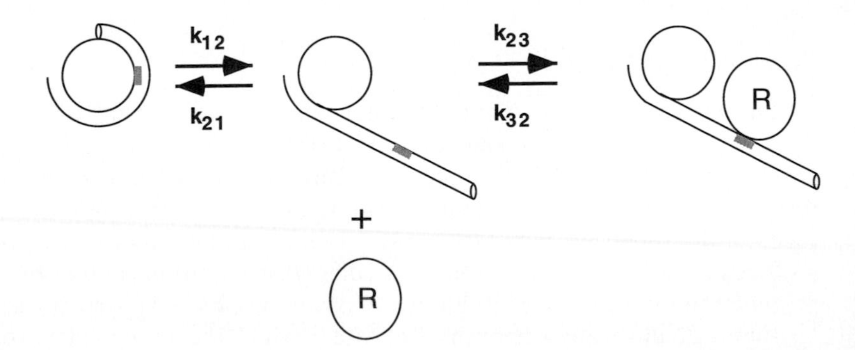

the time; positioning is statistical, not "precise."[2,9,10] Thus, such scenarios do not represent mechanisms for guaranteeing the accessibility of binding sites *in vivo*.) If binding to this site also requires no significant deformation of the DNA, the binding affinity would closely approximate that for naked DNA, i.e., differing from the prediction of the site exposure model by the large factor K_{eq}^{conf}. If binding required a conformational change in the DNA site, as illustrated in Fig. 2b, the binding affinity would be reduced by an amount corresponding to the energetic cost of deforming the DNA. In either case, the site exposure model does not describe the binding event.

Figure 2c illustrates an opposite hypothetical case in which a binding protein (or restriction enzyme) is particularly large or has a particularly unfavorable shape, requiring that substantially more DNA be uncoiled off the histone surface in order for the target site to be sterically accessible. As described earlier, increased uncoiling will generally be accompanied by an increased cost in free energy and will therefore occur with a decreased equilibrium constant for site exposure and a correspondingly suppressed binding affinity or rate of restriction enzyme digestion.

[9] P. T. Lowary and J. Widom, *Proc. Natl. Acad. Sci. U.S.A.* **94,** 1183 (1997).
[10] P. T. Lowary and J. Widom, *J. Mol. Biol.* **276,** 19 (1998).

FIG. 2. (a) Site exposure model for equilibrium binding of a protein (*R*) to a nucleosomal target site compared to binding to naked DNA. Many DNA-binding proteins, like restriction enzymes, contact a large fraction of the DNA circumference at their target sites. Such proteins will be able to bind only after site exposure events make their sites accessible. Exposed DNA behaves similarly to naked DNA in its interactions with *R*. (b) A hypothetical case in which the binding of a protein to a nucleosomal target site can occur with no requirement for prior site exposure, simply because the binding protein is small and can bind to its correspondingly small DNA target site with no requirement for site exposure when the target site happens to face "out." A limiting case is a protein that is small enough to bind with no requirement for site exposure and that binds to a DNA target that has the same local DNA helical parameters as the DNA in that region of the nucleosome. In this case, the affinity for the nucleosomal target site will be identical to that for naked DNA. More generally, however, even very small proteins are likely to require at least some distortion of their DNA target site on binding (e.g., as indicated by the distortions). In this case, binding to the nucleosomal target site will be suppressed by a factor corresponding to the free energy cost of deforming the DNA away from its favored nucleosomal conformation, even though no site exposure is required for the binding to occur. (c) An opposite hypothetical case in which a binding protein (or restriction enzyme) is particularly large, has a particularly unfavorable shape, or is accompanied by a particularly unfavorable deformation of the DNA. In these cases, substantially more DNA than would otherwise be required (e.g., as in a) must be uncoiled off the histone surface in order for the target site to be sterically accessible. Site exposure appears to proceed progressively inward from an end, with increasing energy penalties for increased uncoiling; thus proteins requiring site exposure of greater lengths will have their affinity for nucleosomal target sites suppressed further.

These potential exceptions do not refute this model for the general behavior of nucleosomes. Rather, they simply represent second-order corrections that are inevitably required to account for system-specific details in most physical models.

Site Exposure Model Predicting Arbitrary DNA-Binding Proteins, including Restriction Enzymes, Binding Cooperatively to Sites within Same Nucleosome

The site exposure model has within it a surprising prediction, which is quantitatively upheld by real nucleosomes *in vitro*[4]: if a nucleosome contains sites for two or more DNA-binding proteins, the binding of these proteins may occur cooperatively, with each protein facilitating the binding of the other. This cooperativity occurs as a consequence of mechanical coupling between binding events at two distinct sites, as illustrated in Fig. 3. It is manifested as an increase in the affinity for X in the presence of bound Y, and a linked increase in the affinity for Y in the presence of bound X. Importantly, this novel cooperativity is distinct from, and occurs in addition to, and "conventional" cooperative interactions between X and Y that may also obtain. This novel cooperativity is, in general, substantially larger than previously recognized conventional cooperativities.[4]

Let δG_{XY} equal the amount by which the prior binding of X facilitates the binding of Y, which necessarily also equals the amount by which the prior binding of Y facilitates the binding of X. Our analysis reveals that δG_{XY} (measured in kcal mol^{-1}) is exactly equal to minus one times the energetic cost of exposing the outermost site,

$$\delta G_{XY} = -\Delta G_1^0 \qquad (15)$$

which in turn equals the measured $-\Delta G_{conf}^0$ for the binding of X when there only is one site.

While our previous analysis of cooperativity focused on "coupled binding" studies, i.e., on coupled chemical equilibria describing the cooperative equilibrium binding at pairs of sites (those for X and Y as illustrated), the analysis is expected to apply equally to cases in which one of the two proteins is a restriction enzyme. In this case, cooperativity would be manifested (in the limit of low enzyme concentrations) as an increased accessibility of nucleosomal target sites. There will be a new observed rate constant k_{obs}^{coop} for digestion at nucleosomal target sites with this cooperativity manifested, given by

$$k_{obs}^{coop} = k_{obs} \times e^{-\delta G_{XY}/RT} \qquad (16)$$

where k_{obs} is the rate observed for cleavage at the nucleosomal target site in the absence of other binding interactions [Eq. (7)].

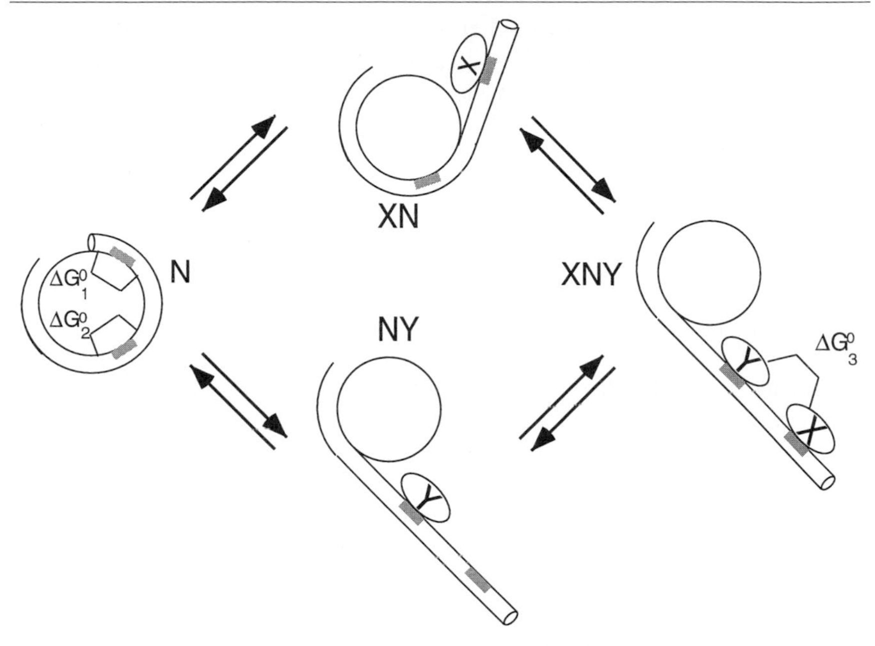

FIG. 3. Cooperativity in the binding of multiple proteins to target sites in a single nucleosome. A nucleosome is shown containing binding sites (shaded) for two proteins: X and Y. X and Y may be two unrelated proteins or two molecules of the same protein. X is defined as the protein binding to the outermost site, and Y is defined as binding to the innermost site. ΔG_1 is the free energy cost for uncoiling enough DNA so as to expose the site for X. ΔG_2 is the additional free energy cost for uncoiling sufficient additional DNA so as to expose the site for Y. In some cases, X and Y may have "conventional" cooperative interactions, which are also detectable in their binding to naked DNA (e.g., from favorable protein–protein contacts between X and Y); these are represented collectively as ΔG_3. Even in the absence of conventional cooperative interactions, the prior binding of X facilitates the binding of Y and vice versa, both changing the free energy for binding by the same amount, given by $\delta G_{XY} = -\Delta G_1^0$. If either X or Y happens to be a restriction enzyme, then binding of the other protein to near saturation will increase the time-averaged accessibility of the target site of the restriction enzyme and hence the rate of digestion by the factor $\exp(-\delta G_{XY}/RT)$.

Thus, for example, suppose that X (referring to the protein binding to the outermost site) in Fig. 3 represents a particular restriction enzyme and that Y is an ordinary site-specific, DNA-binding protein. Then, if binding of Y is driven to near saturation by the addition of sufficient [Y], the model predicts that the rate of cleavage at the X site will be increased by the factor $\exp(\delta G_{XY}/RT)$, at which point it will approximate the rate obtained for naked DNA in identical conditions. Note that in this case, $K_{eq}^{conf} \gg 1$ and Eqs. (8)–(10) are not valid.

Conversely, suppose that Y (which acts at the innermost site) represents a particular restriction enzyme, whereas X is an ordinary site-specific, DNA-binding protein. Then, if binding of X is driven to near saturation by the addition of sufficient [X], the model predicts that the rate of cleavage at the Y site will still be suppressed relative to that for naked DNA, but will be increased above its initial value in the nucleosome by the same factor as for the converse case, $\exp(\delta G_{XY}/RT)$.

Experimental Methods

Design of DNA Templates

The ideal substrate for assays probing nucleosome dynamics is a homogeneous population of particles with well-defined translational and rotational positions. This is best achieved by (i) using a strong nucleosome positioning sequence to bias the positioning to one particular location[9,10] and (ii) keeping the overall length near the minimum 147 bp present in the core particle (which discourages alternative positionings because these would come at a cost of unsatisfied DNA-binding sites on the histone octamer). Our designs utilized the positioning sequence from the sea urchin 5S RNA gene, first characterized by Simpson and colleagues.[11,12] This sequence was used as a template in preparative-scale polymerase chain reactions (PCR). The total length was kept at 150 bp. Desired restriction enzyme sites were incorporated into this sequence using PCR primers containing appropriate base changes. While structural considerations suggest that site exposure will always be required to allow the access of restriction enzymes to nucleosomal target sites (see earlier discussion), we also addressed this issue experimentally, as follows. Rather than incorporating isolated single sites, we nested together several sites within a small patch of about one DNA helical turn in length. This allows the measurement of site exposure at sites all around the periphery of the DNA. If, in separate reactions, one probes distinct sites within the patch and discovers that in each case all of the molecules participate in the reactions and yield comparable apparent equilibrium constants for site exposure, this discriminates real exposure of an entire patch from the possibility that a single site happened to face "out" notwithstanding our structural argument.

Conditions of PCR reactions were as follows: template DNA (50 ng ml^{-1}), *Tfl* polymerase buffer [50 m*M* Tris–HCl (pH 9.0), 20 m*M* ammonium sulfate, 1.5 m*M* $MgCl_2$], primer oligonucleotides (1 μM each), dNTPs (200 μM each), and *Tfl* DNA polymerase [5 units enzyme (Epicentre Technologies, Madison, WI) per 100-μl reaction]. PCR reactions are optimized for

[11] R. T. Simpson and D. W. Stafford, *Proc. Natl. Acad. Sci. U.S.A.* **80,** 51 (1983).
[12] P. C. FitzGerald and R. T. Simpson, *J. Biol. Chem.* **260,** 15318 (1985).

yield and purity (as judged on agarose gels stained with ethidium) by adjusting annealing temperatures and total number of cycles. PCR products are high-performance liquid chromatography (HPLC) purified on a Mono Q HR5/5 anion-exchange column (Amersham Pharmacia Biotech, Piscataway, NJ) using a linear gradient of 0.65 *M* NaCl in TE (10 m*M* Tris, 1 m*M* EDTA, pH 8.0, at room temperature) to 0.8 *M* NaCl in TE over 90 min at a flow rate of 0.25 ml min^{-1}. The purified material is concentrated on Centricon 30 filters (Amicon, Danvers, MA) and resuspended in 0.1× TE. Final concentrations are determined from absorbance measurements at 260 nm; typical yields were 100 μg of DNA from 4.8 ml of PCR reactions (48 100-μl reactions), after HPLC purification.

Core Particle Reconstitution and Purification

Core particles were reconstituted from purified histone proteins[13] and DNA constructs via gradual salt dialysis.[1] The construct of interest is labeled radioactively with ^{32}P and is divided in half for use as naked DNA and reconstitution into core particles. Reconstitution reactions (300 μl) contained 100 ng of labeled construct DNA, 19.2 μg of chicken erythrocyte core particle DNA, 15.5 μg of purified chicken erythrocyte histone octamer (0.8 : 1 molar ratio), 2 *M* NaCl, 1 m*M* Tris (pH 8.0), 0.1 m*M* EDTA, 0.5 m*M* phenylmethylsulfonyl fluoride (PMSF), and 1 m*M* benzamidine (BZA). Prior to loading the reconstitution reactions into dialysis tubing, the tubing is soaked and rinsed extensively in buffer containing 2 *M* NaCl. The reactions are dialyzed successively (minimum 2 hr per step, 4°) into 2.0, 1.5, 1.0, and 0.5 m*M* NaCl followed by an overnight dialysis in 5 m*M* NaCl. All buffers contain 0.5× TE, 0.5 m*M* PMSF, and 1 m*M* BZA.

Reconstituted core particles are purified away from naked DNA and nonnucleosomal aggregates on 5–30% (w/v) sucrose gradients in 0.5× TE. Gradients are spun in a Beckman SW41 rotor at 41,000 rpm for 24 hr, fractionated into 0.5-ml fractions, and quantified by Cerenkov counting. Peak core particle fractions are pooled and concentrated on Centricon 30 microconcentrators and resuspended in 300 μl of 0.5× TE/5 m*M* NaCl. Final concentrations of purified core particles are calculated from absorbance measurements at 260 nm and are typically greater than 50% of the input material.

Restriction Enzyme Digestion Assays

Assays are carried out on naked DNA and reconstituted core particles in parallel so that digestion rates can be compared directly. DNA digestions utilized 100- to 1000-fold lower enzyme concentrations than used for core

[13] H.-P. Feng, D. S. Scherl, and J. Widom, *Biochemistry,* **32,** 7824 (1993).

particle digestions so that they would exhibit similar rates of digestion. Reactions are set up as follows, with the goal of equilibrating them at the desired temperature prior to diluting the enzyme and adding enzyme to substrate. For each substrate, two 50-μl solutions are prepared: one containing the substrate and the other buffer to which enzyme is later added. These solutions are preincubated at the digestion temperature for 5 min, at which point the enzyme is added to the solution containing no substrate. Any necessary dilutions of the enzyme are first made into appropriate buffers over ice. The digestion assay is then initiated by mixing the two half-reactions at the desired temperature. Aliquots are removed as a function of time and quenched on ice in formamide gel-loading buffer (50% formamide, 10 m*M* EDTA, 0.05% xylene cyanol). Each aliquot is then digested (in the formamide) with proteinase K (250 μg ml^{-1}) at 37° for 2 hr.

Accurate determination of digestion rates requires that a significant fraction of the labeled core particles participate in the reaction. As described earlier this is also important for formally proving that digestion is occurring at an exposed patch, rather than at a site that faces out in a subpopulation of particles. For these reasons, digestions should be tailored to each individual enzyme in terms of ionic strength and buffer conditions so that optimum enzyme activity may be achieved. Available enzyme stock concentrations, activity, and cost should therefore be considered in construct design. It is important that digestion conditions are balanced precisely between DNA digestions, using diluted enzymes and core particle digestions, which require concentrated enzyme. Any differences in composition between storage and dilution buffers should be balanced in the digestion. The total concentration of glycerol in the reactions should not exceed 5%, establishing a practical upper limit for enzyme concentration at 10% of the total reaction volume.

Quantitative Analysis of Digestion Kinetics

Digestion products are resolved on denaturing acrylamide gels of appropriate percentage for the product lengths generated. Gels are imaged and results are quantified by a phosphorimager as follows. When adequate resolution is obtained among the digestion products and substrate, all could be quantified separately with appropriate backgrounds subtracted. The fraction uncut is calculated as (counts in substrate)/(counts in substrate + product 1 + product 2). When probing sites near one end of the construct, it is difficult to obtain resolution between the substrate and the larger product without losing the smaller product from the gel. In this case, the shorter product is resolved from the substrate and longer product, and the fraction uncut is calculated as (counts in substrate + longer product) − (counts in shorter product)/(counts in substrate + longer product + shorter product). Although this calculation alleviates the need for difficult resolutions, a problem can arise if the two DNA ends are not labeled with the same efficiency. Unequal labeling can lead to nonzero baselines in the kinetic plots. Baselines can be corrected by independent measurement of DNA digestions carried to completion. This experimentally measured baseline is used in the least-squares fit of data. Both of these calculations are insensitive to variations in gel loading.

Plots of fraction uncut versus time revealed first-order digestion kinetics for most enzymes as expected. However, problems emerged that led to multiphasic kinetics in some assays. This behavior suggests two possible sources: heterogeneity in the substrate, possibly attributable to distinct subpopulations with different site exposure characteristics; or a loss of enzyme activity during the course of the assay. In all cases where multiphasic kinetics were observed with core particle digestions, digestions on naked DNA demonstrated the same behavior, suggesting that the problem is due to the enzyme, not the substrate. In these cases, initial rates were determined from the early times in the digestion where the plots were linear. Because of this problem, it is best to use siliconized tubes and tips for sample manipulation and to include bovine serum albumin in all reactions. High-temperature enzymes typically demonstrate more consistent activity than 37° enzymes. Rate constants obtained from each exponential decay curve defined k_{obs}.

Equilibrium Constants

Equilibrium constants for site exposure were calculated from rates for digestion of naked DNA and reconstituted core particles according to Eq. (10). The initial assays utilized high-temperature enzymes, and results from a number of different temperatures were compiled for analysis. Equilibrium constants calculated from these data showed no indication of systematic temperature dependence beyond random fluctuations due to experimental error. Average values were calculated along with standard deviations for each individual site, from multiple measurements over this modest temperature range.

Equilibrium constants for site exposure differ significantly from zero and can be characterized as follows. For sites just inside the protected nucleosome sequence, values for K_{eq}^{conf} varied from 1 to 4×10^{-2} for a set of sites spanning one full turn of the DNA helix. This is consistent with exposure of a stretch of DNA from the octamer surface rather than cleavage at an individual outward-facing site. For sites 30 to 40 bp in from the end, values for K_{eq}^{conf} varied from 5×10^{-4} to 3×10^{-3}. Sites near the dyad axis of symmetry revealed values for K_{eq}^{conf} of 10^{-4} to 10^{-5}. Thus, restriction enzyme assays reveal a progressive trend of decreasing equilibrium accessibility with distance inward from the core particle end.

Additional Experimental Concerns and Controls

Tests for Rapid Preequilibrium

The rapid preequilibrium limit is expected on theoretical grounds. The available concentrations of restriction enzyme are unlikely to exceed 100 n*M*. For a diffusion-controlled encounter between a target DNA sequence and an enzyme active site, one expects $k_{23} \leq 10^8\ M^{-1}\ \text{sec}^{-1}$, hence we expect $k_{23}(E_0) \leq 10\ \text{sec}^{-1}$. In contrast, simple theoretical models for the site exposure and recapture process, assuming either an activated or a diffusive process for recapture, lead to the expectation that $k_{21} \gtrsim 10^5\ \text{sec}^{-1}$ (see earlier discussion), in which case $k_{21} \gg k_{23}(E_0)$ and digestions will proceed in the rapid equilibrium limit.

An alternative limiting behavior is formally possible in which k_{12} is rate limiting and the rate of digestion simply monitors the direct rate of site exposure (or perhaps nucleosome dissociation).

The two limits are readily distinguished by the dependence of k_{obs} on (E_0). In the expected rapid preequilibrium limit, the dependence of k_{obs} is first order in (E_0), whereas in the alternative (opening-limited) mechanism the dependence is zero order. All assays performed demonstrated qualitatively that k_{obs} depended linearly on the total concentration of added enzyme. Further quantitative measurements of k_{obs} as a function of enzyme concentration demonstrated clear first-order dependence, in that data were well described by a linear fit, which passed through the origin. Analysis of the exponent that related k_{obs} to (E_0) supported this conclusion. For any particular small range of (E_0), we may write

$$k_{obs} = a_1(E_0)^{a_2} \tag{17}$$

The value of a_2 varies between the extreme values of 1 and 0 as the mechanism progresses from the rapid preequilibrium limit to the opening-limited regime. We determined a_2 by plotting $\ln(k_{obs})$ versus $\ln([E_0])$ and obtained the slope $a_2 = 1.07$, confirming that the system obeys the expected rapid preequilibrium limit. Thus, the ratios of cleavage rates yield the desired equilibrium constants K_{eq}^{conf}.

Contamination by Free DNA

We have emphasized the necessity of having homogeneous substrates for the accurate interpretation of digestion results. A common contaminant of unpurified reconstituted core particles is naked DNA. In restriction enzyme digestion assays, this contamination leads to digestion kinetics that are biphasic, consisting of a fast phase for naked DNA digestion and a slow phase for digestion of the core particles. This can be problematic if naked DNA constitutes a large fraction of the total substrate. One could simply measure the rate of digestion for the slower phase and attribute this to the digestion of nucleosomes. However, significant levels of naked DNA can mask other problems that may be occurring in the digestion, such as a loss of enzyme activity. Instead of biphasic kinetics, multiphasic kinetics will be obtained and it will be impossible to distinguish one effect from the other.

Contaminating naked DNA is removed easily by purification on 5–30% sucrose gradients. Native gel analysis of particles purified in this way typically shows less than 3% remaining DNA, an amount too small to contribute to the digestion kinetics in any significant way. Even this small amount may be artificial in that it may be generated by the electrophoresis process itself. In any case, digestions of material at this level of purity should display first-order kinetics; the failure to do so is indicative of other problems.

Stability of Nucleosomes during Assays

The time-dependent digestion of nucleosomal DNA could arise from disruption of the native nucleosome structure and the subsequent digestion of naked DNA [although the dependence of the rate on (E_0) argues against this; see earlier discussion]. To test this possibility further, core particles were incubated in digestion buffer at various temperatures for up to 1 hr, the longest digestion time explored in the assays. Samples were then loaded onto 5% native acrylamide gels running at 15 V cm^{-1} in 1/3× TBE along with a 100-bp DNA ladder as size standards. Even at a digestion temperature of 65°, there was a minimal loss of intact core particles as judged by an electrophoretic mobility shift assay, consistent with the known stability of nucleosomes at such elevated temperatures and ionic strengths.[14] Remarkably, we found that even actual digestion of nucleosomes at a site near the outer end of the core particle sequence led to a negligible loss of nucleosome organization.[1] We conclude that the mechanism of site exposure observed in the digestion assays is independent of core particle disruption.

Substrate Concentrations

Measurement of K_{eq}^{conf} according to Eq. (10) requires that two experimental conditions are met: (1) that K_{eq}^{conf} is small ($\ll 1$), as our results show

[14] J. Bashkin, J. J. Hayes, T. D. Tullius, and A. P. Wolffe, *Biochemistry* **32,** 1895 (1993).

it is, and (2) that [S] $\ll K_m$, so that $[E] \approx [E_0]$. Typical K_m values for restriction enzymes are 1–10 nM, so naked DNA digestions are set up with [DNA] $\ll$ 1 nM. For particle digestions, the total concentration of sites, including sites present statistically in the nucleosomes assembled on carrier DNA, may approach several nanomolars, but these sites are exposed with probabilities of 10^{-2} or less, so that again $[S] \ll K_m$. Thus, we can be confident that the concentration of free enzyme is approximately equal to the concentration of the total enzyme.

Restriction Enzyme Concentrations

If enzyme concentrations were high enough, binding to nucleosomal DNA could drive the system toward site exposure, yielding artifactually high measurements of K_{eq}^{conf}. However, our measurement of the exponent $a_2 = 1.07$ [(Eq. (17)] provides formal proof that our experiments were carried out in the low (E_0) limit. This is demonstrated further by the fact that the equilibrium constants measured were explicitly independent of (E_0).

A separate concern is that nonspecific binding by the restriction enzymes could enhance site exposure, as discussed earlier for cooperative binding in general [Eq. (16) and accompanying discussion]. This behavior is unexpected because the enzyme concentrations used should be small in comparison to typical dissociation constants for nonspecific binding (which may also be further suppressed for the nucleosomal DNA). We tested for this possibility experimentally as follows. Two parallel digestions were set up on reconstituted core particles: one containing an enzyme (*Taq*I) that had a single site in the particle and the other containing *Taq*I plus an equal concentration of another enzyme (*Bsm*AI) for which no site existed in this particular DNA construct. Concentrations of both enzymes were near the upper end of the concentration range used in our assays (i.e., also for *Bsm*AI, an enzyme for which sites *were* present, in other constructs). We found identical digestion rates, and thus identical equilibrium constants for site exposure, using *Taq*I in the presence or absence of *Bsm*AI. We conclude that nonspecific binding by restriction enzymes does not contribute to the site exposure process.

Polynucleosomal Systems

It is important to extend these studies of mononucleosomes to more natural polynucleosomal substrates. Such systems have been studied *in vitro* with restriction enzymes[15] and DNA-binding proteins[16,17] and yielded equilibrium constants for site exposure that are similar to those observed for isolated mononucleosomes.

How could DNA in a central "probe" nucleosome undergo site exposure while remaining connected to a nucleosome filament on either side? One way in which this might be imagined is as follows. With just modest deformation of the linker DNA between nucleosomes, a combined uncoiling coupled to a motion of the uncoiled DNA in a direction parallel to the axis of the nucleosomal disk allows uncoiling beyond the dyad (which is as far as necessary to allow binding anywhere) with no required crossings and little motion of the other nucleosomes.

These increasingly complex systems pose several new technical problems. A major problem concerning polynucleosomal substrates is heterogeneity in the translational positions. DNA constructs used in the studies of nucleosome core particles were kept short so as to provide the histone octamer with no other choices of translational position (except ones that would leave large amounts of the DNA-binding surface of the histone octamer unsatisfied), which is not the case for substrates designed to accommodate multiple nucleosomes. The positioning of histone octamers on long DNA molecules is inherently statistical,[9] and many of the "strong" positioning sequences used to date do not provide the positioning power in these templates that one might expect.

Consider a case where a polynucleosomal construct is designed with a repeat length of 208 bp based on the 5S RNA gene used in our mononucleosomal studies. The probability of finding a nucleosome at the most probable position relative to all possible positions is given by[9]

$$P = \frac{1}{1 + (L - 147)e^{\Delta\Delta G_{\mathrm{HO}}/RT}} \tag{18}$$

where L is the length of the DNA available for binding (208 bp in this example) and $\Delta\Delta G_{\mathrm{HO}}$ is the difference free energy for the reconstitution of histone octamer into nucleosomes, for two DNA molecules having identical lengths but where one contains a single favored position and the other contains no such site. For the 5S positioning sequence, $\Delta\Delta G_{\mathrm{HO}} \approx -1.6$ kcal mol^{-1},[18] yielding a remarkably low ≈19% probability for occupancy at the

[15] C. Logie and C. L. Peterson, *EMBO J.* **16,** 6772 (1997).

[16] T. Owen-Hughes, R. T. Utley, J. Cote, C. L. Peterson, and J. L. Workman, *Science* **273,** 513 (1996).

[17] T. Owen-Hughes and J. L. Workman, *EMBO J.* **15,** 4702 (1996).

[18] T. E. Shrader and D. M. Crothers, *Proc. Natl. Acad. Sci. U.S.A.* **86,** 7418 (1989).

expected position. Although other positioning forces arising from higher order chromatin structure may contribute,[19] the positioning power of the individual sequence in this context is plainly inadequate to provide a homogeneous substrate.

Digestions of such material yield multiphasic kinetics that are difficult to compare directly to results obtained from core particle assays. It is not evident which rates correspond to any given position or what the contributions are, if any, from organization into higher order structures. Such problems may be overcome, at least to some extent, by utilizing much stronger positioning sequences that are now becoming available.[10] These sequences exhibit $\Delta\Delta G_{HO} \approx 2.8$ kcal mol^{-1} relative to the 5S sequence. Taking the value of $\Delta\Delta G_{HO} \approx -1.6$ kcal mol^{-1} for the 5S sequence relative to arbitrary-sequence DNA[18] (although our own measurements yield somewhat lower values), the new positioning sequences yield a net $\Delta\Delta G_{HO} \approx$ 4.4 kcal mol^{-1} relative to random sequence DNA. This would provide a probability of ≈96% occupancy at a specific position on a 208-bp fragment, a large improvement over previously available reagents.

[19] J. Yao, P. T. Lowary, and J. Widom, *Proc. Natl. Acad. Sci. U.S.A.* **90,** 9364 (1993).

[16] Nucleoprotein Gel Assays for Nucleosome Positioning and Mobility

By SARI PENNINGS

Introduction

Nucleosomes form the basic repeating unit of chromatin, the structure in which the genome of eukaryotes is packaged in the nucleus. Traditionally, the nucleosome subunit is defined as the particle consisting of an octamer of histone proteins around which two turns of DNA are wrapped in a left-handed superhelix. The histone core consists of two each of histones H2A, H2B, H3, and H4. The length of the DNA, which includes the linker DNA between nucleosomes, totals one repeat length. In addition to core histones, a linker histone H1 secures the two DNA coils around the core, thus completing the nucleosome.

The nucleosome was an operationally defined particle long before its exact composition and structure were known. Nucleosomes are produced by micrococcal nuclease digestion of chromatin and were first detected as ladders of bands on DNA gel electrophoresis. Digestion eventually trims

0076-6879/99 $30.00

away the linker DNA but is halted when 146 bp of DNA is left around the histone core. This residual particle is termed the core particle. An intermediate stop when the nucleosome is trimmed to just two turns of DNA (168 bp) and still contains a linker histone is called the chromatosome.[1] In practice, the term nucleosome is loosely employed, and length of the DNA and presence of linker histones are specified in context. Moreover, in reality, considerable variation can be found in the components of nucleosomes.

Our current understanding of chromatin structure–function relationships owes much to the analysis of the heterogeneity of the nucleosomes that make up the chromatin fibers. Indeed, histone modifications and histone variants, the presence of linker histones and nonhistone proteins, have all been associated with specific functions in transcription, replication, or the regulation of these processes.[1,2] Polyacrylamide gel electrophoresis has been one of the principal methods used to analyze and distinguish between such particles.

With the more recent advances in knowledge about the separation characteristics of polyacrylamide gels, new applications have emerged, such as in studies of macromolecular conformation. It is this separation capacity that also lies at the basis of the use of the technique for analyzing nucleosome positioning. This in turn has led to the discovery of nucleosome mobility. This article describes the procedures for nucleoprotein gel assays for nucleosome positioning and mobility, as well as their background, development, and interpretation.

Separation of Native Nucleosomes by Polyacrylamide Gel Electrophoresis

Polyacrylamide vertical slab gel electrophoresis has become a standard technology in molecular biology to separate and analyze DNA fragments and proteins.[3] It is also suitable for the preparation of small quantities of these macromolecules. Nucleosomes were one of the earliest protein–DNA complexes to be analyzed by polyacrylamide gel electrophoresis.[4,5] The gels, referred to as nucleoprotein gels or native particle gels, are essentially an adaptation of the standard DNA gel. This mainly involves reducing the buffer strength compared to standard DNA gel conditions to protect the nucleosomes from dissociation, although in most cases this is not an absolute

[1] K. E. van Holde, "Chromatin." Springer-Verlag, New York, 1988.
[2] A. Wolffe, "Chromatin: Structure and Function," 2nd Ed. Academic Press, London, 1995.
[3] F. W. Studier, *J. Mol. Biol.* **79,** 237 (1973).
[4] A. Varshavsky, V. V. Bakayev, and G. P. Georgiev, *Nucleic Acids Res.* **3,** 477 (1976).
[5] R. D. Todd and W. T. Garrard, *J. Biol. Chem.* **252,** 4729 (1977).

requirement. If very low ionic strength buffers are used, buffer recirculation becomes necessary to avoid the formation of pH gradients across the gel.[6] Electrophoresis is usually carried out in the cold to preserve against degradation, as well as for other reasons discussed later.

Similar gels are now also commonly used in the band-shift assays of transcription factor–DNA binding studies, which may include nucleosomes.[7,8] Here, concern about dissociation of the complexes can justify the addition of stabilizing agents. Trails of dissociating components are rarely observed, however, and it is thought a gel "cage effect" favors association inside the gel matrix.[9]

Procedures

Good quality electrophoresis grade reagents are used throughout. Stock solutions are renewed regularly. Acrylamide stock solutions are filtered after dissolving or are purchased ready-made.

Gels consist of 5% (29:1) acrylamide–bisacrylamide, 0.5× TBE [45 m*M* Tris, 45 m*M* boric acid, 1.25 m*M* EDTA (pH 8.3)], 0.1% ammonium persulfate, and 0.15% TEMED to catalyze polymerization. Gels of 1.5 mm thickness are cast between 17- × 20-cm glass plates, inserting a comb forming wide slots. When polymerization is complete, the gel is mounted on a large vertical slab gel apparatus. A cooling jacket keeps the back glass plate in contact with the top buffer.

Electrophoresis is at 4° in 0.5× TBE and is preceded by a prerun of at least an hour (optional). Samples are adjusted to 0.5× TBE, 3% Ficoll, 0.2% xylene cyanol (and optionally, bromphenol blue). To minimize the exposure of nucleosomes to high ionic strength buffer, droplets of concentrated sample solution are pipetted on the walls of the tube and then mixed in by vortexing. At least 100 ng of material per expected nucleosome band is required for ethidium bromide staining, as fluorescence from nucleosomal DNA is less intense than from free DNA.[10]

Electrophoresis is at 5 to 10 V/cm maximum for about 1000 V·hours, depending on particle size. Lower voltages are recommended when nucleosomes containing linker histones or nonhistone proteins are involved.[11]

[6] S.-Y. Huang and W. T. Garrard, *Methods Enzymol.* **170,** 116 (1989).

[7] J. Carey, *Methods Enzymol.* **208,** 103 (1991).

[8] R. T. Utley, T. A. Owen-Hughes, L. J. Juan, J. Côté, C. C. Adams, and J. L. Workman, *Methods Enzymol.* **274,** 276 (1996).

[9] M. G. Fried and D. M. Crothers, *Nucleic Acids Res.* **9,** 6505 (1981).

[10] C. T. McMurray and K. E. van Holde, *Proc. Natl. Acad. Sci. U.S.A.* **83,** 8472 (1986).

[11] S. Pennings, G. Meersseman, and E. M. Bradbury, *Proc. Natl. Acad. Sci. U.S.A.* **91,** 10275 (1994).

Gels are stained in freshly made 0.03 mg/ml ethidium bromide under slow agitation for 20 min and destained in H_2O. They are photographed on a UV transilluminator on ISO 80 Polaroid positive/negative film or equivalent high-resolution sheet negative film using an orange No. 15 Wratten filter. Alternatively, the gels are dried on Whatman (Clifton, NJ) DE81 ion-exchange paper and exposed for autoradiography or phosphor imaging.

Optimization of Procedures

Gel electrophoresis following the protocol just described separates free DNA from mononucleosomes and dinucleosomes, as well as distinguishing among core particles, chromatosomes, and nucleosomes containing longer DNA. These gels also fractionate nucleosomes according to their positioning on the DNA fragments. Furthermore, nucleosomes associated with additional nonhistone or other proteins may be isolated. All these separations give rise to some overlap and bands cannot normally be identified unless the starting material is known or otherwise characterized.

Nucleoprotein gel fractionation offers probably the highest sensitivity to the different characteristics of native nucleosomes of any technique; however, because electrophoresis acts on a combination of parameters with possibly counteracting effects, there are limits to its resolution. Polyacrylamide gel electrophoresis separates particles according to charge, mass, and conformation. Variations on the standard protocol can give advantage to one type of fractionation over another. This is because the parameters for separation are influenced in an independent way by the gel pore size, electric field strength, buffer ionic strength, pH, and temperature. Some examples of this are reviewed later. However, the unsurpassed resolving power of the gels can be fully exploited in studies using reconstituted nucleosomes, where the number of variables can be controlled. This is demonstrated for the case of nucleosomes varying only in their position on DNA fragments.

Despite advances in the complex theories behind polyacrylamide gel electrophoresis, reviewed recently with respect to nucleoprotein gels,[12] methods are still mainly developed empirically. It is difficult to extract unifying principles from these protocols, and procedures should therefore be further optimized with each specific application in mind.

[12] S. Pennings, *Methods* **12,** 20 (1997).

Glycerol: A Stabilizing Agent That Changes Gel Separation Characteristics

Glycerol is often included in gel recipes to protect nucleosomes or other protein–DNA complexes from dissociation during electrophoresis. Because it is added solely as a stabilizing agent, separation is presumed to be indifferent to the presence (up to 30%) or absence of this additive. However, when the gel migrations of various types of nucleosomes and bent DNA were compared as a function of glycerol, a strong effect on the separation characteristics of polyacrylamide gels was revealed. The conformation factor in separation is progressively lost with increasing glycerol concentration, leaving a fractionation largely based on particle mass and charge. Nucleosome positions are no longer resolved, while the difference in electrophoretic mobility between core particles and nucleosomes carrying longer DNA becomes smaller and is eventually lost. The retardation of bent DNA is also much reduced.[13]

Glycerol lowers electrophoretic mobility because it both increases the viscosity and reduces the dielectric constant of the electrolytic medium. This is manifested in the much longer times needed to complete electrophoresis. However, evidence of a more specific action of glycerol was demonstrated by comparison with sucrose, another stabilizing viscosogen, included in the gels at the same viscosity.[13] The physicochemical basis for the stabilizing effect of glycerol (and sucrose) on complexes in solution lies in the preferential hydration of macromolecules in these solutions. This thermodynamically unfavorable hydration layer drives protein stabilization and assembly by minimizing the surface of contact between macromolecules and solvent.[14,15] With increasing glycerol concentration, the sensitivity of polyacrylamide gels to conformational differences is gradually lost. In addition, the more prominent roles of charge and mass in gel separation seem to cancel out the contribution of extra DNA extending from the nucleosome. The net effect of this is that glycerol gels show an insensitivity to the DNA length contained in nucleosomes in a wide range of glycerol concentrations. It is also important to note that glycerol inhibits the polymerization of polyacrylamide, resulting in a different, looser gel matrix structure for the same percentage of gel.[13]

In conclusion, the incorporation of glycerol in the gel is indicated for nucleoprotein analyses of nucleosome-binding proteins or macromolecular composition. Here, the stabilizing effect will also be beneficial. In contrast,

[13] S. Pennings, G. Meersseman, and E. M. Bradbury, *Nucleic Acids Res.* **20,** 6667 (1992).
[14] K. Gekko and S. N. Timasheff, *Biochemistry* **20,** 4667 (1981).
[15] J. C. Lee and S. N. Timasheff, *J. Biol. Chem.* **256,** 7193 (1981).

glycerol has to be omitted from assays for nucleosome positioning, DNA bending, and macromolecular conformation in general.

Separation According to Presence of Linker Histones or Nonhistone Proteins

Earlier use of nucleoprotein gels focused on the identification of subpopulations of nucleosomes carrying varying amounts of linker histones or nonhistone proteins. The methods of Huang and Garrard[6] have been reviewed earlier in this series. These studies involved native nucleosome particles obtained by micrococcal digestion. The DNA length of bulk digested nucleosomes is coupled to the presence or absence of certain additional proteins, and even there some heterogeneity is inevitable. These variations in DNA length constitute only a small proportion of particle mass but contribute significantly to particle charge, obscuring resolution of the nucleosome subpopulations.

Separation according to particle mass seems most efficient in 3.5% polyacrylamide–0.5% agarose gels run at 4° in a recirculating low-ionic strength running buffer.[6] Furthermore, 30% glycerol was included as a stabilizing agent in these studies, which as discussed earlier also reduces the sensitivity of gel electrophoresis to DNA length and conformation. Finally, the complete resolution of nucleosomes of different composition could be achieved by a second-dimension DNA gel electrophoresis in the presence of sodium dodecyl sulfate (SDS). Particles containing either one or two molecules of either histone H1 or HMG14/17 were revealed.[16,17] This system is also sensitive to the level of histone ubiquitination in native nucleosomes.[6]

Separation According to Histone Acetylation

Nucleoprotein gel electrophoresis can detect the histone acetylation of intact nucleosomes.[18] This reversible modification of histone N-terminal lysine residues is correlated with transcriptional activity.[19] Acetylation neutralizes the positive charge of lysine residues, increasing the net negative charge of nucleosomes. However, paradoxically, acetylated nucleosomes migrate slower toward the anode in nucleoprotein gels compared to nonacetylated controls. As the mass of these particles remains virtually the same, this effect has been attributed to a conformational change within the nucleo-

[16] R. D. Todd and W. T. Garrard, *J. Biol. Chem.* **254,** 3074 (1979).
[17] S. C. Albright, J. M. Wiseman, R. A. Lange, and W. T. Garrard, *J. Biol. Chem.* **255,** 3673 (1980).
[18] J. Bode, M. M. Gomez-Lira, and H. Schröter, *Eur. J. Biochem.* **130,** 437 (1983).
[19] E. M. Bradbury, *BioEssays* **14,** 9 (1992).

some.[18] This change has so far proven very difficult to detect with other methods. The retardation of acetylated nucleosomes is noticeable beyond a level of 10 acetylated residues per particle. Residues located in the H3 and H4 N-terminal tails are mainly responsible for this effect.[20]

For these purposes, electrophoresis conditions are similar to those used to fractionate DNA (4 or 5% polyacrylamide in 0.5 or 1× TBE).[18,20] Again, resolution of these subpopulations in native nucleosomes is incomplete in nucleoprotein gels, but can be enhanced by a second-dimension SDS protein gel or acid–urea protein gel.[18]

Nucleoprotein Gels in Combination with Other Assays

Native particle gels can be a very effective tool in any experimental strategy. As reviewed earlier, second-dimension DNA or protein gel electrophoresis can take the nucleoprotein gel analysis further in terms of characterization, but these nucleosomes are lost to subsequent experimentation. In some cases this problem can be circumvented by manipulating the nucleosome sample before separation on the nucleoprotein gel, if the assay does not affect the integrity of the nucleosome. DNase I-digested nucleosomes, for example, will still fractionate according to their position on the DNA, revealing their different footprinting patterns.[21] In addition, the gels themselves do leave some room for experimentation. Smaller restriction enzymes can be diffused into the gel and can be activated *in situ* through buffer changes containing the appropriate cations.[22] Experiments such as these have provided evidence that nucleosomes varying only in the position of the histone octamer relative to a DNA fragment, i.e., positioning isomers, can be separated on polyacrylamide gels in their native state.

Separation of Positioned Nucleosomes on Polyacrylamide Gels

The separation of nucleosomes according to their positioning on DNA is another manifestation of the influence of particle conformation on gel migration: no differences in particle mass or charge are involved. In native nucleosomes, this separation capacity is again obscured by DNA length heterogeneities that produce their own spread of electrophoretic mobilities. Furthermore, it becomes less effective as the DNA is trimmed shorter.[22] Although difficult to recognize with native nucleosomes,[23] this additional separation function came to light when nucleoprotein gels were employed

[20] K. W. Marvin, P. Yau, and E. M. Bradbury, *J. Biol. Chem.* **265,** 19839 (1990).
[21] S. Pennings, G. Meersseman, and E. M. Bradbury, *J. Mol. Biol.* **220,** 101 (1991).
[22] G. Meersseman, S. Pennings, and E. M. Bradbury, *EMBO J.* **11,** 2951 (1992).
[23] W. Linxweiler and W. Hörz, *Nucleic Acids Res.* **12,** 9395 (1984).

for monitoring nucleosome assembly on DNA fragments of defined length. Because these reconstitutions use purified components, a homogeneous population of particles was expected, migrating on the gel as one band.

In contrast, reconstitution on 207-bp fragments of 5S rDNA results in assembled nucleosomes migrating on nucleoprotein gel as three bands, using gel electrophoresis procedures as outlined earlier.[21] The sea urchin 5S rDNA sequence contains a strong positioning site for nucleosomes.[24] However, differences in length of DNA protruding from the positioned histone octamer were demonstrated to be solely responsible for the multiple band pattern.[21] Restriction enzyme cutting to trim away this DNA draws the three bands together to one faster migrating band, showing that no heterogeneities are involved at the level of the histone core. As described earlier, DNase I digestion prior to separation reveals different footprints for nucleosomes migrating as different bands. This is also consistent with restriction enzyme accessibility assays performed prior to separation, which affect certain nucleoprotein bands but leave others intact. Mapping of micrococcal nuclease-generated core particle boundaries on this sequence had indicated that histone octamers reconstitute on 5S rDNA in one dominant position surrounded by weaker overlapping positions 10 bp apart.[25] Thus, by establishing that differently positioned nucleosomes can be separated on a native gel, the noninvasive technique of nucleoprotein gel analysis could substantiate the enzymatic mapping results.

While the observation was unexpected, it hinted at a logical connection with the gel retardation of curved DNA. As is the case for intrinsically curved or protein-induced bent DNA,[26,27] nucleosomes positioned at the ends of the DNA fragment migrated faster through the gel than those positioned near the center.[20] This link was demonstrated convincingly in the analysis of the electrophoretic mobilities of mononucleosomes assembled on a 5S rDNA dimer construct.[22] The many possible positions of a single histone core on this 414-bp fragment give rise to a complex band pattern (Fig. 1B). Restriction cutting of end-labeled nucleosomes was performed inside the gel strip of the first separation to separate the monomeric halves out on a second-dimension nucleoprotein gel. In this way, the complex band pattern derived from mononucleosomes assembled on the dimer construct could be broken down to the simpler and fully characterized pattern for the monomer fragment. As a result, histone octamer positions could be attributed to each of the bands in the pattern from the dimer

[24] R. T. Simpson and D. W. Stafford, *Proc. Natl. Acad. Sci. U.S.A.* **50,** 51 (1983).
[25] G. Meersseman, S. Pennings, and E. M. Bradbury, *J. Mol. Biol.* **220,** 89 (1991).
[26] H.-M. Wu and D. M. Crothers, *Nature* **308,** 509 (1984).
[27] D. M. Crothers, M. R. Gartenberg, and T. E. Shrader, *Methods Enzymol.* **208,** 118 (1991).

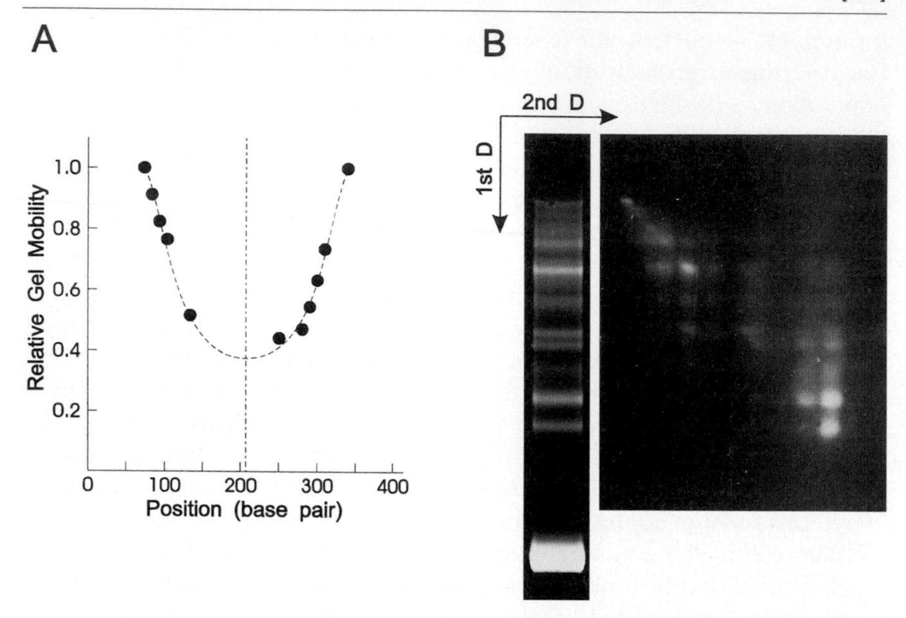

FIG. 1. (A) Graph of the relative gel mobility of positioned nucleosomes in a 5% polyacrylamide gel plotted versus their location on the DNA fragment (taken from the midpoint of the core particle). The gel migration distances of mononucleosomes reconstituted on 414-bp dimers of sea urchin 5S rDNA sequence were measured from the complex band pattern, such as the one shown in (B, 1st D). These bands are formed by nucleosomes differing only in the position of the histone octamer on the DNA fragment, which was determined at high resolution by other methods. (B) Two-dimensional nucleoprotein assay demonstrating mobile nucleosomes on the 5S rDNA sequence. The first-dimension gel band pattern, which shows the nucleosome positioning on this DNA fragment in accordance with (A), was excised and incubated at 37° between runs. Off-diagonal dots represent mobile nucleosomes with an altered electrophoretic mobility (and thus position) during the second gel electrophoresis.

construct. The graph that plots relative gel electrophoretic mobility of nucleosomes against the position of the histone octamer on the 414-bp DNA fragment (Fig. 1A) clearly shows similarities to plots obtained for the gel retardation of bent DNA.[26,27] The symmetry of this graph also makes it clear that this assay requires additional information in order to assign positions to the 5′ or 3′ half of the sequence.

Reconstitution on a variety of sequences has since confirmed these findings of multiple, sometimes overlapping nucleosome positions and their fractionation on nucleoprotein gels. These positions, frequently spaced by 10 bp (the DNA helical repeat) or multiples thereof, were previously often dismissed as artifacts of invasive enzymatic techniques. Native particle gels have therefore been instrumental in establishing the principle of multiple overlapping positioning on DNA sequences.

The gel retardation effect of positioned nucleosomes is due to the frictional drag produced by the extra DNA protruding from the nucleosome and is likely to be linked to the angle at which these DNA strands enter and leave the core particle. This behavior is not dependent on the source of histones or their state of acetylation.[13] At least an extra 40 bp of DNA in addition to core particle length is required for nucleosome positioning to be discernible on a 5% polyacrylamide gel.[22] This was determined by analysis of the DNA lengths contained in untrimmed native nucleosomes after separation of these nucleosomes on a nucleoprotein gel. Second-dimension DNA gel electrophoresis confirmed that the gel retardation of nucleosomes increases with the overall length of DNA. It also revealed the presence of DNA fragments of the same length in nucleosomes with different electrophoretic mobilities, producing a pattern of migration radiating out beyond a length of 190 bp of DNA.[22] This result from bulk nucleosomes is therefore in agreement with the observation that multiple band patterns are characteristic for the assembly of nucleosomes on a variety of DNA sequences.

Preparative Gel Electrophoresis

One of the major advantages of nucleoprotein gels is that they can give information on the positioning of nucleosomes in the native state, albeit at low resolution. This has offered the possibility to study these differently positioned nucleosomes individually.[28–34] For an example of how this technique can be applied,[34] see Fig 2.

Intact nucleosomes can be eluted from excised nucleoprotein gel bands,[29] but preparative applications of nucleoprotein gels have been relatively rare. For larger operations, a few designs of preparative gel apparatus are available, among which the toroidal design has the advantage of concentrating the fractions.[35] The difficulty with nucleosome elution lies precisely in maintaining a workable concentration and avoiding dissociation. Despite

[28] I. Duband-Goulet, V. Carot, A. V. Ulyanov, S. Douc-Rasy, and A. Prunell, *J. Mol. Biol.* **224,** 981 (1992).

[29] M.-F. O'Donohue, I. Duband-Goulet, A. Hamiche, and A. Prunell, *Nucleic Acids Res.* **22,** 973 (1994).

[30] V. M. Studitsky, D. J. Clark, and G. Felsenfeld, *Cell* **76,** 371 (1994).

[31] M. S. Roberts, G. Fragoso, and G. L. Hager, *Biochemistry* **34,** 12470 (1995).

[32] K. Ura, J. J. Hayes, and A. P. Wolffe, *EMBO J.* **14,** 3752 (1995).

[33] A. Hamiche, P. Schultz, V. Ramakhrishnan, P. Oudet, and A. Prunell, *J. Mol. Biol.* **257,** 30 (1996).

[34] C. Davey, S. Pennings, and J. Allan, *J. Mol. Biol.* **267,** 276 (1997).

[35] J. M. Harp, E. L. Palmer, M. H. York, A. Gewiess, M. Davis, and G. J. Bunick, *Electrophoresis* **16,** 1861 (1995).

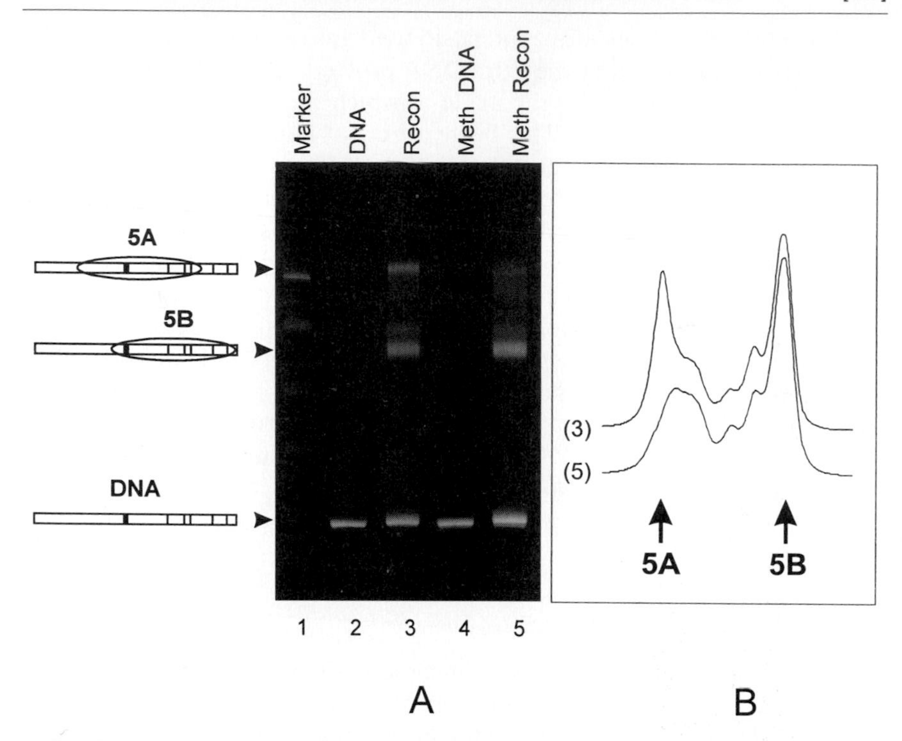

FIG. 2. Example of the use of nucleoprotein gels in the analysis of nucleosome positioning, providing a first demonstration that DNA methylation can alter nucleosome positioning. (A) A 245-bp DNA sequence upstream from the chicken adult β-globin promoter supports two major possible nucleosome positioning sites, 5A and 5B, determined at high resolution by other methods. These nucleosomes separate on the gel as indicated by the diagram. Position 5A incorporates a CpG triplet sequence near the nucleosome dyad (solid bar). Nucleosomes no longer assemble in this position when the DNA is CpG methylated. (B) Intensity traces of lanes 3 and 5 of (A), obtained by phosphorimage analysis of the dried gel. The peak corresponding to position 5A is missing in the trace for reconstituted methylated DNA.

the very high association constant of the histone–DNA interaction, concentrations in the gel or in solution can easily reach the point where nucleosomes start to come apart, especially if one relies on trace radiolabeled particles.[36,37]

An elegant way to avoid elution of the various and often incompletely separated populations of nucleosomes is through the use of two-dimensional gel electrophoresis. The assay to detect mobile nucleosomes is based on this approach.

[36]. J. Ausió, D. Seger, and H. Eisenberg, *J. Mol. Biol.* **176,** 77 (1984).

[37] J. S. Godde and A. P. Wolffe, **270,** 27399 (1995).

Assays for Nucleosome Mobility

The multiple positions for nucleosomes found on a majority of DNA sequences represent a choice of alternative, often overlapping locations in which a histone octamer can reconstitute. The frequency with which these possible sites are actually occupied can differ greatly and reflects their affinity.[38] Nucleoprotein gels can provide a direct visualization of the abundance and distribution of nucleosome positions (Figs. 1B and 2). Consequently, this positioning assay has proven to be a unique tool in following redistributions of nucleosomes on DNA fragments.

Distribution patterns of nucleosomes over the various possible positions were found to be in dynamic equilibrium at 37°, which introduced the concept of "mobile" nucleosomes.[21] To detect this nucleosome movement, a two-dimensional gel assay was developed employing identical nucleoprotein gels run in the same conditions at 4°. Because electrophoresis is the same in both dimensions, the second-dimension nucleoprotein gel displays the first dimension pattern on a diagonal line if no changes in nucleosome position occur. When thin gel strips comprising the full band pattern from the first-dimension gel are incubated at 37°, nucleosomes are found to redistribute between possible positioning sites on the DNA fragment.[21,22] These mobile nucleosomes have an altered electrophoretic mobility when electrophoresed in the second-dimension gel and are therefore observed as off-diagonal dots in the two-dimensional gel pattern (Fig. 1B).

Procedures

The procedure is outlined in Fig. 3. Nucleosomes assembled on DNA fragments of defined sequence are first separated on a nucleoprotein gel as described earlier. The best results are obtained using fast protein liquid chromatography (FPLC)-purified DNA fragments and quantities of reconstituted material sufficient for ethidium bromide staining (at least 100 ng per expected band). Nucleosomes reconstituted with radiolabeled trace fragments are also suitable, however. When different such DNAs are compared, gel migration patterns of the reconstituted unlabeled carrier DNA can serve as a quality control.

Typically, part of the gel will be destined to provide an image of the first-dimension gel. The remaining lanes are used for preparative purposes and will carry excess sample. This part of the gel should not be stained, as ethidium bromide causes the dissociation of nucleosomes.[10] A good resolution first-dimension gel separation is essential. The slots of the gel

[38] C. Davey, S. Pennings, G. Meersseman, T. J. Wess, and J. Allan, *Proc. Natl. Acad. Sci. U.S.A.* **92,** 11210 (1995).

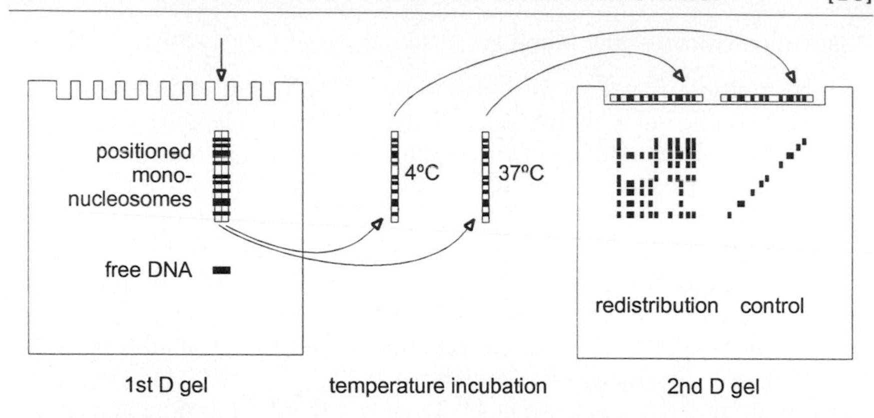

FIG. 3. Schematic diagram of the two-dimensional nucleoprotein gel electrophoresis procedure to detect the mobility of nucleosomes on DNA. Conditions of the first and second-dimension polyacrylamide gel (run at 4°) are identical. Nucleosomes that have redistributed to different locations on the DNA fragment at 37° no longer migrate as a diagonal in the second-dimension gel. The pattern of off-diagonal dots represents the proposed redistributions on the 414-bp dimer of 5S rDNA (Fig. 1) in the absence of end effects.

should be wide, and samples are loaded as small volumes (less than 50 μl) at high concentration. Sample solutions should use a high molecular weight viscosogen such as Ficoll to prevent reverse osmosis effects in the wells that can affect the shape of the bands. Likewise, the final NaCl concentration in the samples should not exceed 50 m*M* to avoid distortion of the lanes.

After the run, the gel plates are opened while still in the cold room. The part of the gel to be stained with ethidium bromide or dried for autoradiography is cut off and removed. The remainder of the gel is left on the glass plate and kept at 0–4° covered with Saran wrap, until the gel strips can be excised. The exact cutting of the strips is decided on the basis of the stained gel or the radioactive signal as measured by a Geiger monitor, as the dimensions of the second gel are size limiting. To assist in locating the region of interest within the lanes, a marked graph paper is placed underneath the glass plate carrying the gel. A new razor blade is pressed down into the gel (rather than dragged) alongside a transparent ruler, avoiding jagged edges. Lanes are usually cut lengthwise to provide two identical strips. Wider strips will produce stripes rather than dots in the second dimension. A higher gel percentage in the second dimension may focus the dots, but it will also distort the diagonal. The strips are cut obliquely at one end to mark the direction of electrophoresis.

Throughout the procedure, some bromphenol blue-containing buffer can aid in contrasting the delicate strips, which are carefully handled using microspatulas and transferred on gel spacers. The strips are sealed in thin plastic bags with a minimum of buffer to keep the samples moist and to improve contact with the water bath. They are completely submerged in the water bath and incubated at 37° or other desired temperature for 1 hr. The control gel strip is kept on ice in the cold room during this time.

Next, control and incubated strips are arranged on top of a second-dimension gel, which has been precooled. This gel has one continuous well, formed by casting with the slot comb upside down. For convenience, the manipulation is performed before the gel is mounted on the apparatus. Some sample solution is added to ease loading of the gel strips. Care is taken to avoid the trapping of air bubbles between the strip and the gel, as it causes streaking. Electrophoresis conditions are identical to the first-dimension gel and start with a precooled apparatus and running buffer. The gel is stained and photographed as described earlier.

Interpretation

Nucleosome mobility is a short-range, temperature-dependent process that occurs in low salt conditions with all histone–DNA contacts intact. It is distinct from the long-range nucleosome sliding that takes place at higher ionic strengths,[39] which may be due to a partial loss of histone–DNA interactions. Although nucleosome mobility is a general behavior observed for many DNA sequences and is visible with bulk nucleosomes, it is not uncommon to find that not all possible nucleosome positions on the diagonal of the two-dimensional gel are engaged in all the redistributions.[22] We have proposed that nucleosomes may be allowed to move as long as the coiling path of the DNA can be continued beyond their present location. This would mean that nucleosome movement is only energetically favorable between positions sharing the same rotational setting of the DNA on the histone core.

Another common observation is that nucleosomes seem to accumulate to some extent on the far side of the two-dimensional pattern, i.e., in the end positions. This can be attributed to the difference between the temperature of the mobility assay incubation and the lower temperature of reconstitution. End positions seem more favored at 37°. If necessary, the effect can be neutralized by preincubation of the sample at 37°.[21]

The 0–4° control gel strip is included to assess the quality of the zero-mobility diagonal. Furthermore, this control also serves to reveal nucleosome dissociation.[29] The resulting free DNA, if observed at all, will migrate ahead of the two-dimension pattern as a horizontal band. Smears can indicate dissociation/mobility during electrophoresis.

[39] P. Beard, *Cell* **15,** 955 (1978).

Summary

Recent recognition of the sensitivity of polyacrylamide gel electrophoresis to macromolecular conformation has provided a source of new applications. In chromatin research, nucleoprotein gel electrophoresis can yield a direct and visual estimate of the number and relative abundance of different positions adopted by the core histone octamer on DNA, as well as their locations relative to the middle of the DNA fragment. It is the only technique available for the fractionation of such nucleosome positioning isomers and leaves them intact. Thus this simple method constitutes a powerful tool to analyze and manipulate populations of variously positioned nucleosomes in their native state. Complementing conventional invasive enzymatic procedures that rely on the analysis of cutting patterns on nucleosomal DNA, these procedures are now revealing that histone octamers can reconstitute to a number of discrete, often overlapping, locations on most DNA sequences.

Further capitalizing on these advantages of nucleoprotein gel analysis, the development of the technique into a two-dimensional assay has permitted a rare view at the dynamics of nucleosome positioning. Nucleosomes can redistribute between possible positions on DNA, with the distribution patterns of nucleosomes along the DNA being in dynamic equilibrium at 37° in relatively low ionic strength conditions.

This mobility of nucleosomes on DNA means that possible positions of nucleosomes can be defined precisely but that the actual locations of the nucleosomes are dynamic. It provides a compelling argument that a nucleosome position should be regarded as a probability rather than a static factor type of binding. This supports a more dynamic view of the nucleosomal organization, which seems more in accordance with the dynamic nature of gene expression. In providing the flexibility for adaptation, multiple positioning and nucleosome mobility could constitute essential ingredients of the mechanisms by which chromatin participates in gene regulation.

Acknowledgments

Nucleosome positioning/mobility nucleoprotein gel assays were developed jointly with Geert Meersseman. The contribution of Morton Bradbury to this work is also gratefully acknowledged. The procedures described benefited from work by Serge Muyldermans in perfecting the nucleoprotein gel technique. DNA methylation experiments were a collaboration with Colin Davey. I thank Richard Meehan and Jim Allan for valuable comments on the manuscript. S.P. is supported by a Wellcome Trust Senior Fellowship in Biomedical Sciences.

[17] Assays for Interaction of Transcription Factor with Nucleosome

By QIAO LI, ULLA BJÖRK, and ÖRJAN WRANGE

Introduction

Gene regulation is controlled by various transcription factors. The first step in gene induction is the binding of these factors to regulatory promoter/enhancer segments. Some of the regulatory DNA segments, however, are known to harbor positioned nucleosomes. One nucleosome contains 146 bp of DNA wrapped as 1.7 negative supercoils around a histone octamer complex. The accessibility of nucleosomal DNA is restricted by the following factors: (i) the shielding of one side of the DNA helix that is facing the histone protein surface, (ii) the strong bending and kinking of the DNA around the histone octamer, (iii) the electrostatic interactions between the histone surface and the phosphate backbone of the DNA; and, perhaps most importantly, (iv) the interaction between the positively charged N-terminal tails of the histones and the DNA. Different DNA-binding proteins are affected differentially by nucleosomal DNA organization.[1–3] In order to understand the regulatory events at a molecular level, it is thus crucial to evaluate the binding affinity of transcription factors in their correct nucleosomal context. This article describes the preparation of oligonucleosomes used as donors of histone octamers, the *in vitro* nucleosome reconstitution by histone exchange, the purification of reconstituted mononucleosomes by glycerol gradient centrifugation, and the characterization of the translational and rotational positions of the nucleosomal DNA. In addition, it describes methods for affinity determination and characterization of protein–DNA interactions that have been useful for studying the interaction of transcription factor with nucleosomes, such as electrophoretic mobility shift assay, quantitative DNase I footprinting, and dimethyl sulfate (DMS) methylation protection. In our case the transcription factor used in nucleosome-binding experiments was the glucocorticoid receptor protein (GR) purified from rat liver[4] or recombinant nuclear factor 1 (NF-1).[3] However, the methods described are general and should be applicable for most other DNA-binding proteins.

[1] Q. Li and Ö. Wrange, *Mol. Cell. Biol.* **15,** 4375 (1995).
[2] J. S. Godde, Y. Nakatani, and A. P. Wolffe, *Nucleic Acids Res.* **23,** 4557 (1996).
[3] P. Blomquist, Q. Li, and Ö. Wrange, *J. Biol. Chem.* **271,** 153 (1996)
[4] Ö. Wrange, J. Carlstedt-Duke, and J.-Å. Gustafsson, *J. Biol. Chem.* **254,** 9284 (1979).

0076-6879/99 $30.00

Methodological Steps

1. Preparation of oligonucleosomes from purified cell nuclei (in our case from rat liver nuclei) to be used as a histone octamer source.
2. Reconstitution and purification of mononucleosomes.
3. Characterization of histone/DNA organization in mononucleosomes (translational and rotational positioning).
4. Transcription factor/DNA binding.

Preparation of Oligonucleosomes

This protocol is essentially as described by Lutter[5] and can probably be applied to most tissues or cell lines provided that a nuclear preparation can be obtained and proteolytic activity is low. Protease inhibitors may be required for histone preparations from certain tissues. We routinely use rat liver nuclei purified according to Gorski *et al.*[6] for the extraction of chromatin.

Solutions

PBS (phosphate-buffered saline)[7]: 137 m*M* NaCl; 2.7 m*M* KCl; 8.1 m*M* Na_2HPO_4; 1.8 m*M* KH_2PO_4, pH 7.0

Homogenization buffer: 2.1 *M* sucrose; 10 m*M* HEPES, pH 7.6; 15 m*M* KCl; 2 m*M* Na_2EDTA; 10% (v/v) glycerol [additions added just before use: 0.15 m*M* spermidine; 0.5 m*M* spermine; 0.5 m*M* dithiothreitol (DTT); 0.5 m*M* phenylmethylsulfonyl fluoride (PMSF) from a 100 m*M* stock solution in 2-propanol]

Buffer A: 15 m*M* Tris–HCl, pH 7.4; 60 m*M* KCl; 15 m*M* NaCl; 0.34 *M* sucrose (additions added just before use: 0.15 m*M* spermine; 0.5 m*M* spermidine; 15 m*M* 2-mercaptoethanol)

MNase stop mix: 100 m*M* EDTA; 10% SDS

Buffer B: 450 m*M* NaCl; 5 m*M* Tris–HCl pH 7.5; 0.2 m*M* 2-mercaptoethanol added just before use

Procedures

Two rats (body weight 250–300 g) are sacrificed by cervical dislocation. The rat livers are removed, rinsed in ice-cold PBS, transferred to a glass

[5] L. C. Lutter, *J. Mol. Biol.* **124,** 391 (1978).

[6] K. Gorski, M. Carneiro, and U. Schibler, *Cell* **47,** 767 (1986).

[7] J. Sambrook, E. F. Fritsch, and T. Maniatis, *in* "Molecular Cloning: A Laboratory Manual," 2nd ed. Cold Spring Harbor Laboratory Press, 1989.

beaker without buffer, and minced with a pair of scissors (all operations are done on ice). Then 60 ml of ice-cold homogenization buffer is added to the minced livers and is followed by homogenization of two sets of three strokes with a Teflon/glass Potter–Elvehjem pestle homogenizer. The extent of homogenization depends on the speed of rotation of the Teflon pestle and on the tightness of fit between the Teflon pestle and the glass cylinder. We routinely use a loose-fitting homogenizer for the preparation of nuclei (as opposed to a tight-fitting one). The glass cylinder of the homogenizer should be held in a plastic beaker filled with ice. The sucrose-containing buffer is viscous at 0° and hence homogenization is heavy. About 30 sec of cooling should be allowed between each of the three strokes of homogenization with the Teflon pestle.

The liver homogenate is transferred to a cold measuring cylinder and diluted to 160 ml with the same homogenization buffer as described earlier. Six centrifuge tubes (for SW 27, Beckman rotor, Palo Alto, CA, or AH 627, Sorvall rotor, Wilmington, DE) are filled with a cushion of 10 ml homogenization buffer, and the diluted liver homogenate, about 26 ml per tube, is poured slowly along the tube wall so that a clear zone is formed between the lower 10 ml of clean buffer cushion and the upper liver homogenate. Centrifugation is performed for 30 min at 24,000 rpm and 0° in an ultracentrifuge. The supernatant is poured off, making sure not to contaminate the white nuclear pellet at the bottom of the tube. The tube walls are dried with tissue paper, and the nuclear pellets are suspended in buffer A, about 13.3 ml per tube or about 40 ml per liver, and transferred to two 50-ml centrifuge tubes. The nuclear suspension is centrifuged at 7000 rpm and 0° for 4 min. The supernatant is discarded and the two nuclear pellets are suspended in a total volume of 5 ml of buffer A and pooled into one 50-ml centrifuge tube. A 2-μl aliquot of the nuclear suspension is transferred to a tube containing 498 μl of 0.1 M NaOH, and the A_{260} units are measured in a spectrophotometer. The remaining nuclear suspension is adjusted by diluting with buffer A to 50 A_{260} units/ml. (Example: the measured A_{260} units in spectrophotometer is 0.7, hence there is 250 $\times$ 0.7 A_{260} units/ml = 175 A_{260} units/ml; the nuclear suspension should be diluted 3.5-fold with buffer A.) To obtain an even nuclear suspension from the nuclear pellets, we recommend using a glass rod with a glass sphere with a diameter 1–2 mm smaller than the diameter of the centrifuge tubes at one end.

Because the aim is to obtain 30–60 nucleosome long chromatin by micrococcal nuclease digestion, we always titrate the amount of enzyme to obtain an optimal distribution of chromatin lengths. For this purpose, a 396-μl volume of the nuclear suspension is mixed with 4 μl of 0.1 M $CaCl_2$. Then 48 μl of this mixture is aliquoted to each of six tubes containing 15, 10, 5, 2.5, 1.25, or 0.625 units of micrococcal nuclease (Pharmacia Biotechnology, Piscataway, NJ) in 2 μl buffer A solution and incubated for 4 min at 20°. After the incubation, 5 μl of MNase stop mix and 45 μl of water are added to each tube, followed by an equal volume of phenol/chloroform (2:1) extraction. Two microliters of each clear water phase after phenol extraction is run on a 1% agarose gel with 0.5 μg/ml of ethidium bromide in must 1× TBE buffer[7] at 10 V/cm. The ethidium bromide-stained DNA in the agarose gel is visualized by UV light (Fig. 1A), and an appropriate concentration of micrococcal nuclease is selected for digestion of the residual nuclear suspension, which is kept on ice after the first titration step. In the experiment displayed in Fig. 1A, we selected a concentration of 10 U of micrococcal nuclease per 50 μl reaction volume. The residual nuclear suspension, also adjusted to 1 m*M* $CaCl_2$, is prewarmed to 20° and then the micrococcal nuclease is added, followed by a 4-min incubation at 20°. The reaction is stopped by adding 0.5 *M* of Na_2EDTA, pH 8.0, to a final

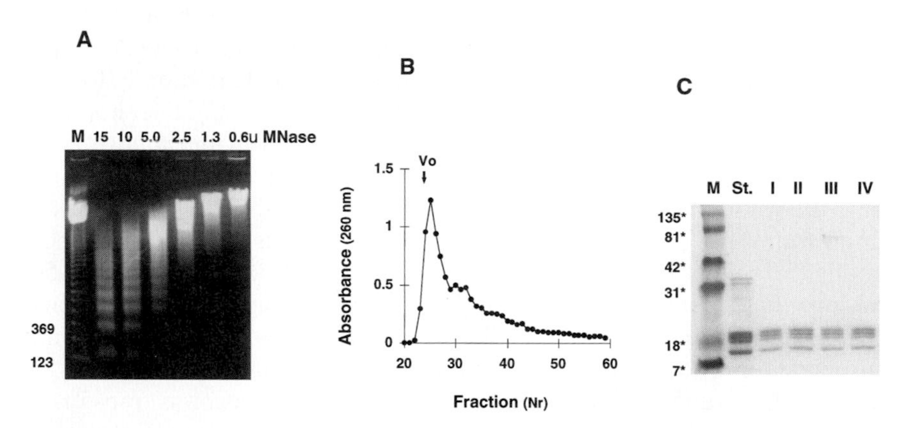

FIG. 1. Purification and analysis of long chromatin from rat liver nuclei. (A) Titration of the amount of micrococcal nuclease required to obtain an optimal length distribution of chromatin fragments. An ethidium bromide-stained agarose gel (1%) showing the indicated amounts of micrococcal nuclease (units of MNase given above each lane) is incubated in 50 μl of nuclear suspension as described in the text. (B) Sepharose CL-4B gel-filtration chromatography of long chromatin after micrococcal nuclease digestion of purified rat liver nuclei. Absorbance at 260 nm per 50 μl is given on the ordinate axis. Fractions from just before the exclusion volume (V_o) until about 0.7 column volumes were analyzed. (C) Electrophoresis analysis of purified histone proteins. Polyacrylamide gel (14%) electrophoresis in SDS shows aliquots (about 4 μg per lane = 0.08 A_{260} units) of pooled fractions (I: fractions 24 and 25; II: 26–29; III: 30–32; and IV: 33–36) from the chromatogram in (B). The molecular weight markers ($\times 10^{-3}$) (Kaleidoscope prestained standards, Bio-Rad) are on the left-hand side. St, histone protein standard (calf thymus type IIA, H-7755, Sigma). The gel is stained with Coomassie Brilliant Blue R. Note the double band of histone H1 visible in the histone standard around the 31-kDa marker but is absent in the chromatin preparations.

concentration of 2.2 m*M*. The nuclear suspension is then centrifuged at 7000 rpm for 4 min at 2°. The supernatant is discarded and the nuclear pellet is suspended in 5 ml of 0.2 m*M* Na_2EDTA, pH 8.0, and recentrifuged for 4 min at 7000 rpm at 2°. At this step the nuclei will lyse, and the pellet formed after centrifugation is viscous and ill-defined. The supernatant, i.e., the nuclear extract, which contains soluble long chromatin, is saved and its volume is estimated, e.g., by taking it up into a pipette. Then 0.127 ml of 4 *M* NaCl per milliliter of nuclear extract is added dropwise with mixing on ice. The final concentration of NaCl should be 0.45 *M*. Usually a precipitate forms at the beginning of the NaCl addition but it redissolves as the ionic strength is further increased.

It is important to keep the ionic strength around 0.45 *M* in order to dissociate histone H1 and most nonhistone proteins from the DNA. The core histones remain bound to the DNA at this ionic strength. In the next step, dissociated proteins are separated from long oligonucleosomes by gel filtration. The clear supernatant is applied onto a 200-ml Sepharose CL-4B (Pharmacia) column (2.5 cm in diameter) that has been equilibrated with buffer B. The column is eluted with a flow rate of about one drop per 6 sec, for 90 drop per fraction, equal to about 2.4 ml of each fraction. Gel-filtration chromatography is conveniently run overnight in the cold room at 2–4°.

To determine the chromatin distribution in the chromatogram, a 50-μl aliquot from each fraction is mixed with 0.5 ml of 0.1 *M* NaOH and analyzed at $A_{260\ nm}$ with spectrophotometry (Fig. 1B). The chromatin peak, which comes just after the exclusion volume, about 0.3 column volumes after sample application, is usually taken for nucleosome reconstitution (see Fig. 1B). As seen from the 14% polyacrylamide gel in SDS (Fig. 1C), this chromatin preparation contains essentially only core histones. The long chromatin is stored at 4° with 0.02% NaN_3. However, we have also stored long chromatin in the same buffer B in the freezer (−110°) and have used this material for nucleosome reconstitution. In our hands this chromatin yields mononucleosomes identical to those obtained when using long chromatin stored at 4° for *in vitro* nucleosome reconstitution (unpublished observation, 1996). The length of DNA fragments in each fraction of oligonucleosomes should be monitored by agarose gel electrophoresis as in Fig. 1. We prefer to use fragments of 30–60 nucleosomes, i.e., about 5.5- to 11-kbp-long DNA fragments, for nucleosome reconstitution. The quality of the histone proteins should be controlled by SDS–PAGE (see Fig. 1C) as they are very sensitive to potentially contaminating proteolytic activity. If chromatin is stored for longer periods, this control should be repeated.

In Vitro Nucleosome Reconstitution

We usually reconstitute nucleosomes by a histone exchange method that is similar to the method described previously by Losa and Brown[8] using long fragments of rat liver chromatin as the histone source (see earlier discussion).

Solutions

10× reconstitution buffer: 150 m*M* Tris–HCl, pH 7.5; 2 m*M* Na_2EDTA; 2 m*M* PMSF (added fresh just before use)
Donor chromatin: about 5 A_{260} units/ml = 250 μg/ml

Procedures

Routinely, about 0.5–1.5 pmol of a ^{32}P 5′ end-labeled DNA fragment with a radioactivity of 2–5 × 10^6 cpm is lyophilized in a Speed-Vac. To the lyophilized radioactive DNA, 8 μl of 5 *M* NaCl, 4 μl of 10× reconstitution buffer, and 28 μl of donor chromatin are added. It thus renders a final donor chromatin concentration of 175 μg/ml (1 A_{260} units = 50 μg of long chromatin) and the buffer finally contains 1 *M* NaCl, 15 m*M* Tris–HCl, pH 7.5, 0.2 m*M* Na_2EDTA, and 0.2 m*M* PMSF. The mixture is incubated for 30 min at 37° and is followed by stepwise dilution during 4 hr at room temperature with 1× reconstitution buffer but lacking NaCl. Dilution protocol: 6× 5 μl every 15 min, 7× 10 μl every 10 min, and 8× 20 μl every 10 min. After completed dilution, the final NaCl concentration is 0.133 *M* and the final volume is 300 μl. Reconstituted nucleosomes are then separated from long chromatin and nonincorporated DNA fragments by glycerol gradient centrifugation in a 7–30% glycerol gradient containing 50 m*M* Tris–HCl, pH 7.5, 1 m*M* Na_2EDTA, and 0.1 mg/ml BSA (bovine serum albumin fraction V, 775827 Boehringer) with a total volume of 5 ml in a Beckman SW 50.1 rotor at 35,000 rpm for 16 hr at 4°. BSA is added to prevent the nonspecific adherence of nucleosomes to the tube walls. After centrifugation we collect 230 μl per fraction for about 21 fractions total. The 11S mononucleosome peak is usually in the middle of the gradient when the conditions just described are applied. Unincorporated labeled DNA is short, about 200 bp or less (see later), and sediments are considerably slower (see Fig. 2). It should be emphasized that the sedimentation rate of DNA depends on the fragment length. Hence unincorporated long chromatin and long unlabeled DNA fragments originating from the long chromatin will sediment to the bottom of the gradient. Occasionally an

[8] R. Losa and D. D. Brown, *Cell* **50,** 801 (1987).

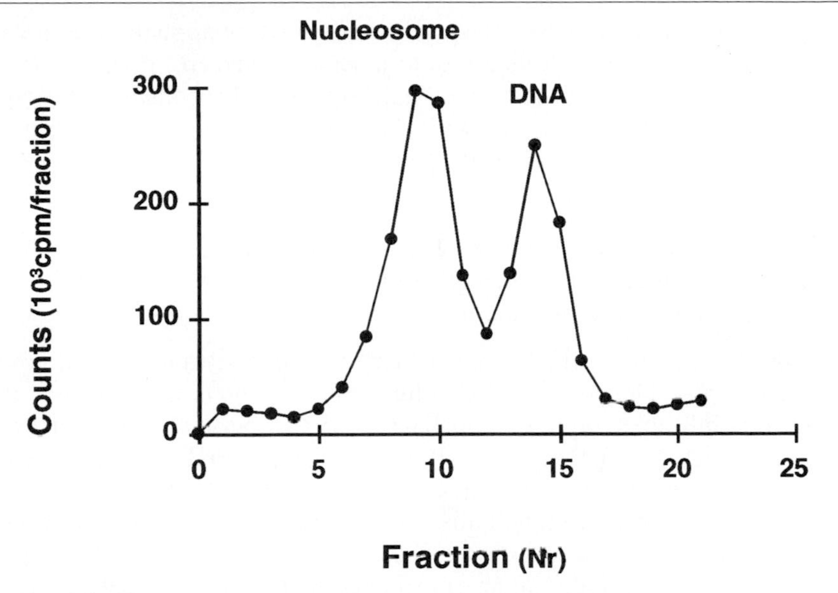

FIG. 2. Purification of *in vitro*-reconstituted mononucleosomes by glycerol gradient centrifugation. The gradient is fractionated from the bottom of the centrifuge tubes and then analyzed by liquid scintillation counting. Radioactivity analysis may be conducted simply by placing snap cap tubes with each of the gradient fractions into liquid scintillation vials followed by Cerenkov counting in a liquid scintillation counter using the setting for tritium detection. The radioactively labeled DNA fragment used here is 197 bp long and harbors the −241/−51 DNA segment of the mouse mammary tumor virus promoter.

additional complex that sediments faster than the 11S mononucleosome complex is observed, which probably represents histone–DNA complexes that contain additional histone protein (Q. Li, U. Björk, and Ö. Wrange, unpublished observation, 1996). The occurrence of this faster-moving complex is dependent on the DNA sequence of reconstituted nucleosome, and such complexes should not be regarded as mononucleosomes.

Purified mononucleosomes are stored at 4°; we do not see detectable DNA dissociation from mononucleosomes even after several weeks at 4°. This may, however, depend on the stability of the DNA fragment used for nucleosome reconstitution (see later). The purity and stability of the reconstituted mononucleosomes may be controlled by electrophoretic mobility shift assay on 5% polyacrylamide gels[9,10] (see also later).

It is valuable to compare the sedimentation rate of reconstituted nucleosomes to that of mononucleosomes generated by the micrococcal nuclease

[9] Y. Lorch, J. W. LaPointe, and R. D. Kornberg, *Cell* **49,** 203 (1987).
[10] T. Perlmann and Ö. Wrange, *EMBO J.* **7,** 3073 (1988).

digestion of long chromatin. This can be achieved by applying the mononucleosome from long chromatin onto a separate glycerol gradient or together with a small amount of radioactively labeled reconstituted mononucleosome on the same gradient.[11] The unlabeled nucleosomes can be detected by $A_{260\ nm}$ analysis of the gradient fractions.

Choice of DNA Fragment for Nucleosome Reconstitution and Characterization of Histone–DNA Organization in Reconstituted Nucleosomes

Histone octamers can be reconstituted onto essentially any double-stranded DNA. The double helix within a nucleosome is strongly bent and forms one full supercoil per 80 bp. The DNA flexibility and curvature are strongly affected by the sequences. It is not surprising that a DNA sequence with a preponderance for a unidirectional bend forms a very stable nucleosome,[12] whereas interruption of this bending sequence destabilizes histone–DNA contacts and increases accessibility for putative DNA-binding proteins.[13] Hence the stability of an entire nucleosome or local histone–DNA contacts within a nucleosome is significantly influenced by the DNA sequence. Consequently the local sequence context around a transcription factor-binding site may have large effects on its DNA-binding affinity in a nucleosome. Therefore it is important to maintain the natural sequence context around a transcription factor-binding site if its binding affinity in a particular chromatin setting is to be evaluated.

However, if the binding of a transcription factor to a reconstituted nucleosome is to be evaluated, as opposed to its binding affinity to free DNA, it is a clear advantage if the reconstituted material is structurally homogeneous. This may be achieved by placing the DNA-binding site in a defined position relative to a DNA-bending sequence.[13] Using this approach, the DNA-bending sequence will usually direct the nucleosome structure such that the rotational and translational positions of the histone octamer relative to the DNA are identical or at least similar for all reconstituted nucleosomes.[1] The reason for taking this approach is to try to avoid heterogeneous populations of nucleosomes that would display a mixture of different affinities caused by differences in rotational or translational positions of the factor-binding element.[1] It should also be mentioned that naturally occurring DNA sequences originating from gene regulatory re-

[11] T. Perlmann and Ö. Wrange, *EMBO J.* **7,** 3073 (1988).
[12] T. E. Shrader and D. M. Crothers, *Proc. Natl. Acad. Sci. U.S.A.* **86,** 7418 (1989).
[13] Q. Li and Ö. Wrange, *Genes Dev.* **7,** 2471 (1993).

gions have been shown to harbor rotational and, to some extent, translational nucleosome positioning sequences.[9,14]

A finding common to many DNA-binding proteins is that if its cognate DNA site is located close to the nucleosome border, then it is more accessible than the same DNA site located close to the nucleosome dyad.[14,15] The glucocorticoid receptor (GR) does not follow this rule because it also binds a glucocorticoid response element (GRE) with high affinity when positioned at the nucleosome dyad.[13] The other extreme is nuclear factor 1 (NF-1), which has a 100- to 300-fold lower affinity for its nucleosomal site compared to free DNA independent of the translational and rotational positions.[3] It should be stressed, however, that a rotationally unfavorable positioned DNA element, where the factor-binding surface forming the base-specific protein contacts is facing toward the histone octamer, in most cases usually has drastically lower accessibility than if this surface is rotationally positioned away from the histone surface.[1]

Analysis of Translational Nucleosome Positioning by Exonuclease III Protection

Because a nucleosome core organizes 146 bp of DNA around the histone octamer, a convenient way to restrict the freedom of its translational positioning is to use DNA fragments only slightly longer than this for nucleosome reconstitution. In our case we have routinely used DNA fragments of 161[1] and 198 bp.[11] Exonuclease III is useful for the analysis of the translational positioning of nucleosomes as this enzyme digests DNA from the 3′ end only. It is important to use parallel reconstituted nucleosomes containing 5′ end-labeled DNA on either strand for comparison of the exonuclease III protection pattern from both strands of the nucleosomal DNA. A histone octamer bound to DNA results in a significantly reduced rate of exonuclease III digestion at the entry of the DNA into the nucleosome and thus generates a band at this position when analyzed on a sequencing gel. However, the histone–DNA contacts do not cause a complete stop but instead only reduce the digestion rate of the exonuclease III. Thus exonuclease III is able to bypass the histone–DNA-contacting point, only to be slowed down again at the next histone–DNA contact along the nucleosome surface, i.e., after another 10 bp. In this way a 10-bp pattern of histone-induced exonuclease III protection is generated. If the first histone-dependent exonuclease III stop occurs 146 bp apart when comparing the two strands of nucleosomal DNA, then the nucleosome has one translational positioning. The digestion pattern obtained for one single

[14] D. Y. Lee, *et al., Cell* **72,** 73 (1993).

[15] H. Chen, B. Li, and J. L. Workman, *EMBO J.* **13,** 380 (1994).

translational positioning as compared to two alternative translational positionings has been published elsewhere.[1,3]

Exonuclease III protection analysis is performed with 1.5 fmol of 5′ end-labeled free or nucleosomal DNA and 15 U/ml of exonuclease III enzyme (Pharmacia) in 100 μl Exo III buffer [20 m*M* Tris–HCl, pH 7.6; 1 m*M* Na_2EDTA; 10% (v/v) glycerol; 5 m*M* DTT; 0.1 mg/ml BSA; 3 m*M* $MgCl_2$] and in a 100-μl reaction volume at 25°. Usually the time of incubation is 2–4 min, but this as well as enzyme concentration may need to be optimized for the substrate used. After incubation the reaction is stopped by adding 15 μl of stop mix (125 m*M* Na_2EDTA; 1% SDS), then 5 μg of *Escherichia coli* tRNA in a 1/10 volume of 3 *M* sodium acetate, pH 5, followed by phenol chloroform (2:1) extraction, ethanol precipitation, and analysis on a 6% polyacrylamide sequencing gel.

Analysis of Rotational Positioning by DNase I Footprinting

The rotational positioning of the DNA fragment on the nucleosome should also be considered. The endonuclease DNase I cuts DNA in the minor groove and with a preference for DNA bent away from the enzyme.[16] It thus cut more frequently on the periphery of the nucleosome and can be excluded from the minor groove facing the histone protein surface. DNA positioned rotationally on a nucleosome with one unique rotational frame will give rise to a 10–bp ladder of DNase I hypersensitive sites with intervening segments protected from DNase I cutting. This pattern is easily detectable when analyzed on a denaturing sequencing gel. DNA lacking any preferred rotational positioning or with several alternative frames of rotational positioning will not show a distinct 10-bp ladder or will, in some cases, even lack a nucleosome-specific pattern when analyzed by DNase I footprinting. If a distinct 10-bp ladder is seen, the rotational positions of the DNA fragment on the histone octamer may be revealed by DNase I footprinting of both DNA strands, e.g., by 5′ end labeling at either end and parallel analysis.[1,11,13] Sequences cut by DNase I are localized by a Maxam–Gilbert sequencing reaction of the naked form of the same DNA fragment[17] (cf. Fig. 5A).

We conclude that the DNA sequence of the fragment used for nucleosome reconstitution is an important determinant of whether it will be positioned rotationally and/or translationally on the histone octamer. If the analysis of nucleosomal DNA access concerns a certain sequence context of natural DNA segment, any information on possible translational or rotational nucleosome positioning *in vivo* should be taken into account

[16] M. Noll, *Nucleic Acids Res.* **1,** 1573 (1974).

[17] A. M. Maxam and W. Gilbert, *Proc. Natl. Acad. Sci. U.S.A.* **74,** 560 (1977).

when designing the DNA fragment to be used for *in vitro* nucleosome reconstitution.[11,18–20] If the capacity of a factor binding to a nucleosome is to be investigated at different nucleosomal contexts, the nucleosomal probe should, in each case, have a reasonably well-defined translational and rotational positioning. The use of a synthetic DNA-bending sequence, as developed by Shrader and Crothers,[12] may be advantageous in order to direct the factor-binding DNA segment into a single rotational frame. The translational positioning can be reasonably well controlled by the use of short, in our case 161 bp, DNA fragments. The strategy for how to use bending DNA to direct a DNA segment into a defined nucleosome positioning *in vitro* is described elsewhere.[1,3,13]

Transcription Factor/DNA Binding: Electrophoretic Mobility Shift Assay

Electrophoretic mobility shift assay (EMSA) is a sensitive, rapid, and well-established method to detect specific interactions between the transcription factor and its DNA-binding site.[21] Application of this method can also reveal the ternary complex of the transcription factor, DNA, and histones. It has been used in several laboratories for the nucleosome binding of many different factors.[13,14,22] The procedure involves two steps: the binding of transcription factor to nucleosomal template and the separation of the DNA-binding factor–nucleosome complex from unbound nucleosomes by agarose or polyacrylamide gel electrophoresis. The optimal binding conditions of a particular protein to its nucleosomal template are case specific, or factor dependent, whereas the electrophoresis conditions are more general. Figure 3 presents a titration experiment utilizing EMSA for evaluating GR binding to two different nucleosomal probes: n3Go1 and n3GoA1.

GR is a glucocorticoid hormone-activated transcription factor. On hormone activation, it regulates gene expression by binding to a GRE in the chromatin-organized promoter.[23] It binds to DNA as a homodimer, which also forms in the absence of DNA.[24] To assess the influence of nucleosome organization on the access of a GRE, we designed two 161-bp DNA fragments: 3Go1 and 3GoA1. Both 3Go1 and 3GoA1 contain an identical

[18] B. Pina, U. Bruggemeier, and M. Beato, *Cell* **60,** 719 (1990).
[19] T. K. Archer *et al., Mol. Cell. Biol.* **11,** 688 (1991).
[20] G. Fragoso, *et al., Genes. Dev.* **9,** 1933 (1995).
[21] M. G. Fried and D. M. Crothers, *J. Mol. Biol.* **172,** 244 (1984).
[22] I. C. A. Taylor, *et al., Genes. Dev.* **5,** 1285 (1991).
[23] K. S. Zaret and K. R. Yamamoto, *Cell* **38,** 29 (1984).
[24] Ö. Wrange, P. Eriksson, and T. Perlmann, *J. Biol. Chem.* **264,** 5253 (1989).

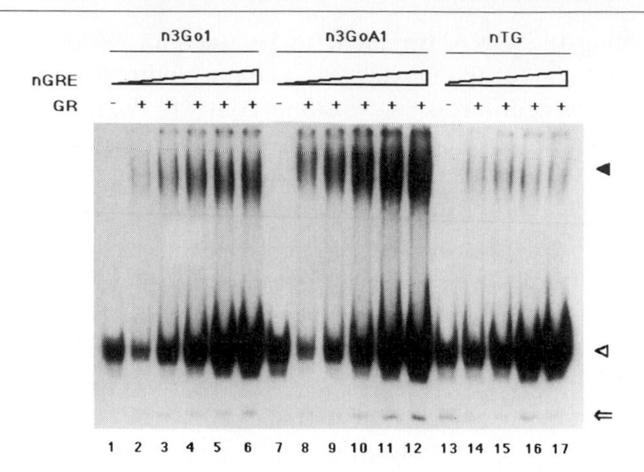

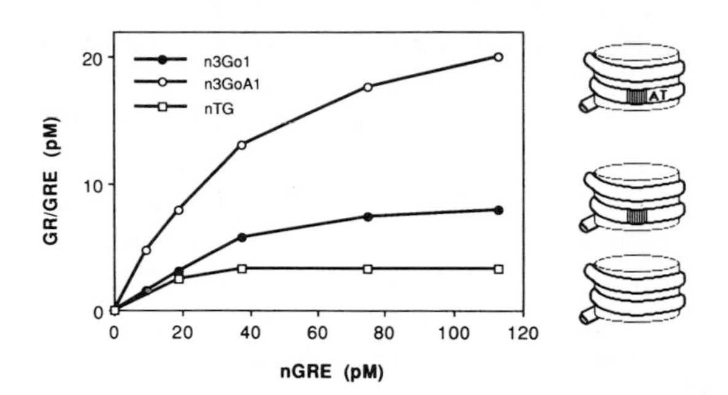

FIG. 3. Binding of GR to nucleosomes analyzed by electrophoretic mobility shift assay. (A) Relative GR affinity to nucleosomal template n3Go1 and n3GoA1. Increasing amounts of indicated nucleosome probe (9.4, 18.8, 37.5, 75.0, and 112.5 p*M*) are incubated with a constant concentration of GR (1.2 n*M*). nTG is used as a control for nonspecific GR binding. Solid arrowhead, the ternary complex of GR and nucleosome probe; open arrowhead, the unbound nucleosomes; open arrow, the position of free DNA visible only after prolonged exposure. (B) Quantitation of the experiment in (A) by PhosphorImager analysis. (Right) Diagrams of nucleosome probes: (top) n3GoA1 (where the GRE, the shadowed segment, is positioned 30 bp from the nucleosome dyad[13] and is flanked by an AT-rich segment, marked AT), (middle) n3Go1 (where the GRE is flanked by a GC triplet in each side), and (bottom) nTG (a nucleosome lacking a GRE and used to measure nonspecific binding of GR).

GRE but with different flanking sequences: in 3Go1 the GRE is flanked by a triplet of GC base pairs on both sides, but in 3GoA1 the GRE is flanked by a GC triplet on one side and by an AT-rich segment on the other side.[13] The DNA fragments are 5′ end labeled using T4 polynucleotide kinase and [γ-^{32}P]ATP and are subjected to nucleosome reconstitution (see earlier) to generate their nucleosomal counterparts n3Go1 and n3GoA1.

The GR used in the binding experiments is purified from rat liver and its concentration is defined as the concentration of active GR monomer based on quantitation of its hormone- and DNA-binding activity.[24] A constant concentration of GR (1.2 n*M*) is incubated with increasing amounts of n3Go1 or n3GoA1 (9.4, 18.8, 37.5, 75.0, and 112.5 p*M*) for 40 min at 25° in a 20-μl reaction in GR-binding buffer [50 m*M* NaCl, 20 m*M* Tris–HCl (pH 7.6), 1 m*M* Na_2EDTA, 10% (v/v) glycerol, 5 m*M* DTT, 0.1 mg pork insulin/pml]. The samples are then separated by electrophoresis on a nondenaturing gel [4% acrylamide of a 1.5-mm gel with 0.03% Nonidet P-40 (NP-40), acrylamide/bisacrylamide: 40/0.5] at 14 V/cm and 4° for 2 hr in 0.25× TBE buffer.[7] Two microliters of loading buffer [0.25% bromphenol blue (BPB) in 50% glycerol] is added to each sample before loading. The inclusion of BPB in the sample is for monitoring the loading and running process without interfering with the stability of the complex. The 0.03% NP-40 in the gel is required to prevent GR from aggregating in the sample well. It is important to prerun the gel at 17.5 V/cm and 4° for 30 min before loading the samples as the buffer front may otherwise disturb the electrophoretic separation. After electrophoresis, the gel is vacuum dried and autoradiographed on X-ray film (Fig. 3A). Quantitation of the shifted bands (Fig. 3B) is performed by PhosphorImager analysis with ImageQuant v3.0 and the Fast Scan System (Molecular Dynamics, Sunnyvale, CA).[25]

If the affinity of a DNA-binding protein is to be evaluated, it is essential to keep its concentration in the binding reaction below the dissociation constant for its binding to the nucleosomal target. In the case of GR this is 1–2 n*M*. With this approach, the affinity of the protein for the nucleosomal site will be driving the complex formation, and hence the amount of ternary complex formed will be correlated directly to the binding affinity. It is also important, however, to generate a signal of enough intensity for the quantitation, so the concentration of the DNA-binding protein in the binding reaction should be titrated carefully. In each binding reaction there is a component of nonspecific binding that needs to be controlled for. Here, nTG, a nucleosome without any GRE, is employed as a control and is used for the subtraction of nonspecific GR binding. As revealed by the experiment, the relative affinity of GR to n3GoA1 is about threefold higher than that of n3Go1. The experiment illustrates that the flanking DNA sequence may have a strong impact on factor accessibility for a nucleosomal DNA site.[13]

[25] R. F. Johnston, S. C. Pickett, and D. L. Barker, *Electrophoresis,* **11,** 355 (1990).

The relative affinity of GR for its nucleosomal target may also be determined by a competition binding assay. In the experiment shown in Fig. 4A, the specific complex of GR and its nucleosomal target, nGo4, is analyzed as a function of increasing amounts of calf thymus (CT) DNA competitor. A constant concentration of GR and nGo4 is incubated with increasing amounts of CT DNA (0, 0.2, 0.8, 3.2, and 12.8 μg/ml). The relative GR affinity is estimated as the concentration of CT DNA required to reduce the specific GR binding to nGo4 site by 50% (Fig. 4B).[21] In this kind of experiment, it is crucial to determine the correct condition for specific GR binding as opposed to nonspecific GR binding. This is influenced by both the concentration of GR and the concentration of the CT

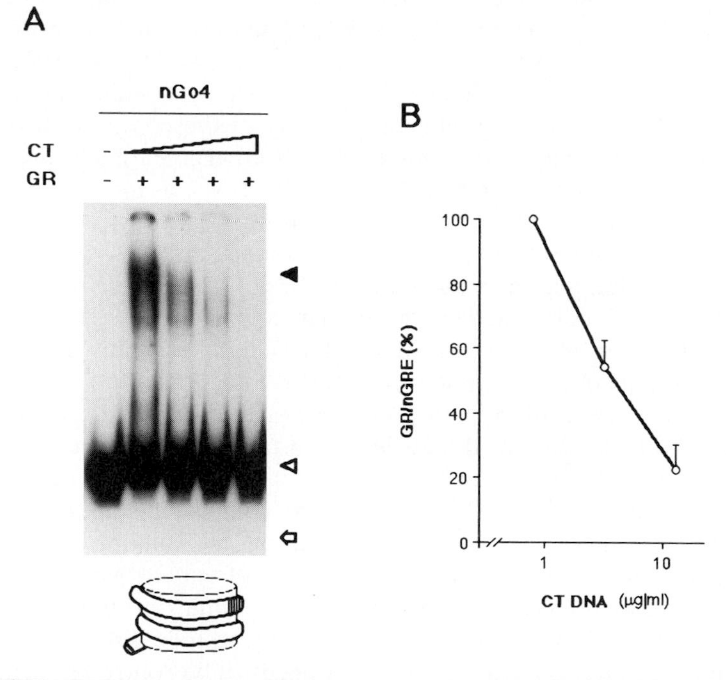

FIG. 4. Competition experiment by electrophoretic mobility shift assay. (A) Relative GR affinity for nucleosome template nGo4. A constant concentration of nGo4 (75 p*M*) and GR (3.6 n*M*) is incubated with increasing amounts of calf thymus DNA competitor (CT) (0, 0.2, 0.8, 3.2, and 12.8 μg/ml). Solid arrowhead, the shifted ternary complex of GR and nGo4; open arrowhead, unbound nucleosomes; open arrow, position of free DNA, only visible after prolonged exposure. (Bottom) Nucleosome probe nGo4 with a GRE, the shadowed segment, located 40 bp from the nucleosome dyad.[13] (B) Quantitation of the experiments as in (A) by PhosphorImager analysis. Experiments performed as in (A) were quantitated (n = 9) by PhosphorImager analysis. Specific binding (100%) is set at 0.8 μg/ml of CT DNA to eliminate the influence of nonspecific binding as inferred from nTG control experiments (see Fig. 3).

DNA in the binding reaction. The contribution of nonspecific binding in a GR–nucleosome complex may be quantitated and subtracted by parallel analysis of a nucleosome with the same DNA sequence but lacking the GRE (see nTG curve in Fig. 3B). Nonspecific binding may also be investigated by methylation interference analysis as described elsewhere.[13,26]

Quantitative DNase I Footprinting

Quantitative DNase I footprinting is a useful method for estimating the dissociation constant of a transcription factor interacting with free DNA.[27] DNase I cleaves nucleosomal DNA at the minor groove facing outward (see earlier discussion). Many transcription factors bind their cognate DNA site through the major groove; GR is one such example. The 10-bp ladder of DNase I cutting, observed on the nucleosome with a rotationally positioned DNA fragment, will be modified on binding of the transcription factor. Hence a DNase I footprint caused by a specific DNA-binding protein on a nucleosome is often much less dramatic in terms of an altered DNase I cutting pattern than the same factor-induced footprint on naked DNA.[1,3,11] In some cases the nucleosome-induced DNase I pattern may in fact overshadow the factor-induced footprint completely, and in such cases the method cannot be used to monitor factor binding to nucleosomes (Q. Li, U. Björk, and Ö. Wrange, unpublished observation, 1995); however, in most cases, quantitative DNase I footprinting is a very useful method for determining the affinity of transcription factor to its nucleosomal site.[1,28]

Solutions

Divalent cation mix: 200 mM $MgCl_2$; 200 mM $CaCl_2$; 6 ng nonspecific DNA/μl

DNase I buffer: 50 mM Tris–HCl, pH 7.5; 0.1 mM $MgCl_2$; 0.1 mM $CaCl_2$

DNase I stop mix: 125 mM EDTA, 1% SDS

Denaturing sample loading buffer: 80% (w/v) formamide, 0.025% (w/v) bromphenol blue, 0.025% xylene cyanol, 1 mM EDTA

Binding of GR to Nucleosome Template

A constant amount of nucleosome template n2Go2 (37.5 pM) is incubated with increasing amounts of GR (0, 2.5, 5.0, 10, and 20 nM) in a 40-μl reaction volume as described for the EMSA.

[26] Q. Li and Ö. Wrange, *Methods* **12,** 96 (1997).

[27] M. Brenowitz. *et al., Methods Enzymol.* **130,** 132 (1986).

[28] T. Perlmann, *Proc. Natl. Acad. Sci. U.S.A.* **89,** 3884 (1992).

DNase I Footprinting

The binding reaction is incubated for 40 min at 25° and then subjected to 30 sec of DNase I treatment at room temperature as the following: first 3 μl of a divalent cation mix and then 3 μl DNase I buffer containing 4–20 ng of DNase I (Boehringer, grade 1) are added. The concentration of DNase I should be titrated out to achieve one cut per strand.[27] DNase I digestion is stopped by adding 15 μl of stop mix. For convenience, the sample is first diluted with 45 μl of water to a final volume of 100 μl, extracted with phenol/chloroform (2:1), precipitated with ethanol, dissolved in 2 μl of denaturing loading buffer, and separated on 6% sequencing gel.[7] The gel is directly vacuum dried and autoradiographed on X-ray film (Fig. 5A).

The DNase I footprinting result is quantitated by PhosphorImager. The GR-induced DNase I footprinting on n2Go2 is plotted as a function of free GR concentration. The binding curve illustrates that the binding is saturable, characteristic of a specific binding reaction (Fig. 5B). Scatchard analysis is carried out as described previously.[29] The dissociation constant is 1.4 n*M*, the maximal GR binding is 62 p*M*, and the correlation coefficient is 0.99 (Fig. 5B, inset).

DMS Methylation Protection Analysis

The specific contact of a transcription factor and a GC base pair in DNA often results in the blocking of DMS methylation of the engaged guanosine, which is methylated readily at the N-7 position by DMS in the absence of bound protein. This feature has been used to study the contact of several transcription factors and their cognate DNA sites.[3,30] We have found this method to be very useful for detection of the specific interaction between transcription factor and its nucleosomal-binding site.[1,3] It should be mentioned that DMS also methylates the N-3 position of adenine, which is located in the minor groove of the DNA and thus can be used to evaluate such protein–DNA contacts. Note that DMS is extremely toxic and that special precautions for its use and waste disposal are strongly recommended in order to avoid any exposure. All work with this chemical should be done in a hood, using gloves and goggles. Small amounts of DMS waste can be disposed of by alkaline hydrolysis in 4 M NaOH solution. Local regulations concerning the storage, use, and disposal of DMS should be consulted before ordering of DMS is even considered.

[29] G. Scatchard, *Proc. Natl. Acad. Sci. U.S.A.* **51,** 660 (1949).

[30] C. Scheidereit and M. Beato, *Proc. Natl. Acad. Sci. U.S.A.* **81,** 3029 (1984).

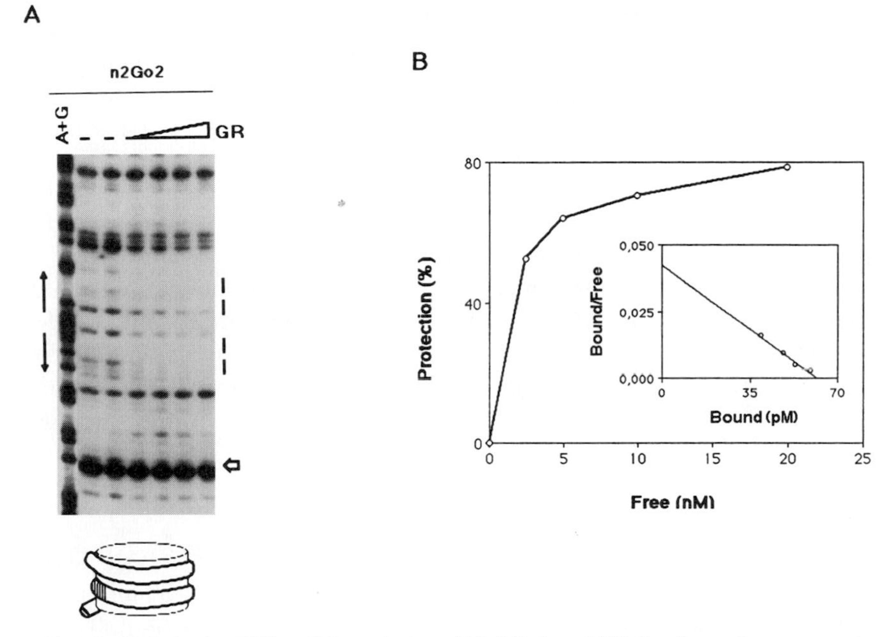

FIG. 5. Quantitative DNase I footprinting. (A) Affinity of GR for the nucleosome probe n2Go2. A constant concentration of the labeled mononucleosome n2Go2 (37.5 p*M*) is incubated with increasing amounts of GR (0, 2.5, 5, 10, and 20 n*M*). Vertical arrows, the position of the partially palindromic GRE; dashed line, the GR-protected region used for the quantitation of GR/n2Go2 binding by PhosphorImager analysis. Open arrow, the reference band, which is not protected by GR and thus is used to correct for variations in sample loading and in the extent of DNase I cutting.[27] (Bottom) Nucleosome probe n2Go2 where the GRE, the shadowed segment, is located at the nucleosome dyad.[13] (B) Quantitation of experiments in (A). GR-induced protection is plotted as a function of free GR concentration according to Scatchard (inset).[29] The apparent dissociation constant, K_d, is 1.4 n*M*; maximal binding is 62 p*M*; and the correlation coefficient is 0.99.

Solutions

DMS stop mix: 1.5 *M* ammonium acetate, 25 m*M* EDTA, 2.8 *M* 2-mercaptoethanol

DNA extraction solution: 20 m*M* EDTA, 0.15% SDS, 100 μg *E. coli* tRNA/ml

Binding of GR to Nucleosomal GRE

GR is incubated at 10 n*M* with 37.5 p*M* of end-labeled nucleosomal DNA n2Go2 in a 20-μl reaction volume as described earlier for EMSA.

Methylation of Nucleosomal DNA by DMS

Two microliters of ice-cold and freshly diluted 10% (v/v) dimethyl sulfate is added to the binding reaction without or with GR for 2 or 4 min, respectively, at 25°. The methylation reaction is inhibited by the addition of 100 μl of ice-cold DMS stop mix and is followed by ethanol precipitation with ice-cold EtOH. Because the DMS stop mix does not completely abolish the DMS methylation reaction, it is essential to work fast and to carry out ethanol precipitation and the following centrifugation as close to 0° as possible. The extent of DMS methylation should be about one methylated guanine per DNA strand. A control for monitoring the extent of DMS methylation occurring after the addition of the DMS stop mix and during ethanol precipitation should also be analyzed. DMS is not added to this control until right before the addition of stop mix. This control may then be used for the correction of background DMS methylation.

Cleavage and Separation of Methylated DNA

The DNA pellet is dissolved in 100 μl of 1 *M* piperidine (freshly diluted with ice-cold water in a glass tube on ice) and incubated for 30 min at 90°. Piperidine is then removed by the addition of 100 μl of DNA extraction solution at ambient temperature and followed by phenol extraction and ethanol precipitation. The DNA pellet is dissolved in 2 μl of denaturing sample loading buffer and finally separated on a 6% polyacrylamide sequencing gel.[7] The extent of protected guanine in the nucleosomal GRE is quantitated using a PhosphorImager. One of the guanines showed a distinct DMS methylation protection as a function of GR binding to the nucleosomal GRE located at the nucleosome dyad (Fig. 6). A reaction without GR was used as reference.

Methodological Concerns and Concluding Remarks

The methods discussed in this article have been found useful for the analysis of binding of transcription factors to nucleosomal DNA-binding sites. These methods are essentially the same as those used routinely to study protein–DNA interaction in the absence of histones. A first concern is whether the *in vitro*-reconstituted material mirrors the correct structure of mononucleosomes. Briefly, this can be addressed by physical characterization of the histone–DNA complex (e.g., by glycerol gradient centrifugation and EMSA) and by the analysis of protein content (e.g., by SDS–PAGE).

The special considerations for using nucleosomal DNA as a template for factor binding are (i) whether the reconstituted nucleosome is homoge-

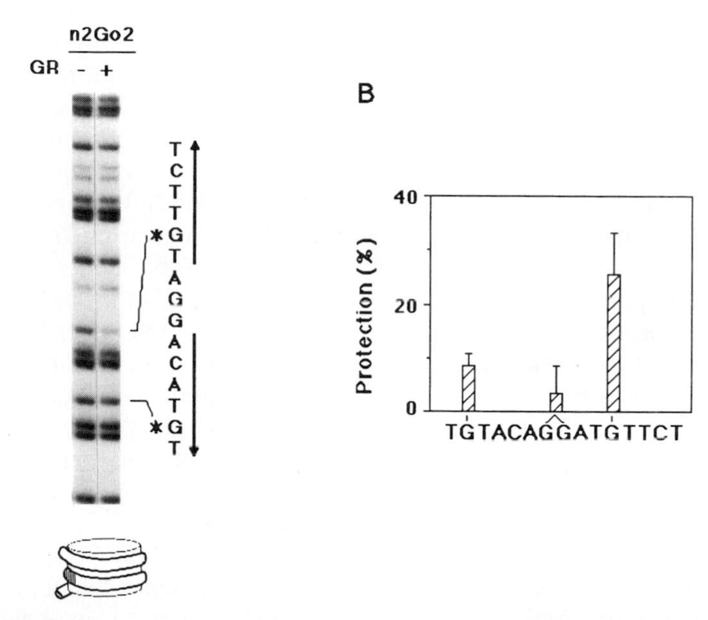

FIG. 6. DMS methylation protection analysis. (A) GR-dependent methylation protection for the top strand of n2Go2. DMS methylation is performed at a constant concentration of n2Go2 (37.5 pM) in the absence (−) or the presence (+, 10 nM) of GR. The two protected guanine residues are marked (*). (Bottom) Nucleosome probe n2Go2 as in Fig. 5. (B) Quantitation of experiments as in (A) by PhosphorImager analysis (n = 7).

neous in terms of translational and rotational positioning (this can be followed by exonuclease III protection analysis or DNase I footprinting, respectively), (ii) how the histone–DNA contacts and nucleosome structure are affected by the factor binding (this can be analyzed by DNase I if the nucleosome harbors a rotationally positioned DNA fragment rendering a distinct 10-bp DNase I ladder), and (iii) to what extent the detection of the ternary complex is disturbed by the nucleosomal or factor-free DNA complex (this can be revealed by EMSA).

The electrophoretic mobility shift assay is a sensitive and fast method and usually is the first method of choice for any protein–DNA-binding study. Potential shortcomings are aggregation and instability of the bound complex during electrophoresis; sometimes the instability of the nucleosome may seriously hamper the evaluation of the results. The aggregation of protein in the sample may be prevented by the use of detergent in the gel, provided that the DNA-binding protein under study is stable in the presence of the detergent. If the DNA has dissociated from the histones or if the nucleosome has been contaminated with naked DNA, it may interfere with the quantitation of the ternary complex when the migration rate of the free DNA–protein complex is the same as that of the nucleosome–protein complex.

DNase I footprinting is a little more time-consuming and technically demanding, but it is usually easy to achieve a complete saturation of the protein-binding site, and quantitation of the results usually generates consistent affinity measurements. Moreover, DNase I footprinting monitors the structure of the nucleosomal DNA, seen as the 10-bp ladder of DNase I hypersensitivity, concomitantly with the analysis of factor binding. This is of particular importance if the rotational position of the factor-binding site is addressed and related to factor access[1] or if the factor binding is causing a nucleosome disruption or any other structural modification of the nucleosomal DNA.[1] When titrating DNase I digestions for affinity studies, it is best to keep the digestion time short, about 30 sec, while varying the enzyme concentration in order to minimize the disturbance of the equilibrium-binding conditions during the DNase I cutting reaction.[27]

DMS methylation protection analysis is a powerful method for studying the binding of a transcription factor to its nucleosome template. This is especially useful when many nonspecific interactions occur or when histone–DNA interactions make the evaluation of specific protein–nucleosome interaction by DNase I footprinting difficult. We have not experienced the occurrence of nonspecific DMS methylation protection or methylation interference, and histone–DNA contacts do not appear to have any influence on DMS methylation. This makes the method especially useful in evaluating specific protein–DNA interactions in nucleosomes; however, as opposed to DNase I footprinting, it does not allow any evaluation of the nucleosome structure. It should also be pointed out that the extent of GR-dependent DMS methylation protection of different guanines within a GRE may vary when the DNA site is in a nucleosomal context compared to in free DNA, which is probably due to the strong bending of the nucleosomal target.[1]

Acknowledgments

This work was supported by the Swedish Cancer Foundation (2222-B97-13XCC). We are grateful to Dr. Tom Klenka for critically reading the manuscript.

[18] Assays for Chromatin Remodeling during DNA Repair

By JONATHAN G. MOGGS and GENEVIÈVE ALMOUZNI

Introduction

The biochemical mechanisms for remodeling chromatin during the repair of DNA damage in eukaryotes are not well understood. The most detailed studies of the interrelationship between chromatin dynamics and DNA repair have focused on the nucleotide excision repair (NER) pathway.[1] Eukaryotic NER enzymes repair a variety of DNA lesions, including UV photoproducts (cyclobutane pyrimidine dimers and 6-4 photoproducts) and bulky chemical adducts.[2,3] Observations of transient changes in the nuclease sensitivity of chromatin undergoing NER in cultured human cells have led to the proposal of a model in which nucleosomes are rearranged during NER.[4]

A simplified approach for investigating the mechanism of nucleosome rearrangements associated with NER involves the use of naked DNA substrates containing lesions that are repaired by NER as a model for the disrupted chromatin organization observed during NER *in vivo*. Incubation of UV-C-irradiated DNA with eukaryotic cell extracts proficient for both NER and chromatin assembly has provided insights into the biochemical mechanism and protein factors involved in the assembly of nucleosomes on NER substrates.[5] Detailed methods for using *Xenopus* egg extracts and UV-C-irradiated DNA to perform a simultaneous analysis of NER and chromatin assembly on the same damaged circular DNA template have been described.[6] The use of *Drosophila* preblastoderm embryo extracts in conjunction with a DNA substrate containing a site-specific cisplatin cross-link has proven to be a powerful model system for the selective and precise

[1] E. C. Friedberg, G. C. Walker, and W. Siede, "DNA Repair and Mutagenesis." ASM Press, Washington, D.C., 1995.

[2] R. D. Wood, *Annu. Rev. Biochem.* **65,** 135 (1996).

[3] A. Sancar, *Annu. Rev. Biochem.* **65,** 43 (1996).

[4] M. J. Smerdon, *in* "DNA Repair Mechanisms and Their Biological Implications in Mammalian cells" (M. W. Lambert and J. Laval, eds.), p. 271. Plenum, New York, 1989.

[5] P.-H. Gaillard, E. M. Martini, P. D. Kaufman, B. Stillman, E. Moustacchi, and G. Almouzni, *Cell* **86,** 887 (1996).

[6] P.-H. Gaillard, D. R. Roche, and G. Almouzni, *in* "Chromatin Protocols" (P. B. Becker, ed.). Humana Press, Totowa, NJ, 1998.

0076-6879/99 $30.00

analysis of nucleosome assembly from a single target site for NER.[7] This article describes specific properties of this system that should be helpful in designing experiments to investigate the chromatin dynamics associated with DNA repair and other DNA transactions.

Covalently Closed Circular DNA Substrates Containing a Single Site-Specific DNA Lesion

We have utilized a covalently closed circular DNA substrate containing a single site-specific DNA lesion formed by the chemotherapeutic drug cisplatin to analyze nucleosome assembly from a single site of NER.[7] Cisplatin reacts with purine bases in DNA to form intrastrand cross-links, interstrand cross-links, and monoadducts. The main repair mechanism for intrastrand cross-links is believed to be NER, and although cisplatin forms primarily 1,2-intrastrand cross-links between adjacent purines, NER enzymes repair the less common 1,3-intrastrand cross-links with a much higher efficiency due to their more distorting structure.[8,9] Furthermore, several biochemical assays are available for the precise analysis of NER of a single 1,3-intrastrand cisplatin cross-link[10–12] (discussed later in section on embryo extracts).

Methods for preparing covalently closed circular DNA substrates containing a single site-specific 1,3-intrastrand cisplatin cross-link have been described in detail[13] and are summarized in Figs. 1A–1C. Briefly, a 24-mer oligonucleotide containing a unique GTG sequence is reacted with cisplatin to form a single 1,3-intrastrand d(GpTpG)–cisplatin cross-link. The formation of this cross-link reduces the mobility of the oligonucleotide in a gel, allowing excision of the desired product. After purification and 5′-phosphorylation, the platinated oligonucleotide is annealed to the complementary sequence in a single-stranded M13 DNA derivative. The 3′ terminus of the oligonucleotide acts as a primer for complementary strand synthesis by T4 DNA polymerase. The newly synthesized DNA strand is

[7] P.-H. Gaillard, J. G. Moggs, D. M. J. Roche, J.-P. Quivy, P. B. Becker, R. D. Wood, and G. Almouzni, *EMBO J.* **16,** 6281 (1997).

[8] D. B. Zamble, D. Mu, J. T. Reardon, A. Sancar, and S. J. Lippard, *Biochemistry* **35,** 10004 (1996).

[9] J. G. Moggs, D. E. Szymkowski, M. Yamada, P. Karran, and R. D. Wood, *Nucleic Acids Res.* **25,** 480 (1997).

[10] J. G. Moggs, K. J. Yarema, J. M. Essigmann, and R. D. Wood, *J. Biol. Chem.* **271,** 7177 (1996).

[10a] G. Almouzni and A. P. Wolffe, *Genes Dev.* **7,** 2033 (1993).

[11] E. Evans, J. Fellows, A. Coffer, and R. D. Wood, *EMBO J.* **16,** 625 (1997).

[12] E. Evans, J. G. Moggs, J. R. Hwang, J.-M. Egly, and R. D. Wood, *EMBO J.* **16,** 6559 (1997).

[13] M. K. K. Shivji, J. G. Moggs, I. Kuraoka, and R. D. Wood, *in* "DNA Repair Protocols: Eukaryotic Systems" (D. S. Henderson, ed.). Humana Press, Totowa, NJ, 1998.

covalently closed with T4 DNA ligase and nicked forms are eliminated by CsCl/ethidium bromide density gradient centrifugation. Control DNA substrates are prepared in parallel by the same method but starting with a nonmodified oligonucleotide. These substrates are stable for more than 1 year when stored in 10 m*M* Tris, 1 m*M* EDTA, pH 8.0, at −80°. The two different DNA substrates employed for the analysis of nucleosome assembly from a single site of NER are shown in Fig. 1D.

Drosophila Embryo Extracts Competent for Nucleosome Assembly and Nucleotide Excision Repair

Detailed methods for the preparation of *Drosophila* preblastoderm embryo extracts have been described previously.[14–16] These extracts contain an endogenous pool of maternal histones as well as high amounts of enzymes and cofactors involved in DNA metabolism and support the assembly of physiologically spaced nucleosomes onto plasmid DNA. Briefly, preblastoderm embryos [0–90 min after egg laying (AEL)] are harvested from a population of approximately 200,000 flies. After dechorionation, embryos are homogenized and successive centrifugation steps are used to obtain a soluble "cytosolic" extract, which is quick frozen and stored as aliquots at −80°. Extracts can also be prepared from *Drosophila* postblastoderm embryos,[15,17] although they need to be supplemented with purified core histones for efficient nucleosome assembly (due to the depletion of maternal histones during the rapid replication cycles of early *Drosophila* development). Postblastoderm embryo extracts used for this study were prepared from embryos collected for 2 hr AEL and then aged outside of fly cages for a further 4 hr.

We have investigated the capacity of both *Drosophila* preblastoderm and postblastoderm embryo extracts to support NER reactions using two different functional assays that measure incision of the damaged DNA strand and DNA repair synthesis, respectively. Detection of short damaged oligonucleotides after the formation of dual incisions flanking a 1,3-intrastrand cisplatin cross-link provides a highly specific assay for NER activity.[10,13,18] With *Drosophila* preblastoderm and postblastoderm embryo extracts, oligonucleotides 27–33 nucleotides long were detected (Fig. 2A)

[14] P. B. Becker and C. Wu, *Mol. Cell. Biol.* **12,** 2241 (1992).

[15] P. B. Becker, T. Tsukiyama, and C. Wu, *Methods Cell Biol.* **44,** 207 (1994).

[16] T. A. Blank, R. Sandaltzopoulos, and P. B. Becker, *Methods* **12,** 28 (1997).

[17] R. T. Kamakaka, M. Bulger, and J. T. Kadonaga, *Genes Dev.* **7,** 1779 (1993).

[18] A. M. Sijbers, W. L. de Laat, R. R. Ariza, M. Biggerstaff, Y.-F. Wei, J. G. Moggs, K. C. Carter, B. K. Shell, E. Evans, M. C. de Jong, S. Rademakers, J. de Rooij, N. G. J. Jaspers, J. H. J. Hoeijmakers, and R. D. Wood, *Cell* **86,** 811 (1996).

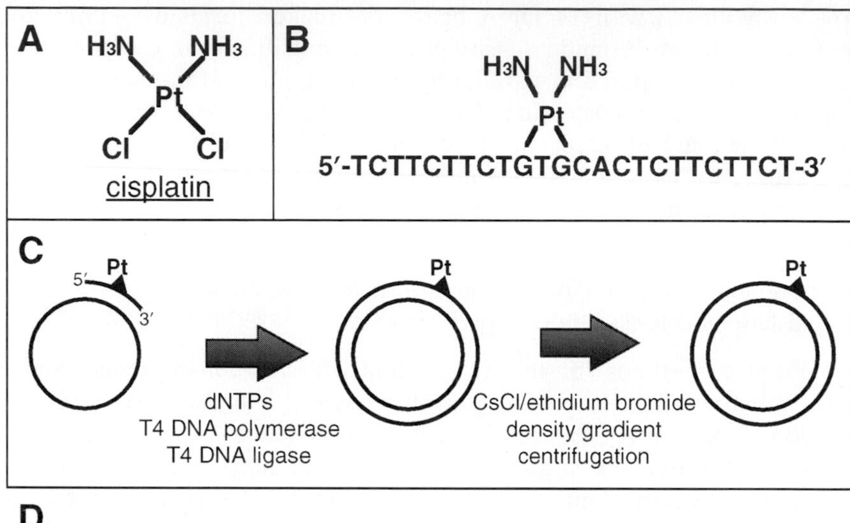
A
H3N
NH3
Pt
Cl
Cl
cisplatin
B
H3N
NH3
Pt
5′-TCTTCTTCTGTGCACTCTTCTTCT-3′
C
Pt
5′
3′
dNTPs
T4 DNA polymerase
T4 DNA ligase
CsCl/ethidium bromide
density gradient
centrifugation

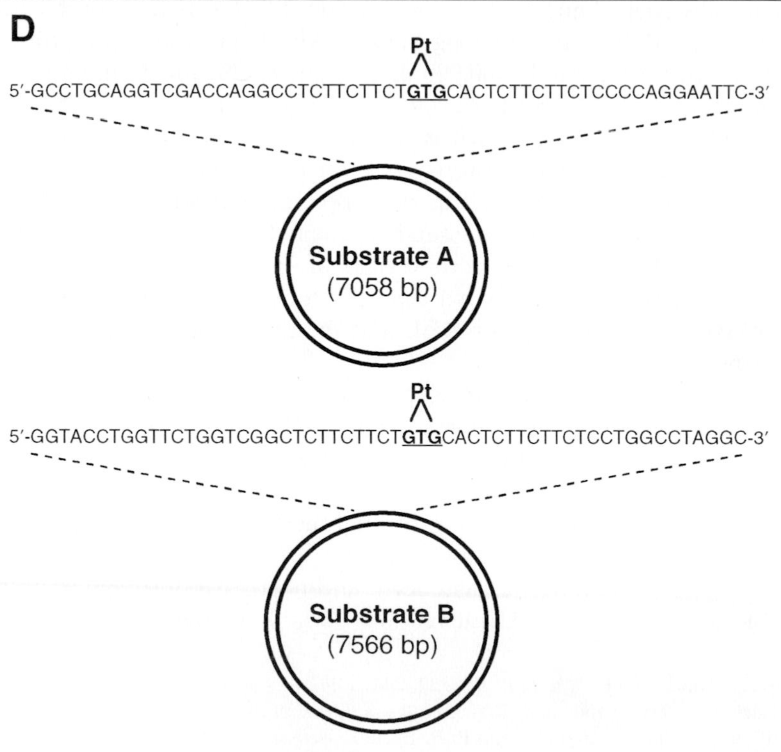
D
Pt
5′-GCCTGCAGGTCGACCAGGCCTCTTCTTCTGTGCACTCTTCTTCTCCCCAGGAATTC-3′
Substrate A
(7058 bp)
Pt
5′-GGTACCTGGTTCTGGTCGGCTCTTCTTCTGTGCACTCTTCTTCTCCTGGCCTAGGC-3′
Substrate B
(7566 bp)

consistent with authentic NER activity. The precise pattern of oligonucleotides observed varies among invertebrate, mammalian, and amphibian cell-free systems[7,19] and is also dependent on the local DNA sequence context of the cisplatin cross-link (see Figs. 1D, 2A, and 4A). DNA repair synthesis is expected to be limited to a short patch at the site of a single 1,3-intrastrand cross-link corresponding to the sizes of excised damaged oligonucleotides and can be monitored by the specific incorporation of ^{32}P-labeled dNMPs into a restriction fragment spanning the site of the lesion.[7,10] Most lesion-dependent radiolabel incorporation was observed within the 40-bp restriction fragment of substrate B after incubation with either *Drosophila* preblastoderm or postblastoderm embryo extracts (Fig. 2B) consistent with the size range of excised damaged oligonucleotides observed in Fig. 2A.

The efficiency of nucleotide excision repair of UV photoproducts and a 1,3-intrastrand cisplatin–DNA cross-link can be estimated using DNA repair synthesis assays.[7,20] The number of femtomoles of dNMP incorporated during a repair synthesis reaction can be calculated by excision and Cerenkov counting of a gel slice containing the repaired DNA, taking into account the specific activity of the ^{32}P-labeled dNTP and a correction for NER-independent DNA synthesis in nondamaged control DNA molecules. The percentage of lesions repaired can then be calculated from the quantity of DNA present in the reaction, the number of DNA lesions per molecule, and by making the assumption that repair patches are 30 nucleotides long. The use of single lesion DNA substrates avoids the need to estimate the number of lesions present per DNA molecule and allows an additional correction for the base composition of the damaged DNA strand within the restriction fragment spanning the lesion site, thus potentially improving the accuracy of the calculated NER efficiency. This approach was used to calculate a repair efficiency of $\approx$40% for DNA containing a single 1,3-intrastrand cisplatin–cross-link (based on the radiolabel incorporation into

[19] J. G. Moggs, P.-H. Gaillard, R. D. Wood, and G. Almouzni, unpublished results (1996).
[20] R. D. Wood, M. Biggerstaff, and M. K. K. Shivji, *Methods* **7,** 163 (1995).

FIG. 1. Covalently closed circular DNA substrates containing a single 1,3-intrastrand cisplatin–DNA cross-link. (A) Structure of cisplatin [*cis*-dichlorodiammineplatinum(II)]. (B) A 24-mer oligonucleotide containing a single 1,3-intrastrand d(GpTpG)–cisplatin cross-link bridging bases 10 and 12. (C) Schematic of synthesis and purification of covalently closed circular DNA containing a single 1,3-intrastrand cisplatin–DNA cross-link. (D) Single lesion DNA substrates A and B were derived from M13mp18[10] and M13mp19E4G5,[10a] respectively, by insertion of a DNA sequence complementary to the platinated 24-mer oligonucleotide shown in (B). The local DNA sequence context of the 1,3-intrastrand cisplatin cross-link is shown in each case.

the 33-bp *Bst*NI restriction fragment of substrate A) after 2 hr of incubation with a *Drosophila* preblastoderm embryo extract.[7]

Subsequent sections describe methods for the selective analysis of nucleosome assembly from a single NER site using *Drosophila* preblastoderm embryo extracts. Under these specific reaction conditions, NER activity in preblastoderm embryo extracts was not significantly sensitive to changes of KCl concentrations within the physiological range (Fig. 2B). This is in contrast to the NER activity in postblastoderm embryo extracts (compare lanes 7 and 9), *Xenopus* egg extracts,[7] or cultured human cell extracts[21] that are inhibited by elevated KCl concentrations. Thus, in addition to subtle species-specific differences in the NER reaction, there may also be important differences between repair enzyme activities derived from different developmental stages. However, we cannot exclude the possibility that the methods for extract preparation from different eukaryotic species and/or developmental stages lead to the observed differences in NER activity.

Selective Analysis of Nucleosome Assembly From Target Site for Nucleotide Excision Repair

We routinely assess the quality of nucleosome assembly reactions on single lesion DNA substrates by partial digestion with micrococcal nuclease (MNase), an enzyme that preferentially cleaves the most accessible regions of nucleosomal DNA (e.g., linker DNA between nucleosomes), resulting in a series of oligonucleosomal fragments. If an array of regularly spaced nucleosomes is assembled on the majority of DNA substrate, then agarose gel electrophoresis of the associated DNA fragments after MNase digestion produces a characteristic ladder of bands whose lengths differ by the size of DNA in a nucleosomal repeat unit. The spacing of nucleosomes assembled with *Drosophila* embryo extracts resembles the physiological nucleosomal repeat lengths observed in *Drosophila* nuclei.[14] MNase-digested DNA can also be subjected to further analysis to provide information on the extent of nucleosome assembly at different positions in the DNA substrate (discussed later in section on analysis of mechanism).

Reaction conditions optimized for efficient nucleosome assembly on plasmid DNA by *Drosophila* preblastoderm embryo extracts[16] led to the formation of regular nucleosomal arrays on both single lesion and control DNA substrates (Fig. 3A). Thus, under these conditions, it was not possible to selectively observe any nucleosome assembly potentially initiated by or

[21] R. D. Wood, P. Robins, and T. Lindahl, *Cell* **53,** 97 (1988).

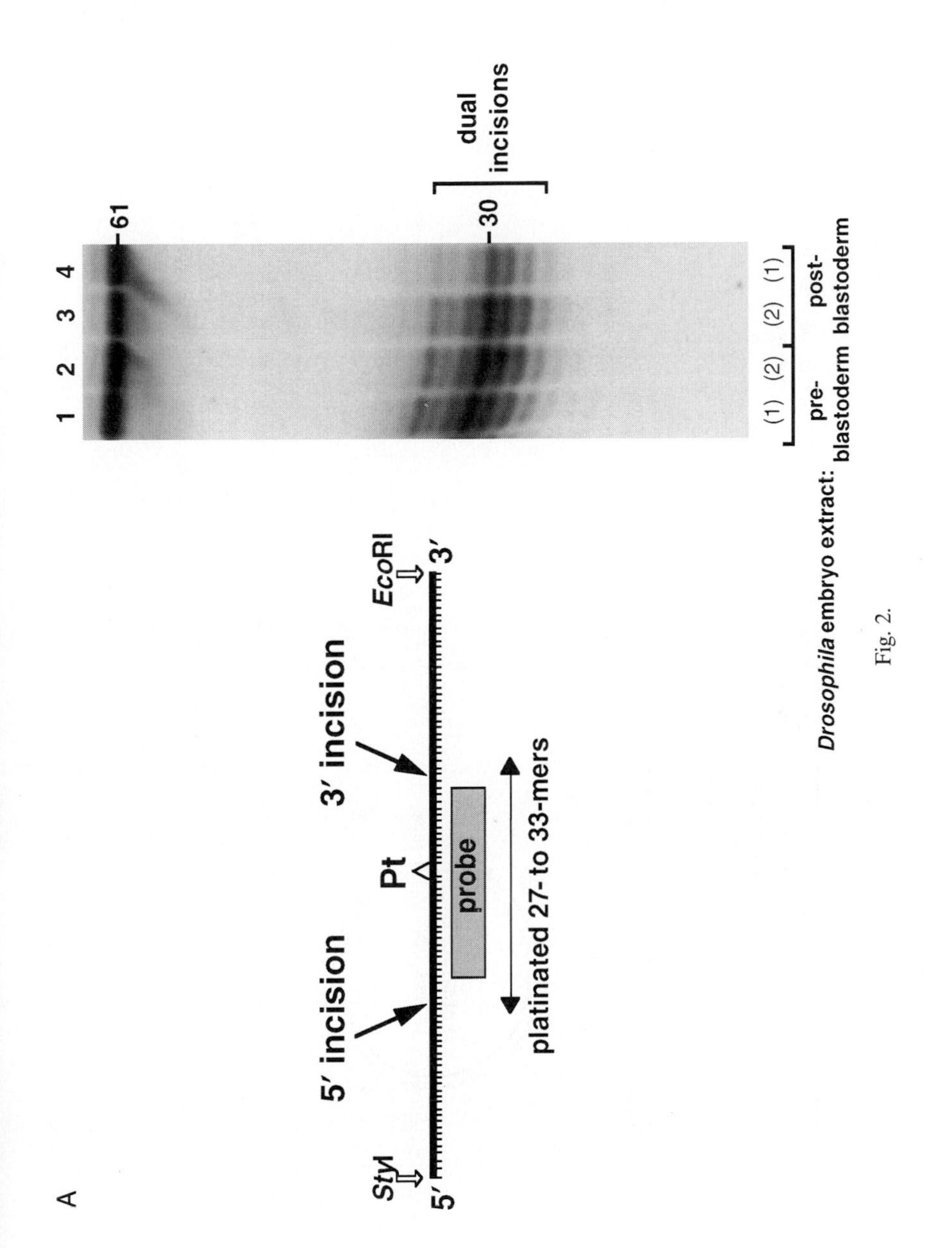
A
StyI
5′
5′ incision
Pt
3′ incision
EcoRI
3′
probe
platinated 27- to 33-mers
1 2 3 4
61
30
dual incisions
(1) (2) (2) (1)
pre-blastoderm
post-blastoderm
Drosophila embryo extract:

Fig. 2.

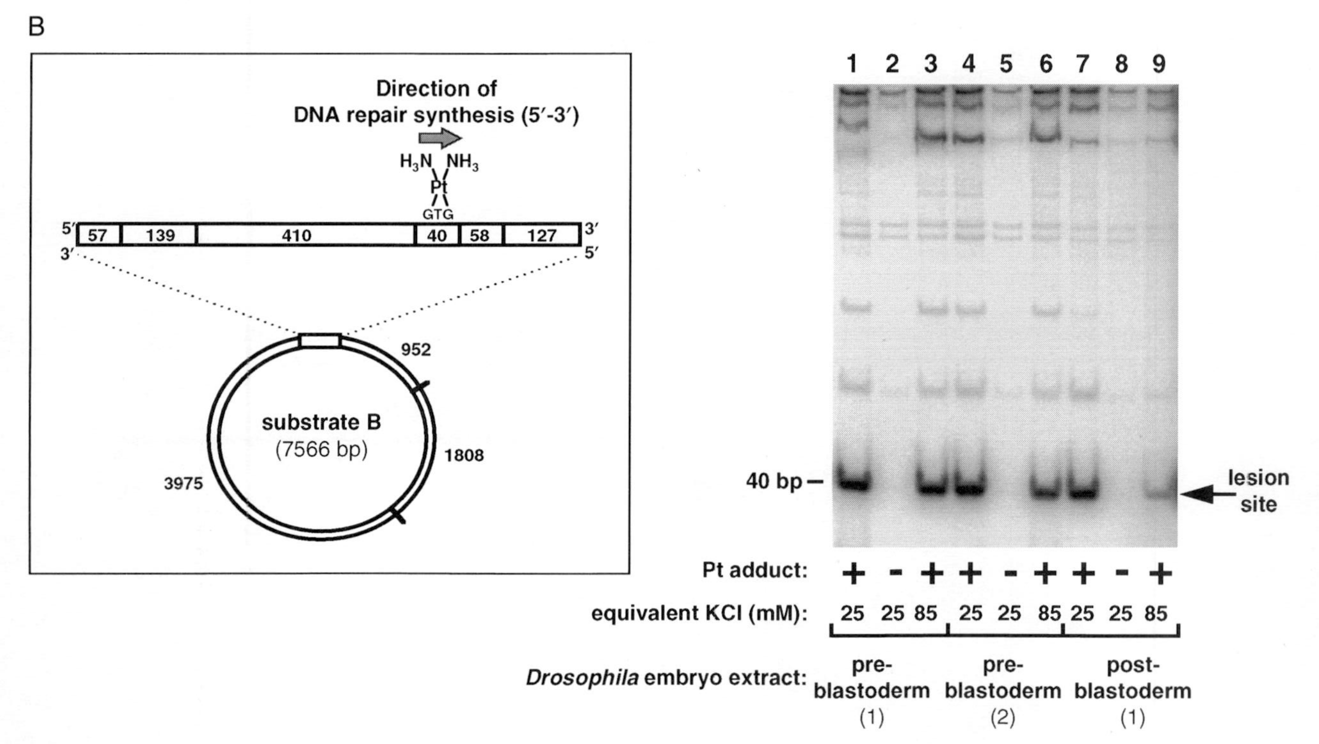

Fig. 2. (*continued*)

coupled to NER. Consistent with previous studies,[22] we observed an increased spacing between the DNA bands within the MNase ladders derived from reactions containing higher ionic strengths. This may reflect altered spacing within assembled nucleosomal arrays via charge repulsion effects on nucleosomes that are translationally adjacent to one another or possibly altered MNase accessibility due to the influence of ionic strength on higher order nucleosomal organization.

Repair-independent assembly of DNA into nucleosomal arrays can, however, be suppressed using the following modified assembly reaction conditions[7] (see Fig. 3B). Generally, *Drosophila* preblastoderm embryo extract and DNA substrate (either $\approx$7 or $\approx$7.5 kbp) are incubated in a 75-μl reaction containing 40 m*M* HEPES–KOH, pH 7.8, 5 m*M* $MgCl_2$, 0.5 m*M* DTT, 4 m*M* ATP (dilithium salt), 5 μCi [α-^{32}P]dCTP (3000 Ci/mmol), 40 m*M* phosphocreatine (disodium salt), and 2.5 μg creatine phosphokinase (type I) at 23° for 6 hr. The optimum amount of DNA, extract, and the final ionic strength (generally $\geq$85 m*M* equivalent KCl) should be determined empirically for each new extract preparation. The contribution of an extract to the final ionic strength of the reaction can be measured (in equivalent KCl) using a conductivity meter (e.g., Model CDM210, Radiometer Copenhagen). The final equivalent KCl concentration of the reaction is then obtained by adding KCl. *Drosophila* preblastoderm embryo extracts are crude and contain endogenous pools of deoxynucleotide triphosphates (estimated to be approximately 50 μM by isotopic dilution[23]). The choice of ^{32}P-

[22] T. A. Blank and P. B. Becker, *J. Mol. Biol.* **252,** 305 (1995).
[23] J. J. Blow and R. A. Laskey, *Cell* **47,** 577 (1986).

FIG. 2. Analysis of nucleotide excision repair catalyzed by *Drosophila* embryo extracts. Single lesion DNA substrate B was incubated for 40 min with *Drosophila* embryo extracts under the reaction conditions described in the text at the equivalent KCl concentrations indicated, except that [^{32}P]dCTP was omitted from the reaction mixture for the incision assay. Pre- and postblastoderm embryo extract reactions contained equal amounts of total protein. DNA purification included extraction with phenol : chloroform : isoamyl alcohol (25 : 24 : 1, v/v) prior to ethanol precipitation. (A) Southern hybridization analysis of NER incisions formed around a 1,3-intrastrand cisplatin–DNA cross-link. Purified DNA was digested with two restriction enzymes (61 nucleotides apart) flanking 3′ and 5′ incision sites. After separation by denaturing polyacrylamide gel electrophoresis, DNA fragments were transferred to a nylon membrane that was hybridized with a radiolabeled oligonucleotide probe complementary to the DNA sequence flanking the cisplatin cross-link, allowing the detection of dual incision events as oligonucleotides 27–33 nt long. Nonrepaired or fully repaired DNA is detected as 61-mers. (B) Analysis of the extent of DNA synthesis associated with NER of single 1,3-intrastrand cisplatin–DNA cross-link. Purified DNA was digested with *Bst*NI prior to separation in a nondenaturing polyacrylamide gel. The sizes of *Bst*NI restriction fragments spanning and adjacent to the repair site in substrate B are indicated.

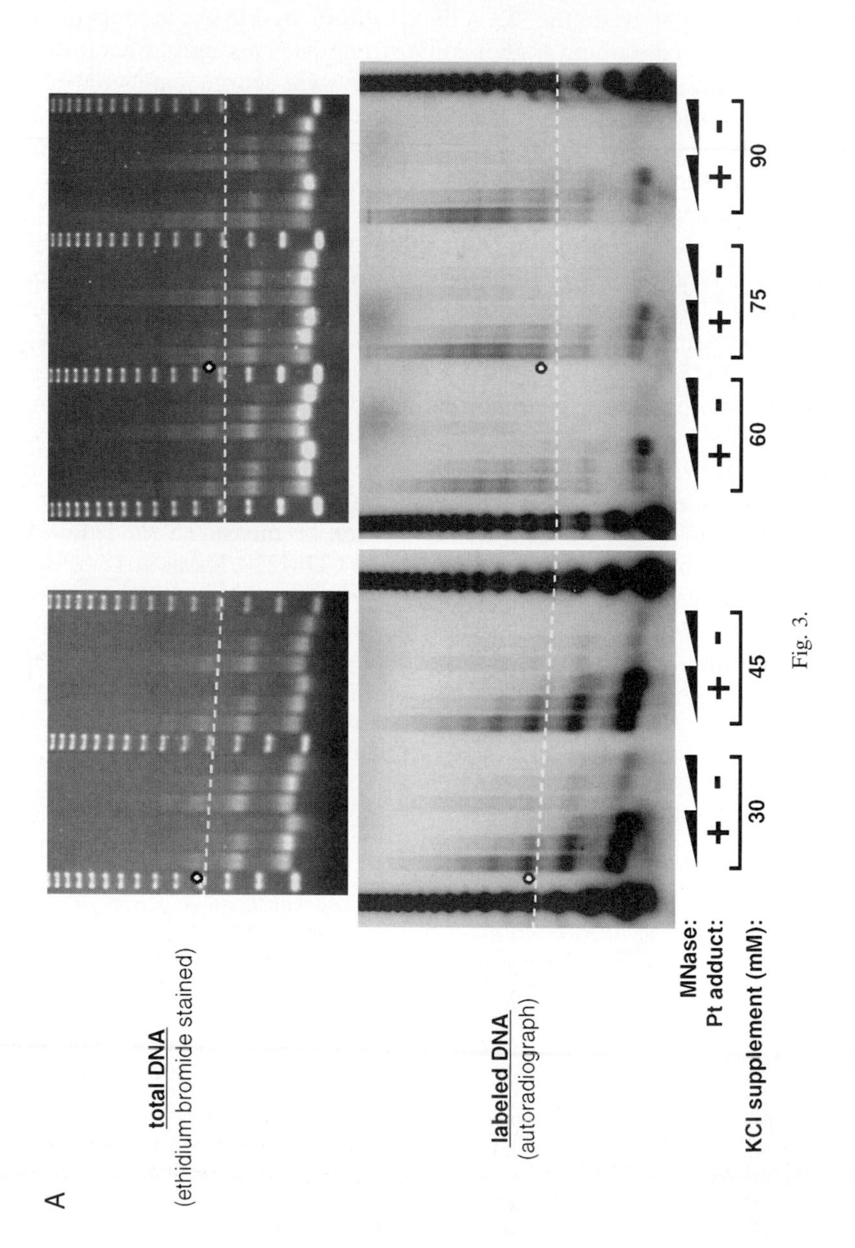
A
total DNA
(ethidium bromide stained)
labeled DNA
(autoradiograph)
MNase:
Pt adduct:
KCl supplement (mM):
+
-
30
45
60
75
90

Fig. 3.

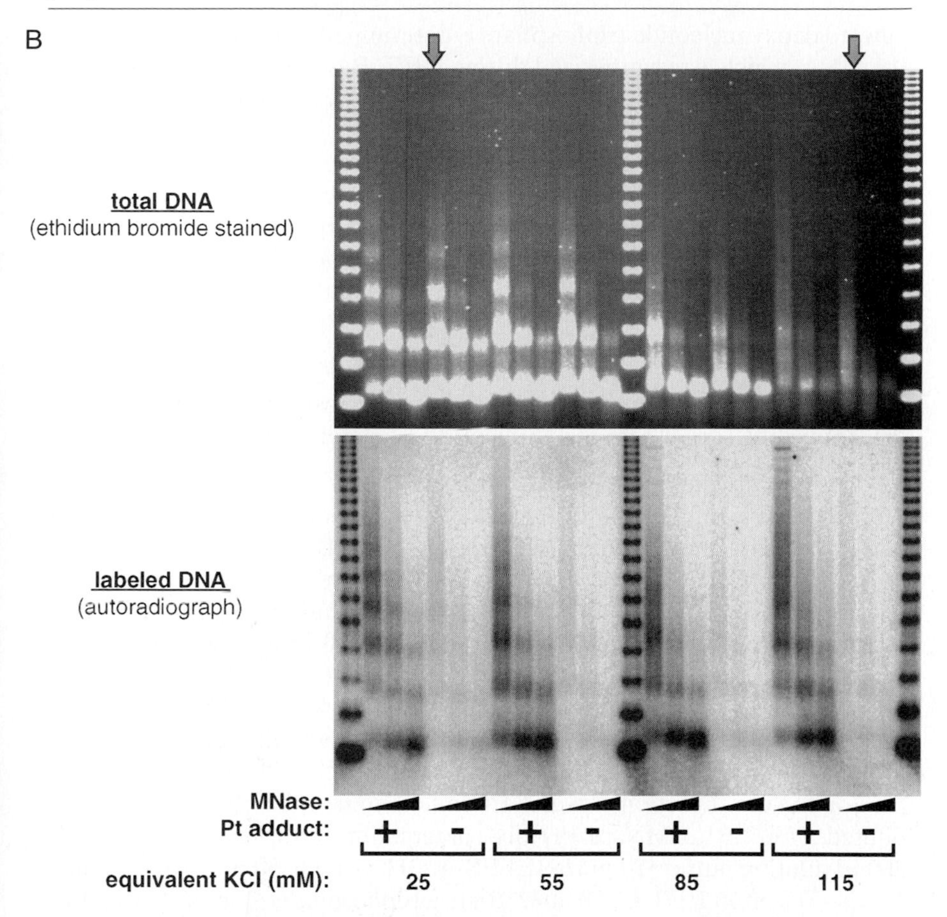

FIG. 3. Micrococcal nuclease analysis of nucleosome assembly/nucleotide excision repair reactions using *Drosophila* preblastoderm embryo extracts. (A) Incubation of either single lesion or control DNA substrates with *Drosophila* preblastoderm embryo extracts under reaction conditions optimized for efficient nucleosome assembly[16] resulted in the formation of regularly spaced nucleosomal arrays over a range of ionic strengths (top, total DNA). Under these reaction conditions the assembly of nucleosomes onto repaired DNA (bottom, labeled DNA) was indistinguishable from those assembled onto unrepaired or control DNA. The effect of ionic strength on the spacing of the DNA bands is indicated by the relative distance of the round symbols (denoting trinucleosomal DNA) from the dashed line drawn between DNA size markers (123-bp ladder, GIBCO-BRL). (B) Selective analysis of nucleosome assembly on DNA containing a single NER site using *Drosophila* preblastoderm embryo extracts. Increasing the equivalent KCl concentration to ≥85 m*M* under the reaction conditions described in the text preferentially suppresses the assembly of nucleosomes onto control or unrepaired DNA (e.g., compare the two lanes highlighted by arrows). Consistent with the specific suppression of repair-independent nucleosome assembly, the quality and extent of assembly of NER-labeled DNA were similar at all equivalent KCl concentrations tested (bottom, labeled DNA).

labeled deoxynucleotide triphosphate is determined by the base sequence of the DNA, which is resynthesized during NER. Base sequences flanking the single cisplatin–DNA cross-link in both substrates A and B are cytosine rich (Fig. 1D).

The biochemical basis for the suppression of repair-independent nucleosome assembly under these specific reaction conditions is not fully understood. However, one potentially important difference is that the overall ionic strength we employ is higher than the reaction conditions described previously for chromatin assembly by *Drosophila* preblastoderm embryo extracts.[16]

Protocols for MNase Analysis of Nucleosome Assembly on DNA Containing Target Site for Nucleotide Excision Repair

The typical composition of a reaction (75 μl) for selective analysis of nucleosome assembly from a single NER site is as follows: 9 μl covalently closed circular DNA containing single cisplatin–DNA adduct (50 ng/μl), 30 μl *Drosophila* preblastoderm embryo extract (contributes ≈28 m*M* equivalent KCl to the reaction), 15 μl 5× reaction buffer (200 m*M* HEPES–KOH, pH 7.8, 25 m*M* $MgCl_2$, 2.5 m*M* DTT, 200 m*M* phosphocreatine), 3 μl ATP (100 m*M*), 0.5 μl [α-^{32}P]dCTP (3000 Ci/mmol), 3 μl creatine phosphokinase (2.5 mg/ml in H_2O), 6 μl KCl (1 *M*) and 8.5 μl H_2O. The extract is allowed to equilibrate with buffer components for 5 min at 23°. DNA is then added and the reaction is incubated at 23° for up to 6 hr.

For MNase digestion, 75-μl NER/nucleosome assembly reactions are diluted with 213 μl MNase premix (preequilibrated to 23°) consisting of 200 μl dilution buffer [10 m*M* HEPES–KOH, pH 7.6, 50 m*M* KCl, 1.5 m*M* $MgCl_2$, 0.5 m*M* EDTA, 10 m*M* β-glycerophosphate, 1 m*M* DTT, 10% (v/v) glycerol], 10 μl $CaCl_2$ (100 m*M*), and 3 μl MNase (15 U/μl in H_2O, Boehringer Mannheim, Mannheim, Germany). Aliquots (90 μl) are removed at 15, 45, and 180 sec and mixed with 22.5 μl nuclease stop buffer [2.5% (w/v) *N*-lauryl sarkosyl, 100 m*M* EDTA]. Arrested MNase reactions are incubated with DNase-free RNase (500 μg/μl) at 37° for 30 min. SDS is added to 0.5% and reactions are incubated with proteinase K (500 μg/μl) at 37° for ≥12 hr. DNA is then precipitated with 2 volumes of ethanol in the presence of 3 *M* ammonium acetate and 10 μg glycogen (Boehringer). DNA pellets are washed with 80% (v/v) ethanol and dried before resuspension in 4 μl buffer (10 m*M* Tris–HCl, pH 8.0, 1 m*M* EDTA). DNA is separated by electrophoresis in a 1.3% agarose gel run in 0.5× TBE buffer at 4 V/cm for 3–4 hr. These conditions usually result in the resolution of DNA fragments corresponding to nucleosomal fragments containing up to ≈8 nucleosomes. A 123-bp DNA ladder (GIBCO-BRL, Gaithersburg, MD) forms a convenient size marker that can also be radiolabeled by incubation with *Escherichia coli* DNA polymerase I (Klenow fragment) and [α-^{32}P]dCTP. It is important to use Orange G as a tracking dye as bromphenol blue migrates at the same position as, and thus will obscure, mononucleosomal DNA. Ethidium bromide staining and fluorography of the gel reveal the quality of nucleosome assembly in the total DNA population. The gel can then be dried onto DE81 chromatography paper (Whatman, Clifton, NJ), which prevents loss of small DNA fragments, and subsequent autoradiography reveals the quality of nucleosome assembly on NER-labeled DNA.

Alternatively, for an analysis of the extent of nucleosome assembly at different sites on the DNA substrate by Southern blotting (discussed later in section on mechanism analysis), gels are soaked in 0.5 *M* NaOH, 1.5 *M* NaCl for 30 min followed by two washes in 1 *M* Tris–HCl, pH 7.5, 1.5 *M* NaCl for 15 min each. DNA is then transferred to a nylon membrane (Hybond N+, Amersham) by vacuum transfer (Appligene apparatus, Appligene Oncor, Illkirch, France) in 2× SSC for 2 hr. DNA is fixed onto the membrane using a Stratalinker (Stratagene, La Jolla, CA) UV source. A site-specific oligonucleotide probe is radiolabeled by incubating 10 pmol of oligonucleotide in a 10-μl reaction containing 70 m*M* Tris–HCl, pH 7.6, 10 m*M* $MgCl_2$, 5 m*M* DTT, 50 μCi [γ-^{32}P]ATP (>5000 Ci/mmol), and 10 units T4 polynucleotide kinase (10 U/μl) at 37° for 30 min. After heat inactivation at 68° for 15 min, free ATP is removed using a Sephadex G-25 spun column (Boehringer). Hybridization is performed for ≥12 hr at 45° in hybridization bottles containing 10 ml of buffer [7% (w/v) SDS, 10% (w/v) polyethylene glycol 8000, 250 m*M* NaCl, and 130 m*M* sodium phosphate buffer (pH 7.0)] and 10 pmol of ^{32}P-labeled oligonucleotide probe. Membranes are washed twice for 5 min in 1× SSC, 0.1% SDS at 30° before autoradiography. Additional site-specific oligonucleotide probes may be used after stripping each preceding probe from the membrane by incubation in 0.4 *M* NaOH at 45° for 30 min followed by 0.2 *M* Tris–HCl, pH 7.5, 0.1% (w/v) SDS, 0.1× SSC at 45° for 15 min.

Analysis of Mechanism of Nucleosome Assembly from Target Site for Nucleotide Excision Repair

A kinetic analysis of nucleosome assembly from a single NER site[7] revealed that at early reaction times, mononucleosomes were formed at the repair site and contained DNA labeled during repair synthesis. The subsequent appearance of labeled polynucleosomal DNA indicated that repair of a single site can lead to the formation of extended nucleosomal arrays. The extent of nucleosome assembly at different regions of the single lesion DNA substrate can be assessed by probing MNase-digested DNA with radiolabeled site-specific oligonucleotide probes. The use of oligonucleotide probes much shorter (20 to 30 nucleotides long) than mononucleosomal DNA fragments is important for the analysis of nucleosome assembly reactions originating from a DNA transaction such as NER, which is limited to a specific site on the DNA substrate. The high NER efficiency of *Drosophila* preblastoderm embryo extracts combined with reaction conditions that suppress repair-independent nucleosome assembly results in similar MNase digestion patterns for NER-labeled DNA and DNA detected with a probe specific for the NER site[7] (see reactions without aphidicolin in the labeled DNA and Pt site probe panels of Fig. 4B). This is an observation that we have routinely observed using various preparations of extract and DNA substrate. The use of several site-specific probes can reveal the extent of propagation of nucleosomal arrays over a given region of DNA. Our interpretation takes into account the fact that we could not observe any specific positioning of nucleosomes on the M13 DNA substrates shown in Fig. 1D,[24] although we cannot formally exclude some sequence modulation in the assembly process. Thus in a population of DNA molecules containing randomly positioned nucleosomal arrays, a single oligonucleotide probe should not be excluded from nucleosomal regions on a significant proportion of DNA molecules. Nucleosomal arrays extending from different sites can be compared conveniently by the densitometric scanning of nucleosomal DNA fragments produced at identical MNase digestion times. The number and distribution of peaks indicate the extent of propagation of regularly spaced nucleosomal arrays on either side of a particular site located by the probe. For example, a change in the distribution of peaks toward mononucleosomal DNA indicates the presence of less extensive regular nucleosomal arrays. Using this approach we have observed a decrease in the extent and quality of assembled nucleosomal arrays moving away from the repair site in either direction.[7] The use of additional intermediate oligonucleotide probes at ± 200 bases from the NER site supports this interpretation.[25] Whether the poor MNase ladder observed at sites distal to the lesion represents a defined limit to the distance of propagation of a nucleosomal array from the NER site remains to be resolved.

These data demonstrate that there is not a unique direction for the propagation of nucleosome assembly from the target site for NER. One simple interpretation of these observations is that there is a bidirectional propagation of nucleosome assembly from the NER site.[7] Alternatively, it is possible that the direction of nucleosome assembly initiated from the

[24] J. G. Moggs, J.-P. Quivy, P.-H. Gaillard, and G. Almouzni, unpublished results (1997).

[25] J. G. Moggs and G. Almouzni, unpublished results (1997).

NER site is random, resulting in roughly equal numbers of DNA molecules possessing nucleosomal arrays in each direction. To differentiate between these two possibilities, it would be necessary to analyze nucleosome assembly events on individual DNA molecules in the population. Electron microscopy potentially represents a powerful approach for resolving the directionality of nucleosome assembly from a single repair site.

The unidirectionality (5′ to 3′) of DNA repair synthesis during NER contrasts with both of the models in which the direction of propagation of nucleosomal arrays from the NER site is either bidirectional or random. We demonstrated that DNA repair synthesis was not essential for nucleosome assembly from a single NER site.[7] This finding raised the possibility that either the helical distortion produced by the lesion itself and/or early NER reaction steps are responsible for the recruitment of nucleosome assembly machinery. Although benzopyrene diol epoxide DNA adducts have been shown to enhance the formation of a positioned nucleosome, UV irradiation of DNA partially inhibited nucleosome formation,[26] probably due to the presence of 6-4 photoproducts.[27] The influence of a 1,3-intrastrand cisplatin cross-link on nucleosome formation remains to be determined. Importantly, aphidicolin did not significantly inhibit the damage recognition and dual incision reactions during NER (Fig. 4A), implicating either a specific DNA structure (perhaps an incised DNA strand) or a component of the NER machinery as the signal for the recruitment of nucleosome assembly machinery. Furthermore, in the presence of aphidicolin, the NER site was particularly sensitive to MNase digestion (Fig. 4B, bottom), possibly reflecting an unligated gap and/or the presence of an incompletely formed nucleosome. Thus, although NER reaction steps preceding DNA repair synthesis appear to lead to the initiation of nucleosome assembly and the propagation of regular nucleosomal arrays, the formation of a regular nucleosomal array spanning the NER site requires the completion of the repair reaction. These results are consistent with changes in the sensitivity of chromatin to DNase I undergoing NER *in vivo* where the lack of a 10.4 nucleotide repeat pattern implied that DNA was not tightly bound to the surface of core histones in a native nucleosome conformation during or immediately after NER.[28] Furthermore, the restoration of canonical nucleosome structures at sites of NER required the completion of both repair synthesis and ligation.[29,30]

[26] D. B. Mann, D. L. Springer, and M. J. Smerdon, *Proc. Natl. Acad. Sci. U.S.A.* **94,** 2215 (1997).
[27] H. Matsumoto, A. Takakusu, T. Mori, M. Ihara, T. Todo, and T. Ohnishi, *Photochem. Photobiol.* **61,** 459 (1995).
[28] M. J. Smerdon and M. W. Lieberman, *Biochemistry* **19,** 2992 (1980).
[29] D. J. Hunting, S. L. Dresler, and M. W. Lieberman, *Biochemistry* **24,** 3219 (1985).
[30] M. J. Smerdon, *J. Biol. Chem.* **261,** 244 (1986).

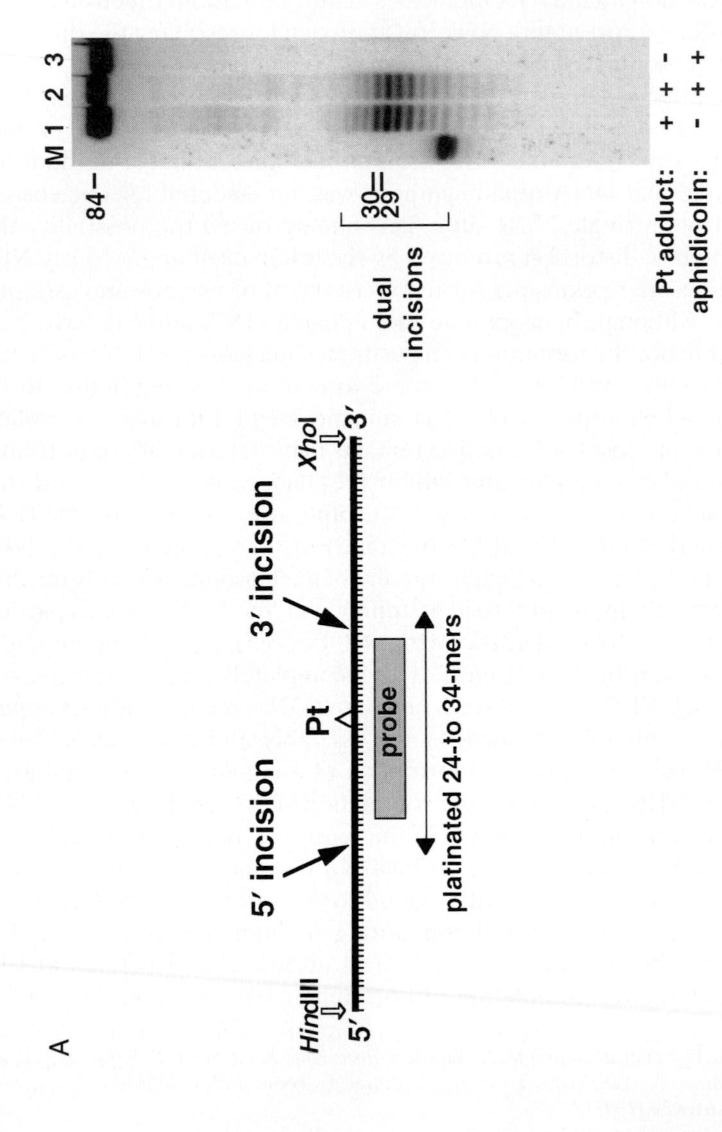
A
HindIII
5′
5′ incision
Pt
probe
3′ incision
XhoI
3′
platinated 24-to 34-mers
M 1 2 3
84–
30
29
dual incisions
Pt adduct: + + -
aphidicolin: - + +

Fig. 4.

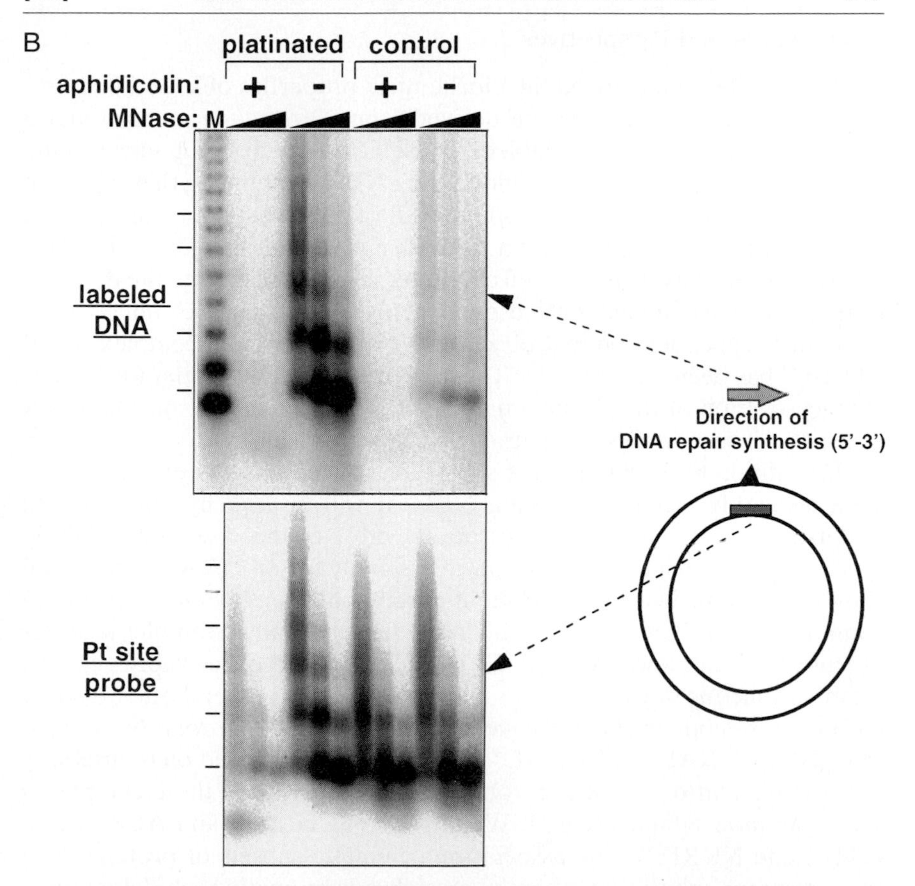

FIG. 4. Effect of inhibiting DNA repair synthesis on nucleosome assembly at a target site for NER. (A) *Drosophila* preblastoderm embryo extracts catalyze dual incision formation in the absence of DNA repair synthesis. DNA repair synthesis was inhibited by preincubating reactions (described in the text) with 200 μg/ml aphidicolin for 20 min. DNA substrate A was then added and reactions were incubated for a further 40 min. Both single lesion and control reactions contained 8% (v/v) dimethyl sulfoxide. Analysis of the formation of dual incisions on substrate A was performed as described in Fig. 2A except that different restriction enzymes (84 nucleotides apart) and oligonucleotide probe were used. The size marker in lane 1 is the 5′-phosphorylated platinated 24-mer oligonucleotide shown in Fig. 1B. (B) Increased sensitivity of the NER site to MNase in the absence of DNA repair synthesis. Extracts were preincubated with aphidicolin as described in (A) before the addition of DNA substrate A and a further 6-hr incubation. (Top) NER-labeled DNA after MNase digestion, agarose gel electrophoresis, and Southern blotting onto a nylon membrane. This membrane was then hybridized with a "Pt site" (5′-CCAGGCCTCTTCTTCTGTGCACTCTTC-3′) oligonucleotide probe (bottom).

Conclusions and Perspectives

This article characterized the biochemical properties of a modified cell-free chromatin assembly system derived from *Drosophila* preblastoderm embryo extracts that has enabled the selective analysis of nucleosome assembly associated with DNA undergoing NER. Combining this approach with DNA containing a single site-specific lesion has led to the precise analysis of nucleosome formation from a single target site for NER.[7] This cell-free system has also been well characterized with respect to the dynamic properties of assembled nucleosomal arrays. For example, nucleosome movement dependent on hydrolyzable ATP and a protein complex called CHRAC has been observed.[31,32] There is thus a great potential for further characterization of the chromatin dynamics associated with specific DNA transactions such as DNA repair.

In addition to elucidating the signal for the recruitment of the nucleosome assembly machinery during NER, it will be equally important to understand which factors lead to the assembly of the observed regularly spaced nucleosomal arrays. Chromatin assembly factor 1 (CAF-1) has been shown to be sufficient to complement chromatin assembly that occurs concomitantly with NER.[5] *Drosophila* CAF-1 preferentially assembles histones H3 and H4 onto newly replicated DNA[33] and could conceivably perform a similar function during NER. In addition to CAF-1, several other distinct chromatin-remodeling factors have been purified from *Drosophila* embryos (e.g., DF 31[34]; NAP-1[35]; CHRAC[32]; NURF[36]; ACF[37]) based on their ability to assemble and/or remodel chromatin *in vitro*. Some of these complexes share common subunits (e.g., ISWI in NURF, CHRAC, and ACF; p55 in CAF-1 and NURF[38]). The association of unique subsets of proteins with these common subunits may confer specialized chromatin remodeling functions during specific DNA transactions such as NER.

Although initial *in vitro* studies focused on the problem of how chromatin organization is restored during and after NER, an equally important

[31] P. D. Varga-Weisz, T. A. Blank, and P. B. Becker, *EMBO J.* **14,** 2209 (1995).

[32] P. D. Varga-Weisz, M. Wilm, E. Bonte, K. Dumas, M. Mann, and P. B. Becker, *Nature* **388,** 598 (1997).

[33] R. T. Kamakaka, M. Bulger, P. D. Kaufman, B. Stillman, and J. T. Kadonaga, *Mol. Cell. Biol.* **16,** 810 (1996).

[34] G. Crevel and S. Cotterill, *EMBO J.* **14,** 1711 (1995).

[35] M. Bulger, T. Ito, R. T. Kamakaka, and J. T. Kadonaga, *Proc. Natl. Acad. Sci. U.S.A.* **92,** 11726 (1995).

[36] T. Tsukiyama, C. Daniel, J. Tamkun, and C. Wu, *Cell* **83,** 1021 (1995).

[37] T. Ito, M. Bulger, M. J. Pazin, R. Kobayashi, and J. T. Kadonaga, *Cell* **90,** 145 (1997).

[38] M. A. Martinez-Balbas, T. Tsukiyama, D. Gdula, and C. Wu, *Proc. Natl. Acad. Sci. U.S.A.* **95,** 132 (1998).

question is how chromatin is remodeled to allow repair enzymes access to DNA lesions. Evidence shows that NER activity in human cell extracts can be suppressed by chromatin structure.[39,40] It may be possible to develop *in vitro* approaches for the elucidation of the biochemical mechanism of chromatin remodeling during the early stages of NER. This would complement *in vivo* approaches involving the analysis of NER rates at specific sites in DNA in small yeast minichromosomes possessing a well-characterized nucleosomal organization.[41–43] Furthermore, the experimental strategies described here may well be applicable to biochemical investigations of chromatin remodeling during other types of DNA repair.

Acknowledgments

We thank all the members of our laboratory for reagents, advice, and support during this study. We are grateful to Jean-Pierre Quivy and Pierre-Henri Gaillard for stimulating discussions and for sharing unpublished data. We also thank Peter Becker and his collaborators for help in the preparation of *Drosophila* embryo extracts. This work was supported by an EMBO long-term fellowship (J.G.M), an ATIPE from the CNRS (G.A.), Institut Curie, and ARC.

[39] Z. Wang, X. Wu, and E. C. Friedberg, *J. Biol. Chem.* **266,** 22472 (1991).
[40] K. Sugasawa, C. Masutani, and F. Hanaoka, *J. Biol. Chem.* **268,** 9098 (1993).
[41] M. J. Smerdon and F. Thoma, *Cell* **61,** 675 (1990).
[42] J. Bedoyan, R. Gupta, F. Thoma, and M. J. Smerdon, *J. Biol. Chem.* **267,** 5996 (1992).
[43] R. E. Wellinger and F. Thoma, *EMBO J.* **16,** 5046 (1997).

[19] Nuclear Run-On Assays: Assessing Transcription by Measuring Density of Engaged RNA Polymerases

By KAZUNORI HIRAYOSHI and JOHN T. LIS

Introduction

The transcriptional level of a gene can be estimated by a variety of methods. Some of the most widely used methods measure the accumulation of specific mRNAs. These include Northern blot analysis, S1 and RNase protection, mRNA dot-blot hybridization, and *in situ* hybridization.[1] Al-

[1] F. M. Ausubel, R. Brent, R. E. Kingston, D. D. Moore, J. G. Seidman, J. A. Smith, and K. Struhl, "Current Protocols in Molecular Biology" (Janssen, K., ed.). Wiley, New York, 1998.

0076-6879/99 $30.00

though these assays are relatively simple and sensitive, the level of RNA measured is a function of not only the amount of transcription, but also RNA processing and stability.

Transcription or promoter activity is also often assessed indirectly at the level of protein or enzymatic activity. β-Galactosidase, luciferase, or chloramphenicol acetyltransferase (CAT) activity of reporter genes is used frequently as a measure of transcriptional activity of a particular promoter that has been altered systematically and/or provided with different combinations of regulatory factors.[1] Although these assays are indirect measurements of promoter activity, they do provide convenient estimates of relative transcriptional activity and are reliable if comparing different constructs that conserve sequences important for reporter RNA processing, stability, and translation efficiency.

A more direct view or snapshot of transcription is provided by protein–DNA cross-linking followed by immunoprecipitation with RNA polymerase antibody. Cross-linking agents such as UV light[2] or formaldehyde[3] can be used to cross-link RNA polymerase to DNA *in vivo*. The immunoprecipitation of RNA polymerase complexes and the assay of particular covalently attached DNAs provide a direct measure of the density and distribution of RNA polymerase on specific DNA sequences *in vivo*. Cross-linking can be done rapidly, with a flash of a laser or xenon flash lamp, allowing the change in RNA polymerase density on sequences to be monitored as a function of time.[4] However, the cross-linking method does not discriminate between different states of RNA polymerase, e.g., elongationally competent versus arrested.

This article describes the use of nuclear run-on assays as a direct measure of the density of elongating RNA polymerases. Assuming a relatively constant rate of elongation (for polymerases that have progressed beyond the promoter region and into a fully competent elongational mode), this density provides a measure of transcription at the moment of nuclei isolation. Transcriptionally engaged RNA polymerases in isolated nuclei are allowed to continue elongation in the presence of labeled nucleoside triphosphate. The labeled RNA generated from a specific gene sequence is then quantified by hybridization to specific unlabeled DNAs bound to filters. The levels of hybridization to these DNAs, which are present in vast excess over specific labeled RNAs, are proportional to polymerase density on these

[2] D. Gilmour, A. Rougvie, and J. Lis, *in* "Methods in Cell Biology, Functional Organization of the Nucleus: A Laboratory Guide" (B. A. Hamkalo and S. C. R. Elgin, eds.), p. 369. Academic Press, San Diego, 1991.

[3] V. Orlando and R. Paro, *Cell* **75,** 1187 (1993).

[4] T. O'Brien and J. T. Lis, *Mol. Cell. Biol.* **13,** 3456 (1993).

sequences in the isolated nuclei. Probing several DNAs simultaneously by simply spotting the single-stranded DNA sequence on a filter and hybridizing the labeled run-on RNA allows the relative density of transcriptionally active polymerase on different genes or on fragments within a single gene to be compared directly.

The run-on method has its historical roots in the 1970s[5,6] and has become increasingly utilized as a direct measure of transcription over the past decade. Run-on reactions with isolated nuclei lead to the incorporation of labeled nucleotide into the existing nascent RNA strands associated with elongating RNA polymerase. The labeling method is a simple reaction containing nuclei, buffer, and nucleotides. To minimize the effects of chromatin structure, the detergent sarkosyl (sodium *N*-lauroylsarcosine) can be added to the reaction. Sarkosyl has been used to dissect the steps involved in transcription initiation by RNA polymerase II *in vitro*.[7] A sarkosyl concentration of 0.05% (w/v) is sufficient to inhibit transcription initiation by RNA polymerase II. However, RNA polymerase II complexes that have initiated transcription prior to sarkosyl addition are resistant to this treatment and will continue transcription if supplied with nucleoside triphosphates.[7,8] Transcribing RNA polymerase II molecules in isolated nuclei are also resistant to concentrations as high as 0.6%, a concentration that strips nontranscribing RNA polymerases and the vast majority of other chromatin-associated proteins, including histones from the DNA.[5,6]

This article describes the standard protocols used by researchers in a variety of cell and animal systems. Nuclear run-on assays consist of three distinct steps: (1) the isolation of nuclei, (2) the run-on reaction to generate labeled RNA, and (3) the hybridization of labeled run-on RNA to specific sequences. We describe first the protocols that we have used and have developed in working with *Drosophila* cell cultures and whole animals. We then provide protocols used with mammalian cells and with yeast (*Saccharomyces cerevisiae*).

Drosophila Nuclear Run-On Assay Using Cell Cultures and Whole Flies

General Considerations

For many of the following protocols, the purity and intact nature of nuclei are critical. Many methods for nuclei preparation have been reported

[5] P. Gariglio, J. Buss, and M. H. Green. *FEBS Lett.* **44,** 330 (1974).
[6] M. H. Green, J. Buss, and P. Gariglio. *Eur. J. Biochem.* **53,** 217 (1975).
[7] D. K. Hawley and R. G. Roeder. *J. Biol. Chem.* **260,** 8163 (1985).
[8] D. K. Hawley and R. G. Roeder. *J. Biol. Chem.* **262,** 3452 (1987).

for different cell or tissue types. The following methods are standard protocols of nuclei preparation for each of several model organisms. In deciding on the number of cells to process, a good rule of thumb is to use a minimum of 10^6 nuclei in a run-on reaction. Usually, 10^7 nuclei work well for a single reaction, although more may be necessary if the sensitivity of the assay is pushed to its limit.

In lysing cells, the number of strokes with a Dounce homogenizer or Teflon homogenizer (Wheaton Glass Co., Millville, NJ) will vary with cell type and should be determined empirically by examining the yield of nuclei. A convenient way to count nuclei is to stain preparations with 1 mg/ml DAPI (4,6-diamidino-2-phenyl-indole) and examine by fluorescence microscopy. Microscopic observation of nuclei stained with 0.1% cresyl violet under phase contrast is also an easy way to estimate the yield of nuclei. Because contamination of intact cells can contribute to background, it is best to check the purity of every nuclear preparation.

Buffers and Reagents

Buffer A: 300 m*M* sucrose, 2 m*M* magnesium acetate, 3 m*M* $CaCl_2$, 10 m*M* Tris–Cl (pH 8.0), 0.1% Triton X-100, 0.5 m*M* dithiothreitol (DTT)
Buffer B: 2 *M* sucrose, 5 m*M*, magnesium acetate, 10 m*M* Tris–Cl (pH 8.0), 0.5 m*M* DTT
Buffer B′: 1.75 *M* sucrose, 5 m*M* magnesium acetate, 10 m*M* Tris–Cl (pH 8.0), 0.5 m*M* DTT
Buffer C: 1.9 *M* sucrose, 5 m*M* magnesium acetate, 10 m*M* Tris–Cl (pH 8.0), 0.5 m*M* DTT
Buffer D: 25% glycerol, 5 m*M* magnesium acetate, 50 m*M* Tris–Cl (pH 8.0), 5 m*M* DTT, 0.1 m*M* EDTA
Run-on reaction buffer: 7.2 m*M* Tris–HCl (pH 8.0), 4.4 m*M* magnesium acetate, 1.4 m*M* $MnCl_2$, 150 m*M* KCl, 3 m*M* each of ATP, CTP, and GTP, 10 μM UTP, 100 μCi [α-^{32}P]UTP (3000 Ci/mmol), 1 μl of RNasin (40 U/μl)
Stop solution: 2% (w/v) sodium dodecyl sulfate (SDS), 7 *M* urea, 0.35 *M* NaCl, 1 m*M* EDTA, 10 m*M* Tris–HCl (pH 8.0)
Hybridization solution: 6× SSC, 50% (w/v) formamide, 1% (w/v) SDS, 10% (w/v) dextran sulfate, 250 μg/ml salmon sperm DNA
Washing solution I: 2× SSC, 0.1% SDS
Washing solution II: 0.2× SSC, 0.1% SDS

Nuclei Isolation from Drosophila Cell Cultures

Harvest cells by transferring suspended cultures to a centrifuge tube and place the tube immediately on ice. [For rapid kinetic analysis of heat shock gene activation, we have used liquid nitrogen to freeze cells.[4]] The cells are collected by brief centrifugation (e.g., 500g for 3 min) at 4°. Cell pellets (ca. 5 × 10^7 cells) are suspended in 7.5 ml of buffer A. Cells are homogenized in a Dounce homogenizer by 25 strokes with pestle B. Homogenates are mixed immediately with an equal volume of ice-cold buffer B. The mixture is layered over a cushion of buffer B in an ultracentrifuge tube. We use 10 ml of buffer B for 15 ml of homogenate mixture. Centrifuge at 11,500 rpm for 25 min at 4° in the SW28 rotor (Beckman).

The following part of the procedure should be performed in a cold room. The supernatant is poured off and residual supernant is removed with a Kimwipe. Nuclei are recovered from the pellet. Each tube of nuclear pellet is resuspended in 0.5 ml of buffer D by pipetting using a 1-ml Pipetman with the end of the tip cut to widen the orifice. Nuclei are transferred to an Eppendorf tube and centrifuged for 20 sec at the maximum speed at 4°. The supernatant is removed and the pellet is resuspended in 100 μl (for 10^7 cells) of buffer D using a 200-μl Pipetman with the tip cut and widened to at least 1 mm to reduce shearing. Nuclei can be stored at −80° after quick freezing in liquid nitrogen.

Nuclei Isolation from Drosophila Adult Flies

One to 1.5 g of adult flies or 10^7–10^8 nuclei are sufficient for a run-on experiment. All surfaces that come into contact with flies or their extracts are prechilled to 0° at the start of the experiment. The flies are collected in a prechilled bottle and weighed. The bottle is left on ice until flies stop moving, and then flies are transferred to a prechilled Omni mixer (Dupont, Wilmington, DE). Fifteen milliliters of ice-cold buffer A is added, and the flies are ground on ice for 75 sec at the maximum speed. The fly mash is filtered through two sheets of 100-μm nylon mesh to remove large debris such as wings and cuticle. Filtrates are transferred to a 30-ml Teflon/glass homogenizer and stroked 40 times on ice. Homogenates are filtered through two sheets of 35-μm nylon mesh twice, and then mixed with 15 ml of buffer B′. The mixture is then divided into two equal portions and is layered over a sucrose gradient consisting of 8 ml of buffer C, over 4 ml of buffer D in tubes of a SW28 rotor (Beckman). The sample is then spun at 14,000 rpm for 15 min at 4°. The supernatant is discarded and residual supernant is removed with a Kimwipe. After centrifugation, all steps are performed in the cold room. Each pellet is resuspended in 500 μl of buffer D with a 1-ml Pipetman using a tip with its end cut and transferred to an Eppendorf tube. The suspension is spun for 1 min in a cold microcentrifuge at maximum speed. Nuclei are recovered as a pellet. Pellets are resuspended in 100 μl of buffer D dissolved by pipetting with a Pipetman using a 200-μl tip with its end cut. Suspended nuclei are frozen in liquid nitrogen and stored at −80°.

This protocol is optimized to be gentle enough to preserve paused RNA polymerase II molecules in their paused state. Paused polymerases have been found near the start of a variety of genes.[9,10] In a standard run-on reaction done in the absence of detergent or high salt, a promoter-paused polymerase will remain slow to enter into elongation. Interestingly, we find that this protocol also gives the best nuclei as assessed cytologically and that these nuclei produce the most RNA from polymerases that are known to be actively elongating. All solutions used in preparing nuclei should be detergent free, and the run-on reaction must be terminated very quickly. Our experience[11] is that these protocols produce a low background of transcription from a promoter-paused polymerase without sarkosyl and a 3- to 10-fold higher signal after addition of sarkosyl. In the presence of 0.6% of sarkosyl, both normally engaged and promoter-paused transcription complexes can elongate. High salt concentrations (e.g., 800 mM KCl) is another condition that can allow paused polymerases to enter elongation. Similar to sarkosyl, these conditions strip histone HI from chromatin[12] and nucleosomes are destabilized.[13]

Run-On Reaction

A frozen or freshly prepared 100-μl aliquot of nuclei is mixed with 114 μl of run-on reaction buffer and incubated at 22°. This is a relatively low temperature compared to mammalian run-on reactions; however, the normal culture temperature for *Drosophila* is between 22 and 25°. If reactions are to be done in the presence of 0.6% sarkosyl, then sarkosyl is added to the nuclei reaction mix on ice and is mixed well with a Pipetman using a tip with its end cut. Tubes are transferred to a water bath or block heater immediately. The reaction is allowed to proceed for a fixed time, usually 2–5 min, and is stopped by transferring the sample to 5 ml of stop solution in a 15-ml tube with a tight-fitting cap. The large volume of stop solution dilutes the nucleotides to ensure that the robust RNA polymerase reaction is stopped and it also facilitates the separation of the RNA fraction from protein in the subsequenct phenol extraction. Eight milliliters of Leder phenol (phenol/chloroform/isoamyl alcohol at 25 : 24 : 1) is added and mixed by vortexing. Tubes are left on ice for 10 min and spun at 3000 rpm in a clinical centrifuge for 10 min at room temperature. This can be repeated, and we often do three extractions with Leder phenol. Following one more

[9] A. E. Rougvie and J. T. Lis, *Mol. Cell. Biol.* **10**(11), 6041 (1990).
[10] A. Krumm, L. B. Hickey, and M. Groudine, *Genes Dev.* **9**(5), 559 (1995).
[11] A. E. Rougvie and J. T. Lis, *Cell* **54**(6), 795 (1988).
[12] K. Tatchell and K. E. Van Holde, *Biochemistry* **16,** 5295 (1977).
[13] W. O. Weischet, *Nucleic Acids Res.* **7,** 291 (1979).

extraction with chloroform, the aqueous phase is transferred to an Oakridge tube and 2.5 volumes of cold ethanol is added. After incubation at −20° for at least 30 min, RNA is pelleted by centrifugation using a SA600 rotor (Sorvall) at 12,000 rpm for 30 min at 4°. Pellets are washed with 70% ethanol. Run-on products are collected as a pellet. The progress of every step can be monitored by a Geiger counter. After drying, pellets are resuspended in 400 μl of water. If pellets include proteins, they are quite difficult to dissolve. Therefore, the complete removal of proteins in the phenol extraction step is important. Usually pellets are dissolved easily by vortexing or pipetting. Dissolved RNA is transferred to an Eppendorf tube. After addition of 2.4 μl each of 1 *M* $MgCl_2$ and 1 *M* $CaCl_2$, 1 μl of RQ1 DNase (Promega, Madison, WI, RNase-free) is added and the reaction is incubated at 37° for 20 min. Ten microliters of 4 *M* NaCl and 1 ml of ethanol are added, incubated at −20° for 10 min, and RNA is recovered by centrifugation at 14,000 rpm for 15 min at 4° in a microcentrifuge. RNAs are washed with 70% ethanol and then dried. Pellets are resuspended in 128 μl of TE (pH 8.0), then incubated at 65° for 5 min to help dissolve the RNA.

In situations where high-resolution mapping of polymerase is required, the run-on RNA can be fragmented by brief treatments with base. Tubes are chilled on ice, and then 32 μl of ice-cold 1 *M* NaOH is added to the tube. A 10-min treatment fragments RNA to lengths of approximately 500 bases. This increases the resolution of the analysis, as the RNAs are elongated by about an additional 300 bases during the run-on reaction. This labeled stretch of run-on RNA is extended from an unlabeled, nascent RNA that was made in cells prior to the run-on reaction. This longer RNA would hybridize to regions upstream of the engaged polymerase being assayed. Fragmenting the RNA can therefore position the transcriptionally engaged polymerase on the transcription unit more accurately and is critical in positioning polymerases that are distributed unevenly on a transcription unit or in analyzing the changes in RNA polymerase distribution on a rapidly induced or repressed gene.[4] The resolution is also improved by reducing the time of the run-on reaction to produce shorter run-on products. The base hydrolysis reaction is terminated by the addition of 50 μl of 1 *M* HEPES (acid form). Twenty-one microliters of 3 *M* sodium acetate and 600 μl of ethanol are added and incubated at −20° for 10 min. RNA is pelleted by centrifugation at 14,000 rpm for 15 min at 4°. Pellets are washed with 70% ethanol and dried. RNAs are resuspended in 50 μl of TE and are ready for hybridization.

Hybridization of Labeled Run-On RNA to Specific Sequences

To detect the run-on products, unlabeled, specific nucleic acid sequences consisting of double-strand DNA, single-strand DNA, RNA, or oligonucleotides can be used. These nucleic acids are transferred to membranes by standard methods used in Southern blots, slot blots, or dot blots. Labeled run-on RNAs are then hybridized to these membranes and the resulting signals are quantified.

A single-strand DNA complementary to the RNA to be assayed is a useful probe for run-on reactions. These are often conveniently generated by cloning a desired fragment into an M13 phage vector and isolating clones for both a sense (control) and an antisense probe. Single-stranded probes lack a complementary DNA strand that can compete for hybridization with run-on RNAs and are specific for detecting transcription from the appropriate strand. In addition, synthetic oligonucleotides of 50 bases or more that correspond to the noncoding strand are convenient to generate and have been shown to detect run-on products.[14]

Membranes containing the probe DNAs are best kept as small as possible. These are placed in a sealed bag or hybridization tube. In our experience, hybridizations in a tube produce stronger signals and lower background. We usually perform the prehybridization with 5 ml of hybridization solution per 10-cm^2 filter at 42° for more than 6 hr. Skimping on the volume of prehybridization solution can produce higher backgrounds. Hybridizations are performed with 2 ml hybridization solution; the smaller volumes produce more complete hybridization and stronger signals. Run-on products in TE are denatured at 100° for 5 min and are added to the hybridization solution. We generally perform hybridizations between 16 and 20 hr, but they can go for as long as 96 hr.

After hybridization, membranes are washed at room temperature until the counts in washing solution I are at background. This is followed by two washes with washing Solution II at 60° for 15 min. The temperature of washing with solution II can be altered and of course depends on the stability of the hybridization product. Membranes are exposed to X-ray film or an imaging screen and signals are quantified by standard scanning protocols.

Nuclear Run-On Using Mammalian Cells

Buffer and Reagents

1× SSC: 150 m*M* NaCl, 15 m*M* sodium citrate (pH 7.0)
NP-40 lysis buffer: 10 m*M* Tris–HCl (pH 7.4), 10 m*M* NaCl, 3 m*M* $MgCl_2$, 0.1 m*M* EDTA, 0.5% (v/v) Nonidet P-40 (NP-40)

[14] T. Albert, J. Mautner, J. O. Funk, K. Hoertnagel, A. Pullner, and D. Eick, *Mol. Cell. Biol.* **17**(8), 4363 (1997).

Freezing buffer: 50 m*M* Tris–HCl (pH 8.3), 40% (v/v) glycerol, 5 m*M* $MgCl_2$, 0.1 m*M* EDTA

Reaction buffer: 10 m*M* Tris–HCl (pH 8.0), 5 m*M* $MgCl_2$, 300 m*M* KCl, 0.5 m*M* each of ATP, CTP, and GTP, and 100 μCi of [α^{32}P]UTP 800 Ci/mmol

Stop buffer: 20 m*M* Tris–HCl (pH 7.4), 2% SDS, 10 m*M* EDTA, 200 μg/ml proteinase K

Procedure Essentially as Described Previously[15,16]

Cells are quick-chilled on ice. All procedures should be done at 4° or less. In the case of adherent cells, cells are washed two or three times with ice-cold 1× SSC, and then cells are recovered from a culture plate by cell scraper with 1× SSC and collected by centrifugation, 500*g* for 5 min. For floating cells, cells are recovered by centrifugation and then washed with 1× SSC. Pellets are suspended in NP-40 lysis buffer by pipetting 10 times with a 1-ml Pipetman (use a 1-ml tip with its end cut to reduce shearing) and are incubated on ice for 5 min. We use 1 ml of buffer for 1×10^7 cells. After incubation, lysates are transferred to a new Eppendorf tube and centrifuged at 3000 rpm for 5 min. Pellets are washed twice with 1 ml of NP-40 lysis buffer and then resuspended in 100 μl (for 10^7 cells) of freezing buffer with a Pipetman (use a 200-μl tip with its end cut). The mixture is frozen in liquid nitrogen and can be kept at −80°. Nuclei are stable for at least 6 months. Alternatively, one can use the same method of nuclear preparation as described earlier with cultured *Drosophila* cells.

For run-on reactions, 100 μl of frozen nuclei (about 10^7 nuclei) is thawed on ice and then mixed with 100 μl of reaction buffer or freshly prepared nuclei can be mixed directly with reaction buffer and incubated at 30°. To ensure that the run-on reaction is as complete as possible, unlabeled UTP should be added to the reaction. Alternatively, if highly labeled shorter RNAs are needed, use UTP with high specific activity and omit unlabeled UTP. GTP, ATP, and CTP can also be used as the labeled nucleotide. The reaction time is usually 5–20 min. After completing the run-on reaction, DNase I (RNase-free) is added to the reaction at a final concentration of 10 mg/ml and incubated at 30° for 10 min. An equal volume of stop buffer is added to the reaction mixture and then incubated at 42° for 30 min. The EDTA and SDS in the stop buffer halt the reaction, and DNase I and other proteins are digested by proteinase K. After incubation, yeast RNA is added to the reaction mixture to a final concentration of 100 μg/ml and then RNA is extracted with Leder phenol (phenol : chloroform : isoamyl

[15] M. Groudine, M. Peretz, and H. Weintraub, *Mol. Cell. Biol.* **1,** 281 (1981).

[16] M. E. Greenberg and E. B. Ziff, *Nature* **311,** 433 (1984).

alcohol at 25:24:1) two or three times. RNA is precipitated twice with ethanol. Unincorporated labeled nucleotide can be removed by using a Sephadex G-50 spin column [Quick Spin G-50 (Boehringer)] and RNA can be recovered by ethanol precipitation. Another simple and convenient way to purify the run-on products is to use commercially available reagents RNAzol (Cinna/Biotecx, Houston, TX) or Trisol (BRL).

Nuclear Run-On Using Yeast

Buffers and Reagents

TMN buffer: 10 m*M* Tris–HCl (pH 7.4), 100 m*M* NaCl, 5 m*M* $MgCl_2$

Run-on reaction mixture: 50 m*M* Tris–HCl (pH 7.9), 100 m*M* KCl, 5 m*M* $MgCl_2$, 1 m*M* $MnCl_2$, 2 m*M* DTT, 0.5 m*M* ATP, GTP, and CTP, 100 μCi [α-^{32}P]UTP, 1 U/μl RNasin, 10 m*M* phosphocreatine, and 1.2 μg/μl creatine kinase

Procedure

In the case of yeast, the isolation of nuclei is not required. The preparation of nuclei from yeast is difficult because of the rigid cell wall and takes precious time during which the *in vivo* state of transcription could be changing. A convenient run-on method without nuclei preparation was developed by Elion and Warner.[17] A culture containing 2–3 $\times$ 10^7 log-phase cells is chilled in crushed ice. Cells are harvested at 5000 rpm for 6 min at 4°. Pellets are washed with TMN buffer. Cells are resuspended in 0.5% sarkosyl in 1 ml of water, transferred to an Eppendorf tube, and kept on ice for 15 min. Sarkosyl permeabilizes the cell and these permeabilized cells are recovered by centrifugation at 6000 rpm for 1 min at 4°. The pellet is now ready for a run-on reaction.

The permeabilized cells (2–3 $\times$ 10^7 cells) are resuspended in 100 μl of the run-on reaction mixture. Phosphocreatine and creatine kinase are not essential for the reaction but seem to increase the amount of labeled RNA synthesis slightly. The reaction is incubated at 25° for 8 min and is terminated by the addition of α-amanitin to 10 μg/ml. The reaction is terminated by DNase I addition and is followed by the addition of proteinase K. The mixture is extracted with Leder phenol, and the RNA is precipitated with ethanol as described for *Drosophila* cells.

[17] E. Elion and J. R. Warner, *Mol. Cell. Biol.* **6,** 2089 (1986).

Concluding Comments on Strengths and Weaknesses of Run-On Assay

The nuclear run-on assay provides a snapshot of the density of transcriptionally engaged RNA polymerases on specific DNA sequences in nuclei. If nuclei are prepared rapidly and carefully, this density of polymerase in isolated nuclei should represent the *in vivo* situation at the point of nuclei isolation and reflects the level of transcription of a particular DNA sequence.

The rate of elongation of RNA polymerase must be factored into the interpretation of run-on results. In cells, the rate of elongation has been estimated by several laboratories in several different ways to be about 1.2 kb/min.[4] If most transcriptionally engaged polymerases move at this rate, as suggested by these independent measurements, then the density of polymerases measured by the run-on assay reflects the frequency of transcription of a gene or the rate of promoter firing.

A strong correlation between the density of RNA polymerase II measured by UV cross-linking/immunoprecipitation and by run-ons assays done in sarkosyl or high salt has been found.[4,9] Engaged polymerases transcribe under these conditions. In the presence of sarkosyl or high salt, the promoter-associated, paused polymerases that we and others have examined are stimulated to transcribe like normal elongationally competent polymerases.[11,18] However, there is a report of an engaged polymerase on adenovirus whose transcription appears to be terminated in the presence of sarkosyl.[19] To evaluate the different forms of engaged RNA polymerase and to avoid missing a particular state of engaged polymerase, we recommend doing run-on assays in varying conditions, particularly under normal physiological conditions and in the presence of sarkosyl or high salt.

The ability to hybridize labeled run-on RNAs to a single filter that contains multiple genes or different segments of a single gene allows for rigorous quantitative comparisons. The relative transcription frequency of different genes can be compared directly. Moreover, if the absolute polymerase density on one of these genes is known, then the absolute transcription frequency of all these genes can be estimated. In the case of *Drosophila hsp70,* the absolute number of RNA polymerase II molecules on the induced gene is known,[20] and thus one knows the density of polymerase on genes that have been compared to it.[9] Likewise, the comparison of RNA polymerase density on different segments of a single gene can reveal control of transcription at the level of elongation. In this regard, the method

[18] A. Krumm, T. Meulia, M. Brunvand, and M. Groudine, *Genes Dev.* **6,** 2201 (1992).
[19] D. K. Wiest and D. K. Hawley, *Mol. Cell. Biol.* **10,** 5782 (1990).
[20] T. O'Brien and J. T. Lis, *Mol. Cell. Biol.* **11**(10), 5285 (1991).

has been extremely useful in characterizing promoter-associated pausing on a variety of genes.[9,10,21,22]

The snapshot quality of the run-on assay provides a means of examining the kinetics of polymerase associations. The rate of increase in polymerase density along sequences of a particular gene can provide insight to possible mechanisms of gene activation. Likewise, the kinetics of loss of RNA polymerase along the length of a gene can provide insight to mechanisms of repression.[4]

In summary, the nuclear run-on assay has become an important tool in measuring transcriptional levels *in vivo*. It has likewise provided important observations that need to be considered in evaluating mechanisms of transcriptional regulation.

[21] L. J. Strobl and D. Eick, *EMBO J.* **11,** 3307 (1992).

[22] K. Yankulov, J. Blau, T. Purton, S. Roberts, and D. L. Bentley, *Cell.* **77**(5), 749 (1994).

Section III

Assays for Chromatin Structure and Function *in Vivo*

[20] Mapping Chromatin Structure in Yeast

By PHILIP D. GREGORY and WOLFRAM HÖRZ

Introduction

The study of gene expression has become increasingly inseparable from that of chromatin structure. Whereas molecular research previously focused on *cis*-acting DNA elements and the factors that bind to them, the roles that such factors and elements play within a chromatin context will become more and more important. In addition, because the fundamental cellular processes of replication, transcription, recombination, and repair are provided with a nucleosomal DNA substrate, techniques that address the structure of chromatin and that identify perturbations in this matrix are of importance to the study of a wide range of cellular functions and activities.

A unique nucleosomal organization often characterizes the regulatory state of a specific gene. Nuclease digestion experiments originally correlated the appearance of hypersensitivity within a promoter with gene activation, whereas inactive genes maintain a less accessible structure, often with a more regular positioned array of nucleosomes. Recent work has served to underscore the importance of these architectural changes,[1-3] and techniques that facilitate the determination of the underlying chromatin structure can provide important information vital to the understanding of the molecular processes and factors required for proper transcriptional regulation *in vivo*.

To this end, we have successfully employed three basic techniques that provide a high degree of reproducibility to determine the nucleosomal organization of yeast chromatin at various loci. The first employs DNase I with yeast nuclei to determine the presence of positioned nucleosomes or hypersensitive sites within a specific region of chromatin. In addition we provide a micrococcal nuclease-based method that assays for the presence or absence of a nucleosome on a particular stretch of DNA. Although these techniques can provide important structural information, the level of hypersensitivity or the extent to which the chromatin structure is perturbed is hard to assess by such methods. Therefore, we also present a complementary technique that employs restriction enzymes with yeast nuclei to provide quantitative accessibility data. The combination of these methods thus

[1] D. G. Edmondson and S. Y. Roth, *FASEB J.* **10,** 1173 (1996).
[2] P. D. Gregory and W. Hörz, *Eur. Biochem.* **251,** 9 (1998).
[3] A. P. Wolffe, *Nature* **387,** 16 (1997).

0076-6879/99 $30.00

presents an effective diagnostic tool for the study of chromatin structure in yeast.

Methods

Preparation of Yeast Nuclei

The following protocol for the isolation of yeast nuclei is based on the original method of Wintersberger *et al.*[4] and has been subsequently refined. There are, however, several, more time-consuming methods for the preparation of highly purified nuclei[5,6] and faster procedures for the treatment of nuclei with nucleases in crude lysates.[7–10] The protocol provided here attempts to balance purity and speed and reproducibly provides identical results with DNase I, micrococcal nuclease, and restriction enzyme analyses for which the following protocols are given.

Solutions for Nuclei Preparation

Preincubation solution: 0.7 *M* 2-mercaptoethanol and 2.8 m*M* EDTA
Sorbitol solution: 1.0 *M* sorbitol
Lysis solution: 1.0 *M* sorbitol and 5 m*M* 2-mercaptoethanol
Ficoll solution: 18% (w/v) Ficoll, 20 m*M* KH_2PO_4, pH 6.8, 1 m*M* $MgCl_2$, 0.25 m*M* EGTA, and 0.25 m*M* EDTA
Zymolyase solution: 20 mg Zymolyase 100T (ICN) dissolved in 1 ml water

Procedure

1. A 1-liter yeast culture is grown to early logarithmic phase ($2–4 \times 10^7$ cells/ml) and the cells are collected by centrifugation (3000*g*/10 min at room temperature). This is approximately 2–4 OD at 600 nm and should provide approximately 2 g of cells (wet weight) or 0.2 mg of DNA. The measurement of cell density may vary between spectrophotometers, thus counting cells to determine the conversion factor for a particular spectrophotometer is advised. It is important

[4] U. Wintersberger, P. Smith, and K. Letnansky, *Eur. J. Biochem.* **33,** 123 (1973).
[5] J. P. Aris and G. Blobel, *Methods Enzymol.* **194,** 735 (1991).
[6] D. Lohr, *in* Yeast: A Practical Approach (I. Campbell and J. H. Duffus, eds.), p. 125. IRL Press, Oxford, 1988.
[7] M. J. Fedor, N. F. Lue, and R. D. Kornberg, *J. Mol. Biol.* **204,** 109 (1988).
[8] M. W. Hull, G. Thomas, J. M. Huibregtse, and D. R. Engelke, *Methods Cell Biol.* **35,** 383 (1991).
[9] J. M. Huibregtse and D. R. Engelke, *Methods Enzymol.* **194,** 550 (1991).
[10] N. A. Kent, L. E. Bird, and J. Mellor, *Nucleic Acids Res.* **21,** 4653 (1993).

to process a culture immediately after taking cells out of the incubator. Excessive delays or storing the culture first at low temperature may impede the lysis of the cells greatly.

2. Wash cells in ice-cold water and suspend in 50 ml water.
3. Transfer into preweighed 50-ml centrifuge tubes and centrifuge (3000*g*/5 min). Determine wet weight.
4. Add 2 volumes preincubation solution (relative to wet weight of cells) and shake for 30 min at 28°. This step facilitates digestion of the cell wall with Zymolyase. During this time it may be prudent to maintain induction or repression during preincubation and Zymolyase treatment by adding the appropriate inducer or repressor or, for temperature-sensitive strains, adjusting the incubation temperature accordingly.[11] It is also possible to treat cells with Zymolyase in medium supplemented with sorbitol and a reducing agent.
5. Collect by centrifugation (3000*g*/5 min at 5°) and wash in 50 ml 1 *M* sorbitol.
6. Collect again (3000*g*/5 min at 5°) and resuspend in 5 ml lysis solution per 1 g of cells (wet weight).
7. Dilute 20-μl aliquots 100-fold in water and read optical density at 600 nm, which should be in the range of between 1 and 2.
8. Add 1/50 volume of a freshly prepared Zymolyase solution to the cells.
9. Incubate with gentle agitation at 28°.
10. Measure optical density at 600 nm after 15 and 30 min as in step 7. As a relative measure of lysis, values should drop to 5–20% of the original measurement. Lysis is, however, strain and growth stage dependent, with stationary cells being more difficult to lyse than logarithmically growing cells. We have successfully used nuclei obtained from cells that continued to give 60% of the starting OD value at the end of the Zymolyase treatment. In such cases it is advisable to monitor a constitutively accessible restriction site in chromatin as a control.
11. Centrifuge (2000*g*/5 min at 5°) and wash in 50 ml 1 *M* sorbitol.
12. Centrifuge (3000*g*/10 min at 5°) and resuspend in 7 ml Ficoll solution per 1 g cells (original wet weight). Cells lyse at this stage, but nuclei are stabilized by the Ficoll.
13. Distribute into as many aliquots as desired and centrifuge (30,000*g*/30 min at 5°). Aliquots equivalent to 0.5 or 1 g wet weight cells are suitable for subsequent digestion experiments. Ten-milliliter polypropylene centrifuge tubes are convenient.

[11] A. Schmid, K. D. Fascher, and W. Hörz, *Cell* **71,** 853 (1992).

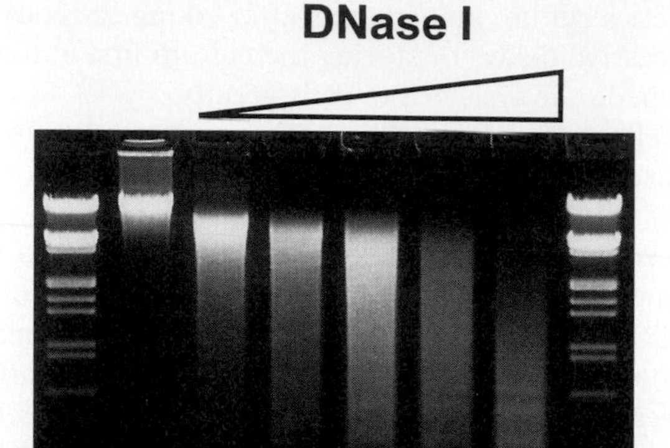

FIG. 1. Monitoring the extent of DNase I digestion of isolated yeast nuclei. Purified yeast nuclei were treated with 0 (lane 1), 0.5 (lane 2), 1 (lane 3), 2 (lane 4), 4 (lane 5), or 8 (lane 6) U/ml DNase I at 37° for 20 min. Marker lanes (M) contain λ DNA digested with *Hin*dIII and *Eco*RI. Suitably digested samples would therefore be in lanes 2–5.

14. Decant supernatant, and freeze the nuclear pellet in liquid nitrogen and store at −70°.

Chromatin Analysis with DNase I

The following protocol is designed to assay for histone–DNA interactions and has been used extensively to study inducible *PHO5*,[12] *PHO8*,[13] and *TDH3*[14] promoters. DNase I has also been used extensively to determine the footprints of factors binding to the DNA in chromatin.[8,9] In contrast to the method provided here, these protocols aim to minimize the dissociation of DNA-bound factors by the immediate addition of nuclease following lysis, the use of low salt buffers, and lower digestion temperatures. Furthermore, because these methods assay single-strand cuts rather than the double-strand cuts scored here (see later), the optimum extent of digestion is also lower. An example of the range of DNase I digestion obtained in a digestion series is given in Fig. 1, and an example of a typical result of this procedure

[12] P. D. Gregory, A. Schmid, M. Zavari, L. Lin, S. L. Berger, and W. Hörz, *Mole. Cell* **1,** 495 (1998).

[13] S. Barbaric, K. D. Fascher, and W. Hörz, *Nucleic Acids Res.* **20,** 1031 (1992).

[14] B. Pavlovic and W. Hörz, *Mol. Cell. Biol.* **8,** 5513 (1988).

with respect to the chromatin transition at the *PHO5* promoter is shown in Fig. 2.

Solutions for DNase I Analysis

Digestion buffer: 15 m*M* Tris–HCl, pH 7.5, 75 m*M* NaCl, 3 m*M* $MgCl_2$, 0.05 m*M* $CaCl_2$, and 1 m*M* 2-mercaptoethanol

DNase I dilution buffer: 10 m*M* Tris–HCl, pH 7.4, 0.1 mg/ml bovine serum albumin (BSA)

Stop solution: 1.0 *M* Tris–HCl, pH 8.8, and 0.08 *M* EDTA, pH 8.0

Proteinase K solution: 10 mg/ml proteinase K in 10 m*M* Tris–HCl, pH 8.0

Chloroform solution: chloroform/isoamyl alcohol (24 : 1, v/v)

RNase solution: 5 mg/ml ribonuclease A (DNase-free) dissolved in 5 m*M* Tris–HCl, pH 7.5, and heated for 10 min/100°

Procedure

1. Suspend pelleted nuclei from approximately 500-mg cells (wet weight) for one experiment in 3 ml digestion buffer by vortexing.
2. Centrifuge (2000*g*/5 min at 5°) and resuspend in 1.2 ml digestion buffer. Transfer 200-μl aliquots to microfuge tubes.
3. Add DNase I at four different concentrations in the range of 0.5 to 20 U/ml and incubate for 20 min at 37°. Keep one sample on ice and one at 37° without nuclease.
4. Terminate digestion by adding 10 μl stop solution, 5 μl 20% SDS, and 20 μl proteinase K solution. Incubate for 30 min at 37°.
5. Add 1/5 volume 5 *M* $NaClO_4$, 1 volume phenol, vortex, and then 1 volume chloroform solution, vortexing well.
6. Centrifuge for 5 min.
7. Take off supernatant, reextract with 1 volume chloroform solution.
8. Take off supernatant and add 2.5 volumes ethanol to precipitate nucleic acids.
9. Collect by centrifugation and resuspend in 125 μl TE.
10. Add 10 μl DNase-free RNase solution and incubate for 1 hr at 37°.
11. Add 5 μl 5 *M* NaCl and 0.6 volume 2-propanol and centrifuge immediately at room temperature.
12. Wash the pellet with 70% ethanol and dissolve in 80 μl TE.
13. Analyze 5-μl aliquots in 1% agarose gels and stain with ethidium bromide. An example of such a gel is given in Fig. 1.
14. Select appropriate samples and use 20 μl for secondary digestion and indirect end labeling. Optimal extents of digestion may differ according to the chromatin structure of the region of interest and as a product of how distant the region of interest is in relation to

A

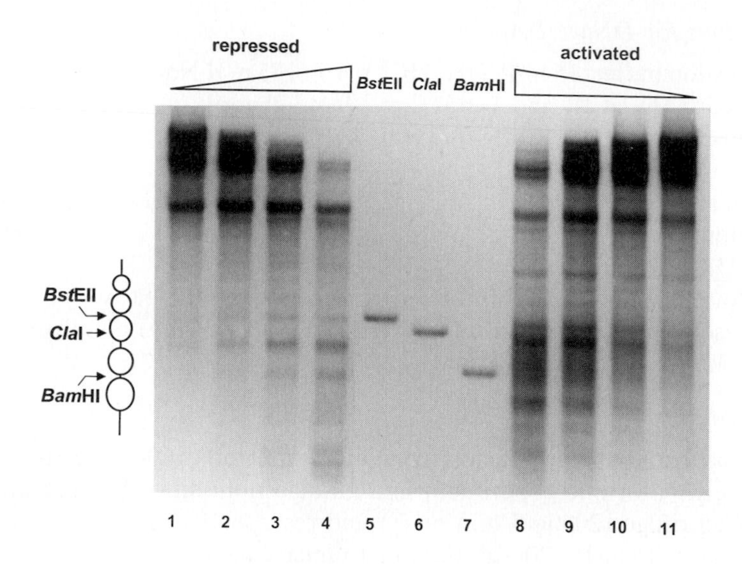

B

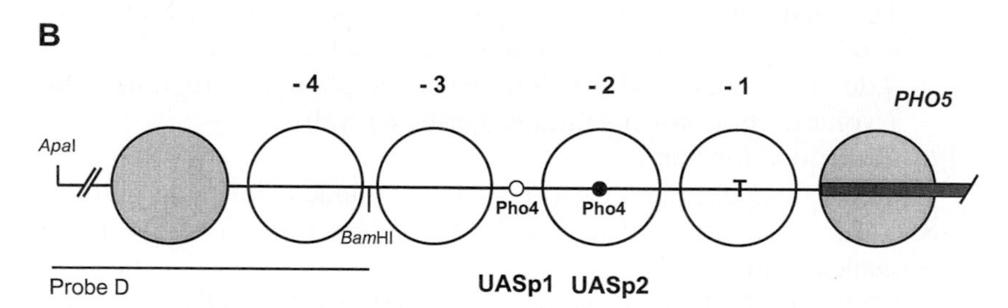

FIG. 2. DNase I analysis of the two chromatin states of the yeast *PHO5* promoter. (A) Nuclei from repressed or activated nuclei were digested with 0.25 (lanes 1 and 11), 0.5 (lanes 2 and 10), 1 (lanes 3 and 9), or 2 (lanes 4 and 8) U/ml DNase I for 20 min at 37°. DNA was isolated, digested with *Apa*I, and analyzed on a 1.5% agarose gel, blotted, and hybridized with probe D.[16] Double digests of purified genomic DNA with *Apa*I/*Bst*EII, *Apa*I/*Cla*I, and *Apa*I/*Bam*HI shown in lanes 5, 6, and 7, respectively, serve as markers. The schematic on the left shows the positions of the restriction sites used for generating the marker fragments together with the derived positions of the nucleosomes. (B) A schematic representation of the chromatin organization of the repressed *PHO5* promoter. The TATA box (T) and the positions of the Pho4-binding sites UASp1 and UASp2 are shown. Nucleosomes removed on activation are shown as open circles. Also shown are probe D and the location of the *Apa*I site used for indirect end labeling.

the secondary digest and probe position. In general, a range of digests are taken for secondary digestion, which should show neither a high proportion of mononucleosome fragments nor no apparent change on digestion. For an example of a suitable extent of digestion, see Fig. 1, lanes 2–5.

15. Results of a typical analysis are given in Fig. 2.

Mononucleosome Analysis with Micrococcal Nuclease

Micrococcal nuclease demonstrates strong DNA sequence specificity and shows preference for DNA in the linker region between positioned nucleosomes.[15] The method described here utilizes the latter to reduce the chromatin into mono-, di-, tri-, and tetranucleosomal DNA fragments. The DNA is then resolved on an agarose gel and is analyzed by Southern blotting and hybridization to short DNA probes (see the schematic in Fig. 3A). The assay asks whether there is a nucleosome on a given stretch of DNA. Because the technique does not utilize indirect end labeling, the assay does not provide information relating to nucleosome position. One advantage of this method, however, is that by direct comparisons between the results of the same blot probed with several different DNA fragments it is possible to effectively "walk" across a region of chromatin determining the presence or absence of nucleosomes. To exemplify this technique, it has been used to study the chromatin of the *PHO5* promoter[11,16] (see Fig. 3B) and the *Gal1-10* promoter.[17]

Solutions

Digestion buffer: 15 m*M* Tris–HCl, pH 8.0, 50 m*M* NaCl, 1.4 m*M* $CaCl_2$, 0.2 m*M* EDTA, 0.2 m*M* EGTA, and 5 m*M* 2-mercaptoethanol

Nuclease dilution buffer: 10 m*M* Tris–HCl, pH 7.4, and 0.1 mg/ml BSA

Procedure

1. Suspend nuclei from approximately 500-mg cells (wet weight) in 3 ml digestion buffer by vortexing.
2. Centrifuge (2000*g*/5 min at 4°), resuspend in 1.2 ml digestion buffer, and transfer 200-μl aliquots to microfuge tubes.
3. Add micrococcal nuclease [Boehringer Mannheim (also listed as nuclease S7)] at four different concentrations in the range of 5 to 100 U/ml and incubate for 20 min at 37°. The extent of digestion is

[15] W. Hörz and W. Altenburger, *Nucleic Acids Res.* **9,** 2643 (1981).

[16] A. Almer, H. Rudolph, A. Hinnen, and W. Hörz, *EMBO J.* **5,** 2689 (1986).

[17] M. J. Fedor and R. D. Kornberg, *Mol. Cell. Biol.* **9,** 1721 (1989).

A 1. Micrococcal nuclease digestion

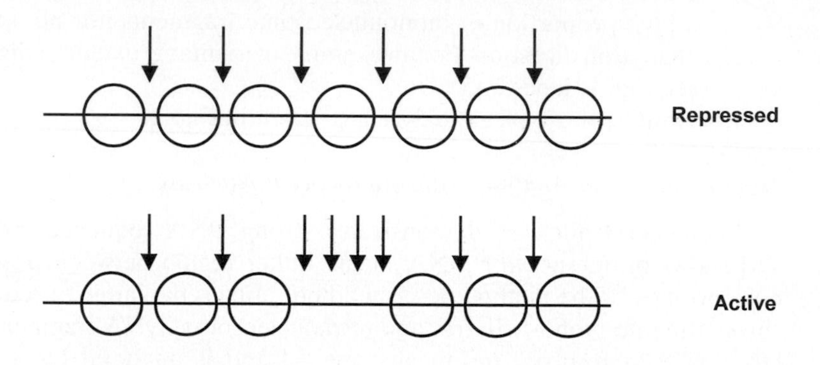

2. Gel electrophoresis, Southern transfer, hybridization, and autoradiography

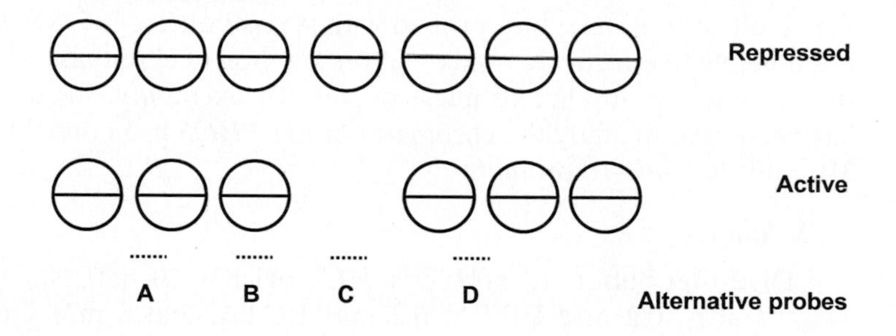

3. Mononucleosome signal

Probe	Repressed	Active
A	+	+
B	+	+
C	+	-
D	+	+

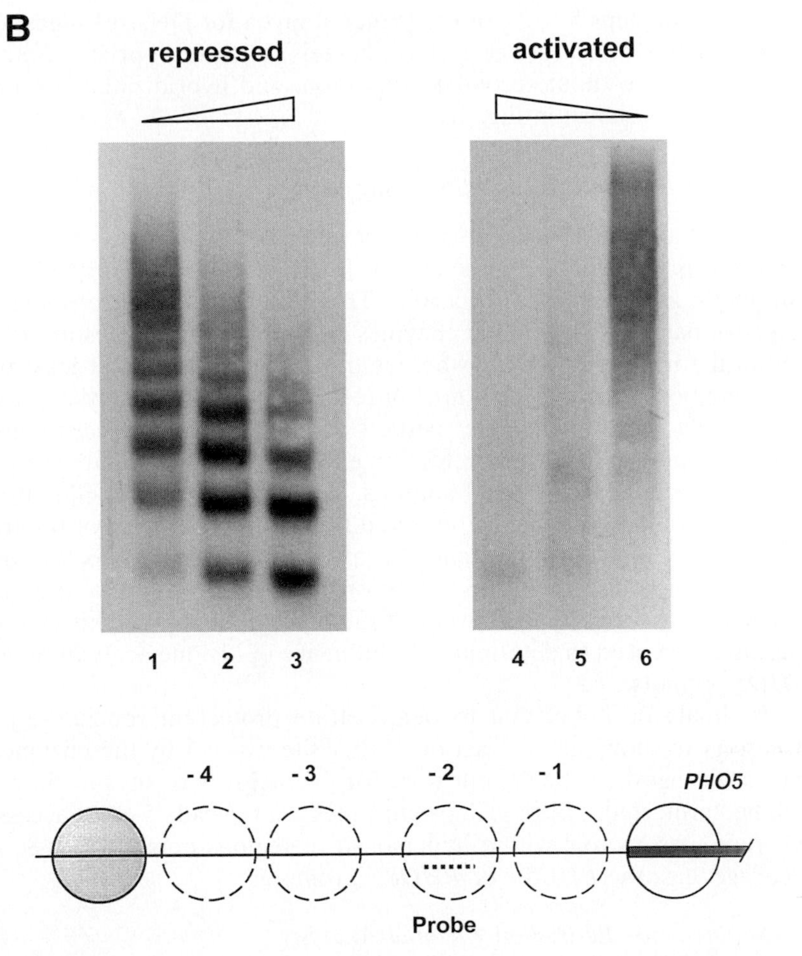

FIG. 3. Assaying for the presence or absence of nucleosomes. (A) A schematic of the protocol. (B) Nuclei isolated from repressed or activated cells were treated with 32 (lanes 1 and 6), 64 (lanes 2 and 5), and 128 (lanes 3 and 4) U/ml micrococcal nuclease. DNA was isolated, separated on a 2% agarose gel, blotted, and hybridized to a nucleosome -2 specific probe. The schematic underneath depicts the chromatin organization of the *PHO5* promoter and the position of the probe employed. Nucleosomes disrupted under activating conditions are shown as dashed circles.

monitored by ethidium bromide-stained gels. Select samples that have mostly mononucleosomes and also samples that still have di-, tri-, and tetranucleosomal DNA.

4. Terminate digestion by adding 10 μl stop solution, 5 μl 20% SDS, and 20 μl proteinase K solution. Incubate for 30 min at 37°.

5. Follow steps 5 to 14 of the protocol given for DNase I digestion.
6. Analyze by agarose gel electrophoresis (2% is appropriate), Southern transfer without secondary digestion, and hybridization. A typical result is shown in Fig. 3B.

Restriction Enzyme Accessibility Assay

A limitation of the use of nonspecific nucleases in the study of chromatin structure is the inherent difficulty in determining the extent to which a chromatin modulation has occurred. The presence of a nucleosome is an effective barrier to restriction enzymes and prevents the digestion of sites within the underlying DNA, whereas transcription factors, at least under the conditions employed, do not. A restriction enzyme analysis provides a relatively easy and reliable method of quantifying the accessibility of nucleosomal DNA and is therefore an excellent complementary technique to the more standard DNase I analysis. This analysis can provide information on the boundaries of the observed transition, the nature of the transition and resultant structure, and the question of what proportion of the cell population undergoes this transition. The strategy employed is shown schematically with a typical result of the analysis in Fig. 4. This technique has also been used to determine the influence of histone acetylation at the *PHO5* promoter.[12]

It should be noted that to demonstrate protection conclusively it is necessary to show that at least one other site was cut by the enzyme in a particular digest, as the default state for the majority of sites in chromatin will be "protected." Intranucleosomal sites are typically 5–10% accessible, whereas sites located within hypersensitive regions demonstrate 80–95% cleavage, e.g., the *PHO5*[16] or *PHO8*[13] promoter.

Solutions for Restriction Enzyme Analysis

Digestion buffer: 10 m*M* Tris–HCl, pH 7.4, 50 m*M* NaCl, 10 m*M* $MgCl_2$, 0.5 m*M* spermidine, 0.15 m*M* spermine, 0.2 m*M* EDTA, 0.2 m*M* EGTA, and 5 m*M* 2-mercaptoethanol

Procedure

1. Suspend pelleted nuclei in digestion buffer by vortexing.
2. Centrifuge (2000*g*/5 min at 5°) and resuspend in digestion buffer. Nuclei from approximately 50-mg cells (wet weight) are used for one experiment and are suspended in 200 μl. Transfer to microfuge tubes. The buffer employed balances the need to preserve the chromatin structure while facilitating enzyme activity. The buffer therefore contains 50 m*M* salt, which appears to be sufficient for most enzymes

1. Nuclei isolation

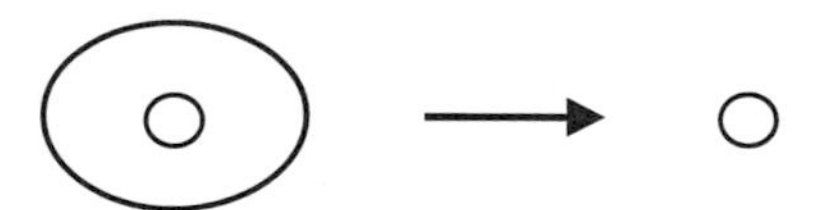

2. Restriction enzyme treatment (*Cla*I)

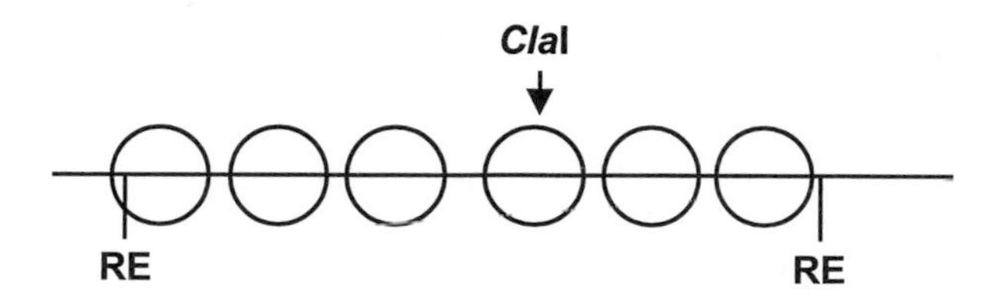

3. DNA isolation and secondary digestion with restriction enzyme (RE)

*Cla*I site:

Protected

Accessible

Probe

4. Gel electrophoresis, Southern transfer, hybridization, and autoradiography

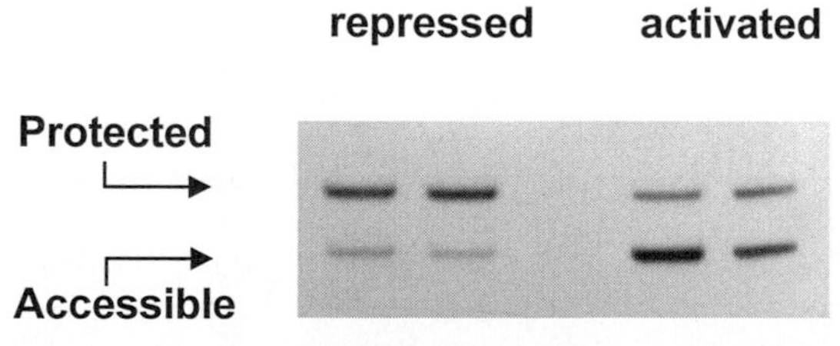

FIG. 4. Restriction site accessibility of the *PHO5* promoter. Nuclei isolated from repressed and activated strains were treated with either 50 (left) or 200 (right) u of *Cla*I and incubated for 30 min at 37°. In order to monitor the extent of cleavage, DNA was isolated, cleaved with *Hae*III, analyzed on a 1.5% agarose gel, blotted, and hybridized with probe D.[16] A schematic representation of the method is provided at the top.

in the chromatin digestion, and the polyamines, spermidine and spermine, which serve to stimulate restriction enzyme activity and (usually) suppress nonspecific cellular nucleases.

3. Add restriction nuclease at two different concentrations (range between 150 and 1500 U/ml). Incubate at 37° for 30 min. The two concentrations of enzyme differing by a factor of 3–4 should provide approximately identical levels of cleavage. This demonstrates that the amount of enzyme was not limiting. Therefore, all accessible sites can be assumed to be cut within the time course of the experiment. It is also advisable to include a 0 and 37° control to monitor the extent to which endogenous nucleases were active during the incubation. In this respect it is preferable if the longer (protected) and shorter (accessible) fragments generated by the analysis are of an approximately similar size and thus presumably equally susceptible to degradation by endogenous nucleases.
4. Terminate digestion as described for DNase I except raise the EDTA concentration to 12 m*M*.
5. Follow steps 5 to 14 of the protocol given for DNase I analysis.
6. Use 10 μl of the DNA from restriction nuclease-digested nuclei for secondary digestion with the appropriate restriction enzyme.
7. Analyze by Southern transfer and hybridization (Fig. 4).
8. Accessibility can be quantified by determining the ratio of the amount of each fragment per lane by phosphoimaging analysis or densitometry.

Acknowledgments

This work was supported by grants from the European Commission Human Capital and Mobility Network (ERBCHRXCT940447), the Deutsche Forschungsgemeinschaft (SFB190), and Fonds der Chemischen Industrie.

[21] Assays for Nucleosome Positioning in Yeast

By MICHAEL P. RYAN, GRACE A. STAFFORD, LIUNING YU, KELLIE B. CUMMINGS, and RANDALL H. MORSE

Introduction

The yeast *Saccharomyces cerevisiae* is an excellent eukaryotic model system to investigate how chromatin functions in biological processes such

0076-6879/99 $30.00

as transcription, replication, and silencing. To understand the relationship between chromatin structure and function, it is critical to define where nucleosomes, the repeating unit of chromatin, are positioned along a DNA sequence. Each nucleosome contains 147 bp of DNA wrapped $1\frac{3}{4}$ turns around a cylindrical-shaped histone core consisting of two molecules each of histones H2A, H2B, H3 and H4 (reviewed in Wolffe[1]). In higher eukaryotes, histones H1 and H5 bind to DNA entering and exiting the nucleosome and promote higher order folding of chromatin. Although a homolog of histone H1 has been identified in yeast, genetic studies suggest that it plays a minor role with respect to chromatin structure and function.[2]

Positions of nucleosomes are defined translationally and rotationally.[3,4] Translational positioning specifies where the histone core begins and ends along the DNA sequence. Rotational positioning refers to the orientation of the DNA helix relative to histone core, e.g., whether the minor groove faces inward to the core or outward. Several techniques have been developed to map both the translational and the rotational positioning of nucleosomes. In general, these techniques utilize enzymes or chemicals that cleave nucleosomal and nonnucleosomal DNA differentially. Differential cleavage is largely due to the wrapping of DNA around the nucleosome core, which makes DNA in chromatin less accessible to cutting.

When mapping the structure of chromatin in yeast, the position of a nucleosome is determined from a population of sequences. Within this population, a given region of DNA may contain nucleosomes that are randomly or well positioned. Well positioned implies that the nucleosome occupies a particular sequence in the majority of the population being examined. Well-positioned nucleosomes have been characterized in the promoter regions of a number of yeast RNA polymerase II-dependent genes in their inactive state, including the *PHO5, GAL1*, and *GAL10* genes, and in yeast minichromosomes.[5–8] In many cases, the mechanism for positioning is unclear. Experimental evidence indicates that nucleosome positioning can be directed by the DNA sequence and by DNA-binding proteins that position nucleosomes (e.g., the $\alpha 2$ repressor protein[7]) or gener-

[1] A. P. Wolffe, "Chromatin: Structure and Function." Academic Press, New York, 1995.

[2] H.-G. Patterton, C. C. Landel, D. Landsman, C. L. Peterson, and R. T. Simpson, *J. Biol. Chem.* **273,** 7268 (1998).

[3] R. T. Simpson, *Prog. Nucleic Acids Res. Mol. Biol.* **40,** 143 (1991).

[4] F. Thoma, *Biochem. Biophys. Acta* **1130,** 1 (1992).

[5] D. Lohr, *J. Biol. Chem.* **272,** 26795 (1997).

[6] J. Svaren and W. Hörz, *TIBS* **22,** 93 (1997).

[7] S. Y. Roth, A. Dean, and R. T. Simpson, *Mol. Cell. Biol.* **10,** 2247 (1990).

[8] R. H. Morse, *Science* **262,** 1563 (1993).

ate nucleosome-free regions that influence positioning in adjacent sequences (e.g., GRF2[9,10]).

This article discusses and outlines methodologies used to assess nucleosome positioning in yeast. The principles and methods for the analysis of chromatin structure in yeast have been covered in previous reviews that provide excellent companions to the methods presented here.[11,12]

Principles

Reagents Used to Cleave DNA

The most common cleavage reagents used to examine chromatin structure are micrococcal nuclease (MNase) and DNase I. MNase rapidly introduces double-stranded cuts in DNA between nucleosomes (i.e., within linker DNA) that initially generates a 180- to 200-bp fragment of DNA per nucleosome. Further digestion cleaves DNA to 147 bp; the DNA remaining includes protected sequences that are wrapped around the nucleosome. MNase will also introduce single-stranded nicks into nucleosomal DNA. This feature must be taken into consideration when performing high-resolution mapping (see later). When chromatin is digested with limiting concentrations of MNase, cleavages occur such that fragments are generated corresponding to mononucleosomes, dinucleosomes, trinucleosomes, and so on. DNA purified from such MNase-digested DNA runs as a "nucleosome ladder" when visualized on an agarose gel (Fig. 1).

DNase I differs from MNase in that it introduces single-stranded cuts in DNA within the nucleosome with a periodicity of 10 bp. This cleavage pattern reflects cutting in the minor groove of DNA, which is exposed once per helical turn as DNA is wrapped around the histone core. Differences in the accessibility of DNA in the nucleosome to DNase I and MNase may be associated with the architecture of DNA in the nucleosome; DNA is bent away from DNase I during cleavage so that nucleosomal DNA may provide a more suitable structure for cleavage by DNase I than for MNase.[13]

Restriction enzymes have also been used to examine the position of nucleosomes.[14,15] Restriction endonucleases, like MNase, fail to cleave

[9] M. J. Fedor, N. F. Lue, and R. D. Kornberg, *J. Mol. Biol.* **204,** 109 (1988).

[10] D. I. Chasman, N. F. Lue, A. R. Buchman, J. W. Lapointe, Y. Lorch, and R. D. Kornberg, *Genes Dev.* **4,** 503 (1990).

[11] S. M. Paranjape, R. T. Kamakaka, and J. T. Kadonaga, *Annu. Rev. Biochem.* **63,** 265 (1994).

[12] F. Thoma, *Methods Enzymol.* **274,** 197 (1996).

[13] R. H. Morse and R. T. Simpson, *Cell* **54,** 285 (1988).

[14] A. Almer and W. Hörz, *EMBO J.* **5,** 2681 (1986).

[15] T. K. Archer, M. G. Cordingley, R. G. Wolford, and G. L. Hager, *Mol. Cell. Biol.* **11,** 688 (1991).

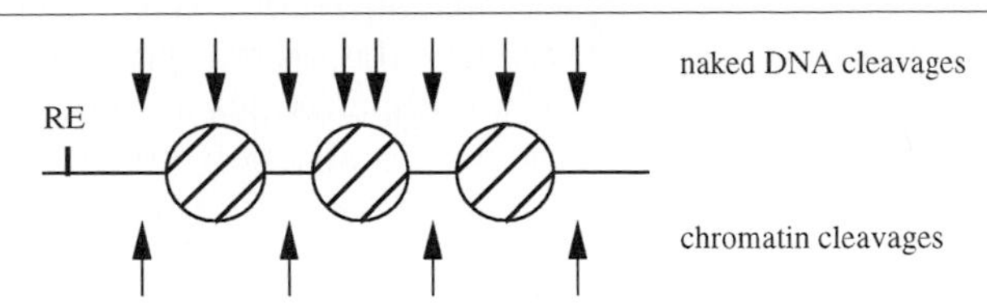

1. Partially digest chromatin and naked DNA with MNase.

2. Purify DNA. Run chromatin samples on agarose gel to confirm DNA is appropriately digested.

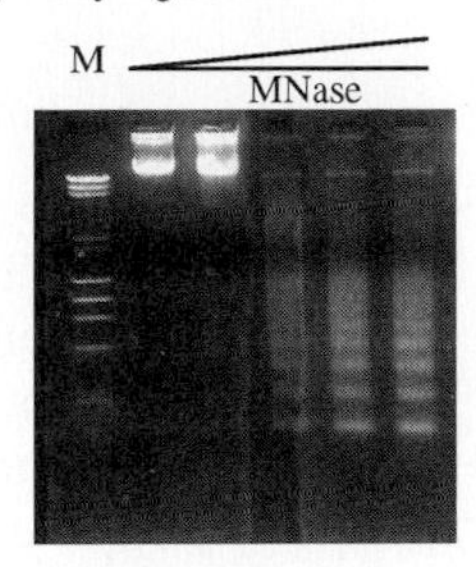

3. Digest with a restriction enzyme.

CHROMATIN

RE

*—— (labeled probe)

NAKED DNA

RE

*—— (labeled probe)

4. Electrophorese on an agarose gel and Southern blot. Hybridize with probe abutting restriction enzyme cleavage site.

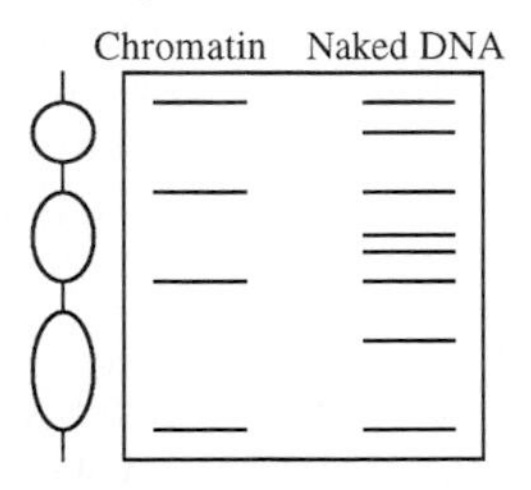

FIG. 1. The indirect end-label technique.

DNA within nucleosomes. Defining the precise position of a nucleosome is likely to be more difficult with these enzymes if appropriate sites are not available for analysis. In this case, restriction enzymes can be used in conjunction with MNase or DNase I to confirm the presence or absence of a nucleosome along a particular sequence. Several other reagents can be used to differentiate nucleosomal and nonnucleosomal DNA, including bacterial methylases (expressed exogenously in yeast),[16,17] methidiumpropyl-EDTA (MPE · Fe[II]),[18] and hydroxyl radicals (hydroxyl radicals are used primarily for *in vitro* studies).[19]

Low-Resolution Mapping of Nucleosomes by Indirect End Labeling

The indirect end-label technique (Fig. 1) is the preferred method to initially assign the position of nucleosomes at low resolution (to ±20 bp).[12,20] In this assay, MNase is added at limiting concentrations to yeast nuclei[21] or detergent-permeabilized spheroplasts[22] (see methods) to digest chromatin, and the cleaved DNA is purified. Naked (deproteinized) DNA, purified from nuclei or permeabilized spheroplasts, is also digested with MNase and generally exhibits some sequence specificity in its cleavage. MNase-digested DNA from chromatin and naked DNA samples is subsequently digested with a restriction enzyme whose site lies near the sequence being examined (0.2–2 kbp away from the restriction site). DNA is fractionated on an agarose gel along with a molecular weight marker. Southern analysis is subsequently performed using a 150- to 200-bp probe complementary to sequences that start from the restriction site toward the region being examined. To assign the position of nucleosomes, cleavage patterns generated with chromatin samples are compared to naked DNA. Regions that are not cleaved in chromatin, but are in naked DNA, that encompass 140–160 bp are provisionally assumed to be positioned nucleosomes (Fig. 1). If nucleosomes are positioned randomly along a sequence, chromatin and naked DNA samples will have similar digestion patterns.

Examples and Interpretation. Figure 2 shows representative indirect end-labeling experiments performed with MNase on yeast harboring the high-copy plasmids TABIC4Δ80 (Fig. 2A) or 1-10/104 (Fig. 2B). TABIC4Δ80,

[16] M. P. Kladde and R. T. Simpson, *Proc. Natl. Acad. Sci. U.S.A.* **91,** 1361 (1994).
[17] M. P. Kladde and R. T. Simpson, *Methods Enzymol.* **274,** 214 (1996).
[18] I. L. Cartwright and S. C. R. Elgin, *Methods Enzymol.* **170,** 359 (1989).
[19] J. Yang and J. Carey, *Methods Enzymol.* **259,** 452 (1995).
[20] S. A. Nedospasov and G. P. Georgiev, *Biochem. Biophys. Res. Commun.* **92,** 532 (1980).
[21] M. Shimizu, S. Y. Roth, C. Szent-Gyorgyi, and R. T. Simpson, *EMBO J.* **10,** 3033 (1991).
[22] N. A. Kent, L. E. Bird, and J. Mellor, *Nucleic Acids Res.* **21,** 4653 (1993).

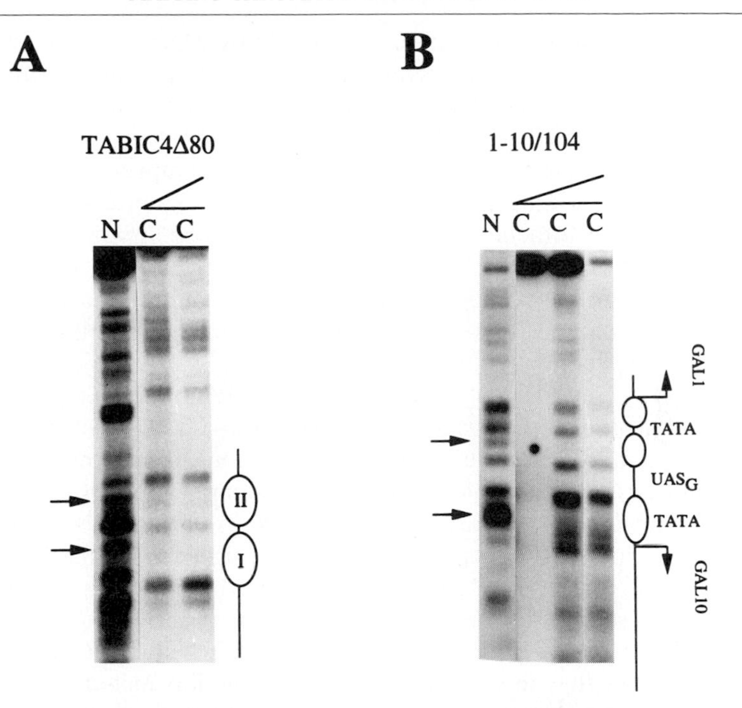

FIG. 2. Mapping nucleosome positions in the episomes TABIC4Δ80 and 1-10/104 using the indirect end-label technique. (A) TABIC4Δ80 contains sequences from the TRP1ARS1 minichromosome, including those of nucleosomes I and II (ovals I and II).[3,4,8] Arrows indicate cleavage sites in naked DNA (N) that are not cleaved (or barely cleaved) in chromatin. In 1-10/104 (B) positions predicted to be nucleosomal[5] (i.e., the ovals at the *GAL10* TATA element, and on either side of the *GAL1* TATA element) are cleaved in chromatin but not in naked DNA [the dot in second lane in (B) is an autoradiography artifact]. Note that the region encompassing GAL4-binding sites (UAS_G) is not cleaved in either chromatin or naked DNA.

a derivative of TA17Δ80,[8] contains sequences of the TRP1ARS1 minichromosome that encompass nucleosomes I and II and four Bicoid-binding sites introduced into sequences normally occupied by nucleosome I. The minichromosome 1-10/104 contains the *GAL1-10* promoter, including the *GAL1* and *GAL10* TATA elements and transcription start sites located on either side of four GAL4-binding sites (UAS_G). Each band in the naked DNA and chromatin lanes represents a population of DNA fragments that is cut on one end with the restriction enzyme and on the other end by MNase. Because the location of the restriction site is known, the distance from this site to MNase-specific cleavage sites can be determined by calculating the molecular weight of each band. Naked DNA samples are used to define sites that are cleaved preferentially by MNase. Some cleavage sites in naked DNA are also seen in the chromatin samples; however, cleavages in naked DNA that are not observed in chromatin are clearly evident in both TABIC4Δ80 and 1-10/104 (Fig. 2, arrows). Based on the size of these protected regions in the chromatin samples, it is predicted that these sequences are occupied by nucleosomes (Fig. 2, ovals). Other regions are more difficult to assign positioned nucleosomes to without additional data. For example, the region above nucleosomes I and II in TABIC4Δ80 chromatin shows clear protection against MNase cleavage, but more than one positioning assignment is compatible with this protection.

The nucleosome is not a "fixed"structure but is more dynamic in nature. Well-positioned nucleosomes may become lost in a fraction of the population being examined as a consequence of nucleosome disassembly or sliding during replication, transcription, or improper sample handling, resulting in only partial protection against cleavage reagents. This tends to be more apparent in chromatin samples that are overdigested with MNase. For this reason, a range of MNase should be used in a given experiment. In TABIC4Δ80, for example, a band is evident in the region designated nucleosome II in both naked DNA and chromatin samples (arrow in Fig. 2A). We expect that this reflects a small population of nonnucleosomal sequences being cleaved as the band is relatively weak and is flanked by bands that are more intense, which is indicative of cutting at the nucleosome edges. If this region were nonnucleosomal in the majority of the population, the relative intensities of the bands would be similar to naked DNA samples.

Because MNase has some sequence specificity, a given region may not be cut even as naked DNA. This can result in uncertainties when assigning nucleosome positions. Under these circumstances, additional cleavage reagents, such as a restriction endonuclease (see later), MPE · Fe[II] or DNase I should be used to determine if sequences are nucleosomal or simply resistant to cleavage. (DNase I will introduce double-stranded cleavages at high concentrations and can be used for indirect end labeling, but the banding pattern tends to be diffuse compared to that observed with MNase.[12,23]) Such resistance to MNase digestion has been observed in the *GAL1-10* promoter. Neither chromatin nor naked DNA is digested with MNase at the UAS_G (Fig. 2B). However, MPE · Fe[II] and DNase I cleave these sequences efficiently, indicating that this region of the *GAL1-10* promoter is not protected by a nucleosome.[24]

[23] D. Lohr and J. E. Hopper, *Nucleic Acids Res.* **13,** 8409 (1985).

[24] D. Lohr, *Nucleic Acids Res.* **12,** 8457 (1984).

Mapping Chromatin with Restriction Endonucleases

Restriction enzymes can be used to substantiate results obtained with MNase and the indirect end-label technique, providing that regions of interest contain an appropriate restriction site or sites. A sequence predicted to be nucleosomal by the indirect end-label technique should not be cleaved by a restriction enzyme whose site is positioned within a nucleosome. Similarly, if a restriction site is located in an MNase-sensitive region, cutting should be evident when chromatin is digested with the restriction enzyme. Like indirect end labeling, restriction enzyme digestion is performed using isolated nuclei or permeabilized spheroplasts as the chromatin substrate. Chromatin is digested for various times, and control samples are incubated without enzyme to monitor endogenous nuclease activity. After chromatin is digested and DNA purified, a secondary digestion is done with an enzyme (or enzymes) that cuts DNA on either side of the sequence being examined. The secondary digestion yields a band of a predicted molecular weight in undigested chromatin samples and a smaller band if the region in chromatin is accessible to digestion. Samples are analyzed by the indirect end-label technique discussed earlier.

Examples and Interpretation. Differential cleavage of nucleosomal and nonnucleosomal DNA with a restriction enzyme is shown with the plasmids 314-17lacZ and 314-17Δ80lacZ[8] (Fig. 3). 314-17lacZ contains TRP1ARS1 sequences that encompass nucleosomes I and II (similar to TABIC4Δ80 discussed earlier) and the β-galactosidase reporter gene. A single GAL4-binding site (UAS_G) was introduced such that it should be positioned at the edge of nucleosome I (Fig. 3A). Eighty base pairs of DNA located within nucleosome I in 314-17lacZ were deleted to generate the plasmid 314-17Δ80lacZ; we expected that this deletion would reposition the GAL4-binding site from the edge of nucleosome I to its center (Fig. 3A). Indirect end labeling confirmed that nucleosome I is well positioned in 314-17lacZ (not shown) and 314-17Δ80lacZ.[8] We took advantage of a unique *Pst*I site located 30 bp away from the UAS_G to determine whether the UAS_G is more accessible in 314-17lacZ compared to 17Δ80lacZ (Fig. 3). Spheroplast lysates isolated from yeast containing 314-17lacZ or 314-17Δ80lacZ were incubated in the absence of enzyme ("0") or were digested for 15 or 30 min with *Pst*I. DNA was purified, cut with *Pvu*II, and visualized by Southern analysis using a plasmid-specific probe. Both 314-17lacZ and 314-17Δ80 lacZ samples contain a band in mock-digested samples ("0") that corresponds to the *Pvu*II fragment containing the *Pst*I site. *Pst*I cleavage results in a lower molecular weight band corresponding to the predicted *Pvu*II–*Pst*I fragment (labeled *Pst*I; Fig. 3B). The extent of digestion in 314-17lacZ is much greater than that observed with 314-17Δ80lacZ, indicating that the *Pst*I site in 314-

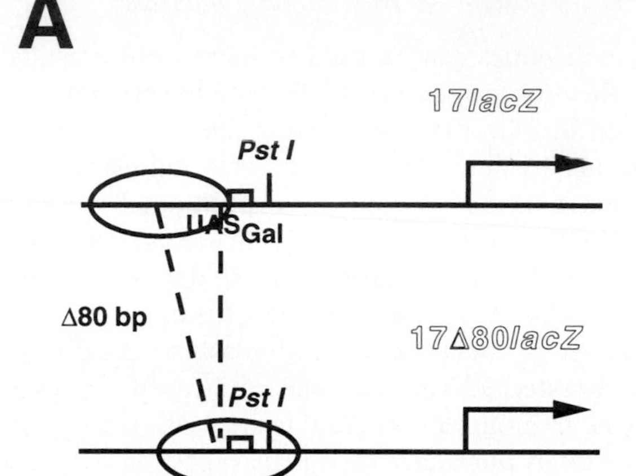

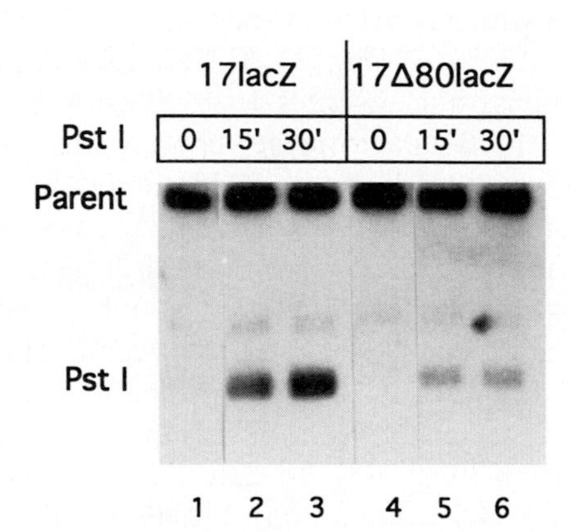

FIG. 3. Accessibility of 314-17lacZ and 314-17Δ80lacZ to *Pst*I. (A) Schematic illustration showing the predicted position of nucleosome I, based on its position in TA17 and TA17Δ80, and a GAL4-binding site in 314-17lacZ (17lacZ) and 314-17Δ80lacZ (17Δ80lacZ). (B) Spheroplast lysates from yeast containing 314-17lacZ or 314-17Δ80lacZ were mock digested ("0") for 30 min or digested with 40 units of *Pst*I for 15 or 30 min as indicated. The parent *Pvu*II fragment ("Parent") and *Pvu*II–*Pst*I fragment are indicated.

17lacZ is much more accessible to cleavage with the restriction enzyme. This, in conjunction with indirect end-labeling analysis, strongly suggests that the UAS_G in 314-17Δ80lacZ is positioned within a nucleosome.

High-Resolution Mapping Using Primer Extension Analysis

Cleavages introduced into chromatin by MNase or DNase I can be mapped at the nucleotide level to precisely define the position of nucleosomes using high-resolution primer extension. In this technique, MNase- or DNase I-digested DNA is denatured and a ^{32}P-end-labeled oligonucleotide complementary to the sequence being examined is annealed to one strand. *Taq* polymerase is used to extend DNA from the primer to the cleavage site. A single denaturation, annealing, and extension reaction is performed when analyzing abundant sequences (i.e., high-copy plasmids). If sequences are in low abundance (i.e., genomic sequences or low copy plasmids), multiple extensions are performed with a thermocycler. Reactions are also done with ddNTPs to sequence the region being analyzed. Extended DNAs are separated on a sequencing gel and visualized by autoradiography.

Primer extension reactions will optimally resolve cleavage sites in sequences 150–250 bp away from the primer and generally provide an acceptable signal-to-noise ratio when analyzing low-copy sequences. Ligation-mediated polymerase chain reaction, which is used more often when analyzing single copy sequences in large genomes, is an alternative approach that can be utilized.[25]

Examples and Interpretation. The *STE6* locus contains multiple positioned nucleosomes in yeast haploid α cells,[21] which are clearly evident as regions of protection in MNase-digested chromatin analyzed by primer extension[2] (Fig. 4). Some sites are weakly cleaved relative to naked DNA and are flanked by sequences highly sensitive to MNase (i.e., at the nucleosome edges). Such sites typically represent single-stranded cleavages that are not seen when the same samples are analyzed by the indirect end-label technique, which reveals only double-stranded cleavages. The precise location of the nucleosome edges are determined using the sequencing reactions. DNA isolated from nucleosome monomers can also be used to define nucleosome edges.[9,26] Oligonucleotides are annealed to sequences protected from MNase digestion and primer extension analysis is performed.

When chromatin is digested with DNase I and examined by primer extension analysis, bands separated by 10 nucleotide intervals are observed on a sequencing gel in regions containing rotationally positioned nucleo-

[25] P. R. Mueller and B. Wold, *Science* **246,** 780 (1989).

[26] G. Fragoso, S. John, M. S. Roberts, and G. L. Hager, *Genes Dev.* **9,** 1933 (1995).

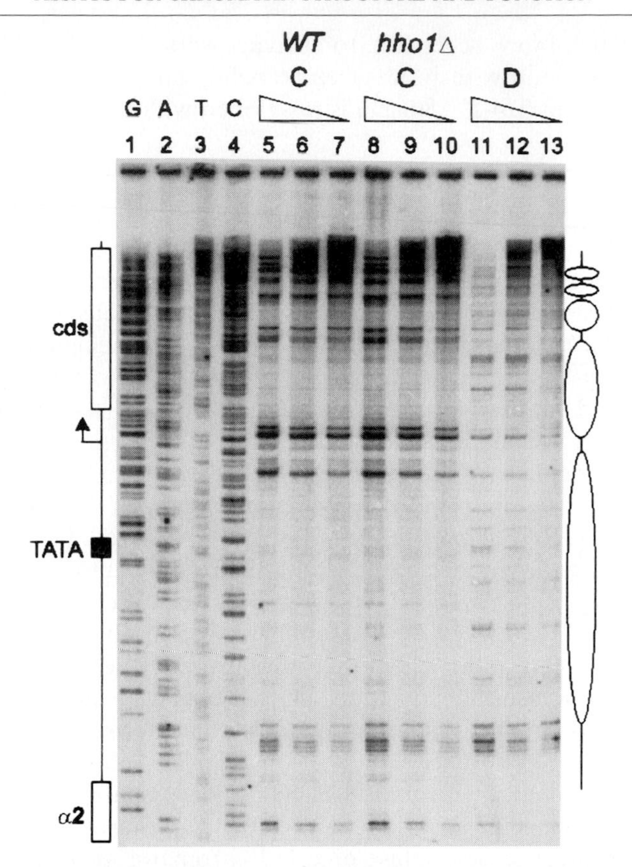

FIG. 4. Mapping positions of MNase cleavage sites in the *STE6* locus by primer extension. Nuclei from two yeast strains (WT, lanes 5–7; and *hho1*Δ, lanes 8–10) were used to digest chromatin (C) with 10 (lanes 5 and 8), 5 (lanes 6 and 9), and 2.5 (lanes 7 and 10) U/ml MNase. Naked DNA (D) was digested with 0.1 (lane 11), 0.05 (lane 12), or 0.025 (lane 13) U/ml MNase. Primer extension was performed using a ^{32}P-labeled oligonucleotide that anneals downstream of the $\alpha 2$ operator. Sequencing reactions are shown in lanes 1–4. The schematic at the left indicates sequences in the *STE6* locus, and the positions of nucleosomes are indicated by ovals on the right. [Reprinted with permission of the American Society for Biochemistry from H.-G. Patterton, C. Church Landel, D. Landsman, C. L. Peterson, and R. T. Simpson, *J. Biol. Chem.* **273,** 7268 (1998).]

somes. This periodicity is observed at the recombination enhancer in yeast α cells[27] (Fig. 5). In α cells, chromatin is highly organized at the recombination enhancer by the $\alpha 2$ protein, as evident by bands cleaved at approximately 10-bp intervals at positions 29,403–29,560 of chromosome III. In **a** cells,

[27] K. Weiss and R. T. Simpson, *EMBO J.* **16,** 4352 (1997).

chromatin is less organized and this pattern is not observed. Based on the characteristic cleavage patterns observed in chromatin with DNase I, the rotational position of nucleosomes can be determined. Cleavage sites determined by primer extension represent minor groove sequences that face outward from the histone core.

Analyzing data obtained with primer extension can be difficult, especially if MNase introduces single-stranded nicks in nucleosomal DNA or if nucleosomes occupy multiple positions. To differentiate single-stranded cuts in nucleosomal DNA from double-stranded cuts introduced into linker DNA by MNase, it is recommended that the opposite strand be analyzed by priming with another oligonucleotide. Data obtained using the indirect end-label technique, restriction enzymes, and primer extension analysis should be combined to come to a consensus of where a nucleosome is positioned.

Methods and Reagents

Comments on Methodologies

Chromatin prepared as isolated nuclei[21] or as permeabilized spheroplasts[10,22] can be used for MNase or DNase I digestion and indirect end-labeling analysis. For primer extension analysis, we recommend using isolated nuclei.[21] Although permeabilized spheroplasts prepared using a method similar to the one outlined later have been used for high-resolution analysis of MNase-accessible sites in the *SUC2* promoter,[28] we have found that such preparations sometimes exhibit single-stranded cleavages by endogenous nucleases, whereas this is rarely seen with nuclei preparations. Samples used for primer extension should be examined on agarose gels to ensure that DNA is digested minimally with either MNase or DNase I, as moderate levels of double-stranded cleavage, as visualized on nondenaturing agarose gels, often herald a substantial amount of single-stranded cleavage of nucleosomal DNA.

Restriction enzymes vary in their ability to cleave accessible regions in chromatin prepared as spheroplast lysates, with nuclei typically yielding better results. It is suggested that an appropriate plasmid or DNA fragment be added in low amounts to chromatin samples being digested with restriction enzymes to ensure that the enzyme is active during the incubation period. Such control DNA can be visualized after Southern blotting by hybridization using an appropriate probe.

[28] L. Wu and F. Winston, *Nucleic Acids Res.* **25,** 4230 (1997).

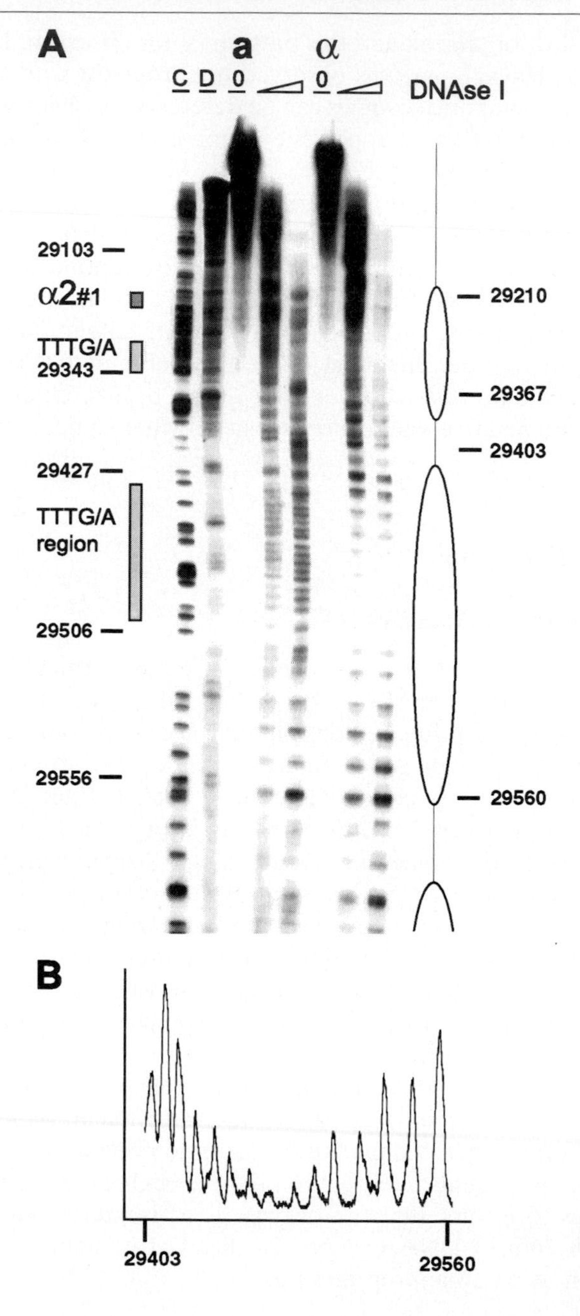
A
a
α
C
D
0
0
DNAse I
29103
α2#1
TTTG/A
29343
29427
TTTG/A region
29506
29556
29210
29367
29403
29560
B
29403
29560

The concentrations of MNase and DNase I given in the following procedure are suggested starting concentrations. Some variability is observed between yeast strains so concentrations may need to be adjusted to achieve appropriate levels of digestion. Higher concentrations of MNase may be required when using permeabilized spheroplasts. To ensure that chromatin and naked DNA samples are digested appropriately, 10 μl of MNase- or DNase-digested DNA (i.e., DNA isolated prior to the secondary restriction digestion in the indirect end-label technique), corresponding to 10–100 ml of yeast culture, depending on the type of preparation, is separated on a 1.4% agarose/TAE gel and DNA visualized with ethidium bromide. Samples digested with MNase for indirect end-label analysis should show a nucleosome ladder (see Fig. 1). If overdigested, samples will primarily contain low molecular weight species. Naked DNA samples have a wide molecular weight distribution and run as a smear.

Isolation of Nuclei

Reagents. 100 m*M* phenylmethylsulfonyl fluoride (PMSF) in 2-propanol (note that PMSF is very toxic and should be handled with great care); 2-mercaptoethanol; 10 mg/ml Zymolyase 100T (Seikagaku America Inc., Ijamsville, MD) prepared in 1 *M* sorbitol, 10 m*M* DTT; 10 U/μl micrococcal nuclease (Worthington Biochemical Corp., Freehold, NJ) prepared in water or 10 U/μl DNase I (Boehringer-Mannheim, Indianapolis, IN) or restriction endonuclease and 10× enzyme reaction buffer; stop buffer (5% SDS/5 mg/ml proteinase K); equilibrated phenol; chloroform; 4 *M* ammonium acetate; 100% ethanol; 70% ethanol; TE buffer (10 m*M* Tris, pH 8.0, 1 m*M* EDTA, pH 8.0); 10 mg/ml RNase A (Boehringer-Mannheim, Indianapolis, IN); microscope slides and coverslips; S buffer (1.4 *M* sorbitol, 40 m*M* HEPES, pH 7.5, 0.5 m*M* $MgCl_2$); F buffer [18% (w/v) Ficoll 400, 20 m*M* PIPES, pH 6.5, 0.5 m*M* $MgCl_2$]; GF buffer [20% (v/v) glycerol, 7% (w/v) Ficoll 400, 20 m*M* PIPES, pH 6.5, 0.5 m*M* $MgCl_2$]; D buffer (10 m*M* HEPES, pH 7.5, 0.5 m*M* $MgCl_2$, 0.05 m*M* $CaCl_2$); D 5/2 buffer (D buffer with 2 m*M* $CaCl_2$, 5 m*M* $MgCl_2$).

FIG. 5. Mapping DNase I cutting sites in the recombination enhancer in **a** and α cells by primer extension. (A) Undigested and DNase I-digested chromatin (two concentrations) from yeast **a** and α cells was analyzed by multiple cycle primer extension with a chromosome III-specific ^{32}P-labeled oligonucleotide. D, naked DNA; C, the dideoxycytosine-terminated sequencing reaction. Ovals indicate the inferred position of nucleosomes in α cells. In α cells, a 10-bp periodicity is observed at sequences 29,403–29,560. (B) Scan showing the cutting pattern produced by DNase I at sequences 29,403–29,560 in α cells. [Reprinted by permission of Oxford University Press from K. Weiss and R. T. Simpson, *EMBO J.* **16,** 4352 (1997).]

Methods

1. Grow 1 liter of cells to OD_{600} 0.5–1.0 in media buffered with 25 m*M* phthalic acid, pH 5.5.
2. Set water baths to 37 and 30°.
3. Harvest cells in 500-ml centrifuge bottles. Weigh one empty bottle without cap, aliquot cell suspension, and centrifuge at 5000 rpm (4225*g*) for 5 min at 4° in a Sorvall GS3 rotor.
4. Pour off media and resuspend pellets in a total of 50 ml S buffer containing 10 m*M* 2-mercaptoethanol (undiluted 2-ME = 14 *M*) and 1 m*M* PMSF (2-ME and PMSF are added just prior to use). Transfer the cell suspension to the weighed centrifuge bottle and spin at 5000 rpm (4225 *g*) for 5 min at 4°.
5. Resuspend pellet in 25 ml S buffer containing 10 m*M* 2-ME, 1 m*M* PMSF and incubate at 30° with gentle shaking (about 60 rpm) for 10 min.
6. Spin at 5000 rpm (4225*g*) for 5 min at 4°. Drain completely and determine the weight of the pellet.
7. Add a volume of S buffer containing 2 m*M* 2-ME, 1 m*M* PMSF equal to four times the weight of the pellet (e.g., 12 ml for a 3-g pellet). Cells are resuspended and spheroplasted by adding 1 ml of 10 mg/ml Zymolyase 100T and incubating at 30° with gentle shaking (about 60 rpm) for 30–90 min. Note that zymolyase is used as a suspension so it should be agitated before adding to the cells to resuspend the enzyme. Be sure that cells are not under- or overspheroplasted. Spheroplasting can be monitored using the squish test; 5 μl of cells is placed on a slide and a coverslip slid back and forth with light pressure two to three times until it sticks. If cells are spheroplasted, cells in the middle are broken and worm-like while cells at the edge are intact. Perform remaining steps on ice using cold buffers.
8. Dilute cells to 30 ml with chilled S buffer containing 1 m*M* PMSF. Transfer to a 30-ml Oak Ridge tube and spin for 5 min at 3500 rpm (2000*g*) in a HB4 Sorvall (swinging bucket) rotor for 5 min. (A SS-34 rotor can be used at 4500 rpm.)
9. Wash pellet twice with 30 ml cold S buffer containing 1 m*M* PMSF.
10. After the second wash, resuspend pellet in 19 ml F buffer containing 1 m*M* PMSF using a large-bore pipette. Transfer to a homogenizer and dounce on ice 10 strokes using an overhead mechanical stirrer (e.g., Con Torque power unit, Eberbach Corp., Ann Arbor, MI).
11. Pipette 19 ml GF buffer containing 1 m*M* PMSF into the bottom of a 30-ml Oak Ridge tube. Layer lysate from homogenizer on top

of the GF buffer and centrifuge at 11,500 rpm in an HB4 rotor (20,000*g*) for 30 min (glycerol must be added to the balance tube to achieve sufficient density to balance).

12. Discard supernatant and resuspend pellet in 20 ml F buffer containing 1 m*M* PMSF using a large-bore pipette. Vortex for 5 min in cold room.
13. Centrifuge for 15 min at 4500 rpm (3000*g*) in a swinging bucket rotor.
14. Transfer *supernatant* to a clean 30-ml Oak Ridge tube and centrifuge at 11,500 rpm (20,000*g*) for 30 min.
15. Pour off supernatant. The pellet, which contains nuclei (volume approximately 0.2 ml), is resuspended in 2–4 ml D buffer and digested with the appropriate enzyme (i.e., MNase, DNase I, or a restriction enzyme).

Digestion of Nuclei with MNase and DNase I

1. Transfer 300-μl aliquots of nuclei into eight, 1.5-ml microfuge tubes (five for digestion of chromatin, three for digestion of naked DNA). Incubate at 37° for 5 min.
2. For chromatin samples, add MNase (0, 2, 5, 20, and 50 U/ml for indirect end labeling; 0, 0.125, 0.25, 0.5, and 1.0 U/ml for primer extension analysis) or DNase I (0, 0.25, 0.5, 1, and 3 U/ml for indirect end labeling; 0, 0.0625, 0.125, 0.25, and 0.5 U/ml for primer extension) to each tube, mix by inversion, and incubate for 10 min at 37°. MNase should be diluted from stock so that 1–10 μl is added per tube. Stop digestion by adding 55 μl of stop buffer and incubate tubes at 37° for >2 hr (overnight incubation is suitable). For naked DNA samples, stop buffer is added immediately and samples may be extracted immediately.
3. Extract chromatin and naked DNA samples once with phenol : chloroform (1 : 1). Remove aqueous phase, add 4 *M* ammonium acetate to a final concentration of 0.8 M, mix, and add 2.5 volumes of 100% ethanol. Precipitate on dry ice for 10 min, centrifuge for 10 min at 12,000 rpm (13,000*g*) in a microfuge, wash pellet with 70% ethanol, and dry pellet. For chromatin samples, resuspend pellet in 100 μl TE buffer containing 3 μl of 10 mg/ml RNase A. For naked DNA samples, resuspend pellet in 300 μl of D 5/2 buffer. Incubate at 37° for 5 min. Digest for 10 min at 37° with MNase (1, 4, and 10 U/ml for indirect end labeling; 0.025, 0.05, and 0.1 U/ml for primer extension) or DNase I (0.05, 0.1, and 0.2 U/ml for indirect end labeling; 0.01, 0.025, and 0.05 U/ml for primer extension). Add 55 μl stop buffer and incubate at 37° for 30 min. Extract with phenol/chloro-

form, precipitate DNA, and resuspend in 100 μl TE/RNase A. Samples are used for indirect end label or primer extension analysis (see later).

Digestion of Nuclei with Restriction Enzymes

1. After resuspending nuclei in D buffer, add the appropriate 10× buffer supplied with the restriction enzyme of choice to a final concentration of 1×. Aliquot 300 μl into three 1.5-ml microfuge tubes and incubate at 37° for 5 min.
2. Digestions are performed with 200–250 U/ml of enzyme for 15 and 30 min at 37°. One sample is mock-digested for 30 min ("0" sample) to monitor endogenous nuclease activity. To ensure that the restriction enzyme is active during the incubation period, 10 ng of an arbitrary plasmid DNA containing the restriction site is added to the 30-min tube after an initial 15-min incubation (i.e., for the last 15 min of digestion). Reactions are terminated using 55 μl of stop buffer and samples are incubated at 37° for >2 hr. Extract with phenol/chloroform, precipitate DNA, and resuspend dried pellet in 100 μl of TE/RNase A as described earlier (step 3 under "Digestion of Nuclei with MNase and DNase I").

Preparation of NP-40 Spheroplasts

Reagents. Yeast media containing 1 *M* sorbitol (i.e., YNB plus carbon source) buffered with 25 m*M* phthalic acid, pH 5.5; 50 m*M* Tris–HCl, pH 7.4, containing 0.1% 2-mercaptoethanol (add 2-ME fresh); 10 mg/ml Zymolyase; 10% SDS; MNase, DNase I, or restriction endonuclease; stop buffer (5% SDS/5 mg/ml proteinase K); equilibrated phenol; chloroform; 4 *M* ammonium acetate; 100% ethanol; 70% ethanol; TE buffer; 10 mg/ml RNase A; buffer A [1 *M* sorbitol, 50 m*M* NaCl, 10 m*M* Tris–HCl, pH 7.4, 5 m*M* $MgCl_2$, 1 m*M* $CaCl_2$, 1 m*M* 2-mercaptoethanol, 0.5 m*M* spermidine (2-ME and spermidine are added just prior to use)]. Buffer B (buffer A containing 0.15% NP-40 added prior to use). D 5/2 buffer (10 m*M* HEPES, pH 7.5, 5 m*M* $MgCl_2$, 2 m*M* $CaCl_2$).

Methods

1. Grow a 100- to 200-ml culture to OD_{600} 0.5–1.2 (5×10^6–1.2×10^7 cells/ml).
2. Collect cells by centrifugation at 5000 rpm (4225*g*) for 5 min in a 500-ml centrifuge bottle in a Sorvall GS3 rotor.

3. Resuspend pellet in 50 mM Tris, pH 7.4/0.1% 2-ME at a concentration of approximately 5×10^7 cells/ml. Incubate in a 30° shaking water bath at 60 rpm for 15 min.
4. Collect cells by centrifugation as described earlier and resuspend pellet in media/sorbitol at a concentration of 1×10^9 cells/ml in a 500-ml centrifuge bottle. Add 1/10 volume of 10 mg/ml Zymolyase 100T and shake in a 30° water bath for approximately 15 min. To determine if cells are spheroplasted, take an aliquot of cells (15–20 μl) prior to, and after addition of, Zymolyase and dilute 100-fold in 1% SDS. Monitor clearing by reading absorbance at OD_{600}. When the absorbance is approximately 20% of the untreated sample, cells are well spheroplasted (samples with values <10 % are overspheroplasted).
5. Dilute cells with 15 volumes of cold media/sorbitol, transfer to a 30-ml Oak Ridge tube, and centrifuge at 3500 rpm (2000g) for 5 min in a Sorvall HB4 (swinging bucket) rotor.
6. Resuspend pellet in 20 ml cold media/sorbitol and centrifuge at 3500 rpm (2,000g) for 5 min in a Sorvall HB4 (swinging bucket) rotor.
7. Resuspend pellet in cold buffer A to a concentration of approximately 2×10^9 cells/ml. Place on ice.

Permeabilizing Spheroplasts and Digesting with MNase or DNase I

1. Place eight 1.5-ml microfuge tubes (five for chromatin samples, three for naked DNA samples) on ice. For chromatin samples to be digested with MNase or DNase I, aliquot enzyme into each tube (refer to protocol for the digestion of nuclei for enzyme concentrations).
2. Aliquot 150 μl of spheroplasts in buffer A into each tube.
3. For MNase and DNase I digestion, add 150 μl buffer B, mix, and incubate at 37° for 5 min. The addition of NP-40 contained in buffer B permeabilizes the spheroplasts and allows digestion to begin. Add 55 μl of stop buffer and incubate at 37° for at least 2 hr. Some clearing should be evident, although samples may still be somewhat cloudy. For naked DNA samples, add 150 μl of buffer B and immediately add 55 μl stop buffer. Incubate at 37° for at least 2 hr.
4. Extract all samples with phenol/chloroform.
5. Collect aqueous phase. For chromatin samples, add 3 μl of 10 mg/ml RNase A to the aqueous phase, incubate at 37° for 30 min, extract with phenol/chloroform, collect the aqueous phase, and add 4 M ammonium acetate to a final concentration of 0.8 M. Add 1 ml 100% ethanol and precipitate on dry ice for 10 min. For naked DNA samples, omit RNase treatment and precipitate as described earlier.

For all samples, collect pellets by centrifugation at 12,000 rpm (13,000*g*) for 10 min, wash with 70% ethanol, and dry pellets. Resuspend chromatin samples in 100 μl TE buffer. For naked DNA samples, resuspend pellet in 300 μl of D 5/2 buffer, preincubate at 37° for 5 min, and digest with MNase or DNase I as described in step 3 for naked DNA samples prepared from nuclei. Examine the extent of digestion as discussed earlier (see Digestion of Nuclei with MNase and DNase I).

Digestion of Permeabilized Spheroplasts with Restriction Enzymes

1. Place three 1.5-ml tubes on ice (for 0, 15, and 30 min digestion) and aliquot enzyme into 15- and 30-min tubes (200–250 U/ml totaling 300 μl).
2. Aliquot 150 μl of spheroplasts in buffer A into each tube.
3. Add 10× reaction buffer (supplied with the enzyme) to buffer B to a final concentration of 2×. Add 150 μl of buffer B/2× reaction buffer to each tube, mix, and incubate at 37° for 0, 15, and 30 min. Add a control plasmid to the 30-min sample as described earlier for nuclei preparations (step 2 under "Digestion of Nuclei with Restriction Enzymes"). Add 55 μl of stop buffer and incubate at 37° for at least 2 hr.
4. Extract with phenol/chloroform, collect the aqueous phase, and add 4 *M* ammonium acetate to a final concentration of 0.8 *M*. Add 1 ml 100% ethanol and precipitate on dry ice for 10 min. Resuspend in 100 μl TE buffer containing 3 μl RNase A.

Mapping Chromatin Structure by Indirect End Labeling

Reagents. Restriction enzyme and 10× reaction buffer; 4 *M* ammonium acetate; 100% ethanol; 70% ethanol; molecular biology grade agarose; 50× TAE buffer [2 *M* Tris–acetate, 0.993 *M* glacial acetic acid (concentrated glacial acetic acid = 17.4 *M*), 0.05 *M* EDTA]; 6× DNA loading buffer [0.25% xylene cyanol FF (bromphenol blue is omitted because it inhibits the transfer of low molecular weight species of DNA during Southern blotting), 15% Ficoll 400].

Reagents for Southern Blotting. 0.5 *N* HCl; 0.5 *N* NaOH; 20× SSC (3 *M* NaCl, 0.3 *M* sodium citrate); 0.9 *M* Tris/1.5 *M* HCl; 1 *M* $Na_2HPO_4 \cdot 7H_2O$, pH 7.2; hybridization buffer (0.525 *M* $Na_2HPO_4 \cdot 7H_2O$, pH 7.2, 7% SDS, 1 m*M* EDTA, pH 8.0, 1% bovine serum albumin); wash buffer 1 (0.04 *M* $Na_2HPO_4 \cdot 7H_2O$, pH 7.2, 5% SDS, 1 m*M* EDTA, pH 8.0); wash buffer 2 (0.04 *M* $Na_2HPO_4 \cdot 7H_2O$, 1% SDS, 1 m*M* EDTA, pH 8.0); wash buffer 3 (0.2× SSC/0.1% SDS); 20× SSC/50 m*M* Tris, pH 8.0; nylon membrane; Whatman 3MM filter paper; paper towels; pH indicator paper; random primed-labeled ^{32}P probe; hybridization bags or bottles.

Methods

1. For low-resolution mapping of chromatin structure, purified MNase-, DNase I-, or restriction enzyme-digested DNA is digested with an appropriate restriction enzyme and DNA visualized by Southern blot analysis. Thirty microliters (for nuclei preparations) or 50 μl (for NP-40 lysed spheroplasts) of chromatin and naked DNA samples are digested in a 1.5-ml tube with a restriction enzyme in a total volume of 200 μl containing the manufacturer's recommended reaction buffer. The restriction enzyme (200–250 U/ml) is added and the reaction is incubated at 37° for 2–4 hr.
2. After DNA is digested, add 4 *M* ammonium acetate to a final concentration of 0.8 *M*.
3. Add 2.5 volumes of 100% ethanol and precipitate on dry ice for 10 min.
4. Collect DNA by centrifuging at 13,000 rpm and wash pellet with 1 ml of 70% ethanol.
5. Dry pellet, resuspend in 12–20 μl TE buffer DNA and add 6× loading dye to a final concentration of 1×.
6. Load DNA onto an agarose/1× TAE gel (use a 1.2% gel to resolve DNA between 300 bp and 1 kbp or a 1% gel to resolve DNA between 1 and 2 kbp) and electrophorese in 1× TAE buffer (20 × 25-cm gel for samples initially digested with MNase or DNase I; 11 × 14-cm gel for samples initially digested using restriction enzymes).

Southern Blot Analysis[29,30]

1. After the gel is run, measure size and incubate twice for 10 min in 0.5 *N* HCl in a glass or plastic tray.
2. Remove HCl and soak twice for 10 min in 0.5 *N* NaOH.
3. Remove NaOH. Neutralize by soaking in 20× SSC containing 1–2 ml 0.9 *M* Tris/1.5 *M* HCl. Soak for 10–20 min until buffer reaches pH 6–8 (check with pH indicator paper).
4. While performing incubations, cut a piece of nylon membrane and three pieces of Whatman (Clifton, NJ) 3MM paper the same size as the gel and three pieces of 3 MM filter paper slightly larger than the gel. Soak nylon membrane briefly in 6× SSC.

[29] I. Sambrook, E. F. Fritsch, and T. Maniatis, "Molecular Cloning: A Laboratory Manual." Cold Spring Harbor Laboratory, Cold Spring Harbor, NY, 1989.

[30] G. Church and W. Gilbert, *Proc. Natl. Acad. Sci. U.S.A.* **81,** 1991 (1984).

5. Lay neutralized gel bottom side up on plastic tray (use tray that was used to pour gel). Lay membrane over gel and roll out gently with a plastic pipette to remove air bubbles.
6. Soak 3MM filter paper larger than gel in 6× SSC and place on top of membrane. Roll out bubbles with pipette. Turn gel/membrane/3MM paper over, being careful not to slide gel and membrane, and place onto a large stack of paper towels. Soak 3MM filter paper that is the same size as the gel in 20× SSC, place on top of gel, and roll out air bubbles with a pipette.
7. Cover with plastic wrap and paper towel and place the gel tray and a light weight on top.
8. After 15–20 min, remove weight, gel tray, paper towel, and plastic wrap. Pick up 3MM filter paper/membrane/gel sandwich and remove wet paper towels from underneath. Replace with dry towels and place sandwiched gel on top. Moisten top of 3MM filter paper (that was soaked in 20× SSC) with 20× SSC/50 m*M* Tris, pH 8.0. Replace plastic wrap, paper towels, gel tray, and weight. Allow to transfer for at least 8 hr.
9. Disassemble blot, making sure to note the orientation of the membrane and rinse membrane in 6× SSC. Cross-link DNA side up using a UV cross-linker.
10. Place membrane into hybridization bag or bottle. Add a suitable volume of hybridization buffer (e.g., 10 ml in a 1-quart bag) and prehybridize for 10–30 min at 65°.
11. Heat random primed-labeled probe to 95° for 3 min. Place on ice. Add chilled probe to 10 ml fresh hybridization buffer. Remove prehybridization buffer and add fresh buffer containing probe to membrane. Hybridize for 12–16 hr at 65°. Prewarm wash buffers to 65°.
12. Remove hybridization buffer. Wash membrane as follows: hot wash buffer 1, 2× 5 min on bench; hot wash buffer 2, 2× 5 min on bench; hot wash buffer 2, 15 min at 65°; and hot wash buffer 3, 30 min at 65°.
13. Remove excess wash buffer by placing membrane on 3MM filter paper, wrap in plastic wrap, and expose to X-ray film with an intensifying screen.

Mapping Chromatin Structure by Primer Extension Analysis[21,31]

Reagents. TE Select D G-50 spin columns (5 prime–3 prime, Inc., Boulder, CO); 0.3 pmol ^{32}P-end-labeled primer (10^6 cpm; primers should have

[31] J. D. Axelrod and J. Majors, *Nucleic Acids Res.* **17,** 171 (1989).

a T_m of 60–65°); 5× *Taq* buffer (50 m*M* Tris, pH 8.3, 250 m*M* KCl, 15 m*M* $MgCl_2$, 0.25% NP-40, 0.25% Tween 20); 2.5 m*M* dNTP mix (2.5 m*M* each dATP, dCTP, dGTP, dTTP); 5 m*M* dNTP mix (5 m*M* each dATP, dCTP, dGTP, dTTP); 0.2 U/μl *Taq* DNA polymerase (diluted in 1× *Taq* buffer); 25× ddG mix (1.25 m*M* ddGTP, 0.25 m*M* dGTP, 2.5 m*M* each of dATP, dTTP, dCTP); 25× ddA mix (2.5 m*M* ddATP, 62.5 μ*M* dATP, 2.5 m*M* each dGTP, dTTP, dCTP); 25× ddT mix (2.5 m*M* ddTTP, 0.25 m*M* dTTP, 2.5 m*M* each dGTP, dATP, dCTP); 25× ddC mix (2.5 m*M* ddCTP, 0.25 m*M* dCTP, 2.5 m*M* each of dGTP, dATP, dTTP); mineral oil; 10 m*M* EDTA, pH 8.0; chloroform; 4 *M* ammonium acetate; 100% ethanol; 70% ethanol; sequencing gel-loading buffer [six parts formamide, one part gel-loading dye (0.25% bromphenol blue, 0.25% xylene cyanol FF, 15% Ficoll 400)].

Method: One Cycle Primer Extension for Mapping Multicopy Plasmids

1. Clean digested samples by passing through a Sephadex G-50 spin column.
2. Combine 10 μl of MNase- or DNase I-treated DNA (containing 10–50 ng of plasmid) with 0.3 pmol ^{32}P-end-labeled primer (about 10^6 cpm), 5 μl of 5× *Taq* buffer, and water to a total volume of 24 μl.
3. Heat at 95° for 5 min (denaturation) and cool at 48° for 20 min (annealing).
4. Add 1.2 μl of 5 m*M* dNTPs and 5 μl 1× *Taq* buffer containing 1 unit of *Taq* polymerase (0.2 U/μl) and incubate at 70° for 5 min (extension). Terminate reaction by chilling on ice.
5. Bring reaction up to 100 μl with 10 m*M* EDTA, pH 8.0, add 4 *M* ammonium acetate to 0.8 *M*, and add 2.5 volumes of 100% ethanol. Precipitate on dry ice for 10 min, wash with 70% ethanol, dry pellet, and resuspend in 4 μl sequencing gel-loading dye. Heat at 95° for 5 min, cool on ice, and load onto a 6% polyacrylamide sequencing gel (see later). ^{32}P-end-labeled *Hae*III-digested ∅X174 DNA can be used as a size standard.

Use the same amount of undigested sample to perform a sequencing reaction as follows.

1. Into four different tubes, combine DNA with the labeled primer, 5× *Taq* buffer, and water as described earlier.
2. Denature, anneal, and instead of adding 5 m*M* dNTPs, add 1.2 μl of 25× ddG, ddA, ddT, or ddC, mix, and perform extension. Terminate reaction, precipitate, resuspend dried pellet in 4 μl sequencing gel-loading dye, heat at 95° for 5 min, and cool on ice.

Method: Multiple Cycle Primer Extension for Mapping Low-Copy Sequences

1. Clean digested samples by passing through a Sephadex G-50 spin column.
2. Combine 10 μl of MNase- or DNase I-treated DNA (containing 5–10 μg of genomic DNA) with 0.3 pmol ^{32}P-end-labeled primer (about 10^6 cpm), 4 μl of 5× *Taq* buffer, 4 μl 2.5 mM dNTPs, and water to a total volume of 20 μl in a 0.5-ml tube and place on ice. Add 1 unit of polymerase in 5 μl 1× *Taq* buffer and layer 25 μl of mineral oil on top.
3. Subject samples to the following cycles using a DNA thermal cycler: 94° for 1 min, 55° for 2 min, and 72° for 2 min, repeated 15×, followed by a soak at 4°.
4. Pipette 75 μl of 10 m*M* EDTA, pH 8.0, through the oil and into the reaction mix. Transfer reaction mix to a new tube containing 100 μl chloroform. Extract and centrifuge at 13,000 rpm.
5. Transfer aqueous phase to a fresh tube containing 20 μl 4 *M* ammonium acetate (final concentration 0.8 *M*), add 2.5 volumes of 100% ethanol, and precipitate on dry ice for 10 min.
6. Centrifuge at 13,000 rpm for 10 min, wash pellet with 70% ethanol, and dry.
7. Resuspend pellet in 4 μl sequencing gel-loading dye, heat at 95° for 5 min, cool on ice, and run on a 6% polyacrylamide sequencing gel (see later).

Use the same amount of undigested sample to perform a sequencing reaction as follows.

1. Into four different 0.5-ml tubes, combine DNA with the labeled primer, 4 μl 5× *Taq* buffer, and 1.2 μl of 25×, ddG, ddC, ddT, or ddA mix, and water to a total volume of 20 μl.
2. Add 5 μl containing 1 unit of *Taq* polymerase in 1× *Taq* buffer and mineral oil, and subject to cycles as described earlier.
3. After the reaction, proceed to step 5 in the previous section.

Denaturing Polyacrylamide Gel Electrophoresis for Primer Extension

Reagents. Urea; 40% acrylamide : bisacrylamide (29 : 1), 10× TBE (0.9 *M* Tris base, 0.02 *M* EDTA, pH 8.0, 0.89 *M* boric acid); 10% ammonium persulfate; TEMED.

Methods

1. For a 6% gel, mix in a 250-ml beaker 33.8 g of urea (final concentration of 7.5 M), 11.25 ml 40% (29 : 1) acrylamide : bisacrylamide, 7.5 ml

10× TBE and 41.25 ml water (place on hot plate at very low setting to dissolve urea).
2. When urea is in solution, add 450 μl 10% ammonium persulfate and 30 μl TEMED.
3. Pour gel and let polymerize for at least 1 hr.
4. Rinse wells with 1× TBE and prerun gel in 1× TBE for 15 min at 60 W.
5. While prerunning gel, heat samples at 95° for 5 min and cool on ice.
6. Rinse wells and load samples. Run at 60 W until second dye front is at the bottom of the gel (approximately 2 hr).
7. Disassemble the apparatus and place an old piece of X-ray film on the gel and lift gel off the glass plate. Cover the gel with plastic wrap and expose to X-ray film with an intensifying screen. Alternatively, gels can be lifted using filter paper, dried, and exposed.

Acknowledgments

We thank Hugh Patterton, Kerstin Weiss, and Robert Simpson for providing Figs. 4 and 5. Work in the authors laboratory is supported by Grant GMS51993 from the NIH.

[22] Mapping DNA Interaction Sites of Chromosomal Proteins Using Immunoprecipitation and Polymerase Chain Reaction

By ANDREAS HECHT and MICHAEL GRUNSTEIN

Introduction

Chromosomal proteins that affect the maintenance, propagation, and expression of the genome often interact with DNA only indirectly through other DNA-binding factors. This is especially true for certain factors and enzymes that interact with nucleosomes to produce chromosomal domains that are functionally and structurally distinct.[1] A complete understanding of the mechanisms controlling DNA replication, repair, or transcription ultimately necessitates the study of factors involved in these processes in their natural chromosomal environment. This article describes a method that allows one to determine where in the genome such factors interact

[1] M. Grunstein, *Nature* **389,** 349 (1997).

0076-6879/99 $30.00

(A) Formaldehyde cross-linking of cells

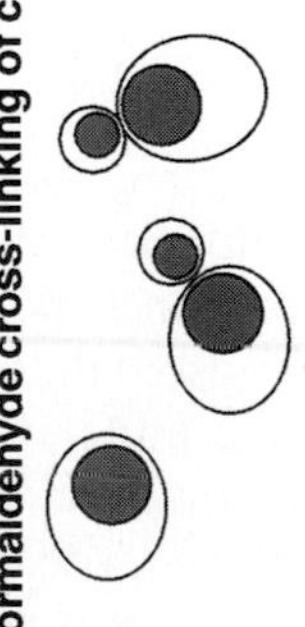

(B) Cell lysis and fragmentation of chromatin

(C) Immunoprecipitation

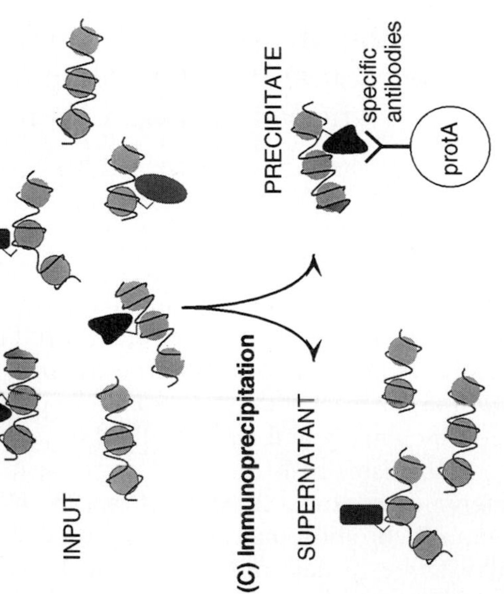

(D) Reversal of cross-link, extraction of DNA

(E) PCR analyses with gene-specific primer pairs

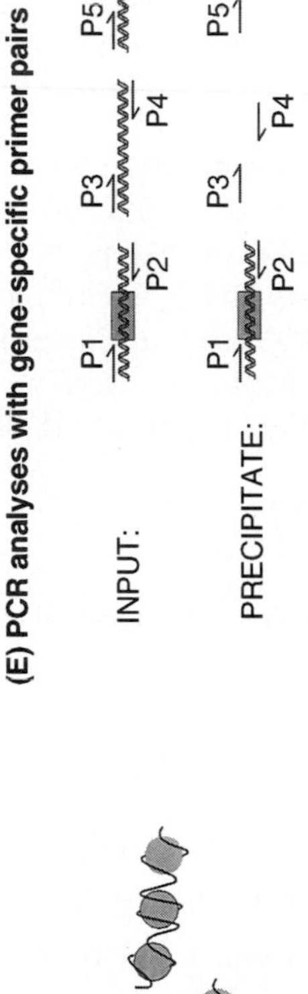

(F) Polyacrylamide gel analyses of PCR products:

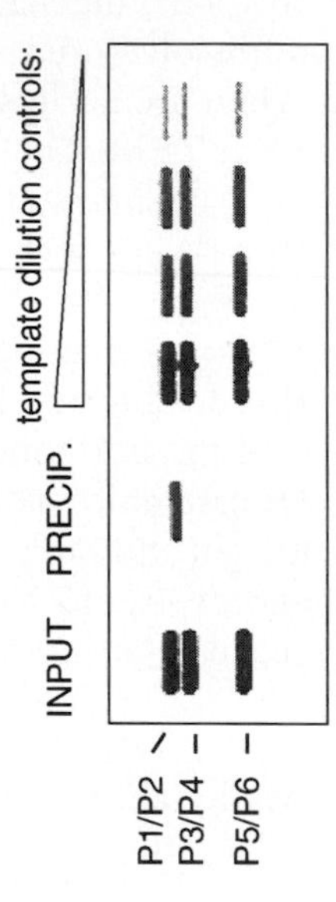

directly or indirectly with DNA. Given its wholly sequenced genome and the ease by which it is possible to introduce targeted mutations in specific genes, we have used the yeast *Saccharomyces cerevisiae* for this purpose. This combination has generated a powerful tool for the spatial and temporal mapping of chromatin-bound factors *in vivo* and for the investigation of the means by which chromosomal regions are remodeled.[2–6]

A schematic outline of the method is presented in Fig. 1. The initial step is the cross-linking of live cells. Formaldehyde (FA) is a reagent particularly useful for this purpose. It has long been used in studies of histone organization in the nucleosome[7] or protein–DNA interactions, e.g., in simian virus 40 (SV40) minichromosomes.[8] The main advantages of formaldehyde are its solubility in water and that it is active over a wide range of buffer conditions and temperatures. Most importantly, formaldehyde penetrates biological membranes readily. Thus, cross-linking can be done with intact cells, which reduces the risk of redistribution or reassociation of chromosomal proteins during the preparation of cellular or nuclear extracts. The chemical targets for formaldehyde are primary amino groups,

[2] A. Hecht, S. Strahl-Bolsinger, and M. Grunstein, *Nature* **383,** 92 (1996).
[3] S. Strahl-Bolsinger, A. Hecht, and M. Grunstein, *Genes Dev.* **11,** 83 (1997).
[4] T. Tanaka, D. Knapp, and K. Nasmyth, *Cell* **90,** 649 (1997).
[5] O. M. Aparicio, D. M. Weinstein, and S. P. Bell, *Cell* **91,** 59 (1997).
[6] K. Ekwall, T. Olson, B. M. Turner, G. Cranston, and R. C. Allshire, *Cell* **91,** 1021 (1997)
[7] V. Jackson, *Cell* **15,** 945 (1978).
[8] M. J. Solomon and A. Varshavsky, *Proc. Natl. Acad. Sci. U.S.A.* **82,** 6470 (1985).

FIG. 1. General outline of the procedure for mapping of the chromosomal distribution of nonhistone chromatin-associated factors. Yeast cells are cross-linked with formaldehyde *in vivo* (A) and subsequently lysed mechanically. Chromatin released from the cells is fragmented by sonication (B) to an average size suitable for the desired accuracy and resolution of the mapping experiment. Protein–DNA complexes containing a particular chromosomal nonhistone protein are recovered from the cell lysate by the addition of specific antibodies and adsorption to protein A–Sepharose beads (C). DNA is isolated from samples of the input and precipitate materials after reversal of the cross-links by heat treatment and proteinase digestion (D). The purified DNA is then used as a template in multiplex PCR reactions with primer pairs derived from different chromosomal regions. To allow for a quantitative interpretation of the results it is important to control that the PCR was still in the exponential phase. Therefore, control reactions with increasing amounts of template are analyzed in parallel (E). (F) PCR products are resolved by polyacrylamide gel electrophoresis and visualized with ethidium bromide. Bound as well as unbound sequences are represented equally in the genome and will be amplified accordingly from the input sample. Only sequences occupied specifically by the factor of interest (symbolized by the shaded area in the P1/P2 region) will give rise to PCR products from the precipitate.

such as the ε amino group of lysine residues, but also side chains present in adenine, guanine, and cytosine. This leads to both protein–protein and protein–DNA cross-links bridging distances of about 2 Å. Both types of cross-links can be reversed selectively. Extended incubation at 65° breaks protein–DNA bonds whereas the reversal of protein–protein cross-links requires temperatures close to boiling.[7,8]

After cross-linking, the cells are lysed mechanically by vortexing in the presence of glass beads. Lysates prepared in this way contain DNA fragments that are roughly 20 kb or more in length, which is too long to determine the precise chromosomal location of most chromatin-associated proteins (see later). Sonication is a rapid and simple way to shear the chromatin fragments and to generate rather uniformly sized pieces of DNA with well-defined lengths as needed for the desired mapping accuracy of the experiment. In general, shorter DNA fragments provide higher mapping resolution.

After preparation of the cellular extract and chromatin fragmentation, specific chromosomal factors, together with cross-linked DNA, are immunoprecipitated. Protein–DNA cross-links in the immunoprecipitated material are then reversed and the DNA fragments are purified (Figs. 1C and 1D). If the protein under investigation is associated with a specific genomic region *in vivo,* DNA fragments from this region should be enriched in the immunoprecipitate (IP) compared to other portions of the genome (see Figs. 1E and 1F). The presence of relevant genomic regions in the IP is determined by polymerase chain reactions (PCR) in which the immunoprecipitated material is used as a template for the simultaneous amplification with gene-specific primers from the region in question and reference regions. Finally the PCR products are analyzed by polyacrylamide gel electrophoresis.

Examples for applications and typical results obtained with this method are shown in Fig. 2. We have used this method to study the association of silent information regulator 3 (SIR3) with heterochromatin-like regions in the genome of *S. cerevisiae.* SIR3 belongs to a group of regulatory proteins required for the transcriptional repression of subtelomeric regions and the silent mating type loci *HMLα* and *HMR*a.[9,10] SIR3 is not believed to bind directly to DNA, rather it interacts with chromatin by binding to the histone

[9] P. Laurenson and J. Rine, *Microbiol. Rev.* **56,** 543 (1992).

[10] J. S. Thompson, A. Hecht, and M. Grunstein, *Cold Spring Harb. Symp. Quant. Biol.* **58,** 247 (1993).

H4 N terminus.[2,11] Yeast strains expressing epitope-tagged SIR3 (SIR3HA) were cross-linked and SIR3HA was immunoprecipitated. DNA coprecipitating with SIR3HA was then used as a template for PCR reactions with primer pairs from *HMLα* and *ADH4;* the latter is integrated approximately 1 kb from the telomere. *ACT1, PHO5*, and *GAL1* were used as controls, as they are not believed to be affected by SIR3. *HMLα* and *ADH4* can be detected readily in the immunoprecipitate, whereas DNA from *ACT1, PHO5*, or *GAL1* genes are much less abundant (Fig. 2A). To demonstrate the stringency and the specificity of the method, we performed control experiments with unfixed cells, cells expressing WT SIR3, which is not recognized by the antibody against the epitope tag (Fig. 2A), and with strains carrying mutations in histone H4 (Fig. 2B). Without cross-linking or in the absence of the epitope tag, *HMLα* and *ADH4* are no longer enriched over the reference regions in the immunoprecipitate. In addition, the immunoprecipitate from strains with the silencing defective H4 mutant G28P[12] shows much reduced levels of *HMLα* whereas a second H4 mutant K20G, which has little effect on silencing,[12] also has no obvious effect on SIR3HA/*HMLα* coprecipitation. In additional experiments we used this method to further analyze the involvement of histone H4 in the association of SIR3 with chromatin[2] to map interactions of SIR2, SIR3, and SIR4 with subtelomeric heterochromatin[3] and to study gene dosage effects on the assembly and spreading of heterochromatic multiprotein complexes.[2,3] More recently, Tanaka and co-workers[4] and Aparicio and colleagues[5] applied the method to analyze the occupation of single chromosomal sites (the ARS1 and ARS305 origins of replication) by ORC, CDC6, CDC45, and MCM proteins in a cell cycle-dependent fashion, which provides additional examples of the power of the method. An adaptation of the protocol for fission yeast in which antibodies to the acetylation sites of H4 have been used has also been published.[6]

Others have previously used similar formaldehyde cross-linking and immunoprecipitation procedures. Braunstein *et al.*[13] described the use of antibodies to tetraacetylated histone H4 to demonstrate that silent chromosomal regions in yeast (HM loci, telomeres) are hypoacetylated. Orlando and Paro[14] utilized a similar protocol to analyze the chromosomal distribu-

[11] A. Hecht, T. Laroche, S. Strahl-Bolsinger, S. M. Gasser, and M. Grunstein, *Cell* **80,** 583 (1995).

[12] L. M. Johnson, G. Fisher-Adams, and M. Grunstein, *EMBO J.* **11,** 2201 (1992).

[13] M. Braunstein, A. B. Rose, S. G. Holmes, C. D. Allis, and J. R. Broach, *Genes Dev.* **7,** 592 (1993).

[14] V. Orlando and R. Paro, *Cell* **75,** 1187 (1993).

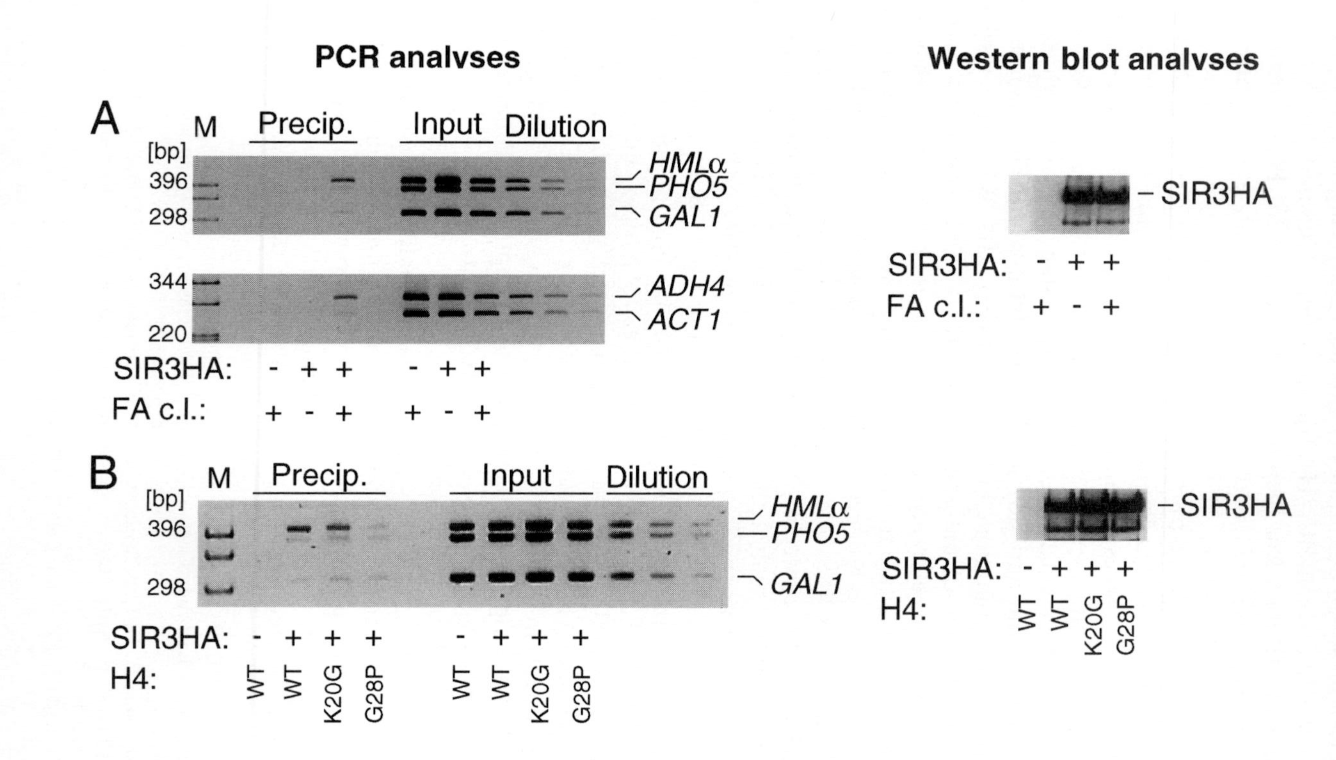
PCR analvses
Western blot analvses
A
M
Precip.
Input
Dilution
[bp]
396
298
344
220
HMLα
PHO5
GAL1
ADH4
ACT1
SIR3HA: - + + - + +
FA c.l.: + - + + - +
SIR3HA
SIR3HA: - + +
FA c.l.: + - +
B
M
Precip.
Input
Dilution
[bp]
396
298
HMLα
PHO5
GAL1
SIR3HA: - + + + - + + +
H4: WT WT K20G G28P WT WT K20G G28P
SIR3HA
SIR3HA: - + + +
H4: WT WT K20G G28P

tion of the Polycomb protein in *Drososphila melanogaster.* One of the differences between the method of Orlando and Paro[14] and our protocol is that we eliminated the CsCl gradient step used to separate non-cross-linked material from the desired protein–DNA adducts. This was possible because of the smaller size of the yeast genome, which provides a better ratio of specific (bound) versus nonspecific (unbound) DNA. Differences between the method used by Braunstein *et al.*[13] and our procedure are the buffer composition during cell lysis and washing of the immunoprecipitates. We also utilize a simplified and more rapid technique for cell breakage. However, the main difference to the other two methods is the use of PCR rather than hybridization to detect immunoprecipitated sequences. The entire yeast genome has been sequenced. This information can be exploited to design PCR primers for virtually any genomic region and thus all of the yeast genome is accessible for investigation. Consequently the hybridization step for analyses of the immunoprecipitated DNA can be replaced by the more rapid and versatile PCR detection method. The advantage of the PCR technique is most evident when multiple primer sets are used simultaneously (multiplex PCR) to determine binding at several locations in comparative studies. By subcloning and sequencing of the immunoprecipitated DNA, it should also be possible to modify the procedure for the identification of unknown interaction sites.

[14] V. Orlando and R. Paro, *Cell* **75,** 1187 (1993).

FIG. 2. Examples of experiments using the method described to analyze the association of SIR3 with silent chromosomal regions. (A) Yeast cells expressing WT or epitope-tagged SIR3 were treated with or without formaldehyde [FA c.l. (cross-linking)]. Cell lysates were prepared and subjected to immunoprecipitation with an antibody directed against the HA epitope in the tagged SIR3 protein. (Left) DNA from input and precipitate was isolated and analyzed in multiplex PCR with primer pairs from *HMLα, PHO5, GAL1, ADH4,* and *ACT1* loci. Only the known targets for SIR3, *HMLα* and telomeric *ADH4* sequences, are enriched in the precipitate. Without cross-linking or without epitope-tagged SIR3, *HMLα* and *ADH4* are no longer detectable in the precipitate. This demonstrates the stringency and specificity of the procedure. (Right) Fractions of the immunoprecipitates were processed for Western blotting with anti-HA antibodies to show that equal amounts of SIR3HA were precipitated from cross-linked and non-cross-linked cells. (B) Cell lysates were prepared from formaldehyde-treated yeast strains expressing either WT or mutant H4 as indicated. Lysates were used for the immunoprecipitation of SIR3HA and subsequent PCR analyses (left) and Western blotting (right). The glycine-to-proline exchange at position 28 of H4 (G28P) abrogates silencing of *HMLα* while the lysine-to-glycine mutation at position 20 (K20G) has no effect on *HMLα.*[12] In accordance with the genetic observations, PCR analyses of the immunoprecipitates reveal an interaction of SIR3HA with *HMLα* in WT or K20G strains, but not in G28P strains.

Methods

Preparation of Cell Lysates from Formaldehyde Cross-Linked Yeast Cultures and Chromatin Fragmentation

Materials and Solutions

Liquid SD media[15]
37% formaldehyde (Merck)
2.5 *M* glycine (Sigma) in double-distilled H_2O
Phosphate-buffered saline (PBS): 140 m*M* NaCl, 2.5 m*M* KCl, 8.1 m*M* Na_2HPO_4, 1.5 m*M* KH_2PO_4, pH 7.5
Lysis buffer: 50 m*M* *N*-(2-hydroxyethyl piperazine-*N'*-2-ethanesulfonic acid) (HEPES)/KOH, pH 7.5, 140 m*M* NaCl, 1 m*M* ethylenediaminetetraacetic acid (EDTA), 1% (v/v) Triton X-100, 0.1% (w/v) sodium deoxycholate
100× stock solutions of protease inhibitors (Sigma): 100 m*M* phenylmethylsulfonyl fluoride (PMSF) freshly prepared in 2-propanol, 100 m*M* benzamidine in H_2O stored at −20°, 5 mg/ml *N*-tosyl-L-phenylalanine chloromethyl ketone (TPCK) in ethanol stored at −20°, 25 m*M* N^{α}-*p*-tosyl-L-lysine chloromethyl ketone (TLCK) in 50 m*M* sodium acetate, pH 5.0, stored at −20°
1000× stock solutions of protease inhibitors (Sigma): 10 mg/ml aprotinin in 10 m*M* HEPES/KOH, pH 8.0, 1 mg/ml Leupeptin in H_2O; 1 mg/ml pepstatin A in ethanol, 2 mg/ml antipain in H_2O; all solutions are stored at −20°
Glass beads (0.45–0.52 mm diameter) (Thomas Scientific). Before use, the glass beads are acid washed by soaking in concentrated nitric acid for 1 hr, followed by extensive rinsing with distilled water and a final wash with 70% (v/v) ethanol and baking at 80° until dry[16]

Cell Growth. Aliquots from overnight cultures of the strains to be analyzed are diluted into 50 ml fresh liquid media so that the cultures will have reached an OD_{600} of approximately 1.5 the next morning. When handling several strains in parallel, it is desirable to work with synchronously growing cultures and to harvest all cultures in the same growth phase, preferably log phase. Wild-type strains and their mutant derivatives often differ in their growth properties. In addition, the type

[15] M. D. Rose, F. Winston, and P. Hieter, *in* "Methods in Yeast Genetics: A Laboratory Manual." Cold Spring Harbor Laboratory Press, Cold Spring Harbor, NY, 1990.
[16] C. S. Hoffman and F. Winston, *Gene* **57,** 267 (1987).

of growth media—complex media such a YEPD or synthetic selective media—influences the growth rate of strains and thus the time needed to obtain a sufficient number of cells. Therefore it is advisable to determine the doubling time of the working strains in pilot experiments. This allows one to calculate the size of the inoculum from the preculture required to achieve the desired cell density after a given period of time. As the entire protocol from cross-linking through cell lysis, immunoprecipitation, and cleanup of the precipitated DNA adds up to a quite lengthy procedure, we usually start our cultures in the late afternoon and adjust them in such a way that the cells are ready for harvesting the next morning.

Cross-Linking. When the cultures have reached the desired density, formaldehyde is added to a final concentration of 1% (avoid skin contact or breathing formaldehyde fumes). Mix rapidly and incubate at room temperature with constant mixing. Our standard time for the cross-linking reaction is 15 min. However, this may vary with the particular protein–DNA or protein–protein interaction under consideration. Another important aspect is to process the same number of cells from different cultures, especially when comparing mapping experiments from different strains. In case not all strains are ready for the cross-linking step at the same time, it is possible to keep the faster growing cultures on ice, while the slower strains continue to grow to the desired density.

Cell Harvest and Lysis. The cross-linking reaction is terminated by the addition of 2.5 *M* glycine to a final concentration of 125 m*M* and the incubation is continued for 5 min. After this the cells are transferred onto ice and in the following steps they are kept at 4° whenever possible. Cells are harvested by centrifugation for 5 min at 1500*g* in a prechilled centrifuge. The supernatant is removed, the cells are washed twice by resuspending in ice-cold PBS, and pelleted as described earlier. The final washing solution is poured off and residual liquid is removed by wiping the walls of each tube with a tissue while the centrifuge tubes are kept inverted. Each cell pellet is then resuspended in 400 μl ice-cold lysis buffer with protease inhibitors by pipeting up and down several times. Cell suspensions are transferred to 1.5-ml Eppendorf tubes containing an equal volume of glass beads (about 500 μl). Cell lysis occurs by vortexing the cell suspension/glass bead mixtures on an Eppendorf shaker (Model 5432) for 45 min at 4°.

Collection of Whole Cell Extract. The following steps serve to separate the cellular extracts from the glass beads and cell debris. First the tubes are spun briefly in a table-top centrifuge to remove any liquid from the top of the tubes. Then, with the lid of the Eppendorf tubes open, the bottoms of the tubes are punctured with a red-hot 26-gauge (0.45 mm diameter) needle. Each tube is then placed into a second (collection) Eppendorf tube and the lids of the upper tubes are closed. Some liquid will leak out. By centrifuging the tube combinations for 5 sec at full speed in a table-top centrifuge, cellular debris and the crude cell lysate will be captured in the bottom tube, while the glass beads remain behind. Before centrifuging it should be checked that the tube assembly does not obstruct rotation. If necessary the lid of the centrifuge may have to be left open or a different centrifuge and/or rotor used.

Chromatin Shearing. Next, the size of the chromatin fragments present in the cell lysate is reduced by sonication. This is a critical step because the extent of chromatin fragmentation will influence the resolution of the mapping experiment. Smaller DNA fragments allow one to better distinguish between bound and adjacent unbound DNA regions of the chromosome. To establish the conditions for sonication with a particular sonicator, some cellular lysate is prepared and subjected to various numbers of sonication cycles. Small aliquots of cellular lysate are removed after each cycle and finally DNA is isolated from all collected aliquots as described later. The DNAs are analyzed by agarose gel electrophoresis and ethidium bromide staining to determine how many cycles of sonication are needed to shear the chromatin to a certain size range. An example of such an experiment is shown in Fig. 3. DNA fragments, the majority of which are between 0.5 and 1.0 kb in size, are generated with two 10-sec pulses using the Model W-375 sonicator (Heat Systems Ultrasonic Inc.) equipped with the microtip and operated at setting 3. Between each pulse the samples are allowed to cool for 20 sec in an ice/water bath. Even smaller DNA fragments (100–500 bp; Fig. 3, lane 10) can be obtained, but this requires much more extensive sonication. When choosing a certain size range for the DNA, one should also keep in mind that the amplification efficiency during the PCR may decrease as the average fragment size approaches the distance between the PCR primers used. DNA slot or dot blots and hybridization techniques may then be the methods of choice as detection systems.[13,14]

After sonication the lysates are centrifuged at 10,000*g* or more for 5 min at 4°. As much as possible of the supernatant is transferred to a fresh tube and spun again at 10,000*g* for 15 min at 4°. The final supernatant, which is the crude cellular extract (WCE), is pipetted into a fresh tube. A 20-μl aliquot of the WCE is set aside as input material and stored on ice until further processing.

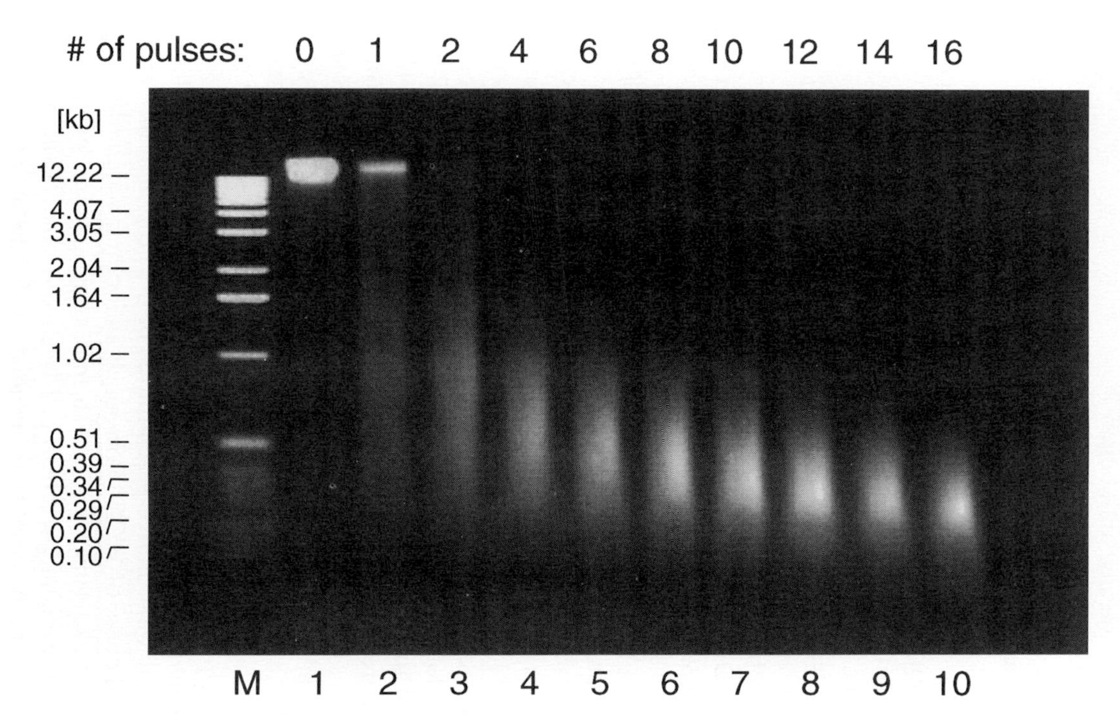

FIG. 3. Adjusting the size distribution of DNA fragments by sonication. A whole cell lysate was prepared from a 100-ml culture of exponentially growing yeast cells as described in the text. Samples of the lysate were taken before and after the number of ultrasound pulses indicated. Each pulse lasted for 10 sec (setting 3 of the Model W-375, Heat Systems Ultrasonic Inc.) followed by a 20-sec interval with cooling on ice. DNA from each sample was purified and analyzed by electrophoresis on a 1.2% agarose gel and ethidium bromide staining. Initial breakdown of the DNA to sizes below 2 kb occurs rapidly, whereas extensive sonication is required to yield DNA fragments primarily in the size range between 0.1 and 0.5 kb.

Immunoprecipitation and DNA Isolation

Materials and Solutions

Affinity-purified, specific antibodies[17]
50% (v/v) suspension of protein A-Sepharose beads (Pharmacia, Piscataway, NJ), swollen according to the manufacturer and equilibrated in lysis buffer
Wash buffer 1: lysis buffer but with 500 m*M* NaCl
Wash buffer 2: 10 m*M* Tris–HCl, pH 8.0, 0.25 *M* LiCl, 0.5% Nonidet P-40 (NP-40), 0.5% sodium deoxycholate, 1 m*M* EDTA
10 m*M* Tris–HCl, pH 8.0, 1 m*M* EDTA (TE)
TE/1% sodium lauryl sulfate (SDS)
1 mg/ml glycogen (Boehringer, Indianapolis, IN)
20 mg/ml proteinase K (Boehringer)
Phenol : chloroform : isoamyl alcohol, 25 : 24 : 1 (PCI solution)
Reagents for DNA precipitation: 5 *M* LiCl, 50 m*M* Tris–HCl, pH 8.0; absolute ethanol; 70% ethanol
10 mg/ml RNase A (DNase-free; Sigma St. Louis, MO)

The outcome of the experiment depends critically on the quality of the antibodies used. They should possess a high affinity for the antigen and should cross-react minimally with other proteins. Thus, before starting an immunization it is advisable to test preimmune sera from a number of animals and to select the ones that show the lowest nonspecific background. To further minimize the risk of potential problems due to the nonspecific reactivity of polyclonal antisera, one should affinity purify specific antibodies by absorption to the antigen used for immunization,[17] but even then one should be aware that polyclonal and even monoclonal antibodies often cross-react. In cases in which cross-reactivity cannot be eliminated, it is still possible to do immunoprecipitations with the available antibody or even serum as long as the antibody is mainly to one species of protein. However, it is important in these cases that lower quantities of antibody are used to minimize the precipitation of cross-reacting species.

Immunoprecipitation. After preparing the WCE the amount of antibodies necessary for the quantitative precipitation of the protein of interest is added to 400-μl aliquots of WCE. How much antibody is required can be determined in pilot experiments in which increasing amounts of affinity-purified antibodies are combined with cell lysates. A control sample that does not receive antibody should be included. A good starting point for the antibody concentration is 0.1 μg/ml as the lowest amount and approxi-

[17] E. Harlow and D. Lane, *in* "Antibodies: A Laboratory Manual." Cold Spring Harbor Laboratory Press, Cold Spring Harbor, NY, 1988.

mately 25 μg/ml as the highest. If necessary, the titration has to be repeated with lower or higher antibody concentrations. All samples are subject to immunoprecipitation as outlined later and finally the precipitated proteins and aliquots of the supernatants are analyzed by Western blotting to monitor at which point the protein of interest is found exclusively in the precipitate. The amount of antibodies that completely removes the protein from the lysate will be used in the actual experiments.

After antibody addition, all samples are incubated on a nutator for 3 hr at 4°. Thereafter 60 μl of a protein A–Sepharose slurry is added and incubation is continued for an additional hour to allow for adsorption of the antibody/antigen complexes. Not all antibodies react efficiently to protein A, however. In cases where the antibodies cannot be recovered with protein A–Sepharose beads, one can try using protein G beads instead.[17] Antibody recovery can also be simplified by covalently coupling the antibodies to the protein A–Sepharose beads before use,[17] which results in less chance for experimental variations. In this manner, the antibody heavy and light chains cannot elute from the beads and therefore will not copurify with the immunoprecipitate. This may be important not only when the DNA but also the protein composition of the IP is to be analyzed because detection of the antibodies used for the immunoprecipitation can obscure weaker signals from the precipitated factors in Western blots.

The protein A–Sepharose beads with the immunoprecipitate are pelleted for 5 sec at 10,000*g*. The supernatant is removed carefully and transferred to a fresh tube. It may be kept on ice for further processing if desired. Even though the goal of these experiments is primarily to characterize the immunoprecipitated DNA, it may be necessary to control efficiency of the precipitation by analyzing the protein content of the WCE (input), the supernatant after immunoprecipitation, and the precipitate itself in Western blots. For this purpose, aliquots from each of these samples are heated to 95° in the presence of 0.5 *M* 2-mercaptoethanol for 30 min to reverse protein–protein cross-links prior to SDS–polyacrylamide gel electrophoresis.[7,8]

The protein A–Sepharose beads with the immunoprecipitates are resuspended in 1 ml lysis buffer, incubated for 5 min on a nutator, and pelleted as described earlier. This washing step is repeated once more with lysis buffer, once with 1 ml of wash buffer 1, once with 1 ml of wash buffer 2, and once with 1 ml TE. The composition of the buffers used for cell lysis, immunoprecipitation, and washing determines the stringency of the analyses. Increasing salt concentrations or including 0.1% SDS in the lysis and wash buffers will increase the stringency and could reduce nonspecific precipitation. However, these agents might also be detrimental to the antibody. Nonetheless, varying the type and concentration of the detergent and salt in the lysis and wash buffers may be helpful in adapting the procedure for a particular application.

After the final wash, as much of the TE supernatant as possible is pipetted off. The tubes are centrifuged again and residual liquid is removed while taking care not to lose any beads (a Hamilton syringe or a glass capillary with a narrow tip are convenient to use for this step). To release the immunoprecipitated material, the beads are resuspended in 100 μl TE/1% SDS and heated to 65° for 10 min. After pelleting the beads by centrifugation for 2 min at 10,000*g*, the TE/1% SDS solution now containing the immunoprecipitated material is transferred to a fresh Eppendorf tube. This sample is the precipitate.

Reversal of the Cross-Link and DNA Purification. The following steps involve the reversal of formaldehyde-induced cross-links and the cleanup of the DNA that was coimmunoprecipitated with the protein of interest. Add 100 μl TE/1% SDS to 20 μl of the precipitate and the input. Incubate at 65° for at least 6 hr up to overnight[8] to reverse the DNA–protein cross-linking. The remainder of the precipitate can be stored at 4° for additional analyses if desired. All samples are allowed to cool after the 65° incubation and the condensate is collected at the bottom of the tubes by brief centrifugation. The proteins in the samples are removed by digestion with 100 μg proteinase K for 2 hr at 37° and subsequent PCI extraction. After the addition of 2 μg glycogen, 1/10 volume 5 *M* LiCl, 50 m*M* Tris–HCl, pH 8.0, and 3 volumes of absolute ethanol, the nucleic acids are precipitated by centrifugation for 20 min at 10,000*g* or more. The pellets are rinsed once with 70% ethanol and then air dried. DNA from the input is resuspended in 200 μl TE, whereas DNA from the precipitate is resuspended in 20 μl TE. Copurifying RNA is degraded by digestion with 10 μg RNase A per sample for 30 min at 37°. Input and precipitate samples can be used immediately or stored at −20° until further use.

PCR Analyses and Gel Electrophoresis

Materials and Solutions

- Oligonucleotides: 24-mers with approximately 50% GC content and similar melting temperatures
- 10× PCR buffer: 0.2 *M* Tris–HCl, pH 8.3, 0.5 *M* KCl, 0.015 *M* $MgCl_2$, 0.5% Tween 20, 1 mg/ml gelatin
- dNTP mixture (10×) containing 2 m*M* dATP, dGTP, dCTP, dTTP each (Pharmacia)
- Recombinant *Taq* polymerase (GIBCO/BRL, Gaithersburg, MD) 5 U/μl
- Supplies and reagents for polyacrylamide gel electrophoresis

40% (19:1) acrylamide/bisacrylamide solution (Accugel, National Diagnostics, Atlanta, GA)
10× Tris–borate–EDTA buffer (TBE)[18]
N,N,N′,N′-Tetramethylethylenediamine (TEMED) (Bio-Rad, Richmond, CA)
10% (w/v) ammonium persulfate in H_2O
10 mg/ml ethidium bromide in H_2O

Our method of choice to detect the presence of specific genomic regions in the immunoprecipitate is multiplex PCR in which several PCR products from different primer pairs are amplified simultaneously in a single test tube. For this purpose we designed most of our oligonucleotide primers as 24-mers with 50% GC content so that the same annealing temperatures could be used for all primer pairs. The length of the PCR products was in the range between 200 and 400 bp with a size difference of approximately 20 bp between individual products. Thus, amplification efficiency should be biased minimally by length differences. Individual PCR products can be distinguished easily from each other on polyacrylamide gels or high-resolution agarose gels. Nonetheless, it is recommended that all primer pairs be tested in separate reactions and that the products are analyzed side by side before setting up multiplex PCR. This allows for the unequivocal identification of individual PCR products. In addition, Southern blot hybridizations can be performed to verify the identity of the PCR products. Another alternative is the isolation of PCR products from gels and sequencing them.

PCR Analyses. PCR amplifications are carried out in 50 μl volume with 50 pmol of each primer, 0.2 m*M* dNTPs, and 2.5 units *Taq* polymerase. Usually we prepare a common PCR premix for all samples containing appropriate amounts of 10× PCR buffer, primers, dNTPs, *Taq* polymerase, and H_2O. The minimal number of samples includes a control reaction without DNA (to check for DNA contaminations by carryover) and one reaction for each input and each precipitate–DNA, as well as four reactions with dilutions from one of the input samples as template.

Place 48-μl aliquots of the PCR premix in thermal cycler tubes and add the DNA templates. The amount of template DNA used is typically 2 μl of a 1:120 dilution of the input–DNA and 2 μl of the undiluted precipitate–DNA. These amounts correspond to 1/13,500 of the input (assuming a total volume of approximately 450 μl of the initial WCE) and 1/50 of the precipitate. Input control dilutions receive 4 μl of serial 2.5-fold dilutions made from the initial 1:120 input dilution. If a thermal cycler with heated

[18] J. Sambrook, E. F. Fritsch, and T. Maniatis, *in* "Molecular Cloning: A Laboratory Manual." Cold Spring Harbor Laboratory Press, Cold Spring Harbor, NY, 1989.

cover is available, the samples are placed directly in the cycler, if not the samples are overlaid with mineral oil before starting the PCR. Our standard cycling program is 2 min initial denaturation at 95° followed by 25 cycles with 30 sec at 95° (denaturation), 30 sec at 55° (annealing), 60 sec at 72° (elongation), and a final extension step of 5 min at 72°.

Gel Electrophoresis. After completion of the PCR amplification, 20 μl from each reaction is loaded along with a size standard on 6% polyacrylamide/0.5× TBE gels (20 cm long). Electrophoresis is carried out with 150-V constant voltage until the bromphenol blue dye has reached the end of the gel. Gels are disassembled and stained in an 0.1-μg/ml aqueous ethidium bromide solution. Stained gels are photographed using a Polaroid camera or other gel documentation system.

To deduce whether a chromosomal protein is associated with a particular genomic section, one compares the relative abundance of PCR products from the region in question relative to a reference. Therefore, it is important that the PCR products are quantified while the PCR is still in the exponential phase. Under these conditions the amount of PCR product should be proportional to the amount of template DNA added to the reaction. The dilution series of the input–DNA serves to control this relationship. Specific applications may require increasing or decreasing the number of PCR cycles. Additional tests to see whether a particular DNA sequence coprecipitates specifically can be omission either of the antibody during immunoprecipitation or of the cross-linking step. Another possibility is to compare precipitation from WT and mutant strains with known phenotypes so that certain predictions as to the precipitation behavior can be made (see Fig. 2 for examples).

[23] Immunological Analysis of Yeast Chromatin

By PAMELA B. MELUH and JAMES R. BROACH

Introduction

Chromosome metabolism underlies a variety of biological problems, including DNA packaging and organization, gene expression, DNA replication, and chromosome segregation. These processes require a number of specific, but often transient, interactions between chromosome-binding proteins and specific domains on chromosomes. Some of these processes, such as transcription, are amenable to *in vitro* reconstruction, permitting analysis of the role of individual components and interactions in the context of the

0076-6879/99 $30.00

overall process. However, many of these processes cannot be reconstituted *in vitro* and, in fact, can best be assessed only in the context of their *in vivo* activity. Accordingly, analysis of these processes depends critically on an ability to access the specific and dynamic interactions of protein–chromosome interactions *in vivo*.

The general association of a protein with chromatin can be detected by the isolation of bulk chromatin followed by immunological analysis of the protein in the various fractions generated during the isolation procedure. Several protocols for the isolation of chromatin have been described.[1,2] With appropriate refinement of such a technique, one can assess not only whether a protein associates with chromosomes, but also whether that association is cell cycle regulated or dependent on particular *trans*-acting factors. In a recent application of this technique, two groups demonstrated that proteins of the origin recognition complex in yeast are chromatin associated throughout the cell cycle, whereas certain initiator proteins cycle on and off bulk chromatin.[1,2]

Another approach, which not only demonstrates that a protein associates with chromatin *in vivo* but also defines the specific locus to which the protein binds, entails *in vivo* cross-linking followed by immunoprecipitation of the cross-linked chromatin with an appropriate antibody. Hybridization or polymerace chain reaction (PCR) analysis can then identify the coprecipitated DNA (Fig. 1). Use of a cross-linking agent precludes artifacts that arise as a result of exchange, sliding, or adventitious binding of proteins to DNA during chromatin isolation. Typically, cross-linking *in vivo* has been achieved by treatment with UV light or formaldehyde.[3] The latter procedure provides several advantages. Formaldehyde introduces protein–protein cross-links in addition to protein–nucleic acid cross-links, which allows detection of indirect as well as direct protein–DNA associations. In addition, formaldehyde cross-links are reversible by mild heat treatment, which facilitates analysis of the immunoprecipitated material.[4–6]

The use of *in vivo* formaldehyde cross-linking in chromatin analysis was pioneered by Varshavsky and colleagues,[4,7] and elaborated by Paro, Orlando, and colleagues working with *Drosophila*[6,8] and by Allis, Gorovsky,

[1] C. Liang and B. Stillman, *Genes Dev.* **11,** 3375 (1997).

[2] S. Donovan, J. Harwood, L. S. Drury, and J. F. X. Diffley, *Proc. Natl. Acad. Sci. U.S.A.* **94,** 5611 (1997).

[3] D. S. Gilmour, A. E. Rougvie, and J. T. Lis, *Methods Cell Biol.* **35,** 369 (1991).

[4] M. J. Solomon and A. Varshavsky, *Proc. Natl. Acad. Sci. U.S.A.* **82,** 6470 (1985).

[5] P. C. Dedon, J. A. Soults, C. D. Allis, and M. A. Gorovsky, *Anal. Biochem.* **197,** 83 (1991).

[6] V. Orlando, H. Strutt, and R. Paro, *Methods* **11,** 205 (1997).

[7] M. J. Solomon, P. L. Larsen, and A. Varshavsky, *Cell* **53,** 937 (1988).

[8] V. Orlando and R. Paro, *Cell* **75,** 1187 (1993).

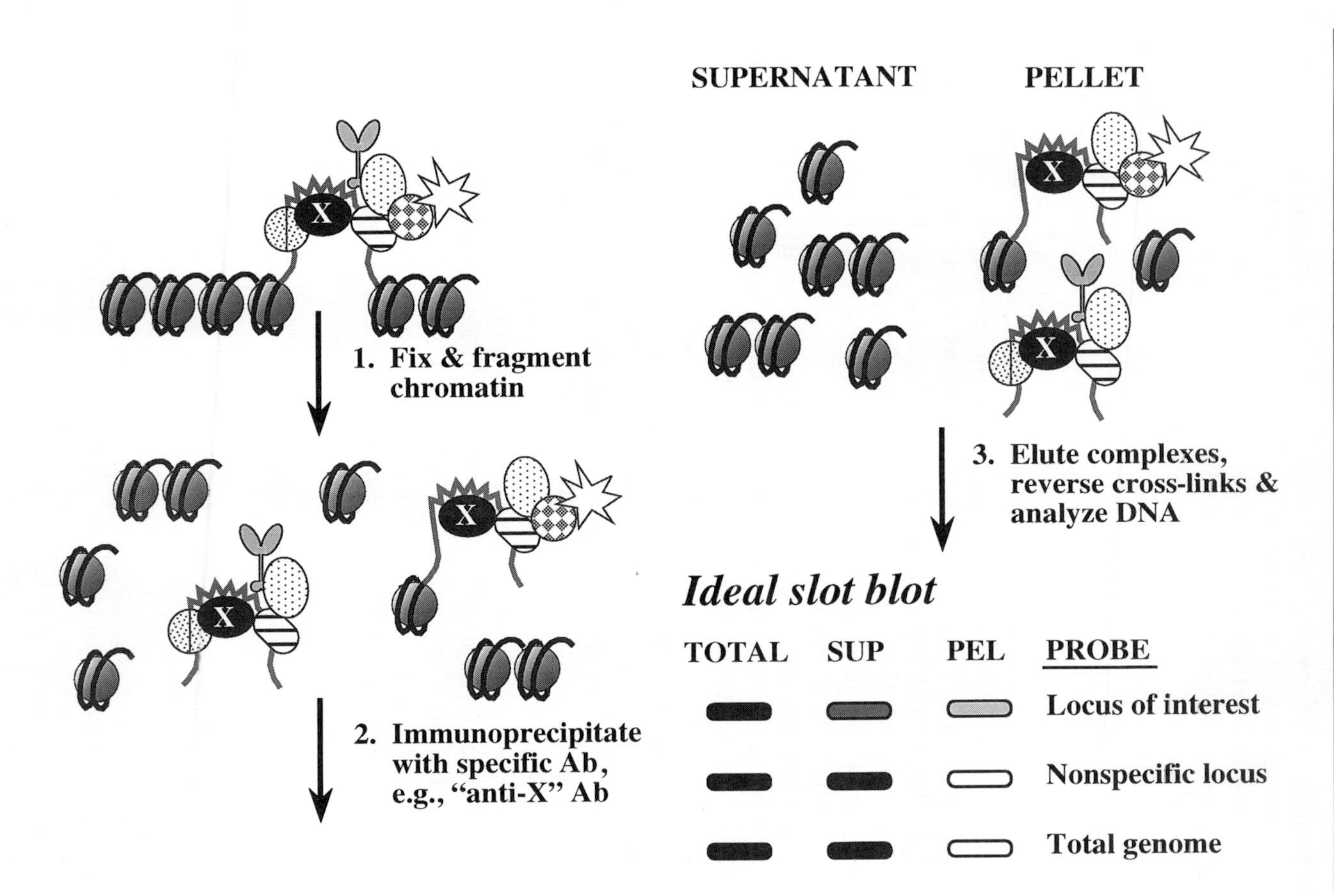
X
1. Fix & fragment chromatin
2. Immunoprecipitate with specific Ab, e.g., "anti-X" Ab
SUPERNATANT
PELLET
3. Elute complexes, reverse cross-links & analyze DNA
Ideal slot blot
TOTAL
SUP
PEL
PROBE
Locus of interest
Nonspecific locus
Total genome

and colleagues working with *Tetrahymena*.[5,9] These studies defined the distribution of specifically modified histones or Polycomb group proteins, which bind to multiple regions of the genome. Immunological analysis of chromatin in the budding yeast *Saccharomyces cerevisiae* was first described by Braunstein *et al.*,[10,11] where they examined the acetylation state of histone H4 at silent loci and telomeres in wild-type strains as well as in strains defective in silencing proteins. This work revealed the potential power of *in vivo* cross-linking when applied to an organism with a compact genome and facile molecular genetics.

Recent years have witnessed a resurgence of the application of *in vivo* cross-linking to budding yeast[12–16] as well as to fission yeast.[17,18] This is due in part to the availability of the complete genomic sequence of the budding yeast as well as to the exquisite sensitivity of PCR as a DNA detection method. Thus, the study of telomeric chromatin has been extended to include the Rap1 and Sir proteins[12,13] and to map the extent of telomeric heterochromatin domains. *In vivo* cross-linking has also been used to define protein–DNA interactions that occur at either origins of replication[14,15] or centromeres.[16–18] Notably, in budding yeast these chromosomal structures form on short (<150 bp) consensus loci and the proteins being monitored most likely exist in single or low-copy number within their respective complex. Thus, *in vivo* cross-linking followed by immunoprecipitation can be

[9] P. C. Dedon, J. A. Soults, C. D. Allis, and M. A. Gorovsky, *Mol. Cell. Biol.* **11,** 1729 (1991).

[10] M. Braunstein, A. B. Rose, S. G. Holmes, C. D. Allis, and J. R. Broach, *Genes Dev.* **7,** 592 (1993).

[11] M. Braunstein, C. D. Allis, B. M. Turner, and J. R. Broach, *Mol. Cell. Biol.* **16,** 4349 (1996).

[12] A. Hecht, T. Laroche, S. Strahl-Bolsinger, and M. Grunstein, *Nature* **383,** 92 (1996).

[13] S. A. H. Strahl-Bolsinger, K. Luo, and M. Grunstein, *Genes Dev.* **11,** 83 (1997).

[14] O. M. Aparicio, D. M. Weinstein, and S. P. Bell, *Cell* **91,** 59 (1997).

[15] T. Tanaka, D. Knapp, and K. Nasmyth, *Cell* **90,** 649 (1997).

[16] P. B. Meluh and D. Koshland, *Genes Dev.* **11,** 3401 (1997).

[17] S. Saitoh, K. Takahashi, and M. Yanagida, *Cell* **90,** 131 (1997).

[18] K. Ekwall, T. Olsson, B. M. Turner, G. Cranston, and R. C. Allshire, *Cell* **91,** 1021 (1997).

FIG. 1. General schematic for formaldehyde *in vivo* cross-linking and chromatin immunoprecipitation. Step 1: Yeast cells in culture medium are treated *in situ* with formaldehyde and then harvested. A whole cell lysate is prepared from the fixed cells and cross-linked chromatin present in the lysate is sheared into small fragments by sonication. Step 2: Cross-linked protein–DNA complexes are immunoprecipitated with protein-specific antibodies (e.g., "anti-X" antibody). Step 3: After harvesting, chromatin immunoprecipitated by the antibody (i.e., in the pellet fraction) is eluted from the affinity matrix and heated to reverse the cross-links. Coimmunoprecipitation of various DNA sequences is assessed by slot blot (shown) or Southern hybridization or by PCR. *Note:* A 300-bp fragment represents less than 0.0025% of the budding yeast genome.

applied with great success to very specific protein–DNA interactions in budding yeast.

This article presents a straightforward protocol for *in vivo* analysis of protein–DNA complexes in budding and fission yeast. The protocol is based largely on that developed by Braunstein *et al.*[10] for budding yeast and allows one to test whether a protein of interest interacts with a particular chromosomal locus *in vivo,* provided protein-specific antibodies and locus-specific probes are available. This protocol has been used to study the distribution of acetylated histone H4 on budding yeast chromosomes as well as to confirm the specific association of several candidate centromere proteins with centromeric DNA. We also point out protocol variations currently in use. Finally, we recommend interested readers to several very useful articles that elaborate on the theory, practice, and general aspects of chromatin immunoprecipitation (ChIP) and that discuss the relative merits and pitfalls of the approach.[4–6,19]

Chromatin Immunoprecipitation

Formaldehyde Fixation of Cells

1. Grow a culture of yeast cells in 120 ml medium (synthetic or YEPD[20] to midlogarithmic phase (OD_{660} = 1.0 to 1.5).
2. Measure culture volume and add formaldehyde directly to the culture to a final concentration of 1%. Formaldehyde stock solutions are generally available as 37% solutions. Fix cells at ambient temperature with gentle agitation. *Note:* The optimal time of fixation that affords the best compromise in the recovery of soluble chromatin without inactivation of immunological determinants on the protein of interest must be determined experimentally. Fixation times reported previously range from 10 min to several hours. Cross-linking reactions can be stopped by harvesting cells or by adding glycine to 0.125 *M* final concentration, followed by incubation for 5 min. This latter step is needed only for short fixation times.
3. Harvest 100 ml formaldehyde-fixed cells by low-speed centrifugation (3000*g* at 4° for 5 min, e.g., Sorvall SS34 rotor at 5000 rpm). *Note:* Cells can be stored on ice for several hours at this point.

[19] L. P. O'Neill and B. M. Turner, *Methods Enzymol.* **274,** 189 (1996).

[20] C. Kaiser, S. Michaelis, and A. Mitchell, *in* "Methods in Yeast Genetics: A Cold Spring Harbor Laboratory Course Manual." Cold Spring Harbor Laboratory, Cold Spring Harbor, NY, 1994.

Preparation of Whole Cell Lysates Containing Cross-Linked Chromatin

Cell lysates can be prepared either by formation and lysis of spheroplasts or by mechanical breakage of cells with glass beads.

The following procedures provide material for 6–10 separate immunoprecipitation reactions. Several different strains can be processed simultaneously.

Preparation of Chromatin by Lysis of Spheroplasts. Spheroplasts are prepared by enzymatic digestion of the yeast cell wall following a high pH treatment. Volumes are for 100 ml fixed cells.

1. Resuspend formaldehyde-fixed cells in 5 ml cold, freshly prepared 0.1 *M* Tris–HCl, pH 9.4, 10 m*M* dithiothreitol (DTT) and transfer to a polycarbonate Oak Ridge centrifuge tube. Incubate on ice for 20 min. Harvest cells by centrifugation at 3000*g* for 5 min at 4°. Decant supernatant.
2. Wash cells once with 5 ml spheroplast buffer A (1.2 *M* sorbitol, 20 m*M* HEPES, pH 7.4).
3. Resuspend cells in 5 ml spheroplast buffer A supplemented with 0.5 m*M* phenylmethylsulfonyl fluoride (PMSF). Add 40 μl 1 mg/ml oxalyticase (ca. 2000 units; Enzogenetics, Corvallis, OR) or 2 mg zymolyase 100T (200 units, ICN Biochemicals, Costa Mesa, CA). Incubate at 30° for 30 min with gentle agitation. Confirm microscopically that cells have been converted to spheroplasts.
4. Add 10 ml spheroplast buffer B (1.2 *M* sorbitol, 20 m*M* PIPES, pH 6.8, 1 m*M* $MgCl_2$) supplemented with 0.5 m*M* PMSF and mix gently. Harvest spheroplasts at 3000*g* for 5 min at 4° and decant supernatant.
5. Gently resuspend spheroplasts in 5 ml cold PBS (10 m*M* KH_2PO_4, 40 m*M* K_2HPO_4, 0.15 *M* NaCl). Harvest spheroplasts at 3000*g* for 5 min at 4° and decant supernatant.
6. Gently resuspend spheroplasts in 5 ml cold Triton/HEPES buffer (0.25% Triton X-100, 10 m*M* EDTA, 0.5 m*M* EGTA, 10 m*M* HEPES, pH 6.5) supplemented with 0.5 m*M* PMSF, 0.8 μg/ml pepstatin, and 0.6 μg/ml leupeptin. Harvest spheroplasts at 7500*g* at 4° for 5 min and decant supernatant. Note the increased centrifugation force.
7. Gently resuspend spheroplasts in 5 ml cold NaCl/HEPES buffer (200 m*M* NaCl, 1 m*M* EDTA, 0.5 m*M* EGTA, 10 m*M* HEPES, pH 6.5) supplemented with 0.5 m*M* PMSF, 0.8 μg/ml pepstatin, and 0.6 μg/ml leupeptin. Harvest spheroplasts at 7500*g* at 4° for 5 min and decant supernatant. Drain well.

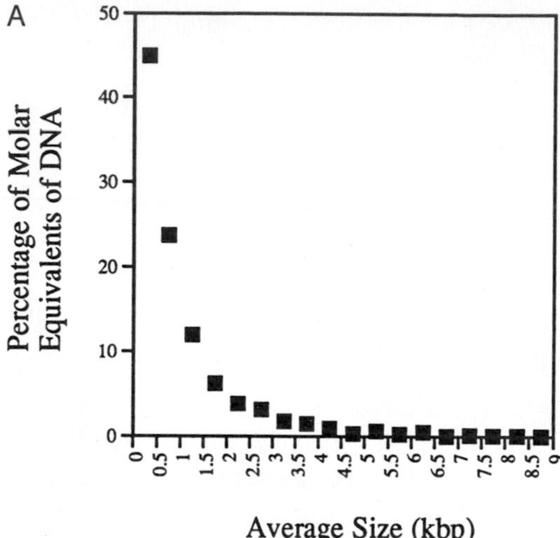

FIG. 2. Fragment size and resolution by chromatin immunoprecipitation. (A) Size distribution of DNA in chromatin following fragmentation by sonication. DNA fragments produced by the protocol through step B-9 were recovered after reversal of cross-links, extraction of protein, and treatment with RNase and fractionated on a 1.7% agarose gel. After staining the gel with ethidium bromide, the staining intensity versus migration distance was determined by densitometry of a Polaroid negative obtained from the stained gel. We then calculated the staining intensity for each 500-bp segment of the gel, divided the staining intensity by size to obtain molar equivalents of DNA in each section, and then plotted the molar fraction of total DNA in each segment versus the segment size range. Taken from M. Braunstein, Ph.D. dissertation, Princeton University (1996), with permission. (B) The influence of average chromatin fragment size on the ability to finely map a detected interaction. For purposes of illustration, we suppose that a hypothetical protein (stippled) interacts *in vivo* with a specific chromosomal locus (white box). After performing chromatin immunoprecipitation, the coprecipitated DNA is analyzed by PCR using primers that amplify overlapping 300-bp fragments that span the region of interest. If the starting average size of the sonicated chromatin fragments were small (300–400 bp in this example; top), such a PCR analysis would localize the detected interaction to a single interval. In contrast, if the average fragment size were somewhat larger (e.g., 600–700 bp; bottom), the interaction would appear to map over an extended region of the chromosome.

8. Resuspend spheroplasts thoroughly in 1 ml sodium dodecyl sulfate (SDS) lysis buffer (1% ultrapure SDS, 10 m*M* EDTA, 50 m*M* Tris–HCl, pH 8.1) supplemented with 0.5 m*M* PMSF, 0.8 μg/ml pepstatin A, and 0.6 μg/ml leupeptin. The buffer should be cool but not ice cold in order to avoid precipitation of SDS. Divide the suspension equally between two microfuge tubes.
9. Sonicate the spheroplast suspension in order to complete lysis and to break the chromatin into small fragments. For a Branson Sonifier 250 (Branson Ultrasonics Corporation, Danbury, CT) fitted with a

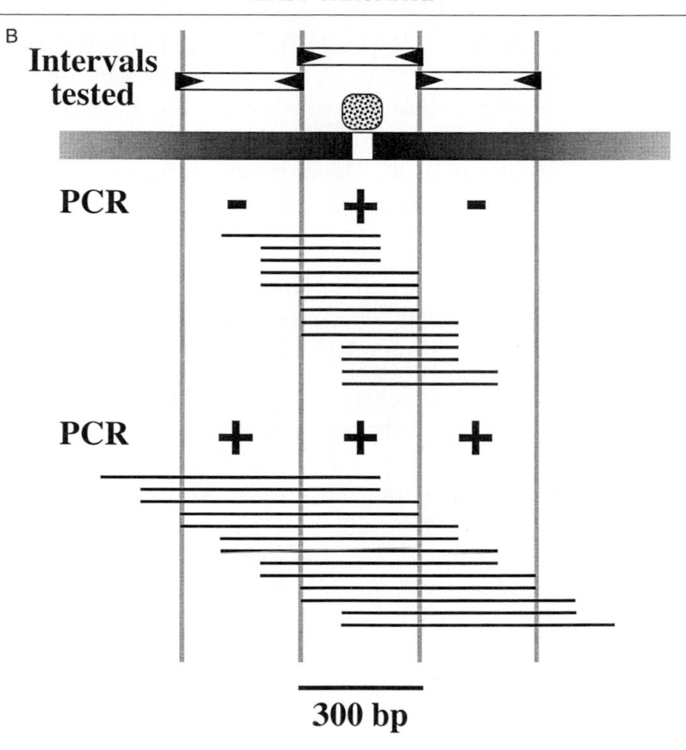

FIG. 2. (*continued*)

microprobe tip, six 10-sec pulses of constant output at 15–20% power is generally sufficient to achieve the desired DNA fragment size range of between 100 and 2000 bp centered around 400–500 bp. Samples should be maintained on ice during the sonication procedure. *Note:* Fragment size after sonication can be assessed by gel electrophoresis and should be determined for each fixation procedure. A small average fragment size is important because it promotes solubility and permits high-resolution mapping of detected interactions (Fig. 2). Prior to gel fractions the sample should be heated to remove cross-links and phenol extracted to remove protein (see section D, step 3 ff.)

10. Clarify sonicated material by centrifugation at 17,000*g* for 10 min at 4° in a microfuge.
11. Transfer both supernatants (ca. 1.1 ml total) to a single prechilled 14-ml round-bottomed polypropylene tube (e.g., Falcon 2059). Lysate may be slightly cloudy due to the SDS. Add 9 volumes (ca. 10 ml) of ice-cold IP dilution buffer (0.01% SDS, 1.1% Triton X-100,

1.2 m*M* EDTA, 16.7 m*M* Tris-HCl, pH 8.1, 167 m*M* NaCl) supplemented with 0.5 m*M* PMSF, 0.8 μg/ml pepstatin, and 0.6 μg/ml leupeptin. Incubate diluted lysate on ice for 15 min to allow any insoluble precipitate to form. Clarify dilute lysate by centrifugation at 8000*g* for 10 min at 4°. Decant supernatant into a fresh 15-ml prechilled tube and store on ice. This is the solubilized chromatin solution.

Preparation of Chromatin by Mechanical Breakage with Glass Beads. Fixed cells are broken directly in SDS lysis buffer, and the whole cell lysate is then sonicated. Volumes are for 100 ml of cells.

1. Wash fixed cells two times with 10 ml each cold PBS. Resuspend cell in 2 ml PBS and divide between two microfuge tubes. Harvest cells by centrifugation.
2. Resuspend each cell pellet in 250 μl of SDS lysis buffer supplemented with protease inhibitors. Add 0.4-g glass beads (0.35–0.50 mm in diameter; acid washed) such that the column of displaced liquid above the beads is ca. 3 mm. *Note:* An alternative lysis buffer, FA lysis buffer (50 m*M* HEPES–KOH, pH 7.5, 140 m*M* NaCl, 1 m*M* EDTA, 1% Triton X-100, 0.1% sodium deoxycholate), was described by Strahl-Bolsinger *et al.*[13] and has been used by several groups in combination with glass beads.[14,15,18] In this case, the immunoprecipitation reactions (see later) are performed under the same buffer conditions. Although we have not compared the efficacy of the two buffers directly, it is likely that any of several detergent-containing buffers can be used to prepare lysates from formaldehyde-fixed cells.
3. Vortex the sample at high speed for 4 min. Typically this is done in intervals of 30 sec each, with incubation on ice in between each cycle for ca. 1 min. It is extremely useful to agitate all samples simultaneously using a microfuge tube mixer to ensure even processing. The extent of cell breakage can be monitored microscopically.
4. Spin briefly to concentrated lysate at bottom of tube. Cell debris may form a pellet. Add an additional 250 μl of SDS lysis buffer. Resuspend any pellet by pipetting gently to generate a homogeneous suspension. Transfer as much lysate as possible (ca. 0.55 ml) to a new microfuge tube, avoiding the glass beads. This can be accomplished either by pipetting or by making a small hole at the bottom of the tube with a red hot needle and centrifuging the suspension into a fresh tube.
5. Sonicate as described in the previous section (step 9) and continue as described. Clarification of the lysate prior to sonication will reduce

the yield of cross-linked chromatin significantly. *Note:* A survey of recent *in vivo* cross-linking applications to yeast indicates that mechanical breakage of fixed yeast cells is the current method of choice for both budding and fission yeast.[17,18] As we have seen only subtle differences in results obtained using glass beads as compared to the spheroplast method, we also recommend this technique. Breakage of intact cells has several advantages, including increased ease of processing multiple cultures and elimination of variation resulting from differences in spheroplasting efficiency between strains.

Immunoprecipitation Reactions

The solubilized chromatin solution can be used in immunoprecipitation reactions to recover specific genomic regions associated with specific proteins. Sufficient material is obtained by the lysis procedures described previously to allow immunoprecipitation with several different antibodies. The solubilized chromatin solution contains 0.1% SDS, which corresponds to buffers used in denaturing immunoprecipitation procedures. Accordingly, the ability of a particular antibody preparation to immunoprecipitate quantitatively a protein of interest from non-cross-linked lysates under the same buffer conditions should be assessed prior to using the solubilized chromatin material.

1. Aliquot the chromatin solution to microfuge tubes. Typically 1–1.5 ml of chromatin solution, corresponding to $2–4 \times 10^8$ cells or 10–20 OD_{660} equivalents, is used per immunoprecipitation reaction.
2. Add antibody and incubate the reactions several hours to overnight at 4° with gentle mixing. *Note:* See comments on controls in section E.
3. Harvest immune complexes by adding 2 μg of sonicated λ genomic DNA (to block nonspecific binding of DNA) and 40 μl protein A–Sepharose CL-4B beads (Pharmacia Biotech, Uppsala, Sweden) to each reaction. Prepare beads in advance as a 50% slurry in buffer containing 10 m*M* Tris, pH 8.0, 1 m*M* EDTA, 0.1% (w/v) BSA, and 0.1% (w/v) azide. Incubate the reactions at ambient temperature for 1–2 hr with gentle mixing. *Note:* The protein–DNA complexes may be immunoprecipitated in a single step using antibody coupled to an appropriate bead resin.
4. Centrifuge for 1 min at 17,000*g* and 4° to concentrate beads. If desired, retain an aliquot of the supernatant (i.e., the unbound fraction) for analysis. Aspirate the remaining supernatant carefully.
5. Wash beads successively with 1–1.5 ml each of the following buffers, incubating for 3–5 min with gentle mixing each time, followed by centrifugation to concentrate beads. Transfer suspension to a fresh

microfuge tube prior to the final TE wash to reduce contamination from nonspecific chromatin that might be stuck to the tube walls.

a. TSE-150 (1% Triton X-100, 0.1% SDS, 2 m*M* EDTA, 20 m*M* Tris–HCl, pH 8.1, 150 m*M* NaCl). This buffer approximates immunoprecipitation conditions.

b. TSE-150 or TSE-500 (1% Triton X-100, 0.1% SDS, 2 m*M* EDTA, 20 m*M* Tris–HCl, pH 8.1, 500 m*M* NaCl). *Note:* Some protein–antibody interactions may be sensitive to the higher NaCl concentration.

c. LiCl/detergent wash (0.25 *M* LiCl, 1% NP-40, 1% DOC, 1 m*M* EDTA, 10 m*M* Tris–HCl, pH 8.1).

d. TE (10 m*M* Tris, pH 8.0, 1 m*M* EDTA).

e. TE.

Note: If the alternative FA lysis buffer described in the previous section is used, then beads should be washed first with FA lysis buffer, then with FA lysis buffer supplemented with 500 m*M* NaCl, before proceeding to the LiCl/detergent wash (step 5c, above).

Purification of Coimmunoprecipitated DNA

To recover coimmunoprecipitated DNA, the immune complexes are first eluted from the protein A beads. The eluted samples are then heat treated to reverse the cross-links, followed by protease treatment and extraction with organic solvents to remove proteins and SDS.

1. Add 250 μl of elution buffer (1% ultrapure SDS, 0.1 *M* $NaHCO_3$) to the protein A–Sepharose beads. Vortex briefly and then incubate for ca. 15 min at ambient temperature with gentle mixing. Centrifuge for 1 min at 17,000*g* and 4° to concentrate beads. Carefully transfer the supernatant, which contains the immunoprecipitated chromatin, to a fresh microfuge tube, avoiding beads.
2. Repeat step 1. Pool the first and second supernatants. Use a fine pipette tip to aspirate liquid remaining in the beads and add to the combined eluate (final volume ca. 500 μl).
3. To facilitate ease of handling, supplement all samples with NaCl to a 0.2 *M* final concentration, i.e., add 20 μl (1/25 volume) 5 *M* NaCl to the eluted immune complexes (500 μl) and 3 μl (1/100 volume) of 5 *M* NaCl to reserved 300-μl aliquots of the solubilized chromatin solution ("total" or "input") and any unbound fractions from the immunoprecipitation reactions. Mix well.
4. Incubate all samples at 65° for 4 to 5 hr to reverse the formaldehyde cross-links. Aggregates may form in total chromatin samples during this period.

5. Spin tubes briefly to remove evaporated liquid from the lid and add 2 volumes absolute ethanol to each sample. Precipitate the DNA at −20° for 2 hr. DNA samples can be stored indefinitely at this point.
6. Centrifuge the samples at 17,000*g* for 20 min at 4°. The precipitate will appear flocculent due to the coprecipitation of SDS. Take care not to lose any of this material while aspirating the supernatant. Wash the pellets with 70% (v/v) ethanol, vortex, and recentrifuge. Dry pellets briefly. Resuspend in 100 μl TE and allow to rehydrate at 4°. Aggregates in the total samples will disperse on addition of proteinase K buffer (see later). *Note:* DNase-free RNase A (20 μg) can be added to each sample at this point, followed by a 30-min incubation at 37°. This step is especially important for input and unbound fractions that will be analyzed by slot-blot hybridization (see later) or by DNA gel electrophoresis.
7. Add 25 μl 5× proteinase K buffer (50 m*M* Tris, pH 8.0, 25 m*M* EDTA, 1.25% SDS) to each sample, mix, and then add 1.5 μl proteinase K solution (ca. 30 μg; Boehringer Mannheim, 1413 783). Incubate at 42° for 1 to 2 hr.
8. Add 175 μl TE to the immunoprecipitated material (300 μl final volume) and 275 μl TE to the total samples (400 μl final volume).
9. Extract the immunoprecipated samples once with an equal volume (300 μl) of freshly prepared PCI (phenol : chloroform : isoamyl alcohol; 25 : 24 : 1) and then with an equal volume of chloroform. Back extraction is not required, provided only a small volume of the aqueous phase is left behind.
10. Extract input and unbound fractions twice with an equal volume (400 μl) of PCI and then twice with an equal volume of chloroform. Back extract the organic phases with 100 μl TE and combine with the primary aqueous phase.
11. Add 5 μg glycogen to all samples, followed by 1/10 volume of 3 *M* sodium acetate and 2 volumes absolute ethanol. Precipitate the DNA for 2 hr at −20°. DNA samples can be stored indefinitely at this point. *Note:* Glycogen serves as a carrier to maximize DNA precipitation.
12. Centrifuge the samples at 17,000*g* for 20 min at 4°. Ethanol-precipitated material corresponding to coimmunoprecipated DNA will likely not be visible, whereas DNA from the total samples and supernatant fractions should form a visible pellet. Wash all pellets with 70% ethanol, vortex, and recentrifuge. Dry pellets briefly under vacuum.
13. Resuspend all samples in TE and allow DNA to rehydrate at 4° prior to analysis. For PCR analysis, we typically resuspend the

coimmunoprecipitated DNA in a volume of TE equivalent to 1/10 the initial volume of chromatin solution used for immunoprecipitation (e.g., 150 μl TE for a 1.5-ml aliquot). We resuspend samples for input DNA and unbound fractions in a volume of TE equivalent to the volume of chromatin solution originally set aside. For other detection methods, the samples can be resuspended in smaller volumes.

Note: The order of execution of the various steps in the DNA purification scheme can be varied. For example, Aparicio and colleagues[14] routinely perform a proteinase K treatment immediately after reversing the cross-links at 65°. For this purpose, they recommend using proteinase K buffer adjusted to 1% SDS to elute immune complexes from the beads. Samples can then be incubated at 65° to reverse the cross-links, diluted with proteinase K buffer lacking SDS to reduce the SDS concentration to 0.25–0.4%, and further incubated with proteinase K at 37° for 2 hrs. The resulting material is PCI extracted and ethanol precipitated.

Analysis of Coimmunoprecipitated DNA

Although only a small amount of coimmunoprecipitated DNA may be obtained by the preceding protocol, the presence of various DNA sequences in the resultant samples can be assessed readily by slot blot or Southern hybridization or by PCR. Slot blot hybridization provides a straightforward method of quantitating the immunoprecipitated DNA in different samples. However, Southern hybridization or PCR provides more stringent assays for the presence of specific sequences. Moreover, in some cases, only a small amount of the input DNA may coprecipitate with the protein of interest. For example, ca. 0.01% of input telomeric DNA was recovered in immunoprecipitates of the Rap1 or Sir proteins.[13] Thus, the sensitivity afforded by PCR analysis may be obligatory for detection.

For analysis of coprecipitated (bound) DNA by either hybridization method, the entire sample may be required in order to detect a signal (i.e., a volume equivalent to 1–1.5 ml of chromatin solution). For input and unbound fractions, however, an amount equivalent to 150 μl of chromatin solution should be sufficient for Southern analysis. A titration series of input material equivalent to 150 μl or less of chromatin solution should be applied to slot blots. For slot blot analysis, the samples must be pretreated with RNase (step 6 in the preceding section) to avoid spurious hybridization signals due to RNA. For Southern analysis, select a restriction enzyme(s) that will yield small diagnostic fragments, as the DNA, having been previously sonicated, will be composed of random small fragments. Add RNase to the samples after digestion to reduce background. Pretreatment of agarose gels with HCl to fragment the DNA is unnecessary.

PCR reactions are typically carried out in 50-μl reaction volumes using standard reaction conditions and a limited number (23–25) of amplification cycles. For immunoprecipitated material, reactions should be programmed with 1/30 to 1/50 of the immunoprecipitated DNA (equivalent to 30–50 μl of chromatin solution). For total input or unbound chromatin, the reactions should be programmed with two- to threefold serial dilutions starting with DNA from 5 μl of chromatin solution. PCR primers should be designed so as to amplify diagnostic fragments between 100 and 350 bp in length. Reaction products are resolved on a polyacrylamide gel. The limited amplication and titration of templates ensure that the yields of product are roughly proportional to the amount of starting template in each reaction. Under such conditions the percentage recovery can be estimated.

Regardless of the method of detection, assaying for several DNA loci in addition to those regions of interest is critical. The degree to which nonspecific DNA sequences are detected in the immunoprecipitates provides insight into the specificity and biological relevance of the interactions being tested.

Strengthening a Positive Result: A Question of Specificity

At least two criteria must be met before concluding that a protein–DNA interaction detected by the method just described reflects an authentic *in vivo* interaction. First, detection of a particular DNA sequence in the immunoprecipitate must be antibody dependent. Little or no signal should be detected in material recovered from mock-treated reactions to which no antibody was added. Any result to the contrary would indicate nonspecific precipitation of chromatin. Second, the antibody must be specific for the protein of interest. The antibody should be well characterized and, if possible, affinity purified. A useful control to confirm this specificity is to include excess purified antigen in the chromatin precipitation reaction. This should block coprecipitation of the target DNA. If it does not, then the possibility has to be entertained that the antibody preparation is precipitating some other antigen.

Chromatin immunoprecipitation analysis of protein–DNA interactions in a genetically tractable organism such as budding yeast affords additional controls. For example, a well-characterized antiepitope antibody (e.g., the antihemagglutinin monoclonal antibody 12CA5) can be used with cross-linked chromatin prepared from a pair of isogenic strains, one of which expresses an epitope-tagged version of the protein as its sole source and the other of which expresses the untagged version. Specificity for the protein would be demonstrated by the recovery of DNA fragments of interest only in immunoprecipitates from the tagged strain. Alternatively, provided the protein of interest is not essential, a pair of isogenic strains can be used in which one of the strains carries a deletion of the target protein gene. In this case, specificity for the protein would be demonstrated by the recovery of DNA fragments of interest in immunoprecipitates from the wild type but not the deleted strain.

While appropriate outcomes in these tests would clearly implicate the protein of interest as a DNA-associated factor, they do not address the DNA sequence specificity of the detected interaction. Interaction of the target protein with the chromosome may occur globally (like histones), may be restricted to a particular chromosomal domain (like acetylated histone H4, Rap1p, or Sir proteins), or may be sequence specific (like centromere or Orc proteins). In addition, the interaction might be specifically localized during one stage of the cell cycle but delocalized at others, as is the case for DNA polymerase. To distinguish among these possibilities, several DNA sequences in the immunoprecipitates, in addition to the ones of interest, must be monitored. In cases of localized interactions, one should test whether the interaction is diminished or abolished when the target DNA sequence is mutated. In addition, as long as the chromatin has been well fragmented, the site of interaction can be mapped using small overlapping probes to analyze the coprecipitated DNA (Fig. 3).

After establishing the existence of a particular protein–DNA interaction, one can use chromatin immunoprecipitation to probe the architecture and assembly of a chromosomal complex. For example, the effect of cell cycle position or of mutations in relevant *trans*-acting factors on the interaction can be tested. Variations in a given signal must be interpreted carefully and in the context of other information, however, as they may reflect actual disruption of the protein–DNA interaction or simply a change in the accessibility of the protein being immunoprecipitated. Similarly, care must be taken with putative *trans*-acting mutants to distinguish a direct effect on the complex from an indirect effect conferred by the mutation.

Finally, a lack of interaction determined by chromatin immunoprecipitation is only meaningful in the context of other positive results. Failure to detect an interaction can reflect any number of contingencies, some of which can be addressed experimentally. First, whether the protein of interest can be immunoprecipitated quantitatively under the conditions of the experiment must be confirmed. Denaturation of the epitope due to high SDS concentrations, inaccessibility of the epitope, or inactivation of the epitope by fixation could preclude immunoprecipitation. Western analysis of appropriate control immunoprecipitations should address this question. In addition, the presence of the target protein in the chromatin immunoprecipitate

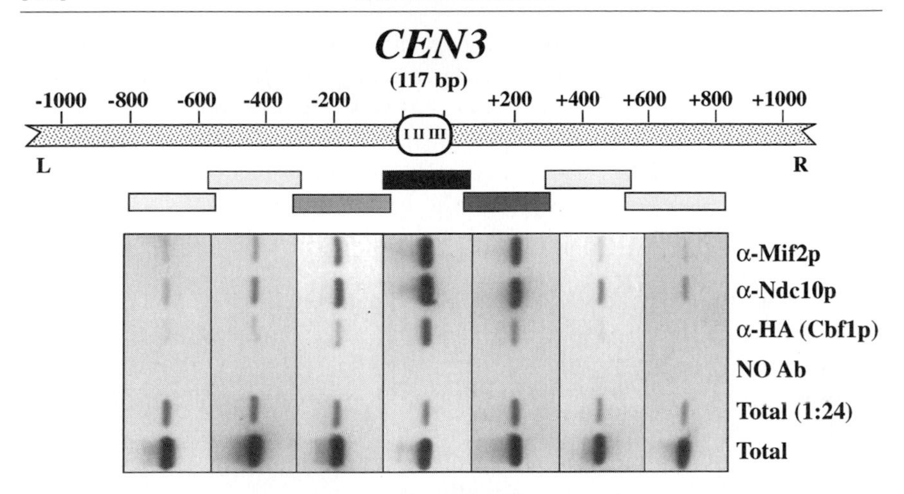

FIG. 3. Interaction of yeast centromere proteins is limited to the centromeric DNA. The interaction of three yeast centromere proteins—Mif2p, Ndc10p, and Cbf1p—with ca. 2 kb encompassing the *CEN3* genomic was investigated using *in vivo* cross-linking and chromatin immunoprecipitation. Cross-linked chromatin (2-hr fixation) prepared from a *CBF1-HA* strain was immunoprecipitated with anti-Mif2p, anti-Ndc10p and anti-HA (specific for epitope-tagged Cbf1-HAp) or was mock-treated (NO Ab). Total input material (3 μl of chromatin solution), a 1:24 dilution thereof, and coprecipitated DNAs (30 μl of chromatin solution) were analyzed by PCR using primers specific for the indicated series of *CEN3*-flanking fragments. Shading indicates the relative enrichment in the immunoprecipitates of DNA corresponding to those intervals and is based on the negative images of ethidium bromide-stained PCR products as shown. Only the *CEN3* DNA-containing region was strongly enriched in the immunoprecipitates, indicating that all three proteins interact specifically with centromeric DNA. If a protein participated in interactions beyond the centromeric DNA, more than one PCR fragment should have been amplified strongly. Taken from P. B. Meluh and D. Koshland, *Genes Dev.* **11,** 3401 (1997), with permission.

can be assessed by Western analysis following SDS–PAGE fractionation of the sample after incubating the precipitate in sample buffer at 100° for 30 min or at 65° for several hours. Second, the interaction may occur only during a certain window of the cell cycle and thus exist only in a subset of cells from an asynchronous population. If this is a likely problem, the assay can be repeated with synchronized cells. Finally, the protein may be present at a particular DNA locus but the geometry of potential reactive groups may preclude cross-linking with formaldehyde. This eventuality would be difficult to rule out.

Conclusions

The chromatin immunoprecipitation assay described in this article is a simple approach to probing the organization of protein–DNA complexes that exist within a cell. For small organisms such as the budding yeast, this provides an approach to the molecular cytology of chromosomes that are too small to be visualized microscopically. With the budding yeast, the availability of a wide array of molecular gentic tools as well as the complete genome sequence makes effective application of this technique straightforward. That is, probes for analysis of the immunoprecipitated DNA can be selected readily and genetic controls to confirm detected interactions can be constructed easily. However, even with organisms with more complex genomes and perhaps less tractable molecular genetics, the technique has been and can be applied to great advantage. PCR analysis of precipitated DNA has extended the limit of sensitivity of the technique, making a small genome less of a constraint. In addition, the growing amount of genomic sequence available for a number of organisms facilitates probe selection. Finally, careful attention to biochemical and immunological controls can obviate the need for extensive genetic controls.

We expect that this technique can be extended beyond simply identifying the *in vivo* association of a protein with a particular region of the genome. The procedure has already been used to examine dynamic events within the cell, such as the outward procession of components of the replication apparatus from replication origins.[14] In addition, the procedure has been used for examining the effects of mutations on the structure and organization of particular chromosomal structures.[10,16] We anticipate that this aspect of the application of the technique will continue to expand. Finally, we and others foresee that the technique can be used to identify the binding sites for proteins of interest.[6] That is, coprecipitated DNA obtained following *in vivo* cross-linking to a protein of interest can be cloned and sequenced or used as a probe to identify the corresponding locus. Accordingly, we anticipate that this technique will become a staple in the repertoire of molecular geneticists examining a wide variety of biological problems in a number of different experimental systems.

Acknowledgments

The authors thank Doug Koshland, in whose laboratory part of this work was performed, and Miriam Braunstein, for comments, insights, and efforts in developing the technique described. We also thank Tomoyuki Tanaka, Kim Nasmyth, Oscar Aparicio, and Steve Bell for sharing with us details of protocols prior to publication and to David Allis for encouragement in developing this technology. This work was supported by grants from the NIH to JRB and by a postdoctoral research fellowship from the Helen Hay Whitney Foundation to PBM.

[24] DNA Methyltransferases as Probes of Chromatin Structure *in Vivo*

By MICHAEL P. KLADDE, M. XU, and ROBERT T. SIMPSON

Introduction

A complementary strategy to conventional methods for probing chromatin organization utilizes the expression of foreign DNA methyltransferases in intact *Saccharomyces cerevisiae.* At current levels of expression of several foreign methyltransferases in yeast, no deleterious effects to cellular metabolism have been detected. The innocuousness of methyltransferase expression and the absence of endogenous adenine or cytosine methylation in the yeast genome[1,2] allow detection of *de novo* DNA modification in DNA that is directly isolated from engineered, methylation-proficient cells. The ability to bypass the isolation of nuclei, thereby avoiding loss of labile constituents or possible reorganization of chromosome structure, has provided the motivation for developing methyltransferase probing strategies.

Background

Brooks *et al.*[3] first reported that the *Escherichia coli dam* gene could be introduced into yeast, where it was expressed and led to N^6-methylation of adenine in genomic DNA. Fehér *et al.*[4] were the first to present data demonstrating expression of a foreign DNA (cytosine-5) methyltransferase, *Bsp*RI methyltransferase (M.*Bsp*RI), in yeast. Monitoring the resistance of isolated genomic DNA to cleavage by *Bsp*RI restriction endonuclease, these authors showed that the majority of target sites in bulk genomic DNA were not protected against *Bsp*RI cleavage (i.e., not methylated), whereas two transcribed loci that were analyzed were somewhat more susceptible to methylation. The differential levels of methylation between bulk DNA and specific genes led Fehér *et al.*[4] to suggest that these enzymes may serve as useful probes of chromosome structure. Subsequent studies confirmed that Dam methyltransferases from *Escherichia coli*[5] and bacteriophage T4[6]

[1] S. Hattman, C. Kenny, L. Berger, and K. Pratt, *J. Bacteriol.* **135,** 1156 (1978).
[2] J. H. Proffitt, J. R. Davie, D. Swinton, and S. Hattman, *Mol. Cell. Biol.* **4,** 985 (1984).
[3] J. E. Brooks, R. M. Blumenthal, and T. R. Gingeras, *Nucleic Acids Res.* **11,** 837 (1983).
[4] Z. Fehér, A. Kiss, and P. Venetianer, *Nature* **302,** 266 (1983).
[5] M. F. Hoekstra and R. E. Malone, *Mol. Cell. Biol.* **5,** 610 (1985).
[6] Z. Fehér, S. L. Schlagman, Z. Miner, and S. Hattman, *Gene* **74,** 193 (1988).

0076-6879/99 $30.00

as well as *Bacillus subtilis* SPR methyltransferase[6] could be expressed in yeast. Incomplete methylation of bulk, genomic DNA was also observed and chromatin structure was mentioned as one potential reason for this phenomenon. These initial experiments were qualitative and the low modification in the genome could be explained by insufficient expression of the enzyme, loss of plasmids expressing the methyltransferases, or efficient removal of modified nucleotides by yeast repair systems. Later studies have concluded that although the yeast excision repair system (e.g., *RAD3, RAD1*) can recognize and remove both N^6-methyladenine[7] and 5-methylcytosine,[6] when quantified in excision repair-deficient strains, levels of modification increased only nominally, about 10–30%, in either bulk DNA[6] or at individual loci.[8,9]

Singh and Klar[10] and Gottschling[8] expressed Dam methyltransferase from *E. coli* in yeast and showed reduced DNA accessibility to the enzyme when genes were transcriptionally repressed versus transcribed[10] or when a gene was inserted near a silenced telomere.[8] Adenine methylation at GATC sites was detected by indirect end-labeling analysis of DNA, which was cleaved secondarily by methylation-sensitive restriction endonucleases. Uncontrolled expression of the probe, leading to relatively high levels of methylation, limited quantification of Dam accessibility to the first GATC site distal to the probe. A similar limitation is also encountered when primer extension is used to assess Dam methylation levels.[11] In addition, the paucity of GATC sites, once every ~300 bp in native yeast sequences, posed an obvious limitation in the resolution of probing. Finally, these data were correlative and did not provide a structural basis for the differential accessibility of methylation sites in the various transcriptional states.

Using detailed knowledge of the chromatin structure of a minichromosome,[12,13] we determined that positioned nucleosomes present one barrier to Dam methylation in yeast.[11,14] Resolution was increased through the insertion of additional GATC sites, minor sequence changes that apparently did not affect the presence or translational positioning of the nucleosome. The problem of quantifying methylation at individual, closely arrayed sites was solved by the hybridization of DNA digested with *Dpn*I or *Dpn*II to a radioactively end-labeled oligonucleotide that spanned each site. While

[7] M. F. Hoekstra and R. E. Malone, *Mol. Cell. Biol.* **6,** 3555 (1986).
[8] D. E. Gottschling, *Proc. Natl. Acad. Sci. U.S.A.* **89,** 4062 (1992).
[9] M. P. Kladde and R. T. Simpson, *Nucleic Acids Res.* **26,** 1354 (1998).
[10] J. Singh and A. J. S. Klar, *Genes Dev.* **6,** 186 (1992).
[11] M. P. Kladde and R. T. Simpson, *Proc. Natl. Acad. Sci. U.S.A.* **91,** 1361 (1994).
[12] S. Y. Roth, A. Dean, and R. T. Simpson, *Mol. Cell. Biol.* **10,** 2247 (1990).
[13] M. Shimizu, S. Y. Roth, C. Szent-Gyorgyi, and R. T. Simpson, *EMBO J.* **10,** 3033 (1991).
[14] M. P. Kladde and R. T. Simpson, *Methods Enzymol.* **274,** 214 (1996).

sites located in the central region of the nucleosome core were refractory to methylation, unexpectedly, the methyltransferase modified sites in the peripheral 20 bp of the nucleosome core equally as well as those localized in a linker region.[11] This distribution of Dam methylation in yeast is somewhat reminiscent of that for naturally occurring N^6-methyladenine in *Tetrahymena thermophila.* In macronuclei isolated from *Tetrahymena,* methyladenine-containing sequences were preferentially digested by micrococcal nuclease, indicating localization predominantly in linker DNA.[15] In addition to histones, the origin recognition complex (ORC), which interacts with *ARS1* sequences in the minichromosome we employed, also limited accessibility of Dam methyltransferase.[11] Consistent with our observation, Dam methylation can also be blocked in *E. coli,* at least transiently, by the replicative complex[16] and by binding of various nonhistone proteins.[17]

Several characteristics limit the general utility of Dam methyltransferases as chromatin probes. As no chemical method for detection of N^6-methyladenine is known, the precise localization of Dam methylation relies on using cleavage with methylation-sensitive restriction endonucleases. *Dpn*I, which reportedly is restricted to cleaving dimethylated GATC sites,[18,19] will also digest unmethylated and hemimethylated substrates at high enzyme concentrations, limiting the ability to quantitate data.[11] Because digestion is confined to unmethylated sites, analysis with *Dpn*II is more precise, but because it produces a negative display of methylation, its use is not amenable to single-hit kinetic analysis. Finally, the 4-bp recognition specificity of Dam methyltransferase limits its utility in probing chromatin structure within native sequences.

The development of a method that allows a positive display of 5-methylcytosine by Frommer and colleagues,[20,21] referred to as bisulfite genomic sequencing (Fig. 1), facilitated a significant advance in the utilization of DNA methyltransferases to investigate chromatin structure. Under appropriate conditions, deamination of denatured DNA with sodium metabisulfite leads to quantitative conversion of cytosine to uracil; 5-methylcytosine is resistant to deamination and is retained in the DNA.[22,23] The procedure

[15] K. I. Pratt and S. Hattman, *Mol. Cell. Biol.* **1,** 600 (1981).
[16] J. L. Campbell and N. Kleckner, *Cell* **62,** 967 (1990).
[17] M. X. Wang and G. M. Church, *Nature* **360,** 606 (1992).
[18] S. Lacks and B. Greenburg, *J. Biol. Chem.* **250,** 4060 (1975).
[19] S. Lacks and B. Greenburg, *J. Mol. Biol.* **114,** 153 (1977).
[20] M. Frommer, L. E. MacDonald, D. S. Millar, C. M. Collis, F. Watt, G. W. Grigg, P. L. Molloy, and C. L. Paul, *Proc. Natl. Acad. Sci. U.S.A.* **89,** 1827 (1992).
[21] S. J. Clark, J. Harrison, C. L. Paul, and M. Frommer, *Nucleic Acids Res.* **22,** 2990 (1994).
[22] H. Hayatsu, Y. Wataya, K. Kai, and S. Iida, *Biochemistry* **9,** 2858 (1970).
[23] H. Hayatsu, *Prog. Nucleic Acids Res.* **16,** 75 (1976).

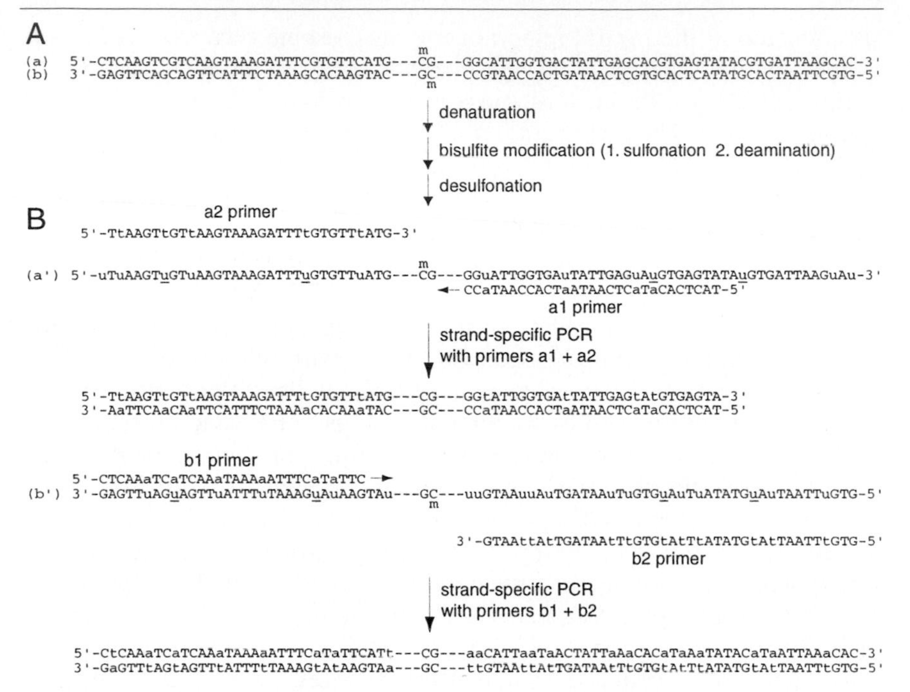

FIG. 1. Detection of 5-methylcytosine by deamination with bisulfite. (A) DNA containing 5-methylcytosine (^{m}C) is denatured to upper (a) and lower (b) single strands and treated with sodium metabisulfite. During bisulfite treatment, cytosines are first sulfonated to produce cytosine sulfonate and then deaminated to produce a sulfonyl adduct of uracil. Alkali treatment effects desulfonation, removing the sulfonyl moiety. (B) In the resulting modified upper (a′) and lower (b′) DNA strands, each cytosine is converted to uracil (indicated in lowercase), whereas 5-methylcytosine is retained. Following deamination, the desired region of the upper or lower DNA strand is amplified in separate PCR reactions by primer pairs a1 and a2 or b1 and b2, respectively. For synthesis of the a2 and b2 primers, all cytosines are converted to thymines (C to t). Conversely, the a1 and b1 primers are synthesized with transitions of each guanine to adenine (G to a). Methylated cytosines, which remain in the population of amplified PCR products, are detected by direct thermal cycle sequencing in the presence of a high concentration of ddGTP using primers a1 or b1 end labeled with ^{32}P. Representative data, using the shown b1 and b2 primers for PCR amplification from deaminated DNA and ^{32}P-end-labeled b1 primer for thermal cycle sequencing of the resulting PCR product, are presented in Fig. 2.

is extremely sensitive, utilizing primer pairs that flank the region under study for amplification by the polymerase chain reaction (PCR). Due to noncomplementarity of the two DNA strands following bisulfite treatment, primer pairs are chosen that specifically amplify only one or the other strand. Cytosines within the PCR product are detected by direct thermal cycle sequencing in a modified ddGTP-termination reaction. Following separation of the extension products on a sequencing gel and autoradiography, intensity of the visualized bands is directly proportional to the level of cytosine present at each potential methyltransferase modification site. The pertinent information for mapping chromatin is obtained by comparison of methylation levels in protein-free DNA, modified *in vitro,* to the same segment of DNA complexed with proteins in cells.

Probing Chromatin Structure with SssI Methyltransferase

To utilize the method of bisulfite genomic sequencing in the *in vivo* analysis of methyltransferase accessibility, we used M.*Sss*I, identified and cloned by others,[24] to introduce 5-methylcytosine at susceptible CG target sites in yeast.[25] Although the CG dinucleotide is somewhat underrepresented in the yeast genome, a site is present about once every 35 nucleotides, providing an 8- to 10-fold increase in resolution over previously employed methyltransferases that recognize 4-bp sites. We cloned M.*Sss*I, with a simian virus 40 (SV40) nuclear localization sequence appended at its amino terminus, into a yeast shuttle vector that expresses the chimeric gene from the regulatable *GAL1* promoter. Using standard transformation methods, the expression vector was integrated as a single copy into the *LYS2* locus of *S. cerevisiae.* Coupling the single-hit kinetic levels of M.*Sss*I expression with the method of bisulfite genomic sequencing allows quantitative, simultaneous comparison of methylation at many sites.

Methods

Although there is no deleterious consequence known to be caused by expression of M.*Sss*I in yeast, as a precaution, we maintain all M.*Sss*I strains on glucose-containing medium to repress methyltransferase expression. Under inducing conditions, in medium containing galactose and excluding glucose, cells accumulate a higher percentage of methylation as cell density increases.[25] The best signal-to-noise ratio for footprinting is obtained with cells grown to high densities. Starter cultures of yeast are grown overnight in galactose medium (10 ml) until they reach an optical density at 600 nm (OD_{600}) of approximately 1. Cells are then pelleted and resuspended in 10 ml fresh galactose medium to OD_{600} of 1 and are incubated at 30° with shaking (300 rpm) for 16 hr. These conditions lead to fully induced expression of M.*Sss*I and a maximal accumulation of methylation within accessible CG sites. However, even under these conditions, the total level of 5-methyl-

[24] P. Renbaum, D. Abrahamove, A. Fainsod, G. Wilson, S. Rottem, and A. Razin, *Nucleic Acids Res.* **18,** 1145 (1990).

[25] M. P. Kladde, M. Xu, and R. T. Simpson, *EMBO J.* **15,** 6290 (1996).

cytosine accumulation is relatively low. We estimate one modification every 2–3 kb producing no known phenotype and meeting criterion for single-hit kinetic analysis over the 600- to 1000-bp ranges we usually analyze. Reasonably good footprinting results are obtained with cultures resuspended at densities as low as OD_{600} of 0.5; however, a significant decrease in the amount of methylation occurs in cultures seeded at $OD_{600} \leq 0.1$.[25] It is necessary to include a control sample of protein-free DNA, either isolated genomic DNA or a plasmid or PCR product that contains the relevant region, that is methylated *in vitro*. In initial experiments, to identify M.*Sss*I specific changes, it is also recommended to include a control of cells grown in glucose.

Total nucleic acid is isolated rapidly from yeast cells by the glass bead method[26] as modified by Kladde and Simpson[9] to ensure that methylation levels are indicative of those present *in vivo*. It is not necessary to remove RNA from the samples. Resuspend the isolated nucleic acid in 50 μl sterile 0.1× TE (1.0 m*M* Tris–HCl, pH 8.0, 0.1 m*M* EDTA). To obtain structural information for a covalently closed circular molecule, e.g., an episomal plasmid, the DNA must be linearized with an appropriate restriction endonuclease as deamination requires complete denaturation of duplex DNA. Digest 16.5 μl total yeast DNA (ca. one-third of that isolated) in a final volume of 20 μl. The selected enzyme must cleave outside the region to be amplified by PCR following deamination (see later). Omit the endonuclease digestion if the region of interest is chromosomal.

Preparation of the reagents for deamination of the DNA is critical. Avoid exposure of solid sodium metabisulfite to the atmosphere until immediately prior to use; oxidation of the reactive sulfite ion to the inactive sulfate species is a common source of difficulty in the procedure.[23] We purchase 0.5 kg (Aldrich, Milwaukee, WI) and aliquot it in an oxygen- and water-free glove box into 5-g capacity glass vials. Precision in dispensing the solid chemical is not essential as a saturated metabisulfite solution is to be employed. With the aid of a funnel, fill each vial near capacity, secure the vial lids tightly, and store the capped vials in the dark in a vessel that contains Drierite. It is not necessary to flood the chamber with an inert gas. The reagent is stable for at least 2 years when aliquoted and stored under these conditions. Alternatively, single-use, 5-g quantities of sodium metabisulfite can be purchased.

The solutions used for deamination are prepared as follows. Protective eyeglasses and gloves should be worn as the solutions are caustic, mutagenic, or suspected carcinogens. Boiling distilled water is used to fill a bottle to

[26] M. D. Rose, F. Winston, and P. Hieter, "Methods in Yeast Genetics: A Laboratory Course Manual." Cold Spring Harbor Laboratory Press, Cold Spring Harbor, NY, 1990.

full capacity, which is then sealed air-tight. Following cooling to room temperature, the degassed distilled water is used to make up fresh solutions of 3 *N* NaOH (0.4-g NaOH pellets make 3.3 ml) and 100 m*M* hydroquinone (0.05 g makes 10 ml), which serves as an antioxidant. Avoid aeration through excessive vortexing. Next, make an appropriate volume (depending on the number of samples being deaminated) of 3× sample denaturation buffer by mixing the following in the ratio of 3 μl freshly prepared 3 *N* NaOH : 0.7 μl 3 mg/ml sheared, calf thymus DNA : 0.5 μl 0.5 *M* EDTA (pH 8.0) : 5.8 μl degassed, distilled water (i.e., 10 μl 3× denaturation buffer is added to each DNA sample). Save the remainder of 3 *N* NaOH; following deamination, it will be needed to make the desulfonation solution. Just prior to deamination and before making the saturated sodium metabisulfite solution, add 10 μl 3× sample denaturation buffer to a 0.5-ml microcentrifuge tube that contains 20 μl of each sample of total genomic DNA or control DNA and vortex. If the DNA was digested by a restriction enzyme, the 3× sample denaturation buffer can be added directly to the reactions. The samples can safely be left at room temperature for 10–15 min while preparing the saturated sodium metabisulfite solution.

Make the saturated sodium metabisulfite by dumping approximately 5 g from a previously unopened vial into a 20-ml capacity, glass scintillation vial that contains a small stir bar and 100 μl of freshly made 100 m*M* hydroquinone. Degassed, distilled water (7 ml) is then added, stirred immediately, and followed by the quick addition of 1.2 ml of the freshly made 3 *N* NaOH solution. Adjust the pH at room temperature to 5.0 with additional 3 *N* NaOH. Prewarm the solution to 50°. A variable (depending on the weight of sodium metabisulfite in a particular vial) amount of undissolved solid will usually remain in the final solution.

Complete denaturation of the DNA samples is attained by incubation at 98° for 5 min. Maintaining the samples at 98°, add 0.2 ml of stirring, prewarmed saturated sodium metabisulfite solution directly into the sample and vortex immediately. After bisulfite has been added to each tube, overlay each reaction with ~100 μl mineral oil (add 2 drops from a dropper), and incubate at 50° in the dark for 6 hr. Deamination is terminated by transferring the metabisulfite solution to a 1.5-ml microcentrifuge tube and recovering the DNA by a desalting resin and column from the Wizard PCR Prep protocol (Promega Corp., Madison, WI). Briefly, resin purchased in bulk is diluted twofold with 6 *M* guanidine thiocyanate and 1 ml is added directly to each deamination reaction and vortexed. The resin is recovered by vacuum filtration through the minicolumn and then is washed three times with 1 ml 80% (v/v) 2-propanol. Evacuate the volume *completely* before adding each subsequent wash and dispose of the filtrate solution in the manifold properly. Residual isopropanol is removed from the minicolumn by pressing it into a 1.5-ml microcentrifuge tube and centrifugation at 14,000 rpm for 2 min in a microcentrifuge. The minicolumn is transferred to a new 1.5-ml microcentrifuge tube and the deaminated DNA is recovered from the resin by applying 52 μl 0.1× TE, preheated to 95°, to the resin in the minicolumn. After incubation at room temperature for 5 min, the eluate is collected in the bottom of the microcentrifuge tube by microcentrifugation at 14,000 rpm for 20 sec.

To effect desulfonation of the eluted DNA, the sample is transferred, carefully avoiding any pelleted resin that passed through the minicolumn, to a new 1.5-ml microcentrifuge tube containing 8 μl desulfonation solution. After vortexing, incubate the samples at 37° for 15 min. Desulfonation solution is made fresh before use and contains the following in the ratio of 7 μl 3 *N* NaOH : 1 μl 3 mg/ml sheared, calf thymus DNA. The DNA is recovered for PCR amplification by ethanol precipitation, following the addition of 18 μl 10.0 *M* ammonium acetate (pH ~7.8) and 0.2 ml absolute ethanol, vortexing, incubation at −70° for at least 5 min to overnight, and microcentrifugation at 14,000 rpm for 20 min at room temperature. The DNA pellet is washed with 0.3 ml 70% (v/v) ethanol : 30% 0.1× TE, dried slightly, and resuspended in 50 μl sterile 0.1× TE. Store the samples at −20°.

Because the deamination procedure results in denatured DNA, we find that "hot start" PCR[27] greatly improves the amplification reactions by eliminating nonspecific bands and substantially increasing the yield of the desired product. A high-quality PCR product can be purified directly by Wizard PCR Preps, thereby avoiding the cumbersome techniques of gel purification. The lower PCR mixture of each reaction contains 3.25 μl distilled water, 1.25 μl 10× *Taq* buffer [100 m*M* Tris–HCl, pH 8.3, 500 m*M* KCl, 30 m*M* $MgCl_2$, 0.5% (v/v) Nonidet P-40 (NP-40), 0.5% (v/v) Tween 20], 4 μl dNTP mix (2.5 m*M* each dATP, dCTP, dGTP, and TTP), and 2 μl of each primer at a 20 μ*M* concentration. A wax bead (Ampliwax PCR Gem 50, Perkin-Elmer, Norwalk, CT) is added to the lower PCR mixture, melted, and cooled prior to layering 35.5 μl upper PCR mixture (30.25 μl distilled water, 5 μl 10× *Taq* buffer, 0.25 μl 5 U/μl *Taq* DNA polymerase), and 2 μl deaminated DNA template. The amplification protocol is preheating at 94° for 3 min; followed by 30 cycles of 94° for 45 sec, 5° below the calculated T_m (see later) for 45 sec, and 72° for 1 min; and a final extension cycle of 72° for 4 min. Usually, the PCR reaction will yield a single amplification product of the correct size and such products are purified directly by the Wizard PCR Prep protocol. If an amplified product is not homogeneous, particularly if smaller, contaminating products are present, the full-length product must be gel purified. Occasionally, higher

[27] Q. Chou, M. Russell, D. Birch, J. Raymond, and W. Bloch, *Nucleic Acids Res.* **20,** 1717 (1992).

molecular weight amplification products are observed. However, they will not compromise the experiment.

Careful design of primers for PCR amplification from deaminated DNA is essential for obtaining high-quality PCR products and footprinting results. In our experience, in order of most to least importance, in designing primers: (1) choose primers that will amplify fragments <1 kbp in size, (2) distribute base transitions in the primers evenly (i.e., G to a or C to t transitions) to effect selective amplification of deaminated DNA and discrimination against wild-type, untreated DNA (i.e., avoid grouping transitions in one region, especially at the 5′ end of the primer), (3) the T_m should be close to 60°, based on nearest-neighbor thermodynamic quantities calculated according to Breslauer *et al.*,[28] (4) avoid long stretches of A or T (>7 nucleotides), (5) include a C (a1 and b1 primers) or G (a2 and b2 primers) at the 3′ priming position, and (6) avoid target CG sites, particularly at the 3′ end of the primer. Conform to as many of these guidelines as possible. The last parameter can usually be disregarded because the total percentage of methylated sites at any given CG will be in the minority (even if the site is in an accessible region of chromatin) and, consequently, cytosines at that position will be deaminated in a majority of the template molecules. Typical primers will be longer than customary, approximately 30 bp in length, to achieve suitable thermal stabilities for PCR amplification from deaminated DNA.

Purified PCR products are sequenced directly by a modified version of thermal cycle sequencing that is necessitated by the low level of resident cytosines. The modified reactions (final volume of 8–10 μl) contain 50–100 fmol purified PCR product (for a 600-bp PCR product, 100 fmol ~ 40 ng DNA), 1× Sequitherm buffer (50 m*M* Tris–HCl, pH 9.3, 2.5 m*M* $MgCl_2$), 5 μM each dATP, dCTP, and dTTP, 50 μM ddGTP, 1.0–1.2 pmol oligonucleotide primer end labeled with ^{32}P to a specific activity of $1–2 \times 10^8$ cpm/μg, 1.25 units Sequitherm DNA polymerase, and sterile, distilled water to volume. High-purity nucleotides are essential. We use the ultrapure grade deoxynucleotide kit (Pharmacia) and ddGTP (Pharmacia). Because the bulk of the reaction volume will be occupied by the DNA and the radiolabeled primer, nucleotides are added to the sequencing mixture from high-concentration stocks. The dGTP is deliberately omitted, and a high concentration of ddGTP is included to achieve a high efficiency (>96%) of termination at cytosines. Dideoxy-A, -C, and -T ladders are obtained using one of the PCR products as template DNA exactly as described in the Sequitherm cycle sequencing protocol, except that, due to the high content of thymine

[28] K. J. Breslauer, R. Frank, H. Blocker, and L. A. Markey, *Proc. Natl. Acad. Sci. U.S.A.* **83,** 3746 (1986).

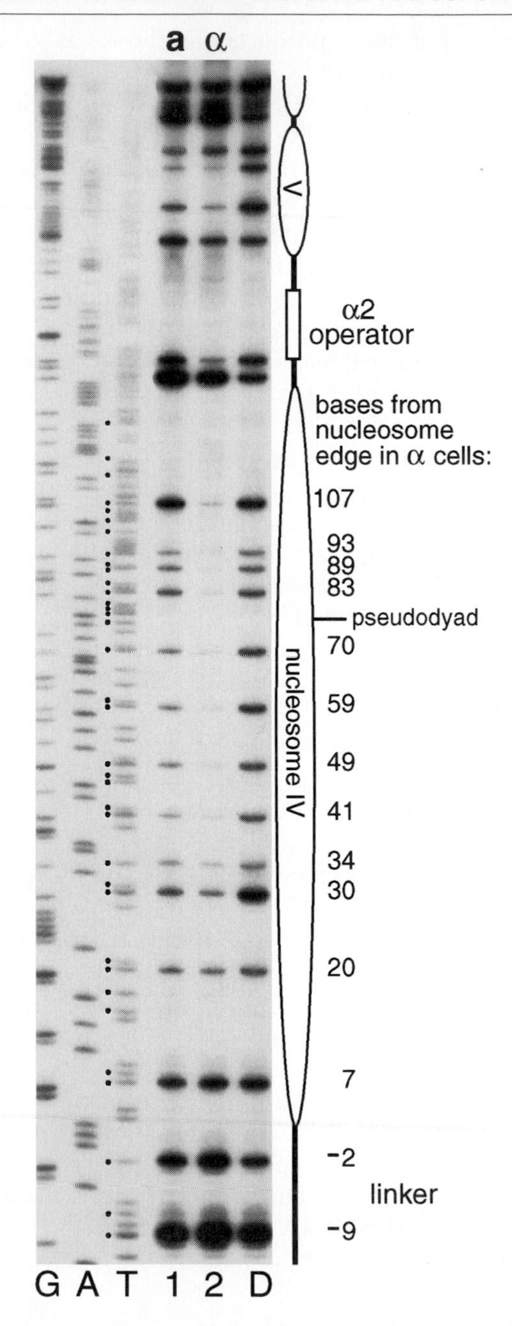
a
α
V
α2
operator
bases from
nucleosome
edge in α cells:
107
93
89
83
pseudodyad
70
59
49
41
34
30
20
7
-2
linker
-9
nucleosome IV
G A T 1 2 D

in the DNA, only half as much "A" termination mix is employed. Reactions are stopped by adding 0.5 volume reaction stop solution [95% (v/v) deionized formamide, 10 m*M* EDTA, pH 7.6, 0.025% each xylene cyanole FF and bromphenol blue]. Samples (3 μl) are then separated on a denaturing 6% polyacrylamide [19 : 1 (w/w) acrylamide : bisacrylamide], 50% (w/v) urea sequencing gel and analyzed by exposure of the dried gel to X-ray film or a phosphorimager screen.

Results

To validate the methodology, we transformed strains of yeast expressing M.*Sss*I with a minichromosome that has a well-characterized chromatin structure. The minichromosome contained an *ARS1* element, conferring multicopy number, extrachromosomal maintenance, a *TRP1* selectable marker, and an α2 operator sequence. In yeast α-cells, nucleosomes flank the α2 operator.[12] When analyzed by high-resolution primer extension following micrococcal nuclease and DNase I digestion of isolated nuclei, these nucleosomes were found to have precise translational and rotational positions.[13] Chromatin flanking the operator in **a** cells lacked this precise organization of nucleosomes.

The accessibility of the DNA region encompassing the α2 operator in the minichromosome to M.*Sss*I in both **a** and α cells is shown in Fig. 2. The bisulfite genomic sequencing protocol identifies all the cytosines in CG sites that are methylated by M.*Sss*I in naked DNA, modified *in vitro* (Fig. 2, lane D). The high efficiency of the deamination reaction is demon-

FIG. 2. Accessibility of the TALS4 minichromosome to M.*Sss*I. The locations of positioned nucleosomes in α cells, inferred from micrococcal nuclease digestions of purified nuclei, are indicated by ellipses flanking the α2 operator. Modification of the minichromosome by M.*Sss*I under control of the *GAL1* promoter is shown for isogenic **a** and α cells grown in galactose-containing medium. In intact α cells (lane 2), CG sites located 30 bp or further from the linker region were highly refractory to methylation relative to protein-free DNA, methylated *in vitro* (lane D), indicating inaccessibility of M.*Sss*I to the central 86–106 bp of a positioned nucleosome. In comparison, the protection against methylation in **a** cell chromatin (lane 1) relative to naked DNA and α cell chromatin (lane 2) is not as pronounced, indicating the presence of nucleosomes that are not positioned as precisely. Also note protection of the CG site within the α2 operator in α cells, which express Matα2p, relative to control DNA and **a** cells, where the repressor is absent. The sequencing ladders were generated using the PCR product amplified from the deaminated α cell DNA as template. In the region of the gel where the extension products are clearly resolved, cytosines from wild-type DNA that were converted to thymines in deaminated DNA are demarcated by filled dots to the left of the "T" ladder. Termination products in the "T" ladder that correspond to products in lanes 1, 2, and D are present because the majority of the cytosines in the PCR product are converted to thymines.

strated by the absence of cytosines located between these known CG sites. Deamination efficiency can also be ascertained by including a control sample of cells grown in glucose-containing medium, which represses expression of M.*Sss*I from its *GAL1* promoter. Lack of methylation will lead to an accumulation of extension products of high molecular weight that have "run off" the end of the PCR product.[25]

The methylation pattern generated by M.*Sss*I in α cell chromatin (Fig. 2, lane 2) as compared to that of **a** cell chromatin (lane 1) and modified, naked DNA (lane D) is indicative of a positioned nucleosome. Note that to increase the density of methylation sites, five additional M.*Sss*I targets were manipulated into the DNA of nucleosome IV. These minor sequence alterations did not affect the location of the nucleosome.[11,25,29] Accessibility is highest in the linker region (sites −2 and −9, bottom of gel), decreases gradually in the region of DNA entering the nucleosome (sites 7 to 30), and is almost eliminated at more internal sites, where DNA is most tightly sequestered by the histones. This continuum of accessibility was also observed when probing with Dam methyltransferase[11,14] and may be an important structural characteristic of nucleosomes that affects DNA replication,[30] transcriptional activation,[29] and *trans*-acting factor binding, in general.[31,32]

Analysis of methylation by M.*Sss*I at the $\alpha 2$ operator demonstrates that methyltransferases are also effective at detecting DNA-bound, nonhistone, regulatory proteins. The Matα2p repressor is expressed in α cells and binds to its operator sequence. This is evidenced by the protection of guanines, one of which is located within a CG site in one half-site of the operator, against methylation by dimethyl sulfate.[33] Similarly, when compared to naked DNA, methylation by M.*Sss*I at the CG site within the $\alpha 2$ operator is blocked substantially by the bound repressor. As a control, in **a** cells, which lack the Matα2p repressor, accessibility to the methyltransferase is not inhibited. Due to the short half-life of the repressor,[34] prior studies that mapped Matα2p binding in isolated nuclei required a rapid DNase I footprinting protocol as well as a yeast strain that overexpressed the repressor and was deficient in its degradation.[33] In contrast, the *in vivo* methyltransferase methodology both locates nucleosomes and identifies factor binding in otherwise wild-type cells.

[29] M. Xu, R. T. Simpson, and M. P. Kladde, *Mol. Cell. Biol.* **18,** 1201 (1998).

[30] R. T. Simpson, *Nature* **343,** 387 (1990).

[31] K. J. Polach and J. Widom, *J. Mol. Biol.* **254,** 130 (1995).

[32] K. J. Polach and J. Widom, *J. Mol. Biol.* **258,** 800 (1996).

[33] M. R. Murphy, M. Shimizu, S. Y. Roth, A. M. Dranginis, and R. T. Simpson, *Nucleic Acids. Res.* **21,** 3295 (1993).

[34] M. Hochstrasser, M. J. Ellison, V. Chau, and A. Varshavsky, *Proc. Natl. Acad. Sci. U.S.A.* **88,** 4606 (1991).

Probing Chromatin Structure with *Cvi*PI Methyltransferase

The frequency of natural CG sites in the yeast genome, about once per 35 bp, should lead to the occurrence, on average, of four possible modification sites in each nucleosome, making it feasible to determine whether sequences are wrapped as positioned nucleosomes. However, the current level of resolution precludes determination of the precise translational location of a positioned nucleosome. In addition, most proteins that bind to DNA with a high degree of specificity have cognate binding sites of about 20–30 bp or less, limiting the use of M.*Sss*I in all situations. To increase the utility of the methyltransferase probing strategy, a major need is to identify, clone, and express additional enzymes with reduced target site size specificity.

A family of double-stranded DNA viruses that infect *Chlorella,* a unicellular green alga, provides a particularly attractive source for additional DNA (cytosine-5) methyltransferases.[35] Among the several dozen virus strains whose DNA has been analyzed, the fraction of 5-methylcytosine ranges from 0.1 to 47% of the total cytosine content. One (cytosine-5) methyltransferase, M.*Cvi*JI, has been cloned from *Chlorella* virus IL-3A, one of the viruses with a lower 5-methylcytosine content. This enzyme is thought to recognize (G/A)GC(T/C/G) sites.[36] Characterization of five viruses with highly methylated genomes by digestion with a panel of methylation-sensitive restriction endonucleases led to the deduction that methyltransferases were present that modified CC and RCY sites.[35] In the yeast genome, which has a G + C content of 38%, on average, a single CC and RCY site would be expected to occur once per 13.9 and 10.7 bp, respectively. The cloning and expression of such enzymes, or others with di- or partially ambiguous trinucleotide specificity, would enhance the methyltransferase mapping technique greatly, increasing resolution to levels comparable to nonspecific nucleases.

Taking advantage of the high degree of conservation among DNA (cytosine-5) methyltransferases at the amino acid level, we have utilized degenerate primer-based PCR to clone a novel enzyme from *Chlorella* virus NYs-1, designated M.*Cvi*PI. The gene for M.*Cvi*PI encodes a polypeptide of 362 amino acids with a predicted molecular mass of 41.9 kDa. The protein exhibits extensive amino acid homology to the conserved domains of other cytosine methyltransferases and more than 70% overall similarity with M.*Cvi*JI, previously cloned from *Chlorella* virus IL-3A.

[35] M. Nelson, Y. Zhang, and J. L. Van Etten, *in* "DNA Methylation: Molecular Biology and Biological Significance" (J. P. Jost and H. P. Saluz, eds.), p. 186. Birkhäuser Verlag, Basel, 1993.

[36] S. L. Shields, D. E. Burbank, R. Grabherr, and J. L. Van Etten, *Virology* **176,** 16 (1990).

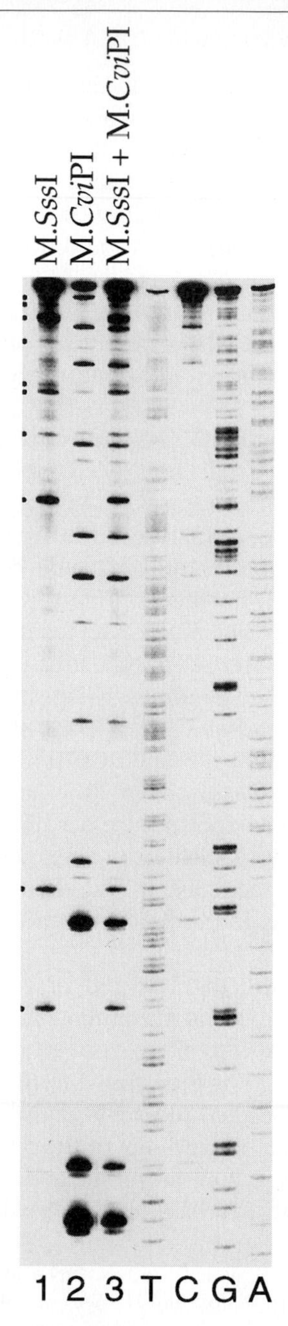

M.*Sss*I
M.*Cvi*PI
M.*Sss*I + M.*Cvi*PI
1 2 3 T C G A

Methods

The gene encoding M.*Cvi*PI, with the SV40 nuclear localization sequence fused to its amino terminus, was cloned into the same integration vector used for the transformation of M.*Sss*I. This gene, under control of the *GAL1* promoter, is integrated as a single copy at the *LYS2* locus. M.*Cvi*PI- or M.*Sss*I-expressing yeast cells were grown in medium containing 2% galactose and DNA was rapidly isolated and processed for the detection of 5-methylcytosine exactly as outlined earlier for M.*Sss*I.

Results

Figure 3 compares *in vivo* methylation by M.*Sss*I or M.*Cvi*PI at the 3′ end of the single copy *STE6* gene on chromosome IX of *S. cerevisiae.* As seen earlier, methylation by M.*Sss*I occurs at CG sites (Fig. 3, lane 1). To determine the specificity of M.*Cvi*PI (lane 2), the methylated residues were aligned with the sequencing ladders, generated from a PCR product amplified from deaminated DNA, to demonstrate that each of the methylated cytosines is contained within a GC site. In other experiments analyzing the *in vivo* methylation pattern in both yeast and *E. coli* that express M.*Cvi*PI, we have determined that the enzyme has a novel dinucleotide specificity, modifying GC sites that are embedded in all of the 16 possible tetranucleotide sequence contexts. With its novel GC dinucleotide specificity, M.*Cvi*PI is complementary to M.*Sss*I and provides a powerful new reagent for probing chromatin structures *in vivo.*

Interpretation of Methyltransferase Probing Results

The degree of methylation at a particular target site is dependent on at least four factors: (1) the inherent preference of the enzyme for the site, (2) the presence of histones in nucleosomes, (3) the precise location of the site in a positioned nucleosome, e.g., periphery of the nucleosome versus

---

FIG. 3. Methylation of the *STE6* gene region by *Sss*I and *Cvi*PI methyltransferases. Yeast with either M.*Sss*I (lane 1) or M.*Cvi*PI (lane 2) under control of the *GAL1* promoter were grown in galactose to induce methylation. Genomic DNA was isolated and then processed to identify 5-methylcytosines. An equal mixture of genomic DNA isolated from the two methyltransferase-expressing strains was mixed prior to deamination for data shown in lane 3. The locus analyzed by the deamination method is on chromosome IX at the 3′ end of *STE6,* a region that does not contain positioned nucleosomes (Y. Tsukagoshi and RTS, unpublished results). CG sites modified by M.*Sss*I are indicated by filled circles. Analysis of the *in vivo* methylation pattern generated by the expression of M.*Cvi*PI indicates that all GC sites were modified.

pseudodyad, and (4) the presence of nonhistone proteins that may bind to, or in the vicinity of, the particular target site. Removal by DNA excision-repair (*RAD1, RAD3,* etc.) exerts only a nominal effect on the steady-state levels of 5-methylcytosine and therefore has little effect on methyltransferase approaches to chromatin structure.[6,8,9]

The combined use of the methyltransferase probes outlined in this article will permit investigations at a resolution; statistically, of about one natural site every 15 bp, making data quite amenable to interpretation in terms of chromatin structure. This is a conservative estimate of resolution because the relatively large sizes of the methyltransferases (43.1 kDa for M.*Sss*I and 41.9 kDa for M.*Cvi*PI) lead to an operational increase in resolution; target sites are protected in the vicinity of DNA-bound proteins as well as when they contribute to a binding site. Thus, on average, five to six sites would be expected to be available in the central, inaccessible 100 bp of a nucleosome. At first blush, protected regions spanning this distance can be equated with nucleosomes. It is naive to purport that inhibition of methylation at a single site is due to a nucleosome, as the presence of a bound nonhistone protein could produce a similar result. It is equally naive to conclude that accessibility of a single site, or a short 20-bp cluster of sites, is indicative of a nucleosome-free region; such a site may be located within the two helical turns of DNA entering/exiting a positioned nucleosome. However, protection of such short regions, or of isolated sites, is likely to reflect site-specific binding of non histone proteins. As in any chromatin mapping experiment, proper evaluation of *in vivo* methyltransferase footprinting data requires comparison of the ratio of methylation in chromatin to that in protein-free DNA, modified *in vitro.* The effect of higher order chromatin structure, matrix attachments, or noncanonical structures of DNA on methyltransferase accessibility remains a matter only for conjecture at this time.

Summary and Prospects

The approach of expressing methyltransferases *in vivo* that have no apparent phenotypic consequences has proven an effective one for complementing conventional techniques that rely on probing chromatin structure with chemical reagents or nonspecific endonucleases. To date, we have not encountered any situation where the strategy failed to detect a nucleosome or bound factor that, from its known biology, was expected to be present. In fact, methyltransferases have been used successfully to detect DNA-bound regulatory factors where other techniques have failed.[25] The high efficiency of the *in vivo* strategy likely results from significant steric clashes of the methyltransferases with DNA-binding factors.

Currently, as outlined in this article, investigations of chromatin structure with DNA methyltransferases can be carried out with a resolution of ~15 bp. If necessary, resolution can be increased through the introduction of additional reporter sites. However, mutagenesis of native sequences is undesirable, especially in the absence of detailed knowledge regarding the locations of *cis*-acting regulatory sequences. Although the methyltransferase footprinting strategy has thus far been limited to yeast, the general approach should be exportable to some other species. In vertebrates, the occurrence of endogenous CG methylation and its resulting effects on gene expression present obvious barriers to the use of the CG methyltransferase M.*Sss*I, and *Cvi*PI, where one methyltransferase-susceptible cytosine in four will occur in the sequence GCG. For this reason, as well as to achieve further increases in resolution, we are continuing our search for additional (cytosine-5) methyltransferases.

Acknowledgments

We extend our gratitude to Bill Jack (New England Biolabs) for supplying the cloned M.*Sss*I gene. We also thank Jim Van Etten at the University of Nebraska for providing the *Chlorella* viral DNA that was used for cloning M.*Cvi*PI. This work was supported by NIH Grants GM52908 and GM56907 to R.T.S.

[25] Mapping Cyclobutane–Pyrimidine Dimers in DNA and Using DNA–Repair by Photolyase for Chromatin Analysis in Yeast

By BERNHARD SUTER, MAGDALENA LIVINGSTONE-ZATCHEJ, and FRITZ THOMA

Introduction

Protocols are described on how to measure ultraviolet light (UV)-induced cyclobutane–pyrimidine dimers (CPDs) and their repair in yeast chromatin and how to use DNA photolyase as a molecular tool to monitor chromatin structures in living yeast cells.

All organisms have developed repair systems to protect their genome against deleterious effects of DNA damage formation. Cyclobutane–pyrimidine dimers and pyrimidine (6-4) pyrimidone photoproducts (6-4PP) are two major classes of stable DNA lesions generated by UV (predominantly 254 nm). Pyrimidine dimers are removed by two pathways: nucleo-

0076-6879/99 $30.00

tide excision repair (NER) and photoreactivation. NER is a ubiquitous multistep pathway in which more than 30 proteins are involved to sequentially execute damage recognition, excision of an oligonuceotide with the pyrimidine dimer, and gap repair synthesis.[1,2] As an additional pathway, many organisms, including bacteria, yeast, plants, invertebrates, and many vertebrates, can revert CPDs by CPD photolyase in the presence of photoreactivating light (of wavelengths 350 to 450 nm), restoring the bases to their native form.[3–5]

In eukaryotic cells, genomic DNA is packaged by histone proteins in an array of nucleosomes connected by linker DNA. This nucleosome filament is further condensed into compact chromatin fibers and higher order structures.[6] The nucleosomal array is interrupted by regions in which the DNA is readily accessible to nucleases [nuclease-sensitive regions (NSRs), also called hypersensitive sites]. NSRs are found at active promoters, regulatory regions, or origins of replication and are thought to be generated by factors that bind to DNA, disrupt nucleosomes, or prevent nucleosome formation.[7] Protein–DNA interactions and packaging of DNA into nucleosomes alter the structure of DNA and restrict its accessibility to proteins and drugs. These properties directly affect DNA damage formation and repair. Moreover, damage formation and repair are modulated by the location of histone octamers (the protein core of nucleosomes) on the DNA sequence (nucleosome positions) and by structural and dynamic properties of nucleosomes.[8,9] To understand DNA repair and mutagenesis, we need to know how DNA lesions are recognized and repaired in chromatin. Conventional approaches to characterize chromatin structures include footprinting of proteins and histone octamers on the DNA sequence by nucleases [micrococcal nuclease (MNase), DNase I] (for a protocol, see Thoma[10]) or chemical cleavage reagents (hydroxyl radicals, methidium propyl-EDTA-iron, copper phenanthroline).[11] The cleavage sites are moni-

[1] E. C. Friedberg, G. C. Walker, and W. Siede, "DNA Repair and Mutagenesis." ASM Press, Washington, DC, 1995.

[2] R. D. Wood, *Annu. Rev. Biochem.* **65,** 135 (1996).

[3] A. Sancar, *Science* **272,** 48 (1996).

[4] A. Sancar and G. B. Sancar, *Annu. Rev. Biochem.* **57,** 29 (1988).

[5] A. Yasui, A. P. Eker, S. Yasuhira, H. Yajima, T. Kobayashi, M. Takao, and A. Oikawa, *EMBO J.* **13,** 6143 (1994).

[6] A. Wolffe, "Chromatin." Academic Press, San Diego, 1995.

[7] L. L. Wallrath, Q. Lu, H. Granok, and S. C. R. Elgin, *Bioessays* **16,** 165 (1994).

[8] S. Tornaletti and G. P. Pfeifer, *Bioessays* **18,** 221 (1996).

[9] M. Smerdon and F. Thoma, *in* "DNA Damage and Repair: Molecular and Cell Biology" (M. F. Hoekstra and J. A. Nickoloff, eds.). Humana Press, Clifton, NJ, 1998.

[10] F. Thoma, *Methods Enzymol.* **274,** 197 (1996).

[11] I. L. Cartwright and S. C. R. Elgin, *Methods Enzymol.* **170,** 359 (1989).

tored using low- and high-resolution protocols.[10] By investigating photoreactivation in yeast, we have discovered that photolyase is tightly modulated by chromatin structure and transcription. Repair was slow in nucleosomes, but fast in linker DNA and in nuclease-sensitive regions, and photolyase was inhibited on the transcribed strand possibly by RNA polymerases stalled at CPDs.[12,13] These observations showed that photolyase can be used as a tool to monitor structural and dynamic properties of chromatin in a living cell. This article describes how to map CPDs in the yeast genome and in minichromosomes and how to do photoreactivation experiments in yeast.

Approach

Figure 1 illustrates the generation of CPDs and their repair by photolyase in yeast cells (Fig. 1A) and the indirect end-labeling protocol for mapping CPDs along the DNA sequence (Fig. 1B). This approach allows a direct comparison with the traditional way of mapping nucleosomes by micrococcal nuclease footprinting (Fig. 1C). (1) DNA damage is induced by the irradiation of yeast cultures with UV light of a germicidal lamp (predominantly 254 nm). This generates CPDs throughout the genome. Polypyrimidine regions, such as T tracts in promoter regions of yeast genes, are "hot spots" of CPD formation and accumulate high yields of CPDs.[12] CPD yields can be modulated by protein–DNA interactions, chromatin structure, and DNA sequence.[8,14–16] (2) Repair of CPDs by photolyase (photoreactivation) is done by limited exposure of the cell suspension to photoreactivating light. Aliquots are taken at different repair times. If photoreactivation is used for chromatin analysis, the NER pathway of yeast should be inactivated to avoid competition between the two repair processes. This can be done by deletion of the *RAD1* gene, which is essential for the incision step of NER.[17] (3) In order to analyze CPD yields and CPD distribution along the DNA sequence, DNA is purified and treated with T4 endonuclease V (T4-endoV) (Fig. 1B). T4-endoV is a CPD-specific enzyme that introduces single-strand cuts at CPDs.[18] The cutting sites on each strand of the DNA are displayed by indirect end-labeling from a

[12] B. Suter, M. Livingstone-Zatchej, and F. Thoma, *EMBO J.* **16,** 2150 (1997).

[13] M. Livingstone-Zatchej, A. Meier, B. Suter, and F. Thoma, *Nucleic Acids Res.* **25,** 3795 (1997).

[14] J. M. Gale, K. A. Nissen, and M. J. Smerdon, *Proc. Natl. Acad Sci. U.S.A.* **84,** 6644 (1987).

[15] J. R. Pehrson, *Proc. Natl. Acad. Sci. U.S.A.* **86,** 9149 (1989).

[16] U. Schieferstein and F. Thoma, *Biochemistry* **35,** 7705 (1996).

[17] A. J. Bardwell, L. Bardwell, A. E. Tomkinson, and E. C. Friedberg, *Science* **265,** 2082 (1994).

[18] L. K. Gordon and W. A. Haseltine, *J. Biol. Chem.* **255,** 12047 (1980).

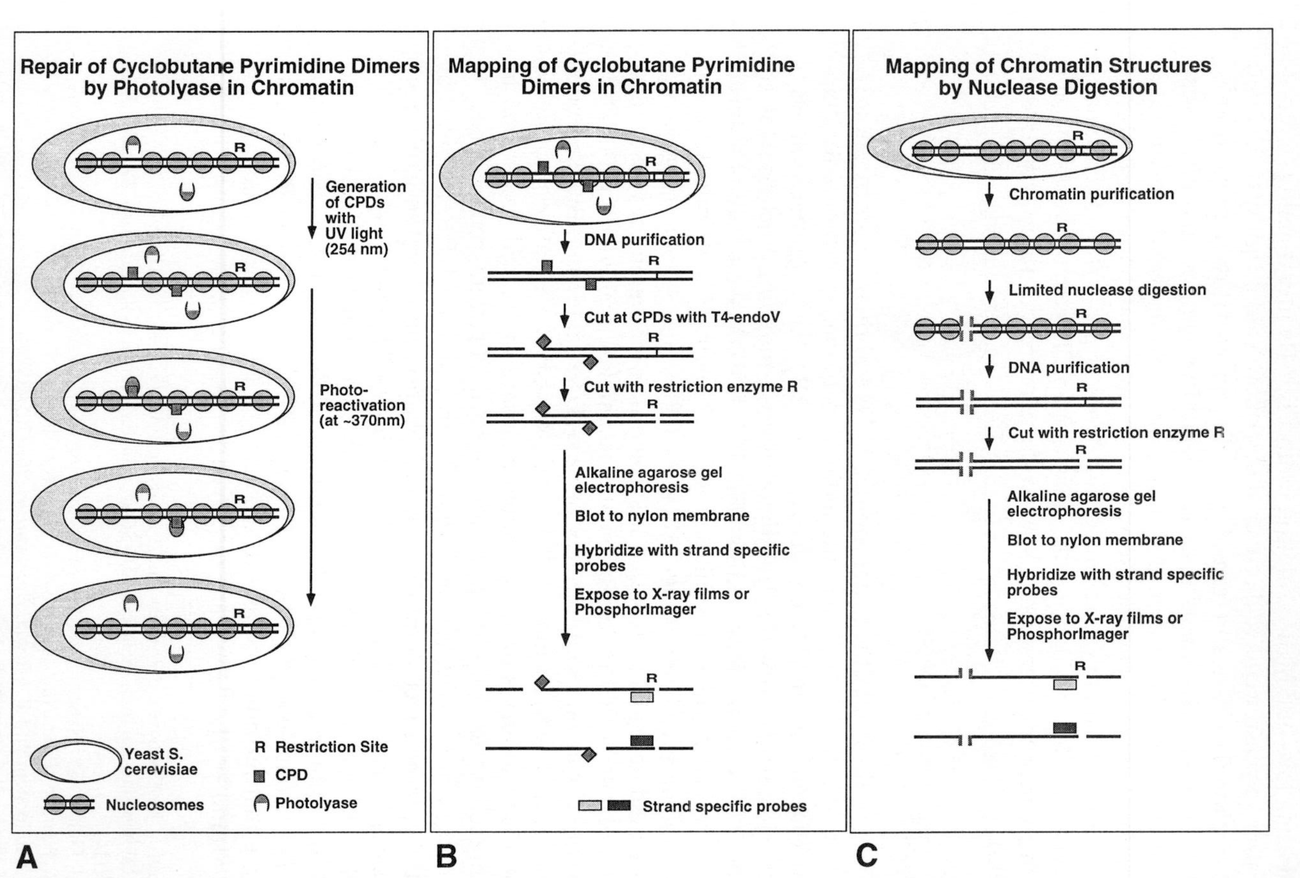

FIG. 1. Comparison of chromatin analysis by MNase footprinting and by CPD repair using photolyase. (A) The generation of CPDs by UV light and site-specific repair by photolyase; (B) mapping of CPDs using T4-endoV and an indirect end-labeling assay; and (C) mapping of nuclease cleavage sites by indirect end-labeling.

restriction site using strand-specific probes.[19] For that purpose, DNA is cut with a restriction enzyme that provides the reference point for mapping (R, Fig. 1). The DNA fragments are fractionated by electrophoresis on denaturing alkaline agarose gels and transferred to a membrane using a Southern blot protocol. Strand-specific radioactive probes that abut the restriction site are used for hybridization. The membranes are exposed to X-ray films or PhosphorImager screens.

The results of such an experiment are shown in Figs. 2B and 2C (lanes 4–10). The top band indicates the fraction of restriction fragments that were not damaged. The other bands represent DNA fragments that have the restriction site at one end and the CPD cut at the other end. The length of the fragments is measured using molecular size markers (lanes 11, Fig. 2) and is used to calculate the site of CPD (distance from the restriction site). The intensity of the bands reflects the CPD yields at that particular site. Decreasing band intensities with increasing photoreactivation time reflect DNA repair (lanes 5–8, Fig. 2).

For chromatin analysis (Fig. 1C), chromatin is partially purified and cut with MNase. MNase introduces double-strand cuts in linker DNA and in NSRs, whereas nucleosomal DNA is protected against MNase digestion (for a detailed protocol, see Thoma[10]). For comparison, deproteinized–free DNA is also digested with MNase to a limited extent. Because T4-endoV and MNase induce DNA cuts, chromatin analysis by the photoreactivation of CPDs *in vivo* can be compared side by side on the same gel with chromatin analysis by MNase *in vitro* (Fig. 2). The protocol described here differs from classical chromatin mapping protocols[10] only by using denaturing alkaline agarose gels and strand-specific probes.

A comparison among chromatin structure, DNA damage formation, and CPD repair by photolyase is illustrated for a yeast strain FTY117 that is deficient in NER (*rad1*Δ) (Fig. 2). FTY117 contains a well-characterized minichromosome (YRpCS1) as a model substrate (Fig. 2A). Indirect end-labeling was done from the *Xba*I site (X) in a counterclockwise direction using strand-specific RNA probes. Chromatin analysis is shown in Figs. 2B and 2C (lanes 1–3). The free DNA lanes (lane 1, Fig. 2) display all cutting sites of MNase. Intensities and positions of bands are similar on both strands (compare lanes 1 in Figs. 2B and 2C), which shows the preference of MNase to generate double-strand breaks. A comparison of the chromatin pattern (lanes 2 and 3, Fig. 2B/C) and DNA pattern (lane 1, Fig. 2B/C) reveals protected regions of about 150 bp, which are consistent with positioned nucleosomes (indicated as white boxes). The positioned nucleosomes are

[19] M. J. Smerdon and F. Thoma, *Cell* **61,** 675 (1990).

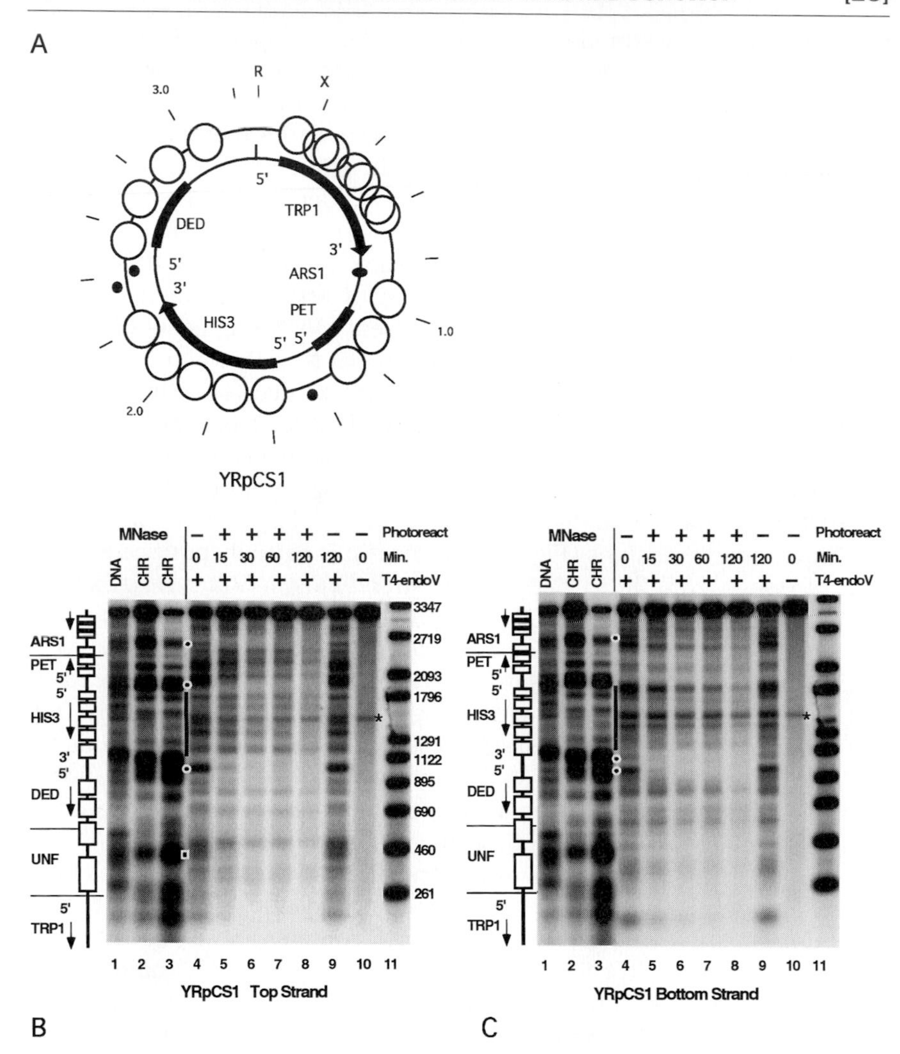

FIG. 2. CPD repair by photolyase correlates with chromatin structure. (A) The minichromosome YRpCS1 of *S. cerevisiae* strain FTY117 contains the *HIS3* gene (arrow) and the truncated *DED1* and *PET56* genes (decapitated arrows) inserted in the TRP1ARS1 circle. Multiple positions of nucleosomes in the *TRP1* gene (overlapping circles) and precisely positioned nucleosomes (white circles) are as described[35]; the promoter regions (5′), the 3′ ends of the genes (3′), the ARS1 origin of replication (ARS1), and the *Eco*RI (R) and *Xba*I (X) restriction sites. Map units in base pairs (bp) are indicated in 0.2-kb steps. Dots refer to polypyrimidine regions, which are hot spots of CPD formation and which are fast repaired by photolyase (outside is top strand; inside is bottom strand). (B and C) Chromatin structure and CPD repair by photolyase in the top strand and bottom strand, respectively. The bottom strand is the transcribed strand of *TRP1* and *HIS3* genes. FTY117 (*rad1*Δ) cells were grown in SD medium (0.67% yeast nitrogen base without amino acids, 2% dextrose, 20 mg/liter uracil) and

separated by linker DNA and nuclease-sensitive regions [5′ and 3′ ends of genes (*DED1, HIS3, PET56*); *ARS1*, origin of replication].

CPD analysis is shown in lanes 4 to 10 (Fig. 2). Cells were irradiated with 100 J/m^2 and photoreactivation was done for up to 120 min. Unirradiated DNA (not shown) and mock-treated DNA showed a background smear due to nicking of DNA during preparation (−T4-endoV; lane 10, Fig. 2). In contrast, T4-endoV-treated DNA revealed numerous bands of different intensities (+T4-endoV; lanes 4 to 9, Fig. 2). These bands can be assigned to dipyrimidines and polypyrimidine tracts in the DNA sequence. Many strong bands correspond to CPDs in T tracts in the promoter regions of the *DED1, HIS3,* and *PET56* genes, demonstrating that these tracts are "hot spots" of CPD formation.

No repair was observed during incubation in the dark for 120 min (dark control, lane 9, Fig. 2). However, more than 90% of CPDs were removed from both strands during exposure to photoreactivating light for 120 min (lane 8, Fig 2). This demonstrates that CPDs were repaired by photolyase. Site-specific inspection reveals two classes of repair: fast repair, when CPDs are removed within 15 to 30 min (dots in Fig. 2), and slow repair, when CPDs remain detectable for up to 60 to 120 min. The comparison of CPD repair with the chromatin analysis shows that fast repair correlates strictly with the accessibility of DNA to MNase (bands in chromatin lanes), and that slow repair corresponds to inaccessibility to MNase (no bands in chromatin lanes). Hence, chromatin structure regulates the accessibility of photolyase

---

UV irradiated with 100 J/m^2 in SD medium without uracil. Uracil was added after irradiation. The chromatin structure was analyzed by micrococcal nuclease digestion (MNase) of DNA (lane 1) and chromatin (CHR, lanes 2 and 3) which were extracted from irradiated cells. Photoreactivation (+ Photoreact) was for 15 to 120 min (lanes 5 to 8). CPD distribution and repair were analyzed by T4-endoV cleavage (+ T4-endoV, lanes 4 to 9). Lane 10 is irradiated DNA (same as lane 4) without T4-endoV cleavage. An aliquot of cells was kept in the dark for 120 min (lane 9). Cleavage sites for MNase and T4-endoV are shown by indirect end-labeling from the *Xba*I site using strand-specific RNA probes. RNA probes were generated from the *Eco*RI–*Xba*I restriction fragment subcloned in a Bluescript vector. A schematic interpretation of chromatin structure is shown (left side). Chromatin regions of 140 to 200 bp that are protected against MNase cleavage represent footprints of positioned nucleosomes (rectangles), cutting sites between nucleosomes represent linker DNA, and long regions with multiple cutting sites represent nuclease-sensitive regions (ARS1; 5′PET-5′HIS3; 3′HIS3-5′DED; 5′TRP1). The 5′ and 3′ ends of the genes, direction of transcription (arrows), are indicated. Dots and squares indicate fast repair in nuclease-sensitive regions and linker DNA, respectively. CPD bands that correspond to the footprints of positioned nucleosomes are repaired slowly. Bars represent the transcribed region of the *HIS3* gene. Asterisks denote cross-hybridization with genomic DNA. Size markers (in bp) are shown in lane 11 and indicate the distance from the *Xba*I site (counterclockwise). Adapted from B. Suter, M. Livingstone-Zatchej, and F. Thoma, *EMBO J.* **16**, 2150 (1997), with permission of Oxford Univ. Press.

to CPDs. The fact that CPDs in nucleosomes are repaired can be taken as evidence for dynamic properties of nucleosomes that make lesions accessible.[12] This result illustrates that photolyase can be used as a tool to monitor chromatin structure and dynamics in living cells.

The indirect end-labeling approach displays and measures whole yeast genes as well as individual regions, such as nuclease-sensitive promoter regions (e.g., *HIS3*, Fig. 2). When CPD repair by photolyase was measured in the transcribed region of the *HIS3* gene (excluding promoter and 3′ end), repair of the transcribed strand was slower than repair of the nontranscribed strand.[12] This strand bias of repair is probably due to the inhibition of photolyase by RNA polymerase II, which is stalled at CPDs on the transcribed strand.[13] Hence, the photorepair approach can also be used to obtain information on the transcriptional state of a chromatin region.

Methods

UV Irradiation and Photoreactivation of Yeast cells in Suspension

Materials. Dark room equipped with gold fluorescent light (Sylvania, type GE; "safety light" to prevent photoreactivation). A UV irradiation box (homemade) containing six germicidal lamps (Sylvania, Type G15 T8), arranged parallel, with peak emission at 254 nm. An illumination stage of a 55 cm × 55 cm area at a distance of 39 cm from the lamps. The average flux used for the experiments is about 0.40 mW/cm^2 (using two lamps). A photoreactivation box (homemade) containing six lamps (Sylvania, Type F15 T8/BLB), arranged parallel with peak emission at 375 nm. An illumination stage (55 cm × 55 cm). The surface of cell suspension is ca. 8.5 cm from the lamps. The illumination stage is connected to a water bath for temperature control. The average flux used in the experiments is about 1.4 mW/cm^2 (using four lamps). A UVX radiometer with a UVX-25 and a UVX-36 photocell to measure 254 and 366 nm, respectively (UVP, San Gabriel, CA).

Procedure

1. Grow three to four 1-liter yeast cultures in the appropriate medium and temperature to an absorbance at 600 nm of about 0.8–1.2 corresponding to approximately 10^7 cells/ml.[20]
2. Harvest cells by centrifugation in a GS3 rotor (Sorvall) at 6000*g* and 4° for 8 min.

[20] F. Sherman, G. R. Fink, and J. B. Hicks, "Laboratory Course Manual for Yeast Genetics." Cold Spring Harbor Laboratory Press, Cold Spring Harbor, NY, 1986.

3. Resuspend cells and pool in minimal medium without amino acids and adjust to about 2.5–3.5 × 10^7 cells/ml at room temperature.
4. Do all the following steps from irradiation to lysis of spheroplasts in a dark room equipped with gold fluorescent light to avoid photoreactivation.
5. Split cell suspension into aliquots of 250 ml and pour into plastic trays (21.9 cm × 31.2 cm) to produce a 4 mm thin layer.
6. Irradiate at room temperature with 100 to 150 J/m^2 of UV light (predominantly 254 nm) generated by germicidal lamps.
7. Immediately after irradiation, pool cells and supplement the medium by adding the appropriate amino acids or uracil.
8. For photoreactivation, split the cell suspension into aliquots. Put a 250-ml aliquot on ice (0′ repair). This aliquot records the initial damage. Incubate another 250-ml aliquot at room temperature for 120 min in the dark. This aliquot records DNA repair by NER (dark repair). Put two aliquots (500 ml) into two plastic trays. Place the trays on the illumination stage connected to a water bath for cooling. Irradiate with photoreactivating light (peak emission at 375 nm) at 1.4 mW/cm^2. The temperature of the cell suspension during photoreactivation is between 23 and 26°. Remove aliquots of 250 ml at 15, 30, 60, and 120 min and chill on ice.
9. Collect cells by centrifugation in a GS3 rotor at 6000*g*, 4° for 8 min.

DNA Purification

Materials, Buffers, and Enzymes. Proteinase K (Boehringer Mannheim, Germany): 10 mg/ml in 50 m*M* Tris–HCl, pH 8.0, stored at −20°. RNase A (Boehringer Mannheim): 10 mg/ml in 10 m*M* Tris, pH 7.5, 15 m*M* NaCl; stored at −20°. Zymolyase 100T (Seikagaku Corporation, Tokyo, Japan): 10 mg/ml suspended in 40 m*M* potassium phosphate, pH 7.5, 1 *M* sorbitol, stored at −20°. Zymolyase reaction buffer: 250 m*M* EDTA, pH 8.0, 1 *M* sorbitol, 20 m*M* 2-mercaptoethanol, 1 m*M* phenylmethanesulfonyl fluoride (PMSF) (mixed before use). PMSF (Merck, Darmstadt, Germany): 0.5 *M* dissolved in DMSO (methyl sulfoxide; Fluka, Ronkonkoma, NY). 2-Mercaptoethanol (Fluka); 20% SDS; 1 *M* sorbitol, 10 m*M* EDTA; 50T20E (50 m*M* Tris–HCl, 20 m*M* EDTA, pH 8.0); 10T1E (10 m*M* Tris–HCl, 1 m*M* EDTA, pH 8.0); 5 *M* potassium acetate (pH 5.5); P (phenol saturated with 0.1 *M* Tris, pH 8.0); D/I [mixture of dichloromethane/isoamyl alcohol (24 : 1)]; P/D/I [mixture of phenol/dichloromethane/isoamyl alcohol (25 : 24 : 1)]; 3 *M* sodium acetate, pH 4.8; isopropanol; ethanol (99.9%); Elutip-d columns (Schleicher & Schuell, Keene, NH); low salt solution: 0.2 *M* NaCl, 20 m*M* Tris–HCl, pH 7.4, 1.0 m*M* EDTA; high salt solution:

1.0 *M* NaCl, 20 m*M* Tris–HCl, pH 7.4, 1.0 m*M* EDTA; SS34, HB4 rotors (Sorvall; DuPont, Wilmington, DE).

Procedure. This procedure should be done in rooms equipped with a safety light to avoid photoreactivation.

1. Resuspend pellets of UV-irradiated and photoreactivated cells in 15 ml of zymolyase reaction buffer and transfer the solutions to 50-ml Falcon tubes.
2. Add 180–300 μl zymolyase 100T (10 mg/ml) and convert the cells to spheroplasts by incubation at 30°.
3. To test spheroplasting, mix 50-μl aliquots with 950 μl 1% SDS. Spheroplasting is complete when A_{600} drops from the initial ~0.8–1.0 to below 0.1. Spheroplasting takes 10 to 45 min.
4. Collect spheroplasts by centrifugation in a table-top centrifuge at 3900*g* and 4° for 2 min.
5. Resuspend pellets in 12 ml 1 *M* sorbitol, 10 m*M* EDTA, centrifuge as described earlier, and resuspend in 7.5 ml 50T20E.
6. Add 375 μl 20% SDS and 75 μl proteinase K (10 mg/ml). Incubate at 65° for 2 hr and at room temperature overnight.
7. Add 3 ml of 5 *M* potassium acetate, mix well, and incubate on ice for 1 hr.
8. Transfer the suspension into 14-ml Falcon tubes and centrifuge in a SS34 rotor (Sorvall) at 12,000*g* and 4° for 30 min.
9. Collect the clear supernatant containing DNA in 50-ml Falcon tubes.
10. Purify DNA by sequential extraction once with P, once with P/D/I, and once with D/I.
11. Collect the aqueous phase and precipitate DNA with 0.7 volumes of 2-propanol at room temperature for about 1 hr followed by centrifugation in an HB4 rotor (Sorvall) at 16,000*g* and 4° for 30 min.
12. Dry the pellets *in vacuo* and dissolve DNA in 1.9 ml 10T1E, pH 8.0.
13. Digest RNA with 19 μl RNase A (10 mg/ml) at 37° overnight.
14. Purify DNA by sequential extractions with P, P/D/I, and D/I.
15. Transfer the aqueous phase to Falcon tubes and precipitate the DNA with 0.1 volume sodium acetate (3 *M*) and 3 volumes of cold ethanol. Incubate at −20° for 1 hr and centrifuge in an HB4 rotor (Sorvall) at 16,000*g* and 4° for 30 min.
16. Dry the pellets *in vacuo* and dissolve the DNA in 300 μl–1 ml low salt solution at 37° for 20–30 min.
17. Bind the DNA on Elutip-d (Schleicher & Schuell) and elute with high salt solution following the instructions of the supplier.
18. Precipitate DNA and dissolve it in 300 μl 10T1E, pH 8.0.

19. Check DNA content by agarose gel electrophoresis and ethidium bromide staining. The yield is roughly 8 μg per 10^{10} cells.

Mapping of CPDs by Indirect End-Labeling

Materials, Buffers, and Enzymes. T4 endonuclease V (Epicentre Technologies, Madison, WI). *In vitro* transcription kit (Stratagene, La Jolla, CA); appropriate DNA fragments subcloned in a Bluescript vector (Stratagene). Restriction enzymes and appropriate 10× buffers. P/D/I [mixture of phenol/dichloromethane/isoamyl alcohol (25:24:1)]; D/I [mixture of dichloromethane/isoamyl alcohol (24:1)]; ethanol; 10T1E (10 m*M* Tris–HCl, 1 m*M* EDTA, pH 8.0). 10× T4RB [T4-endoV reaction buffer: 200 m*M* Tris–HCl (pH 7.4), 100 m*M* EDTA (pH 8.0), 1 *M* NaCl, 1/20 volume of bovine serum albumin (BSA) (20 mg BSA/ml in 50 m*M* Tris–HCl, 100 m*M* NaCl, 0.25 m*M* EDTA, 1 m*M* 2-mercaptoethanol, 50% glycerol, pH 7.5; Boehringer Mannheim)]; agarose (GIBCO-BRL, Gaithersburg, MD, ultrapure); 5× ALB [alkaline loading buffer: 12.5% Ficoll 400, 5 m*M* EDTA, 0.125% bromocresol green, 250 m*M* NaOH (added just before use)[21]]; alkaline electrophoresis buffer (50 m*M* NaOH, 1 m*M* EDTA); Zeta-Probe GT (genomic tested blotting membrane; Bio-Rad Laboratories, Richmond, CA); 3MM Whatman (Clifton, NJ) paper; 0.4 *N* NaOH; 20× SSC (3 *M* NaCl, 0.3 *M* sodium citrate, pH 7.0); 0.5 *M* sodium phosphate, pH 6.5; 100× Denhardt's solution [1 g Ficoll, 1 g polyvinylpyrrolidone K30, 1 g BSA in 50 ml H_2O]; 20% SDS. Herring sperm DNA (4 mg/ml sonicated, stored at −20°). Prehybridization solution (4× SSC, 50 m*M* $NaPO_4$, pH 6.5, 10× Denhardt's solution, 1% SDS, 0.5 mg/ml sonicated herring sperm DNA, boiled and snap cooled on ice just before use); size markers prepared by a combination of restriction digests of genomic or plasmid DNA that hybridize to the probe. Alternatively, commercial size markers might be used, but they need to be hybridized separately. Gel apparatus (for large agarose gels, 20 × 25 cm, Horizon, BRL).

Procedure

1. Digest DNA to completion with the appropriate restriction enzyme. Purify DNA once with P/D/I and once with D/I.
2. Precipitate the DNA by the addition of three volumes of ethanol at −20° for 1 hr followed by centrifugation in an HB4 rotor (Sorvall) at 16,000*g* and 4° for 20 min.
3. Dry the pellet *in vacuo* and redissolve it in 10T1E, pH 8.0, to a DNA concentration of about 50–100 ng/μl (plasmid-containing DNA) and 150–300 ng/μl (genomic DNA).

4. Digest with T4-endoV in a final reaction volume of 40 μl adjusted to 1× T4RB at 37° for 2.5 hr. For mock-treated samples, replace T4-endoV by 1× T4RB. For T4-endoV digestion, use about 100 to 150 ng of restricted plasmid-containing DNA or about 2.5 μg of restricted genomic DNA.
5. Add 10 μl 5× ALB to stop the reaction and to denature the DNA.
6. Prepare two horizontal agarose gels (A and B, for hybridization with the top and bottom strand probes, respectively) (20 × 25 cm, BRL; 300 ml 1.5% agarose in 50 m*M* NaCl, 1 m*M* EDTA).[21] Soak the gels in 1× alkaline electrophoresis buffer (for up to 3 hr). Prerun the gels at 50–60 V (constant voltage) and 180–270 mA for about 3 hr at 4° using a peristaltic pump and magnetic stirrers at both electrodes to circulate the buffer.
7. Load 24 μl DNA samples per slot and allow the DNA to enter the gel at 60–65 V and 210–340 mA for 10–20 min (without buffer circulation). Cover the gel with a glass plate (to stop the gel from floating) and continue electrophoresis with buffer circulation at 50–55 V (constant voltage) and 270–110 mA for 15–17 hr at 4° until the dye (bromocresol green) migrates about 14–16 cm.
8. For Southern transfer, soak the gel briefly in 0.4 *N* NaOH. Lay two sheets of 3MM Whatman paper presoaked in 0.4 *N* NaOH over a plastic plate bridging a tray with 0.4 *N* NaOH. The paper remains in contact with 0.4 *N* NaOH on both sides. Carefully invert the gel and place it on the paper. Cover it with a Zeta-Probe GT membrane (presoaked first in H_2O and then in 0.4 *N* NaOH). Cover the membrane with four 3MM Whatman papers presoaked in 0.4 *N* NaOH, two layers of dry 3MM Whatman paper, 10 cm of paper towels, a glass plate, and a weight of about 250 g. Transfer DNA to the membrane for about 5 to 20 hr. Replace wet towels after a few hours with fresh ones.
9. After transfer, neutralize the membranes by rinsing them twice for about 15 min in 2× SSC. Make sure that the pH of the 2× SSC solution falls to seven; if necessary, rinse the membranes again. Bake the membranes at 80° for 1 to 2 hr.
10. Place the membranes in rotary cylinders with the DNA-bound side facing inside and incubate them in 20 ml prehybridization solution per 20 cm × 24 cm membrane at 65–67° for 4–16 hr in a hybridization oven.

[21] T. Maniatis, E. Fritsch, and J. Sambrook, "Molecular Cloning." Cold Spring Harbor Laboratory Press, Cold Spring Harbor, NY, 1982.

11. Prepare strand-specific RNA probes using an *in vitro* transcription kit and the appropriate DNA templates subcloned in Bluescript vectors (Stratagene).
12. Replace prehybridization solution with the same volume of freshly boiled and chilled prehybridization solution. Add the strand-specific RNA probe (approximately 3.3–7.5 $\times$ 10^7 cpm in 100–400 μl). Hybridize at 65–67° overnight.
13. Pour off the hybridization solution and wash the membranes six times for 20 min each in 0.6× SSC, 0.1% SDS at 65–67°.
14. Dry the membranes briefly between two sheets of 3MM Whatman paper and expose them to X-ray films (Fuji, Tokyo, Japan) using enhancer screens or to PhosphorImager storage plates (Molecular Dynamics, Sunnyvale, CA).
15. Measure bands and calculate the distance from the restriction sites.
16. Quantify overall damage (CPDs/restriction fragment) using a PhosphorImager (Molecular Dynamics). At each repair time, measure the signal in the intact restriction fragment (IRF) and divide by the signal of the whole lane (T4 +, Fig. 2) to obtain a signal normalized with respect to the overall DNA in that lane [$IRF_{(T4+)}$]. Proceed the same way to obtain the normalized signal of the mock-treated DNA (T4 lanes) [$IRF_{(T4-)}$]. Calculate the CPD content in the restriction fragment using the Poisson expression [$-\ln(IRF_{(T4+)}/IRF_{(T4-)})$.[22]
17. Quantification of repair in a defined region. Define the region of interest by indirect end-labeling. Measure the CPD content in that region at each repair time as the signal in that region divided by the signal of the whole lane (+T4). Determine and subtract the background from the −T4 lanes. To generate repair curves, normalize the values with respect to the initial damage (0 min repair).

Comments

1. UV dose. The cell density in the liquid culture and the absorption properties of the medium need to be considered for DNA damage induction as well as for photoreactivation. The UV dose should be optimized by quantification of the CPD content in DNA. Good conditions for photoreactivation experiments are about 0.5 CPDs per 1 kb (or 100 to 200 J/m^2 under conditions described earlier[12]). Using gold fluorescent light as a "safety light" is highly recom-

[22] I. Mellon, G. Spivak, and P. C. Hanawalt, *Cell* **51,** 241 (1987).

mended, as photoreactivation is a fast process and could be promoted by traces of daylight.

2. Purification of DNA. Spheroplasting may depend on the strain and growth conditions. In the case of NER-proficient strains, spheroplasting needs to be fast in order to reduce the risk of CPD removal by NER. This risk of nucleotide excision repair is eliminated by using *rad1*Δ mutants. Photoreactivation is avoided by working in safety light. For purification of genomic DNA as well as for purification of DNA from plasmid-containing cells, we use conventional purification by organic extractions (phenol) and protocols and materials from Qiagen (Chatsworth, CA). Both procedures yield DNA that is suitable for T4-endoV cleavage and indirect end-labeling. The Qiagen procedure is faster and gives higher yields.
3. Mapping of CPDs by indirect end labeling. For optimal display of the region of interest, a restriction site should be selected that places the region of interest at a distance of 0.3 to 2.5 kb. To generate strand-specific probes it is recommended to use short DNA fragments (about 150 to 200 bp) subcloned in a Bluescript vector (Stratagene). Alternatively, DNA sequences can be amplified directly from the yeast genome using strand-specific primers and the polymerase chain reaction. Radioactive probes can be generated using small DNA fragments as substrates and primer extension by *Taq* polymerase with radioactive dNTPs. Several rounds of primer extension can be used to increase the amount of radioactive DNA. It is important to check the strand specificity of indirect end-label probes. Because CPD distribution is strand specific, each strand will produce a characteristic CPD pattern (compare Figs 2B and 2C).[12]
4. Comparison of chromatin structures and CPD repair by indirect end-labeling. A protocol to footprint chromatin structures in yeast genomes and minichromosomes by indirect end-labeling has been described previously.[10] This material can be used to display chromatin structures side by side with CPD mapping data and, hence, to correlate repair and damage formation with chromatin structures.[12] It is important to realize that chromatin analysis by nucleases represents an averaged structure of the whole population of cells. However, the fraction of molecules with a CPD at a specific site is very small (below 1%). Therefore, it is impossible to analyze by nuclease digestions whether a CPD at a particular site results in an altered chromatin structure. However, the tight correlation of MNase accessibility and photoreactivation strongly suggests that CPD induction does not grossly alter chromatin structure *in vivo*. This is also consistent with

the observation that reconstituted nucleosomes irradiated *in vitro* can tolerate distortions imposed by DNA damage.[16,23]

5. Alternative protocols. The protocol described here allows analysis of CPD distribution and repair in a range from a few base pairs to several kilobases. In order to investigate the effect of protein binding on CPD formation and repair or the effect of repair within nucleosomes, high-resolution protocols have been developed. For yeast, linear primer extension protocols[24] have been used to footprint chromatin structures.[10,25,26] Because CPDs and 6-4 PP block *Taq* polymerase,[27] the same primer extension protocol can be used for quantitative repair analysis of both classes of pyrimidine dimers.[27–29] Protocols for mapping specifically CPDs at high resolution have been published.[30,31] For higher eukaryotes with large genomes, the indirect end-labeling protocol has not yet been applied to CPD repair, but DNA lesions and repair were detected at nucleotide resolution using linker-mediated polymerase chain reactions.[32–34]

Acknowledgments

We thank Dr. M. Smerdon for discussions and initiation of DNA repair projects in our group. We thank Dr. A. Aboussekhra and A. Meier for corrections of the manuscript and Dr. U. Suter for continuous support. This work was supported by grants from the Swiss National Science Foundation and by the ETH-Zürich (to FT).

[23] U. Schieferstein and F. Thoma, *EMBO J.* **17,** 306 (1998).
[24] J. D. Axelrod and J. Majors, *Nucleic Acids Res.* **17,** 171 (1989).
[25] M. C. Marsolier, S. Tanaka, M. Livingstone-Zatchej, M. Grunstein F. Thoma, and A. Sentenac, *Genes Dev.* **9,** 4, 10 (1995).
[26] S. Tanaka, M. Livingstone-Zatchej, and F. Thoma, *J. Mol. Biol.* **257,** 919 (1996).
[27] R. E. Wellinger, Ph.D. Thesis, ETH-Zürich, No. 11931, 1996.
[28] R. E. Wellinger and F. Thoma, *EMBO J.* **16,** 5046 (1997).
[29] A. Aboussekhra and F. Thoma, *Genes Dev.* **12,** 411 (1998).
[30] M. Tijsterman, J. G. Tasseron-de Jong, P. van de Putte, and J. Brouwer, *Nucleic Acids Res.* **24,** 3499 (1996).
[31] Y. Teng, S. Li, R. Waters, and S. H. Reed, *J. Mol. Biol.* **267,** 324 (1997).
[32] G. P. Pfeifer, R. Drouin, A. D. Riggs, and G. P. Holmquist, *Proc. Natl. Acad. Sci. U.S.A.* **88,** 1374 (1991).
[33] G. P. Pfeifer, R. Drouin, A. D. Riggs, and G. P. Holmquist, *Mol. Cell. Biol.* **12,** 1798 (1992).
[34] S. Tornaletti and G. P. Pfeifer, *J. Mol. Biol.* **249,** 714 (1995).
[35] R. Losa, S. Omari, and F. Thoma, *Nucleic Acids Res.* **18,** 3495 (1990).

[26] Analysis of *Drosophila* Chromatin Structure *in Vivo*

By IAIN L. CARTWRIGHT, DIANE E. CRYDERMAN, DAVID S. GILMOUR, LORI A. PILE, LORI L. WALLRATH, JANET A. WEBER, and SARAH C. R. ELGIN

Introduction

Gene activation *in vivo* is a complex process. Previous results have led to an increased appreciation for the functional role of the chromatin template, particularly as components of the protein machinery that modifies chromatin structure to facilitate activation have been identified. Similarly, new results pointing to significant organization and compartmentalization of the nucleus have led to a heightened awareness of the role of higher order structure. While much useful work has been done in *in vitro* systems, there is a constant need for "reality checks," experiments that test ideas against observations based on chromatin assembly and organization in the cell. Because of the ability to utilize a combination of biochemical, genetic, and cytological approaches, work with *Drosophila melanogaster* is particularly advantageous in this regard. This article discusses in detail approaches for mapping chromatin structure, both at the level of nucleosome arrays and at the level of base pair resolution, using cells or nuclei isolated from *Drosophila* at different stages of the life cycle.

A necessary step for the initial work on chromatin structure was to develop culture methods that provided grams (if needed, kilograms) of starting materials. D. Hogness and J. Peacock developed a well-organized system for maintaining a population of 400,000 flies (in eight cages), capable of producing 100 g of embryos per day (described in Shaffer *et al.*[1]). Embryos of 6–18 hr age proved most advantageous, having a high yield of nuclei per embryo, while avoiding contamination with ingested yeast typical of larval or adult stages. Embryos can be used immediately or frozen and stored for several months at −80°. Nuclear preparations from embryos were used in pioneering studies to examine the nucleosome array of a given gene (generated using micrococcal nuclease) and to establish the presence and map the position of DNase I hypersensitive sites.[2,3] The information obtained from chromatin mapping experiments can be increased substantially by the utilization of chemical cleavage reagents, e.g., methidiumpro-

[1] C. D. Shaffer, J. M. Wuller, and S. C. R. Elgin, *Methods Cell. Biol.* **44,** 99 (1994).
[2] C. Wu, P. M. Bingham, K. J. Livak, R. Holmgren, and S. C. R. Elgin, *Cell* **16,** 797 (1979).
[3] C. Wu, *Nature (London)* **286,** 854 (1980).

0076-6879/99 $30.00

pyl-EDTA · Fe(II), which show significantly less sequence bias in cleaving DNA than most enzymes.[4,5] Basic protocols for mapping the chromatin structure of specific genes in *Drosophila* using these reagents with the indirect end-labeling technique are presented in the first part of this article.

Having determined characteristics of the nucleosome array, one often wishes to explore further by analyzing protein–DNA interactions and/or by mapping alterations in DNA structure at base pair resolution. This can be done using the indirect end-labeling approach, blotting, and probing a sequencing gel; excellent results can be obtained, although the technique requires some dexterity.[6,7] Primer extension can also be used to map protein–DNA interactions revealed by DNase I digestion at the sequence level. The advent of ligation-mediated polymerase chain reaction (LM-PCR) techniques to amplify a set of fragments with one end unspecified has provided an alternative, although care is required to obtain a reliable quantitative analysis.[8] Use of LM-PCR has allowed us to work with smaller amounts of starting material, facilitating the characterization of chromatin structure in specific tissues of *Drosophila*.[9] An example using LM-PCR to assay the $KMnO_4$ sensitivity of a gene in salivary glands is given in the second part of this article.

An *in vivo* analysis of chromatin structure and its functional role in regulating the expression of a given gene requires the exploitation of genetic tools, whereby the effects of sequence alterations within the putative regulatory region, and of mutations in the proposed *trans*-acting regulatory proteins, can be examined. The extensive genetic information available[10] and the ability to return an altered gene to the genome by P-element transformation[11] make *Drosophila* an excellent system for these types of functional studies. A large number of loci that encode chromosomal proteins (regulatory factors, enzymes, and structural proteins) have been identified, mutants characterized, and (in many cases) the genes cloned; screens for modifiers

[4] I. L. Cartwright, R. P. Hertzberg, P. B. Dervan, and S. C. R. Elgin, *Proc. Natl. Acad. Sci. U.S.A.* **79,** 5470 (1983).

[5] I. L. Cartwright and S. C. R. Elgin, *Mol. Cell. Biol.* **6,** 779 (1986).

[6] G. H. Thomas and S. C. R. Elgin, *EMBO J.* **7,** 2191 (1988).

[7] M. W. Hull, G. Thomas, J. M. Huibregtse, and D. R. Engelke, *Methods Cell Biol.* **35,** 383 (1991).

[8] P. R. Mueller, P. A. Garrity, and B. Wold, *in* "Current Protocols in Molecular Biology" (F. M. Ausubel, R. Brent, R. E. Kingston, D. D. Moore, J. G. Seidman, J. A. Smith, and K. Struhl, eds.), Vol. 2, p. 15.5.1. Wiley, New York, 1992.

[9] J. A. Weber, D. J. Taxman, Q. Lu, and D. S. Gilmour, *Mol. Cell. Biol.* **17,** 3799 (1997).

[10] D. L. Lindsley and G. G. Zimm, "The Genome of *Drosophila melanogaster*." Academic Press, New York, 1992.

[11] G. M. Rubin and A. C. Spradling, *Science* **218,** 348 (1982).

of position effect variegation,[12,13] or for modifiers of a weakened promoter,[14] have been particularly useful in this regard. When dissecting the regulatory region of a gene, it can be useful to create deletions, point mutations, or substitutions *in vitro;* the modified gene can then be returned to the genome by P-element-mediated germ line transformation for chromatin structure analysis. By mobilizing a chromosomally situated P-element containing the gene under study to new locations, or by creating rearrangements, it is possible to examine the responses of the gene to different chromosomal environments. Such approaches are likely to require a comparative analysis of a large number of different genetic stocks. One analysis that has proven useful is a quantitative assessment of access to key regulatory sites, established by digesting the chromatin with an excess of restriction enzyme.[15,16] This assay, described in the third section, can be performed using relatively small numbers of animals taken at the embryonic, larval, or adult stages.

Finally, it should be mentioned that efficient *in vitro* transcription[17] and chromatin assembly[18,19] systems can be prepared from *Drosophila* embryonic extracts; the definition of critical components within these has become increasingly precise. Thus hypotheses generated by genetic analysis can also be tested in a well-defined biochemical system. These various tools have made *Drosophila* an extremely useful organism for studies designed to describe and analyze the mechanisms by which chromatin structure controls gene regulation.

1. Identifying Major Embryonic Chromatin Structural Features at Specific Loci: Perspective

Nuclear Substrate

Because most probes of chromatin structure are cell membrane impermeant, it is usual to isolate nuclei in the first step of analysis. Procedures for the preparation of nuclei from *Drosophila* embryos are fairly standard, and although there are some variations in the buffers that can be used, the results obtained are not affected by such differences for the most part. In the embryonic disruption step, a standard "physiological" ionic strength

[12] T Grigliatti, *Methods Cell. Biol.* **35,** 588 (1991).
[13] L. L. Wallrath, *Curr. Opin. Genet. Dev.* **8,** 147 (1998).
[14] F. Sauer, D. A. Wassarman, G. M. Rubin, and R. Tjian, *Cell* **87,** 1271 (1996).
[15] Q. Lu, L. L. Wallrath, H. Granok, and S. C. R. Elgin, *Mol. Cell. Biol.* **13,** 2802 (1993).
[16] L. L. Wallrath and S. C. R. Elgin, *Genes Dev.* **9,** 1263 (1995).
[17] R. T. Kamakaka and J. T. Kadonaga, *Methods Cell Biol.* **44,** 225 (1994).
[18] P. B. Becker, T. Tsukiyama, and C. Wu, *Methods Cell Biol.* **44,** 208 (1994).
[19] R. T. Kamakaka, M. Bulger, and J. T. Kadonaga, *Genes Dev.* **7,** 1779 (1993).

is used because chromatin compaction (and hence probe accessibility) is strongly influenced by salt conditions. Because release of nuclei from cells might lead to deregulation of endogenous metal-dependent nucleases, it is usual for the isolation buffer to contain divalent metal-chelating agents such as EDTA and EGTA. However, because the native higher order structure of chromatin is sensitive to the presence of divalent cations, it is common to add compensating cations in the form of organic polyamines (such as spermine and/or spermidine) in order to preserve the chromatin structural integrity. As it turns out, embryonic nuclei from *Drosophila* seem not to have a very high content of nucleases, and we have successfully isolated nuclei for chromatin analysis in Mg^{2+}-containing buffers that are devoid of EDTA, spermine, and spermidine.[6]

Cleavage Reagents

Structural information is derived by the use of approaches that allow the visualization of differential accessibility of the DNA in chromatin compared to that in its naked, unadorned form. Nucleases such as deoxyribonuclease I (DNase I) and micrococcal nuclease (MNase) are utilized extensively; more recent approaches have included the use of chemical reagents such as the ferrous complex of methidiumpropyl-EDTA [MPE · Fe(II)] and potassium permanganate ($KMnO_4$).

DNase I is the classic reagent for detecting structural alterations in the nucleosomal array of chromatin. DNase I hypersensitive (DH) sites, representing major local perturbations of the order of 50–200 bp, reflect either a complete disruption of local histone–DNA contacts (i.e., nucleosome loss) or a substantial perturbation or rearrangement thereof. DH sites frequently demarcate regions in which high-affinity-specific DNA–protein interactions are occurring, e.g., transcription factors binding to their cognate recognition sequences. MNase has been used most frequently in studies aimed at mapping the precise translational location (usually known as the "position"), if any, of nucleosomes on a particular DNA sequence of interest. However, MNase has distinct sequence preferences in A/T-rich sequences,[20,21] and the periodicity with which such sequence-preferential cleavages can occur in eukaryotic DNA has, in some instances, shown a remarkable similarity to the typical nucleosome repeat length.[22] Situations have arisen in which the MNase-derived chromatin and naked DNA patterns were either quite similar or virtually identical, making interpretation difficult. Under such circumstances, an alternative chemical approach using

[20] C. Dingwall, G. P. Lomonosoff, and R. A. Laskey, *Nucleic Acids Res.* **9,** 2659 (1981).
[21] W. Hörz and W. Altenburger, *Nucleic Acids Res.* **9,** 2643 (1981).
[22] M. A. Keene and S. C. R. Elgin, *Cell* **27,** 57 (1981).

the chemical nuclease MPE · Fe(II) can be of great value in helping unravel specific details of the chromatin structure.

MPE consists of the DNA intercalator methidium joined by a three carbon linker arm to the divalent metal chelator EDTA. When complexed with certain reduced transition metal ions [in particular the Fe(II) ion], the compound generates highly reactive hydroxyl radicals that are localized, because of DNA intercalation, to the immediate neighborhood of the double-stranded helical backbone.[23,24] In this regard the chemistry is identical to that of the DNA structural probe, iron-EDTA, used by Tullius and coworkers.[25] Hydroxyl radical abstraction of sugar protons leads, via successive stages, to the cleavage of phosphodiester linkages through an eliminative reaction, thereby producing nicks in the DNA backbone and ultimately double-stranded cleavage. Studies with psoralens had shown a highly preferential localization of intercalators in the internucleosomal linkers of chromatin.[26,27] As anticipated, therefore, when added to eukaryotic chromatin, MPE · Fe(II) efficiently liberated oligonucleosomal arrays.[4] Beyond this, however, MPE · Fe(II) is a particularly useful reagent because, being of such small size, it can disclose information about all accessible regions of DNA in chromatin that can accommodate an intercalative event. Thus DH sites, among others, are strongly cleaved. Given its extremely low level of sequence-specific cleavage, it is apparent that MPE · Fe(II) is important for nucleosome mapping in situations where MNase data are compromised by its DNA sequence preference. MPE · Fe(II) is now available commercially at a reasonable cost and it is not difficult to use; it should be given strong consideration for use in any mapping study. There are numerous instances where obtaining high-quality information on nucleosome positioning and other structural features has been dependent on the unique capabilities of this reagent.[5,28–33]

Mapping Features of Chromatin Structure

The typical method used to map chromatin-specific cleavages has been the indirect end-labeling approach.[3,34] The general strategy (Fig. 1) is to

[23] R. P. Hertzberg and P. B. Dervan, *J. Am. Chem. Soc.* **104,** 313 (1982).
[24] R. P. Hertzberg and P. B. Dervan, *Biochemistry* **23,** 3934 (1984).
[25] M. A. Price and T. D. Tullius, *Methods Enzymol.* **212,** 194 (1992).
[26] C. V. Hanson, C.-K. J. Shen, and J. E. Hearst, *Science* **193,** 62 (1976).
[27] T. Cech and M. L. Pardue, *Cell* **11,** 631 (1977).
[28] I. L. Cartwright and S. C. R. Elgin, *EMBO J.* **3,** 3101 (1984).
[29] H. Richard-Foy and G. L. Hager, *EMBO J.* **6,** 2321 (1987).
[30] M. F. Norman, T. N. Lavin, J. D. Baxter, and B. L. West, *J. Biol. Chem.* **264,** 12063 (1989).
[31] J. C. Hansen, K. E. van Holde, and D. Lohr, *J. Biol. Chem.* **266,** 4276 (1991).
[32] M. Truss, J. Bartsch, A. Schelbert, R. J. Hache, and M. Beato, *EMBO J.* **14,** 1737 (1995).
[33] S. A. Stanfield-Oakley and J. D. Griffith, *J. Mol. Biol.* **256,** 503 (1996).
[34] S. A. Nedospasov and G. P. Georgiev, *Biochem. Biophys. Res. Commun.* **92,** 532 (1980).

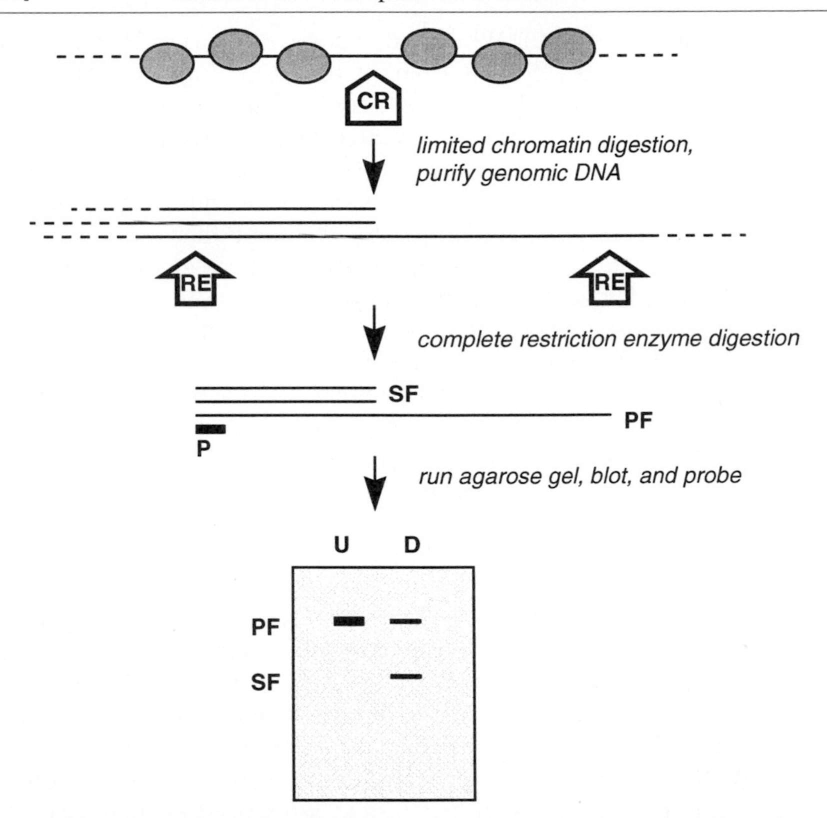

FIG. 1. The indirect end-labeling technique. In this schematic, a region missing a specific nucleosome from the normal chromatin array is depicted. Limited chromatin digestion with the cleavage reagent (CR) of choice is applied; only those genomic fragments resulting from this treatment that can hybridize to the probe (P) are shown. Restriction enzyme (RE) cleavage of the purified population of DNA molecules leads to the production of a parent fragment (PF) defined by the two flanking restriction sites or a subfragment (SF) defined by the left-most restriction site and the site of reagent cleavage. Lane U depicts the result derived from an undigested control; lane D the result after limited digestion with the particular cleavage reagent.

partially digest the nuclei with the reagent of choice, isolate and purify the genomic DNA, cut it to completion with a restriction enzyme that releases a fragment spanning the genomic region of interest, and then subject the sample to size separation by agarose electrophoresis, followed by Southern blotting and subsequent hybridization to a labeled DNA probe. The particular probe chosen will ideally be a relatively short one that directly abuts (but does not extend beyond) the restriction site at one end of the region of interest. In such a situation the labeled probe only detects DNA fragments on one side of the restriction site (therefore achieving end labeling indirectly, hence the name). Where no cuts have occurred in chromatin, the only band revealed will be the parent restriction fragment. However, wherever the chosen cleavage reagent has cut within this restriction fragment in nuclei, the site(s) will be revealed as a shorter DNA fragment on the autoradiogram, its size directly revealing the accessible chromatin location relative to the known restriction site. If the indirect end-labeling probe is relatively small (200–400 bp), the likelihood that it will hybridize across a region (or regions) of chromatin accessibility (and hence cutting) is not high. With larger probes, such an overlap is more likely, with the result that two (or more) fragments will hybridize to the probe, only one of which will correctly indicate a site of cutting by virtue of its size. Mapping confusion can ensue in such a situation. If there is suspicion that a chromatin feature of interest falls within the labeled probe region itself, its position can usually be confirmed by independently mapping its location with a labeled probe abutting the restriction site at the other end of the parent fragment.

In a collection of nuclei the goal is to have the cleavage reagent produce, on average, slightly less than one cut per DNA template in the general genomic region of interest. Because of stochastic considerations under these conditions, some nuclei will sustain no cuts in this region, whereas others may have one, two, or more cuts. Irrespective of the actual number of accessible chromatin sites occurring within the region of interest, however, such an idealized digestion point will reveal maximum information from one sample. With the luxury of a large number of nuclei for experimentation, an extensive digestion series can be performed using fixed aliquots of nuclei with increasing amounts of enzyme for a given time period. In this way, all samples (including the important "no reagent" control) are exposed equally to any endogenous nuclease activity. Samples can then be compared side by side on the gel to disclose which one provides maximum information content. When the number of nuclei available per experiment is small, some trial and error may be necessary in order to find conditions that reveal the most compelling data. The following protocols give suggested conditions that have worked well in our hands for both large and small preparations of embryonic nuclei.

Experimental Protocols

A. Large-scale Nuclei Preparation

1. Fresh or frozen (thawed to room temperature) embryos (approximately 30 g) are dechorionated by magnetic stir bar suspension in 200 ml of 50% (v/v) bleach/1% (v/v) Triton X-100 solution for approx-

imately 2 min (making sure all clumps are dissipated quickly), followed by collection on a fine mesh nylon filter (usually a square of 100-μm Nitex material) and subsequent copious washing with cold water, 70% ethanol (to eliminate the last traces of Triton), and cold water again. Embryos are blotted dry through the Nitex by use of a pad of paper towels and resuspended in a fivefold volume (150–200 ml) of ice-cold buffer A–1 *M* sucrose. (Note that buffer recipes are listed in section G.) All subsequent steps are performed in a cold room or in an ice bucket as appropriate.

2. Homogenization of embryos is achieved through 10–20 strokes with a Teflon pestle in either a custom-milled stainless steel homogenizer or a regular commercial glass homogenizer (Potter–Elvehjem type, but take extra care in case of breakage). The progress of embryo disruption can be monitored microscopically. When most of the preparation is disrupted adequately, the homogenate is filtered through two layers of Miracloth (Calbiochem, La Jolla, CA), excess filtrate is squeezed out from the Miracloth using gloved hands, and the combined filtrate is poured into 50-ml nylon (or other plastic) centrifuge tubes. Low-speed centrifugation (480*g*) for 5 min eliminates cellular debris; the supernatant is poured carefully into fresh tubes and Nonidet P-40 [10% (w/v) stock)] is added to each tube to achieve a final concentration of 0.25%.
3. Each tube is sealed carefully with multiple layers of stretched Parafilm and then vortexed at top speed for approximately 3 × 45 sec, allowing the tube contents to equilibrate for a minute in the ice bucket between each vortexing. Crude nuclei are pelleted for 10 min at 4300*g*, the supernatant is discarded, and the pellet is allowed to drain briefly, after which the sides of the tube are wiped with Kimwipe tissue.
4. Resuspend nuclei in 10–20 ml of buffer A–1 *M* sucrose and distribute suspension evenly on to the top of 20 ml of buffer A*–1.7 *M* sucrose (buffer A* is identical to buffer A except for the omission of EDTA and EGTA) in each of two 30-ml glass (Corex) centrifuge tubes. Swirl the interface between the two layers briefly using the end of a Pasteur pipette to achieve a partial mixing, and then centrifuge the tubes at 16,500*g* in a swinging bucket rotor for 20 min. Nuclei should sediment as a creamy-colored pellet. The supernatant is discarded, the pellet drained briefly, the sides of the tube wiped with a Kimwipe, and nuclei resuspended in a small volume (1 ml is used initially) of the appropriate digestion buffer (see later) and kept on ice until needed. Resuspension is conveniently performed by pipetting up and down through the blue tip of a 1000-μl micropipettor.

Experimental Note. There are several possible variations on this basic procedure; one that was mentioned earlier involved the use of a Mg^{2}-containing buffer.[6] This "nuclear buffer" was substituted in place of buffer A/A* in the just-described protocol, with uniformly good results for a genomic footprinting study. Another variation omits the pelleting step through 1.7 *M* sucrose; crude nuclei obtained from the 4300*g* centrifugation are dispersed directly in buffer A*–1 *M* sucrose, repelleted at 4300*g*, and then resuspended in a small volume of the relevant digestion buffer and kept on ice.

B. Small-Scale Nuclei Preparation

It is not trivial to find the ideal digestion point for maximum information extraction when the number of nuclei available is limited, but there are occasions when it is not convenient to set up several population cages. For example, when *in vivo* functional and structural analyses are explored through the creation of transgenic flies by P-element-mediated transformation, embryo collections from several different lines may be needed. Expanding each of these for short periods into single population cages is quite realistic, and 1- to 2-g collections of embryos are achieved rather readily (although it may be necessary to pool the frozen embryos obtained from several independent collections). The following nuclear preparation protocol has given good results for DNase I, MNase, and restriction enzyme analyses.

1. Remove embryos (2.5 g) from −80° storage and dechorionate as described earlier. Rinse well in embryo wash solution [0.7% (w/v) NaCl, 0.04% Triton X-100], then with water, and pat dry. All steps from this point on are conducted at 4°. Place embryos in a 15-ml glass-Teflon homogenizer, add 3.5 ml buffer A^{+}, and grind with five up and down strokes of a motorized pestle (we place the pestle in the chuck of a variable-speed electric drill).
2. Transfer the homogenate to a 7-ml Wheaton (Millville, NJ) homogenizer and hand grind with 10 strokes of pestle A. Filter the homogenate through two layers of Miracloth and squeeze cloth with gloved hands to obtain the maximum amount of liquid. Rinse the homogenizer with 2.5 ml buffer A^{+} and filter this through the same two layers of Miracloth. Carefully layer the filtered homogenate over 1.5 ml buffer AS and spin at 2600*g* in a swinging bucket rotor for 6 min at 4°.
3. Remove the supernatant and resuspend pelleted nuclei in 3 ml buffer A^{+}. Place the resuspended nuclei in the 7-ml Wheaton homogenizer

and disrupt further with five strokes using pestle B. Carefully layer over 1.5 ml buffer AS and spin at 2,000*g* for 5 min at 4°.

4. Remove the supernatant and resuspend pelleted nuclei in 3 ml buffer A. Place the nuclear suspension in the 7-ml Wheaton homogenizer and disrupt pellet with five strokes using pestle A. Remove to a centrifuge tube and spin at 4000*g* for 5 min at 4° in order to generate a tight nuclear pellet.
5. Pour off the supernatant, invert the tube, and wipe with a Kimwipe tissue to remove as much extraneous liquid as possible. Resuspend the pellet in appropriate digestion buffer (see later) and keep on ice.

C. Large-Scale Nuclease Digestions

Nuclear digestions are commonly performed on several aliquots at a fixed nuclear concentration and with varying concentrations of enzyme, i.e., DNase I or MNase. Enzymes are stored at high concentration in a −80° freezer in 25-μl aliquots: DNase I at 20,000 U/ml [dissolved in 100 m*M* NaCl, 10 m*M* Tris–HCl (pH 7.4), 100 μg/ml gelatin] and MNase at 15,000 U/ml [dissolved in 10 m*M* NaCl, 10 m*M* Tris–HCl (pH 7.4), 100 μg/ml gelatin]. Note that we have routinely used enzymes from Worthington Biochemical Corporation (Lakewood, NJ) and that these units are those quoted by the manufacturer. It is *very* important to be aware that other suppliers frequently quote their nuclease units according to different activity criteria, which must be taken into account if the guidelines quoted here are being followed.

1. Large-scale nuclear preparations (see earlier) are resuspended in 1 ml of the appropriate digestion buffer on ice, i.e., DNase I digestion buffer or MNase digestion buffer. Prior to dispensing into aliquots, nuclei are counted microscopically in a Petroff–Hausser bacterial counting chamber, and the suspension is adjusted to either 5×10^8 or 1×10^9 nuclei/ml (preferably the latter if sufficient nuclei have been obtained) in the relevant digestion buffer.
2. Enzymes are diluted serially into tubes immediately prior to use. For DNase I, a 25-μl aliquot of stock solution is diluted with 375 μl of DNase I digestion buffer, and then twofold serial dilutions into DNase I digestion buffer are made into a series of six to eight tubes. For MNase, a 25-μl aliquot of stock is diluted with 175 μl of MNase digestion buffer, and then twofold serial dilutions are made into MNase digestion buffer. Do not vortex the tubes hard during this dilution process, as the enzyme may be partially inactivated.
3. Separately dispense 475-μl aliquots of nuclear suspension into six to eight fresh tubes and incubate at 25° for 2–3 min to achieve tempera-

ture equilibration. To each tube add 25 μl of the appropriate enzyme dilution, vortex very briefly, and incubate at 25° for 3 min. At the end of this time, add 12.5 μl of 0.5 *M* EDTA, vortex briefly, followed by 12.5 μl of 20% (w/v) SDS with subsequent stronger vortexing. It is very easy to tell at this stage if the digestion series has worked, as SDS lysis of the nuclei and denaturation of chromatin will lead to the release of genomic DNA in the tube. In a good digestion series the aliquot digested at the highest enzyme concentration will show virtually no viscosity on vortexing, whereas the sample receiving the least concentrated enzyme should show a pronounced viscosity. It is also a good idea to run a no enzyme-added control, as this will provide a useful indication of the integrity of the nuclear DNA after exposure to endogenous nucleases.

4. DNA is purified from each sample by fairly standard procedures. First, 2.5 μl of proteinase K [from a 10-mg/ml stock in 10 m*M* Tris–HCl (pH 8.0), 1 m*M* $CaCl_2$] is added and incubated at 37° overnight (or at 45–50° for 2 hr), and the sample is extracted twice with phenol/chloroform/isoamyl alcohol (25 : 24 : 1, v/v), with the top aqueous phase from each step being recovered after microfuge separation. RNase A (2.5 μl from a 10-mg/ml stock) is added and incubated at 37° for 2 hr, and then each sample is extracted once more with phenol/chloroform/isoamyl alcohol and once with chloroform alone. Samples are brought to 0.3 *M* in sodium acetate, and DNA is precipitated by the addition of 3 volumes of 100% ethanol. DNA pellets are collected by 15 min of centrifugation in the microfuge, washed in 70% (v/v) ethanol, repelleted, and dried very briefly under vacuum.
5. DNA is resuspended in a small volume of sterile deionized water (typically 100–200 μl) and, once it is solubilized completely, brought up to 1× TE from a 20× stock [1× TE is 10 m*M* Tris–HCl (pH 7.5), 1 m*M* EDTA], and its concentration measured spectrophotometrically. For ease of use subsequently, each DNA sample is then adjusted to a standard final concentration of 1 mg/ml and stored at 4°.

Experimental Note. The interpretation of particular cutting patterns as chromatin-specific features necessarily implies that such patterns are not characteristic of naked DNA. Implicit in this approach therefore is the requirement to perform the relevant controls with purified, protein-free DNA digested to nearly identical extents as the chromatin samples under investigation. Experimental evidence has shown that in locations where nucleosome linker positions overlap DNA sequence-preferential sites, are not fixed rigidly, or consist of multiple, overlapping positions, it can become extremely difficult to interpret what data derived from MNase digestions of chromatin actually mean. Under these circumstances, when the patterns appear virtually identical (or even quite similar), it is definitely worthwhile to consider using the chemical nuclease MPE·Fe(II) (see section E) to help unravel details of the chromatin structure.

D. Small-Scale Nuclease Digestions

1. Resuspend a small-scale nuclear preparation in either 900 μl DNase I digestion buffer or 900 μl MNase digestion buffer (we have sometimes used this buffer with $CaCl_2$ at 2 m*M* instead of 1 m*M* as in the standard large-scale digestion buffer). By taking the volume of pelleted nuclei into consideration, this will give a total volume of approximately 1 ml. Dispense 250-μl aliquots into each of four microfuge tubes, keeping everything on ice.
2. For DNase I digestion, add 0, 10, 30, and 60 U of enzyme per sample and incubate on ice for 3 min with occasional gentle mixing. For MNase digestion, add 0, 3, 6, and 30 U of enzyme per sample and incubate at 25° for 5 min with occasional gentle swirling of the samples.
3. Stop each reaction with 5 μl 0.4 *M* EDTA (pH 8.0). Follow with SDS/proteinase K digestion as in section C and subsequently purify DNA using procedures basically identical to those given for the large-scale nuclease digestion.

Experimental Note. In these small-scale nuclear digestions we find it is easier to obtain consistent DNase I digestion by performing the incubation on ice as opposed to 25°. In general, we suggest the actual amounts of enzyme given be viewed as guidelines only, and optimal digestions be determined empirically.

E. Methidiumpropyl-EDTA·Fe(II) Digestions

MPE is a maroon solid that can be obtained commercially in 5× 100-μg samples (Sigma Chemical Company, St. Louis, MO). We store it as a 1 m*M* aqueous solution shielded from light at −80° (concentration calculated from its absorbance at 488 nm using a molar extinction coefficient of 5994 M^{-1} cm^{-1}). Because MPE works at quite low concentrations in our hands, 500 μg should prove adequate for several analyses. Undoubtedly MPE can be used for digestion of small-scale nuclear preparations, but the protocol given here is optimized for the larger scale preparation.

1. Nuclei are prepared as described in section A and resuspended at 5×10^8 per ml in nuclear digestion buffer (NDB). It is important to

note, however, that MPE will work well in the presence of EDTA and EGTA, thereby minimizing the action of endogenous nucleases at every stage. Thus it is possible to centrifuge the nuclei through buffer A–1.7 *M* sucrose (rather than buffer A*) and resuspend the drained nuclear pellet at the required concentration in NDB for subsequent digestion. As desired, EDTA and EGTA can be added to 1 and 0.1 m*M* respectively. We have also used MPE successfully on a crude nuclear preparation obtained after pelleting nuclei through buffer A–1 *M* sucrose at 4300*g* (see Lowenhaupt *et al.*[35]) followed by resuspension in NDB.

2. The MPE · Fe(II) complex is prepared and activated immediately prior to use. Although the complex will work in the presence of nuclear suspensions containing EDTA and EGTA, it needs to be initially prepared in their absence. Using a freshly made stock (5 m*M*) of aqueous ferrous ammonium sulfate, a 1 : 1 equimolar complex of ferrous ion and MPE is prepared at 2×10^{-4} *M*. Thus, to make 500 μl of MPE · Fe(II) complex, 20 μl of ferrous ammonium sulfate stock and 100 μl of MPE stock are mixed together with 380 μl of NDB. In order to fully activate the complex and ensure that the iron is recycled back to the +2 valence state throughout the ensuing hydroxyl radical-generating redox reaction, 5 μl of freshly prepared 1 *M* dithiothreitol (DTT) is now added to the complex. Prepare the complex literally a few moments before addition to nuclei and keep shielded from light during use.
3. Remove the nuclear suspension in NDB from ice and allow to equilibrate to 25° for 2 or 3 min. Because the MPE redox reaction requires activated oxygen species for the efficient generation of hydroxyl radicals, we now generally add hydrogen peroxide (from a freshly prepared 100 m*M* stock) to the nuclear suspension to give a final concentration of 0.5 m*M* or 1 m*M* H_2O_2. Digestion is initiated by adding 1 volume of activated MPE · Fe(II) complex to 9 volumes of nuclear suspension [i.e., 2.5×10^{-5} *M* final concentration of MPE · Fe(II)], followed by brief vortexing and incubation at 25°.
4. At suitable time intervals (e.g., every 5 min), withdraw aliquots from the digestion, dispense into a tube containing 0.1 volume of 50 m*M* bathophenanthroline disulfonate (Sigma), and vortex briefly to stop the reaction. This compound is a very strong iron chelator and literally displaces the iron from the grasp of MPE, signified by development of a bright red color in solution. The subsequent addition of SDS/

[35] K. Lowenhaupt, I. L. Cartwright, M. A. Keene, J. L. Zimmerman, and S. C. R. Elgin, *Dev. Biol.* **99,** 194 (1983).

EDTA and DNA purification is now performed in identical fashion to the procedure described in section C. The red coloration remains in the aqueous phase during the organic extraction steps, but stays in the supernatant during the final ethanol precipitation and washing steps. If the DNA precipitate retains some slight red color it can be redissolved in TE and reprecipitated via sodium acetate/ethanol if desired. When DNA samples produced by this procedure are visualized on agarose gels by ethidium bromide staining, the bulk chromatin nucleosomal pattern revealed is basically indistinguishable from that produced by MNase digestion (Fig. 2).

Experimental Note. For MPE · Fe(II) we have routinely added a fixed concentration of chemical to the complete nuclear suspension and withdrawn equal aliquots of nuclei from the digestion at different time points. This approach has been necessitated by the relative scarcity of the reagent in the past and could now potentially be revised given its commercial availability. There is, however, a secondary consideration. Because MPE is an intercalator, it is possible that, if used at substantially higher concentrations than described, it might distort chromatin structure by virtue of the base pair unwinding that accompanies intercalation. Under the conditions given here, we estimate that this induced distortion should not be a major factor, with probably less than one intercalated molecule per 200 bp of DNA.

F. Indirect End-Label Mapping

1. After choosing a suitable restriction enzyme that spans the region of interest, DNA samples are digested to completion. We generally take anywhere between 2 and 10 μg of DNA per sample (depending on availability, the more the better) and digest with a 5- to 10-fold excess of restriction enzyme. Reactions are terminated by the addition of an SDS/EDTA mix, bromphenol blue/xylene cyanol running dye is added, and samples are loaded onto an agarose gel for electrophoresis. Gels are run until the bromphenol blue has migrated to a position of about 80% of the gel length, and the gel is prepared for Southern blotting by standard procedures. Whereas regular 15- to 20-cm gels are quite adequate, we have been able to get excellent resolution of structural details ($\pm$20 bp) by running long agarose gels (40–45 cm) cast in TAE in the submarine mode; these are kept cool by running in the cold room for up to 40 hr at 2.5 V/cm with constant buffer recirculation. We have achieved excellent results either by blotting to regular nitrocellulose membranes in a standard 20× SSC transfer buffer or by using nylon membranes (either neutral or posi-

FIG. 2. Nucleosomal pattern obtained by digestion with either MPE · Fe(II) or micrococcal nuclease (MNase). *Drosophila* 6- to 18-hr embryonic nuclei were digested under conditions described in the text, DNA was purified, fractionated on a 1.2% agarose gel, and stained with ethidium bromide. Adapted, with permission, from I. L. Cartwright and S. C. R. Elgin, *EMBO J.* **3,** 3101 (1984).

tively charged) with standard alkaline transfer buffer. In the case of nitrocellulose or neutral nylon membranes, it is important to either bake the membrane in a vacuum oven at 80° for 2 hr or immobilize the DNA by UV cross-linking.

2. Probes for mapping are labeled routinely by standard random primer procedures using [α-^{32}P]dATP. Commercially available kits are perfectly adequate. The cleanest results are usually obtained when the probe of interest is purified away from any vector DNA prior to labeling and the labeled probe is separated from unincorporated nucleotides on a spin column. We have performed hybridizations by a variety of methods using either classical SSC-containing buffers (with or without 10% dextran sulfate) or the SDS–phosphate buffer system of Church and Gilbert.[36] Containers for hybridization have included specially constructed low-volume slit-well chambers, Seal-a-Meal bags, or rotisserie oven cylindrical bottles. All give good results, although slit-well chambers routinely give the lowest backgrounds of all. Visualization of the signal is performed by standard film autoradiography methods or can be quantified as required on a phosphor imaging system.

Experimental Note. The highest resolution of all can be achieved by running sequencing-type polyacrylamide gels (see Thomas and Elgin[6]), although handling and blotting these is significantly more irksome in general. Most genomic footprinting (for that is what it is at this level of resolution) is currently performed by ligation-mediated polymerase chain reaction approaches, described in detail in section 2.

G. Buffers and Reagents Used in Embryonic Chromatin Structure Analysis

Buffer A (store at −20°): 60 m*M* KCl, 15 m*M* NaCl, 15 m*M* Tris–HCl (pH 7.4), 1 m*M* EDTA, 0.1 m*M* EGTA, 0.5 m*M* spermidine trihydrochloride, 0.15 m*M* spermine tetrahydrochloride, 0.5 m*M* DTT, 0.1 m*M* phenylmethylsulfonyl fluoride (PMSF). Note that each time buffer A is thawed DTT and PMSF are added fresh (from concentrated stocks of 1 *M* and 25 m*M*, respectively, stored at −20°) to the volume of buffer to be used. Other buffer A stocks contain various concentrations of sucrose (e.g., buffer A–1 *M* sucrose); these are also stored at −20°. Buffer A is identical to buffer A, but lacks EDTA and EGTA.

Nuclear buffer (store at −20°): 60 m*M* KCl, 15 m*M* NaCl, 15 m*M* Tris–HCl (pH 7.4), 5 m*M* $MgCl_2$, 0.5 m*M* DTT, 0.1 m*M* PMSF

[36] G. M. Church and W. Gilbert, *Proc. Natl. Acad. Sci. U.S.A.* **81,** 1991 (1984).

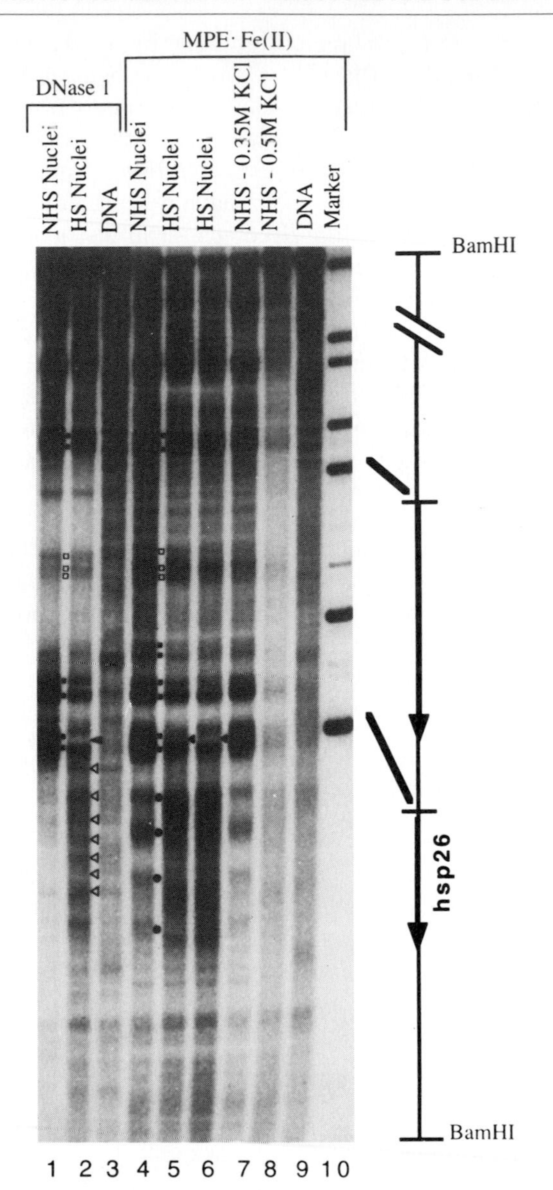
MPE· Fe(II)
DNase 1
NHS Nuclei
HS Nuclei
DNA
NHS Nuclei
HS Nuclei
HS Nuclei
NHS - 0.35M KCl
NHS - 0.5M KCl
DNA
Marker
BamHI
hsp26
BamHI
1 2 3 4 5 6 7 8 9 10

Buffer A+ (store at −20°): 60 m*M* KCl, 15 m*M* NaCl, 15 m*M* Tris–HCl (pH 7.4), 1 m*M* EDTA, 0.1 m*M* EGTA, 0.5 m*M* spermidine, 0.15 m*M* spermine, 0.5 m*M* DTT, 0.5% Triton X-100

Buffer AS (store at −20°): 60 m*M* KCl, 15 m*M* NaCl, 15 m*M* Tris–HCl (pH 7.4), 1 m*M* EDTA, 0.1 m*M* EGTA, 0.5 m*M* spermidine, 0.15 m*M* spermine, 0.5 m*M* DTT, 0.3 *M* sucrose

DNase I digestion buffer (store at −20°): 60 m*M* KCl, 15 m*M* NaCl, 15 m*M* Tris–HCl (pH 7.4), 3 m*M* $MgCl_2$, 0.05 m*M* $CaCl_2$, 0.5 m*M* DTT, 0.25 *M* sucrose

MNase digestion buffer (store at −20°): 60 m*M* KCl, 15 m*M* NaCl, 15 m*M* Tris–HCl (pH 7.4), 1 m*M* $CaCl_2$, 0.5 m*M* DTT, 0.25 *M* sucrose

Nuclear digestion buffer (store at −20°): 60 m*M* KCl, 15 m*M* NaCl, 15 m*M* Tris–HCl (pH 7.4), 0.25 *M* sucrose

Note. Add DTT, PMSF, and Triton X-100 fresh to each assay.

An Example From the *Drosophila hsp26* Gene

Through the use of these procedures we were able to derive a significant amount of information at the small heat shock gene locus of *Drosophila.*[5] A typical result using DNase I and MPE · Fe(II) to map chromatin structure at the *hsp26* gene and promoter region both before and after heat shock (i.e., pre- and post-gene activation) is shown (Fig. 3). DNase I analysis reveals the presence of two prominent, compound hypersensitive sites upstream of the gene promoter prior to activation (lane 1, Fig. 3), each of which spans the location of a DNA heat shock sequence element (HSE). The gene body itself is quite DNase I resistant. MPE analysis both confirms

FIG. 3. Chromatin structure in the vicinity of the *Drosophila hsp26* gene. Nuclei were derived from either non-heat-shocked (NHS) or heat-shocked (HS) embryos to facilitate a comparison of structural features prior to and during transcription. After digestion with either DNase I or MPE · Fe(II), DNA was purified, cut to completion with *Bam*HI, and fractionated on a 40-cm 1.2% agarose gel. Blotting to nitrocellulose was followed by probing with a small labeled fragment directly abutting a downstream *Bam*HI site. The genomic organization of the locus across the probed region is shown to the right of the autoradiogram (gene R is a developmentally regulated transcript of unknown function). Lanes marked DNA (lanes 3 and 9) were naked DNA controls digested to similar levels as the nuclear DNA. Salt extraction of NHS nuclei prior to digestion was performed in lanes 7 (0.35 *M* KCl) and 8 (0.5 *M* KCl). Black squares denote the DH sites, and black arrowheads denote regions of induced inaccessibility on heat shock. Black circles denote nucleosomal linker regions (lane 4) within *hsp26*, whereas open arrowheads show a novel DNase I cutting pattern within the *hsp26* transcribed region on gene activation (lane 2). Adapted, with permission, from S. C. R. Elgin, H. Granok, Q. Lu, and L. L. Wallrath, *Cold Spring Harbor Symp. Quant. Biol.* **58,** 83 (1993).

these data and extends the structural information further, as it is clear that an array of nucleosomes is positioned on the gene body prior to transcription (lane 4, Fig. 3). On gene activation, marked resistance to digestion appears in the center of each hypersensitive region (lanes 2, 5, and 6, Fig. 3), consistent with the strong binding of active heat shock transcription factor to the HSE elements at these locations. In addition, the nucleosomal pattern detected by MPE on the gene becomes quite indistinct (lanes 5 and 6, Fig. 3), suggesting a strong perturbation of the nucleosome array coincident with passage of the RNA polymerase. This latter region now appears very interesting from the DNase I analysis: not only has the gene body itself become very sensitive to the enzyme, but a series of distinct cleavages, occurring at approximately 100-bp intervals, is revealed (lane 2, Fig. 3). These might arise as a result of a discrete nucleosomal unfolding concomitant with transcription rather than a total relaxation of histone–DNA contacts. Salt extraction of the nuclei with 0.5 *M* KCl prior to MPE digestion reveals that loss of the hypersensitive site pattern (possibly due to extraction of the GAGA protein factor known to bind in this region[37]) is accompanied by an apparent randomization of the nucleosomal array across the gene (lane 8, Fig. 3): under these conditions it is expected that the nucleosomes themselves are still strongly associated with DNA and that their apparent positioning seen in lane 4 (Fig. 3) is due to their juxtaposition to the DH site region. Thus this would be a good example of statistical positioning of a nucleosomal array.[38]

2. Genomic Footprinting using *Drosophila* Salivary Glands: Perspective

The salivary gland polytene chromosomes have long offered a way to visualize changes in chromatin structure that accompany gene activation. We have found salivary gland cells to be useful for analyzing protein–DNA interactions.[9] Because of their polytene state, a small number of glands easily provides enough material to perform a genomic footprinting analysis.[39] This has been very useful for analysis of a variety of different fly lines carrying transgenes with altered promoters stably introduced into the genome by P-element-mediated transformation. One bottle of flies will easily provide enough larvae for several reactions. Most of our analyses have been performed using potassium permanganate; this reagent has allowed us to monitor the interaction of TFIID at the TATA element and of RNA

[37] D. S. Gilmour, G. H. Thomas, and S. C. R. Elgin, *Science* **245,** 1487 (1989).
[38] R. D. Kornberg and L. Stryer, *Nucleic Acids Res.* **16,** 6677 (1988).
[39] P. R. Mueller and B. Wold, *Science* **246,** 780 (1989).

polymerase II at the transcription start site of the *hsp70* gene. Permanganate penetrates the plasma membrane and reacts preferentially with thymine residues. In the case of the TATA box, the presence of TFIID protects thymines in the TATA element from permanganate oxidation. In the case of RNA polymerase, thymines located within the single-stranded transcription bubble are hyperreactive to permanganate (see Fig. 5). Unfortunately, permanganate does not detect the interaction of all proteins known to be present at the *hsp70* promoter. GAGA factor, the presence of which can be observed readily with DNase I in isolated nuclei,[37] fails to be disclosed by our permanganate footprints. We anticipate that there will be situations where other reagents, such as dimethyl sulfate, are likely to be suitable for monitoring particular protein interactions. Dimethyl sulfate, however, has not been informative in these *hsp70* studies.

Experimental Protocols

A. Dissection of Salivary Glands and Permanganate Treatment

1. Before starting, prepare a fresh stock of 40 m*M* $KMnO_4$ and $KMnO_4$ stop solution (the composition of all buffers used is given in section F). Prepare the 40 m*M* $KMnO_4$ solution by dissolving the magenta-colored crystals in dissection buffer and store on ice. Keep the $KMnO_4$ stop solution at room temperature so that the SDS does not precipitate.
2. Place six or more larvae on moist filter paper in a small petri dish. As they crawl around, debris attached to the larvae should transfer onto the filter paper. (If the larvae are to be heat shocked, preheat three petri dishes with moist paper in the 37° incubator. At 5-min intervals, place two to four larvae in a dish. Heat shock for 15 to 20 min and then dissect. By setting up three dishes, it is possible to dissect six larvae while keeping the heat shocks within the 15- to 20-min period.)
3. Place each larva in approximately 100 μl of dissection buffer under a dissecting scope and, using fine tweezers, remove the salivary glands and transfer them to 100 μl of ice-cold dissection buffer held in a siliconized 1.5-ml microfuge tube. Continue dissecting glands until 10 to 12 glands (five to six pairs) have been transferred to the tube. This should take no more than 20 min. While collecting salivary glands, also transfer other soft body tissue (avoid the cuticle) to a second tube that contains 200 μl of dissection buffer. DNA from this soft body tissue will be used to prepare naked DNA and molecular weight markers.

4. Treat only the salivary glands at this time with permanganate. Add 100 μl of 40 m*M* permanganate to the tube of salivary glands and gently tap the sides of the tube to promote mixing. Avoid vigorous mixing as the glands can stick to the sides of the tube above the permanganate solution. Put the glands back on ice for 2 min. After 2 min, add 200 μl of $KMnO_4$ stop solution to both the salivary glands and the soft body tissue. Tap the sides of the tube to mix. The magenta-colored solution should quickly change color to yellow and then clear. The glands will remain a brownish color for about 10 sec. Shake all samples vigorously for a few minutes. Add 50 μg of proteinase K to each sample. Digest for 1 hr at 37°. A 10-mg/ml stock of proteinase K can be prepared by dissolving solid enzyme in water to a concentration of 20 mg/ml and then diluting this with an equal volume of 100% glycerol. Store the stock at −20°.
5. Extract the sample with a sequence of Tris-equilibrated phenol (pH 8), phenol/chloroform (1:1, v/v), and chloroform alone. Perform each extraction with 300 μl of the organic solution. Shake samples for 5 min and then centrifuge for 5 min to separate phases. For both of the extractions containing phenol, transfer the aqueous phase to a fresh tube. For chloroform extraction, remove the chloroform layer by pipetting it out from underneath the aqueous phase. Add 40 μl of 3 *M* sodium acetate (pH 7) and 1 ml of cold ethanol. Mix thoroughly by inverting the tube, then place on ice for at least 15 min. Microfuge in the cold for at least 20 min. A small pellet should be evident. Discard the supernatant and add 100 μl of cold 75% ethanol. Microfuge in the cold for 5 min and discard the supernatant. Allow the samples to air dry.
6. Dissolve the nucleic acid pellet from the salivary glands in 20 μl of TE [10 m*M* Tris–HCl (pH 7.5), 1 m*M* EDTA] and the nucleic acid pellet from the soft body tissue in 50 μl of TE. Determine the concentration of DNA by measuring 1 μl of the sample with a fluorometer. DNA concentrations for samples derived from 12 glands are typically around 10 ng/μl and those from the soft body tissue are around 50 ng/μl. Note that the samples contain RNA, which probably accounts for why a pellet can be seen in the previous ethanol precipitation. The RNA will not interfere with the fluorometer measurements or with the LM-PCR. Store samples in the refrigerator for short term or in a −20° freezer for long term.

B. Permanganate Treatment of Naked DNA

In order to interpret the pattern of permanganate reactivity that occurs in cells, the pattern must be compared to that which occurs in naked genomic DNA. Prepare a tube containing 300 ng of the genomic DNA purified from soft body tissue and diluted to a final volume of 100 μl in dissection buffer. Place on ice. Add 100 μl of ice-cold 40 m*M* $KMnO_4$ (dissolved in dissection buffer) and incubate for 60 sec. If necessary, this time can be varied to obtain patterns of modification suitable for comparison to the DNA from salivary glands. Stop the permanganate reaction with 200 μl of $KMnO_4$ stop solution. Add 40 μl of 3 *M* sodium acetate (pH 7) and 1 ml of ethanol. Place the sample on ice for 15 min, microfuge for 20 min, discard the supernatant, and wash the pellet with cold 75% ethanol. Air dry the DNA and then dissolve it in 20 μl of TE. Quantify the DNA with a fluorometer.

C. Formic Acid Treatment of Naked DNA to Generate G/A Markers

Formic acid depurinates the DNA, and subsequent heating of this DNA with piperidine cleaves the DNA backbone. Formic acid treatment of 1 μg of DNA should yield enough material for 10 gels. In a siliconized 1.7-ml tube, combine 1 μg of genomic DNA in 10 μl of TE with 10 μl of H_2O and 50 μl of 99% formic acid. Incubate at 15° for 5 min. Add 200 μl of cold 0.3 *M* sodium acetate (pH 7) and 50 μg/ml tRNA to stop depurination. Add 750 μl of cold ethanol and put on ice for 15 min. Microfuge the sample in the cold for 15 min to collect the DNA. Wash the pellet with 200 μl of 75% cold ethanol and air dry. Dissolve the DNA in 15 μl of TE and quantify with a fluorometer. Store samples in a −20° freezer.

D. Piperidine Cleavage of Permanganate-Treated and Formic Acid-Treated DNA

1. Piperidine cleaves the DNA at thymine residues that have been oxidized by permanganate and at apurinic sites in formic acid-treated DNA. Put 100 ng of DNA dissolved in a final volume of 15 μl of TE into siliconized 1.7-ml tubes. The collection of samples should include naked DNA treated for 60 sec with permanganate, DNA from permanganate-treated salivary glands, naked DNA that has not been treated with permanganate (derived from soft tissue), and DNA partially depurinated with formic acid (derived from soft tissue).
2. Add 75 μl of H_2O and 10 μl of piperidine to each tube and mix thoroughly. Heat the samples at 90° for 20 min. Add 300 μl of H_2O to each sample and extract three times with 800 μl portions of isobutanol (2-methyl-1-propanol). Separate the phases after each extraction by microfuging for 2 min. The isobutanol, which is the top phase, should be removed. Ether extract each sample one time. Adjust the volume of the DNA to approximately 100 μl by adding

water and then add 10 μl of 3 *M* sodium acetate (pH 7) and 250 μl of ethanol. Mix each sample thoroughly, place on ice for 15 min, and microfuge for 20 min in the cold to collect the DNA precipitate. Wash each precipitate with 75% ethanol. Air dry the DNA precipitate and dissolve in 5 μl of TE. *Note:* Transfer the DNA to a fresh siliconized 600-μl tube or other vessel suitable for use with a thermocycler. This transfer is essential as there is something about the tube where the piperidine reaction is performed that completely inhibits the Vent polymerase used subsequently. Samples are now ready for LM-PCR.

E. LM-PCR Analysis

This protocol is derived from a procedure originally described in detail by Mueller *et al.*[8] based on earlier seminal work.[39,40] A schematic of the procedure is provided in Fig. 4, and an example of the use of the technique in permanganate footprinting at the endogenous *hsp70* locus, as well as a copy of *hsp70* transduced into the germ line by P-element-mediated transformation, is shown in Fig. 5. Five primers are required. Primers 1, 2, and 3 are target specific and are designed so that primer 1 has the lowest annealing temperature and primer 3 has the highest. Note that the annealing temperatures used are given in the legend to Fig. 5 and that these vary depending on the particular primers used. Linker A′ and linker B are annealed together to provide a substrate suitable for the ligation reaction. We have successfully used oligonucleotides that have been subjected to varying levels of purification ranging from a crude, deprotected state first recovered from an oligonucleotide synthesizer to material purified using the SurePure oligonucleotide purification kit (USB). Note that linker A′ is modified from the original linker A described by Mueller and Wold[39] so that it better matches the annealing temperature of primer 2.

1. Combine 5 μl of DNA (100 ng in TE) with 25 μl of first-strand synthesis mix in a 600-μl siliconized tube. Do not overlay the samples with mineral oil. In a thermocycler, denature DNA at 95° for 5 min, anneal the primer for 30 min, and extend the primer at 76° for 10 min. Spin down condensate and place tube on ice. Add 20 μl of ligation dilution solution and 25 μl of ligation mix. Incubate overnight at 15°.
2. Add 9.4 μl of salt precipitation mix and 220 μl cold ethanol. Mix well and place on ice for at least 15 min. Microfuge the samples in the cold for 20 min. Wash one time with 100 μl of 75% ethanol,

[40] P. A. Garrity and B. Wold, *Proc. Natl. Acad. Sci. U.S.A.* **89,** 1021 (1992).

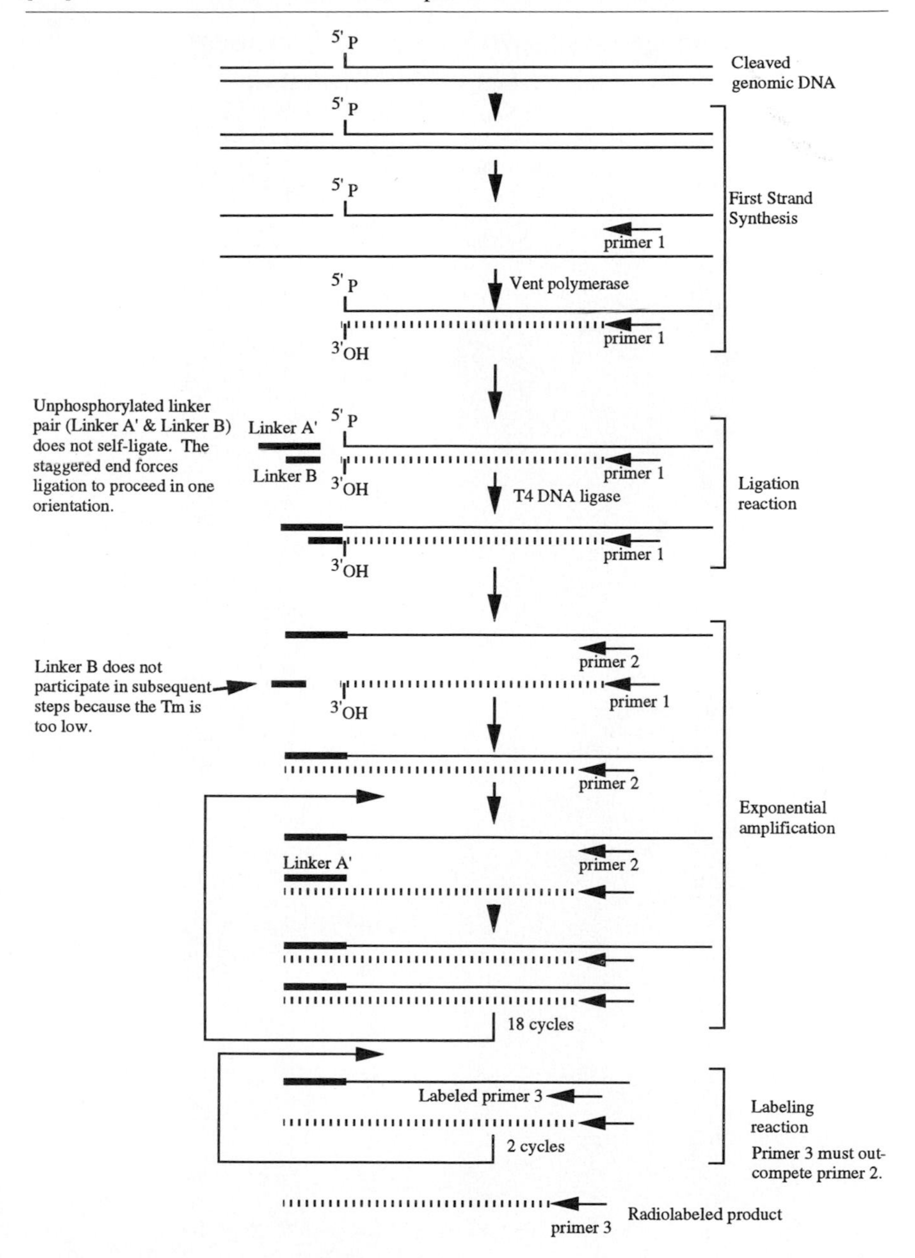

FIG. 4. Schematic of the LM-PCR procedure. This example is for detection of the location of a single nick. Typically, a population of cleaved DNA molecules is analyzed. See text for further discussion.

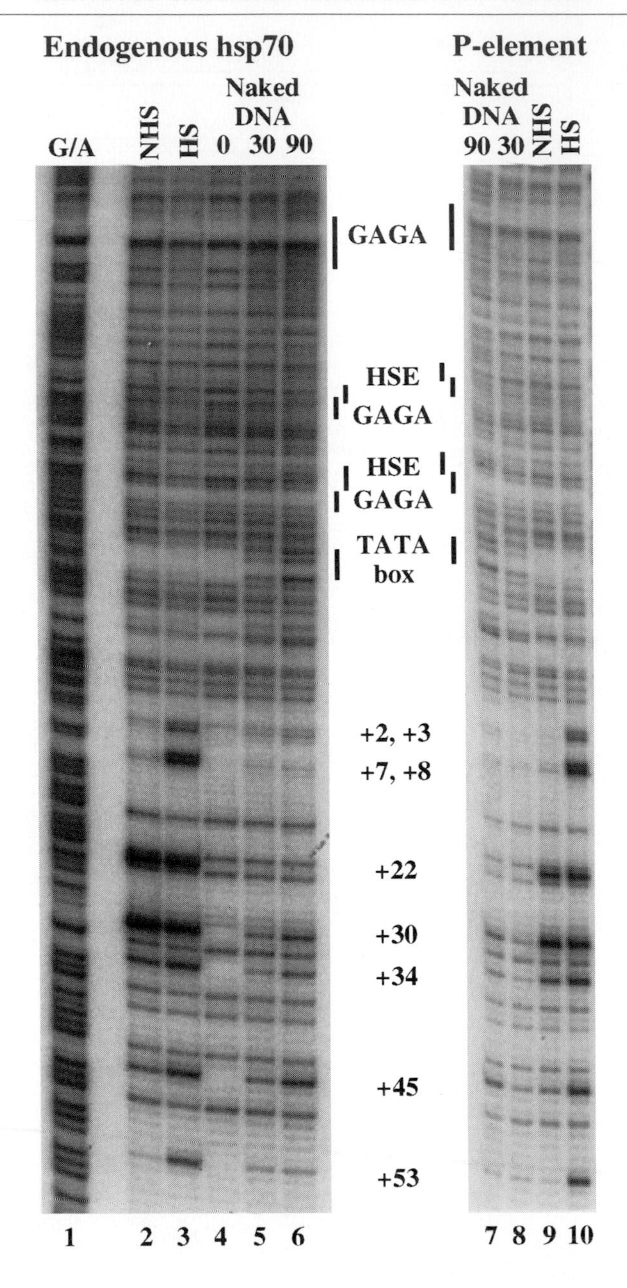

Endogenous hsp70
P-element
G/A
NHS
HS
Naked DNA
0 30 90
90 30
NHS
HS
GAGA
HSE
GAGA
HSE
GAGA
TATA box
+2, +3
+7, +8
+22
+30
+34
+45
+53
1 2 3 4 5 6
7 8 9 10

microfuging to secure the precipitate. Air dry the pellet and dissolve in 30 μl of H_2O. Place sample on ice. Add 70 μl of Vent amplification mix and cover with 3 drops of mineral oil. Perform 18 amplification cycles. The first cycle should have a 3-min denaturing step at 95° and the remaining denaturing steps should be for 1 min. All cycles have annealing steps of 2 min. The first 9 cycles have extension steps of 3 min and the last 9 cycles have extension steps of 4 min. All extension steps are perfomed at 76°. Following amplification, place the tubes on ice.

3. During the PCR reaction, prepare radiolabeled primer 3. Each sample uses approximately 1 to 2 pmol of radiolabeled primer so the following reaction will yield enough primer for at least 20 samples. Combine 2.8 μl of H_2O, 1 μl of 10× kinase buffer, 4 μl of 10 μM primer 3, 2 μl [γ-^{32}P]ATP (7000 Ci/mmol, 150 μCi/μl), and 0.2 μl polynucleotide kinase (30 U/μl). Incubate at 37° for 30 min. Purify the oligonucleotide on a NucTrap column (Stratagene). At least 20 million Cerenkov counts should be recovered in a total volume of 100 to 150 μl.
4. After the 18 cycles of amplification, proceed with the labeling reaction. Add 20 μl of labeling mix to each sample. Perform two amplifi-

---

FIG. 5. Permanganate footprinting of the *hsp70* promoter. (Left) The permanganate pattern for the endogenous *hsp70* promoter. (Right) The pattern for a version that has been transformed into the fly on a P-element (sequences extending from −194 to +84). Lane 1 is the G + A cleavage pattern for the endogenous promoter; the pattern for the transformed promoter is identical (not shown). The permanganate pattern derived from salivary glands from non-heat shocked (lanes 2 and 9) and from heat-shocked (lanes 3 and 10) larvae is shown. Lanes 5, 6, 7, and 8 show patterns for naked DNA samples that were treated for 30 or 90 sec. Lane 4 is the background pattern of cleavage that is detected in a sample that has not been treated with permanganate. The strong bands detected in lane 4 correspond to cutting at guanine residues. Note the protection of thymines at the TATA box that occurs in glands from heat-shocked and non-heat-shocked larvae: these correspond to the association of TFIID. Also note the hyperreactivity of thymines in the region downstream from the transcription start. These sites are due to the transcription bubble associated with RNA polymerase. Detection of the endogenous promoter used the following primers (the annealing temperatures are provided in parentheses): primer 1, CAGTAGTTGCAGTTGATTTAC (53°); primer 2, GCAGTTGATTTACTTGGTTGC (56°); and primer 3, GCAGTTGATT TACTTGGTTGCTGG (60°). These primers span the region from +155 to +124. Detection of the transformed promoter used the following primers: primer 1, ACGACGTTGTAAAAC-GAC (50°); primer 2, GTAAAACGACGGGATCGATTCC (58°); and primer 3, CGACGG GATCGATTCCAATAGGC (62°). These primers span the region from +142 to +106 in the transformed promoter. In both situations, the same combination of linker A′ and linker B was used in the ligation step: linker A′, GCGGTGATTTAAAAGATCTGAATTC and linker B, GAATTCAGATC. Linker A′ was used in combination with primers 2 and 3 in subsequent amplification and labeling steps.

cation cycles. The first cycle is initiated with a 3-min denaturation at 95°, and the second cycle is for 1 min. Each annealing is done for 2 min, and each extension is done for 10 min at 76°. Following the labeling cycles, transfer 100 μl of each reaction to a fresh 1.7-ml tube. Leave behind the oil and 20 μl of the aqueous phase. Add 120 μl stop solution to the transferred samples. Extract one time with 100 μl of chloroform/phenol mixture. Transfer the aqueous phase to a fresh tube and add 550 μl of ethanol. Chill on ice for 20 min and then microfuge for 20 min. Remove supernatant and wash pellet with 100 μl of 75% ethanol, microfuge, and discard supernatant. Air dry the samples. Dissolve precipitates in 10 μl of sequencing loading buffer and analyze 5 μl on a sequencing gel.

F. Buffers and Reagents Used for LM-PCR/Permanganate Analysis

Dissection buffer (store at 4° for periods of several days or at −20° for longer periods of time): 130 m*M* NaCl, 5 m*M* KCl, 1.5 m*M* $CaCl_2$

$KMnO_4$ stop solution (store frozen for only a few days): 20 m*M* Tris–HCl (pH 7.5), 20 m*M* NaCl, 40 m*M* EDTA, 1% SDS, 0.4 *M* 2-mercaptoethanol

First strand synthesis mix (prepare fresh and store on ice): for one sample, combine 16.75 μl H_2O, 6 μl 5× first-strand buffer, 1.5 μl of 0.2 pmol/μl primer 1, 0.5 μl 10 m*M* dNTPs, 0.25 μl Vent polymerase (2 U/μl, NEB No. 254L)

5× first-strand buffer (stock stored at −20°): 200 m*M* NaCl, 50 m*M* Tris–HCl (pH 8.9), 25 m*M* $MgSO_4$, 0.05% gelatin

Ligation dilution solution (prepare fresh and keep on ice): For one sample, combine 13.1 μl H_2O, 2.2 μl 1 *M* Tris–HCl (pH 7.5), 1.2 μl 300 m*M* $MgCl_2$, 1 μl 1 *M* DTT, 2.5 μl 1 mg/ml BSA

Ligation mix (prepare fresh and keep on ice): For one sample, combine 7.9 μl H_2O, 0.85 μl 300 m*M* $MgCl_2$, 0.5 μl 1 *M* DTT, 7.5 μl 10 m*M* ATP, 1.25 μl 1 mg/ml BSA, 5 μl linker pair (A′ : B at 20 pmole/μl), 2 μl T4 DNA ligase (1 Weiss unit/μl)

Linker pair: The linker pair (A′ and B) are annealed prior to use in the ligation mix. Linkers A′ and B are combined in 250 m*M* Tris–HCl (pH 7.7) so that the final concentration of each oligonucleotide is 20 μ*M*. The mixture is heated at 95° for 5 min and transferred to a beaker of water adjusted to 70°. The mixture is allowed to cool to room temperature for 1 hr and then left for at least another hour at room temperature. The beaker of water harboring the oligonucleotide mix is then placed on ice for slow cooling to 4°. The oligonucleotide mixture is left on ice overnight and then stored at −20°.

Salt precipitation mix: For one sample, combine 8.4 μl 3 *M* sodium acetate (pH 7), 1 μl 10 mg/ml yeast tRNA

Vent amplification mix (prepare fresh and keep on ice): For one sample, combine 45.5 μl H_2O, 20 μl 5× Vent amplification buffer, 1 μl of 10 pmol/μl primer 2, 1 μl of 10 pmol/μl linker primer A′, 2 μl 10 m*M* dNTPs, 0.5 μl Vent polymerase (2 U/μl)

5× Vent amplification buffer (stock stored at −20°): 200 m*M* NaCl, 100 m*M* Tris–HCl (pH 8.9), 25 m*M* $MgSO_4$, 0.05% gelatin, 0.5% Triton X-100

Labeling mix: For one sample, combine 9.5 μl H_2O, 4 μl 5× Vent amplification buffer, 1 μl 10 m*M* dNTPs, 5 μl radiolabeled primer 3 (1 to 2 pmol), 0.5 μl Vent polymerase (2 U/μl). Note that one can increase or decrease the amount of radiolabeled primer and compensate for these changes by varying the amount of water. The intensity of the signal will vary in proportion to the amount of radiolabeled primer.

Stop solution: For one sample, combine 95.8 μl of TE, 22 μl 3 *M* sodium acetate (pH 7), 2.2 μl 0.5 *M* EDTA

3. Quantitative Measurements of Chromatin Accessibility: Perspective

General cleavage reagents such as DNase I or MNase provide only qualitative measurements of the accessibility of a given region. Restriction enzymes can be used to quantitatively measure the accessibility of a given site within the genome. Typically nuclei are isolated and incubated with an excess amount of a restriction enzyme that cleaves within a desired region. The DNA is isolated from the nuclei and cleaved to completion with a restriction enzyme that flanks the region of interest. The DNA is fractionated on an agarose gel, transferred to membrane, and probed with a DNA fragment abutting one of the flanking restriction sites (i.e., indirect end labeling as in Fig. 1). The amount of cleavage can be determined by measuring the amount of hybridization of a probe to particular fragments on the membrane. Both radioactive and nonradioactive labeling can be used for the probes.[15,16]

The amount of restriction enzyme to be used must be determined empirically for each enzyme. It is essential that the enzyme be in excess; a limiting amount of enzyme may cause the accessibility of a site to be grossly underestimated. To determine the correct amount, a titration experiment is performed. Identical amounts of nuclei are incubated with increasing amounts of restriction enzyme for a given period of time. A plateau in which additional units of enzyme do not produce further cleavage should be achieved.

Typically 10% more than the lowest amount of enzyme at which no further cleavage is observed should be used for experimental reactions. We have found that 200 to 400 units of restriction enzyme is sufficient for an aliquot of nuclei containing approximately 25 μg of genomic DNA.

Restriction enzyme analysis has been used for a variety of purposes in analyzing the *in vivo* chromatin structure of a gene.[16,41–43] Sites that are relatively nucleosome free, based on other criteria, are 50–80% accessible; sites within nucleosomal-associated DNA are typically less than 10% accessible.[15,43–46] Here we describe applications using the *hsp26* gene of *Drosophila* as a model system. One convenient feature of *hsp26* is that the gene is heat shock inducible in almost all tissues at any stage of development; therefore, whole larvae are a suitable material for chromatin structure studies. The *hsp26* promoter has two nucleosome-free regions that map to the location of the heat shock elements.[5,6] Sequences within the HSE contain recognition sites for the restriction enzyme *Xba*I (Fig. 6). Thus, the extent of *Xba*I cleavage can be quantified to measure the accessibility of the regulatory region.

Several studies have shown quantitative differences in the chromatin structure of the *hsp26* promoter in comparisons between wild-type and mutant transgenes.[15,45,46] Analyses of transgenes with deletions of the $(CT)_n$ regions, which bind GAGA factor,[47] demonstrated that these regions are essential for establishing the open chromatin structure of the *hsp26* promoter prior to heat shock. For *hsp26* mutant transgenes, formation of the appropriate chromatin structure correlates with the ability of the gene to be heat shock inducible.[15,45] In a similar study, deletion of the region known to be associated with a specifically positioned nucleosome had little effect on the chromatin structure and heat shock inducibility of the gene.[46]

To learn more about the nature of chromatin domains and how they influence gene expression, we have mobilized an *hsp26* transgene present on a P-element transposon to new genomic locations.[16] Stocks with inserts at pericentric, telomeric, and fourth chromosome locations were recovered, regions known to be heterochromatic (reviewed by Weiler and Wakimoto[48]). At these new locations, the *hsp26* transgene was not heat shock

[41] C. Straka and W. Hörz, *EMBO J.* **10,** 361 (1991).

[42] T. K. Archer, M. G. Cordingly, R. G. Wolford, and G. L. Hager, *Mol. Cell. Biol.* **11,** 688 (1991).

[43] E. Verdin, P. Paras, Jr., and C. Van Lint, *EMBO J.* **12,** 3249 (1993).

[44] R. S. Jack, P. Moritz, and S. Cremer, *EMBO J.* **9,** 2603 (1991).

[45] Q. Lu, L. L. Wallrath, B. D. Allan, R. L. Glaser, J. T. Lis, and S. C. R. Elgin, *J. Mol. Biol.* **225,** 985 (1992).

[46] Q. Lu, L. L. Wallrath, and S. C. R. Elgin, *EMBO J.* **14,** 4738 (1995).

[47] H. Granok, B. A. Leibovitch, C. D. Shaffer, and S. C. R. Elgin, *Curr. Biol.* **5,** 238 (1995).

[48] K. S. Weiler and B. T. Wakimoto, *Annu. Rev. Genet.* **29,** 577 (1995).

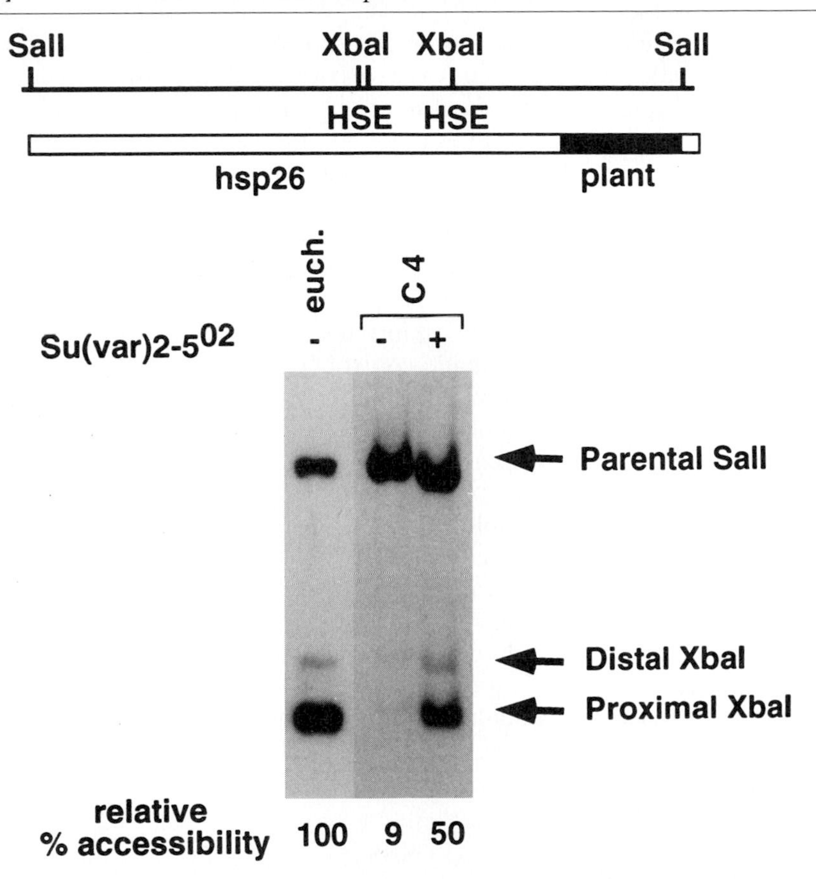

FIG. 6. Accessibility of the *hsp26* transgene promoter region in the background of an HP1 mutant. (Top) Schematic diagram of the *hsp26* transgene showing the location of restriction enzyme sites used and the heat shock elements (HSE). The transgene is marked with a fragment of plant (barley) DNA. (Bottom) Nuclei were isolated from non-heat-shocked third instar larvae and treated with an excess amount of the restriction enzyme *Xba*I. The DNA was purified, cleaved to completion with *Sal*I, separated by size on an agarose gel, transferred to membrane, and probed with the plant cDNA fragment. The transgene in stock 39C-X is located in euchromatin (euch.). The transgene in stock 118E-10 is located near the centromere of the mostly heterochromatic fourth chromosome (C 4). The analysis of 118E-10 was performed on stocks with (+) and without (−) *Su(var)2-5*^{02}, a mutation in the gene encoding HP1. The percentage cleavage at the proximal *Xba*I site was calculated by measuring the hybridization signal intensity of the fragment generated by cleavage at this proximal site divided by the total intensity of the hybridization signals in a given lane. Relative values are shown after setting the value for 39C-X at 100%. Adapted, with permission from D. E. Cryderman, M. H. Cuaycong, S. C. R. Elgin, and L. L. Wallrath, *Chromosoma* **107,** 277 (1998).

inducible to wild-type levels. Chromatin structure analyses revealed that heterochromatic *hsp26* transgenes were less accessible to restriction enzyme digestion and were packaged in a more regular nucleosome array.[16] Thus, a wild-type gene undergoes alterations in chromatin structure and gene expression in response to its location along a chromosome.

We have asked whether the altered chromatin structure of pericentric *hsp26* transgenes could be "reversed" by mutations in a protein that is involved in heterochromatin formation. Figure 6 shows the inaccessible chromatin structure of a transgene at a pericentric location on chromosome 4 (stock 118E-10); this transgene exhibits 9% cleavage at the proximal *Xba*I site compared to the euchromatic control (stock 39C-X). A more open chromatin structure, 50% cleavage compared to the euchromatic control, is formed in the presence of a mutation in the gene encoding heterochromatin protein 1 (HP1).[49,50] An increase in *hsp26* heat shock-inducible expression is also observed in the mutant background.[16] This powerful combination of genetics and molecular biology in *Drosophila* provides insight into the functional role of chromosomal proteins.

The following section describes methods and protocols to quantitatively assess the chromatin structure of a given region within *Drosophila* larval cells. When dealing with a large number of transformed stocks, we find it easiest to collect larvae as a starting material. However, similar procedures can be used with small amounts of embryos or adults as starting material.[51]

Experimental Protocols

A. Collection of Larvae

Typically four or more culture bottles containing cornmeal media[1] are filled with approximately 800 adult flies (5 days postemergence). The bottles are cultured at 25° at approximately 65% relative humidity for 4 days; the flies are then transferred to a second set of bottles. This can be repeated at least one additional time to provide three sets of staged bottles for larval collection. As larvae begin to crawl out of the food and coat the bottle walls, they are collected with a spatula and immediately dropped into liquid nitrogen. Frozen larvae are weighed in a prechilled weighboat, grouped into 1-g amounts, wrapped in foil, and placed at −80°. Each bottle typically

[49] J. C. Eissenberg, T. C. James, D. M. Foster-Hartnett, T. Hartnett, V. Ngan, and S. C. R. Elgin, *Proc. Natl. Acad. Sci. U.S.A.* **87,** 9923 (1990).

[50] J. C. Eissenberg, G. D. Morris, G. Reuter, and T. Hartnett, *Genetics* **131,** 345 (1992).

[51] L. L. Wallrath, M. J. Swede, and S. C. R. Elgin, *in* "Chromatin: A Practical Approach" (H. Gould, ed.), p. 59. Oxford Univ. Press, Oxford, 1998.

yields 1 to 3 g of larvae. Variations in the amount of larvae occur due to the age and fecundity of the adults and freshness of the food. Frozen samples of larvae can be stored for several years with no detectable change in the outcome of the following assay.

B. Isolation of Nuclei from Whole Larvae and Restriction Enzyme Analysis

1. Pulverize 1 g of larvae in liquid nitrogen using a prechilled mortar and pestle until the material has a fine grain-like consistency.
2. Transfer the ground larvae to a beaker containing 3 ml of buffer A^+ (buffer recipes are given in section F) and mix throughly. Transfer the suspension into a 7-ml ground-glass homogenizer and disrupt with five strokes of the pestle. Transfer the homogenate to a 15 ml-Dounce homogenizer, disrupt with 10 strokes using the type B pestle, and filter through two layers of Miracloth. Rinse the ground-glass homogenizer and the Dounce homogenizer with 3 ml of buffer A^+ and filter the suspension through the same layers of Miracloth. Squeeze the cloth with gloved fingers to complete the filtration.
3. Pipette the filtrate onto 1 ml of buffer AS and centrifuge in a swinging bucket rotor at 1500*g* for 5 min at 4°. Decant the supernatant and resuspend the pellet in 3 ml of buffer A^+. Place the suspension into a 7 ml-Dounce homogenizer, and disperse with five strokes of the type A pestle. Pipette the suspension onto 1 ml of buffer AS and centrifuge at 1500*g* for 5 min at 4°. Decant the supernatant and resuspend the pellet in 3 ml of buffer A, transfer the suspension to the 7-ml Dounce homogenizer, and disperse with five strokes of the type B pestle. Centrifuge at 700*g* for 5 min at 4°.
4. Decant the supernatant and resuspend the pellet in 1.47 ml of a particular restriction enzyme digestion buffer. Add protease inhibitors PMSF, leupeptin, and aprotinin to final concentrations of 0.5 m*M*, 5 mg/ml, and 0.1 unit/ml, respectively. Add the appropriate units of restriction enzyme (see previous discussion) to 250-μl aliquots of nuclei and pipette to mix. Incubate for 45 min at the appropriate temperature (see manufacturer's specifications). At the conclusion of the incubation, add 5 μl of 0.4 *M* EDTA, pH 8.0, to terminate the reaction.

C. Purification of DNA from Larval Nuclei

1. Collect nuclei from the terminated restriction enzyme digestion by centrifugation for 30 sec in a microcentrifuge. Resuspend the pellet in 250 μl of sarkosyl lysis buffer, add 3 μl of proteinase K (20 mg/

ml in distilled H_2O), and incubate at 37° for at least 3 hr or overnight. At the conclusion of the incubation add 83 μl of 2 M NaCl, 400 μl of phenol : chloroform : isoamyl alcohol (25 : 24 : 1, v/v) and vortex for 5 min. Spin for 5 min in a microcentrifuge at room temperature. Remove the aqueous phase and place it in a new tube. Add 400 μl of 2-propanol to the aqueous phase, mix by inversion, and place on dry ice for 15 min or at $-20°$ overnight.

2. Spin in a microcentrifuge for 10 min at 4°. At this step, either a visible pellet or a gelatinous liquid layer may be present at the bottom of the tube. If a pellet is visible, decant the 2-propanol and add 200 μl of 70% ethanol. If a gelatinous layer is apparent, pipette off the top 2-propanol layer using a micropipettor and add 70% ethanol. Invert the tube to mix the layers; a white precipitate should form. In either case, repeat the centrifugation for 10 min at 4°, decant the 70% ethanol, and vacuum dry the pellet.
3. Resuspend the pellet in 120 μl of RNase A digestion buffer. Add 3 μl of RNase A [10 mg/ml in 10 mM Tris–HCl (pH 7.5), 15 mM NaCl] and incubate for 1 hr at 37°. At the conclusion of the incubation, add 40 μl of 2 M NaCl and extract with phenol : chloroform : isoamyl alcohol as described earlier. Add 400 μl of chloroform : isoamyl alcohol (24 : 1, v/v) to the aqueous phase, vortex for 5 min, spin in a microcentrifuge for 5 min at room temperature, and place the aqueous phase in a new tube. Add 1 ml of water-saturated ethyl ether to the aqueous phase, vortex for 5 min, spin in a microcentrifuge for 5 min at room temperature, and discard the organic phase. Place the tubes with their lids open at 65° for 5 min to eliminate residual ether. Add 400 μl of 100% ethanol and precipitate for 15 min on dry ice or at $-20°$ overnight.
4. Spin in a microcentrifuge for 10 min at 4°. Decant the ethanol and add 200 μl of 70% ethanol. Recentrifuge for 10 min at 4°. Vacuum dry the pellet and resuspend the DNA in 20 μl of TE buffer. Use at least one-half of the sample for restriction enzyme digestion and Southern blot analysis.

D. Isolation of Nuclei from Larval Salivary Glands and Restriction Enzyme Digestion

Unlike the general expression of heat shock genes, many genes are tissue specific and/or developmentally regulated. This section describes the isolation of nuclei from hand-dissected salivary gland tissue. This protocol can be used for the isolation of nuclei from other dissected tissues. For quantitative results, one must use sufficient amounts of material to allow

Southern blot analysis. Techniques such as primer extension and LM-PCR can be applied to DNA from nuclei treated with restriction enzymes, but careful controls must be included to place confidence in the quantitation.[52,53]

1. Manually dissect 100 pairs of salivary glands from third instar larvae in dissection buffer (see recipe in section 2,F). Transfer glands to 100 μl of buffer M in a 1.5-ml Eppendorf tube. Add 2.5 μl of 20% Nonidet P-40 (0.5% final concentration) and gently homogenize with 50 strokes of a plastic pestle while keeping the tube on ice. Incubate on ice for 15 min, vortexing (or flicking) gently every 3 min.
2. Pipette the suspension of salivary glands up and down five times through a yellow pipette tip that has its tip cut off to make the opening approximately 1 mm in diameter. Spin at 2100*g* for 5 min at 4°. Carefully remove the supernatant using a pipette.
3. Add 200 μl of the appropriate restriction enzyme buffer with PMSF, leupeptin, and aprotinin (same concentrations as used in section B). Add the appropriate amount of enzyme (see earlier discussion) and incubate at 37° for 45 min.
4. Stop the digestion by adding 200 μl of buffer S and add proteinase K to 200 μg/ml final concentration. Incubate for 3 hr at 37°.

E. Purification and Analysis of DNA from Salivary Gland Nuclei

1. Extract the reaction mixture from section D, step 4, three times with 400 μl of phenol : chloroform : isoamyl alcohol. Add 20 μl of 3 *M* sodium acetate and 1 ml of 95% ethanol and place at −20° for 1 hr.
2. Spin in a microcentrifuge for 10 min at 4°. Decant the ethanol and add 200 μl of 70% ethanol. Recentrifuge for 10 min at 4°. Decant the ethanol, vacuum dry the pellet, and dissolve in 23 μl of distilled H_2O.
3. Digest the DNA with the desired restriction enzyme(s) for indirect end-labeling analysis in a total volume of 30 μl, including 1 μl of RNase A (10 mg/ml), for 3 hr at the appropriate temperature for the enzyme. Load the entire sample into one well of an agarose gel for subsequent Southern blot analysis.

F. Buffers Used in Quantitative Analyses

Buffer A^+: 60 m*M* KCl, 15 m*M* NaCl, 13 m*M* EDTA, 0.1 m*M* EGTA, 15 m*M* Tris–HCl (pH 7.4), 0.15 m*M* spermine, 0.5 m*M* spermidine, 0.5 m*M* DTT, 0.5% Nonidet P-40. DTT is added fresh from a 1 *M* stock just prior to use.

[52] J. Schlossherr, H. Eggert, R. Paro, S. Cremer, and R. S. Jack, *Mol. Gen. Genet.* **243,** 45 (1994).
[53] M. Hershkovitz and A. D. Riggs, *Methods* **11,** 253 (1997).

Buffer AS: 60 m*M* KCl, 15 m*M* NaCl, 1 m*M* EDTA, 0.1 m*M* EGTA, 15 m*M* Tris–HCl (pH 7.4), 0.15 m*M* spermine, 0.5 m*M* spermidine, 0.5 m*M* DTT, 0.3 *M* sucrose. DTT is added fresh from a 1 *M* stock just prior to use.

Buffer A: 60 m*M* KCl, 15 m*M* NaCl, 1 m*M* EDTA, 0.1 m*M* EGTA, 15 m*M* Tris–HCl (pH 7.4), 0.15 m*M* spermine, 0.5 m*M* spermidine, 0.5 m*M* DTT. DTT is added fresh from a 1 *M* stock just prior to use.

Sarkosyl lysis buffer: 50 m*M* Tris–HCl (pH 8.0), 100 m*M* EDTA, 0.5% sodium laurylsarcosine

RNase A digestion buffer: 50 m*M* Tris–HCl (pH 8.0), 10 m*M* EDTA, 100 m*M* NaCl

Buffer M: 10 m*M* HEPES (pH 7.6), 25 m*M* KCl, 5 m*M* $MgCl_2$, 5% glycerol, 0.5 m*M* PMSF, 0.5 m*M* DTT. PMSF and DTT are added fresh just prior to use.

Buffer S: 20 m*M* Tris–HCl (pH 7.4), 200 m*M* NaCl, 2 m*M* EDTA, 2% SDS

Acknowledgments

We thank Susan Gerbi and Fyodor Urnov for the *Sciara coprophila* salivary gland nuclei isolation protocol that we adapted for use with *Drosophila melangaster.* Work in S.C.R.E's laboratory is supported by NIH (R01-GM31432 and R01-HD23844); that in I.L.C.'s laboratory by NIH (R01-ES07543 and P30-ES06096); that in D.S.G.'s laboratory by NIH (R01-GM47477); and that in L.L.W.'s laboratory by the American Cancer Society (RPG-97-128-01-GMC) and the March of Dimes.

[27] Ultraviolet Cross-Linking Assay to Measure Sequence-Specific DNA Binding *in Vivo*

By MARK D. BIGGIN

Introduction

DNA replication, gene transcription, and chromosome structure are regulated by proteins that bind specific DNA sequences. The DNA-binding specificities of many of these proteins have been characterized extensively *in vitro*. However, to determine how they function in the context of a whole organism, it is essential to also understand the range of DNA sequences that they bind *in vivo*. Do conditions in the cell, such as altered chromatin structure or cooperative interactions with other proteins, radically alter the DNA-binding specificities of many proteins? *In vivo,* do proteins only bind

0076-6879/99 $30.00

to a limited subset of the DNA sequences that they recognize *in vitro*? These questions are especially pressing in plants and animals. These organisms have large genomes and complex regulatory networks that include families of related transcription factors, often showing broad and similar DNA sequence specificities *in vitro*. This makes it impossible to infer which DNA sequences a protein binds *in vivo* if the only data available derive from *in vitro* and molecular genetic experiments. This article describes a method that can directly quantitate binding by sequence-specific DNA-binding proteins *in vivo*. It also discusses the results obtained and suggests possible differences between transcription factor DNA binding in prokaryotes and higher eukaryotes.

In Vivo DNA-Binding Assays

There are three classes of *in vivo* DNA-binding assays: footprinting, immunolocalization, and cross-linking. Footprinting assays provide high-resolution mapping of DNA sequences that are protected from chemical or enzymatic attack by the binding of a protein.[1,2] However, this approach cannot identify which protein(s) occupies a particular DNA site and cannot determine the range of genes that a protein binds. Immunolocalization of proteins on polytene chromosomes in *Drosophila* can identify chromosomal regions at which proteins are found at their highest concentrations. Importantly, for some proteins these chromosomal regions include expected target genes.[3,4] The resolution of the polytene method is usually only ±100 kb, but if binding is assayed to P-element constructs inserted into chromosomes, it is possible to map binding to regions of 100–500 bp in length.[5] This technique has not been used to quantitate relative levels of DNA binding and has only been used effectively in *Drosophila*. Cross-linking assays employ either formaldehyde or ultraviolet (UV) light to covalently couple endogenous proteins to DNA.[6–10] Because antibodies are used to immunoprecipitate proteins coupled to their target DNAs, this approach can, unlike

[1] G. P. Pfiefer and A. D. Riggs, *Genes Dev.* **5,** 1102 (1991).
[2] P. Bossard, C. E. McPherson, and K. S. Zaret, *Methods* **11,** 180 (1997).
[3] L. D. Urness and C. S. Thummel, *Cell* **63,** 47 (1990).
[4] J. T. Westwood, J. Clos, and C. Wu, *Nature* **353,** 827 (1991).
[5] L. S. Shopland, K. Hirayoshi, M. Fernades, and J. T. Lis, *Genes Dev.* **9,** 2756 (1995).
[6] D. S. Gilmour and J. T. Lis, *Mol. Cell. Biol.* **6,** 3984 (1986).
[7] M. Solomon and A. Varshavsky, *Cell* **53,** 937 (1988).
[8] A. Hecht, S. Strahl-Bolsinger, and M. Grunstein, *Nature* **383,** 92 (1996).
[9] H. Strutt, G. Cavalli, and R. Paro, *EMBO J.* **16,** 3621 (1997).
[10] A. Carr and M. D. Biggin, *in* "Methods in Molecular Biology: Chromatin Protocols" (P. B. Becker, ed.). Humana Press, 1999.

footprinting assays, identify which proteins are bound to a specific region. For some proteins, binding can be mapped to 1- to 3-kb regions of DNA that encompass multiple DNA-binding sites.[11,12] However, for proteins that cross-link with high efficiency, it is possible to map binding to regions as short as 50–200 bp.[6,13,14]

In Vivo Ultraviolet Cross-Linking

In vivo UV cross-linking has been used to study a variety of proteins, including RNA polymerase II, topoisomerase I, and sequence-specific transcription factors such as Eve, Zeste, and GAGA.[6,11,15,16] In many cases, the pattern of DNA binding discovered differs greatly from that predicted by earlier indirect approaches, leading to fundamental reassessments of how these proteins act. For example, it was commonly assumed that most eukaryotic promoters would be regulated at the level of transcription initiation. However, using *in vivo* UV cross-linking, Gilmour and Lis[6] discovered that an RNA polymerase molecule was paused just 3′ of the start site of the *hsp70* promoter. Further experiments suggest that the pause is a rate-limiting step used to regulate the *hsp70* promoter[17] and that this mode of transcriptional control is common on many, but not all, promoters.[18,19] In a second example, studies of the Zeste protein indicate that, contrary to one model, some enhancers do not function by increasing the binding of activators to proximal promoter regions. Instead, activators can bind near the RNA start site in the absence of the enhancer, suggesting that synergy must result from some other mechanism.[13]

In vitro experiments show that UV light can induce covalent coupling between thymine bases in DNA and amino acids in proteins.[20–23] Ultraviolet light is a zero-length cross-linking agent that only induces covalent coupling

[11] J. Walter, C. A. Dever, and M. D. Biggin, *Genes Dev.* **8,** 1678 (1994).
[12] J. Walter and M. D. Biggin, *Proc. Natl. Acad. Sci. U.S.A.* **93,** 2680 (1996).
[13] J. D. Laney and M. D. Biggin, Proc. *Natl. Acad. Sci. U.S.A.* **94,** 3602 (1997)
[14] P. B. Meluh and D. Koshland, *Genes Dev.* **11,** 3401 (1997).
[15] D. S. Gilmour, G. Pflugfelder, J. C. Wang, and J. T. Lis, *Cell* **44,** 401 (1986).
[16] T. O'Brien, R. C. Wilkins, C. Giardina, and J. T. Lis, *Genes Dev.* **9,** 1098 (1995).
[17] C. Giardina, M. Perez-Riba, and J. T. Lis, *Genes Dev.* **6,** 2190 (1992).
[18] A. E. Rougvie and J. T. Lis, *Mol. Cell. Biol.* **10,** 6041 (1990).
[19] A. Krumm, L. B. Hickey, and M. Groudine, *Genes Dev.* **9,** 559 (1995).
[20] J. W. Hockensmith, W. L. Kubasek, W. R. Vorachek, E. M. Evertz, and P. H. von Hippel, *Methods Enzymol.* **208,** 211 (1991).
[21] K. R. Williams and W. H. Konigsberg, *Methods Enzymol.* **208,** 516 (1991).
[22] E. E. Blatter, Y. W. Ebright, and R. H. Ebright, *Nature* **359,** 650 (1992).
[23] J. Walter and M. D. Biggin, *Methods* **11,** 215 (1997).

between molecules that are in intimate contact with one another. Therefore, this method should not detect cross-linking of proteins that do not contact DNA directly. This contrasts with *in vivo* formaldehyde cross-linking, which is thought to induce high levels of protein/protein cross-linking and to detect proteins that are not directly in contact with DNA because proteins become coupled indirectly via other proteins.[9,14] *In vitro* experiments also show that UV cross-linking gives a quantitative measure of the level of binding of a protein to different DNA fragments.[12,23] The quantitative nature of this assay is important because some regulators bind *in vivo* at significant levels to most genes tested.[11,16] Thus, a description of the degree of binding to different genes is necessary, and it is not possible to describe such binding in a qualitative, bound-or-not-bound, manner. However, it should be cautioned that some proteins may not cross-link with equal efficiency to all DNA sequences and that proteins whose recognition sites do not include thymine may not UV cross-link to DNA at all.[20,22] Thus, *in vitro* experiments should be conducted to determine how quantitative UV cross-linking is for each protein studied (for details, see Walter and Biggin[23]).

Briefly, the *in vivo* UV cross-linking assay involves the following steps (Fig. 1). First, cells or intact embryos are irradiated with 254-nm UV light to cross-link endogenous proteins to DNA. Nuclei are then isolated. The chromatin is extracted by treatment with 2% Sarkosyl, which strips noncovalently bound proteins from the DNA. The cross-linked chromatin is then separated from the free proteins by buoyant density ultracentrifugation. After dialysis and restriction digestion, the protein of interest, together with the DNAs that are cross-linked to it, is immunoprecipitated. The cross-linked protein is then removed from the DNA by proteolysis. Finally, the immunoprecipitated DNA is analyzed on Southern blots to determine which of a variety of potential target DNAs are bound by the protein under study.

Sequence-specific DNA-binding proteins UV cross-link to DNA with much lower efficiency than proteins such as RNA polymerase II.[11] Consequently, it was necessary to develop a procedure that is considerably more sensitive than the original method of Gilmour and Lis.[6,10,11] A version of this highly sensitive protocol is described in the following section. This has been specifically adapted to study binding in *Drosophila* embryos. For modifications required to study binding in tissue culture cells and other organisms, see Walter and Biggin[23] and Boyd and Farnham.[24]

[24] K. E. Boyd and P. J. Farnham, *Mol. Cell. Biol.* **17,** 2529 (1997).

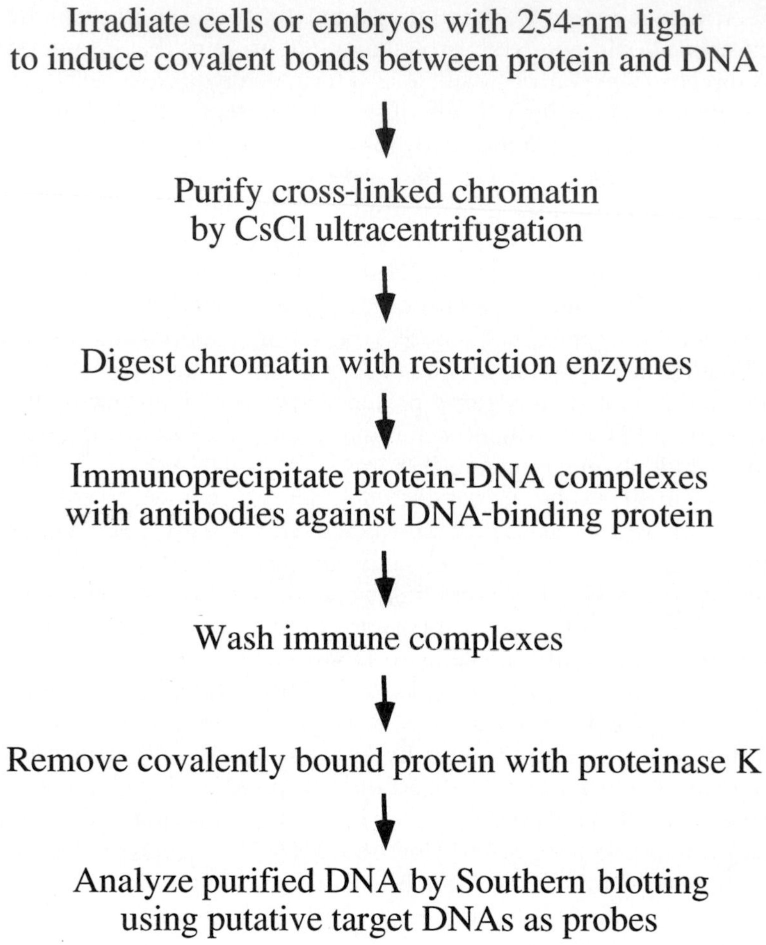

Irradiate cells or embryos with 254-nm light
to induce covalent bonds between protein and DNA

Purify cross-linked chromatin
by CsCl ultracentrifugation

Digest chromatin with restriction enzymes

Immunoprecipitate protein-DNA complexes
with antibodies against DNA-binding protein

Wash immune complexes

Remove covalently bound protein with proteinase K

Analyze purified DNA by Southern blotting
using putative target DNAs as probes

FIG. 1. Flow diagram for the *in vivo* UV cross-linking method.

Detailed Protocol

A. *UV Irradiation of Embryos*

Solutions

50% (v/v) Clorox bleach (2.6% sodium hyperchlorite solution)
0.1% (v/v) Tween 20 (Sigma, St. Louis, MO)

Method. Embryos are collected from population cages that each contain 90 ml of well-fed flies that are maintained at 25° and 50% relative humidity.

Embryos collected for between 1 and 2 hr are aged at 25° until they have reached the desired stage of development. They are then harvested with deionized water using a fine Nitex mesh (Tetko), blotted dry, weighed, dechorionated by submersion in 50 ml of 50% Clorox bleach for 2 min, and then rinsed thoroughly with deionized water to remove all traces of bleach. Six cages of wild-type flies typically lay 3 g of embryos per hour.

The following steps are carried out in a cold room. To irradiate the dechorionated embryos, 3–6 g of embryos is resuspended in 40 ml of ice-cold 0.1% Tween 20 and distributed between two 9 × 13-cm plastic trays. The trays are placed on ice, 3 cm below the four 15-W, 254-nm bulbs of a Fotodyne DNA transfer lamp (Model 2-1500). Embryos are irradiated for a total of 30 min and are shaken gently every 5 min to reorient them. After irradiation, embryos are collected onto a fine Nitex mesh, blotted dry, transferred to disposable polypropylene centrifuge tubes (Sarstedt), and frozen in liquid nitrogen. Cross-linked embryos can be stored at −70° for up to 2 years without loss of signal.

Comments. The UV irradiation time given is optimal for *Drosophila* embryos. Irradiation of tissue culture cells requires a total of only 4–6 min.[23,24]

B. Purification of Chromatin

Solutions

1 *M* dithiothreitol (DTT); stored at −20° for no more than 3 months

200 m*M* phenylmethylsulfonyl fluoride (PMSF) (Sigma) in 100% ethanol; stored at −20° for no more than 1 month

Nuclear incubation buffer (NIB): 0.3 *M* ultrapure sucrose, 15 m*M* Tris–HCl, pH 7.5, 15 m*M* NaCl, 60 m*M* KCl, 5 m*M* $MgCl_2$, 0.1 m*M* EDTA, pH 8.0, 0.1 m*M* EGTA; filter sterilized using a 0.22-μm Nalgene filter and stored at 4°. Directly prior to use, DTT is added to 0.5 m*M* and PMSF is added to 1 m*M* from the just-described stock solutions

20% (v/v) Triton X-100 (Sigma); filter sterilized using a 0.22-μm Nalgene filter and stored at room temperature

Nuclear lysis buffer: 10 m*M* Tris–HCl, pH 8.0, 100 m*M* NaCl, 1 m*M* EDTA, pH 8.0, 0.1% NP-40; filter sterilized using a 0.22-μm Nalgene filter and stored at 4°. Directly prior to use, PMSF is added to 1 m*M* from the stock described previously

20% (w/v) *N*-lauroylsarcosine (Sarkosyl) (Sigma); filter sterilized using a 0.22-μm Nalgene filter and stored at room temperature

CsCl buffer: 0.5% (w/v) Sarkosyl, 1 m*M* EDTA, pH 8.0; filter sterilized using a 0.22-μm Nalgene filter and stored at room temperature

CsCl gradient solutions:

1.75 g/ml is obtained by dissolving 400 g of CsCl in 300 ml of CsCl buffer
1.5 g/ml is obtained by dissolving 66.7 g of CsCl in 83.3 ml of CsCl buffer
1.3 g/ml is obtained by dissolving 40 g of CsCl in 90 ml of CsCl buffer

The three solutions are filter sterilized using a 0.22-μm Nalgene filter and stored at room temperature. Directly prior to use, PMSF is added to 1 m*M* from the above stock.

10× dialysis buffer, pH 8.0: 0.5 *M* Tris, 20 m*M* EDTA; solid Tris base and disodium EDTA are dissolved in water and the pH is adjusted using HCl. The solution is autoclaved and then stored at 4°

1× dialysis buffer: 10× dialysis buffer is diluted in autoclaved, deionized water. Directly prior to use, PMSF is added to 1 m*M* from the stock described previously

Method. All chromatin purification steps are carried out at 4°, except for buoyant density ultracentrifugation. The amount of embryos to be processed should be calculated from Table I, and varies with the developmental stage of the embryos. Batches of 5–10 g of frozen embryos are disaggregated in 35 ml of NIB buffer with one stroke of a motorized Teflon pestle tissue homogenizer (Thomas, Swedesboro, NJ) at 8000 rpm. Embryos are then homogenized further at 7000 rpm with two more strokes. The homogenate is then passed through prewetted Miracloth (Calbiochem, La Jolla, CA). Remaining batches of the same stage frozen embryos are homogenized and then these homogenates are pooled. Cells are lysed by the addition of Triton X-100 to a final concentration of 0.3% while the pooled, filtered homogenate is stirred continuously in a beaker for 1 min. Then, nuclei are pelleted at 4000 rpm for 15 min at 4° in a SS34 rotor (Sorvall, Newtown, CT).

The supernatant is removed by aspiration. The amount of nuclei to be loaded on one SW28 gradient (see Table I) is resuspended in 8.1 ml of nuclear lysis buffer and then completely disaggregated with a glass B

TABLE I
PARAMETERS FOR CHROMATIN PURIFICATION

Stage embryo[a]	Weight (g) of embryos processed per SW28 tube	Chromatin yield (μg) per gram of embryo
4–5a	25–35	50
5b–8	15–25	150–200
9–10	10–15	400–500
11	5–8	600–800

[a] From J. A. Campos-Ortega and V. Hartenstein, "The Embryonic Development of *Drosophila melanogaster*." Springer-Verlag, Berlin, 1997.

Dounce (Bellco, Vineland, NJ). Nuclei are lysed by the addition of 1/10 volume of 20% Sarkosyl, followed by rapid vortexing. The chromatin is sheared by passing the viscous solution twice through an 18-gauge, $1\frac{1}{2}$ inch needle and twice through a 25-gauge, $1\frac{1}{2}$ inch needle. This solution is stored on ice while the CsCl gradients are prepared.

The CsCl gradients are prepared in 25 × 89-mm, polyallomer, SW28 ultracentrifuge tubes (Beckman, Fullerton, CA) and contain 18.5 ml of 1.75 g/ml CsCl, 6.0 ml of 1.5 g/ml CsCl, and 3.5 ml of 1.3 g/ml CsCl. Nine milliliters of nuclear lysate is layered on top of each gradient. These centrifuge tubes are then spun for 40 hr at 25,000 rpm in a SW28 rotor at 20°. The purified chromatin is dripped from the gradient by inserting an 18-gauge, $1\frac{1}{2}$ inch needle 1.5 cm from the bottom of the centrifuge tubes. One-milliliter fractions are collected, and 2-μl aliquots from each fraction are electrophoresed on a 1% agarose gel containing 0.3 μg/ml ethidium bromide to identify fractions containing the bulk of the chromatin. Typically, four to six fractions are pooled, which have an average density of 1.66 g/ml. CsCl is removed by dialysis in Spectra/por Spec 2 dialysis tubing (Spectrum, Houston, TX) against three changes of 2 liters of 1× dialysis buffer, each dialysis lasting 2 hr at 4°. Any insoluble material is removed by centrifugation at 2000 rpm for 10 min in a clinical centrifuge. Most preparations contain only trace amounts of RNA. Thus, the DNA concentration can be determined from the optical density of the solution at 260 nm. If significant amounts of RNA are present, the DNA concentration is instead estimated by comparing the intensity of ethidium bromide staining of the DNA in an agarose gel to that of DNA of a known concentration. The chromatin is then quick frozen in liquid nitrogen and stored at −70°, where it is stable for up to 2 years.

Comments. Because the chromatin purification and immunoprecipitation procedures include prolonged incubations at room temperature or at 37° in the presence of detergents, trace amounts of proteases can degrade some sensitive proteins. These proteases probably derive from low-level bacterial or fungal growth in stock solutions. To prevent this problem, all solutions are sterilized where possible, all glassware is cleaned rigorously, plasticware is kept dust free, and PMSF is added to solutions as directed. To test for the presence of proteases, the protein of interest is incubated at 10 μg/ml in various solutions overnight. (Because many proteases are more active in the presence of detergents, buffers should be tested with any detergent included.) The degree of proteolysis is determined the following day by SDS–PAGE. With these simple precautions, this problem can be eliminated.

The buoyant density of UV cross-linked chromatin is virtually indistinguishable from that of uncross-linked DNA on CsCl gradients because the level of cross-linking achieved is very low. Uncross-linked protein has a buoyant density of 1.3 g/ml.

C. Restriction Digestion and Immunoprecipitation of Chromatin

Solutions

10× restriction enzyme buffer; buffers supplied with commercial enzymes are not used. Aliquots are stored frozen at −20°

5 mg/ml bovine serum albumin (BSA), Fraction V, radioimmunoassay (RIA) grade (BSA) (Sigma); stored at −20°

20% (v/v) Triton X-100 (Sigma); filter sterilized using a 0.22-μm Nalgene filter and stored at room temperature

200 m*M* PMSF (Sigma) in 100% ethanol; stored at −20° for no more than 1 month

0.5 *M* EDTA, pH 8.0; autoclaved and then stored at room temperature

20%(w/v) Sarkosyl (Sigma); filter sterilized using a 0.22-μm Nalgene filter and stored at room temperature

Sonicated UV cross-linked chromatin; 400–800 μg/ml of chromatin purified by the procedure described earlier is sonicated to an average size of less than 1 kb

5 mg/ml leupeptin (Boehringer Mannheim); dissolved in water and stored at −20°

0.5 mg/ml pepstatin (Boehringer Mannheim); dissolved in ethanol and stored at −20°

2.0 mg/ml aprotinin (Sigma); supplied at this concentration. Stored at 4°

Wash solution 1: 50 m*M* Tris–HCl, pH 8.0, 2 m*M* EDTA, pH 8.0, 0.2% Sarkosyl; filter sterilized using a 0.22-μm Nalgene filter and stored at 4°. Directly prior to use in the immunoprecipitation procedure, PMSF is added to 1 m*M*, leupeptin to 4.0 μg/ml, pepstatin to 0.4 μg/ml, and aprotinin to 1 μg/ml; all from the stock solutions described previously

20% (w/v) *Staphylococcus aureus* protein A-positive cells (Staph A cells): 1 g of lyophilized Staph A cells (Boehringer Mannheim) is resuspended with 10 ml of wash solution 1 (protease inhibitors are not added to this solution for this procedure), pelleted by centrifugation, and washed one more time with 10 ml of wash solution 1. Cells are then resuspended to 20 ml of 1× phosphate-bufferd saline (PBS), 3% (w/v) SDS, 10% (v/v) 2-mercaptoethanol, heated in a boiling water bath for 30 min, washed twice with 20 ml wash solution 1, resuspended to 20% (w/v) with 5 ml of fresh wash solution 1, frozen in liquid nitrogen in 100 μl-aliquots, and stored at −70°. These aliquots are stable for at least 2 years. They should be thawed only once. Any excess cells should be discarded after one set of experiments

Wash solution 2: 100 m*M* Tris–HCl, pH 9.0, 500 m*M* LiCl, 1% (v/v) NP-40, 1% (w/v) deoxycholic acid (Sigma); filter sterilized using a 0.22-μm Nalgene filter and stored at 4°. Directly prior to use, PMSF is added to 1 m*M* from the stock solution, described previously

Elution buffer: 50 m*M* $NaHCO_3$–NaOH, pH 10.0, 1% SDS, 1.5 μg/ml sonicated calf thymus DNA (of average size less than 1 kb); filter sterilized using a 0.22-μm Nalgene filter and stored at room temperature

Proteinase K solution: 50 m*M* Tris–HCl, pH 7.5, 10 m*M* EDTA, pH 8.0, 0.3% SDS; filter sterilized using a 0.22-μm Nalgene filter and stored at room temperature. Immediately before use, proteinase K (Boehringer Mannheim) is added to 1 mg/ml

3 *M* sodium acetate, pH 5.3; autoclaved and then stored at room temperature

20 mg/ml yeast RNA (Sigma, type VI from *Torula* yeast); RNA is dissolved in deionized water and insoluble material is removed by centrifugation. The solution is extracted three times against an equal volume of phenol/chloroform, once against 1 volume of chloroform, ethanol precipitated, and resuspended to 20 mg/ml

5× loading dye: 20% (w/v) ultrapure sucrose, 5× TBE, 0.1% (w/v) bromphenol blue, 5 m*M* EDTA, pH 8.0; filter sterilized using a 0.22-μm Nalgene filter and stored at room temperature

1× loading dye: Made fresh on the day of use. 5× loading dye is diluted with water and RNase (Boehringer Mannheim, DNase-free) is added to a final concentration of 0.5 μg/ml

Method. Chromatin is thawed by placing polypropylene storage tubes in water at 4°. (Any excess chromatin is refrozen in liquid nitrogen and stored at −70°.) Restriction digestions are carried out in 1.7 ml and contain 450 μg of DNA, 1× restriction enzyme buffer, 100 μg/ml BSA, 0.01% (v/v) Triton X-100, 1 m*M* PMSF, 1 μl (0.5 μg) RNase (Boehringer Mannheim, DNase free), and 450 units of restriction enzyme (Biolabs). Reactions are set up in 2.0-ml microfuge tubes using the stock solutions described earlier. After overnight digestion at 37°, another 250 units of restriction enzyme is added and the digestion is allowed to continue for another 2 hr. A 2-μl aliquot of the reaction is examined by electrophoresis on a 1% agarose gel to ensure that the digestion is complete. The reaction is terminated by the addition of 68 μl of 0.5 *M* EDTA, pH 8.0 (final concentration 20 m*M*), and then 25.5 μl of 20% Triton X-100 (final concentration 0.3%) and 4 μl of 20% Sarkosyl (final concentration 0.05%) are added. Any insoluble material is removed by centrifugation at full speed in a microfuge for 15 min at 4°. The supernatant is then transferred to a fresh 2-ml microfuge tube.

To immunoprecipitate DNA cross-linked to the protein of interest, 1–2 μg of affinity-purified rabbit antibody is added to each 1.7-ml reaction mixture. The antibody is incubated with chromatin on a Nutator (Clay Adams, Parsippany, NJ) for 3 hr at 4°. While this is underway, 100 μl of 20% Staph A cells is blocked by incubation with 200 μl of 400–800 μg/ml sonicated UV cross-linked chromatin for 3 hr on a Nutator at 4°. The Staph A cells are pelleted by centrifugation for 1 min in a microfuge and the supernatant is removed by aspiration. The cells are resuspended in 500 μl of wash solution 1, repelleted, and resuspended in 100 μl of wash solution 1. This suspension of cells is referred to as the "blocked 20% Staph A cell solution."

After the 3-hr incubation of antibody with the 1.7-ml chromatin samples, aggregated material is removed by centrifugation for 15 min at 4°. The remaining steps are carried out at room temperature. Remaining particulate matter is removed by passing the supernatants through a 3-ml syringe fitted with a 0.22-μm Millex-GP filter (Millipore, Bedford, MA). Twenty-five microliters of the "blocked 20% Staph A cell solution" is added to each 1.7-ml immunoprecipitation reaction, and this mixture is nutated at room temperature for 15 min. The antibody/cross-linked–protein/DNA complexes are pelleted by centrifugation for 1 min in a microfuge at room temperature. (The supernatant is saved to provide standards for the total amount of DNA in the immunoprecipitation reaction. This sample is referred to as "total DNA"; see later.)

Staph A cells and the attached immunoprecipitated complex are washed by successive rounds of resuspension and centrifugation. Cells are washed twice with 1.4 ml of wash solution 1 and four times with 1.4 ml of wash solution 2. Each time, pellets are resuspended in 200 μl of wash solution by repeated pipetting up and down. Any Staph cells adhering to the tip are flushed out with a further 200 μl of wash solution. A further 1 ml of wash solution is added to a final volume of 1.4 ml and the suspension is mixed briefly. Cells are collected by centrifugation for 1 min in a microfuge. Before the first and last washes, the initial 200 μl of resuspended cells is transferred to a new 1.5-ml microfuge tube to remove any chromatin that may be bound to the tubes.

After the wash steps, the immunoprecipitated chromatin is eluted from the Staph A cells by resuspending these cells in 100 μl of elution solution and vortexing them for 10 min at room temperature. Cells are pelleted by centrifugation and the supernatant is saved for proteolytic digestion. The pelleted cells are reextracted two more times with 100 μl of elution solution.

The three elutions are pooled into one 300-μl sample. The extracted cells are discarded.

To remove protein from the DNA, 200 μl of proteinase K solution is added to each 300-μl sample of eluted, immunoprecipitated complex and the mixture is allowed to digest overnight at 65°. At the same time, 1% of the "total DNA" sample (typically 17 μl) is diluted to a volume of 300 μl with elution solution, 200 μl of proteinase K solution is added, and this mixture is also digested overnight at 65°. DNA from the digestions of both immunoprecipitated complexes and "total DNA" samples is recovered by the addition of 40 μg of yeast RNA, 50 μl of 3 *M* sodium acetate, pH 5.3, and 1.25 ml of 100% ethanol. After mixing, the ethanol precipitation mixtures are left at −70° until they are frozen (typically 45 min). The DNA is pelleted in a microfuge for 15 min at 4°. After the pellets have been washed in 75% ethanol and dried, DNA from the immunoprecipitated samples is resuspended in 20 μl of 1× loading dye and samples of 1% of "total DNA" are resuspended in 1 ml of 1× loading dye.

Comments. It is often convenient to set up single large volume restriction digestions of DNA for multiple immunoprecipitation reactions, rather than multiple 1.7-ml digestions, then to divide the solution into 1.7-ml aliquots after digestion, which are then ready for separate immunoprecipitation reactions. Restriction enzymes are chosen to produce 0.5–6 kb DNA fragments within the gene locus under study.

Affinity-purified polyclonal antibodies are much preferred for two reasons: they reduce the chances of aberrant cross-reaction with proteins other than the one under investigation and less antibody needs to be added to immunoprecipitations. Consequently, less Staph A cells have to be added, reducing any nonspecific association of chromatin with the cells. The optimum amount of antibody to add to immunoprecipitations varies and should be determined by titration. If antibodies other than rabbit polyclonal antibodies are used, it may be necessary to employ a secondary antibody to ensure efficient binding to the Staph A cells (see Boehringer Mannheim catalogue for details). Any secondary antibody is added after the incubation with the primary antibody and the reaction is allowed to proceed for 1 hr at 4°.

D. Electrophoresis and Southern Blot

Solutions

20× SSPE, pH 8.0: 0.2 *M* NaH_2PO_4, 3.6 *M* NaCl, 20 m*M* EDTA; after the dry ingredients are dissolved, the pH is adjusted with NaOH

Denaturation solution: 1.5 *M* NaCl, 0.5 *M* NaOH

Neutralization solution, pH 7.2: 0.5 *M* Tris, 1.5 *M* NaCl, 1 m*M* EDTA; after the dry ingredients are dissolved, the pH is adjusted with HCl

50× Denhardt's: 1% Ficoll 400 (Sigma), 1% polyvinylpyrrolidone (Sigma), 10 mg/ml BSA (Sigma); stored at −20°

1 mg/ml calf thymus DNA (Sigma); sonicated to less than 1 kb in length. Half of the sample is denatured by heating in a boiling water bath for 15 min and then snap cooled on ice. The native and denatured DNAs are recombined and stored at −20°

1 mg/ml *Escherichia coli* DNA (Sigma); sheared through an 18-gauge needle and stored at −20°

Prehybridization solution: 6× SSPE, 50% formamide (Fluka, Ronkonkoma, NY), 5× Denhardt's, 90 μg/ml calf thymus DNA, 10 μg/ml *E. coli* DNA, 10% (w/v) dextran sulfate (Pharmacia, Piscataway, NJ), 5% SDS, 1% Sarkosyl; this is prepared just before use, using the just-described stock solutions. Solid dextran sulfate, SDS, and Sarkosyl are added last while the solution is stirred rapidly in a beaker. The solution is heated for 15–20 min to dissolve the dextran sulfate, then filtered through a 5.0-μm Millex SV filter (Millipore)

4 *M* ammonium acetate, pH 4.5; stored at room temperature

5×10^6 cpm/ng DNA probe. Restriction fragments of 1–7 kb are purified from their associated plasmid DNAs by two successive rounds of agarose gel electrophoresis. Twenty nanograms of restriction fragment is labeled with 150 μCi of 6000 Ci/m*M* dCTP in 38 μl using a random prime label kit (Amersham, Arlington Heights, IL). After the labeling reaction, unincorporated nucleotides are removed by two successive ethanol precipitations in the presence of ammonium acetate. (For the first precipitation, add 38 μl of 4 *M* ammonium acetate, pH 4.5, and 60 μg of yeast RNA, followed by 150 μl of ethanol.) After the second precipitation, the pelleted DNA is resuspended in 150 μl of TE buffer, pH 7.9, and denatured by heating in a boiling water bath for 5 min followed by snap cooling on ice

Methods. A 0.7% agarose, 1× TBE gel is poured with 1× 6-mm wells. Dilutions of the "total DNA" sample are prepared that contain 0.005, 0.0025, 0.001, 0.0005, 0.00025, and 0.0001% of the total DNA, each in 20-μl aliquots. These samples are loaded into adjacent wells of the gel, next to their associated immunoprecipitated DNA samples (see Fig. 2). DNA size markers are also loaded in a separate well. The gel is run at 5 V/cm until the bromphenol blue dye has traveled 6–8 cm. The gel is then stained by incubating for 5 min with 3–4 gel volumes of 1× TBE containing 1.5 μg/ml ethidium bromide. The background staining of the agarose is reduced by soaking the gel for another 10 min in 3–4 volumes of 1× TBE. Excess

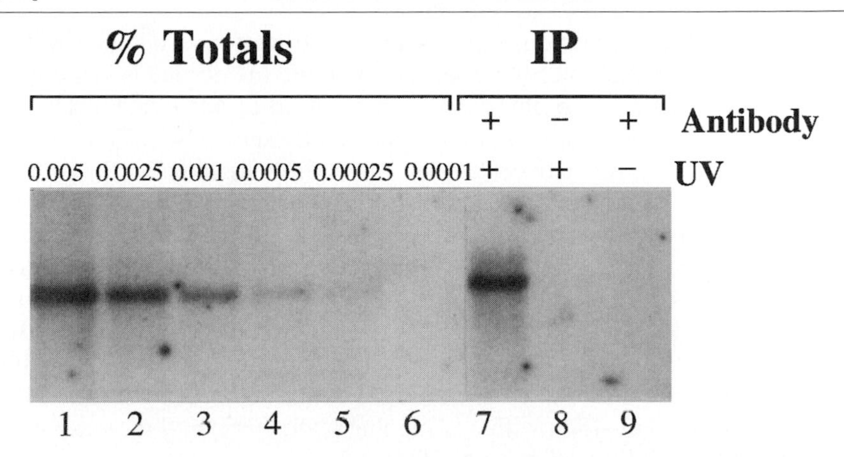

FIG. 2. Eve protein binds to the *eve* promoter in *Drosophila* embryos (adapted from Walter *et al.*[11]). Results of an *in vivo* UV cross-linking experiment carried out on UV-irradiated 4- to 5-hr-old embryos (lanes 7 and 8) or unirradiated 4- to 5-hr embryos (lane 9). Chromatin from embryos was isolated and digested with *BglI*, and 450 μg of chromatin was immunoprecipitated with affinity-purified anti-Eve antibodies (lanes 7 and 9) or mock precipitated (lane 8). Immunoprecipitated DNA was analyzed on a Southern blot using *eve* promoter sequences as a probe detecting a 7.3-kb *eve* promoter fragment extending from −0.3 to −7.6 kb. For quantitation of the immunoprecipitated material, lanes 1–6 contain a known percentage of the DNA present in a single immunoprecipitation reaction prior to the addition of antibody (referred to as % total DNA).

gel is removed, and the gel is photographed alongside a ruler in such a way that the distance that a DNA has migrated can be used to estimate its length (see Sambrook *et al.*[25]).

DNA is denatured by soaking the gel in 3–4 volumes of denaturation solution for 30 min while rocking. The pH of the gel is returned to neutral by incubating twice with 3–4 volumes of neutralization solution, each time for 15 min. An uncharged Hybond-N nylon membrane (Amersham) is prewetted in distilled/deionized water for 10 min and then equilibrated with 20× SSPE for 10 min. The agarose gel and nylon membrane are assembled into a Southern blot and the DNA is transferred using 20× SSPE for at least 12 hr as described by Sambrook *et al.*[25] After transfer, the blot is disassembled and the nylon membrane is baked at 80° for 30 min in a vacuum oven. The DNA is then cross-linked to the dry membrane by UV irradiating the blot for 5 sec with a Fotodyne DNA transfer lamp.

To hybridize the blot, the nylon membrane is placed in a 35 × 225-mm

[25] J. Sambrook, E. F. Fritsch, and T. Maniatis, "Molecular Cloning: A Laboratory Manual." Cold Spring Harbor Press, Cold Spring Harbor, NY, 1989.

bottle (Integrated Separation Systems/Owl) and wetted in distilled water for 1–2 min. After the water has been drained, the membrane is incubated overnight with 10 ml of prehybridization solution at 42° in a UniTherm 6/12 hybridization oven (Integrated Separation Systems/Owl). The following day, denatured, radioactively labeled DNA probe is added to the prehybridization solution to $1–2 \times 10^7$ cpm/ml. The blot is incubated with the probe for 18 hr at 42°. Then, the excess solution is drained. The blot is placed in a large plastic box and is washed twice for 15 min in 200 ml of 2× SSPE, 0.1% SDS at room temperature. The blot is further washed with 200 ml 0.1× SSPE, 0.1% SDS for 15 min at 65° and then for at least 1 hr in 200 ml 0.1× SSPE, 0.1% SDS at 65°. If background radiation on the membrane is more than 1–5 cps, washing is continued overnight. The blot is first exposed for 12–15 hr with an Imaging plate (Fuji) and then exposed for 7–10 days with autoradiographic film.

Binding of a protein to different DNA fragments can be determined from one immunoprecipitation reaction by stripping and reprobing blots with different DNAs. Five hundred milliliters of 0.1% SDS is boiled and then removed from the heat. The blot is immediately placed in this solution and shaken until it has cooled to room temperature (2 hr). The blot is then hybridized with another DNA probe, as described earlier. Although the sensitivity of blots gradually decreases, blots can generally be stripped and reprobed three to six times.

Comments. In our hands, the optimum time for UV irradiation of the nylon membrane does not vary between batches of membrane, but the manufacturer recommends determining this time for each batch. The Southern blot protocol described can detect either 10 fg of a 4-kb restriction fragment or a restriction fragment from a digest of 225 pg of total *Drosophila* genomic DNA. It is strongly recommended that the protocol described here be used as it has been found to work with high reproducibility and to give consistently low backgrounds.

Expected Results and Controls

The amount of DNA immunoprecipitated is dependent on the cross-linking efficiency of the protein under study, the number of sites the protein binds in a given DNA fragment, the degree to which each binding site is occupied, and the percentage of cells that express the protein. A single molecule of RNA polymerase II gives an immunoprecipitation signal equivalent to 0.2% of the total amount of a DNA fragment in an immunoprecipitation reaction.[6,17] However, sequence-specific transcription factors such as Eve and Zeste cross-link much less efficiently and give signals equivalent to only 0.005% of the total amount of a DNA fragment in an immunoprecipitation reaction, even though they bind to multiple functional sites within the DNA fragments precipitated (Fig. 2).[11–13] However, because these signals are 100-fold higher than the smallest amount of DNA that can be detected, they are still well above background.

It is important to include a number of controls to ensure that the immunoprecipitation signals obtained are due to the protein/DNA interaction under investigation. These controls are (1) mock immunoprecipitations that contain a nonspecific antibody, such as a rabbit antimouse immunoglobulin antibody (Promega, Madison, WI). These should give no detectable signal (Fig. 2, lane 8). (2) Immunoprecipitation reactions using chromatin purified from embryos that have not been UV irradiated. Again, these should not give a signal (Fig. 2, lane 9). (3) Antibodies should be used that recognize separate regions of the protein being studied. These antibodies should give similar results to each other.[11,16] (4) Cross-linked chromatin derived from cells that do not express the protein of interest should be assayed. No DNA should be detected in immunoprecipitations of these reactions.[11] It is also important to repeat immunoprecipitations at least three to five times and to test different batches of chromatin. The variation between duplicate experiments should be small; values within 1 SD lie between ±20% of the mean signal.

Binding should be tested both to probable target DNAs and to DNAs to which there is no reason to suspect that the protein may bind. Some proteins have been shown to bind highly specifically *in vivo*. For example, Zeste is detected only on a cluster of high-affinity, functional sites in the *Ubx* promoter and is not found on other genes tested (Fig. 3).[11,13] Other proteins have been found to bind a much broader range of DNA sequences. For example, the two homeoproteins Eve and Ftz bind at significant levels to all DNA fragments tested (Fig. 4). They bind most strongly to DNA fragments throughout the length of three genes that they are known to regulate, the *eve, ftz,* and *Ubx* genes, but they also bind at only 2- to 10-fold lower levels to four genes initially chosen as unlikely targets, the *hsp70, rosy, actin 5C,* and *Adh* genes. Subsequent experiments indicated that the four unexpected target genes are regulated by Eve, Ftz, and other closely related homeoproteins.[26] This surprising result suggests that these proteins act much more broadly than previously assumed and has led to the proposal of a new model for how homeoproteins control animal development (reviewed by Biggin and McGinnis[27]).

The broad cross-linking of Eve and Ftz raises the question of whether much of the cross-linking observed results from molecules bound nonspe-

[26] Z. Liang and M. D. Biggin, *Development* **125,** 4471 (1998).
[27] M. D. Biggin and W. McGinnis, *Development* **124,** 4425 (1997).

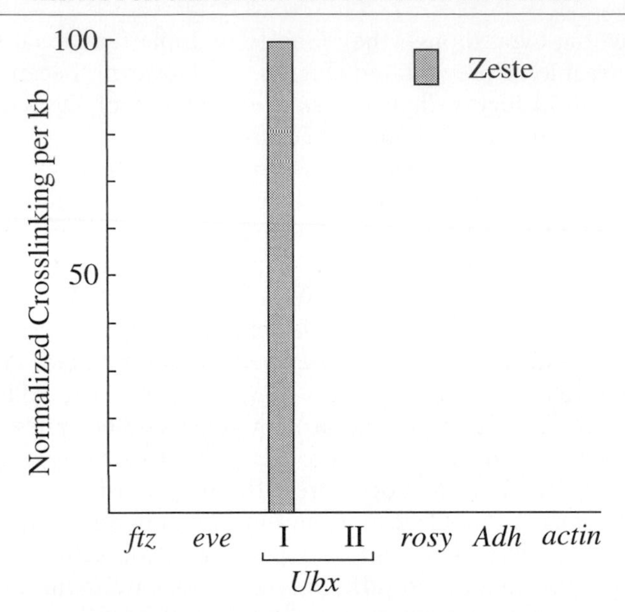

FIG. 3. Zeste binds highly selectively to the proximal promoter of the *Ubx* gene (adapted from Walter *et al.*[11]). Data are expressed as the relative level of zeste cross-linking divided by the length of the DNA fragment and thus represent the mean density of binding across each DNA. The DNA fragments, ranging between 1 and 7 kb in length, contain promoter elements from the genes indicated. Binding is strongly detected on functional DNA sites at the proximal promoter of *Ubx* (*Ubx* I), but not on the adjacent 3108 *Ubx* DNA fragment (*Ubx* II) or on other genes.

cifically to DNA (i.e., interacting in a sequence-independent manner, largely through electrostatic interactions with the phosphate backbone of DNA). This is unlikely for the following reasons. First, nonspecific binding of Zeste is not detected *in vivo* (Fig. 3). Second, all of the genes bound by Eve and Ftz *in vivo* contain high- and moderate-affinity homeoprotein recognition sites at a density of 5–10 binding sites per kilobase.[12,28] Third, at physiological salt concentrations of 150–180 m*M* KCl, *in vitro* UV cross-linking does not detect nonspecific DNA binding by Eve or other sequence-specific DNA-binding proteins.[12,23,29] (It should be noted that this is not true at nonphysiological salt concentrations of 10 m*M* KCl[29].) Thus, it is most probable that the *in vivo* UV cross-linking method will principally detect only sequence-specific DNA binding for most transcription factors.

[28] D. D. Dalma-Weiszhausz and M. D. Biggin, unpublished data.

[29] S.-Y. Lin and A. D. Riggs, *Proc. Natl. Acad. Sci. U.S.A.* **71,** 947 (1974).

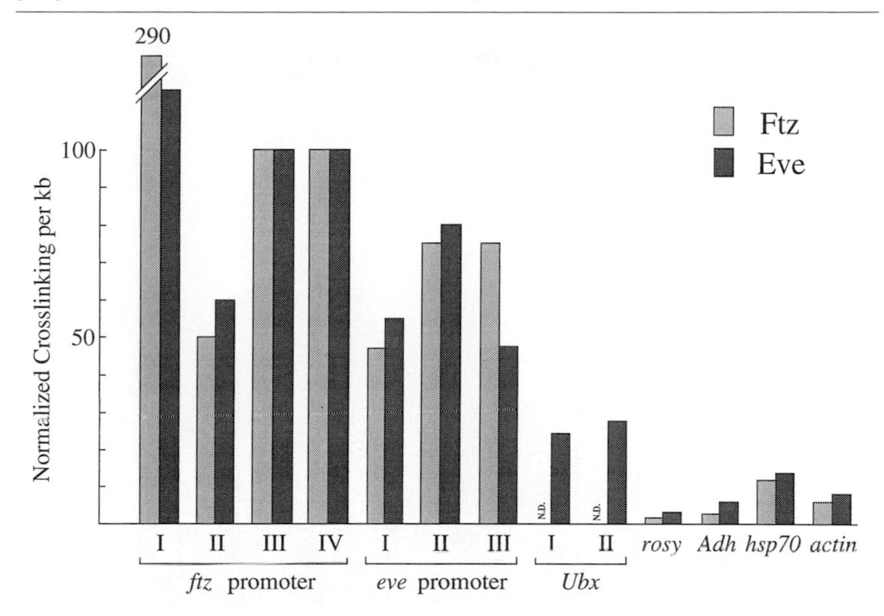

FIG. 4. Eve and Ftz homeoproteins bind with similar specificity to a broad array of DNA sites in *Drosophila* embryos. Relative levels of Eve protein (dark shading) and Ftz protein (light shading) UV cross-linking to various DNA fragments *in vivo* (adapted from Walter *et al.*[11]). Data are expressed as the relative level of cross-linking divided by the length of the DNA fragment, and thus represent the mean density of binding across each DNA. The DNA fragments, ranging between 1 and 7 kb in length, contain promoter elements from the genes indicated. *eve, ftz,* and *Ubx* genes are known to be regulated by Eve and Ftz. *actin 5C, adh, hsp70,* and *rosy* genes were chosen as they were initially thought to be unlikely targets of Eve or Ftz. Data include several contiguous DNA fragments from *ftz, eve,* and *Ubx* promoters. Binding of Ftz protein to the *Ubx* promoter has not been examined (N.D.).

Speculations on Transcription Factor DNA Binding in Higher Eukaryotes

In bacteria, transcriptional regulators such as *lac* and λ repressor are present at between 15 and 150 molecules per cell.[30,31] These proteins generally recognize specific 16- to 20-bp DNA sequences with dissociation constants of 10^{-10}–10^{-12} *M,* and can also bind DNA in a sequence-independent manner with dissociation constants of 10^{-4}–10^{-5} *M.* Typically, each regulator binds to only two to six high-affinity sites in the genome through which they regulate transcription of one to four operons. The majority of the

[30] S.-Y. Lin and A. D. Riggs, Cell 4, 107 (1975).

[31] M. Ptashne, "A Genetic Switch: Phage Lambda and Higher Organisms." Blackwell Scientific, Cambridge, MA, 1992.

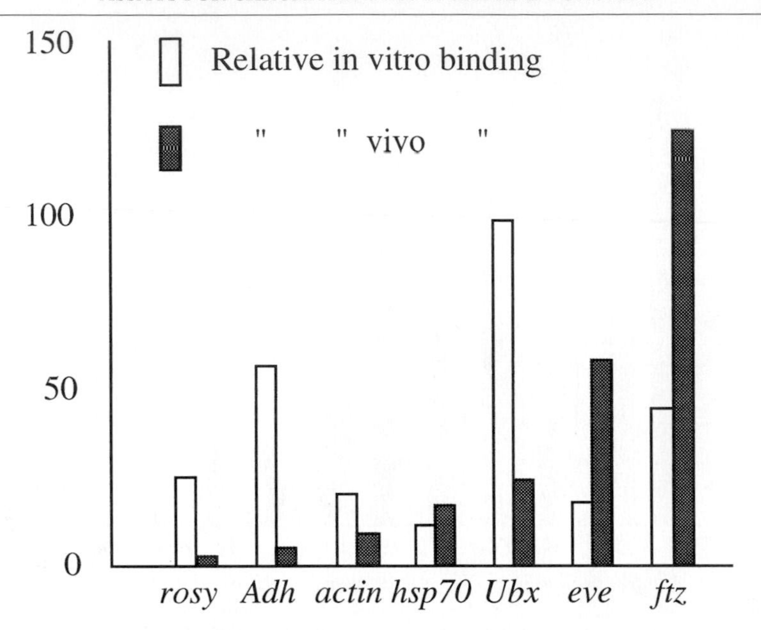

FIG. 5. Conditions in the embryo alter the relative occupancy of Eve protein to a range of genes. Relative levels of Eve protein cross-linking *in vivo* (dark shading) and binding *in vitro* (no shading) to various DNA fragments (adapted from Walter and Biggin[12]). Data are expressed as the relative level of binding or cross-linking divided by the length of the DNA fragment. The DNA fragments, ranging between 1 and 7 kb in length, contain promoter elements from the genes indicated.

remaining regulatory molecules are thought to bind in a sequence-independent manner throughout the rest of the genome at a density of one molecule for every 30–300 kb.[30–33] In *Drosophila*, each homeoprotein is present at 50,000 molecules per cell, or more, and *in vitro* they bind to highly degenerate six nucleotide sequences with dissociation constants of 10^{-9}–$10^{-;10}$ *M*.[27] Because of the high frequency of these specific sites in the *Drosophila* genome, it is suggested that, unlike prokaryotic regulators, the majority of homeoprotein molecules may not be bound to DNA in a sequence-independent manner, but, instead, the bulk of the molecules might occupy specific DNA sites. This should represent a minimum density of one molecule for every 4 kb of the genome, but on the most strongly bound genes, the density could be up to 10 times higher than this.

[32] P. H. von Hippel, A. Revzin, C. A. Gross, and A. C. Wang, *Proc. Natl. Acad. Sci. U.S.A.* **71,** 4808 (1974).

[33] S.-W. Yang and H. A. Nash, *EMBO J.* **14,** 6292 (1995).

In the Introduction, the question was raised of how conditions *in vivo* might alter the DNA-binding specificities of proteins. In this regard, it is very revealing to compare *in vitro* and *in vivo* binding data quantitatively. Figure 5 compares the relative level of Eve protein binding *in vitro* and cross-linking *in vivo* to a number of different genes. Data show that the overall specificity *in vivo* differs from that *in vitro*. For example, Eve cross-links *in vivo* more strongly to the *eve* gene than it does to the *rosy* or *Adh* genes, but *in vitro* the opposite preference is seen. Because UV cross-linking accurately reflects the relative levels of DNA binding,[12] differences between *in vitro* and *in vivo* data presumably reflect differences in relative occupancy. A similar difference is also seen for Zeste. This protein binds very selectively *in vitro* and *in vivo*, but on strongly bound DNA fragments, 5- to 10-fold differences in preferences exist between *in vitro* and *in vivo* data.[12] Although speculative, perhaps the following generalization may apply: For many proteins, conditions *in vivo*, i.e., the combined effect of chromatin inhibiting DNA binding at some sites and cooperative interactions with other transcription factors increasing DNA binding at other sites, may not radically alter DNA-binding specificities, but may modulate relative levels of occupancy at different sites some 2- to 10-fold.

Our knowledge of the range of DNA sequences that proteins bind *in vivo* is limited. Mechanisms determining binding site selection, especially in complex multicellular organisms, are understood even less. Although existing methods have yielded important data, none can provide all of the information desired. In the future, it will be important to develop approaches that will allow both very high-resolution mapping and quantitation of DNA binding. This should permit a better understanding of this fundamental problem.

Acknowledgments

I am grateful to Janann Ali, Alan Carr, Joseph Toth, Johannes Walter, and Trevor Williams for detailed advice on the UV cross-linking method and for corrections to this manuscript.

[28] DNA–Protein Cross-Linking Applications for Chromatin Studies *in Vitro* and *in Vivo*

By DMITRY PRUSS, IGOR M. GAVIN, SVETLANA MELNIK, and SERGEI G. BAVYKIN

Introduction

Covalent protein–nucleic acid cross-linking is a powerful tool for the analysis of protein–DNA and protein–RNA interactions in various biological systems. It allows a researcher to "freeze" protein–nucleic acid contacts within the context of a nucleoprotein complex and analyze cross-linked species by a variety of methods such as two-dimensional denaturing gel electrophoresis.[1,2] This article describes some of the more recent methodical advances in the field of protein–nucleic acid covalent cross-linking applications. The innovations discussed here include a fast, relatively noninvasive method of cross-linking by radical-producing chemicals, such as bleomycin–iron and phenanthroline–copper complexes[3]; applications of chemical and photochemical cross-linking techniques to nucleosome-containing complexes following *in vitro* nucleosome reconstitution on specific DNA sequences[4–6]; and a time-resolved cross-linking technique for studies of post-replicative chromatin assembly, based on immunoenrichment of newly replicated DNA.[7]

Radical-Induced Protein–Nucleic Acid Cross-Linking *in Vitro* and *in Vivo*

The major limitation for the use of cross-linking techniques is that most of them are utilized for *in vitro* experiments and only UV irradiation has

[1] A. D. Mirzabekov, S. G. Bavykin, A. V. Belyavsky, V. L. Karpov, O. V. Preobrazhenskaya, and K. K. Ebralidse, *Methods Enzymol.* **170,** 386 (1989)

[2] D. Pruss and S. G. Bavykin, *Methods* **12,** 36 (1997).

[3] I. M. Gavin, S. M. Melnik, N. P. Yurina, M. I. Khabarova, and S. G. Bavykin, *Anal. Biochem.* **263,** 26 (1998).

[4] D. Pruss and A. P. Wolffe, *Biochemistry* **32,** 6810 (1993).

[5] J. J. Hayes, D. Pruss, and A. P. Wolffe, *Proc. Natl. Acad. Sci. U.S.A.* **91,** 7817 (1994).

[6] D. Pruss, B. Bartholomew, J. Persinger, J. J. Hayes, G. Arents, E. N. Moudrianakis, and A. P. Wolffe, *Science* **274,** 614 (1996).

[7] S. Bavykin, L. Srebreva, T. Banchev, R. Tsanev, J. Zlatanova, and A. Mirzabekov, *Proc. Natl. Acad. Sci. U.S.A.* **90,** 3918 (1993).

0076-6879/99 $30.00

been applied to a living cell.[8] However, exposure of the probes with extensive doses of UV light to obtain sufficient amounts of protein–nucleic acid cross-links results in a high level of protein degradation and DNA modification that may interfere with the subsequent analysis of cross-linked complexes.[9] Moreover, UV irradiation induces the cross-linking of proteins, mainly to thymine bases that also limits the application of this method.[10] The dimethyl sulfate (DMS)-mediated cross-linking[2] has broader chemical selectivity and does not damage DNA as extensively as UV irradiation, but it is a prolonged process and, therefore, it has only been applied to analysis of protein–DNA interactions *in vitro.* We have demonstrated[3] that radical-producing reagents create irreversibly protein-bound molecular species of the nucleic acids and applied this fast technique to cross-linking *in vivo.*

The methodology of protein–nucleic acid cross-linking by radical-producing systems is based on the chemistry of DNA modification by bleomycin–Fe(III) (BLM-Fe) or 1,10-phenanthroline–Cu(II) (OP-Cu) complexes. Both reagents may generate DNA abasic sites without strand scission in the presence of hydrogen peroxide under anaerobic conditions (resulting in abstraction of H-4′).[11] The aldehyde functional groups of the resulting abasic sites react with histidine or lysine amino acid residues of proteins, resulting in protein–DNA cross-linking by a mechanism similar to the one described for the dimethyl sulfate cross-linking protocol (Fig. 1, II vs IIa). Experiments *in vitro* showed that the OP-Cu complex effectively cross-links proteins to both DNA and RNA without protein degradation or formation of protein–protein cross-links.[3] The radical-induced cross-linking methodology has several advantages over other cross-linking techniques that make this technique unique for studies of protein–nucleic acid interactions: simplicity, applicability to cross-linking of proteins to both DNA and RNA *in vitro* or *in vivo,* and a high yield of cross-linked complexes without extensive protein and nucleic acid damage. Mild reaction conditions and a short cross-linking time also decrease the possibility of redistribution of proteins along DNA during cross-linking.

[8] I. G. Pashev, S. I. Dimitrov and D. Angelov, *Trends Biochem. Sci.* **16,** 323 (1991).

[9] H. Gorner, *Photochem. Photobiol.* **26,** 117 (1994).

[10] J. W. Hockensmith, W. L. Kubasek, W. R. Vorachek, E. M. Evertsz, and P. H. von Hippel, *Methods Enzymol.* **208,** 211 (1991).

[11] W. K. Pogozelski and T. D. Tullius, *Chem. Rev.* **98,** 1089 (1998).

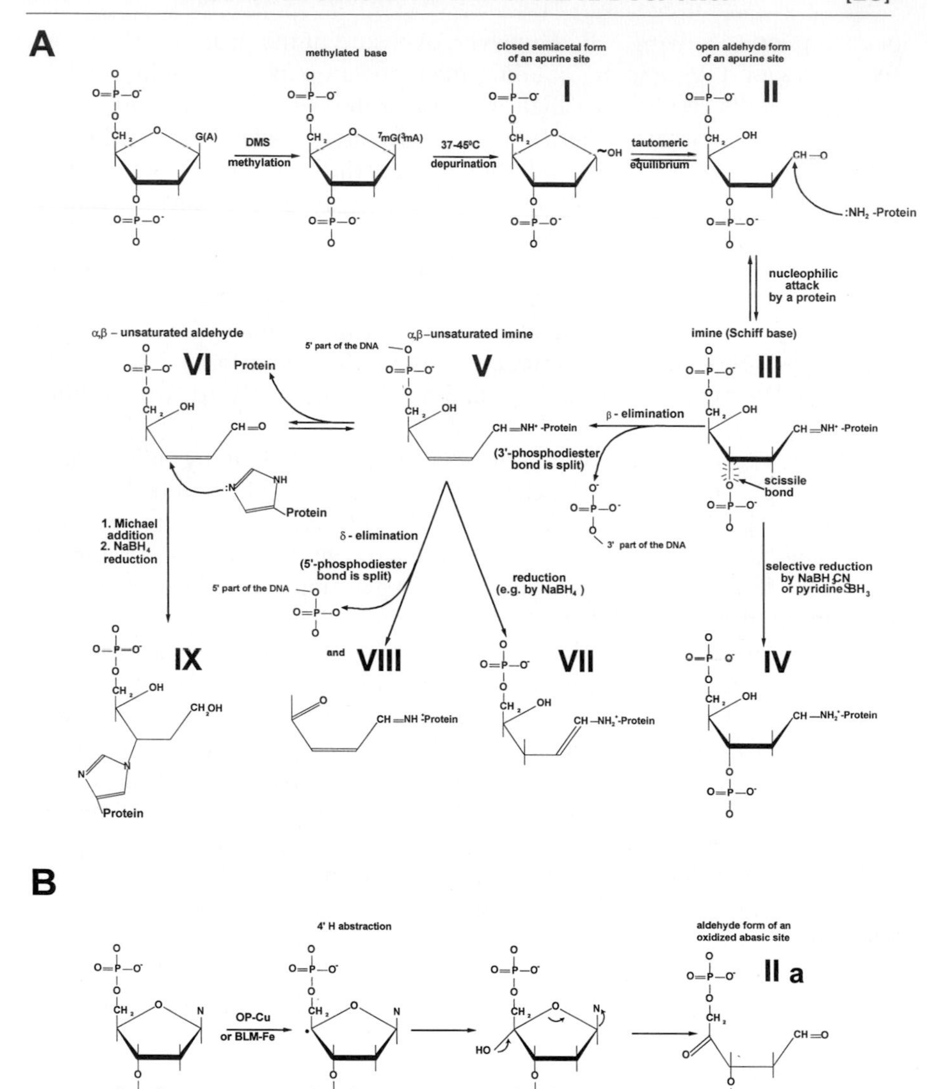

FIG. 1. The chemistry of dimethyl sulfate-induced and radical-induced cross-linking reactions. (A) The classical Mirzabekov cross-linking reaction takes advantage of the fact that the N-glycosidic bonds that form by methylated purine bases are unstable even at neutral pH at room temperature. Depurination of methylated guanines and adenines accelerates at slightly increased temperatures. A free glycosilic center of the resulting apurinic DNA site exists in tautomeric equilibrium between a semiacetal form (**I**) and an aldehyde form (**II**). An aldehyde group may be attacked by a nearby nucleophilic moiety (such as N-terminal α-amino groups, ε-amino groups of lysines, and possibly imidazole rings of histidines), which

Experimental Procedures

OP-Cu and BLM-Fe-Mediated Protein–DNA Cross-Linking within Isolated Chromatin

With argon, deoxidize all stock solutions and solubilize chromatin or whole organelles (nuclei, chloroplasts, etc.), isolated by an appropriate method,[12] in 20 m*M* sodium phosphate, pH 7.4, at 37°. Chemicals that quench free radicals such as alcohols, glycerol, and so on should be excluded from the cross-linking buffer. For a OP-Cu reaction, mix *o*-phenanthroline hydrochloride (Fluka, Ronkonkoma, NY) and $CuSO_4$ to a final concentration of 2 and 0.2 m*M*, respectively (20× concentrated OP-Cu). Add the OP-Cu complex to a probe, initiate cross-linking by adding O_2-depleted H_2O_2 to 1 m*M*, and incubate the mixture under strong anaerobic conditions (argon bubbling) at 37° for 30 min. For a BLM-Fe reaction, make a complex of 2 m*M* bleomycin hydrochloride ("Bleocin," Nippon Kayaku) and 0.2 m*M* $Fe(NH_4)(SO_4)_2$ (20× BLM-Fe) with argon-purged water and adjust the pH to 7.0 with sodium hydroxide. Activate the BLM-Fe complex by adding H_2O_2 to 20 m*M* and incubate at room temperature for 30 sec. Initiate cross-linking by adding the activated BLM-Fe complex to chromatin and incubate at 37° for 30 min with argon bubbling. Stop the reaction by adding EDTA to 5 m*M*. Add HEPES, pH 7.4, to 100 m*M* and $NaBH_4$ to 20 m*M* and reduce cross-linked complexes by incubation on ice in the dark

[12] R. D. Kornberg, J. W. LaPointe, and Y. Lorch, *Methods Enzymol.* **170,** 3 (1989).

creates a reversible covalent bond between the protein and the DNA strand. The resulting imines, also known as Schiff bases (**III**), are prone to fast hydrolysis. They may, however, be converted into stable covalent adducts in the presence of a reducing agent, e.g., sodium borohydride (**IV**). Imine-specific reducers, such as sodium cyanoborohydride or pyridine borane complex, are particularly useful, as they may be present continually during the reaction to selectively trap the imine adducts (**III**) from equilibrium. In the absence of reducing agents, abasic site-derived Schiff bases undergo additional spontaneous conversions. First, the 3′-phosphodiester bond is cleaved by the mechanism of β-elimination, with the formation of an α,β-unsaturated aldehyde (**VI**) via the α,β-unsaturated imine intermediate (**V**). The unsaturated imines can also be stabilized by reducing agents, and some end products of cross-linking may emerge from this pathway (e.g., **VII**). Another potential mechanism of stable adduct formation is electrophilic alkylation of proteins by β-elimination products (Michael addition), resulting in His and Cys side chain attachment via the 3′ position (**IX**). Finally, a portion of α,β-unsaturated intermediates decay without forming stable cross-links by the mechanism of δ-elimination (**VIII**). (B) Radical-induced cross-linking starts from 4′-hydrogen atom abstraction. The resulting 4′-deoxyribose radical decays with the loss of the DNA base and formation of an oxidized abasic site (**IIa**), which then undergoes chemical conversions analogous to the conversions of the compound **II** on (A).

for 30 min. Precipitate with ethanol and enrich by Cetavlon precipitation and phenol–chloroform extraction as described,[1] with the following modifications:

1. For complete Cetavlon DNA precipitation, the Cetavlon:DNA (w/w) ratio in solution should be 4:1 or higher and ionic strength 0.4 *M* NaCl or lower. For the best results, precipitate cross-linked nuclei with Cetavlon three times. First, lyse nuclei (20 A_{260} U/ml) in NUTE buffer (1 *M* NaCl, 7 *M* urea, 1 m*M* EDTA, 10 m*M* Tris–Cl, pH 8.0), add 10% Cetavlon to a final concentration of 0.5%, and precipitate by adding 4 volumes of distilled water. For the successful second and third precipitation, the amount of Cetavlon in the solution should not exceed double DNA weight. For this purpose, dissolve the pellet in NUTE buffer at a concentration 30 A_{260} U/ml, add Cetavlon to 0.3%, and precipitate by adding 4 volumes of water.
2. Ethanol precipitation of the cross-linked complex from the organic layer during phenol–chloroform treatment is performed without adding tRNA as a carrier. After precipitation, dilute the cross-linked complex in water at concentration 50–100 A_{260} U/ml, centrifuge for 5 min at 14,000*g*, discard supernatant from an insoluble gelatinous pellet, and precipitate it again with ethanol.

Analyze DNA–protein cross-links in the "DNA" version of a two-dimensional (2D) gel electrophoresis.[1,2] Two-dimensional gels of chromatin cross-linked with OP-Cu, BLM-Fe, and DMS *in vitro*[3] are shown in Fig. 2A. Two additional diagonals corresponding to DNA cross-linked to core histones and linker histones appear in all three gels and are absent in the gel corresponding to the uncross-linked probe (Fig. 2A, IV). The level of DNA fragmentation is proportional to the yield of protein–DNA cross-links (1 to 5% of the input DNA) and is similar for all three cross-linking procedures.

Cross-Linking of Proteins to DNA in Ehrlich Ascites Tumor Cells in Vivo

The OP-Cu complex can be applied to cross-link proteins to DNA *in vivo*. The following protocol describes a cross-linking procedure for 7-day-old mouse ascites carcinoma cells.

Mix ascitic fluid collected from two mice with an equal volume of Dulbecco's modified Eagle's medium (DMEM), 50 m*M* HEPES, pH 7.4, and 0.1 mg/ml heparin prewarmed to 37° and filter through several layers of cheesecloth. Add OP to 1.5 m*M* and $CuSO_4$ to 0.15 m*M* and initiate the reaction by adding H_2O_2 to 10 m*M*. Incubate at 37° for 30 min and stop

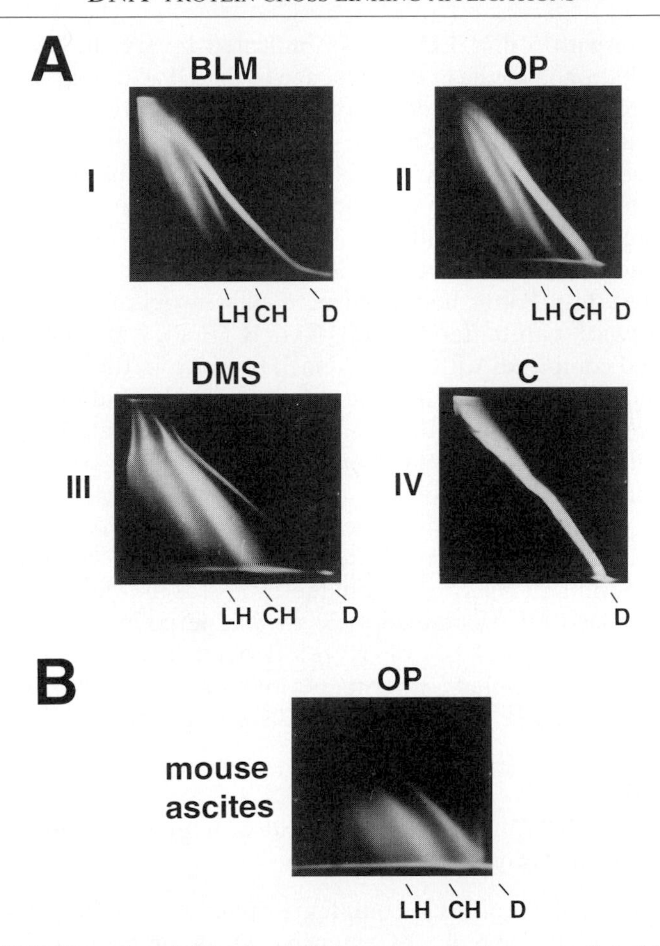

FIG. 2. "DNA" version of 2D gels of DNA–protein complexes cross-linked *in vitro* and *in vivo*.[3] (A) Complexes are purified either from isolated chromatin cross-linked with BLM-Fe (I), OP-Cu (II), and DMS (III) or from uncross-linked chromatin (IV). A strip of a denaturing gel of the first dimension is turned by 90° and subjected to electrophoretic separation of free DNA components of the parental cross-linked adducts in the second dimension, following an exhaustive protease digestion of the protein components of the cross-linked adducts by pronase directly in the gel. Because both of the dimensions use the same SDS–urea gel formulation, the uncross-linked DNA fragments migrate the same distance in both directions and form a diagonal. Covalently bound proteins slow down DNA chains in the first-dimension gel; the more bulky the protein molecule, the larger the resulting deflection from the free DNA diagonal. As a result, covalently bound proteins give rise to separate diagonal arrays of DNA spots, with the slopes of the diagonals proportional to the molecular weight of proteins cross-linked to DNA. D, uncross-linked DNA fragments. CH and LH, DNA cross-linked to either core or linker histones. (B) Two-dimensional gel of chromatin cross-linked in Ehrlich ascites tumor cells *in vivo*.

the reaction with 5 m*M* EDTA. As indicated by trypan blue exclusion assay, incubation in media in the presence of the OP-Cu complex and H_2O_2 under anaerobic conditions does not decrease the number of living cells. However, some cells, e.g., Chinese hamster ovary (CHO) cells, have an increased sensitivity to the OP-Cu/H_2O_2 reaction and may die during incubation. Transfer cells to ice, wash twice with 0.25 *M* sucrose, 10 m*M* HEPES, pH 7.4, and 3 m*M* $CaCl_2$ (buffer 1), resuspend them in buffer 1 containing 0.1 m*M* phenylmethylsulfonyl fluoride (PMSF) and 0.5% Nonidet P-40 (NP-40), and lyse in a Potter homogenizer. Pellet nuclei at 4000*g* for 10 min and wash twice with buffer 1. Add 100 m*M* HEPES, pH 7.4, and reduce cross-linked complexes with 20 m*M* $NaBH_4$ on ice in the dark for 30 min. Wash nuclei twice with buffer 1 and lyse in NUTE buffer. Shear cross-linked chromatin by sonication and clear by centrifugation at 8000*g* for 20 min. Enrich DNA–protein complexes by Cetavlon precipitation and phenol–chloroform extraction and analyze in the 2D gel as described.[1,2]

As shown in Fig. 2B, this procedure effectively cross-links DNA to either core histones or linker histones. The level of DNA fragmentation in these experiments is surprisingly higher, whereas the yield of cross-linked DNA is similar to that obtained in *in vitro* experiments (Fig. 2A), which may be duc to the presence of oxygen-derived metabolic species in the cell. However, cross-linking experiments on yeast cells *in vivo* and on intact pea chloroplasts show the low level of DNA fragmentation.

Application of OP-Cu Cross-Linking Methodology to "Protein Shadow" Hybridization Assay

The ability of the OP-Cu complex to penetrate cellular and nuclear membrane and the absence of extensive DNA or protein degradation, as well as protein–protein cross-linking, makes it possible to apply this methodology to the "protein shadow" hybridization assay described earlier.[2,7] This procedure is used for the identification of proteins interacting with different DNA loci within intact nuclei or whole cells. It is based on the fixation of protein–DNA contacts by cross-linking and analysis of cross-linked complexes by a 2D gel electrophoresis followed by hybridization to probes specific to a particular DNA region. Figure 3 shows an example of the analysis of the proteins interacting with DNA in intact pea chloroplasts.[3] Three diagonals in an ethidium bromide-stained gel corresponding to cross-linked proteins (Fig. 3A) also appear in the blot, hybridized to a OP-Cu 5.7-kDa probe corresponding to *5′-psbD* locus, *trnE, trnY, trnD* genes, and *5′-psaA* locus of chloroplast genome (Fig. 3B).

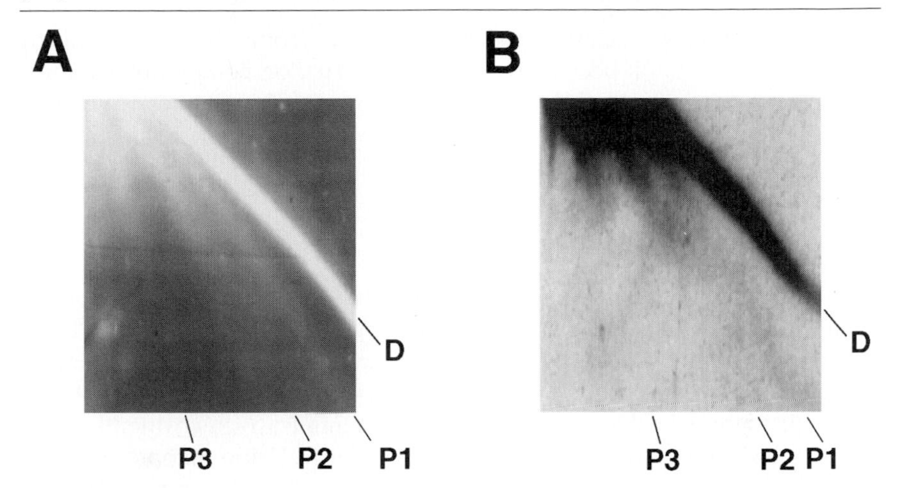

FIG. 3. "DNA" version of 2D gel of DNA–protein complexes cross-linked with OP-Cu within intact pea chloroplasts. (A) Two-dimensional gel of bulk cross-linked chloroplast chromatin. The positions of diagonals corresponding to uncross-linked DNA (D) or cross-linked complexes (P1–P3) are indicated. (B) Interaction of P1–P3 proteins with *5′-psbD* locus, *trnE, trnY, trnD* genes, and *5′-psaA* locus of chloroplast genome. Chloroplast DNA fractionated in the 2D gel shown in (A) is transferred to a nylon membrane and hybridized to the 5.7-kb probe cloned into *Pst*I and *Kpn*I sites of pBR322.[3]

Chemical and Photochemical Cross-Linking Applications to *in Vitro*-Assembled Nucleosomes.

Although it has long been demonstrated by genetic studies that histones play an essential role in the gene-specific regulation of transcription,[13,14] the precise molecular mechanisms of such an influence remained unclear. Among the probable mechanisms of histone-dependent transcription regulation are the modulation of transcription factor access by steric hindrance and direct protein–protein interactions resulting in the formation of ternary DNA–histone-transcription factor complexes (e.g., reviewed by Wolffe *et al.*[15] and Grunstein[16]). The high resolution of protein–DNA contact mapping by cross-linking analysis makes this an important tool for addressing the detailed roles played by chromatin components in gene regulation.

Since it was shown by cross-linking that nucleosomes can be reconstituted from purified histones and naked DNA so that the internal structure

[13] M. Grunstein, *Annu. Rev. Cell. Biol.* **6,** 643 (1990).
[14] G. Felsenfeld, *Nature* **355,** 219 (1992).
[15] A. P. Wolffe, J. Wong, and D. Pruss, *Genes Cells* **2,** 291 (1997).
[16] M. Grunstein, *Cell* **93,** 325 (1998).

of the resulting particles is indistinguishable from chromatin-derived nucleosome cores,[17] it became possible to embark on comparative studies of nucleosomes assembled *in vitro* on particular DNA sequences. We have gained particularly interesting insights from our nucleosome reconstitution and cross-linking studies of the internal structural details of 5S rRNA gene nucleosomes, where it has been shown that histone "tails" modulate the access of transcription factor TFIIIA to the internal promoter of the 5S rRNA genes of *Xenopus laevis.*[4,18]

The chemical mechanism of DMS-induced cross-linking (Fig. 1A) has been discussed extensively elsewhere.[2] Following the loss of dimethyl sulfate-methylated purine bases, the resulting apurinic sites bind covalently to amino groups and histidine side chains of proteins. Therefore, protein–sugar–phosphate backbone interactions are being trapped covalently. This classic technique[1,2] allows for single-nucleotide precision mapping of protein–nucleic acid interactions in nucleosome reconstituted on the fragment of the 5S rRNA gene of *X. laevis,*[1,4] but it also has several drawbacks. First, it requires a purine to be present in the right spot. The elevated reaction temperature leads to the risk of nucleosome mobility on the templates, and postcross-linking checkup of the intactness of nucleoprotein complexes is, therefore, a must. As a rule, long templates with low nucleosome densities, as well as elevated salt concentrations, cannot be used, as at elevated temperatures, the nucleosomes may be expected to change their position to a significant extent. For protein–DNA interactions, which, unlike histones, either display sensitivity to the major groove methylation of guanine N7 positions or protect these major groove positions from methylation, use of DMS may not result in cross-linking. Consequently, only a subset of nonhistone protein–DNA interactions may be detected.

The advantages of photoreactive DNA derivatives include broad chemical specificity of cross-linking, high yields, and a possibility to change the length of linkers to optimize yields of cross-linking for different types of proteins. For photochemical cross-linking, we use specifically modified, photoactive base derivatives to induce photochemical DNA–protein cross-linking at specific DNA positions with mild UV irradiation.[6,19] After cross-linking, DNA is subjected to exhaustive chemical hydrolysis. The residual radioactively labeled oligonucleotide blocks remain covalently attached to proteins, which can be resolved by sodium dodecyl sulfate (SDS) gel electrophoresis. The photoreactive arylazido groups are attached to the bottom of the major groove by a ca. 13-Å linker to achieve the optimal

[17] S. G. Bavykin, S. I. Usachenko, and A. D. Mirzabekov, *Mol. Biol.* (*Moscow*) **22,** 531 (1984).
[18] D. Y. Lee, J. J. Hayes, D. Pruss, and A. P. Wolffe, *Cell* **72,** 73 (1993).
[19] B. Bartholomew, G. A. Kassavetis, and E. P. Geiduschek, *Mol. Cell. Biol.* **11,** 5181 (1991).

cross-linking efficiency for the proteins that, like histones, interact largely with the sugar–phosphate backbone rather than with the grooves of DNA (Fig. 4A). If desired, a linker length can be fine-tuned by altering the number of glycil linkages in the reagent molecule. Strategies for linker

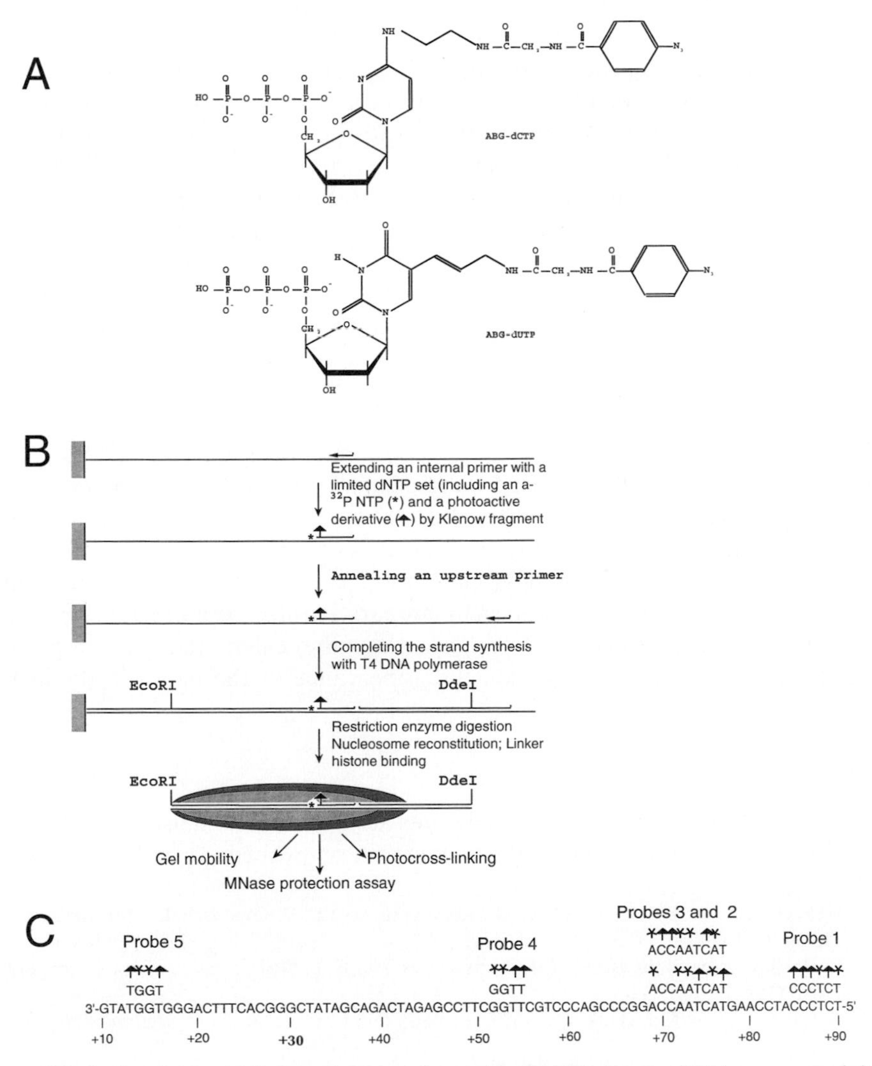

FIG. 4. Chemistry and design of photoactive probes. (A) Photoactive DNA precusors. (B) Synthesis of photoactive probes. (C) Positions of photoactive bases (arrows) and ^{32}P-labeled nucleotides (asterisks) within the bottom strand of the 5S RNA gene.

length selection have been discussed elsewhere.[19–22] Photoreactive base derivatives are introduced into specific loci of template DNA by limited primer extension (Fig. 4B). The rest of the template strand is then polymerized using "cold" dNTPs. Note that a nick remains in the template DNA after the successive primer extensions. Because histone–DNA interactions turn out to be insensitive to the nicks, a ligation step[22] could be omitted, thus making the preparation of the specifically labeled templates easier.

Photochemical cross-linking is very fast and efficient, and nucleosome relocations do not pose a problem. However, a major drawback of the approach is that a separate reconstitution has to be performed for every DNA site studied. Only pyrimidine positions could be studied with the ABG-substituted triphosphates shown in Fig. 4A. Finally, one has to protect the photoreactive compounds from UV exposure at all times.

Reconstituted nucleosomes can be used to study interactions of nucleosomes with various proteins. In studies described in more detail later, we concentrated on 5S rRNA gene nucleosome interactions with linker histones[5] and transcription factor TFIIIA.[4,18]

Experimental Procedures

DNA templates for reconstitution, containing the sequence of the *Xenopus borealis* 5S rRNA gene and 5′ labeled at a single end, are prepared from plasmids pXP-10,[23] pXbs-1,[24] or pJHX1[25] by the first restriction enzyme digestion, followed by dephosphorylation using *Escherichia coli* alkaline phosphatase, thorough phenol–chloroform extraction, second restriction enzyme digestion, and purification by preparative electrophoresis in agarose gel. The concentration of the fragments was either calculated spectrophotometrically or estimated based on the appearance of the bands in the agarose gels.

Nucleosome Reconstitution

Random-sequence nucleosome cores are prepared and used as a source of histones for the salt exchange reconstitution procedure of Tatchell and

[20] B. Bartholomew, D. Durkovich, G. A. Kassavetis, and E. P. Geiduschek, *Mol. Cell. Biol.* **13,** 942 (1993).

[21] B. Bartholomew, B. R. Braun, G. A. Kassavetis, and E. P. Geiduschek, *J. Biol. Chem.* **269,** 18090 (1994).

[22] B. Bartholomew, R. L. Tinker, G. A. Kassavetis, and E. P. Geiduschek, *Methods Enzymol.* **262,** 476 (1995).

[23] A. P. Wolffe, E. Jordan, and D. D. Brown, *Cell* **44,** 381 (1986).

[24] R. C. Peterson, J. L. Doering, and D. D. Brown, *Cell* **20,** 131 (1980).

[25] K. Ura, A. P. Wolffe, and J. J. Hayes, *J. Biol. Chem.* **269,** 27171 (1994).

van Holde[26] as described[4,5] with the total DNA concentrations in the dialysis bags of approximately 0.1 μg/μl and the donor chromatin/naked acceptor DNA ratios between 5:1 and 25:1. The volumes of samples prior to the step dialysis procedure should be between 60 and 300 μl. In the case of photocross-linking experiments, the acceptor DNA mixture consists of naked nonspecific DNA (1 μg) and the photoactive-labeled 5S fragment (50 fmol).

Linker Histones and Linker Histone Fragments

Linker histones are known to reduce the gel mobility of nucleosomes on stoichiometric (1:1) binding, as well as to protect about 168 bp of nucleosomal DNA from micrococcal nuclease (so-called "chromatosome barrier" phenomenon). These observations are used to determine the optimal ratios of linker histone and nucleosome preparations for subsequent cross-linking experiments. Chicken erythrocyte histone H5 is prepared by 5% $HClO_4$ extraction from chromatin, and the histone H5 globular domain is excised by limited trypsin digestion in a buffer containing 0.75 *M* NaCl, 20 m*M* potassium phosphate, 0.2% Triton X-100, and 250 m*M* Tris–HCl buffer (pH 8.0), followed by protease inhibition by diisopropyl fluorophosphate (Sigma), as described previously.[5,27]

An expression plasmid pET 3dH1°A, derived from the pET 3d vector by inserting the *Xenopus* histone H1°A cDNA sequence[28] between *Nco*I and *Bam*HI restriction sites, was kindly donated by J. J. Hayes (University of Rochester). We chose to retain both the N-terminal "tail" and a portion of the C-terminal "tail" of similar size in the deletion mutant tested for nucleosome binding, based on the observation[29] that the globular domains of linker histones, whenever "tailed" from only one side, tend to produce weak or ambiguous results in the chromatosome barrier assay. Allan *et al.*[29] hypothesized that the two linker histone tails interact strongly with both of the DNA linkers outside of the nucleosomes and that whenever only one tail is present, the imbalance of the resulting linker DNA interactions may shift the globule away from the nucleosome.

For *in vitro* mutagenesis, we use an adapter primer 3′-tcaacggttcggg tttAttcCTAGGttcagagg-5′ (bases positions not matching the sequence of the pET 3dH1°A insert are shown in uppercase), substituting a Lys codon at amino acid position 126 in the sequence of *Xenopus* histone H1°A by a

[26] K. Tatchell and K. E. van Holde, *Biochemistry* **24,** 52 (1977).

[27] A. D. Mirzabekov, D. V. Pruss, and K. K. Ebralidse, *J. Mol. Biol.* **211,** 479 (1989).

[28] D. Rousseau, S. Khochbin, C. Gorka, and J. J. Lawrence, *Eur. J. Biochem.* **208,** 775 (1992).

[29] J. Allan, T. Mitchell, N. Harborne, L. Bohm, and C. Crane-Robinson, *J. Mol. Biol.* **187,** 591 (1986).

stop codon (i.e., all but the first 23 amino acid residues of the C-terminal tail were removed), as well as introducing a *Bam*HI site immediately downstream. A polymerase chain reaction product derived from the use of this primer is *Bam*HI and *Nco*I cut, gel purified, inserted into pET3d, and used to prepare the recombinant H1°Δ126-C protein.

Linker Histone Binding

Incubate approximately 25 ng of reconstituted nucleosomes (including, in the case of photochemical cross-linking experiments,[6] 0.25 fmol of a photoprobe-containing nucleosome) with or without various amounts of linker histones in 10 μl of binding buffer containing 10 m*M* Tris–HCl (pH 8.0), 50 m*M* NaCl, 0.1 m*M* EDTA, 5% (v/v) glycerol, and 0.05% Triton X-100 at room temperature for 15 min and analyze the products by gel mobility shift and "chromatosome barrier" assays.[5,30] Gel mobility shift studies are performed in 0.7% agarose gels. To avoid potential complications with nonspecific linker histone binding, we recommend using linker histones (H1°, H5)/nucleosome ratios of about 0.5, as confirmed by the titration experiment described earlier. Binding to the nucleosomes of the deletion mutant H1°126-C and the globular domain of histone H5 does not result in clear gel mobility shifts, so to select the optimal amounts of these linker histone derivatives, one should use roughly 25% of the amounts sufficient to cause extensive DNA aggregation in the samples, based on the observation that full-length linker histones cause DNA aggregation at a 2:1 ratio to the nucleosomes.

For "chromatosome barrier" micrococcal nuclease (MNase) digestion assays, dilute the chromatin samples, containing ca. 200 ng of DNA (including internally radioactively labeled probes), by 4 volumes of 10 m*M* Tris–HCl (pH 8.0), and 2 m*M* $CaCl_2$ and digest with 2 ng of MNase for 5–60 min at room temperature. Stop digestion with the addition of EDTA (5 m*M* and SDS (0.25%, w/v), deproteinize DNA by proteinase K (1 μg/ml) (37°, 30 min), phenol extract, and ethanol precipitate. After electrophoresis, excise core particle and chromatosome-sized bands from the 9% nondenaturing acrylamide gel. Electroelute (90 V, 25 min) and resolve on a denaturing gel.

Analyze the nucleoprotein complexes either by chemical cross-linking, followed by two-dimensional electrophoresis, or by photocross-linking and then perform the label-transfer assay, as described later.

[30] J. J. Hayes and A. P. Wolffe, *Proc. Natl. Acad. Sci. U.S.A.* **90,** 6415 (1993).

Depurination-Induced Cross-Linking and Two-Dimensional Electrophoretic Mapping of Nucleosomes

Nucleosomes' reconstitution at specific DNA sequences should be performed as described by Mirazbekov *et al.*,[1] with the following modifications: DNA–protein cross-link enrichment is achieved by the ethanol precipitation of cross-linking products in the presence of 0.5% SDS and 0.5 mg/ml uncross-linked nucleosome cores as a carrier and subsequent phenol extraction in the presence of 0.25% SDS (which results in the retention of protein-bound DNA in the organic layer), followed by ethanol precipitation.[4] Note that the low concentration of cross-linked DNA species in the phenol phase may require using a carrier during precipitation. We recommend using random-length, single-stranded DNA fragments as a carrier, at or above the final concentration of 0.1 mg/ml.

Photoactive Probes

This section describes the procedure, using an example of 5S gene probe synthesis.[6] Probes are synthesized by a solid-phase modification of a previously described method[19,20] using streptavidin–agarose to immobilize the 5-biotinylated template strand of 5S DNA prior to primer extensions. The single-stranded biotinylated 5S DNA template (top strand between positions −113 and +218 relative to the transcription start site) is synthesized with a 5-biotinylated primer (−113 ÷ −96, 5-ATTTCACACAGG AAACAG-3) and plasmid pXP10 (12), linearized by *Hin*dIII. Anneal a probe-specific oligonucleotide [(+110 ÷ +91 (5′-CACCTGGTATTCC CAGGCGG-3′) for 5S DNA probe 1, +95 ÷ +78 (5′-GGCGGTCT CCCATCCAAG-3′) for 5S DNA probes 2 and 3, +72 ÷ +56 (5′-CCA GGCCCGACCCTGC-3′) for probe 5S DNA 4, and +41÷ +17 (5′-CA GACGATATCGGGCACTTTCAGGG-3′) for 5S DNA probe 5] to a single-stranded DNA template. Extend primer with ABG-dUTP or ABG-dCTP and [^{32}P]dATP, [^{32}P]dGTP, or [^{32}P]dTTP, using exonuclease-free Klenow fragment DNA polymerase I (U.S. Biochemical). Spin down the resin to wash out labeled dNTPs and resume DNA synthesis by adding all four deoxynucleoside triphosphates (to 200 μM), bovine serum albumin (BSA; to 100 μg/ml), and 1 U of exonuclease-free Klenow fragment DNA polymerase. After 5 min at 37°, terminate the reaction by adding 0.2–0.4% SDS and heating at 65° for 10 min. Wash out SDS, nucleotides, and inactivated DNA polymerase and complete probe synthesis by adding a fivefold molar excess of the second (upstream) oligonucleotide primer [165 ÷ 150 (5′-GGAGGAGACTGCCCCC-3′) for 5S DNA probes], annealing for 30 min at 37°, and extension in the presence of 1–5 U of T4 DNA polymerase

(GIBCO/BRL) with 500 μM each of all four dNTPs at 37° for 10 min. Cleave the labeled DNA by appropriate restriction enzymes and gel purify as described.[19]

Photochemical Cross-Linking and Label-Transfer Cross-Linking Assay

Adjust the samples to 15 mM NaCl and conduct photocross-linking by mild ultraviolet irradiation[19,22] using a chilling bath and an appropriate UV source (e.g., a hand-held UV lamp with removed filter). To digest cross-linked DNA chemically, adjust to 70% formic acid, 2% diphenylamine and incubate at 70° for 20 min. Following chemical digestion, only 5,3-phosphorylated oligopyrimidine blocks remain attached covalently to the histones.[27] The cross-linking events are now manifested by the ^{32}P-labeled residual oligonucleotide tag that is transferred onto proteins in the course of cross-linking. Extract formic acid and diphenylamine by 10 volumes of ether (twice) and freeze-dry the samples prior to loading on a 15% SDS gel. Stain the gel with Coomassie blue to reveal unlabeled protein positions, dry, and autoradiograph.

Probing Protein–DNA Interactions in Replicating Chromatin

The assembly of eukaryotic chromatin during the postreplication period is a complex, multistep process. The structure of nascent chromatin during the first 10–20 min after DNA replication is distinguishable from that of bulk chromatin. Newly assembled chromatin is highly sensitive to nonspecific nucleases. Newly made nucleosomes are unstable and slide easily along DNA during nuclease digestion. The histones themselves, especially newly synthesized ones, are bound weakly to nascent DNA. Importantly, short periods of maturation, a low percentage of nascent chromatin in the nucleus, and difficulties in selective labeling of newly replicated chromatin limit the application of the most common method of analysis of protein–DNA interactions (reviewed in Bavykin *et al.*[7]).

However, newly replicated DNA may be detected easily after BrdUrd labeling by an anti-BrdUrd antiserum. We developed a novel immunochemical technique for analyzing the protein content of nascent BrdUrd-labeled chromatin[7] that includes reversible fixation of nuclei with formaldehyde to provide quick "freezing" of preexisting chromatin structure,[31] followed by irreversible protein–DNA cross-linking by dimethyl sulfate (see earlier discussion), two-dimensional electrophoretic fractionation,[1,2] immunode-

[31] G. A. Nacheva, D. Y. Guschin, O. V. Preobrazhenskaya, V. L. Karpov, K. K. Ebralidse, and A. D. Mirzabekov, *Cell* **58,** 27 (1989).

tection of the nascent BrdUrd-labeled DNA with anti-BrdUrd antiserum, and quantitative analysis of two-dimensional gel images by scanning/volume integration. This method allows us to estimate the molecular weight of proteins cross-linked to DNA, their stoichiometry, rigidity of protein–DNA interactions and to differentiate "tail"–DNA and "globule"–DNA contacts of histones in chromatin. The same method may also be applied for studies of protein–nascent DNA interaction in prokaryotic cells.

Fixation of chromatin with formaldehyde creates protein–protein as well as protein–DNA cross-links,[32] which makes it unsuitable for identifying proteins interacting with newly replicated DNA by 2D gel electrophoresis.[1,2] However, these cross-links are created relatively rapidly and can be reversed easily,[33] which makes them useful for preventing protein redistribution and loss that may occur during DMS cross-linking in immature chromatin regions close to the replication fork.[7] After the formation of covalent cross-links between proteins and DNA by the DMS method,[1,2] which never gives rise to protein–protein cross-links, the formaldehyde-formed cross-links have to be destroyed so as to permit 2D gel analysis of the protein–DNA complexes.

Experimental Procedures

Labeling of Newly Replicated Ehrlich Ascites Carcinoma DNA by BrdUrd

Ascitic fluid containing 5- to 7-day-old Ehrlich ascitic tumor (EAT) cells is filtered through two layers of cheesecloth and diluted with an equal volume of Medium 199 containing 50 m*M* HEPES, pH 7.4, and 5 U/ml of heparin. Incubate cells for 15 min at 37° under gentle shaking and add BrdUrd to a final concentration of 100 μg/ml. Terminate incorporation of the drug by adding 9 volumes of ice-cold buffer I (0.25 *M* sucrose, 10 m*M* Tris–HCl, pH 7.5, 2 m*M* $MgCl_2$).

We emphasize that excess BrdUrd may inhibit the incorporation of drug into the DNA. The optimal concentration of BrdUrd may vary for different cells. For an excellent overview of principles and methods used for nuclei isolation, readers are referred to Bellard *et al.*[12,34] For isolation of nuclei from EAT cells, see Bavykin *et al.*[7]

[32] V. Jackson, *Cell* **15,** 945 (1978).
[33] V. Jackson and R. Chalkley, *Biochemistry* **24,** 6930 (1985).
[34] M. Bellard, G. Dretzen, A. Giangrande, and P. Ramain, *Methods Enzymol.* **257,** 317 (1995).

Fixation of Nuclei with Formaldehyde and Cross-Linking of Proteins to DNA

Fix isolated nuclei (20–50 A_{260} U/ml) with 1% formaldehyde, 30 m*M* sodium phosphate, pH 7.4, and 5 m*M* $MgCl_2$ for 18 hr at 0° and cross-link proteins to DNA with DMS as described.[1] Methylate nuclei (20–50 A_{260} U/ml) with 5 m*M* dimethyl sulfate in 50 m*M* cacodylate, pH 7.0, 0.5 m*M* EDTA, 0.5 m*M* diisopropyl fluorophosphate (DFP), and 1% DMS for 18 hr at 4°. For partial (15%, approximately) depurination of methylated bases, wash nuclei twice with 80 m*M* NaCl, 5 m*M* EDTA, and 20 m*M* sodium phosphate, pH 6.8, at 4° and incubate in the same buffer with 0.5 m*M* DFP and 1% dimethyl sulfoxide for 8 hr at 45°. Wash pellet at 4° in 100 m*M* sodium phosphate and reduce Schiff base by incubating nuclei in the same solution with 25 m*M* of freshly prepared sodium borohydride at 4° for 30 min in the dark. To cleave formaldehyde cross-linking, wash and sonicate nuclei in 0.5 *M* Tris–HCl, pH 8.0, at 44 kHz in 10-sec bursts in an ice–water-cooled jacket, add SDS to a final concentration of 1%, and boil for 30 min. To prevent unspecific cross-linking of proteins to DNA via the aldehyde groups that appear on complete depurination of methylated bases during boiling, a second reduction with sodium borohydride should be done followed by precipitation with ethanol.

It should be noted that for nuclei with a high level of nuclease activity, $MgCl_2$ should be substituted for a monovalent salt at physiological concentration and 5 m*M* of EDTA should be introduced to the buffer for formaldehyde fixation. For cells with low levels of protease activity, DFP may be substituted for PMSF. Because of their instability in water solutions, both DFP and PMSF must be dissolved in dimethyl sulfoxide and added to the cross-linking buffers just before the experiments.

Purify cross-linked DNA–protein complexes by Cetavlon precipitation and phenol–chloroform extraction and analyze in a 2D gel electrophoresis as described earlier for radical-induced cross-linking. To increase the sensitivity of the method, we recommend minimizing the size of 2D gels (5 × 4 cm and smaller) and increasing the amount of purified cross-linked complexes loaded onto the gels to 0.25–0.5 mg of DNA per gel.

Immunoblotting

Transfer DNA fragments from the 2D gel to nitrocellulose filters (BA-85, Schleicher & Schull) by electroblotting, wash blot in PBS buffer (130 m*M* NaCl, 10 m*M* potassium phosphate, pH 7.5) containing 0.25% gelatin for 2 hr at room temperature and incubate in the same buffer containing 1% Tween 20 and antiserum against BrdUrd diluted 1:3000 for 3 hr at 37°. Wash blot five times with PBS–Tween buffer at room temperature and incubate in the same solution containing 2×10^6 dpm of ^{125}I-labeled protein A per milliliter for 2 hr. After three final washes with PBS–Tween buffer, dry blot and expose to X-ray film for autoradiography. Specific activity of protein A should be 30×10^6 dpm/mg or higher.

Acknowledgment

We thank Jodi Flax for help in the preparation of this manuscript.

[29] Chromatin Immunoprecipitation Assays in Acetylation Mapping of Higher Eukaryotes

By COLYN CRANE-ROBINSON, FIONA A. MYERS, TIM R. HEBBES, ALISON L. CLAYTON, and ALAN W. THORNE

Acetylation of specific lysine residues in the N-terminal domains of core histones is a biochemical marker of active genes. To determine the spatial and temporal distribution of this reversible posttranslational modification, affinity-purified polyclonal antibodies recognizing acetylated core histones (principally H4), and also the epitope ε-acetyllysine, have been used in chromatin immunoselection procedures (CHIP assays) with mononucleosomes and salt-soluble chromatin fragments generated by micrococcal nuclease. The DNA of the antibody-selected chromatin was slot-blotted and probed with a variety of gene sequences: an enhanced hybridization signal, with respect to that from the DNA of the input chromatin, demonstrated elevated acetylation levels on histones associated with the probed sequences. Using chicken embryonic erythrocytes as the chromatin source and probes from the tissue-specific β-globin locus, it was shown that both embryonic ε and adult β genes are acetylated at 5 and 15 days and that the acetylation uniformly covers the whole of the locus, precisely comapping with the 33 kb of open chromatin structure. These results indicate that globin locus acetylation is not a consequence of transcription but rather a prerequisite and may be responsible for either generating or maintaining the open structure of poised and active genes. In contrast, acetylation at the chicken thymidine kinase housekeeping gene is restricted to an ~800-bp region at the 5′ end that encompasses the CpG island. The difference between these two genes may represent a distinction between wide-scale acetylation that primarily plays a structural role and directed acetylation that facilitates transcriptional initiation.

0076-6879/99 $30.00

In 1964, Allfrey and colleagues[1] proposed that acetylation of the ε-amino groups of lysine residues of core histones promoted increased transcription. Over the following 20 years, a large body of correlative evidence accumulated in support of this proposal. For example, it was shown that chromatin released rapidly from nuclei by DNase I is enriched both in active gene sequences[2] and in acetylated core histones.[3,4] It remained uncertain, however, whether the gene-rich and acetylated histone-rich chromatins were one and the same subset or came from two different segments of the genome, both sensitive to digestion by DNase I. Demonstration of overlap between these subsets required either selection of chromatin fragments on the basis of active gene sequences, followed by the observation of enhanced histone acetylation or, alternatively, selection on the basis of enhanced acetylation and observation of the enrichment of active gene sequences. The second approach was realized in 1988 by the selection of chromatin fragments having enhanced acetylation using an antibody raised against chemically acetylated histone H4: with chicken erythrocytes as the source, the immunoselected chromatin was shown to be enriched in α-globin sequences but not in those of the inactive gene ovalbumin.[5] This antibody has subsequently been used to map acetylation in various cell types, and this article describes the technology that has been used for the work in Portsmouth with chicken embryo erythrocytes, together with some results.

Immunoselection is a powerful biochemical approach used to define the immediate cellular neighbors of the immunogen, with one major proviso: the ability to immunoselect without breaking contacts with those neighbors. In most cases, this means that immunoselection must be preceded by cross-linking, which can bring its own specific problems. In the case of histone acetylation, the core histones are so well embedded in the nucleosome that an initial preparation of oligonucleosomes or even mononucleosomes that is representative of the total chromatin is a suitable input material for immunoselection, without any cross-linking. The same is not true of chromatin containing linker histones or HMG proteins, components known to show considerable mobility and therefore requiring prior cross-linking.

The methodologies described here are certainly not the only ones that can be adopted. They use relatively large amounts of affinity-purified antibody (50–100 μg per experiment) and, consequently, select large amounts

[1] V. G. Allfrey, R. Faulkner, and A. E. Mirsky, *Proc. Natl. Acad. Sci. U.S.A.* **51,** 786 (1964).
[2] H. Weintraub and M. Groudine, *Science* **193,** 848 (1976).
[3] L. Sealy and R. Chalkley *Nucleic Acids Res.* **5,** 1863 (1978).
[4] G. Vidali, L. C. Boffa, E. M. Bradbury, and V. G. Allfrey, *Proc. Natl. Acad. Sci. U.S.A.* **75,** 2239 (1978).
[5] T. R. Hebbes, A. W. Thorne, and C. Crane-Robinson, *EMBO J.* **7,** 1395 (1988).

of acetylated histone-rich chromatin (typically ~10 μg from 400 μg). The advantages are that a protein gel can be obtained from the immunoprecipitated chromatin to demonstrate the true levels of histone modification in the selected chromatin fraction and also that slot-blot hybridization, rather than polymerase chain reaction (PCR), can be used to quantify the enrichment of different gene sequences (thereby avoiding the problems of quantitative PCR). Alternative protocols using cross-linking, e.g., with formaldehyde,[6] have also been very successful, using lower levels of antibody and with smaller amounts of immunoselected chromatin being available for hybridization or PCR amplification. The overriding criterion for success in this approach is probably the quality of the antibody, i.e., its specificity, affinity, and purity.

Antibody Raised against Chemically Acetylated H4

Mammalian histone H4 is fully acetylated at all lysines with acetic anhydride and is used in a complex with tRNA as immunogen.[7] Rabbits are immunized using 200–500 μg of acetylated H4 mixed with Freund's complete adjuvant and spread over five sites of subcutaneous injection. Boost injections use Freund's incomplete adjuvant. Rabbit serum is initially screened by ELISA against several antigens: the immunogen and chemically acetylated H3, H2A, and H2B. All acetylated core histones are recognized by the antiserum and this is then screened against chemically acetylated bovine serum albumin (BSA) and chemically acetylated histone H1, both of which give a good response (0.2 OD at 10,000-fold serum dilution). It is clear that the very limited epitope ε-acetyl lysine is being recognized.[8] Western analysis of histones from butyrate-treated HeLa cells, however, shows that tri- and tetraacetylated histone H4 are recognized by the serum to a greater extent than the acetylated forms of the other core histones. The serum thus contains a mixture of affinities. Experience has shown that the response of rabbits to this immunogen is slow in appearing and it typically takes 3 months to observe a good ELISA response to chemically acetylated H4. The response to acetylated BSA or H1 takes longer and a rabbit is not considered an appropriate serum donor until this latter response has also risen. Typically, no more than one-third of New Zealand White rabbits produce an adequate response. We have no knowledge as to whether outbred or other strains of rabbits would respond differently.

[6] M. Braunstein, A. B. Rose, S. G. Holmes, C. D. Allis, and J. R. Broach, *Genes Dev.* **7,** 592 (1993).

[7] B. D. Stollar and M. Ward, *J. Biol. Chem.* **245,** 1261 (1970).

[8] T. R. Hebbes, C. H. Turner, A. W. Thorne, and C. Crane-Robinson, *Mol. Immun.* **26,** 865 (1989).

Affinity Purification

We have always used affinity-purified antibody for immunoselection of chromatin, prepared as follows. Ammonium sulfate (45% saturation) is used to precipitate a crude total immunoglobulin (Ig) fraction, freeing it from the bulk of the albumin. This is then loaded onto a Whatman (Clifton, NJ) DEAE-cellulose column in 30 m*M* phosphate, pH 7.3, and eluted in this buffer. The total Ig constitutes the bulk of the run-through peak. In order to affinity purify antibodies, a column of immobilized acetylated histone H4 is prepared as follows: histone H4 is chemically acetylated in 50 m*M* bicarbonate buffer to a level of ~70% by adding acetic anhydride at a 0.7 molar ratio to total lysine residues (i.e., assuming quantitative reaction). In practice, we frequently conduct the reaction in D_2O and monitor the disappearance of free lysine by nuclear magnetic resonance using the ε-CH_2 resonance at 3.05 ppm so as to achieve an appropriate level of modification. This modified H4 is then coupled via its remaining free ε amino groups to bromoacetyl-derivatized Sepharose 4B-CL agarose beads having a hydrophilic and anionic spacer arm. The Sepharose beads are first reacted with maleic anhydride in 50/50 acetone/water to introduce an activated double bond, which is then reacted with β-mercaptoethylamine. The terminal amino group is then reacted with the NHS ester of bromoacetic acid (prepared by carbodiimide coupling of *N*-hydroxysuccinimide to bromoacetic acid). The final spacer arm is thus

$$\text{Resin}-O-\underset{\underset{O}{\|}}{C}-CH_2-\underset{\underset{\underset{OH}{|}}{C=O}}{\underset{|}{CH}}-S-CH_2-CH_2-NH-\underset{\underset{O}{\|}}{C}-CH_2-Br$$

Reaction of the activated halide to amino groups of modified H4 is carried out at 37° overnight. The procedure is described in greater detail in Hebbes *et al.*[8] In a typical preparation of affinity-purified antibodies, 20 ml of serum gives ~80 mg of total Ig, which is passed twice through an affinity Sepharose column of 4 ml bed volume that has ~20 mg of coupled partially acetylated H4. After washing with phosphate buffered saline (PBS)/1 *M* NaCl, antibodies are eluted with 3.5 *M* potassium thiocyanate (glycine buffer, pH 2 is ineffective) and immediately desalted on a Sephadex G25 column in bicarbonate buffer. The final yield is typically ~500 μg of affinity-purified antibodies. Affinity purification of antibodies carries the risk of losing those having the highest affinity for antigen, although in the present case the strong elution procedure probably avoids that problem. Immediate renaturation of antibodies following elution is, however, essential.

Effectiveness of Antibodies for Chromatin Fractionation

This is tested by comparing bulk nucleosomes from chicken erythrocytes, which have very low levels of core histone acetylation (~1.5 acetyl groups per nucleosome) with nucleosomes from HeLa cells treated with butyrate (~7 acetyl groups per nucleosome). Increasing amounts of both types of mononucleosome are treated separately with a fixed amount of antiserum and then the excess, unreacted antibodies are back titrated by enzyme-linked immunosorbent assay (ELISA) against chemically acetylated H4 as antigen. Under conditions where 50% of the antibodies are removed from the serum by HeLa butyrate nucleosomes, only 3% are removed by chicken erythrocyte nucleosomes.[5,8] We conclude that these antibodies are indeed able to recognize nucleosomes carrying acetylated core histones.

Preparation of Chromatin for Immunoselection

We have always used micrococcal nuclease digestion of nuclei to prepare chromatin for immunoselection, and an important requirement is that this chromatin is representative of total chromatin. In particular, it is important to establish that the input chromatin for immunoselection is not severely depleted of the genes of interest due to their enhanced accessibility to MNase. It is therefore advisable to check the digestion conditions by probing a time course MNase digestion with the sequences of interest. Reducing the bulk of the chromatin to short oligonucleosomes or monosomes is desirable for high-resolution mapping and for yield, but overdigestion can result in the sequences of interest being almost entirely removed. Sensitivity to MNase is gene dependent, i.e., one must not assume that conditions appropriate for one gene in the cell type being studied can necessarily be transferred to another gene.

Salt-Soluble Chromatin

A nuclear suspension of 5 mg/ml DNA digested with 200/U/ml of MNase releases a first supernatant. Nuclei are then pelleted and lysed in 0.25 m*M* EDTA to release a second supernatant; all procedures are carried out in 10 m*M* sodium butyrate. NaCl is then added to these pooled supernatants to 150 m*M* final concentration to precipitate H1-containing chromatin, with the remaining supernatant being about 85% mononucleosomes for both chicken erythrocytes and human culture cells. Such supernatant chromatin has been found to be representative of total chromatin as judged by hybridization of its DNA to both active and inactive sequences (see later).

Preparation of Mononucleosomes

The resolution of core histone acetylation mapping is the length of the gene probe used or the length of the chromatin fragments selected, whichever is longer. It is therefore ideal to use mononucleosomes and genomic (not cDNA) probes 200–300 bp long (any shorter and the efficiency of hybridization deteriorates). For studying the acetylation of the chicken β-globin locus, a representative mononucleosome population could be prepared by taking the combined supernatants described earlier, depleting in H1-containing material by the addition of 150 m*M* NaCl (or by adding only 50 m*M* NaCl together with a cation-exchange resin such as Sephadex CM-25), and then fractionating on a sucrose gradient. For some other genes this approach results in loss of the sequences of interest, but the situation is remedied by digesting a nuclear suspension at only 0.5 mg/ml DNA with 150 U/MNase per ml at 37° for 5 min. Under these conditions, a reasonable representation of active gene sequences is found in monomer, dimer, and trimer nucleosomes, which can be depleted of histone H1 using either of the two procedures. Depending on the resolution sought in the mapping, this material can be fractionated on sucrose gradients.

Immunoselection of Chromatin

In a typical fractionation, 400 μg of input chromatin (as DNA) is mixed with 100 μg of affinity-purified antibodies in a Tris–HCl buffer, pH 7.5, containing 1 m*M* EDTA and 10 m*M* butyrate in a total volume of 800 μl and incubated for 2 hr at 4° with constant agitation. The fractionation has also been carried out frequently at half this scale. To immobilize immunocomplexes, 50 mg of washed protein A–Sepharose is added and the suspension is incubated for a further hour at 4°. Immunocomplexes are collected by centrifugation and the supernatant is also retained. After repeat washing with incubation buffer, the pellet is resuspended in 150 μl of incubation buffer containing 1.5% SDS to release complexes. Sepharose beads are removed by centrifugation and reextracted with 150 μl of buffer containing 0.5% SDS, combining the two supernatants. The histones and DNA in this fraction represent the antibody-bound chromatin whereas the initial supernatant represents unbound chromatin.

Protein Analysis

It is critically important to be certain that the antibodies truly immunoselect chromatin fragments only on the basis of the specificities revealed in prior ELISA screening or Western blotting. Because the aim of the experiment is to look for enrichments or depletions of particular gene sequences, one needs to establish the modification levels of all the histones in the antibody-bound chromatin. It is not sufficient to know only that the most highly acetylated histones are largely or entirely within this fraction because if a significant amount of poorly acetylated chromatin is also present, this will dilute any gene enrichments/deficiencies subsequently observed. Using 400 μg of input salt-soluble chromatin, we typically immunoselect about 10 μg, i.e., 2.5%. This represents sufficient total histone for a single lane of an acetic acid/urea/Triton (AUT) gel. When mononucleosomes are used as the input, the yield is typically 5–6 μg of bound chromatin, which, with care, will also yield a satisfactory protein gel. To achieve this, the bound chromatin is extracted twice with an equal volume of phenol/chloroform and once with chloroform, recovering the DNA by ethanol precipitation. The proteins are recovered from the phenol phase of the first extraction by adding 1/100 the volume of 10 *M* HCl, followed by 12 volumes of acetone and allowing proteins to precipitate overnight at −20°. The protein pellet is washed twice with acidified acetone, three times with dry acetone, and finally vacuum dried.

DNA Analysis

With 5–10 μg of bound DNA available, this is quantitated by UV absorption to permit accurate known amounts to be used in slot blotting with 0.5–2.0 μg/slot. The DNA is denatured in 0.5 *M* NaOH/1.5 *M* NaCl for 10 min at 37°, then for 1 min at 100° and applied through a manifold to a Biodyne B membrane (Pall, Portsmouth, UK) prewetted in 10× SSC. The filters are immersed in the NaOH/NaCl solution for 5 min and then in a 50 m*M* Tris/EDTA neutralizing solution for 30 sec before blotting dry and fixing the DNA by baking at 80° for 30 min. We have also used Hybond-N membranes (Amersham, UK) with essentially the same protocol.

For sequence analysis of the DNA on the membranes it is essential to use genomic DNA probes rather than cDNA, particularly if mononucleosomes have been fractionated. Plasmid-borne probes are excised and labeled by random priming. In some cases, probes are generated by the restriction of cosmids and cutting the required band from an agarose gel. The membranes are prehybridized for 60 min and then hybridized for 2 hr using Quick Hyb (Stratagene, La Jolla, CA) at 68° in bottles in a hybridization oven. Final washing of the membranes is typically in 0.2× SSC and 0.1% SDS for 20 min at 68° in an oven. When comparisons are required between probes of a significantly different GC/AT ratio, it is advisable to vary the SSC concentrations in the final wash conditions to achieve equivalent T_m values for the different probes.

In a typical experiment, equal amounts of DNA from antibody-bound chromatin, the unbound chromatin, and the input chromatin are immobilized in adjacent slots. An equal weight of genomic DNA, sonicated to the limit level of ~600 bp, is also immobilized to check the gene representation in the input chromatin. To assist quantitation, several increased loadings of the input, unbound, and genomic DNA samples are also immobilized and, as a check on the effectiveness of hybridization, several loadings of the plasmid carrying the probe are immobilized, all quantitated by their UV absorption.

Illustrative Results Using Chicken Embryo Erythrocytes

Figure 1 shows results from the immunofractionation of mononucleosomes from 15-day chicken embryo erythrocytes. The probe is a genomic fragment of β^A, which is the adult gene transcriptionally active in this tissue. The AUT gel of extracted proteins shows that for histone H4, the antibody-bound material has about equal quantities of Ac4, Ac3, and Ac2 with very reduced amounts of Ac1 and very little Ac0, whereas the input chromatin

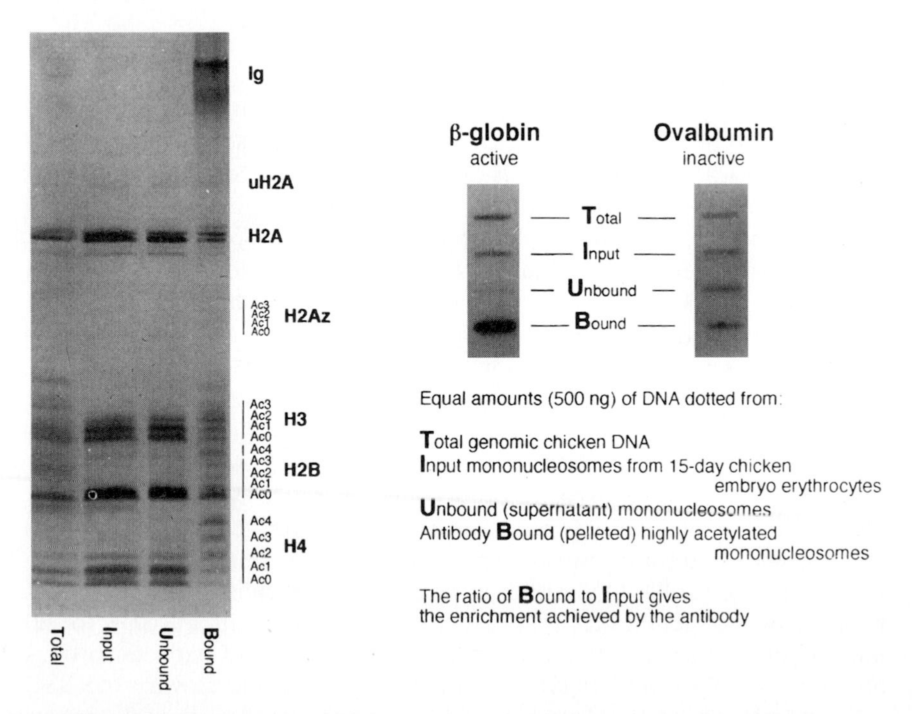

FIG. 1. Analysis of histones and DNA from an antibody fractionation of mononucleosomes obtained from 15-day chicken embryo erythrocytes by micrococcal nuclease digestion.

contains largely Ac0, Ac1, and a little Ac2. For histone H2B the distribution of acetylated species is rather bipartite, with significant amounts of Ac4 and Ac3, together with a large proportion of Ac0. For histone H3, the situation is complicated by the presence of primary sequence variants having different mobilities: two-dimensional gels with greater loadings show that the distribution of acetylated H3 species is intermediate between that of H4 and H2B. In the case of histone H2A, having only a single site of acetylation, there seems little enrichment of Ac1 in the bound fraction. It is striking that multiply acetylated forms of the replacement histone H2AZ can be seen in the bound lane: this is to be expected on the premise that replacement core histones are deposited only onto transcriptionally active chromatin (which is enriched in the bound mononucleosomes, as seen from hybridization results). It has been proposed that ubiquitinated H2A and H2B are features of active chromatin[9] and so we looked to find an enrichment of uH2A (or uH2B) in the bound material: no such enrichment was ever observed, even in samples showing a 30-fold enrichment of active sequences, such as β^A globin in the present example. The conclusion for ubiquitinated histones is that highly acetylated nucleosomes (the basis for selection) do not contain elevated levels of uH2A: we cannot exclude the fact that there might be regions of active genes that are not highly acetylated but which are enriched in uH2A/H2B.

The hybridization analysis in Fig. 1 shows a ~30-fold enrichment of β^A sequences in the bound chromatin relative to the input (which is about 2-fold depleted in β^A, relative to total genomic DNA). With such a large enrichment in the bound fraction, a significant depletion of β^A sequences in the unbound supernatant can be seen (~0.2× input). It is important to appreciate the enrichments and depletions that could be expected: if ~3% of the input is immunoselected and displays a ~30-fold enrichment of active sequence, then the unbound should be almost completely depleted. If, however, an enrichment of 10-fold is found (as is often the case), then only 30% of the active sequences are carried into the bound fraction and the hybridization signal from the unbound will be 70% that of the input, i.e., a barely detectable difference. Depletion of the unbound chromatin is thus much more difficult to observe when only a small fraction of the chromatin is immunoselected than is enrichment in the bound chromatin. If 20% of the input was immunoselected, then an enrichment of only 5-fold would be sufficient to fully deplete the unbound supernatant. However, we have found in practice that immunoselecting 20% of the input never results in significant enrichments of the bound chromatin or depletions of the unbound supernatant.

[9] L. Levinger and A. Varshavsky, *Cell* **28,** 375 (1982).

Analysis of Chicken β-Globin Locus

This locus contains two embryonic genes, β^{ρ} and β^{ε}, together with the hatching gene β^{H} and the adult gene β^{A}. In 5-day embryo erythrocytes (the primitive cell line), only β^{ρ} and β^{ε} are active, whereas in 15-day embryo erythrocytes (the definitive cell line), these embryonic genes are inactive and the adult β^{A} (and β^{H}) genes are active: mononucleosomes were therefore prepared from both 5- and 15-day erythrocytes, the highly acetylated fraction immunoselected, and the extracted DNA probed with genomic sequences from both β^{ρ} and β^{A}. As previously, the ovalbumin gene was used as the negative control for a gene inactive in erythrocytes. From the enrichments in the bound chromatin (Fig. 2), it can be seen that the embryonic gene β^{ρ} carries hyperacetylated histones both in 5-day erythrocytes (as expected since it is active then) and in 15-day erythrocytes where it is not active. In the same way, the adult β^{A} gene carries hyperacetylated histones in 15-day erythrocytes where it is active (as also seen in Fig. 1) and also in 5-day erythrocytes in which it is inactive. This conclusion is even more striking because the primitive cell line (in which the inactive β^{A} carries hyperacetylated histones) dies out by 15 days, when it is replaced by the definitive cell line. Thus the β^{A} gene in the primitive cell line will *never* be transcribed. Conversely, in the definitive line, the embryonic β^{ρ} gene (which carries hyperacetylated histones) has never and will never be transcribed. We therefore concluded[10] that both poised and actively transcribed genes carry the histone modification and thus the acetylation at the chicken β-globin genes is not a consequence of the transcriptional event but is a prerequisite.

Locus-Wide Mapping

Following the use of the two gene probes described previously, several more single-copy gene probes (P1–P10) were generated from a 45-kb region covering the whole of the chicken β-globin locus and used with immunoselected mononucleosomes obtained from 15-day erythrocytes.[11] A broad region of ~30 kb was found to be hyperacetylated (Fig. 3). We also reinvestigated the extent of open chromatin using the criterion of relative sensitivity to DNase I. This was done by treating nuclei with increasing levels of DNase I, restricting the extracted DNA, and probing Southern blots simultaneously with at least two probes, one of which was known to be outside the locus (probe P1, which detects fragment D). The rate of disappearance of the

[10] T. R. Hebbes, A. W. Thorne, A. L. Clayton, and C. Crane-Robinson, *Nucleic Acids Res.* **20,** 1017 (1992).

[11] T. R. Hebbes, A. L. Clayton, A. W. Thorne, and C. Crane-Robinson, *EMBO J.* **13,** 1823 (1994).

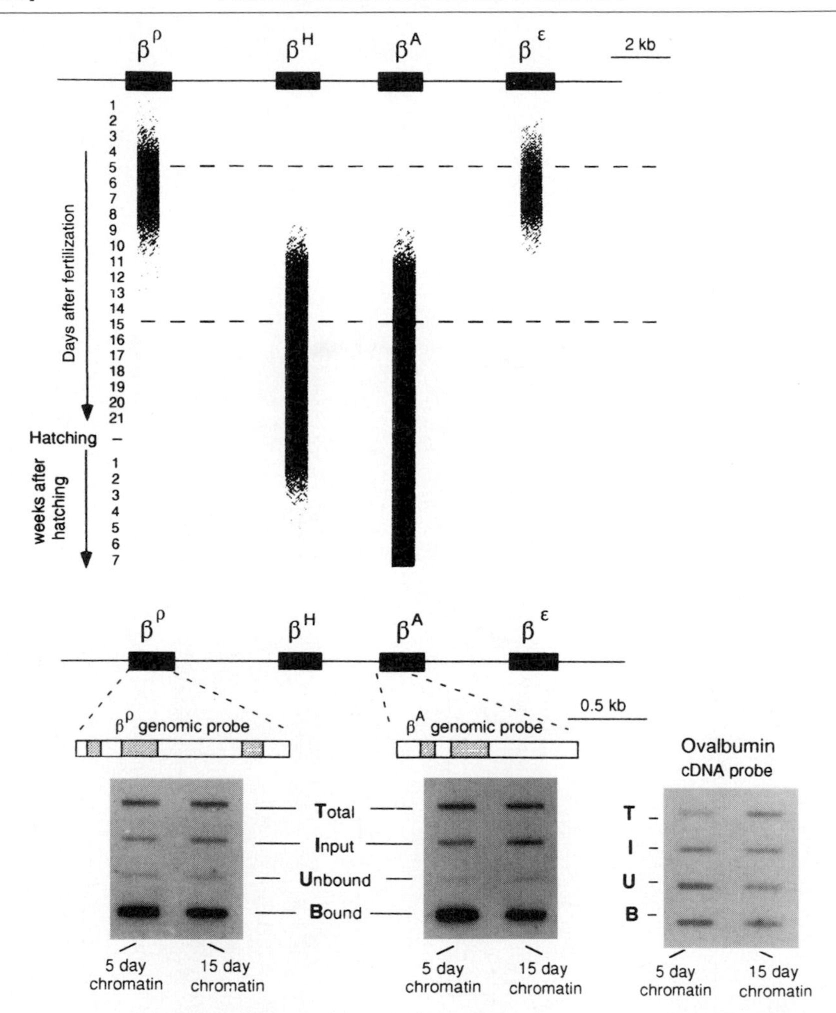

FIG. 2. Analysis of whether globin gene switching is linked to changes in histone acetylation. (Top) The time dependence of β-globin expression in the chicken embryo. (Bottom) Hybridization probings of mononucleosomes from 5- and 15-day embryo erythrocytes with genomic probes covering the embryonic β^{ρ} and adult β^{A} genes. Probing of an identical filter with inactive ovalbumin gene sequences is also shown. Five hundred nanograms of DNA is loaded in each slot. Data from Hebbes *et al.*[10]

hybridizing restriction fragment, relative to that of fragment D, was quantitated and the results are shown in Fig 3. After correcting for the target size of the fragment and ignoring fragments known to contain hypersensitive sites, the extent of open chromatin was determined as ~33 kb. A close correspondence was observed of the upstream and downstream limits of open chromatin and the limits of hyperacetylation. This was made more

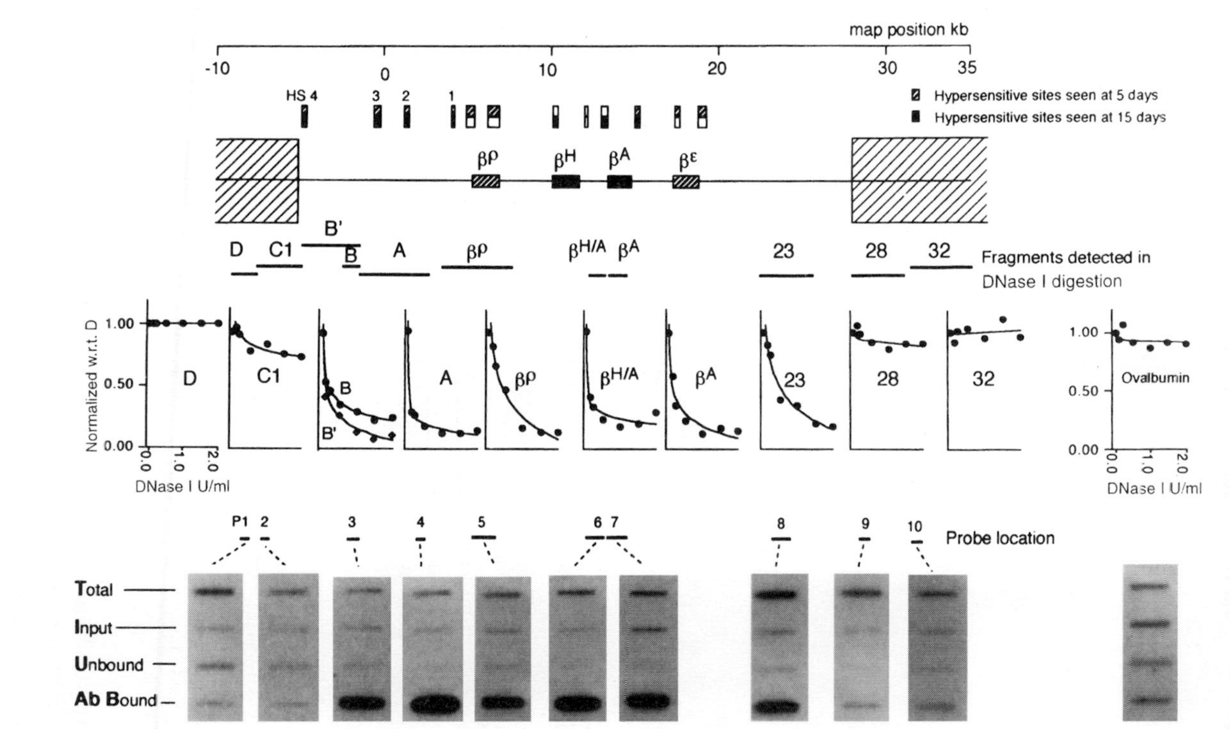

FIG. 3. Comparison of the distribution of acetylation and open chromatin at the chicken β-globin locus shows a close correspondence between the two. (Top) Location of hypersensitive sites and various restriction fragments. (Center) Intensity of restriction fragments C1 to 32 (relative to D) in hybridizations (using probes P1 to P10) of Southern transfers of DNA from DNase I-digested nuclei from 15-day chicken embryo erythrocytes. (Bottom) Hybridizations of slot blots (500 ng/slot) of DNA from antibody fractionations of mononucleosomes from 15-day chicken embryo erythrocytes. Single-copy probes P1 to P10 are positioned correctly with respect to the β-globin locus. Corresponding data for the inactive ovalbumin gene are also given. Data from Hebbes *et al.*[11]

precise at the upstream end of the locus using an additional probe (P2a) centered ~500 bp upstream of HS4, the constitutive DNase I hypersensitive site, and immediately adjacent to the insulator element defined by Chung *et al.*[12] A strong enrichment of these sequences was observed in the antibody-bound chromatin, whereas probe P2, centered ~1.4 kb further upstream, showed no enrichment (data not shown, see Hebbes *et al.*[11]). This narrowed the boundary of hyperacetylation to a segment containing ~7 nucleosomes, i.e., one turn of the chromatin supercoil. The close comapping of DNase I sensitivity and hyperacetylation suggests that this histone modification is responsible for either generating or maintaining the open chromatin structure. The observation that acetylation is found throughout the locus, not only in transcribed or promoter/enhancer regions, reinforces the view that it is not exclusively concerned with transcription itself or the immediate preparation for it. This conclusion is not in accordance with observations showing that acetylation facilitates the access of transcription factors to their recognition sequences in nucleosomes[13,14] and the expectation (based on the recruitment of acetyltransferases to induced promoters, reviewed in Wade and Wolffe[15] and Grant *et al.*[16]) that acetylation is directed to promoter/enhancer sequences. A concentration of acetylation at the 5′ end of genes has been demonstrated directly[17] by excising chromatin from CpG island regions (present in all housekeeping genes and in about half of the tissue-specific genes). Whether the uniform acetylation seen over a wide region of the chicken β-globin locus is typical of active loci is investigated later for housekeeping genes.

Structural Changes in Acetylated Nucleosomes

If nuclei are treated with a biochemical probe capable of yielding structural information and acetylated nucleosomes are then immunoselected, comparison with nucleosomes in the unbound supernatant might reveal structural differences consequent on acetylation. This has been done using chemically induced histone DNA cross-linking and analysis of the sites of cross-linking using histone peptide analysis.[18] Significant differences in the

[12] J. Chung, M. Whiteley, and G. Felsenfeld, *Cell* **74,** 505 (1993).
[13] D. Y. Lee, J. J. Hayes, D. Pruss, and A. P. Wolffe, *Cell* **72,** 73 (1993).
[14] M. Vettese-Dadey, P. A. Grant, T. R. Hebbes, C. Crane-Robinson, C. D. Allis, and J. L. Workman, *EMBO J.* **15,** 2508 (1996).
[15] P. Wade and A. P. Wolffe, *Curr. Biol.* **7,** R82 (1997).
[16] P. A. Grant, D. E. Sterner, L. J. Duggan, J. Workman, and S. L. Berger, *Trends Cell Biol.* **8,** 193 (1998).
[17] J. Tazi and A. Bird, *Cell* **60,** 909 (1990).
[18] K. K. Ebralidse, T. R. Hebbes, A. L. Clayton, A. W. Thorne, and C. Crane-Robinson, *Nucleic Acids Res.* **21,** 4734 (1993).

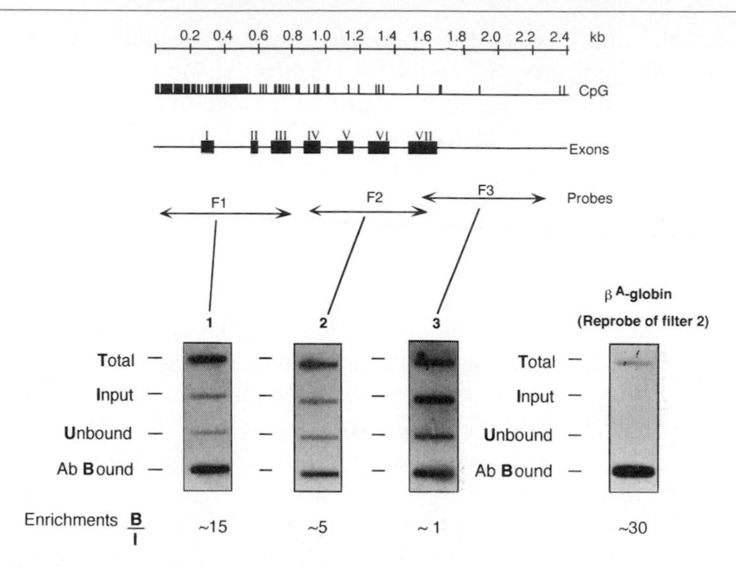

FIG. 4. Mapping acetylation at the chicken thymidine kinase gene. Use of the three probes F1, F2, and F3 with the same antibody fractionation of di/trinucleosomes shows an enrichment of ~15 in the region of the CpG island that decreases to unity at the 3′ end of the gene.

pattern of cross-linking were found, implying a restructuring in hyperacetylated nucleosomes, although the persistence of several cross-links indicated that the N-terminal tails of the core histones (primary sites of cross-linking) are not displaced totally from the DNA in hyperacetylated nucleosomes.

Housekeeping Genes

The acetylation pattern in the thymidine kinase (TK) gene was investigated in 15-day chicken erythrocytes by the methods described in this article using a di/trinucleosomal fraction purified on sucrose gradients. Although this gene consists of seven exons, it is quite short and covers about 2.5 kb of DNA, with the CpG island covering about 800 bp of the known 5′ sequence. Three genomic probes were used that together covered the whole of the gene and Fig. 4 shows the results. The striking finding is that while an enrichment of ×15 was observed for the 5′ sequences that cover the CpG island, this had fallen off to about ×5 in the region of exons IV to VIII and to unity just downstream of this. To be certain that this tail off of the enrichments was genuine, the second filter was reprobed with β^{A} globin sequences and shown to give an enrichment (~×30) typical for that gene. The finding of considerable acetylation in the region of the CpG island is in accord with the findings of Tazi and Bird[17] but is at considerable variance with the broad distribution of the modification found at the β-globin locus.

Directed vis-a-vis General Acetylation at Active Genes

The understanding that several transcriptional coactivators/repressors have either histone acetyltransferase (HAT) activity or deacetylase (HDAC) activity has led to a model of "directed acetylation" in which gene activation/induction is accompanied by the recruitment of protein complexes containing HAT activity, whereas on gene repression these are replaced by complexes containing HDAC activity.[15,16] This implies the deposition and removal of acetylation from specific nucleosomes in the region of promoters/enhancers. Such a model is in accord with the general observation of a very high level of H4 modification within CpG island chromatin.[17] The most striking feature of the observations reported here for chicken embryo erythrocytes is the very much more spacially restricted region over which the acetylation is found in the TK gene as compared to the β-globin locus. This does not appear to be due to an intrinsic property of CpG islands as the β^{ρ} globin gene contains a CpG island but the β^{A} globin gene does not. Housekeeping genes are active constitively and not regulated developmentally, whereas in erythrocytes the β-globin locus is present constitutively as open chromatin but is regulated developmentally. Neither is in accord with the externally inducible/repressible genes that represent the typical target of a putative directed acetylation/deacetylation cycle.

The ability of immunoselection to map the biochemical modifications of active gene loci is very considerable, particularly regarding core histone acetylation, as cross-linking is not always essential. The current challenge is to generate antibodies having finer specificities, sufficiently high affinities, and be capable of reproducible production in sufficient quantities to permit a more detailed picture to be elaborated than has been possible with the antibodies used here. The use of antibodies recognizing acetylated histone H4 subspecies to define their participation in the silencing of yeast genes[19,20] shows the way forward. It remains to apply this approach to higher eukaryotes.

Acknowledgments

The earlier phase of this work was supported by the Cancer Research Campaign and the later phases by the Wellcome Trust. We are very grateful to both these organizations for their support.

[19] M. Braunstein, R. E. Sobel, C. D. Allis, B. M. Turner, and J. R. Broach, *Mol. Cell. Biol.* **16,** 4349 (1996).

[20] S. E. Rundlett, A. A. Carmen, N. Suka, B. M. Turner, and M. Grunstein *Nature* **392,** 831 (1998).

[30] Chromatin Structure Analysis by Ligation-Mediated and Terminal Transferase-Mediated Polymerase Chain Reaction

By G. P. PFEIFER, H. H. CHEN, J. KOMURA, and ARTHUR D. RIGGS

Introduction

Several protocols for chromatin analysis are described here, each of which depends on the ligation of a linker (or adapter) oligonucleotide to DNA prior to polymerase chain reaction (PCR). These ligation-mediated PCR assays provide single-nucleotide resolution and sensitivity adequate for the study of mammalian cells. The standard method, which has been used for many years, is called ligation-mediated PCR (LM-PCR).[1,2] Over 100 publications have now reported the use of LM-PCR for studies of DNA structure, DNA adducts, *in vivo* footprinting of DNA-binding proteins, and nucleosomes. For information beyond that provided here, we recommend several reviews[3–5] and a special issue of the journal *Methods.*[6] In addition to LM-PCR, a protocol is included here for a technique that uses tailing with terminal transferase before ligation to an oligonucleotide. This method[7] is called terminal transferase-dependent PCR (TD-PCR). Both methods can be used to obtain information on minimally perturbed *in vivo* chromatin structure, and each requires only a microgram or less of total genomic DNA per lane of a sequencing gel. Good quality sequence-type ladders are obtained that contain information on DNA structure and DNA-bound proteins.

As illustrated in the left-hand side of Fig. 1, a key step in the LM-PCR procedure is the ligation of a double-stranded oligonucleotide linker onto the 5′ end of each target DNA molecule. This provides a common sequence at each 5′ end and permits a family of fragments of different length to be amplified by PCR. The first step of LM-PCR is cleavage of DNA to produce

[1] P. R. Mueller and B. Wold, *Science* **246,** 780 (1989).

[2] G. P. Pfeifer and A. D. Riggs, *Genes Dev.* **5,** 1102 (1991).

[3] M. M. Becker and G. Grossman, *in* "Footprinting of Nucleic Acid–Protein Complexes" (A. Revzin, ed.), p. 129. Academic Press, New York, 1993.

[4] I. K. Hornstra and T. P. Yang, *Anal Biochem* **213,** 179 (1993).

[5] A. D. Riggs and G. P. Pfeifer, Advances in Molecular Cell Biology, Vol. 21, p. 47. JAI Press, Greenwich, CT, 1997.

[6] K. S. Zaret, ed., *Methods* **11,** 149 (1997).

[7] J. Komura and A. D. Riggs, *Nucleic Acids Res.* **26,** 1807 (1998).

0076-6879/99 $30.00

5′-phosphorylated molecules. This can be achieved, for example, by use of DNA sequencing (Maxam–Gilbert) chemicals or by cutting with enzymes such as DNase I. Primer extension of a gene-specific oligonucleotide (primer 1) creates molecules that have a blunt end on one side. Linkers are ligated to these blunt ends, and then an exponential PCR amplification of the linker-ligated fragments is done using the longer oligonucleotide of the linker (linker–primer) and a second, nested gene-specific primer (primer 2). The PCR-amplified fragments are separated on sequencing gels, electroblotted onto nylon membranes, and hybridized with a gene-specific single-stranded probe. Single-stranded hybridization probes can be made from PCR products that flank these LM-PCR primers (see Fig. 1B).

LM-PCR has been used for direct sequencing of genomic DNA and for determining complete DNA cytosine methylation patterns.[8,9] Methylation analysis by LM-PCR is not discussed further here but has been described in a previous protocol.[10] Methods for chemical [dimethyl sulfate (DMS) and potassium permanganate], enzymatic (DNase I), and UV photofootprinting are described in detail.[1,2,8,10–12]

Genomic Footprinting with Dimethyl Sulfate

Dimethyl sulfate footprinting has been the most commonly used reagent for the detection of sequence-specific transcription factors. Both hyporeactivity and hyperreactivity of specific guanines after DMS treatment of cells indicate protein binding at or near these sites. The type of data obtained is shown in Fig. 2.

Treatment of Cells with DMS and DNA Isolation

1. Replace cell culture medium with medium containing 0.2% DMS (freshly prepared).
2. Incubate at room temperature for 5–10 min.
3. Take out the DMS-containing medium (centrifuge quickly for cells in suspension).
4. Wash with 10–20 ml of phosphate-buffered saline (PBS).
5. Remove cells by trypsinization, dilute with ice-cold PBS, and centrifuge.

[8] G. P. Pfeifer, S. D. Steigerwald, P. R. Mueller, B. Wold, and A. D. Riggs, *Science* **246,** 810 (1989).
[9] S. Tornaletti and G. P. Pfeifer, *Oncogene* **10,** 1493 (1995).
[10] G. P. Pfeifer, J. Singer-Sam and A. D. Riggs, *Methods Enzymol.* **225,** 567 (1993).
[11] G. P. Pfeifer, R. L. Tanguay, S. D. Steigerwald, and A. D. Riggs, *Genes Dev.* **4,** 1277 (1990).
[12] S. Tornaletti and G. P. Pfeifer, *J. Mol. Biol.* **249,** 714 (1995).

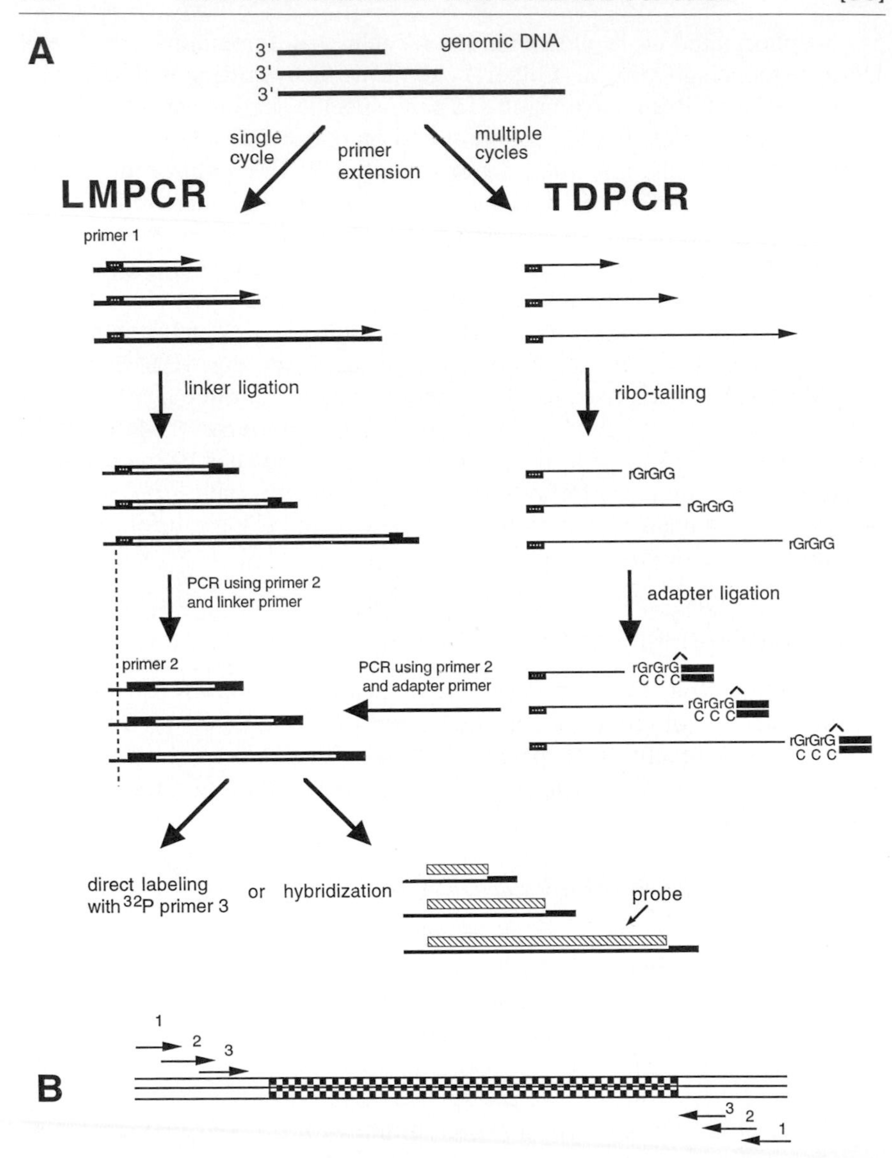

FIG. 1. (A) Outline of the LM-PCR and TD-PCR procedures. After cleavage and denaturation of genomic DNA, a gene-specific primer is annealed and extended once (LM-PCR) or multiple times (TD-PCR). Then a double-stranded linker is blunt end ligated (LM-PCR, left) or an adapter with a deoxycytidine overhang is ligated after tailing the 3′ ends with ribo dGTP (rG) (TD-PCR, right). All fragments are PCR amplified using a second nested gene-specific primer and a primer specific for the linker or adapter. Visualization of the sequence ladders can be done either by use of a labeled primer 3 and doing primer extension or by hybridization with a single-stranded probe that abuts on primer 2. (B) Arrangement of

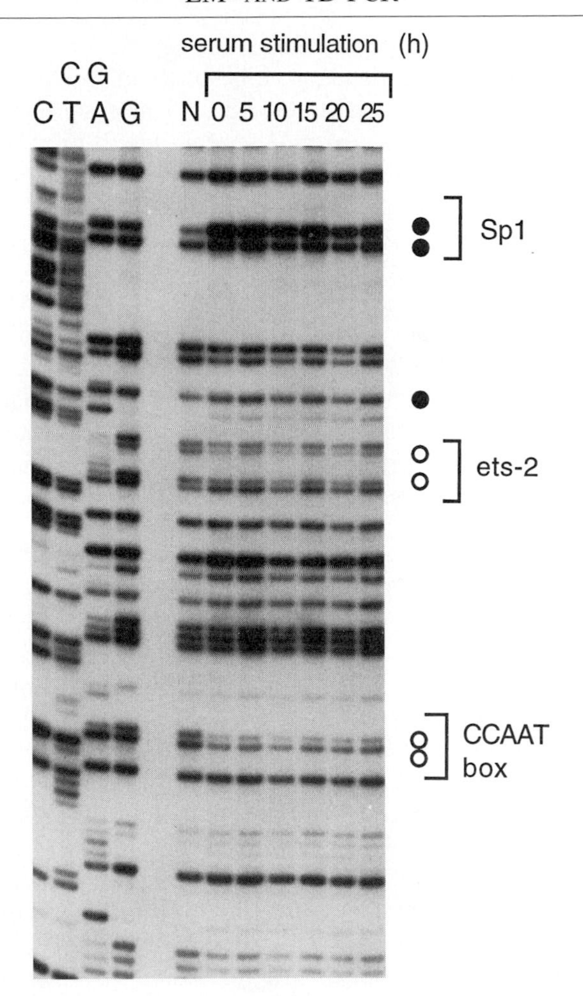

FIG. 2. Dimethyl sulfate (DMS) genomic footprinting by ligation-mediated PCR. The promoter of the human cdc2 gene was analyzed by DMS footprinting of human fibroblasts that were serum starved and then restimulated with serum for various time periods.[21] The footprints indicate binding sites for transcription factors Sp1 and ets-2 and a CCAAT box-binding protein. Protections (○) and hyperreactivities (●) can be seen when naked DNA controls (N) and DMS-treated cells are compared. Lanes C, CT, GA, and G are Maxam–Gilbert sequencing ladders obtained from HeLa cell DNA.

LM-PCR/TD-PCR primers to reveal both strands of a 200-bp region (checkered area). Primer 1, first primer used in primer extension; primer 2, PCR primer. Primers 3 are used to make initially a double-stranded PCR product; then, one of primers 3 is used to make a single-stranded hybridization probe from the double-stranded template.

6. Wash cell pellet with cold PBS.
7. Isolate nuclei by suspending cells in 10 ml of buffer A (0.3 *M* sucrose, 60 m*M* potassium chloride, 15 m*M* sodium chloride, 60 m*M* Tris–HCl, pH 8.0, 0.5 m*M* spermidine, 0.15 m*M* spermine, 2 m*M* EDTA) containing 0.5% Nonidet P-40. This step will remove most of the cytoplasmic RNA. Incubate on ice for 5 min.
8. Centrifuge at 1000*g* for 5 min at 4°.
9. Wash nuclei once with 15 ml of buffer A.
10. Resuspend thoroughly in 2–5 ml of buffer B (150 m*M* NaCl, 5 m*M* EDTA, pH 8.0), and add 1 volume of buffer C [20 m*M* Tris–HCl, pH 8.0, 20 m*M* NaCl, 20 m*M* EDTA, 1% sodium dodecyl sulfate (SDS)] containing 600 μg/ml of proteinase K. Incubate for 2 hr at 37°. It is important to keep the time the DNA is incubated at 37° to a minimum to prevent the depurination of 3-methyladenine. This would otherwise be scored as an *in vivo* hyperreactivity of adenine bases.
11. Add DNase-free RNase A to a final concentration of 100 μg/ml. Incubate for 1 hr at 37°.
12. Extract with 1 volume of buffer-saturated phenol. Extract with 0.5 volumes of phenol and 0.5 volumes of chloroform. Repeat this step until there is no interface. Finally, extract with 1 volume of chloroform.
13. Add 0.1 volume of 3 *M* sodium acetate, pH 5.2, and precipitate the DNA with 2.5 volumes of ethanol at room temperature.
14. Centrifuge at 2000*g* for 1 min at room temperature. Wash the pellet with 75% ethanol and air dry briefly.
15. Dissolve the DNA in TE buffer to a concentration of approximately 0.2 μg/μl. Keep at 4° overnight. If all of the DNA is to be used for piperidine treatment, redissolve it directly in 100 μl 1 *M* piperidine and continue as described for naked DNA controls, below, step 17.

Naked DNA Controls

As an essential control, naked DNA from the same cell type used for *in vivo* footprinting is modified with DMS *in vitro* as follows.

1. In a ventilated chemical safety hood, mix, on ice, 5 μl genomic DNA (20–100 μg), 200 μl DMS buffer (50 m*M* sodium cacodylate, 1 m*M* EDTA, pH 8), and 1 μl DMS (99+% from Aldrich, Milwaukee, WI). Dispose of pipette tips and supernatants containing DMS in a solution of 5 *M* NaOH.

2. Incubate at 20° for 2 min.
3. Add 50 μl of DMS stop solution (1.5 *M* sodium acetate, pH 7, 1 *M* 2-mercaptoethanol).
4. Add 750 μl precooled ethanol (−70°) and mix well.
5. Keep samples in a dry ice/ethanol bath for 10–20 min.
6. Spin for 15 min at 14,000*g* at 0–4°.
7. Take out supernatant and respin to remove all liquid.
8. Resuspend pellet in 225 μl water.
9. Add 25 μl 3 *M* sodium acetate, pH 5.2.
10. Add 750 μl precooled ethanol (−70°) and mix well.
11. Put on dry ice for 10 min.
12. Spin for 10 min at 14,000*g* at 0–4°.
13. Take out supernatant and respin.
14. Wash with 1 ml of 75% ethanol.
15. Dry pellet in a Speed-Vac or air dry.
16. Dissolve DNA in 100 μl 1 *M* piperidine (Fluka, Ronkonkoma, NY, freshly diluted).
17. Heat at 90° for 30 min in a heat block; use lid locks and Teflon tape to prevent evaporation.
18. Add 10 μl 3 *M* sodium acetate, pH 5.2, and 2.5 volumes of ethanol.
19. Keep on dry ice for 20 min.
20. Spin at 14,000*g* for 15 min to remove supernatant.
21. Wash twice with 1 ml of 75% ethanol.
22. Remove traces of remaining piperidine by drying the sample overnight in a Speed-Vac. Dissolve DNA in water to a concentration of about 0.5 μg/μl.
23. Determine the approximate modification frequency by running 1 μg of the samples on a 1.5% alkaline agarose gel. There should be fragments below the 200 nucleotide size range. Fragment sizes should be similar between DMS-treated naked DNA and DNA from cells treated *in vivo*.
24. Process the samples by LM-PCR (see later).

Genomic Footprinting with Potassium Permanganate

Single-stranded DNA reacts preferentially with potassium permanganate. Treatment of intact cells with this reagent can be used to identify melted or partially melted DNA structures, such as those associated with a paused RNA polymerase.[13]

[13] S. A. Brown, A. N. Imbalzano and R. E. Kingston, *Genes Dev.* **10,** 1479 (1996).

1. Freshly prepare a solution of $KMnO_4$ (5–20 m*M*) in PBS and add this to cell monolayers after the removal of medium and washing in PBS.
2. Treat the cells with $KMnO_4$ for 2 min at 20°, remove the solution, and quench the reaction by adding 1 *M* 2-mercaptoethanol in 150 m*M* NaCl, 5 m*M* EDTA.
3. Remove cells from the plate, e.g., by scraping or trypsinization, and isolate DNA as described earlier.
4. After purification, break the DNA at the sites of oxidized bases by heating in 1 *M* piperidine for 30 min at 90°. Then precipitate the DNA in ethanol and dry overnight in a Speed-Vac concentrator. The piperidine-cleaved fragments are used directly for LM-PCR (see later).

Naked DNA is treated *in vitro* with $KMnO_4$ in PBS under the same conditions as described for cells, the reaction is stopped with 2-mercaptoethanol, and the DNA is precipitated and then cleaved with piperidine. If there is a brown precipitate of MnO_2 at any step when the DNA is in solution, remove it by centrifugation and continue with the supernatant.

Genomic Footprinting with DNase I

DNase I footprinting can be used to detect sequence-specific transcription factors as well as to characterize other aspects of chromatin structure, e.g., the positioning of nucleosomes. A cell permeabilization procedure[2,14] is described here for *in situ* digestion of chromatin with DNase I. Specific DNase I footprints of single-copy genes are then obtained by LM-PCR amplification of the nicked genomic DNA. Digestion of isolated nuclei with DNase I is often done, but care needs to be taken to prevent loss of footprints, presumably due to the detachment of factors from their binding sites during certain nuclear isolation procedures.[2] Cell permeabilization can be used for both monolayer and suspension culture cells,[15] and, with minor modifications, the method should be adaptable for homogenized tissue samples.

Procedure

1. Grow cells as monolayers to about 80% confluency or grow them in suspension. Treatment of suspension culture follows centrifugation of the cells.

[14] S. Ymer and D. A. Jans, *Bio Techniques* **20,** 834 (1996).
[15] P. L. Chin, J. Momand and G. P. Pfeifer, *Oncogene* **15,** 87 (1997).

2. To permeabilize the cells, treat cells (about 4×10^6) with 0.05% (w/v) lysolecithin (type I, Sigma, St. Louis, MO) in prewarmed solution I [150 m*M* sucrose, 80 m*M* KCl, 35 m*M* HEPES, 5 m*M* K_2HPO_4, 5 m*M* $MgCl_2$, 0.5 m*M* $CaCl_2$ (pH 7.4)] for 1 min at 37°. Use enough solution to cover the cells.
3. Remove lysolecithin and wash with 5–10 ml of solution I. Incubate the cells with DNase I (2–10 μg/ml, grade I, Boehringer Mannheim, Indianapolis, IN) in solution II [150 m*M* sucrose, 80 m*M* KCl, 35 m*M* HEPES, 5 m*M* K_2HPO_4, 5 m*M* $MgCl_2$, 2 m*M* $CaCl_2$ (pH 7.4)] at room temperature for 5 min. Lysolecithin and DNase I concentrations and incubation times need to be adjusted for different cell types. During DNase I treatment, less than 20% of the cells should become detached from the plastic surface.
4. Stop the reaction and lyse the cells by removal of the DNase I solution and the addition of 2.5 ml of stop solution (20 m*M* Tris–HCl, pH 8.0, 20 m*M* NaCl, 20 m*M* EDTA, 1% SDS, 600 μg/ml proteinase K). Add 2.5 ml of 150 m*M* NaCl, 5 m*M* EDTA, pH 8.0, and incubate the solution for 3 hr at 37°.
5. Purify DNA by phenol/chloroform extraction and ethanol precipitation. Remove RNA by digestion with RNase A (50 μg/ml in TE buffer, 1 hr at 37°). Extract with phenol/chloroform and ethanol precipitate.
6. Naked DNA controls with a similar distribution of fragment sizes are obtained by DNase I digestion of purified DNA. Digest 40 μg of genomic DNA in 400 μl of 40 m*M* Tris–HCl, pH 7.7, 10 m*M* NaCl, 6 m*M* $MgCl_2$ for 5 min at room temperature with 0.2–1.6 μg/ml of DNase I.
7. Check the size of the DNase I-derived DNA fragments on alkaline 1.5% agarose gels. Prepare alkaline agarose gels (1.5%) in 50 m*M* NaCl, 4 m*M* EDTA. The running buffer is 30 m*M* NaOH, 2 m*M* EDTA. The gels are soaked in running buffer for at least 2 hr. Samples are denatured prior to running by the addition of 1 volume of 2× loading buffer (50% glycerol, 1 *M* NaOH, 0.05% bromocresol green). After the run, gels are neutralized for 1 hr in 0.1 *M* Tris–Cl, pH 7.5, stained for 30 min in ethidium bromide (1 μg/ml), and destained in water.
8. Process the sample by LM-PCR using extension product capture (see later).

Notes. Identification of *in vivo* protein–DNA contacts requires a comparison of the *in vivo*-treated sample with a naked DNA control. Keep in mind that even purified DNA samples might react differently with DNase

I depending on cell type. This is due to heterogeneous cytosine methylation patterns, which can give altered DNase I cleavage patterns.[2] Micrococcal nuclease can be used with the LM-PCR procedure,[2] but it should be noted that many intranucleosomal nicks are made, generating a high background that can obscure double-strand breaks made preferentially in internucleosomal DNA. For this reason, McPherson *et al.*[16] successfully used ligation of linker prior to primer extension to reveal nucleosomes over a liver-specific enhancer.

For optimum molecule usage in LM-PCR, the average fragment size, as measured by ethidium staining, should be about 250 nucleotides. DNase I digestion often results in a very broad distribution of fragment sizes (50 to 1500 nucleotides). Estimate the amount of DNA to be used in LM-PCR from the relative amount of DNA fragments within the lower size range to obtain similar band intensities on the sequencing gel in all lanes.

Genomic Footprinting with Ultraviolet Light

Altered reactivity with UV light is a useful and precise indicator of protein–DNA interactions.[3,12] Two types of UV-induced DNA photoproducts can be analyzed by LM-PCR: cyclobutane–pyrimidine dimers and (6-4) photoproducts. Cells growing as monolayers in petri dishes are UV irradiated after removal of the medium and washing with PBS. Suspension culture cells can be irradiated in suspension in PBS. UV doses are measured with a UV radiometer (Ultraviolet Products, Upland, CA). Typical UV doses for UV footprinting are 1000 to 2000 J/m^2 of 254-nm light; 1000 J/m^2 corresponds to approximately 1 min at a distance of 20 cm from five germicidal UV lamps. For controls, purified genomic DNA from the same cell type is irradiated in water, or in TE buffer, in small 5-μl droplets at a concentration of 0.2–0.5 μg/ml. It is usually necessary to use a UV dose twice as high for purified DNA in order to achieve the same frequency of UV damage as in irradiated cell monolayers.

Procedure

1. Irradiate cells with UV.
2. Isolate DNA (see earlier discussion).
3. To analyze (6-4) photoproducts, heat the UV-irradiated DNA in 1 *M* piperidine. This will destroy the photolesion and create ligatable strand breaks. Dissolve 10–50 μg of UV-irradiated DNA in 100 μl of 1 *M* piperidine. Continue as described in step 17 of the DMS footprinting procedure.

[16] C. McPherson, E. Shim, D. Friedman, and K. Zaret, *Cell* **75,** 387 (1993).

4. To analyze cyclobutane–pyrimidine dimers, incubate DNA with T4 endonuclease V (obtained from R. S. Lloyd, University of Texas, Galveston, or from Epicentre Technologies, Madison, WI) and then with *Escherichia coli* photolyase (kindly supplied by A. Sancar, University of North Carolina at Chapel Hill; also available from Pharmagen, Inc., San Diego, CA) to create fragments with 5′-phosphate groups and ligatable ends. Mix the UV-irradiated DNA (10 μg in 50 μl) with 10 μl of 10× T4 endonuclease V buffer [500 m*M* Tris–HCl, pH 7.6, 500 m*M* NaCl, 10 m*M* EDTA, 10 m*M* dithiothreitol (DTT), 1 mg/ml bovine serum albumin (BSA)] and a saturating amount of T4 endonuclease V in a final volume of 100 μl. Incubate at 37° for 1 hr. Add 3 μg of *E. coli* photolyase under yellow light. Irradiate the samples in 1.5-ml tubes from two 360-nm black lights (Sylvania 15W F15T8) for 1 hr at room temperature at a distance of 3 cm. Extract once with phenol–chloroform and once with chloroform and precipitate the DNA.
5. Determine the frequency of (6-4) photoproducts and CPDs by running 1–2 μg of the samples on a 1.5% alkaline agarose gel.
6. Continue with LM-PCR.

LM-PCR Procedures

Procedure 1 is the standard LM-PCR protocol; procedure 2, which is based on extension product capture on magnetic beads,[17] gives increased specificity and lower background for DNase I footprinting experiments.

Procedure 1: Standard LM-PCR

a. First Primer Extension. A gene-specific primer is used on genomic DNA fragments for primer extension with Sequenase (United States Biochemical Corp., Cleveland, OH). The primers we have used as Sequenase primers are 15- to 20-mer oligonucleotides with a T_m of 48° to 54°, calculated by the Oligo 4.0 computer program (National Biosciences, Plymouth, MN) with default settings (50 m*M* salt and 250 p*M* oligonucleotide concentrations).

1. In a siliconized tube, mix 0.5–2 μg of cleaved genomic DNA, 0.6 pmol of primer 1, and 3 μl of 5× Sequenase buffer (250 m*M* NaCl, 200 m*M* Tris–Cl, pH 7.7) in a final volume of 15 μl.
2. Denature at 95° for 3 min and incubate at 45° for 30 min.
3. Cool on ice and spin for 5 sec.

[17] V. T. Törmänen, P. M. Swiderski, B. E. Kaplan, G. P. Pfeifer, and A. D. Riggs, *Nucleic Acids Res.* **20,** 5487 (1992).

4. Add 7.5 μl cold, freshly prepared Mg-DTT-dNTP mix (20 m*M* $MgCl_2$, 20 m*M* DTT, 0.25 m*M* of each dNTP) and 5 units of Sequenase 2.0 (US Biochemicals, Cleveland, OH).
5. Incubate at 48° for 15 min and then cool on ice.
6. Add 6 μl 300 m*M* Tris–Cl (pH 7.7).
7. Heat inactivate Sequenase at 67° for 15 min.
8. Cool on ice and spin for 5 sec.

Note. Other polymerases, such as Vent or Pfu and their exonuclease minus derivatives, can be used for first primer extension, and excellent results often can be obtained. Sequenase has the advantage of being inactivated easily by heating prior to ligation.

b. Ligation. The primer-extended, blunt-ended molecules (which must have a 5′ phosphate) are ligated to an unphosphorylated synthetic double-stranded linker. All oligonucleotides, including linker primers, are dissolved in TE buffer as a stock solution of 50 pmol/μl. Linkers are prepared in 250 m*M* Tris–Cl (pH 7.7) by annealing a 25-mer (5′-GCGGTGACCCGG GAGATCTGAATTC, final concentration 20 pmol/μl) to an 11-mer (5′-GAATTCAGATC, final concentration 20 pmol/μl). This mixture is heated to 95° for 3 min and gradually cooled to 4° over a time period of at least 3 hr. Linkers are stored at −20°. After thawing, they are kept on ice.

1. Add 45 μl of freshly prepared ligation mix [13.33 m*M* $MgCl_2$, 30 m*M* DTT, 1.66 m*M* ATP, 83 μg/ml BSA, 3 units/reaction T4 DNA ligase (Promega, Madison, WI), and 100 pmol linker/reaction (= 5 μl linker solution)].
2. Incubate overnight at 18°.
3. Heat inactivate at 70° for 10 min.
4. Precipitate the DNA by adding 8.4 μl 3 *M* sodium acetate, pH 5.2, 10 μg *E. coli* tRNA, and 220 μl ethanol and cool samples on dry ice for 20 min.
5. Centrifuge for 15 min at 4° at 14,000*g*.
6. Wash pellets with 1 ml of 75% ethanol.
7. Dry samples in a Speed-Vac or air dry.
8. Dissolve pellets in 50 μl H_2O and transfer to 0.5-ml siliconized tubes.

c. PCR. Gene-specific fragments are amplified with a second, nested gene-specific primer and the common linker primer, LP25, the longer oligonucleotide of the linker. The primers used in the amplification step (primer 2) are 21- to 28-mers. They are designed to extend 3′ to primer 1 and primer 2 should overlap several bases with primer 1. The annealing temperature in the PCR is chosen to be at the T_m of the gene-specific primer (usually between 60 and 68°).

1. Add 50 μl of freshly prepared 2× *Taq* polymerase mix (20 m*M* Tris–HCl, pH 8.9, 80 m*M* NaCl, 0.02% gelatin, 4 m*M* $MgCl_2$, 0.4 m*M* of each dNTPs). This mix also contains 10 pmol of the gene-specific primer (primer 2), 10 pmol of the linker-primer (5′-GCGGTGACCCGGGAGATCTGAATTC), and 3 units of *Taq* polymerase (Perkin-Elmer, Norwalk, CT or Boehringer Mannheim) for each sample.
2. Cover samples with 50 μl mineral oil and spin briefly.
3. Cycle 18 to 20 times at 95° for 1 min, 60 to 68° for 2 min, and 76° for 3 min.
4. To extend completely all DNA fragments and uniformly add an extra nucleotide through the 3′ terminal transferase activity of *Taq* polymerase, an additional *Taq* polymerase extension step is performed. Add 1 unit of fresh *Taq* polymerase per sample together with 10 μl reaction buffer. Incubate for 10 min at 74°.
5. Stop the reaction by adding sodium acetate, pH 5.2, to 300 m*M*, EDTA to 10 m*M*, and add 10 μg tRNA.
6. Extract with 70 μl of phenol and 120 μl chloroform (premixed).
7. Add 2.5 volumes of ethanol and put on dry ice for 20 min.
8. Centrifuge samples for 15 min at 14,000*g* at 4°.
9. Wash pellets in 1 ml of 75% ethanol.
10. Dry pellets in a Speed-Vac.

d. Gel Electrophoresis. PCR-amplified fragments are separated on 0.4-mm-thick, 60-cm-long sequencing gels usually consisting of 8% polyacrylamide and 7 *M* urea in 0.1 *M* TBE. The gel is run until the xylene cyanol marker reaches the bottom. Fragments below the xylene cyanol dye hybridize only very weakly.

Dissolve pellets in 1.5 μl of water and add 3 μl formamide loading dye (94% formamide, 2 m*M* EDTA, pH 7.7, 0.05% xylene cyanol, 0.05% bromphenol blue). Heat the samples to 95° for 2 min prior to loading and load one-half of the sample with an elongated, flat tip.

e. Electroblotting. We have been using electroblotting and hybridization instead of directly extending a ^{32}P-labeled primer followed by gel electrophoresis[1] because longer single-stranded probes provide a higher specific activity than end-labeled oligonucleotides and there is less exposure of the worker to radiation. Electroblotters for the transfer of sequencing gels are available from Hoefer Scientific (San Francisco, CA) and Owl Scientific (Cambridge, MA).

1. After the run, transfer the lower part of the gel (length 40 cm) to Whatman (Clifton, NJ) 3MM paper and cover it with Saran wrap.

2. Pile two layers of Whatman 17 paper, 43 × 19 cm, presoaked in 100 m*M* TBE onto the lower electrode. Squeeze the paper with a rolling bottle to remove air bubbles between the paper layers.
3. Place the gel piece covered with Saran wrap onto the paper and remove the air bubbles between gel and paper by wiping the Saran wrap with a soft tissue.
4. When all air bubbles are squeezed out, remove the Saran wrap and cover the gel with a GeneScreen nylon membrane cut somewhat larger than the gel and presoaked in 100 m*M* TBE.
5. Put two layers of Whatman 17 paper presoaked and cut as just described onto the nylon membrane.
6. Place the upper electrode onto the paper and perform electroblotting at 1.6 A.
7. After 45 min, remove the nylon membrane and mark the DNA side.

f. Hybridization

1. Dry the nylon piece briefly at room temperature. Irradiate the membrane at a UV dose of 1000 J/m^2 (approximately 1 min at a distance of 20 cm from five germicidal tubes).
2. Perform the hybridization in rotating 250-ml glass cylinders in a hybridization oven. Wet the nylon membrane in TBE and roll it into the cylinders so that the membrane sticks completely to the walls of the cylinders.
3. Prehybridize with 10 ml hybridization buffer (0.25 *M* sodium phosphate, pH 7.2, 1 m*M* EDTA, 7% SDS, 1% BSA) for 10 min at 62°.
4. Remove the prehybridization solution. Dilute the labeled probe into 7 ml hybridization buffer. Hybridize overnight at 62°.
5. After hybridization, wash each nylon membrane with a total of 2 liters of prewarmed washing buffer (20 m*M* sodium phosphate, pH 7.2, 1 m*M* EDTA, 1% SDS) at 60 to 65°. Dry the membranes briefly at room temperature, wrap them into Saran wrap, and expose to Kodak (Rochester, NY) XAR-5 films. Probes can be stripped by soaking the filters in 0.2 *M* NaOH for 30 min at 45°.

g. Preparation of Single-Stranded Hybridization Probes. The most convenient method to prepare labeled single-stranded probes is repeated primer extension with *Taq* polymerase with a single primer (primer 3) on a cloned, double-stranded template DNA, which can be a plasmid containing the sequences of interest or a PCR product containing sequences just 3′ of primer 2. The length of these probes can be controlled easily by an appropriate restriction cut or by the size of the PCR product. The primer used to make this probe (primer 3) should be on the same strand just 3′

to the amplification primer (see Fig. 1) and should have a T_m of 60 to 66°. It can overlap a few bases with primer 2.

1. Mix 50 ng of the respective restriction-cut plasmid DNA (or 10 ng of a PCR product) with 20 pmol of primer 3, 100 μCi of [^{32}P]dCTP, 20 μM of the other three dNTPs, 10 mM Tris–Cl (pH 8.9), 40 mM NaCl, 0.01% gelatin, 2 mM $MgCl_2$, and 3 units of *Taq* polymerase in a volume of 100 μl. Cover with mineral oil.
2. Do linear amplification using 30 cycles of 95° for 1 min, 60–66° for 2 min, and 75° for 3 min.
3. Recover the probe by phenol/chloroform extraction, addition of ammonium acetate to a concentration of 0.7 M, ethanol precipitation at room temperature, and centrifugation. Alternatively, the probe may be recovered by centrifugation through a Sephadex G-50 spin column.

Procedure 2 (for DNase I footprinting)

This procedure is identical to the standard procedure (procedure 1) except for the following steps.

1. Use 2 μg of DNase I-treated genomic DNA per reaction. A biotinylated primer, specific for the gene of interest, is annealed to the heat-denatured (95°, 3 min) DNase I-cleaved DNA at 45° for 30 min in 40 mM Tris–HCl, pH 7.7, 50 mM NaCl.
2. Carry out a primer extension reaction at 48° with Sequenase 2.0 as in procedure 1. Inactivate the polymerase for 15 min at 67°.
3. Ligate the universal linker to the primer extended molecules as described earlier.
4. Streptavidin-coated magnetic beads (Dynabeads, Dynal, Inc., Great Neck, NY) are prepared immediately before use by washing twice in 10 mM Tris–HCl, pH 7.7, 1 mM EDTA, 2 M NaCl (washing and binding buffer) and resuspending in the same buffer at a concentration of 5 mg beads/ml.
5. Add the ligation mixture to 1 volume of the prewashed beads, and incubate the beads at 37° for 30 min.
6. Remove the supernatant on a magnetic separation stand and wash the beads once with 75 μl of washing and binding buffer.
7. Elute the nonbiotinylated strand from the beads in 37.5 μl of 0.15 N NaOH by incubating at 37° for 10 min.
8. Transfer the supernatant to a new tube, neutralize by adding 3.75 μl of 2 M Tris–HCl, pH 7.7, and 37.5 μl of 0.15 N HCl.

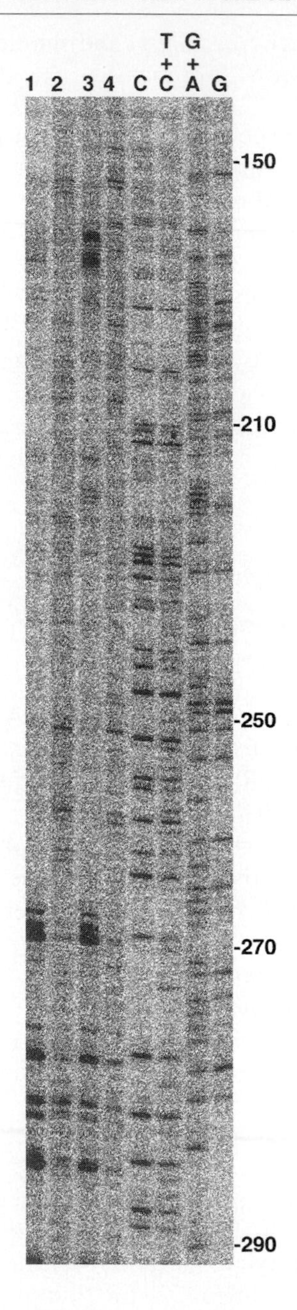
1
2
3
4
C
T + C
G + A
G
-150
-210
-250
-270
-290

9. Precipitate the DNA by adding 10 μg tRNA, 0.1 volumes of 3 *M* sodium acetate, pH 5.2, 2.5 volumes of ethanol and incubating on dry ice.
10. Centrifuge, wash in 75% ethanol, and dissolve in 50 μl of water.
11. Continue with PCR and the following steps as for the standard procedure.

Short 2-Day Procedure with Direct Primer Labeling

The following short procedure, which is adapted from a procedure described by Cairns and Murray,[18] allows LM-PCR data to be obtained in 2 days or less, rather than the 4 or 5 days required for the standard procedure. In this simplified procedure, most purification and transfer steps are omitted (all enzymes and primers remain in the same reaction), and, in its shortest version, a labeled third primer is used, thereby eliminating the Southern transfer and hybridization steps. As was done originally by Mueller and Wold,[1] and by many investigators since, the gel-separated PCR bands can be visualized by direct extension of a ^{32}P-labeled primer 3. The simplified direct labeling method given here gives good results with the majority of primers and does not require special equipment for transfer to a membrane. Figure 3 shows the quality of data that can be obtained, even for DNase I samples, which tend to have a high background. For most work, though, we prefer the long procedure, with transfer and hybridization, because it is about 10 times more sensitive and it is more specific if stringent hybridization conditions are used. Often, though, the sensitivity of LM-PCR is so great that signal strength is not limiting and, with good primers, extra specificity is not needed. The short procedure described here not only saves time but also uses less of expensive reagents, as smaller volumes are

[18] M. J. Cairns and V. Murray, *Biochemistry* **35,** 8753 (1996).

FIG. 3. LM-PCR by a short procedure and direct labeling. The DNase I and short procedure protocols described here were used with mouse DNA from cells treated *in vivo* by permeabilization with lysolecithin in the presence of DNase I. Dynabeads were not used. The mouse Pgk1 promoter region is shown, with nucleotide position upstream from the major transcriptional start site[22] indicated. A 16:1 mixture of Pfu (exo$^-$)/Pfu was used. After 18 cycles of PCR, labeling was done using the direct labeling protocol. For detection, a PhosphorImager (Molecular Dynamics) was used, with overnight exposure. Lanes 1 and 3: *in vivo* DNase I treatment; average single-strand size 3 and 10 kb, respectively, as determined by alkaline agarose electrophoresis. Lanes 2 and 4: *in vitro* DNase I-treated DNA; average size 3 and 10 kb, respectively. Lanes labeled C, T + C, G + A, and G are Maxam–Gilbert-treated mouse DNA analyzed by the same short procedure.

used. Although not described in detail, this short protocol can also be used with a 5′ biotinylated primer 1 and Dynabeads M-280 (Dynal, Inc.) (J. LeBon and A. D. Riggs, unpublished results, 1998), similar to the use of beads in the DNase I procedure described earlier.

First Primer Extension

Most gene-specific first primers that work in the standard procedure will also work in the short procedure, even though they have a low T_m appropriate for Sequenase. For reasons not yet fully understood, there is a broad temperature optimum (H. H. Chen and A. D. Riggs, unpublished data).

1. In a siliconized tube, mix 1 μl of cleaved genomic DNA (0.5–2 μg) and 4 μl of *Pfu* reaction mix consisting of 1 μl of H_2O, 0.5 μl of 10× *Pfu* buffer (supplied by the manufacturer), 0.5 μl of 2.5 m*M* dNTP mix, 0.5 μl of 2 μM gene-specific primer 1, and 1 μl of a 16:1 *Pfu* enzyme mixture [dilute 1 μl of *Pfu* DNA polymerase with 15 μl of TE, pH 8.0, then mix 1:1 (v/v) with *Pfu* (exo^-) DNA polymerase]. Mix well and spin briefly.
2. Add 10 μl of mineral oil and spin briefly.
3. Perform primer extension by denaturing at 95° for 2–3 min, annealing at 0–5° above the calculated T_m of primer 1 for 15 min, and incubation at 72° for 5 min.
4. Cool on ice and spin briefly. Leave the samples on ice while preparing the ligation mix.

Notes. Pfu generates blunt ends and generally works well for LM-PCR. It does, however, have editing and 3′-exonuclease activities that cause significant degradation of the primer from the 3′ end. We have found that a mixture of exonuclease negative (exo^-) Pfu and normal Pfu works well (H. H. Chen and A. D. Riggs, unpublished results, 1998). A mixture of $Vent_R$ (exo^-) and normal $Vent_R$ (New England Biolabs, Beverly, MA) also works well for the short LM-PCR procedure, as it does for standard LM-PCR.

Ligation

The linker is the same as for the standard procedure.

1. For each sample, prepare a 7.2-μl ligation mix consisting of 49 m*M* Tris–HCl, pH 7.5, 13.6 m*M* $MgCl_2$, 33 m*M* DTT, 1.66 m*M* ATP, 83 μg/ml BSA, 24 pmol linker, and 4.8 units of T4 DNA ligase (Promega, Madison, WI, 20 units/μl).
2. Add 7.2 μl of the ligation mix directly to the 5-μl reaction resulting from the primer extension step. Mix by pipetting and spin briefly.

3. Incubate at 17° for 2 hr or overnight, then keep the tubes on ice while preparing the PCR mix.

Notes. By using high concentration T4 DNA ligase (such as can be obtained from Promega, Inc.), the volume of the ligation mix can be kept small, which may be important. Ligation time course and linker concentration studies have shown that under the conditions just described, a 2-hr incubation time at 17° is adequate, but overnight is often convenient.

Polymerase Chain Reaction

The T_m of primer 2 primer should be close to that of the linker primer, which is 66° as calculated by the Oligo 4.0 program.

1. Prepare 8 μl of a PCR mix consisting of 2 μl of 10× *Pfu* buffer, 0.2 μl of 25 m*M* dNTP, 0.2 μl of 20 μ*M* primer 2, 0.2 μl of 20 μ*M* linker primer, 2 μl of 16:1 *Pfu* mix (see first primer extension, step 1), and 3.4 μl H_2O.
2. Add 8 μl of the PCR mix to the ligation reaction mixture, making the total volume about 20 μl. Mix well and spin briefly.
3. Add 20 μl of mineral oil.
4. Denature at 95° for 3 min, then perform 18 thermocycles of 45 sec at 95°, 2 min at the annealing temperature, and 3 min at 72°. Add a final cycle of 1 min at 95°, 2 min annealing, and 10 min at 72°.
5. Cool on ice and spin briefly. Store at 4°.

At this point, either or both of the following can be done: (i) All or part (we generally use only half) of the PCR product can be alcohol precipitated and dried down, then used for gel electrophoresis, Southern blotting, and hybridization with single-stranded probes, as described in the conventional LM-PCR protocol or (ii) 10 μl of the PCR product can be used for direct labeling with end-labeled primer 3 (as described in the following section).

Direct Labeling by Extension of End-Labeled Primer 3

1. Transfer 10 μl of the PCR product to a new tube.
2. Add 1 μl of [γ-^{32}P]ATP-labeled primer 3, mix well, and spin briefly.
3. Add 10 μl of mineral oil on top.
4. Denature at 95° for 3 min, then do three to five thermocycles of primer extension for 45 sec at 95°, 2 min at T_m + 0–5°, and 3 min at 72°. Allow 5 min at 72° at the end of cycling.
5. Cool on ice and store at −20° until gel electrophoresis.

Notes. The thermostable polymerases are still quite active after 18 cycles of PCR and remain so if stored at 4°. (Do not freeze.) The labeling step can be done immediately after PCR or on the next day while preparing the sequencing gel. If the short procedure is being used, there is no purification of the PCR product prior to the labeling reaction, as generally this is not necessary. However, until this is demonstrated experimentally for a given primer set, it would be prudent to purify the PCR product by phenol extraction and ethanol precipitation before labeling, as described for the standard LM-PCR procedure.

Preparation of End-Labeled Primer 3

1. Mix 1 μl of 10× T4 DNA kinase buffer with 6.0 μl of H_2O and then add 1 μl of 20 μ*M* primer 3.
2. Add 1 μl of [γ-^{32}P]ATP (150 mCi/ml, Amersham, Arlington Heights, IL).
3. Add 1 μl of T4 polynucleotide kinase (PNK, 10 units/μl, New England Biolabs, Beverly, MA), mix well, and spin briefly.
4. Incubate at 37° for 1 hr and then heat-inactivate the enzyme at 65° for 15 min.
5. Purify the labeled primer 3 by using a Sephadex G-25 spin column (5 prime–3 prime, Inc., Boulder, CO). Store at −20°.

Notes. For direct labeling with an end-labeled primer 3 after PCR, use a gene-specific primer 3 that has a T_m greater than that of primer 2. If this is not the case, then it would be prudent to purify the PCR product to reduce the amount of primer 2, which will compete with primer 3.

Gel Electrophoresis and Detection of Bands

Gel electrophoresis is done as for the standard procedure. After the run, transfer the lower part of the gel (about 40 cm) to Whatman 3MM paper, as is normally done for radioactive DNA sequencing gels, and cover it with Saran wrap. Dry the gel under vacuum at 80° for 30–60 min using a gel dryer. Expose the dried gel to an X-ray film overnight at −70°, usually with intensifying screens.

TD-PCR: Ligation-Mediated Polymerase Chain Reaction with Ribonucleotide Tailing by Terminal Deoxynucleotidyltransferase

The key difference between TD-PCR and standard LM-PCR is that an oligonucleotide is ligated to the 3′ end of molecules that result from repeated primer extension and then the addition of 3 rGs by use of terminal transferase (see right hand side of Fig. 1). This method is quite sensitive and versatile, allowing detection of any DNA lesion or strand break that stops primer extension. Thymidine dimers, for example, can be detected[7] without needing special enzymes (see Fig. 4). TD-PCR has proven to be quite robust and can be more sensitive than LM-PCR.[7,19]

First Primer Extension

1. In a siliconized tube, mix 10 μl of DNA sample (0.5–1 μg) and 20 μl Vent reaction mix containing 30 m*M* Tris–HCl, pH 8.8, 15 m*M* KCl, 15 m*M* $(NH_4)_2SO_4$, 0.15% Triton X-100, 1 pmol of primer 1, 375 μM each dNTP, and 1 μl of 16:1 $Vent_R$ (exo^-)/$Vent_R$ DNA polymerase mix.
2. Add 30 μl mineral oil and spin briefly.
3. Denture at 95° for 3 min, start 10 cycles of 1 min at 95°, 3 min at 0–5° above the T_m of primer 1, and 2 min at 72°, and denature again at 95° for 2 min after cycling and immediately chill the sample on ice.
4. Transfer the solution beneath the oil to a new chilled tube and transfer it again to another tube containing chilled 80 μl salt solution 1 (57.6 μl of H_2O, 20 μl of 10 *M* ammonium acetate, 0.4 μl of 0.5 *M* EDTA, pH 8.0, and 2 μl of 10 mg/ml glycogen).
5. Add 260 μl of 100% ethanol and leave on dry ice for 15 min.
6. Centrifuge at maximum speed for 15 min and remove the supernatant.
7. Wash DNA pellets with 0.5 ml 80% (v/v) ethanol and spin for 5 min to remove supernatant. Quick spin and remove as much as possible of the residual alcohol.
8. Air dry DNA pellets for 10–15 min.

Notes. Usually 10 cycles of extension are performed. A 16:1 mix of $Vent_R$ (exo^-) and $Vent_R$ DNA polymerase works well, probably because the presence of some exo^+ enzymes lessens the addition of 3′ nucleotides and provides some editing of extension-stopping misincorporated bases. A similar mixture of *Pfu* (exo^-) and *Pfu* also works well. More than 10 cycles can be done, although the background does begin to increase.[7]

Terminal Deoxynucleotidyl Transferase (TdT) Tailing

1. Dissolve the DNA pellets with 10 μl of 0.1× TE, pH 7.5 (1 m*M* Tris–HCl, 7.5 and 0.1 m*M* EDTA).

[19] M. F. Denissenko, T. B. Koudriakova, L. Smith, T. R. O'Connor, A. D. Riggs, and G. P. Pfeifer, *Oncogene,* **17,** 3007 (1998).

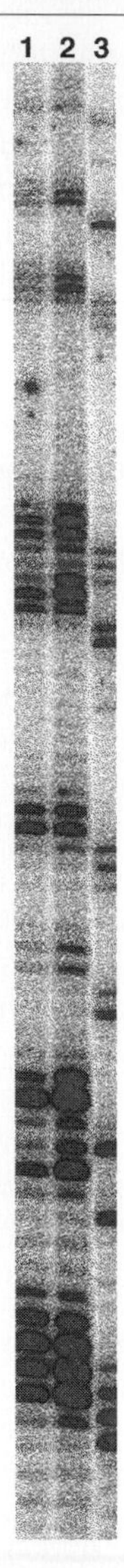

Fig. 4. Terminal transferase-dependent PCR (TD-PCR) and comparison with LM-PCR. TD-PCR was done using DNA that had been UV irradiated *in vitro* (lane 1) or *in vivo* (lane 2). A primer set specific for the mouse Snrpn promoter was used. For comparison, LM-PCR (lane 3) was done using the same *in vivo*-irradiated DNA and primer set. DNA used for lanes 2 and 3 was isolated from UV-irradiated cells (254 nm, 1000 J/m^2) and was used without any additional treatment (lane 2, TD-PCR) or was pretreated with T4 endonuclease IV and photolyase (lane 3, LM-PCR) to generate ligatable ends at positions of pyrimidine dimers.

2. Add 10 μl TdT mix consisting of 10 units of TdT (GIBCO-BRL, Grand Island, NY) in 2× TdT buffer supplied by the manufacturer. Mix well and spin briefly.
3. Incubate at 37° for 15 min and chill on ice.
4. Add 80 μl of salt solution 2 (59.6 μl of H_2O, 20 μl of 10 *M* ammonium acetate, 0.4 μl of 0.5 *M* EDTA, pH 8.0) and ethanol precipitate as in steps 5–8 described in the previous section.

Notes. Ribonucleotide triphosphates are not the preferred substrates for TdT; thus the reaction is self-limited to the addition of only about 3 riboGs.[7,20]

Adapter Preparation

After TdT treatment, most molecules have rG tails that can be ligated to double-stranded DNA adapters with a complementary 3′ overhang (CCC) as described by Schmidt and Mueller[20] for cDNA cloning. The lower strand of the adapter is a 24-mer with an aminopentyl blocking group at the 3′ end (5′AATTCAGATCTCCCGGGTCACCGC3′ blocked, see Komura and Riggs[7]). This oligonucleotide must be 5′ phosphorylated before use. The upper strand of the adapter (5′GCGGTGACCCGGGAGAT CTGAATTCCC3′) is a 27-mer that gives a CCC 3′ overhang when annealed with the lower strand.[7]

1. Dissolve both upper and lower strands of the adapter in water as stock solutions of 200 μM (200 pmole/μl).
2. To phosphorylate the lower strand, prepare a 100-μl reaction containing 22.2 μM of the oligonucleotide, 50 units of T4 polynucleotide kinase (PNK, 10 units/μl from New England Biolabs), and 1 m*M* ATP in 1× buffer supplied by the manufacturer.
3. Incubate at 37° for 2 hr and then heat inactivate the enzyme at 65° for 15 min.
4. Add 11.1 μ1 of the upper primer (200 μM) to the phosphorylated lower primers.
5. Heat the reaction at 95° for 3–5 min with a lid lock, cool gradually to room temperature, and store at −20°. The adapter is thawed and kept on ice before use.

[20] W. M. Schmidt and M. W. Mueller, *Nucleic Acids Res.* **24,** 1789 (1996).
[21] S. Tommasi and G. P. Pfeifer, *Mol. Cell Biol.* **15,** 6901 (1995).
[22] S. Tommasi, J. M. LeBon, A. D. Riggs, and J. Singer-Sam, *Somat. Cell Mol. Genet.* **19,** 529 (1993).

Adapter Ligation

1. Dissolve the air-dried sample resulting from the tailing reaction in 15 μl of 0.1× TE, pH 7.5 (1 m*M* Tris–HCl, pH 7.5, 0.1 m*M* EDTA).
2. Prepare a ligation mix (15 μl/sample) by mixing 7.95 μl H_2O, 1.5 μl 1 *M* Tris–HCl, pH 7.5, 0.3 μl 1 *M* $MgCl_2$, 0.3 μl 1 *M* DTT, 0.3 μl 100 m*M* ATP, 0.15 μl 10 mg/ml BSA, 3 μl of 20 μ*M* TD-PCR adapter, and 1.5 μl of T4 DNA ligase (Promega, Madison, WI, 3 U/μl).
3. Add 15 μl ligation mix to each sample. Mix well and spin briefly.
4. Incubate overnight at 17°.

PCR, Labeling, and Electrophoresis

The molecules that have participated in ligation to the adapter will function as templates for PCR with a second gene-specific primer (primer 2) and the same linker–primer (LP25) that is used in the LM-PCR protocol. Generally no purification is done and the ligation reaction (30 μl) is used directly for a PCR reaction (100 μl final volume). After 18–20 thermocycles, the product is phenol extracted, ethanol precipitated, and dissolved in 5 μl of water for use in either (i) electrophoresis, blotting, and hybridization with single-stranded probes as described for standard LM-PCR or (ii) direct labeling with an end-labeled primer 3.

To perform direct labeling of the TD-PCR product:

1. Prepare a labeling cocktail (5 μl/sample) by mixing 0.75 μl of 10× Vent buffer (New England Biolabs), 0.1 μl of 25 m*M* each dNTP, 1.5 μl of end-labeled primer 3, 1 μl of $Vent_R$ (exo^-)/$Vent_R$ DNA polymerase 16:1 mix, and water to a total of 5 μl.
2. Mix half of the DNA sample (2.5 μl) with 5 μl labeling cocktail and spin briefly.
3. Add 10 μl of oil on top and spin briefly.
4. Denature at 95° for 2 min and then start three to six cycles of 45 sec at 95°, 3 min at annealing temperature, 2 min at 72°, and an extra 5 min at 72°.
5. Cool on ice and store at −20°.

After labeling, the samples are ready for gel electrophoresis, gel drying, and autoradiography as described for the short procedure.

Notes. Excellent results have been obtained with Vent (exo^-) for primer extension and Vent for PCR.[7] Pfu or a 16:1 mixture of the exonuclease minus and normal versions of Pfu works well for both primer extension and PCR. For the PCR reaction, Amplitaq polymerase (Perkin-Elmer, Norwalk, CT) can be used in place of the 16:1 Vent (exo^-)/Vent mixture. Dynabeads M-280 (Dynal, Inc.) and a 5′-biotinylated first primer can be used, similar to the procedure described above for LM-PCR (J. Komura, H. H Chen, and A. D. Riggs, unpublished data, 1998). If Dyanbeads are used, the alcohol precipitation steps before PCR can be replaced by TE buffer washes of the beads. Tailing, ligation, and PCR are done using the beads; as for TD-PCR it is the biotinylated strand that participates in tailing, ligation, and the first cycle of PCR.

Choice of Procedure

We have made numerous comparisons of enzymes and procedures over a period of years and have consistently found that the standard LM-PCR procedure described here gives the most reliably strong signal and the lowest background. This is especially true for genomic samples with a low density of lesions. The extra purification, membrane transfer, and hybridization steps in the standard procedure add an extra margin of reliability and sensitivity, as well as minimizing radiation exposure and contamination. However, the standard procedure is tedious and takes several days. End-labeled primers, as used for the direct labeling procedures described here, usually provide a signal strength about one-tenth that of the standard method, but often this is not of great importance because good data nevertheless can be obtained after overnight exposure. With adequately pure genomic DNA and favorable primers, the short procedure gives excellent data (Fig. 3). However, this method is not quite as robust as the standard method as all reagents—DNA, oligonucleotides, enzymes, and impurities—stay in the same reaction. The TD-PCR procedure is still quite new, but it seems quite robust and has great potential for footprinting and chromatin structure analysis. It has already been shown to work well for the detection of UV-induced footprints[7] and for aflatoxin adducts.[19]

Acknowledgments

We thank Chia-Wei Cheung, Dr. Stella Tommasi, and Dr. Jeanne LeBon for valuable contributions. This work was supported by NIH Program Project Grant CA69449 (G. Holmquist, P.I.) and NIH Grants GM50575 to A.D.R and ES06070 to G.P.P.

[31] Guanine–Adenine Ligation-Mediated Polymerase Chain Reaction *in Vivo* Footprinting

By ERICH C. STRAUSS and STUART H. ORKIN

Introduction

Hematopoietic cells represent an attractive biological model system for the study of molecular mechanisms that control lineage and developmental specific gene expression. To generate the various hematopoietic lineages, common pluripotent stem cells express distinct sets of genes during cellular commitment and maturation. Activation of lineage-specific programs of gene expression in hematopoietic cells is presumed to be mediated by the interaction of cell-specific and ubiquitous transcriptional factors with their cognate *cis* elements. The identification of regulatory motifs and the characterization of DNA-binding proteins that recognize these sequences provide a basis for understanding the mechanisms involved in cell-specific gene expression.

The expression of α- and β-like globin genes in developing erythroid cells is dependent on the integrity of distant, upstream regulatory elements referred to as locus control regions (LCRs).[1] LCRs serve to maintain chromatin in an open configuration and supersede inhibitory actions on gene expression.[2] The major activity of the α-LCR is coincident with a single erythroid-specific deoxyribonuclease I hypersensitive site located 40 kb upstream of the embryonic zeta gene.[3] The β-LCR corresponds to four erythroid-specific, developmentally stable hypersensitive sites, referred to as HS-1 to HS-4 that lie 6–18 kb upstream of the embryonic epsilon-globin gene[4,5]; the primary activity of the β-LCR has been associated with HS-2 and HS-3. Current models suggest that high-level developmentally appropriate transcription of individual globin genes is accomplished through

[1] S. H. Orkin, *Cell* **63,** 665 (1990).
[2] G. Felsenfeld, *Nature* **355,** 219 (1992).
[3] D. R. Higgs, W. G. Wood, A. P. Jarman, J. Sharpe, J. Lida, I.-M. Pretorius, and H. Ayyub, *Genes Dev.* **4,** 1588 (1990).
[4] D. Tuan, W. Solomon, Q. Li, and I. M. London, *Proc. Natl. Acad. Sci. U.S.A.* **82,** 6384 (1985).
[5] W. C. Forrester, S. Takegawa, T. Papayannopoulou, G. Stamatoyannopoulos, and M. Groudine, *Nucleic Acids Res.* **15,** 10159 (1987).

METHODS IN ENZYMOLOGY, VOL. 304
0076-6879/99 $30.00

an interaction between gene promoters or enhancers and LCR elements.[1,6,7]

The functional activity of LCR elements is thought to be mediated through their interaction with cell-specific and ubiquitous regulatory proteins. The properties of LCR elements and the basis of cell-specific gene expression have been investigated by *in vitro* studies.[8–15] These analyses have revealed multiple binding sites for several types of *cis* elements that include the erythroid transcription factor GATA-1[11] or closely related family members,[16] an AP-1 and/or erythroid AP-1-like activity,[8–12,15] termed NF-E2,[17] and proteins that recognize GGTGG/CACCC sequences.[12,14] However, *in vitro* studies have several inherent limitations. First, *in vitro* assays may reveal protein binding to sites that are unavailable in native chromatin. Second, in the presence of factors with overlapping specificities, these studies may fail to distinguish appropriate occupancy of regulatory motifs as a consequence of protein abundance. Third, proteins present at low concentrations in nuclear extract may preclude *in vitro* detection. Furthermore, *in vitro* techniques are insensitive to chromatin structure. For these reasons, we have used dimethyl sulfate (DMS) *in vivo* footprinting[18] as a complementary method for the dissection and characterization of active regulatory elements within the human α-LCR,[19] an HS-3 of the human β-LCR.[20]

[6] R. R. Behringer, T. M. Ryan, R. D. Palmiter, R. L. Brinster, and T. M. Townes, *Genes Dev.* **4,** 380 (1990).

[7] T. Enver, N. Raisch, A. J. Ebens, T. Papayannopoulou, F. Costantini, and G. Stamatoyannopoulos, *Nature* **344,** 309 (1990).

[8] A. P. Jarman, W. G. Wood, J. A. Sharpe, G. Gourdon, H. Ayyub, and D. R. Higgs, *Mol. Cell. Biol.* **11,** 4679 (1991).

[9] P. Moi and Y. W. Kan, *Proc. Natl. Acad. Sci. U.S.A.* **87,** 9000 (1990).

[10] P. A. Ney, B. P. Sorrentino, C. H. Lowrey, and A. W. Nienhuis, *Nucleic Acids Res.* **18,** 6011 (1990).

[11] P. A. Ney, B. P. Sorrentino, K. T. McDonagh, and A. W. Nienhuis, *Genes Dev.* **4,** 993 (1990).

[12] S. Philipsen, D. Talbot, P. Fraser, and F. Grosveld, *EMBO J.* **9,** 2159 (1990).

[13] S. Pruzina, O. Hanscombe, D. Whyatt, F. Grosveld, and S. Philipsen, *Nucleic Acids Res.* **19,** 1413 (1991).

[14] D. Talbot and F. Grosveld, *EMBO J.* **10,** 1391 (1991).

[15] D. Talbot, S. Philipsen, P. Fraser, and F. Grosveld, *EMBO J.* **9,** 2169 (1990).

[16] M. Yamamoto, L. J. Ko, M. W. Leonard, H. Beug, S. H. Orkin, and J. D. Engel, *Genes Dev.* **4,** 1650 (1990).

[17] V. Mignotte, L. Wall, E. deBoer, F. Groveld, and P.-H. Romeo, *Nucleic Acids Res.* **17,** 37 (1989).

[18] A. Ephrussi, G. M. Church, S. Tonegawa, and W. Gilbert, *Science* **227,** 134 (1985).

[19] E. C. Strauss, N. C. Andrews, D. R. Higgs, and S. H. Orkin, *Mol. Cell. Biol.* **12,** 2135 (1992).

[20] E. C. Strauss and S. H. Orkin, *Proc. Natl. Acad. Sci. U.S.A.* **89,** 5809 (1992).

Generally, *in vivo* footprinting has involved the use of DMS, an alkalating agent that methylates genomic DNA of intact cells at the N-7 position of guanines and the N-3 position of adenines.[21] The reactivity of guanine residues has been used exclusively for *in vivo* footprinting of complex genomes as adenines methylate less efficiently than guanines[22] and the N-7 position of guanines resides in the major grove of DNA, a common protein-binding site. As a requisite of our analysis of LCR elements, we modified the ligation-mediated polymerase chain reaction (PCR) *in vivo* footprinting procedure to permit the assessment of protein–DNA interactions at both guanine and adenine residues rather than merely at guanines. This modification, referred to as GA-LMPCR *in vivo* footprinting, provides the opportunity to evaluate protein contacts at the purine component of each base pair of DNA. GA-LMPCR *in vivo* footprinting revealed protein binding at GATA-1, AP-1/NF-E2, and CACC/GGTGG motifs in erythroid cells and specific differences compared with *in vitro* studies.[19,20] At several GATA-1 motifs, protein contacts were observed exclusively at adenine residues within the core sequence of this *cis* element.[19,20] Moreover, GA-LMPCR *in vivo* footprinting provided a more extensive profile of protein contacts at GATA-1 and AP-1/NF-E2 regulatory sites. Our investigation also detected erythroid-specific patterns of protein–DNA interactions in LCR regions that are not associated with previously characterized regulatory motifs. The sequences coincident with these contacts fail to reveal specific protein binding *in vitro*. Therefore, these regions may reflect a local chromatin structure consequent to interactions of proteins assembled on their respective core motifs. Our composite analyses suggest that three sites—GATA-1, AP-1/NF-E2, and CACC/GGTGG motifs—are minimally required for the distinctive properties of globin LCR elements. This article presents a comprehensive discussion of the GA-LMPCR *in vivo* footprinting technique and provides representative analyses from our studies to demonstrate the enhanced detection and resolution obtained with this procedure.

Methods

Methylation and Isolation of Genomic DNA from Cultured Cells

Cell Line and Culture. The interspecies human–mouse somatic cell hybrid line J3-8B has been established by the method of Deisseroth and

[21] P. B. Becker and G. Schutz, *Genet. Eng.* **10,** 1 (1988).
[22] A. M. Maxam and W. Gilbert, *Proc. Natl. Acad. Sci. U.S.A* **74,** 560 (1977).

Hendrick[23] as modified by Zeitlin and Weatherall.[24] Cells are cultured in Ham's F12 medium supplemented with 15% fetal calf serum; J3-8B cells containing human chromosome 16 are selected with methotrexate (10 μM), adenine (0.1 mM), and thymidine (30 μM). The interspecies human–mouse somatic cell hybrid line Hu-11[25] is cultured in RPMI 1640 medium supplemented with 10% fetal bovine serum. The presence of human chromosome 11 β-globin sequences is maintained by selection in hypoxanthine/aminopterin/thymidine medium in which 10 μM methotrexate is used in place of aminopterin. Induced maturation of J3-8B and Hu-11 cells is achieved by the addition of 1.5% dimethyl sulfoxide (DMSO) to their respective culture media 48 hr prior to *in vivo* methylation. HepG2 cells are cultured in Dulbecco's modified Eagle's medium with 10% fetal calf serum. K562 cells[26] are cultured in RPMI 1640 medium containing 10% fetal calf serum.

In Vivo Methylation. Approximately 4×10^8 cells are pelleted and resuspended in 4 ml of complete culture medium (with fetal calf serum) in a 50-ml Falcon tube (Becton Dickinson Labware, Franklin Lakes, NJ). The mixture is incubated on ice for 5 min while a 20° water bath is prepared (e.g., a beaker with water adjusted to 20° with ice). Twenty microliters of DMS (5 μl/ml of culture medium) is added and incubated at 20° for 4 min. The reaction is terminated by adding ice-cold phosphate-buffered saline (PBS) to 50 ml. Cells are pelleted and resuspend in 10 ml of cold PBS; an additional 40 ml PBS is added. The cells are pelleted again and resuspended in 2 ml PBS. Protein-free (naked) control DNA is prepared by omitting the DMS steps.

Preparation of Genomic DNA. In a separate 50-ml Falcon tube, 1.5 mg of proteinase K is added to 4 ml lysis buffer [300 mM Tris (pH 7.8), 150 mM EDTA, 1.5% SDS]. The 2-ml cell/PBS mixture is added to the 4 ml lysis solution, swirled gently to mix, and incubated at 37° for 3–4 hr. One milliliter of 6 M $NaClO_4$ is added and mixed gently. The lysis mixture is extracted four times with phenol/chloroform (1:1); phenol is saturated with 100 mM Tris, pH 8. To prevent shearing of genomic DNA, polyethylene pipettes with several millimeters of the tip removed are used to transfer the aqueous phase to a new 50-ml Falcon tube. The genomic DNA is precipitated with 2 volumes ethanol and mixed by inversion. A Pasteur pipette with a solid, curved tip, created by exposing the tip briefly to a flame, is used to collect and transfer the genomic DNA. Residual ethanol is removed by gently contacting the spooled DNA with the dry, inner

[23] A. Deisseroth and D. Hendrick, *Proc. Natl. Acad. Sci. U.S.A* **76,** 2185 (1979).
[24] H. C. Zeitlin and D. J. Weatherall, *Mol. Biol. Med.* **1,** 489 (1983).
[25] V. Dhar, A. Nandi, C. L. Schildkraut, and A. I. Skoultchi, *Mol. Cell. Biol.* **10,** 4324 (1990).
[26] T. R. Rutherford, J. B. Clegg, and J. D. Weatherall, *Nature* **280,** 164 (1979).

surface of a Falcon tube. Genomic DNA is transferred to another 50-ml tube containing 40 μg RNase in 4 ml TE (10 m*M* Tris, pH 7.5; 1 m*M* EDTA) and incubated overnight at 4°. The digestion mixture is extracted once with phenol/chloroform (1 : 1) and precipitated in 0.3 *M* sodium acetate with 2 volumes of ethanol. Genomic DNA is again collected by the spool method, and residual ethanol is removed as described previously and transferred to a 1.5-ml tube containing 1 ml of TE. Genomic DNA is allowed to resuspend in TE for several days before checking the OD_{260}.

In Vitro Methylation of Protein-Free (Naked) DNA. To generate *in vitro* protein-free control DNA, 30 μg of genomic DNA is digested with an enzyme (2 units/μg DNA) that cleaves outside the target region for footprinting analysis. Digests are performed at 37° overnight. The digest mixture is precipitated in 0.3 *M* sodium acetate with 2 volumes of ethanol. The pellet is resuspended in 180 μl water and 20 μl 0.5 *M* sodium cacodylate (stored at 4°). DMS is added and incubated at room temperature. Conditions for *in vitro* methylation of protein-free genomic DNA (DMS concentration, incubation time) need to be determined empirically to match the distribution of band intensities obtained with *in vivo*-methylated DNA. Overmethylation of genomic DNA *in vitro* will produce a banding pattern that is characterized by a general overrepresentation of lower molecular weight bands and an underrepresentation of higher molecular weight bands; the converse occurs with undermethylation of genomic DNA *in vitro*. The *in vitro* DMS methylation reaction is terminated with 50 μl stop solution (1 *M* 2-mercaptoethanol, 1.5 *M* sodium acetate). Reaction products are precipitated with 750 μl ethanol, pelleted, and resuspended in water at approximately 1 μg/μl.

Guanine–Adenine-Specific Cleavage Reaction

Guanine–adenine-specific cleavage of methylated DNA is modified from the G > A chemistry described by Maxam and Gilbert.[22] Water is mixed with 30 μg of methylated genomic DNA to a final volume of 100 μl in a PCR tube (to avoid shearing of genomic DNA, pipette tips with several millimeters removed are used). An equal volume (100 μl) of 20 m*M* sodium phosphate (pH 7.0) is added and mixed well. Reactions are incubated at 90° for 15 min in a thermal cycler, collected by brief centrifugation, and transferred to ice. Twenty microliters of 1 *M* NaOH is added to the solution, mixed well, and incubated at 90° for 30 min in a thermal cycler. The reactions are collected by brief centrifugation, transferred to a 1.5-ml tube, neutralized with 1 *M* HCl, and precipitated in 0.3 *M* sodium acetate with 2 volumes of ethanol. The cleavage products are resuspended in water at approximately 1 μg/μl.

5′-GCAGTGACTCGAGAGATCTGAATTC-3′ 25-mer

3′-CTAGACTTAAG-5′ 11-mer

FIG. 1. Structure and sequence of the linker used in GA-LMPCR *in vivo* footprinting. The underlined nucleotides refer to changes introduced to the Mueller and Wold linker.[27]

GA-LMPCR in Vivo Footprinting

GA-LMPCR *in vivo* footprinting is performed essentially as described by Mueller and Wold.[27] Modifications that we have introduced to this technique and suggestions to optimize GA-LMPCR *in vivo* footprinting are discussed in the following sections.

Annealing Reaction. To provide an appropriate founder population for the analysis of adenine residues, the annealing reaction is performed with at least 5 μg of cleaved genomic DNA.

Ligation Reaction. The sequence of the 25-mer linker primer is modified (Fig. 1) to be compatible with primer 2 (region-specific, amplification reaction primer) T_m values of 60°. This has been suitable for the positioning and design of all primer sets used for our studies. The sequence of the 11-mer component remains the same as that described by Mueller and Wold.[27]

Amplification Reaction. The amplification and labeling reactions have been separated; this modification from the Mueller and Wold[27] procedure appears to be important for obtaining appropriate band intensity at adenine residues. The annealing step of the PCR amplification reaction is performed at 60° with a region-specific primer and the 25-mer component of the linker shown in Fig. 1. After 17 cycles of exponential amplification of the ligation products, the 100-μl reaction volume is transferred to a 1.5-ml tube containing 295 μl *Taq* stop buffer [260 m*M* sodium acetate (pH 7.0), 10 m*M* Tris–HCl (pH 7.5), 4 m*M* EDTA (pH 8.0), 35 μg/ml tRNA]. The mixture is extracted once with phenol/chloroform (1 : 1) and precipitated with 2.5 volumes of ethanol. Reaction products are resuspended in 100 μl water.

Labeling Reaction. Fifty microliters of the amplification reaction is used in the labeling reaction; the remaining 50 μl is stored at −20° for future labelings, if necessary. For each labeling reaction, 50 μl of amplification products is combined with 20 μl of 5× *Taq* buffer [200 m*M* NaCl, 50 m*M* Tris–Cl, pH 8.9, 17.5 m*M* $MgCl_2$, 0.05% (w/v) gelatin], 8 μl of dNTP mix (prepared by adding equal volumes of 10 m*M* dATP, dCTP, dGTP, and

[27] P. R. Mueller and B. Wold, *in* "Current Protocols in Molecular Biology" (F. M. Ausubel, R. Brent, R. E. Kingston, D. D. Moore, J. A. Smith, and J. G. Seidman, eds.). Greene/Wiley, New York, 1991.

dTTP stock solutions), 18 μl H_2O, 4 μl (3 pmol) 5′-end-labeled primer 3, and 2.5 units of *Taq* DNA polymerase. The contents are mixed well and three cycles are used for the labeling reaction. The 100-μl reaction volume is transferred to a 1.5-ml tube containing 295 μl *Taq* stop buffer, extracted once with phenol/chloroform (1:1), and precipitated with 2.5 volumes of ethanol. The labeled products are resuspended in 10 μl formamide–dye mix; 1–2 μl of each reaction is applied to denaturing polyacrylamide gels. The gels are dried and exposed to Kodak (Rochester, NY) X-AR film with an intensity screen at −70° for 15–45 hr.

Applications

Of the various techniques available to investigate mechanisms of gene regulation, only *in vivo* footprinting assesses protein–DNA interactions as they occur in living cells. We have used the DMS/guanine-based *in vivo* footprinting to complement functional and *in vitro* studies of the GATA-1 gene promoter.[28] To enhance the detection of protein binding at GATA motifs and expand the resolution of protein contact at other regulatory elements, we have modified the chemistry used with DMS-based LMPCR *in vivo* footprinting to permit the analysis of protein interactions at both guanine and adenine residues rather than at guanines alone.[19] This modification was required for a comprehensive analysis of the human α-LCR[19] and subregions of the human β-LCR.[20]

As shown by the examples in Fig. 2, the use of GA cleavage chemistry with DMS-based LM PCR *in vivo* footprinting significantly augments the ability to detect protein binding at regulatory motifs in which guanines are not contacted. With G cleavage chemistry alone, no *in vivo* footprint is observed in either J3-8B cells or nonerythroid HepG2 cells at a potential GATA motif in the human α-LCR element (Fig. 2A). However, as indicated in Fig. 2B, GA cleavage chemistry reveals protection of two adenines in J3-8B cells but not in either K562 cells, which exhibit erythroid, megakayocytic, and myeloid properties,[26,29] or HepG2 cells. Two inferences can be made from these data. First, the use of GA cleavage chemistry with LMPCR *in vivo* footprinting provides information that cannot be obtained with G cleavage chemistry. Second, although partially erythroid in phenotype, K562 cells may be inadequate for *in vivo* footprinting analysis of some regulatory regions. The absence of *in vivo* footprints in HepG2 cells is consistent with a chromatin structure that renders these sites inaccessible in nonerythroid and nonglobin expressing cells. The complete analysis of

[28] S.-F. Tsai, E. Strauss, and S. H. Orkin, *Genes Dev* **5,** 919 (1991).
[29] N. L. Lumelsky and B. G. Forget, *Mol. Cell. Biol.* **11,** 3528 (1991).

A G-cleavage chemistry

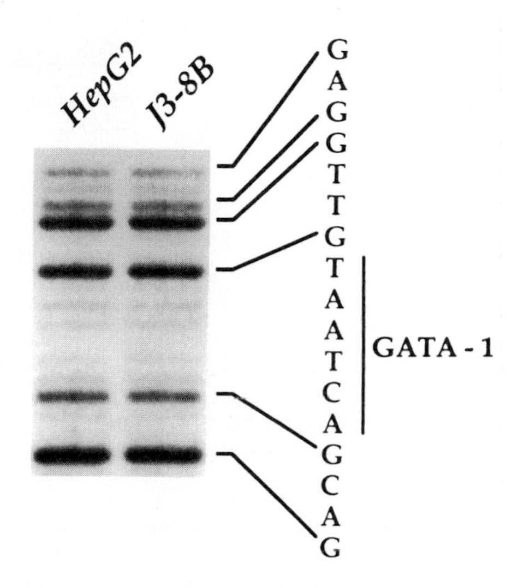

B GA-cleavage chemistry

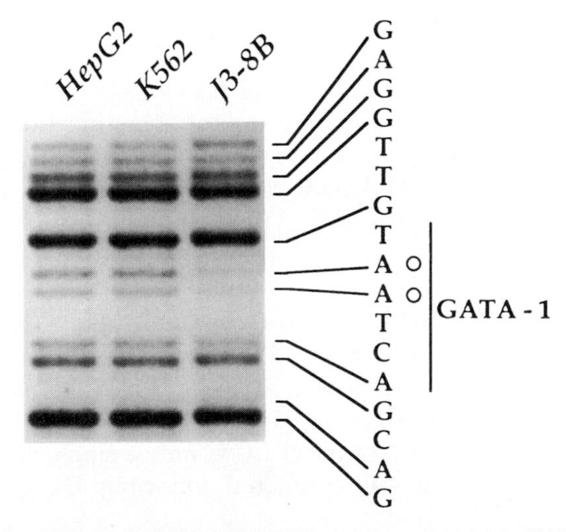

FIG. 2. Analysis of *in vivo* protein–DNA interactions at the nonconsensus GATA-1 binding site in the human α-LCR element[19] using guanine (A) and guanine–adenine cleavage chemistry (B) with LMPCR *in vivo* footprinting. Expressing cell lines include *in vivo*-methylated J3-8B and K562; *in vivo*-methylated HepG2 cells were used as a nonexpressing control. K562 cells were treated with 30 μM hemin for 2 days prior to *in vivo* methylation; J3-8B cells were treated with DMSO as described in the text. The same preparations of methylated genomic DNA were used for these studies.

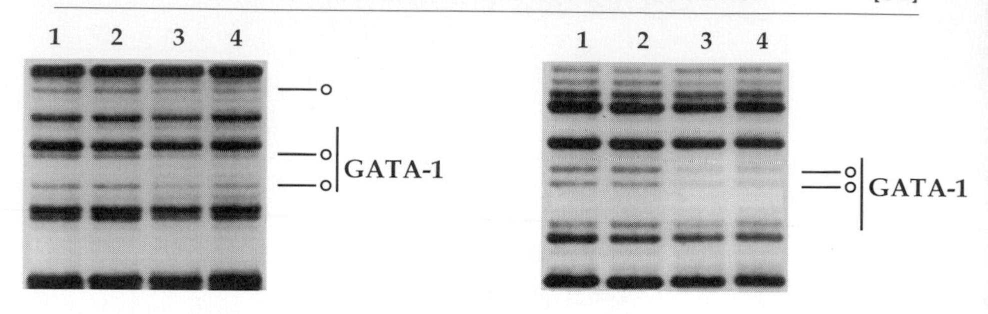

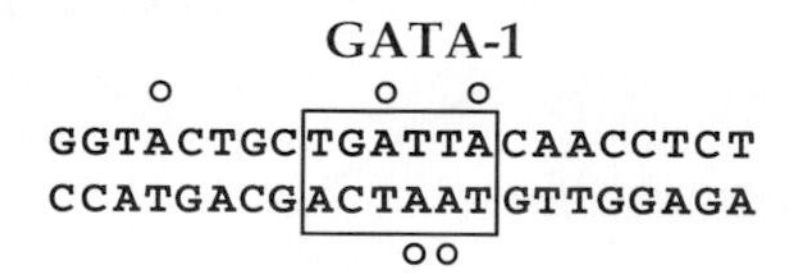

FIG. 3. GA-LMPCR *in vivo* footprinting of the top (left) and bottom (right) strands of the nonconsensus GATA-1 motif in the human α-LCR element. Lane 1: *in vitro*-methylated protein-free K562 DNA; lane 2: *in vivo*-methylated HepG2 DNA (nonexpressing control); lane 3: *in vivo*-methylated J3-8B DNA (expressing line, uninduced cells); and lane 4: *in vivo*-methylated J3-8B DNA (expressing line, DMSO-induced cells). Protections are indicated by open circles.

this nonconsensus GATA site (TGATTA vs T/AGATAA/G) is shown in Fig. 3. As indicated in the summary of altered methylation reactivities, only adenines were contacted at this motif. Two additional GATA sites, one each in the α-LCR and HS-3 of the β-LCR, also displayed adenine exclusive contacts in the core sequence.[19,20] The nonconsensus GATA sequence has subsequently been shown to be a high-affinity target for GATA-1 protein binding.[30]

GA-LMPCR *in vivo* footprinting was also critical for the analysis of protein binding to AP-1/NF-E2 sites in human LCR elements. As illustrated in Fig. 4, the top strand analysis of an AP-1/NF-E2 element in the α-LCR reveals a more extensive profile of protein contacts with GA-LMPCR *in vivo* footprinting. In HS-3 of the β-LCR, only a single adenine in the core AP-1/NF-E2 sequence was contacted following DMSO-induced maturation.[20]

Our *in vivo* footprinting studies of HS-3 of the β-LCR demonstrated that protein complexes formed in this region are regulated developmentally both with respect to cell type and stage of erythroid maturation. First, no *in*

[30] M. Merika and S. H. Orkin, *Mol. Cell. Biol.* **13,** 3999 (1993).

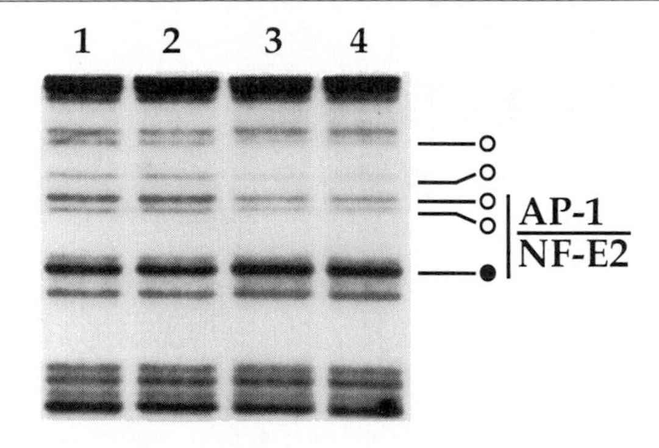

AP-1/NF-E2

GGGCCAACCATGACTCAGTGCTTCTGG
CCCGGTTGGTACTGAGTCACGAAGACC

FIG. 4. GA-LMPCR *in vivo* footprinting of the top strand of an AP-1/NF-E2 motif in the human α-LCR element. Lane 1: *in vitro*-methylated protein-free K562 DNA; lane 2: *in vivo*-methylated HepG2 DNA (nonexpressing control); lane 3: *in vivo*-methylated J3-8B DNA (expressing line, uninduced cells); and lane 4: *in vivo*-methylated J3-8B DNA (expressing line, DMSO-induced cells). Protections are represented with open circles; enhancements are indicated by solid circles.

vivo footprints were detected in a nonerythroid environment, e.g., HepG2. Second, HS-3 is not formed in K562 cells, a line of mixed hematopoietic lineages expressing embryonic and fetal hemoglobins.[4,31] Third, and most important, our results reveal only limited *in vivo* contacts in uninduced Hu-11 cells, which subsequently become more extensive and complex after DMSO-induced maturation.[20] As shown in Fig. 5, protein contacts are not observed in uninduced cells (lane 3) that do not express appreciable β-globin; however, protein binding is readily apparent following DMSO-induced maturation (Fig. 5, lane 4). These results indicate that complexes formed at HS-3 are altered during induced maturation. As the interaction of LCRs with promoters or enhancers of individual globin genes is presumed to mediate transcriptional competence,[1] this type of dynamic interaction may influence the complex developmental pattern of globin gene expression.

[31] T. Ikuta and Y. W. Kan, *Proc. Natl. Acad. Sci. U.S.A.* **88,** 10188 (1991).

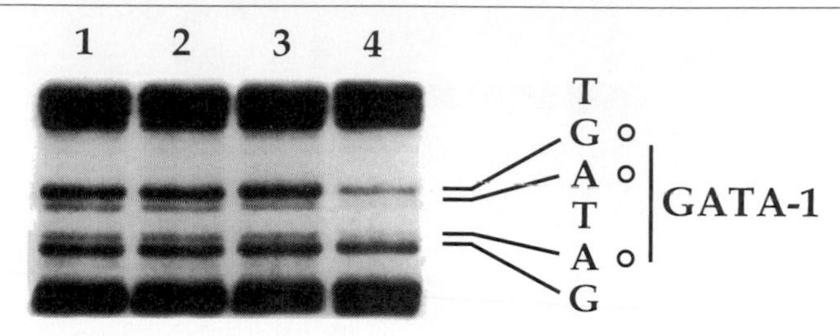

FIG. 5. GA-LMPCR *in vivo* footprinting of the bottom strand of a GATA-1 site in HS-3 of the human β-LCR. Lane 1: *in vitro*-methylated protein-free K562 DNA; lane 2: *in vivo*-methylated HepG2 DNA (nonexpressing control); lane 3: *in vivo*-methylated Hu-11 DNA (expressing line, uninduced cells); and lane 4: *in vivo*-methylated Hu-11 DNA (expressing line, DMSO-induced cells). Protections are indicated by open circles.

An objective of our *in vivo* footprinting studies was the identification of novel protein–DNA interactions that may provide insights into the relative position independence and potent enhancer activity of LCR regions. Several possibilities may explain the unique properties of these elements. First, a chromatin domain might be opened with DNA binding of specific factors, distinct from proteins that contact GATA-1, AP-1/NF-E2, and CACC/GGTGG motifs. Alternatively, protein binding to these regulatory elements may act in concert to open a closed chromatin domain during hematopoietic differentiation and prevent closure during development. Altered chromatin structures generated during this process might be revealed by *in vivo* footprinting as protections or enhancements in regions not recognized by nuclear proteins *in vitro*. We have consistently detected a set of protections and/or enhancements that do not localize to established protein-binding sites or to sequences that bind nuclear extract proteins *in vitro*.[19,20] In the central region of HS-3, GA-LMPCR *in vivo* footprinting reveals erythroid-specific protections exclusively at adenines.[20] The top strand analysis of this region is displayed in Fig. 6. This site does not contain a characterized protein-binding motif and has failed to reveal specific *in vitro* binding of nuclear proteins. We have detected similar patterns of altered DMS reactivity *in vivo* at distinct sequences in the α-LCR and HS-2 of the β-LCR.[19,20] As in HS-3, these sites do not resemble regulatory motifs or bind nuclear proteins *in vitro*. Although we cannot exclude the possibility that *in vitro* assays fail to detect factors that interact with these sites as a result of protein instability or trace concentration levels, we favor the view that these *in vivo* contacts reflect local chromatin structure induced by proteins bound to GATA-1, AP-1/NF-E2, and CACC/GGTGG motifs interacting with each other and proteins that do not contact DNA directly. These findings may facilitate resolution of the paradox that the same repertoire

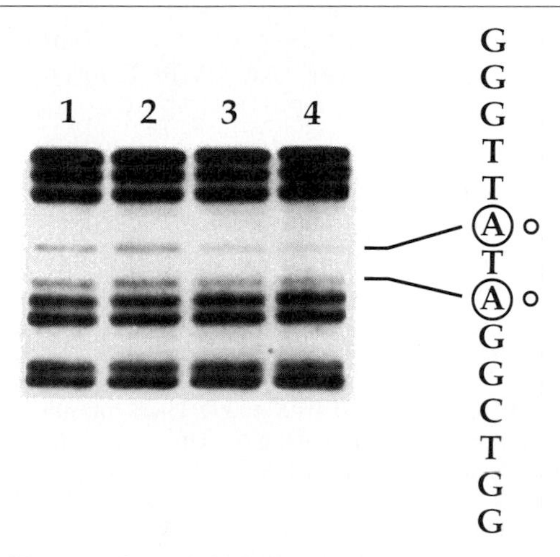

FIG. 6. GA-LMPCR *in vivo* footprinting of the top strand of a central region of the human HS-3. Lane 1: *in vitro*-methylated protein-free K562 DNA; lane 2: *in vivo*-methylated HepG2 DNA (nonexpressing control); lane 3: *in vivo*-methylated Hu-11 DNA (expressing line, uninduced cells); and lane 4: *in vivo*-methylated Hu-11 DNA (expressing line, DMSO-induced cells). Protections are represented by open circles.

of protein–DNA interactions is found in cells poised for terminal differentiation and those subjected to maturation-inducing agents. Subtle alterations in protein–protein interactions or posttranslational modifications of preexisting nuclear factors may transduce signals for differentiation and high-level globin gene expression. These changes may activate chromatin without modifying existing protein–DNA interactions.

Summary

The analysis of functional DNA regulatory sequences involved in transcriptional control is critical to establishing which proteins mediate cell-specific gene expression. The organization of erythroid LCRs is complex, consisting of multiple, interdigested *cis* elements. As *in situ* binding to these sites is determined by the accessibility of these regulatory regions in native chromatin and the availability of relevent cell-specific and ubiquitous factors, *in vivo* footprinting was used to define protein DNA interactions in human globin LCRs.[19,20] To further enhance the detection of protein contacts with this technique, we have modified the dimethyl sulfate-based ligation-mediated PCR *in vivo* footprinting procedure to permit the assessment of protein binding at guanine and adenine resides, rather than exclusively at guanines. This modification, termed GA-LMPCR *in vivo* footprinting, was essential for the analysis of GATA-1 motifs in the α-LCR and HS-3 of the β-LCR. Moreover, GA-LMPCR *in vivo* footprinting provided high-resolution analysis of AP-1/NF-E2 elements and revealed protein contacts at sequences that are not coincident with previously described regulatory motifs. A comprehensive discussion of this modification and sample illustrations from our studies have been presented to demonstrate the enhanced detection and resolution obtained with this procedure.

Acknowledgments

We are indebted to W. G. Wood, J. Sharpe, and D. Higgs for establishing the interspecific somatic cell hybrid line J3-8B and for providing the Hu-11 line. This work was supported in part by a grant from the National Institutes of Health to S.H.O. and a grant from the Johnson and Johnson Research Awards to E.C.S. and S.H.O. through the Harvard-Massachusetts Institute of Technology Division of Health Sciences and Technology Program. S.H.O. is an Investigator of the Howard Hughes Medical Institute.

[32] Exonuclease III as a Probe of Chromatin Structure *in Vivo*

By TREVOR K. ARCHER and ANDREA R. RICCI

Introduction

Developments during the last 25 years have placed chromatin structure as a fundamentally important area in understanding nuclear functions.[1] Much of this work, both biochemical and genetic, has been focused at the level of the nucleosome with particular emphasis on examining the interaction of *trans*-acting factors with regulatory sequences.[2-4] Our ability to define the precise positions or boundaries of nucleosomes as well as the proteins that interact with them in living cells has afforded us great insight into the transcriptional process.[5,6] While many experimental tools and approaches have been important in arriving at this understanding of chromatin

[1] K. E. van Holde, "Chromatin." Springer-Verlag, Heidelberg, 1988.
[2] A. P. Wolffe, "Chromatin Structure and Function." Academic Press, London, 1995.
[3] R. E. Kingston, C. A. Bunker, and A. N. Imbalzano, *Genes Dev.* **10,** 905, (1996).
[4] C. Wu, *J. Biol. Chem.* **272,** 28171 (1997).
[5] T. K. Archer and C. E. Watson, *in* "Molecular Biology of Steroid and Nuclear Hormone Receptors" (L. P. Freedman, ed.), p. 209. Birkhäuser, Boston, 1998.
[6] G. L. Hager, T. K. Archer, G. Fragoso, *et al., Cold Spring Harb. Symp. Quant. Biol. DNA Chrom.* **58,** 63 (1993).

0076-6879/99 $30.00

structure, studies with exonucleases have been particularly important for assessing the mechanistic implications of protein DNA interactions within living cells.[7–10] This article provides a detailed methodology of the approach we have taken to map chromatin structure and DNA protein interactions within human and mouse cells using the mouse mammary tumor virus (MMTV) promoter as a model template.[11] The methodology presented will focus on using exonuclease III (exo III) as a probe for chromatin structure *in vivo.*[8–10,12] However, exo III is also well established for defining nucleosome positions and protein–DNA interactions *in vitro,* and detailed protocols for this *in vitro* approach have been published previously.[13–15]

Opening Considerations

Exo III may be used as a probe for chromatin structure with or without preliminary information with respect to the chromatin structure of the gene of interest.[8,16,17] The presence of a region of constitutive or inducible hypersensitivity, indicative of chromatin remodeling and protein binding, and adjacent restriction enzyme cleavage sites is sufficient to permit analysis.[18,19] However, if there is information as to the organization of the chromatin structure, then exo III can be a powerful tool for the structural and functional characterization of the promoter.[9,20,21] Thus in the case of the MMTV promoter, precise nucleosome positioning across the promoter has been described with respect to micrococcal nuclease and DNase I assays.[20,22] These low resolution studies provide an approximate position of nucleo-

[7] H. P. Saluz and J.-P. Jost, *Crit. Rev. Eukary. Gene Exp.* **3,** 1 (1993).
[8] C. Wu, *Nature* **317,** 84 (1985).
[9] M. G. Cordingley, A. T. Riegel, and G. L. Hager, *Cell* **48,** 261 (1987).
[10] T. K. Archer and H.-L. Lee, *Methods* **11,** 235 (1997).
[11] T. K. Archer and J. S. Mymryk, *in* "The Nucleus" (A. P. Wolffe, ed.), p. 123. JAI Press, Greenwich, CT, 1995.
[12] Y. W. Kow, *Biochemistry* **28,** 3280 (1989).
[13] B. Neubauer and W. Hörz, *Methods Enzymol.* **170,** 630 (1989).
[14] M. S. Roberts, G. Fragoso, and G. L. Hager, *Biochemistry* **34,** 12470 (1995).
[15] T. K. Archer, M. G. Cordingley, R. G. Wolford, and G. L. Hager, *Mol. Cell. Biol.* **11,** 688 (1991).
[16] W. J. Feaver and R. E. Pearlman, *Curr. Genet.* **18,** 17 (1990).
[17] C. Wu, *Nature* **309,** 229 (1984).
[18] N. Tourkine, N. Mechti, M. Piechaczyk, P. Jeanteur, and J.-M. Blanchard, *Oncogene* **4,** 973 (1989).
[19] S. S. Chen, E. C. Ruteschouser, S. N. Maity, and B. De Crombrugghe, *Nucleic Acids Res.* **25,** 3261 (1997).
[20] H. Richard-Foy and G. L. Hager, *EMBO J.* **6,** 2321 (1987).
[21] H.-L. Lee and T. K. Archer, *Mol. Cell. Biol.* **14,** 32 (1994).
[22] K. S. Zaret and K. R. Yamamoto, *Cell* **38,** 29 (1984).

somes that can then be coordinated with the positions of various transcription factor binding sites located within the promoter (Fig. 1). This information is of importance as the use of exo III is dependent on an initial restriction enzyme cleavage that affords exo III an entry site to begin digestion.[15,23] Once provided with an entry site, exo III will digest the DNA in a 3′ to 5′ direction until it encounters an obstacle.[8,13,24] This obstacle may represent the intimate contact between the histone octamer and DNA, thus providing a boundary for the nucleosome at the first DNA–protein contact.[14,25,26] Alternatively, at higher concentrations, the enzyme will digest through the DNA protein contacts between the DNA and the histone octamer until it encounters a bound *trans*-acting factor as seen for the MMTV promoter, c-*myc,* HIS4, and the proα1 and proα2 collagen genes.[16,18,19,27] These high-affinity binding proteins then provide a second type of obstacle or block to progress by the enzyme. The result is that exo III will generate a series of partially digested DNA fragments that correspond to the initial restriction enzyme cleavage, the predominant contact between the DNA and histone octamer, and then subsequent bound *trans*-acting factors present on the chromatin (Fig. 2). These exo III-dependent DNA fragments can then be visualized by indirect end labeling using ^{32}P-labeled oligonucleotides.[21,27,28] A flow chart of the procedure is presented in Fig. 3 and the protocol is broken up into its five fundamental steps (A–E).

Detailed Protocol for Cell Culture

For our experiments, we have used human and mouse cells that have been transformed stably to contain multiple copies of the mouse mammary tumor virus long terminal repeat driving a series of reporter genes that include the bacterial chloramphenicol acetyltransferase (CAT), firefly luciferase (luc), and/or the *ras* oncogene.[29–32] These cell lines contain multiple

[23] J. D. Hoheisel, *Anal. Biochem.* **209,** 238 (1993).

[24] W. Metzger and H. Heumann, *in* "DNA–Protein Interactions: Principles and Protocols" (G. G. Kneale, ed.), p. 11. Humana Press, Totowa, NJ, 1994.

[25] R. U. Protacio, K. J. Polach, and J. Widom, *J. Mol. Biol.* **274,** 708 (1997).

[26] M. Buttinelli, E. Di Mauro, and R. Negri, *Proc. Natl. Acad. Sci. U.S.A.* **90,** 9315 (1993).

[27] T. K. Archer, P. Lefebvre, R. G. Wolford, and G. L. Hager, *Science* **255,** 1573 (1992).

[28] M. Shimizu, S. Y. Roth, C. Szent-Gyorgyi, and R. T. Simpson, *EMBO J.* **10,** 3033 (1991).

[29] T. K. Archer, M. G. Cordingley, V. Marsaud, H. Richard-Foy, and G. L. Hager, *in* "The Steroid/Thyroid Hormone Receptor Family and Gene Regulation" (J. A. Gustafsson, H. Eriksson, and J. Carlstedt-Duke, eds.), p. 221. Birkhäuser Verlag, Basel, 1989.

[30] G. L. Hager, *Prog. Nucleic Acid Res. Mol. Biol.* **29,** 193 (1983).

[31] C. J. Fryer, S. K. Nordeen, and T. K. Archer, *J. Biol. Chem.* **273,** 1175 (1998).

[32] S. K. Nordeen, B. Kühnel, J. Lawler-Heavner, D. A. Barber, and D. P. Edwards, *Mol. Endocrinol.* **3,** 1270 (1989).

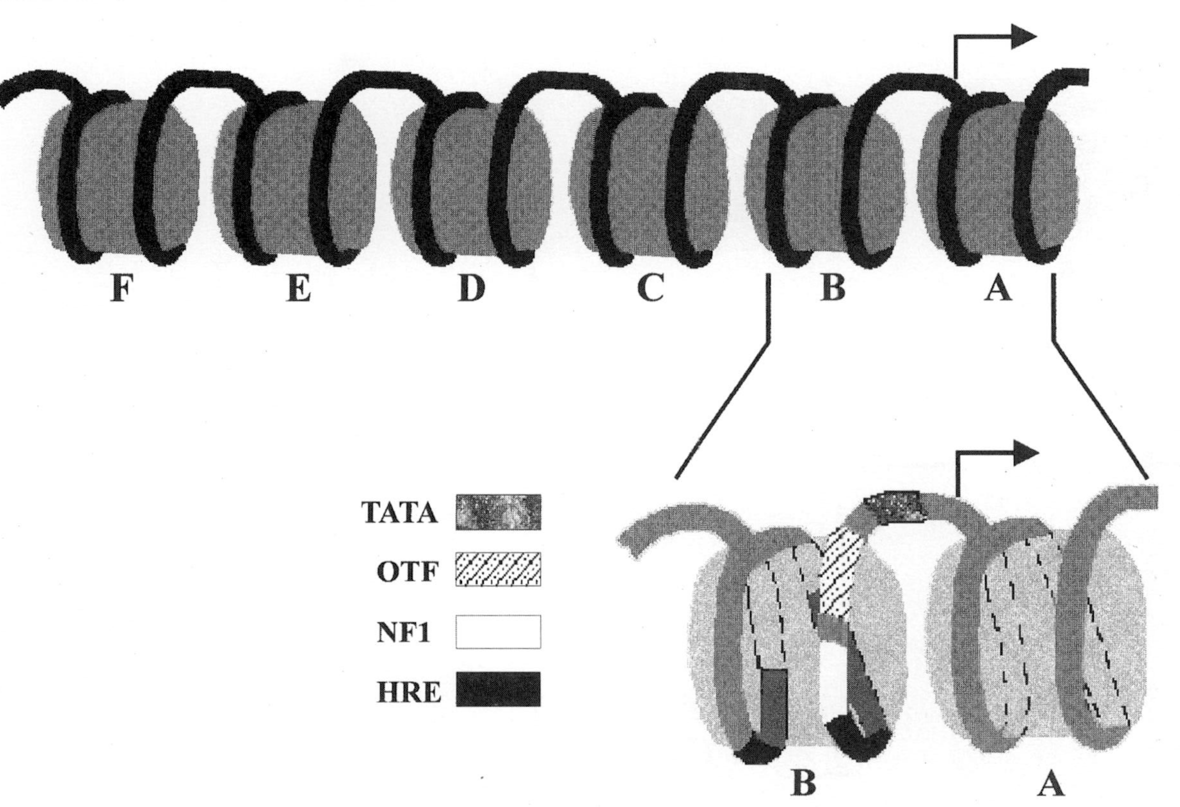

FIG. 1. The mouse mammary tumor virus (MMTV) promoter chromatin structure. A schematic of the MMTV promoter showing the positions of the six phased nucleosomes (A–F) and the start site of transcription.[20] The proximal promoter encompassed by nucleosomes A and B and the recognition elements for *trans*-acting factors are indicated. HRE, hormone response element; NF1, nuclear factor 1; OTF, octamer factor and TATA-binding site for TBP.

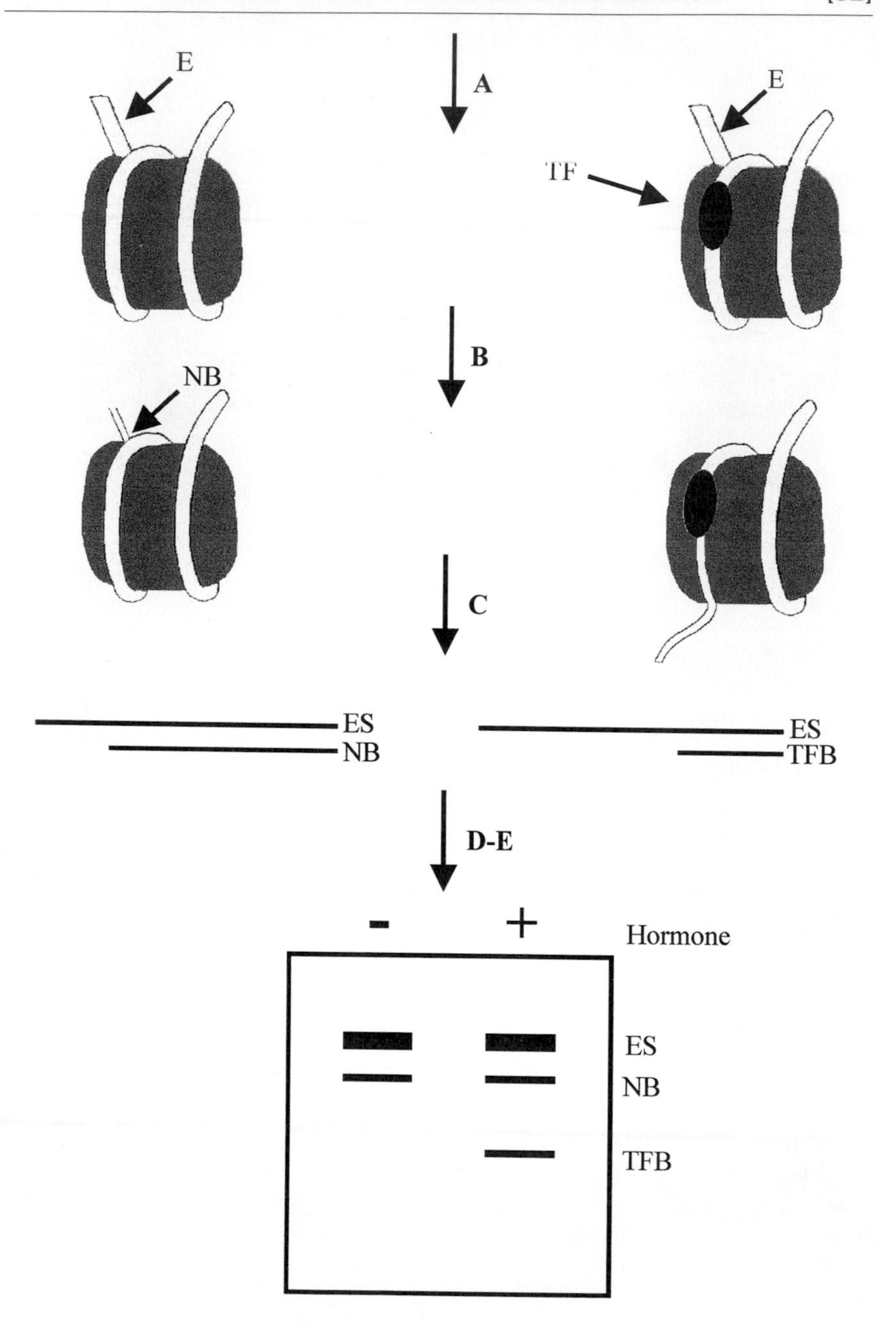
E
A
E
TF
B
NB
C
ES
NB
ES
TFB
D-E
-
+
Hormone
ES
NB
TFB

copies of the MMTV sequences ranging from 10 to 1000 copies per cell for the individual cell line.[11] The use of multicopy cell lines has greatly enhanced our ability to map the chromatin structure as it provides a very strong "signal-to-noise" ratio for the MMTV sequences.[33] Cells are maintained in 150-mm tissue culture plates in 25 ml of Dulbecco's modified minimal essential media supplemented with L-glutamine (2 m*M*), HEPES (10 m*M*), penicillin (100 μg/ml), streptomycin (100 μg/ml), and fetal bovine serum while grown in a humidified incubator at 37° with 5% (v/v) CO_2. For the purposes of our experiments, cells were untreated or treated with hormone for 1 hr prior to nuclei isolation.

A. Nuclei Isolation

Reagents

Homogenization buffer: 10 m*M* Tris–HCl, pH 7.4, 15 m*M* NaCl, 60 m*M* KCl, 1 m*M* EDTA, 0.1 m*M* EGTA, 0.1% Nonidet P-40 (NP-40), 5% sucrose, 0.15 m*M* spermine,* and 0.5 m*M* spermidine* (store at 4°)

Sucrose pad: 10 m*M* Tris–HCl, pH 7.4, 15 m*M* NaCl, 60 m*M* KCl, 10% sucrose, 0.15 m*M* spermine,* and 0.5 m*M* spermidine.* Filter sterilize (0.45 μm) (store at 4°)

Wash buffer: 10 m*M* Tris–HCl, pH 7.4, 15 m*M* NaCl, 60 m*M* KCl, 0.15 m*M* spermine,* and 0.5 m*M* spermidine* (store at 4°)

Note: The homogenization buffer, wash buffer, and sucrose pad are generally not maintained at 4° for more than 2–3 months. The following buffer stocks may be used in the preparation of these solutions.

10× buffer salt stock: 100 m*M* Tris–HCl, pH 7.4, 150 m*M* NaCl, and 600 m*M* KCl (store are room temperature)

[33] T. K. Archer, *Ann. N.Y. Acad. Sci.* **684,** 196 (1993).

* 1000× spermine (0.15 *M*) and spermidine (0.5 *M*) stocks (stored at −20°) should be prepared and added to an aliquot of each buffer just prior to nuclei isolation.

FIG. 2. Exo III as a probe for transcription factors and nucleosome boundaries *in vivo.* Schematic of the exo III footprinting assay of a hormone-inducible nucleosome template. Steps A–E correspond to the detailed protocol presented in the text. Illustrations to the left of the arrows demonstrate the use of exo III footprinting to determine the nucleosome boundary (E, entry site restriction enzyme; NB, nucleosome boundary; ES, entry site). Illustrations to the right of the arrows depict the use of this technique to determine transcription factor binding in addition to nucleosome boundaries (TF, transcription factor; TFB, transcription factor boundary). In both cases the DNA is analyzed on a denaturing polyacrylamide gel.

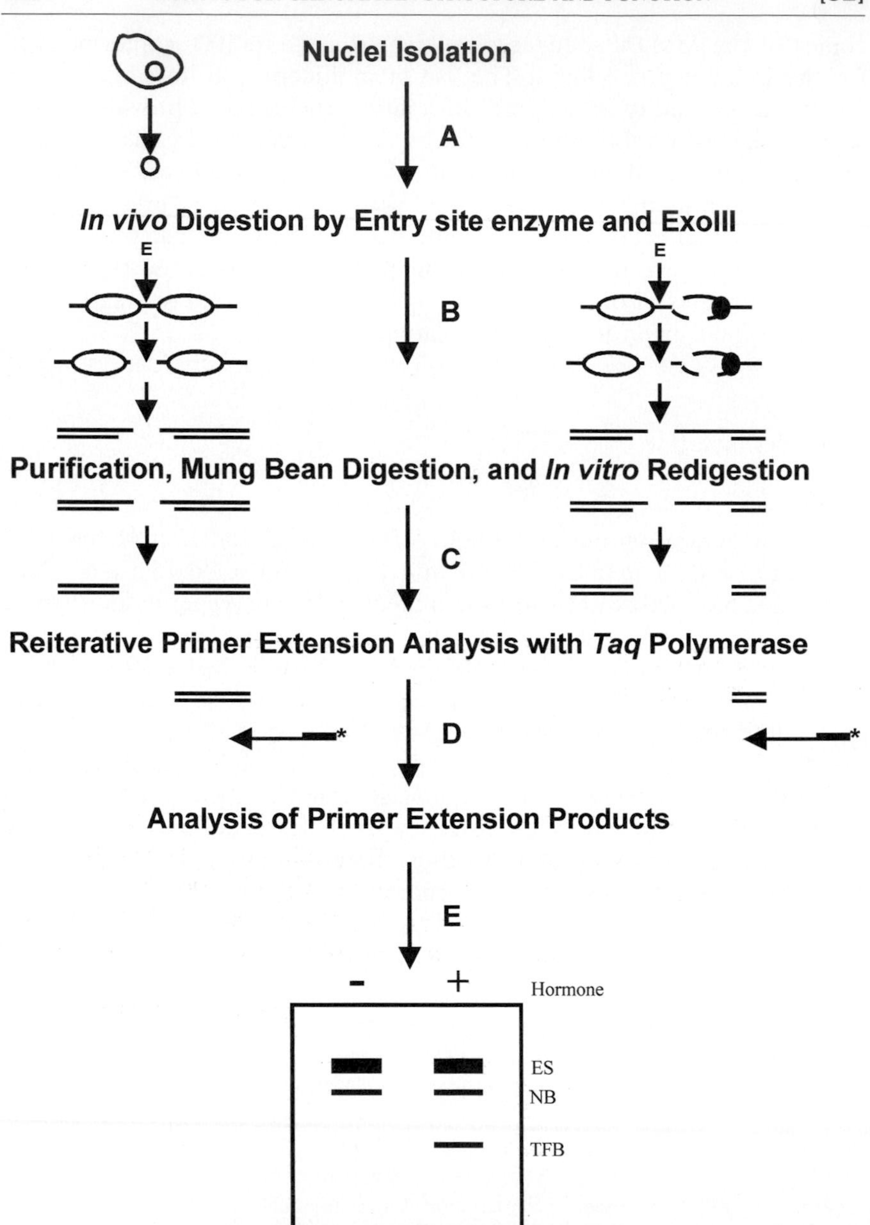

FIG. 3. Flow chart for the use of exo III as a probe for chromatin structure.

100× EDTA/EGTA stock: 100 m*M* EDTA (pH 8.0) and 10 m*M* EGTA (store at room temperature)

Protocol. The following protocol is carried out on ice and all solutions must be at 4°.

1. Rinse cells with cold 1× phosphate-buffered saline (PBS), detach them from the plates with a rubber policeman into 10 ml of cold 1× PBS, and transfer to a prechilled 50-ml centrifuge tube.
2. Pellet cells in a Beckman GS-6R centrifuge (Beckman, Sunnyvale, CA) at 1900 rpm (750*g*) for 5 min at 4° and remove excess PBS.
3. Gently resuspend cells in 5 ml of cold homogenization buffer and transfer to a prechilled 7.5-ml Dounce homogenizer (VWR, Missassauga, Ontario, Canada) for 2 min.
4. Lyse cells with five strokes of a Dounce A pestle and transfer the solution to a prechilled 15-ml conical centrifuge tube. The number of strokes of the Dounce homogenizer pestle required to effectively break the cells and maintain nuclei intact are determined empirically for each cell line, and in initial experiments the degree to which the nuclei have been liberated from cytoplasm is examined under a light microscope.
5. Add 1 ml of sucrose pad directly to the bottom of the tube with a P1000 micropipette (Gilson, Guelf, Ontario, Canada). Nuclei are isolated after sedimentation through this pad at 2700 rpm (~1400*g*) for 20 min at 4°.
6. Carefully remove the supernatant with a pipette and discard. To remove the remaining NP-40, gently resuspend the nuclei in wash buffer (5 ml) and centrifuge at 1900 rpm (750*g*) for 5 min at 4°. Carefully remove all traces of wash buffer.

B. *In Vivo* Digestion by Entry Site Enzyme and Exonuclease III

Reagents

Enzyme digestion buffer: 10 m*M* Tris–HCl, pH 7.4, 15 m*M* NaCl, 60 m*M* KCl, 0.1 m*M* EDTA, 5 m*M* $MgCl_2$, 5% glycerol, and 1 m*M* dithiothreitol (DTT)

Proteinase K buffer: 10 m*M* Tris–HCl, pH 7.6, 10 m*M* EDTA, 0.5% sodium dodecyl sulfate (SDS), and 0.2 mg/ml proteinase K

Note: We have purchased exonuclease III from GIBCO/BRL (Gaithersburg, MD) and New England Biolabs (NEB; Beverly, MA) with similar results. DTT should be prepared as a 1000× stock (1 *M*) and added to the enzyme digestion buffer just prior to use. Proteinase K should be prepared as a 100× stock and added to the buffer just prior to use.

Protocol

1. Gently resuspend nuclei in cold restriction enzyme digestion buffer and transfer aliquots of 50–100 μl to prechilled 1.5-ml microfuge tubes.

2. Partial digestion of nuclei is carried out with entry site restriction enzyme (300–1000 U/ml) and exo III (1000–3000 U/ml) at 30° for 15 min. Stop the reaction by adding 10 volumes of proteinase K buffer and then transfer tubes to 37° for 5–24 hours.

CHOICE OF ENTRY SITE RESTRICTION ENZYME. Because exo III requires an entry site in order to digest DNA, it is very important to determine which restriction enzyme will be used. For example, in MMTV, the choice of enzyme is determined experimentally as it relies on the location of the nucleosomes and the location of unique restriction enzyme sites that lie within the linker regions. As our promoter is affected by hormone activation, we use *Hae*III as our entry site enzyme as its cleavage is unaffected by hormone treatment. A titration of entry site enzyme may also be carried out to determine the optimum level of *in vivo* cleavage prior to carrying out the experiment with exo III. The initial digest should be located far enough away from the putative binding site of the protein in question such that the resulting fragment can be separated electrophoretically from the parental subfragment (Fig. 2).

Exo III is a small monomeric protein with a molecular mass of approximately 28,000.[12,23,34] The 3′ → 5′-exonuclease activity, which is most important with respect to its mapping of chromatin, releases deoxyribose 5′-monophosphates from the 3′ terminus of duplex DNA.[24,35] However, the enzyme possesses multiple enzymatic activities; in addition to its 3′ → 5′-exonuclease activity, it is a class II apurinic/apyrimidinic endonuclease, a 3′-phosphomonoesterase, and a 3′-phosphodiesterase.[12,23] Despite its multiple activities, exo III is thought to have a single active site and these activities are all dependent on the presence of magnesium.[12,34] The processivity of the enzyme is temperature dependent such that it is highly processive at temperatures below 20° and distributive at higher temperatures.[36,37]

CHOICE OF RESTRICTION ENZYME DIGESTION BUFFER. We have selected an enzyme digestion buffer that is compatible with the digestion of the restriction enzyme that provides the entry site as well as the digestion by exo III. Our approach is to use a single buffer instead of individual buffers tailored either to the restriction enzyme choice or the exo III. This, in our

[34] C. D. Mol, C.-F. Kuo, M. M. Thayer, R. P. Cunningham, and J. A. Tainer, *Nature* **374,** 381 (1995).

[35] D. Shalloway, T. Kleinberger, and D. M. Livingston, *Cell* **20,** 411 (1980).

[36] K. R. Thomas and B. M. Olivera, *J. Biol. Chem.* **253,** 424 (1978).

[37] R. Wu, G. Ruben, B. Siegel, E. Jay, P. Spielman, and C. P. Tu, *Biochemistry* **15,** 734 (1976).

view, provides greater flexibility with respect to the choice of enzyme used for the entry site as well as providing an easier comparison of exo III digestion with multiple restriction enzymes. Clearly the alternative to use specific buffers with respect to enzyme and exo III is possible; however, this would require additional steps with respect to the extent of cleavage by either enzyme or exo III with regard to the specific buffer.

The quantities of both the restriction enzyme and exo III used for each experiment are derived empirically. The choice of restriction enzyme and the selection of the number of units per restriction enzyme used are such to provide a significant degree of cleavage within the nuclei as this is a rate-limiting step with respect to the ability of exo III to digest DNA. Initially the quantities of exo III are determined by a broad titration with the understanding that at high concentrations of exonuclease, the exo III will digest through the nucleosome as well as transcription factor blocks.[10,24] Consequently, it is important to establish control experiments using purified DNA and exo III to determine if there are any endogenous pause sites for exonuclease on the DNA sequence being analyzed.

Note: In addition to exo III, we have also made use of the λ exonuclease as an alternative to determining the boundaries of protein–DNA interactions *in vivo*.[38] λ exonuclease provides complementary information with respect to DNA–protein interactions by detecting protein interactions with the complementary strand of the DNA duplex. In addition, digestion by λ exonuclease is significantly more processive than digestion by exo III under similar temperature and reaction conditions.[36] Thus one is able to use significantly lower concentrations of enzymes to achieve a similar degree of digestion.[39] Second, because λ exonuclease hydrolyzes DNA in a 5′ and 3′ direction, when coupled with a *Taq* polymerase assay, one does not require mung bean or S1 nucleases to remove an overhang as does exo III.[38]

C. Purification, Mung Bean Digestion, and *in Vitro* Redigestion

Purification Protocol

1. Transfer *in vivo*-digested DNA to 5-ml polypropylene tubes and purify with four to six extractions with phenol/chloroform/isoamyl alcohol (PCI; 25:24:1, v/v) and two extractions with chloroform. The first extraction is carried out with 2× volume of PCI and each subsequent extraction with 1× volume. Samples should be mixed

[38] J. S. Mymryk and T. K. Archer, *Nucleic Acids Res.* **22,** 4344 (1994).
[39] C. M. Radding, *J. Mol. Biol.* **18,** 235 (1966).

vigorously prior to centrifugation at 9000 rpm (10,400*g*) for 5 min at room temperature in a Beckman J2-MI centrifuge (J20.1 rotor).
2. Precipitate the DNA in 2–3 volumes of 95% (v/v) ethanol (1/10 volume of 1 *M* NaCl may be added) at −80° for at least 1 hr. DNA is pelleted after centrifugation at 9000 rpm (10,400*g*) for 30 min at 4°.
3. Wash the pellet with cold 70% (v/v) ethanol (500 μl) and centrifuge at 9000 rpm (10,400*g*) for 5 min at 4°. Dry the pellet briefly in a Sorval Speed-Vac (Fisher, Pittsburgh, PA) (do not over dry) and resuspend in 100 μl of water. The DNA should now be transferred to a 1.5-ml microfuge tube.

Mung Bean Digestion

Because the exo III digestion generates 5′ single-stranded overhangs, these must be trimmed back in order to determine the protein boundaries. This is carried out using mung bean nuclease which digests single-stranded DNA (and RNA) but leaves duplex regions intact. The resulting fragments are now double stranded at the point where exo III digestion was impeded.

Note: While we have predominantly used mung bean nuclease as a way of removing the 5′ strand, S1 nuclease is equally effective in our hands at removing the overhang.[27]

Reagents

10× mung bean digestion buffer: 1 *M* sodium acetate, pH 5, 10 m*M* zinc acetate, 100 m*M* L-cysteine, 5 *M* NaCl, and 50% glycerol (store at −20°)

Protocol

1. To remove single-stranded overhangs, incubate DNA in 1× mung bean digestion buffer and 45 units of mung bean nuclease. Digestion is carried out at 30° for 30 min.
2. Place samples on ice and add 5 μl of 5 *M* NaCl to stop the reaction.
3. Purify the DNA further with two to three rounds of phenol/chloroform/isoamyl alcohol extraction (250–500 μl) and one round of chloroform extraction and precipitate in 2–3 volumes of 95% ethanol for 1 hr at −80°.
4. Pellet the DNA in a microcentrifuge at 14,000 rpm for 15 min at 4°. Wash the pellet with 70% ethanol (250 μl) and spin for 5 min at 4°. The pellet may be dried briefly (do not overdry) and then resuspended in 100 μl of water.

In Vitro Redigestion Protocol

Once the DNA has been purified following the removal of the overhang by mung bean nuclease, we carry out a second restriction enzyme digestion using an enzyme that is 5′ to the initial entry site. The cutting by this enzyme serves as an internal control that will allow us to determine the extent to which the initial *in vivo* restriction enzyme digest was successful.[40] This *in vitro* cleavage is useful as an internal loading control, when comparing different experimental conditions, as well as indicating the number or percentage of templates that are subject to analysis by exo III. An example is provided in Fig. 4 for the MMTV promoter in cells treated with and without hormone (see later).

Protocol

1. Digest the DNA to completion with the entry site enzyme (4 U/μg) for 4 hr at 37° using the commercially supplied restriction enzyme buffer.
2. Purify the redigested DNA with two rounds of PCI (250 μl) and one round of chloroform (250 μl).
3. Precipitate the DNA with 2–3 volumes of 95% ethanol and place at −80° for 1 hr. Spin the samples at 14,000 rpm for 15 min at 4° in a microcentrifuge.
4. Wash the pellet with 70% ethanol (250 μl) and spin at 14,000 rpm for 5 min at 4°. Dry the pellet briefly (do not overdry) and resuspend in water. The amount of water used for resuspension is determined empirically based on the size of the pellet.
5. Quantify the concentration of DNA by UV absorbance and visualize on a 1% agarose gel (1× TBE) prior to primer extension analysis to compare the amount of DNA.

D. Reiterative Primer Extension Analysis with *Taq* Polymerase

Primer Selection

The primer selected should be downstream from the transcription factor-binding site, approximately 200–300 bp from the cleavage site of the entry site restriction enzyme. Keep in mind that the primer length and sequence are very important in designing the parameters of a successful amplification. The melting temperature (T_m) of a nucleic acid duplex increase both with its length and with increasing (G + C) content. The annealing temperature for the primer extension should be 5–10% above the T_{m50} of the primer chosen. Because the amplification takes place in the presence of a large excess of genomic DNA, we have generally used primers greater than 18 bases long and with a T_{m50} between 45 and 70°.

[40] J. S. Mymryk, D. Berard, and G. L. Hager, and T. K. Archer, *Mol. Cell. Biol.* **15,** 26 (1995).

End Labeling of Oligonucleotides

1. In a 1.5-ml microfuge tube, combine 2 μl of 5 *pM* oligo, 5 μl of [^{32}P]ATP (5000 ci/mmol), 2 μl of 10× T4 polynuclcotidc kinase buffer (supplied with enzyme), 10 μl of water, and 20 U T4 polynucleotide kinase.
2. Incubate the mixture for 10 min at 37°.
3. Prepare Sephadex columns (e.g., G-50) as directed by the manufacturer's guidelines.
4. Pass labeling reaction over the column, as directed by the manufacturer's guidelines.
5. Determine the cpm/μl of the labeled primer.

Reiterative Primer Extension Analysis

Reagents

5× PCR buffer: 50 m*M* Tris–HCl, pH 8.3, 250 m*M* KCl, 15–25 m*M* $MgCl_2$, and 0.25% Tween 20

Note: It is well known that the concentration of magnesium (Mg^{2+}) can influence the yield as well as the specificity of the *Taq* polymerase.[28] Consequently, a major consideration for primer extension analysis involves the titration of the Mg^{2+} used in the extension reaction. It has been our experience that elevated Mg^{2+}, while producing a higher quantity of the desired PCR product, also promotes the appearance of artifactual bands. These are of course controlled for by digestions that lack either restriction enzyme or the exonuclease as well as by primer extension experiments with plasmid and genomic DNA. A titration with genomic DNA is strongly recommended in order to determine the optimal Mg^{2+} concentration. For MMTV, low Mg^{2+} concentrations of 2–3 m*M* provide specific results.

PCR stop buffer: 200 m*M* sodium acetate, pH 7.0, 10 m*M* Tris–HCl, pH 7.5, 5 m*M* EDTA, and 0.1 μg/μl yeast tRNA

Loading buffer: 80% formamide, 0.01 *M* NaOH, 1 m*M* EDTA, 0.04% bromphenol blue, and 0.04% xylene cyanol

Protocol

1. Amplify purified DNA (10–20 μg) with 2.5 units of *Taq* polymerase in 1× PCR buffer with 200 μm deoxyribonucleotides and a ^{32}P-labeled primer specific for the template of interest in a final volume of 30 μl.
2. Program a thermocycler for a denaturation cycle of 3 min at 94° and 2 min at the annealing temperature of the primer, followed by 2 min at 72° for primer extension. An additional 29 cycles should be carried out as follows: 2 min at 94°, 2 min at the annealing temperature of the primer, and 2 min at 72°, with a final extension carried out at

72° for an additional 10 min. The precise conditions for this portion of the protocol are of course dependent on the specific themocycler used and its ramping times. Thus these times should be modified for each machine using *in vitro*-digested genomic DNA as a control for specificity and yield.

3. After the primer extension, add 100 μl of PCR stop buffer.
4. Purify the samples with one round each of PCI and chloroform. Precipitate with 2–3 volumes of 95% ethanol at $-80°$ for 1 hr and spin at 14,000 rpm for 15 min at 4°. Remove ethanol and dry the pellet briefly. Resuspend in 7 μl of loading buffer.

Analysis of Primer Extension Products

Reagents. Our laboratory uses premixed sequencing solutions, but the following denaturing polyacrylamide gel solutions may be substituted.

Acrylamide stock: 38% acrylamide and 2% bisacrylamide (w/v)

For 8% gel: 25.2 g urea (7 *M*), 6 ml 10× TBE, and 12 ml acrylamide stock. Add H_2O to 60 ml. Filter solution. Add 10 μl TEMED and 200 μl ammonium persulfate (10%). Pour gel immediately.

1. Heat samples for 5 min at 95° and apply to a 8% denaturing polyacrylamide sequencing gel (1× TBE; 16 or 45 cm).

2. Electrophorese samples to allow for maximal separation between the band corresponding to the transcription factor-binding site and the parental band derived from the initial restriction enzyme digest that provides the entry for exo III. After electrophoresis, transfer the gel to Whatman (Clifton, NJ) filter paper, dry, and expose the dried gel to a PhosphorImager screen (Molecular Dynamics, Sunnyvale, CA) or expose to film at $-80°$.

Note: We have used both 45- and 16-cm gels and the analysis of NF1 binding is visualized effectively in both systems. However, the detection of nucleosome boundaries and multiple protein-binding sites is visualized more effectively using the longer denaturing gel.

Application of Technique to Mouse Mammary Tumor Virus Promoter

The MMTV promoter is a well-established model system that has been used to study the mechanisms by which steroid hormone receptors activate transcription within chromatin[5] (Fig. 1). When introduced into human and rodent cells stably, the MMTV promoter adopts a phased array of nucleosomes as monitored by micrococcal nuclease and restriction enzyme hypersensitivity assays.[20,22,29] Analysis of DNase I and restriction enzymes will reveal a region of hormone-dependent hypersensitivity that encompasses

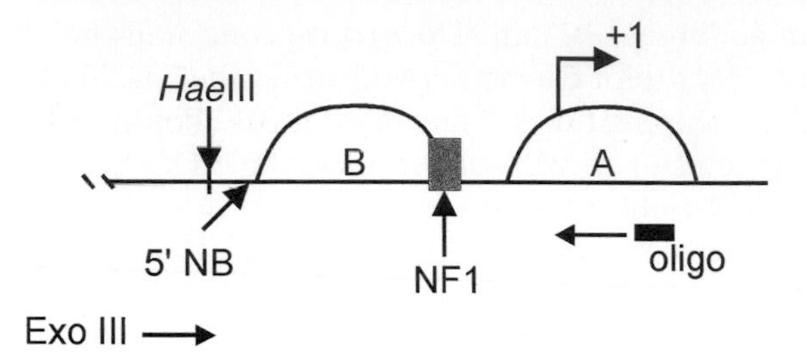

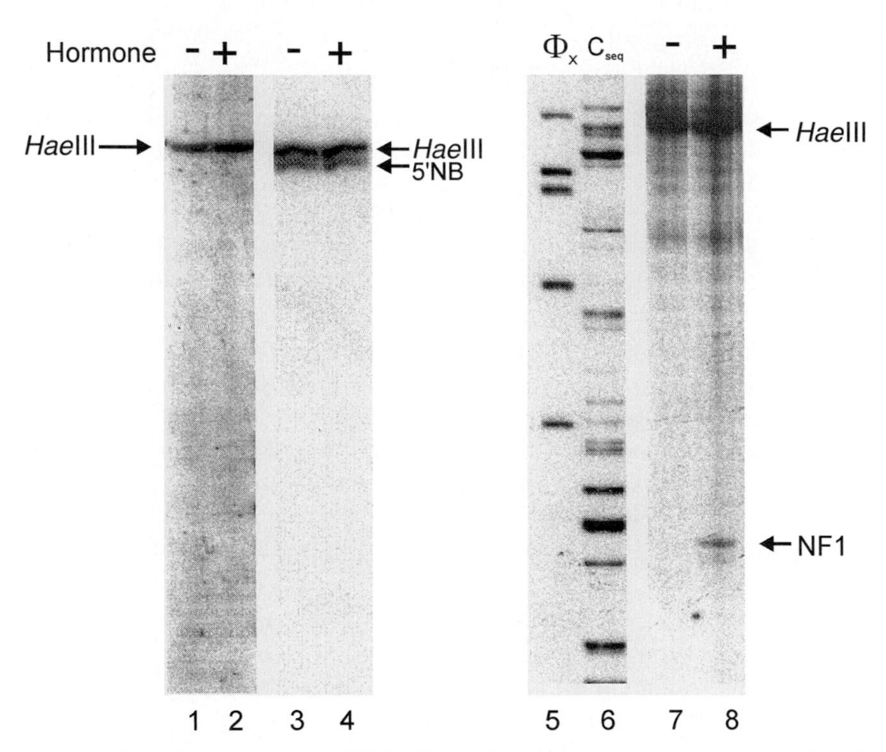

FIG. 4. Exo III as a probe for NF1 loading and nucleosome boundaries in MMTV. The illustration depicts the use of exo III footprinting of the MMTV promoter. Human (lanes 1–4) and mouse (lanes 7 and 8) cell lines transformed stably with the MMTV promoter attached to the CAT gene were grown in the absence or presence of hormone, dex 10^{-7} *M* or R5020 10^{-8} *M*, and assayed as outlined in the text. *Hae*III is the entry site enzyme; 5′ nucleosome boundary (5′NB) and NF1 (transcription factor) sites are shown. To detect boundaries, nuclei were digested with *Hae*III alone (lanes 1 and 2) or *Hae*III plus exo III (lanes 3 and 4). In the experiment shown in lanes 7 and 8, nuclei were digested with *Hae*III plus exo III. The confirmation that the shorter fragment corresponds to the NF1 border is by comparison to the sequencing tract (C_{seq}, lane 6) and size markers (Φ ×174, lane 5) and previous experiments using purified NF1 and MMTV DNA *in vitro*.[43]

the region including and adjacent to nucleosome B in the promoter.[15,41] Assembly of these sequences as chromatin *in vitro* and subsequent analysis using exo III reveal a position of nucleosomes that is highly reminiscent of that observed *in vivo*.[15] Exo III digestions of the MMTV promoter *in vivo* detect an exo III boundary that is consistent with similar experiments carried out *in vitro* as well as footprinting by micrococcal nuclease.[14,15,20,42] Interestingly, the results with exo III suggest the presence of a single predominant boundary for nucleosome B, consistent with *in vitro* restriction enzyme and exo III assays, but differing from micrococcal nuclease data that suggest multiple translational frames for nucleosome B (Fig. 4, lanes 1–4).[14,15,42] Experiments in mouse cells demonstrate the ability of exo III to detect the hormone-dependent binding of the NF1 protein to the MMTV promoter. In the absence of hormone, the exo III digests through the NF1-binding site, indicating that no protein is bound. In cells exposed to hormone, exo III digestion results in the appearance of a novel DNA fragment that extends to the 5′ border of the NF1-binding site (Fig. 4, lanes 7 and 8).

In summary, exo III represents an excellent method by which to determine the precise positioning of the nucleosomes both *in vivo* and *in vitro*. In addition, it allows one to capture the image of DNA–protein interactions that occur within chromatin on complex promoters such as the MMTV promoter.[21,27] As more regulatory regions become characterized, both with respect to nucleosome positioning by micrococcal nuclease and hypersensitivity by DNase I, exo III will prove to be an invaluable tool in determining a precise position of the nucleosome and the proteins that interact to control transcription.

Acknowledgments

We thank past and present members of the Archer laboratory who have contributed to the development of the exonuclease assays, Christy Fryer for comments on the manuscript, and D. Power for help in its preparation. This work was supported by grants to T.K.A. from the National Cancer Institute of Canada (NCIC), the Canadian Breast Cancer Research Initiative, and the Medical Research Council of Canada. T.K.A. is a Scientist of the NCIC supported by funds from the Canadian Cancer Society and A.R.R. is supported by a Natural Sciences and Engineering Research Council (NSERC) of Canada Scholarship.

[41] E. H. Bresnick, M. Bustin, V. Marsaud, H. Richard-Foy, and G. L. Hager, *Nucleic Acids Res.* **20,** 273 (1991).
[42] G. Fragoso, S. John, M. S. Roberts, G. L. Hager, *Genes Dev.* **9,** 1933 (1995).
[43] M. G. Cordingley and G. L. Hager, *Nucleic Acids Res.* **16,** 609 (1988).

[33] Protein Image Hybridization: Mapping and Positioning DNA–Protein Contacts along DNA

By S. Belikov, O. Preobrazhenskaya, and V. Karpov

Introduction

DNA–protein cross-linking provides a useful approach for the study of chromatin structure and function.[1–5] The method discussed in this article includes the following steps: (a) DNA–protein cross-linking performed on isolated nuclei or whole cells; (b) purification/enrichment of cross-linked DNA–protein complexes; and (c) analysis of cross-linking by two-dimensional (2D) electrophoresis/hybridization. The method was named "protein image" hybridization because the cross-linked proteins are no longer present at the final stage of detection of their binding along the DNA. The efficiency of this approach has been demonstrated in a number of papers concerning protein-DNA interactions in chromatin.[6–12]

Method

The schematic representation of the "protein image" hybridization assay is given in Fig. 1. Proteins are first cross-linked to DNA. We performed cross-linking either on whole cells by UV irradiation or on isolated nuclei by mild, partial depurination of DNA with or without preliminary formaldehyde fixation. This method is referred to here as the dimethyl sulfate

[1] A. Mirzabekov, *Gene* **135,** 111 (1993).
[2] V. Orlando, H. Strutt, and R. Paro, *Methods* **11,** 205 (1997).
[3] J. Walter and M. D. Biggin, *Methods* **11,** 215 (1997).
[4] T. Moss, S. Dimitrov, and D. Houde, *Methods* **11,** 225 (1997).
[5] D. Pruss and S. Bavykin, *Methods* **12,** 36 (1997).
[6] G. Nacheva, D. Guschin, O. Preobrazhenskaya, V. Karpov, K. Ebralidse, and A. Mirzabekov, *Cell* **58,** 27 (1989).
[7] S. Belikov, A. Dzherbashyajan, O. Preobrazhenskaya, V. Karpov, and A. Mirzabekov, *FEBS Lett.* **273,** 205 (1990).
[8] Y. Postnikov, V. Shick, A. Belyavsky, K. Khrapko, K. Brodolin, T. Nikolskya, and A. Mirzabekov, *Nucleic Acids Res.* **19,** 717 (1991).
[9] S. Belikov, A. Belgovsky, O. Preobrazhenskaya, V. Karpov, and A. Mirzabekov, *Nucleic Acids Res.* **21,** 1031 (1993).
[10] S. Belikov, A. Belgovsky, M. Partolina, V. Karpov, and A. Mirzabekov, *Nucleic Acids Res.* **21,** 4796 (1993).
[11] A. Pemov, S. Bavykin, and J. L. Hamlin, *Biochemistry* **34,** 2381 (1995).
[12] D. Papatsenko, I. Priporova, S. Belikov, and V. Karpov, *FEBS Lett.* **381,** 103 (1996).

0076-6879/99 $30.00

(DMS) method. Two different protocols of DMS cross-linking are used. The rationale for this is that they target different sets of amino acids involved in cross-linking, either mostly histidines ("histidine" protocol) or lysines ("lysine" protocol). Following cross-linking, nuclei are lysed and DNA–protein adducts are separated from uncross-linked proteins. The easiest way to achieve this would be by centrifugation in CsCl gradients followed by dialysis and/or concentration. DNA is then sheared nonspecifically by sonication or DNase I treatment or specifically by a restriction endonuclease(s). Specific shearing by restriction endonuclease cleavage "unifies" the DNA component of DNA–protein complexes and can be accompanied by DNA trimming with 3′-exonucleases (Exo III), which allows precise localization of protein cross-linking sites along DNA. Uncross-linked DNA released after DNA fragmentation was removed by phenol extraction of DNA–protein cross-links, leaving DNA–protein complexes ready to be resolved by two-dimensional gel electrophoresis. In the first dimension, the complexes are separated according to the size of both DNA and protein, following which proteins are digested quantitively by pronase directly in the gel, and the resulting DNA is resolved in the second dimension. Because the same buffer system is utilized in the gels for both dimensions, randomly sheared uncross-linked DNA forms a diagonal. The cross-linked protein retards migration of the DNA molecule (relative to an uncross-linked DNA molecule of the same length), thus shifting the diagonal: the higher the molecular weight of the protein, the larger the shift of the diagonal. As a result, each cross-linked protein appears as a separate diagonal or diagonal array of spots (see Fig. 1). The specific cross-linking pattern for each sequence of interest can be visualized ("developed") by hybridization (Fig. 1). Signals obtained do not contain any protein but rather "image" of the molecular weight and the relative abundance of the proteins along the DNA sequence of interest. An elegant alternative to DNA hybridization was described by Bavykin and co-workers,[13] where antiserum against BrdUrd incorporated into newly replicated DNA was used to "develop" the cross-linking pattern between the latter and histones.

Techniques

UV Cross-Linking and Isolation of Nuclei[7–10,12]

UV cross-linking can be performed easily using an inverted transilluminator (without filter) emitting 254 nm UV light. The working dose deter-

[13] S. Bavykin, L. Srebreva, T. Banchev, R. Tsanev, J. Zlatanova, and A. Mirzabekov, *Proc. Natl. Acad. Sci. U.S.A.* **90,** 3918 (1993).

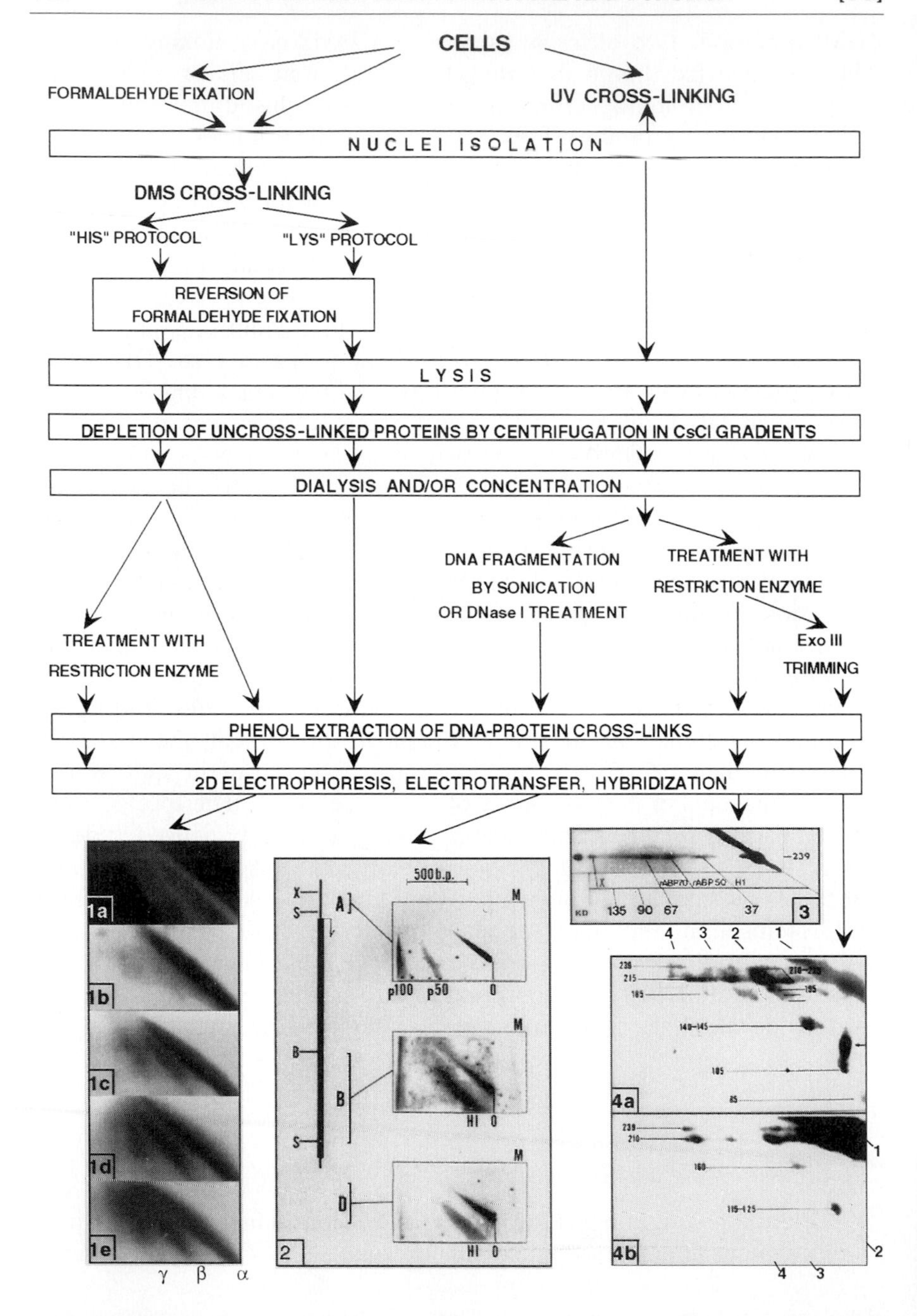
CELLS
FORMALDEHYDE FIXATION
UV CROSS-LINKING
NUCLEI ISOLATION
DMS CROSS-LINKING
"HIS" PROTOCOL
"LYS" PROTOCOL
REVERSION OF
FORMALDEHYDE FIXATION
LYSIS
DEPLETION OF UNCROSS-LINKED PROTEINS BY CENTRIFUGATION IN CsCl GRADIENTS
DIALYSIS AND/OR CONCENTRATION
DNA FRAGMENTATION
BY SONICATION
OR DNase I TREATMENT
TREATMENT WITH
RESTRICTION ENZYME
Exo III
TRIMMING
TREATMENT WITH
RESTRICTION ENZYME
PHENOL EXTRACTION OF DNA-PROTEIN CROSS-LINKS
2D ELECTROPHORESIS, ELECTROTRANSFER, HYBRIDIZATION
1a
1b
1c
1d
1e
γ β α
500 b.p.
M
p100 p50 0
HI 0
2
KD 135 90 67 37
ABP70 ABP50 H1
239
3
4a
4b

mines the amount of DNA–protein cross-linking, but is restricted by photodegradation of DNA. We found that a wide range of proteins could be cross-linked to DNA at irradiation doses of 0.2–0.5 J/cm^2 and absorption of the layer of cell suspension 0.5–0.7 A_{260} units. This article presents a protocol for *Drosophila melanogaster* tissue culture cells; however, it can be adapted easily for other organisms and tissues (adaptation for yeast is described in Papatsenko *et al.*[12]). The optimum working dose can be estimated by irradiating tissue for increasing periods of time with subsequent examination of the amount of cross-linking of the protein of interest.[3] DNA degradation can be measured by, for example, inhibition of cleavage by restriction enzymes (such as *Eco*RI). Walter and Biggin[3] recommend not to exceed the inhibition level of 10%.

Protocol

1. Place a 5-mm-thick suspension of tissue culture cells (*D. melanogaster*, subline K_c, 0.5–1 × 10^7 cells/ml in Echalier medium) in a dish with a cooling jacket and irradiate under an inverted transilluminator tipped with six 15-W low-pressure mercury bulbs without filter (UVP

FIG. 1. Flow diagram of "protein image" hybridization assay. (1) DNA–histone interactions within DHFR locus as assayed by "histidine" DMS cross-linking. Diagonal α corresponds to free DNA and diagonals β and γ to DNA cross-linked to the core histones and histone H1, respectively. (1a) Ethidium bromide-stained pattern of bulk histone–DNA cross-linking. (1b–e) Hybridization with different regions of DHFR locus. See Pemov *et al.*[11] for details. Adapted, with permission, from A. Pemov, S. Bavykin, and J. L. Hamlin, *Biochemistry* **34,** 2381 (1995). (2) DNA–histone interactions within the active *hsp-70* gene (A, B) and nontranscribed type II ribosomal insertion (D) of *D. melanogaster* revealed by UV cross-linking. The diagonals represent uncross-linked DNA (0), DNA cross-linked to histone H1 (H1), and proteins p50 (p50) and p100 (p100). M, internal standard applied onto the gel of second dimension. Reproduced, with permission, from S. Belikov, A. Belgovsky, O. Preobrazhenskaya, V. Karpov, and A. Mirzabekov, *Nucleic Acids Res.* **21,** 1031 (1993). (3) Mapping proteins on Alu repeat of ribosomal genes of *D. melanogaster* by UV cross-linking. Diagonal lines under autoradiographs indicate positions of "protein images" for histone H1 (H1), two nucleosome-binding proteins (rABP 50, rABP 70), and an unknown protein (X). Bottom diagonal lines indicate positions of "protein images" of marker proteins (37, 67, 90, and 135 kDa). M, *D. melanogaster* DNA digested with *Alu*I. See Belikov *et al.*[10] for details. Adapted, with permission, from S. Belikov, A. Belgovsky, M. Partolina, V. Karpov, and A. Mirzabekov, *Nucleic Acids Res.* **21,** 4796 (1993). (4) Positioning of proteins on Alu repeat by UV cross-linking followed by Exo III trimming. Hybridization with Alu repeat probe [coding (a) and noncoding (b) strand]. Numbers 1–4 above and below designate diagonals of uncross-linked DNA (1), DNA cross-linked to histone H1 (2), rABP 50 (3), and rABP70 (4). Adapted, with permission, from S. Belikov, A. Belgovsky, M. Partolina, V. Karpov, and A. Mirzabekov, *Nucleic Acids Res.* **21,** 4796 (1993).

Inc., San Gabriel, CA) placed 10 cm above the cells for 10–15 min at 0–4° while stirring.

2. *D. melanogaster* nuclei are isolated following the protocol of Gilmour and Lis,[14] with modifications. All procedures are done at 0–4°. Collect cooled cells by centrifugation at 750*g* for 10 min, suspend in buffer A [100 m*M* KCl, 50 m*M* NaCl, 5 m*M* $MgCl_2$, 10 m*M* Tris–HCl, pH 7.4, 0.5 m*M* diisopropyl fluorophosphate (DFP; Sigma, St. Louis, MO)] supplemented with 0.2 *M* sucrose at about 10^8 cells/ml. Add one-tenth volume of freshly prepared 10% Nonidet P-40 (Sigma) and vortex the suspension of cells thoroughly for 5 min. Collect nuclei by centrifugation at 4400*g* for 10 min. Go to section entitled "Depletion of Uncross-linked Proteins by Centrifugation in CsCl Gradients."

Formaldehyde Fixation[6,8,11] *and DMS Cross-Linking*[6,8,11]

To prevent artifactual redistribution of proteins along the DNA, we performed formaldehyde fixation to "freeze" the native conformation of chromatin. Formaldehyde very quickly and efficiently produces both protein–DNA and protein–protein zero-length cross-links *in vivo* (via 2-Å-long methylene "bridges") that can be reversed under mild conditions.

"DMS cross-linking" within nuclei is induced by the mild methylation of purines at position N-7 of guanosine and N-3 of adenine by DMS followed by the depurination of methylated bases. Aldehyde groups thus formed react with neighboring amino groups of proteins: α-amino groups, ε-amino groups of lysines, and imidazole groups of histidines, forming cross-links. Subsequent stabilization of the cross-links is achieved by their reduction with sodium borohydrate. Under these conditions, proteins cross-link to DNA mostly via histidines ("histidine" protocol). The "lysine" protocol differs from the "histidine" one in that cross-linking and stabilization/reduction are performed simultaneously with sodium cyanoborohydride or pyridine borane complex.

An excellent description of the "chemistry" of DMS DNA–protein cross-linking can be found in Pruss and Bavykin.[5]

Protocol

1. Add formaldehyde to 1% (from 37% formaldehyde stock solution; Merck, Darnstadt, Germany) to the suspension of nuclei at 20 A_{260} units/ml in 30 m*M* sodium phosphate, pH 7.4, 5 m*M* $MgCl_2$, 0.5 m*M*

[14] D. S. Gilmour and J. T. Lis, *Mol. Cell. Biol.* **5,** 2009 (1985).

DFP and incubate at 4° for 18 hr. Wash fixed nuclei three times in the same buffer.

2. Suspend formaldehyde-fixed nuclei at 20 A_{260} units/ml in 30 m*M* sodium phosphate, pH 7.4, 5 m*M* $MgCl_2$, 0.5 m*M* DFP. Add 1/2 volume of 20 m*M* dimethyl sulfate (Merck) in the same buffer and incubate nuclei at 4° for 18 hr.
3. Pellet nuclei by centrifugation at 3000*g* for 10 min at 4°; resuspend in 30 m*M* sodium phosphate, pH 7.4, 2 m*M* EDTA, 0.5 m*M* DFP and incubate at 37° for 20 hr to perform depurination of methylated bases and cross-linking.
4. Chill the suspension of nuclei in an ice–water bath and perform reduction of $>C{=}N-$ bonds in either of two ways:
   a. "Histidine" protocol: add 1/10 volume of freshly prepared 200 m*M* $NaBH_4$ in 30 m*M* sodium phosphate, pH 7.4, 2 m*M* EDTA and incubate the mixture in an ice–water bath for 30 min in the dark.
   b. "Lysine" protocol: add 1/10 volume of freshly prepared 200 m*M* $NaCNBH_3$ during the depurination step to the suspension of nuclei in 30 m*M* sodium phosphate, pH 7.4, 2 m*M* EDTA. Reduction and depurination are continued simultaneously at 37° for 20 hr. To complete reduction, add 1/10 volume of freshly prepared 200 m*M* $NaBH_4$ in 30 m*M* sodium phosphate, pH 7.4, 2 m*M* EDTA and incubate the mixture in an ice–water bath for 30 min in the dark.
5. To reverse formaldehyde-induced fixation after DMS cross-linking, nuclei are sonicated briefly at 0° in 2% sarcosyl (Sigma). The sample is adjusted to 1 *M* in NaCl and 0.5 *M* in Tris–HCl, pH 8.0, followed by incubation for 5 hr at 80°. Proceed to the next section.

Depletion of Uncross-Linked Proteins by Centrifugation in CsCl Gradients[13,15]

This method relies on the difference in floating density of DNA and proteins. Two points are crucial for this procedure: (i) using guanidine hydrochloride as a strong dissociating agent and (ii) loading of high molecular weight DNA. Under a low degree of cross-linking (one cross-linked protein per 3–10 kb of DNA), the influence of the protein component on the density is negligible and so the DNA–protein complex behaves almost as uncross-linked DNA in the CsCl gradient.

[15] D. Papatsenko, S. Belikov, O. Preobrazhenskaya, and V. Karpov, *Methods Mol. Cell. Biol.* **5,** 169 (1995).

Protocol

1. Suspend cross-linked nuclei in medium containing 50 m*M* Tris–HCl, pH 8.0, 5 m*M* EDTA, and 0.5 m*M* DFP, and/or add 8 *M* guanidine hydrochloride to a 6 *M* solution. Adjust the concentration of DNA in this solution until 1 mg/ml by 6 *M* guanidine hydrochloride. Gently grind lysate in Dounce homogenizer and spin at 3000*g* for 5 min to remove insoluble debris.
2. Add 6 ml CsCl (ρ = 1.63 g/cm^3) solution containing 5 m*M* EDTA, 50 m*M* Tris–HCl, pH 8.0, 0.5 m*M* DFP to a 10-ml ultracentrifuge tube (type "Ultroclean"). On top of this, load 0.2 ml of 8 *M* guanidine hydrochloride and then 3.8 ml of lysate and centrifuge in a Beckman Ti-50 rotor (100,000*g*; 40 hr; 20°). Collect the DNA-containing fraction from the bottom by monitoring UV absorbance (usually one-third of the tube volume) using a capillary tube connected to a peristaltic pump. Important: do not overload gradients as this leads to aggregation of the material.
3. Add 3 volumes of water and recover the cross-linked complexes by ethanol precipitation. Collect DNA–protein complexes by centrifugation for 20 min at 12,000*g*.

Optional: dialyze samples extensively against 5 m*M* Tris–HCl, pH 8.0, 0.5 m*M* EDTA, 0.1% NP-40, 0.5 m*M* DFP and proceed to "DNA Shearing" or store at −20°. This is only required for UV cross-linking or prior to restriction endonuclease treatment.

DNA Shearing[10,12] *and Exo III Trimming*[10]

In order for cross-linked protein–DNA complexes to be separated by gel electrophoresis, the size of the DNA component should be in the range of 50–500 bp. Besides, the size of DNA fragments may also be crucial for high-resolution mapping. It is worth nothing that in the "histidine protocol," DNA splitting is induced at the sites of cross-linking[16] and that degree of DNA "self-shearing" depends on the degree of DNA methylation. If treatment with restriction enzyme is to be done (see Fig. 1), the amount of DMS added to generate cross-links should be reduced two- to threefold to increase the average size of DNA (see "Formaldehyde Fixation and DMS Cross-Linking" section).

UV irradiation during cross-linking leaves DNA almost intact, so to shear DNA randomly, we recommend sonication; this generates a statistical

[16] E. Levina, S. Bavykin, V. Shick, and A. Mirzabekov, *Anal. Biochem.* **110,** 93 (1981).

set of DNA fragments with no sequence specificity. However, it is difficult to obtain fragments shorter than ca. 200 bp. Enzymatic fragmentation could be used as well. Endonucleases such as DNase I or micrococcal nuclease are good enough, but one should keep in mind that most enzymes display distinct sequence preference for DNA cleavage.

In order to increase the hybridization signal and locate the protein cross-linking site(s) within a short DNA fragment, we recommend treatment with a suitable restriction endonuclease(s) that converts the diagonal corresponding to the cross-linked protein into a spot or diagonal array of spots.[7,10] Subsequent treatment with exonuclease III trims 3′ terminus of the DNA fragment up to the point of cross-linking, thus allowing precise localization of the cross-linked protein on the DNA fragment.[10]

Protocol

DNA-Sonication. We routinely use Branson Sonifer Cell Disruptor B15 (Branson Ultrasonics Corp., Danbury, CT), equipped with a microtip. However, almost every available sonicator can be used if it produces DNA fragments of an average size of about 300–400 bp as revealed by denaturating gel electrophoresis. If this level is achieved, longer sonication would not improve shearing. Orlando and co-workers[2] recommend the addition of microglass beads (0.1–0.5 mm diameter) during sonication to improve sonication efficiency.

Shearing with DNase I.[9] After dialysis, add $MgCl_2$ to 5 m*M* and then DNase I (0.01 mg per 50 mg of DNA). Incubate for 5–10 min at 37°, according to the preliminary analytical experiment to generate DNA fragments about 100–500 bp long. Stop the reaction by adding EDTA to 10 m*M* and SDS to 0.5%.

Treatment with Restriction Endonucleases.[7,10] DNA–protein complexes were digested with an appropriate restriction enzyme(s) in the buffer recommended by the manufacturer supplemented with 0.1% NP-40 and 0.5 m*M* DFP. The exact number of units of enzyme to be added and incubation time should be determined in preliminary experiments.

Exonuclease III Trimming.[10] After completion of restriction endonuclease treatment add Exo III (Amersham) (5 units per 1 mg of DNA) and DFP to 0.5 m*M* and incubate for 30 min at 37°. Proceed to the following section.

Phenol Extraction of DNA–Protein Complexes[15]

The phenol extraction method is based on the well-known fact that free DNA remains in the aqueous phase, whereas proteins and DNA–protein complexes are transferred into the organic and interphases.

Protocol

1. Dissolve cross-linked complexes at approximately 0.5 mg DNA/ml in 1% SDS and 100 m*M* Tris–HCl, pH 8.0, or add SDS and Tris–HCl to 1% and 100 m*M*, respectively.
2. Denature DNA by boiling for 2 min, cool to room temperature, and add an equal volume or at least 1 ml of phenol–chloroform–isoamyl alcohol (80 : 20 : 1, v/v/v) if dialyzed, nonconcentrated DNA–protein complexes are to be assayed.
3. Mix by vortexing for 60 sec and centrifuge for 2 min at 15,000*g*. Discard the upper aqueous layer.
4. Add an equal volume of 100 m*M* Tris–HCl, pH 8.0, and repeat steps 3 and 4 twice.
5. Add 50 mg of dextran (moleculer weight 500,000, Sigma) to the organic phase and interphase, mix, and recover the cross-linked complex by ethanol precipitation by adding 3 volumes of ethanol containing 0.3 *M* sodium acetate and then incubating at −20° for at least 30 min.
6. Centrifuge for 15 min at 15,000*g*, resuspend the pellet in a small volume of water, and precipitate the complex by ethanol again.
7. Wash pellets twice with cold 80% ethanol.

Two-Dimensional Gel Electrophoresis

Electrophoresis[17] is performed in a SDS–polyacrylamide system[18] for both dimensions. The exact percentage of the acrylamide in the gels is dictated by the molecular weight of cross-linked proteins and the length of cross-linked DNA.

Protocol

1. Suspend the pellet in 8–9 *M* urea with heating. Add SDS to 1%, DTT to 1 m*M*, and bromphenol blue to 0.025%. Heat for 2 min at 95° and load onto a gel of first dimension (3.75% stacking gel) containing 8 *M* urea.
   Resolving gel buffer: 0.375 *M* Tris–HCl, pH 8.8, 0.1% SDS, 8 *M* urea[17]
   Stacking gel buffer: 0.125 *M* Tris–HCl, pH 6.8, 0.1% SDS, 8 *M* urea[17]
   Reservoir buffer: 0.192 *M* glycine, 0.025 *M* Tris, pH 8.3, 0.1% SDS[17]
2. Run the electrophoresis (10 V/cm), excise the strip, and equilibrate it in 0.125 *M* Tris–HCl, pH 6.8, 0.1% SDS for 15 min (for 1-mm gel thickness). Polymerize the strip on top of the stacking gel of the

[17] A. Mirzabekov, S. Bavykin, A. Belyavsky, V. Karpov, O. Preobrazhenskaya, V. Shick, and K. Ebralidse, *Methods Enzymol.* **170,** 386 (1989).

[18] U. Laemmli, *Nature* **277,** 680 (1970).

second dimension (5%, without urea) using 5% acrylamide in stacking gel buffer without urea. On top of that load pronase-containing buffer (0.063 *M* Tris–HCl, pH 6.8, 0.1% SDS, 10% Ficoll 400, and 0.025% bromphenol blue). In a separate well, load the deproteinized DNA that serves as an internal control (see "Comments and Applications," step 7). Run the electrophoresis (10 V/cm).
3. After electrophoresis, soak the gel in water for 5 min.

Electrotransfer of DNA from Polyacrylamide Gel[6] *and Hybridization with Radioactive Probes*[19,20]

Any commercial or homemade apparatus for electrotransfer (e.g., Trans-Blot Electrophoretic Transfer Cell, Bio-Rad, Richmond, CA) and nylon hybridization membrane (e.g., HybondN+, Amersham) should be adequate. The exact way of labeling is dictated by the specific requirements of each particular study and/or researcher's own preferences. We recommend protocols for labeling and hybridization described in Church and Gilbert[19] and Belikov *et al.*[20] Blots can be stripped and rehybridized up to 10 times.

Protocol

Electrotransfer of DNA from Polyacrylamide Gel. Wash the gel in transfer buffer (10 m*M* Tris–acetate, pH 7.8, 5 m*M* sodium acetate, 0.5 m*M* EDTA) twice for 15 min under gentle shaking. Transfer to nylon membrane at 5 V/cm overnight, according to supplier's instructions.

Hybridize with ^{32}P-labeled probe and wash membrane as described in Church and Gilbert.[19] Seal membrane in plastic bag and expose it.

Probe Stripping and Rehybridization. Wash the membrane twice for 15 min in boiling solution containing 0.1× SSC and 0.5% SDS just before rehybridizaton. Check membrane by overnight exposure. *Important:* while reprobing, do not allow membrane to dry between hybridizations.

Comments and Applications

1. Work with DNA–protein complexes requires protease-free conditions. The addition of 0.5 m*M* phenylmethylsulfonyl fluoride or 0.5 m*M* diisopropyl fluorophosphate to all solutions is a minimum requirement; the addition of other protease inhibitors is highly advisable. All solutions should be filter sterilized before use.

[19] G. Church and W. Gilbert, *Proc. Natl. Acad. Sci U.S.A.* **81,** 1991 (1984).

[20] S. Belikov, D. Papatsenko, and V. Karpov, *Anal. Biochem.* **240,** 152 (1996).

2. Two cross-linking methods, DMS cross-linking and UV cross-linking, have shown themselves to be feasible, reliable, and informative. We can also recommend DNA–protein cross-linking by *cis*-dichlorodiammineplatinum (II) to study the distribution of histone H1 along the chromation. Histone H1/H5 is cross-linked to DNA in significant amounts whereas core histones remain practically unattached. The yield of DNA–protein complexes is high enough to allow "protein image" hybridization analysis immediately after cross-linking without CsCl purification/phenol enrichment. A detailed protocol can be found in Belikov and Karpov.[21]
3. We had shown that "histidine" cross-links involve mostly globular domains of histones. In contrast, "lysine" cross-links are formed predominantly via N- and C-terminal tails of histones that are highly enriched in positively charged lysine resides.[6] The simultaneous use of two protocols provides a unique opportunity to study the dynamics of histone–DNA interactions *in situ*.

   DMS cross-linking was applied successfully to the analysis of histone–DNA interactions along the *hsp-70* gene,[6] ribosomal genes,[7] chicken erythrocyte chromatin,[8] and constitutively expressed DHFR genes.[11] Analysis of the chromatin structure of the DHFR gene is presented in Fig. 1 (1a–e) as an example. An ethidium bromide-stained 2D gel is presented in Fig 1 (1a); the blot of this gel was hybridized in succession with probes representing different genomic regions [Fig 1 (1b–e)]. Note the absence of histone cross-links in the region covering transcription initiation site [Fig. 1 (1b)].
4. The restriction enzyme cleavage of DMS cross-linked DNA–proteins complexes (Fig. 1) "unifies" the 5′ end of the DNA fragment in the complex. As the formation of cross-links is accompanied by a single-stranded splitting at the site of cross-linking, the sites of protein cross-linking can be localized easily by hybridization with a strand-specific ^{32}P-labeled probe to the region adjacent to the restriction enzyme site. Note the analogy with the indirect end-labeling technique. In this assay, the concentration of DMS should be reduced three- to fourfold.
5. For UV-induced DNA–protein cross-linking, we use a transilluminator without a filter which is available in any molecular biology laboratory. Unlike conventional UV light sources, lasers provide a stable, high-energy source of UV radiation.[4] Furthermore, the use of lasers (i) allows a dramatic increase in the efficiency of cross-linking in a significantly shortened irradiation time and (ii) leads to formation

[21] S. Belikov and V. Karpov, *Biochem. Mol. Biol. Int.* **38,** 997 (1996).

of a different, although overlapping, subset of DNA–protein cross-links. The only limitation could be the cost of the laser equipment.

6. A wide spectrum of nuclear proteins can be cross-linked to DNA using UV irradiation; however, the major products are histone H1–DNA adducts. The diagonal corresponding to histone H1 can be visualized simply by staining 2D gels with ethidium bromide. Cross-linking of core histones is negligible, but these core histone–DNA cross-links can still be detected, although the corresponding diagonal is much weaker than that for histone H1. Cross-linking of histones occurs predominantly via N and C tails.[22,23]
7. Figure 1 (2) shows the use of UV-induced DNA–protein cross-linking for probing DNA–protein interactions within the *D. melanogaster hsp-70* gene. Purified cross-linked DNA–protein complexes were digested with DNase I to shorten the average DNA fragment length. "Protein image" hybridization analysis shows the presence of histone H1 in the highly transcribed *hsp-70* gene [Fig 1 (2B)], documented by subsequent hybridization with the probe representing inactive chromatin [Fig. 1 (2D)]. Hybridization with a short probe covering the promoter revealed an absence of histone H1 cross-links and the presence of proteins p50 and p100 associated with the active *hsp-70* promoter [Fig. 1 (2A)].[10] For quantitative evaluation of cross-linking efficiencies, an internal standard containing deproteinized sheared DNA from the same cross-linked sample was applied to gel for the second dimension [lane M in Fig. 1 (2)].
8. To estimate an approximate molecular weight of the cross-linked proteins, we recommend DMS cross-linking *in vitro* of proteins with a known molecular weight to randomly labeled DNA followed by 2D gel electrophoresis resulting in a "fan" of diagonals analogous to a molecular weight ladder [Fig. 1 (3)]. Note that histone H1 has abnormal mobility in SDS-containing polyacrylamide gels, its apparent molecular mass of 36–37 kDa (data not shown). The corresponding protocol is described in Mirzabekov *et al.*[17]
9. The identification of sites of UV-induced DNA–protein cross-linking can be achieved by restriction enzyme cleavage/Exo III trimming with a subsequent hybridization with a short strand-specific ^{32}P-labeled probe adjacent to the 5′ end of the fragment. Using this method, we revealed two new nucleosome-binding proteins and histone H1

[22] V. Stefanovsky, S. Dimitrov, D. Angelov, and I. Pashev, *Biochem. Biophys. Res. Commun.* **164,** 304 (1989).

[23] V. Stefanovsky, S. Dimitrov, V. Russanova, D. Angelov, and I. Pashev, *Nucleic Acids Res.* **17,** 10069 (1989).

positioned with respect to the histone octamer within so-called Alu repeats in the *D. melanogaster* ribosomal spacer [Fig. 1 (4)].[10]

Acknowledgments

We express deep gratitude to Professor A. Mirzabekov for introducing us to the fascinating field of DNA–protein cross-linking. We thank Drs. M. Isaguliants and T. Klenka for critical comments on the manuscript.

[34] *In Vivo* Analysis of Chromatin Structure

By KENNETH S. ZARET

Introduction

Chromatin structure *in vivo* covers a wide range of cell biological and biochemical perspectives on the physical state of genes in the nucleus. The field is further broadened by the highly dynamic aspect of chromatin structure, which involves changes in nucleosomes, large chromatin domains, and attachments to the nuclear scaffold. Such changes occur during gene regulation, DNA repair, and cell division. This article provides a critical overview of many of the techniques being used to identify chromatin structure changes *in vivo* or *in situ*. At the end of the article, a protocol is provided for mapping nucleosome positions in complex genomes at the nucleotide level of resolution.

An *in vivo* analysis of chromatin structure can provide several kinds of information regarding gene function. First, the analysis can be used as a primary investigative tool to identify genetic regulatory sequences. For example, sequences spanned by nuclease hypersensitive sites or unusual nucleosome arrangements often correspond to genetic regulatory elements.[1] Second, certain experimental criteria can be used to define physiological states of chromatin, such as "open" or "closed," nucleosomes "present" or "absent," and factor "bound" or "not bound." With enough knowledge about these chromatin states, the experimental criteria may have predictive value with regard to gene function in novel physiological conditions. Of course, all such analysis is critically dependent on the integrity of the native chromatin substrate used in the experiment.

[1] D. S. Gross and W. T. Garrard, *Annu. Rev. Biochem.* **57,** 159 (1988).

0076-6879/99 $30.00

Native chromatin is a relative term, which may be confusing or dismaying to casual observers of the field. To a biophysicist interested in chromatin folding, native chromatin may mean purified nucleosome arrays with a conformation reflecting that in a cell; to a molecular biologist interested in nucleosome positions or transcription factor binding, it may mean the state of DNA in an isolated nucleus; and to a cell biologist, it may mean the organization of chromosomes within intact cells. This article addresses chromatin that is found within intact cells, permeabilized cells, and isolated nuclei, as all of these conditions are relevant to the goal of *in vivo* analysis of chromatin structure. Note that results obtained from studying chromatin in nonintact cells must acknowledge that the substrate was not *in vivo* or "native" during the experiment.

Three parameters govern the ability to obtain useful information from a chromatin experiment: The quality of the chromatin substrate, the structural specificity of the chromatin probe, and the sensitivity of the assay. Each of these parameters is discussed separately in terms of the benefits and hazards of relevant techniques, and specific experimental protocols are referred to or provided. Only a few references are given for each method; these either provide a clear and generally accepted description of experimental details or illustrate novel ways the techniques can be used. The list of techniques is not intended to be comprehensive, but rather inclusive of the more commonly used methods and those that may deserve more use. Summaries of findings obtained from the methodologies can be obtained from the many excellent reviews that have appeared on the topic.[1–6]

Obtaining High-Quality Native Chromatin Substrates

The ideal chromatin substrate exists in the intact cell, and there are certain cell-permeable probes, such as dimethyl sulfate (DMS), that diffuse into the nucleus and target DNA. Also, strategies have been developed to genetically express, within a cell, DNA-modifying enzymes such as methyltransferases, whose activities are sensitive to chromatin structure.[7,8] These reagents can provide a dynamic view of chromatin structural changes under different physiological conditions. The modifying enzymes recognize specific DNA sequences, which occur every few dozen to few hundred base

[2] D. S. Pederson, F. Thoma, and R. T. Simpson, *Annu. Rev. Cell Biol.* **2,** 117 (1986).

[3] R. D. Kornberg and Y. Lorch, *Annu. Rev. Cell Biol.* **8,** 563 (1992).

[4] G. Felsenfeld, *Nature* **355,** 219 (1992).

[5] M. Grunstein, *Nature* **389,** 349 (1997).

[6] A. Wolffe, "Chromatin Structure and Function." Academic Press, San Diego, 1992.

[7] D. E. Gottschling, *Proc. Natl. Acad. Sci. U.S.A.* **89,** 4062 (1992).

[8] M. P. Kladde, M. Xu, and R. T. Simpson, *EMBO J.* **15,** 6290 (1996).

pairs, whereas the chemical probes recognize single bases; thus there are many more targets for the latter. However, a problem with small molecule probes is that it can be difficult to stop their reaction with DNA. Simply lysing the cells and initiating proteolysis, the typical first step in DNA preparation, can be insufficient because the reagent remains chemically reactive while the relevant binding proteins have been stripped from the DNA. It therefore may be necessary to expose the intact cells to the chemical probe for a short period of time and then isolate cell nuclei.[9] Nuclei are washed to remove remaining amounts of reactive probe, and then nuclei are lysed and DNA is purified. To facilitate these procedures, and to facilitate uniform access of the chemical probe to the cell population, it is usually best to start with a cell suspension. However, it is also possible to deliver a DMS probe uniformly to cells in intact animal tissues by perfusing the probe solution via the endogenous venous system.[10] This is about as close as the field has come to using biochemical probes for analyzing the chromatin structure of specific genes in an animal.

A nearly ideal chromatin substrate exists in the permeabilized cell, and nonionic detergents such as lysolecithin[11] and Nonidet P-40[12] have been found to permeabilize the cell membrane sufficiently to allow the entry of enzymatic probes of chromatin such as deoxyribonuclease I (DNase I) and micrococcal nuclease (MNase). A convenience of this assay is that the cells can be either in a suspension or in a monolayer. A concern is that permeabilized cells will lyse after a certain amount of time in detergent; thus care must be taken to ensure cell integrity by performing microscopy within the time course of the experiment. A further difficulty with the permeabilization technique is that for each cell type there appears to be a relatively narrow detergent concentration range over which the assay can be performed. Considering that chromatin probe concentrations must also be calibrated for each cell type or gene being assayed, the permeabilization approach may not be feasible if cell sources are limiting. However, cell permeabilization has been an underexplored approach to studying chromatin structure and further work in the area may simplify some of the technical concerns. The retention of cytoskeletal architecture around the nucleus seems highly likely to maintain the integrity of chromatin, and therefore cell permeabilization may be particularly useful for studies of higher-order chromatin structure.

Historically, the typical source of chromatin is within isolated nuclei.

[9] P. Bossard, C. E. McPherson, and K. S. Zaret, *Methods* **11,** 180 (1997).

[10] C. E. McPherson, E.-Y. Shim, D. S. Friedman, and K. S. Zaret, *Cell* **75,** 387 (1993).

[11] L. Zhang and J. D. Gralla, *Genes Dev.* **3,** 1814 (1989).

[12] G. Rigaud, J. Roux, R. Pictet, and T. Grange, *Cell* **67,** 977 (1991).

Clearly there are numerous parameters to consider before concluding that chromatin in an isolated nucleus represents native chromatin in a cell. Definitive experiments to address the integrity issue, at least for higher-order structures, are not in hand, and Pfeifer and Riggs[13] found a striking depletion of transcription factors bound to chromatin within isolated nuclei compared to that within permeabilized cells. Kornberg *et al.*[14] have published an excellent discussion of the technical parameters for preparing nuclei, with special emphasis on avoiding nuclear lysis and chromatin degradation during isolation. The reader is urged to stringently monitor these latter parameters for each preparation of isolated nuclei. The efficiency of cell breakage and nuclear integrity are easily checked by phase-contrast microscopy of a portion of the crude cell lysate after cell breakage; diluting nuclei into phosphate-buffered saline may be necessary if the preparation is excessively dense. Cell lysis should be virtually complete before proceeding to purify nuclei further. The extent of chromatin cleavage during isolation and the extent of contamination by endogenous nucleases are readily monitored by including in the experiment an incubation of nuclei in the absence of the relevant chromatin probe for the same time duration as for samples that are treated with the probe. Establishing conditions to maintain nuclear integrity is essential before concluding that a probe is causing a signal in the assay.

The primary means for maintaining nuclear stability is by appropriately adjusting the buffers used to isolate and assay nuclei. A proper divalent cation concentration is necessary to maintain chromatin structure, yet cations can activate endogenous nucleases. Indeed, the endogenous nuclease problem can be fierce with certain cell types, which has led to the use of spermine and spermidine as counterions to replace divalent cations as chromatin stabilizers.[14] However, there is some evidence that these agents can perturb chromatin-binding factors.[13] Another way to significantly reduce the effects of endogenous nucleases is to be sure that all test tubes, pipettes, pipette tips, and so on are kept cold and that the nuclear isolation procedure is performed as quickly as possible. Finally, nuclei should be warmed up the minimum time necessary before adding enzymatic or chemical probes. The benefits of cold and speed cannot be overstated. These parameters also help reduce the amount of diffusion of proteins from the nuclei; such proteins are likely to maintain relevant chromatin structures.

Burch and Weintraub[15] developed a rapid method for preparing crude nuclei that preserves nuclease hypersensitive sites in chromatin, but which

[13] G. Pfeifer and A. D. Riggs, *Genes Dev.* **5,** 1102 (1991).

[14] R. D. Kornberg, J. W. LaPointe, and Y. Lorch, *Methods Enzymol.* **170,** 3 (1989).

[15] J. B. E. Burch and H. Weintraub, *Cell* **33,** 65 (1983).

is often ineffective for eliminating endogenous nucleases. That is, nuclease hypersensitive sites appear quite strong on Southern blots, but the cleavages are due to endogenous nucleases. However, many endogenous nuclease cleavages seen with the Burch and Weintraub[15] technique correspond precisely to cleavages caused by DNase I, when the latter is used with a nuclear isolation procedure that suppresses endogenous nuclease activity.[16]

A strategy for initially characterizing the chromatin structure of a gene would be to first use the relatively simple Burch and Weintraub[15] procedure. Once hypersensitive regions have been mapped and it becomes important to characterize them in greater detail, then use the more involved protocol of Kornberg *et al.*[14] It remains incumbent on the investigator to establish whether chromatin cleavages are due to the exogenous probe or to endogenous nucleases.

Structural Specificity of Native Chromatin Probes

Cell-Permeable Probes

Dimethyl sulfate (DMS) adds a methyl group to the N-7 position of guanine residues in the DNA major groove and to the N-3 position of adenines in the minor groove. Proteins in intimate contact with the DNA may protect G's and A's from DMS, and the extent of protection of these bases is a diagnostic footprint of the factor.[17] By comparing a DMS protection pattern *in vivo* with that given by a purified protein on DNA *in vitro,* one can identify a protein bound to DNA in native chromatin with some degree of confidence. A caveat with the analysis is that different members of the same transcription factor family may exhibit identical DMS footprints. Also, the core histone proteins do not elicit DMS protections of nucleosomal DNA,[18] so the DMS probe is restricted to analyzing interactions with nonhistone proteins. A detailed description of the *in vitro* and *in vivo* DMS protection methodology is provided by Bossard *et al.*[9] Strauss *et al.*[19] describe a method to more efficiently reveal cleavages at A residues.

Potassium permanganate is another DNA-modifying agent that is permeable to intact cells. $KMnO_4$ interacts preferentially with single-stranded DNA, most prominently with thymines. It has been useful for analyzing changes in the conformation of DNA in and around seqences that are bound by transcription factors or that are adjacent to nucleosomes and for

[16] J.-K. Liu, Y. Bergman, and K. S. Zaret, *Genes Dev.* **2,** 528 (1988).
[17] U. Siebenlist and W. Gilbert, *Proc. Natl. Acad. Sci. U.S.A.* **77,** 122 (1980).
[18] J. D. McGhee and G. Felsenfeld, *Proc. Natl. Acad. Sci. U.S.A.* **76,** 2133 (1979).
[19] E. C. Strauss, N. C. Andrews, D. R. Higgs, and S. H. Orkin, *Mol. Cell. Biol.* **12,** 2135 (1992).

studying the dynamics of promoter melting during transcriptional activation.[11]

Ultraviolet light has also been used to monitor changes in DNA conformation in native chromatin *in vivo*.[20,21] The nature of the conformational change that is detected as UV sensitivity is unclear, but in some contexts, bound transcription factors can block the UV light.[22] Importantly, the ability to deliver discrete pulses of light allows time-resolved views of chromatin structure changes during gene regulation, and some DNA sequences exhibit dramatic changes in UV sensitivity.[23] The original "photo footprinting" studies were performed in the "pre-LM-PCR" era and few investigators have pursued the approach recently. Now that LM-PCR can sensitively detect single base pair cleavages in complex genomes (see later), UV-based footprinting should be revisited.

Endogenously Expressed Probes

An exciting new development has been to conditionally express enzymes such as DNA methyltransferases that probe chromatin structure within a living cell.[7] Methyltransferase can be induced in different physiological states so that chromatin transitions can be monitored in time. The relative methylation sensitivity of certain DNA regions, compared to others, indicates the state of chromatin compaction and, in some cases, positioned nucleosomes.[24] The original DAM methyltransferase, which was used for the assay, recognizes target sequences of 4 bp, which occur randomly every several hundred base pairs. Kladde *et al.*[8] have utilized the *Sss*I methyltransferase, which recognizes CpG sites and increases the number of potential target sequences greatly. However, the *Sss*I enzyme probe is limited primarily to organisms such as yeast, which lack endogenous CpG methylases, or to mammalian genes in cell types where the target sequences have clearly been shown to be unmethylated.

Cross-Linking Reagents

The chromatin probes just described, which use DNA as a primary target, can detect protein bound to a specific DNA sequence, but they do not definitively establish the identity of the protein. The latter information now can be addressed by protein–DNA cross-linking technology. That is,

[20] M. M. Becker and J. C. Wang, *Nature* **309,** 682 (1984).

[21] S. B. Selleck and J. Majors, *Nature* **325,** 173 (1987).

[22] Z. Wang and M. M. Becker, *Proc. Natl. Acad. Sci. U.S.A.* **85,** 654 (1988).

[23] S. B. Selleck and J. Majors, *Proc. Natl. Acad. Sci. U.S.A.* **85,** 5399 (1988).

[24] M. P. Kladde and R. T. Simpson, *Proc. Natl. Acad. Sci. U.S.A.* **91,** 1361 (1994).

chromatin-bound proteins are cross-linked to DNA, the DNA is reduced to small-sized fragments, the protein–DNA complexes are subjected to immunoprecipitation with specific antibodies, and the DNA sequences that are cross-linked to the primary antigen are identified by hybridization.[25] The two primary reagents for cross-linking are formaldehyde, which is used on intact cells and isolated nuclei, and ultraviolet light, which is used on intact cells. Formaldehyde reacts with primary amines and thus HEPES-based buffers, rather than Tris, are used when performing the reactions.[26] Formaldehyde can result in extensive cross-linking and has the benefit of being reversible by extensive incubation of the sample at 68°.[27,28] However, excessive formaldehyde treatment may mask epitopes that are critical for the immunoprecipitation reaction, and thus it can be useful to set up control reactions where cross-linking conditions are tested with purified or defined antigen. Ultraviolet light is less efficient in cross-linking, but recent developments in laser-based UV sources have led to much more efficient reactions and extremely short pulse times.[28] Most of the chromatin proteins that have been detected by cross-linking bind multiple sites over large regions of DNA because the immunoprecipitation and hybridization steps have been less sensitive than that required to observe a single protein bound to a single DNA site. Thus, chromatin cross-linking and immunoprecipitation studies have analyzed the core histone proteins.[27] and their acetylated variants,[29] linker histones,[30] RNA polymerase,[31] and regulatory factors that appear to bind repeatedly at target loci,[32,33] The use of carefully calibrated, PCR-based approaches has greatly increased the sensitivity with which immunoprecipitated material can be detected, allowing much smaller regions of DNA to be assayed.[34,35]

4,5′,8-Trimethylpsoralen is a DNA interstrand cross-linking agent that is inhibited by the histone octamer and therefore has been used to map

[25] M. J. Solomon, P. L. Larsen, and A. Varshavsky, *Cell* **53,** 947 (1988).

[26] V. Jackson, *Cell* **15,** 954 (1978).

[27] V. Jackson, *Biochemistry* **29,** 719 (1990).

[28] S. I. Dimitrov, V. Yu. Stefanovsky, L. Karagyozov, D. Angelov, and I. G. Pashev, *Nucleic Acids Res.* **18,** 6393 (1990).

[29] T. R. Hebbes, A. L. Clayton, A. W. Thorne, and C. Crane-Robinson, *EMBO J.* **13,** 1823 (1994).

[30] E. H. Bresnick, M. Bustin, V. Marsaud, H. Richard-Foy, and G. L. Hager, *Nucleic Acids Res.* **20,** 273 (1991).

[31] D. S. Gilmour and J. T. Lis, *Proc. Natl. Acad. Sci. U.S.A.* **81,** 4275 (1984).

[32] V. Orlando and R. Paro, *Cell* **75,** 1187 (1993).

[33] H. Strutt, G. Cavalli, and R. Paro, *EMBO J.* **16,** 3621 (1997).

[34] A. Hecht, S. Strahl-Bolsinger, and M. Grunstein, *Nature* **383,** 92 (1996).

[35] M.-H. Kuo, J. Zhou, P. Jambeck, M. E. A. Churchill, and C. D. Allis, *Genes Dev.* **12,** 627 (1998).

linker regions between nucleosomes.[36] Psoralen is permeable to the cell and its cross-linking is activated by exposure to ultraviolet light. One method to visualize the position of psoralen cross-links is to purify the DNA from the cell and then employ electron microscopy under denaturing conditions for the DNA. Single-stranded DNA "bubbles" occur between regions where double strandedness is maintained by cross-linking. This approach is especially useful when analyzing episomal DNAs that can be purified from the bulk chromatin.[37] Another method for revealing the extent to which a region has been cross-linked by psoralen is based on the ability of psoralen cross-linking to retard the migration of DNA under nondenaturing gel electrophoresis. After DNA purification, genomic sites flanking the region of interest are cleaved with restriction enzymes and the electrophoretic migration of the DNA fragment is visualized by Southern blot hybridization.[38] The more psoralen cross-links over a region, the more the mobility of the gel fragment may be retarded. The extent to which psoralen access is inhibited *in vivo* has been used to distinguish nucleosomes of different stabilities.[37] The method has also been used to monitor the extent of local domains of supercoiling in native chromatin.[39,40]

Cell-Impermeable/Nucleus-Permeable Probes

Both enzymatic and chemical probes are available for mapping the presence of nucleosomes and nucleosome organization in specific regions of chromatin. Micrococcal nuclease makes double-stranded cleavages in the linker DNA between nucleosomes and thus has been used extensively for nucleosome characterization. Partial MNase digestion of polynucleosomal DNA yields a "ladder" of DNA fragment sizes that are multiples of 160–200 bp, depending on the nucleosomal repeat length in the source of chromatin.[41] If extensively digested chromatin is subjected to a Southern blot hybridization experiment with a specific gene sequence probe, and a mononucleosome-sized band and larger multiples is seen, it implies that the DNA region being probed was nucleosomal during the MNase digestion. If the DNA purified from a *partial* MNase digestion is further treated with a restriction enzyme prior to Southern blot analysis, then by using the appropriate "indirect end-label" probe, it is possible to determine if the gene sequences in question may have existed in phased or positioned

[36] C. V. Hanson, C. J. Shen, and J. E. Hearst, *Science* **193,** 62 (1976).
[37] R. Gasser, T. Koller, and J. M. Sogo, *J. Mol. Biol.* **258,** 224 (1996).
[38] A. Conconi, R. M. Widmer, T. Koller, and J. M. Sogo, *Cell* **57,** 753 (1989).
[39] E. R. Jupe, R. R. Sinden, and I. L. Cartwright, *Bichemistry* **34,** 2628 (1995).
[40] G. Cavalli, D. Bachmann, and F. Thoma, *EMBO J.* **15,** 590 (1996).
[41] K. E. van Holde, "Chromatin." Springer-Verlag, New York, 1989.

nucleosomes.[42] For indirect end-label probing, it is critical that the target gene sequences are only partially digested with MNase. Overly digested DNA will yield subfragments of about 180 bp periodicity with MNase-generated end points that artifactually give the appearance of phased nucleosomes. The investigator should also note that even low levels of MNase can elicit single-stranded cleavages on nucleosomes[43] and that high levels of enzyme can elicit double-stranded cleavages on the particles, particularly at A+T-rich regions.[44] These potential artifacts should be kept in mind, especially when using methods to map MNase cleavages at the nucleotide level of resolution (see later). With all such analyses, it is critical to assay an MNase digestion of protein-free DNA side by side with chromatin samples. MNase cleaves DNA sequences nonrandomly and it is important to distinguish between a banding pattern that is specific to nucleotide sequence versus a pattern that is intrinsic to chromatin. This "free DNA" control is also essential for the other DNA-cleaving reagents discussed later. Despite these caveats, MNase analyses have been invaluable in determining the nucleosomal organization of chromatin, and an MNase assay of permeabilized cells or isolated nuclei, coupled with an indirect end-label analysis of a Southern blot of the DNA, represents an excellent first step in characterizing the chromatin of a genomic domain.

Methidiumpropyl–EDTA–Fe(II) (MPE) is a relatively small molecular probe in comparison to the 16.8-kDa MNase, yet like MNase, MPE cleaves double-stranded DNA in chromatin preferentially in the linker regions between nucleosomes.[45] Unlike MNase, MPE works in the presence of metal chelators, thereby making MPE a better choice for cell types with high levels of endogenous nucleases. Furthermore, MPE exhibits much less DNA sequence specificity than MNase, thus in principle making MPE a more generally applicable reagent for probing nucleosome position. However, there are reports of steroid hormone-induced MPE hypersensitivities that occur within apparent positioned nucleosomes, where the nucleosome boundaries are defined by MNase cleavages that remain stable with and without hormone.[46] These hormone-induced hypersensitive sites are also DNase hypersensitive.[46] In conclusion, MPE can provide an independent view of nucleosome boundaries from that obtained with MNase, but in some contexts, MPE may cleave within a perturbed nucleosome or locally altered DNA structure that is not sensitive to MNase. Unfortunately, we

[42] C. Wu, *Nature* **286,** 854 (1980).

[43] M. Cockell, D. Rhodes, and A. Klug, *J. Mol. Biol.* **170,** 423 (1983).

[44] J. D. McGhee and G. Felsenfeld, *Cell* **32,** 1205 (1983).

[45] I. L. Cartwright, R. P. Hertzberg, P. B. Dervan, and S. C. R. Elgin, *Proc. Natl. Acad. Sci. U.S.A.* **80,** 3213 (1983).

[46] H. Richard-Foy and G. L. Hager, *EMBO J.* **6,** 2321 (1987).

do not yet know the molecular basis for these "perturbations" and "altered structures."

Another small molecule probe for nucleosome boundaries is the 1, 10-phenanthroline–cuprous complex,[47] which cleaves DNA in a reaction involving hydroxyl radicals. The copper phenanthroline complex has a highly nonrandom DNA sequence preference that is remarkably similar to that of MNase[47,48]; both reagents cleave A + T-rich regions and perhaps subtle deviations from the B form of DNA.[48]

Restriction enzymes cleave nucleosomal DNA weakly and thus they have been used to determine whether nucleosomes are present at specific DNA sites under particular physiological conditions.[49,50] Restriction enzyme assays are useful when the existence of positioned nucleosomes has been established first by other means (e.g., MNase, MPE). Not all nucleosomal sites are strongly inhibited from cleavage and not all restriction enzymes appear to exhibit strong sensitivity to the nucleosomal status of the DNA, thus some degree of empirical testing is required to set up the assay. Furthermore, DNA cleavage by a restriction enzyme may be strongly inhibited by a site-specific binding factor that occupies otherwise nucleosome-free DNA in a nucleus. Nonetheless, restriction enzyme cleavage can be a convenient means to assess DNA availability once the basic chromatin parameters are characterized.

Deoxyribonuclease I has been used in multiple ways as a chromatin probe. First, DNase I has been used to define the general nuclease sensitivity of relatively large genomic domains (several kilobases to several hundred kilobases).[51] General DNase I sensitivity has been taken to imply that the genomic domain being assayed is either in "open" or "closed" chromatin, and often there is a good correlation with the potential to express a gene and its existence in DNase I sensitive chromatin (e.g., see Pikaart *et al.*[52]). The general sensitivity assay typically involves quantitating the disappearance of a full-length genomic restriction fragment on a Southern blot as a function of increasing DNase I concentration or increasing time incubation with enzyme. Data must be normalized to the rate of bulk chromatin digestion or preferably to the rate of digestion of another genomic region probed on the same blot. Care must also be taken to avoid DNase I hypersensitive sites within the region being assayed, as these can lead to

[47] B. Jessee, G. Gargiulo, F. Razvi, and A. Worcel, *Nucleic Acids Res.* **10,** 5823 (1982).

[48] I. L. Cartwright and S. C. R. Elgin, *Nucleic Acids Res.* **10,** 5835 (1982).

[49] J. D. McGhee, W. I. Wood, M. Dolan, J. D. Engel, and G. Felsenfeld, *Cell* **27,** 45 (1981).

[50] A. Schmid, K.-D. Fascher, and W. Hörz, *Cell* **71,** 853 (1992).

[51] W. C. Forrester, E. Epner, C. Driscoll, T. Enver, M. Brice, T. Papayannopoulou, and M. Groudine, *Genes Dev.* **4,** 1637 (1990).

[52] M. Pikaart, J. Feng, and B. Villeponteau, *Mol. Cell. Biol.* **12,** 5785 (1992).

depletion of the main genomic band, despite a relative DNase I resistance of flanking sequences. More recently, a PCR-based assay has been used to map discrete regions of DNase accessibility across a genomic domain.[52]

DNase I has also been used to map nuclease hypersensitive sites and regulatory sequences in chromatin. While not all DNase I hypersensitive sites correspond to regulatory elements, virtually all active regulatory elements in chromatin are hypersensitive to DNase I and endogenous nucleases.[1] Thus, the hypersensitivity assay can be used to expediently localize the position of regulatory sequences within a large chromatin domain. Furthermore, the hypersensitivity assay can be used to measure rapid transitions in local chromatin structure during gene regulation.[53] An unresolved issue in the field is the nature of the chromatin structure that causes DNase I hypersensitivity. The original view was that hypersensitive sites correspond to nucleosome-free regions of DNA.[49] In cases where other nucleosome-sensitive probes yield similar results[49] or where genomic footprinting shows that the DNase I cleavages over the region in chromatin are similar to that seen on free DNA, it is convincing that a DNase I hypersensitive site is nucleosome free. However, as more regulatory regions have been subjected to scrutiny, there are data consistent with the possibility that a DNase I hypersensitive site may be nucleosomal, and where the underlying nucleosome exhibits structural perturbations of the DNA on the particle or a loss of some but not all of the core histone proteins.[10,30,54,55] DNase I hypersensitive sites are mapped by the indirect end-label assay with all of the controls described earlier for MNase.

Finally, DNase I can be used to detect transcription factor footprints in nuclei or permeabilized cells using LM-PCR as described later.[10,11] DNase I intranuclear footprinting works best when a region has already been characterized for DNase I hypersensitivity at the Southern blot level of resolution. Because bands on a Southern blot are due to double-stranded DNase cleavage, and the LM-PCR assay detects single-stranded cleavages, which are far more frequent, the best DNase I-treated chromatin substrate for LM-PCR is that which has been lightly treated with enzyme and which just begins to exhibit a subband over the relevant region in an indirect end-label assay. Essential control samples include DNase I-treated free DNA and DNase I-treated chromatin from cells that lack binding of the transcription factor(s). The assay can be highly diagnostic for DNA site occupancy by a specific transcription factor if the factor has already been characterized

[53] K. S. Zaret and K. R. Yamamoto, *Cell* **258,** 1780 (1984).

[54] G. Fragoso, S. John, M. S. Roberts, and G. L. Hager, *Genes Dev.* **9,** 1933 (1995).

[55] M. Truss, J. Bartsch, A. Schelbert, and M. Beato, *EMBO J.* **14,** 1737 (1995).

to elicit DNase I hypersensitivity, on binding DNA, to specific nucleotides within or flanking the binding site.[12]

Exonuclease III degrades one strand of double-stranded DNA in the 3′ to 5′ direction, and its movement is impeded by nucleosomes and bound transcription factors. A strategy to use exo III as a chromatin probe is to first treat isolated nuclei with a restriction enzyme that is known to cleave DNA in the region of interest; this generates a free 3′ end that can serve as an exo III substrate when the enzyme is subsequently added to the nuclei.[56] Clearly, it is necessary to first demonstrate nearly quantitative access of the restriction enzyme. In such cases, the exo III assay can be a highly sensitive indicator of the frequency with which particular DNA sequences in chromatin are occupied by protein in a nuclear population.[57]

Ligation-Mediated PCR Analysis of DNA Cleavages in Chromatin

The development of the ligation-mediated PCR assay by Mucller and Wold[58] was a major advance for being able to detect DNA cleavages in the chromatin of complex genomes, i.e., *in vivo* footprinting. There have been subsequent papers that provide useful experimental details of the protocol[10, 12] and further optimizations.[9,59] All of these references describe ways to map single-stranded cleavages generated from DMS, DNase I, and other chromatin probes.

For LM-PCR, the nuclear DNA is purified from protein, and for experiments with DMS, the DNA is further treated with piperidine to cleave modified G residues. The first steps of LM-PCR involve denaturing the DNA, annealing a primer to one DNA strand, and extending the primer with T7 DNA polymerase (Sequenase 1.0; United States Biochemical Corp., Cleveland, OH). The primer extends across the desired target region until a DNA cleavage is encountered, at which point the enzyme generates a blunt end. Considering that the population of DNA templates will have, on average, cleavages at different sites along the region being extended, the initial polymerase extension reaction generates a collection of double-stranded target regions with blunt ends containing 5′ phosphates at the original chromatin cleavage sites. An asymmetric double-stranded linker is then ligated to the blunt ends. The linker is asymmetric in that the "top" strand is longer at its 5′ end than the "bottom" strand to which it is annealed;

[56] C. Wu, *Nature* **317,** 84 (1985).
[57] H.-L. Lee and T. K. Archer, *Mol. Cell. Biol.* **14,** 32 (1994).
[58] P. R. Mueller and B. Wold, *Science* **246,** 780 (1989).
[59] J.-P. Quivy and P. B. Becker, *Nucleic Acids Res.* **21,** 2779 (1993).

thus, only the 3′ end of the top strand can be ligated to the 5′ phosphate of the blunt-ended genomic substrate. The ligated molecules are subjected to PCR with one primer consisting of the same sequence as the linker top strand and a second primer that hybridizes to a genomic sequence within the region extended by the first primer. Finally, a third ^{32}P-labeled primer specific to the genomic region is used in a linear amplification reaction, and the reaction products are displayed on a DNA sequencing gel. The resulting bands will migrate according to DNA sizes equal to the distance from the third primer to the original DNA cleavage site, plus the length of the top primer used in ligation and PCR. Thus, experiments that display DNase I cleavages, for example, must include control LM-PCR reactions with DMS-cleaved, protein-free DNA to provide a sequence ladder with which the position of DNase I-generated bands can be assigned unambiguously. The use of several different primers of increasing T_m for the target genomic region provides the specificity required to observe clear signals without artifacts due to nonspecific hybridization.

The main parameters to adjust for LM-PCR are the amount of cleavage of chromatin template, the upper limits of PCR annealing temperatures, and the minimum number of PCR cycles necessary to observe clear DMS cleavage ladders. A DMS cleavage ladder that solely reflects the expected sequence of G's indicates that the priming specificity is suitable for examining unpredictable cleavage patterns such as that from DNase I or MNase. For the amount of chromatin cleavage, as described earlier, the best signals with LM-PCR are achieved when relatively low amounts of template cleavage are used, in comparison with that required to observe hypersensitive site subbands on Southern blots. The upper limits for primer annealing temperatures must be defined empirically in pilot PCR reactions with plasmid and genomic DNA. Try varying the magnesium concentration in 0.5 m*M* increments, from 1 to 2.5 m*M* at each temperature tested, and choose an annealing temperature that gives the maximal PCR product about 5° below the temperature at which priming is abolished. Excess PCR cycles can result in the appearance of a "free DNA" pattern and a loss of footprints; try working out conditions so that only 20–22 cycles are necessary with 1 μg of input genomic DNA. Performing footprint reactions with purified proteins *in vitro* and comparing the specific protection patterns with that observed *in vivo* provides compelling evidence for having identified a specific protein–DNA interaction in native chromatin.[60]

[60] R. Gualdi, P. Bossard, M. Zheng, Y. Hamada, J. R. Coleman, and K. S. Zaret, *Genes Dev.* **10,** 1670 (1996).

Mapping Nucleosome Boundaries with MNase at Nucleotide Level of Resolution

Mapping MNase cleavages in chromatin at the nucleotide level of resolution allows the apparent positions of nucleosome boundaries to be compared with genetic landmarks such as the positions of regulatory protein-binding sites. For some genes, this sort of analysis has led to the surprising conclusion that certain transcription factors occupy their binding sites on nucleosome-like particles in chromatin, not in the linker DNA between particles.[10,54,55] A problem with mapping MNase cleavages by conventional LM-PCR is that the initial Sequenase primer extension step uses single-stranded DNA as a template and, as described earlier, MNase can make single-stranded cleavages on nucleosomes.[43] A way to avoid this problem, and thereby selectively map double-stranded MNase cleavages that occur in linker DNA, is to eliminate the initial primer extension step of LM-PCR and directly ligate the asymmetric linker to the MNase-cleaved, double-stranded DNA.[10] Because MNase cleavage leaves a 5′-hydroxyl, it is necessary to first phosphorylate the cleaved, double-stranded substrate so that the top linker strand can be ligated to it. As with DNase I, chromatin that has been lightly cleaved with MNase, so that the nucleosome ladder just becomes visible in an indirect end-label Southern blot, makes the best substrate for this modified LM-PCR protocol. Also, a DMS cleavage pattern with genomic DNA, run on the same gel as the final MNase LM-PCR products, is necessary to map MNase cleavages along the sequence. The following protocol modification is for mapping MNase cleavages in mammalian chromatin by LM-PCR.

MNase cleaved nuclear chromatin should be extracted once with phenol (adjusted to pH 8.0) and three to five times with chloroform. After ethanol precipitation, spooling out the DNA, and rinsing the DNA with 70% ethanol, the DNA is resuspended in a solution of 10 m*M* Tris–HCl (pH 7.8), 1 m*M* EDTA and is suitable for LM-PCR. The DNA is phosphorylated by combining the following in a microfuge tube: 5 μg of MNase-cleaved genomic DNA, 5 μl of 10× kinase buffer supplied by the manufacturer (e.g., New England Biolabs), 0.5 μl of 10 m*M* ATP, distilled H_2O to 50 μl final volume, and 1 μl T4 polynucleotide kinase. The mixture is incubated at 37° for 1 hr, then 1 μl of 0.5 *M* EDTA (pH 8) is added and the mixture is incubated at 68° for 10 min. DNA is precipitated with sodium acetate and ethanol, rinsed with 70% ethanol, and resuspended into distilled H_2O at 0.1–0.2 μg/μl and stored at −20° until use. The linker ligation step and subsequent PCR procedures are those used for conventional LM-PCR; see Bossard *et al.*[9] and the other references cited earlier[10,12,58,59] for the details. The only other variation for MNase is to use a few extra cycles in the PCR reaction (e.g., 22 cycles instead of 20). Clusters of MNase cleavages of linker DNA at the nucleotide level of resolution should occur in the same genomic segments as cleavages observed in nucleosome ladders on Southern blots by indirect end-label analysis.[10]

Acknowledgments

Clifton McPherson and Pascale Bossard contributed to technical developments described in this article. Chromatin work in the author's laboratory is supported by a grant from the National Institutes of Health (GM47903).

[35] Analysis of Nucleosome Positioning in Mammalian Cells

By GORDON L. HAGER and GILBERTO FRAGOSO

Introduction

Remodeling of chromatin structure is now recognized as a major component of gene regulation.[1,3] Not all transcription factors can interact with DNA sequences present in replicated nucleoprotein templates; reorganization of the native chromatin structure is a first step in activating many genes. The nature of these remodeling processes is now the subject of intensive investigation. Nucleosome position is one potential determinant in regulating factor interaction with chromatin, and there are several examples of regulatory regions with nonrandomly positioned nucleosomes.[4–8] It is therefore important to accurately determine intrinsic phasing of nucleosomes *in vivo* and to evaluate the extent to which these positions are duplicated during reassembly of chromatin *in vitro*. This article describes current methodology to map nucleosome positions, both *in vivo* and *in vitro*.

[1] C. L. Smith and G. L. Hager, *J. Biol. Chem.* **272,** 27493 (1997).
[2] P. A. Wade and A. P. Wolffe, *Curr. Biol.* **7,** R82 (1997).
[3] C. L. Peterson, *Curr. Opin. Genet. Dev.* **6,** 171 (1996).
[4] G. Fragoso and G. L. Hager, *Methods* **11,** 246 (1997).
[5] G. Fragoso, S. John, M. S. Roberts, and G. L. Hager, *Genes Dev.* **9,** 1933 (1995).
[6] M. S. Roberts, G. Fragoso, and G. L. Hager, *Biochemistry* **34,** 12470 (1995).
[7] H. Richard-Foy and G. L. Hager, *EMBO J.* **6,** 2321 (1987).
[8] J. Svaren and W. Horz, *Curr. Opin. Genet. Dev.* **3,** 219 (1993).

0076-6879/99 $30.00

Mapping Nucleosome Positions *in Vitro*

The preparation of histones and DNA fragments suitable for reconstitution experiments will not be covered here. We will first briefly mention various reconstitution procedures that are commonly utilized to prepare defined nucleosomal substrates.[9–11] These can be subdivided into histone transfer and high-salt histone-binding procedures. For the transfer protocols, a histone carrier such as nucleoplasmin or a polyglutamic acid–histone complex (or RNA–histone) is incubated with DNA fragments under physiological salt conditions. An advantage of using histone carriers is that additional components can be included during the reconstitutions in low salt ($\mu \approx 0.15$) if so desired. One disadvantage, however, of using polyglutamic acid is that if intermediate histone–DNA ratios are utilized in reconstitutions of long templates, cooperative assembly of nucleosomes results in subpopulations of molecules differing in the histone–DNA ratio.[12]

The second set of reconstitution procedures involves the direct exposure of histones to the DNA fragment at sufficiently high concentrations of salt so that the mixture does not precipitate, followed by dialysis or stepwise dilution to physiological or subphysiological salt concentrations.[9,10] Such a protocol can be performed in either the presence or the absence of urea.[13] For specific experiments the high to low salt reconstitution approach may be desirable, as the resulting reconstitute does not have to be purified away from histone carriers; such components could affect the nucleosomal status of the reconstitute when, for instance, challenged by transcription factors.

Which procedure is used to prepare nucleosomal reconstitutes depends on the specific aims of the experiment. In a case where the position of nucleosomes reconstituted by transfer from polyglutamic acid was compared against that obtained by a salt dialysis procedure, there were no significant differences in the rotational positioning.[14] It has been reported that the number of positions obtained by salt dialysis can be minimized if the reconstitution temperature is kept at 0°.[15] This might be an important parameter to control if population effects are to be minimized *in vitro*.

[9] D. Rhodes and R. A. Laskey, *Methods Enzymol.* **170,** 575 (1989).
[10] A. Stein, *Methods Enzymol.* **170,** 585 (1989).
[11] L. Sealy, R. R. Burgess, M. Cotten, and R. Chalkley, *Methods Enzymol.* **170,** 612 (1989).
[12] A. Stein, J. P. Whitlock, Jr., and M. Bina, *Proc. Natl. Acad. Sci. U.S.A.* **76,** 5000 (1979).
[13] L. Kleiman and R. C. Huang, *J. Mol. Biol.* **64,** 1 (1972).
[14] J. R. Jackson and C. Benyajati, *Nucleic Acids Res.* **21,** 957 (1993).
[15] A. Flaus and T. J. Richmond, *J. Mol. Biol.* **275,** 427 (1998).

Rotational Positions and Dyads Mapped with Endonucleolytic Reagents

Because the DNase I cutting sites of DNA on a surface are maximally exposed at multiples of 10 bases,[16–18] the presence of a 10 base ladder in a digest of an end-labeled reconstitute is indicative of rotational positioning.[19] Initial test digests with DNase I (Worthington, DPRF) can be performed in 2 m*M* magnesium acetate, 0.1 m*M* $CaCl_2$, 100 μg/ml bovine serum albumin (BSA), 10 m*M* Tris–Cl, pH 7.5, with DNase I at concentrations ranging from 0.1 to 10 units/ml. The digestion cocktail is prepared as a 2× concentrate and mixed with the reconstitute by pipetting. Reactions are conducted for 1–5 min at room temperature and stopped by the addition of EDTA to a final concentration of 10 m*M*. If multiple nucleosomal species are present in the reconstituted population, the digest can be resolved by electrophoresis at 4° in native (0.5× Tris–borate–EDTA) acrylamide gels prior to extraction of the DNA.

Additional factors determining the intensity of cleavage by DNase I are the sequence of the DNA substrate and the position of the cleavage site relative to the end of the core particle.[16–20] More even cutting can be obtained by using the hydroxyl radical[21]; the main factor modulating cleavage is the orientation of the DNA on the surface of the histone octamer. General considerations for the use of this reagent are described elsewhere.[22] Iron(II)–EDTA reagent is prepared immediately prior to use by mixing equal volumes of 600 μM ferrous ammonium sulfate and 1.20 m*M* EDTA. The reagent, 1 μl of Fe(II)–EDTA, together with 1 μl of 10 m*M* sodium ascorbate, 1 μl of 0.3% H_2O_2, and 2 μl of H_2O, is applied to the walls of a microcentrifuge tube containing 5 μl of end-labeled reconstitute (in 10 m*M* HEPES or Tris–Cl, pH 7.4) and is spun to combine and mix. The reaction is allowed to proceed for 1–3 min at room temperature and is then quenched by the addition of 1/10 volume 0.1 *M* thiourea in 0.2 *M* EDTA. The DNA can be extracted at this point or, as described earlier, the digested reconstitute can be resolved by native gel electrophoresis. The extracted DNA is then electrophoresed in denaturing sequencing gels.

Because the pitch of the DNA changes in the vicinity of the dyad,[23,24]

[16] M. Noll, *J. Mol. Biol.* **116,** 49 (1977).

[17] L. C. Lutter, *J. Mol. Biol.* **117,** 53 (1977).

[18] D. Rhodes and A. Klug, *Nature* **286,** 573 (1980).

[19] R. T. Simpson and D. W. Stafford, *Proc. Natl. Acad. Sci. U.S.A.* **80,** 51 (1983).

[20] L. C. Lutter, *J. Mol. Biol.* **124,** 391 (1978).

[21] J. J. Hayes, D. J. Clark, and A. P. Wolffe, *Proc. Natl. Acad. Sci. U.S.A.* **88,** 6829 (1991).

[22] W. J. Dixon, J. J. Hayes, J. R. Levin, M. F. Weidner, B. A. Dombroski, and T. D. Tullius, *Methods Enzymol.* **208,** 380 (1991).

[23] J. J. Hayes, T. D. Tullius, and A. P. Wolffe, *Proc. Natl. Acad. Sci. U.S.A.* **87,** 7405 (1990).

[24] J. J. Hayes, J. Bashkin, T. D. Tullius, and A. P. Wolffe, *Biochemistry* **30,** 8434 (1991).

this procedure can be used to map very precisely the translational position of a nucleosome in addition to determining its rotational phase. The position of the dyad can be assigned to the center point of the pitch transition; the end points of the core particle are then taken as the position of the dyad ± 73 bases.

Core Particle Ends Generated by Micrococcal Nuclease

In case multiple translational frames are present in an unfractionated reconstitute, this approach yields the number of frames in addition to their translational location. Because of sequence preferences in the cleavage by micrococcal nuclease (MNase), the translational position of the core particle can be determined with perhaps an accuracy of ±3 bases. Reconstituted material is digested with MNase with the aim of generating core particles (containing ~147 bp of DNA). The reconstitute is mixed with an equal volume of 100 m*M* Tris–Cl, pH 7.4–7.5, 50 m*M* KCl, 8 m*M* magnesium acetate, 2 m*M* $CaCl_2$ (and protease inhibitors if necessary)[25] and digested with MNase for 5–10 min at 37°. Test reactions can be performed with MNase at 1–50 U/ml. The extent of digestion can be monitored by terminating the reaction with EGTA to 10 m*M*, treating aliquots with sarkosyl (0.1% final) and proteinase K (100 μg/ml) for 10 min at 37°, electrophoresis in a high concentration agarose (2–2.5%) or a low concentration acrylamide (4–6%) gel in Tris–borate–EDTA, and visualizing the resolved material by UV transillumination after staining with ethidium bromide. DNA is then extracted from the appropriate digests (it may be necessary to gel purify the 147-bp band) and used for the mapping.

The ends of the core particle are mapped relative to positions internal to the putative location of the reconstituted nucleosome: either restriction sites[26,27] or a priming site to be used in a primer extension assay.[28] The DNA is end labeled, restricted, electrophoresed in 6% acrylamide gels, and visualized by autoradiography. In case multiple translational positions are represented, pairs of fragments are matched according to the sum of their sizes, which equal ~147 bases. If the ends are mapped by primer extension, it is best to end label the oligonucleotide and perform the DNA synthesis with cold deoxynucleotides. In addition, because DNA polymerase will stop at the sites of UV-induced DNA lesions,[29] the core DNA must be

[25] C. von Holt, W. F. Brandt, H. J. Greyling, G. G. Lindsey, J. D. Retief, J. D. Rodrigues, S. Schwager, and B. T. Sewell, *Methods Enzymol.* **170,** 431 (1989).
[26] F. Dong, J. C. Hansen, and K. E. van Holde, *Proc. Natl. Acad. Sci. U.S.A.* **87,** 5724 (1990).
[27] J. C. Hansen, J. Ausio, V. H. Stanik, and K. E. van Holde, *Biochemistry* **28,** 9129 (1989).
[28] M. J. Fedor, N. F. Lue, and R. D. Kornberg, *J. Mol. Biol.* **204,** 109 (1988).
[29] J. D. Axelrod and J. Majors, *Nucleic Acids Res.* **17,** 171 (1989).

band purified from lanes that have not been irradiated. The primer extension products are resolved in 6–8% sequencing acrylamide gels.

Another procedure has been devised for mapping the positions of multiple nucleosomes reconstituted on long DNA fragments.[30] This procedure is particularly well suited for the analysis of nucleosomal arrays. After digestion with MNase and isolation of the ~147-bp core particle DNA, the DNA is end labeled at the 5′ end and hybridized to single-stranded plasmid DNA containing the entire region of interest. The labeled monomer DNA is then extended with the Klenow fragment of DNA polymerase in the presence of a restriction enzyme that cuts downstream of the region analyzed. The products are electrophoresed in sequencing gels and visualized by autoradiography. The extension products indicate the position of nucleosomes reconstituted throughout a fragment of up to 1 kb long. The reader is referred to the original publication for the specific conditions of the monomer extension assay.

3′ Core Particle Ends Detected with Exonuclease III

Exonuclease III (Exo III) will digest one strand of free DNA in a 3′ → 5′ direction until the substrate is no longer double stranded. When DNA is complexed with a protein, Exo III will pause or stop at the position where the protein binds the DNA. Although this enzyme has been used to map nucleosome positions, its use should be considered carefully as it penetrates nucleosomes and pauses at 10 base intervals.[31] In an examination of gel-purified mouse mammary tumor virus (MMTV) nucleosome B reconstitutes, Exo III consistently produced a series of bands spaced 10 bases apart.[6] Importantly, the major bands were not at the edge of the core, but 10 bases inside. Furthermore, the major fragment did not necessarily correspond to the correct nucleosome position; it is clear that there are local sequence effects on the frequency of Exo III pausing at a given position. This technique must therefore be utilized with great caution.

Nucleosomes are reconstituted with 5′-end-labeled DNA. In case multiple species are obtained because of translational heterogeneity, the reconstitute is first resolved by electrophoresis in native gels. Gel slices are equilibrated in 10 m*M* Tris–Cl, pH 8.0, 3 m*M* $MgCl_2$, 5 m*M* 2-mercaptoethanol (2-ME) (15 min at room temperature), and treated with Exo III at 10–100 U/ml for 30 min. The digested DNA is eluted, extracted, and electrophoresed in sequencing gels.

[30] A. Yenidunya, C. Davey, D. Clark, G. Felsenfeld, and J. Allan, *J. Mol. Biol.* **237,** 401 (1994).
[31] A. Prunell and R. D. Kornberg, *Phil. Trans. R. Soc. Lond. B Biol. Sci.* **283,** 269 (1978).

Mapping Nucleosome Dyads with Histone H4 Cleavage Reagent

A cleavage procedure has been developed utilizing a histone H4 with an engineered Ser-47 → Cys.[32] This Cys-47 is derivatized with cysteaminyl-EDTA, which on binding Fe^{2+} and exposure to H_2O_2 will generate hydroxyl radical in its vicinity. The dyads of nucleosomes reconstituted with this reagent can thus be mapped very precisely; the main hydroxyl radical cutting sites along one DNA strand occur three to four bases upstream and six to seven bases downstream of the dyad. From the symmetrical cuts on the opposite strands the dyad can be assigned at base pair resolution. The reader is referred to the original publication for specific protocols.

This procedure can unambigously determine the presence of multiple translational frames in crude reconstitutes prepared with the modified H4 and allows the rotational orientation of the nucleosomal DNA in each frame to be inferred from the position of the dyad. Mapping MMTV nucleosome B reconstitutes with this technique[15] confirmed the position of the dyads of the predominant nucleosomes determined previously with Fe(II)–EDTA.[6] However, the H4 cleavage reagent failed to uncover alternative rotational orientations that were detected and mapped in that study.[6] Although it might seem unlikely, it is not yet clear whether the reactive moiety influenced the position adopted by the octamer during reconstitution. Complicating the comparison is the fact that the temperature during the salt dialysis step of the reconstitutions was different.[6,15] At least initially during the course of a study, it might be prudent to map reconstitutes by additional techniques.

Mapping Nucleosome Positions *in Vivo*

Low-Resolution Mapping by Indirect End-Labeling Analysis

An assessment of whether nucleosomes are positioned in a given genetic region should be attempted *in vivo*. This can be accomplished at low resolution by indirect end-labeling analysis of Southern blots of DNA obtained from MNase digestions of nuclei. However, this is technically difficult in low copy number systems from high complexity genomes because of the low intensities of the hybridization signals. The amount of DNA digested, electrophoresed, and probed must be high enough to yield a reasonable signal spread throughout an electrophoretic lane without overly influencing the mobility of the DNA fragments. In addition, the hybridization conditions must yield very low nonspecific hybridization and membrane back-

[32] A. Flaus, K. Luger, S. Tan, and T. J. Richmond, *Proc. Natl. Acad. Sci. U.S.A.* **93,** 1370 (1996).

grounds. Furthermore, the investigator must be cautious of highly repeated genomic sequences found interspersed throughout the genome, e.g., Alu elements in humans. These sequences may contain similarities to a specific probe and could produce false hybridization signals. The following set of protocols have been utilized successfully to map *in vivo* nucleosome positions by indirect end labeling. We are most experienced in analyses of mammalian cell lines carrying elements ranging in copy number from about 10 to 1000. We will mention the steps that merit additional caution to enable analysis of single copy sequences in mammalian cells (about 3 pg DNA per haploid genome).

Isolation of Nuclei. We have modified a nuclear isolation procedure described by others[33] to make nuclei for transcription experiments, restriction enzyme access assays, and nucleosome positioning studies with native and formaldehyde-fixed cells.[4,5,34,35] Growing adherent cells are rinsed twice with cold (4°) phosphate-buffered saline (PBS) lacking calcium and magnesium, harvested by scraping into cold PBS, and centrifuging for 3–5 min at 1000*g*. Cells growing in suspension are washed by centrifugation prior to homogenization. The cell pellet is resuspended and disrupted in a Dounce homogenizer with an "A" pestle in 5–10 ml of 0.3 *M* sucrose, 2 m*M* magnesium acetate, 3 m*M* $CaCl_2$, 0.1–1% Triton X-100, 0.5 m*M* dithiothreitol (DTT), 0.5 m*M* phenylmethylsulfonyl fluoride (PMSF), and 10 m*M* HEPES, pH 7.8. Homogenates are diluted 1 : 1 with 5–10 ml of 25% glycerol, 5 m*M* magnesium acetate, 0.1 m*M* EDTA, 5 m*M* DTT, and 10 m*M* HEPES, pH 7.8 (pad buffer, PB), centrifuged at 1000*g* for 15 min in a tabletop centrifuge through a 10–20 ml layer of PB, and resuspended in PB containing 0.5 m*M* DTT (nuclei storage buffer, NSB), using approximately 150–200 μl per 150-mm culture dish. An aliquot is taken to measure the nucleic acid concentration; if the preparation is not used within 15 min, it is aliquoted, frozen with liquid nitrogen, and stored in liquid nitrogen.

The OD_{260} is measured after disrupting the nuclei with 1% SDS and diluting 20-fold with 1 *M* NaCl in 10 m*M* Tris–Cl, pH 8, 0.1 m*M* EDTA. Note that the bulk of the OD_{260} is from RNA; if the MNase digestions need to be standardized using DNA instead of the total nucleic acid content, DNA can be determined by the fluorescence of Hoechst 33258 measured in 2 *M* NaCl[36] or with diphenylamine by the method of Burton.[37] Alternatively, nuclei can be counted in a hemocytometer. Some cell lines require more

[33] W. F. Marzluff, Jr., E. C. Murphy, Jr., and R. C. Huang, *Biochemistry* **12,** 3440 (1973).
[34] W. D. Pennie, G. L. Hager, and C. L. Smith, *Mol. Cell. Biol.* **15,** 2125 (1995).
[35] G. Fragoso, W. D. Pennie, S. John, and G. L. Hager, *Mol. Cell. Biol.* **18,** 3633 (1998).
[36] C. Labarca and K. Paigen, *Anal. Biochem.* **102,** 344 (1980).
[37] K. Burton, *Biochem. J.* **62,** 315 (1956).

Triton than others to obtain nuclei free of cytoplasmic tags; we have not required concentrations higher than 1% with our current cell lines. Although initially we included spermine–spermidine in our solutions,[7] we currently avoid their use.[38,39]

Digestion of Nuclei with MNase. Components are assembled and mixed at 0°. Nuclei are diluted to 500–600 μg/ml into 100–200 μl of buffer containing 1 m*M* $CaCl_2$ and are digested with MNase (Worthington) at 0, 10, 20, and 40 U/ml for 5 min at 25° without a temperature preequilibration step. Various buffering conditions have been utilized without a noticeable difference in the results. The buffer of Hewish and Burgoyne,[40] 10 m*M* Tris–Cl, pH 7.5, 15 m*M* NaCl, 60 m*M* KCl, 5 m*M* $MgCl_2$, 0.1 m*M* EDTA, and 1 m*M* DTT, has been used with and without spermine–spermidine; other buffers have been tested as well. The concentration of MNase is decreased for digestions containing lower concentrations of nuclei or naked DNA controls. Control reactions are performed either with genomic DNA or with an appropriate restriction enzyme-linearized plasmid in the presence of nonspecific carrier DNA. The reaction is terminated with 1/10 volume of 0.11 *M* EGTA, 0.11 *M* EDTA, 5.5% sarkosyl, 1.1 mg/ml proteinase K and incubation for 1 hr at 37°. The mixture is diluted to about 600 μl with H_2O, the DNA is then extracted with 1 volume of 25 : 24 : 1 phenol : chloroform : isoamyl alcohol (PCI), and recovered by precipitation with 1 volume of isopropanol after adjusting the concentration of NaCl to 0.2 *M*. After washing the precipitate with 80% (v/v) ethanol and drying lightly, the DNA is dissolved in TE containing 10 m*M* EGTA and 0.1% sarkosyl and treated with RNase T1 at 1000 U/ml for 1 hr at 37°, aliquots are set aside for analysis of the extent of cleavage and comparison against restricted samples (later). The DNA is reextracted with PCI, adjusted to 0.2 *M* NaCl, precipitated with 1 volume of 2-propanol, washed and dried as before, and dissolved in TE containing 10 m*M* EGTA.

Aliquots containing 5 μg (high copy number elements) to 50 μg (low copy number) of DNA are digested with the appropriate restriction enzyme for the subsequent indirect end-labeling analysis in the buffer suggested by the manufacturer (but also including EGTA in the buffer). Restrictions are conducted at enzyme/DNA ratios of 2 to 5 units/μg DNA, for 2 to 16 hr. Digests containing large amounts of DNA can be concentrated by precipitating with alcohol at this point. Analysis of single copy elements may require digesting and electrophoresing up to 80–100 μg of DNA[41] to

[38] G. P. Pfeifer and A. D. Riggs, *Genes Dev.* **5,** 1102 (1991).
[39] P. Colson and C. Houssier, *FEBS Lett.* **257,** 141 (1989).
[40] D. R. Hewish and L. A. Burgoyne, *Biochem. Biophys. Res. Commun.* **52,** 504 (1973).
[41] Y. L. Sun, Y. Z. Xu, M. Bellard, and P. Chambon, *EMBO J.* **5,** 293 (1986).

increase the hybridization signal. The MNase digests ± restriction can be examined at this step by electrophoresing 0.5- to 1-μg aliquots in 1–1.5% agarose minigels. The best samples for analysis display a nucleosomal ladder but still contain a large proportion of the DNA in a high molecular weight form. The completeness of the restriction is harder to judge, although a difference in the appearance of the samples should be observable. The investigator may wish to include phage λ DNA in some samples prior to digestion.

Electrophoresis, Southern Blotting, and Hybridization. Electrophoresis is conducted in horizontal 1.1–1.5% agarose gels, 20–25 cm long and up to 6 mm thick (dictated by the amount of DNA), in the buffer of choice. Southern blotting[42] is conducted after a 0.25 *N* HCl treatment of the gel to increase transfer of the larger DNA species.[43] We use nylon membranes for the transfers (Hybond-N, Amersham). Blots using positively charged membranes have produced variable backgrounds; however, extensive comparisons have not been performed. After transfer, the orientation of the gel on the membrane is marked, the membrane is rinsed in 2× SSC to remove gel particles, and DNA is cross-linked by UV irradiation. After irradiation, the membrane is rinsed thoroughly with H_2O; if stored and allowed to dry, wet with H_2O before prehybridizing. Prehybridizations are performed overnight (>12 hr) with 50–100 ml of 3× SSC, 20 m*M* sodium phosphate, pH 6.8, 0.02% PVP 360, 0.02% Ficoll 400, 0.1% SDS, and 100–500 μg/ml carrier RNA or DNA at 60–65°. Stock solutions are filtered and BSA is omitted (from the Denhardt's); carrier nucleic acid at 1 mg/ml is used in the analysis of low copy number cell lines. Probes are up to 200 bp long and end at the selected restriction site. The probes are prepared from PCR or restriction fragments by random priming,[44] boiled for 2–5 min after removing the unincorporated nucleotide by sephadex G-25 spin column chromatography, and utilized at a concentration no higher than 10^7 cpm/ml. Hybridization volumes of 10–20 ml are utilized for a typical 10–15 × 20-cm^2 membrane. We also deareate the hybridization solution by boiling 5–10 min in an attempt to decrease the number of bubbles formed on the membrane during prolonged high temperature incubations. Avoid introducing folds in the membrane during the prehybridization and hybridization steps. After a 14- to 18-hr hybridization, the probe is removed, and the membrane is rinsed with 1× SSC, 0.1% SDS. It is then washed twice with 0.5× SSC, 10 m*M* sodium phosphate, 0.1% SDS, 30 min each

[42] E. M. Southern, *J. Mol. Biol.* **98,** 503 (1975).
[43] G. M. Wahl, M. Stern, and G. R. Stark, *Proc. Natl. Acad. Sci. U.S.A.* **76,** 3683 (1979).
[44] A. P. Feinberg and B. Vogelstein, *Anal. Biochem.* **132,** 6 (1983).

time, and twice with 0.1× SSC, 10 m*M* phosphate, 0.1% SDS, for 15–30 min each time, rocking every so often. The washes are conducted at 60–65°.

These rather stringent hybridization and washing conditions affect the membrane background, nonspecific hybridization, and the specific hybridization signal. The signal can be increased by relaxing the hybridization conditions with an increased salt concentration (5× SSC). If longer hybridization times are desired, lower the incubation temperature (40–45°, typically 42°) and include 50% formamide to maintain the desired stringency. Finally, 10% dextran sulfate can be included during the hybridization step (works only with double-stranded probes),[43] and the stringency of the last wash decreased (increase the salt to 0.2× SSC instead of decreasing the temperature). Denhardt's (0.02% PVP, 0.02% Ficoll, and 0.02% BSA)[45] is used to maintain low backgrounds; it is normally included at a 5× concentration. Large prehybridization volumes are recommended; if backgrounds are still too high, increase the hybridization volume 50–100% and decrease the probe concentration to $1–2 \times 10^6$ cpm/ml. An additional discussion of membranes and hybridization conditions can be found elsewhere.[46]

It is very important that the specificity of the hybridization be tested with MNase-untreated restricted material. Another indication that the conditions may be too relaxed is if the hybridization pattern looks like the bulk nucleosomal ladder, i.e., similar to the ethidium bromide-stained pattern. No difference in the results before and after restriction could result from hybridization to repeated elements. Mapping should be conducted from both the upstream and the downstream directions, preferably with multiple probes at different distances from the target region.

Alternatives and Improvements. The use of permeabilized cells instead of isolated nuclei simplifies experiments by eliminating a preparative step and greatly facilitates specific studies where multiple samples need to be analyzed. Also, potential artifacts introduced during nuclei isolation are minimized or eliminated altogether.[38] However, one preparation of nuclei can be examined for aspects other than nucleosome positioning. At some point during the course of a study, perhaps after successful blot hybridization conditions have been established, an analysis of cleavage by MPE · Fe(II) should be conducted.[47] This reagent exhibits less cutting specificity than MNase and thus provides a more accurate picture of low resolution nucleosome positioning; it is now available commercially (Sigma).

[45] D. T. Denhardt, *Biochem. Biophys. Res. Commun.* **23,** 641 (1966).

[46] G. M. Church and W. Gilbert, *Proc. Natl. Acad. Sci. U.S.A.* **81,** 1991 (1984).

[47] I. L. Cartwright, R. P. Hertzberg, P. B. Dervan, and S. C. Elgin, *Proc. Natl. Acad. Sci. U.S.A.* **80,** 3213 (1983).

A mapping strategy utilizing psoralen cross-linking has been introduced.[48] Although it is too early to review adequately, we believe it offers definite advantages over the classical procedure involving cleavage followed by indirect end-labeling analysis. It is based on the fact that psoralen does not bind nucleosomal DNA as well as linker DNA, thus linker DNA will contain the highest density of cross-links. Following extraction, the DNA is restricted, treated with a 5′-exonuclease, and subjected to multiple cycles of primer extension with a thermostable DNA polymerase, which does not progress beyond the position of the cross-link(s). The resulting DNA is electrophoresed in alkali gels, blotted, and analyzed by indirect end labeling with a complementary single-stranded probe. The psoralen binding and photoactivation steps are performed at the level of the living cell. However, we consider the linear amplification step the most significant improvement: (1) it increases the mass of the DNA eventually probed by indirect end labeling and (2) it reduces the amount of total DNA electrophoresed and Southern blotted. This contributes substantially to the decrease in nonspecific hybridization signals. Consequently, this procedure can be expected to facilitate the analysis of single copy elements in mammals, and perhaps even in organisms with a higher genomic complexity.

High Resolution Mapping

Three methodologies can be utilized to examine chromatin structure at high resolution *in vivo*: DNase I footprinting analysis, core-linker boundary determinations, and double-stranded MNase footprinting. These methods all incorporate a DNA amplification step and examine the location of nuclease cuts dependent on specific chromatin features at base pair resolution.

DNase I Footprinting. Ligation-mediated polymerase chain reaction (LM-PCR) was introduced to enable the footprinting analysis of low copy elements in high complexity genomes.[49,50] Such analyses have been performed with a variety of agents, including UV light, DMS, $KMnO_4$, DNase I, and MNase, and a number of reviews have been published.[51–58] With

[48] R. E. Wellinger and J. M. Sogo, *Nucleic Acids Res.* **26,** 1544 (1998).
[49] G. P. Pfeifer, S. D. Steigerwald, P. R. Mueller, B. Wold, and A. D. Riggs, *Science* **246,** 810 (1989).
[50] P. R. Mueller and B. Wold, *Science* **246,** 780 (1989).
[51] P. R. Mueller and B. Wold, *Methods* **2,** 20 (1991).
[52] M. Hershkovitz and A. D. Riggs, *Methods* **11,** 253 (1997).
[53] A. Dey and K. Ozato, *Methods* **11,** 197 (1997).
[54] G. P. Pfeifer and S. Tornaletti, *Methods* **11,** 189 (1997).
[55] P. Bossard, C. E. McPherson, and K. S. Zaret, *Methods* **11,** 180 (1997).
[56] J. P. Quivy and P. B. Becker, *Methods* **11,** 171 (1997).

regard to chromatin structure, the observation of a 10 base DNase I cleavage periodicity over an extended stretch can be diagnostic of a rotationally phased nucleosome(s) in a particular chromatin region. Interpretation of 10 base ladders is difficult, however (for a discussion, see Fragoso *et al.*[5,35] and Roberts *et al.*[6]), and caution must be exercised in drawing simple conclusions from these patterns.

We have performed DNase I digestions of nuclei under a variety of conditions without appreciable differences in the results (in the case of transcription factor footprinting this is likely incorrect). In addition to buffer (10–50 m*M* Tris–Cl or HEPES, pH 7.5–8.0), salt (0–60 m*M* KCl or <40 m*M* ammonium sulfate), glycerol (0–7%), and detergent (0–1% Triton X-100), we include Mg^{2+} at 2 m*M* and $CaCl_2$ at 0.1 m*M*. Digestions are performed at 25–37° for 2–6 min with DNase I at 0.5–40 U/ml and nuclei at 100 OD_{260}/ml, terminated with SDS (final 1%), EDTA (20 m*M*), and proteinase K (100–500 μg/ml), and the DNA extracted.

The entire set of reactions is processed and analyzed, and then suitable naked DNA and chromatin samples are selected for side-by-side comparisons. Samples are selected according to the extent of digestion, with samples displaying 50–70% of the signal at distances >200 bases from the oligonucleotide primer preferred over more digested samples. We have utilized linear amplification in studies of high copy number elements; the reader is referred to other articles for details on LM-PCR for analysis of low copy number targets. Because of the increased complexity of *in vivo* systems, such as the potential presence of multiple translational positions and template-bound nonhistone proteins, the cleavage patterns could be difficult to interpret. A restriction enzyme treatment step, introduced to provide a primer extension end point, would facilitate comparisons and quantitative analysis.[59]

Core-Linker Boundaries. This is the only method yet to be performed with LM-PCR. As described earlier *in vitro,* this approach involves the isolation of mononucleosomal material, followed by the determination of core-linker boundaries, i.e., the ends of the DNA protected by the histones in the nucleosome core particle. It was designed explicitly to detect the presence of heterogeneity in nucleosome positions,[28] and its utility has been demonstrated with both native and cross-linked chromatin. The method

[57] E. C. Strauss and S. H. Orkin, *Methods* **11,** 164 (1997).

[58] T. Grange, E. Bertrand, M. L. Espinas, M. Fromont-Racine, G. Rigaud, J. Roux, and R. Pictet, *Methods* **11,** 151 (1997).

[59] M. Brenowitz, D. F. Senear, M. A. Shea, and G. K. Ackers, *Methods Enzymol.* **130,** 132 (1986).

and its strengths and weaknesses have been reviewed previously[4] and will not be covered here.

A recent article[15] prompts us to comment that this procedure could be conducted with chromatosomal DNA as well as with core particle material. However, no information on the status of H1 in the undigested target nucleosome can be derived from an analysis of core DNA obtained from fixed cells. Nucleosomal material containing 167 bp of contiguous DNA (chromatosomes in unfixed samples) is an intermediate in the MNase digestion and it is not an end product, even when utilizing fixed chromatin substrates (e.g., see Fragoso *et al.*[5]).

Double-Stranded MNase Footprinting. To select against intranucleosomal cuts, the LM-PCR technique was modified to detect only flush ends generated during MNase digestion.[60] The first primer extension step of LM-PCR was eliminated and double-stranded linkers ligated directly to the MNase blunt ends. This technique therefore allows the detection of a subset of the generated cuts that are present predominantly in the internucleosomal linker DNA.

This technique can provide information similar to that from a core boundary assay, but in a shorter period of time. Although it might be expected that assessments of nucleosome frame occupancy (in the case of heterogeneity in nucleosome positions) would differ between both techniques, insufficient data are available to compare the results derived from each.[61–63] However, analyses of the MMTV promoter have been performed by different groups, and the blunt MNase ends detected by LM-PCR in the Nuc-A-Nuc-B region[62] matched very closely in both position and number the 5′ ends of the Nuc-A nucleosomes detected in core-linker boundary assays.[5]

[60] C. E. McPherson, E. Y. Shim, D. S. Friedman, and K. S. Zaret, *Cell* **75,** 387 (1993).
[61] G. F. Sewack and U. Hansen, *J. Biol. Chem.* **272,** 31118 (1997).
[62] M. Truss, J. Bartsch, A. Schelbert, R. J. Hache, and M. Beato, *EMBO J.* **14,** 1737 (1995).
[63] S. T. Okino and J. P. Whitlock, Jr., *Mol. Cell. Biol.* **15,** 3714 (1995).

[36] Measurement of Localized DNA Supercoiling and Topological Domain Size in Eukaryotic Cells

By PHILLIP R. KRAMER, OLGA BAT, and RICHARD R. SINDEN

Introduction

Advantages and Rationale for Using Psoralen to Measure Localized Supercoiling and Topological Domain Sizes

Psoralen has distinct advantages as an *in vivo* probe to measure accurately the level of unrestrained supercoiling.[1,2] The main advantages of psoralen are that it is readily permeable to eukaryotic cells,[3,4] it has little effect on cells in the absence of longwave (360 nm) UV light, it will not displace nucleosomes,[5,6] and cells can survive high numbers of psoralen cross-links.[7,8] Supercoiling measurements have been made accurately in the whole chromosome of plants,[9] *Escherichia coli,*[3] *Bacillus brevis,*[10] and plasmids.[11] Furthermore, using psoralen as a probe, the level of supercoiling has been measured in specific chromosomal locations in *Rhodobacter capsulatus,*[12] *Drosophila,*[13,14] mouse (P. R. Kramer *et al.,* unpublished results), and human cells.[15–17]

Psoralen can also be used to estimate a localized topological domain size within living cells. The existence of unrestrained negative supercoiling

[1] R. R. Sinden and D. W. Ussery, *Methods Enzymol.* **212,** 319 (1992).
[2] D. W. Ussery, R. W. Hoepfner, and R. R. Sinden, *Methods Enzymol.* **212,** 242 (1992).
[3] R. R. Sinden, J. O. Carlson, and D. E. Pettijohn, *Cell* **21,** 773 (1980).
[4] G. D. Cimino, H. B. Gamper, S. T. Isaacs, and J. E. Hearst, *Annu. Rev. Biochem.* **54,** 1151 (1985).
[5] W. De-Bernardin, T. Koller, and J. M. Sogo, *J. Mol. Biol.* **191,** 469 (1986).
[6] T. Cech and M. L. Pardue, *Cell* **11,** 631 (1977).
[7] R. R. Sinden and R. S. Cole, *J. Bacteriol.* **136,** 538 (1978).
[8] M. E. Zolan, C. A. Smith, and P. C. Hanawalt, *Biochemistry* **23,** 63 (1984).
[9] R. J. Thompson and G. Mosig, *Nucleic Acids Res.* **18,** 2625 (1990).
[10] A. Bohg and H. Ristow, *Eur. J Biochem.* **170,** 253 (1987).
[11] G. Zheng, T. Kochel, R. W. Hoepfner, S. E. Timmons, and R. R. Sinden, *J. Mol. Biol.* **221,** 107 (1991).
[12] D. N. Cook, G. A. Armstrong, and J. E. Hearst, *J. Bacteriol.* **171,** 4836 (1989).
[13] E. R. Jupe, R. R. Sinden, and I. L. Cartwright, *EMBO J.* **12,** 1067 (1993).
[14] E. R. Jupe, R. R. Sinden, and I. L. Cartwright, *Biochemistry* **34,** 2628 (1995).
[15] M. Ljungman and P. C. Hanawalt, *Nucleic Acids Res.* **23,** 1782 (1995).
[16] M. Ljungman and P. C. Hanawalt, *Proc. Natl. Acad. Sci. U.S.A.* **89,** 6055 (1992).
[17] P. R. Kramer and R. R. Sinden, *Biochemistry* **36,** 3151 (1997).

0076-6879/99 $30.00

is dependent on the partitioning of DNA into individual topological domains that have traditionally been envisioned as loops. The loops may be organized from folding the chromosome, as proposed for *E. coli,* or from the periodic attachment of DNA to a nuclear scaffold (or nuclear matrix), as suggested in eukaryotes, or attachment to the bacterial membrane in bacterial cells. Attachment-forming loops delimit the boundaries of topologically closed domains.[18] Topological domain size measurements have been made in *E. coli,*[19] bacteriophage T4,[20] human cells,[17] and mouse cells (P. R. Kramer *et al.,* unpublished results). Psoralen-based techniques establish the domain size(s) of a particular chromosomal region without disruption of the chromatin structure. Significantly, the critical photobinding step is performed in intact, living cells. Thus, this technique can determine an *in vivo* domain size on chromosomal DNA with its nucleosome organization and higher order organizational structure intact.

The assay to measure unrestrained supercoiling relies on a comparison of the rate of psoralen cross-linking of duplex DNA containing a certain level of unrestrained supercoiling and the rate of cross-linking to nicked DNA with no unrestrained supercoiling. For DNA with a superhelical of about $\sigma = -0.065$, as found in naturally supercoiled DNA from bacteria, the rate of photobinding *in vitro* in TEN buffer is about 1.7 times higher for supercoiled DNA than relaxed DNA.[3] A near-linear relationship exists between the rate of psoralen photobinding to supercoiled DNA relative to relaxed DNA as a function of superhelical density. The level of psoralen photobinding is also dependent on the accessibility of DNA, which is defined by its extent of association with nucleosomes or other proteins that prevent trimethylpsoralen binding.[21,22] For measurements of supercoiling *in vivo,* the rate of psoralen photobinding is dependent on both supercoiling and protein organization. Nicking the DNA releases unrestrained supercoiling by allowing one strand to swivel around the other. A decrease in supercoiling should lead to a decrease in the rate of cross-linking, reflecting the inherent affinity of that chromosomal location to the intercalation of trimethylpsoralen. Provided there is no change in protein association with DNA on nicking, the decrease in photobinding reflects only a change in supercoiling. Control experiments by Jupe *et al.*[13] showed that no major global changes in the nucleosomal organization occurred upon nicking DNA in *Drosophila* cells.

[18] L. A. Freeman and W. T. Garrard, *Crit. Rev. Eukary. Gene Exp.* **2,** 165 (1992).

[19] R. R. Sinden and D. E. Pettijohn, *Proc. Natl. Acad. Sci. U.S.A.* **78,** 224 (1981).

[20] R. R. Sinden and D. E. Pettijohn, *J. Mol. Biol.* **162,** 659 (1982).

[21] R. R. Sinden, D. E. Pettijohn, and B. Francke, *Biochemistry* **21,** 4484 (1982).

[22] J. M. Sogo, P. J. Ness, R. M. Widmer, R. W. Parish, and T. Koller, *J. Mol. Biol.* **178,** 897 (1984).

The measurement of the size of a topological domain containing a specific DNA sequence, or locus, requires determination of the level of unrestrained supercoiling before and during the introduction of nicks into the DNA molecule. The rate of cross-linking decreases to a minimum after all supercoils are lost from a population of topological domains after a nick is introduced into each domain encompassing the particular locus on the chromosome.

Experimentally, the number of nicks within a local chromosomal region must be measured and related to the fraction of the total unrestrained superhelical tension present. Plotting the percentage of the relaxation of unrestrained supercoils (F_r) against the number of nicks in the DNA allows theoretical curves to be fitted to experimental data. Theoretical curves are created for different domain sizes (m) using the Poisson formula ($P_0 = F_r = e^{-(x/m)}$), where F_r represents the fraction of nonnicked domains (zero term of the Poisson). As discussed later, a theoretical model that fits experimental data can explain the distribution and size(s) of domains within that chromosomal location. Thus far, the rate of decay of supercoiling as nicks are introduced is nearly first order[3,17,19,20] and the fraction of supercoiling remaining following introduction of a defined number of nicks will fit a Poisson distribution. The distribution of nicks among independent topological domains is described by a Poisson distribution, where x is the number of nicks introduced into the genome and m is the number of independent topological domains (see later equations). This approach has been used to measure domain sizes in *E. coli*,[19] bacteriophage T4,[20] and human cells.[17]

Procedures for Measuring Supercoiling and Topological Domains

Cell Growth, DNA Nicking, and Cross-Linking

To measure unrestrained supercoiling, the rate of 4,5′,8-trimethylpsoralen photobinding to DNA must be measured in cells before and after the introduction of nicks in DNA (to relax supercoils). Nicks can be introduced by ionizing radiation[3,19,21] or photolysis by bromodeoxyuridine incorporated into DNA.[13,20] Because radiation sources are not generally available, we will describe nicking using 5-bromo 2-deoxyuridine photolysis.

Unrestrained superhelical density was analyzed in human and mouse cell lines that attach to the surface of tissue culture plates, but a method using eukaryotic cells in suspension has been discussed previously.[1,13] The human or mouse cell cultures are started from a frozen culture and grown to a density of 5×10^6 cells/100-mm plate. The cells are washed twice in phosphate-buffered saline (PBS: 0.137 M NaCl, 2.7 mM KCl, 0.5 mM $MgCl_2$, 0.9 mM $CaCl_2$, 8.1 mM Na_2HPO_4, and 1.5 mM KH_2PO_4, pH 7.5) and then

0.5 ml of trypsin (0.25%) per plate is added. The human or mouse cells are resuspended to a density of 1×10^5 cells/ml in minimal essential medium (MEM) or Dulbecco's modified Eagle's medium (DMEM), respectively, with serum and 10 m*M* bromo-2-deoxyuridine (BrdUrd). Ten milliliters of media is added to each plate (1×10^6 cells/plate). Using this protocol in hamster cells, 45% of the thymidines are substituted with BrdUrd.[23] Cells are grown to a density of 5×10^6 cells/plate. Cells (attached to a 100-mm plate) are washed twice with 20 ml of PBS buffer, 2 ml of PBS buffer is added, and the plates are chilled to 0–4° by placing the dish in ice (such that the top edge of the petri plate is flush with the top of the ice layer). A light filter with a peak transmittance of 313 nm is placed over the 100-mm plate containing the cells; the filter consists of a sealed, polyethylene petri dish (150 mm diameter by 15 mm height) containing a 1-cm solution of 0.1 m*M* K_2CrO_4 in 0.1 *M* NaOH. A Blak-Ray Model 100A ultraviolet lamp (Ultraviolet Products, Inc., Upland, CA) with a 100-W mercury flood bulb at a distance of 10 cm from the top of the light filter is used to introduce nicks into the DNA using 313-nm light.[1,20] To the nicked and nonnicked samples, 20 μl of a saturated solution of 4,5′,8-trimethylpsoralen in ethanol (1% v/v) is added and, after a 5-min incubation on ice (0–4°), the plates are exposed to various doses of 360-nm light from two General Electric BLB20 black light bulbs at an incident light intensity of about 1.2 kJm^{-2} per minute. Treatment with psoralen has been described in detail previously.[1,2] Following nicking and cross-linking, all procedures are performed at 4°, and the cells are kept under minimal light conditions, being subjected to only nondirect, nonfluorescent lighting.

The nonnicked chromosomal sample should be performed in duplicate for each separate experimental batch performed. The level of psoralen photobinding in the nonnicked chromosomal sample is measured most accurately due to the sharpest contrast between the peak and the baseline level of radioactive signal detected on PhosphoImager analysis. It is a good gauge of the experimental technique and will indicate any changes due to experimental variables (i.e., growth conditions) in the level of cross-linking that may occur. A very similar level of cross-linking in nonnicked samples between experiments would indicate similar conditions and that a similar technique was used for each repetition of the experiment. After treatment, cells are washed twice with PBS and 0.5 ml of trypsin (0.25%) is added; after 5 min at room temperature, gentle pipetting with 10 ml of PBS removes cells fixed to the plate. The cell mixture is added to a 13-ml polypropylene tube, pelleted at 3000*g* for 10 min, and resuspended in 3 ml of extraction buffer (200 m*M* Tris, pH 8.0, 200 m*M* NaCl, 100 m*M* EDTA, 4% SDS).

[23] E. R. Kaufman and R. L. Davidson, *Proc. Natl. Acad. Sci. U.S.A.* **74,** 4982 (1978).

Fifty microliters of RNase A (10 mg/ml) is added, and the mixture is incubated at 37° for more than 8 hr. One hundred twenty-five microliters of proteinase K (20 mg/ml) is then added and following an 8-hr incubation at 55°, the protein is removed by one extraction with 3 ml phenol equilibrated with 10 m*M* Tris, pH 7.6, 50 m*M* NaCl, 1 m*M* EDTA (TEN) and one extraction with 3 ml of 24:1 (v/v) chloroform:isoamyl alcohol. Each extraction requires vigorous shaking, centrifugation at 4000*g*, and removal of the top layer to a fresh tube. DNA is precipitated by adding 1/10th volume 3 *M* potassium acetate and 2.5 volumes of 95% ethanol, mixing, and centrifuging at 3000*g*. DNA samples are air dried overnight resuspended in 10 m*M* Tris, pH 7.0, 5 m*M* EDTA (TE) at a concentration of approximately 300 μg/ml.

Southern Analysis of Photocross-Linked DNA

To measure the number of psoralen cross-links per kilobase of a given region of chromosomal DNA, DNA containing cross-links is cut with a restriction enzyme to generate a restriction fragment of interest. This DNA fragment is denatured, the cross-linked and noncross-linked DNAs are separated on a glyoxal gel, the gel is then probed by Southern hybridization, and the fraction of cross-linked and noncross-linked DNA is determined using densitometry on a PhosphorImager. The procedure for Southern hybridization described by Lupski *et al.*[24] has been used successfully. Briefly, 7 μg of chromosomal DNA is digested to completion at 37° with 100 units of a restriction endonuclease in a 60-μl volume. Quantitative analysis of the restriction fragment is improved by a high signal to background ratio, and one of the simplest ways to increase this ratio is to analyze a large restriction fragment in the Southern analysis. An optimum size range of DNA fragments is between 2 and 6 kb. Fragments larger than 2.0 kb can be analyzed easily, while analysis of fragments less than 1 kb is difficult. This is due to the weaker hybridization signal of smaller DNA fragments and to the higher number of nicks required in the DNA to relax supercoiling in small domains. After digestion, 300 μl of TE is added to the digests and the sample is extracted twice with 100 μl of phenol saturated with TEN and twice with 100 μl of chloroform:isoamyl alcohol (24:1). The samples are ethanol precipitated, pelleted, dried, and then resuspended in 6 μl of H_2O plus 3 μl of 100 m*M* sodium phosphate buffer, pH 7.0, and vortex mixed to vigorously resuspend the sample. Complete resuspension is necessary. The samples are boiled for 1 min to denature the DNA, quick cooled

[24] J. R. Lupski, R. M. de Oca-Luna, S. Slaugenhaupt, L. Pentao, V. Guzzetta, B. J. Trask, O. Saucedo-cardenas, D. F. Barker, J. M. Killian, C. A. Garcia, A. Chakravarti, and P. I. Patel, *Cell* **66,** 219 (1991).

on ice, and then 5 μl of 40% ultrapure glyoxal aldehyde (Fisher, keep frozen in aliquots at $-80°$) 6% final concentration (v/v) and 21 μl dimethyl sulfoxide (DMSO) final concentration 60% (v/v) are added for a total volume of 35 μl. The samples are incubated for 1 hr at 50° (the sample volume can be reduced under vacuum to a volume of 20 μl), 3 μl of a 40% sterile glucose solution is added, and the samples are separated on a 1% agarose gel in 10 m*M* sodium phosphate (pH 7.0) as described.[25] Once electrophoresis is complete, the gel is treated for 3 hr in denaturing solution (0.5 *M* NaOH, 1.5 *M* NaCl). The DNA is then transferred to a nylon membrane (Genescreen plus, NEN, Dupont, Boston, MA) and dried at 80° under vacuum for 2 hr.

A restriction endonuclease fragment or polymerase chain reaction (PCR) product can be isolated and used as a probe for detecting the chromosomal location of interest. The quality of experimental data requires a high radioactive signal to background ratio, and thus it is important to maximize the percentage of the probe region homologous to the chromosomal region of interest. The membrane is rinsed in H_2O for 5 min, prehybridized for 2–3 hr with 8 ml of hybridization solution (1 *M* NaCl, 1% SDS, 10% dextran sulfate 200,000 g/mol, 1.5 mg of herring sperm DNA denatured by boiling for 5 min). A probe is labeled by the random hexamer labeling method (Boehringer Mannheim) using [α-^{32}P]dCTP 3000 Ci/mmol (DuPont, NEN). In 2 ml hybridization solution (with an additional 1.5 mg of herring sperm), $1–2 \times 10^7$ cpm/ml of probe is added and the sample is boiled for 5 min. This is incubated at 65° for 2–3 hr (longer is fine) and is then added to the 8 ml of solution prehybridizing the membrane. Hybridization is performed at 65° for 8–20 hr. The membrane is then washed twice at 65° (1 hr each time) in a prewarmed solution of 2× SSC, 0.1% SDS, dried, and exposed to film. The Southern blot will have a slower migrating band, which is the cross-linked product (designated XL), and a faster migrating band, which represents the noncross-linked fragment. The cross-links per kilobase (Xl/kb) are calculated from the intensities of the cross-linked (Xl) and noncross-linked bands. Quantitation is completed using a Molecular Dynamics (Sunnyvale, CA) PhosphorImager with ImageQuant software.

The fraction of DNA cross-linked (*Fx*) is calculated by dividing the area of cross-linked peak by the sum of the area for the cross-linked and noncross-linked peaks. Cross-links per kilobase (Xl/kb) values are calculated using the formula $\text{Xl/kb} = -\ln(1 - Fx)/S$, as described previously,[13] where (S) is the size of the restriction fragment in kilobases. The ratio of the mean cross-linking rate in intact versus relaxed domains ($R_{I/N}$ = Xl/

[25] G. K. McMaster and G. G. Carmichael, *Proc. Natl. Acad. Sci. U.S.A.* **74,** 4835 (1977).

$kb_I/XI/kb_N$) reflects the level of unrestrained supercoiling.[13] To determine an $R_{I/N}$ value, an average of two Southern blots are performed for a minimum of four separate experiments for most analyses. A ratio of $R_{I/N} = 1$ indicates completely relaxed DNA with a lack of unrestrained torsional tension. A ratio of $R_{I/N} > 1$ indicates the presence of negative unrestrained torsional tension. It is not yet known if an $R_{I/N} < 1$ would represent positively supercoiled DNA. An unpaired, two-tailed *t* test can be performed with these $R_{I/N}$ values to determine if a significant difference in $R_{I/N}$ values is indicated.

Measurement of Number of Single-Strand Breaks in DNA of Eukaryotic Cells

To measure the frequency of single-strand breaks introduced by BrdUrd photolysis within the whole genome, 5–9 $\times 10^6$ cells/100-mm plate are grown for three to four generations in the presence of 0.02 mCi/ml of [^{3}H]thymidine before the DNA is nicked as described earlier. Following the nicking protocol, cells are washed in 3 ml of PBS buffer and pelleted. Cells are resuspended in 1 ml of PBS and stored on ice. Cells (1–5 $\times 10^4$) and 2000 cpm ^{14}C-labeled T7 phage are added to 0.1 ml of 1 *M* NaOH, 10 m*M* EDTA layered on an alkaline sucrose gradient (5–20% sucrose in 0.3 *M* NaOH, 2 *M* NaCl, 10 m*M* EDTA). After loading, the gradients are put in the dark for 3 hr at 20° before sedimentation at 30,000 rpm in a Beckman SW55Ti rotor for 2–4 hr at 20°. Thirty equal-volume fractions are collected from the bottom of the tube onto 2-cm sections of a 0.5-inch strip of Whatman (Clifton, NJ) 3MM filter paper. Once dry, the paper is washed twice in 5% trichloroacetic acid and once in 95% ethanol. After drying, 2-cm sections are placed into vials with scintillation fluid and counted for radioactivity. Molecular weights of the single-strand DNA fragments are calculated,[26] and the number of single-strand breaks per genome (nicks per genome equivalent) is calculated using a molecular weight of 338 g/mol base and the number of bases per genome. Alkaline sucrose gradient analysis generates the average molecular weight of the single-stranded DNA fragments made from the entire genome (Z). The number of bases per fragment (Y) = $Z/338$ g/mol. Nicks per genome equivalent = (bp/genome)/(number of bp/nick) where the number of bp/nick = $Y/2$, where Y is the size of the single-strand DNA (number of bases per fragment). (The number of breaks per length of duplex DNA is twice that of the single-strand spacing of the nicks.[26])

[26] W. D. Rupp and P. Howard-Flanders, *J. Mol. Biol.* **31,** 291 (1968).

Measurement of Single-Stranded Breaks at Specific Chromosomal Locations

The measurement of single-strand breaks introduced into a specific region or locus of the genome by BrdUrd photolysis is similar to the procedure described earlier with the addition of Southern analysis. Following the nicking protocol, the cells are washed in 3 ml of PBS buffer and pelleted. The cells are resuspended in 1 ml of PBS and stored on ice. Cells (5×10^4 to 1.5×10^5) and 2000 cpm ^{3}H-labeled T7 phage are layered onto an alkaline gradient and incubated as described previously. For fragments sizes of 5000–80,000 bases, sedimentation is performed at 35,000 rpm in a Beckman SW55Ti rotor for 4 hr at 20°. Thirty equal-volume fractions are collected from the bottom of the tube onto 3-cm^2 sections of nylon membrane. After baking at 80° for 2 hr, the membrane is hybridized with an [^{32}P]dCTP, random prime-labeled restriction endonuclease fragment or PCR product. The membrane is washed twice with 2× SSC at 65° for 1 hr each time. The membrane is allowed to air dry and then 3-cm^2 sections are placed into vials with scintillation fluid and counted for radioactivity. Molecular weights of the single-strand DNA fragments (Z) analyzed from a specific chromosomal location are taken from the alkaline sucrose gradient Southern analysis (described earlier).[26] The term, chromosomal location, is the region of DNA in which both supercoiling measurements and nicking measurements are taken. The number of single-strand breaks per chromosomal location is calculated as described previously. Nicks per chromosomal location = (bp/chromosomal location)/(number of bp/nick) where the number of bp/nick = $Y/2$, where Y is the size of the single-strand DNA (number of bases per fragment).

Measuring the nicks within a chromosomal location in which the level of unrestrained supercoiling was measured has several advantages. First, it does not assume the nicking of the whole genome is the same within the chromosomal location where the level of supercoiling was analyzed. Second, by plotting the fraction of supercoils remaining as a function of the number of nicks introduced (both data obtained from a specific location), a theoretical curve can be drawn to fit experimental data more accurately as described later. The theoretical curve calculated to fit experimental data should describe the distribution and domain sizes within a chromosomal location and whether it is a first-order (or some other order) rate of decay (single domain size) as described previously or a chromatin region with topological domains of varying sizes, as described later.

Estimation of Topological Domain Size in Chromosomal DNA: Theoretical Analysis of F_R as Function of Topological Domain Sizes

From a probabilistic point of view, an experiment in which we introduce nicks into the genome can be considered a series of Bernoulli trials for each topological domain. A Bernoulli trial is one that has only two possible outcomes; in this case the introduction of a nick or no occurrence of a nick. We assume that a Bernoulli trial for each domain genome occurs when a single nick is introduced at random, somewhere in the genome (whether by γ radiation or by 313-nm light-induced photolysis of BrdUrd). If a break occurs in the domain under consideration, then this Bernoulli trial has a positive outcome. If a break occurs somewhere else in the genome, the Bernoulli trial for that particular domain has a negative outcome. Also, the Bernoulli trials must be identical and independent. These conditions are also met by irradiation with γ rays or 313-nm light. Thus, the number of breaks (successes) introduced into each domain following x trials fits the definition of binomial random variable. The probability of n successes after x trials for the binomial random variable is described by the formula

$$P(N = n) = \binom{x}{n} p^n (1 - p)^{x-n}, \qquad n = 0, 1, 2, \ldots$$

where p is a probability of a success after one trial. The mean of the binomial variable is defined as $\mu = xp$. In our case, $p = 1/m$, where m is the number of domains. It implies that $\mu = x/m$. If the number of trials $x \to \infty$ and $p \to 0$ in such a way that the mean $\mu = xp$ is constant, then

$$\lim_{\substack{x\to\infty,\, p\to 0\\ \mu = xp}} P(N = n) = \lim_{\substack{x\to\infty,\, p\to 0\\ \mu = xp}} \binom{x}{n} p^n (1 - p)^{x-n} = e^{-\mu}\mu^n/n!$$

This implies that the probability of n breaks (successes) in one domain in x trials is defined by the Poisson distribution:

$$P(n) = e^{-\mu}\mu^n/n!$$

where $\mu = x/m$. In practice the conditions under which one can switch from a binomial distribution to the Poisson distribution without losing accuracy are mathematically not very strict, namely $x \geq 20$ and $p \leq 0.5$. These conditions are met in our experiment: $1 \times 10^6 \leq x \leq 6 \times 10^6$ and $3.1 \times 10^{-5} \leq p \leq 1.25 \times 10^{-4}$. (Given $p = 1/m$ and $8 \times 10^3 \leq m \leq 32 \times 10^3$.)

The probability for a domain to get zero breaks and remain supercoiled is the zero term of the Poisson:

$$P(0) = e^{-\mu} = e^{-x/m}$$

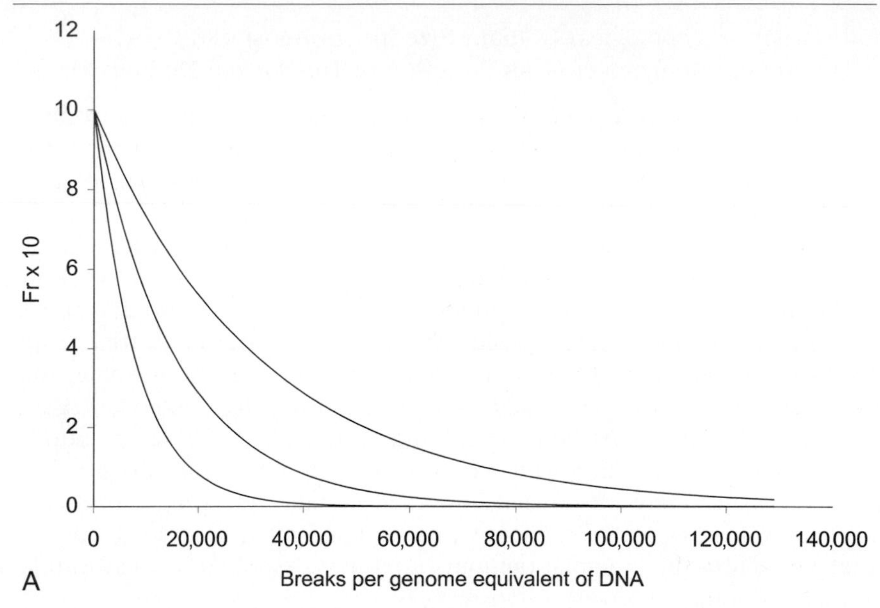

FIG. 1. *Fr* is the fraction of supercoils remaining in a DNA population after introduction of a defined number of breaks, described by the equation: $Fr_x = [(R_{I/Nc}$, at complete relaxation) $- (R_{I/Nx}$, at number of nicks $x)] \div [(R_{I/Nc}$, at complete relaxation) $- (R_{I/N0}$, in the nonnicked sample)], where $R_{I/N}$ values represent the ratio of the mean cross-linking rate in intact versus relaxed domains for a given number of nicks (x) ($R_{I/NO}$ is the ratio in control cells with no nicking and $R_{I/NO} = 1$). (A) Theoretical Fr_x curves for a situation where the domain size is 32 kb (bottom curve), 16 kb (middle curve), or 8 kb (top curve). (B) Five different curves (solid lines) for the Fr_x values for situations where the population of topological domains is composed of two domain sizes, 8 and 32 kb, which represent 10 and 90%, 30 and 70%, 50 and 50%, 70 and 30%, or 90 and 10%, respectively, of the total domains (representing the curves from top to bottom). These curves are superimposed on the three curves shown in (A) (dotted lines).

(given $0! = 1$). In terms of the entire genome, the same equation describes the fraction of domains containing no breaks:

$$Fr = e^{-\mu} = e^{-x/m}$$

Experimentally, Fr_x is the fraction of supercoils remaining in a DNA population after introduction of a defined number of breaks, described by the equation:

$$Fr_x = [(R_{I/Nc}, \text{at complete relaxation}) - (R_{I/Nx}, \text{at number of nicks } x)] \div [(R_{I/Nc}, \text{at complete relaxation}) - (R_{I/N0}, \text{in the nonnicked sample})]$$

where $R_{I/N}$ values represent the ratio of the mean cross-linking rate in intact versus relaxed domains for a given number of nicks (x) ($R_{I/NO}$ is the

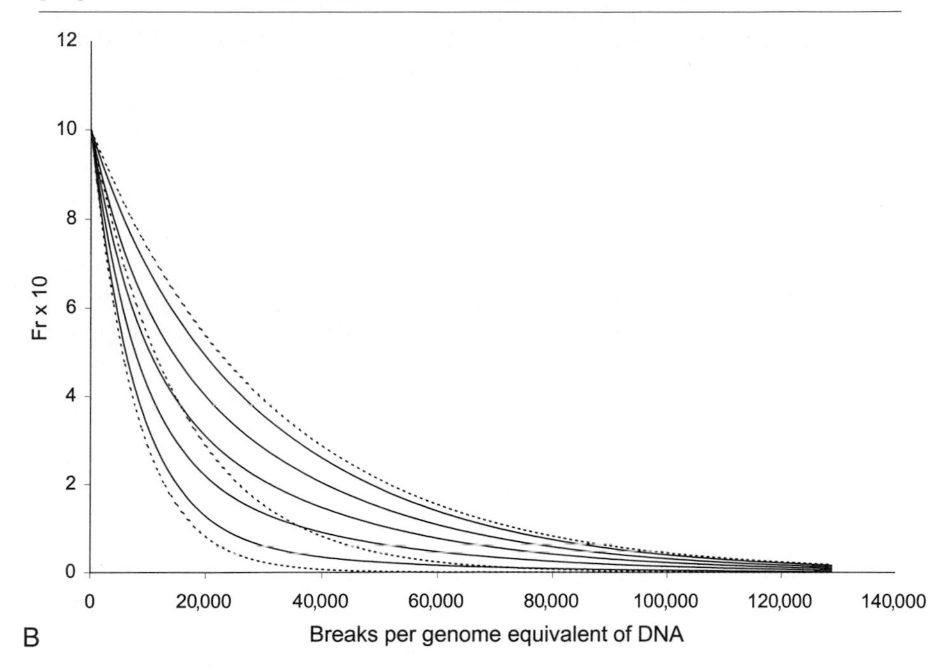

FIG. 1. (*Continued*)

ratio in control cells with no nicking and $R_{I/NO} = 1$). Effectively, within the ability to detect supercoiling, $R_{I/Nc}$ represents about >95% of the domains containing at least one nick. For a single-sized domain, an exponential decrease in Fr_x occurs as nicks are introduced into the DNA. A reasonable fit to an exponential decrease was observed in *E. coli* and bacteriophage T4 following λ radiation or photolysis by BrdUrd and 313-nm light.[19,20] However, the decrease in Fr_x observed in human or mouse cells did not seem to fit an exponential decay as well as that in bacterial cells.[17]

We have analyzed supercoiling and topological domains at a locus in a mouse cell line that contains 200 tandem repeats of a 9-kb region (P. R. Kramer *et al.,* unpublished results). In this system the domain size changes from about 32 to 16 to 8 kb on the activation of transcription (under different conditions) throughout the region. Thus the locus may contain a mixture of two (or more) domain sizes. The deviation from a perfect fit to the Poisson distribution is expected when *m*, the number of domains varies (or when the domain size is a collection of two or more different size classes). Figure 1A shows the theoretical Fr_x curves for a situation where the domain size is 32, 16, or 8 kb. Figure 1B shows five different curves for the Fr_x values for situations where the population of topological domains is composed of two domain sizes, 8 and 32 kb, which represent 10 and 90%, 30 and 70%, 50 and 50%, 70 and 30%, or 90 and 10% of the total domains (representing the curves from top to bottom). These curves (solid lines) are superimposed on the three curves from Fig. 1A (dotted lines). It is clear that the theoretical Fr_x curves for mixed populations do not fit the exponential decay expected for a single population. If two different domain sizes are present, then a best fit to these can be estimated. However, the situation becomes complicated if there are multiple different domain sizes (especially more than two), as by judicious selection of the domain sizes and the proportion of domains, nearly identical Fr_x curves can be calculated from rather different values of domain size and proportion. This consideration must be kept in mind when estimating the size of a topological domain *in vivo*.

[37] Fluorescence *in Situ* Hybridization Analysis of Chromosome and Chromatin Structure

By WENDY BICKMORE

Introduction

The level of complexity being studied in the analysis of chromatin structure is increasing. Rather than analyzing the interaction of a single protein, e.g., a transcription factor, with specific binding sites on a DNA molecule, we now want to understand how multimeric protein complexes interact with each other, and with the DNA, to form chromatin fibers and even whole chromosomes. Unraveling the complexity of eukaryotic chromosomes is daunting. In a mammalian cell, the 2 m of DNA double helix is compacted through hierarchical levels of chromatin structure to form an interphase chromosome and further chromosome condensation occurs as cells enter mitosis. One advantage that such a large and compact structure affords, however, is size. Metaphase chromosomes from vertebrate cells are isolated readily and they are large enough to be seen with the light microscope (chromosome 1, the largest human chromosome, is approximately 8 μm in length in conventional cytogenetic preparations). This means that visual means of analysis can make significant and immediate impacts on our understanding of chromosome structure.

The DNA within chromosomes can be denatured and so hybridized with complementary DNA probes. These probes can be chosen to highlight specific chromosomal loci such as centromeres, telomeres, or specific gene

0076-6879/99 $30.00

loci, but they can also comprise more complex mixtures of DNA molecules that have been prepared according to chosen criteria (e.g., base composition, extent of CpG methylation, attachment to the chromosome scaffold)[1–3] or they can even be prepared so as to highlight a specific chromosome, chromosome arm, or subregion.[4,5] Probes can be labeled, either directly or indirectly, with differently colored fluorochromes, allowing sites of hybridization to be discerned by fluorescence microscopy. The resulting procedure—fluorescence *in situ* hybridization (FISH) and its application to the study of vertebrate chromosome structure—is the subject of this article.

FISH can contribute to an understanding of chromatin and chromosome structure in two basic ways. In the first, standard cytogenetic preparations of condensed chromosomes from cells arrested in metaphase are used as the template on which to hybridize complex probe mixtures that have been prepared from chromatin according to a specified biochemical characteristic (e.g., their ability to bind to the nuclear matrix).[3] The resulting hybridization pattern across the karyotype is used to assess the nature of the DNA sequences that are contained in the probe. In this instance, chromosomes spread on the glass slide act as a way of displaying the entire genome in a visually recognizable way.

In the second application, it is the chromosomes or nuclei on the microscope slide that are manipulated prior to hybridization (e.g., by extraction of soluble proteins),[6,7] and FISH with specific probes is used to reveal something of the structure of the manipulated chromatin. This article describes examples of both of these types of application.

Chromosome Painting with Biochemical Fractions of the Genome

The first experiments that used fractions of the genome as probes for FISH to investigate chromosome organization looked at the distribution of different families of interspersed repeats, sequences of differing base composition, and CpG islands along chromosome length.[1,2,8] This was ex-

[1] S. Saccone, A. De Sario, J. Wiegant, A. K. Raap, G. Della Valle, and G. Bernardi, *Proc. Natl. Acad. Sci. U.S.A.* **90,** 11929 (1993).

[2] J. M. Craig and W. A. Bickmore, *Nature Genet.* **7,** 376 (1994).

[3] J. M. Craig, S. Boyle, P. Perry, and W. A. Bickmore, *J. Cell Sci.* **110,** 2673 (1997).

[4] P. Rabbits, H. Impey, A. Hepell-Parton, C. Langford, C. Tease, N. Lowe, D. Bailey, M. Ferguson-Smith, and N. Carter, *Nature Genet.* **9,** 369 (1995).

[5] X.-Y. Guang, H. Zhang, M. Bittner, Y. Jiang, P. Meltzer, and J. Trent, *Nature Genet.* **12,** 10 (1996).

[6] W. A. Bickmore and K. Oghene, *Cell* **84,** 95 (1996).

[7] M. G. Gerdes, K. C. Carter, P. T. Moens, and J. B. Lawrence, *J. Cell Biol.* **126,** 289 (1994).

[8] J. R. Korenberg and M. C. Rykowski, *Cell* **53,** 391 (1988).

tended to chromatin fractions by examining the chromosomal distribution of sequences assembled into hyperacetylated histones and fractions of the genome associated with the nuclear/chromosome scaffold, matrix, and skeleton.[3,9]

Production of Metaphase Chromosome Spreads

Principle of Method

The first requirement for this approach is the production of long, straight, and well-spread metaphase chromosomes. In the case of human chromosomes, these are easiest to attain from peripheral blood cultures. Human peripheral blood lymphocytes or cells isolated from mouse spleens are stimulated to divide with phytohemagglutinin (PHA) or lipopolysaccharide (LPS), respectively. Mitotic cells can be accumulated using the spindle poison Colcemid. The time of incubation with Colcemid increases the percentage of mitotic cells within a population (mitotic index), but also increases the degree of chromosome condensation. For this application it is important to keep the time of incubation short (<40 min). Blood cultures can also be used to prepare metaphase chromosomes with minimal use of Colcemid by blocking cells in the S phase prior to mitosis with methotrexate.

After harvest, cells are collected by centrifugation and then swollen in a hypotonic solution of KCl before fixation with 3 : 1 methanol : acetic acid. For rodent cells, a hypotonic solution of trisodium citrate/KCl gives better preparations than KCl alone. The 3 : 1 fixative should be freshly prepared with absolute methanol and glacial acetic acid.

Method

1. Add 1 ml peripheral blood to 10 ml RPMI 1640 medium supplemented with 15% (w/v) fetal calf serum (FCS), 100 U/ml penicillin, 100 μg/ml streptomycin, and 1% PHA. Incubate for 72 hr at 37° with 5% (v/v) CO_2.
2. Add methotrexate to a final concentration of 45 ng/ml for 17 hr and then spin cells at 200*g* for 5 min at room temperature. Resuspend in fresh, prewarmed RPMI/FCS to release the block.
3. Incubate at 37° for 4.5 hr and then add Colcemid to a final concentration of 0.1 μg/ml for 30 min.
4. Spin down cells at 200*g* for 5 min at room temperature and remove culture medium, wash cells in PBS, repeat spin, and resuspend cell pellet in residual PBS.

[9] J. W. Breneman, P. M. Yau, R. R. Swiger, R. Teplitz, H. A. Smith, J. D. Tucker, and E. M. Bradbury, *Chromosoma* **105,** 41 (1996).

5. Slowly add 10 ml 0.075 *M* KCl hypotonic solution with continual mixing (the concentration of cells in hypotonic should be $<2 \times 10^7$/ml). For rodent cells, use 0.017 *M* trisodium citrate, 0.033 *M* KCl as hypotonic.
6. Incubate at room temperature for 20 min.
7. Spin down cells at 200*g* for 5 min, remove hypotonic, and resuspend cell pellet in residual liquid.
8. Add a few drops of freshly made 3:1 (v/v) methanol:acetic acid, mixing continually. Add further fix to bring volume to 10 ml.
9. Leave on ice for 20 min, then spin down cells at 200*g* for 5 min at room temperature.
10. Repeat step 8, add fresh fix, and leave at 4° overnight.
11. The next day, spin cells again and resuspend in desired volume of fresh fix. The final cell suspension should have a slightly milky appearance. Drop onto clean microscope slides. Leave overnight and then store in a vacuum desiccator for 3–7 days before use.

As cells are released from the methotrexate block, 5-bromo-2′-deoxyuridine (BrdU) can be added to the cultures (to 0.1 m*M*). This mildly inhibits chromosome condensation and allows sites of late replication to be detected by immunofluorescence in parallel with FISH.[2]

Labeling of DNA Fractions

Principle of Method

DNA probes useful for the analysis of chromosome structure by FISH are likely to be complex mixtures of sequences. Depending on the criteria by which they were purified (e.g., by immunoprecipitation with an antibody directed against a chromatin protein or by binding to the nuclear matrix),[3,9] the average fragment size may be either large (>1 kb) or small (<1 kb). What is important is that, after labeling, the average probe size should be ~600 bp. Probes larger than this give spotty background on the slides after hybridization. Very small probes hybridize inefficiently. It is these considerations that determine the method of labeling.

DNA fragments that are >1 kb in length are usually labeled by nick translation. Probes that are smaller than this can be labeled by random priming. However, if probes are amplifiable by polymerase chain reaction (PCR), then it is convenient for label to be incorporated during the PCR reaction, as long as the amplification products remain in the 400- to 1000-bp size range. If in doubt, the size range of labeled products can be assessed by alkaline agarose gel electrophoresis. Probe mixtures can be converted to an amplifiable form by the attachment of suitable catch linkers.

Method

Nick Translation

1. To 1 μg DNA add 4 μl 10× nick translation buffer [0.5 *M* Tris, pH 7.5, 0.1 *M* $MgSO_4$, 1 m*M* dithiothreitol (DTT), 0.5 mg/ml bovine serum albumin (BSA) fraction V] and 4 μl each of 2 m*M* dATP, dCTP, and dGTP. Add 3 μl 0.5 m*M* dTTP and 3 μl of either 1 m*M* biotin-16-dUTP or digoxigenin-11-dUTP.

2. Adjust volume to 37 μl with distilled H_2O. Add 1 U DNA polymerase I and 2 μl of a fresh 1:500 dilution DNase I (10 U/μl stock concentration). (If probe length after labeling is not correct, adjust DNase I accordingly.)

3. Incubate at 14° for 90 min. Transfer reaction to −20°.

Random Priming. This is best achieved using a commercially available kit (e.g., BCL) and substituting half of the dTTP with biotin- or digoxigenin-conjugated dUTP.

Labeling by PCR

1. To 1–2 μl of amplifiable DNA, add 5 μl 10× Tricine buffer (200 m*M* Tricine, pH 8.4, 20 m*M* $MgCl_2$, 100 m*M* 2-mercaptoethanol, 0.1% gelatin), 5 μl each of 2 m*M* dATP, dGTP, and dCTP, 3 μl 1 m*M* biotin-16-dUTP or digoxigenin 11-dUTP, 3 μl of 0.5 m*M* dTTP, 400 ng appropriate primers, and 2 U *Taq* polymerase. Adjust volume to 50 μl with distilled H_2O.
2. Amplify for 30 cycles using suitable conditions for the primers, keeping extension times long enough to maintain average size of products >500 bp.

PCR buffers based on Tricine rather than Tris seem to be better at maintaining the size range of complex probes during amplification cycles.

Method

Quantifying Label Incorporation. Unincorporated labels must be removed from probes by molecular sieving before use.

1. Plug a 1- or 2-ml syringe barrel with siliconized glass wool and pack with Sephadex G-50 (fine) preswollen in TE.

2. Load into 13-ml centrifuge tube. Spin at 135*g* for 1 min at room temperature.

3. Put 1.5-ml screw-cap microfuge tube into centrifuge tube to collect probe. Load labeling reaction onto column and add 60 μl TE to column.

4. Spin as in step 2 and remove labeled probe in microfuge tube.

5. Make 1:100 and 1:1000 dilutions of probe in TE and spot 1 and 2 μl of each dilution together with 1-, 2-, 10-, and 20-pg standards of

appropriately λ DNA onto a gridded nitrocellulose circle, presoaked in distilled H_2O and 20× SSC and air dried.

6. Cross-link filter with UV (50 mJ) and wash the filter for 5 min in buffer 1 (0.1 *M* Tris, pH 7.5, 0.15 *M* NaCl).

7. Incubate for 30 min at 37° in buffer 1 + 3% BSA fraction V.

8. Incubate in 5 ml of buffer 1 containing 5 μl of either streptavidin–alkaline phosphatase or antidig–alkaline phosphatase for 30 min at room temperature on a shaking platform.

9. Wash 2× 15 min in 200 ml buffer 1 and then for 5 min in 0.1 *M* Tris, pH 9.5.

10. Place filter in hybridization bag. Add 5 ml of 0.1 *M* Tris, pH 9.5, containing 2 drops each from one, two, and three bottles of a Vector BCIP (5-bromo-4-chloro-3-indoylphosphate)/NBT (4-Nitro blue tetrazolium chloride) kit. Incubate in the dark for 2–4 hr. Estimate concentration of labeled DNA (blue color) against standards.

Hybridization

Before hybridization of complex probes to chromosomes fixed in 3:1 (v/v) methanol : acetic acid, RNA must be removed from the slide by RNase digestion. The amount of probe applied to the slide depends on probe complexity. For most applications relevant to chromosome structure, 150–300 ng of each chromatin fraction is needed per slide, and hybridization from repetitive sequences present in these fractions must be suppressed with 10–15 μg *Cot*I.[3,10,11] Complex probes also benefit from extended hybridization times (48 hr).[11]

Method

1. To probe (approximately 250 ng each) add 25 μg of *Cot*I DNA, 5 μg of sonicated salmon sperm DNA, and 2 volumes of ice-cold absolute ethanol.

2. Place at −20° for 30 min and then precipitate in a Speed-Vac. For each slide, resuspend probe well in 10 μl of hybridization mix [50% (v/v) deionized formamide, 2× SSC, 1% (v/v) Tween 20, 10% (w/v) dextran sulfate]. Leave at 4°.

3. Put slides into metal slide rack and place into glass trough containing 2× SSC, 100 μg/ml RNase, at 37° for 1 hr. (This step is not necessary for hybridization to extracted metaphase chromosomes.)

[10] P. Lichter, C.-J. Chang, K. Call, G. Hermanson, G. A. Evans, D. Housman, and D. C. Ward, *Science* **247,** 64 (1990).

[11] S. de Manoir, M. R. Speicher, S. Joos, E. Schrock, S. Popp, H. Dohner, G. Kovacs, M. Robert-Nicoud, P. Lichter, and T. Cremer, *Hum. Genet.* **90,** 590 (1993).

4. Put formamide denaturant (70% formamide, 2× SSC, pH 7.5) into glass trough and prewarm to 45° and then transfer to 70°. Put some 70% ethanol on ice. Clean 22 × 22-mm coverslips in ethanol/HCl.

5. Rinse slides in 2× SSC and then dehydrate through 70, 90 and 100% ethanol for 2 min each and dry under vacuum.

6. Warm slides in 70° oven for 5 min and then denature for 3 min at 70° (if hybridizing to extracted chromosomes reduce this to 2 min).

7. Transfer slides quickly to ice-cold 70% ethanol for 2 min and then through 90 and 100% ethanol as before and dry under vacuum.

8. While slides are drying, denature probe mix at 70° for 5 min. Transfer to 37° water path to preanneal repeats for 15 min.

9. Pipette 10 μl of preannealed probe onto warmed coverslip and pick up with slide. Seal with rubber solution glue. Incubate in covered tray in 37° water bath for 48 hr.

Signal Detection

Principle of the Method

After hybridization, slides are washed once in 2× SSC at room temperature to loosen the coverslips and to wash off the hybridization mixture. The slides are then washed in 2× SSC twice at 45° followed by three washes at the same temperature in 50% formamide, 2× SSC. The final series of washes is at high stringency (0.1× SSC, 60°). Slides are then ready for detection and must not be allowed to dry out at any point.

Experiments to analyze chromosome structure using FISH will often involve the use of two differently (biotin or digoxigenin) labeled probes simultaneously (e.g., sequences that are associated with the nucleoskeleton vs those released from this nuclear structure).[3] Biotin is usually detected with avidin–Texas Red conjugate and the signal can be amplified with biotinylated antiavidin antibody. Digoxigenin is detected with antidigoxigenin fluorescein isothiocyanate (FITC) (made in sheep) and amplified with antisheep antibody conjugated to FITC. Other fluorochrome/antibody combinations can be used, but care must be taken to avoid cross-reactivity between antibodies. The following method details the detection system used most commonly for the simultaneous detection of biotin- and digoxigenin-labeled probes.

Method

1. Peel rubber solution glue from slides and place them into a glass rack. Wash in 2× SSC at room temperature for 3 min with occasional agitation. Carefully discard coverslips that have floated off.

2. Wash slides twice, for 3 min each, in 2× SSC at 45° and then 4 × 3 min in 50% formamide, 2× SSC at 45°. The final high stringency wash is 4 × 3 min in 0.1× SSC at 60°. Transfer slides to 4× SSC, 0.1% Tween 20. Make sure that slides do not dry out at any time.

3. Incubate each slide with 40 μl blocking buffer (4× SSC, 5% dried milk powder) under a 22 × 40-mm coverslip for 5 min at room temperature.

4. Drain excess blocking buffer from slide and add 40 μl first antibody(s), e.g., FITC antidigoxigenin (sheep), avidin–Texas Red diluted in blocking buffer. All antibody dilutions should be spun in a refrigerated microfuge for 10 min at 4° before use.

5. Incubate slides with antibodies, under a coverslip, in a humidified chamber at 37° for 30 min.

6. Wash slides 3 × 2 min in 4× SSC, 0.1% Tween 20 at 37°.

7. Add next antibody, e.g., FITC antisheep (rabbit). Incubate and wash as in steps 5 and 6.

8. Incubate with biotinylated antiavidin (goat). After washing this layer, the final incubation is with avidin–Texas Red. Thus both biotin and digoxigenin labels have had one layer of amplification.

9. Wash slides as in step 6. Drain excess fluid from slide and mount in 40 μl antifadent (Vectashield, Vector Laboratories, is best in our experience) containing 2 μg/ml 4,6-diamidino-2-phenylindole (DAPI). Seal with rubber solution glue.

The simple counterstaining just described can be used to identify human and mouse chromosomes, especially after electronic enhancement of banding (see later). However, improved banding can be achieved by using actinomycin D in combination with DAPI.[12]

Image Analysis

Unnormalized, original gray-scale images for each fluorochrome should be collected with a cooled CCD camera (Photometrics 1400) so that the full dynamic range of information on pixel intensity is retained. It is important that the exposure times for each fluorochrome do not result in signal saturation. Chromosomes are identified from the inverted DAPI banding pattern, which can be enhanced digitally using commercially available algorithms (e.g., SmartCapture, Digital Scientific/Vysis).

To compare hybridization signals from chromatin fractions, commercially available comparative genomic *in situ* hybridization (CGH) programs can be used. Alternatively, use an algorithm to remove background from

[12] D. Schweizer, *Hum. Genet.* **57,** 1 (1981).

FITC and Texas Red images and then segment the chromosomes using the DAPI signal. Pixels from FITC and Texas Red images can be quantitated along chromosome length using data from the full width of the chromosomes and then compared.[3] Total DNAs from two normal genomes hybridized to metaphase chromosomes and detected with red and green fluorochromes produce the same hybridization pattern, painting the chromosomes yellow. Deviation from the yellow signal indicates that one of the genomes is either missing or has extra copies of that chromosome region. This concept is extended to analyze the relative proportions of DNA from the same genome that has partitioned into different biochemical fractions averaged over any given chromosome or resolvable chromosome region. One limitation of this approach is that the analysis is restricted by the resolution limit of metaphase chromosomes and the measured changes in hybridization signal are only the average value over ~10 Mb.[13]

Hybridization to Metaphase Chromosomes Decondensed by Extraction of Soluble Proteins

The organization of eukaryotic genomes into topologically constrained domains is an intriguing but poorly understood level of chromosome structure. It is illustrated graphically by the halo of DNA loops that emanate from a "metaphase scaffold" at the core or axis of mitotic chromosomes that have been swollen or extracted.[14] Normal cytogenetic preparations of metaphase chromosomes, prepared for FISH, are too condensed to allow resolution of the topological folding of DNA within the chromosomes. However, because swollen or extracted chromosomes are an order of magnitude wider (they are also longer), they can be used as substrates for FISH analysis to determine the nature of sequences located close to the chromosome core or in the DNA loops.[6]

Isolation of Suspensions of Unfixed Metaphase Chromosomes

Principle of the Method

The overnight Colcemid treatment of exponentially growing cells produces a high mitotic index. Because the mammalian nuclear membrane breaks down during mitosis, detergents that are selective for the cell membrane, e.g., digitonin, allow the isolation of a suspension of unfixed meta-

[13] S. de Manoir, E. Schrock, M. Bentz, M. R. Speicher, S. Joos, T. Reid, and P. Lichter, *Cytometry* **19,** 27 (1995).

[14] J. R. Paulson and U. K. Laemmli, *Cell* **12,** 817 (1977).

phase chromosomes away from intact interphase nuclei from such cultures.[15] Chromosomes may be purified further over a sucrose step gradient. Chromosomes are best prepared from large-scale cultures (>10^8 cells).

Method

1. Harvest cells from culture blocked overnight with Colcemid (for monolayer cultures, mitotic cells should have rounded up and can be detached selectively by mitotic shake off, thus minimizing contamination with interphase cells). Centrifuge cells at 450*g* for 5 min at room temperature.
2. Resuspend cells in 50 ml PBS. Disperse any cell clumps and count cells.
3. Centrifuge cells as in step 1 and resuspend in 0.075 *M* KCl at 2×10^6 cells/ml.
4. Incubate for 10 min at 37° and centrifuge at 275*g* for 5 min.
5. Resuspend the pellet in freshly prepared ice-cold polyamine (PA) buffer (15 m*M* Tris–HCl, pH 7.2, 0.2 m*M* spermine, 0.5 m*M* spermidine, 0.5 m*M* EGTA, 2 m*M* EDTA adjusted to pH 7.2 with 1 *M* KOH, 80 m*M* KCl, 20 m*M* NaCl) at 8×10^6 cells/ml.
6. Withdraw a 0.5-ml aliquot, fix in 3:1 methanol:acetic acid, and spread onto slides to assess the mitotic index.
7. Spin the rest of the cells at 200*g* at 4° for 5 min.
8. Resuspend the pellet at 10^7 cells/ml in cold PA buffer containing 1 mg/ml digitonin. [Because batches of digitonin differ in their solubility, add digitonin to PA buffer, heat to 37° for 15–20 min, filter (0.22 μm) to remove any undissolved digitonin, and transfer to ice.]
9. Vortex for 2×15 sec to burst the cell membranes and hence release metaphase chromosomes from mitotic cells.
10. Spin at 200*g* for 10 min at 4° to pellet the nuclei (these can be frozen in PA, 40% glycerol for use in other experiments). Reserve the supernatant containing the chromosomes.
11. Add another 1 ml of PA buffer with digitonin to the nuclear pellet and respin, as many chromosomes get trapped in among the nuclei.
12. Pool the two supernatants and load onto a sucrose step gradient prepared in 16-ml polyallomer tubes with 3 ml 60% (w/v) sucrose in PME (5 m*M* PIPES, pH 7.2, 5 m*M* NaCl, 5 m*M* $MgCl_2$, 1 m*M* EGTA), 2 ml 50% sucrose in PME, 2 ml 40% (w/v) sucrose in PME, and 2 ml 30% (w/v) sucrose in PME.
13. Spin at 4000*g* in a swing-out rotor at 4° for 15 min with no brake. Aspirate until the middle of the 40% step and collect the flocculent white

[15] A. B. Blumenthal, J. D. Diedan, L. N. Kapp, and J. W. Sedat, *J. Cell Biol.* **81,** 255 (1979).

material at the 40–50% and 50–60% interfaces using a Pasteur pipette. Small aliquots of chromosomes can be frozen at −70°.

Extraction of Isolated Metaphase Chromosomes and FISH

Principle of Method

Isolated metaphase chromosomes can be swollen and soluble proteins extracted from them with a variety of agents. The chelation of divalent cations during chromosome preparation induces a degree of swelling[16]; however, progressively more extensive decondensation of the chromosomes is achieved by incubating the chromosomes in buffers containing high salt, polyanions, or the ionic detergent lithium 3,5-diiodosalicylate (LIS).[14] In practice, we have found that although chromosomes extracted with 2 *M* NaCl, 25 m*M* LIS, or 4 *M* NH_4SO_4 retain a good morphology when stained with DAPI, the resulting decondensed chromosomes are too fragile to retain their morphology after the harsh conditions of FISH.[6] We routinely examine chromosomes that have been extracted with concentrations of NaCl up to 1.8 *M*. Salt extraction is performed by lowering the slides horizontally, onto a perforated platform, into buffers with progressively increased concentrations of NaCl. Agitation of the chromosomes prior to fixation must be minimized to prevent streaking of chromatin fibers. We do not recommend the use of poly(L-lysine)-coated slides as this inhibits chromosome decondensation. After extraction and fixation the chromosomes are ready for analysis by standard FISH protocols with the following modifications to minimize chromosome damage: no RNase treatment of the slides is necessary, it is important that the pH of the slide denaturant is checked before use (pH 7.5) and that denaturation times are kept to a minimum (1.5–2 min). Additionally, hybridization is only allowed to proceed for 24 hr.

Because of the decondensed nature of chromosomes that have been extracted with high salt, relatively high concentrations (10 μg/ml) of DAPI are needed in order to see the halo of DNA loops that surround the chromosome core.

Method

1. Take a 10-μl aliquot of isolated chromosome suspension (~10^6 chromosomes) and smear across a microscope slide using the side of a Gilson tip. Cover and allow chromosomes to settle onto the slide overnight.

[16] M. P. F. Marsden and U. K. Laemmli, *Cell* **17,** 849 (1979).

2. Lay the slides onto a wire mesh and gently lower into chromosome isolation buffer (CIB): 10 mM Tris, pH 8.0, 10 mM EDTA, pH 8.0, 0.1% Nonidet P-40, 0.1 mM $CuSO_4$, 20 μg/ml PMSF. Leave for 5 min.

3. Transfer the slides on their carrier into CIB containing 0.5 M NaCl and leave for 5 min.

4. Repeat step 3, increasing the NaCl concentration until the desired concentration is reached. Remove slides and carrier from the high salt solution and drain off excess liquid on a kitchen towel.

5. Transfer the slides and carrier into fresh 3:1 (v/v) methanol:acetic acid. Leave for 10 min. Repeat with fresh fix.

6. Remove the slides from fix and air dry.

After signal detection the position of hybridization signals is analyzed with respect to the chromosome core, which is visible by DAPI staining. The extent of extraction and the preservation of chromosome morphology usually vary across the slide. The best chromosomes to analyze are those with a good level of decondensation, but where the symmetry of the sister chromatids is preserved.[6] Signals from bacterial artificial chromosomes (BACS), PI-based artificial chromosomes (PACS), and yeast artificial chromosomes (YACS) should be strung out as a series of dots rather than the single spots of hybridization seen with these probes on fully condensed chromosomes.[6] Another key indicator of chromosome morphology is the appearance of centromeric regions; these should remain condensed and stain brightly by DAPI as the DNA of this part of the chromosome remains tightly packaged under the residual kinetochore after extraction, and many of the kinetochore proteins are known to be scaffold components.[6,14,17]

Analyzing Nuclear Matrix by FISH

Many methods have been developed for decondensing chromatin for FISH analysis. However, most of these were designed to allow high resolution mapping of DNA sequences with respect to each other and so preservation of nuclear structure was not a consideration.[18–20] The protein composition of scaffolds prepared from mitotic chromosomes is simple, whereas the nuclear "matrix" is complex, containing components of the peripheral lamina, nuclear pores, and residual nucleoli, as well as an ill-defined internal network.[21] Just as the organization of specific sequences with respect to

[17] W. C. Earnshaw, B. Halligan, C. A. Cooke, and N. Rothfield, *J. Cell Biol.* **98,** 352 (1984).

[18] H. H. Q. Heng, J. Squire, and L.-C. Tsui, *Proc. Natl. Acad. Sci. U.S.A.* **89,** 9509 (1992).

[19] I. Parra and B. Windle, *Nature Genet.* **5,** 17 (1993).

[20] T. Haaf and D. C. Ward, *Hum. Mol. Genet.* **3,** 629 (1994).

[21] J. Mirkovitch, S. M. Gasser, and U. K. Laemmli. *Phil. Trans. Roy. Soc. Lond. B* **317,** 563 (1987).

the "core" or "scaffold" of the metaphase chromosome can be analyzed by FISH to chromosome preparations from which soluble proteins have been removed, so can similarly treated interphase nuclei be used as substrates for FISH to investigate the partitioning of specific sequences into the nuclear halo or their retention within the residual extracted nucleus.[22]

Nuclei can be prepared for this purpose using any established protocol and can be deposited onto slides either by incubation on the bench (as for metaphase chromosomes in the preceding protocol) or by cytocentrifugation.[22] Extraction can be done by incubating slides for 5 min in suitable buffers such as CIB (preceding protocol) supplemented with NaCl (up to 2 *M*) or 25 m*M* LIS, in 2 *M* NaCl buffer (2 *M* NaCl, 10 m*M* PIPES, pH 6.8, 10 m*M* EDTA, 0.1% digitonin, 0.05 m*M* spermine, 0.125 m*M* spermidine),[22,23] or in LIS buffer (5 m*M* HEPES/KOH, pH 7.4, 0.25 m*M* spermine, 2 m*M* KCl, 0.1% digitonin, 25 m*M* LIS, 2 m*M* EDTA).[22] Results obtained using these different extraction buffers appear to be comparable.[22]

After extraction, extracted nuclei can be fixed by baking[22] or with 3:1 methanol : acetic acid or 4% formaldehyde. Hybridization conditions should be as gentle as possible, as with extracted metaphase chromosomes. After hybridization and detection it is determined whether the signal is located within the residual nucleus or released into the surrounding halo of DNA loops.

Concluding Remarks

This article has served to illustrate how FISH, in its different guises, can be used to investigate the structure of both interphase and mitotic chromatin/chromosomes. These approaches both complement and augment more classical biochemical approaches to chromatin studies. In particular, chromosome painting with chromatin fractions gives a picture of the global properties of the sequences partitioning through a specific fractionation procedure. This can be used to assess whether individual sequences being studied using the same fractionation are typical of the whole, thus ensuring that any conclusions reached from these studies are likely to be generally applicable.

FISH experiments where the chromatin template (chromosome or nucleus) on the microscope slide is itself manipulated prior to hybridization open up new views of the higher order chromatin structure that have been difficult to study by other techniques. In particular, these approaches should help resolve the enduring controversies about the existence/nature of the operationally defined scaffold, matrix, and skeleton.

[22] M. G. Gerdes, K. C. Carter, P. T. Moen, and J. B. Lawrence, *J. Cell Biol.* **126,** 289 (1994).
[23] B. Vogelstein, D. M. Pardoll, and D. S. Coffey, *Cell* **22,** 79 (1980).

[38] Analysis of Nuclear Organization in *Saccharomyces cerevisiae*

By M. GOTTA, T. LAROCHE, and S. M. GASSER

Introduction

To study the three-dimensional organization of the nucleus, it is necessary to visualize specific DNA sequences or proteins while preserving the native nuclear structure. Yeast provides specific problems. Unlike higher eukaryotes, in which the nuclear lamina provides a rigid support for the shape of the nucleus, yeast nuclei collapse in nonionic detergents.[1-3] Morphological work in yeast therefore requires that cells or nuclei be well fixed by chemical cross-linking prior to permeabilization of the nuclear envelope. An alternative approach is to localize endogenous proteins in yeast by fusion with green fluorescent protein (GFP)[4] or derivatives thereof or to localize specific chromosome domains by targeting a DNA-binding factor fused to GFP (e.g., *lac* or *tet* repressors) to clusters of its recognition consensus inserted at the chromosomal site of interest.[5-7] However, these methods are not particularly well adapted to labeling multiple targets, nor do they allow localization of a specific protein relative to a given DNA sequence or a given structure, such as the nuclear envelope. The combined immunofluorescence (IF) fluorescent *in situ* hybridization (FISH) protocol described in this article has been optimized to analyze the subnuclear localization of several proteins at once and to localize proteins relative to DNA sequences.[8,9]

Earlier FISH protocols for budding yeast included treatments with protease after fixation or extraction with sodium dodecyl sulfate (SDS) and

[1] M. E. Cardenas, T. Laroche, and S. M. Gasser, *J. Cell Sci.* **96,** 439 (1990).
[2] F. Klein, T. Laroche, M. E. Cardenas, J. F. Hofmann, D. Schweizer, and S. M. Gasser, *J. Cell Biol.* **117,** 935 (1992).
[3] F. Palladino, T. Laroche, E. Gilson, L. Pillus, and S. M. Gasser, *Cold Spring Harb. Symp. Quant. Biol.* **58,** 733 (1993).
[4] S. L. Shaw, E. Yeh, K. Bloom, and E. D. Salmon, *Curr. Biol.* **7,** 701 (1997).
[5] C. Michaelis, R. Ciosk, and K. Nasmyth, *Cell* **91,** 35 (1997).
[6] A. F. Straight, A. S. Belmont, C. C. Robinett, and A. W. Murray, *Curr. Biol.* **6,** 1599 (1996).
[7] A. F. Straight, W. F. Marshall, J. W. Sedat, and A. W. Murray, *Science* **277,** 574 (1997).
[8] F. Palladino, T. Laroche, E. Gilson, A. Axelrod, L. Pillus, and S. M. Gasser, *Cell* **75,** 543 (1993).
[9] M. Gotta, T. Laroche, A. Formenton, L. Maillet, H. Scherthan, and S. M. Gasser, *J. Cell Biol.* **134,** 1349 (1996).

0076-6879/99 $30.00

Triton X-100 to enhance hybridization efficiency.[10,11] Protocols presented here omit protease and harsh detergent treatments. Cells are fixed after spheroplasting in osmotically buffered growth medium to efficiently preserve nuclear structure. In this way, the cell wall does not itself become a target of the cross-linking reagents, a phenomenon that can prevent efficient diffusion of the fixative to the nucleus. At the same time, we obtain high efficiency labeling with both antibodies and DNA probes. The described protocol can also be performed on cells that are fixed prior to spheroplasting if preservation of the cell shape is desired.[9]

The double *in situ*/immunofluorescence staining described here provides a powerful tool for the characterization of nuclear protein localization and the positioning of specific chromosomal regions. Confocal microscopy of the fixed cells demonstrates that nuclei maintain their three-dimensional organization throughout the procedure.[8,9] We have found that immunolabeling with antibodies specific for the nuclear pore complex is a sensitive means to monitor both the integrity of the nuclear envelope and the size of the nucleus. The diameter of an intact diploid nucleus is about 2 μm, whereas that of a flattened chromatin mass is ~6–8 μm. However, by aiming to maximize the maintenance of nuclear integrity, we limit the sensitivity of the FISH assay, and unique sequence probes need to contain from 4 to 6 kb for reproducible detection. Moreover, the combination of IF and FISH can result in a reduced sensitivity for immunolocalization, as not all antibodies resist the hybridization conditions. For specific applications, such as immunostaining alone, immunostaining on cells fixed prior to spheroplasting, or double GFP immunolocalization, this protocol can be modified, as indicated at the end of the article.

Yeast Strains and Media

Diploid yeast cells are preferred for these studies because nuclei are nearly twice the size of haploid nuclei. For cell synchronization experiments, a/a diploids can be used after release from α factor arrest.[12] There is a significant variation in the efficiency with which different strains are converted to spheroplasts, probably reflecting differences in the cell wall composition. Whenever mutant and wild-type strains are compared, it is preferable that they share the same genetic background. Moreover, it is important to note that the efficiency of spheroplasting can be affected by growth

[10] V. Guacci, E. Hogan, and D. Koshland, *J. Cell Biol.* **125,** 517 (1994).

[11] B. M. Weiner and N. Kleckner, *Cell* **77,** 977 (1994).

[12] M. D. Rose, F. Winston, and P. Hieter, "Methods in Yeast Genetics." Cold Spring Harbor Laboratory Press, Cold Spring Harbor, NY, 1990.

conditions, i.e., carbon source, rate of growth, temperature, and cell density at the time of harvest. The best results are obtained with cells grown on rich medium (YPD)[12] at 30° to mid logarithmic phase (1–2 × 10^7 cells/ml).

Antibody Purification and Specificity

Polyclonal antibodies are a good tool for immunofluorescence because they recognize a variety of different epitopes and are likely to recognize at least one present on the denatured protein in fixed cells. However, because most rabbit sera react with a variety of yeast proteins in addition to the specific antigen, we strongly recommend that all antibodies used for yeast immunofluorescence be affinity purified against recombinant antigen as follows.

1. Perform a Western transfer of the recombinant antigen onto a nitrocellulose filter. After staining with Ponceau red (0.05% in 3% TCA), cut out the strip containing the protein band.
2. Wash the nitrocellulose strip 3 × 10 min in 1× TEN (20 m*M* Tris–Cl, pH 7.5, 1 m*M* EDTA, 140 m*M* NaCl), 0.05% Tween 20. Block excess protein-binding sites by incubating in 1× TEN + 0.05% Tween 20 + 1% dry milk powder at room temperature for 30 min.
3. Incubate the strip with 10–50 μl of serum (depending on the titer of the antibody) in 1 ml of 1× TEN, 0.05% Tween 20, 1% dry milk powder overnight at 4° with constant agitation (rocker or wheel).
4. Remove the supernatant and wash the strip 3 × 10 min in 1× TEN, 0.05% Tween 20 at room temperature.
5. Elute the bound immunoglobulin (Ig) with 300 μl of 100 m*M* glycine, pH 3.0, for 2 min.
6. Immediately raise the pH to 7.0 by adding 1 *M* Tris base (the volume needed should be determined before starting) and place on ice.
7. Repeat the elution once or twice and pool the elutions. The number of elutions required depends on the antibody and the first time elutions should be checked separately. It may be necessary to use glycine at a lower pH (pH 1.9, e.g.) to elute the antibody.
8. Once purified, the antibodies can be stored as aliquots at −80°. Stabilization is enhanced by the addition of 1–2% ovalbumin and 20% glycerol. The antibody is used at a dilution of 1:2 or more for immunofluorescence. The specificity of the purified antibodies should be demonstrated by Western blot and immunofluorescence on strains lacking the protein in question.

Monoclonal antibodies have the advantage of recognizing a single epitope, thereby decreasing the probability of nonspecific background staining.

Therefore they do not have to be affinity purified before use. However, because this same epitope may be shared by other proteins, it is essential to test the specificity of the antibody on Western blots. The obvious disadvantage of monoclonal antibodies is that the unique epitope may be masked or denatured under the conditions used to fix and permeabilize cells.

Secondary antibodies coupled to fluorophore are used to detect primary antibodies. They should be tested for a lack of cross-reactivity with yeast proteins by performing immunofluorescence in the absence of the primary antibody. To eliminate nonspecific background, it is advisable to preabsorb the secondary antibody on fixed yeast cells as described in part I.

Choice of Fluorophore

The choice of the fluorophore depends on the number of different probes that need to be visualized on each sample and on the filter sets available for the microscope. It is important to choose those that can be excited and visualized independently. If there is overlap between emission spectra, some signal must be attenuated in order to avoid "bleed through." Some of the more commonly used fluorophores include FITC (A = 494 nm, E = 518 nm), DTAF (A = 494 nm, E = 518 nm), Cy3 (A = 554 nm, E = 566 nm), Texas Red (A = 595 nm, E = 615 nm), and Cy5 (A = 649 nm, E = 666 nm).

In Situ Hybridization Probes

To label probes for *in situ* hybridization, plasmids containing the target sequence can be used or the DNA can be produced by polymerase chain reaction (PCR) using appropriate primers. Probes for FISH are labeled by a nick-translation protocol in which dTTP is replaced by a 1:1 mixture of digoxigenin-derivatized dUTP (dig-dUTP, Boehringer Mannheim, Rotkreuz, Switzerland) and dTTP. The optimal probe size is 4–6 kb.

Reagents

10× dNTP mix: 500 m*M* Tris–HCl, pH 7.8, 50 m*M* $MgCl_2$, 100 m*M* 2-mercaptoethanol, 0.2 m*M* dCTP, dGTP, and dATP, and 0.1 m*M* dTTP
DNA polymerase I
DNase I
Nuclease-free bovine serum albumin (BSA)
Digoxigenin (dig)-dUTP (Boehringer Mannheim)
[α-^{32}P]dATP

EDTA–NaOH, pH 7.4
ssDNA (sheared and heat denatured)
Dextran sulfate (Pharmacia, Piscataway, NJ)
Deionized formamide
20× SSC,[13] pH 7.0
3 *M* sodium acetate, pH 7.0
100 and 75% ethanol

1. Mix on ice: 2 μg probe fragment (phenol free, RNA free), 10 μl of 10× dNTP mix, 1 μl (1 nmol/μl) dig-dUTP, 1 μl [α-^{32}P]dATP diluted 1/1000 in H_2O, 1 μg of nuclease-free BSA, 0.1 ng DNase I, 10 units of DNA polymerase I, and H_2O to 100 μl.
2. Incubate for 3 hr at 16°.
3. Add 5 μl of 300 m*M* EDTA–NaOH, pH 7.4.
4. Denature the probe for 5 min at 98°.
5. Chill on ice.
6. Add 20 μg sonicated salmon sperm DNA, 12 μl of 3 *M* sodium acetate, and 2 volumes of chilled 100% ethanol.
7. Leave 30 min to 1 hr at −70°.
8. Centrifuge for 10 min at 4°.
9. Wash the pellet twice in 1 ml of chilled 75% ethanol.
10. Check the reaction by monitoring the [α-^{32}P]dATP incorporation.
11. Dry the pellet 10 min using a Speed-Vac.
12. Add the probe mix (50% formamide, 10% dextran sulfate, 2× SSC), dissolve the pellet at 37° for 20 min, and keep on ice before the hybridization. The final concentration of the probe should be between 10 and 100 ng/μl. The probe can be stored either in hybridization mix or as a dried pellet at −20°.

Part I: Immunofluorescence

1. Grow the cells overnight to about 0.5–1 × 10^7 cells/ml in 50 ml YPD or selective media.[12]
2. Harvest the cells at 1200*g* for 5 min at room temperature in preweighed 50-ml polypropylene tubes (keep the Erlenmeyer flasks).
3. Decant the supernatant and weigh the cell pellet.
4. Resuspend the cells in 1 ml/0.1 g of cells in 0.1 *M* EDTA–KOH (pH 8.0) and 10 m*M* dithiothreitol (DTT).
5. Incubate at 30° for 10 min with gentle agitation.
6. Collect the cells by centrifugation at 800*g* for 5 min at room temperature.

[13] J. Sambrook, E. F. Fritsch, and T. Maniatis, "Molecular Cloning: A Laboratory Manual," 2nd ed. Cold Spring Harbor Laboratory Press, Cold Spring Harbor, NY, 1989.

7. Carefully resuspend the cell pellet in 1 ml/0.1 g cells YPD + 1.2 *M* sorbitol (mix 22 g sorbitol with 100 ml YPD). To resuspend evenly, suspend the cell pellet first in 500 μl.
8. Add lyticase (β-glucanase[14]) to 1000 U/ml and Zymolyase (20T, Seikagaku) to 400 μg/ml.
9. Incubate at 30° in the original Erlenmeyer flask with gentle agitation and monitor spheroplast formation in the microscope at 5, 10, 15, and 20 min. Cells become round, dark gray, and lose the dark peripheral ring. It is best to harvest prior to completion of spheroplasting for optimal maintenance of the nuclear structure.
10. Transfer to the polypropylene tube and add paraformaldehyde to a final concentration of 4% from a stock of 20%. The paraformaldehyde should be freshly prepared before the experiment begins by mixing 5 g of paraformaldehyde, 15 ml H_2O, and 25 μl of 10 *N* NaOH. Dissolve at 62° in a closed bottle for about 30 min with occasional shaking. Add 5 g of sorbitol and H_2O up to 25 ml.
11. Fix for 20 min at room temperature (keep the tubes horizontal with occasional gentle shaking).
12. Dilute with YPD + 1.2 *M* sorbitol to 40 ml. Centrifuge for 5 min at 800*g* at room temperature.
13. Wash twice in 40 ml YPD + 1.2 *M* sorbitol, resuspending gently. Centrifuge for 5 min at 800*g* at room temperature.
14. Resuspend 0.5 g spheroplasts in 0.8 ml YPD. The concentration of the cells in this suspension should be such that only one layer of nonconfluent cells will adhere to the slide. Leave a drop on each spot of the slide (Polylabo, Geneva, Switzerland; super-Teflon slides) for 1–2 min to allow the spheroplasts to adhere on the surface of the slide, and take away as much liquid as possible using a pipette. Superficially air dry for 2 min. Perform all the following washes by immersing the slide in a Coplin jar containing the appropriate buffer.
15. Put the slides in methanol at −20° for 6 min (prechill the methanol).
16. Transfer the slides to acetone at −20° for 30 sec (prechill the acetone).
17. Air dry for 3–5 min.
18. Incubate slides in 1× phosphate buffered saline (PBS)[13] + 1% ovalbumin + 0.1% Triton X-100 for 10 min or more. Shake gently two or three times at room temperature. After this step, the cells appear transparent and the nucleus can be seen as a dark spot. This is an indication of good spheroplasting. If this is not the case, it may help

[14] S. H. Shen, P. Chretien, L. Bastien, and S. N. Slilaty, *J. Biol. Chem.* **266,** 1058 (1991).

to leave the slides for a longer time in PBS + 1% ovalbumin + 0.1% Triton X-100 (go to part II for FISH alone).

19. Dry the black surface and bottom surface of the slides with a paper towel.
20. Cover each spot on the slide with 10 μl of the appropriate primary antibody diluted as required in PBS containing 0.1% Triton X-100. If affinity-purified antibodies are used, they should be diluted two- or threefold in 0.5× PBS + 0.1% Triton X-100 to avoid high salt concentrations.
21. Incubate for 1 hr at 37° in a humid chamber or overnight at 4°. In the latter case the slides should be covered with a coverslip to avoid drying of the antibody solution.
22. Preabsorb the secondary antibody on yeast cells. For this purpose, use the remaining fixed spheroplasts by washing them three times in PBS and resuspending them in 1 ml of PBS. Dilute the secondary antibody (stock is usually 1 mg/ml) 1:50 in this spheroplast suspension and incubate for 1 hr on a rotating wheel at 4° in the dark. Centrifuge at top speed, collect the supernatant, and add Triton X-100 to a final concentration of 0.1%. Store on ice until use.
23. After the primary antibody incubation, wash the slides 3 × 5 min in PBS + 0.1% Triton X-100 at room temperature.
24. Dry the black surface and bottom surface of the slides. Cover each slide with 10 μl/spot of the fluorescent secondary antibody prepared as described in step 22 and incubate for 1 hr at 37° in the humid chamber in the dark.
25. After the secondary antibody, wash the slides 3 × 5 min in PBS + 0.1% Triton at room temperature.
26. Postfix the cells in 4× SSC and 4% paraformaldehyde for 20 min at room temperature after the last wash. Rinse 3 × 3 min in 4× SSC. This step is important when continuing with FISH, as the secondary and/or primary antibodies tend to dissociate under the harsh conditions of *in situ* hybridization. It is not required when performing only immunofluorescence.

Part II: FISH

1. Immerse cells in 4× SSC, 0.1% Tween 20, and 20 μg/ml preboiled RNase A. Incubate overnight at room temperature (in the dark if immunofluorescence has been performed previously).
2. Wash in H_2O.
3. Dehydrate in 70, 80, 90, and 100% ethanol consecutively at −20°, 1 min each bath.

4. Air dry.
5. Add 10 μl per spot of 2× SSC and 70% formamide. Cover with a coverslip. Leave for 5 min at 72°. (Place the slide on top of an aluminum block that is partially submerged in a 72° water bath. On the narrow edges of the slide, place a few drops of water, which will spread between the aluminum block and the slide, improving the heat conductance.)
6. Dehydrate in 70, 80, 90, and 100% ethanol consecutively at −20°, 1 min each bath.
7. Air dry.
8. Apply hybridization solution: 3 μl for each spot. The optimal concentration of probe depends on the sequence and must be determined empirically. Place a coverslip on top, avoiding air bubbles, and seal with nail polish.
9. Incubate for 10 min at 72°.
10. Incubate for 24–60 hr at 37°.
11. Remove the coverslip and wash twice in 0.05× SSC for 5 min at 40°.
12. Incubate in BT buffer (0.15 *M* $NaHCO_3$, 0.1% Tween 20, pH 7.5) 0.05% BSA, 2 × 30 min at 37° in the dark.
13. Add sheep antidigoxigenin diluted 1:50 in BT buffer without BSA + the secondary goat–antimouse or rabbit antibody 1:50 (for refreshing the immunofluorescence signal, if necessary) (Boehringer Mannheim). Stock solutions are usually 1 mg/ml.
14. Incubate for 1 hr at 37° in a humid chamber.
15. Wash 5 × 3 min in BT buffer.
16. Add 15 μl per spot of antifading (PBS, 50% glycerol, 24 mg/ml diazabicyclo[2.2.2.]octane or DABCO, pH 7.5). Cover with a coverslip, avoiding air bubbles, and seal with nail polish. Keep the slides at 4° in the dark.

To visualize the DNA, various DNA-staining agents such as ethidium bromide (diluted to 1 μg/ml in antifading), DAPI (1 μg/ml), or POPO-3 (Molecular Probes, Eugene, OR) can be used, depending on the filters of the microscope and on the secondary antibodies that have been used for FISH and IF. It is important to check that the wavelengths of absorbance and emission do not interfere with the fluorescent probes already used.

Variations

A. Fluorescence *in situ* hybridization only. If the goal is simply to localize DNA sequences, follow the protocol to part I, step 17, and then skip to part II, step 1. We have had equivalent success using biotin-dUTP to derivatize FISH probes. The labeling protocol can be modified to use biotin-16-dUTP instead of dig-dUTP, and the secondary detection can be achieved by streptavidin conjugated to a fluorescent molecule (Enzo Pharmaceuticals, Farmingen, NY). In principle, two differently labeled probes can be combined for double detection.

B. Immunofluorescence only. If the goal is simply to localize proteins, follow only part I of the protocol to step 26 and mount the slide in DABCO with the appropriate DNA stain. It is possible to study the localization of more than one protein at a time if using primary antibodies from different species (e.g., mouse, rabbit, sheep) with species-specific secondary antibodies. Mixtures of the primary antibodies or mixtures of the secondary antibodies can be made to reduce incubation times. It is essential to test the secondary antibody with each of the primary antibodies separately to ensure that they do not cross-react. In case of cross-reactivity, different secondary antibodies should be tested.

C. When maintenance of the cell shape and cytosolic structures is required, this same IF/FISH protocol can be performed by first fixing the cells in growth medium for 10 min at 30° in 4% paraformaldehyde prior to spheroplasting. The conversion to spheroplasts is done identically as described earlier, except with 3000 U/ml of lyticase and 0.6–1.2 mg/ml of Zymolyase (20T). The cells should be checked microscopically every 5 min, and spheroplasting is completed when they appear transparent. Buds should remain attached and the cell shape is usually maintained.

D. An alternative way to localize proteins both in living and in fixed cells is to fuse the gene with that encoding GFP.[4] However, proteins fused to GFP must be assayed to ensure that they retain as many of their physiological attributes as can be measured. For very abundant or overexpressed protein, the GFP fluorescence will survive the immunofluorescence protocol. When this is not the case, antibodies raised against GFP can be used successfully in combination with antibodies raised against the second antigen of interest.

E. There are specific applications in which it is not necessary to preserve nuclear integrity to draw some conclusions about the localization of DNA sequences, e.g., assays for the pairing of mitotic or meiotic chromosomes.[10,11] However, one must be careful not to draw conclusions about nuclear architecture from results obtained with flattened preparations.

Evaluation of the Method and Troubleshooting

As an assay for nuclear integrity, dimensions of the nucleus can be calculated from ethidium bromide or DAPI fluorescence. When nuclei are well preserved, diameters of 2.0 ± 0.2 μm in the *XY* plane and 2.4 ± 0.2 μm in the *Z* axis are measured. This 20% distortion along the *Z* axis is also observed when cells are fixed prior to spheroplasting and reflects an integral *Z*-stretch function of the confocal microscope and software program. However, if cells are treated with detergents prior to fixation or if spheroplasts are protease treated prior to FISH, cells become flattened, no longer maintaining this spherical shape,[2,3,9] and the number of *Z* sections possible is reduced to one or two, indicating that the flattened nucleus is less than 0.5 μm in height.

As an independent assay for nuclear integrity, a mouse monoclonal antibody raised against the human nuclear pore protein p62, a homolog of yeast Nsp1p,[15] can be used. Immunofluorescence of Nsp1p shows a ring-like staining at the nuclear periphery typical of nuclear pore staining. This staining is lost if cells are spheroplasted to completion prior to formaldehyde fixation or if the nuclear envelope is disrupted by detergents. This confirms that the presence of a punctate/ring-like staining with anti-p62 correlates with a degree of nuclear integrity. In addition, this antibody allows monitoring of the relationship of the immunofluorescence signals to the nuclear periphery as defined by the nuclear pore signal.

The resolution of the light microscope depends on the fluorochrome used. Under optimal conditions with DTAF or FITC, a resolution of 0.2 μm can be achieved in the *XY* axis with optimal optics and alignment. In the *Z* axis, resolution is much less efficient (0.6 μm). Quantitation of colocalization of two signals can be carried out on computer graphic representations of the fluorescent signals from one focal section taken as near as possible to the equator of the field of cells. A threshold for contour tracing can be set using Adobe Photoshop v4.0 software after a normalization of each filter channel independently to give the same maximum signal. To control whether a given localization is statistically significant, a simulation by computer randomization for a given size and number of signals has been developed (contact Philippe Bucher, ISREC, e-mail pbucher@eliot.unil.ch). Under standard imaging conditions, it is essential to check that no signal from one fluorochrome can be detected on the other filter set and that image capture and background subtraction are done uniformly on all images to allow direct comparison.

[15] C. V. Wimmer, V. Doye, P. Grandi, U. Nehrbass, and E. C. Hurt, *EMBO J.* **11,** 5051 (1992).

Section IV

Chromatin Remodeling Complexes

[39] Histone Acetyltransferases: Preparation of Substrates and Assay Procedures

By CRAIG A. MIZZEN, JAMES E. BROWNELL, RICHARD G. COOK, and C. DAVID ALLIS

Introduction

Reversible, posttranslational acetylation of ε-amino groups of evolutionarily conserved lysine residues within the amino-terminal domains of core histones has long been correlated with chromatin replication and transcription.[1-5] Acetylation is catalyzed by enzymes referred to collectively as histone acetyltransferases (HATs), which promote the transfer of acetate from the substrate acetyl-CoA to histone and possibly other protein substrates (see later). Acetylation converts these primary ε-amines, which are positively charged at physiological pH, to uncharged secondary amines. This decrement in charge, and possibly the presence of the nonpolar acetyl moiety itself, is thought to alter functional interactions of histone amino termini with DNA and/or other proteins. Much of the information available regarding histone acetylation comes from studies employing radioisotopic acetate and acetyl-CoA to monitor acetylation *in vivo* and *in vitro*. This article describes an activity gel assay and a liquid assay that employ radioisotopic acetyl-CoA to detect acetyltransferase activity in cell extracts and preparations of recombinant proteins. Because experimental evidence demonstrates that various HATs display distinct histone/chromatin substrate specificities *in vitro*, we discuss the preparation of these substrates. In addition, methods associated with the identification of sites acetylated by HATs *in vitro* are described. Methods for the preparation of HATs are not described here but are available in many of the references cited.

Selection of Substrates

Historically, two types of HATs have been recognized based on their subcellular localization and (presumed) substrate specificity.[6] Type A HATs

[1] S. Y. Roth and C. D. Allis, *Cell* **87,** 5 (1996).
[2] P. A. Wade, D. Pruss, and A. P. Wolffe, *Trends Biochem. Sci.* **22,** 128 (1997).
[3] M. Grunstein, *Nature* **389,** 349 (1997).
[4] C. A. Mizzen and C. D. Allis, *Cell. Mol. Life Sci.* **54,** 6 (1998).
[5] J. T. Kadonaga, *Cell* **92,** 307 (1998).
[6] J. E. Brownell and C. D. Allis, *Curr. Opin. Gen. Dev.* **6,** 176 (1996).

0076-6879/99 $30.00

are nuclear enzymes commonly thought to affect gene transcription by promoting the accessibility of the DNA template through the modification of nucleosomal histones. Type B HATs have been described as predominantly cytoplasmic enzymes (however, see Verreault *et al.*[7]), which acetylate newly synthesized histones H3 and H4 prior to nuclear import. Before initiating assays, workers should consider findings that illustrate that these distinctions are arbitrary in some instances and that selection of the most appropriate protein substrate to use in assays for a given HAT may have to be made empirically. Furthermore, the appropriate use of different substrates may facilitate the characterization of a given HAT and may also permit specific detection in crude preparations containing other HAT activities with distinct substrate specificities.

Replication-associated acetylation, presumably catalyzed by type B HATs, has only been described for histones H3 and H4.[1] However, yeast and human HAT B activities described to date have been assayed successfully using crude mixtures of histones.[7–9] Synthetic peptides mimicking the amino terminus of histone H4 have also been employed.[9,10] Note that these enzymes do not acetylate nucleosomal histones *in vitro* under the conditions employed.

Analyses of transcriptionally active or competent chromatin have detected acetylation of all four core histones.[11] However, the type A HATs described to date differ significantly in their ability to acetylate nucleosomal histones compared to free histones and also differ with respect to the identities of the histones they acetylate *in vitro*. Some enzymes, including recombinant p300/CBP, PCAF, SRC-1, and, to a lesser extent, $TAF_{II}250$, are active with nucleosomal substrates *in vitro*,[12–15] but other HATs such as yeast GCN5 require the presence of other proteins in multiprotein com-

[7] A. Verreault, P. D. Kaufman, R. Kobayashi, and B. Stillman, *Curr. Biol.* **8,** 96 (1997).

[8] M. R. Parthun, J. Widom, and D. E. Gottschling, *Cell* **87,** 5 (1996).

[9] L. Chang, S. S. Loranger, C. Mizzen, S. G. Ernst, C. D. Allis, and A. Annunziato, *Biochemistry* **36,** 469 (1997).

[10] S. Kleff, E. D. Andrulis, C. W. Anderson, and R. Sternglanz, *J. Biol. Chem.* **270,** 24674 (1995).

[11] A. Csordas, *Biochem. J.* **265,** 23 (1990).

[12] V. V. Ogryzko, R. L. Schiltz, V. Russanova, B. H. Howard, and Y. Nakatani, *Cell* **87,** 953 (1996).

[13] X.-J. Yang, V. V. Ogryzko, J. Nishikawa, B. H. Howard, and Y. Nakatani, *Nature* **382,** 319 (1996).

[14] T. E. Spencer, G. Jenster, M. M. Burcin, C. D. Allis, J. Zhou, C. A. Mizzen, N. J. McKenna, S. A. Onate, S. Y. Tsai, M.-J. Tsai, and B. W. O'Malley, *Nature* **389,** 194 (1997).

[15] C. A. Mizzen, X.-J. Yang, T. Kokubo, J. E. Brownell, A. J. Bannister, T. Owen-Hughes, J. Workman, L. Wang, S. L. Berger, T. Kouzarides, Y. Nakatani, and C. D. Allis, *Cell* **87,** 1261 (1996).

plexes for activity with nucleosomal substrates.[16] Although nucleosomal HATs such as p300/CBP, PCAF, SRC-1, and $TAF_{II}250$ are also active with free histone substrates, this may not be a property of all type A HATs, as an activity acting exclusively on nucleosomal histones has been detected in *Tetrahymena.*[17] Thus, putative acetyltransferases should be assayed using free histone and nucleosomal substrates in parallel whenever possible. In addition, certain enzymes identified initially as type A HATs by virtue of their abilities to acetylate histone substrates are also able to acetylate nonhistone proteins,[18] demonstrating that substrates other than histones and chromatin can be employed *in vitro*. However, the ease of preparation of nucleosomal and free histone substrates, combined with the fact that the histones modified by a given HAT are identified readily by fluorography of SDS–PAGE gels, makes these two substrates good choices for initial experiments.

Synthetic peptides mimicking the amino termini of histones or putative nonhistone acetyltransferase substrates have also been used to monitor type A acetyltransferase activities. This approach has the advantage that series of homologous peptides with ε-*N*-acetyllysine incorporated at different positions during synthesis can be employed to determine the site specificity of HATs *in vitro.*[12,15] Although it has not been demonstrated that site specificities displayed by a specific HAT with peptide substrates are equivalent to those with other substrates *in vitro* or *in vivo*, these assays can permit discrimination of different HAT activities and represent a convenient method for the initial characterization of HATs.

Preparation of Substrates

Substrate Characteristics

Substrates for HAT assays should be free of contamination by endogenous acetyltransferase activities and should not contain adventitious inhibitors of HAT activity. In addition, levels of acetyllysine in substrates should be as low as possible or their sequence context known. Chromatin substrates should be free of subnucleosomal particles, as these may dissociate to release free histones or otherwise possess distinct substrate properties in assays. This section describes methods to prepare free histone and soluble

[16] P. A. Grant, L. Duggan, J. Cote, S. M. Roberts, J. E. Brownell, R. Candau, R. Ohba, T. Owen-Hughes, C. D. Allis, F. Winston, S. L. Berger, and J. L. Workman, *Genes Dev.* **11,** 1640 (1997).

[17] R. Ohba, D. J. Steger, J. E. Brownell, C. A. Mizzen, R. G. Cook, J. Côté, J. L. Workman, and C. D. Allis, *Mol. Cell. Biol.* **19,** 2061 (1999).

[18] A. Imhof, X.-J. Yang, V. V. Ogryzko, Y. Nakatani, A. P. Wolffe, and H. Ge, *Curr. Biol.* **7,** 689 (1997).

chromatin substrates in accordance with these criteria. We assume the reader is familiar with methods for isolating nuclei and acid extraction of histones. These techniques have been described previously (e.g., this series, Vol. 170, [1], [23], [25]).

Preparation of Free Histone and Synthetic Peptide Substrates

To avoid bias in acetyltransferase assays and acetylation site mapping procedures, histone preparations with low contents of acetylated forms should be employed. In practice, this is not difficult to achieve because the overall level of acetylated histones in bulk chromatin of most cell types is low and also because histone ε-*N*-acetyl groups are easily lost if inhibitors of deacetylases (e.g., sodium butyrate, trichostatin, trapoxin) are omitted during the isolation of nuclei. When nuclei prepared in this fashion are extracted with 0.4 *N* H_2SO_4, a solvent that solubilizes all five classes of histones nearly quantitatively regardless of whether they are acetylated or not, the level of acetylated forms in the resulting crude histones is extremely low. We have successfully employed crude calf thymus histone obtained commercially (type II-A; Sigma, St. Louis, MO) and also crude histone prepared in the laboratory by acid extraction of nuclei isolated from a number of sources, including *Tetrahymena,* cultured human cells (HeLa), and chicken erythrocytes. We have shown previously that, as in yeast, steady-state levels of acetylated histones are high in macronuclei of growing *Tetrahymena,* making it necessary to induce deacetylation by incubating isolated macronuclei in an EDTA solution prior to extracting histones for use as substrate.[19] In contrast, the content of acetylated histones in mature chicken erythrocytes appears to be quite low and these histones have been employed in HAT assays and acetylation site mapping experiments without evidence of bias due to preexisting acetylation. The content of acetylated forms can be determined by analyzing preparations on one-dimensional acetic acid–urea polyacrylamide gels (with or without Triton X-100) that resolve acetylated forms of the core histones.[20] Two-dimensional procedures can be employed for greater resolution if necessary (this series, Vol. 170, [25]).

Due possibly to their high content of basic residues, histones and histone peptides recovered from acidic solutions may be associated with sufficient acid so as to lower the pH of HAT assay buffers on addition of substrate. This can be avoided by the thorough washing of histone precipitates with an appropriate solvent such as pure acetone following acid extraction. Alternatively, dialysis may be used to remove excess acid or other small

[19] K. J. Vavra, C. D. Allis, and M. A. Gorovsky, *J. Biol. Chem.* **257,** 2591 (1982).

[20] F. Zweidler, *Methods Cell Biol.* **17,** 223 (1978).

solutes. Contaminants of synthetic peptide preparations are best removed by reversed-phase high-performance liquid chromatography (RP-HPLC) purification of peptides. Peptide fractions from RP-HPLC should be lyophilized to complete dryness at least once to remove excess trifluoroacetic acid.

Chromatin

Mononucleosomes and oligonucleosomal fragments for use in HAT assays are prepared readily by the digestion of isolated nuclei with micrococcal nuclease. To minimize the potential selective solubilization of nucleosomes enriched in acetylated core histones by micrococcal nuclease,[11] extensive digestion is performed on nuclei isolated in the absence of deacetylase inhibitors. Because the kinetics of micrococcal nuclease digestion, the rate of generation of subnucleosomal species, and the potential for proteolysis all vary significantly according to the origin of nuclei and details of their preparation, investigators must optimize digestion conditions according to their own experimental system. The following procedure provides relatively concentrated solutions (0.5–2.0 mg/ml with respect to DNA) of short chromatin oligomers from nuclei of chicken erythroctyes appropriate for use in HAT assays.

Micrococcal Nuclease Digestion of Chicken Erythrocyte Nuclei

All steps are performed at 4° unless otherwise noted. Chicken blood is collected into an equal volume of ice-cold 150 m*M* NaCl, 20 m*M* EDTA, 1.0 m*M* phenylmethylsulfonyl fluoride (PMSF), pH 8.0, at a poultry packer, and nuclei are prepared by detergent lysis of erythrocytes as described previously.[21] All buffers contain 1.0 m*M* PMSF but deacetylase inhibitors are omitted. Nuclei are stored as loosely packed pellets equivalent to 25 ml of blood in a minimal volume of 10 m*M* Tris–HCl, 50 m*M* NaCl, 5 m*M* $MgCl_2$, 50% (v/v) glycerol, pH 8.0, at −70° until used. Prior to digestion, an aliquot of nuclei is thawed rapidly and washed twice gently with digestion buffer (10 m*M* Tris–HCl, 1 m*M* $CaCl_2$, 1 m*M* PMSF, pH 7.4). Nuclei are then resuspended in digestion buffer to give A_{260} (measured in 2% SDS) = 100, equivalent to a DNA concentration of approximately 5 mg/ml. The suspension is warmed quickly to room temperature and micrococcal nuclease (EC 3.1.31.1, Worthington Biochemical, Freehold, NJ) is added to give a final concentration of 100 U/ml. Digestion is allowed to proceed for 60 min at room temperature and is terminated by making the suspension 2 m*M* in EDTA (i.e., add 1/50th volume of 100 m*M* EDTA, pH 8.0). Nuclei are sedimented by centrifugation at 600*g* for 5 min, and supernatant S1,

[21] A. L. Olins, R. D. Carlson, E. B. Wright, and D. E. Olins, *Nucleic Acids Res.* **3,** 3271 (1976).

consisting largely of mononucleosomes and short oligonucleosomes, is transferred to a new tube and kept on ice. Nuclei are lysed using periodic trituration over 20 min on ice to resuspend the pellet in 6.0 ml of 1 m*M* EDTA, 1 m*M* PMSF. This suspension is clarified by centrifugation at 10,000*g* for 20 min, and supernatant S2, enriched in short oligonucleosomes but containing significant amounts of higher oligomers, is transferred to a new tube and kept on ice. The A_{260} of the S1 and S2 fractions are measured, and aliquots equivalent to 10–20 μg of DNA are phenol extracted and analyzed in 2% (w/v) agarose or 5% (w/v) polyacrylamide gels in Tris–borate–EDTA (TBE) buffer as described previously.[22] Using a suitable DNA molecular weight marker (e.g., 123-bp ladder, GIBCO-BRL, Gaithersburg, MD), the two fractions are inspected for the presence of subnucleosomal DNA (less than 165 bp) and the range of DNA sizes is noted. If subnucleosomal DNA is not detected, then the S1 and S2 fractions can be used in HAT assays individually or pooled following dialysis against TE (10 m*M* Tris–HCl, 1 m*M* EDTA, pH 8.0) containing 1 m*M* PMSF and concentration by ultrafiltration if desired. If significant amounts of subnucleosomal DNA are detected, then this material should be removed by density gradient ultracentrifugation or the digestion repeated using a shorter digestion time and/or less nuclease.

Sucrose Gradient Purification of Chromatin

We routinely purify soluble chromatin on sucrose density gradients in order to reduce contamination by HATs and other potential sources of error in assays. Furthermore, defined populations of mononucleosomes and oligomers can be recovered if desired. In our experience, HAT activity does not cosediment with soluble chromatin fractions of any size from mature chicken erythrocytes. However, this may not be the case for chromatin prepared from other sources. For example, although HAT activity does not cosediment with mononucleosomes prepared from *Tetrahymena* macronuclei, it is detected readily in gradient fractions containing higher oligomers.[23] Thus, when preparing chromatin substrates from nuclei with high levels of acetyltransferase activity, such as *Tetrahymena* macronuclei, it may be necessary to select only mononucleosome fractions for use in assaying exogenous HATs.

We prepare 5–20% (w/v) linear sucrose gradients in TE using a binary gradient mixing apparatus (Model 150, GIBCO-BRL) and a magnetic stirrer. Polyallomer tubes (14 or 38 ml) for use in the SW40 and SW28 rotors

[22] T. Maniatis, E. F. Fritsch, and J. Sambrook, *in* "Molecular Cloning: A Laboratory Manual." Cold Spring Harbor Laboratory, Cold Spring Harbor, NY, 1982.

[23] L. G. Chicoine, R. Richman, R. G. Cook, M. A. Gorovsky, and C. D. Allis, *J. Cell Biol.* **105,** 127 (1987).

(Beckman Instruments) are used depending on the amount of chromatin to be purified. Gradients are formed using room temperature solutions, and the tubes are cooled to 4° in an ice bucket or cold room before use, taking care not to disturb the gradients. Samples are loaded using a 1-ml syringe fitted with a 16-gauge needle, and the gradients are centrifuged at 25,000 rpm for 24 hr (either rotor). Fractions are collected manually using capillary tubing to siphon 0.5- and 1.0-ml fractions from the bottom of 14- and 38-ml tubes, respectively. Sedimentation profiles are determined by measuring the A_{260} of the fractions from one or more tubes, and aliquots corresponding to 10 μg of deproteinized DNA are analyzed on TBE gels. After appropriate fractions have been pooled, the material is dialyzed against TE containing 1 m*M* PMSF and concentrated by ultrafiltration if necessary. Purified chicken erythrocyte chromatin has been found to be stable for up to 6 months at 4°, but aliquot portions of large preparations for long-term storage at −70°.

The maximum amount of chromatin that can be loaded on each gradient tube depends on the composition of the sample and the degree of resolution desired. For general use, we pool chicken erythrocyte S1 and S2 fractions with the goal of preparing large amounts of "low molecular weight chromatin" containing mono-, di-, and trinucleosomes. In this case, up to 1 mg of pooled S1 and S2 can be loaded on each 14-ml gradient and up to 4 mg can be loaded onto each 38-ml gradient in maximum volumes of 0.5 and 2.0 ml, respectively. If we wish to prepare purified mononucleosomes, the maximum loads are reduced by approximately 50%. Note that it can be difficult to completely separate subnucleosomal material from mononucleosomes and, in this case, maximum loads should be reduced further.

Chromatin fractions prepared as just described contain the linker histones H1 and H5. If desired, these proteins can be extracted and separated during ultracentrifugation by including 0.65 *M* NaCl in the density gradient solutions. All steps are performed as described earlier except that solubilized chromatin is made 0.65 *M* NaCl by drop-wise addition of 5 *M* NaCl with constant mixing prior to loading on gradients containing 0.65 *M* NaCl. Following dialysis against TE, linker histone-depleted chromatin fractions can be used as HAT substrates or to prepare nucleosome core particles as described previously (this series, Vol. 170, [1], [5]). Note that this procedure also extracts numerous nonhistone chromatin proteins and thus may be useful for the preparation of chromatin substrates from nuclei containing high levels of HATs or other activities affecting assays.

Acetyl-CoA

Acetyl-CoA, with ^{3}H or ^{14}C incorporated exclusively in the acetyl group, is available in purified form commercially. Because of the lower cost, we have employed ^{3}H-labeled material exclusively to date, but the use of ^{14}C-labeled material is not precluded. Regardless of the radioisotope employed, we recommend obtaining the highest specific activity material available to maximize the sensitivity of assays, particularly the activity gel assay. We routinely obtain [^{3}H]acetyl-CoA with specific activity between 4.0 and 6.0 Ci/mmol. Aliquots containing 10 μCi in the aqueous buffer as supplied (pH 4.5–4.8) can be stored in 0.5-ml vials at $-20°$ for at least 6 months without apparent degradation as judged by the magnitude of acetate transfer in assays. Alternatively, it is possible to prepare radioisotopic acetyl-CoA using radioisotopic sodium acetate and CoA obtained commercially.[24] For liquid assays, it is possible to include an acetyl-CoA generating system utilizing acetyl-CoA synthetase in assay mixtures.[9]

Liquid HAT Assay

This assay measures HAT activity by detecting the amount of acetate transferred from radioisotopic acetyl-CoA to protein substrates. Following the acetylation reaction, aliquots of reactions are spotted onto phosphocellulose filters that are washed to remove unreacted acetyl-CoA prior to scintillation counting. The assay format allows a wide variety of potential substrates and reaction conditions to be examined easily with a minimal consumption of expensive or scarce substrates. Moreover, because all aspects of the assay can be quantified, it is possible to calculate the specific activities of different HATs with a variety of substrates and to determine kinetic parameters under defined conditions. Finally, proteins acetylated in reactions can be identified readily by fluorography of standard protein electrophoretic gels. The basic procedure outlined here has been used to characterize HAT activities from several cell types, including one extracted from *Tetrahymena* macronuclei in our laboratory.[25] The conditions described represent a useful starting point for investigating HATs from other sources, but should be optimized for each case.

Procedure for Free Histone and Chromatin Substrates

Reagents

Buffer A. 50 mM Tris-HCl (pH 8.0), 10% glycerol (v/v), 10 mM butyric acid (included as a histone deacetylase inhibitor), 1 mM PMSF, 1 mM

[24] L. Attisano and P. N. Lewis, *J. Biol. Chem.* **265,** 3949 (1990).

[25] J. E. Brownell and C. D. Allis, *Proc. Natl. Acad. Sci. U.S.A.* **92,** 6364 (1995).

dithiothreitol (DTT). For work with dilute enzyme and substrate samples, a 2× preparation of buffer A is appropriate.

Free Histone Substrates. Prepare a stock solution of acid-extracted histones at 10 mg/ml in distilled H_2O (store at −20°).

Chromatin Substrates. Purified mononucleosomes or low molecular weight chromatin (prepared as described earlier) in TE buffer at a concentration of approximately 1 mg/ml (with respect to DNA). See Comments.

Histone Acetyltransferase. The specific activity of different HAT preparations varies significantly and the amount of enzyme to include in assays must be determined empirically. For early experiments, a range of enzyme concentrations should be tested to determine the amount of enzyme required for the linear incorporation of radioactivity into substrate proteins over the duration of the assay. Recombinant yeast GCN5 is available commercially (Upstate Biotechnology, Lake Placid, NY) and serves as a convenient positive control for assays with free histone substrates. Crude sonicates of bacteria expressing putative HATs frequently contain inhibitory substances and we routinely test these using several dilutions. In our experience, high salt concentrations (i.e., greater than 100 m*M*, see Fig. 2) reversibly inhibit HAT activity. Samples prepared in high ionic strength buffers should be diluted or desalted such that the final concentration of salt in the assay mixture is less than 100 m*M*. Also avoid exposing enzyme preparations to compounds that react with sulfhydryl groups (i.e., cysteine protease inhibitors) as these appear to be potent inhibitors of several HATs studied to date.

[^{3}H]Acetyl-CoA. Immediately prior to use, dilute an aliquot of the radiolabeled stock with the appropriate amount of unlabeled acetyl-CoA in buffer A to provide a 10.0 μM working solution with a specific activity of 0.5 Ci/mmol.

Phosphocellulose Filter Paper (P-81, Whatman, Clifton, NJ). Cut into pieces of approximately 15 × 15 mm and number with a No. 2 pencil. For each reaction, prepare and label two pieces of filter paper so that samples can be collected in duplicate for liquid scintillation counting.

Procedure

Reactions are assembled in 500-μl microcentrifuge tubes on ice. To each tube, add 25 μl buffer A, 5 μl histone stock (50 μg histone), 10 μl enzyme preparation, and 10 μl [^{3}H]acetyl-CoA (final concentration 2.0 μM; 50 nCi ^{3}H). Reactions are incubated in a water bath maintained at 30°. Negative control assays should be included in which either the histone substrates or the enzyme sample is omitted or in which a heat-denatured enzyme sample is included in the reaction mixture to determine background levels of counts.

After 10 min at 30°, terminate reactions by spotting a portion of each onto a piece of P-81 filter paper using a micropipette; it is recommended that two replicate aliquots of 20 μl be spotted for each reaction. (*Note:* If gel electrophoretic analysis of reaction product proteins is desired, add 2 μl of 5× concentrated sample buffer to the remaining 10 μl of the reaction mixture and place on ice.)

Immediately wash the filters to remove unbound reactants in 50 m*M* $NaHCO_3$ (pH 9.0) (five washes, 5 or more minutes each wash). Then air-dry the filters before counting in 2 ml of an aqueous-based scintillation cocktail. Determine the background levels of nonenzymatic labeling from negative control assays; radiolabel incorporated into substrate proteins under these conditions is not attributable to enzymatic activity and should be subtracted from experimental cases.

Comments

Assay Conditions

As shown in Fig. 1, the HAT activity of the *Tetrahymena* macronuclear extract with free histone substrate was proportional to time for more than 10 min when the final concentration of acetyl-CoA in the assay was 2.0 μM as described earlier. When the concentration of acetyl-CoA employed was 10-fold lower, the linear range of the assay was shorter, even though both assays employed equivalent amounts of [^{3}H]acetyl-CoA. These data illustrate that investigators should establish the appropriate concentrations of substrates (both histones and acetyl-CoA) and enzymes before attempting to compare different HAT preparations. It has been reported that the HAT activity of human GCN5 and PCAF was stabilized by incubating these enzymes with acetyl-CoA prior to the addition of protein substrates in assays performed at 37°.[26] Investigators should ascertain whether similar phenomena influence the activity of other HATs under the assay conditions selected.

Figure 2 illustrates the effect of pH and ionic strength on the background signal and sensitivity of the liquid assay. As shown in Fig. 2A, nonenzymatic acetylation of free histone substrates is detected readily in a 10-min assay when the pH exceeds 8.0. We have not examined mechanisms of nonenzymatic acetylation in detail but suggest that the pH dependence reflects

[26] J. E. Herrera, M. Bergel, X.-J. Yang, Y. Nakatani, and M. Bustin, *J. Biol. Chem.* **272,** 27253 (1997).

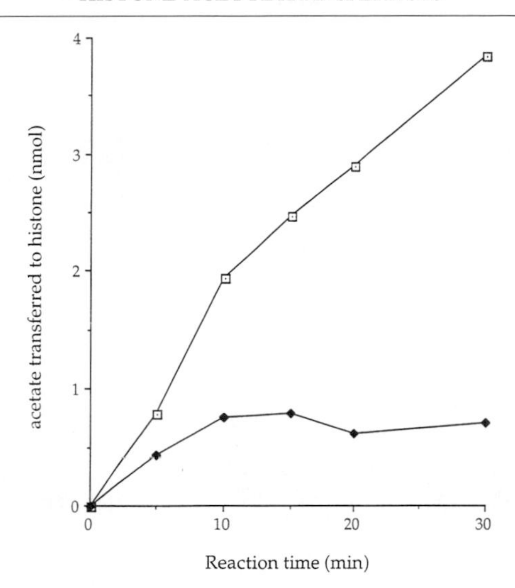

FIG. 1. Liquid assay of histone acetyltransferase activity from *Tetrahymena* macronuclei (for enzyme preparation, see Brownell and Allis[25]). The time course of incorporation of acetyl groups onto free histones in reactions containing 50 μg histones, 10 μl enzyme preparation, and ^{3}[H]acetyl-CoA at final concentrations of either 2.0 μM (0.5 Ci/mmol, ⊡) or 0.2 μM (5.0 Ci/mmol, ◆) is shown; assay volumes were adjusted to 50 μl with buffer A as described in the text. Reactions were incubated at 30°. At the times indicated, duplicate samples were spotted onto phosphocellulose filters, washed, and subjected to liquid scintillation counting. Average values of duplicate samples are plotted. The number of nanomoles of acetyl groups transferred to histones was calculated from the mean value of sample counts after the subtraction of nonspecific counts determined from negative controls (see text). Under these conditions, HAT activity is directly proportional to time for at least 10 min.

decreased stability of the thioether linkage in acetyl-CoA at high pH. Note also that substrates differ in their propensity to react with acetyl-CoA in the absence of acetyltransferases. In our experience, the rate of nonenzymatic reaction of free histone substrates is appreciably greater than that of histone N-terminal peptide and chromatin substrates. Figure 2B shows that high ionic strengths inhibit the HAT activity detected for both *Tetrahymena* macronuclear extract and *Drosophila* $TAF_{II}230$ ($dTAF_{II}230$). We have not investigated the mechanism of this inhibition, but it appears to be reversible and is common to all the HATs we have studied to date.

Free Histone Substrates

We have not established whether native histone secondary structure is reconstituted on the addition of acid-extracted histones dissolved in water to our assay buffer. However, we are not aware of evidence suggesting that

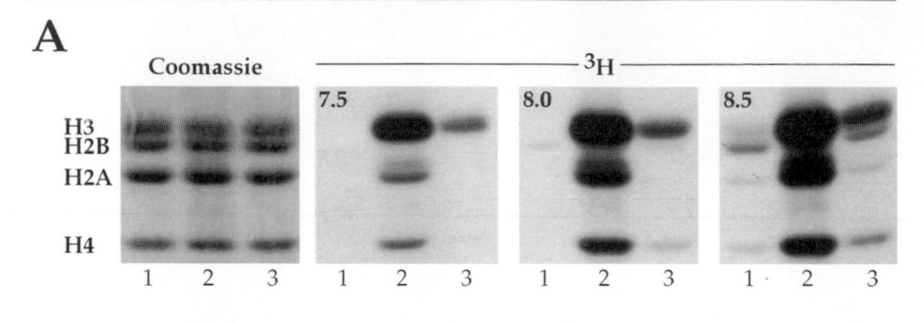

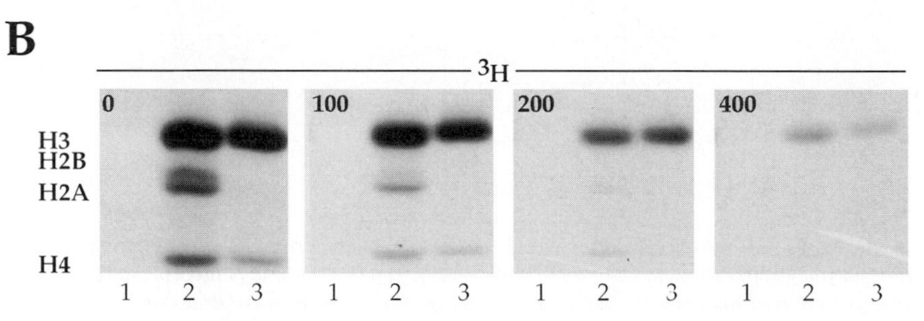

FIG. 2. Influence of pH and ionic strength on the liquid assay for histone acetyltransferase activity. Free histones, prepared by acid extraction of chicken erythrocyte nuclei, were used as substrates in a series of assays designed to test the effect of varying pH or ionic strength on HAT activity *in vitro*. Portions of these reactions were analyzed by SDS–PAGE and fluorography. (A) Fluorograms of reactions performed in buffer A at pH 7.5, pH 8.0, and pH 8.5 are shown. The core histone region of the stained gel corresponding to the reaction performed at pH 8.0 is shown after fluorography in the left-most panel. In each panel, the source of HAT activity was (1) none (i.e., a negative control), (2) *Tetrahymena* macronuclear extract and (3) purified recombinant dTAF$_{II}$230. Note the nonenzymatic labeling of core histones, particularly H2B, in the reaction performed at pH 8.5, which is not apparent in the reactions performed at lower pH. (B) Fluorograms of reactions performed in buffer A (pH 8.0) supplemented with 0, 100, 200, or 400 m*M* KCl are shown. Lane designations are as in (A). Note that the inhibition of the HAT activity of both the *Tetrahymena* extract and dTAF$_{II}$230 is proportional to the final ionic strength in the reaction mixture and is apparent at 100 m*M* KCl. In both (A) and (B), all reactions contained 10 μg histones and were incubated for 10 min at 30° with 0.33 μM [^{3}H]acetyl-CoA (6.1 Ci/mmol). Fluorograms represent exposures of 18 days.

histone conformation has a marked effect on HAT activity *in vitro*. This may derive from the fact that the amino termini of core histones possess little or no regular secondary structure in nucleosome crystals.[27] HAT activities of human GCN5 and dTAF$_{II}$230 with chicken erythrocyte histones prepared by acid extraction and subsequently dissolved in water were simi-

[27] K. Luger, A. W. Mader, R. K. Richmond, D. F. Sargent, and T. J. Richmond, *Nature* **389,** 251 (1997).

lar to those obtained with the same histones renatured by slow dialysis from 6.0 *M* guanidine hydrochloride into buffered saline following acid extraction.[28] Furthermore, even though convenient protocols for the preparation of (presumably) "native" histones from nuclei or solubilized chromatin have been reported,[29] it should be emphasized that the selective solubilization of histones by acid, combined with the denaturing conditions employed during histone recovery, is likely to minimize the contamination of substrate preparations by catalytically active HAT polypeptides. In this regard, new preparations of histones and synthetic peptides should be assayed for endogenous HAT activity before use in assays with valuable experimental samples.

Another factor that may influence assays performed with free histone substrates is the propensity of histones to assemble into multisubunit complexes, notably the H2A–H2B dimer and the $(H3–H4)_2$ tetramer, under conditions similar to those employed here.[30,31] Investigators should consider the influence of pH, ionic strength, protein concentration, and buffer composition on the associative behavior of histones when interpreting data.

Chromatin Substrates

In some cases, nucleosomes are more likely to be representative of *in vivo* substrates. Because of the labor involved in preparing these substrates, we reduce the reaction volume to reduce the amount of substrate required, using 5–10 μg of nucleosomes in a final assay volume of 25 μl. For dilute chromatin samples, it may be necessary to assemble reactions using 2$\times$ buffer A in order to attain standard reaction conditions. To assess whether chromatin substrates remain intact during the reaction, we suggest that a reaction containing recombinant GCN5 be performed in parallel. Although GCN5 is a potent HAT in assays with free histone substrates, it shows no detectable activity with nucleosomes under the conditions employed here.[16,32] Thus, evidence of acetate transfer in the GCN5-containing reaction may indicate the release of free histones secondary to nucleosome dissociation or degradation.

Analysis of Histone Reaction Products

Because histones are resolved readily in SDS–PAGE and acetic acid–urea gels, the extent to which individual histones have been acetylated in

[28] C. A. Mizzen and C. D. Allis, unpublished data, (1996).

[29] R. H. Simon and G. Felsenfeld, *Nucleic Acids Res.* **6,** 689 (1979).

[30] V. Karantza, A. D. Baxevanis, E. Freire, and E. N. Moudrianakis, *Biochemistry* **34,** 5988 (1995).

[31] V. Karantza, E. Freire, and E. N. Moudrianakis, *Biochemistry* **35,** 2037 (1996).

[32] C. A. Mizzen, J. E. Brownell, and C. D. Allis, unpublished data (1996).

assays containing mixtures of histones can be assessed using fluorography to detect the incorporation of radioisotopic acetate (e.g., Fig. 2). Assessing the pattern of histone labeling is essential for initial studies in that it provides a direct indication of specific histones that are acetylated. For example, GCN5 is known to acetytate free H3 and H4 (but not H2A or H2B) in reactions with free histone substrates.[33]

Procedure for Peptide Substrates

We employ sets of peptides that mimic unacetylated histone N termini and the same N termini acetylated at distinct positions in assays to establish preferred sites of acetylation for HATs under investigation. This assay is designed to compare the activity of a particular HAT with a number of different substrates. The final assay volume is 20 μl from which two to three replicates of 5 μl can be spotted onto P81 filters for LSC counting. Each reaction consists of 10 μl of peptide solution plus 10 μl of "enzyme mix." Peptides are dissolved in water alone and stored at $-70°$. The enzyme mix contains the protein of interest and all other necessary components at $2\times$ concentration.

Generally, we employ peptides at a working concentration of 100 ng/μl. For assays, 10 μl of each peptide working stock is pipetted into labeled tubes on ice. Therefore, each assay tube contains 1 μg of peptide and 250 ng is subsequently counted in each 5-μl replicate spotted onto P81 paper. These conditions were determined empirically to give a good signal-to-noise ratio for purified recombinant HATs and crude extracts.[15]

"Enzyme mixes" are assembled on ice. One tube contains enough putative enzyme and [^{3}H]acetyl-CoA to test all the peptides. The appropriate amount of each enzyme to use per reaction should be determined empirically in initial trials. For example, approximately 2 pmol of affinity-purified $dTAF_{II}230$ typically transfers >1000 dpm [^{3}H]acetate to 250 ng of H3 peptide in 20 min. Much less GCN5 will give an equivalent signal. As with free histone substrates, several dilutions of crude sonicates of bacteria expressing putative HATs should tested to compensate for the presence of inhibitory substances. Because the enzyme mix is added to peptides dissolved in water alone, the remaining ingredients are present in the enzyme mix at twice the final concentations listed here: 2.0 nCi/μl [^{3}H]acetyl-CoA, 50 m*M* Tris–HCl, pH 8.0, 10% (v/v) glycerol, 0.1 m*M* EDTA, 1.0 m*M* DTT, 0.5 m*M* PMSF.

[33] M.-H. Kuo, J. E. Brownell, R. E. Sobel, T. A. Ranalli, R. G. Cook, D. G. Edmonson, S. Y. Roth, and C. D. Allis, *Nature* **383,** 269 (1996).

Reactions are initiated by adding 10 μl of enzyme mix to 10 μl of peptide solution. Tubes are incubated at 30° for 20 min. Aliquots (5 μl) are spotted onto P81 filters and washed as described earlier prior to liquid scintillation counting.

Activity Gel Assay

The methodology employed in HAT activity gels is essentially the same as that employed in kinase activity gels (e.g., this series, Vol. 200, [33]) except that radioisotopic acetyl-CoA is substituted for [γ-^{32}P]ATP in the assay portion of the procedure. Samples containing putative HATs are resolved on substrate-containing SDS–PAGE gels. After electrophoresis is completed, activity gels are treated sequentially to remove SDS and then to fully denature and gradually renature sample polypeptides and assay HAT activity *in situ*. This enables identification of catalytic polypeptides according to their approximate molecular weight and can be used to monitor specific HAT activities during the course of purification.[25] The following procedure, as performed in this laboratory, has been described in detail.[34]

Procedure

The buffer volumes and wash times listed here have been determined for minigels approximately 6 × 8 cm and 0.75 mm thick. Recovery of activity in different gel formats may require adjustment of these parameters. The resolving portion of the activity gel containing 1 mg/ml (final) calf thymus histone is prepared and allowed to polymerize at room temperature for at least 1 hr. A control activity gel containing 1 mg/ml (final) BSA (a protein not known to be ε-*N*-acetylated *in vivo*) is prepared in parallel. Conventional stacking gels prepared without substrate proteins are polymerized on top of both gels. Electrophoresis is conducted at room temperature using standard SDS–PAGE conditions. To replenish substrate proteins in the gel matrix as they are depleted during the course of electrophoresis, 0.1 mg/ml histones or BSA can be included in the appropriate upper reservior buffer. This addition may enhance the detection of high molecular weight HATs.[15]

Following electrophoresis, the gels are transferred to separate plastic containers and washed as outlined. All steps are performed at room temperature with gentle agitation on an orbital shaker unless otherwise noted.

[34] J. E. Brownell, C. A. Mizzen, and C. D. Allis, *in* "Methods in Molecular Biology: Chromatin Protocols" (P. Becker, ed.). Humana Press, Totowa, NJ, 1999.

1. SDS is removed by washing gels in buffer 1 [50 m*M* Tris–HCl, pH 8.0, 20% (v/v) 2-propanol, 0.1 m*M* EDTA, 1 m*M* DTT]. Four washes of 15 min duration and employing 100 ml buffer 1 are performed.
2. Denature sample proteins by washing gels in buffer 2 (50 m*M* Tris–HCl, pH 8.0, 8 *M* urea, 0.1 m*M* EDTA, 1 m*M* DTT) as described earlier.
3. Renature sample proteins slowly by incubating gels with buffer 3 (50 m*M* Tris–HCl, pH 8.0, 0.005% Tween 40, 0.1 m*M* EDTA, 1 m*M* DTT) at 4° without agitation. Rinse gels once in 100 ml buffer 3 for 15 min at 4° followed by a 12-hr (overnight) incubation in 100 ml buffer 3 at 4°. The following morning, gels are washed a third time in 100 ml buffer 3 for 30 min at 4°. A fourth and final 30-min wash in 100 ml buffer 3 is performed at room temperature without agitation to allow the gel and buffer to equilibrate at room temperature.
4. Equilibrate each gel in 100 ml buffer 4 [50 m*M* Tris–HCl, pH 8.0, 10% (v/v) glycerol, 0.1 m*M* EDTA, 1 m*M* DTT] at room temperature for 15 min with gentle agitation on an orbital shaker.

The acetylation reaction is conducted in a minimal volume of buffer to keep the concentration of isotopic acetyl-CoA as high as possible. Place each gel in a heat-sealable plastic bag that is just slightly larger than the gel itself and add 3 ml of fresh buffer 4 containing 5 μCi of [^{3}H]acetyl-CoA. Remove as much air as possible from the bag and seal it. Distribute thoroughly the [^{3}H]acetyl-CoA-containing reaction buffer about the gel. Incubate at 30° for 1 hr to allow the acetylation reaction to proceed (optimal reaction times will vary depending on the amount of enzyme activity recovered during the renaturation step).

Terminate the acetylation reaction by placing the gels in Coomassie Blue R250 staining solution. Despite the background staining due to the presence of substrate proteins throughout the gel, in most cases sample and molecular weight marker proteins can be visualized. Gels should be destained exhaustively (overnight) with several changes of destain solution to reduce the levels of background staining and to remove unbound radiolabel. In lieu of staining, gels can be washed using 5% trichloroacetic acid, again with several changes. In either case, thorough washing prior to fluorography is recommended to reduce the background signal.

Process gels for fluorography (e.g., Entensify; NEN Life Science Products, Boston, MA) and expose to film. The exposure time required is a function of the amount of activity loaded onto the gel and the degree to which activity has been renatured, as well as the quality of the acetyl-CoA used.

Comments

1. Stock solutions of calf thymus histones (type II-A; Sigma, St. Louis, MO and BSA (Sigma) at 10 mg/ml in water are used to prepare activity gels. These stocks can be stored for months at −20° without degradation.
2. The highest specific activity of [^{3}H]acetyl-CoA or [^{14}C]acetyl-CoA available should be used because the concentration of acetyl-CoA in the assay is low. The addition of unlabeled acetyl-CoA, even if only in a small molar excess over radioisotopic acetyl-CoA, is not recommended as the specific signal for less active HATs will be reduced and may not be detectable.
3. Virtually any type of sample can be assayed for HAT activity using activity gels, provided that the method of preparation is compatible with SDS–PAGE and detection of acetyltransferase activity. The assay is particularly useful for recombinant proteins that form inclusion bodies during expression as inclusion bodies generally are solubilized readily in SDS–PAGE sample buffer. Recombinant yeast GCN5p (available commercially from Upstate Biotechnology Inc.) is a convenient positive control in that as little as 100 ng of the purified 6His-fusion protein gives a positive signal after 6 days of autoradiography.

Interpretation of Results

It should be noted that while the HAT activity gel method is useful for identifying catalytically active HAT polypeptides, in some instances polypeptides that are not HATs may also incorporate low levels of [^{3}H]acetyl groups under the assay conditions. Sample polypeptides must therefore also be tested in negative control gels such as those that contain BSA (or other nonsubstrate protein) or in which no protein substrate has been included in the gel. Incorporation of [^{3}H]acetyl groups by sample polypeptides under these circumstances can be judged to be artifactual due to either nonenzymatic protein acetylation or autoacetylation of sample polypeptides. Moreover, minor radiolabeled bands detected in histone gels should also be viewed with caution. However, histone-dependent, major radiolabeled bands are likely to be HATs, given that known HATs demonstrate high specificity for histone substrates. Particularly promising potential HATs are those polypeptides whose autoradiographic band intensity on a histone-containing activity gel is disproportionately greater than the band intensity in Coomassie staining.

Acetylation Site Mapping

To identify sites acetylated by specific HATs *in vitro,* substrates are labeled as in the liquid assay procedure but using more [^{3}H]acetyl-CoA. Proteins of interest are then resolved from other components of the acetylation reaction by electrophoretic or chromatographic methods prior to N-terminal microsequence analysis. Histone fractions are prepared readily for microsequencing by electrophoretic transfer from PAGE gels to polyvinylidene difluoride (PVDF) membrane. However, in our experience, histone fractions prepared by liquid chromatography sequence more efficiently, allowing acetylation sites to be assigned without the ambiguity that can arise due to the more pronounced sequencing lag encountered with electroblotted proteins. Furthermore, the removal of α-*N*-acetyl groups by incubation with concentrated TFA (commonly referred to as "deblocking") that is necessary to sequence the N-termini of histones H2A and H4 in most species also appears to be more efficient when performed with protein in solution. The following section describes the preparation of histones by RP-HPLC for acetylation site mapping.

Preparation of Substrates for N-Terminal Microsequencing

An important consideration in preparing material for acetylation site mapping is that sites acetylated weakly by a given HAT will only be detected unambiguously in histones labeled to high specific activities. This requires that acetylation reactions are performed using high acetyl-CoA to protein substrate ratios. Given the expense of isotopic acetyl-CoA, it is desirable to select the minimal amount of histone or chromatin substrate for labeling that will provide sufficient material for sequencing (e.g., 10–100 pmol) following purification and deblocking (if required).

Free Histones

Individual histones purified previously or crude mixtures can be used as appropriate. The minimum amount of sample required is determined in part by whether deblocking is necessary. As a general guideline, we suggest that experiments involving sequencing of only H2B and H3 start with at least 4.0 μg of these proteins. Experiments involving sequencing of H2A and H4 should employ 8.0 μg, or more, of these proteins as deblocking is usually required. Acetylation reactions are conducted for 10 to 20 min at 30° using up to 2.5 μCi [^{3}H]acetyl-CoA per 50 μl reaction. Radioisotopic acetyl-CoA is employed exclusively. At the end of the reaction, samples are taken to determine the extent of labeling and the remainder of the reaction is placed on ice, made 20% (final) in TCA, and allowed to stand 1 hr. Precipitated proteins are pelleted by centrifugation, and the supernatant, containing unincorporated label, is discarded. The protein pellet is washed once with cold acetone/0.1% HCl, twice with cold acetone, and dried. The sample is then dissolved in a small volume of 0.1% TFA and chromatographed as described next.

Chromatin Substrates

The considerations regarding minimal amounts of chromatin samples to label are similar to those for free histones. However, preparation of chromatin samples differs in two significant ways. First labeling reactions can be allowed to proceed for long intervals due to the lower rate of nonenzymatic reaction of acetyl-CoA with nucleosomal histones. We have incubated chromatin samples for up to 1 hr at 30° in 50-μl reactions containing 2.5 μCi [^{3}H]acetyl-CoA without detecting nonenzymatic histone labeling. Second, histones should be separated from DNA prior to RP-HPLC to avoid fouling of the column by DNA. After aliquots have been taken to quantitate the extent of labeling at the end of the reaction, samples are made 0.4 *N* H_2SO_4 (add 12 μl concentrated H_2SO_4 per milliliter) and allowed to stand on ice for 1 hr. Precipitated DNA is sedimented by centrifugation, and the supernatant is transferred to a new tube. Histones can be separated from unincorporated label by TCA precipitation as described earlier or injected directly following adjustment of the sample to approximately pH 2.5 by the addition of 50 mg of anhydrous Na_2HPO_4 per milliliter of sample.

High-Performance Liquid Chromatography

For good recovery, it is recommended that a narrow bore RP-HPLC column (e.g., 2.0 mm i.d.) and an HPLC system capable of providing a flow rate of 200–250 μl/min be used. If a previously purified histone fraction was labeled, it can then be recovered using a relatively rapid gradient that increases acetonitrile content by 1–2% per minute. However, if crude core histone or chromatin substrates were labeled, then a method capable of resolving the histone fractions of interest is required. Optimal methods for resolving histones from various organisms on particular columns vary somewhat and workers should refer to the original literature. The following methodology can be used to resolve core histones from humans and chickens.

All chromatography is performed at room temperature. Protein elution is monitored by UV absorption at 214 nm. We use the same solvents and gradient profiles for human and chicken histones but use different columns and flow rates for the two species. Under these conditions, each of the core histones from chicken erythrocytes elutes as a single peak regardless of acetylation state. Similarly, H2B and H4 from HeLa cells elute as single peaks regardless of acetylation state. However, HeLa histones H2A and H3 elute in two and three peaks, respectively. Acetylation does not influence the rctention of these proteins under the conditions employed and we presume the peaks represent different amino acid sequence variants of these proteins as described previously.[35]

Human (HeLa) Histones

Column: 2.1 × 250 mm Aquapore RP-300 (C8). Part numbers 0711-0060 (column cartridge) and 0711-0090 (guard cartridge; Perkin-Elmer Applied Biosystems, Foster City, CA)
Flow rate: 250 μl/min.

Chicken Histones

Column: 2.1 × 250 mm Vydac C_{18}. Part numbers 218TP52 (column) and 218GCC52 (guard cartridge; The Separations Group, Hesperia, CA)
Flow rate: 200 μl/min.

Separation Method (both species)

Buffer A: 0.1% (v/v) Trifluoroacetic acid (TFA) in 5% (v/v) CH_3CN
Buffer B: 0.090% (v/v) TFA in 95% (v/v) CH_3CN

Gradient:

Time (min)	%B
0–2	0
2–7	0–27 (linear)
7–97	27–52 (linear)
97–102	52–100 (linear)
102–107	100

Microsequencing

Purified histones are analysed by conventional microsequencing using an inline split to divert a portion (approximately 50%) of each cycle to a fraction collector to detect the elution of ε-*N*-[^{3}H]acetyllysine) by liquid scintillation counting while residue identification is performed with the

[35] K. W. Marvin, P. Yau, and E. M. Bradbury, *J. Biol. Chem.* **265,** 19839 (1990).

remainder.[36] Note that histones H2A and H4 from many species must be deblocked by incubation in concentrated TFA prior to sequencing.[36,37]

Interpretation of Results

Two factors affecting the sensitivity of this method should be considered when interpreting data. One is the influence of repetitive yield. In the absence of significant amino acid sequence-dependent factors, the fraction of polypeptide molecules continuing to sequence decreases with each cycle dependent on the repetitive yield (efficiency) of the sequencer. Thus, the signal for ^3H cpm eluted during sequencing of a substrate acetylated equivalently at multiple sites will decrease when plotted as a function of amino acid sequence.

A second factor is the effect that preexisting patterns of acetylation have on the ability of HATs to transfer isotopic acetate to substrates *in vitro*. Using the strategy described here, the ability of a HAT to acetylate one or more specific sites will not be recognized if one or more of those sites are fully or largely acetylated in the substrate prior to incubation with isotopic acetyl-CoA. This occurrence can be minimized by preparing substrates as outlined earlier. In addition, careful assessment of the levels of ε-*N*-acetyllysine released in each cycle, in comparison to ^3H cpm data, can be used to monitor for this and provide a rational basis for the interpretation of disparate results for a particular HAT with seemingly similar substrates.[33]

Concluding Remarks

Together, the assays described in this article represent methods that enable both the identification and the characterization of histone acetyltransferases. As work in this field progresses, we anticipate that these methods will be increasingly complemented by immunochemical and mass spectroscopic techniques. Reports that certain enzymes, initially described as HATs, are also able to acetylate specific nonhistone proteins *in vitro*[18,38] suggest the possibility that the activities of many nonhistone chromatin proteins may be regulated by acetylation. With appropriate modifications, the methods descibed here should be efficacious for the detection of acetyltransferases specific for nonhistone substrates and facilitate investigation of the role(s) of protein ε-N-acetylation in nuclear function.

[36] R. E. Sobel, R. G. Cook, C. A. Perry, A. T. Annunziato, and C. D. Allis, *Proc. Natl. Acad. Sci. U.S.A.* **92,** 1237 (1995).

[37] D. Wellner, C. Paneerselram, and B. L. Horecker, *Proc. Natl. Acad. Sci. U.S.A.* **87,** 1947 (1990).

[38] W. Gu and R. G. Roeder, *Cell* **90,** 595 (1997).

Note Added in Proof

Since this work was originally submitted, it has been reported that although recombinant yeast GCN5 does not acetylate nucleosomal histones when assayed in low ionic strength buffers as described here, acetylation of nucleosomal histones occurs in buffers containing optimum concentrations of NaCl or $MgCl_2$,[39] and that GCN5 activity with free histone substrates is not inhibited at elevated ionic strengths, in contrast to that reported here for *Tetrahymena* macronuclear extract and recombinant $dTAF_{II}230$. These differences in enzyme characteristics underscore the importance of optimizing assay conditions for the particular enzyme under investigation.

Acknowledgments

We thank Drs. Y. Nakatani and X.-J. Yang for providing recombinant $dTAF_{II}230$. This work was supported by a grant to C.D.A. (NIH-GM53512).

[39] C. Tse, E. I. Georgieva, A. B. Ruiz-Garcia, R. Sendra, and J. C. Hansen, *J. Biol. Chem.* **273,** 32388 (1998).

[40] Analysis of Activity and Regulation of hGcn5, a Human Histone Acetyltransferase

By NICKOLAI A. BARLEV, JERRY L. WORKMAN, and SHELLEY L. BERGER

Introduction

In eukaryotes, nuclear DNA is complexed with histone and nonhistone proteins to form higher order chromatin. The basic repeating unit of chromatin is the nucleosome, composed of ~146 bp of DNA wrapped around histone octamers. DNA-related biological processes are inhibited in chromatin, including transcription, replication, and DNA repair. Nucleosomal inhibition may be caused by physical blocking of cognate-binding sites for specific factors. Chromatin structures undergo remodeling prior to or during any of these processes. Mechanisms that are involved in chromatin remodeling include posttranslational modifications of core histones, such as ubiquitinization, phosphorylation, and acetylation. Acetylation neutralizes the charge of lysine residues located in the amino-terminal tails of histones,

0076-6879/99 $30.00

which may reduce the ability of the histone octamer to associate with DNA and with nonhistone repressor proteins.[1]

It has been well documented that acetylation of histones correlates with the transcriptional state.[2] Hyperacetylated histones are found associated with transcriptionally active chromosomal domains, including the mammalian globin locus and the male X chromosome in *Drosophila melanogaster*. In contrast, hypoacetylated histones are detected at silenced regions in *Saccharomyces cerevisiae,* including the telomeric regions and the mating type loci. Genetic manipulation of lysines in the tails of histone H3 and H4 in the yeast *S. cerevisiae* results in increased or decreased levels of transcription.[3,4] In addition, binding of transcriptional activators to nucleosomal DNA substrates *in vitro* is potentiated by acetylation of histone tails.[5] A significant breakthrough has been the revelation that the enzymes responsible for histone acetylation and deacetylation are previously known transcriptional regulatory factors (p55, yGCN5,[6] hGcn5,[7] P/CAF-CBP/p300,[8] $TAF_{II}250$,[9] RPD3[10]). Thus, evidence shows that alterations in the acetylation state of core histones provide an important mechanism of gene regulation.

The finding that histone acetyltransferases (HATs) and histone deacetylases (HDACs) are involved in transcriptional activation or repression raises several questions regarding their own regulation. It is likely that the dynamic interplay between acetylation and deacetylation defines levels of gene expression. Thus, according to this model, the activation of transcription is characterized by shifting of the acetylation/deacetylation balance toward acetylation. In contrast, when transcription is shut down, histones become deacetylated. Several potential mechanisms alter HAT/HDAC activity. For example, modulation of HAT/HDAC accessibility to histones may be achieved by selective physical targeting to specific genomic loci via protein–protein interactions. Thus, the native yeast Gcn5-dependent histone acetylation complex, SAGA, may be recruited to promoters by

[1] J. Brownell and C. Allis, *Curr. Opin. Genet. Dev.* **6,** 176 (1996).
[2] T. R. Hebbes, A. W. Thorne, and C. Crane-Robinson, *EMBO J.* **7,** 1395 (1988).
[3] R. Mann and M. Grunstein, *EMBO J.* **11,** 3297 (1992).
[4] J. S. Thompson, X. Ling, and M. Grunstein, *Nature* **369,** 245 (1994).
[5] M. Vettese-Dadey *et al., EMBO J.* **15,** 2508 (1996).
[6] J. E. Brownell *et al., Cell* **84,** 843 (1996).
[7] L. Wang *et al., Mol. Cell Biol.* **17,** 519 (1997).
[8] X.-J. Yang *et al., Nature* **382,** 319 (1996).
[9] C. A. Mizzen *et al., Cell* **87,** 1261 (1996).
[10] J. Taunton, C. A. Hassig, and S. L. Schreiber, *Science* **272,** 408 (1996).

interaction with activation domains, while HDAC complexes may be recruited by interaction with repression domains.[11]

A second more speculative mechanism by which acetylation may be regulated is via the modulation of intrinsic enzymatic activity. For example, posttranslational modification(s), such as phosphorylation of HAT/HDAC catalytic subunits, may alter activity and ultimately result in either increased or decreased levels of histone acetylation.

Many HATs and HDACs are found in high molecular weight complexes.[12,13] These complexes may contain various enzymatic activities, potentially including protein kinases. One interesting case is the largest TBP (TATA-binding protein) associated factor $TAF_{II}250$, which possesses both kinase and HAT activity,[9,14] although it has not been shown that these activities have any mutual effect.

We have detected repression of the HAT activity of hGcn5 via phosphorylation mediated by DNA-dependent protein kinase (DNA–PK).[15] The DNA–PK holoenzyme consists of a 450-KDa catalytic serine/threonine kinase subunit (DNA-PKcs)[16] and a DNA-binding component known as Ku autoantigen.[17] The effect of phosphorylation on HAT activity is reversible, as dephosphorylation of hGcn5 leads to the restoration of HAT activity. Moreover, we demonstrated that the HAT activity of hGcn5 differs in matched human cell lines possessing or lacking DNA–PKcs, suggesting DNA–PK-dependent regulation of hGcn5 *in vivo*. These findings suggest that the HAT activity of hGcn5 may be regulated by phosphorylation.

There are several approaches to determine whether HAT or HDAC activity is regulated by posttranslational modifications, such as phosphorylation. Initially it should be determined whether the protein of interest is, in fact, phosphorylated *in vitro*. The next step is to test whether phosphorylation affects the histone-modifying activity (HAT or HDAC) of the protein. Finally, the specificity of effects observed *in vitro* should be tested *in vivo*.

This article describes a variety of methods that have proven useful in analyzing the activity and regulation of hGcn5. These include methods to prepare recombinant hGcn5 or endogenous hGcn5 fractionated from HeLa nuclear extract. We also describe protocols for assaying the phosphorylation of hGcn5 by DNA–PK and for determination of the effect on HAT activity of hGcn5 both *in vivo* and *in vitro*. These approaches may be generally

[11] L. Alland *et al., Nature* **387,** 49 (1997).
[12] P. A. Grant *et al., Genes Dev.* **11,** 1640 (1997).
[13] S. E. Rundlett and M. Grunstein, *Proc. Natl. Acad. Sci. U.S.A.* **93,** 14503 (1996).
[14] R. Dikstein, S. Ruppert, and R. Tjian, *Cell* **84,** 781 (1996).
[15] N. A. Barlev *et al., Mol. Cell. Biol.* **18,** 1349 (1998).
[16] T. Carter *et al., Mol. Cell Biol.* **10,** 6460 (1990).
[17] T. Gottlieb and S. Jackson, *Cell* **72,** 131 (1993).

useful in analogous studies of the steadily increasing number of HATs and HDACs that have now been identified, as well as to analyze other posttranslational modifications that may alter activity.

1. Expression and Purification of Recombinant hGcn5

Most of the proteins known to date that possess HAT activity are difficult to express in bacteria as recombinant proteins. Similarly, we have found hGcn5 protein difficult to express at high levels and problematical to purify in soluble form. This section describes methods to prepare enzymatically active and soluble recombinant hGcn5 either as a hexahistidine (His_6)-tagged protein or as a glutathione (GST) fusion protein.

a. His_6-Tagged hGcn5

We have obtained large amounts of recombinant hGcn5 from transfected *Escherichia coli* (JM109 cells) followed by superinfection with phage that produces T7 polymerase (Invitrogen, Carlsbad, CA). Due to insolubility of recombinant hGcn5 protein, it is necessary to use denaturing conditions of purification with subsequent stepwise dialysis against decreasing amounts of urea.

The cDNA coding for hGcn5 was cloned into the pRSET vector (Invitrogen) and the plasmid construct was transformed into JM109.

1. Inoculate a single colony into 5 ml of M9 synthetic medium containing 100 mg/ml of ampicillin.
2. The next day dilute the overnight culture into 100 ml of SOB, supplemented with ampicillin. Grow the culture at 37° until OD_{600} of 0.3 (approximately 1–2 hr).
3. Add isopropylthiogalactoside (IPTG) to a final concentration of 1 m*M* and incubate for 1.5 hr. At this point, save 1 ml for a negative control to monitor the level of induction of protein expression.
4. Determine the OD_{600} and then calculate the amount of M3 phage required for infection. Cells are infected with phage at a multiplicity of infection (MOI) of 5 pfu (plague-forming units) per cell based on phage titer (pfu/ml) and assuming that an OD_{600} of 1.0 is equivalent to 10^9 cells/ml. After addition of phage, continue growing the culture for 3 hr at 37°. Formula for amount of phage:

$$\text{phage (ml)} = OD_{600} \times 10^9 \times 5 \times (\text{volume of culture})/(\text{pfu/ml})$$

5. Before harvesting, reserve 1 ml of the culture to check expression levels of hGcn5 by SDS–PAGE.

6. Harvest cells at 6000 rpm in a GS-3 (Sorvall) rotor at 4° for 10 min.

7. Discard the supernatant and either snap-freeze the pellet in dry ice and store at −70° or proceed with purification.

8. Thaw cell pellet on ice and lyse in 10 ml of lysing buffer for 1 hr while rotating at 4°.

9. Centrifuge lysed cells at 10,000 rpm in a SS-34 rotor at 4° for 10 min.

10. Save the supernatant for binding to a Ni^{2+}–agarose column.

11. Wash 0.75 ml of Ni^{2+}–NTA resin slurry (Qiagen, Santa Clarita, CA) three times with 10 ml of Qiagen buffer A. To do this, centrifuge the resin at low speed (1000 rpm for 1 min) in a tabletop centrifuge, decant the supernatant, and add a fresh portion of buffer A.

12. Incubate the resin with precleared cell extract for 1 hr at 4° with rotation.

13. Gently spin the resin as in step 11, decant supernatant, and wash the resin three times with buffer A as in step 11.

14. Wash the resin three times with buffer B (at room temperature).

15. Transfer Ni^{2+}–NTA resin to 1 ml minicolumn (Bio-Rad, Richmond, CA).

16. Wash the column as follows: 2 CV (column volume) Qia buffer B, pH 8.0; 2 CV Qia buffer B, pH 6.3; and 3 CV Qia buffer B, pH 5.3.

17. Elute the hGcn5 protein from the column at pH 4.5 (3 CV) and pH 3.75 (2 CV) and collect as 0.5-ml fractions. Check the protein amount by the Bradford assay.

18. Pool fractions containing hGcn5 and dilute four times with dialysis buffer C without urea.

19. Dialyze the sample for 3–4 hr against several stepwise dilutions of urea in buffer C (2, 1, 0.5, and 0.25 *M* urea).

20. The last step is dialysis against storage buffer D. Large amounts of the hGcn5 will be insoluble and should be removed by centrifugation. Collect soluble material and analyze by SDS–PAGE.

21. Quick freeze samples in dry ice/methanol. The yield of soluble hGcn5 varies, but typically is 20–40 μg/ml.

Solutions

Qia Buffer A: 6 *M* guanidine hydrochloride/0.1 *M* NaH_2PO_4/0.01 *M* Tris–HCl, pH 8.0

Qia Buffer B: 8 *M* urea/0.1 *M* NaH_2PO_4/0.01 *M* Tris–HCl, pH 8.0

Qia Buffer C: 20 m*M* HEPES–KOH, pH 7.9/20% glycerol/0.2 m*M* EDTA/100 m*M* NaCl/5 m*M* dithiothreitol (DTT)/1 m*M* phenylmethylsulfonyl fluoride (PMSF)

b. Preparation of GST–hGcn5

A second method used to obtain relatively pure recombinant protein is to purify as a GST fusion on glutathione-Sepharose beads. Because GST–hGcn5 is also largely insoluble, ionic detergent is used to solubilize the protein, and, before binding to beads, Gcn5 is renatured by dilution with nonionic detergent. Using these methods, GST–hGcn5 protein and its derivatives (e.g., GST–BrD, containing the bromo domain derived from hGcn5) are prepared for use in HAT assays, phosphorylation reactions, and for competition experiments.

1. Inoculate a single colony of newly transformed *E. coli* into 5 ml LB-amp (ampicillin).
2. Dilute overnight culture 1:40 with LB-amp and grow an additional 1.5 to 2 hr at 37°.
3. Induce protein expression by adding IPTG to a final concentration of 0.5 m*M* followed by continued incubation for 3.5 hr at 32°.
4. Harvest bacteria and resuspend in 3 ml of PBS (do not freeze pellet).
5. While on ice, add egg white lysozyme (0.1 mg/ml final concentration), DTT (5 m*M* final), and PMSF (1 m*M* final) and incubate for 15 min.
6. To solubilize GST–Gcn5, add laurylsarcosine to a final concentration of 4%.
7. Vortex the cell lysate briefly and sonicate two to three times on ice for 40 sec (microprobe tip set at 30% output).
8. Remove cellular debris by centrifugation at room temperature (SS-34, 10,000 rpm, 10 min).
9. Add Triton X-100 to the supernatant at a final concentration of 5%. This is an important step to neutralize the denaturing effect of the ionic detergent.
10. Vortex briefly.
11. Wash glutathione–Sepharose beads (400 μl of 50:50 slurry) three times with phosphate-buffered saline (PBS) by resuspending the resin in 10 ml of PBS, followed by gentle centrifugation (1000 rpm, 1 min) and decanting the supernatant.
12. Incubate the washed resin with cell lysate for 1 hr with rotation at 4°.
13. Wash the loaded resin extensively (six to seven times) with cold PBS supplemented with 1 m*M* PMSF as described in step 11.
14. Boil 15–20 ml of beads in Laemmli buffer and electrophorese on SDS–PAGE.
15. Stain the gel with Commassie blue to visualize GST–hGcn5.

GST–hGcn5 fusion protein can be stored at 4° in PBS and is stable for 3–4 weeks. We have used GST–hGcn5 in experiments testing functional association with DNA–PK holoenzyme and have used GST–BrD as a competitor in experiments described later.

Either method of purification described yields high amounts of soluble, renatured recombinant protein with minimal proteolytic degradation. Correct refolding and enzymatic activity of hGcn5 is verified by two methods. First, hGcn5 is shown to have correct structure by demonstrating specific interaction with hAda2 protein. For example, GST–hGcn5 interacts physically with *in vitro*-translated [^{35}S]hAda2 (Fig. 1A). Second, His$_6$-hGcn5 possesses enzymatic HAT activity on free histone substrates (Fig. 1B).

2. Phosphorylation and HAT Assay of Recombinant hGcn5

The following protocols allow determination of whether a HAT is a substrate of a particular protein kinase and whether potential phosphorylation affects histone-modifying activity.

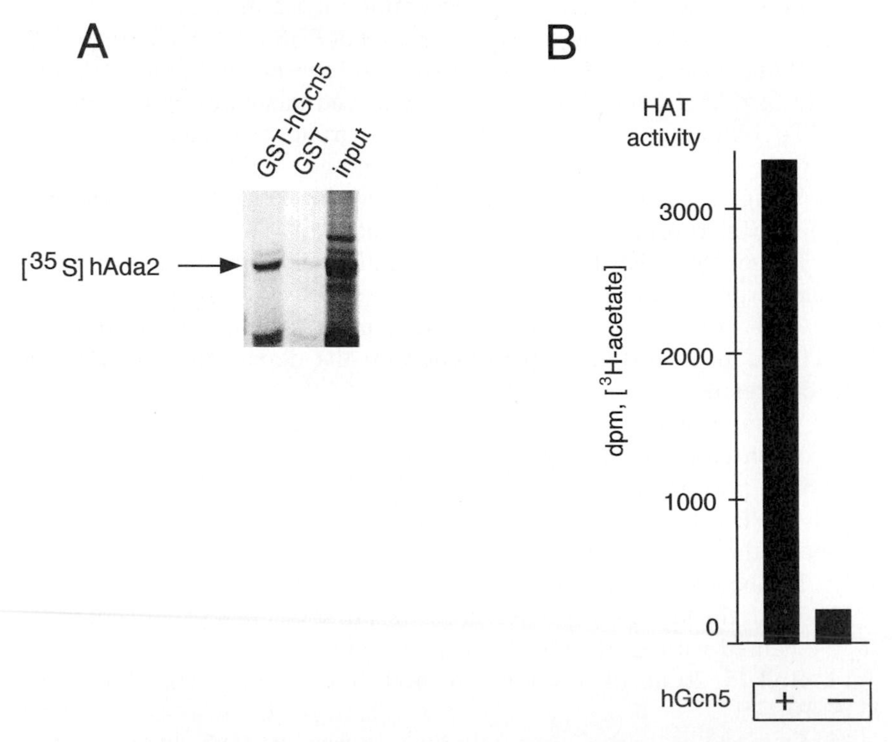

FIG. 1. Analysis of purified recombinant hGcn5. (A) Interaction of GST–hGcn5 or GST with *in vitro*-translated hAda2. Input represents one-third of amount used in each binding reaction. (B) HAT activity assay of His$_6$-hGcn5 on purified core histones. One hundred nanograms of His$_6$-hGcn5 is used in the reaction. See text for details.

a. In Vitro Kinase Assay

As an example of testing phosphorylation by a known protein kinase, the following protocol describes an assay for analysis of phosphorylation of hGcn5 by DNA–PK.

1. To a prechilled Eppendorf tube, add 5 μl of 5× reaction buffer, 100 ng of purified recombinant hGcn5, and 0.5 μg of sonicated salmon sperm DNA. The DNA is an important cofactor required to activate DNA–PK.
2. Add 1.5 μl Ku and 1.5 μl DNA–PKcs obtained from HeLa cells.[15] Semipurified DNA–PKcs[16] and Ku heterodimer [each at (100 ng/μl)] are used. Mix by pipetting.
3. Add 12.5 μM [γ-^{32}P]ATP to test phosphorylation or 0.5 mM nonradiolabeled ATP to determine HAT activity. Mix with an appropriate amount of $MgCl_2$ to yield 10 mM final concentration of magnesium.
4. Incubate at 30° for 30 min.
5. Mix reaction with Laemmli buffer and load on SDS–PAGE (to determine phosphorylation state) or stop reaction by adding 1 ml of RIPA buffer supplemented with 50 mM NaF (to assay HAT activity).
6. In the case of determination of HAT activity, add 5 μl of antigen-purified hGcn5 antisera or preimmune serum and incubate on ice for 2 hr.
7. Add protein A–Sepharose beads prewashed in RIPA buffer and incubate for an additional 30 min.
8. Wash loaded beads five to six times with RIPA buffer (1 ml each wash).
9. Renature hGcn5 protein by washing three times in cold PBS with the addition of 50 mM NaF for subsequent HAT assays.

Solutions

Reaction buffer: HEPES, pH 7.6, 50 mM NaCl, 8% glycerol, 10 mM $MgCl_2$, 1 mM DTT, 1 mM PMSF

RIPA buffer: 10 mM Tris, pH 7.5, 150 mM NaCl, 1% Triton X-100, 1% SDS, 1% DOC

We have observed that either GST–hGcn5 (Fig. 2A) or His$_6$-hGcn5 (Fig. 2B) is phosphorylated by semipurified DNA–PK. The phosphorylation is dependent on both Ku70/80 (Fig. 2B) and exogenous DNA (Figs. 2A and 2B), which are both hallmarks of DNA–PK-dependent phosphorylation.

b. HAT Assay of Purified Recombinant Gcn5

Following the phosphorylation reaction, enzymatic activity of the protein is determined on histone substrates. In the case of hGcn5, HAT activity is determined on core histones. Other HATs, such as recombinant CBP/

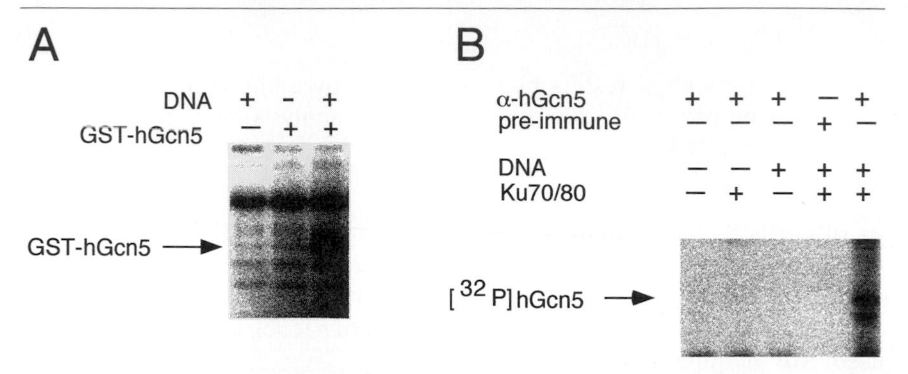

FIG. 2. Phosphorylation of purified recombinant hGcn5. (A) Phosphorylation of GST–hGcn5 by semipurified DNA–PK. Sonicated salmon sperm DNA was added where indicated. (B) Phosphorylation of His_6-hGcn5 by semipurified DNA–PK. His_6-hGcn5 was immunoprecipitated by α-hGcn5 immune sera or, as a negative control, preimmune sera. Sonicated salmon sperm DNA and Ku70/80 were added where indicated.

p300, acetylate nucleosomal histones,[18] thus making nucleosomes the preferred substrate for analysis. The histone acetylation assay is performed as described[19] with modifications for testing HAT activity of immunoprecipitated hGcn5.

1. Incubate 1 μl (20 μg/ml) of purified recombinant hGcn5 or 20 μl of protein A–Sepharose beads loaded with immunoprecipitated hGcn5 with DNA–PKcs, ATP, Ku70/80, and DNA to phosphorylate the protein (as described in Section 2a).
2. Dilute phosphorylated Gcn5 with an appropriate volume of 5× HAT buffer. If using GST–hGcn5 on beads, wash the resin three times with 1× HAT buffer to equilibrate the beads in reaction buffer. If bead-bound hGcn5 is being used, then wash beads three times with PBS/0.1% NP-40 buffer to remove DNA–PK and unincorporated ATP prior to incubation in 1× HAT buffer.
3. While on ice, add 2 μl of free calf thymus type IIA histones (Sigma, St. Louis, MO) (10 μg/μl) and 0.1 μCi of [^{3}H]acetyl-CoA (Sigma). Use histones and [^{3}H]acetyl-CoA as a negative control.
4. Incubate for 30 min at 30° (with rotation if beads are used).
5. Run samples containing acetylated histones on 15% SDS–PAGE.
6. Fix the gel in 20% methanol/10% acetic acid (v/v).
7. Soak in intensify liquid scintillant (NEN Life Science Products, Du Pont, Boston, MA), dry, and perform autoradiography.

[18] V. V. Ogryzko *et al., Cell* **87,** 953 (1996).
[19] J. Brownell and C. D. Allis, *Proc. Natl. Acad. Sci. U.S.A.* **92,** 6364 (1995).

8. In parallel, spot 10-μl aliquots of reactions on P-81 Whatman (Clifton, NJ) filters, wash four times (15 min each wash) in 50 m*M* sodium bicarbonate buffer (pH 9.0), air dry, and count in a liquid scintillation counter.

Solutions

5× HAT buffer: 250 m*M* Tris–HCl, pH 8.0, 0.25 *M* NaCl, 50% glycerol, 5 m*M* DTT, 0.5 m*M* EDTA, 5 m*M* PMSF, 50 m*M* butyric acid

Figure 3 illustrates the effect of DNA–PK-mediated phosphorylation on HAT activity of purified His_6-hGcn5. The level of histone acetylation by hGcn5 is both concentration and time dependent. The inhibitory effect of DNA–PK on HAT activity is most apparent at higher levels of acetylation.

Effect of Phosphatase Treatment on Recombinant hGcn5

To determine whether the effect of phosphorylation on HAT activity of the recombinant protein is reversible, phosphatase treatment is performed.

1. Phosphorylate hGcn5 as described in Section 2a.
2. Immunoprecipitate hGcn5 as described in Section 2a.
3. Wash beads in PBS five to six times.
4. Divide protein A beads bearing hGcn5 into equal portions (40 μl of beads) and either treat or mock treat with 50 units of CIP (Boehringer Mannheim, Indianapolis, IN) at 37° for 30 min with rotation.
5. Test immunoprecipitates either for HAT activity (20 μl of loaded beads; nonradiolabeled ATP) or analyze phosphorylation state (radiolabeled ATP) by SDS–PAGE in 10% gel and expose to X-ray film.

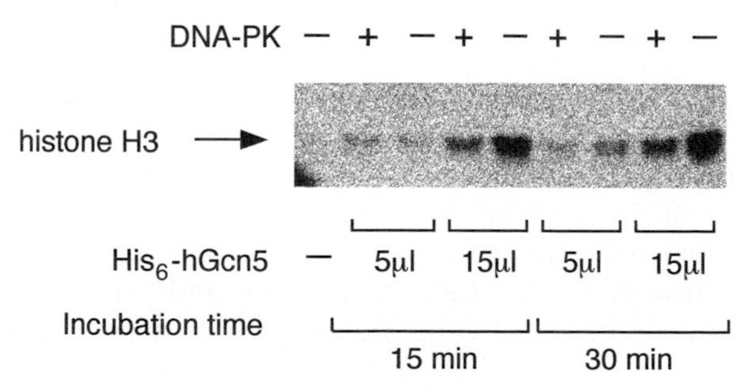

FIG. 3. Effect of DNA–PK-mediated phosphorylation on HAT activity of His_6-hGcn5. As shown, recombinant hGcn5 preferentially acetylates free histone H3. The DNA–PK holoenzyme is added where indicated. The concentration of hGcn5 and time of HAT reaction are indicated.

3. Biochemical Analysis of Effects Caused by Phosphorylation on Endogenous hGcn5 Activity

a. Fractionation of HeLa Nuclear Extract and Western Analysis

If phosphorylation alters HAT activity of the recombinant protein, then a next step would be to determine whether the endogenous protein is phosphorylated and how the phosphorylation state affects the HAT activity of the endogenous protein. Ideally, it is desirable to employ a relatively crude cellular fraction that is enriched in the specific HAT activity as well as the specific kinase activity. As described in this section, the flow-through (FT) fraction after phosphocellulose P11 column chromatography of HeLa nuclear extract contains both enzymatically active hGcn5 and DNA–PK holoenzyme.

1. HeLa nuclear extract is prepared as described previously.[20]
2. Load the nuclear extract (100 ml = 0.5 g) onto a 100-ml phosphocellulose (P11) column (Whatman) preequilibrated in 0.1 *M* KCl DB.
3. Collect the FT (100 ml).
4. Perform SDS–PAGE electrophoresis and analyze the fraction for the presence of hGcn5 and DNA–PK by Western blot analysis using specific antibodies. The antibody dilutions used for Western blot analysis were as follows: α-hGcn5 rabbit polyclonal sera 1:4000, α-Ku70 monoclonal antibody (N3H10) 1:700,[21] α-Ku80 monoclonal antibody (III) (1:700),[21] and α-DNA-PKcs monoclonal antibody (25-4) 1:300.[16]

Solutions

DB: 20 m*M* HEPES, pH 7.9, 10% glycerol, 0.2 m*M* EDTA, 0.5 m*M* DTT, 0.2 m*M* PMSF, 1 μg/ml leupeptin, 1 μg/ml pepstatin A, 5 μg/ml apoprotinin

b. In Vitro Kinase Assay

The kinase assay on the crude fraction is performed essentially as described for purified recombinant hGcn5 (Section 2a). Because hGcn5 and DNA–PK holoenzyme are present in the same fraction, kinase activity can be activated directly in the fraction. If HAT and protein kinase are in separate fractions, then they combine and kinase activity would be activated.

[20] S. M. Abmayr and J. L. Workman, *in* "Current Protocols in Molecular Biology" (F. M. Ausbel, ed.).
[21] W. H. Reeves and Z. M. Sthoeger, *J. Biol. Chem.* **264,** 5047 (1989).

1. Dialyze 50-μl aliquots of P11 FT containing hGcn5 and DNA–PK holoenzyme against buffer A for 2 hr at 4°.
2. Add 0.5 μg of sonicated salmon sperm DNA to activate DNA–PK.
3. Add 12.5 μM [γ-^{32}P]ATP or 0.5 mM nonradiolabeled ATP (for subsequent HAT assays).
4. Incubate at 30° for 30 min.
5. Proceed to HAT assay (next section) or stop reaction by adding 1 ml RIPA buffer supplemented with 50 mM NaF for subsequent immunoprecipitation and analysis of the phosphorylation state by SDS–PAGE electrophoresis (see Section 2a).

Phosphorylation of endogenous hGcn5 from the HeLa nuclear P11 FT is shown in Fig. 4A. [^{32}P]hGcn5 is detected in α-hGcn5 immunoprecipitate (lane 2, Fig. 4), but not in preimmune immunoprecipitate (lane 1, Fig. 4).

c. Histone Acetyltransferase Assay

HAT assays are performed using either the untreated P11 FT fraction or the kinase-activated fraction to determine how phosphorylation affects total HAT activity. In addition, to test how hGcn5 in particular is affected by phosphorylation, the HAT activity of immunoprecipitated hGcn5 is determined, either untreated or after stimulation of kinase activity. An example of this experiment is shown in Fig. 4B, which illustrates that the HAT activity of phosphorylated hGcn5 is reduced compared to the HAT activity of nonphosphorylated hGcn5. The detailed protocol is described in Section 2b.

d. Effect of Phosphatase Treatment

To determine whether the effect of phosphorylation on HAT activity of the endogenous protein is reversible, phosphatase treatment is performed on bead-bound hGcn5 as described earlier for the recombinant protein.

1. Phosphorylate hGcn5 as described in Section 2a.
2. Immunoprecipitate hGcn5 as described earlier from 100 μl of P11 FT fraction either immunodepleted for DNA–PKcs or mock treated with α-HA monoclonal antibody.
3. Follow steps 3, 4, and 5 from the protocol described in Section 2c.

An example of this experiment is shown in Fig. 4, where the phosphorylation of endogenous hGcn5 is reversed by treatment with calf intestinal phosphatase (Fig. 4A, lane 3 compared to lane 2). Moreover, phosphatase treatment is seen to restore HAT activity of hGcn5 (Fig. 4B).

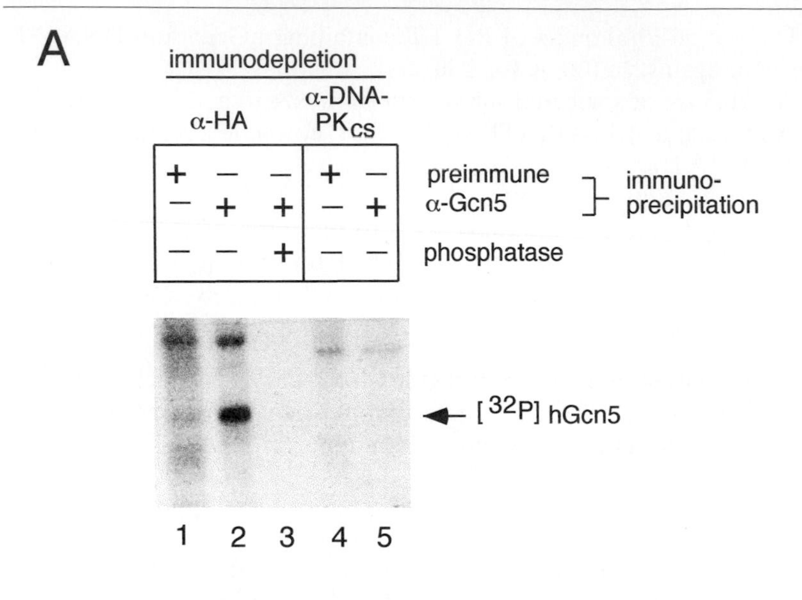

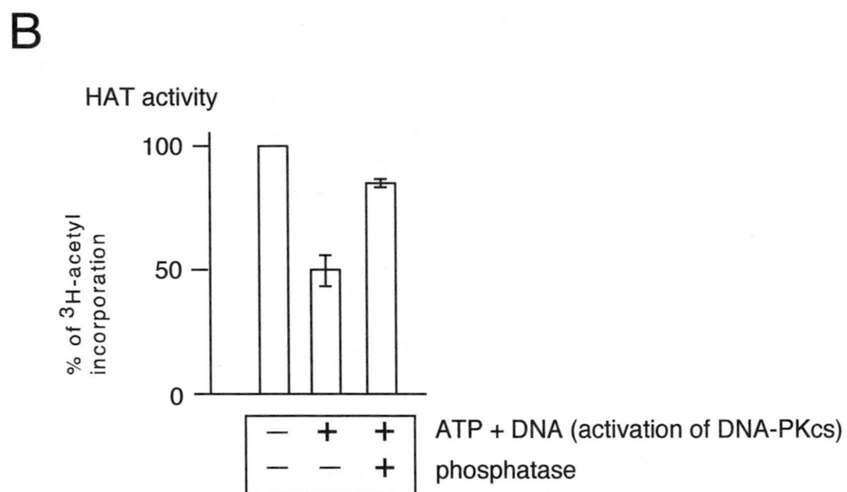

FIG. 4. DNA–PK holoenzyme-mediated effects on endogenous hGcn5. (A) Reversible phosphorylation of hGcn5 by the DNA–PK holoenzyme. Also shown is the absence of phosphorylation in samples immunodepleted of DNA–PKcs. See details in text. Samples were immunoprecipitated with α-hGcn5 or preimmune sera, as indicated. (B) Reversible effect of phosphorylation on hGcn5-mediated HAT activity. See text for details. In (A) and (B) calf intestinal phosphatase was added where indicated.

e. Immunodepletion

To test whether phosphorylation affects the HAT activity of the endogenous protein, it is important to know both what portion of total HAT activity is due to the specific activity of the HAT of interest and what portion of total kinase activity is due to the kinase of interest. This is achieved by comparison of HAT or kinase activities of the intact fraction vs the fraction following immunodepletion of the specific HAT or kinase. This section describes a protocol for the immunodepletion of hGcn5 or DNA–PK from the HeLa-derived crude P11 FT fraction.

To determine the contribution of hGcn5 to total HAT activity in the P11 FT, the sample is immunodepleted with α-hGcn5 polyclonal antisera or, as a negative control, preimmune sera. The amount of Ab is titrated to determine a minimum concentration of immune sera, but not preimmune sera, that lowers activity. In the case of hGcn5 in the P11 FT fraction, this amount was determined to be 6 μl of antisera.

To determine whether DNA–PK phosphorylates hGcn5 in the P11 FT and, if so, what effect the phosphorylation has on the HAT activity of hGcn5, immunodepletion of DNA–PKcs is performed. The sample is immunodepleted with DNA–PKcs monoclonal antibodies or, as a negative control, an unrelated monoclonal antibody.

1. Add either 6 μl of immune α-Gcn5 (or preimmune) antisera of 4 μl of α-DNA–PKcs (200 ng/ml) or 1 μl of α-HA (1 μg/μl) monoclonal antibody to 50-μl portions of P11 FT fraction. Incubate overnight on ice with rotation.
2. The next day add the same amounts of fresh antibodies to the samples and incubate for an additional 4–5 hr with rotation.
3. Add 40 μl of protein A–Sepharose bead slurry preblocked in 5% BSA and incubate with rotation for 2 hr at 4°.
4. Remove beads by centrifugation and save supernatants.
5. Examine the efficiency of immunodepletion using Western blot analysis of the supernatants from step 4.

Following immunodepletion, the phosphorylation state and HAT activity of the immunoprecipitated hGcn5 are examined. Figure 4A shows an example of the effect of DNA–PKcs immunodepletion on phosphorylation of endogenous hGcn5. Immunodepletion of DNA–PK prior to the kinase assay abolishes phosphorylation of hGcn5 (Fig. 4A, compare lanes 2 and 5).

f. Affinity Competition

Of the domain of the HAT protein responsible for interaction with the kinase is known, it is important to demonstrate that this interaction is required for phosphorylation of the HAT. In the crude fraction, this is done by pretreatment with a competitor composed of GST fused to a peptide encompassing the interaction domain.

1. To elute GST fusion proteins from glutathione–Sepharose beads, add 5 bead volumes of elution buffer to beads (use approximately 5 μg of bead-bound protein).
2. Incubate overnight with rotation at room temperature.
3. Centrifuge beads for 1 min and save supernatants.
4. Analyze supernatants by SDS–PAGE for the presence of eluted proteins by staining the gel in Coomassie blue.
5. Dialyze equal amounts of protein eluted from glutathione beads (either GST or GST–peptide) against the kinase buffer overnight. Approximately 2 μg of protein will be obtained after dialysis.
6. Add 1/5 of total volume of dialyzed samples (~0.5 μg) to 100-μl portions of FT and incubate for 2 hr at 4°, following 30 min of incubation at room temperature.
7. Resulting samples are used in the kinase assay as described in Section 3C.

In the case of hGcn5 we had identified an interaction between hGcn5's bromo domain and Ku70. Thus, because Ku70/80 is required to recruit DNA–PKcs, GST-BrD is used as a competitor for interaction with Ku70 and, hence, lowers phosphorylation (Fig. 5). GST alone is used as a negative control for the level of phosphorylation in the absence of a specific competitor (Fig. 5).

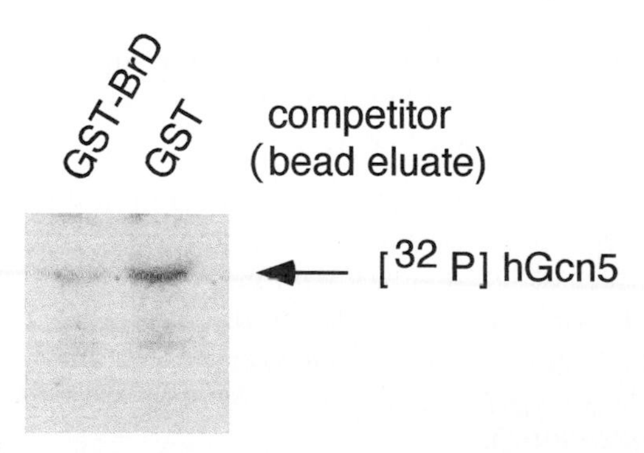

FIG. 5. Effect of affinity competition on DNA–PK-mediated phosphorylation of endogenous hGcn5. GST or GST-BrD competitor was added prior to the kinase reaction where indicated. See text for details.

Solutions

Elution buffer: 100 m*M* Tris–HCl, pH 8.0, 250 m*M* NaCl, 0.1% NP-40, 20 m*M* reduced glutathione

Kinase buffer: 20 m*M* HEPES, pH 7.6, 50 m*M* NaCl, 10 m*M* $MgCl_2$, 8% glycerol, 1 m*M* DTT, protease inhibitors

The experiments described in the previous section reveal whether a particular HAT identified in a crude fraction is enzymatically active, whether it is phosphorylated by a kinase, and whether this modification alters HAT activity.

4. In Vivo Analysis of Effects Caused by Phosphorylation on hGcn5-Mediated Histone Acetylation

It is crucial to determine whether phosphorylation occurs *in vivo* and how the modification affects HAT activity. To address the question of whether the protein of interest is a phosphoprotein *in vivo* and whether its activity is regulated by phosphorylation, ^{32}P labeling of cells expressing the protein of interest is performed. The role of a specific kinase can be tested using selective inhibitors and/or cell lines that lack the potential kinase activity. For example, we used two neuroblastoma cell lines that differ in their DNA–PKcs expression[22] and found that hGcn5 exhibits a higher phosphorylation level and possesses lower HAT activity in the DNA–PK^{+}_{cs} cell line compared to the DNA–PK^{-}_{cs} cell line.

To analyze the phosphorylation state of hGcn5 in COS cells, cells are transfected with plasmid containing hGcn5 fused to hemeagglutinin (HA) tag. This allows immunoprecipitation of the protein with α-Gcn5 antibody or α-HA monoclonals. To monitor the level of expression of hGcn5, cells are labeled with [^{35}S]methionine followed by immunoprecipitation. Labeling with ortho[^{32}P]phosphate and immunoprecipitation indicates whether the protein is phosphorylated.

DEAE-Dextran-Mediated Transfection of hGcn5

DEAE-dextran is used for the transfection of plasmid DNA bearing the hGcn5 gene fused to the HA tag. Similar results are obtained using lipofectin-mediated transfection.

1. On day 0, seed 35-mm culture dishes with 2×10^5 cells. For nuclear extracts, plate 10^6 cells per 100-mm dish, scale up buffer volumes and DNA amounts fivefold, and maintain the same incubation times.

[22] S. P. Lees-Miller *et al., Science* **267,** 1183 (1995).

2. On day 1, prepare transfection cocktail
3. Aliquot 1 ml of cocktail in a sterile tube.
4. Add 1–2.5 μg plasmid DNA (CsCl or Qiagen purified).
5. Incubate for 15 min at room temperature
6. Wash cells twice with IMDM without serum.
7. Add 1 ml DNA/transfection cocktail per 35-mm dish.
8. Incubate at 37° for 2.5 hr.
9. Add 1 ml 10% DMSO in IMDM with 10% fetal bovine serum (FBS).
10. Incubate for 2 min at room temperature.
11. Aspirate DMSO and immediately wash two times with 10% FBS–IMDM.
12. Add 2 ml 10% FBS–IMDM complete media.
13. Incubate at 37° for 40–48 hr.

Solutions

Transfection cocktail: 10.1 ml Iscove's modified Dulbecco's medium (-) serum (IMDM), 100 μl Fungizone (250 μg/ml) filtered through a 0.2-μm membrane, 100 μl 200 m*M* glutamine, 500 μl DEAE-dextran 10 mg/ml filtered through a 0.2-μm membrane (e.g., Pharmacia, Piscataway, NJ, molecular weight approximately 500,000), 30 ml chloroquine (0.125 g/5 ml PBS, make fresh filtered through a 0.2-μm membrane, 100 μl Nutridoma SP (Boehringer Mannheim)

b. ^{35}S Labeling and Immunoprecipitation

Because we are unable to detect expression of hGcn5 by Western analysis in extracts from transfected COS cells, [^{35}S]Met labeling of total proteins is used with subsequent immunoprecipitation of hGcn5.

1. Wash two times with (-)-methionine, (-)-cysteine DMEM (ICN, Costa Mesa, CA).
2. Add 1 ml (-)-Met, (-)-Cys DMEM. Incubate at 37° for 30 min to starve.
3. Aspirate and add 1 ml (-)-Met, (-)-Cys DMEM (plus glutamine) with 15 ml Tran^{35}S-label (1 mCi/ml; ICN, Costa Mesa, CA).
4. Incubate at 37° for 2 hr.
5. Remove media containing radiolabel.
6. Add 1 ml lysis buffer (e.g., RIPA plus protease inhibitors) and lyse for 10 min at room temperature.
7. Clarify at 100,000*g* for 30 min at 4° (40,000 rpm in Ti90 rotor).
8. Collect supernatant and use 0.5 ml of extract per immunoprecipitation (diluted with RIPA buffer to 1 ml total).

9. Add 2 μl of preimmune, 1 μl of α-HA, or 4 μl of α-GCN5 antibody. Incubate on ice for 2 hr.

10. Add 50 μl of 10% slurry of prewashed protein A–agarose beads (Pharmacia).

11. Incubate for 40 min on ice.

12. Wash beads extensively (six to seven times) with RIPA buffer (1.5 ml each wash) by centrifugation for 20 sec and aspirating supernatant.

13. Transfer beads to fresh tubes before the last wash.

14. Aspirate supernatant, add 40 μl of phosphate buffered saline (PBS), add 10 μl of 5× Laemmli, boil, and load the supernatant on 10% SDS–PAGE gel. Run gel.

15. Dry gel and perform autoradiography.

Labeling of transfected hGcn5 in COS cells is shown in Fig. 6. Labeling of hGcn5 with [^{35}S]Met is shown by immunoprecipitation with α-hGcn5 antibody.

c. Phosphorylation in Vivo

Once the expression of the protein is confirmed by either Western blot analysis or immunoprecipitation of [^{35}S]Met-labeled material, the phosphorylation state of the protein is tested by labeling with radioactive inorganic phosphate.

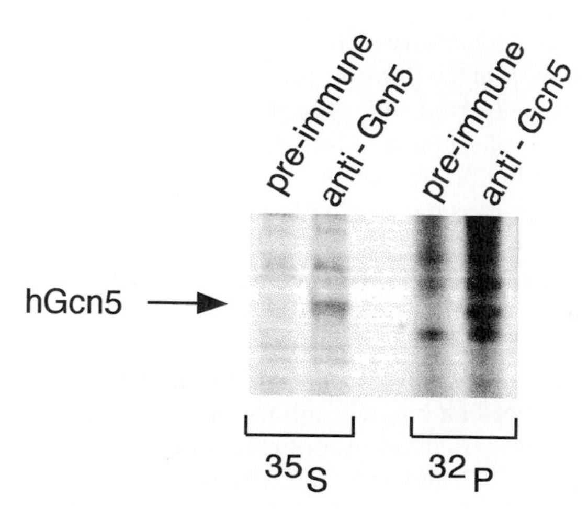

FIG. 6. *In vivo* labeling of transfected hGcn5 in COS cells. Cells were labeled metabolically with [^{35}S]Met or [^{32}P]Pi as indicated. Samples were immunoprecipitated with α-hGcn5 or preimmune sera as indicated. See text for details.

For labeling with inorganic phosphate, a 10-cm dish of COS cells transfected with hGcn5 (described in Section 4A) at 70–80% confluence is used. Cells are cultivated as described earlier.

1. Aspirate medium and wash once with TBS.
2. Add 4 ml of "starvation medium" [DMEM medium supplemented with 10% of dialyzed fetal bovine serum (FBS)] and incubate for 3 hr to exhaust for endogenous phosphate.
3. Aspirate medium and add 4 ml of fresh "starvation medium" supplemented with 1 mCi of [^{32}P]orthophosphate. Incubate for 5–6 hr and remove radioactive medium by pipetting.
4. Wash cells twice with TBS.
5. Add 2 ml of RIPA buffer with protease/phosphatase inhibitors as described in Section 3a.
6. Lyse cells in RIPA buffer for 10 min at room temperature with gentle rocking.
7. Collect the lysate by pipetting and transfer to centrifuge tubes.
8. Clear the extract by centrifugation (40,000 rpm in Ti90 rotor for 30 min at 4°).
9. Perform immunoprecipitation as described in Section 2a using double amounts of antibodies and protein A–Sepharose beads.
10. Check one-third of the beads on SDS–PAGE for autoradiography.

Labeling of transfected hGcn5 in COS cells is shown in Fig. 6. Labeling of hGcn5 with [^{32}P]Pi is shown by immunoprecipitation with the α-hGcn5 antibody.

To demonstrate specificity of phosphorylation, several methods can be used. First, to show that the phosphorylation is reversible, the immunoprecipited protein can be treated with phosphatase, as described earlier. Second, to show that a specific kinase is involved in the phosphorylation, where available, specific inhibitors can be used. In the case of DNA–PK, OK-1035 has been shown to be specific for DNA–PK kinase activity *in vitro* and possibly *in vivo*.[23] Third, as described earlier, if matched cell lines are available that possess or lack the specific kinase activity, these should be tested for the state of phosphorylation of the protein.

Finally, a set of parallel experiments should be done to determine whether HAT activity is affected by inhibition of specific kinase activity, again, in the presence of a kinase inhibitor or comparing matched cell lines. In this case the protein is immunoprecipitated from cells treated with inhibitor or not treated, and HAT activity is determined.

[23] Y. Take *et al., Biochem. Biophys. Res. Commen.* **215,** 41 (1995).

Acknowledgments

We thank T. Owen-Hughes for fractionation of HeLa extract; C. Ying for assistance in the preparation of recombinant His_6-hGcn5; V. Poltoratsky and T. Carter for samples of semipurified Ku70/80 and DNA–PK; and T. Carter and W. Reeves for monoclonal antibodies against DNA–PK and Ku, respectively. Research was supported by grants from the National Institutes of General Medical Sciences to J.L.W. and from the American Cancer Society, the National Science Foundation, and the Council for Tobacco Research to S.L.B.

[41] Purification of a Histone Deacetylase Complex from *Xenopus laevis:* Preparation of Substrates and Assay Procedures

By PAUL A. WADE, PETER L. JONES, DANIELLE VERMAAK, and ALAN P. WOLFFE

Introduction

Since its discovery in the 1960s, histone acetylation has been correlated with changes in transcription.[1,2] Enzymes that catalyze the deposition and removal of acetate from the ε-amino groups of core histone lysines have been identified.[1–3] Currently, two structural families of histone deacetylase are known. The first consists of homologs of the yeast Rpd3 protein. There are five Rpd3-like histone deacetylases in yeast and at least three in mammalian cells.[2] These enzymes are nuclear, and mutation of their genes in yeast and *Drosophila* results in transcriptional defects.[2] A subset of the Rpd3-like deacetylases are found in a large macromolecular complex of megadalton size that contains both the histone-binding protein RbAp48/p46 and the known transcriptional regulator Sin3.[4] This complex has been implicated in transcriptional regulation by members of the nuclear receptor family and the Myc family of transcription factors.[1,2] It is also clear that this large complex is not the only entity containing an Rpd3-like deacetylase both in yeast and in mammalian cells.[1,4] The second structural class of histone deacetylases currently contains a single member, the maize HD2 protein.[5] This phosphoprotein is found exclusively in the nucleolus and appears to

[1] K. Struhl, *Genes Dev.* **12,** 599 (1998).
[2] M. Grunstein, *Nature* **389,** 349 (1997).
[3] P. A. Wade and A. P. Wolffe, *Curr. Biol.* **7,** R82 (1997).
[4] C. A. Hassig, J. K. Tong, T. C. Fleischer, T. Owa, P. G. Grable, D. E. Ayer, and S. L. Schreiber, *Proc. Natl. Acad. Sci. U.S.A* **95,** 3519 (1998).
[5] A. Lusser, G. Brosch, A. Loidl, H. Haas, and P. Loidl, *Science* **277,** 88 (1997).

0076-6879/99 $30.00

lack accessory proteins of the type identified with the nuclear Rpd3-like enzymes. The role, if any, of this molecule *in vivo* is currently unknown. At the current time, little is known regarding the identity of polypeptides associated with the Rpd3-like deacetylases or their roles in modulating enzymatic activity. This article outlines a protocol for assaying histone deacetylase activity in cell extracts and for purifying a large, nuclear macromolecular complex containing Rpd3-like deacetylases and associated proteins.

Preparation of Acetylated Histone Substrates and Deacetylase Assay

Purification of proteins from cell and tissue extracts is highly dependent on the sensitivity and reliability of the detection methodology. In the case of enzymes, preservation of biological activity as well as recovery can be monitored through the use of an appropriate activity assay. For histone deacetylases, several assays exist that differ primarily in the type of substrate. The first, and oldest, assay utilizes histones acetylated *in vivo* in the presence of a radioactive acetyl coenzyme A precursor and deacetylase inhibitor.[6] Hyperacetylated core histones are then purified by acid extraction or chromatographically. An obvious benefit of these substrates is that they are obtained from a natural source. Acetylation occurs at physiologically relevant sites and the substrates can retain their native structures. The major drawbacks of this method are that relatively low specific activity substrates are obtained and that there is no control over histone and lysine specificity. A second related type of substrate is obtained through the chemical acetylation of purified histones.[6] This can yield substrates with very high specific activity, but acetylation occurs randomly throughout the molecule. As lysine acetylation is highly specific *in vivo* and restricted to amino termini, this lack of sequence specificity is problematic. The final assay substrate involves the chemical acetylation of synthetic or naturally derived histone peptides.[7] The resulting substrate can have very high specific activity with acetylation occurring at physiologically relevant residues. However, these small substrates typically lack the histone fold domain, a region that may provide crucial structural information for the binding of histone modification enzymes.

We present here an alternative substrate for deacetylation assays. We have opted to use purified recombinant histone acetyltransferases to deposit radiolabeled acetyl groups onto purified histones *in vitro*. There are several

[6] A. Inoue and D. Fujimoto, *Biochem. Biophys. Res. Commun.* **36,** 146 (1969).

[7] D. E. Krieger, G. Vidali, B. W. Erickson, V. G. Allfrey, and R. B. Merrifield, *Bioorgan. Chem.* **8,** 409 (1979).

specific advantages to this technique. First, it is possible to obtain very high specific activity substrates. Of course, the limits of detection of any assay are related directly to the specific activity of the substrate. Second, as intact histones are used, native structures are preserved. As a previously described deacetylase subunit, RbA p48/p46, is known to contact histone H4, at least in part, through amino acids in the histone fold motif,[8] this may offer increased binding information over synthetic peptide substrates. Finally, through the use of unique histone acetyltransferases or through the purification of individual histones, one can obtain deacetylase substrates with both histone and lysine specificity. The method presented here utilizes yeast Hat1p for substrate production, but any reasonably pure histone acetyltransferase could be used with similar results.

Recombinant Hat1p Production

1. A plasmid for expression of yeast Hat1p in *E. coli* (gift of Dr. M. Parthun, Ohio State University[9]) under T7 RNA polymerase control is introduced into *Escherichia. coli* strain BL21 (DE3). A single colony is propagated overnight in 5 ml LB plus 50 μg/ml ampicillin at 37°.
2. The following day, 500 ml LB (with 50 μg/ml ampicillin) is innoculated with 500 μl of the overnight culture and grown at 37° to an absorbance of 0.5 to 1.0 at 600 nm. Protein production is induced by the addition of isopropylthiogalactoside (IPTG) to a final concentration of 0.5 m*M* and the culture is incubated an additional 3 hr.
3. Bacteria are harvested by centrifugation for 5 min at 5000 rpm in a Sorvall GS3 rotor (4°). The bacterial pellet may be processed immediately or frozen at −20° for later use.
4. The bacterial pellet is resuspended in 25 ml buffer B(100) [buffer B: 25 m*M* Tris–HCl, pH 8.0, 10% (v/v) glycerol, 1 m*M* EDTA, 0.2 m*M* phenylmethylsulfonyl fluoride (PMSF)] (the number in parentheses indicates millimolar concentration of sodium chloride) and cells are lysed by sonication.
5. Debris is removed by centrifugation at 10,000 rpm for 20 min in a Sorvall SA600 rotor.
6. The resulting lysate (approximately 2 mg protein/ml) is applied to a DE-52 column preequilibrated in buffer B(100) at a ratio of 10 mg protein/ml packed bed volume. The column is washed with 5 column volumes of B(100) followed by a 5 column volume linear gradient from B(100) to B(400) and a 2 column volume wash in B(400).

[8] A. Verreault, P. D. Kaufman, R. Kobayashi, and B. Stillman, *Curr. Biol.* **8,** 96 (1997).
[9] M. R. Parthun, J. Widom, and D. E. Gottschling, *Cell* **87,** 85 (1996).

Fractions (one-fourth column volume) are checked for Hat1p by SDS–PAGE and pooled. The DE-52-purified Hat1p is dialyzed versus buffer B(50), and protein concentration is determined [Bio-Rad (Richmond, CA) protein assay using bovine serum albumin (BSA) as standard] and frozen in aliquots of about 100 μg.

In Vitro Acetylation

Purified histone octamers from chicken erythrocyte nuclei[10] are acetylated as follows.

1. One milligram of core histones is incubated with 100 μg recombinant Hat1p and 100 μl [^{3}H]acetyl coenzyme A (CoA, Amersham, Piscataway, NJ) in a final volume of 8.8 ml. Final concentrations of buffer components are 25 m*M* Tris–HCl, pH 8.0, 100 m*M* NaCl, 0.1 m*M* EDTA, 0.2 m*M* PMSF, and 10% glycerol. It should be noted that the NaCl concentration is crucial to efficient acetylation with this enzyme and that the addition of salt with the purified histones and rHat1p must be accounted for in calculating the proper amount of NaCl to add to the reaction buffer. Acetylation proceeds at 37° for 30 min. The reaction is then chased by the addition of 100 nmol cold acetyl-CoA and a further incubation for 30 min.
2. Acetylated histones are purified from the reaction components by immediate application to a 1-ml Bio-Rex 70 column preequilibrated with buffer C(200) (buffer C: 10 m*M* Tris–HCl, pH 8.0, 0.1 m*M* EDTA; the number in parentheses indicates millimolar concentration of NaCl). Prolonged incubation under these conditions results in histone proteolysis. The column is washed with at least 10 ml buffer C(200) and eluted with buffer C(2000). Integrity of the acetylated histones is monitored by SDS–PAGE, and the extent of acetylation is assessed by Triton acid urea gel and fluorography. Using this protocol, we routinely achieve better than 95% acetylation with approximately 50% of the H4 molecules receiving two acetyl groups. The specific activity achieved using this method is normally in the range of 2000–5000 dpm per picomole histone H4. This specific activity can be increased by altering the molar ratio of [^{3}H] acetyl-CoA to histones in the initial pulse-labeling period. A typical substrate preparation is depicted in Fig. 1.

[10] S. Chandler and A. P. Wolffe, "Methods in Molecular Biology: Chromatin Protocols." Humana Press, Clifton, NJ, 1998.

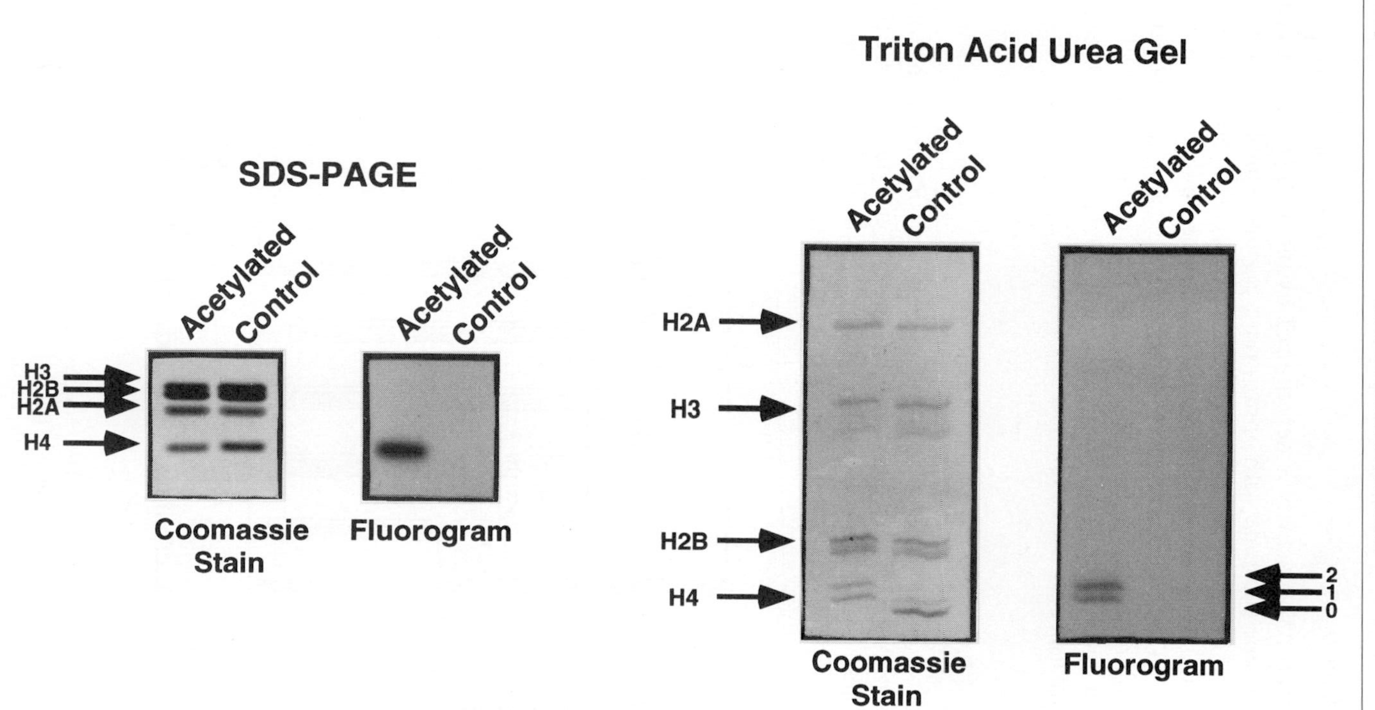

FIG. 1. Histone substrates for deacetylase assay. Three micrograms of histones (chicken erythrocyte histones with or without acetylation as described in the text) per lane was electrophoresed on 15% SDS or Triton acid urea gels and stained with Coomassie Blue. After photography, the stained gels were soaked in Amplify (Amersham) and fluorograms were made.

Deacetylase Assay

Histone deacetylase activity is determined by incubating 1 μg of acetylated free histone substrate with appropriate enzyme samples in a final volume of 200 μl of 25 m*M* Tris–HCl, pH 8.0, 10% glycerol, 50 m*M* NaCl, 1 m*M* EDTA at 30° for 30 min. The reaction is stopped by the addition of 50 μl 0.1 *M* HCl, 0.16 *M* acetic acid. Acetate is subsequently extracted from the reaction with 600 μl ethyl acetate and 75% of the organic phase is counted in a liquid scintillation counter. Figure 2 depicts a typical result from this assay using an unfractionated egg extract.

Preparation of Protein Extracts and Fractionation

We present a method for purification of a nuclear Rpd3-like histone deacetylase enzyme complex from *Xenopus* egg and oocyte extracts. Extracts from these cells have been used previously for *in vitro* studies of chromatin assembly, transcription, and other aspects of nuclear physiol-

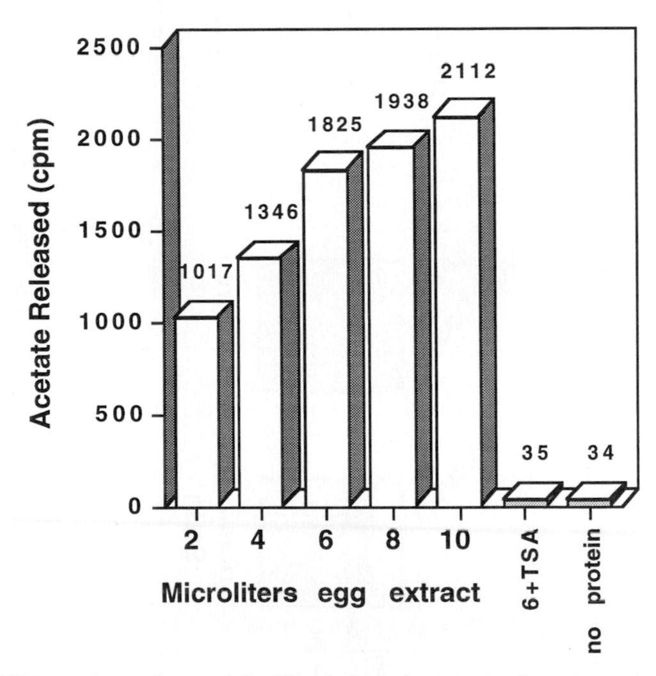

FIG. 2. Histone deacetylase activity. The indicated amounts of egg extract were assayed for histone deacetylase activity as described in the text. The deacetylase inhibitor trichostatin-A was added at a concentration of 300 n*M* in the lane marked +TSA.

ogy.[11] As the egg must provide sufficient materials for the early embryo to develop to the 4000 cell stage in the absence of zygotic transcription, many crucial biomolecules are present at high levels. Histone deacetylation, in the context of *in vitro* chromatin assembly, has been reported previously for both oocyte[12] and egg extracts.[13] A major advantage offered by these cells is the presence of chromatin components as storage forms. This allows for simple extraction at low salt, potentially preserving protein–protein interactions that could be disrupted in the process of stripping these species from chromatin. The protocol presented here (summarized in chart form in Fig. 3A) results in isolation of a multisubunit histone deacetylase complex containing stoichiometric quantities of at least six polypeptides. Notably, this enzyme contains *Xenopus laevis* homologs of Rpd3 and RbAp48/p46 (see Fig. 3C).

Egg Extract Production

1. Eggs are obtained from mature female *X. laevis* by pretreatment with 50 units of human chorionic gonadotrophin (Sigma Chemical, St. Louis, MO). Following a period of 7 to 10 days, the animals are treated with 500 units HCG (human chorionic gonadotropin).
2. The following day (roughly 12 hr after HCG treatment), eggs are collected over an approximately 8-hr period in 0.2 *M* NaCl, dejellied in 2% cysteine (pH 8.0), and washed briefly in buffer A: 20 m*M* HEPES, pH 7.5, 5 m*M* KCl, 1.5 m*M* $MgCl_2$, 1 m*M* EGTA, 10% glycerol, 10 m*M* β-glycerophosphate, 0.5 m*M* dithiothreitol, 1 m*M* PMSF, 2 μg/ml pepstatin A, and 1 μg/ml leupeptin.
3. Eggs are then transferred gently to SW-41 polyallomer tubes (filling slightly more than half the tube with packed eggs) and washed three times with buffer A. Tubes are subsequently filled with buffer A and centrifuged in an SW-41 rotor at 38,000 rpm for 2 hr at 4°.
4. Following centrifugation, the tubes are punctured with a 21-gauge needle and the clear straw-colored supernatant is removed carefully. The protein concentration is determined by the Bio-Rad protein assay using bovine serum albumin as standard.

Oocyte Extract Production

Oocyte extracts are prepared by dissection of ovaries from mature female *X. laevis*.

[11] G. Almouzni and A. P. Wolffe, *Exp. Cell Res.* **205,** 1 (1993).
[12] A. Shimamura and A. Worcel, *J. Biol. Chem.* **264,** 14524 (1989).
[13] S. Dimitrov, M. C. Dasso, and A. P. Wolffe, *J. Cell Biol.* **126,** 591 (1994).

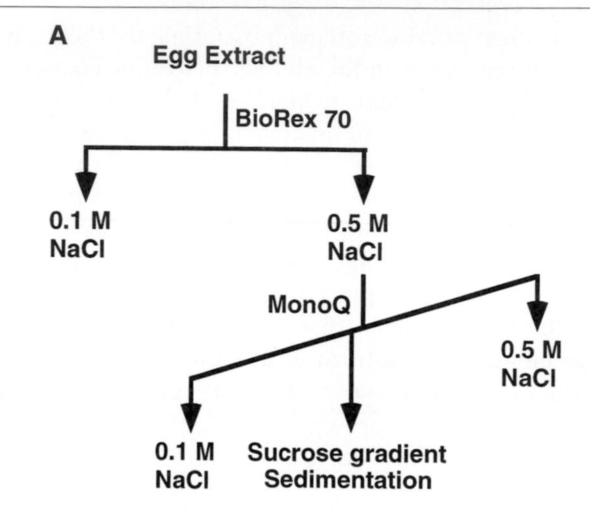

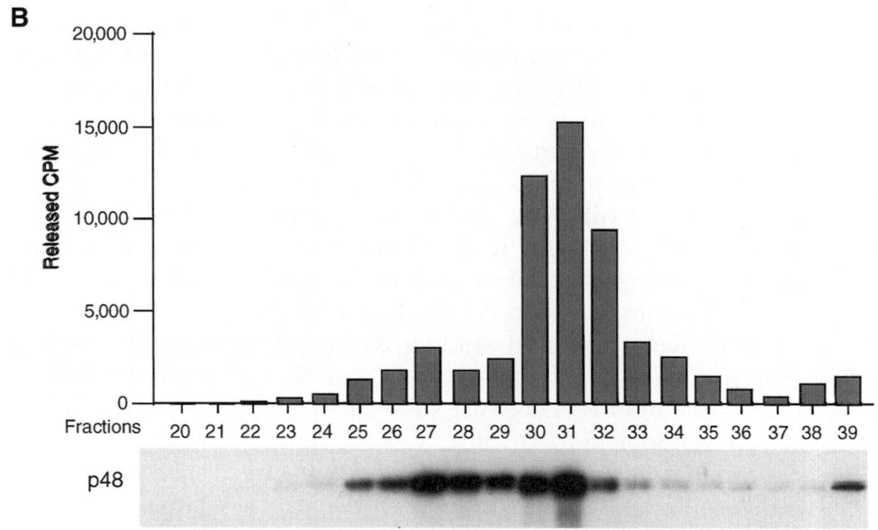

FIG. 3. Histone deacetylase purification. (A) Flow chart. (B) Mono Q gradient profile. Indicated column fractions from the Mono Q gradient were assayed for deacetylase activity and analyzed for RbAp48/p46 by immunoblot. (C) Sucrose gradient. Indicated fractions from the sucrose gradient were electrophoresed on 10% SDS polyacrylamide gels and assayed for deacetylase activity. Enzymatic activity exactly cosediments with six polypeptides: 240 kDa, 80 kDa, a 62/63/64-kDa triplet, a 58/59-kDa doublet, a 54/55-kDa doublet, and a 35/36-kDa doublet.

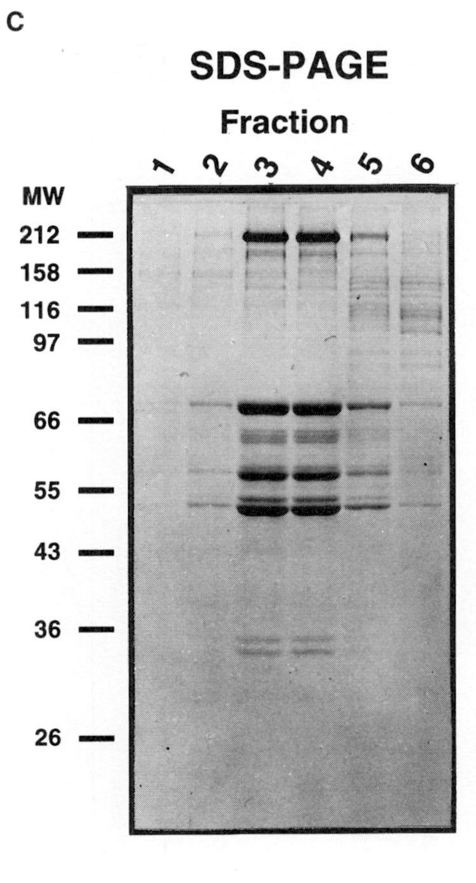

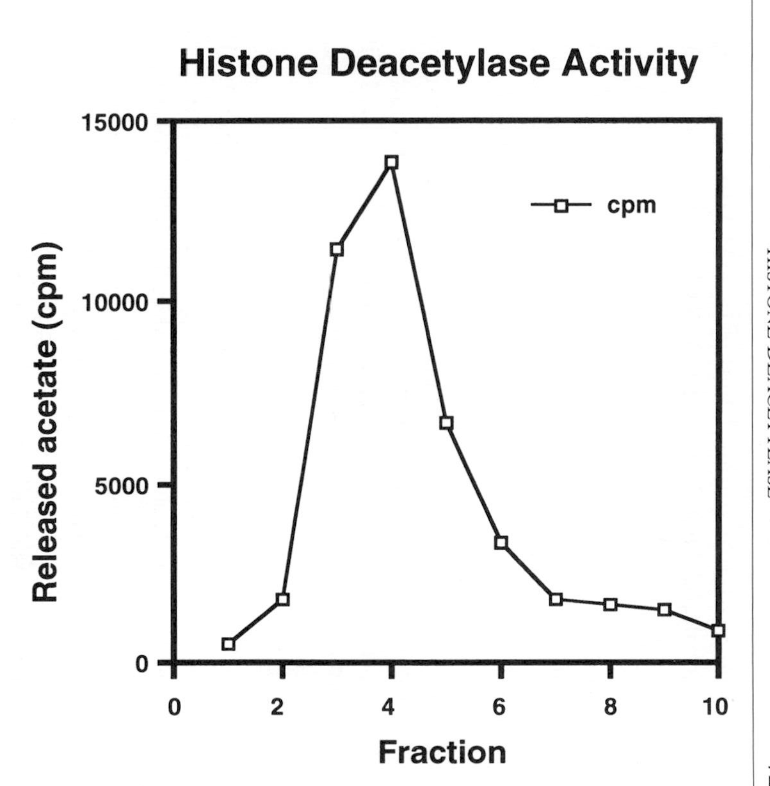

FIG. 3. (*Continued*)

1. Tissue is washed in OR-2 buffer (5 mM HEPES, 1 mM sodium phosphate, 82.5 mM NaCl, 2.5 mM KCl, 1 mM $CaCl_2$, 1 mM $MgCl_2$) to remove blood.
2. Ovaries are carefully cut into small pieces with a pair of scissors. Approximately 20 ml of ovary tissue is placed in 50-ml conical tubes, 25 mg collagenase type II (Sigma Chemical) is added, and tubes are filled to approximately 45 ml with OR-2, placed on a platform shaker, and agitated vigorously for 45 min to 1 hr until the oocytes are clearly dispersed.
3. Oocytes are washed extensively in OR-2 buffer with rapid buffer decanting to facilitate the removal of immature oocytes.
4. Mature oocytes are transferred to SW-41 tubes, washed with buffer A, and centrifuged as described for eggs.

Extract Fractionation

Bio-Rex 70. Extracts are initially fractionated on Bio-Rex 70 (Bio-Rad) cation-exchange columns as follows. All procedures are carried out at 4°. The fractionation protocol is summarized in flow chart form in Fig. 3A.

1. Bio-Rex 70 (sodium form, 100–200 mesh) is equilibrated in buffer A(100); the number in parentheses indicates millimolar NaCl concentration.
2. Extract protein is applied (10 mg protein/ml packed bed volume) and the column is washed with at least 5 column volumes of buffer A(100).
3. Bound protein is eluted with 5 column volumes A(500).
4. Fractions containing approximately 80% of the protein from each step are pooled.

Mono Q

1. The 500-mM NaCl fraction from the Bio-Rex 70 column is dialyzed versus 200 to 500 volumes of buffer A until the conductance is equal to or less than 0.1 M NaCl, and insoluble material is removed by centrifugation for 20 min at 10,000 rpm in a Sorvall SA-600 rotor.

TABLE I
PURIFICATION FROM EGG EXTRACT

Fraction	Volume (ml)	Protein concentration (mg/ml)	Total protein
Egg extract	136	6.9	940 mg
Bio-Rex70 pool	30	3.5	105 mg
Mono Q 10/10	8	0.46	3.7 mg
Sucrose pool	5	≤0.01	~50 μg

2. The clear supernatant is loaded on a Mono Q 10/10 column (Pharmacia) equilibrated in buffer A(100).
3. The column is eluted by a 5 column volume wash with A(100), a linear gradient (in 20 column volumes) from A(100) to A(500), and a 5 column volume wash in A(500). Fractions (4 ml) are collected and analyzed for histone deacetylase activity. The elution of histone deacetylase activity as well as the deacetylase associated polypeptide RbAp48/p46 from a typical Mono Q gradient is depicted in Fig. 3B.

Sedimentation. Peak activity fractions from the Mono Q column are purified further by sedimentation.

1. Sucrose gradients (5–20% sucrose) are formed in buffer A(150) using a Biocomp Gradient Master (SW-41 tubes).
2. Gradients are chilled to 4° and loaded with up to 1 ml Mono Q-purified histone deacetylase and centrifuged in SW-41 rotors for 28 hr at 41,000 rpm at 4°.
3. The gradients are subsequently fractionated into 15 to 20 fractions and assayed for histone deacetylase activity. Figure 3C depicts the protein and deacetylase activity profiles following sedimentation of the major deacetylase peak from the Mono Q column. This large deacetylase complex sediments with an apparent molecular mass of 1–1.5 MDa. The doublet at 58/59 kDa reacts with *Xenopus* Rpd3 antisera.

Table I presents protein recovery data for a typical purification from an egg extract. The final purified protein is robustly active using free histone substrates as seen in Fig. 3C. We have never detected significant differences in protein profile, activity, or chromatographic properties when comparing enzyme preparations from eggs or oocytes. However, care should be taken with an enzyme prepared from oocyte extracts to avoid residual proteases from collagenase treatment. Careful washing of the dispersed oocytes with copious quantities of buffer is the best protection from this potential problem. The enzyme prepared according to this protocol is stored at −70° and is stable to multiple rounds of freeze–thaw.

[42] Purification and Biochemical Properties of Yeast SWI/SNF Complex

By COLIN LOGIE and CRAIG L. PETERSON

Introduction

Saccharomyces cerevisiae SWI/SNF complex is a 2-MDa multiprotein assembly that contains 11 distinct polypeptides.[1-3] Genetic and biochemical evidence indicate that this macromolecular machine participates in the induction of genes otherwise repressed at a transcriptional level via nucleosome-dependent mechanisms. *In vitro,* on mononucleosome substrates, SWI/SNF uses the energy of ATP hydrolysis to perturb histone–DNA interactions, which results in enhanced binding of recombinant transcription factors to their DNA sites.[1] In the context of nucleosome arrays, SWI/SNF also enhances nucleosomal DNA binding by Gal4 protein, which can result in the persistent disruption of a nucleosome.[4]

The stimulation of transcription factor binding is not readily amenable to kinetic analysis as it depends on the detection of a reversible event. However, the restriction enzyme cleavage of DNA is more amenable because it involves an enzymatic covalent modification that is easy to detect. Furthermore, site-specific cleavage of DNA has the virtue of quantitatively reporting the degree of DNA occlusion mediated by histone octamers.[5] Because nucleosome arrays are more likely to mimic the natural substrate of the SWI/SNF complex, we developed a coupled enzyme assay where the cleavage of a unique site within a well-defined nucleosome array is enhanced by the nucleosome remodeling activity of the SWI/SNF complex.

This article describes our current SWI/SNF purification scheme and a brief SWI/SNF ATPase assay. We describe the methods that we use to assemble radiolabeled nucleosome arrays and to determine the degree of saturation of the reconstituted arrays. Finally we discuss the considerations that dictated our choice of experimental parameters for measuring SWI/

[1] J. Cote, J. Quinn, J. L. Workman, J. L, and C. L. Peterson, *Science* **265,** 53 (1994).

[2] I. Treich, B.R. Cairns, T. De Los Santos, E. Brewster, and M. Carlson, *Mol. Cell. Biol.* **15,** 4240 (1995).

[3] B. R. Cairns, K. J. Kim, M. H. Sayre, B. C. Laurent, R. D. Kornberg, *Proc, Natl. Acad. Sci. U.S.A.* **91,** 1950 (1994).

[4] T. Owen-Hughes, R. T. Utley, J. Cote, C. L. Peterson, and J. L. Workman, *Science* **273,** 513 (1996).

[5] K. J. Polach and J. Widom, *J. Mol. Biol.* **254,** 130 (1995).

0076-6879/99 $30.00

SNF activity in the coupled SWI/SNF–restriction enzyme assay using reconstituted nucleosome arrays as substrates.

SWI / SNF Complex Purification

The SWI/SNF complex is a low abundance protein found at approximately 100 to 200 copies per cell. In order to enable nickel affinity purification of the complex, a yeast strain was modified by the insertion of DNA encoding six histidines and an HA epitope tag into the 3′ end of the SWI2/SNF2 open reading frame.[6] The purification scheme consists of a nickel affinity purification step followed by DNA cellulose chromatography, ion-exchange chromatography, and glycerol gradient sedimentation. Fractionation of a whole cell extract from a 20-liter yeast culture routinely yields 0.1 mg of purified SWI/SNF complex.

Solutions

- E buffer: 20 m*M* HEPES (pH 8.0), 350 m*M* NaCl, 10% glycerol, and 0.1% Tween 20; store at 4° (prepare 2.5 liter)
- Imidazole stock: 1 *M* imidazole in water, adjust pH to 7.2 with 1 *N* NaOH (store 40-ml aliquots at −20°)
- Imidazole buffers: 20 and 500 m*M* imidazole in 75 m*M* NaCl, 10% glycerol, and 0.1% Tween 20) (prepare 50 ml of each)
- DNA-binding buffer: 25 m*M* Tris–HCl (pH 7.5), 100 m*M* NaCl, 5 m*M* $MgCl_2$, 10% (v/v) glycerol, 0.1% (v/v) Tween 20, and 0.5 m*M* (DTT); store at 4° (prepare 4 liter, also prepare 250 ml of 1 *M* NaCl–DNA-binding buffer)
- Mono Q buffer A: 50 m*M* Tris–HCl (pH8.0), 10% glycerol, 0.1% Tween, and 0.5 m*M* DTT; Mono Q buffers should be filtered through a 45-μm filter unit and are stored at room temperature (prepare 500 ml)
- Mono Q buffer B: Prepare 500 ml as Mono Q buffer A, but containing 1 *M* NaCl

Materials

The following materials are used for this SWI/SNF purification method: Yeast strain CY396 or its equivalent. Glass beads and a 250-ml bead beater equipped with a stainless-steel chamber (Biospec products, Bartlesville, OK). An ultracentrifuge and Ti45 and SW28 rotors or their equivalents. Ni^{2+}-NTA agarose beads (Qiagen, Chatsworth, CA). DNA cellulose resin

[6] C. L. Peterson, A. Dingwall, and M. P. Scott, *Proc. Natl. Acad. Sci. U.S.A.* **91,** 2905 (1994).

(Sigma, St. Louis, MO). Centricon-30 concentrators (Amicon, Danvers, MA). Ion-exchange chromatography is carried out on a fast protein liquid chromatography (FPLC) instrument (Pharmacia, Piscataway, NJ) using a 5/5 Mono Q column (Pharmacia) (note that SWI/SNF does not bind well to other types of anion-exchange resins, such as Econo Q or Q-Sepharose).

Procedure

To obtain submilligram quantities of the SWI/SNF complex, a 20-liter YEPD (2% yeast extract, 1% Bacto-peptone with 2% glucose) culture of strain CY396 is grown to an OD_{600} of 2.0 to 3.0. In our experience, cultures grown to higher densities result in lower yields of SWI/SNF complex. Cells are harvested in 1-liter bottles by centrifugation at 4550*g* for 10 min at 4°. The cell pellets are then washed and pooled in 1 liter E buffer and sedimented as described earlier.

The yeast pellet is suspended in 100 ml ice-cold E buffer containing the protease inhibitors pepstatin (1 μg/ml), leupeptin (0.5 μg/ml), aprotinin (2 μg/ml), and phenylmethylsulfonyl fluoride (1 m*M*). The cell suspension is poured into the stainless-steel bead beater that has been filled to the ridge with glass beads, and the beads are stirred to remove air bubbles. The bead beater is then filled to the rim with E buffer. The bead beater is packed tightly in ice and the cells are beaten five to eight times for 30 sec at 2-min intervals. The extent of lysis is verified under a microscope. The lysate is poured off into an ice-cold beaker, the beads are rinsed with 100–150 ml of chilled E buffer containing protease inhibitors, and the rinses are pooled with the lysate. The lysate is then centrifuged at 4550*g* for 5 min at 4° to eliminate the remaining glass beads. The lysate is then transferred to four 50-ml ultracentrifuge tubes and centrifuged at 43,000 rpm for 45 min at 4° in a Beckman Ti45 rotor.

The ultracentrifuge supernate is mixed with 20 ml of Ni^{2+}-NTA agarose beads that have been preequilibrated in chilled E buffer. The mixture is then rocked gently at 4° for 2–3 hr. After binding, the nickel bead suspension is centrifuged for 7 min at 3500*g* at 4°. The resin is suspended in 1 liter of ice-cold E buffer and sedimented as before. The Ni^{2+}-NTA agarose beads are then suspended in 50 ml E buffer and poured into a 2.5 × 30-cm column. The Ni^{2+}-NTA column is washed twice with 50 ml E buffer, once with 20 m*M* imidazole buffer, and the nickel bound proteins are eluted with 50 ml of 500 m*M* imidazole buffer. The 50 ml eluate is then dialyzed for 4 hr at 4° against 4 liters of DNA-binding buffer.

After dialysis the nickel column eluate is clarified by centrifugation at 4550*g* for 10 min and loaded onto a preequilibrated DNA cellulose column (5 ml bed volume at 4°) at a rate of 0.5 ml/min at 4°. The DNA cellulose column is then washed with 20 ml DNA-binding buffer, and the DNA cellulose-bound proteins are eluted with DNA-binding buffer containing 400 m*M* NaCl and 1-ml fractions are collected. The peak DNA cellulose fractions (as assayed by the Bradford assay or by absorbance at 280 nm) are pooled and diluted fourfold in Mono Q buffer A and then loaded at a flow rate of 1 ml/min onto a FPLC Mono Q 5/5 column preequilibrated in Mono Q buffer (90% A:10% B) at room temperature. The column is washed with 5 ml Mono Q buffer (90% A:10% B), a 20-ml gradient is applied from 10 to 50% Mono Q buffer B, and 500-μl fractions are collected. The SWI/SNF complex elutes from the column at about 340 m*M* NaCl. The presence of the SWI/SNF complex in the Mono Q fractions is determined by Western analysis of 10 μl of every second fraction using a monoclonal antibody (BABCO) to detect the HA-tagged SW12/SNF2 subunit. Fractions containing SW12/SNF2-HA are pooled and concentrated to 0.5 ml with a Centricon-30 concentrator.

The concentrated Mono Q pool is then applied to the top of a 16-ml linear 13–30% glycerol gradient in E buffer and sedimented for 16–24 hr at 33,000*g* in a SW-28 rotor (Beckman) at 4°. The gradient is then cut into 0.5-ml fractions from the top. After 16 hr, the peak of the SWI/SNF complex is found in the 8th fraction and after 24 hr in the 10th fraction. The peak usually spans 5 fractions. Fractions containing HA immunoreactivity are pooled and concentrated as before to a final volume of 200 to 500 μl and stored as 15- to 50-μl aliquots at −70° after rapid freezing in liquid nitrogen.

The concentration of SWI/SNF complex is determined by comparative Western blot analysis using an antibody against the SNF5 subunit of the complex and dilutions of recombinant GST–SNF5 protein. Typically, this purification protocol yields a 100–300 n*M* solution of the SWI/SNF complex. This represents an approximate 150,000-fold purification with a 25–50% overall yield.[1]

SWI/SNF-Specific ATPase Activity

SWI/SNF-mediated nucleosome remodeling requires ATP hydrolysis.[1] The ATPase activity of SWI/SNF is stimulated by cofactors such as DNA and nucleosomes.[1] Cofactor-stimulated ATPase activity can therefore be employed to determine the specific activity of purified SWI/SNF preparations. ATPase activity is determined routinely in the presence of DNA.

In the past we have monitored ATP hydrolysis by measuring the amount of [^{32}P]pyrophosphate released from [γ-^{32}P]ATP by binding the remaining ATP to activated charcoal in 20 m*M* phosphoric acid.[1] The disadvantages of this method are twofold and lie in the fact (1) that only the amount of radioactive pyrophosphate is measured and (2) that a standard background value caused by residual quantities of unbound ATP must be subtracted from the measured signal before multiplication by the dilution factor.

In contrast, thin-layer chromatography (TLC) allows simultaneous measurements of radiolabeled ATP and the released pyrophosphate. This method also yields more reproducible results and lends itself well to detailed kinetic analysis. The following TLC protocol is what we have used to evaluate the kinetic parameters of the ATPase activity of SWI/SNF.

Solutions

ATPase buffer: 0.1% Tween, 20 m*M* Tris, pH 8.0, 5% glycerol, 0.2 m*M* DTT, 5 m*M* $MgCl_2$, 100 μg/ml bovine serum albumin (BSA). Prepare 1 ml of a 10-fold concentrated stock without BSA and store at −20°,

Developer solution: 0.75 *M* monobasic potassium phosphate solution acidified to a pH of 3.5 with concentrated phosphoric acid

Materials

SWI/SNF complex; SWI/SNF complex isolated as described earlier.

Nucleic acid cofactor: Linear (>200 bp in length) DNA, supercoiled plasmid DNA, or reconstituted nucleosome arrays (see later)

[γ-^{32}P]ATP: [γ-^{32}P]ATP with a specific activity of 3000 mCi/mmol is obtained commercially (NEN, Boston, MA)

PEI-cellulose: polyethyleneimine-cellulose can be obtained from a variety of sources, including Em Science (Gibbstown, NJ). Usually the 20 × 20-cm sheets are cut in halves and used as such.

Procedure

Cofactor-stimulated ATPase activity is determined routinely in the presence of 0.02 μg/μl of DNA (e.g., a final concentration of 12 n*M* of the 2400-bp nucleosome array template described later) using a final concentration of 5 n*M* of SWI/SNF complex, 100 μ*M* ATP, and 0.2 μCi of [γ-^{32}P]ATP in ATPase buffer plus 10 μg/μl BSA in a final volume of 20 μl at 37°. The rate of the reaction is linear for 5–10 min under these conditions. To stop the reaction, 1 μl of the reaction is spotted onto a PEI-cellulose sheet about 10 mm from the bottom of the PEI sheet. The proportion of hydrolyzed [^{32}P]pyrophosphate is determined by resolution in the 0.75 *M* KPO_4 (pH 3.5) solvent. It is important to determine the proportion of [^{32}P]pyrophosphate already present in the [γ-^{32}P]ATP stock solution, as this varies from batch to batch as well as with the conditions and time of storage.

Chromatography is started by dipping the PEI sheet in 2 mm of KPO_4

solution and is stopped by removing the PEI-cellulose sheet from the KPO_4 solution when the front of the solvent has migrated about 9 cm. The sheet is then allowed to dry for 20 min. Under these conditions, ATP is clearly resolved from the origin and from the pyrophosphate. Quantitation can be carried out on a phosphorimager or by exposing the sheet to film overnight, cutting out the radioactivity spots, and counting in a scintillation counter. We have found that the K_m and V_{max} kinetic parameters of the ATPase activity of the SWI/SNF complex are best estimated by titrating the ATP concentration from 1 to 300 μM. We usually plot the obtained reaction velocities using the graphic direct linear plot method of Eisentahl and Cornish-Bowden[7] because this method does not require a dedicated computer program to estimate K_m and V_{max}, and the distribution-free statistics render it insensitive to aberrant measurements.

Using both the charcoal separation and the PEI chromatography methods, we have determined a V_{max} in the presence of 12 nM naked DNA of 1000 ± 200 ATP per minute per complex and a K_m of $10^{-4} \pm 5 \times 10^{-5}$ M. Therefore, at 100 μM ATP, about 500 ATP molecules should be hydrolyzed per minute per complex (giving a specific activity of 1 μmol ATP hydrolyzed per minute per 0.004 mg SWI/SNF complex). We believe this value is larger than the previously published estimate[1] because we replaced the Superose 6 purification step with a glycerol gradient and because of variations in the specific activity from one purification to another.

Nucleosome Array Reconstitution and Characterization.

Because nucleosome arrays are more likely to mimic the natural substrate of nucleosome-remodeling complexes, we have developed a novel nucleosome array remodeling assay.[8] The DNA template that we have used is composed of 11 head-to-tial repeats of a *Lytechinus variegatus* 5S rRNA gene (the 208-11 template). Each repeat can rotationally and translationally position a nucleosome after *in vitro* reconstitution with purified histone octamers, yielding a homogeneous, positioned array of nucleosomes.[9–11] The sixth repeat of our nucleosome array template[8] contains a *L. variegatus* 5S sequence engineered by Polach and Widom[5] that bears a unique restriction site (GTCGAC; *Sal*I, *Hinc*II) close to the dyad axis of symmetry of the positioned nucleosome.

[7] R. Eisentahl and A. Cornish-Bowden, *Biochem. J.* **139,** 715 (1974)

[8] C. Logie and C. L. Peterson, *EMBO J.* **16,** 6772 (1997).

[9] R. T. Simpson, F. Thoma, and J. M. Brubaker, *Cell* **42,** 799 (1985).

[10] J. C. Hansen, J. Ausio, V. H. Stanik, and K. E. van Holde, *Biochemistry* **28,** 9129 (1989).

[11] J. C. Hansen, K. E. van Holde, and D. Lohr, *J. Biol. Chem.* **266,** 4276 (1991).

Nucleosome Array Reconstitution Protocol

The present salt dialysis reconstitution protocol has been described by Richards and Pardon.[12] It has been used successfully in our laboratory with histone octamers purified from chicken erythrocytes as well as from HeLa cells. The degree of nucleosome saturation of the array is controlled simply by varying the ratio of 5S rDNA positioning sequences to histone octamers prior to dialysis.

Milligram quantities of plasmid DNA bearing a given 5S rDNA array template are purified routinely by alkaline lysis of *E. coli* cells harboring the plasmid. The DNA template for reconstitution is then isolated by restriction enzyme cleavage. Care has to be taken not to overdigest the DNA as this results in nonspecific cleavage and nicking of the DNA template. The plasmid is digested with *Hha*I when the nucleosome reconstitution template exceeds 1.5 kbp (cleavage sites for *Hha*I are located 60 to 400 bp apart on pBR322 and pUC vectors, resulting in the appearance of only one high molecular weight DNA species) and/or by cleavage with appropriate restriction enzymes (e.g., *Not*I and *Hin*dIII) to release the array template from the vector sequences (the 5S array templates generated in our laboratory are cloned into the *Sma*I site of pBluescript SK+). The 5S concatemers are then isolated by chromatography over an exclusion matrix (Bio-rad A-150m) at rates close to gravity flow[13] or by low melting point agarose electrophoresis followed by gel purification (β-agarase purification, New England Biolabs) and concentrated by ethanol precipitation to 1 mg/ml in TE (0.25 mM EDTA, 10 mM Tris, pH 8.0).

The DNA template is radiolabeled by the Klenow fill-in reaction in 100 mM NaCl, 10 mM $MgCl_2$, 1 mM DTT, and 100 μg/ml BSA. Labeling reactions typically contain 10–20 μg of DNA template, 100–150 μCi [α-^{32}P]dCTP, a final concentration of 0.25 mM dATP, dGTP, and dTTP, and 10 units of Klenow in a final volume of 60 μl. After 20 min at room temperature, the labeled DNA is purified by phenol extraction and passed over a spin column (Sephadex G-25) to eliminate the unincorporated nucleotides. The specific activity of the labeled DNA template should be between 2×10^6 and 2×10^7 cpm/μg DNA.

A method to purify milligram quantities of histone octamers from chicken erythrocytes has been described.[14,15] This yields histone octamer

[12] B. M. Richards and J. F. Pardon, *Exp. Cell Res.* **62,** 184 (1970).
[13] J. C. Hansen and H. Rickett *Anal. Biochem.* **179,** 167 (1989).
[14] T. D. Yager, C. T. McMurray, and K. E. van Holde, *Biochemistry* **28,** 2271 (1989).
[15] C. von Holt, W. F. Brandt, H. J. Greyling, G. G. Lindsey, J. D. Retief, J. D. Rodrigues, S. Schwager, and B. T. Sewell, *Methods Enzymol.* **170,** 431 (1989).

preparations containing 0.5–10 mg/ml octamer in 2.5 *M* NaCl and 100 m*M* KPO_4 (OD_{230} multiplied by the dilution factor and divided by 4.3 gives the concentration of octamers milligrams per milliliter). Histone octamers are stored on ice and are stable for over 1 year. However, because octamers tend to aggregate on prolonged storage on ice, it is prudent to centrifuge an aliquot of the octamers in a bench-top centrifuge at full speed for 15 min and to transfer the supernate to a new tube prior to concentration determination and inclusion in dialysis reactions.

Reconstitution reactions are assembled on ice and contain a final concentration of 100 μg/ml BSA, 10 m*M* TrisCl (pH 8.0), 0.25 m*M* EDTA, 2.0 *M* NaCl, and 100 μg/ml of array DNA template. Typically, 1×10^7 cpm-labeled template DNA is included in 200-μl reconstitution reactions. The molecular mass of a single 208 5S rDNA nucleosome positioning sequence is about 140 kDa and that of a histone octamer is 108 kDa. Therefore, to obtain a saturated array (r = 1.0) for every microgram of array template, 0.77 μg of histone octamers has to be included. Dialysis is carried out against 30 ml of NaCl–TE buffers in a Spectra/Por microdialyzer (Spectrum, Houston, TX) using dialysis membranes with a 10,000 MW exclusion limit. The initial buffer is 1.5 *M* NaCl–TE [the TE solution is 10 m*M* Tris–Cl (pH 8.0) and 0.25 m*M* EDTA]. The buffer is exchanged every 4 hr, first with 1.0 *M* NaCl–TE and subsequently with 0.8 *M* NaCl–TE, 0.6 *M* NaCl–TE, and then either 0.1 *M* NaCl–TE or TE. This procedure yields a solution of 60 n*M* for an 11-mer array. The reconstituted arrays are stored as such on ice and are stable for 2 weeks to 1 month.

Experimental Determination of Degree of Nucleosome Saturation of Reconstituted Nucleosome Arrays

Although deposition of histone octamers onto arrays of 5S repeats is a saturatable process, histone deposition is not; when the concentration of histone octamers exceeds that of DNA repeats, excess histones associate with the nucleosome array to from larger molecular weight complexes.[16] Oversaturation of the reconstituted nucleosome arrays should be avoided as it is less likely to yield nucleosome arrays of a physiological nature. A traditional means to assess the integrity of nucleosome arrays is micrococcal nuclease digestion. This assay, however, only reports nucleosome positions and is insensitive to the oversaturation of arrays.[16] In our hands, arrays that are grossly oversaturated still yield a typical repeating pattern of micro-

[16] J. C. Hansen and D. Lohr, *J. Biol. Chem.* **268,** 5840 (1993).

coccal nuclease digestion. The most reliable method to assess the degree of array saturation is analysis by analytical ultracentrifugation. However, this instrument is not readily available and requires substantial quantities of material. An alternative method is a simple restriction enzyme assay developed by Hansen and colleagues.[10]

Every 208-bp *L. variegatus* 5S rDNA nucleosome positioning sequence within the array templates is flanked by two *Eco*RI recognition sites. These sites lie outside the sequences that position the histone octamers, and therefore digestion of reconstituted nucleosome arrays with *Eco*RI should release either mononucleosomes or naked DNA depending on whether a given positioning sequence is occupied by a histone octamer. Resolution of the free *Eco*RI 5S DNA fragments from the nucleosomal 5S DNA is achieved by native polyacrylamide gel electrophoresis [in 1× TBE, 4% (v/v, 20:1) acrylamide:bisacrylamide]. Restriction digestions are performed with 10 units of *Eco*RI at 37° for 10–45 min using twofold diluted reconstituted arrays in final concentrations of 125 m*M* NaCl, 2.5 m*M* $MgCl_2$ and 1 m*M* DTT. As control, naked DNA is used. Prior to loading on the gel, glycerol is added to the *Eco*RI restriction reaction to reach a final concentration of 5%. A 20 × 20 × 0.1-cm gel is typically electrophoresed at 50–100 V for 4–6 hr. Mononucleosomes will migrate in such a native acrylamide gel with an apparent mobility of 350–400 bp. Comparison of the amount of free DNA (195 bp) with the amount of histone octamer associated DNA provides an accurate measurement of the level of occupancy of the positioning sequences. (Note that octamer bound DNA absorbs ethidium bromide about 2.5 times less efficiently than naked DNA.[17]) Direct comparison of data obtained with the *Eco*RI assay and analysis by analytical ultracentrifugation indicate that the presence of about 5% naked DNA repeats corresponds to a fully saturated array. Less than 5% free repeats is indicative of oversaturation of the arrays, whereas more than 5% free repeats indicates that the arrays are subsaturated.

The *Eco*RI assay also reports the proportion of alternative positions that the octamers occupy on the 5S array by virtue of the degree of occlusion of the *Eco*RI sites. Alternative nucleosome positions that result in occlusion of *Eco*RI sites are therefore indicated by the appearance of higher mobility complexes (di-, tri-, and polynucleosomes) in the native acrylamide gel. Oversaturation of arrays, of course, will result in a higher proportion of these complexes and loss of the free DNA.

[17] C. T. McMurray and K. E. van Holde, *Proc. Natl. Acad. Sci. U.S.A.* **83,** 8472 (1986).

Choice of Experimental Conditions for SWI/SNF–Restriction Enzyme Coupled Assay

Glycerol and Salt Effects on Restriction Enzyme Activity

To couple a nucleosome-remodeling reaction to a restriction enzyme reaction, salt conditions have to be optimized so that both enzymes function and so that the nucleosome arrays maintain a physiological structure. The activity of SWI/SNF, as well as a restriction enzyme, is optimal in the presence of magnesium as well as sodium. In coupled assays it is also necessary to ensure that the activity of the coupling enzyme (restriction enzyme) exceeds that of the enzyme under study (SWI/SNF). This can be achieved by the addition of a vast excess of the coupling enzyme. However, commercially available restriction enzymes are stored in 50% glycerol, a known inhibitor of restriction enzyme activity. We therefore undertook the task of measuring the effect of titrating NaCl, Mg, and glycerol concentration on *Sal*I activity. The result of a glycerol titration is shown in Fig. 1.

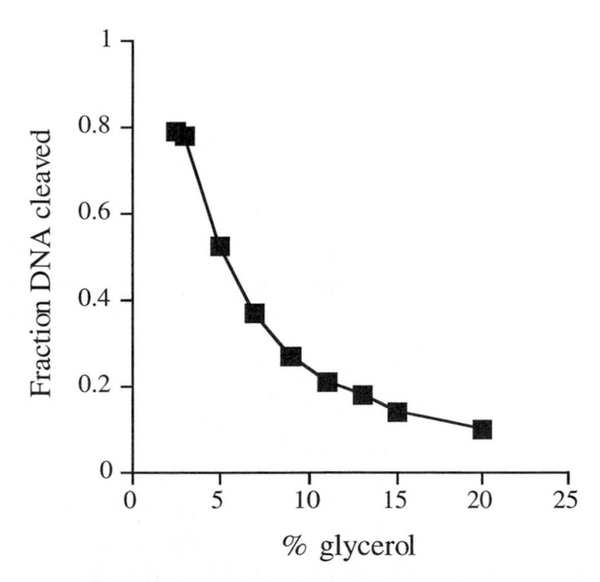

FIG. 1. Effect of increasing glycerol concentration from 2.5 to 20% on *Sal*I restriction enzyme activity. A 3 n*M* solution of the radiolabeled DNA template used for nucleosome array reconstitution was made in 2.5 m*M* $MgCl_2$, 125 m*M* NaCl, 1 m*M* DTT, and 10 m*M* Tris–Cl (pH 8.0) and was then exposed to 100 units/ml of *Sal*I (0.2 n*M*) for 45 min. Reactions were stopped by vigorous mixing with 2 volumes of a 1 : 1 solution of phenol and chloroform. The amount of cleaved DNA was obtained by gel electrophoresis of the DNA and dividing the cut DNA signal by the sum of the cut and uncut signals.

Even at 5% glycerol, we find that *Sal*I activity is decreased substantially; thus we generally use a 1/20 dilution of a restriction enzyme that is stored in 50% glycerol.

Next, we determined the effect of mono- and divalent cations on restriction enzyme digestion of naked DNA. These control reactions contained a 208-11 DNA template (3 n*M*) and 100 U/ml *Sal*I (equivalent to 0.2 n*M* of *Sal*I) or 5 U/ml *Hin*cII. Data displayed in Fig. 2a indicate that *Sal*I enzyme activity on naked DNA is reduced at concentrations of NaCl below 125 m*M*. Notably, below 125 m*M* NaCl, lowering the $MgCl_2$ concentration leads to a further decrease in activity (Fig. 2a). In contrast, *Hin*cII, a degenerate isoschizomer of *Sal*I, was able to digest naked DNA with 50% efficiency in the absence of any added NaCl (Fig. 2b). *Hin*cII activity is further stimulated by the presence of 50 m*M* NaCl but-inhibited by 125 m*M* NaCl (Fig. 2b). Furthermore, conversely to *Sal*I, increasing the concentration of $MgCl_2$ has a deleterious effect on *Hin*cII activity even in the presence of 50 or 125 m*M* NaCl. These two restriction enzymes therefore display complementary magnesium and sodium dependencies and thus should permit the study of nucleosome array accessibility under a variety of salt conditions.

Salt Effects on Nucleosome Array Conformation and Self-Association

Analysis of reconstituted arrays by electron microscopy and analytical ultracentrifugation have shown that monovalent and divalent salts can affect the shape and deformability of nucleosome arrays. Saturated model 208-11 and 208-12 arrays are dynamic macromolecular assemblies that undergo reversible folding events, much like physiological chromatin fibers (for review Fletcher and Hansen.[18] In very low ionic strength buffers (i.e., TE), 208-12 arrays exist as an extended, flexible fiber that sediments in the analytical centrifuge as a 29S species. When monovalent (Na^+) or divalent (Mg^{2+}) cations are introduced, the arrays undergo two distinct folding equilibria. Moderate levels of NaCl (20–100 m*M*) establish an intermediate level of folding; these structures sediment at a maximum of 40S in the analytical ultracentrifuge. In contrast, 1–2 m*M* $MgCl_2$ induces a rigid, compact structure that sediments at 55S. The 55S structure is consistent with the extent of folding expected for a 30-nm interphase chromatin fiber. Unfortunately, ultracentrifuge data on 11- or 12-mer arrays under conditions where both magnesium and sodium are present are still pending. Based on previous work, however, it is hypothesized that in buffers containing 125 m*M* NaCl and 2.5 m*M* $MgCl_2$, the arrays are in a dynamic equilibrium between the 29S, 40S, and 55S conformations.[18]

[18] T. M. Fletcher and J. C. Hansen, *Crit. Rev. Eukary. Gene Expr.* **6,** 149 (1996).

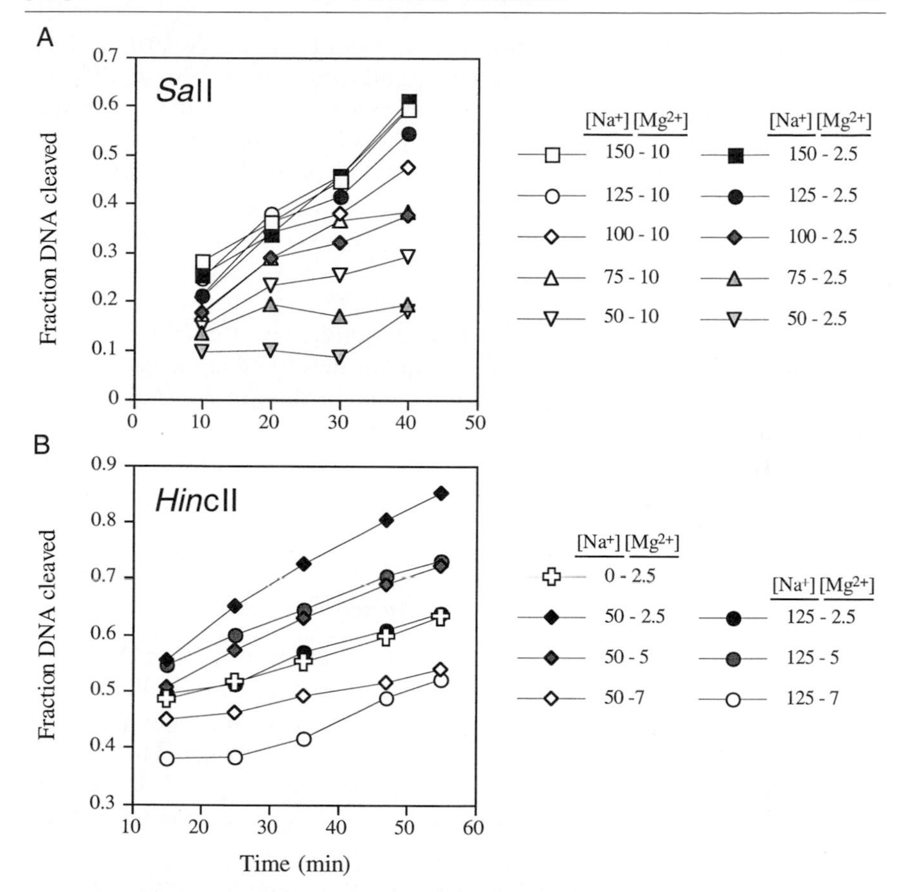

FIG. 2. Effect of titrating magnesium and sodium concentrations on restriction enzyme activity. (a) Fifty units per milliliter of *Sal*I solutions was made in the indicated salt conditions and the reactions were started by adding radiolabeled array DNA template to a final concentration of 2.5 n*M*. At the indicated time points, aliquots of the reactions were mixed vigorously with two volumes of a 1:1 solution of phenol and chloroform. Quantification was carried out as for Fig. 1. (b) Same as (a) but using 5 units/ml of *Hinc*II under the indicated salt conditions

Concentrations of divalent cations, such as Mg^{2+}, above 2.5 m*M* can also induce interarray association (also termed self-association). Moreover, interarray association induced by divalent cations is antagonized by monovalent cations, such as sodium.[19] We have determined the degree of self-association of nucleosome arrays in salt concentrations normally used for

[19] P. M. Schwarz, A. Felthauser, T. M. Fletcher, and J. C. Hansen, *Biochemistry* **35,** 4009 (1996).

coupled restriction enzyme-remodeling assays (Table I). A saturated 11-mer array (30 n*M*) does not associate significantly after 2 hr in 125 m*M* NaCl/2.5 m*M* $MgCl_2$ (Table I). Interarray association does occur in 5 m*M* magnesium alone but this is reversed by the addition of sodium to 50 m*M* (Table I). At 7.5 m*M* magnesium, sodium is only partially inhibitory (Table I). Interarray association is an intermolecular phenomenon,[19] and the rate of association is therefore expected to be proportional to a power of the concentration of arrays. Consistent with this, when the concentration of arrays was reduced to 3 n*M*, even 10 m*M* magnesium did not result in the appearance of detectable amounts of array aggregates after 2 hr of incubation on ice (Table I). The SWI/SNF–*Sal*I assay described in the next section is performed routinely using arrays at a concentration of 3 n*M* in 125 m*M* NaCl plus 2.5 m*M* $MgCl_2$.

Coupling *Sal*I Restriction Endonuclease Reaction to SWI/SNF Complex Reaction

SalI Cleavage Kinetics on Reconstituted Nucleosome Arrays

When reconstituted nucleosome arrays are exposed to 10 n*M* *Sal*I restriction enzyme (5000 U/ml), the kinetics of cleavage are biphasic. The first phase is very rapid, where up to 25–50% of the nucleosome arrays are cleaved at a rate close to the rate of naked DNA cleavage. The fraction of arrays cleaved in this first phase varies proportionally with the degree of saturation of the array template (Fig. 3). The second phase is much

TABLE I
INTERARRAY ASSOCIATION AS FUNCTION OF MAGNESIUM, SODIUM, AND ARRAY CONCENTRATION[a]

$[MgCl_2]$ (m*M*)	60 n*M* array no NaCl added % self-associated	60 n*M* array +50 m*M* NaCl % self-associated	60 n*M* array +125 m*M* NaCl % self-associated	3 n*M* array no NaCl added % self-associated
TE	<1	<1	<1	
2.5	<1	<1	<1	
5	64	<1	<1	
7.5	88	69	60	
10	95	86	82	<1

[a] Array self-association was measured by the method of P. M. Schwarz, A. Felthauser, T. M. Fletcher, and J. C. Hansen, *Biochemistry* **35,** 4009 (1996). Reconstituted arrays (60 n*M*) were diluted 2- or 20-fold into the indicated salt solutions, incubated on ice for 2 hr, and centrifuged in a bench-top Eppendorf centrifuge at 13,000*g* for 10 min. The optical density at 260 nm (OD_{260}) was then measured and compared to the OD_{260} of the original array solution.

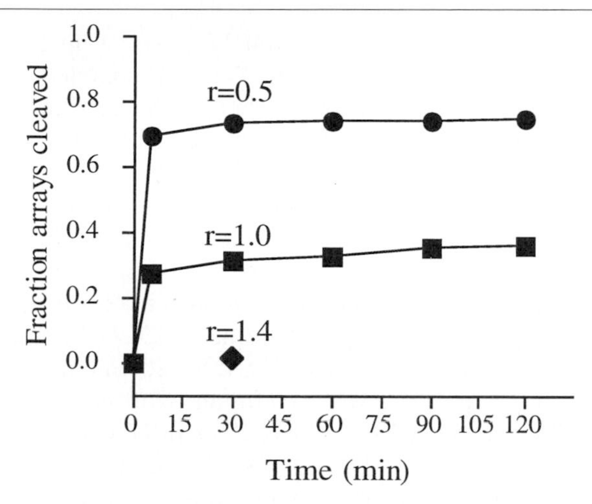

FIG. 3. Effect of varying the ratio of histone octamers to 5S positioning sequences on the amplitude of the first phase of the restriction kinetics. Nucleosome arrays were reconstituted with 0.5, 1.0, or 1.4 purified histone octamers per 5S positioning sequence. The reconstituted arrays were diluted 20-fold into 125 m*M* NaCl, 2.5 m*M* $MgCl_2$, 1 m*M* DTT, and 10 m*M* Tris–Cl (pH 8.0) to yield a final concentration of arrays of 3 n*M*. *Sal*I restriction enzyme (10 n*M*, 5000 units/ml) was added, and the fraction of cleaved arrays was measured at the indicated time points as described in Figs. 1 and 2.

slower and represents the occlusion of the *Sal*I site by a positioned nucleosome. The biphasic kinetics are most consistent with the adoption of alternative positions by 25–50% of the histone octamers resulting in an accessible *Sal*I site.[10]

Coupled SWI/SNF–Restriction Enzyme Reaction

$$\text{Nucleosome array} \xrightarrow[\text{SWI/SNF}]{1} \text{Remodeled 6th nucleosome} \xrightarrow[\textit{Sal}\text{I}]{2} \text{Digested array (readout)}$$

In a coupled two-enzyme assay, care has to be taken to ensure that the activity of the second enzyme (*Sal*I) is not limiting. To verify that this condition is met, one must titrate the second enzyme and choose concentration windows within which twofold changes in the concentration of the second enzyme do not affect the readout of the coupled assay. Such an experiment was carried out for the SWI/SNF–*Sal*I coupled assay and is described in detail.

Final concentrations of 2, 6, or 10 n*M* *Sal*I (corresponding, respectively, to 1000, 3000, or 5000 units/ml of New England Biolabs high concentration *Sal*I) were added to a solution containing 100 μg/ml BSA, 1 m*M* ATP,

125 m*M* NaCl, 2.5 m*M* $MgCl_2$, and 1 m*M* DTT and a final concentration of 2.5% glycerol. Then, 4, 2, or 0.5 n*M* SWI/SNF complex was added by serial dilution. The reaction tubes were then prewarmed at 37° for 5 min. At time zero, [α-^{32}P]dCTP-labeled nucleosome arrays (3×10^5 cpm total) were added to each reaction tube to obtain a final array concentration of 3 n*M* (equivalent to 33 n*M* nucleosomes). Ten-microliter aliquots were removed after 5 min and every 15 min thereafter for 2 hr and mixed vigorously for 10 sec with 50 μl 1 : 1 phenol : chloroform equilibrated against 10 m*M* Tris–Cl (pH 8.0). The resulting aqueous phase was then loaded on 1% TAE agarose gels containing 0.5 μg/ml ethidium bromide. The gels were electrophoresed at 100 V for 2 hr, fixed for 20 min in 10% acetic acid, and dried at 50° for 5 hr onto 3MM Whatman (Clifton, NJ) paper using a gel drier equipped with a dry ice cold trap. The dried gel was exposed onto a phosphorimager screen for 12 hr and read into the ImageQuant software. Radioactive signal intensities were obtained using the volume integration function. The fraction uncut array DNA was determined by taking the ratio of the signals emanating from the uncut DNA template and the sum of the signals from the cut and uncut bands. The results presented in Fig. 4 indicate that the amount of array cleaved in the 10 and 6 n*M* *Sal*I reactions are equivalent and vary with the concentration of the SWI/SNF complex. The fraction of arrays cleaved by 2 n*M* *Sal*I is slightly reduced, more so at the lowest concentration of SWI/SNF. A concentration of 10 n*M* of *Sal*I is therefore adequate for the concentration range of SWI/SNF used here.

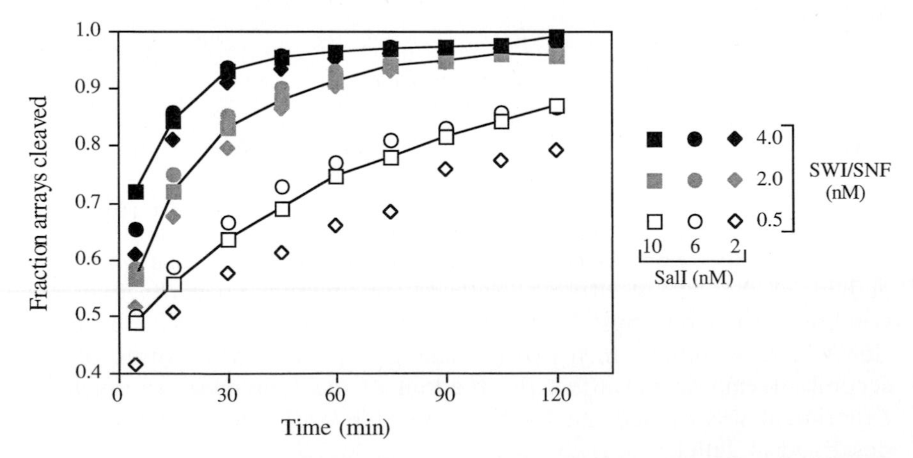

FIG. 4. Concomitant titration of the SWI/SNF complex and of the *Sal*I restriction endonuclease. The indicated concentrations of SWI/SNF and *Sal*I were used in reactions containing 3 n*M* of reconstituted nucleosome array. The fraction of cleaved nucleosome arrays was measured as described in the text.

Applications of Coupled SWI/SNF–Enzyme Assay

We have exploited the coupled array assay to measure several kinetic parameters of the SWI/SNF-remodeling reaction.[8] We have calculated an approximate turnover number for nucleosome remodeling and have shown that the reaction is highly reversible. Using a variety of assays, including restriction enzyme accessibility, it was found that the remodeled state is persistent when mononucleosomes were used as substrates, although some reversal could be seen 2 hr after apyrase treatment.[20,21] The discrepancy between the result obtained with nucleosome arrays and mononucleosome substrates indicates that mononucleosomes lack a type of interaction that takes place within nucleosome arrays to promote reversal from the SWI/SNF-induced state.

The coupled array assay is well suited to study the targeting and specificity of nucleosome-remodeling enzymes. For example, the DNA array template can be modified so that one or more binding sites for sequence-specific transcription factors are added at different locations within the array relative to the reporter nucleosome (*Sal*I site bearing). In addition, arrays can be created that contain two nucleosomes that harbor a *Sal*I site. These templates have the potential to distinguish between models whereby a remodeling enzyme perturbs individual nucleosomes sequentially or whether multiple nucleosomes are remodeled simultaneously.

We have found that nucleosome arrays assembled with histones that lack their N-terminal domains (a situation that mimics histone hyperacetylation) lead to an enhanced rate of *Sal*I digestion vs digestion of arrays containing intact histones.[22] This result suggests that the coupled array assay will prove useful for quantitating the activity of histone acetyltransferase and deacetylase complexes. Given that SWI/SNF and GCN5 acetyltransferase show genetic interactions *in vivo*,[23] the array assay described here may prove useful for investigating such functional interactions *in vitro*.

Acknowledgments

We thank present and past members of the Peterson laboratory for developing the current SWI/SNF purification protocol; very special thanks to J. Hansen (University of Texas Health Sciences Center, San Antonio) for performing analytical ultracentrifugation analysis of arrays and in particular for advice and patience concerning our nucleosome array studies. SWI/SNF studies have been supported by grants from the NIH to C.L.P. and from the Human Frontiers Science Program Organization to C.L. C.L.P. is a Scholar of the Leukemia Society of America.

[20] A. N. Imbalzano, G. R. Schnitzler, and R. E. Kingston, *J. Biol. Chem.* **271,** 20726 (1996).
[21] J. Cote, C. L. Peterson, and J. L. Workman, *Proc. Nat. Acad. Sci. U.S.A.* **95,** 4947 (1998).
[22] C. Logie, C. Tse, J. Hansen, and C. L. Peterson, *Biochemistry* **38,** 2514.
[23] K. J. Pollard and C. L. Peterson, *Mol. Cell. Biol.* **17,** 6212 (1997).

[43] Analysis of Modulators of Chromatin Structure in *Drosophila*

By PATRICK D. VARGA-WEISZ, EDGAR J. BONTE, and PETER B. BECKER

Introduction

The chromatin organization of eukaryotic DNA is an important determinant of all functions in the nucleus, notably the regulated transcription of genes.[1] Unraveling the highly condensed and hence repressive chromatin fiber to allow regulatory factors and enzymatic machineries to access the DNA substrate is an important first step in the series of events that lead to gene activity. The complex winding of the chromatin fiber is built on the ever-repeating fundamental unit of organization: the nucleosome. Structural alterations at this level influence not only the local organization of the DNA, but is likely to have an impact on as yet ill-defined higher order structures as well. Such structures need to be opened for transcription to occur. However, even the nucleosome, alone or in context of a nucleosomal array, frequently poses a considerable hurdle for the sequence-specific interaction of proteins with DNA. It is likely that the interaction of transcription factors with chromatin is highly regulated by modulation of the accessibility of particular sites.[2] A number of protein factors have been identified that are good candidates for regulators of accessibility in chromatin.[3] Such an activity was first identified in yeast: the SWI/SNF complex, a paradigm for multisubunit, energy-dependent remodeling machines.[4] However, three new and distinct protein complexes that are highly active for chromatin assembly and nucleosome remodeling, NURF,[5] CHRAC,[6] and ACF,[7] were purified from *Drosophila* embryo extracts.[8,9] The common denominator between these diverse factors appears to be the targeting of nucleosomes followed by an ill-defined, energy-consuming process referred

[1] G. Felsenfeld, *Cell* **86,** 13 (1996).
[2] T. Tsukiyama and C. Wu, *Curr. Opin. Genet. Dev.* **7,** 182 (1997).
[3] R. Kingston, C. Bunker, and A. N. Imbalzano, *Genes Dev.* **10,** 905 (1996).
[4] C. L. Peterson and J. W. Tamkun, *TIBS* **20,** 143 (1995); M. Pazin and J. T. Kadonaga, *Cell* **88,** 737 (1997).
[5] T. Tsukiyama and C. Wu, *Cell* **83,** 1011 (1995).
[6] P. D. Varga-Weisz, M. Wilm, E. Bonte, K. Dumas, M. Mann, and P. B. Becker, *Nature* **388,** 598 (1997).
[7] T. Ito, M. Bulger, M. J. Pazin, R. Kobayashi, and J. T. Kadonaga, *Cell* **145,** (1997).
[8] P. B. Becker and C. Wu, *Mol. Cell. Biol.* **12,** 2241 (1992).
[9] R. T. Kamakaka, M. Bulger, and J. T. Kadonaga, *Genes Dev.* **7,** 1779 (1993).

0076-6879/99 $30.00

to as "nucleosome remodeling." Although the physiological role for these distinct ATP-dependent remodeling factors in extracts from highly metabolically active fly embryos is presently unclear, it is tempting to speculate that the diversity of remodeling machines may reflect their involvement in diverse nuclear functions such as transcription, DNA replication, DNA repair, and chromatin assembly.[2] It is quite likely that more remodeling machineries are yet to be discovered.

This article describes methodology that led to the identification and preliminary characterization of one specific remodeling machine, the chromatin accessibility complex (CHRAC).[6] CHRAC activity was initially identified in chromatin reconstituted in a crude preblastoderm *Drosophila* embryo extract (S150 extract).[6,10] The activity mediates ATP-dependent accessibility of chromatin to restriction endonucleases, which were used as tools to probe for general changes in the accessibility of chromatin. We used restriction enzymes as a fast and simple assay to monitor the CHRAC activity qualitatively. Restriction enzymes have also been employed to quantify the catalytic action of the SWI/SNF complex on nucleosomal arrays.[11] CHRAC activity is not only abundant in cytoplasmic extracts of preblastoderm embryos, but also in nuclei of actively transcribing embryos once an inhibitory fraction was removed from the extracts.[12] Monitoring the ATP-dependent increase in accessibility of chromatin toward restriction enzymes in the presence of CHRAC, we purified the complex to homogeneity.[6] The purification involved a methanol precipitation, which preferentially precipitates proteins that require a high salt concentration to be soluble, anion exchange (CM-Sepharose) and cation exchange (Mono Q) chromatography and gel filtration (Superose-6; all columns from Pharmacia, Piscataway, NJ).[6]

In addition to granting access to chromatin, CHRAC also functions as an ATP-dependent nucleosome spacing factor during chromatin assembly, perhaps by facilitating the relocation of nucleosomes on DNA.[6] CHRAC shares this "nucleosome spacing" activity with ACF, a remodeling machine that was isolated in the Kadonaga laboratory during the fractionation of the chromatin assembly system.[7]

This article describes general methods required to establish a restriction enzyme accessibility assay that may be used to purify CHRAC or similar activities from other biological sources and species. We further detail three additional assays that may be employed to analyze remodeling complexes:

[10] P. D. Varga-Weisz, T. A. Blank, and P. B. Becker, *EMBO J.* **14,** 2209 (1995).

[11] C. Logie and C. L. Peterson, *EMBO J.* **16,** 6772 (1997).

[12] The inhibitor fraction consisted mainly of histone H1 (P. V. Weisz, Katia Dumas, and P. B. Becker, unpublished observation, 1997).

a nucleosome spacing assay, an ATPase assay, and a strategy to assess interactions of nucleosome-remodeling machines with nucleosomes that involves single nucleosomes immobilized onto paramagnetic beads. In closing we refer the reader to various other assays that have been used by other researchers to characterize nucleosome-remodeling complexes.

Establishing Restriction Enzyme Accessibility Assay

As substrate for nucleosome-remodeling activities may either serve a single nucleosome[13] or an array of nucleosomes reconstituted by salt gradient dialysis[14] or more complex chromatin reconstituted under physiological conditions in extracts such as those from *Drosophila* embryos. Because of the potential of nucleosome–nucleosome interactions and the presence of linker DNA, nucleosomal arrays are better models for physiological chromatin than single nucleosomes.[15]

Nucleosome Reconstitution Using Drosophila Embryo Extracts

Drosophila embryos contain large maternal stocks of chromatin constituents, such as histones and assembly factors. Extracts from those embryos form the basis of very powerful chromatin assembly systems.[8,9] Chromatin assembled from such extracts on recombinant DNA resembles native chromatin in many respects. It displays physiological nucleosome repeat lengths with accessible linker DNA in between the nucleosomes. Linker histone H1 is absent in these chromatin assembly extracts, but can be added exogenously and incorporated efficiently into the chromatin.[8]

There are essentially three variants of *Drosophila* chromatin assembly systems available: One uses extracts from early (0–2 hr old, preblastoderm) embryos.[8] This type of extract assembles chromatin from the endogenous pool of histones. Another strategy is based on extracts from older embryos (0–6 or 3–6 hr after egg laying), but because the histone pools have been depleted during the development of the embryo, exogenous histones have to be supplemented.[9] Finally, the Kadonaga group described an assembly system with purfied components.[7] This system has just been established and it may take a while until it is widely available. This section describes the assembly system derived from extracts from preblastoderm embryos for chromatin reconstitution, which we use routinely in the laboratory because of its ease and reliability: the extract already comes with the right amount of histones and assembles transcriptionally repressive chromatin

[13] J. Côté, J. Quinn, J. L. Workman, and C. L. Peterson, *Science* **265,** 53 (1994).
[14] T. Owen-Hughes and J. L. Workman, *EMBO J.* **15,** 4702 (1996).
[15] T. M. Fletcher and J. C. Hansen, *Crit. Rev. Eukary. Gene Expr.* **6,** 149 (1996).

with high efficiency. Populations of about 200,000 flies allow the routine collection of sufficient amounts of preblastoderm embryos (0–90 min after egg laying) to yield about 10–15 ml of extracts per day. The maintenance of such a population and the collection of embryos have been described.[16] The performance of such a population depends greatly on temperature (26°) and humidity, which has to be optimized for each particular setting. The extract preparation described deviates slightly from earlier versions.[17] Clearly, active extracts can be obtained under a variety of conditions.

Extract Preparation

1. Harvest 0- to 2-hr embryos in several (three to five) successive collections on apple juice agarose plates (a variant of the classical grape juice agar plates). Wash embryos off the plates each time with a paintbrush and water, wash through two sieves (of 0.71- and 0.355-mm mesh), and collect them on a fine sieve (0.125-mm mesh). Accumulate embryos in embryo wash buffer [0.7% (w/v) NaCl, 0.05% (v/v) Triton X-100] on ice. It is not necessary to purge older embryos retained by females during the overnight period in a precollection, as described earlier.
2. Allow the embryos to settle through fresh embryo wash at room temperature to warm them up. For dechorionation, add 60 ml of 13% hypochloric acid to the 200-ml embryo suspension in embryo wash and stir vigorously for 3 min. Wash embryos extensively on a fine sieve (0.125-mm mesh) under a sharp stream of tap water.
3. Settle embryos once in 500 ml embryo wash and then in 500 ml embryo wash without Triton and then twice in 250 ml cold EX buffer [10 m*M* HEPES–KOH, pH 7.6, 10 m*M* KCl, 1.5 m*M* $MgCl_2$, 0.5 m*M* EGTA, 10% (v/v) glycerol, 10 m*M* β-glycerophosphate].
4. Transfer the embryos in the EX buffer to a glass homogenizer. Tight packing of the embryos is achieved by allowing them to settle for extended times (30–45 min). Aspirate off the supernatant until about 2 ml of buffer covers the surface of the embryos. Add $MgCl_2$ to an additional 5 m*M,* phenylmethylsulfonyl fluoride (PMSF; from of 0.2 *M* stock) to 0.2 m*M*, and dithiothreitol (DTT) to 1 m*M*. Homogenize embryos by six complete strokes at 2000–3000 rpm with a Teflon pestle connected to a motor-driven drill press.
5. Centrifuge the extract for 10 min at 10,000 rpm (17,000*g*) in a chilled HB4 rotor in Corex tubes at 4°. Discard the tight white lipid layer with the help of a pipette tip and collect the supernatant.

[16] C. Shaffer, J. M. Wuller, and S. C. R. Elgin, *Methods Cell Biol.* **44,** 99 (1994).

[17] P. B. Becker, T. Tsukiyama, and C. Wu, *Methods Cell Biol.* **44,** 207 (1994).

6. Centrifuge the supernatant for 2 hr at 4° at 150,000*g* (45,000 rpm in a SW55 Ti rotor; Beckman, or equivalent). Paraffin of low viscosity is used to top the centrifuge tubes.
7. The homogenate splits into four fractions after centrifugation: a solid pellet, a loose cloudy white pellet above the solid one, a clear supernatant, and a floating lipid layer. Collect the clear supernatant by puncturing the tube with a syringe and needle above the pellets. Store the extracts in aliquots at −80°. Protein concentration is between 20 and 30 mg/ml.

Chromatin Assembly, Sarkosyl Strip, and Spin-Column Purification of Chromatin. CHRAC, NURF, and other chromatin-remodeling machines associate with reconstituted chromatin.[5,6] To establish an assay to test for these activities in biological samples, one needs to remove the endogenous activities. In the case of CHRAC and NURF, this can be done by washing the chromatin with the detergent *N*-lauroylsarcosine (sarkosyl) followed by its purification over a gel-filtration column. We usually prepare about 2.5 ml of the sarkosyl-stripped chromatin 1 day before the assay and store the chromatin on ice until use. We have used successfully chromatin that was stored for more than a week on ice and chromatin that was prepared fresh.

1. Prepare the creatine phosphokinase (CPK) mix. The energy generation mix is crucial for efficient chromatin assembly and has to be prepared fresh every time. Some of the components are unstable even when stored at −80°. Special care should be taken with the CPK. Resuspend one vial of 20 mg lyophilized CPK (Boehringer Mannheim, Germany) in 1000 μl 100 m*M* imidazole (pH 6.6). Divide the solution in 20-μl aliquots and store at −80° for maximally 2 months. Prior to use, add 380 μl 100 m*M* imidazole (pH 6.6) to one 20-μl aliquot and keep on ice.
2. Prepare 300 μl of 10× energy regeneration mix, consisting of 30 m*M* $MgCl_2$, 10 m*M* DTT, 300 m*M* creatine phosphate (Sigma, St. Louis, MO), and 30 m*M* ATP (Sigma, grade I, pH of 1 *M* stock solution adjusted to pH 8 with Tris).
3. Set up the assembly reaction. Thaw the embryo extract on ice and spin for 5 min in a cooled tabletop centrifuge to remove precipitate. To 1200 μl extract add 1280 μl EX50 (EX buffer as described earlier containing 50 m*M* KCl), 280 μl 10× energy regeneration mix, 2.8 μl creatine kinase dilution, and 40 μl plasmid DNA at 500 ng/μl (the source of DNA is irrelevant). Incubate at 26° for about 6 hr (several hours longer is fine).
4. Prepare 20 spin columns in parallel. Add 2 ml of S 300 HR slurry (Pharmacia; final settled bed volume about 1.8 ml) to Pharmacia spin columns (recycled Pharmacia cDNA spun columns). Wash the

columns by gravity flow at least four times with 2 ml of EX120 (as described earlier, but containing 120 m*M* KCl, 5 m*M* $MgCl_2$, 1 m*M* DTT, no 1 m*M* PMSF included). After the washes, drain the columns by spinning them at 1100*g* (2500 rpm in 18 cm rotors) for 2 min at room temperature.

5. To the assembly reaction add sarkosyl to a final concentration of 0.075% [from a 7.5% stock solution of *N*-lauroylsarcosine (Sigma) in water] and incubate for 10 min at room temperature. Carefully load 135-μl aliquots of the sarkosyl-treated chromatin to the center of each drained spin column and spin through at 1100*g* for 2 min at room temperature. Pool all the chromatin aliquots again and store on ice until use.

Nucleosome Assembly from Purified Histones by Salt Gradient Dialysis. Chromatin reconstituted in *Drosophila* extracts is very complex, containing a multitude of nonhistone proteins. While this enabled the identification of chromatin modulators such as CHRAC and NURF in the first place, the complexity of the system complicates the identification of the targets for the remodeling factors. In order to establish that histones are the targets for CHRAC, we modified a salt dialysis reconstitution protocol to assemble nucleosomes from pure histones and DNA. Methods to achieve this have been described in earlier volumes of this series.[18] Depending on the application, nucleosomes may be assembled on plasmid DNA, long linear DNA [e.g., λ phage DNA] or polymerase chain reaction (PCR) fragments, short enough to accommodate only a single histone octamer. We purified histones from *Drosophila* embryos essentially as described earlier.[19]

1. Dialysis reconstitution is performed in microcollodion bags (Sartorius, Göttingen) held by Styrofoam floaters at 4°. Prior to the first use, the bags have to be rinsed for 1 hr in water. To minimize the loss of histones and DNA through adsorption to the membrane walls, the bags are precoated by dialyzing histones (a side fraction from the purification will do) and 0.1 mg/ml bovine serum albumin (BSA) against dialysis buffer (DB: 10 m*M* HEPES–KOH, pH 7.6; 1 m*M* EDTA; 1 m*M* 2-mercaptoethanol; 0.05% NP-40) containing 2 *M* NaCl (DB2000) followed by rinsing with 2 *M* NaCl. Collodion bags can be re used up to five to six times. After each use, rinse and store the bags in 2 *M* NaCl/dialysis buffer.
2. The input mixture contains, in a volume of 50 μl, 2–10 μg DNA, slightly less (in mass) of purified histones, 0.5 mg/ml BSA (Sigma), and 0.05% NP-40; the reaction is filled up with DB2000. The optimal

[18] B. Neubauer and W. Hörz, *Methods Enzymol.* **170,** 630 (1989).

[19] R. H. Simon and G. Felsenfeld, *Nucleic Acids Res.* **6,** 689 (1979).

ratio between histones and DNA maximizing the nucleosome reconstitution while avoiding ill-defined aggregates must be determined empirically by mobility shift assays using radiolabeled DNA.[18]

3. Add the histone/DNA mixture to a collodion bag and dialyze the sample twice for 1 hr each against 1 liter of DB2000.
4. Dialyze against 1 liter of DB1200 (1.2 *M* NaCl) for 1 hr. Transfer floater into a new beaker with 500 ml DB1200.
5. Reduce the salt of the dialysis buffer gradually to about 0.55 *M* NaCl by pumping (during the course of about 15 hr) 1.8 liter of DB550 into the beaker while pumping a corresponding volume of the mixed dialysis buffer out of the beaker (pump with 2 ml/min).
6. Dialyze at least 3 hr against DB (no NaCl).
7. Collect the samples from the collodion bags and store them on ice in siliconized Eppendorf tubes. During the dialysis reconstitution the volumes of the samples may increase, depending on the age of the collodion bags.

Restriction Enzyme Accessibility Test. Restriction enzymes are used to monitor accessibility in chromatin, which can be increased by factors such as CHRAC or the SWI/SNF complex.[6,11] The CHRAC assay can be performed with sarkosyl-stripped chromatin assembled with the *Drosophila* extract or nucleosomes reconstituted from DNA and pure histones by salt gradient dialysis. Representative results of such assays are illustrated in Figs. 1 and 2.

For each sample to test, a pair of reactions is set up, with or without ATP. Monitoring the ATP dependence of the reaction is crucial as accessibility in chromatin may be generated by energy-independent processes, e.g., histone stripping by high concentrations of polyanions (such as RNA).

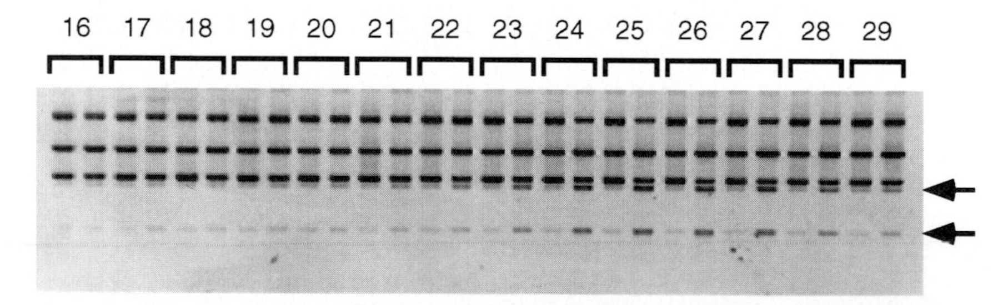

FIG. 1. Monitoring the fractionation of CHRAC activity over a gel filtration column by the restriction enzyme accessibility assay. One-microliter aliquots of the Superose 6 fractions (a late column during the CHRAC purification[6]) were assayed for ATP-dependent *Hinc*II accessibility on chromatin reconstituted with a *Drosophila* embryo extract. Each fraction was tested in the absence (left sample of each pair) and in the presence of ATP (right). Arrows indicate an increased cleavage efficiency in the presence of ATP.

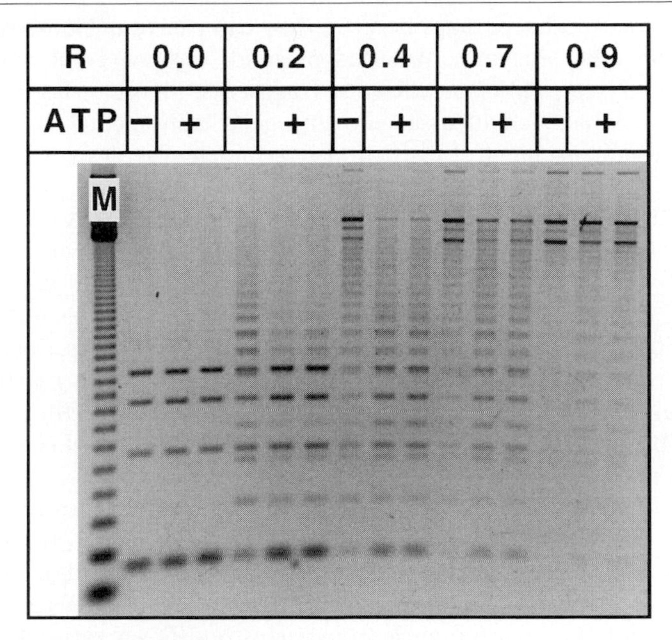

FIG. 2. CHRAC increases the access of *Dra*I to nucleosomal DNA in an ATP-dependent way. Chromatin was reconstituted by salt gradient dialysis with increasing mass ratios of histones to DNA indicated in the figure (R).

1. Set up a reaction containing (in 50 μl) 10 μl sarkosyl-stripped chromatin, 1 μl 100 m*M* ATP (if applicable), 36–37.5 μl EX120 (containing 0.2 mg/ml BSA, 5 m*M* $MgCl_2$, 1 m*M* DTT, but no PMSF), 0.5–2 μl of the CHRAC fractions to be tested, and 1 μl restriction enzyme (e.g., *Dra*I; 30–50 units). The final salt concentration of the reaction should be between 120 and 150 m*M* NaCl. Setting up many reactions can be facilitated by preparing premixes (e.g., of chromatin, restriction enzyme, and ATP).
2. Incubate at 26° for 30–60 min.
3. Stop reaction by adding 20 μl stop mix (1.6% sarkosyl, 64 m*M* EDTA, 0.7% SDS), followed by 5 μl proteinase K (10 mg/ml).
4. Digest at 37° overnight.
5. Add 1 μl glycogen (Boehringer-Mannheim) and precipitate DNA in 2.5 *M* ammonium acetate and 2 volumes of 100% ethanol.
6. Wash pellet with 75% (v/v) ethanol.
7. Analyze DNA by electrophoresis on a suitable agarose gel.

For restriction enzyme, we routinely used *Dra*I or *Hinc*II (both from Fermentas, MBI, St. Leon Rodt, Germany). Some enzymes (*Eco*RV) did not give good results, perhaps because they can cleave nucleosomal DNA better than other enzymes. We used plasmid/ enzyme combination that created both many (5) or only few (2) fragments. Figure 1 shows a typical profile of CHRAC activity assayed over a gel-filtration column. The ATP-dependent enhancement of restriction enzyme cleavage (arrows) peaks in fractions 25 and 26.

In order to assay CHRAC activity with nucleosomal DNA reconstituted by salt gradient dialysis we mixed 2.3 μg of a plasmid containing 12 repeats of the 5S gene from *Lytechinus variegatus* (p207-12)[20] with 0, 0.5, 1, 1.5, and 2 μg of purified *Drosophila* histones in 2 *M* salt dialysis buffer (final volume 50 μl) and subjected it to the various dialysis steps described earlier. Two microliters of the nucleosomal DNA was added to a reaction containing EX120, 5 m*M* $MgCl_2$, 1 m*M* DTT, 1.5 m*M* ATP, 200 μg/ml BSA, 0.5 μl of purified CHRAC (about 4 ng), and 1 μl (50 U) *Dra*I (Fermentas). Incubation was for 30 min at 26°.

Figure 2 shows the result of this experiment. Whereas in the absence of nucleosomes there was no ATP-dependent enhanced cleavage, CHRAC activity was obvious at histone/DNA ratios between 0.2 and 0.7. CHRAC activity was less observable under conditions where the DNA was fully occupied with close-packed nucleosomes (histone/DNA ratio of 0.9).

Other Assays for Characterization of Chromatin-Remodeling Factors

Nucleosome Spacing Assay

The process of nucleosome assembly in *Drosophila* embryo extracts can be divided into two steps. In a first, energy-independent step, nucleosomes assemble on DNA, but the chromatin structure is irregular, giving rise to a continuum of fragments on digestion with micrococcal nuclease. The regular spacing of nucleosomes, a hallmark of physiological chromatin, is achieved in a second step at the expense of ATP. The nature of the ATP-consuming spacing factors has been revealed by the demonstration that two of the *Drosophila* nucleosome-remodeling factors, CHRAC[6] and ACF,[7] function as nucleosome spacing factors. If added to an irregular array of nucleosomes (e.g., assembled in the absence of CHRAC and/or ATP), they will reorganize the nucleosomal array to become regular (see Fig. 3). This catalytic ability[7,21] is an interesting contrast to the ability of CHRAC and ACF to increase the accessibility of sequences in chromatin toward incoming proteins. It has been suggested that both functions are achieved by the

[20] R. T. Simpson, F. Thoma, and J. M. Brubaker, *Cell* **42,** 799 (1985).
[21] P. D. Varga-Weisz and P. B. Becker, unpublished observations, 1997.

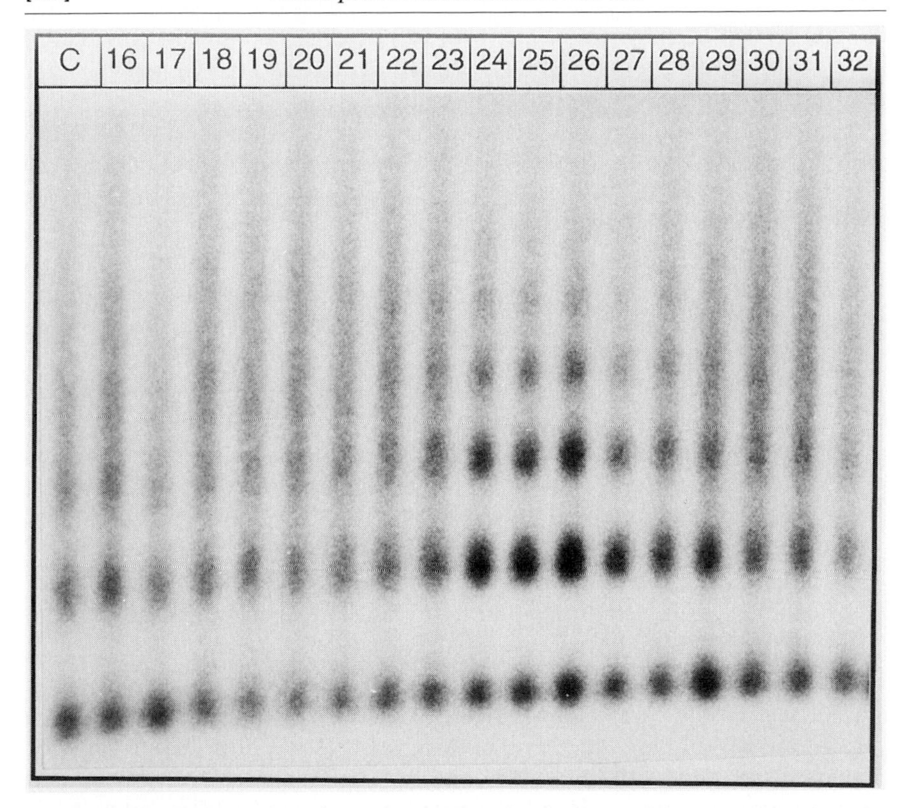

FIG. 3. Monitoring the fractionation of CHRAC activity over a gel-filtration column by the nucleosome spacing assay. Of the fractions already tested in Fig. 1, 0.5 μl was assayed for their ability to create a regular nucleosomal array in an ATP-dependent manner. Chromatin was assembled in the absence of ATP and endogenous CHRAC was inactivated. Upon addition of CHRAC-containing fractions and in the presence of ATP a regular array of nucleosomes is visualized by the regular MNase digestion pattern.

ability of the chromatin-remodeling machines to facilitate the relocation of the nucleosomes, their movement on DNA. It is unclear yet whether nucleosome spacing is a general property of remodeling machines or a feature of a small subset of factors, although it is already clear that not all remodeling machines function in this assay, as the nucleosome-remodeling factor (NURF)[5] is inefficient in this respect.[6]

The following assay can be used to test for nucleosome spacing activities.

1. To remove endogenous ATP, dialyze the *Drosophila* chromatin assembly extract twice for 4 hr against 300 ml EX120, 5 m*M* $MgCl_2$, 1 m*M* DTT, 0.2 m*M* PMSF, and 1 m*M* sodium metabisulfite (pH 7). Use 3.5-kDa cutoff SPECTRA/POR membrane tubing for dialysis.

2. Mix 240 μl dialyzed extract with 280 μl EX120, 5 m*M* $MgCl_2$, 1 m*M* DTT, and 4 μg plasmid DNA (in 40 μl 10 m*M* Tris–Cl, pH 8, 1 m*M* EDTA) and incubate for 3–4 hr at 26°.
3. Remove endogenous CHRAC activity with sarkosyl and gel filtration as detailed earlier. This irregular chromatin can be used right away or stored overnight on ice. Overnight storage may further reduce endogenous CHRAC activity.
4. To 20 μl irregular chromatin, add 34 μl EX50/0.2 mg/ml BSA, 6 μl energy regeneration mix (see section on chromatin assembly), and a small volume of the samples to be tested. Mix gently and incubate for 30–60 min at 26°.
5. Add 70 μl of micrococcal nuclease mix [400 units micrococcal nuclease/ml (Boehringer Mannheim), 5 m*M* $CaCl_2$ in EX120].
6. After 1 and 2 min, remove aliquots of 60 μl and add them to 24 μl stop mix (see earlier). Add 2.5 μl proteinase K (10 mg/ml) and digest overnight at 37°. Separate the purified DNA on an agarose gel and visualize the pattern by Southern blotting as described previously[22] using either the entire random-primed plasmid as probe or the end-labeled oligonucleotides that hybridize to a specific site.

ATPase Assay

All nucleosome-remodeling factors that have been analyzed for the regulation of ATPase activity show a nucleosome-stimulated ATPase activity[3–6,11,23] and, with the exception of NURF,[5] also a DNA-stimulated ATPase activity. We analyzed the regulation of the ATPase activity of CHRAC[6] with an assay that has been adapted from Tsukiyama and Wu.[5]

1. Combine in 9 μl: 60 ng DNA or the equivalent amount of mononucleosomes, 6.6 m*M* HEPES–KOH, pH 7.6, 0.66 m*M* EDTA, 0.66 m*M* 2-mercaptoethanol, 0.033% NP-40, 1.1 m*M* $MgCl_2$, 33 μM cold ATP, 5 μCi [γ-^{32}P]ATP (3000 Ci/mmol, Amersham, without "Redivue" dye), and a small aliquot of the sample to be tested.
2. Incubate for up to 4 hr at 26°.
3. Separate ATP and free phosphate by thin-layer chromatography on PEI-cellulose plates (Schleicher & Schuell) in 0.5 *M* LiCl/1 *M* formic acid in a closed chromatography glass chamber.
4. Quantify the radioactivity retained on the origin (nonhydrolyzed ATP) and that corresponding to the freed phosphate with a PhosphoImager and ImageQuant software.

[22] T. Tsukiyama and C. Wu, *Methods Enzymol.* **273,** 291 (1996).

[23] B. R. Cairns, Y. Lorch, Y. Li, M. Zhang, L. Lacomis, B. H. Erdjument, P. Tempst, J. Du, B. Laurent, and R. D. Kornberg, *Cell* **87,** 1249 (1996).

Interaction of Remodeling Machines with Immobilized Nucleosomes

Nucleosomes can be coupled efficiently to small paramagnetic beads (Dynabeads M280, Dynal, Oslo) via a biotin–streptavidin linkage. Once the nucleosomes are coupled to the beads, changes of buffers and washes of the nucleosomes are performed quickly.[24] This allows a whole range of elegant experiments, such as "pull-down" studies, to investigate the interaction of factors with nucleosomes. A change in histone composition of the nucleosomes after a remodeling step can be investigated easily, but the DNA structure around the nucleosome can also be studied by, e.g., DNase I footprinting.

The preparation of immobilized nucleosomes consists of four steps, which will be described here: (A) preparation of the biotinylated DNA fragment, (B) salt dialysis reconstitution of nucleosomes on this DNA, (C) coupling the nucleosomes to the paramagnetic beads, and (D) monitoring the integrity of the coupled nucleosomes.

Preparation of Biotinylated DNA. A fragment of suitable size (146–180 bp) is created by PCR amplification of a piece of DNA (the nature of which is irrelevant for our purposes here) using one biotinylated and one unmodified oligonucleotide primer. For binding studies, neither of the primers needs to be labeled radioactively. However, for footprint analysis, the unmodified primer needs to be kinased with [^{32}P]phosphate. A trace of radiolabeled PCR fragment mixed into the cold PCR fragment will allow to monitor the efficiencies and losses during the various steps of the procedure.

1. Set up a standard PCR reaction with 10 pmol of each primer and 2 ng of input DNA.
2. After the reaction is completed, add 10 μg of glycogen (Boehringer Mannheim) as a precipitation carrier and precipitate the DNA by adding 0.7 volumes of 7.5 *M* ammonium acetate and 2.5 volumes of 100% ethanol. Wash the pellet with 70% ethanol.
3. Separate the PCR products on a 1.3% agarose gel and gel purify the product from the gel [we use the Quiaquick gel extraction kit (Qiagen) according to the manufacturer's specifications].

Nucleosome Reconstitution. A salt gradient dialysis reconstitution is performed essentially as described earlier with 200 ng of biotinylated DNA fragment (this may be either unlabeled or radiolabeled due to the use of a radiolabeled primer during PCR or essentially unlabeled with a trace amount of labeled fragment) mixed into 2 μg carrier DNA (sheared and

[24] R. Sandaltzopoulos, T. Blank, and P. B. Becker, *EMBO J.* **13,** 373 (1994); R. Sandaltzopoulos and P. B. Becker, *Mol. Cell. Biol.* **18,** 361 (1998).

sonicated herring or salmon sperm DNA). The presence of carrier DNA allows one to establish the optimal histone/DNA ratio more reproducibly.

Coupling of Nucleosomes to Dynabeads

1. Pipette 40 μl of paramagnetic bead suspension per each 200 ng of biotinylated fragment reconstituted into nucleosomes (Dynabeads M280, Dynal Oslo) into a siliconized Eppendorf tube.
2. Wash the beads three times with 100 μl of phosphate-buffered saline (PBS)/0.05% NP-40/0.1 mg/ml BSA according to the manufacturer's instructions. A magnetic particle concentrator (e.g., MPC, Dynal, Oslo) is used to separate the beads rapidly from the wash solutions.
3. After the final wash, resuspend the beads in the nucleosome solution derived from the salt gradient dialysis procedure and incubate overnight on a rotating wheel at 4°.
4. Wash the beads three times with EX120/0.05% NP-40/0.5 mg/ml BSA and resuspend the coupled nucleosomes at a final concentration of 1–10 ng DNA/μl in EX120/0.05% NP-40/0.5 mg/ml BSA.

The coupling efficiency is normally >90%. This can be monitored by following the distribution of the radioactive "tracer" fragment between beads and supernatant either by liquid scintillation counting or by gel electrophoresis of an aliquot of the nucleosome solution before and after the coupling procedure ("missing band test"). Nonnucleosomal histone/DNA aggregates that form during the reconstitution will not be coupled efficiently to the beads. The proper histone stoichiometry of the immobilized nucleosomes can be verified by analyzing the proteins by SDS–PAGE (Fig. 4A). Using the silver-staining procedure of Wray *et al.*,[25] 50 ng of histones is detectable easily. The immobilized nucleosomes are quite stable. When stored in siliconized Eppendorf tubes on ice, no obvious loss of the histones was observed after 2 weeks.

DNase I Footprinting of Immobilized Nucleosomes. The wrapping of DNA around a nucleosome diminishes its accessibility to interacting proteins, such as DNase I. If nucleosomal DNA is incubated with DNase I, the enzyme will have access and cleave only those DNA sequences that are facing outward from the nucleosome. The digestion pattern of the DNA will be a reflection of the helical periodicity of the DNA molecule; DNA at 10- to 11-bp intervals will be cleaved preferentially and can be mapped with respect to one labeled fragment end.[26] If the nucleosome is phased with respect to the underlying DNA sequence, which is usually the case

[25] W. Wray, T. Boulikas, V. P. Wray, and R. Hancock, *Anal. Biochem.* **118,** 197 (1981).
[26] L. C. Lutter, *J. Mol. Bio.* **124,** 391 (1978).

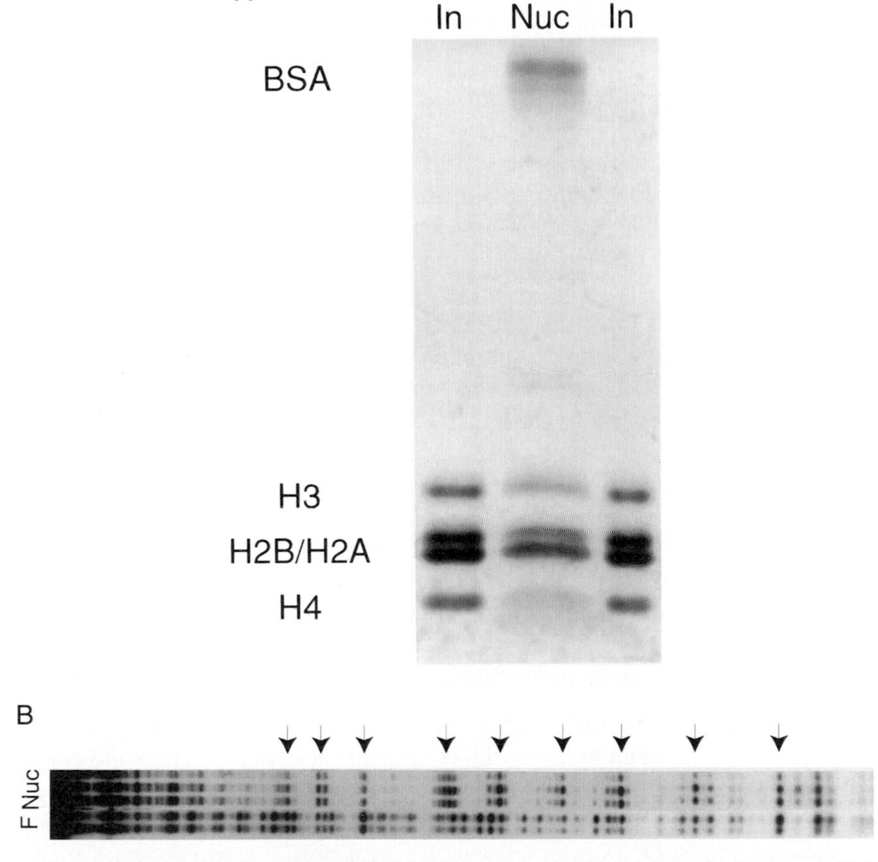

FIG. 4. Characterization of immobilized nucleosomes. (A) Monitoring the histone stoichiometry of immobilized nucleosomes: 50 ng of the input histones (In) and histones from immobilized nucleosomes (Nuc; a calculated equivalent of 80 ng) were analyzed by SDS–PAGE and silver staining. (B) Characterization of immobilized nucleosomes by DNase I digestion. Immobilized free DNA (F) or nucleosomes (Nuc) were analyzed by "solid-phase footprinting." Arrows highlight the 10-bp modulation of DNase I cleavage that characterizes the DNA bent over a nucleosomal surface.

when single nucleosomes are reconstituted on small DNA fragments, the 10-bp periodicity of cleavage will become apparent. This characteristic 10- to 11-bp periodicity is a signature of the nucleosome particle and of the subnucleosomal H3/H4 tetramer and can be used, in combination with the direct analysis of histone stoichiometry (see earlier discussion), to characterize the immobilized nucleosome. If the nucleosome to be analyzed is immobilized on a paramagnetic bead, the DNase I cleavage pattern can be visualized by "solid-phase footprinting"[27] (Fig. 4B).

1. Dissolve DNase I (Boehringer Mannheim) in 5 m*M* $CaCl_2$, 10 m*M* $MgCl_2$ at a concentration of 10 units/μl. Store solution in aliquots of 20 μl at −20°. Discard a thawed aliquot after use. Dilute a DNase I aliquot in 20 μl EX120/0.05% NP-40/0.5 mg/ml BSA.
2. Resuspend 5–10 ng of immobilized nucleosomal DNA (in terms of input DNA) per DNase I digestion point (or the corresponding amount of immobilized naked DNA as control) in 20 μl EX120/0.05% NP-40/0.5 mg/ml BSA.
3. Digest naked DNA by adding 0.01–0.03 units DNase I in 20 μl for 1 min at room temperature. Digest nucleosomes by adding 0.05–0.1 units DNase I in 20 μl for 1 min at room temperature.
4. Stop the digestion by adding 50 μl 4 *M* NaCl, 100 m*M* EDTA.
5. Concentrate the beads on the MPC and remove the supernatant.
6. Wash the beads with 100 μl 2 *M* NaCl, 20 m*M* EDTA and remove the supernatant.
7. Wash the beads with 100 μl 10 m*M* Tris–HCl, pH 7.5, 1 m*M* EDTA and remove the supernatant.
8. Resuspend the beads in 4 μl formamide loading buffer[28] and heat for 5 min at 76°.
9. Analyze the digestion products on an 8% denaturing sequencing gel.

The interaction of nucleosome remodeling factors frequently leads to a distortion of the regular 10- to 11-bp repeat pattern.[23,29] The solid-phase approach therefore allows the simultaneous analysis of the interaction of a factor with a nucleosome in a "pull-down" experiment, the visualization of altered histone/DNA interactions on factor binding, and the monitoring of histone loss through direct analysis of the histone stoichiometry.

Further Options for Analysis of Remodeling Machines

The assays described in this article were particularly useful during the purification and characterization of the chromatin accessibility complex.[6] However, a whole panel of other assays has been described by others that characterize other aspects of nucleosome remodeling, including the analysis

[27] R. Sandaltzopoulos and P. B. Becker, *Nucleic Acids Res.* **22**, 1511 (1994); R. Sandaltzopoulos and E. Bonte, *BioTechniques* **18**, 775 (1995); R. Sandaltzopoulos and P. B. Becker, *Methods Mol. Cell Biol.* **5**, 176 (1995).

[28] J. Sambrook, E. F. Fritsch, and T. Maniatis, "Molecular Cloning: A Laboratory Manual." Cold Spring Harbor, NY, 1989.

[29] A. N. Imbalzano, H. Kwon, M. R. Green, and R. E. Kingston, *Nature* **370**, 481 (1994).

of changes in plasmid topology,[30] of site-specific chromatin transitions on plasmids using a micrococcal nuclease assay,[31] the visualization of superhelical stress,[32] the effect of remodeling machines on transcription initiation,[33] and on promoter clearing of a nucleosome-inhibited polymerase.[34]

Acknowledgment

Patrick Varga-Weisz was supported by a grant from the Deutsche Forschungsgemeinschaft and is currently supported by the Marie Curie Cancer Care, Great Britain.

[30] H. Kwon, A. Imbalzano N., P. A. Khavari, R. E. Kingston, and M. R. Green, *Nature* **370,** 477 (1994).

[31] T. Tsukiyama, P. B. Becker, and C. and Wu, *Nature* **367,** 525 (1994).

[32] J. Quinn, A. M. Fyrberg, R. W. Ganster, M. C. Schmidt, and C. L. Peterson, *Nature* **379,** 844 (1996).

[33] T. Ito, M. Bulger, M. J. Pazin, R. Kobayashi, and J. T. Kodonaga, *Cell* **145,** 1 (1997); G. Mizuguchi, T. Tsukiyama, J. Wisniewski, and C. Wu, *Mol. Cell* **141,** (1998).

[34] S. A. Brown, A. N. Imbalzano, and R. E. Kingston, *Genes Dev.* **10,** 1479 (1996).

[44] Purification of *Drosophila* Nucleosome Remodeling Factor

By Raphael Sandaltzopoulos, Vincent Ossipow, David A. Gdula, Toshio Tsukiyama, and Carl Wu

Introduction

Vital biochemical processes in the cell, such as DNA replication, transcription, repair, and recombination, take place in the environment of chromatin. The efficiency of these biochemical processes, which require access to DNA by very large molecular machines, is not compromised by the compaction of DNA in nucleosomes and higher order chromatin structures. Recent advances indicate a substantial contribution toward the regulation of chromatin accessibility by energy-consuming multisubunit protein complexes that modify, remodel, and rearrange chromatin, rendering it dynamic and accessible. The identification and purification of enzymatic complexes that regulate chromatin dynamics, such as the yeast SWI/SNF[1] and RSC[2] complexes and the *Drosophila* NURF,[3] ACF,[4] and

[1] B. R. Cairns, Y. J. Kim, M. H. Sayre, B. C. Laurent, and R. D. Kornberg, *Proc. Natl. Acad. Sci. U.S.A.* **91,** 1950 (1994).

[2] B. R. Cairns, Y. Lorch, Y. Li, M. Zhang, L. Lacomis, H. Erdjument-Bromage, P. Tempst, J. Du, B. Laurent, and R. D. Kornberg, *Cell* **87,** 1249 (1996).

[3] T. Tsukiyama and C. Wu, *Cell* **83,** 1011 (1995).

[4] T. Ito, M. Bulger, M. J. Pazin, R. Kobayashi, and J. T. Kadonaga, *Cell* **90,** 145 (1997).

0076-6879/99 $30.00

CHRAC[5] complexes, have modified our view of chromatin. The availability of these complexes in a pure form is fundamental for deciphering the mechanism of their functions. This article describes a detailed protocol for the purification of the *Drosophila* nucleosome remodeling factor (NURF).

NURF is a sarkosyl-sensitive, four-subunit complex that remodels nucleosomes in an ATP-dependent manner. Under standard conditions, nucleosome remodeling takes place only in the presence of a sequence-specific DNA-binding factor and is restricted to the nucleosomes in its vicinity.[3] NURF facilitates a step preceding RNA polymerase II preinitiation complex assembly resulting in transcriptional activation of a chromatin template reconstituted *in vitro*,[6] although the detailed mechanism of function and the nature of the remodeling that takes place remain to be elucidated. The stimulation of the ATPase activity of NURF by nucleosomes rather than free DNA or histones suggests that NURF has a special ability to recognize nucleosome structure, important elements of which include the flexible histone tails.[7]

Several of the NURF subunits have been identified and cloned. NURF-140 (140 kDa) is the ISWI ATPase[8] that was cloned on the basis of homology to the ATPase domain of yeast SWI2 protein. NURF-55 (55 kDa) is a WD repeat protein[9] known to be involved in histone metabolism. NURF-38 (38 kDa) has been cloned and expressed as a recombinant polypeptide.[10] The cloning and characterization of the largest subunit, NURF-215 (215 kDa), is currently in progress. Once cDNAs for all subunits become available it will be of interest to reconstitute the NURF complex with recombinant proteins and rigorously compare its activities with native NURF purified from *Drosophila*.

Outline of Purification Scheme

The purification of NURF from *Drosophila* nuclear extracts is a procedure of mild chromatographic steps followed by glycerol gradient centrifugation. The protocol is reliable and reproducible, and the yield of highly pure NURF complex is quite sufficient for biochemical analysis. NURF can be traced along the stages of purification either by testing fractions for

[5] P. D. Varga-Weisz, M. Wilm, E. Bonte, K. Dumas, M. Mann, and P. B. Becker, *Nature* **388,** 598 (1997).

[6] G. Mizuguchi, T. Tsukiyama, J. Wisniewski, and C. Wu, *Mol. Cell* **1,** 141 (1997).

[7] P. T. Georgel, T. Tsukiyama, and C. Wu, *EMBO J.* **16,** 4717 (1997).

[8] T. Tsukiyama, C. Daniel, J. Tamkun, and C. Wu, *Cell* **83,** 1021 (1995).

[9] M. A. Martinez-Balbas, T. Tsukiyama, D. Gdula, and C. Wu, *Proc. Natl. Acad. Sci. U.S.A.* **95,** 132 (1998).

[10] D. A. Gdula, R. Sandaltzopoulos, T. Tsukiyama, V. Ossipow, and C. Wu, *Genes Dev.* **12,** 3206 (1998).

nucleosome remodeling activity or by detection of NURF subunits in the fractions by Western analysis, using a mixture of antisera recognizing p38, p55, and ISWI. The nucleosome remodeling assay is described in detail by Mizuguchi and Wu.[11] For all separation methods tested, the peak of NURF activity coincides with fractions containing p38, p55, and ISWI. Because p55 and ISWI are components of other complexes,[5,9] it is important to trace the codistribution of all NURF subunits to avoid the inadvertent purification of other complexes.

The outline of the purification scheme is shown in Fig. 1. The purity of NURF containing fractions after each purification step is shown in Fig. 2: 1.25 μl (lanes 1, 2, and 3), 5 μl (lanes 4 and 5), or 10 μl (lane 6) of the indicated step was analyzed by 10% SDS–PAGE. Proteins were visualized either by silver staining or by immunoblotting with a mixture of three antisera specific for the indicated NURF subunits. Only four major bands are present after the glycerol gradient. Interestingly, all four of these proteins can be immunoprecipitated by any of the three antibodies from relatively crude fractions.[9,10]

The yield of highly purified NURF is 100–200 μg per 200 g of embryos.

Preparation of Starting Material

The starting material for NURF purification is a nuclear extract prepared from *Drosophila* embryos according to the procedure described by Wampler *et al.*[12] with slight modifications. For a typical purification, approximately 200 g of moist, 12- to 16-hr *Drosophila* embryos are dechorionated and resuspended in 3 ml/g of buffer I (15 m*M* HEPES–KOH, pH 7.6, 10 m*M* KCl, 5 m*M* $MgCl_2$, 0.1 m*M* EDTA, pH 8.0, 0.5 m*M* EGTA, pH 8.0, 350 m*M* sucrose) supplemented with 1 m*M* dithiothreitol (DTT) and 0.2 m*M* 4-(2-aminoethyl)benzenesulfonyl fluoride hydrochloride (AEBSF). The embryos are homogenized with two passes through a Yamato motor-driven glass/Teflon homogenizer at 1500 rpm. The homogenate is filtered through Miracloth, and nuclei are pelleted by centrifugation for 10 min in a Beckman JA 14 rotor at 8000 rpm (10,000*g*). All centrifugations are at 4°. The supernatant is discarded, and nuclei are resuspended again in the original volume of buffer I. While resuspending the nuclei by swirling, it is critical to avoid the much tighter yellow yolk pellet. Nuclei are then dispersed by briefly homogenizing two to three times with a loose glass pestle (Dounce) and immediately pelleted again as described previously. After the second centrifugation there should be hardly any yolk at the bottom of the nuclear pellet. All of it should be avoided as nuclei are

[11] G. Mizuguchi and C. Wu, "Chromatin Protocol." Humana Press, Clifton, NJ, in press.

[12] S. L. Wampler, C. M. Tyree, and J. T. Kadonaga, *J. Biol. Chem.* **265,** 21223 (1990).

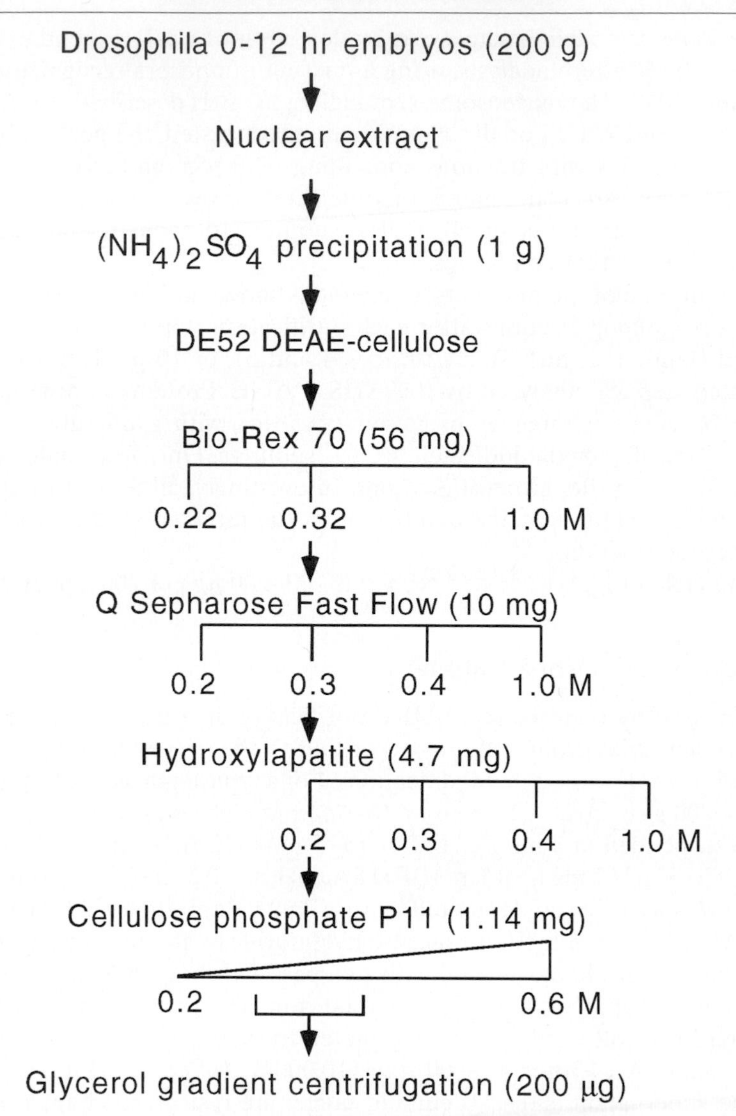

FIG. 1. Schematic outline of the purification protocol. The amount of total protein in the NURF containing fractions after each purification step is indicated.

resuspended by swirling in 1 ml/g of embryos of buffer II (15 m*M* HEPES–KOH, pH 7.6, 110 m*M* KCl, 5 m*M* $MgCl_2$, 0.1 m*M* EDTA, supplemented with 1 m*M* DTT and 0.2 m*M* AEBSF). Nuclei are dispersed well in a Dounce homogenizer (loose pestle). The volume of the nuclear suspension

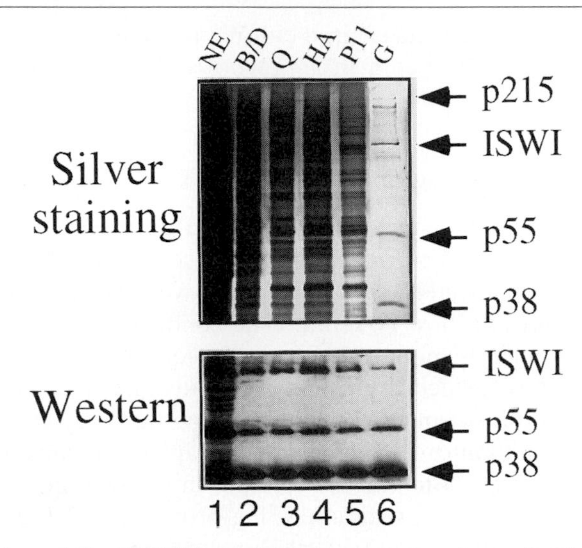

FIG. 2. Protein analysis of NURF containing fractions along the purification procedure.

is measured and divided equally into screw-cap tubes for the Beckman Type 35 rotor. Nuclei are lysed by adding 1/10 volume of 4 *M* $(NH_4)_2SO_4$, pH 8.0, in each tube. The viscous lysate is mixed vigorously for 20 min on a fast rotating wheel and then centrifuged in a precooled ultracentrifuge for 1 hr at 35,000 rpm (90,000*g*). The clear supernatant is collected by plunging a pipette beneath the lipid top layer. The volume is measured again and poured into a beaker. An equal volume of 4 *M* $(NH_4)_2SO_4$, pH 8.0, is added slowly (over a period of 5 min) and the mix is stirred vigorously for another 10 min at 4°. Proteins are precipitated by centrifugation in a JA-14 rotor at 11,000 rpm (18,600*g*) for 30 min. The pellets may be flash frozen in liquid nitrogen and kept at −80° until needed. The pellet is allowed to thaw and is resuspended in 0.2 ml/g of embryos of HEGN-40 [25 m*M* HEPES–KOH, pH 7.6, 1 m*M* EDTA, 10% (v/v) glycerol, 0.02% (v/v) Nonidet P-40 (NP-40), 40 m*M* KCl; the indicator 40 stands for the concentration of KCl in m*M*] supplemented with 1 m*M* DTT and 0.2 m*M* AEBSF by pipetting up and down. It is important to ensure thorough solubilization of the pellet at this step as severe reprecipitation may occur during subsequent dialysis. The extract is then placed in dialysis bags (Spectra/Por 2, molecular weight cutoff 12,000–14,000) and dialyzed twice for 1 hr against 1 liter of HEGN-40 (with 1 m*M* DTT and 0.2 m*M* AEBSF). The dialysis buffer is changed once more and the dialysis continues until the conductivity of the extract equals that of HEGN-150. The conductivity is monitored every 15 min. Finally the dialyzed extract is harvested and cleared before proceeding to chromatography by centrifugation at 8000 rpm in a precooled JA-14 rotor (10,000*g*) for 10 min.

Ion-Exchange Chromatography

All chromatographic steps are performed on an FPLC (fast protein liquid chromatography) system (Pharmacia, Piscataway, NJ). Every column is packed 1 day in advance and equilibrated overnight with at least 5 bed volumes of buffer at a slow flow rate. The conductivity, as well as the pH of the flow through, is monitored to verify equilibration. The estimated volumes of buffers needed are prepared in advance and the appropriate loading superloop is mounted on the system so that they equilibrate to 4° overnight. The first ion-exchange chromatography step consists of a DE52 DEAE-cellulose and a Bio-Rex 70 column, in tandem. For each gram of extract protein (expect approximately 1 g protein per 200 g of embryos) use 20 ml of packed DE52 and 50 ml Bio-Rex 70 resins.

An appropriate amount of preswollen Whatman DE52 resin is resuspended in at least 6 ml of HEG-150 (as HEGN-150 without NP-40) per gram of resin with gentle agitation on a shaking platform. The swollen resin is allowed to settle by gravity and the supernatant containing fines is aspirated away. The resin is then resuspended in fresh HEG-150 buffer and allowed to settle once more to remove ultrafine particles that need a longer time to settle. Then the resin is transferred into a XK16 column (Pharmacia) and allowed to pack.

In parallel, a sufficient amount of Bio-Rex 70 resin (200–400 mesh; Bio-Rad, Richmond, CA) is resuspended in 5 ml HEG-150 per gram of resin. The swollen resin is transferred in a XK26 (Pharmacia) column. Once packed by gravity, the columns are mounted on the FPLC system such that the bottom of the DE52 column is connected to the top of the Bio-Rex 70 column. We use HEGN-0 (containing no KCl) as buffer A of the FPLC and HEGN-1000 (containing 1 *M* KCl) as buffer B, both supplemented with 1 m*M* DTT and 0.2 m*M* AEBSF. The columns are equilibrated with 15% buffer B (containing 150 m*M* KCl) overnight (at least 5 bed volumes). The next day, the flow rate is increased to 2 ml/min for 15 min to ensure tight packing, then the top lid of each column is readjusted to minimize the dead volume.

The extract is loaded at 1 ml/min with 15% buffer B. NURF does not bind efficiently to DE52 under these conditions but it is retained on the Bio-Rex 70. As soon as the UV monitor signal drops to the baseline, the DE52 column is disconnected. The flow rate is increased to 2 ml/min, and the Bio-Rex 70 column is washed with 22% buffer B until the UV signal drops to the baseline. Then NURF is eluted in a single step by raising the relative concentration of buffer B to 32%. The entire peak (about 40 ml for 1 g of loaded protein) is collected. Although NURF is eluted as a broad peak between approximately 280 and 380 m*M* KCl, the elution at 320 m*M* is a good compromise between yield and purification factor.

The majority of free NURF subunits not associated in the complex is removed (as indicated by filtration methods) and NURF activity is extremely high. This first step is highly reproducible. However, as NURF proteins bind avidly to DE52 resin at pH greater than 8.0, careful adjustment of the buffer pH is imperative. NURF containing fractions are then dialyzed against HEGN-100 until conductivity equals that of HEGN-150. The sample is now ready for the second purification step.

Sufficient Q Sepharose fast flow (Pharmacia) slurry is dispensed in a beaker to give 7–8 ml of packed resin. The resin is allowed to settle. The ethanol containing supernatant is aspirated away and the resin is resuspended in an equal volume of starting buffer (HEG-150, Nonidet P-40 is avoided again during packing). When the resin settles again, the supernatant is decanted and is replaced by fresh buffer. The slurry is degassed and poured into an HR 10/10 column (Pharmacia). HEGN-0 is used as buffer A and HEGN-1000 as buffer B. The column is equilibrated with 5 volumes of HEGN-150 (with 1 m*M* DTT and 0.2 m*M* AEBSF freshly added). The flow rate is adjusted at 2 ml/min for 5 min. The sample is centrifuged in a JA-20 rotor for 15 min at 12,000 rpm (17,400*g*). The supernatant is filtered through a 0.22-μm Millex-GV low protein binding filter (Millipore, Bedford, MA) and loaded onto the column at 1 ml/min in HEGN-150. When the UV signal drops to baseline, NURF is eluted in steps with 20, 30, 40, and 100% buffer B. Three-milliliter fractions are collected. NURF elutes in the 0.3 *M* peak (30% of buffer B).

NURF containing fractions are pooled and dialyzed against PPEGN-100 [0.1 m*M* EDTA, 10% (v/v) glycerol, 0.02% (v/v) Nonidet P-40, 100 m*M* K_2 HPO_4/KH_2PO_4, pH 7.6] for 2.5 hr.

Hydroxylapatite and Cellulose Phosphate Chromatography

Hydroxylapatite resin (Bio-Rad) is resuspended in PPEGN-100. The slurry is poured into a HR 10/10 (Pharmacia) column and allowed to pack by gravity. A sufficient amount of slurry to yield 4 ml of packed bed volume is added. PPEGN-0 (as PPEGN-100 without K_2HPO_4/KH_2PO_4) is used as buffer A and PPEGN-800 (contains 0.8 *M* K_2HPO_4/KH_2PO_4, pH 7.6) as buffer B. The column is equilibrated with at least 5 bed volumes of PPEGN-100 (i.e., 12.5% buffer B). The sample is filtered through a 0.22-μm Millex-GV filter and loaded at 0.5 ml/min. When the UV signal drops to baseline, proteins are eluted with steps of PPEGN-200, PPEGN-300, PPEGN-400, and PPEGN-800 collecting 1.5-ml fractions. NURF elutes mainly in the PPEGN-200 peak. Fractions containing NURF are pooled (expect about 12 ml total volume of pooled NURF containing fractions). Although some NURF is also present in the PPEGN-300 peak, these fractions are not pooled because they contain far more impurities. In order to minimize losses due to nonspecific protein sticking to the dialysis bag, pure pH neutralized insulin (Boehringer) is added to a final concentration of 0.2 mg/ml and dialyzed against HEGN-150 until the conductivity is equal to HEGN-200 (about 2 hr).

The last chromatographic step is cellulose phosphate P11 (Whatman). P11 resin comes in a fibrous form that needs precycling before use (see manufacturer's specifications). P11 can be stored at 4° in 0.5 *M* phosphate buffer at pH 7.0. Precycled cellulose phosphate P11 is transfered into 10–20 volumes of HEGN-200. The resin is mixed without vigorous stirring to avoid the generation of fines and is allowed to settle. The supernatant is discarded and the resin is resuspended in 3–5 volumes of HEGN-200. The resin is degassed and the equivalent of 1 ml resin volume is poured into a HR 5/5 column. After the resin has settled by gravity, the top is refilled with resin (to compensate for the reduction of volume after close packing) and the column is closed. HEGN-0 and HEGN-1000 buffers are used as buffers A and B respectively. The column is equilibrated in HEGN-200 with 20 bed volumes at 0.2 ml/min. The flow rate is increased to 0.4 ml/min for 3 min. Then the sample is loaded at 0.4 ml/min for 3 min. Then the sample is loaded at 0.4 ml/min in HEGN-200. When the UV signal drops back to baseline, the column is developed with a shallow linear salt gradient (HEGN-200 to HEGN-600 in 20 column volumes). Four hundred-microliter fractions are collected. NURF elutes as a very broad peak, spanning from HEGN-250 to HEGN-450, probably because P11 is a bifunctional cation exchanger containing both strong and weak acid groups. It is essential not to pool the NURF containing fractions; each one is processed further separately.

Glycerol Gradient Centrifugation

The ultimate purification step is a glycerol gradient centrifugation. Glycerol gradients are formed in SW50Ti tubes (swinging bucket rotor tubes; Beckman) by carefully applying 450-μl aliquots of each of 10 HEGN-150 solutions (supplemented with a cocktail of protein inhibitors; the 100× concentrated inhibitor mix contains 100 m*M* phenylmethylsulfonyl fluoride, 200 μM pepstatin A, 60 μM leupeptin, 200 m*M* benzamidine, and 200 μg/ml chymostatin in ethanol) containing 35 to 17% (v/v) glycerol in increments of 2%, i.e., the first layer (bottom) is 35% (v/v) and the last layer (top) is 17% (v/v) glycerol. All buffers as well as the rotor and the centrifugation chamber must be precooled to 4° because the viscosity of the gradient is a function of the temperature. Four hundred microliters of the NURF containing P11 fractions is applied on top. The gradients are centrifuged at 49,000 rpm (160,000*g*) for 20 hr at 4° (no brake). Five hundred-microliter fractions are collected from the bottom of each tube. NURF is always found in the third fraction. Glycerol gradient centrifugation is a very effective and reproducible purification step that conditions the purified NURF fractions for prolonged storage. Early NURF containing fractions of the P11 column yield very pure NURF preparations when applied to glycerol gradient centrifugation. After this step, only four major protein bands corresponding to the four NURF subunits are visible on a silver-stained SDS–PAGE gel (Fig. 2, lane 6). Late NURF containing P11 fractions usually contain impurities that are not completely removed after glycerol centrifugation.

Acknowledgments

R.S. was supported by an EMBO postdoctoral fellowship. V.O. was a recipient of a long-term fellowship from the Human Frontier Science Program Organization. D.A.G. is a Leukemia Society of America postdoctoral fellow.

Author Index

Numbers in parentheses are footnote reference numbers and indicate that an author's work is referred to although the name is not cited in the text.

A

Abmayr, S. M., 706
Abrahamove, D., 435
Ackers, G. K., 637
Adachi, K., 194
Adachi, Y., 165
Adams, C. C., 300
Adrian, M., 191, 192, 193(8), 199, 200(18), 211(8)
Aeibi, U., 220
Agard, D. A., 205
Akita, M., 80
Albert, T., 358
Alberts, B. M., 80
Albright, S. C., 303
Alfieri, J. A., 35
Alfonso, P. J., 133, 134, 134(4), 138(3), 139(3), 140(3), 142(5, 10), 143(5, 10), 147(10), 148(10), 149(10)
Allan, B. D., 490
Allan, J., 232, 307, 309, 527, 630
Alland, L., 698
Allen, D. J., 216
Allfrey, V. G., 80, 534, 716
Allis, C. D., 50, 85, 87, 87(43a), 88, 88(42), 89, 90, 90(43), 93, 94, 95(54), 97, 97(54), 98, 232, 389(25), 403, 405(13), 415, 417, 417(5), 418(5, 10), 430(10), 535, 545, 547, 618, 675, 676, 676(1), 677, 677(15), 678, 680, 682, 682(9), 687, 688, 688(16), 689, 689(15), 690(15), 695, 697, 704
Allshire, R. C., 401, 405(6), 417, 422(18), 423(18)
Almer, A., 370(16), 371, 375(16), 376(16), 378
Almouzni, G., 50, 63, 75, 76, 82, 143, 146, 146(23), 333, 334, 337, 337(7), 338(7), 341(7), 345(7), 346, 346(7), 347(7), 350(5), 721
Altenburger, W., 371, 465
Amati, B., 668
Amos, W. B., 64, 68(15)
Anderson, C. W., 99, 676
Anderson, J. N., 80, 81(13)
Andrews, N. C., 573, 574(19), 578(19), 579(19), 580(19), 582(19), 583(19), 616
Andrulis, E. D., 99, 676
Angelov, D., 517, 611, 618
Annunziato, A. T., 76, 80, 81, 85, 87, 88, 88(42), 89, 90, 90(43), 93, 94, 95(54), 97, 97(54), 98, 676, 682(9), 695
Aparicio, O. M., 401, 405(5), 417, 422(14), 426(14), 430(14)
Archer, T. K., 323, 378, 490, 584, 585, 586, 586(15, 21), 589, 589(11), 593, 593(10), 594(27), 595, 597(29), 599(15, 21, 27), 623
Arents, G., 231, 232, 516, 524(6), 529(6)
Aris, J. P., 366
Ariza, R. R., 335
Armstrong, G. A., 639
Attisano, L., 682
Ausio, J., 20, 23, 23(6), 34, 136, 139(13), 232, 308, 629, 731, 734(10), 739(10)
Ausubel, F. M., 65, 74(23), 169, 172(37), 176(37), 351, 352(1)
Aviles, F. X., 232
Axelrod, A., 663, 664(8)
Axelrod, J. D., 396, 461, 629
Ayer, D. E., 715
Ayyub, H., 572, 573

B

Bachmann, D., 619
Bailey, D., 651
Bakayev, V. V., 126, 299
Baker, S. M., 40
Bakken, A. H., 216
Balaeff, A., 103
Baldwin, J. P., 239

Banchev, T., 264, 516, 522(7), 530(7), 531(7), 601, 605(13)
Bannister, A. J., 676, 677(15), 689(15), 690(15)
Barbaric, S., 368
Barber, D. A., 586
Bardwell, A. J., 449
Bardwell, L., 449
Barker, D. F., 643
Barker, D. L., 325
Barlev, N. A., 696, 698
Barnues, J., 223
Barr, P. A., 157, 161(17)
Barris, A., 223
Bartholomew, B., 232, 516, 524, 524(6), 526, 526(19), 529(6, 19, 20), 530(19, 22)
Barton, M. C., 63, 64, 66(7, 18), 75, 75(7)
Bartsch, J., 466, 622, 625(55), 638
Bashkin, J., 295, 628
Bat, O., 639
Batson, S. C., 134, 143(6)
Baumeister, W., 229
Bavykin, S. G., 264, 265, 265(1), 267, 267(6), 268(10), 275, 276(5, 6, 10), 516, 517(2, 3), 520(1, 2), 522(1, 2, 7), 524, 524(1, 2), 529(1), 530(1, 2, 7), 531(1, 2, 7), 532(1), 600, 601, 603(11), 604(5, 11), 605(13), 606, 608, 610(11), 611(17)
Baxevanis, A. D., 100, 101(12), 102(12), 235
Baxter, J. D., 466
Bazett-Jones, D. P., 103, 118(19), 221
Bear, D. G., 216
Beard, P., 311
Beato, M., 323, 328, 466, 622, 625(55), 638
Beatty, B. R., 216
Becker, D. M., 41
Becker, M. M., 548, 556(3), 617
Becker, P. B., 50, 53, 63, 65, 66(20), 68(20), 72(20), 76, 334, 335, 337(7), 338(7, 14, 16), 341, 341(7), 343(16), 344(16), 345(7), 346(7), 347(7), 350, 464, 574, 623, 625(59), 636, 742, 743, 743(6), 745, 748(6), 750, 750(6), 751(6), 752(6), 753, 756, 756(6), 757, 758, 759(5)
Bednar, J., 20, 191, 201(6), 203, 203(6), 205(6), 206(6), 207, 208, 210(6, 22), 211(6)
Bedoyan, J., 351
Beer, M., 229
Behringer, R. R., 573
Belgovsky, A., 600, 601(9, 10), 603, 603(10), 606(10), 607(9, 10), 611(10), 612(10)
Belikov, S., 600, 601(7, 9, 10, 12), 603, 603(10, 12), 605, 606(10, 12), 607(7, 9, 10, 15), 609, 610, 610(7), 611(10), 612(10)
Bell, S. P., 401, 405(5), 417, 422(14), 426(14), 430(14)
Bellard, M., 257, 276, 277(20), 531, 633
Bellare, J. R., 199
Belmont, A. S., 663
Beltrame, M., 109(64), 132
Belyavsky, A. V., 264, 265, 268(10), 276(5, 10), 516, 520(1), 522(1), 524(1), 529(1), 530(1), 531(1), 532(1), 600, 601(8), 604(8), 610(8)
Bentley, D. L., 362
Bentz, M., 658
Benyajati, C., 627
Beppo, T., 80
Berard, D., 595
Bergel, M., 684
Berger, L., 431
Berger, S. L., 368, 374(12), 545, 547(16), 676, 677, 677(15), 688(16), 689(15), 690(15), 696
Bergman, L. W., 38, 49(11)
Bergman, Y., 616
Berman, J., 99
Bernardi, G., 651
Berriman, J., 192, 213(7)
Bertrand, E., 637
Beug, H., 573
Bewley, C. A., 156, 161(11), 170(11), 172(11), 173(11), 177(11)
Bianchi, M., 100, 109(64), 132
Bickmore, W. A., 650, 651, 652(3), 653(2), 655(3), 656(2), 658(3, 6), 660(6), 661(6)
Biggerstaff, M., 335, 337
Biggin, M. D., 496, 497, 498, 499(11, 12, 23), 511, 511(11–13, 23), 512, 512(11, 12), 514(12, 27), 515(12), 600, 603(3)
Bina, M., 627
Bingham, P. M., 462
Birch, D., 438
Bird, A., 545, 547(17)
Bird, L. E., 366, 380, 387(22)
Bittner, M., 651
Bjork, U., 313, 319, 327
Bjorklid, E., 93
Blackman, M., 193

Blanchard, J.-M., 585, 586(18)
Blank, T., 65, 66(20), 68(20), 72(20), 335, 338(16), 341, 343(16), 344(16), 350, 743, 753
Blatter, E. E., 498, 499(22)
Blau, J., 362
Blobel, G., 366
Bloch, W., 438
Blocker, H., 439
Blomquist, P., 313, 321(3), 322(3), 323(3), 327(3), 328(3)
Bloom, K. S., 80, 81(13), 663, 671(4)
Blow, J. J., 63, 64, 68(15), 341
Blumenthal, A. B., 659
Blumenthal, R. M., 431
Bock, C.-T., 191
Bode, J., 303, 304(18)
Boeke, J. D., 40
Boffa, L. C., 80, 534
Bohg, A., 639
Böhm, L., 3, 271, 275(16), 527
Boissel, J.-P., 163
Bolshoy, A., 203, 210(22)
Bonnefoy, E., 101, 118
Bonner, J., 223
Bonner, W. M., 83
Bonte, E., 53, 350, 742, 743(6), 748(6), 750(6), 751(6), 752(6), 756, 756(6), 758, 759(5)
Boothby, M., 183, 186(50)
Bortner, C., 218
Bossard, P., 497, 614, 616(9), 623(9), 624, 625(9), 636
Boulikas, T., 142, 754
Bouloukos, C., 223
Boyd, K. E., 499
Boyle, S., 651, 652(3), 655(3), 658(3)
Bozzola, J. J., 192, 193(11)
Bradbury, E. M., 20, 26(7), 257, 264, 267, 275, 300, 302, 303, 304, 305, 305(20–22), 307(13, 22), 309(21, 22), 311(21, 22), 534, 652, 694
Bram, S., 218
Brandt, W. F., 23, 629, 732
Braunstein, M., 403, 405(13), 417, 418(10), 430(10), 535, 547
Breneman, J. W., 652
Brenowitz, M., 107, 112, 327, 328(27), 329(27), 332(27), 637
Brent, R., 65, 74(23), 169, 172(37), 176(37), 351, 352(1)
Breslauer, K. J., 439
Bresnick, E. H., 599, 618, 622(30)
Brewster, E., 726
Briand, G., 271, 275(16)
Brice, M., 621
Brinster, R. L., 573
Broach, J. R., 37, 39(9), 40, 40(9), 403, 405(13), 414, 417, 418(10), 430(10), 535, 547
Brodolin, K., 600, 601(8), 604(8), 610(8)
Brooks, J. E., 431
Brosch, G., 715
Brouwer, J., 461
Brown, D. D., 75, 318, 526
Brown, S. A., 553, 757
Brownell, J. E., 232, 389(25), 675, 676, 677, 677(15), 682, 688, 688(16), 689, 689(15), 690(15), 697, 704
Brubaker, J. M., 11, 20, 22(5), 208, 731, 750
Bruggemeier, U., 323
Brunvand, M., 361
Brynoff, K., 92
Buard, H., 222
Buchman, A. R., 378, 387(10)
Bujalowski, W., 121
Bulger, M., 53, 97, 335, 350, 464, 742, 743(7), 750(7), 757
Bunick, G. J., 307
Bunker, C., 584, 742, 752(3)
Burbank, D. E., 443
Burch, J. B. E., 615, 616(15)
Burcin, M. M., 676
Burgess, R. R., 627
Burglin, T. R., 64
Burgoyne, L. A., 44, 216, 633
Burhans, W. C., 77
Burlingame, R. W., 231
Burton, K., 632
Bushman, F. D., 156
Buss, J., 353
Bustamante, C., 213
Bustin, M., 84, 85, 88, 100, 101(5), 102(63), 132, 133, 134, 134(4), 137(2), 138(3), 139(3), 140, 140(3), 141(2), 142(2, 5, 10), 143, 143(5–7, 10, 11), 144(21), 145(21), 147(10, 21), 148(10), 149(10), 150, 150(19), 151(31), 152(31), 153(31), 155, 156(1), 157(1), 158(1), 161(1), 170(1), 180(1), 599, 618, 622(30), 684

Butler, P. J. G., 3, 11(2)
Buttinelli, M., 586

C

Cairns, B. R., 726, 752, 756(23), 757
Cairns, M. J., 563
Call, K., 655
Calladine, C. R., 26, 57
Campbell, J. L., 433
Campos-Ortega, J. A., 502
Candau, R., 677, 688(16)
Cantor, C. R., 104(40), 105(40), 113, 116(40)
Cardenas, M. E., 663, 672(2)
Carey, J., 300, 380
Carlson, J. O., 639, 640(3), 641(3)
Carlson, M., 726
Carlson, R. D., 679
Carlsson, J., 253
Carlstedt-Duke, J., 313
Carmen, A. A., 547
Carmichael, G. G., 644
Carneiro, M., 74, 314
Carot, V., 307
Carr, A., 497
Carruthers, L. M., 19, 20
Carshavsky, A., 299
Carter, K. C., 335, 651, 662
Carter, N., 651
Carter, T., 698, 703(16)
Cartwright, I. L., 380, 448, 462, 463, 466, 466(4, 5), 474, 476, 479(5), 490(5), 619, 620, 621, 635, 639, 640(13), 641(13), 644(13), 645(13)
Cary, P. D., 101
Cavalli, G., 497, 499(9), 618, 619
Cavallo, L., 216
Cech, T., 466, 639
Chafin, D. R., 231, 250
Chakravarti, A., 643
Chalkley, R., 84, 165, 223, 531, 534, 627
Chambon, P., 257, 633
Chandler, S., 718
Chang, C.-J., 655
Chang, J. J., 192, 193(8), 211(8)
Chang, L., 76, 85, 90, 90(43), 93, 676, 682(9)
Changela, A., 99, 107, 118(32), 120(32)
Chasman, D. I., 378, 387(10)
Chau, V., 442
Chen, H., 205, 321, 548, 564, 571
Chen, S. S., 585, 586(19)
Chen, Y., 233, 242(21), 244(21), 252, 253(11)
Chicoine, L. G., 680
Child, R., 168, 184(33), 185(33), 186(33)
Chimiak, A., 253
Chin, P. L., 554
Chou, Q., 438
Christiansen, G., 216, 226(13)
Chroneos, Z. C., 95
Chui, D.-S., 161
Chung, J., 545
Church, G., 395, 433, 477, 573, 609, 635
Churchill, M. E. A., 99, 101, 103, 103(18), 107, 107(21), 109(18, 21), 110(21), 118(32), 120(32), 121(18), 122(18), 139, 618
Church Landel, C., 386
Chuvpilo, S., 185
Cimino, G. D., 639
Ciosk, R., 663
Citterich, M., 100
Clark, D. J., 35, 48(31), 49, 82, 139, 143, 146, 213, 240, 307, 628, 630
Clark, S. J., 433
Clayton, A. L., 533, 542, 543(10), 544(11), 545, 545(11), 618
Clegg, J. B., 575, 578(26)
Clore, G. M., 156, 161(11), 170(11), 172(11), 173(11), 177(11)
Clos, J., 497
Cockell, M., 620, 625(43)
Coffer, A., 334
Coffey, D. S., 662
Cohen, D. R., 61
Cohen, L. H., 84
Cole, L. S., 168, 184(35), 185(35), 186(35), 187(35)
Cole, R. S., 639
Coleman, J. R., 624
Collis, C. M., 433
Colson, P., 633
Conconi, A., 619
Cook, D. N., 639
Cook, R. G., 85, 87(43a), 675, 680, 688, 695
Cooke, C. A., 661
Cooper, J., 200
Cordingley, M. G., 378, 585, 586, 586(15), 597(29), 598(43), 599, 599(15)
Cordingly, M. G., 490
Cornish-Bowden, A., 731

Costantini, F., 573
Cote, J., 297, 300, 677, 688(16), 726, 729(1), 731(1), 741, 744
Cotten, M., 627
Cotterill, S., 350
Cousens, L. S., 80
Cox, B. S., 40
Cox, L. S., 64
Craig, J. M., 651, 652(3), 653(2), 655(3), 656(2), 658(3)
Crane-Robinson, C., 3, 85, 88(40), 101, 232, 239, 271, 275(16), 527, 533, 534, 535, 536(8), 537(5, 8), 542, 543(10), 544(11), 545, 545(11), 618, 697
Cranston, G., 401, 405(6), 417, 422(18), 423(18)
Cremer, S., 490, 495
Cremer, T., 655
Crevel, G., 350
Crippa, M. P., 133, 134, 134(4), 138(3), 139(3), 140(3), 142(5, 10), 143(5, 10), 147(10), 148(10), 149(10)
Crothers, D. M., 100, 101(4), 103, 107(24), 112(4), 223, 224, 224(35), 297, 298(18), 300, 305, 306(26, 27), 320, 323, 323(12), 326(21)
Crowe, A. J., 63, 75
Cryderman, D. E., 462
Csordas, A., 676, 679(11)
Cuenoud, B., 253
Cummings, K. B., 376
Cunningham, R. P., 592
Cusick, M. E., 77
Cyrklaff, M., 193, 199, 200(18), 211

D

Dadd, C. A., 85, 87, 88, 88(42), 89, 90
Dalma-Weiszhausz, D. D., 512
Dang, T., 64, 66, 68(25)
Daniel, C., 350, 758
Darley-Usmar, V. M., 44
Dasso, M. C., 721
Dattagupta, N., 224
Davey, C., 307, 309, 630
Davidson, N., 214, 215(2)
Davidson, R. L., 642
Davie, J. R., 81, 84, 431
Davis, H. T., 199
Davis, M., 307
Davis, R. W., 214, 215(2)
Dean, A., 36, 38(7), 377, 432, 441(12)
De-Bernardin, W., 639
deBoer, E., 573
Debruin, D., 99
De Crombrugghe, B., 585, 586(19)
Dedon, P. C., 89, 415, 417, 417(5), 418(5)
deHaseth, P. L., 113
Deisseroth, A., 575
de Jong, M. C., 335
de Laat, W. L., 335
Delcuve, G. P., 84
Delfino, J. M., 253
Della Valle, G., 651
De Los Santos, T., 726
Delphech, M., 163
de Manoir, S., 655, 658
Demeler, B., 21, 22, 31(24, 27, 28)
Denhardt, D. T., 635
Denissenko, M. F., 567, 571(19)
de Oca-Luna, R. M., 643
DePamphilis, M. L., 76, 77, 93
De Robertis, E. M., 64
de Rooij, J., 335
Dervan, P. B., 252, 463, 466, 466(4), 620, 635
De Sario, A., 651
Dever, C. A., 498, 499(11), 511(11), 512(11)
Dey, A., 636
Dhar, V., 575
Diedan, J. D., 659
Diffley, J. F. X., 64, 415
Dignam, J. D., 57, 74
Dikstein, R., 698
Di Mauro, E., 586
Dimitrov, S., 76, 100, 129(8), 517, 600, 610(4), 611, 618, 721
Ding, H. F., 134, 143(6, 7)
Dingwall, A., 727
Dingwall, C., 465
Disney, J. E., 168, 169(32), 181(32), 182(32)
Dixon, W. J., 628
Dobson, M. J., 40
Doering, J. L., 526
Dohner, H., 655
Dolan, M., 621, 622(49)
Domas, K., 53
Dombroski, B. A., 628

Dombrowski, B. A., 139
Donald, G., 41
Dong, F., 20, 23, 136, 139(13), 232, 252, 629
Donovan, S., 415
Dorbic, T., 85
Douc-Rasy, S., 307
Dow, L. K., 99, 107, 118(32), 120(32)
Doye, V., 672
Dranginis, A. M., 442
Dresler, S. L., 347
Dretzen, G., 276, 277(20), 531
Drew, H. R., 26, 57, 257
Driscoll, C., 621
Drouin, R., 461
Druckmann, S., 85, 88, 150
Drury, L. S., 415
Du, J., 752, 756(23)
Du, W., 161
Duband-Goulet, I., 307, 311(29)
Dubendorff, J. W., 5, 119
Dubochet, J., 191, 192, 193, 193(8), 195(15), 199, 200(18), 201(6, 15), 203, 203(6), 205(6), 206(6), 209, 210(6, 22), 211, 211(6, 8), 221
Ducommun, M., 221
Duggan, L., 545, 547(16), 677, 688(16)
Dujon, B., 44
Dumas, K., 350, 742, 743, 743(6), 748(6), 750(6), 751(6), 752(6), 756(6), 758, 759(5)
Dumuis-Kervabon, A., 272, 273(18), 275(18)
Dunn, J. J., 5, 119, 161, 162(21)
Durkovich, D., 526, 529(20)
Dustin, I., 209
Dworkin, M. B., 146
Dworkin-Rastl, E., 146
Dzherbashyajan, A., 600, 601(7), 607(7), 610(7)

E

Earnshaw, W. C., 661
Ebens, A. J., 573
Ebralidse, K. K., 265, 268(10), 271, 275, 276(10), 516, 520(1), 522(1), 524(1), 527, 529(1), 530, 530(1, 27), 531(1), 532(1), 545, 600, 604(6), 608, 609(6), 610(6), 611(17)
Ebright, R. H., 233, 242(21), 244(21), 252, 253(11), 498, 499(22)
Ebright, Y. W., 252, 253(11), 498, 499(22)
Edmondson, D. G., 365
Edmonson, D. G., 232, 688
Edwards, D. P., 586
Egelman, E., 209
Eggert, H., 495
Egly, J.-M., 334
Eick, D., 358, 362
Eickbush, T. H., 24
Einck, L., 195, 219, 229(21)
Eisenberg, H., 308
Eisentahl, R., 731
Eissenberg, J. C., 492
Eker, A. P., 448
Ekwall, K., 401, 405(6), 417, 422(18), 423(18)
Elgin, S. C. R., 100, 101(6), 125(6), 129, 130, 380, 448, 462, 463, 464, 465, 465(6), 466, 466(4, 5), 470(6), 474, 476, 479, 479(5), 480, 481(37), 489(15, 16), 490, 490(5, 6, 15, 16), 492, 492(16), 620, 621, 635, 745
Eliasson, R., 92
Elion, E., 360
Ellison, M. J., 442
Elton, T. S., 156, 157, 158, 161(17), 163(15), 164(19), 165(15), 171(5), 177, 181
Emerson, B. M., 64, 66(7, 18), 75(7)
Encontre, I., 272, 273(18), 275(18)
Engel, J. D., 573, 621, 622(49)
Engelke, D. R., 366, 368(8, 9), 463
Enomoto, S., 99
Enver, T., 573, 621
Ephrussi, A., 573
Epner, E., 621
Erdjument, B. H., 752, 756(23)
Erhart, E., 40
Erickson, B. W., 716
Eriksson, P., 323, 325(24)
Erk, I., 213
Ermacora, M. R., 253
Ernst, S. G., 85, 90, 90(43), 93, 676, 682(9)
Escaig, J., 218
Espel, E., 223
Espinas, M. L., 637
Essigmann, J. M., 334, 335(10), 337(10)
Etienne, G., 272, 273(18), 275(18)
Evans, E., 334, 335
Evans, G. A., 655

Evans, J. N. S., 156, 161(11), 170(11), 172(11), 173(11), 177(11)
Evertsz, E. M., 517
Evertz, E. M., 498, 499(20)

F

Faberge, A. C., 229
Fainsod, A., 435
Falciola, L., 100
Fangman, W. L., 36
Farnet, C. M., 156
Farnham, P. J., 499
Fascher, K. D., 367, 368, 371(11), 621
Fashena, S. J., 166, 168(31), 184(31), 185(31), 186(31)
Faulkner, R., 534
Feaver, W. J., 585, 586(16)
Fedor, M. J., 366, 371, 378, 385(9), 629, 637(28)
Fehér, Z., 431, 432(6), 446(6)
Feinberg, A. P., 634
Feldman, M. Ya., 223
Fellows, J., 334
Felsenfeld, G., 23, 57, 213, 231, 239, 252, 307, 523, 545, 572, 613, 616, 620, 621, 622(49), 630, 687, 742, 747
Felthauser, A., 20, 22(16), 33(16), 737, 738(19)
Feng, H.-P., 291
Feng, J., 170(38), 171, 172(38), 621, 622(52)
Ferentz, A. E., 103, 107(21), 109(21), 110(21)
Ferguson-Smith, M., 651
Fernades, M., 497
Finch, J. T., 97, 218
Fink, G. R., 40, 454
Finkel, S. E., 170(38), 171, 172(38)
Fischer, W. H., 74
Fisher-Adams, G., 50, 403
FitzGerald, P. C., 290
Fjose, A., 65
Flaus, A., 3, 11, 17, 17(13), 251, 252, 253(14, 22), 256, 257, 260(22), 263(14, 22), 627, 631, 631(15), 638(15)
Flavo, J. V., 177
Fleischer, T. C., 715
Fletcher, A. B., 40
Fletcher, T. M., 19, 20, 22, 22(16), 26, 27, 31(24, 25), 32(41, 43), 33(16), 736, 737, 738(19), 744
Fogel, S., 39
Forbes, D. J., 64, 66, 68(25), 72(26)
Forget, B. G., 578
Formanek, H., 229
Formenton, A., 663, 664(9), 672(9)
Forrester, J., 157
Forrester, W. C., 572, 621
Foster-Hartnett, D. M., 492
Fotedar, R., 82, 94
Fox, R. O., 253
Frado, L. L., 195, 219, 229(21)
Fragoso, G., 252, 307, 323, 385, 584, 585, 586(14), 599, 599(14), 622, 625(54), 626, 630(6), 631(6), 632, 632(4, 5), 637(6, 35), 638(4, 5)
Francke, B., 640, 641(21)
Frank, J., 229
Frank, R., 439
Fraser, J. M. K., 93, 94(51)
Fraser, P., 573
Fredrickesen, M., 157
Freeman, L. A., 640
Freeman, R., 192
Freire, E., 235, 687
Fried, M. G., 100, 101(4), 112(4), 300, 323, 326(21)
Friedberg, E. C., 333, 351, 448, 449
Friedman, D. S., 314, 622(10), 623(10), 625(10), 626(10), 638
Fritsch, E. F., 23, 119, 121(51), 234, 251(22), 268, 270(12), 314, 316(7), 325(7), 328(7), 330(7), 395, 413, 458, 509, 667, 668(13), 680, 756
Frommer, M., 53, 54(12), 55(12), 56(12), 58(12), 59(12), 61(12), 62(12), 433
Fromont-Racine, M., 637
Fryer, C. J., 586
Fujimoto, D., 716
Fukami, A., 194
Funk, J. O., 358
Furrer, P., 203, 209, 210(22)
Fyrberg, A. M., 757

G

Gaillard, P.-H., 333, 334, 337, 337(7), 338(7), 341(7), 345(7), 346, 346(7), 347(7), 350(5)

Gale, J. M., 449
Gamper, H. B., 639
Ganster, R. W., 757
Garcia, C. A., 643
Garcia-Ramirez, M., 20, 34, 232
Gargiulo, G., 621
Gariglio, P., 353
Garrard, W. T., 126, 299, 300, 303, 303(6), 612, 622(1), 640
Garrity, P. A., 463, 484, 484(8)
Gartenberg, M. R., 305, 306(27)
Gasser, R., 619
Gasser, S. M., 50, 403, 661, 663, 664(8, 9), 668, 672(2, 3, 9)
Gavin, I. M., 264, 267, 267(6), 275, 276(6), 516, 517(3)
Gdula, D., 350, 757, 758, 759(9)
Ge, H., 100, 677, 696(18)
Geiduschek, E. P., 524, 526, 526(19), 529(19, 20), 530(19, 22)
Gekko, K., 302
Georgel, 21
Georgel, P. T., 758
Georgiev, G. P., 126, 299, 380, 466
Geraminejad, M., 99
Gerber, M., 230
Gerchman, S. E., 232
Gerdes, M. G., 651, 662
Germond, J.-E., 257
Gerwig, R., 185
Gewiess, A., 307
Giangrande, A., 276, 277(20), 531
Giardina, C., 498, 499(16), 511(16)
Giese, K., 100, 102(11)
Gilbert, W., 121, 122(53), 322, 395, 477, 573, 574, 576(22), 609, 616, 635
Gill, S. C., 8(15), 17
Gilmour, D. S., 89, 352, 415, 462, 463, 480, 480(9), 481(37), 497, 498, 498(6), 604, 618
Gilson, E., 663, 664(8), 672(3)
Gingcras, T. R., 431
Glaser, R. L., 490
Glauert, A. M., 223
Glibetic, M., 229
Glikin, G. C., 51
Godde, J. S., 308, 313
Goguadze, E. G., 264
Gomez-Lira, M. M., 303, 304(18)
Goodhew, P. J., 219
Goodwin, G. H., 158
Gordon, C. N., 230
Gordon, J., 84
Gordon, L. K., 449
Gorka, C., 527
Gornall, A. G., 95
Gorner, H., 517
Gorovsky, M. A., 89, 415, 417, 417(5), 418(5), 678, 680
Gorski, K., 74, 314
Gotta, M., 663, 664(9), 672(9)
Gottlieb, T., 698
Gottschling, D. E., 99, 432, 446(8), 613, 617(7), 676, 717
Gould, H., 232
Gourdon, G., 573
Goytisolo, F. A., 232
Grabel, P. G., 715
Grabherr, R., 443
Grachev, S. A., 275
Gralla, J. D., 614, 617(11)
Grand, P. A., 698
Grandi, P., 672
Grange, T., 614, 623(12), 625(12), 637
Granok, H., 448, 464, 479, 489(15), 490, 490(15)
Grant, P. A., 545, 547(16), 677, 688(16), 698
Grantcharova, V., 103, 107(21), 109(21), 110(21)
Gray, C. W., 222
Graziano, V., 232
Green, G. R., 195, 219, 229(21)
Green, M. H., 353
Green, M. R., 756, 757
Greenberg, M. E., 359
Greenburg, B., 433
Gregory, P. D., 365, 368, 374(12)
Greyling, H. J., 23, 629, 732
Griffith, J., 214, 215, 216, 218, 219, 220, 220(22), 225, 225(18), 226, 226(13, 22), 466
Griffiths, D. E., 41
Grigg, G. W., 433
Grigliatti, T., 464
Grigoryev, S. A., 207(35), 210, 213, 232
Grilley, M., 216
Gronenborn, A., 156, 161(11), 170(11), 172(11), 173(11), 177(11)
Groom, M., 193, 195(15), 201(15)
Gross, C. A., 514

Gross, D. S., 612, 622(1)
Gross, H., 211
Grosschedl, R., 100, 102(11)
Grossman, G., 548, 556(3)
Grosveld, F., 573
Groudine, M., 356, 359, 361, 362(10), 498, 534, 572, 621
Grummt, F., 185
Grunstein, M., 42, 50, 231, 232(4), 399, 401, 402, 403, 403(2), 405(3), 417, 422(13), 426(13), 461, 497, 523, 547, 613, 618, 675, 697, 698, 715
Gruss, C., 76, 134, 143(11)
Gruss, R., 80
Gu, W., 696
Guacci, V., 664, 671(10)
Gualdi, R., 624
Guang, X.-Y., 651
Guarente, L., 41
Gupta, R., 351
Guschin, D. Y., 271, 530, 600, 604(6), 609(6), 610(6)
Gustafsson, J.-Å., 313
Gutteridge, J. M. C., 252
Guzzetta, V., 643

H

Haaf, T., 661
Haas, H., 715
Hache, R. J., 466, 638
Hager, G. L., 252, 307, 378, 385, 466, 490, 584, 585, 586, 586(14, 15), 594(27), 595, 597(20, 29), 598(43), 599, 599(14, 15, 20, 27), 618, 620, 622, 622(30), 625(54), 626, 630(6), 631(6), 632, 632(4, 5), 633(7), 637(6, 35), 638(4, 5)
Hahne, S., 36
Hainfeld, J. F., 221, 229
Halligan, B., 661
Halliwell, B., 252
Hamada, Y., 624
Hamana, K., 100, 101(6), 125(6)
Hamiche, A., 191, 307, 311(29)
Hamkalo, B. A., 181
Hamlin, J. L., 600, 603(11), 604(11), 610(11)
Han, S., 81
Hanaoka, F., 351
Hanawalt, P. C., 459, 639
Hancock, R., 142, 754
Hanoaoka, F., 97
Hanscombe, O., 573
Hansen, J. C., 19, 20, 22, 22(13, 16), 23(6, 18), 24(9), 26, 26(12, 13), 27, 27(24, 25), 30(18), 31(12, 24, 27, 28), 32(18, 22, 41, 43), 33(13, 16), 232, 252, 466, 629, 731, 732, 733, 734(10), 736, 737, 738(19), 739(10), 744
Hansen, U., 134, 143(6, 7), 638
Hanson, C. V., 466, 619
Harauz, G., 229
Harbone, N., 232, 527
Hardison, R., 84
Harford, A. G., 229
Harlow, E., 410, 411(17)
Harp, J. M., 307
Harris, J. R., 227, 229(49), 230, 230(49)
Harrison, J., 433
Hartenstein, V., 502
Hartl, P., 64, 66, 68(25)
Hartman, P. G., 232
Hartnett, T., 492
Harwood, J., 415
Haseltine, W. A., 449
Hassig, C. A., 697, 715
Hattman, S., 431, 432(6), 433, 446(6)
Hawley, D. K., 119, 125(48), 353, 361
Hayatsu, H., 433, 436(23)
Hayes, J. J., 133, 137, 138(3), 139, 139(3, 15), 140(3), 172, 176(39), 231, 232, 235, 239, 239(24), 240(27), 246, 250, 250(19), 251, 295, 307, 516, 524, 524(6), 526, 526(18), 527(5), 528, 528(5), 529(6), 545, 628
Haykinson, M. J., 103
Heald, R., 65, 66(20), 68(20), 72(20)
Hearst, J. E., 466, 619, 639
Hebbes, T. R., 85, 88(40), 533, 534, 535, 536(8), 537(5, 8), 542, 543(10), 544(11), 545, 545(11), 618, 697
Hecht, A., 50, 399, 401, 402, 403, 403(2), 405(3), 417, 497, 618
Hefner, H. E., 107, 118(32), 120(32)
Hegerl, R., 229
Heinfling, A., 185
Hendrick, D., 575
Heng, H. H. Q., 661
Hepell-Parton, A., 651
Herera, J. E., 150, 151(31), 152(31), 153(31)
Herfort, M., 103, 118(19)

Herman, T., 94
Hermanson, G., 655
Herrera, J. E., 684
Hershkovitz, M., 495, 636
Hertzberg, R. P., 252, 463, 466, 466(4), 620, 635
Hess, G. P., 255
Heumann, H., 586, 592(24), 593(24)
Heuser, J., 218, 225(18)
Hewish, D. R., 44, 216, 633
Hickey, L. B., 356, 362(10), 498
Hicks, J. B., 454
Hieter, P., 406, 436, 664, 665(12), 667(12)
Higgs, D. R., 572, 573, 574(19), 578(19), 579(19), 580(19), 582(19), 583(19), 616
Hill, A. V., 104(37), 113
Hill, D. A., 159, 179(9), 180(9)
Hill, J., 41
Himes, S. R., 168, 184(35), 185(35), 186(35), 187(35)
Hinnen, A., 370(16), 371, 375(16), 376(16)
Hirayoshi, K., 351, 497
Hirosumi, J., 97
Hochstrasser, M., 442
Hock, R., 150, 151(31), 152(31), 153(31)
Hockensmith, J. W., 498, 499(20), 517
Hoeijmakers, J. H. J., 335
Hoekstra, M. F., 431, 432
Hoepfner, R. W., 639
Hoertnagel, K., 358
Hoffman, C. S., 406
Hofmann, J. F., 663, 672(2)
Hogan, E., 664, 671(10)
Hogan, M., 224
Hoheisel, J. D., 586, 592
Hollenberg, C. P., 40
Holmes, S. G., 403, 405(13), 417, 418(10), 430(10), 535
Holmgren, R., 462
Holmquist, G. P., 461
Holth, L. T., 168
Holtlund, J., 157
Homo, J. C., 192, 193(8), 211(8)
Honda, B. M., 97
Hoopes, B., 119, 125(48)
Hopper, J. E., 382
Hopwood, D., 223
Horecker, B. L., 695
Horne, R. W., 227, 229(49), 230(49)
Hornstra, I. K., 548
Horowitz, R. A., 191, 200, 201(6), 203(6), 205(6, 20), 206(6), 210, 210(6), 211(6, 20), 212(20), 227, 229(50), 232
Hörz, W., 304, 365, 367, 370(16), 371, 371(11), 374(14), 375(16), 376(16), 377, 378, 465, 490, 585, 586(13), 621, 626, 747, 748(18)
Houde, D., 600, 610(4)
Hough, P. V. C., 221
Housman, D., 655
Houssier, C., 633
Howard, B. H., 676, 677(12)
Howard-Flanders, P., 645, 646(26)
Hsieh, M., 112
Huang, R. C., 627, 632
Huang, S., 126
Huang, S.-Y., 300, 303(6)
Huberman, J. A., 93, 94(51)
Huibregtse, J. M., 366, 368(8, 9), 463
Hull, M. W., 366, 368(8), 463
Hunter, C., 223
Hunting, D. J., 347
Hurt, E. C., 672
Huth, J. R., 156, 161(11), 170(11), 172(11), 173(11), 177(11)
Hwang, J. R., 334
Hyman, A., 65, 66(20), 68(20), 72(20)
Hyman, L., 134, 143(9)

I

Ian, K. A., 41
Ihara, M., 347
Iida, S., 433
Ikuta, T., 581
Imbalzano, A. N., 553, 584, 741, 742, 752(3), 756, 757
Imhof, A., 677, 696(18)
Impey, H., 651
Inoue, A., 716
Isaacs, S. T., 639
Isenberg, I., 42
Ishimi, Y., 97
Ito, T., 53, 97, 350, 742, 743(7), 750(7), 757
Iwai, K., 223, 224(36)

J

Jack, R. S., 490, 495
Jackson, J. R., 627

Jackson, S., 698
Jackson, V., 401, 405(7), 531, 618
Jacobsen, G., 132, 133(62)
Jaehning, J. A., 40
Jakubowski, U., 229
Jambeck, P., 618
James, T. C., 492
Jans, D. A., 554
Jarman, A. P., 572, 573
Jaspers, N. G. J., 335
Jauregui-Adell, J., 272, 273(18), 275(18)
Jay, E., 592
Jeanteur, P., 585, 586(18)
Jencks, W. P., 117
Jenster, G., 676
Jessee, B., 621
Jett, S. D., 216
Jiang, Y., 651
Johansen, B. V., 220
John, E. W., 157
John, S., 168, 184(33), 185(33), 186(33), 385, 599, 622, 625(54), 626, 632, 632(5), 637(5, 35), 638(5)
Johns, E. W., 157, 158, 158(14), 163(14)
Johnson, K., 156
Johnson, K. D., 157, 161, 161(17)
Johnson, K. R., 156, 168, 169(32), 171(5), 177, 181(32), 182(32)
Johnson, L., 42, 403
Johnson, R. C., 103, 170(38), 171, 172(38)
Johnston, R. F., 325
Jones, D. N. M., 102(49), 119
Jones, E. W., 38, 49(11)
Jones, K. A., 74
Jones, P. L., 715
Joos, S., 655, 658
Jordan, E., 526
Jost, J.-P., 585
Juan, L. J., 300
Jupe, E. R., 619, 639, 640(13), 641(13), 644(13), 645(13)

K

Kadonaga, J. T., 53, 97, 134, 142(8), 335, 350, 378, 464, 675, 742, 743(7), 750(7), 752(4), 757, 759
Kahn, J. D., 103, 107(24)
Kai, K., 433
Kaiser, C., 418
Kam, L., 139
Kamakaka, R. T., 335, 350, 378, 464, 742
Kan, Y. W., 573, 581
Kang, J., 93
Kaplan, B. E., 557
Kapp, L. N., 659
Karagyozov, L., 618
Karantza, V., 235, 687
Karpov, V., 264, 265, 268(10), 271, 276(10), 516, 520(1), 522(1), 524(1), 529(1), 530, 530(1), 531(1), 532(1), 600, 601(7, 9, 10, 12), 603, 603(10, 12), 604(6), 605, 606(10, 12), 607(7, 9, 10, 15), 608, 609, 609(6), 610, 610(6, 7), 611(10, 17), 612(10)
Karran, P., 334
Karsenti, E., 65, 66(20), 68(20), 72(20)
Kassavetis, G. A., 524, 526, 526(19), 529(19, 20), 530(19, 22)
Kaufman, E. R., 642
Kaufman, P. D., 76, 99, 333, 350, 350(5), 676, 717
Kaufmann, P., 219
Keene, M. A., 465, 474
Keiler, K. C., 5
Kellenberger, E., 221
Kenny, C., 431
Kent, N. A., 36, 366, 380, 387(22)
Khabarova, M. I., 516, 517(3)
Khavari, P. A., 757
Khochbin, S., 527
Khrapko, K., 600, 601(8), 604(8), 610(8)
Kijima, M., 80
Kikuchi, A., 97
Killian, J. M., 643
Kim, J., 183, 186(50)
Kim, K. J., 726
Kim, Y. J., 757
King, C.-Y., 118
Kingston, R., 169, 172(37), 176(37), 351, 352(1), 553, 584, 741, 742, 752(3), 756, 757
Kinston, R. E., 65, 74(23)
Kirschner, M. W., 64
Kiselev, N. A., 226
Kiss, A., 431
Kladde, M. P., 380, 431, 432, 433(11), 435, 436(9, 25), 442, 442(11, 14, 25), 446(9, 25), 613, 617, 617(8)
Klar, A. J. S., 432, 460(10)
Kleckner, N., 433, 664, 671(11)

Kleff, S., 99, 676
Kleiman, L., 627
Klein, F., 663, 672(2)
Kleinberger, T., 592
Kleinschmidt, J. A., 55
Klinschmidt, A. K., 214
Klug, A., 203, 207(23), 218, 231, 620, 625(43), 628
Knapp, D., 401, 405(4), 417, 422(15)
Knippers, R., 82
Ko, L. J., 573
Kobayashi, R., 53, 97, 99, 350, 676, 717, 742, 743(7), 750(7), 757
Kobayashi, T., 448
Kochel, T., 639
Kodonaga, J. T., 757
Kokubo, T., 676, 677(15), 689(15), 690(15)
Koller, T., 203, 207(23), 215, 219, 222, 229, 231, 619, 639, 640
Komura, J., 548, 567(7), 569(7), 571, 571(7)
Konigsberg, W. H., 498
Korenberg, J. R., 651
Kornberg, R. D., 84, 99, 126, 151, 319, 321(9), 366, 371, 378, 385(9), 387(10), 480, 519, 531(12), 613, 615, 616(14), 629, 630, 637(28), 726, 752, 756(23), 757
Koshland, D., 417, 429, 430(16), 498, 499(14), 664, 671(10)
Koster, A. J., 200, 205, 205(20), 211(20), 212(20)
Koudriakova, T. B., 567, 571(19)
Kouzarides, T., 676, 677(15), 689(15), 690(15)
Kovacs, G., 655
Kow, Y. W., 585, 592(12)
Kozarich, J. W., 517
Kramer, P. R., 639, 640, 640(17), 641(17)
Kreider, I. K., 22, 31(24, 25)
Krieg, A. J., 99
Krieger, D. E., 716
Krishnan, U., 27, 32(43)
Krokan, H., 93
Krude, T., 82
Kruh, J., 163
Krumm, A., 134, 142(8), 356, 361, 362(10), 498
Kubasek, W. L., 498, 499(20), 517
Kuhlbrandt, W., 193
Kühnel, B., 586
Kuo, C.-F., 592
Kuo, M.-H., 618, 688
Kuraoka, I., 334, 335(13)
Kwon, H., 756, 757

L

Labarca, K., 632
Labhart, P., 222
Lacks, S., 433
Lacomis, L., 752, 756(23)
LaCroute, F., 40
Lade, B., 161
Laemmli, U. K., 165, 168, 180(34), 181(34), 608, 658, 660, 660(14), 661, 661(14)
Lagan, T. A., 161
Laland, S. G., 157
Landel, C. C., 377, 385(2)
Landgraf, R., 170(38), 171, 172(38)
Landsman, D., 85, 88, 100, 101(12), 102(12, 63), 132, 150, 377, 385(2), 386
Lane, D., 410, 411(17)
Laney, J. D., 498, 511(13)
Langan, T., 181
Lange, R. A., 303
Langford, C., 651
Langowski, J., 209
LaPointe, J. W., 126, 151, 319, 321(9), 378, 387(10), 519, 531(12), 615, 616(14)
Laroche, T., 50, 403, 417, 663, 664(8, 9), 672(2, 3, 9)
Larsen, P. L., 89, 415, 618
Larson, D. E., 229
Laskey, R. A., 26, 63, 64, 68, 68(15), 83, 97, 257, 341, 465, 627
Lauger, K., 231
Laurenson, P., 402
Laurent, B. C., 726, 752, 756(23), 757
Lavin, T. N., 466
Lawler-Heavner, J., 586
Lawrence, J. B., 651, 662
Lawrence, J. J., 527
Leblanc, B., 103, 118(19)
LeBlanc, J., 119, 125(48)
LeBon, J. M., 563(22), 564, 569
Lebovitz, R. M., 57, 74
Lebowitz, J., 22, 31(27)
Lee, D. Y., 321, 323(14), 524, 526(18), 545
Lee, H.-L., 585, 586(21), 593(10), 599(21), 623
Lee, J. C., 302

Lee, K.-M., 231, 235, 239(24), 246
Lee, K. S., 223, 224(35)
Lee, S., 216
Lee, Y. K., 229
Lees-Miller, S. P., 711
Lefebvre, P., 586, 594(27), 599(27)
Lehn, D., 156, 157, 161, 161(17), 171(5), 177
Leibovitch, B. A., 490
Leiden, J. M., 168, 184(33), 185(33), 186(33)
Lennox, R. W., 84
Leno, G. H., 64, 68
Leonard, M. W., 573
Leonard, W. J., 168, 184(33), 185(33), 186(33)
Leone, J. W., 85, 87(43a)
Lepault, J., 191, 192, 193(8), 211(8), 213, 218
Letnansky, K., 366
Leuba, S. H., 34, 213
Levin, J. R., 628
Levina, E., 264, 265(1), 606
Levinger, L., 541
Levy, J., 94
Levy-Favatier, F., 163
Lewis, P. N., 682
Li, B., 321
Li, Q., 313, 319, 320, 321(1, 3, 13), 322(1, 3, 13), 323(1, 3, 13), 324(13), 325(13), 326(13), 327, 327(1, 3, 13), 328(1, 3), 329(13), 332(1), 572, 581(4)
Li, S., 461
Li, Y., 752, 756(23)
Liang, C., 415
Liang, Z., 511
Libertini, L. J., 175
Lichter, P., 655, 658
Lida, J., 572
Lieberman, M. W., 347
Liew, C. C., 95
Lin, B., 229
Lin, J.-X., 168, 184(33), 185(33), 186(33)
Lin, L., 368, 374(12)
Lin, R., 85, 87(43a)
Lin, S., 161
Lin, S.-Y., 512, 513, 514(30)
Lindahl, T., 338
Lindsey, G. G., 23, 629, 732
Lindsley, D. L., 463
Ling, X., 697
Linxweiler, W., 304
Lippard, S. J., 103, 109(23, 65), 132, 334
Lis, J. T., 89, 351, 352, 356, 357(4), 361, 361(4, 9, 11), 362(4, 9), 415, 490, 497, 498, 498(6), 499(16), 511(16), 604, 618
Lisgarten, N. D., 193
Lishanskaya, A. I., 264, 276(5)
Liu, J.-K., 616
Livak, K. J., 462
Livingston, D. M., 36, 592
Livingstone, M., 461
Livingstone-Zatchej, M., 447, 449, 453, 454(12, 13), 459(12), 460(12), 461
Ljungman, M., 639
Lloberas, J., 223
Lnenicek-Allen, M., 101
Loening, U. E., 83
Logie, C., 297, 726, 731, 741(8), 743, 748(11), 752(11)
Lohka, M. J., 63
Lohman, T. M., 107, 109, 113, 113(30, 34), 121
Lohr, D., 20, 24(9), 26(12), 31(12), 366, 377, 382, 466, 731, 733
Loidl, A., 715
Loidl, P., 715
Lomonosoff, G. P., 465
London, I. M., 572, 581(4)
Longley, T., 65
Loranger, S. S., 85, 90, 90(43), 93, 676, 682(9)
Lorch, Y., 99, 126, 151, 319, 321(9), 378, 387(10), 519, 531(12), 613, 615, 616(14), 752, 756(23)
Losa, R., 318, 452(35), 461
Love, J. J., 118
Love, W., 231
Lowary, P. T., 298
Lowe, N., 651
Lowenhaupt, K., 474
Lowrey, C. H., 573
Lu, Q., 448, 463, 464, 479, 480(9), 489(15), 490, 490(15)
Lucocq, J. M., 64
Lue, N. F., 366, 378, 385(9), 387(10), 629, 637(28)
Luger, K., 3, 5, 8(7), 11, 13(7), 17(7, 13), 206, 252, 253(14), 256, 263(14), 264, 283, 631, 687
Lumelsky, N. L., 578
Lund, T., 157
Luo, K., 417, 422(13), 426(13)
Lupski, J. R., 643
Lusser, A., 715
Lutter, L. C., 131, 314, 628, 754

M

MacDonald, L. E., 433
Mader, A. W., 206, 231, 264, 283, 687
Maeder, A., 3, 252
Magnuson, N. S., 168, 169(32), 181(32), 182(32)
Maillet, L., 663, 664(9), 672(9)
Maity, S. N., 585, 586(19)
Majors, J., 396, 461, 617, 629
Makhov, A. M., 214, 216
Malone, R. E., 431, 432
Mandelkern, M., 223, 224(35)
Maniatis, T., 23, 119, 121(51), 161, 177, 185(25), 186(25), 234, 251(22), 268, 270(12), 314, 316(7), 325(7), 328(7), 330(7), 395, 413, 458, 509, 667, 668(13), 680, 756
Manley, J. L., 221
Mann, D. B., 347
Mann, M., 350, 742, 743(6), 748(6), 750(6), 751(6), 752(6), 756(6), 758, 759(5)
Mann, R. K., 50, 697
Markey, L. A., 439
Markl, J., 230
Marsaud, V., 586, 597(29), 599, 618, 622(30)
Marsden, M. P. F., 660
Marsh, P., 230
Marshall, W. F., 663
Marsolier, M. C., 461
Martinez, P., 223
Martinez-Balbas, M. A., 350, 758, 759(9)
Martinez-Campa, C., 36
Martini, E. M., 333, 350(5)
Marvin, K. W., 304, 305(20), 694
Marzluff, W. F., Jr., 632
Mascotti, D. P., 107, 109, 113(30, 34)
Massover, W. H., 230
Mastrangelo, I. A., 221
Masui, Y., 63
Masutani, C., 351
Matsumoto, H., 347
Mautner, J., 358
Mavalio, F., 100
Maxam, A. M., 121, 122(53), 322, 574, 576(22)
May, R., 42, 43(24)
McCunezierath, P. D., 99
McDonagh, K. T., 573
McDowall, A. W., 191, 192, 193(8), 211(8)
McEwen, B. F., 229
McGhee, J. D., 107, 113(27), 116(27), 117(27), 252, 616, 620, 621, 622(49)
McGinnis, W., 511, 514(27)
McKenna, N. J., 676
McMaster, G. K., 644
McMurray, C. T., 23, 31, 300, 309(10), 732, 734
McPherson, C. E., 314, 497, 556, 614, 616(9), 622(10), 623(9, 10), 625(9, 10), 626(10), 636, 638
Mechali, M., 50, 75, 82, 143, 146
Mechti, N., 585, 586(18)
Meersseman, G., 20, 26(7), 257, 300, 301, 302, 304, 305, 305(21, 22), 307(13, 22), 309, 309(21, 22), 311(21, 22)
Meier, A., 449, 454(13)
Mellon, I., 459
Mellor, J., 36, 366, 380, 387(22)
Melnik, S. M., 516, 517(3)
Meltzer, P., 651
Meluh, P. B., 414, 417, 429, 430(16), 498, 499(14)
Mendelson, E., 85, 88, 150
Merika, M., 580
Merrifield, R. B., 716
Mery, J., 272, 273(18), 275(18)
Mesnier, D., 272, 273(18), 275(18)
Metzger, W., 586, 592(24), 593(24)
Meulia, T., 361
Michaelis, C., 663
Michaelis, S., 418
Michalowski, S., 214
Mignotte, V., 573
Millar, D. S., 433
Miller, O. L., Jr., 216
Mills, A. D., 64, 68, 68(15), 83, 97
Miner, Z., 431, 432(6), 446(6)
Mirkovitch, J., 661
Mirsky, A. E., 534
Mirzabekov, A., 264, 265, 265(1), 268(10), 271, 275, 276(5, 10), 516, 520(1), 522(1, 7), 524, 524(1), 527, 529(1), 530, 530(1, 7, 27), 531(1, 7), 532(1), 600, 601, 601(7, 8, 10), 603, 603(10), 604(6, 8), 605(13), 606, 606(10), 607(9, 10), 608, 609(6), 610(6–9), 611(10, 17), 612(10)
Mitchell, A., 418
Mitchell, T., 527
Mizuguchi, G., 757, 758, 759
Mizuno, S., 165
Mizzen, C. A., 85, 90, 90(43), 93, 675, 676, 677,

677(15), 682(9), 687, 688, 689, 689(15), 690(15), 697, 698(9)
Modrich, P., 216
Moen, P. T., 662
Moens, P. T., 651
Moggs, J. G., 333, 334, 335, 335(10, 13), 337, 337(7, 10), 338(7), 341(7), 345(7), 346, 346(7), 347(7)
Moi, P., 573
Mol, C. D., 592
Molloy, P. L., 433
Momand, J., 554
Monson, E. K., 99
Moore, C. L., 215
Moore, D. D., 65, 74(23), 351, 352(1)
Moore, S. C., 20
Mori, T., 347
Moritz, P., 490
Morris, G. D., 492
Morris, N. R., 68, 97
Morse, R. H., 376, 377, 378, 381(8), 383(8)
Mosig, G., 639
Moss, T., 103, 118(19), 600, 610(4)
Moudrianakis, E. N., 24, 231, 232, 235, 516, 524(6), 529(6), 687
Moulder, S., 65
Moustacchi, E., 333, 350(5)
Mu, D., 334
Mueller, M. W., 569
Mueller, P. R., 385, 463, 480, 484(8), 548, 549, 549(1), 563(1), 577, 623, 625(58), 636
Muller, S., 84
Muller-Neuteboom, S., 193, 195(15), 201(15)
Murphy, E. C., Jr., 632
Murphy, M. R., 442
Murray, A. W., 64, 663
Murray, V., 563
Myers, F. A., 533
Mymryk, J. S., 585, 589(11), 593, 595

N

Nacheva, G. A., 271, 530, 600, 604(6), 609(6), 610(6)
Nakatani, Y., 313, 676, 677(12, 15), 684, 689(15), 690(15)
Nallaseth, F. S., 77
Namork, E., 220
Nandi, A., 575
Nash, H. A., 514
Nasmyth, K., 401, 405(4), 417, 422(15), 663
Nedospasov, S. A., 380, 466
Negri, R., 586
Nehrbass, U., 672
Neil, K. J., 229
Nelson, D. A., 95
Nelson, M., 443
Nelson, R. G., 36
Ner, S. S., 100, 101, 101(7), 121(7), 125(7)
Ness, P. J., 640
Neubauer, B., 585, 586(13), 747, 748(18)
Newmeyer, D. D., 64
Newport, J., 64, 66(5, 17), 68(5), 71(5)
Ney, P. A., 573
Ng, K. W., 61
Ngan, V., 492
Nicholas, R. H., 158
Nienhuis, A. W., 573
Nightingale, K., 100, 129(8)
Nigro, J. M., 38, 49(11)
Nishikawa, J., 676
Nissen, K. A., 449
Nissen, M. S., 155, 156, 158(6), 159, 161, 161(11), 163(6), 168(7), 169(10), 170(10, 11), 171(5, 6, 24), 172(11), 173, 173(6, 11, 24), 175(6), 177(4, 10, 11), 178(4), 179(10), 180(7), 181, 181(10), 183(7)
Nokolskya, T., 600, 601(8), 604(8), 610(8)
Noll, M., 270, 322, 628
Nordeen, S. K., 586
Norman, M. F., 466

O

O'Brien, T., 352, 357(4), 361, 361(4), 362(4), 498, 499(16), 511(16)
O'Connor, T. R., 567, 571(19)
O'Donohue, M.-F., 307, 311(29)
Oghene, K., 651, 658(6), 660(6), 661(6)
Ogryzko, V. V., 676, 677, 677(12), 696(18), 704
Ohba, R., 677, 688(16)
Ohndorf, U., 109(65), 132
Ohnishi, T., 347
Oikawa, A., 448
Okhema, P. G., 40
Okino, S. T., 638
Olins, A. L., 215, 679

Olins, D. E., 134, 175, 215, 679
Oliver, R. M., 229
Olivera, B. M., 592, 593(36)
Olson, E., 64, 66, 68(25)
Olson, T., 401, 405(6)
Olsson, T., 417, 422(18), 423(18)
O'Malley, B. W., 676
Omari, S., 452(35), 461
Onate, S. A., 676
O'Neill, L. P., 84, 85, 88(41), 89(37), 418
Orkin, S. H., 572, 573, 573(1), 574(19), 578, 578(19, 20), 579(19), 580, 580(19, 20), 581(1, 20), 582(19, 20), 583(19, 20), 616, 637
Orlando, V., 352, 403, 405, 415, 418(6), 430(6), 600, 607(2), 618
Ossipow, V., 757
Ottensmeyer, F. P., 221
Oudet, P., 191, 257, 307
Owa, T., 715
Owen-Hughes, T., 297, 300, 676, 677, 677(15), 688(16), 689(15), 690(15), 726, 744
Ozato, K., 636

P

Pagel, J., 100, 102(11)
Paigen, K., 632
Palladino, F., 663, 664(8), 672(3)
Palmer, E. L., 307
Palmiter, R. D., 573
Pan, C. Q., 170(38), 171, 172(38)
Paneerselram, C., 695
Panyim, S., 84, 165
Paonessa, G., 109(64), 132
Papatsenko, D., 600, 601(12), 603(12), 605, 606(12), 607(15), 609
Papavassiliou, A. G., 252
Papayannopoulou, T., 572, 573, 621
Paranjape, S. M., 134, 142(8), 378
Paras, P., Jr., 490
Pardoll, D. M., 662
Pardon, J. F., 732
Pardue, M. L., 466, 639
Parello, J., 272, 273(18), 275(18)
Parish, R. W., 640
Paro, R., 352, 403, 405, 415, 418(6), 430(6), 495, 497, 499(9), 600, 607(2), 618
Parra, I., 661
Parsell, D. A., 5
Parthun, M. R., 99, 676, 717
Partolina, M., 600, 601(10), 603, 603(10), 606(10), 607(10), 611(10), 612(10)
Pashev, I. G., 517, 611, 618
Pastuzak, J. J., 253
Patel, P. I., 643
Paton, A. E., 134, 175
Patterton, H.-G., 377, 385(2), 386
Paul, C. L., 433
Paule, M. R., 21
Paull, T. T., 103
Paulson, J. R., 658, 660(14), 661(14)
Pavlovic, B., 368, 374(14)
Payet, D., 103, 109(25)
Pazin, M. J., 53, 350, 742, 743(7), 750(7), 752(4), 757
Pearlman, R. E., 585, 586(16)
Pederson, D. S., 35, 36, 38(7), 613
Pehrson, J. R., 449
Pemov, A., 600, 603(11), 604(11), 610(11)
Pendergrast, P. S., 233, 242(21), 244(21)
Pennie, W. D., 632, 637(35)
Pennings, S., 20, 26(7), 257, 298, 300, 301, 302, 304, 305, 305(21, 22), 307, 307(13, 22), 309, 309(21, 22), 311(21, 22)
Pentao, L., 643
Peretz, M., 359
Perez-Riba, M., 498
Perlmann, T., 319, 320, 321(11), 322(11), 323, 323(11), 325(24), 327, 327(11)
Perry, C. A., 80, 85, 87, 88, 88(42), 89, 90, 94, 95(54), 97, 97(54), 98, 695
Perry, P., 651, 652(3), 655(3), 658(3)
Persinger, J., 232, 516, 524(6), 529(6)
Peterson, C. L., 297, 377, 385(2), 386, 626, 726, 727, 729(1), 731, 731(1), 741, 741(8), 742, 743, 744, 748(11), 752(4, 11), 757
Peterson, M. L., 65
Peterson, R. C., 526
Petri, V., 112
Pettijohn, D. E., 609, 610, 611(19–21), 639, 640, 640(3), 641(3), 642(20)
Pfeifer, G. P., 448, 449(8), 461, 497, 548, 549, 549(2), 554, 554(2, 11), 556(2, 12), 557, 567, 569, 571(19), 615, 633, 635(38), 636
Pflugfelder, G., 498
Philipsen, S., 573
Philpott, A., 68
Pickett, S. C., 325

Pictet, R., 614, 623(12), 625(12), 637
Piechaczyk, M., 585, 586(18)
Pikaart, M., 621, 622(52)
Pil, P., 103, 109(23)
Pile, L. A., 462
Pillus, L., 663, 664(8), 672(3)
Pina, B., 323
Polach, K. J., 278, 280, 283, 284, 285(1, 4), 288(4), 291(1), 295(1), 442, 586, 726, 731(5)
Pollard, K. J., 741
Pollard, T. D., 220
Pollock, R., 101, 121(16)
Popp, S., 655
Portmann, R., 219
Postnikov, Y. V., 133, 137, 137(2), 141(2), 142(2), 150, 151(31), 152(31), 153(31), 600, 601(8), 604(8), 610(8)
Powell, R. D., 229
Pratt, K., 431, 433
Prendergrast, S., 252, 253(11)
Preobrazhenskaya, O., 264, 265, 268(10), 271, 276(10), 516, 520(1), 522(1), 524(1), 529(1), 530, 530(1), 531(1), 532(1), 600, 601(7, 9), 603, 604(6), 605, 607(7, 9, 15), 608, 609(6), 610(6, 7), 611(17)
Preston, N. S., 101
Pretorius, I.-M., 572
Price, M. A., 466
Priporova, I., 600, 601(12), 603(12), 606(12)
Proffitt, J. H., 431
Protacio, R. U., 284, 586
Prunell, A., 191, 307, 311(29), 630
Pruss, D., 232, 264, 516, 517(2), 520(2), 522(2), 523, 524, 524(2, 4, 6), 526(4, 18), 527(4, 5), 528(5), 529(6), 530(2), 531(2), 545, 600, 604(5), 675
Pruss, D. V., 527, 530(27)
Pruzina, S., 573
Prydz, H., 93
Ptashne, M., 513, 514(31)
Pullner, A., 358
Purton, T., 362

Q

Querol, E., 223
Quinn, J., 726, 729(1), 731(1), 744, 757
Quivy, J.-P., 65, 334, 337(7), 338(7), 341(7), 345(7), 346, 346(7), 347(7), 623, 625(59), 636

R

Rabbits, P., 651
Radding, C. M., 593
Rademakers, S., 335
Radic, M. Z., 181
Ragnhildstveit, E., 65
Raisch, N., 573
Raju, N. L., 109(65), 132
Ramain, P., 276, 277(20), 531
Ramakhrishnan, C., 307
Ramakrishnan, V., 191, 232
Ranalli, T. A., 688
Rapp, A. K., 651
Rau, D. C., 232
Raymond, J., 438
Razin, A., 435
Razvi, F., 621
Read, C. M., 101, 239
Reardon, J. T., 334
Rechsteiner, T., 3, 5, 8(7), 13(7), 17(7), 256
Record, M. T., 113
Reed, S. H., 461
Reed, S. I., 38, 49(11)
Rees, C., 232
Reeves, K., 156, 171(5), 177(4), 178(4)
Reeves, M., 133
Reeves, R., 100, 101(5), 129(8), 155, 156, 156(1), 157, 157(1), 158, 158(1, 6), 159, 161, 161(1, 11, 17), 163(6, 15), 164(19), 165(15), 166, 168, 168(7), 169(10, 32), 170(1, 7, 8), 171(6, 24), 172(11), 173, 173(6, 8, 11), 175(6), 176(8), 177, 177(10, 11), 179(10), 180, 180(1, 7, 8), 181, 181(10, 32, 46), 183, 183(7), 184(31, 33–35), 185, 185(31, 33, 35), 186(31, 33, 35, 50), 187(35)
Reeves, W. H., 706
Reichard, P., 92
Reid, T., 658
Renbaum, P., 435
Retief, J. D., 23, 629, 732
Reuter, G., 492
Revzin, A., 107, 514
Rhodes, D., 26, 239, 257, 627, 628
Ricci, A. R., 584

Richard-Foy, H., 466, 585, 586, 597(20, 29), 599, 599(20), 618, 620, 622(30), 626, 633(7)
Richards, B. M., 732
Richman, R., 680
Richmond, R. K., 3, 206, 231, 252, 264, 283, 687
Richmond, T. J., 3, 5, 8(7), 11, 13(7, 12), 17, 17(7, 13), 206, 231, 251, 252, 253(14, 22), 256, 257, 260(22), 263(14, 22), 283, 627, 631, 631(15), 638(15), 687
Rickers, A., 133, 137(2), 141(2), 142(2)
Rickett, H., 22, 732
Rickwood, D., 44
Ridgway, P., 61
Ridsdale, J. A., 81
Ridsdale, R. A., 229
Riegel, A. T., 585
Riffe, A., 163
Rigaud, G., 614, 623(12), 625(12), 637
Riggs, A. D., 461, 495, 497, 512, 513, 514(30), 548, 549, 549(2), 554(2, 11), 556(2), 557, 563(22), 564, 567, 567(7), 569, 569(7), 571, 571(7, 19), 615, 633, 635(38), 636
Rimsky, S., 134, 143(6)
Rine, J., 402
Ristow, H., 639
Roberge, M., 668
Robert, C., 213
Robert-Nicoud, M., 655
Roberts, B., 585, 586(14), 599(14)
Roberts, J. M., 82, 94
Roberts, M. S., 252, 307, 385, 599, 622, 625(54), 626, 630(6), 631(6), 632(5), 637(5, 6), 638(5)
Roberts, S., 362, 677, 688(16)
Robinett, C. C., 663
Robins, P., 338
Robinson, H., 118
Rocchini, C., 20
Roche, D. M. J., 334, 337(7), 338(7), 341(7), 345(7), 346(7), 347(7)
Roche, D. R., 333
Rodrigues, J. D., 629, 732
Rodrigues, J. de A., 23
Roeder, R. G., 57, 74, 100, 353, 696
Romeo, P.-H., 573
Roos, N., 211
Rose, A. B., 40, 403, 405(13), 417, 418(10), 430(10), 535
Rose, M. D., 406, 436, 664, 665(12), 667(12)
Rosenberg, A. H., 5, 119, 161, 162(21)
Roth, S. Y., 36, 38(2), 42, 49(2), 50, 232, 365, 377, 380, 385(21), 387(21), 396(21), 432, 441(12, 13), 442, 586, 596(28), 675, 676(1), 688
Rothblum, L. I., 229
Rothfield, N., 661
Rothman, P., 183, 186(50)
Rottem, S., 435
Rougvie, A. E., 352, 356, 361(9, 11), 362(9), 415, 498
Rousseau, D., 527
Rouviere-Yaniv, J., 101, 118
Roux, J., 614, 623(12), 625(12), 637
Ruben, G., 592
Ruberti, I., 51
Rubin, G. M., 463, 464
Ruddle, N. H., 166, 168(31), 184(31), 185(31), 186(31)
Rudolph, H., 370(16), 371, 375(16), 376(16)
Ruiz, T., 213
Rundlett, S. E., 547, 698
Rupp, W. D., 645, 646(26)
Ruppert, S., 698
Russanova, V., 611, 676, 677(12)
Russel, L. D., 192, 193(11)
Russell, M., 438
Ruteschouser, E. C., 585, 586(19)
Rutherford, B., 229
Rutherford, T. R., 575, 578(26)
Ryan, C. A., 76
Ryan, M. P., 376
Ryan, T. M., 573
Rydén, L., 253
Rykowski, M. C., 651

S

Saber, H., 22, 31(28)
Sacone, S., 651
Safer, D., 229
Saitoh, S., 417, 423(17)
Saitoh, Y., 168, 180(34), 181(34)
Salmon, E. D., 663, 671(4)
Saluz, H. P., 585
Sambrook, J., 23, 119, 121(51), 234, 251(22), 268, 270(12), 314, 316(7), 325(7), 328(7),

330(7), 395, 413, 458, 509, 667, 668(13), 680, 756
Samori, B., 213
Sancar, A., 333, 334, 448
Sancar, G. B., 448
Sandaltzopoulos, R., 65, 66(20), 68(20), 72(20), 335, 338(16), 343(16), 344(16), 753, 756, 757
Sanders, M. A., 99
Santarius, U., 229
Sargent, D. F., 3, 206, 231, 252, 264, 283, 687
Saroff, H. A., 102(54), 123
Sato, W., 97
Saucedo-Cardenas, O., 643
Sauer, F., 464
Sauer, R. T., 5
Sautiere, P., 271, 275(16)
Sayre, M. H., 726, 757
Scatchard, G., 113, 117(38), 328, 329(29)
Schaich, K. M., 259
Scheer, U., 150, 151(31), 152(31), 153(31)
Scheidereit, C., 328
Schekman, R., 42
Schelbert, A., 466, 622, 625(55), 638
Schepartz, A., 253
Scherl, D. S., 291
Scherthan, H., 663, 664(9), 672(9)
Schibler, U., 74, 314
Schieferstein, U., 449, 461, 461(16)
Schild, C., 143, 146(24)
Schildkraut, C. L., 575
Schimmel, P. R., 104(40), 105(40), 113, 116(40), 117(39)
Schlagman, S. L., 431, 432(6), 446(6)
Schlehuber, C., 223
Schlitz, R. L., 676, 677(12)
Schlossherr, J., 495
Schmid, A., 367, 368, 371(11), 374(12), 621
Schmidt, M. C., 757
Schmidt, W. M., 569
Schneider, C. A., 76
Schnitzler, G. R., 741
Schomberg, C., 185
Schrader, T. E., 297, 298(18)
Schreiber, S. L., 697, 715
Schreier, A. A., 113, 117(39)
Schrenk, M., 191
Schrock, E., 655, 658
Schröter, H., 303, 304(18)
Schulten, K., 103
Schultz, P., 191, 192, 193(8), 211(8), 307
Schulze, E., 109, 112(35)
Schutz, G., 574
Schwager, S., 23, 629, 732
Schwarz, P. M., 20, 22(13, 16), 26(13), 33(13, 16), 737, 738(19)
Schweizer, D., 657, 663, 672(2)
Scott, J. F., 37, 38(8)
Scott, J. H., 42
Scott, M. P., 727
Scriven, L. E., 199
Seale, R. L., 81, 92, 94, 94(48)
Sealy, L., 534, 627
Searles, M. A., 11, 13(12)
Searles, S., 101
Sedat, J. W., 205, 659, 663
Seed, B., 140
Segel, I. H., 173
Seger, D., 308
Seidman, J. G., 65, 74(23), 169, 172(37), 176(37), 351, 352(1)
Seiter, A., 55
Selleck, S. B., 617
Senear, D. F., 107, 637
Sentenac, A., 461, 668
Sera, T., 20, 32(22)
Serfling, E., 185
Serwer, P., 26, 27, 32(41, 43)
Sewack, G. F., 638
Sewell, B. T., 23, 223, 629, 732
Shaffer, C., 129, 130, 462, 745
Shaffer, S. C. R., 490
Shalitin, C., 36
Shalloway, D., 592
Shannon, M. F., 168, 184(35), 185(35), 186(35), 187(35)
Sharpe, J. A., 572, 573
Shaw, O. J., 27
Shaw, S. L., 663, 671(4)
Shea, M. A., 637
Sheehan, J. C., 255
Sheehan, M. A., 64
Sheen, J. Y., 140
Sheer, U., 192
Sheflin, L. G., 103
Shell, B. K., 335
Shelton, E. R., 93
Shen, C. J., 619
Shen, C.-K. J., 466
Sherman, F., 454

Sherman, M. B., 226
Shick, V., 264, 265, 265(1), 268(10), 276(5, 10), 600, 601(8), 604(8), 606, 608, 610(8), 611(17)
Shields, S. L., 443
Shim, E. Y., 314, 556, 622(10), 623(10), 625(10), 626(10), 638
Shimamura, A., 50, 53(8), 57(8), 721
Shimizu, M., 42, 380, 385(21), 387(21), 396(21), 432, 441(13), 442, 586, 596(28)
Shimkus, M., 94
Shivji, M. K. K., 334, 335(13), 337
Shopland, L. S., 497
Shrader, T. E., 305, 306(27), 320, 323(12)
Siebenlist, U., 616
Siede, W., 333, 448
Siegel, B., 592
Sigman, D., 170(38), 171, 172(38)
Siino, J. S., 173
Sijbers, A. M., 335
Sikorski, R., 38, 49(11)
Silber, K. R., 5
Simon, M., 214, 215(2), 221, 229
Simon, R. H., 3, 23, 57, 239, 687, 747
Simpson, R. T., 11, 13(12), 20, 22(5), 35, 36, 38, 38(2, 7), 42, 49(2, 11), 208, 290, 305, 377, 378, 380, 385(3, 21), 386, 387(21), 389, 396(21), 431, 432, 433(11), 435, 436(9, 25), 441(12, 13), 442, 442(11, 14, 25), 446(9, 25), 586, 596(28), 613, 617, 617(8), 628, 731, 750
Sinden, R. R., 609, 610, 611(19–21), 619, 639, 640, 640(3, 13, 17), 641(1, 3, 13, 17), 642(1, 20), 644(13), 645(13)
Singer-Sam, J., 549, 563(22), 569
Singh, J., 432, 460(10)
Skoultchi, A. I., 575
Slaugenhaupt, S., 643
Sleeman, A. M., 64
Small, E. W., 175
Smerdon, M. J., 333, 347, 351, 448, 449, 451
Smith, C. A., 639
Smith, C. L., 626, 632
Smith, H. A., 652
Smith, J. A., 65, 74(23), 169, 172(37), 176(37), 351, 352(1)
Smith, L., 567, 571(19)
Smith, M. M., 232
Smith, P., 366
Smith, R. C., 146
Smith, R. D., 81
Smith, S., 97
Smythe, C., 64, 66(17)
Sobel, R. E., 547, 688, 695
Sogo, J. M., 76, 77, 215, 219, 222, 619, 636, 639, 640
Solomon, M. J., 89, 401, 402(8), 412(8), 415, 418(4), 497, 618
Solomon, W., 572, 581(4)
Sommerville, J., 192
Sorrentino, B. P., 573
Soults, J. A., 89, 415, 417, 417(5), 418(5)
Southern, E. M., 634
Spaulding, S. W., 103
Spear, B. T., 65
Speicher, M. R., 655, 658
Spencer, T. E., 676
Spielman, P., 592
Spirin, A. S., 46
Spivak, G., 459
Spradling, A. C., 463
Springer, D. L., 347
Squire, J., 661
Srebreva, L., 264, 516, 522(7), 530(7), 531(7), 601, 605(13)
Staehelin, T., 84
Stafford, D. W., 290, 305, 628
Stafford, G. A., 376
Stalder, R., 668
Stamatoyannopoulos, G., 572, 573
Stanfield-Oakley, S. A., 466
Stanik, V. H., 20, 23(6), 629, 731, 734(10), 739(10)
Stark, G. R., 634, 635(43)
Stasiak, A., 203, 209, 210(22)
Stefanovsky, V., 611, 618
Steigerwald, S. D., 549, 554(11), 636
Stein, A., 24, 627
Steitz, T. A., 100
Stern, M., 634, 635(43)
Sterner, D. E., 545, 547(16)
Sternglanz, R., 99, 676
Sthoeger, Z. M., 706
Stillman, B., 82, 90(19), 97, 99, 333, 350, 350(5), 415, 676, 717
Stockley, P. G., 239
Stokrova, J., 115, 118(42)
Stollar, B. D., 535
Strahl-Bolsinger, S., 50, 401, 403, 403(2), 405(3), 417, 422(13), 426(13), 497, 618

Straight, A. F., 663
Straka, C., 490
Strauss, E. C., 572, 573, 574(19), 578, 578(19, 20), 579(19), 580(19, 20), 581(20), 582(19, 20), 583(19, 20), 616, 637
Strobl, L. J., 362
Strokov, A. A., 264, 276(5)
Stros, M., 115, 118(42)
Struhl, K., 65, 74(23), 169, 172(37), 176(37), 351, 352(1), 715
Strutt, H., 415, 418(6), 430(6), 497, 499(9), 600, 607(2), 618
Stryer, L., 480
Stubbe, J., 517
Studier, F. W., 5, 119, 161, 162(21), 299
Studitsky, V. M., 213, 307
Suda, M., 223, 224(36)
Sugasawa, K., 97, 351
Sun, Y. L., 633
Suter, B., 447, 449, 453, 454(12, 13), 459(12), 460(12)
Svaren, J., 377, 626
Swede, M. J., 492
Swiderski, P. M., 557
Swiger, R. R., 652
Swinton, D., 431
Sylvester, S. R., 168, 169(32), 181(32), 182(32)
Symkowski, D. E., 334
Szent-Gyorgyi, C., 42, 380, 385(21), 387(21), 396(21), 432, 441(13), 586, 596(28)

T

Tainer, J. A., 592
Takahashi, K., 417, 423(17)
Takahashi, M., 118
Takakusu, A., 347
Takao, M., 448
Take, Y., 714
Takegawa, S., 572
Talbot, D., 573
Talmon, Y., 199
Tamkun, J., 350, 742, 752(4), 758
Tan, S., 252, 253(14), 263(14), 631
Tanaka, S., 461
Tanaka, T., 401, 405(4), 417, 422(15)
Tanguay, R. L., 549, 554(11)
Tasseron-de Jong, J. G., 461
Tatchell, K., 356, 527
Taunton, J., 697
Taxman, D. J., 463, 480(9)
Taylor, I. C. A., 323
Taylor, J., 216
Taylor, L., 229
Tazi, J., 545, 547(17)
Tease, C., 651
Tempst, P., 752, 756(23)
Teng, Y., 461
Teplitz, R., 652
Terpening, C., 21
Thanos, D., 161, 177, 185(25), 186(25)
Thayer, M. M., 592
Thodes, D., 620, 625(43)
Thoma, F., 11, 20, 22(5), 35, 38, 49, 49(11), 203, 207(23), 208, 231, 251, 252, 351, 377, 378, 380(12), 382(12), 385(4), 447, 448, 449, 449(10), 451, 451(10), 452(35), 453, 454(12, 13), 459(12), 460(12), 461, 461(16), 613, 619, 731, 750
Thomas, C. B., 168, 184(33), 185(33), 186(33)
Thomas, G., 366, 368(8), 463
Thomas, G. H., 463, 465(6), 470(6), 480, 481(37), 490(6)
Thomas, J. O., 3, 11(2), 84, 115, 118(42), 183, 232, 239, 240
Thomas, K. R., 592, 593(36)
Thompson, J. S., 402, 697
Thompson, R. J., 639
Thorlacius, A. E., 168
Thorne, A. W., 85, 88(40), 533, 534, 535, 536(8), 537(5, 8), 542, 543(10), 544(11), 545, 545(11), 618, 697
Thresher, R., 215, 216
Thummel, C. S., 497
Tijsterman, M., 461
Timasheff, S. N., 302
Timmons, S. E., 639
Tinker, R. L., 526, 530(22)
Tjian, R., 464, 698
Todd, R. D., 126, 299, 303
Todo, T., 347
Toh-e, A., 40
Tomkinson, A. E., 449
Tommasi, S., 563(22), 569
Tonegawa, S., 573
Tong, J. K., 715
Törmänen, V. T., 557
Tornaletti, S., 448, 449(8), 461, 549, 556(12), 636

Tourkine, N., 585, 586(18)
Tournebize, R., 65, 66(20), 68(20), 72(20)
Towbin, H., 84
Townes, T. M., 573
Trask, B. J., 643
Travers, A. A., 100, 101, 101(7), 103, 109, 109(25), 121(7), 125(7)
Treich, I., 726
Treisman, R., 101, 121(16)
Tremethick, D., 50, 53, 53(8), 54(12), 55(12), 56(12), 57(8), 58(12), 59(12), 61, 61(6, 12), 62(12), 134, 143(9)
Trent, J., 651
Trieschmann, L., 133, 134, 137(2), 141(2), 142(2, 5, 10), 143, 143(5, 10), 144(21), 145(21), 147(10, 21), 148(10), 149(10)
Trinick, J., 200
Truss, M., 466, 622, 625(55), 638
Tsai, S.-F., 578
Tsai, S. Y., 676
Tsanev, R., 264, 516, 522(7), 530(7), 531(7), 601, 605(13)
Tse, C., 19, 20, 23(18), 30(18), 32(18, 22)
Tsui, L.-C., 661
Tsukagoshi, Y., 445
Tsukiyama, T., 50, 76, 335, 350, 464, 742, 745, 751(5), 752, 752(5), 757, 758, 759(9)
Tsuprun, V. C., 226
Tu, C. P., 592
Tuan, D., 572, 581(4)
Tucker, J. D., 652
Tullius, T. D., 137, 139, 139(15), 239, 251, 258, 295, 466, 628
Turner, B. M., 84, 85, 88(41), 89(37), 401, 405(6), 417, 418, 422(18), 423(18), 547
Turner, C. H., 535, 536(8), 537(8)
Tyler, J. K., 97
Tyree, C. M., 759

U

Ullman, K. S., 66, 72(26)
Ulyanov, A. V., 307
Undritsov, I. M., 264, 276(5)
Unwin, N., 192, 213(7)
Ura, K., 307, 526
Urness, L. D., 497
Usachenko, S. I., 264, 267, 267(6), 275, 276(5, 6), 524
Ussery, D. W., 639, 641(1), 642(1)
Utley, R. T., 297, 300, 726

V

van de Putte, P., 461
Van Etten, J. L., 443
van Holde, H. E., 99, 100(2)
van Holde, K. E., 19, 20, 21, 22, 23, 23(6), 24(9), 28(26), 31, 31(26), 76, 136, 139(13), 147, 207, 213, 231, 232, 252, 299, 300, 309(10), 356, 466, 527, 584, 619, 629, 731, 732, 734, 734(10), 739(10)
Van Lint, C., 490
Van Regenmortel, M. H., 84
Varga-Weisz, P. D., 53, 350, 742, 743, 743(6), 748(6), 750, 750(6), 751(6), 752(6), 756(6), 758, 759(5)
Varshavsky, A., 89, 126, 401, 402(8), 412(8), 415, 418(4), 442, 497, 541, 618
Vassilev, L. T., 77
Vavra, K. J., 678
Venetianer, P., 431
Venkatesan, M., 35
Verdier, J. M., 668
Verdin, E., 490
Verdine, G. L., 103, 107(21), 109(21), 110(21)
Vermaak, D., 715
Verreault, A., 676, 717
Vestner, B., 134, 143(11)
Vettese-Dadey, M., 545, 697
Vidali, G., 534, 716
Villeponteau, B., 621, 622(52)
Vishlizky, A., 36
Vogelstein, B., 38, 49(11), 634, 662
Volker, S., 229
Volkert, F. C., 37, 39(9), 40(9)
Vollenweider, H. J., 215
Vologodskii, A. V., 210
von Hippel, P. H., 8(15), 17, 107, 113(27), 115(28), 116(27), 117(27), 498, 499(20), 514, 517
von Holt, C., 3, 23, 223, 629, 732
Vorachek, W. R., 498, 499(20), 517
Vu, Q. A., 95

W

Wade, P., 50, 232, 545, 547(15), 626, 675, 715
Wagner, C. R., 100, 101(6), 125(6)
Wahl, G. M., 634, 635(43)
Wakimoto, B. T., 490
Walker, G. C., 333, 448
Walker, J. F., 223
Walker, K. P. III, 19
Wall, J. S., 221, 229
Wall, L., 573
Wallrath, L. L., 448, 462, 464, 479, 489(15, 16), 490, 490(15, 16), 492, 492(16)
Walter, C. A., 192
Walter, J., 498, 499(11, 12, 23), 511(11, 12, 23), 512(11, 12), 514(12), 515(12), 600, 603(3)
Walz, J., 200, 205(20), 211(20), 212(20)
Wampler, S. L., 759
Wan, J., 50
Wang, A. C., 514
Wang, B. W., 231
Wang, J. C., 498, 617
Wang, L., 676, 677(15), 689(15), 690(15), 697
Wang, M. X., 433
Wang, S. L., 65
Wang, Z., 351, 617
Ward, D. C., 655, 661
Ward, M., 535
Warner, J. R., 360
Wassarman, D. A., 464
Wassarman, P. M., 76, 77, 93
Wataya, Y., 433
Waterman, M. L., 74
Waters, R., 461
Watson, C. E., 584
Watt, F., 433
Weatherall, J. D., 575, 578(26)
Weber, J. A., 462, 463, 480(9)
Wei, Y.-F., 335
Weidner, M. F., 628
Weigmann, N., 143, 144(21), 145(21), 147(21)
Weiler, K. S., 490
Weiner, B. M., 664, 671(11)
Weinstein, D. M., 401, 405(5), 417, 422(14), 426(14), 430(14)
Weintraub, H., 359, 534, 615, 616(15)
Weischet, W. O., 22, 28(26), 31(26), 356
Weiss, K., 386, 389
Weiss, M. A., 118
Weist, D. K., 361
Weisz, P. V., 743
Welch, J. W., 39
Wellinger, R.-E., 351, 461, 636
Wellner, D., 695
Wernicke, W., 230
Wess, T. J., 309
West, B. L., 466
Westwood, J. T., 497
Whitaker, N., 210, 232
White, J. G., 64, 68(15)
Whitehead, J. P., 109(65), 132
Whiteley, M., 545
Whitlock, J. P., Jr., 627, 638
Whyatt, D., 573
Wickner, R. B., 39, 40
Widmer, R. M., 619, 640
Widom, J., 19, 99, 278, 280, 283, 284, 285(1, 4), 287(2), 288(4), 291, 291(1), 295(1), 298, 442, 586, 676, 717, 726, 731(5)
Wiegant, J., 651
Wilcock, D., 64, 68(15)
Wilkins, R. C., 498, 499(16), 511(16)
Wilkinson, S. E., 134
Wilkinson-Singley, E., 175
Williams, K. R., 498
Williams, R. C., 221
Wilm, M., 53, 350, 742, 743(6), 748(6), 750(6), 751(6), 752(6), 756(6), 758, 759(5)
Wilson, G., 435
Wimmer, C. V., 672
Windle, B., 661
Winston, F., 387, 406, 436, 664, 665(12), 667(12), 677, 688(16)
Wintersberger, U., 366
Wiseman, J. M., 303
Wisniewski, J., 109, 112(35), 757, 758
Wittig, B., 85
Wold, B., 385, 463, 480, 484, 484(8), 548, 549, 549(1), 563(1), 577, 623, 625(58), 636
Wolfe, S. A., 103, 107(21), 109(21), 110(21)
Wolffe, A. P., 19, 20, 32(22), 50, 63, 75, 76, 82, 99, 100, 100(3), 129(8), 134, 137, 139, 139(15), 142(5, 10), 143, 143(5, 10), 146, 146(23, 24), 147(10), 148(10), 149(10), 159, 170(8), 172, 172(8), 173(8), 175(8), 176(8, 39), 179(9), 180(8, 9), 231, 232, 232(2), 239, 240(27), 251, 264, 295, 299, 307, 308, 313, 334, 365, 377, 448, 516, 523, 524, 524(4, 6), 526, 526(4, 18), 527(4, 5),

528, 528(5), 529(6), 545, 547(15), 584, 613, 626, 628, 675, 677, 696(18), 715, 718, 721
Wolford, R. G., 378, 490, 585, 586, 586(15), 594(27), 599(15, 27)
Wong, J., 523
Wong, M. L., 81
Wood, R. D., 333, 334, 335, 335(10, 13), 337, 337(7, 10), 338, 338(7), 341(7), 345(7), 346(7), 448
Wood, W. G., 572, 573
Wood, W. I., 621, 622(49)
Woodcock, C. L., 20, 84, 191, 192, 195, 200, 201(6), 203(6), 205(6, 20), 206(6), 207, 207(35), 208, 210, 210(6), 211(6, 20), 212(20), 213, 219, 227, 229, 229(21, 50), 232
Woodcock, H., 227, 229(50)
Worcel, A., 50, 51, 53(8), 56, 57(8), 61(6), 81, 621, 721
Workman, J. L., 297, 300, 321, 545, 547(16), 676, 677, 677(15), 688(16), 689(15), 690(15), 696, 706, 726, 729(1), 731(1), 741, 744
Wrange, Ö., 313, 319, 320, 321(1, 3, 11, 13), 322(1, 3, 11, 13), 323, 323(1, 3, 11, 13), 324(13), 325(13, 24), 326(13), 327, 327(1, 3, 11, 13), 328(1, 3), 329(13), 332(1)
Wray, V. P., 142, 754
Wray, W., 142, 218, 754
Wright, E. B., 679
Wu, C., 50, 76, 335, 338(14), 350, 462, 464, 466(3), 497, 585, 586(8), 620, 623, 742, 745, 751(5), 752, 752(5), 757, 758, 759, 759(9)
Wu, H.-M., 224, 305, 306(26)
Wu, J., 77
Wu, L., 387
Wu, R., 592
Wu, X., 351
Wuller, J. M., 129, 130, 462, 745

X

Xu, M., 435, 436(25), 442, 442(25), 446(25), 613, 617(8)
Xu, Y. Z., 633

Y

Yager, T. D., 23, 732
Yajima, H., 448
Yamada, M., 334
Yamada, M.-A., 97
Yamamoto, K. R., 323, 585, 597(22), 622
Yamamoto, M., 573
Yanagida, M., 417, 423(17)
Yang, G., 213
Yang, J., 380
Yang, S.-W., 514
Yang, T. P., 548
Yang, X.-J., 676, 677, 677(15), 684, 689(15), 690(15), 696(18), 697
Yankulov, K., 362
Yao, J., 298
Yarema, K. J., 334, 335(10), 337(10)
Yasuhira, S., 448
Yasui, A., 448
Yau, P., 304, 305(20), 652, 694
Yeh, E., 663, 671(4)
Yenidunya, A., 630
Ymer, S., 554
Yokota, S., 97
York, M. H., 307
Yoshida, M., 80, 165
Yu, L., 376
Yu, X., 232
Yurina, N. P., 516, 517(3)

Z

Zahn, R. K., 214
Zakian, V. A.~~, 37, 38(8)~~, 99
Zalenskaya, I. A., 264, 276(5)
Zamble, D. B., 334
Zappavigna, V., 100
Zaret, K. S., 314, 323, 497, 548, 556, 585, 597(22), 612, 614, 616, 616(9), 622, 622(10), 623(9, 10), 624, 625(9, 10), 626(10), 636, 638
Zatchej, M., 49
Zavari, M., 368, 374(12)
Zeitlin, H. C., 575
Zentgraf, H., 191
Zhang, D., 95
Zhang, H., 651

Zhang, L., 614, 617(11)
Zhang, M., 752, 756(23)
Zhang, Y., 443
Zheng, G., 639
Zheng, M., 624
Zhou, J., 618, 676
Ziff, E. B., 359
Zimm, G. G., 463
Zimmerman, J. L., 474
Zlatanova, J., 213, 232, 264, 516, 522(7), 530(7), 531(7), 601, 605(13)
Zolan, M. E., 639
Zollinger, M., 221
Zucker, K., 50, 56, 61(6)
Zweidler, A., 84
Zweidler, F., 678

Subject Index

A

Acetylation mapping, *see* Histone
AFP, *see* α-Fetoprotein
Alcian blue, transmission electron microscopy support treatment, 221–222
Ammonium molybdate, chromatin staining for transmission electron microscopy, 229
Amylamine, transmission electron microscopy support treatment, 221
Aurothioglucose, chromatin staining for transmission electron microscopy, 229

B

Bleomycin–iron, cross-linking of chromatin complexes
  advantages, 517
  cross-linking reaction conditions
    Ehrlich acites tumor cells, *in vivo*, 520, 522
    isolated chromatin, 519–520
  radical production, 516–517
  two-dimensional gel electrophoresis analysis, 520

C

CAF-1, *see* Chromatin assembly factor
CHRAC, *see* Chromatin accessibility complex
Chromatin, *see also* Histone; Minichromosome; Nucleosome; Nucleosome core particle; Solid-phase nucleus
  assembly in *Xenopus* extracts
    advantages of system, 50–51
    anion-exchange chromatography of chromatin assembly components
      chromatin assembly conditions, 58–60
      chromatography conditions, 55–56
    ATP requirement, 53, 58
    extract preparation, 51–52
    histones H2A/H2B, purification, 57–58
    micrococcal nuclease analysis, 62–63
    N1,N2–(H3,H4) complex
      isolation, 60–61
      nucleosome assembly, 61–62
    nucleosome array formation
      DNA concentration optimization, 55
      DNA supercoiling assay, 53, 62–63
    time required for assembly, 53
    topoisomerase treatment of DNA, 53
  cross-linking, *see* Cross-linking, chromatin complexes
  electron microscopy, *see* Electron microscopy
  immunoprecipitation
    antibodies, 84, 88
    fixed chromatin
      formaldehyde fixation, 89–91
      incubation conditions, 91–92
    unfixed chromatin
      incubation conditions, 87
      protein A–agarose bead preparation, 85, 87
      solution preparation, 85
  linker histone H5 incorporation into model nucleosome arrays, 34
  mapping nonhistone chromatin-associated factors
    applications, 402–403, 405, 415, 417–418, 429–430
    cell growth and lysis, 402, 406–407
    chromatin shearing, 408, 422–423
    confirmation of positive results, 427–429
    cross-linking with formaldehyde, 401–402, 407, 415, 418
    DNA purification, 412, 424–426
    extract collection, 407–408
    gel electrophoresis analysis of amplification products, 414, 426–427
    immunoprecipitation, 402, 410–412, 423–424

materials and solutions, 406, 410, 412–413
overview, 401–402, 415
Polycomb mapping, 405, 417
polymerase chain reaction, 402, 405, 412–414, 426–427, 430
silent information regulator 3 mapping, 402–403
spheroplast lysis, 419–422
mapping with restriction endonucleases, 383, 385, 392
native chromatin
definition, 613
high-quality substrate preparation for structural studies, 613–616
remodeling during nucleotide excision repair
DNA substrate preparation, 334–335
Drosophila embryo extract preparation, 335, 338
efficiency of repair, assay, 337–338
extracts for study, 333–334, 338, 350
micrococcal nuclease assay of nucleosome assembly
assembly reaction conditions and repair-independent suppression, 341, 344
digestion reaction, 344
gel electrophoresis and analysis, 344–345
mechanism of nucleosome assembly from target site for repair, 345–347, 351
principle, 338, 341
Southern blot analysis, 345
remodeling factors and functions, 350
remodeling proteins, *see* Chromatin accessibility complex; Chromatin assembly factor 1; Histone acetyltransferase; Histone deacetylase; Nucleosome remodeling factor; SWI/SNF
replication-coupled assembly
cell culture, 77–78
chromatin preparation with micrococcal nuclease digestion, 80–81
dependence on factors, 97, 99
DNA replication in isolated nuclei
labeling conditions, 94–95
overview, 92–93
reagents and solutions, 93–94
histone acetylation in isolated nuclei, 95–97
nuclei preparation from HeLa cells, 79
overview, 76–77
polyacrylamide gel electrophoresis
histone electrophoresis and transfer to membranes, 84
nucleosomal DNA, 82–83
radiolabeling
DNA, 78
histone, 78–79
liquid scintillation counting, 82
metabolic inhibitors, 79–80
structural overview, 19, 50, 231–232, 377, 612, 696
structure probing with DNA methyltransferases
*Cvi*PI methyltransferase probing
cell maintenance and growth, 445
characteristics of methyltransferase, 443
specificity of methylation, 445
Dam methyltransferase probing, 431–433
interpretation of results, 445–446
principle, 431–432
resolution, 447
sensitivity, 446
*Sss*I methyltransferase probing
bisulfite genomic sequencing and positive display of methylcytosine, 433–435, 437–439
cell maintenance and growth, 435–436
deamination of DNA, 436–437
DNA isolation, 436
expression in yeast, 435
gel electrophoresis analysis, 441
polymerase chain reaction, 438–439
validation of methodology, 441–442
transcription, *see* Nuclear run-on assay
Chromatin accessibility complex
ATPase assay, 752
discovery, 743
immobilized nucleosomes, interaction assay
biotinylated DNA preparation, 753
DNase I footprinting, 754–756
magnetic bead coupling of nucleosomes, 754

nucleosome reconstitution, 753–754
overview, 753
nucleosome spacing assay, 750–752
overview of function, 742–743
restriction enzyme accessibility assay
gel electrophoresis, 749
incubation conditions, 748–750
nucleosome reconstitution
chromatin assembly, sarkosyl stripping, and purification, 746–747
dialysis conditions, 747–748
Drosophila embryo extract preparation, 744–746
Chromatin assembly factor 1, role in remodeling during nucleotide excision repair, 350
Chromatosome, definition, 299
Chromosome, structure
fluorescence *in situ* hybridization analysis, *see* Fluorescence *in situ* hybridization
overview, 650
Cross-linking, chromatin complexes
cyclobutane–pyrimidine dimer mapping in chromatin using photolyase
DNA purification and digestion, 449, 451, 460
DNA repair, 447–448
high-resolution techniques, 461
indirect end-labeling, 449, 451, 453–454, 457–461
interpretaton of gels, 451, 453
micrococcal nuclease digestion of chromatin, 451, 453
principle, 448–449
ultraviolet irradiation and photoreactivation repair, 449, 454–455, 459–460
cysteine-substituted histones
binding to reconstituted nucleosomes, 240
gene cloning, 234
modification of cysteines with chemical probes
cross-linking reagent, 236–237
reduction of histones, 236
nucleosome dyad mapping with cysteine-substituted H4, 631
purification
cell lysis, 235
chromatography, 235
expression in *Escherichia coli*, 234–235
rationale for construction, 233
site-directed mutagenesis, 233–234
dimethylsulfate cross-linking in replicating chromatin
cross-linking reaction, 532
fixation of nuclei with formaldehyde, 532
labeling newly replicating DNA with bromodeoxyuridine, 531
overview, 530–531
Western blot analysis, 532–533
high mobility group proteins
HMG14/17 immunofractionation with cross-linking, 153–154
HMG-I(Y) cross-linking to nucleosomes *in vivo*, 182–183
mapping nonhistone chromatin-associated factors, 401–402, 407, 415, 418
methylation of DNA and cross-linking to histones
application to non-histone systems, 278
cross-linking reaction mechanism, 518, 524
H2A contacts with DNA, 275
H4 contacts with DNA, 275
histidine linking to DNA, 271
nuclei cross-linking studies, 276–277
overview, 264–265
proteolytic digestion of histones, 271–272
two-dimensional gel electrophoresis analysis
DNA versus protein gels, 265, 267
electrophoresis conditions, 269–271
interpretation of bands, 273, 275
linker histone binding on reconstituted nucleosomes, 529
peptide resolution on gels, 272–273
resolution in first dimension, 267
short preliminary fractionation for separation time estimation, 267–269
nucleosome reconstitution *in vitro* for cross-linking
dialysis conditions, 239–240, 527
radiolabeling of DNA fragments, 237–238
ratios of components, 239

photoactive 5S probe synthesis and label-transfer cross-linking assay, 529–530
protein image hybridization, *see* Protein image hybridization
radical-induced protein–DNA cross-linking
  advantages, 517
  cross-linking reaction conditions
    Ehrlich acites tumor cells, *in vivo*, 520, 522
    isolated chromatin, 519–520
  protein shadow hybridization assay with phenanthroline–copper, 522
  radical production
    bleomycin–iron, 516–517
    phenanthroline–copper, 516–517
  two-dimensional gel electrophoresis analysis, 520
RNA polymerase to DNA, 352
site-directed photo-cross-linking
  H2A cross-linking to DNA, 245–246
  linker histones and fragments
    binding to reconstituted nucleosomes, 528
    preparation, 527–528
  nucleosome particles with random DNA sequences, 242–244, 246, 526–527
  nucleosomes reconstituted with defined DNA sequences, 241–242, 245–246
  overview, 240–241, 524–526
  site of cross-linking, identification, 244–245, 250–251
4,5′,8-trimethylpsoralen cross-linking of DNA, *see* 4,5′,8-Trimethylpsoralen
ultraviolet cross-linking of protein–DNA complexes *in vivo*
  applications, 498
  chromatin purification
    cesium chloride gradient centrifugation, 502–504
    nuclei pelleting, 502
    protease inhibition, 503
    solution preparation, 501–502
    starting material, 502
  comparison with *in vitro* assays, 515
  controls, 511–512
  density of protein bound to DNA, 513–514
  Drosophila embryo irradiation, 500–501
  efficiency of cross-linking, 499, 510–511
  immunoprecipitation of complexes, 504–507
  limitations, 517
  overview, 499–500, 618
  resolution, 497–498
  restriction digestion of chromatin, 504–507
  Southern blot analysis, 507–510
  specificity of binding, 498–499, 511–512
Cryoelectron microscopy, chromatin and nucleosomes
  advantages, 191–192, 213
  appearance of samples
    chromatin, 206–208
    compaction, 207–208
    linker histone removal, 208
    nucleosomes, 206
  artifacts
    cooling effects, 10
    interface effects, 211–212
  chromatin preparation
    buffers, 202–203
    glutaraldehyde fixation, 203
    micrococcal nuclease digestion, 202–203
    overview, 201–202
    starting concentrations of DNA, 202
  disadvantages, 192, 213
  film thickness and specimen distribution, 212
  frozen hydrated suspension preparation of chromatin
    apparatus, 198–199
    blotting and film thickness, 199–200
    cryogen removal, 199
    overview, 192–193
    plunger automation, 199
    properties of specimen solution, 201
    safety, 201
    water evaporation, 200–201
  grids and support films, preparation
    bare grids, 193
    carbon coating of film, 195
    film removal from slide, 195

grids with perforated film, overview, 193–194
hole formation, 197–198
platinum coating of film, 195
reproducibility, 197–198
slide coating, 194, 197
imaging conditions
accelerating voltage, 204
contrast transfer function correction, 204–205
cryotransfer device, 203
initial survey, 204
temperature equilibration, 203–204
underfocusing, 204–205
recording of images, 205–206
temporal resolution, 213
three-dimensional images
reconstruction, 208–210
tilt angles, 205
Cyclobutane–pyrimidine dimer, mapping in chromatin using photolyase
DNA purification and digestion, 449, 451, 460
DNA repair, 447–448
high-resolution techniques, 461
indirect end-labeling, 449, 451, 453–454, 457–461
interpretaton of gels, 451, 453
micrococcal nuclease digestion of chromatin, 451, 453
principle, 448–449
ultraviolet irradiation and photoreactivation repair, 449, 454–455, 459–460

D

Dimethyl suberimidate, chromatin fixation for transmission electron microscopy, 224
Dimethyl sulfate
cross-linking reaction mechanism, 518, 524
DNA–protein cross-linking in replicating chromatin
cross-linking reaction, 532
fixation of nuclei with formaldehyde, 532
labeling newly replicating DNA with bromodeoxyuridine, 531
overview, 530–531
Western blot analysis, 532–533
footprinting
genomic footprinting
cell treatment with dimethyl sulfate, 549, 552
DNA isolation, 552
naked DNA controls, 552–553
human–mouse somatic cell hybrid footprinting
culture, 574–575
genomic DNA preparation, 575–576
guanine–adenine-specific cleavage reaction, 576–577
methylation reaction, 575
naked DNA methylation, 576
polymerase chain reaction analysis, *see* Polymerase chain reaction
transcription factor binding to nucleosomes
advantages, 332
binding conditions, 329
cleavage and separation of methylated DNA, 330
methylation reaction, 330
principle, 328
safety, 328
solution preparation, 329
nucleosomal DNA methylation and cross-linking to histones
application to nonhistone systems, 278
H2A contacts with DNA, 275
H4 contacts with DNA, 275
histidine linking to DNA, 271
nuclei cross-linking studies, 276–277
overview, 264–265
proteolytic digestion of histones, 271–272
two-dimensional gel electrophoresis analysis
DNA versus protein gels, 265, 267
electrophoresis conditions, 269–271
interpretation of bands, 273, 275
linker histone binding on reconstituted nucleosomes, 529
peptide resolution on gels, 272–273
resolution in first dimension, 267

short preliminary fractionation for separation time estimation, 267–269
permeability of cell membranes, 613
protein image hybridization, *see* Protein image hybridization
uniformity of DNA reactions, 614
DNA bending
HMG-D assay with ligase-mediated circularization, 101, 103, 107, 123–124
HMG-I(Y) assays, 177
DNA methyltransferase, chromatin structure probing
*Cvi*PI methyltransferase probing
cell maintenance and growth, 445
characteristics of methyltransferase, 443
specificity of methylation, 445
Dam methyltransferase probing, 431–433, 617
interpretation of results, 445–446
principle, 431–432, 617
resolution, 447
sensitivity, 446
*Sss*I methyltransferase probing
bisulfite genomic sequencing and positive display of methylcytosine, 433–435, 437–439
cell maintenance and growth, 435–436
deamination of DNA, 436–437
DNA isolation, 436
expression in yeast, 435
gel electrophoresis analysis, 441
polymerase chain reaction, 438–439
validation of methodology, 441–442
DNA photolyase, *see* Cyclobutane–pyrimidine dimer
DNA repair, *see* Nucleotide excision repair
DNA replication
chromatin, replication-coupled assembly
cell culture, 77–78
chromatin preparation with micrococcal nuclease digestion, 80–81
dependence on factors, 97, 99
DNA replication in isolated nuclei
labeling conditions, 94–95
overview, 92–93
reagents and solutions, 93–94
histone acetylation in isolated nuclei, 95–97
nuclei preparation from HeLa cells, 79
overview, 76–77
polyacrylamide gel electrophoresis
histone electrophoresis and transfer to membranes, 84
nucleosomal DNA, 82–83
radiolabeling
DNA, 78
histone, 78–79
liquid scintillation counting, 82
metabolic inhibitors, 79–80
dimethylsulfate cross-linking in replicating chromatin
cross-linking reaction, 532
fixation of nuclei with formaldehyde, 532
labeling newly replicating DNA with bromodeoxyuridine, 531
overview, 530–531
Western blot analysis, 532–533
solid-phase nucleus
assays, 70–72
characteristics, 64, 68, 70
restriction digest analysis, 70–71
DNA supercoiling, psoralen assay
applications, 639
calculations, 641
cell culture, 641–642
cross-linking reaction, 642
DNA topological domain determination
calculations, 647–650
principle, 639–640
extraction of DNA, 642–643
nicking of DNA, 642
principle, 640
single-strand breaks, measurement
eukaryotic cells, 645
specific chromosomal loci, 646
Southern blot analysis of cross-linked DNA, 643–645
DNase I footprinting
chromatin structure probing, applications
hypersensitive site mapping, 622
sensitive sites and chromatin conformation, 621–622
transcription factor footprinting, 622
Drosophila embryo chromatin
buffer preparation, 477, 479
digestion

large-scale, 471–473
small-scale, 473
extent of digestion, 468
hsp26 gene mapping, 479–480
indirect end-labeling, 466–468, 475–477
nuclei preparation
buffers, 464–465
large-scale preparation, 468–470
small-scale preparation, 470–471
overview, 465
genomic footprinting of cell monolayers, 554–556
glucocorticoid receptor binding to nucleosomes, analysis
binding conditions, 327
digestion conditions and analysis, 328
principle, 327
solution preparation, 327
high-resolution mammalian nucleosome positioning, 636–637
HMG14/17, chromatin interaction analysis, 134–135
HMG-I(Y) and directional substrate binding, 169–172
immobilized nucleosomes, 754–756
mammalian nucleosome positioning *in vitro*, 628–629
periodicity of cleavage, 378
permeabilization of cell membranes for nuclear entry, 614
polymerase chain reaction analysis, *see* Polymerase chain reaction
rotational positioning analysis of DNA fragment on nucleosomes
DNA sequence selection, 322–323
principle, 322
yeast chromatin mapping
agarose gel electrophoresis, 369, 371
digestion conditions, 369
inducible promoters, 368
nuclei preparation
cell growth and lysis, 366–367
centrifugation, 367–368
solution preparation, 366
principle, 365
solution preparation, 369
yeast nucleosome positioning assays
digestion conditions, 389, 391–392
indirect end-labeling
digestion reaction, 395
interpretation, 381–382
principle, 380
reagents, 394–395
Southern blot analysis, 395–396
nuclei isolation
cell growth and lysis, 390–391
centrifugation, 391
reagents, 389
primer extension analysis
denaturing polyacrylamide gel electrophoresis, 398–399
interpretation, 385–387
multiple cycle primer extension for mapping low-copy sequences, 398
one-cycle primer extension for mapping multicopy plasmids, 397
principle, 385
reagents, 396–397
spheroplast preparation
cell growth and spheroplasting, 392–393
centrifugation, 393
digestion conditions, 393–394
permeabilization, 393
reagents, 392
Drosophila melanogaster
chromatin accessibility, quantitative measurements with restriction enzymes
buffer preparation, 495–496
digestion conditions, 489–490, 493, 495
DNA purification, 493–495
hsp26 gene, 490, 492
larva collection, 492–493
nuclei isolation
salivary glands, 494–495
whole larvae, 493
overview, 489–490
chromatin footprinting, *see* Dimethyl sulfate; DNase I footprinting; Methidiumpropyl-EDTA–Fe(II) footprinting; Micrococcal nuclease; Permanganate footprinting; Ultraviolet footprinting
embryo production, large-scale, 462
nuclei preparation from embryos
buffers, 464–465
large-scale preparation, 468–470
small-scale preparation, 470–471
P-element transformation, 463–464

E

EDTAcyst(NPS), *see* *S*-(Nitrophenylsulfenyl)-cysteaminyl-EDTA
Electron microscopy, *see* Cryoelectron microscopy; Transmission electron microscopy
Electrophoretic mobility shift assay
  HMG14/17, nucleosome core particle interaction analysis
    binding conditions, 137
    core particle preparation, 136–137
    electrophoresis conditions, 137
    overview, 134–136
  HMG-D
    binding reaction conditions, 123
    cooperativity of DNA binding, analysis, 113, 115–117
    data extraction and analysis, 124–125
    DNA-binding affinities, 107, 109, 112
    DNA-binding specificity analysis, 101
    electrophoresis conditions, 119
    kinetic analysis of DNA binding, 112–113, 124
    limitations, 117–118
    protein mutation effects on DNA binding, 112
  HMG-I(Y)
    nucleosome core particle complexes, 175
    supershift/ablation assays, 183–185
  nucleosome core particle repositioning assay, 17–19
  transcription factor binding to nucleosomes
    affinity analysis, 325–327
    binding conditions, 325
    DNA sequence design, 323–325
    glucocorticoid receptor binding analysis, 323–327
    limitations, 331–332
    principle, 323
EMSA, *see* Electrophoretic mobility shift assay
Ethidium bromide, transmission electron microscopy support treatment, 222
Eve, ultraviolet cross-linking of protein–DNA complexes *in vivo*, 510–512
Exonuclease III
  chromatin structure probing
    cell culture, 586, 589
    digestion reaction, 591–593
    entry site enzyme digestion, 591–592
    mouse mammary tumor virus promoter analysis, 585–586, 589, 597, 599
    mung bean nuclease digestion, 594
    nuclei isolation, 589, 591
    principle, 585–586, 623
    purification of digested DNA, 593–594
    redigestion, *in vitro*, 594–595
    reiterative primer extension analysis
      amplification reaction, 596–597
      end-labeling of oligonucleotides, 596
      primer selection, 595
      product analysis, 597
      reagents, 596
  mammalian nucleosome positioning *in vitro*, 630
  protein image hybridization, DNA trimming, 607–608, 611–612
  translational nucleosome positioning assay
    digestion conditions and analysis, 322
    principle, 321–322

F

α-Fetoprotein, transcription in solid-phase nuclei
  chromatin assembly, 66–68
  coupling with replication, 72, 74–75
  gene constructs, 65
  hepatoma factor effects, 75
  methylation of gene, 70
  repressed template transcription, 74–75
FISH, *see* Fluorescence *in situ* hybridization
Fluorescence *in situ* hybridization
  applications, 651
  chromosome painting
    DNA probe labeling
      nick translation, 654
      overview, 653
      polymerase chain reaction labeling, 654
      quantification of label, 654–655
      random priming, 654
    hybridization, 655–656
    image analysis, 657–658

metaphase chromosome spread preparation, 652–653
overview, 651–652
signal detection, 656–657
decondensed metaphase chromosomes
chromosome isolation, 658–660
overview, 658
protein removal, 660–661
staining and visualization, 661
nuclear matrix analysis, 661–662
principle, 650–651
yeast nuclear organization with immunofluorescence staining
antibody purification and specificity, 665–666
difficulty in analysis, 663
fixation of cells, 671
fluorophore selection, 666, 672
green fluorescent protein applications, 663, 671
hybridization conditions, 669–670
immunofluorescence conditions, 667–669
nuclear integrity assays, 672
overview of assays, 663–664
probe preparation, 666–667
troubleshooting, 672
variations of technique, 670–671
yeast strains and media, 664–665
Footprinting, *see* Dimethyl sulfate; DNase I footprinting; Hydroxyl radical footprinting; Methidiumpropyl-EDTA–Fe(II) footprinting; Micrococcal nuclease; Permanganate footprinting; Ultraviolet footprinting
Formaldehyde
chromatin fixation for transmission electron microscopy, 223–224
DNA–protein cross-linking, 618
mapping nonhistone chromatin-associated factors, 401–402, 407, 415, 418
Ftz, ultraviolet cross-linking of protein–DNA complexes *in vivo*, 511–512

G

Gcn5
human enzyme
affinity competition assay, 709–711
assay of histone acetyltransferase, 703–705, 707
expression and purification of recombinant proteins
glutathione *S*-transferase fusion proteins, 701–702
histidine-tagged enzyme, 699–700
expression quantification by sulfur-35 labeling and immunoprecipitation, 712–713
immunodepletion analysis, 709
phosphorylative regulation
DNA–PK phosphorylation *in vitro*, 703, 706–707
immunodepletion of DNA–PK, 709
overview, 698
phosphatase effects on activity, 705, 707
phosphate-32 labeling *in vivo*, 711, 713–714
Western blot analysis, 706
transfection of COS cells, 711–712
recruitment in yeast, 697–698
α-Globin, locus control region
binding proteins
dimethylsulfate footprinting and ligation-mediated polymerase chain reaction analysis, 580, 582
novel protein identification, 582–584
overview, 573
DNase I hypersensitivity, 572
guanine–adenine ligation-mediated polymerase chain reaction and dimethyl sulfate footprinting, 578, 580–583
β-Globin
gene and histone acetylation analysis in chick embryo, 542, 547
locus control region
binding proteins
dimethylsulfate footprinting and ligation-mediated polymerase chain reaction analysis, 580–582
novel protein identification, 582–583
overview, 573
DNase I hypersensitivity, 572
guanine–adenine ligation-mediated polymerase chain reaction and dimethyl sulfate footprinting, 578, 580–584
Glucocorticoid receptor, DNA-binding to nucleosomes

dimethyl sulfate methylation protection assay
advantages, 332
binding conditions, 329
cleavage and separation of methylated DNA, 330
methylation reaction, 330
principle, 328
safety, 328
solution preparation, 329
DNase I footprinting
binding conditions, 327
digestion conditions and analysis, 328
principle, 327
solution preparation, 327
electrophoretic mobility shift assay
affinity analysis, 325–327
binding conditions, 325
DNA sequence design, 323–325
limitations, 331–332
principle, 323
site of binding, 321
Glutaraldehyde, chromatin fixation
cryoelectron microscopy, 203
transmission electron microscopy, 223–224
GR, *see* Glucocorticoid receptor
Guanine–adenine ligation-mediated polymerase chain reaction, *see* Polymerase chain reaction

H

HAT, *see* Histone acetyltransferase
Hat1p, recombinant enzyme production, 717–718
HDAC, *see* Histone deacetylase
High mobility group proteins
DNA binding
motif, sequence conservation, 100
specificity, 100–101
HMG14/17
DNase I footprinting of nucleosome interactions, 134–135
electrophoretic mobility shift assay of nucleosome core particle interactions
binding conditions, 137
core particle preparation, 136–137
electrophoresis conditions, 137
overview, 134–136
functions, 133–134, 142–143
hydroxyl radical footprinting analysis of chromatin interactions
chromatin particle purification, 139
footprinting conditions, 139
gel electrophoresis, 139–140
nucleosome complex preparation, 139
overview, 134, 137–138
nucleosome core particle binding, 133
organization in cellular chromatin
clustering and quantification, 150–151
gel electrophoresis and analysis, 154–155
immunfractionation, 150–151
immunofractionation with cross-linking, 153–154
oligonucleosome preparation, 151
reconstitution into chromatin
assembly reaction, 146
functions in *in vivo* versus *in vitro*-assembled chromatin, 143
micrococcal nuclease digestion studies, 147–148, 150
sedimentation velocity centrifugation analysis, 146–148
topoisomer analysis of assembly effects, 144, 146
Xenopus egg extract preparation, 143–144, 146
thermal denaturation assay of nucleosome interactions, 134
two-dimensional gel analysis of nucleosome core particle interactions
antibody preparation and immunofractionation, 141–142
electrophoresis conditions, 142
overview, 134, 140–141
HMG-D
DNA bending assay with ligase-mediated circularization, 101, 103, 107, 123–124
DNA preparation for binding studies
extraction, 122
oligonucleotide synthesis, 122–123
restriction digestion, 122
electrophoretic mobility shift assay

binding reaction conditions, 123
cooperativity of DNA binding, analysis, 113, 115–117
data extraction and analysis, 124–125
DNA-binding affinities, 107, 109, 112
DNA-binding specificity analysis, 101
electrophoresis conditions, 119
kinetic analysis of DNA binding, 112–113, 124
limitations, 117–118
protein mutation effects on DNA binding, 112
immunoprecipitation, amplification, and sequencing of bound DNA, 121
purification of recombinant protein from *Escherichia coli*
chromatography, 120
expression system and cell growth, 119
isoforms from oxidation of intercalating methionine, 118–120
lysis and ammonium sulfate fractionation, 120
mass spectrometry, 120–121
quantification, 121
structure, 101
Western blot analysis of chromatin interactions
blocking and probing of membranes, 133
Drosophila embryo collection, 129
gel electrophoresis, 131
membrane transfer, 131–132
micrococcal nuclease digestion of chromatin, 130–131
nuclei isolation, 130
principle, 126–127
rationale, 125–126, 129

HMG-I(Y)
binding site mutagenesis and antisense transfection assays, 185–188
cross-linking to nucleosomes *in vivo*, 182–183
DNA bending assays, 177
DNA-binding specificity, 156, 169–171
electrophoretic mobility shift assay
nucleosome core particle complexes, 175
supershift/ablation assays, 183–185
fluorescence competition binding assays, 172–173
fluorescence *in situ* immunolocalization, 181–182
footprinting
DNase I footprinting and directional substrate binding, 169–172
hydroxyl radical footprinting, 172, 179
reconstituted nucleosomes with defined sequence DNAs, 176
four-way junction binding assays, 179
functions, 155–156, 179–180
nucleosome-binding assays, 173, 175
purification
acid extraction, 157–160
cation-exchange chromatography, 166
cell starting materials, 158–159
gel electrophoresis, 164–165
lysis, 159
polybuffer exchanger purification, 165
quantification by absorbance, 163–164
recombinant protein from *Escherichia coli*, 161–162, 166–167
reversed-phase chromatography, 163, 166–167
structure, 156–157
supercoiling assay with two-dimensional gel electrophoresis, 177, 179
types, 155
Western blot analysis
antibody production, 167–169
incubations and washings, 169
overview of types and function, 100, 133, 155

Hill plot, cooperativity of high mobility group protein binding to DNA, 113, 115, 117

Histone, *see also* Chromatin; Nucleosome; Nucleosome core particle
acetylation mapping, *see* Histone acetyltransferase; Histone deacetylase
cross-linking, *see* Cross-linking, chromatin complexes
cysteine-substituted histones

binding to reconstituted nucleosomes, 240
gene cloning, 234
modification of cysteines with chemical probes
cleavage reagent, 236–237
cross-linking reagent, 236–237
reduction of histones, 236
nucleosome dyad mapping with cysteine-substituted H4, 631
purification
cell lysis, 235
chromatography, 235
expression in *Escherichia coli*, 234–235
rationale for construction, 233
site-directed hydroxyl radical mapping
5S nucleosome studies, 249–250
agarose gel electrophoresis, 247, 249
binding reaction, 246–247
cleavage reaction, 246
sequence analysis of cleaved DNA, 250–251
site-directed mutagenesis, 233–234
functions in chromatin, 50, 298
H2A/H2B, purification from *Xenopus* extracts, 57–58
hydroxyl radical footprinting, *see* Hydroxyl radical footprinting
linker histones
H5 incorporation into model nucleosome arrays, 34
removal, 208
structure, 232
N1,N2–(H3,H4) complex from *Xenopus* extracts
isolation, 60–61
nucleosome assembly, 61–62
nucleosome core particle preparation from recombinant histones
advantages, 3
dialysis conditions for reconstitution, 11, 13
DNA fragment preparation, 11
histone purification
anion-exchange chromatography, 8–9
expression in *Escherichia coli*, 5–6
gel filtration, 7–8
inclusion body preparation, 6–7
yield, 5
octamer refolding, 9–10
overview, 3–4
purification
anion-exchange chromatography, 13–16
modified complexes, 14
preparative gel electrophoresis, 13–14, 16–17
octamer components, 19
polyacrylamide gel electrophoresis and transfer to membranes, 84
radiolabeling, 78–79
reconstitution of nucleosome arrays from purified components
dialysis conditions for reconstitution, 24, 26
DNA template purification, 22–23
histone octamer purification, 23–24
histone saturation of reconstituted arrays
enrichment of saturated arrays, 32–34
importance of determination, 26
overview of assays, 26–27
quantitative agarose gel electrophoresis assay, 27, 30
restriction enzyme digestion assay, 30–31
sedimentation velocity ultracentrifugation assay, 31–32
states of saturation, 20
instrumentation, 22
materials and reagents, 21–22
Histone acetyltransferase, *see also* Gcn5; Hat1p
acetylation mapping of histones
antibody production against chemically acetylated H4
affinity purification, 536–537
enzyme-linked immunosorbent assay of effectiveness, 537
immunization, 535
chicken embryo erythrocyte analysis
β-globin locus analysis, 542, 547
housekeeping gene acetylation, 546–547
hybridization analysis, 541–542

locus-wide mapping, 542–545
probe, 540–541
structural changes in acetylated nucleosomes, 545–546
chromatin preparation for immunoselection
micrococcal nuclease digestion, 537–538
mononucleosomes, 538
salt-soluble chromatin, 537–538
directed acetylation at active genes, 547
immunoselection
DNA analysis, 539–540
immunoprecipitation, 538
principle, 533–535
protein analysis, 539
isolated nuclei, 95–97
N-terminal microsequencing
chromatin substrate preparation, 693
high-performance liquid chromatography to resolve core histones, 694–695
histone substrate preparation, 693
interpretation of results, 695
overview, 692–693
sequencing, 695
acetyl-CoA labeling for assays, 681–682
activity gel assay
acetylation reaction, 690
controls, 691
electrophoresis, 690
gel preparation, 689–690
interpretation of results, 691–692
overview, 689
staining and fluorography, 691
classification and types, 675–677
function, overview, 675, 696–697
liquid assay for free histone or chromatin substrates
filter spotting and washing, 684
incubation conditions, 683–684
overview, 682
pH and ionic strength effects, 684, 686–687
product analysis, 688
radiolabel quantification, 684
reagents, 682–683
structure of substrates in assay
chromatin, 687–688
histone, 687
liquid assay for peptide substrates, 688–689
regulation, 697–698
substrates
chromatin preparation
micrococcal nuclease digestion of chicken erythrocyte nuclei, 679–680
overview, 679
sucrose gradient centrifugation, 680–681
free histone preparation, 678
requirements for assay, 677–678
synthetic peptide preparation, 679
types, 675–677
Histone deacetylase
acetylated substrate preparation for assay
histone acetyltransferase modification of purified histones
acetylation reaction, 718
overview, 716–717
recombinant Hat1p production, 717–718
in vivo, 716
assay conditions, 720
phosphorylative regulation, 698
purification of Rpd3-like complex from *Xenopus* extracts
advantages of *Xenopus* system, 720–721
egg extract production, 721
ion-exchange chromatography, 724–725
oocyte extract production, 721, 724
storage, 725
sucrose gradient density centrifugation, 725
yield, 725
recruitment, 697
structural classification, 715–716
HMG proteins, *see* High mobility group proteins
Hydroxyl radical footprinting
HMG14/17, chromatin interaction analysis
chromatin particle purification, 139
footprinting conditions, 139
gel electrophoresis, 139–140

nucleosome complex preparation, 139
overview, 134, 137–138
HMG-I(Y), 172, 179
radical reactivity with DNA, 252
site-directed hydroxyl radical mapping, histones
cysteine-substituted histones
5S nucleosome studies, 249–250
agarose gel electrophoresis, 247, 249
binding reaction, 246–247
cleavage reaction, 246
modification of cysteines with cleavage reagent, 236–237
sequence analysis of cleaved DNA, 250–251
histone octamers
H4 cysteine modification with cleavage reagent, 253, 256–257
iron ion loading, 258–259
S-(nitrophenylsulfenyl)-cysteaminyl-EDTA synthesis, 253, 255
nucleosome assembly and gel electrophoresis analysis, 257–258
rate of reaction, 259
reaction conditions, 259–260
sequence analysis of cleaved DNA and mapping, 260–263
translational position determination, 252
ultraviolet cross-linked DNA–protein complexes, 504–507
Indirect end-labeling
cyclobutane–pyrimidine dimer, mapping in chromatin using photolyase, 449, 451, 453–454, 457–461
DNase I footprinting of *Drosophila* embryo chromatin, 466–468, 475–477
low-resolution mammalian nucleosome positioning
micrococcal nuclease digestion, 633–634
nuclei isolation, 632–633, 635
optimization, 635–636
overview, 631–632
Southern blot analysis, 634–635
methidiumpropyl-EDTA–Fe(II) footprinting of *Drosophila* embryo chromatin, 466–468, 475–477
micrococcal nuclease
chromatin structural probing, 619–620
Drosophila embryo chromatin analysis, 466–468, 475–477
4,5′,8-trimethylpsoralen, nucleosome positioning, 636
yeast nucleosome positioning assays
digestion reaction, 395
interpretation, 381–382
principle, 380
reagents, 394–395
Southern blot analysis, 395–396

I

Immunoblot, *see* Western blot
Immunoprecipitation
acetylation mapping, 538
chromatin
antibodies, 84, 88
fixed chromatin
formaldehyde fixation, 89–91
incubation conditions, 91–92
unfixed chromatin
incubation conditions, 87
protein A–agarose bead preparation, 85, 87
solution preparation, 85
HMG-D and bound DNA, 121
mapping nonhistone chromatin-associated factors, 402, 410–412, 423–424

L

lac repressor, density of binding on DNA, 513–514
λ repressor, density of binding on DNA, 513–514
LCR, *see* Locus control region
Ligation-mediated polymerase chain reaction, *see* Polymerase chain reaction
Locus control region, *see* α-Globin; β-Globin

M

Methidiumpropyl-EDTA–Fe(II) footprinting

cleavage
reaction mechanism, 620, 666
sequence specificity, 462–463, 466, 620
Drosophila embryo chromatin
buffer preparation, 477, 479
digestion reaction, 473–475
extent of digestion, 468
hsp26 gene mapping, 479–480
indirect end-labeling, 466–468, 475–477
hormone-induced hypersensitive sites, 620–621
Methylamine tungstate, chromatin staining for transmission electron microscopy, 229
Micrococcal nuclease
DNA cleavage characteristics, 378, 619–620
Drosophila embryo chromatin analysis
buffer preparation, 477, 479
digestion
large-scale, 471–473
small-scale, 473
extent of digestion, 468
indirect end-labeling, 466–468, 475–477
nuclei preparation
buffers, 464–465
large-scale preparation, 468–470
small-scale preparation, 470–471
overview, 465
indirect end-labeling probing of chromatin structure, 619–620
mammalian nucleosome positioning
high-resolution, 638
in vitro, 629–630
nucleosome assembly assay
assembly reaction conditions and repair-independent suppression, 341, 344
digestion reaction, 344
gel electrophoresis and analysis, 344–345
mechanism of nucleosome assembly from target site for repair, 345–347, 351
principle, 338, 341
Southern blot analysis, 345
nucleosome boundary mapping at high resolution, 625–626
permeabilization of cell membranes for nuclear entry, 614
sequence specificity, 371, 380, 382
yeast mononucleosome analysis
agarose gel electrophoresis, 374
digestion conditions, 371, 373
nuclei preparation
cell growth and lysis, 366–367
centrifugation, 367–368
solution preparation, 366
principle, 365, 371
solution preparation, 371
yeast nucleosome positioning assays
digestion conditions, 389, 391–392
indirect end-labeling
digestion reaction, 395
interpretation, 381–382
principle, 380
reagents, 394–395
Southern blot analysis, 395–396
nuclei isolation
cell growth and lysis, 390–391
centrifugation, 391
reagents, 389
primer extension analysis
denaturing polyacrylamide gel electrophoresis, 398–399
interpretation, 385–387
multiple cycle primer extension for mapping low-copy sequences, 398
one-cycle primer extension for mapping multicopy plasmids, 397
principle, 385
reagents, 396–397
spheroplast preparation
cell growth and spheroplasting, 392–393
centrifugation, 393
digestion conditions, 393–394
permeabilization, 393
reagents, 392
Minichromosome
intactness assay, 49
isolation from yeast
endogenous plasmid elimination, 39–41
growth of cells, 42
nuclei
lysis, 45–48
preparation from spheroplasts, 43–45
overview of protocols, 35–36
plasmid design, 36–38

spheroplast preparation, 42–43
transformation by electroporation, 41–42
yeast strain selection, 38–39
yield, 49
quantitative analysis, 49
MMTV, *see* Mouse mammary tumor virus
MNase, *see* Micrococcal nuclease
Mouse mammary tumor virus, promoter analysis by exonuclease III probing, 585–586, 589, 597, 599
MPE–Fe(II), *see* Methidiumpropyl-EDTA–Fe(II) footprinting

N

NCP, *see* Nucleosome core particle
NER, *see* Nucleotide excision repair
NF-1, *see* Nuclear factor 1
S-(Nitrophenylsulfenyl)-cysteaminyl-EDTA
histone modification, 253, 256–257
site-directed hydroxyl radical mapping, histones
iron ion loading, 258–259
nucleosome assembly and gel electrophoresis analysis, 257–258
rate of reaction, 259
reaction conditions, 259–260
sequence analysis of cleaved DNA and mapping, 260–263
synthesis
cysteine blocking, 253
linkage to EDTA, 255
mass spectrometry characterization, 255
2-nitrophenylsulfenyl exchange, 255
Nuclear factor 1, DNA-binding site on nucleosomes, 321
Nuclear run-on assay
comparison to other transcription assays, 351–352
Drosophila assays
buffers and reagents, 354
hybridization of labeled run-on RNA to specific sequences, 357–358
nuclei preparation
adult flies, 355–356
cell cultures, 354–355
lysis of cells, 354
overview, 353–354
run-on reaction, 356–357
elongation rate and interpretation of results, 361
kinetic analysis of polymerase associations, 362
mammalian cell assays
buffers and reagents, 358–359
nuclei preparation, 359
run-on reaction, 359–360
principle, 352–353
quantitative comparison of different genes, 361–362
yeast assays
buffers and reagents, 360
permeabilized cell preparation, 360
run-on reaction, 360
Nucleosome, *see also* Chromatin; Histone; Nucleosome core particle
array formation from *Xenopus* extracts
DNA concentration optimization, 55
DNA supercoiling assay, 53, 62–63
array reconstitution from purified components
dialysis conditions for reconstitution, 24, 26
DNA template purification, 22–23
histone octamer purification, 23–24
histone saturation of reconstituted arrays
enrichment of saturated arrays, 32–34
importance of determination, 26
overview of assays, 26–27
quantitative agarose gel electrophoresis assay, 27, 30
restriction enzyme digestion assay, 30–31
sedimentation velocity ultracentrifugation assay, 31–32
states of saturation, 20
instrumentation, 22
materials and reagents, 21–22
DNA accessibility, restricting factors, 313
DNA pitch, 251–252
electron microscopy, *see* Electron microscopy
footprinting, *see* Dimethyl sulfate; DNase I footprinting; Hydroxyl radical foot-

printing; Methidiumpropyl-EDTA–Fe(II) footprinting; Micrococcal nuclease; Permanganate footprinting; Restriction enzyme accessibility mapping; Ultraviolet footprinting
hydroxyl radical footprinting, *see* Hydroxyl radical footprinting
linker DNA in array formation, 19, 298
linker histone H5 incorporation into model nucleosome arrays, 34
oligonucleosome preparation from rat liver nuclei
gel filtration chromatography, 317
liver homogenization, 314–315
nuclei isolation and micrococcal nuclease digestion, 315–317
solution preparation, 314
polyacrylamide gel electrophoresis, *see also* Electrophoretic mobility shift assay
advantages, 312
native nucleosomes
cage effect in stabilization of complexes, 300
casting of gels, 300
comparison to standard DNA gel electrophoresis, 299–300
effectiveness, 304
electrophoresis conditions, 300
glycerol inclusion, 302–303
histone acetylation detection, 303–304
optimization, 301
particle mass separations, 303
staining, 301
nucleosome mobility assays
interpretation of bands, 311
principle, 309, 312
two-dimensional gel electrophoresis, 309–311
positioned nucleosomes
preparative gel electrophoresis, 307–308
principle of separation, 304–307
positioning
exonuclease III assay
digestion conditions and analysis, 322
principle, 321–322
high-resolution mammalian mapping
DNase I footprinting, 636–637
double-stranded micrococcal nuclease footprinting, 638
ligation-mediated polymerase chain reaction, 637–638
importance of studies, 626
low-resolution mammalian mapping
indirect end-labeling, overview, 631–632
micrococcal nuclease digestion, 633–634
nuclei isolation, 632–633, 635
optimization, 635–636
Southern blot analysis, 634–635
mammalian mapping, *in vitro*
core particle end detection with exonuclease III, 630
core particle end generation with micrococcal nuclease, 629–630
nucleosome dyad mapping with cysteine-substituted H4, 631
reconstitution of nucleosomes, 627
rotational positions and dyads mapped with DNase I, 628–629
rotational versus translational, 377
yeast assay
cleavage reagents in assays, 378, 380, 387, 389
digestion reactions, 391–392
indirect end-labeling for low-resolution mapping, 380–382, 394–396
nuclei isolation, 389–391
positioning mechanisms, 377–378
primer extension analysis for high-resolution mapping, 385–387, 396–399
restriction endonuclease mapping, 383, 385
spheroplast preparation and digestion, 392–394
reconstitution by histone exchange
DNA sequence selection for transcription factor interaction assays, 320–321
glycerol gradient centrifugation, 318–320
incubation conditions, 318
solution preparation, 318
structure, overview, 231–232, 298, 313, 377

transcription factor binding, *see* Electrophoretic mobility shift assay; Dimethyl sulfate, footprinting; DNase I footprinting
Nucleosome core particle
definition, 299
gel shift assay and repositioning of particles, 17–19
high mobility group protein binding, *see* High mobility group proteins
preparation from recombinant histones
advantages, overview, 3–4
dialysis conditions for reconstitution, 11, 13
DNA fragment preparation, 11
histone purification
anion-exchange chromatography, 8–9
expression in *Escherichia coli*, 5–6
gel filtration, 7–8
inclusion body preparation, 6–7
yield, 5
octamer refolding, 9–10
purification
anion-exchange chromatography, 13–16
modified complexes, 14
preparative gel electrophoresis, 13–14, 16–17
Nucleosome remodeling factor
overview of function, 742, 757–758
purification
cellulose phosphate chromatography, 764
Drosophila extract preparation, 759–762
glycerol gradient centrifugation, 764–765
hydroxylapatite chromatography, 763–764
ion-exchange chromatography, 762–763
overview, 758–759
yield, 759
subunits, 758
Nucleotide excision repair
chromatin remodeling during repair
DNA substrate preparation, 334–335
Drosophila embryo extract preparation, 335, 338
efficiency of repair, assay, 337–338
extracts for study, 333–334, 338, 350
micrococcal nuclease assay of nucleosome assembly
assembly reaction conditions and repair-independent suppression, 341, 344
digestion reaction, 344
gel electrophoresis and analysis, 344–345
mechanism of nucleosome assembly from target site for repair, 345–347, 351
principle, 338, 341
Southern blot analysis, 345
remodeling factors and functions, 350
lesion targets, 333, 447–448
Nucleus, *see* Solid-phase nucleus
NURF, *see* Nucleosome remodeling factor

P

PAGE, *see* Polyacrylamide gel electrophoresis
PCR, *see* Polymerase chain reaction
Pentylamine, transmission electron microscopy support treatment, 221
Permanganate footprinting
Drosophila salivary glands, genomic footprinting
buffers and reagents, 488–489
hsp70 gene, 481, 484
ligation-mediated polymerase chain reaction analysis, 484, 487–488
naked DNA
formic acid treatment, 483
permanganate treatment, 482–483
piperidine cleavage, 483–484
overview, 480–481
salivary gland
dissection, 481
permanganate treatment, 482
genomic footprinting of cell monolayers, 553–554
modification of DNA, 616
permeability of cell membranes to potassium permanganate, 616
polymerase chain reaction analysis, *see* Polymerase chain reaction
thymine reactivity, 481

Phenanthroline–copper, cross-linking of chromatin complexes
advantages, 517
cross-linking reaction conditions
Ehrlich acites tumor cells, *in vivo*, 520, 522
isolated chromatin, 519–520
DNA cleavage reaction, 521
protein shadow hybridization assay, 522
radical production, 516–517
two-dimensional gel electrophoresis analysis, 520
Phosphotungstic acid, chromatin staining for transmission electron microscopy, 227, 229
Polyacrylamide gel electrophoresis, *see also* Electrophoretic mobility shift assay; Western blot
advantages in nucleoprotein analysis, 312
histone acetyltransferase activity gel assay
acetylation reaction, 690
controls, 691
electrophoresis, 690
gel preparation, 689–690
interpretation of results, 691–692
overview, 689
staining and fluorography, 691
native nucleosomes
comparison to standard DNA gel electrophoresis, 299–300
cage effect in stabilization of complexes, 300
casting of gels, 300
electrophoresis conditions, 300
staining, 301
optimization, 301
glycerol inclusion, 302–303
particle mass separations, 303
histone acetylation detection, 303–304
effectiveness, 304
nucleosome mobility assays
interpretation of bands, 311
principle, 309, 312
two-dimensional gel electrophoresis, 309–311
positioned nucleosomes, principle of separation, 304–307
preparative gel electrophoresis
linker histones, 13–14, 16–17
nucleosome core particles, 13–14, 16–17
positioned nucleosomes, 307–308
replication-coupled assembly assay
histone electrophoresis and transfer to membranes, 84
nucleosomal DNA, 82–83
two-dimensional gel analysis
high mobility group protein interactions with nucleosome core particle
antibody preparation and immunofractionation, 141–142
electrophoresis conditions, 142
overview, 134, 140–141
methylation of DNA and cross-linking to histones
DNA versus protein gels, 265, 267
electrophoresis conditions, 269–271
interpretation of bands, 273, 275
linker histone binding on reconstituted nucleosomes, 529
peptide resolution on gels, 272–273
resolution in first dimension, 267
short preliminary fractionation for separation time estimation, 267–269
protein image hybridization
electroblotting, 609
electrophoresis conditions, 608–609
probe hybridization, 609
radical-induced protein–DNA cross-linking, 520
supercoiling assay, 177, 179
Polycomb, mapping of chromatin interactions, 405, 417
Polylysine, transmission electron microscopy support treatment, 221
Polymerase chain reaction
exonuclease III mapping of chromatin, reiterative primer extension analysis
amplification reaction, 596–597
end-labeling of oligonucleotides, 596
primer selection, 595
product analysis, 597
reagents, 596
fluorescence *in situ* hybridization probe labeling, 654
HMG-D associated DNA, 121
ligation-mediated polymerase chain reac-

tion analysis of *Drosophila* chromatin, 484, 487–488
ligation-mediated polymerase chain reaction in footprinting analysis
direct primer labeling protocol
amplification reaction, 565
direct labeling reaction, 565–566
end-labeling of primer, 566
first primer extension, 564
gel electrophoresis and band detection, 566
ligation, 564–565
overview, 563–564
DNase I footprinting modifications, 561, 563
guanine–adenine ligation-mediated polymerase chain reaction and dimethyl sulfate footprinting
amplification reaction, 577
annealing reaction, 577
applications, 578, 580–584
guanine–adenine-specific cleavage reaction, 576–577
labeling reaction, 577–578
ligation reaction, 577
overview, 574
high-resolution mammalian nucleosome positioning, 637–638
nucleosome boundary mapping at high resolution, 625–626
optimization, 624
overview, 548–549, 623–624
selection of protocol, 571
standard protocol
amplification reaction, 558–559
electroblotting, 559–560
first primer extension, 557–558
gel electrophoresis of products, 559
hybridization analysis, 560
ligation, 558
single-stranded probe preparation, 560–561
transcription factor footprinting, 622–623
mapping nonhistone chromatin-associated factors, 402, 405, 412–414, 426–427, 430
terminal transferase-dependent polymerase chain reaction in footprinting analysis
adapter analysis, 569
adapter ligation, 570
amplification reaction, 570
direct labeling of product, 570
first primer extension, 567
gel electrophoresis and analysis, 570–571
overview, 548, 566–567
tailing reaction, 567, 569
Potassium permanganate, *see* Permanganate footprinting
Protein image hybridization
cesium chloride gradient centrifugation, depletion of free protein, 605–606
cross-link types, 610–611
dimethylsulfate cross-linking reaction, 604–605, 610
DNA shearing, 606–607
exonuclease III trimming, 607–608, 611–612
formaldehyde fixation, 604–605
hsp70 probing in *Drosophila*, 611
molecular weight estimation of cross-linked proteins, 611
overview, 600–601
phenol extraction of DNA–protein complexes, 607–608
protease inhibition, 609
two-dimensional gel electrophoresis
electroblotting, 609
electrophoresis conditions, 608–609
probe hybridization, 609
ultraviolet cross-linking
cross-linking reaction, 603–604
dose, 601, 603
nuclei isolation, 604
sources of light, 610–611
Protein shadow hybridization assay, phenanthroline–copper utilization, 522
Psoralen, *see* 4,5′,8-Trimethylpsoralen

R

Restriction enzyme accessibility mapping
chromatin remodeling assays, *see* Chromatin accessibility complex; SWI/SNF
Drosophila chromatin accessibility, quantitative measurements

buffer preparation, 495–496
digestion conditions, 489–490, 493, 495
DNA purification, 493–495
hsp26 gene, 490, 492
larva collection, 492–493
nuclei isolation
salivary glands, 494–495
whole larvae, 493
overview, 489–490
nucleosome stability and dynamics, analysis
concentrations
substrate, 295–296
restriction enzyme, 296
contamination, free DNA, 294–295
digestion conditions, 291–292
DNA template design, 290–291
equilibrium constants, 293–294
kinetics of digestion, quantitative analysis, 292–293
neutrality of probes, 279
nucleosome core particle reconstitution and purification, 291
polynucleosomal system analysis, 296–298
rapid preequilibrium, testing, 294
site exposure
cooperative binding by arbitrary proteins, modeling, 288–290
definition, 279
digestion kinetics, linking to simple equilibrium-binding reactions, 284–285
kinetic modeling, 280, 282–283
spontaneous versus rapid equilibrium, 283–284
structural model, 283
system-specific, second-order effects, 285, 287–288
specificity and sensitivity, 278–279
stability controls during assay, 295
steric hindrance at target site, 278
yeast chromatin mapping
digestion conditions, 376
nuclei preparation
cell growth and lysis, 366–367
centrifugation, 367–368
solution preparation, 366
principle, 365, 374
solution preparation, 374
Southern blot analysis, 376
Rpd3, *see* Histone deacetylase

S

Scatchard plot, cooperativity of high mobility group protein binding to DNA, 115–117
Sedimentation velocity ultracentrifugation
assay of nucleosome array saturation, 31–32
HMG14/17 reconstitution into chromatin, 146–148
Silent information regulator 3, mapping of chromatin interactions, 402–403
SIR3, *see* Silent information regulator 3
SNF, *see* SWI/SNF
Solid-phase nucleus
applications, 64, 75–76
components, 63–64
DNA replication
assays, 70–72
characteristics, 64, 68, 70
restriction digest analysis, 70–71
DNA template preparation, 65–66
extract preparation from *Xenopus* eggs, 66
α-fetoprotein gene transcription
chromatin assembly, 66–68
coupling with replication, 72, 74–75
gene constructs, 65
hepatoma factor effects, 75
methylation of gene, 70
repressed template transcription, 74–75
Southern blot
chromatin remodeling during nucleotide excision repair, 345
cross-linked DNA, 507–510,643–645
indirect end-labeling and low-resolution mammalian nucleosome positioning, 634–635
micrococcal nuclease and yeast mononucleosome analysis, 395–396
restriction enzyme accessibility mapping of chromatin, 376
yeast nucleosome positioning assay, 395–396

SWI/SNF
  ATPase assays
    activated charcoal binding of phosphorous-32, 729–730
    kinetic parameter determination, 731
    thin-layer chromatography assay
      chromatgraphy, 730–731
      incubation conditions, 730
      materials and solutions, 730
  nucleosome array remodeling assay by restriction enzyme coupling
    applications, 741
    glycerol effects on restriction enzyme activity, 735–736
    incubation conditions, 739–740
    nucleosome array reconstitution
      dialysis conditions, 733
      DNA template, 732
      histone octamer preparation, 732–733
      nucleosome saturation, determination, 733–734
    overview, 731
    *Sal*I cleavage kinetics, 738–739
    salt effects
      nucleosome array conformation and self-association, 736–738
      restriction enzyme activity, 735–736
  nucleosome disruption, overview, 726
  purification of histidine-tagged protein
    DNA cellulose chromatography, 728–729
    glycerol gradient centrifugation, 729
    ion-exchange chromatography, 729
    materials, 727–728
    nickel affinity chromatography, 728
    overview, 727
    yeast growth and lysis, 728
    yield, 729
  subunits, 726

T

TEM, *see* Transmission electron microscopy
Terminal transferase-dependent polymerase chain reaction, *see* Polymerase chain reaction
Transmission electron microscopy, chromatin and DNA–protein complexes
  direct mounting, 216, 218
  fixation
    artifacts, 225–226
    dimethyl suberimidate, 224
    functions, 222
    glutaraldehyde and formaldehyde, 223–224
    optimization, 224–225
  freeze-fracture, 218
  negative staining
    advantages, 226
    ammonium molybdate, 229
    aurothioglucose, 229
    DNA, 226–227
    methylamine tungstate, 229
    phosphotungstic acid, 227, 229
    reaction conditions, 227
    resolution, 226
    surfactant utilization, 230
    uranyl acetate, 227
  support preparation
    carbon supports, 218–220
    treatment to facilitate sample adherence
      Alcian blue, 221–222
      amylamine, 221
      ethidium bromide, 222
      glow charging, 220
      ions, 220
      pentylamine, 221
      polylysine, 221
  surface spreading
    Kleinschmidt method, 214–215
    Miller method, 215–216
4,5′,8-Trimethylpsoralen
  linker region mapping between nucleosomes, 618–619
  nucleosome positioning by indirect end-labeling, 636
  permeability of cell membranes, 639
  supercoiling assay
    applications, 639
    calculations, 641
    cell culture, 641–642
    cross-linking reaction, 642
    DNA topological domain determination

calculations, 647–650
principle, 639–640
extraction of DNA, 642–643
nicking of DNA, 642
principle, 640
single-strand breaks, measurement
eukaryotic cells, 645
specific chromosomal loci, 646
Southern blot analysis of cross-linked DNA, 643–645
Two-dimensional gel electrophoresis, *see* Polyacrylamide gel electrophoresis

U

Ultraviolet cross-linking, *see* Cross-linking
Ultraviolet footprinting
genomic footprinting, reaction conditions, 556–557
polymerase chain reaction analysis, *see* Polymerase chain reaction
principle, 617
protein image hybridization, *see* Protein image hybridization
Uranyl acetate, chromatin staining for transmission electron microscopy, 227

W

Western blot
dimethylsulfate cross-linking in replicating chromatin, 532–533
Gcn5 phosphorylation, 706
HMG-D and chromatin interactions
blocking and probing of membranes, 133
Drosophila embryo collection, 129
gel electrophoresis, 131
membrane transfer, 131–132
micrococcal nuclease digestion of chromatin, 130–131
nuclei isolation, 130
principle, 126–127
rationale, 125–126, 129
HMG-I(Y)
antibody production, 167–169
incubations and washings, 169

Z

Zeste, ultraviolet cross-linking of protein–DNA complexes *in vivo*, 498, 510–512, 515